ENVIRONMENTAL PHYSIOLOGY

ENVIRONMENTAL PHYSIOLOGY

Edited by

N. Balfour Slonim, M.D., Ph.D.

Director, Cardiopulmonary Diagnostic Laboratory,
Denver, Colorado

With twenty-two contributors

with 287 illustrations

THE C. V. MOSBY COMPANY
SAINT LOUIS 1974

Printed in the United States of America

Distributed in Great Britain by Henry Kimpton, London

Library of Congress Cataloging in Publication Data

Slonim N Balfour, 1923-
Environmental physiology.

1. Man—Influence of environment. I. Title.
[DNLM: 1. Adaptation, Physiological. 2. Environment.
QT140 S634e 1974]
QP82.S56 612′.0144 73-17078
ISBN 0-8016-4659-6

CB/CB/B 9 8 7 6 5 4 3 2 1

CONTRIBUTORS

Paul T. Baker, Ph.D.

Head, Department of Anthropology, The Pennsylvania State University, University Park, Pennsylvania

Madeleine F. Barnothy, Ph.D.

Professor of Physics, University of Illinois at the Medical Center, College of Pharmacy, Department of Chemistry, Chicago, Illinois

Jeno M. Barnothy, Ph.D.

Biomagnetic Research Foundation, Evanston, Illinois

Albert R. Behnke, Jr., M.D.

San Francisco, California

N. Karen Bender

Chief Technologist, Cardiopulmonary Diagnostic Laboratory, Denver, Colorado

Ralph W. Brauer, Ph.D.

Professor of Marine Physiology; Head, Department of Marine Bio-medical Research, University of North Carolina at Wilmington, Wilmington, North Carolina

Frank A. Brown, Jr., Ph.D.

Morrison Professor of Biology, Department of Biological Sciences, Northwestern University, Evanston, Illinois

†Konrad J. K. Buettner, Ph.D.

Professor of Atmospheric Science, Department of Atmospheric Sciences, University of Washington, Seattle, Washington

†Deceased.

†Loren D. Carlson, Ph.D.

Division of the Sciences Basic to Medicine, Department of Human Physiology, School of Medicine, University of California, Davis, Davis, California

Farrington Daniels, Jr., M.D.

Professor of Medicine; Head, Dermatology Division, The New York Hospital–Cornell Medical Center, New York City, New York

Frances L. Estes, Ph.D.

Houston, Texas

Ashton Graybiel, M.D.

Naval Aerospace Medical Institute, Naval Aerospace Medical Center, Pensacola, Florida

Austin F. Henschel, Ph.D.

Chief, Physiology and Ergonomics Branch, National Institute for Occupational Safety and Health, Department of Health, Education, and Welfare, Public Health Service, Health Services and Mental Health Administration, Cincinnati, Ohio

Arnold C. L. Hsieh, M.D., D.Sc.

Division of the Sciences Basic to Medicine, Department of Human Physiology, School of Medicine, University of California, Davis, Davis, California

Ulrich C. Luft, M.D.

Head, Physiology Department, Lovelace Foundation for Medical Education and Research, Albuquerque, New Mexico

†Deceased.

A. T. Miller, Jr., M.D., Ph.D.

Professor of Physiology, Department of Physiology, School of Medicine, University of North Carolina, Chapel Hill, North Carolina

Donald E. Parker, Ph.D.

Professor of Psychology, Miami University, Oxford, Ohio

John L. Patterson, Jr., M.D.

Professor of Medicine, College of Medicine, Medical College of Virginia, Health Sciences Division, Richmond, Virginia

Sid Robinson, Ph.D.

Department of Anatomy and Physiology, University of Indiana, Bloomington, Indiana

N. Balfour Slonim, M.D., Ph.D.

Director, Cardiopulmonary Diagnostic Laboratory, Denver, Colorado

Henning E. Von Gierke, Ph.D.

Yellow Springs, Ohio

David L. Wiegman, Ph.D.

Lecturer, Department of Anatomy and Physiology, University of Indiana, Bloomington, Indiana

To Life on Earth

PREFACE

This book is designed primarily as a text for advanced undergraduate and early postgraduate students of the biologic sciences. Sections of this book will interest bioengineers, research scientists, physicians, and others concerned with the environmental systems discussed. The reader will need a working knowledge of general physiology. However, relevant facts, principles, and concepts are reviewed briefly where necessary.

Environmental physiology is emerging as an important branch of the science of physiology. It is the branch concerned with the physiologic responses of the healthy intact organism to environmental change. It focuses on the environment-organism interface, describing, analyzing, and interpreting the dynamic interaction of environmental stimulus with biologic system that produces a physiologic response. Environmental physiology is thus an extension of general physiology—the study of the function of an organism in its usual, natural environment. In this text we are concerned with the present state of adaptation of an organism to its usual natural environment, with the modus operandi of presently existing adaptive response mechanisms, and, to a lesser extent, with the process by which these adaptations evolved. Our emphasis is mammalian and, more precisely, human.

Consideration of the impact of environmental stressors on pathologic states or on the course of preexisting disease is beyond the scope of this book. This subject is treated in numerous works on environmental medicine and environmental health. However, I must reiterate the time-honored dictum that physiology is fundamental to, and provides the rational basis for, clinical medicine.

An environmental system contains a set of elements that are of three general kinds—matter, energy, and force fields. For example, the usual natural environment of man contains gaseous, aqueous, and solid matter; chemical, mechanical, electric, and thermal energy; electromagnetic radiant energy; gravitational, electric, and magnetic force fields; and may be characterized in terms of pressure and temperature.

Biorhythmicity (Chapter 2) is the rhythmic oscillation of biologic processes as a function of time. Derived from, and relating the organism to, its fluctuating natural environment, it is an intrinsic property of all living organisms. As a fundamental biologic phenomenon, it is also basic to environmental physiology. Architects design environments for people to live in and legislators play with the clock, shifting time to suit their convenience. Are they ignorant of biorhythmicity and its essential geophysical cues? Did the ancient architects of Stonehenge, whoever they were, have better intuition? What is the biologic significance of "daylight saving" time? What is being saved and at what physiologic cost?

Biometeorology (Chapter 3) examines climate and its natural changes as an environmental influence. Man has been changing climate inadvertently for some time; however, weather control programs of the future will deliberately change certain aspects of the earth's climate. We need to understand the biologic significance of natural climate and its modification.

A wide spectrum of environmental systems are considered, ranging from the usual to the most extreme and hostile, and from the terrestrial to the extraterrestrial (Chapter 10). Also discussed are new natural, as well as new artificial, environments; and, extrapolating into the future, that tour de force of environmental physiology, artificial closed ecologic systems (Chapter 14). Certain contrived and experimental environments are considered for the light they cast on physiologic regulatory mechanisms (Chapter 12).

Some of the environmental stressors considered in this text are high or low levels, sometimes extremes, of a usual environmental element (Chapters 4 and 5). Magnetobiology (Chapter 8) discusses the biologic significance of our magnetosphere and the present state of our knowledge regarding the biologic effects of increased and decreased magnetic field strength. Hostile environments involve one or more unusual or extreme elements. Clearly, as man asserts and extends his inevitable mastery over environment, environmental physiology is, and will be, the essential bioscience.

In contrast to natural elements, environmental pollution concerns unique exposures to by-products of our twentieth century technology. This pressing contemporary problem is discussed in Chapter 13. Although, in a certain sense, noise is a form of environmental pollution, it is treated in Sound, Vibration, and Impact (Chapter 5).

The contributors have tried throughout to define terms and concepts where they first appear. A brief glossary of key terms is at the end of each chapter. Wherever possible information is presented within the framework of a relevant conceptual model. Feedback control concepts are used where applicable. We have attempted to identify, formulate, and propound the current relevant problems of environmental physiology and to describe the present state of the leading edge of the science. We intend this approach to impart a dynamic, realistic quality to the presentation and, hopefully, to inspire the reader to further inquiry, study, and, possibly, research. By design the chapters are self-sufficient and can thus be studied in any desired sequence. Selected references for additional study are listed at the end of each chapter.

In editing this volume I have tried to preserve the rich diversity of approach and opinion that derives from multiple authorship. I have attempted to integrate the separate topics, establishing continuity where possible, while eliminating unnecessary repetition. However, as in sewing a good seam, some overlapping of the subject matter of adjacent fields is inevitable and desirable. Coming from different sources, it enriches the text, examining an interface from both sides. Thus, some overlapping has been retained where it is the result of rounded presentations of adjacent areas by different contributors. This was not a simple task.

The biologic whole is clearly greater than the sum of its organic parts; and an environmental system contains a set of interacting elements. These facts recommend a holistic viewpoint. Just as we are concerned with all of the responses of the intact organism to a given environment, we are also concerned with the combined effects of the entire set of simultaneously acting elements of that environment. Admittedly, this is a goal for the future.

The developmental sequence of a science begins with static description in terms of entities, evolving later to a stage of thinking in terms of dynamics—processes and mechanisms. There are thus two stages of conceptualization—the early entity-thinking and the later process-thinking. The application of feedback control theory and biologic systems science to physiology are forward steps toward the latter stage. Physiology, traditionally an empiric-experimental science, awaits its theorists and model makers.

The human significance of environmental physiology ranges from the objective and realistic to the subjective and personal. We sense that environment has shaped us and seek, in turn, to understand, predict, and control it. There is thus a highly personal gratification involved in the study, mastery, and application of the subject matter of environmental physiology.

It is a pleasure to acknowledge the competent and dedicated editorial assistance of N. Karen Bender in preparation of the manuscript. Her assistance with the illustrations was also invaluable. I also acknowledge with gratitude the assistance of Laura M. Swigart, Diane S. Miyahara, Suzanna Hutcherson, and Constance E. Devlin in preparing the manuscript.

N. Balfour Slonim

CONTENTS

TABLES

ENVIRONMENTAL PHYSIOLOGY

1 INTRODUCTION

N. BALFOUR SLONIM

WHAT IS ENVIRONMENTAL PHYSIOLOGY?

Environmental physiology is the branch of the science of physiology concerned with the nature and mechanism of the physiologic responses of presently existing forms of life to environmental change. In the language of classic biology, an environmental change is a *stimulus* and the resulting change in an organism is a *response*. In the simplest case a changing environmental element, acting as a stressor, displaces the steady state balance of a biologic system within an organism.

As an experimental science, environmental physiology elucidates the homeostatic regulatory mechanisms that are the essence of all physiology. It derives further significance from its relationship to other fields of science and from its many applications. It is the bioscience basic to survival in, exploration of, and colonization of extreme and hostile environments.

As in other fields of science there are those who teach, some who do research, and bioengineers who apply basic knowledge in the solution of problems. Environmental physiology is essential for design of artificial closed ecologic systems. As the study of the organism-environment interface and interaction, environmental physiology is a vital link in the ecologic chain.

Interaction is the essence of the organism-environment relationship. A living organism always modifies its environment. Some changes, such as depletion, spoilation, or enrichment, are more obvious than others, such as thermal or force field effects. By interaction an organism buffers environmental change. Although a reciprocal aspect of the ecologic loop, study of the effect of organism on environment is largely outside the scope of this book.

Environmental physiology has an as yet undefined relationship to exobiology, the science of extraterrestrial life. In all probability strange forms of life inhabit unknown environments in yet undiscovered worlds. Would such life find our ordinary terrestrial environments extreme and hostile? One can imagine other (contraterrene?) physiologies evolving in and sustained by strange environments. What *are* the universal environmental constraints for the existence of life?

On the organization of this book

It is not a simple matter to select a set of topics for a textbook on the rapidly unfolding science of environmental physiology. For one thing, choosing always involves giving up alternatives. For another, it is difficult to draw the line, if indeed one exists, between the *physiologic* and *psychologic* responses to environmental change. Where should sensory deprivation, a topic usually relegated to the domain of psychology, be discussed? What is psychophysiology? Infant "failure to thrive" may reflect the *emotional quality* of environment. Is this a topic in environmental physiology?

As we learn more, the classic attempts to divide nature among various arbitrarily defined scientific compartments become less and less meaningful. Having served their initial purpose, the traditional artificial boundaries between these departmentalized disciplines blur and then disappear as the inevitable process of interdisciplinary integration proceeds.

The objective of this text is to describe, analyze,

discuss, and interpret the physiologic responses of presently existing animal species to a set of selected environmental stressors. Our emphasis is mammalian and, more precisely, human. We hope to communicate our fascination with environmental physiology to the student and to stimulate further inquiry. Our reader will need a working knowledge of the sciences basic to biology and a good grasp of the fundamental concepts of general physiology. We will refer often to the present state of biologic adaptation to environment and to the evolutionary processes that produced these adaptive mechanisms.

Biorhythmicity is a fundamental property of all terrestrial life. It is an environment-induced bio-oscillation that modulates all physiologic processes. We thus discuss it immediately, in Chapter 2. We next consider the effects of natural climate and climatic extremes. This is followed by presentation of the effects of various physical factors as single environmental stressors. We then discuss altitude, aerospace, marine, and hyperbaric environments. Chapters on the responses to carbon dioxide–containing atmospheres and environmental pollution reflect the current interest in these subjects. This is followed by a chapter on Artificial Closed Ecologic Systems, a futuristic tour de force of environmental physiology, and a final chapter entitled Perspective.

The basis of environmental physiology

Physiology is the science of biologic function. *Analytically,* both physiologic processes (function) and anatomic organization (structure) resolve into the basic biologic sciences of biochemistry, biophysics, and physical biochemistry. *Synthetically,* well-regulated, highly organized molecular events manifest themselves as physiologic processes and anatomic organization. At molecular level the apparent dichotomy between function and structure disappears as these become merely different aspects of the same thing. Pharmacologic tools are often used as molecular dissecting needles to unravel physiologic processes. All physiologic explanations involve the common elements of cell function and biologic control mechanisms that reduce, on analysis, to sequences of physiochemical events. Is it true that we still do not know the fundamental principles of biology that govern the function of a living organism? Is it true that physiology has had its Galileis and Newtons, but that its Einsteins, Plancks, and Heisenbergs are still to come?

Life appears to develop in defiance of the second law of thermodynamics. Biologic systems dam up and collect matter and energy, increasing body size in a positive feedback process that is like a prairie fire. A single organism is thus a relatively transient entity through which matter and energy flow, returning inevitably to the environment. Past environments have left their indelible stamp on both cell structure and cell function. Every living cell contains the record of a billion years of terrestrial evolution, each step of which is a new invention.

The capacity of a living organism for adaptation to changing environment is termed *self-organization.* This process of responsive adaptive change is set in motion by the interaction of organism with environment. Internal reorganization requires the expenditure of free energy that the organism must obtain from body stores or from absorption and metabolism of nutrient materials. The living organism gains information from its successes. Evolution of adaptive mechanisms involves trial and success (preferred by nature to error), random mutation, natural selection, investing gains in further mutation, gene recombination, and further selection. By virtue of adapted structure and function, life exploits environment; however, this specialization involves a loss of some degrees of freedom.

Fig. 1-1 represents the organism-environment interaction during the course of evolutionary time. As fluctuating environments operated on primitive forms of life, more complex and specialized forms evolved that were highly adapted to particular environments. The figure depicts arbitrary evolutionary stages in the development of organisms. One may study the relationship between an organism and its natural environment at any point in evolutionary time. It is important to distinguish between the effects of environmental stressors on presently existing forms of life and the evolutionary process of adaptation to environment. In this text we are concerned primarily with the former—events that occur during the lifespan of an individual organism. Studies of adaptation to natural environments discuss the continuous complex evolutionary interaction between organism and environment.

We are all subjectively familiar with the effects of brief or mild exposure to some of the environmental stressors discussed in this book. Such exposures include the dry heat of the sauna bath; the humid heat of the steam bath; the deafening, high-intensity noise of overamplified music; the hyperbaria of skin and scuba diving; the air pollu-

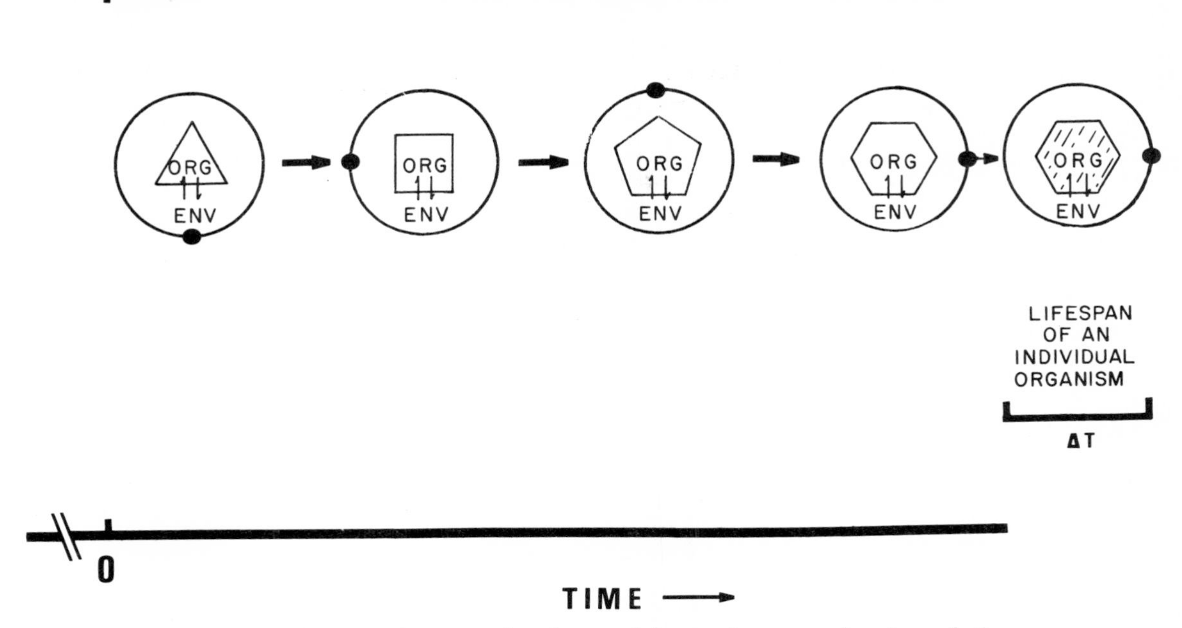

Fig. 1-1. The organism-environment interface and interaction as a function of time.

tion of a smoke-filled room or atmospheric inversion over an industrial city; the acceleration and near weightlessness of the trampoline, skydiving, or the amusement park ride; the electric vibrator; the altitude of mountaineering; the infrared or ultraviolet radiation of heat and suntan lamps; and the brief exposure to cold during an unexpected change in the weather. In this book we are concerned with the effects of relatively intense or prolonged exposures to environmental stressors.

As technology advances man discovers and creates new environments. Aggressive curiosity drives him to explore in every direction to the limit of tolerance and sometimes beyond. In this era, despite the misgivings of the less adventuresome, man will study and explore every accessible environment and eventually colonize strange and hostile environments. The success of these endeavors will require the skillful application of engineering technology and the science of environmental physiology.

In certain respects modern man is estranged from his own human nature. Perhaps he is overdomesticated. He worships comfort unrealistically. He may even attempt to deny the natural instinct to explore and to conquer that is implicit in all human history. Exploratory behavior, unlike specific adaptive behavior, is need-free and results in searching and learning. Such a drive may be involved in the human quest for knowledge and in research. Exploratory behavior is not only human but is also seen in other mammals and even in birds.

Although the drive to explore may be rationalized, it is not a rational formulation. As an aspect of inner reality, it needs no justification. It is a part of our primitive heritage that deserves acceptance. To the extent that it involves courage, it merits admiration. Exploration will bring important scientific and material benefits to the human race as it gratifies basic human instinct.

Population, pollution, and ecology

The spaceship Earth is a semiclosed natural ecologic system containing a finite fixed mass of matter but freely exchanging radiant energy with its interplanetary macroenvironment. Its overpopulation is surely one of the urgent contemporary problems of the human race. The current world population of 3.5 to 4 billion human beings, growing at the present annual rate of about 2%, will produce an estimated world population of 7 billion by the year 2000 and 14 billion by the year 2015.

As populations increase steadily and without

limit, environmental quality inevitably deteriorates. Overpopulation threatens to deplete the finite natural resources that constitute our biospheric life-support system. It threatens to outstrip the food supply, to defile the earth, to amplify noise, to spread violence, and to diminish further personal freedom.

The only possible permanent solution is to achieve an overall equilibrium between the degradative effects of the human population and the earth's resistive and recuperative capacity. Control measures must be designed to achieve and maintain population levels consistent with *optimal* environmental quality, not at the maximum possible number of mouths that the earth can feed nor at the maximum possible number of human beings that the biosphere can tolerate. What is the optimum population density? There is sufficient awareness of the need to curb the so-called population explosion but not yet enough knowledge of ways to enhance the earth's natural capacity to resist and recover from human assault. Overpopulation menaces mankind; education is our hope.

Our biosphere is a complex natural balance of ecologic systems and subsystems. This relatively thin life-supporting layer of our planet is now subject to the overconfident use of a wide variety of potent chemical agents. We are experimenting with our environment on a large scale. An organism has always been at the mercy of its environment; now, uniquely in the case of man, it is also vice versa.

Environmental pollution is objectively the addition of any undesirable, offensive, or harmful substance to an environment; or objectively, the production of any significant change of the natural composition, characteristics, or properties of an environment; or subjectively, any impairment of the quality of an environment. The cumulative effect of environmental pollutants, their interactions and amplifications, is damaging the complex fabric of our biosphere. By enhancing and favoring certain natural ecologic elements while depressing and retarding others, we are displacing ecologic equilibria. In extreme cases we are removing or destroying some natural ecologic elements while introducing others into unaccustomed environments. Our unprecedented power to transform the natural environment far exceeds our control and understanding. Indeed, the assault on the biosphere has reached a magnitude that rivals the scale of the natural processes themselves.

We urgently need more knowledge of the environmental impact of industrial technology and for that matter of all human activity. The ultimate effects of perturbations already set in motion will require decades to evaluate. If unchecked, environmental pollution could render this planet unfit for human life. Even now a biologist looking at contemporary man (through the smog) sees radiostrontium in his bones, radioiodine in his thyroid gland, DDT in his fat, carbon monoxide in his blood, and asbestos in his lungs.

Strictly speaking, overpopulation and noise both conform to the definition of environmental pollution. Noise is sound without value. Although such unwanted sound is technically pollution, we discuss noise as acoustic energy in Chapter 5, Sound, Vibration, and Impact. What are the acute and chronic physiologic effects of noise? How does noise affect human health? What are the effects of incessant background music?

The planet Earth is the only home for the human race for at least several centuries to come. Unfortunately, much of modern industrial technology is inimical to our natural environment. Can we invent technologies that are not? How will the dramatic confrontation between industrial technology and ecology evolve? Will long-range preventive environmental design supplant the present patchwork of retrospective pollution control? What will be the condition of the earth 50 years hence? Will new technologies with compatible artificial ecologic systems replace the present ones? In theory biospheric pollution can be totally eliminated, except for waste heat, the most likely sink for which is interplanetary space.

THE NATURE OF ENVIRONMENT

Environment (nurture) is often contrasted with heredity (nature) as an influence in the development of living things. The *environment of a living thing* may be defined as an arbitrarily limited region of space containing the matter, energy, and force fields that interact with the organism directly or indirectly at any level of organization—physiocochemical, biologic, or psychologic. The concept of environment thus includes a set of elements that may be thought of as parameters, factors, variables, stressors, conditions, and circumstances comprising the surroundings and affecting or influencing the development or function of a living thing.

An environment may be a closed, semiclosed, or open system or subsystem. Its matter may vary in nature, concentration, and distribution; its energy may be of different types, potentials, and capacities; and its force fields may vary in

Table 1-1. Classification and properties of the component elements of an environmental system

MATTER		**Gas**	**Liquid**	**Solid**
Concentration				
Distribution				
Density				
Electric properties				
Chemical properties				
Physical properties				
Thermal properties				
Conductivity				
Heat capacity				
Mechanical properties				
Viscosity				
Convective properties				
Fluid dynamic properties				
Water content				
ENERGIES	**Mechanical (static)**	**Thermal**	**Electric**	**Chemical**
Capacity factor	Volume	Entropy	Electric charge	Mass
Potential factor	Pressure	Temperature	Electric potential	Chemical potential
	Mechanical (dynamic)	**Acoustic**	**Vibrational**	**Impactive**
Amplitude				
Frequency				
Phase				
	Electromagnetic	**Ionizing radiation**		**Nonionizing radiation**
		Cosmic rays		Visible light
		Gamma rays		Infrared
		X-rays		Radiofrequency
		Ultraviolet		Microwave
Intensity				
Frequency				
FIELD FORCES	**Gravitation**	**Electric**		**Magnetic**
Strength				
Distribution				
ACCELERATIVE FORCES				
Strength				
Orientation				
OTHER ORGANISMS				
Number				
Distance				
Physical effects				
Chemical effects				
Emotions				
ENVIRONMENTAL POLLUTANTS				

kind, strength, and pattern of distribution. An environment may be *natural* or *artificial, hostile, favorable,* or *optimal,* a *native* environment (habitat) or an *unusual* one. We may be concerned with *immediate* or *remote* environment. What is an *ideal* environment?

Environmental variables

An environmental system consists of a set of physical elements. Any physical element may remain constant (parameter) or change (variable). Any element may affect an organism. Each element is describable in terms of its properties or characteristics. Biologic relevance depends on interaction. Table 1-1 is a catalog and classification of biologically relevant elements of an environmental system. Note that any element may vary as a function of time. This table provides a means for describing, analyzing, classifying, or quantifying an environmental system. Are there biologically irrelevant environmental variables?

Besides the average value of a given environmental variable, rate of change is often important. Thus description of environmental variables involves consideration of average value; rate, frequency, and amplitude of change; phase; and combined action with other stressors. Some environmental variables are continuous in a mathematical sense. In this text we consider the effects of environmental stressors operating singly.

The separate physiologic effects of two or more simultaneously acting stressors interact, in turn, within an organism. One may imagine that these interactions produce secondary changes of several different classes—additive, multiplicative, and so on. Analysis of the combined effects of two or more simultaneously acting environmental variables is much more complex; indeed, knowledge of such combined effects is still fragmentary.

Some environmental elements, such as ambient temperature, pressure, and humidity, are common to all of the individuals of an exposed group. Each organism of a group is a part of another's environment. However, in some respects environment, and certainly the environment-individual interface and interaction, is unique for each individual organism. Thus it is only meaningful to discuss *the* environment of an individual organism.

In the science of environmental physiology an observer becomes a part of the environment, perturbing other environmental elements and the organism being observed. This effect of an observer on the phenomenon being observed is a universal problem of empiricism.

We are already obtaining useful environmental data from the technology of remote sensing. In the future such technics will be used increasingly to collect data for monitoring familiar environments, as well as for study and analysis of remote and unusual environments.

What is environmental quality?

An environment has *qualitative* as well as *quantitative* aspects. These two aspects of environment may remain relatively constant or vary independently. The qualitative aspects are to be understood in terms of the particular sensory, psychologic, and physiologic reactions of a given organism. An organism samples, or perceives, certain elements of its environment by means of special sensory organs. Even the individuals of a given species differ with regard to sensory perception (for example, human color blindness and PTC nontasters). After detection, sensory information is processed and interpreted. Individual differences in the sequential processes of sensory perception, recognition, association, interpretation, and psychophysiologic reaction produce the unique response of a given individual organism to a given environment. Thus, whereas the quantitative aspects of environment are characteristic of the environment itself, the qualitative aspects depend upon the response of an organism. Although a subjective abstraction, the importance of environmental quality cannot be overestimated.

Where is the line between the physical aspects of environment and the psychologic ones? If such a line exists it lies within the sensing, perceiving, recognizing, interpreting organism; beauty is in the eye of the beholder and pain is in his cerebral cortex.

Perceived versus physical worlds of a species

As sensory physiologists investigate nonhuman biologic systems, we find that there are other perceived worlds than our own. These worlds are partially superimposed, or overlapping. Thus, although all organisms live in the same *physical* world, different species, and indeed individuals, live in different *perceived* worlds and cannot be understood only in terms of their existence in the former. In relating his perceived world to the physical world, man interprets primarily, if not solely, in terms of the models of the world constructed from information generated by his own particular sensory apparatus. The perceived world differs even more from the physical world as a result of the *interpretation* of sensory input.

The ideal environment

From a purely physiologic point of view, the ideal mammalian environment is probably inside the maternal womb. It is our most dependably protective environment. Here favorable environmental conditions are held remarkably constant. Protection against change and hostile physical forces is never more complete. The basic needs of respiration, circulation, nutrition, and excretion are virtually assured. Small wonder we wish to return. Unfortunately, this utopian existence is short-lived and terminates abruptly in the cataclysm of birth as we are expelled into a new and challenging environment of change and competition. The cataclysm of birth is a separation from this constancy and a sudden expulsion into an environment for which biologic evolution has prepared us rather well but an environment that involves uncertainty and the possibility of exposure to unusual and unaccustomed physical forces or chemical action. Indeed, change is the only constant feature of postnatal life. When the environment changes significantly an organism must either adapt or die.

Holistic view

Organism, environment, and their interaction should be viewed holistically. Functionally, a biologic whole is greater than the sum of its parts. At higher levels of organizational complexity, biologic systems manifest new functions that cannot be understood solely in terms of the component subsystems. The elements of an environmental system do not exist in a vacuum either; they interact with each other. Each element of a given environmental set exerts an effect that interacts within an organism with the effects of every other simultaneously acting environmental element, producing a unique homeostatic baseline. We can never completely isolate a living organism from its physical environment, nor can we move it without taking some of its environment with it.

It is thus difficult to isolate a single changing environmental element as a pure stressor; it is even more difficult to isolate the pure *effect* of a given environmental stressor on an organism that is in a steady state of dynamic interaction with a complex welter of environmental influences. To further complicate the problem, although the immediate environmental system of a given organism may contain the most potent stressors, a microenvironment is usually affected by a larger macroenvironment that extends to the sun and even beyond. Experiments in environmental physiology are attempts to isolate the effects of single environmental elements or stressors and are thus always a contrivance and an oversimplification.

BIOLOGIC SYSTEMS SCIENCE

A living biologic system is an open system in a steady state. Fig. 1-2 is a general model for the action of an environmental stressor on such a system. A changing environmental element, or stressor, displaces the existing steady state balance within a biologic system. A homeostatic regulatory mechanism, represented by a negative feedback control system, detects certain effects of the displacement and responds by changing the actual output. The success of this counterprocess of restoration or reparation varies from none to complete, depending on the nature and extent of the displacement. If restoration fails or is only partially successful, residual changes persist that usually constitute damage. If counterregulation succeeds in correcting the displacement fully, then the initial steady state balance is restored (transient stressor) or a new steady state balance is achieved (continuous stressor). Sometimes a counterreaction is biologically disadvantageous. Thus books have been written on the "wisdom" or "stupidity" of the body.

The essential feature of a *feedback control system* is the provision for a more or less continuous flow of information from the variable quantity that is controlled to the mechanism that controls it. This feature permits adjustment in accordance with the degree of error. A feedback system compensates for the effects of *disturbances* (displacements) and follows changes in the *control signal.* The controlled variable is the *actual output.* A *transducer* (detector or sensor) is located where it can measure the actual output of the controlled system, converting the measurement into a *feedback signal.* This signal is relayed to an *error detector* where it is compared with the control signal. The error detector computes the difference between control signal and feedback signal; this difference is termed the *error signal.*

Just as a fire can burn a thermostat, so displacements of the steady state balance within an organism may also directly affect any or all of the components of a homeostatic feedback control system. Imagine the interactions set in motion within an organism by two or more simultaneously acting environmental stressors. Biologic systems and subsystems contain hierarchies of superimposed feedback control systems. Imagine the interactions set in motion by two or more simultaneously acting stressors if each involves hierarchies of superimposed feedback control systems. At death an orga-

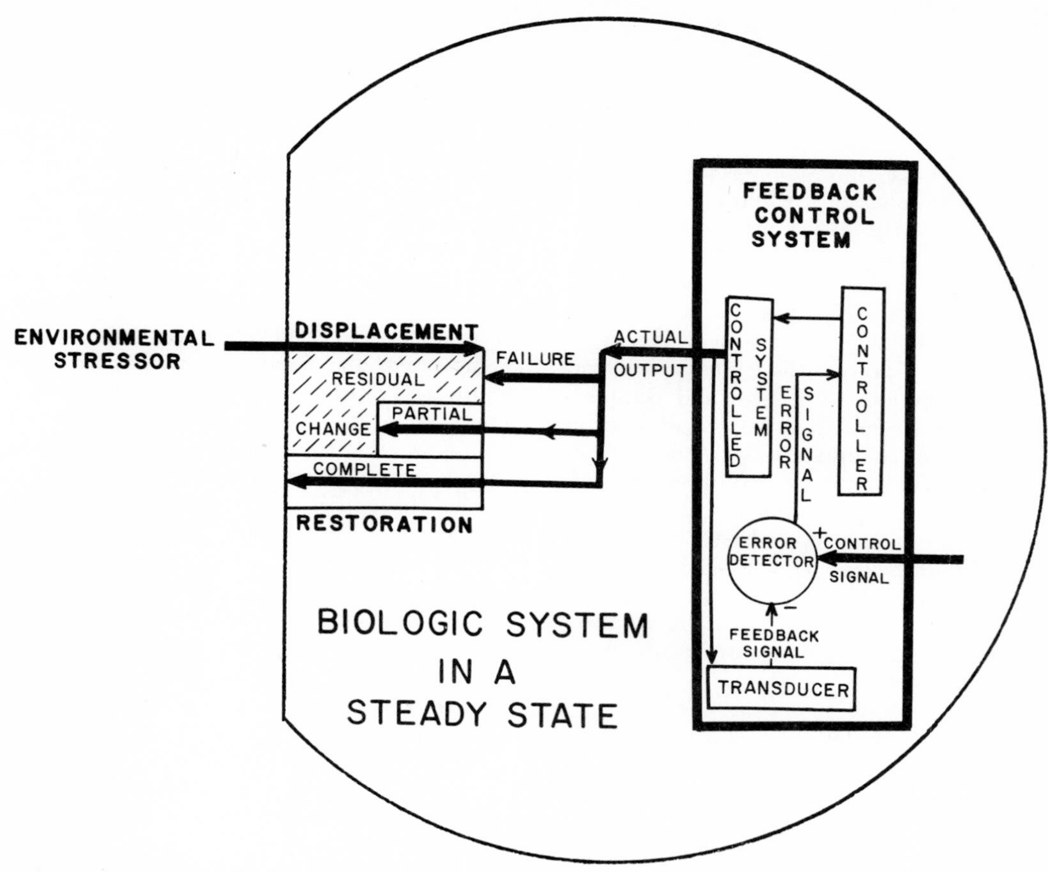

Fig. 1-2. Model for action of environmental stressor on a biologic system in a steady state.

nism approaches the condition of a closed system in an equilibrium state.

Law of initial value

The biologic scientist is often concerned with the stimulus-response relationship. This relationship involves the following set of elements: (1) an activity level of a physiologic process that fluctuates spontaneously about its average homeostatic value, varying between limits as a function of time, (2) a stimulus of either activity-increasing or activity-decreasing nature that may vary with respect to intensity and time of application, and (3) a response that may vary with respect to magnitude and direction of effect on activity. A response to a stimulus is a change from an initial activity value.

The *law of initial value* (LIV) is an empirical-statistical rule that interrelates the intensity of a stimulus, the initial activity level of the physiologic process, and the magnitude of the response. For activity-increasing stimuli of a given intensity, the higher the initial activity value at the instant of stimulation, the less the response. If the activity level has already traversed its intermediate range and reached a higher value prior to stimulation, then a stimulus may produce no response or even a paradoxic response (reversal of effect). The higher the initial value, the more likely a paradoxic response. For activity-decreasing stimuli of a given intensity, the higher the initial activity value at the instant of stimulation, the greater the response; the lower the initial activity value, the more likely a paradoxic response. The more active a physiologic process, the less likely an increase and the more likely a decrease of that activity. Thus initial value may be a more important determinant of response magnitude than is the strength of the stimulus. A biologic system appears to have energetic boundaries on both sides of its homeostatic average at which the maxima of excitation coincide with the minima of excitability. The study and application of the law of initial value in the biologic sciences is termed basimetry.

Is the law of initial value, obviously relevant for biology, simply an expression of probability, or does it reflect a more profound biologic truth? Do some biologic systems involve the principle of a differential computer, feeding back information about the magnitude of changes rather than absolute level of activity? How does this law help us to measure physiologic response? What is the equation for this law?

Like basimetry, biorhythmicity is presently more concerned with describing "what" than with explaining "how." How is the law of initial value related to biorhythmicity? A nonrhythmic stimulus—whether transient or continuous, constant or variable—may set a rhythmic response in motion.

Besides the traditional experimental approach to the science of biology, it is time to begin a second complementary approach that might be termed theoretic biology or biologic systems science. This second approach requires a general systems paradigm or model based on a set of biologic principles. An isomorphic model would be a logical physically based scientific system, suggesting relationships that must agree in form and function with experimental biologic information. To what extent is such an isomorphism possible in biology? The theoretic biologist must always keep in mind the real properties of the living system; in brief, he must remain aware of biologic reality.

2 BIORHYTHMICITY

FRANK A. BROWN, Jr.

A living organism is an open system in a steady state of dynamic equilibrium. Diverse regulatory mechanisms are continuously at work to maintain that mean physiologic state, which is essentially the condition that holds the organism as a viable entity differentiated from its external environment. Each and every environmental stressor that displaces the balanced regulatory systems of an organism from their mean steady state must be met by appropriate countermeasures. Every environmental change evokes an alteration in the character or strength of the reactions of the organism. Homeostasis is a state of relative constancy, a consequence of successful physiologic and behavioral counterreactions, to environmental change.

Continuous fluctuation or change characterizes both living organisms and the geophysical environment with which they steadily interact. A significant number of these fluctuations contain no regularly recurring patterns of change with time and are termed *aperiodic*. Examples of aperiodic variations are the irregular variations in temperature, light, humidity, wind velocity, and numerous other factors of the geophysical environment associated with the passage of weather systems. Similarly aperiodic are the biologic variations resulting from responses to them.

Aperiodic environmental challenges to organisms may also result from chance encounters with predators or prey or changes in the environment as a consequence of the organism's own movements through it. All of these must elicit adequately adaptive responses from living things. The only organisms that survive are those able to deal successfully with all such encounters, encounters whose general aperiodicity and hence substantial unpredictability precludes all but a generalized preparedness to handle them.

In contrast, a significant number of the fluctuations are *periodic;* these are relatively regular changes or patterns of change repeating at essentially constant intervals. Importantly periodic, for example, are the natural fluctuations of daylight, the ebb and flow of ocean tides, and the annual passage of the seasons. Similarly periodic are the multifold variations in a host of phenomena in plants, animals, and microorganisms as they adaptively regulate or adjust to these periodic environmental changes. Phenomena that recur with such regularity of period are termed *rhythmic*.

Because these variations are rhythmic, or periodic, they are predictable. The predictability of rhythmic variations is an advantage to an organism; it provides an opportunity to anticipate and prepare for specific kinds of environmental change that are to occur. The periodic environmental variations may have great biologic impact. An organism well adjusted for daytime activity may be unsuited for nocturnal activity; a marine animal adapted for swimming freely in the sea or for obtaining food and oxygen from sea water is endangered when left to dry by the ebbing tide. In the temperate zones, wintertime is difficult compared to the lush warmer months. The predictability of environmental variations allows an organism to apply a far broader range of its adaptive capacity than would otherwise be possible. It may be invaluable for the physiologic states related to an organism's means of successfully meeting changing environmental conditions to fluctuate with the same periods as these environmental variations. Rhythmically varying physiologic states can systematically improve an organism's ability to cope with difficult conditions and exploit the more hospitable ones to its advantage.

However, the capacity of an organism to predict environmental change as it interacts with its environment does not depend completely on the periodicity of a phenomenon. Aperiodic phenomena may also have limited predictability, which favors a degree of organismic preparedness for them. Predictability results from the almost invariable and often innumerable temporal correlates of environmental change. Prey may be forewarned of the arrival of a predator by odors, sounds, or sight, to any of which appropriate responses are systematically directed toward preparing the potential victim to cope with the encounter. Familiar to geophysicists, meteorologists, and even nonbiologists are many systematically changing, subtle, physical correlates preceding by minutes, hours, or even days weather factors of great immediate significance for organismic survival. An organism is not a physical system responding simply and solely to immediate environmental change; rather, it is the repository and integrator of a complex continuous informational inflow from which it has evolved the capacity to react adaptively, in terms of biologic survival. An organism possesses a temporal organization from which systematic environmental changes, periodic or aperiodic, can elicit significant adaptive physiologic and behavioral responses despite the biologic requirement of finite, and often relatively long, time intervals for their achievement.

In view of the foregoing considerations one must be cautious in interpreting the results of contrived experimental manipulation of single environmental factors. For example, the effect of abrupt changes of light or temperature cannot be equated with the effects produced by the natural gradual systematic changes of these environmental factors. The isolation of an organism and a selected environmental factor from the context of their natural temporal relationship and occurrence within a concurrently changing complex of correlated physical changes duplicates neither the natural condition nor the responses of the organism to natural changes of these same variables.

Periodic biologic variations fall into two rather distinct categories—those with and those without periodic geophysical correlates. Biorhythmic processes without geophysically correlated periods include components of numerous physiologic and behavioral phenomena. Such biorhythms have a very wide range of periods. High-frequency, or short-period, ones include insect wing beats with frequencies ranging from 20 to 2,000 Hz, electroencephalographic variations in brains ranging from 1 to 60 Hz, scratching, walking, and chewing rhythms of 1 to 8 Hz, cardiac frequencies of 20 to 1,000 cycles/min, and breathing frequencies of 4 to 250 cycles/min. Other examples are ciliary and flagellar beats, sound pulses of echo-locating bats, environment-exploring electric pulses of certain fish, electric potential variations in plants, rhythmic oscillations of oxidative metabolic processes in yeast cells, and variations of smooth muscle contraction producing rhythmic variations of blood pressure, intestinal peristalsis, and uterine contractions.

The frequencies of such rhythms without external geophysical correlates change readily to meet the requirements of an organism responding to the demands of its immediate environment. Cardiac and breathing rhythms, for example, change with organismic metabolic requirements, and locomotor rhythms respond to sensory stimuli by changes in character and frequency. The frequency of all these rhythms can also be readily changed by drugs and other chemical substances and, as with metabolic rate, change frequency as body temperature changes. We will not deal further with such rhythms in this chapter; their nature and roles are usually treated in relation to the various specific physiologic and behavioral phenomena of which they are integral parts.

The second category of biologic rhythms and the one that we will discuss in this chapter are those correlated with environmental periodicities such as the ocean tides, days, months, and years; they clearly relate organismic activities to the rhythmic fluctuation of their geophysical environment. These rhythms have special mechanisms, characteristics, and roles and are one of the most extraordinary biologic adaptations to the rhythmic terrestrial environment. These geophysically correlated rhythms have properties that clearly distinguish them from those without geophysical correlates. Unlike the latter, these rhythmic variations have relatively fixed periods, are remarkably resistant to frequency change, and have frequencies normally synchronized with known rhythmic changes of the geophysical environment.

There are two major kinds of biologic rhythms with external periodic correlates. One kind results from direct organismic responses to obvious rhythmic variation of the geophysical environment—for example, responses to light with its 24-hour periodic intensity variation, to monthly variations of moonlight, and to annual variation of the relative lengths of day and night. Also obvious are responses to the diurnal and annual temperature changes and to the rhythmic submergence and emergence of organisms at the seashore effected by the moon- and sun-regulated ocean tides. All these natural variations elicit parallel periodic alterations in many aspects of the physiology and behavior of animals, plants, and microorganisms. In addition to their responses to the most obvious geophysical rhythms, organisms also respond to such less obvious rhythms as environmental variations of humidity and evaporation rate.

Less well known are organismic periodicities arising in response to the weak ambient background radiation or other natural geoelectromagnetic parameters. Natural environmental force fields and certain electromagnetic radiation permeate and penetrate all our carefully controlled constant laboratory conditions; these environmental factors are known to have both aperiodically fluctuating components as well as rhythmic fluctuations at all the major natural geophysical frequencies. These same mean frequencies persist in organisms maintained under the most carefully controlled conditions of laboratory constancy. Such geophysical frequencies include the solar day (24.00 hours), sidereal day (23.95 hours), lunar tidal and lunar day (12.4 and 24.8 hours), lunar month (29.53 days), and solar year (365.25 days). These periodisms have been found in a wide variety of organisms by careful quantitative study of metabolic and activity rates. It has been found that the activity of an organism, other factors remaining equal, varies systematically with local time of day (hour angle of the sun), time of lunar day (hour angle of the moon), time of month (angle of elongation of the moon), and time of year (celestial longitude). A long study of a single potato species conducted under carefully controlled conditions revealed a systematic sidereal-day metabolic variation (period of rotation of the earth relative to the stars). All of these mean organismic periodisms are not only period-stable but also phase-stable; they apparently cannot be reset, or phase-shifted. Thus the mean values about which organismic activity varies fluctuate systematically with all the major geophysical periods. There is no evidence to suggest that these rhythmic changes are adaptive.

Numerous studies under so-called constant conditions relating both hour-by-hour and day-by-day fluctuations of metabolism or spontaneous organismic activity to detailed concurrent and supposedly random variations of meteorologic and other subtle geophysical factors indicate that the exogenous information that normally influences organismic activity continues to do so under these conditions. Extensive experimental studies to discover how such information pervades controlled artificial laboratory environments suggest that organisms are extremely sensitive and responsive to electromagnetic fields similar to the continuously fluctuating natural one. Responsiveness to the horizontal vectors of magnetic and electrostatic fields and to background radiation has been demonstrated. These environmental variables have periodic components with the same frequencies as those found in phase- and period-stable organismic periodicities; these environmental factors also fluctuate aperiodically. Such findings strongly suggest that the observed organismic periodisms result from continuing direct response to one or more of these physical factors. Day-by-day statistically significant correlations of aperiodic biologic fluctuations have been made with concurrent aperiodic fluctuations of background radiation, primary cosmic radiation, and terrestrial magnetism; these correlations suggest that geoelectromagnetic forces are responsible.

The exogenous origin of these organismic periodisms is further suggested by the fact that these geophysically dependent rhythms are present with either the same detailed rhythmic patterns or mirror images of them in all kinds of organisms, unlike those biologic rhythms not correlated with geoelectromagnetic factors that display wide interspecies differences and may be experimentally modified in form and phase. This suggests that all organisms respond to the same specific geophysical variables and that they have only the freedom to determine the sign, positive or negative, of their responses to them.

While evidence is strong that the period lengths of these observed mean periodisms having frequencies similar to those of the geophysical environment also depend on the geophysical environment, the details of the rhythmic pattern are obviously determined by the nature of the responding biologic system. The widespread similarity of response among very diverse plants and animals suggests that it is generated at a fundamental level of protoplasmic organization, yet a level that obviously still reflects the organism as a biochemical and metabolic continuum whose composition and reactivity at any given time are functionally related to past and future. Thus the organism is not completely passive in the generation of these exogenous metabolic periodisms. Indeed, it has a potential capacity for effective filtration of the stimulus, which is the complex fluctuating environmental field.

The second major category of biologic rhythms having geophysical correlates are the endogenous biologic clock–timed rhythms. It was demonstrated more than two centuries ago that plants display a daily sleep movement rhythm, raising their leaves in the daytime and lowering them at night; this is not simply a response to daily light variations. This rhythm persists when plants are shielded from light changes and kept at relatively constant temperature. Since then rhythmic daily fluctuations in a wide variety of living organisms, from unicellular forms to mammals and flowering plants, have also been shown to continue under conditions of constant light and temperature.

Rhythmic activity patterns of many littoral organisms associated with the ocean tidal periods persist after removal from their normal tidal environment. Similarly, semimonthly, monthly, and annual rhythmic variations of many widely different animals and plants continue even after the organisms are deprived of all obvious environmental cues of these longer time periods. A living organism seems to be able to measure closely all the major natural time periods of this planet without cues from rhythmic variations of obvious environmental factors such as light and temperature to which it is known to be responsive.

That organisms often, even when transferred to conditions of continuous darkness or light and unvarying temperature, repeat the rhythmic pattern they displayed during the last few 24-hour exposures to an artificial light-dark or to a natural environmental cycle indicates that an organism can not only definitely time the length of the cycle but even determine points and measure durations of relatively detailed intervals within single cycles. The remarkable stability of the period suggests that a highly specialized timing system is involved, not a physiologic system based upon metabolic rate. Just as any reliable timer, the biologic one is essentially unaffected by temperature change over a wide range or by any common drugs and chemical substances that alter metabolic rate. Clearly, living biologic systems behave as if they contain a unique physiologic timing mechanism. Thus far, biologic clocks have defied direct investigation; they have been studied only indirectly by observing the conventional biochemical, physiologic, and behavioral processes that they time.

The adaptive advantage to organisms of rigorously retained periodicities is obvious; it constitutes a dependable temporal frame of reference enabling plants and animals to regulate rhythmic variations of their processes for continuous maximal adaptation to their rhythmically fluctuating natural environment. Daily and lunar-tidal organismic cycles enable organisms to anticipate and prepare for favorable and unfavorable times of the day, tide, and year. Timed rhythms enable members of a species to synchronize their breeding readiness with one another. Solar-, lunar-, and sidereal-day biorhythms are clocks enabling organisms to utilize the sun, moon, and stars as celestial references for geographic orientation for navigation and homing. The changing seasons of the year are identified using the biologic solar-day timer to measure the changing relative lengths of daylight and darkness. This timing capacity, geared to the rhythmic variations of the geophysical environment, is a most useful mechanism.

The classification of other reported biologic rhythms with or without geophysical correlates is less clear. These involve several recurrences per day, several-day to several-week periods, and even several- to many-year ones. Many biologic rhythms of a few cycles per day have been reported. Examples include variations of activity of very young infants, depth of sleep of adults, liver glycogen content of chick embryos, spiraling of plant tendrils (circumnutation), and several body processes in patients with cancer and diseases of the liver or kidneys. The several-day recurrences of malarial fever and chills and the few-day to several-month reproductive rhythms of mammals are well known. Other low-frequency rhythms include few- to many-day periodic psychoses and edemas and the several- to many-year reproductive cycles of the cicada, fluctuations of animal population density, business activity, and the occurrence of wars.

As a consequence of their continuous fluctuations, both aperiodic and periodic, living creatures change continuously as a function of time. These changes are reflected at every organizational level—molecular, cell, tissue, organ, and organism. They are evident biochemically, physiologically, and behaviorally; they affect an organism's response to physical, chemical, and biotic environmental components.

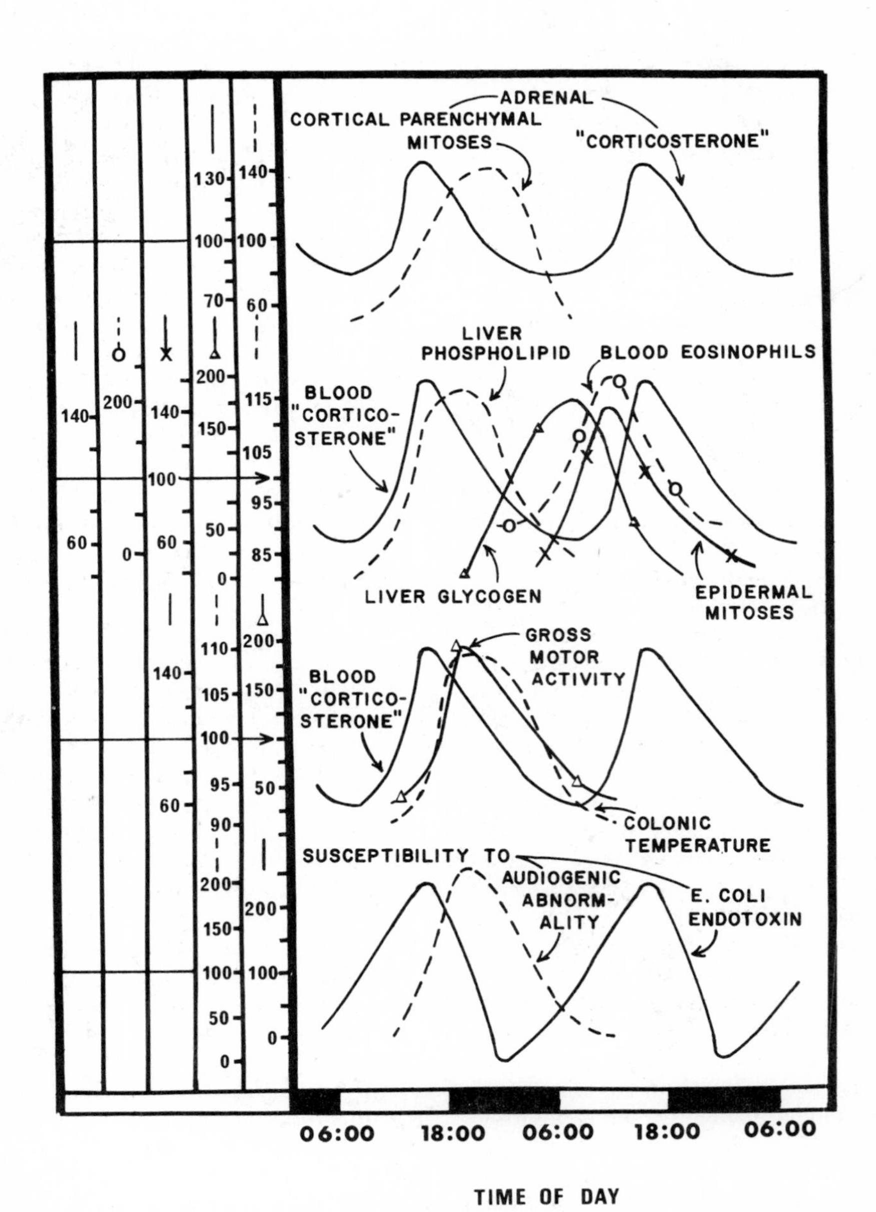

Fig. 2-1. Circadian variations, expressed as percent of mean values, in mice *(Mus musculus)* synchronized to a 24-hour light-dark cycle. The phase differences that contribute to a circadian phase-map are evident. (Modified from Halberg, F.: Physiological rhythms. In Hardy, J. D., editor: Physiological problems in space exploration, Springfield, Ill., 1964, Charles C Thomas, Publisher.)

SOLAR-DAY RHYTHMS

In nature 24-hour organismic rhythms are the rule; these have been variously termed circadian, diel, daily, diurnal, 24-hour, and nycthemeral rhythms. This rhythmicity includes behavioral and physiologic patterns of variation that functionally adapt or adjust each species to the day-night fluctuation of its environment. Life on earth has evolved toward maximal utilization of every habitable niche; specialized species have developed to survive and compete efficiently in virtually every kind of terrestrial environment. This specialization relates not only to habitation of almost every kind of site but even to specific times within a given site. Nocturnal organisms are adapted to the darker, cooler, and more humid hours; diurnal organisms are active only during the lighter, warmer, dryer daytime; whereas others, the crepuscular organisms, are active during the transitional twilight periods. Thus the environment is in use at all times, the biota working in "shifts." The "night shift" includes such animals as mice, bats, skunks, opossums, raccoons, owls, cockroaches, scorpions, and moths; the "day shift" includes man, songbirds, hawks, lizards, butterflies, and honeybees. Green plants turn their leaves toward sunlight by day as they photosynthesize foodstuffs, while emphasizing growth and assimilation at night. Plants also exhibit 24-hour rhythms in their capacity to synthesize chlorophyll and in the efficiency with which they use light for photosynthesis; these are maximal during the daylight hours.

Many plants, including the much-studied bean seedling, exhibit daily sleep movements, their leaves drooping at night and rising by day. Some plants blossom only at night, others only by day. Indeed, many plants blossom at such specific hours that the distinguished naturalist, Linnaeus, constructed a "flower clock." Man himself has a daily sleep-wakefulness rhythm. Thus each species displays its characteristic 24-hour activity pattern adapted to the day-night environmental cycle (Fig. 2-1).

Numerous chemical and physiologic processes that support the organism's circadian, or 24-hour, activity pattern also change rhythmically. It is safe to conclude that, at least in small measure, 24-hour periodicities are manifested in each and every aspect of organismic function. In man, for example, the rhythmic sleep-wakefulness cycle is accompanied by many significant changes. These include the activity of the nervous and endocrine systems, the liver, and the kidneys. Body temperature, cardiac rate, breathing frequency, blood pressure, and blood composition display 24-hour variations, as do cell division in various tissues and enzyme production and activity. Such rhythmically recurring phenomena are temporally interrelated, optimizing functional efficiency and collectively contributing to a characteristic 24-hour "phase map" of the individual. Twenty-four-hour rhythmicity thus provides a temporal coordination termed *endodiurnal* organization.

A striking example of these fundamental 24-hour variations in man and other mammals is shown by the changing susceptibility to toxins and to physical damage. For example, x-irradiation that would kill every individual if administered at a given time during the night will kill few or none if given at a certain time during the day. Doses of such potent poisons as *Escherichia coli* endotoxin or ouabain that are always lethal at one time of day are relatively harmless at another time. White noise of an intensity that induces audiogenic convulsions and a high mortality rate in mice at one time of day fails to do so at another. Similarly, the phlogistic effect of histamine, the intoxicant effect of ethanol, the hypoglycemic effect of insulin, and even the antipyretic and analgesic actions of aspirin vary systematically with the time of day. The resistance of insects to insecticides depends upon the time of application. The resistance of both insects and plants to thermal stress also exhibits a diurnal variation.

Many other biologic phenomena exhibit circadian rhythms. Daily variations in the selection of light versus dark environments are exhibited by fiddler crabs, canaries, and man. Eye pigment movement and eye sensitivity vary rhythmically in insects, crustaceans, and fish. The unicellular alga *Euglena* develops pyrenoids in light that disappear at night; the cristae of its mitochondria are completely disordered during the light period but well formed during darkness. Paramecia display a daily sex reversal. Birds, fish, amphibians, and reptiles exhibit a conspicuous daily variation in their response to the hormone prolactin, fattening in response to midday injections but failing to respond to those given in the early morning.

Extensive daily vertical migrations occur in both lake and ocean plankton, which rise at night and descend by day. This migration involves not only phytoplankton but also the zooplankton that feed upon them and larger members of the marine community such as shimp and fish.

The interdependence of daily rhythmic patterns of organisms within a community plays an important role in community integration. There are rhythmic interdependences of flowers and their pollinators, parasites and their hosts, and predators and their prey. Phase differences between the activity rhythms of the two sexes of a species contribute to sex-assembling and the timing of sex-specific behavior. Indeed, at different times of day there are different food chains in the same biotope.

The functional state of an organism at any given moment reflects two important kinds of influence. The first is the response to the complex of all the physical and biotic factors that directly impinge upon it. The second involves endogenously regulated, clock-timed rhythmic fluctuations of physiologic function and behavior. The organism's rhythmic response pattern to the recurring environmental variations of the past few days tends to repeat itself. Thus, other factors remaining equal, an organism tends to repeat today, hour by hour, what it did yesterday and the day before.

When an organism is placed in constant illumination or in darkness, all variations that were direct responses to light change disappear and, at constant temperature, all variations that were direct responses to temperature change also disappear. Removing an organism from natural environmental variations to such a controlled constant laboratory condition obviously eliminates some of the natural patterns of fluctuation. Observations under such experimental conditions are useful in attempting to differentiate the contributions of specific environmental variables from those of endogenous clock-dependent regulators and those of pervasive physical fields.

The properties of the biologic clocks that regulate solar-day rhythmic activities have been extensively studied. Intensive investigation of circadian rhythms in a wide spectrum of microorganisms, animals, and plants deprived of light and temperature variations have established a number of common properties. Evidence suggests that, except possibly for bacteria, the same kind of clock operates in every living organism. Even single-celled plants and animals, such as *Euglena, Paramecium,* and *Gonyaulax,* possess comparable clocks as accurate as those of the most highly differentiated multicellular organisms. Studies of photosynthesis by oxygen balance measurements in single individuals of the unicellular alga *Acetabularia* suggest that the cytoplasm and nucleus have separate clock-timed diurnal rhythms; rhythmic activity was observed in enucleated cells for as long as 40 days. Rhythmicity also persists in tissue culture preparations of biologic systems as different as carrots and mouse adrenal cortex. Clock rhythms appear indissociable from life itself.

When organisms are placed in an artificial environment of constant illumination and temperature, the rhythms subsequently observed do not usually display an exact mean 24-hour period but instead display regular periods that are either a little longer or a little shorter than 24 hours. In diurnal vertebrates the activity cycles usually occur a little earlier each day, whereas in nocturnal vertebrates they occur a little later. However, many exceptions to this pattern of change are known for vertebrates and there is no comparable generalization for organisms other than vertebrates. In all organisms observed, however, the periods of these rhythms rarely exceed the range of 20 to 28 hours and are usually within an hour of 24 hours. This commonly observed period deviation under these conditions led to the coining of the term "circadian" (*L. circa,* about; *diem,* day).

When organisms are returned to a 24-hour light-dark cyclic environment, they again adopt an accurate mean 24-hour rhythm within a few cycles that involves appropriate adjustments of the times of the organisms' activities to the light and dark phases. Organisms appear to use the light-to-darkness and darkness-to-light changes as the principal clues for adaptively adjusting their daily physiologic cycles. For example, a nocturnal mammal adjusts its time of activity to the period of darkness, a diurnal songbird, to the time of light; the times of light and darkness in a 24-hour rhythm evoke the appropriate response irrespective of their actual temporal relationship to local day and night.

A rhythmic variation of organismic susceptibility to light as a resetting factor facilitates the process of resetting the time of its activity phase within its daily pattern to the appropriate time of day. This daily variation is termed a *response curve.* A light change near the beginning of the activity period of the biologic cycle slightly advances the phase of the cycle to an earlier time of day, whereas a comparable light change near the end of the activity period delays the cycle slightly. Light changes at other times during the cycle have little or no phase-shifting effect. Although response curves have a common general form regardless of the kind of organism—whether alga, flowering plant, crab, bird, or flying squirrel—each individual organism has its own detailed modification of the circadian response curve that facilitates adjustment of its activity pattern to environmental light-dark cycles (Fig. 2-2).

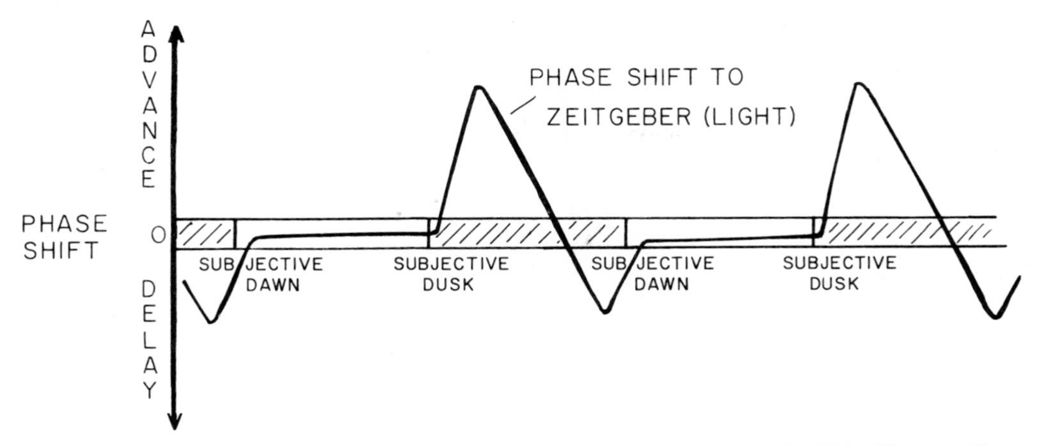

Fig. 2-2. Generalized diagram of a "response curve" of a nocturnal mammal. This diagram illustrates how phase shifts in response to an imposed light stimulus varying systematically with time of day. In a free-running rhythm phase varies with the animal's time of "nocturnal" activity, or subjective night.

Factors that reset, or phase-adjust, rhythmic patterns are variously termed daily clues, synchronizers, phase-setters, and Zeitgeber. Light is the dominant factor in the geophysical environment for adaptively adjusting the phases of clock-timed rhythms to the natural day-night cycles. Although less important than light, temperature is effective for many organisms. Many organisms adjust their activity patterns to 24-hour temperature rhythms under conditions of constant light; lower temperature is the equivalent of lower illumination, and higher temperature is the equivalent of more intense light. Temperature response curves are known to exist but are less well studied than those for light. Other phasing factors operating under certain conditions are sound, feeding time, social interaction, and electromagnetic field changes.

The circadian rhythms of organisms maintained under conditions of constant light and temperature have been the subject of much study. Under such constant conditions the observed rhythms, which usually deviate a little from 24 hours, are termed *free-running* rhythms. They reveal the period of the particular individual organism under the given constant experimental conditions, which include deprivation of the normal fluctuating 24-hour Zeitgeber. The length of the free-running period of an organism can usually be altered a little by changing ambient light intensity or temperature. For example, in nocturnal mice whose free-running period length in constant illumination is typically longer than 24 hours, the brighter the illumination, the longer the period. Reportedly, period length varies directly with the logarithm of ambient light intensity (Fig. 2-3). For diurnal finches or lizards, whose periods are typically shorter than 24 hours, the brighter the light, the shorter the period.

In many organisms the length of the free-running period can also be altered slightly by temperature change. In some organisms, for example, the algae *Oedogonium* and *Gonyaulax,* period length increases with temperature; for most organisms the relationship is the inverse. The changes of free-running period length produced by light intensity or temperature changes are about equal in magnitude and very small; the effect of a 10° C temperature increase seldom exceeds 10% (Q_{10} = 1.10) and is usually less. However, deprived of light, fiddler crabs, whose free-running period length is exactly 24.0 hours, maintain a constant period length despite temperature changes between 6° C and 26° C. The rates of ordinary biochemical and physiologic processes generally increase by a factor of 2 to 3 for each 10° C temperature increase (Q_{10} = 2 to 3).

On reexposure to 24-hour light-dark cycles, chaffinches that have longer free-running circadian periods tend to delay activity initiation somewhat relative to the time of light onset, whereas those with shorter free-running periods tend to commence activity earlier, anticipating light onset.

Under constant ambient conditions free-running rhythms may retain a relatively constant period length for days or weeks. Sometimes they spontaneously undergo relatively abrupt changes of period length; in other instances, period length changes gradually as time passes. Semimonthly and monthly variations of period length have

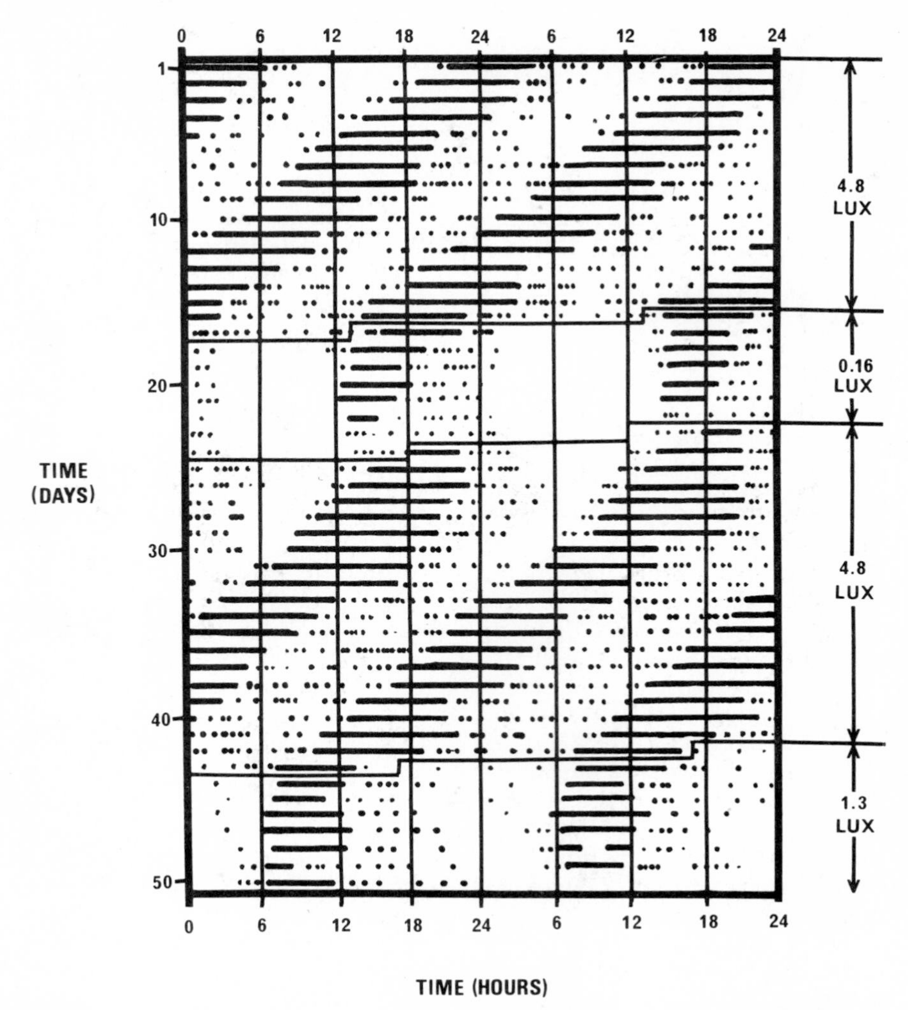

Fig. 2-3. Circadian patterns of activity in the chaffinch *(Fringilla coelebs)* under different intensities of constant illumination. This figure illustrates the relationship between light intensity and rhythmic period. (Modified from Aschoff, J.: Jap. J. Physiol. **16**:363-370, 1966.)

been observed. In a given individual organism the free-running period length may, to some degree, also reflect its particular previous illumination history. Circadian rhythms sometimes suddenly and inexplicably cease, even for several cycles, only to commence again exactly at the scheduled time as if something had maintained the rhythm during the silent period. It appears as if a given physiologic process within an organism may be sometimes coupled to, and at other times uncoupled from, a biologic clock. Free-running circadian rhythms sometimes fade out during the course of a few cycles; in other cases they continue indefinitely, sometimes even with increasing amplitude and regularity.

Many drugs, chemical substances, and metabolic inhibitors have been tested in attempts to alter the periods of free-running rhythms. Although rhythm amplitude is often depressed or even abolished, in most instances it is not possible to change the period length of a free-running rhythm. Some chemical substances alter the observed period for one or two cycles, permitting the rhythm then to return to its former period; this indicates that it is the rhythm *phase* rather than the rhythm *period* that is influenced. Powerful general metabolic inhibitors, such as cyanide, applied for many hours can abolish all overt rhythmic variations. However, upon removal of such a chemical substance the rhythm returns in its expected phase relationships as if some intrinsic mechanism had kept the rhythm. Near-freezing temperatures also inhibit all overt rhythmic changes in fiddler crabs and cockroaches, but with different consequences. If they are rewarmed after a few hours, the restored rhythms are phase-shifted by a time interval that is the approximate duration of the time exposed to low temperature; the rhythm resumes from the point at which temperature fell.

The fundamental nature of the circadian organization of living things is further emphasized by the fact that circadian periodicity does not depend upon previous exposure to environmental cycles of light or temperature, nor to the rhythms of any other agent known to affect or determine phase or period. Beans, fruitflies, chicks, and rats grown or reared from seeds or eggs under conditions of constant light and temperature spontaneously display circadian rhythms or may be induced to do so by a single brief light or temperature stimulus that itself conveys no information regarding period length. Furthermore, circadian rhythms may persist in a population of cells that are undergoing more than one division per day, in a population of fruitflies that are undergoing rapid growth and a series of molts, and even during the period when the insects are undergoing metamorphosis from the larval stage to the adult form.

Bean plants selected and interbred for longer or shorter free-running periods produce offspring with longer or shorter free-running periods, respectively. Thus the length of the free-running period is, in part, genetically determined. The influence of deuterium oxide (D_2O), or heavy water, on circadian rhythms is interesting; treatment of organisms as different as the unicellular alga *Euglena* and the mouse lengthened free-running circadian period. Heavy water also alters the phase relationships of the organismic activity pattern to imposed daily light-dark cycles.

Circadian rhythms have been studied after rapid eastward or westward translocation of organisms to other time zones. The most critical investigations of this type were performed with fiddler crabs and honeybees. No significant change of circadian period resulted from the translocation, despite the fact that it traversed several time zones. Fiddler crabs, transported by airplane from Massachusetts to California, arrived with their free-running color-change rhythm still synchronized with the light cycles of their former location. Shielded from the local natural light and temperature cycles, they retained their Massachusetts setting during 6 days of observation. To rephase their cycles to the new local time required a few days of exposure to local environmental phase-setting factors. This phenomenon is quite familiar in these days of rapid global travel; the physiologic disturbances that result from the inertia of our circadian organization and the process of gradual readjustment at the end of a trip are often called the "jet syndrome."

The multitude of biochemical and physiologic processes within an organism that manifest circadian rhythms do not all attain maximum or minimum activity values at the same time of day. In a healthy organism that is well adjusted to the daily environmental rhythms, the maxima of various physiologic processes occur in a certain temporal sequence and at certain intervals during the day; a description of this characteristic sequence is a *circadian phase map.* To a large extent the sequence and intervals reflect the temporal order of cause and effect and the interactions among the various intraorganismic regulatory processes. Phase maps reveal a transient disturbance while an organism is making a rapid phase readjustment as, for example, after a person moves quickly to a new geographic longitude. Under these circumstances the individual's many 24-hour circadian components that comprise the phase map do not readjust to the new local time at the same rate. On the contrary, during this transient state they become temporally dissociated from each other. Such dissociation of rhythmic circadian components is probably a major cause of the fatigue and decreased efficiency of the "jet syndrome," which commonly requires from 2 to 10 days to abate after a long trip such as one from Chicago to Rome.

Phase map dissociation occurs occasionally in organisms under constant conditions permitting free-running period lengths different from 24-hours. Different rhythmic components may simultaneously display different free-running frequencies. When reexposed to daily light-dark cycles, all components typically revert to 24-hour periodicity, gradually readopting their normal phase map relationships.

Circadian rhythmicity has been studied in animals and plants at the south geographic pole. Free-running circadian rhythms are observed that are not significantly different from those observed in the same species at other geographic latitudes. In the continuous daylight of arctic summer, wolves, foxes, wolverines, certain rodents, and man exhibit 24-hour rhythms. Arctic summer experiments in which men lived by watches that indicated spurious day lengths of 21 or 27 hours resulted in dissociation of some circadian components. Whereas many physiologic processes adopted the same period as the artificially timed 21- or 27-hour activity cycles, renal potassium excretion continued to display a 24-hour cycle.

A kind of dissociation of the circadian components of rats was produced by training the animals to feed by day instead of by night at their normal time. The 10% diurnal blood sugar concentration fluctuation that normally peaks during the night readjusted gradually during a 5- to 14-day interval to follow the feeding phase of the dissociated feeding and activity cycles.

Wild types of the mold *Neurospora* do not usually exhibit circadian rhythmicity. However, some wild types can be made rhythmic by appropriate chemical modification of the culture medium. Furthermore, certain *Neurospora* mutants, two of which were aptly named "timex" and "clock," do exhibit circadian rhythmicity of either growth or conidiation.

As previously described, under constant conditions the circadian photosynthetic rhythm of *Acetabularia* is extremely persistent. Two individuals of this unicellular alga were phase-set to light-dark cycles in opposite phase relationships. When one of them was enucleated and the nucleus of the other transplanted, the phase relationships of the transplanted nucleus gradually reset the rhythmic phase relationships of the host cytoplasm.

Studies of fiddler crabs reveal other interesting aspects of circadian rhythms. When groups of these crabs are kept together in constant darkness, they continue to display a large-amplitude diurnal skin-color rhythm. However, if the crabs are isolated from each other in separate containers, rhythm amplitude diminishes rapidly and the rhythm may even cease. Placing the crabs together again, even only two, rapidly restores the original high-amplitude rhythm. Although this phenomenon suggests that interindividual interactions affect the biologic rhythms of fiddler crabs, other experiments with these crabs indicate independence of the circadian rhythms of the members of a group; when placed together crabs that have different circadian phase-settings do not phase-synchronize each other. Such results illustrate how precarious it is to attempt to resolve the properties of the biologic clock by observing the rhythms timed by it. The nature of the obviously labile linkage between biologic clocks and clock-timed rhythmic phenomena remains largely unknown.

Another remarkable phenomenon is the persistence, in organisms deprived of light and temperature fluctuations, of circadian rhythms with precise 24-hour periods; in some cases, the active period is even set to the usual time of day in the absence of any obvious solar-day cues. The free-running periods of fiddler crabs, gila monsters, kangaroo rats, and mice display 24-hour precision. Chicks and lizards incubated under conditions of constancy with respect to all clues known to signal the time of day to an organism can adjust their circadian activity pattern to local time. Unidentified subtle period cues and phase-setting variables must exist. The naturally favored phase relationship between the environmental fluctuations that function as cues and the organismic circadian rhythm appears to be genetically determined.

Circadian patterns differ from species to species even under the same environmental conditions. Interactions of the nervous and endocrine systems and the body regions that they coordinate contribute to these pattern differences. Under ordinary circumstances, temporal maps of the recurring activity patterns are largely adaptations to fluctuating environmental demands as determined by previous experience. Changes in the functional state of the organism, or any part of it, modify cycle form; however, cycle period length, reflecting that of the geophysical environment, remains unchanged. The rhythms of blinded mice run free, their period length deviating from the 24-hour period length of normal mice kept in constant darkness. Light, operating through the visual system, is the essential synchronizer. The blinded mice ultimately regain 24-hour periodicity, probably by adoption of an alternative Zeitgeber.

The role of an organism's total functional complex in determining circadian cycle form is illustrated by comparing the changing daily respiratory patterns of unhatched chick embryos during early morphologic development with the changing activity patterns of human infants during the first few weeks of life. With temperature constant and continuous low illumination, chick embryos less than 6 days old show a daily pattern of variation with several peaks per day. By the seventh day, as the sensorimotor neural complex approaches full development, the daily pattern shows a single broad daytime peak with a single nocturnal minimum. During the first few weeks

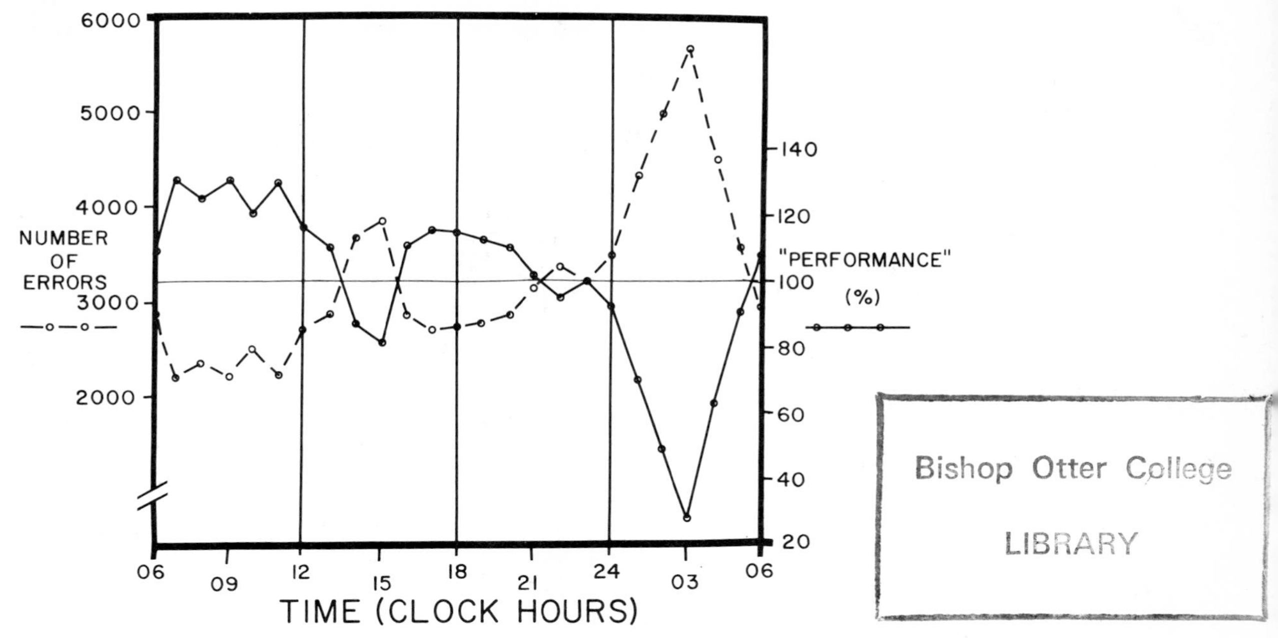

Fig. 2-4. The 24-hour variation in human performance. Data are from a study of 175,000 routine gas meter readings scattered throughout the clock-hours of the day. (Redrawn from Halberg, F.: Physiological rhythms. In Hardy, J. D., editor: Physiological problems in space exploration, Springfield, Ill., 1964, Charles C Thomas, Publisher.)

of life infant sleep-activity patterns exhibit several cycles per day; later the pattern simplifies to wakefulness by day and sleep by night, a change assumed to reflect adjustment to the schedule of household activity. In the embryonic chick, however, the change of pattern with time is genetically determined.

As mentioned earlier, species have evolved that are adapted to every niche of the environment at a specific time of day. Usually, species are rigorously held to their schedules by the activity of other species better adapted to the other times. However, certain large African game animals have been observed to alter their activity patterns from daytime to nocturnal as the population was decimated by man and then to revert to daytime activity as the population increased. Similar adaptive phase shifts may be more common in nature than we know. The circadian patterns of trout and sculpin in arctic Sweden exhibit seasonal variation. The sculpin is diurnal in winter and nocturnal in summer; trout undergo an opposite change.

Man has imposed some special rhythmic behavioral patterns upon himself. The 8-hour work schedule is an artificial activity cycle accurately timed to particular hours of the day. Evidence suggests that an 8-hour work period at one time of day is not the exact equivalent of that at another time. Circadian variations of human efficiency peak between 7:00 AM and 11:00 AM and decline to a minimum at 3:00 AM (Fig. 2-4). This variation and other daily rhythmic changes clearly underlie the socially imposed ones.

Geophysically dependent solar-day periodisms that continue under constant conditions were first characterized in detail for the potato *(Solanum tuberosum)*. The oxygen consumption of sprouting plugs cut from tubers was monitored continuously for 11 years in automatic recording respirometers under conditions of constant light, temperature, pressure, humidity, oxygen, and carbon dioxide. Even under these particular conditions potato metabolism varied substantially and systematically with time. Using 30-day blocks of hourly data to eliminate any lunar-day component, mean oxygen consumption rate is seen to vary systematically with solar-day hour; mean solar-day patterns vary in a systematic manner from month to month, displaying annual modulation of the mean daily cycle form.

The mean annual cycle form is the same year after year, retaining the same general amplitude, characteristic pattern, and phase relationship to local time of day. Every day there are definite maxima close to 7:00 AM, 12:00 noon, and 6:00 PM. The reproducibility of the specific pattern of the mean cycles and their phase relationship to solar-day hours while holding the potato under constant conditions with no known circadian component operative strongly suggests that the mean 24-hour pattern is exogenous. It appears that the observed rhythmicity is a direct response of the potato to fluctuating geophysical factors, the variations correlating with solar-day atmospheric tides. In Chicago these tides peak between 9:00 and 10:00 AM during all seasons of the year, dropping to a minimum in the afternoon; the time of this minimum changes regularly throughout the year. In June, the month with the longest daylight period, the lowest point occurs between 7:00 and 8:00 PM and in December, the month with the shortest daylight period, it occurs at about 2:00 PM. Despite constant ambient pressure, the metabolic rate of the potato during the 5:00 to 7:00 AM and 5:00 to 7:00 PM periods correlate independently day by day with the 2:00 to 6:00 AM and 2:00 to 6:00 PM mean rate of barometric pressure change, respectively. Although, on the average, barometric pressure rises between 2:00 and 6:00 AM, between 2:00 and 6:00 PM the pressure, on the average, falls in summer and rises in winter. Day-to-day differences in the rates of barometric pressure increase and decrease that correlate with potato metabolism result from large, systematic, but aperiodic rises and falls of barometric pressure as weather systems pass. Thus the irregularities in the metabolic patterns of the potato reflect the day-to-day weather-induced irregularities in the atmospheric solar-day tide. The periodic nature of the informational inflow to the organism is evident from the fact that *only* at these two times of day, 12 hours apart, are there correlations with rates of barometric pressure change.

Other indications of a continuing inflow of pervasive geophysical information for the potato is evident from other cross-correlations. The range of daily variation from midnight to noon correlates well day by day with the range of variation of background radiation for the same day, which itself exhibits a mean diurnal variation. Potato metabolic rate also correlates highly with an environmental variable from which it is fully shielded—mean daily temperature. Indeed, the annual modulation of mean solar-day patterns of the potato may be largely the result of these correlating factors together with the well-known annual variations of the solar-day tides and atmospheric temperature. The organism steadily acquires information, not only about the 24-hour geophysical period but also about the proportion of that period during which there is daylight.

Geophysically dependent fluctuations with the same detailed 24-hour patterns occur in the oxygen consumption of bean sprouts *(Phaseolus multiflorus),* mealworm larvae *(Tenebrio molitor),* carrot slices *(Daucus carota),* and 4- and 5-day-old chick embryos *(Gallus domesticus)* (Fig. 2-5). The same 24-hour pattern of variation occurs in the changing nature and degree of response of mice *(Mus musculus)* to a five-fold background radiation increase while being exposed to natural day-night illumination changes. The same pattern underlies a free-running 25¼-hour activity rhythm of the white rat *(Rattus norvegicus albinus).* As the circadian activity pattern moves systematically across the 24-hour day, the mean activity level for the various times of day varies systematically. Thus when the circadian system calls for activity at a given time of day, the activity level varies with the characteristic 24-hour pattern.

All geophysically dependent 24-hour mean periodicities do not exhibit the same pattern even though they appear to be responses to the same periodic geophysical variables. Unlike the potato, the 24-hour pattern of the metabolic rate of the fiddler crab *(Uca pugnax)* displays an early morning maximum and a late afternoon minimum (Fig. 2-6). Although crab metabolic rate, like that of the potato, correlates day by day with the 2:00 to 6:00 AM and 2:00 to 6:00 PM pressure changes, and at no other time, that of the crab correlates *positively* in the afternoon whereas that of the potato correlates *negatively.* For the crab, the more rapid the afternoon pressure decrease, the lower the metabolic rate that afternoon. For the potato, the more rapid the decrease, the higher the rise that afternoon. The factors that determine the sign of the response to the relevant bidaily environmental variables remains unknown. There is evidence for the crab that inversion of the daily light cycle reverses the sign of the afternoon correlation but not that for the morning, producing two metabolic peaks—morning and late afternoon. This is similar to the metabolic pattern of the potato. The correlation sign for potato metabolic rate in the morning undergoes an annual change; it is positive during the warmer months when the mean 2:00 to 6:00 PM pressure change is a decrease, but negative during the colder months

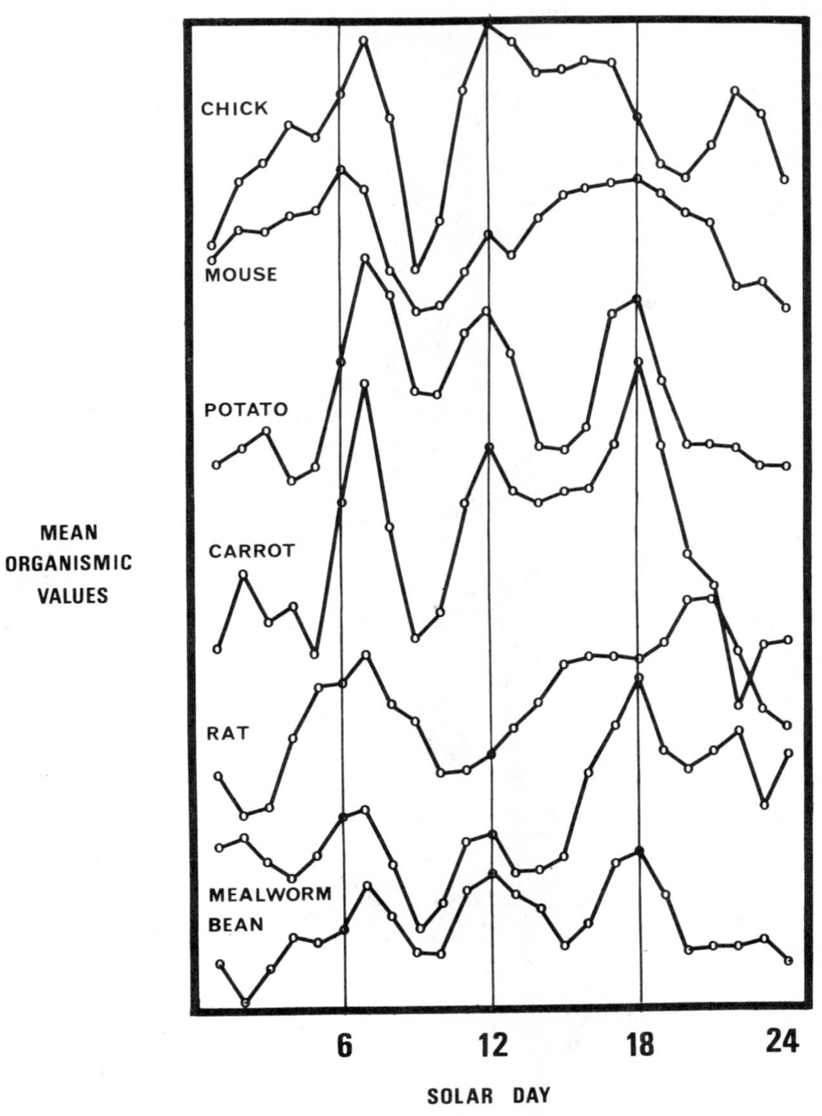

Fig. 2-5. Comparison of mean 24-hour (solar day) patterns of various physiologic processes in organisms held for extended periods in "constant" conditions. These include oxygen consumption in 4-day-old chick embryos *(Gallus domesticus)* in the spring, sprouting potato tubers *(Solanum tuberosum)*, carrot slices *(Daucus carota)*, mealworm larvae *(Tenebrio molitor)*, and germinated bean seeds *(Phaseolus multiflorus)*. Spontaneous rat *(Rattus norvegicus albinus)* activity coexistent with a 25¼-hour free-running rhythm and the response of mice *(Mus musculus)* to a five-fold increase in background radiation are included. (Curves constructed from data obtained from several sources.)

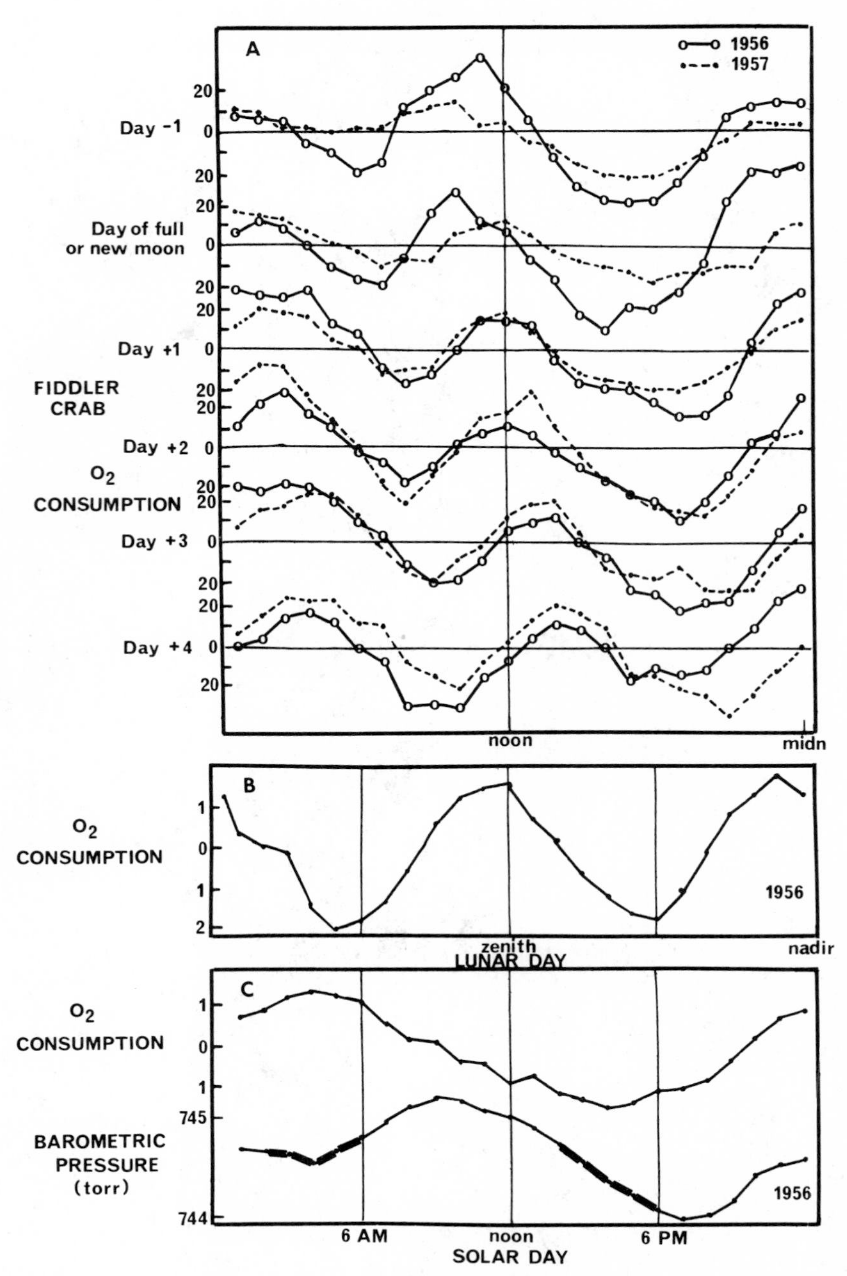

Fig. 2-6. A, Changing 24-hour patterns of oxygen consumption in fiddler crabs *(Uca pugnax)* under "constant" conditions during the summers of 1956 and 1957, as they relate to the times of new and full moon. These patterns result from the simultaneous presence of lunar day (**B**) and solar day (**C**) rhythmic components. **C** also illustrates the form of the solar-day tide of the atmosphere with thickened segments indicating the hours when cross-correlations occur between rates of pressure change and rates of crab respiration. (Modified from Brown, F. A., Jr.: Symp. Quant. Biol. Cold Spring Harbor 25:57-71, 1960.)

when the mean 2:00 to 6:00 PM pressure change is an increase.

Persistent mean 24-hour systematic variations of spontaneous activity also (1) occur in the blind cave crayfish, which exhibits no circadian rhythm, (2) occur in the house sparrow *(Passer domesticus)*, whose circadian activity pattern ceases after prolonged exposure to constant conditions, and (3) coexist with the free-running rhythm components of mice kept in constant low illumination. These patterns can all be explained in terms of two daily responses, 12 hours apart, to the same geophysical variables, each organism determining the correlation sign of its morning and afternoon response.

Correspondence of mean monthly 24-hour patterns of metabolism or of spontaneous activity of several species of animals and plants to the concurrent mean monthly 24-hour patterns of primary cosmic radiation intensity, either in parallel or in mirror image, during a time interval when radiation cycles were anomalous suggested that these organismic variations are responses to fluctuation of some aspect of geomagnetism. It it not possible that the organisms are responding directly to radiation intensity because primary cosmic radiation does not reach ground level; it terminates high in the atmosphere. However, variations in the intensity of this radiation reaching the earth's atmosphere are influenced by the geomagnetic field. The stronger the field, the less primary cosmic radiation arrives, and vice versa. An organism might appear to respond to this distant radiation when, in fact, it is actually responding to the fluctuating geomagnetic field. Simple direct experiments revealed an extraordinary specialized receptiveness for the very weak geoelectromagnetic forces; organisms distinguish strength and direction of very weak horizontal magnetic, electrostatic, and gamma radiation fields. Organismic response to all three of these vector variables changes with time of day.

The probable natural role of the geomagnetic field in biorhythmicity is further suggested by the fact that the activity of gerbils, maintained in light from 6:00 AM to 6:00 PM and in darkness from 6:00 PM to 6:00 AM, displaying an average 24-hour pattern with one activity peak at about 7:00 AM and another at about 6:00 PM, correlates highly on a day-by-day basis month after month with a geomagnetic component that has a highly regular 24-hour variation. Gerbil activity level between 3:00 and 6:00 PM correlates highly with the simultaneous rate of change of the horizontal geomagnetic vector strength. There is no relationship between gerbil activity level and the rate of change of horizontal geomagnetic vector strength at any other time of day or in any other temporal relationship.

Geophysically dependent 24-hour periodisms are found in widely diverse animals and plants and probably occur in all living organisms. Even bacteria, for which a typical pattern and phase-labile, temperature-independent circadian system has not been demonstrated, exhibit diurnal variation of reproductive rate. Correlation of the rate of bacterial cell division with barometric pressure changes under conditions of constant light and temperature suggest the existence of mean 24-hour and other geophysical periodisms.

The living biologic system exhibits a 24-hour periodic component whether it lacks an overt circadian component, whether it has a circadian component with a 24-hour period under natural conditions, or whether, under conditions of constant light and temperature, it is simultaneously exhibiting a free-running period deviating from 24 hours. It derives this periodism by direct response to subtle variations of the geophysical environment that permeate the "constancy" of the traditional experimental microenvironment.

LUNAR-DAY AND TIDAL RHYTHMS

The ebb and flow of ocean tides is another major environmental periodism. Tides subject seashore plants and animals to rhythmic changes, typically with two high and two low tides during each lunar day of about 24 hours and 50 minutes. Ocean tides are produced by rhythmic gravitational changes that result chiefly from the rotation of the earth relative to the moon. Solar gravitation also affects ocean tides, modifying or modulating the tidal rhythm. Solar and lunar influences are additive at the new and at the full moon, producing the greatest tidal ranges of each month, the spring tides; at the moon's quarters the sun and moon oppose each other maximally, producing the smallest tidal ranges, or neap tides. Thus seashore organisms that inhabit the highest and the lowest reaches of the intertidal region are subject to rhythmic brief tidal submergence or exposure once every lunar semimonth, or fortnight. One species of marine snail that inhabits the uppermost level of the beach has a parallel semimonthly activity rhythm.

The details of the patterns of tidal variation differ greatly for different seashores of the world. Most commonly, as for the eastern coast of the

United States, there are two nearly equal tides spaced almost equally within the lunar day. The two lunar-day tides may be very unequal as on sections of the Pacific coast, including a high-high, a low-high, a low-low, and a high-low tide during each lunar day, or a tidal rhythm may be a complex mixture of 24-hour and lunar-tidal components with recurring lunar semimonthly patterns of change characteristic for each coastal area, as in the Gulf of Mexico. Tidal range also varies greatly from one coast to another; for example, tides are more than 50 ft in the Bay of Fundy but nearly absent in the Mediterranean Sea. The direct response of intertidal organisms to the local tidal patterns produces biologic variations that have a correspondingly wide range of patterns.

Certain intertidal organisms, such as crabs, barnacles, clams, oysters, snails, and sea anemones, are most active while submerged at flood tide. Fiddler crabs and most shore birds feed on the beaches exposed at ebb tide. Certain single-celled algae, *Euglena,* diatoms, and the flatworm *Convoluta,* which contains symbiotic algae, migrate down into the sand and mud when the beach is submerged, returning to the illuminated surface for photosynthesis when the tide recedes. *Euglena,* diatoms, worms, and most shore birds are active only when beaches are exposed during the daylight hours. Certain sea anemones, for example *Actinia,* open only when submerged at night. Thus all these organisms simultaneously display both 24-hour and tidal rhythms whose phase relationship changes continuously, repeating a given phase relationship exactly once every lunar semimonth. Indeed, all intertidal organisms react continuously and simultaneously to both daily light and temperature changes and to tidal changes. For example, fiddler crabs simultaneously display an overt daily rhythm of skin color change adaptively adjusted to day and night and a lunar tidal locomotor activity rhythm set to the local tidal schedule.

As in the case of geophysical solar-day rhythms, lunar tidal rhythms are associated with corresponding biorhythms; these involve systematically varying biologic processes and organismic responsiveness to many environmental factors. Furthermore, any living creature that simultaneously exhibits tidal and solar-day biorhythms will also exhibit systematically changing patterns from one day to the next corresponding to the peaks and troughs of the tidal cycles that occur at progressively later times of the solar day. The particular pattern of variation for any given day is repeated 1 lunar semimonth later.

The biologic clock also times lunar tidal rhythms. Plants and animals whose activities on their native shores vary rhythmically with the tides continue to display the same tidal rhythms after removal from the seashore to the laboratory. The period length of these rhythms is about 12.4 hours or 24.8 hours for double tidal patterns. These persistent tidal patterns commonly show two nearly equal peaks during a lunar day, although they may have two quite unequal peaks, unequally spaced. Tidal rhythms may persist in constant light, in constant darkness, in natural day-night light cycles, or in artificial 24-hour light-dark cycles.

The period length of clock-timed tidal or lunar-day rhythms resembles that of circadian rhythms in exhibiting relatively great independence of temperature and drug effects. However, there are distinct differences between these rhythms and circadian ones, the chief of which is that tidal or lunar-day rhythms will not synchronize with a 24-hour light-dark rhythm. Exposed to 24-hour light-dark cycles, they continue to move across the day at the tidal rate. Failure to synchronize with light-dark cycles is obviously adaptive because on their native beaches the phase relationship of tides to solar day time changes systematically.

When one shifts the circadian cycle phase in an organism such as the fiddler crab by changing the time at which light is turned on and off, the tidal cycles shift in the same direction by the same number of hours. If one phase-shifts a circadian rhythm regularly, day after day, using a 24.8-hour light-dark cycle to convert the circadian rhythm to one with this period, the tidal rhythm is corresponding delayed an extra 0.8 hour each day, producing a 25.6-hour "tidal rhythm." Thus tidal cycles are coupled to circadian cycles while the latter are undergoing a phase-shift, but not while the circadian rhythm remains a 24-hour rhythm. This characteristic of tidal rhythm explains the fact that in some cases the precision of the tidal periods seems greater while exposed to a 24-hour light-dark cycle than while deprived of light fluctuation. Under conditions that allowed the 24-hour rhythms of fiddler crabs to run free and to have periods longer than 24 hours, the tidal rhythm also runs free with a period slightly longer than the natural 24.8-hour double tidal period.

Like circadian rhythms, tidal rhythms can be adaptively phase-shifted to correspond to the local tidal schedule of a particular beach. Fiddler crabs and mussels, transported to a different beach where the tides differed by several hours, gradually reset the phase of their tidal cycles within a few days to correspond with the tides of their new location. Thus tidal rhythms shifts phase relative to circadian rhythm. The environmental factor that determines phase remains unknown, although factors such as mechanical wave action, periodic drying, and periodic hypoxia have been suggested. However, mussels, which filter water at a maximum rate at high tide, continue this activity rhythm in synchrony with the tides after removal from tidal changes. They adjust to the times of the local high tide whether they are periodically exposed by ebbing tides, are living in water 30 ft below low tide level, or are attached to the underside of a float. Fiddler crabs time their tidal activity rhythm to the particular time in the tidal cycle that the burrows at their level of the beach are exposed. Crabs that inhabit higher levels of the beach, after removal to "constant" laboratory conditions, become active earlier than crabs that normally inhabit a lower level of the same beach.

The tidal biorhythms of fiddler crabs can also be changed to match a different environmental rhythm. If crabs are collected in the Gulf of Mexico and exposed for a few days to 24-hour tides with a single high tide each day, they continue to display a 24-hour tidal activity cycle after removal to the laboratory. Crabs translocated from the Gulf of Mexico to a seashore on Cape Cod, Massachusetts, which has two nearly equal tides during each lunar day, had a semidiurnal lunar-tidal pattern like that of the native crabs when examined in the laboratory about 2 weeks later. Thus tidal patterns are labile, adaptable, and not genetically rigid.

Clock-timed tidal biorhythms differ from circadian biorhythms in another way; in constant light and temperature and removed from their tidal environment, their phase tends to drift until they have a characteristic relationship to the actual time of lunar day. This phenomenon was first observed for the shell opening of oysters, which in constant light and temperature and removed from tides gradually reset the time of maximum opening to the time of upper and lower lunar transits. Similarly, fiddler crabs from two different beaches with different tidal time removed from ocean tides and isolated from each other in the laboratory gradually, during the course of about 1 week, synchronize their tidal biorhythms with maximum activity at the times of upper and lower lunar transits; it is as if they were substituting some subtle synchronizer associated with the lunar-day tides of the atmosphere for the missing tidal cues.

Lunar-day clocks do not contribute to the timing of tidal rhythms in all cases. Whereas in certain species, such as *Hantzia* (diatom), *Sesarma* (crab), the green crab, and the fiddler crab, both solar-day and lunar-day or tidal variational components continue to evoke rhythmic patterns of spontaneous activity, oxygen consumption, or color changes for long periods of time under "constant" environmental conditions, in other species such rhythmic patterns cease if they are deprived of light or tidal changes. In the alga *Euglena* removed from the tides, either in 24-hour light-dark cycles or in continuous light, only a 24-hour rhythm of migration to the surface persists. The usual tidal and 24-hour variations of activity cease when *Actinia,* the Mediterranean sea anemone, is deprived of tidal, light, and temperature variations. For *Actinia,* the geophysical tidal rhythms are not naturally coupled to a biologic tidal clock.

Geophysically dependent lunar-day variations also occur. Using 30-day blocks of hourly data we can determine the form of any mean biologic variation that is related to the hour-angle of the moon, or hour of the lunar day, regardless of coexisting mean solar-day rhythmic components, whether of geophysically dependent type or phase-labile, circadian type that is being held to a 24-hour period. To determine the form of the lunar-day pattern for organisms under conditions that produce a free-running rhythm whose periods deviate from 24 hours, one needs blocks of hourly data that are simple integral multiples of both lunar months and of the exact length of time required for the free-running rhythm to scan the 24-hour period.

Just as there are solar-day geophysically dependent periodisms, so also are there lunar-day rhythms that similarly fall into a group of temporally related patterns, having a given form, or its mirror image, for a whole day or fraction thereof.

The lunar-day variation of oxygen consumption of the fiddler crab has been extensively studied (Fig. 2-7). Crabs native to a beach where low tide occurs very soon after the times of upper and lower lunar transits continue, under "constant" laboratory conditions, to display a bimodal lunar-day variation with maxima at upper and lower lunar transit times. The lunar-day variations of the opening of the shells of oysters is similar; removed from the ocean tides and after a 2-week period in the laboratory during which rhythms stabilized, the oyster also synchronizes maximum valve opening with upper and lower lunar transit times. During a 9-month study,

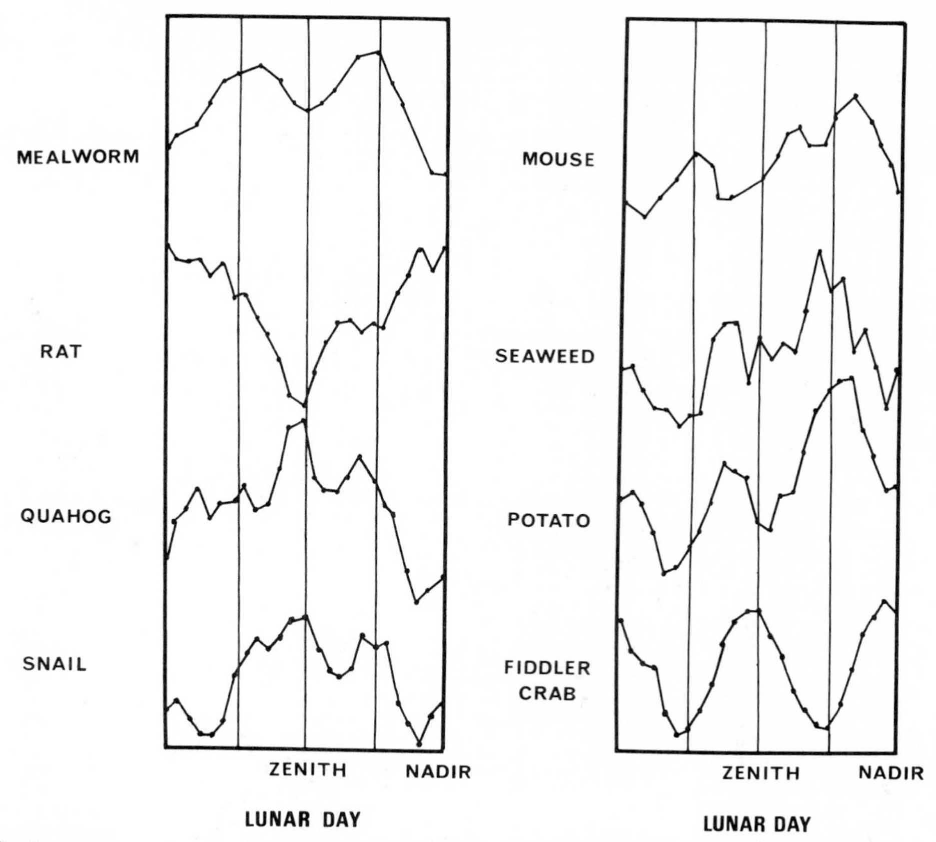

Fig. 2-7. Comparison of mean 24.8-hour (lunar-day) pattern of various physiologic processes in organisms held for extended periods of time in "constant" conditions. These include oxygen consumption in fiddler crabs *(Uca pugnax)*, seaweed *(Fucus vesiculosus)*, potatoes *(Solanum tuberosum)*, mealworms *(Tenebrio molitor)*, and snails *(Nassarius obsoletus)*; spontaneous shell opening in quahogs *(Venus mercenaria)*; and spontaneous running activity in mice *(Mus musculus)* and a rat *(Rattus norvegicus albinus)* in which there coexisted a large amplitude free-running circadian rhythm. (Curves constructed from data obtained from several sources.)

Tenebrio, the mealworm, displayed a lunar-day cycle that is a mirror image of the two preceding examples. In continuous dim illumination organisms as different as *Solanum tuberosum* the potato; salamander; *Lumbricus terrestris,* earthworm; carrot; and mouse have mean lunar-day patterns similar to each other. However, they differ from that of the fiddler crab and oyster in that the half cycle that begins at the time of the moon's upper transit is inverted.

Another type of lunar-day pattern occurs in mudsnails, quahogs, and black bass, whose lunar-day cycles are unimodal with maxima synchronous with those of crabs and oysters at the time of the upper lunar transit but, unlike the latter, their minima occur at the time of lower lunar transit. Still another type of pattern that underlies the free-running circadian rhythm of the white rat and gerbil is almost a mirror image of the unimodal clam and snail cycles, the minima occurring at upper lunar transit and the maxima at lower lunar transit.

Because geophysically dependent solar-day patterns differ from year to year and systematic changes of form occur during a several-year period, it is reasonable to ask whether lunar-day patterns display a similar change with time. The mean solar- and lunar-day patterns of change of many potentially effective geophysical variables also vary in diverse ways with time. It is thus probable that the simultaneously observed lunar-day patterns of different organisms correspond more closely to each other, directly or as mirror images. than do patterns observed at

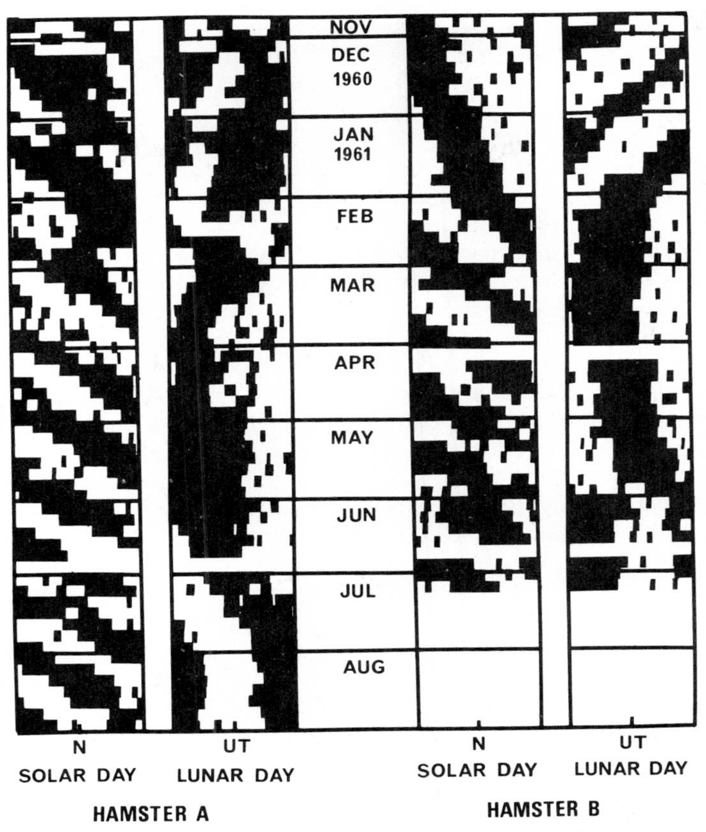

Fig. 2-8. Changing circadian frequencies of wheel-running of two male hamsters *(Mesocricetus auratus)* held for several months in constant dim illumination. Every hour containing animal activity is blacked. The data are plotted on two time coordinates—hours of solar day and hours of lunar day. The figure illustrates the common variabiltiy of circadian period in unchanging conditions and the propensity for adoption of a lunar period. (Modified from Brown, F. A., Jr.: Proc. Soc. Exp. Biol. Med. **120**:792-797, 1965.)

different times. Furthermore, lunar-day patterns probably reflect the longer cycle, systematically varying lunar declination to some degree.

How does an organism determine the sign of its response to the effective periodic lunar factors? The lunar-day patterns of spontaneous activity of two gerbil populations, one of which had undergone a phase-shift of its circadian pattern of motor activity of 180 degrees with respect to the other, concurrently displayed relatively large-amplitude lunar-day patterns that were mirror images of each other. The correlation coefficient for the spontaneous activity of the two gerbil populations during a lunar day was $r = -0.94$.

Geophysically dependent lunar-day fluctuations may be of biologic significance. As described earlier, fiddler crabs and oysters deprived of light, temperature, and tidal fluctuations will, if the tidal times on their home shores deviate from the times of upper and lower lunar transits by 3 or 4 hours, gradually change the times of their maximum activity until this activity synchronizes with the times of upper and lower lunar transits, after which phase and period stabilize. A similar phenomenon occurs with the phase-labile circadian patterns of the white rat and hamster kept in continuous low-intensity illumination. For 2 months the circadian pattern of the rat persistently exhibited not only a precise mean lunar-day frequency but also a large-amplitude circadian pattern that even paralleled in phase and form the lunar-day variation that coexisted with the free-running 25¼-hour circadian rhythm of another rat. The geophysically dependent lunar component can, under certain conditions, capture and even determine the details of a pattern that was originally a solar-day adaptation.

Another example of the lunar-day component entraining a typical circadian rhythm was observed in two hamsters kept for 16 months in continuous dim illumination (Fig. 2-8). The free-running circadian patterns, which varied from 24 to more than 26 hours, adopted lunar-day periodicities for time intervals of a month or more. For each hamster independently, the prolonged period of entrainment began when the initiation of the activity phase in its circadian cycle occurred at lower lunar transit and at moonset. The hamsters thus behaved as if an ambient lunar periodicity were being mistaken for a solar-day synchronizer.

Lunar-day entrainment of a component of human circadian rhythm has been reported. Maintained for a long time under conditions in which the sleep-wakefulness rhythm ran free with a period length of about 33 hours, the free-running period length for the body temperature rhythm dissociated and continued for the duration of the study with a 24.8-hour, or lunar-day, period length (Fig. 2-9).

SIDEREAL-DAY RHYTHMS

The existence of sidereal-day (23 hours and 56 minutes) biorhythms would be of considerable interest for two reasons. First it is difficult to credit the sidereal period with any adaptive significance unless it functions in conjunction with the 24-hour solar-day rhythm to produce annual beats and thus an annual modulation of the solar-day patterns. Second, there are sidereal-day variations of such factors as cosmic radiation caused by the combined effects of the earth's rotation and the intragalactic magnetic fields; this interaction contributes a very low-amplitude sidereal-day component to the ambient terrestrial radiation field. It is theoretically possible that a living organism may respond to this component.

Hourly data for the metabolic rate of potatoes continuously monitored during the years 1956 and 1957 while traditional environmental factors were held constant showed a systematic, sinusoidal variation of about 1% in oxygen consumption, with a maximum near the time of lower transit of the galactic center and a minimum near the time of the upper transit. Subsequent analysis of comparable data for the succeeding 8 years, 1958 to 1965, confirmed the existence of this small variation with respect to both phase and amplitude. The month-to-month, 2-hour shift of this geophysically dependent rhythmic component across the solar day is seen in the annual modulation of the solar-day, geophysically dependent rhythmic pattern of potato metabolism, a modulation that amounts to 4%.

MONTHLY AND SEMIMONTHLY RHYTHMS

The period from one new moon to the next, or synodic month, averages 29.53 days; it results from the changing phase relationship of solar to lunar days. It is thus the period of the beats between these two nearly equal rhythmic period lengths. The synodic month and semimonth are evident to organisms in the times of occurrence of high and low tides. Another obvious synodic monthly environmental rhythm is the amount of nighttime moonlight.

Among the biologic activities linked to the synodic month are the reproductive cycles of many different kinds of marine animals and plants. An example is the monthly reproductive rhythm of *Eunice viridis* (palolo worm) in the southwest Pacific. Despite the nearness to the equator of Samoa, where the reproductive swarming of these worms has been observed for more than a century, there is a rather precise annual, as well as monthly, rhythmic component. The worms swarm during 2 or 3 days at the moon's third quarter between midnight and dawn, but never before October 8 nor after November 23. Swarming is greatest and most certain at the third quarter when this lies between October 18 and November 17 This degree of precision with respect to both year and lunar quarter produces a 19-year metonic rhythm, the period between years when the moon's quarters occur on the same calendar dates.

Another example of highly precise timing of a synodic monthly reproductive rhythm is *Comanthus japonica,* a deepsea lily found near Japan. It liberates its gametes into the seawater once each year between September 28 and October 22 at about 3:00 PM on the day of one of the moon's quarters. In successive years it alternates between the moon's two quarters in successive 3-year periods, first-third-first, repeating this triplet series in 3-year periods to progressively earlier dates in October until late September is reached. The following year it jumps abruptly, occurring at the first quarter nearest the end of the annual limit of its breeding period to start the progression again. The result is an 18-year cycle that is essentially the period of regression of the moon's orbital plane.

The lunar breeding rhythm of the grunion,

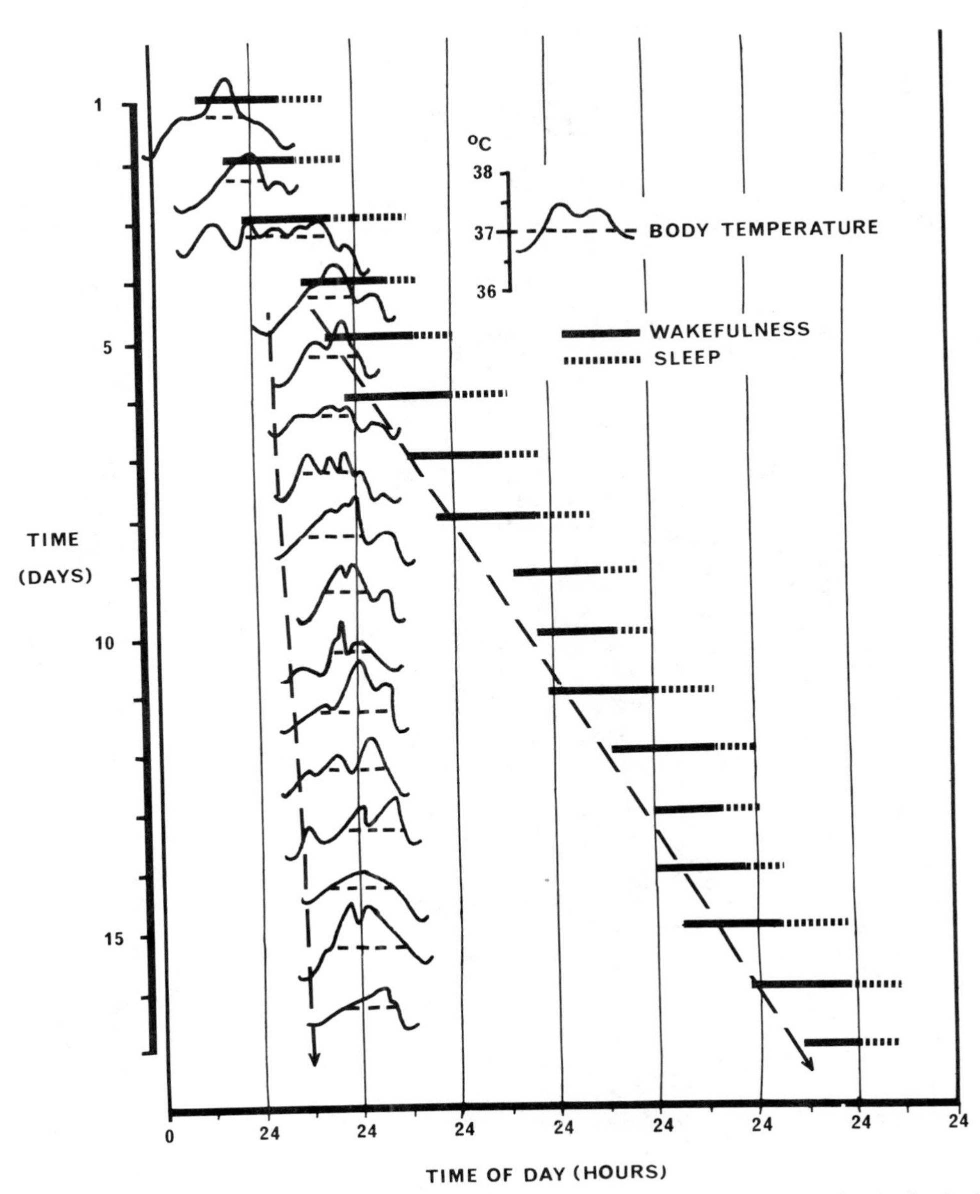

Fig. 2-9. Circadian rhythms of activity and body temperature in a human subject deprived of all time cues. The two rhythms dissociate, one adopting a 33.2-hour period, the other a 24.8-hour one. (Redrawn from Aschoff, J. In International space science symposium, life sciences and space research, ed. 7, Amsterdam, 1967, North-Holland Publishing Co.)

a small pelagic fish of the California coast, is timed to the monthly modulation of the height of the tides. The swarming of these fish to the sandy beaches occurs during the months of April through June within a few minutes of the highest nocturnal spring tides. Such precise timing ensures that the eggs are fertilized and buried in sand at a level of the beach where they will be left undisturbed by wave action until the next spring tides.

The brown alga *Dictyota* liberates gametes according to a lunar rhythm. The time of this activity during the lunar month varies from one location to another, the relationship to lunar phase apparently being determined by the times of occurrence in the day of the local high tides. *Odontosyllis* (Atlantic fireworm) swarms at Bermuda in its monthly breeding rhythm a few minutes after sunset on nights of the full moon. The hatching, or emergence, of *Clunio* (intertidal chironomid) has been extensively studied. On coasts with high tidal ranges there is a semimonthly rhythm synchronized to the times of the spring and neap tides. However, populations of these midges in the Norwegian arctic display only tidal rhythms. In regions where the tidal ranges are small a diurnal rhythmic component becomes relatively more prominent; under conditions where populations are continuously submerged, only a diurnal component is present. The rhythmic pattern is thus adaptable to the conditions of diverse habitats.

Other marine polychaetes display synodic monthly rhythms. An example is the reproduction of the Mediterranean species *Platynereis dumerilii*. Shell deposition of the clams *Mercenaria* and *Tridacna* display a monthly rhythm. Indeed, in comparing the number of days in the month as reflected in fossil clam shells from the Pennsylvanian, Cretaceous, Eocene, and Miocene to the present, a gradual shortening of the month during the course of geologic time is seen.

Synodic monthly biorhythms are known for the physiologic processes of many freshwater and terrestrial species. In the Lake Victoria region of Africa, mayflies swarm only within 5 days of the full moon, with a peak on the second night after the full moon. Mayflies have been observed to swarm simultaneously in areas more than 50 miles apart. Although *Bufo melanostictus* (Javan toad) ovulates and the toads are found in amplexus throughout the year, in almost all instances these activities occur between the times of the new and full moon; they are almost never noted during the waning of the moon.

The light-responsiveness of *Lebistes reticulatus* (guppy), *Dugesia* (planarian), and *Calandra* (beetle) vary with lunar phase, having a maximum at new moon and a minimum at full moon. Human perception of the relative brightness of spectral colors also varies with lunar phase. The tendency to postoperative hemorrhage is maximum at full moon and minimum at new moon. The monthly menstrual cycles of some primates commence at the new moon. Analysis of the time of more than one-half million human births indicates that significantly more births occur during the waning than the waxing of the moon.

The biologic clock times the synodic month of 29.53 days. Synodic monthly rhythms persist even after organisms are shielded from the more obvious environmental cues that provide information about the length of this natural period. The biologic clock that times this period resists alteration as do those that time the circadian and tidal periods. Indeed, semimonthly and monthly rhythms may be timed by interactions between 12.4-hour tidal rhythms and 24-hour ones, or between 24.8-hour lunar-day and 24-hour solar-day ones. Evidence for such a basis for monthly clock-timed rhythms is seen in the semimonthly color-change rhythm of fiddler crabs and the breeding activities of seaweed and of *Clunio* (midge). Each of these species may exhibit semimonthly rhythm phase differences between populations that are accounted for in terms of the differing phase relationships between their local tides and solar-day time.

In many instances synodic monthly biorhythms can be phase-shifted relative to the actual lunar phase. In addition to phase-setting by changes of some unidentified variable associated with the local tidal environment, moonlight variation has been suggested as a phase-setting factor. Three or four nights of low-intensity illumination in a sequence of otherwise 24-hour light-dark cycles shifts the phase of the semimonthly breeding rhythm of *Platynereis* (polychaete) and also the emergence rhythm of some, but not all, populations of *Clunio*. Light, operating in this manner to determine phase, is called a *photoperiodic Zeitgeber*.

It is difficult to establish, but impossible to exclude, 14.8-day and 29.5-day biologic clock-timing that does not depend upon the interaction of two cycles with shorter periods.

Although little is known about the mechanism that times the human menstrual cycle, three hypotheses are suggested to explain the approximately semimonthly estrus cycle of domesticated animals such as sheep, the 6-month cycle of dogs, and the 3-week cycle of cattle and pigs: (1) the cycles are determined by fully independent internal metabolic timers; (2) because most cycles are approximately simple multiples of a quarter-month, they depend upon periodic interference between 24-hour and lunar-day biologic clocks; and (3) particularly for the primate, the menstrual cycle depends upon monthly variations of moonlight.

Experimental evidence suggests that the human menstrual cycle is regulated by the moon. A study of nineteen women who had highly variable menstrual periods revealed that not only did supplementary artificial illumination of the bedroom from the fourteenth through the seventeenth nights after the onset of menstruation result in a substantial regularization of the periods but that the period length approached an average of 29.5 days, the length of the natural synodic month. That this natural period is a significant regulator for the human being is further suggested by the fact that the mean duration of human pregnancy, from ovulation to birth, is precisely 9 29.53-day months. One formula for prediction of the expected date of birth is to "count back 3 months from, and add 7 days to, the first day of the last menstrual period." By either means a gestation period of approximately 266 days is computed.

Variations of many body functions correlate with the human menstrual period; these include body temperature, blood sugar concentration, body weight, heart rate, blood cell counts, blood pH, and pain threshold. All these variations pass through maxima and minima at different times in such a characteristic temporal relationship to each other that phase maps for the menstrual rhythm, comparable to those for the circadian rhythm, can be made.

The week, with its 1 or 2 days of rest, reinforces a weekly rhythm, or periodism, that is evident in human activity. The week may be a natural period for man and perhaps also for other organisms; it may be related to the quarterly lunar phases as the nearest integral number of days.

Geophysically dependent mean synodic monthly biorhythms have been described. Synodic monthly variations occur in the running activity of rats and hamsters, in the spontaneous motor activity of crayfish, in the metabolic rate of potatoes, and in the light-responsiveness of many animals. They persist, locked to lunar phase, in organisms shielded from moonlight and the more obvious solar-day and tidal cues. Thus living organisms must respond to other, subtle, pervasive, geophysical rhythmic variations. A wide variety of subtle electromagnetic geophysical factors undergo synodic monthly variations, and, as discussed earlier, living organisms are very sensitive to such changes.

Despite the apparent lack of adaptive significance, different members or samples of populations of organisms as diverse as *Volvox,* mice, hamsters, crayfish, and planaria all display, concurrently and independently, mean synodic monthly variations. Even under "constant" conditions they often display the same relatively detailed pattern, related in the same manner to lunar phase. Widely different kinds of animals and plants studied under "constant" illumination and temperature exhibit concurrent cyclic patterns that correspond to one another either with a parallel or mirror-image relationship.

Under "constant" conditions lunar monthly biorhythms commonly display maxima and minima at the new and full moon, or maximum or minimum rates of change at these two lunar phases. That a lunar monthly biorhythm depends, at least to some extent, upon pervasive subtle geophysical variables is seen in the light response of *Dugesia* (flatworm). The strength of the negative phototactic response of these worms to a light field displays a definite monthly variation, with a maximum response at the new moon and a minimum at full moon. A 180-degree geographic rotation of the test field immediately resets the phase of the monthly rhythm by 180 degrees; the responses formerly characteristic for full moon and third quarter now occur at new moon and first quarter. That the resetting of the phase of the biorhythm is caused by the changed relationship of the light to the direction of the earth's horizontal magnetic vector is shown by the fact that reversing the magnetic field relative to the test field with a weak horizontal bar magnet also produces a 180-degree phase-shift.

The very existence of solar-day and lunar-day geophysically dependent biorhythms assures the existence of a mean synodic monthly pattern of biologic variation. A bimodal lunar-day and a unimodal solar-day variation produces a semimonthly pattern. The existence of a potentially inverting bimodal lunar day provides a quarter-monthly pattern, just as would a 24-hour pattern with two maxima separated by 6 hours. All these types of geophysically dependent biorhythmic variations exist. Clearly, many fractional periods of a month can be synthesized by periodic interference between uniphasic or polyphasic solar-day and lunar-day patterns derived by the organism from its

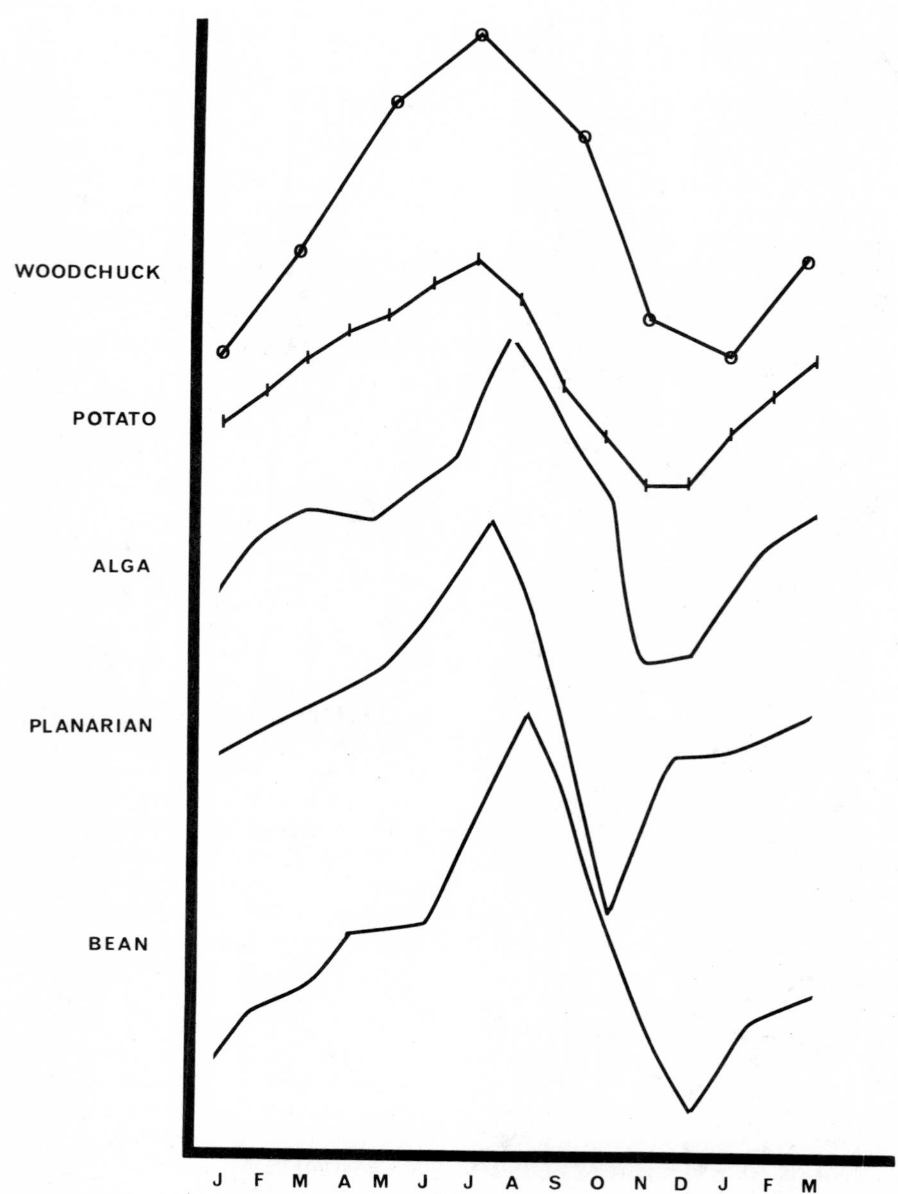

Fig. 2-10. Comparison of annual patterns of diverse physiologic processes in widely different kinds of organisms held for more than a year in "constant" conditions. These include oxygen consumption of potatoes *(Solanum tuberosum)* and beans *(Phaseolus multiflorus)*, nitrate-reducing capacity of algae *(Ankistrodesmus* sp.), food intake of woodchucks *(Marmota monax)*, and orientational responses of planaria *(Dugesia dorotocephala)* to a horizontal gamma radiation beam. (Curves constructed from data obtained from several sources.)

rhythmically fluctuating geophysical environment.

ANNUAL RHYTHMS

The year is another major biorhythmic temporal period. The substantial seasonal light and temperature changes that occur over most of the world correlate with rhythmic biologic phenomena such as activity, growth, and reproduction. In the temperate zones, the growth, flowering, and seed production of higher plants typically occur during the warmer months. The blooming of plankton in the oceans and fresh waters displays an annual rhythm. Animal reproductive rhythms are typically timed for the season of year that is optimal for the birth and rearing of young. The annual biologic cycle may include hibernation during the coldest months of the year or aestivation during hot summer droughts. The superficially observable systematic annual activity cycles of plants and animals are associated with many less obvious physiologic changes that have maxima and minima at particular seasons of the year with sufficient regularity that annual phase maps can be drawn.

Man exhibits many seasonal variations. These include birth rate, birth weight, and child growth rate. There are rhythmic annual variations of body temperature, heart rate, basal metabolic rate, and blood cell counts as well as annual variation in the occurrence of many noninfectious diseases of the skin and nervous, circulatory, respiratory, and gastrointestinal systems. Even mortality shows annual variations.

Many infectious diseases of man exhibit a peak incidence at characteristic times of the year. Even diseases without characteristic seasonal peaks of incidence probably have an annual rhythm with regard to the occurrence of complications and mortality. Test mice show an annual variation in susceptibility to routine pharmaceutical assay.

Clock-timed annual rhythms have been reported. Under constant light and temperature the physiologic processes of certain organisms display persistent annual rhythms (Fig. 2-10). Under such conditions *Ankistrodesmus* and *Chlorella* (algae) and the laboratory rat display annual variations of enzyme activity; seeds display annual variations of capacity to sprout; germinating plants exhibit annual variations of metabolic rate; certain birds show persistent annual, or approximately annual, reproductive cycles; and the food intake of woodchucks varies systematically with the time of year. The biologic clock-timing of the 1-year period, as that of the other periods, is independent of temperature and uninfluenced by chemically induced metabolic rate changes.

Although both pattern and phase of many annual biorhythmic phenomena can be experimentally altered, little is known about the phase lability of the clock-timed annual biorhythms. The annual time of occurrence of many plant and animal activities can be changed by changing the relative lengths of daylight and dark. This fact suggests that the phase of clock-timed annual rhythms can be shifted by photoperiod changes in a manner analogous to the phase-resetting of clock-timed circadian rhythms by changing the time of light onset and termination. However, the annual biorhythms of the woodchuck, which have been studied for several years under "constant" conditions, cannot be phase-shifted by manipulation of either light or temperature. The existence of a true phase- and pattern-labile annual biorhythm has not been established. To do so would involve not only the demonstration of environmentally induced alterations of the annual pattern but also the discovery of evidence that the free-running rhythm thus induced recurs. The interesting possibility of phase-shifting human annual rhythms by manipulation of environmental variables such as photoperiod and temperature is insufficiently studied.

There are good reasons for thinking that living organisms display an annual variation in direct response to annual variations of pervasive geophysical factors. The amplitude of the annual fluctuations of the geophysical variations responsible for the geophysically dependent 24-hour biorhythms is far greater. Furthermore, it is difficult to attribute the observed precision of certain annual biorhythms to anything else. Dried seeds stored under a variety of conditions that greatly affect their biologic processes continue to exhibit an annual variation of germination rate; the 1-year period persists whether the seeds are stored under "constant" conditions at –22° C or at +45° C. There are other examples of the persistence of annual biorhythms under "constant" conditions and of the maintenance of the 1-year period under a wide variety of environmental conditions. The oxygen consumption of sprouted bean seeds continues to show an annual rhythm even after the seeds have been stored in the dark at constant temperature for as long as 7 years.

Also suggesting the exogenous origin of many annual bioperiodisms is the remarkable similarity of the annual pattern and phase relationships for widely different kinds of biologic phenomena. The meteorologic variables with which organisms show day-by-day correlations, particularly mean daily temperature and mean rate and direction of barometric pressure change between 2:00 and 6:00 PM, exhibit definite annual patterns of variation themselves. The effectiveness of a favorable photoperiod for inducing gonadal development in the lizard varies systematically with the time of year; the cycle form of this relationship strikingly resembles what appears from the generality of its occurrence to be a characteristic environmentally dependent pattern.

Phase stability is implied by the fact that the annual biorhythms of samples of a population that are isolated and held under "constant" conditions for several years display synchronous annual maxima and minima when later observed.

For the potato, the only species for which adequate data are available to demonstrate the simultaneous presence of both solar-day and sidereal-day biorhythms, it is evident that periodic interference between these two nearly equal periods accurately defines the year.

OTHER NATURAL-PERIOD RHYTHMS

There are many natural geophysical periodicities besides those that we have discussed in this chapter. One with a very short period is the 10-Hz electromagnetic oscillation that arises in the sun. This solar field resembles the 10-Hz alpha rhythm of the human brain in frequency and also displays a diurnal variation. It is claimed that this 10-Hz rhythm decreases the dissociation of the components of free-running human circadian rhythm.

The best-known periods longer than the year are the 11- and 22-year sunspot cycles associated with variations of climate, rainfall, temperature, and light. There are also the 18.6-year cycle of lunar nutation and the 19-year metonic cycle. Longer periods are the 26,000-year precession of the earth's rotational axis and the even longer rhythmic period of geomagnetic field reversal. Many periods relate to the movements of the other planets of the solar system and interactions among them, the sun, the moon, and the earth. Periodisms arise in the galaxy beyond our solar system. The longest period that has been suggested is the approximately quarter-billion-year period of rotation of our galaxy; it is suggested that this period is correlated with our terrestrial glacial periods.

Numerous several- to many-year biologic periodicities involving a wide variety of phenomena are known. Examples include the 3- to 4-year cycles of lemming migration and the 9- to 10-year cycles of abundance of the Canadian lynx, the varying hare, and the ruffed grouse. Still other periods are found by study of tree rings, business cycles, and the dates of wars. Cycles of about 9, 11, 18 to 19, and 22 years are common. Although there is little reason to doubt that many such long-cycle biologic periodicities exist, the length of their periods, their irregularities, and insufficient data complicate the precise determination of period length and the causal relationship of the periodicities to the fluctuating geophysical environment.

BIOLOGIC RHYTHMS AS CLOCKS

Solar- and lunar-day biorhythms, and possibly also sidereal-day biorhythms, may be involved in the phenomenon of celestial navigation by animals. Birds, fish, turtles, spiders, insects, and crustaceans navigate, or home, using the azimuthal angle of the sun or moon. Birds even use constellations for geographic directional guidance. Navigation by such a sun-compass, moon-compass, or star-compass requires that an animal systematically change its orientational angle at a rate just adequate to compensate for the earth's rotation relative to each of these celestial references; thus, solar-day, lunar-day, and sidereal-day chronometers are required.

There is evidence that the chronometer used in sun-compass navigation is a biologic clock-timed rhythm that is phase-labile; it can be experimentally phase-shifted by changing the times of light and darkness and it can shift spontaneously under conditions favorable for free-running circadian rhythms. Such circadian clocks, which become reset to times other than the natural local sun time, cause the animal to misinterpret geographic direction and hence to err predictably from the direction in which they were trained or otherwise oriented to go.

The biologic clock-timed circadian clock presumably provides for nonlinear variation of the rate of change of the angle of the animal body relative to the sun and to geographic direction. Nonlinearity is necessitated by the fact that the solar azimuthal angle changes relatively slowly at certain times of day and more rapidly at other times, as the sun appears to move from east to west. Furthermore, the rate of angle change varies with terrestrial latitude and time of year.

Circadian clocks also time the relative lengths of daylight and darkness, a measurement used in

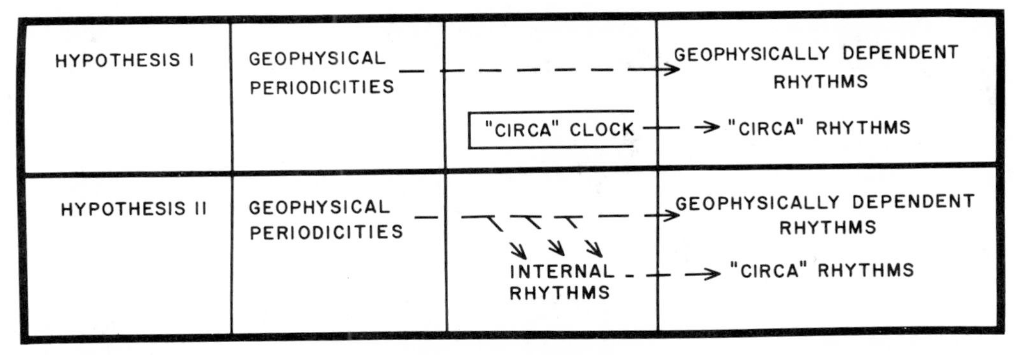

Fig. 2-11. Two biologic clock hypotheses. *I,* A clock, independent of all geophysical rhythms, generates periods only approximating the natural ones. The natural periods are derived by continuous correction of the periods by such rhythmic phasing agents as light and temperature. *II,* A clock, dependent upon geophysical rhythms, provides accurate information concerning environmental periods for use in timing the rhythms. "Circa" periods observed in free-running rhythms in "constant" conditions result from a frequency transformation through regular phase-shifting.

adjusting the annual activities of organisms to the appropriate time of year. Many kinds of plants and animals display this phenomenon termed *photoperiodism.*

NATURE OF THE BIOLOGIC CLOCK

Whereas the natural environmental period-giver for the geophysically dependent biorhythms is obvious, the mechanism of the biologic clock that times the labile behavioral and physiologic circadian, tidal, monthly, and annual biorhythms adapting an organism to the rhythmically changing demands of its environment remain an unsolved problem. Two different hypotheses have been proposed to explain these biorhythms and their characteristics (Fig. 2-11), and neither of these can be excluded at present. The true mechanism of biologic clocks is probably somewhere between the extremes of these two hypotheses and may include parts of each. Meanwhile, biologic research designed and interpreted in terms of each is highly profitable, advancing our knowledge of the phenomenon of biorhythmicity.

The *internal timer hypothesis* suggests that living things have evolved the capacity to oscillate and now do so without any external environmental stimulus or cue at approximately, but not exactly, all major natural environmental frequencies. It suggests that circadian rhythms are timed by circadian clocks whose period lengths are almost never exactly 24 hours. Thus the free-running periods directly reflect the natural periods of the internal clock. Exact organismic solar-day, tidal, monthly, and annual periodicities result from entrainment, synchronization, or day-to-day resetting of the approximately accurate, internally timed biorhythms by geophysical environmental rhythms such as those of light, temperature, tides, and photoperiods. The period length of these internal, self-sustained oscillations increases or decreases slightly in response to different levels of experimentally constant ambient temperature and light, or to chemical substances, such as deuterium oxide, which alter free-running period length.

According to the internal timer theory the biologic clock is temperature-compensated, not temperature-independent, and the lack of significant response to ambient temperature change results from sequences of biochemical reactions that are mutually self-regulatory or from physiochemical processes for which diffusion, with its very low temperature-dependence, is rate-limiting. Oscillator period lengths are inherited, the close approximation to the natural environmental period lengths having resulted from natural selection during the long process of biologic evolution. The "jet syndrome" is interpreted quite simply in terms of the fully autonomous endogenous oscillation theory.

We can consider biorhythms as the oscillations of an internal clock that are analogous to the higher-frequency oscillations of well-known physical systems; these oscillations can be entrained by or synchronized with other slightly different environmental geophysical frequencies. By mathematical manipulation of such a model, good descriptions of the observed dynamics of biologic rhythmic systems are derived, accurate predictions made, and relevant experiments suggested.

However, the autonomous endogenous oscillator hypothesis has several weaknesses. At present, there is no way, except by arbitrary assumption, to distinguish between a biochemical reaction that is part of a clock and one that is only timed by a clock. Why don't two frequencies, as close as 24.0 and 24.8 hours, mutually entrain each other as would be expected in terms of the simple physical analog? In fact, they remain distinct indefinitely. We cannot account satisfactorily for the temperature-independence of the biologic clock or for the fact that its period is slightly light-dependent. There is no convincing demonstration of the existence of any biochemical or physiologic process that exhibits the requisite clock properties. Thus although clock-timed rhythms are omnipresent in organisms and all their major living parts, the search for the mechanism underlying the biologic clock has been unsuccessful.

The *external timer hypothesis* suggests that the periods of the biologic clock are continuously derived by the organism from rhythmic natural geophysical environmental changes to which the organisms have always had access. According to this theory, organisms have evolved the capacity to use a rhythmic geophysical input as the fundamental timing information for their geophysically dependent biorhythms. This information provides a temporal frame of reference for the synthesis of adaptive biorhythmic patterns. Thus the basic clocks are not the inaccurate circadian, circa-tidal, circa-monthly, and circa-annual biologic ones; they are the accurate 24.0-hour, 24.8-hour, 29.5-day, and 365¼-day geophysical ones. As long as the organism remains in its natural rhythmic environment, the basic geophysical clocks and the biorhythms timed by them are synchronous with such natural environmental rhythms as those of light, temperature, and the tides. The biorhythms do not require continuous correction by the rythmic variations of an obvious environmental Zeitgeber; Zeitgeber determine only phase, not period.

According to this theory, when an organism is placed under constant conditions with respect to the major phasing factors, its 24-hour rhythmic pattern shifts phase slightly each day as a result of the specific properties described in its response curve to the normal phasing factors. Phase advances alternate with phase delays in a cyclic manner as the constant light and temperature act at different points on the response curves. The phase-labile rhythmic cycles gradually move toward earlier hours of the day if the advance responses are stronger than the delay responses, or they shift toward later hours of the day if the delay responses are the stronger. These responses to constant light and temperature generate free-running rhythms whose period lengths commonly differ somewhat from the natural environmental ones. Other constant levels of light or temperature would quantitatively alter the interaction between the organism and the artificially constant phasing factors, changing the period length. Response curves differ, even from one individual to another of the same species; this explains the variety of period lengths that exist among the individuals of any given population; all are generated by a clock that has a single natural geophysical period. The inheritance of the period length of free-running rhythms and the alteration of period length by chemical substances such as deuterium oxide are explained as effects that determine or modify the response curves. The postulated daily self-resetting while the normal Zeitgeber is held constant is termed *autophasing*.

The external timer hypothesis explains the temperature-independence of biorhythms, including the synodic monthly and the annual ones. It also readily explains the self-setting to specific clock hours of the solar or lunar day or to the time of the synodic month or the year that is sometimes observed in organisms held under "constant" conditions. In terms of this hypothesis the continuing accuracy of the clocks during rapid east-west geographic translocation requires either (1) the existence of a universal-time geophysical variable, such as the rhythmically varying electric charge of the ionosphere, or (2) that the organism shift the phase of its 24-hour rhythmic pattern relative to an abruptly shifted period-giving environmental cycle. Such a phase-shift could theoretically result from the inertia of the biologic rhythmic component inherent in an ongoing living system.

The controversy over the nature of the biologic clock has gone on for nearly a century. The major difficulty in resolving the two clock hypotheses—internal versus external timing of the clock-timed biorhythms—is inherent in the temporal nature of the problem. If biologic events recur with a relatively constant period length close to that of a natural geophysical period, one cannot distinguish between two possibilities: (1) that the organism has wholly within itself the capacity to measure the duration of the period, a capacity that includes all the requisite clock-like properties, or (2) that the organism recognizes the arrival of a point in time by means of a sensing mechanism that can distinguish points along the 360-degree temporal cycles of the rhythmically varying geophysical environmental complex. One cannot establish the first alternative without excluding the second, and vice versa.

Sensing subtle pervasive geophysical variables by an unknown mechanism (magnetoreception), an organism can distinguish, remember, recognize, and recall specific points along a 360-degree geophysical environmental cycle and relate light-associated events to such points. To circumvent the ambiguity inherent in the temporal cycle, and because geophysical factors vary continuously in both time and space, the fundamental problem was transposed from temporal to spatial coordinates. After thus eliminating temporal ambiguity, it was demonstrated that an organism can distinguish points around a 360-degree horizontal geographic compass cycle and that this capacity is related to biologic rhythmicity; a 180-degree rotation of the geomagnetic field, which the organism uses to determine geographic direction, immediately shifted the temporal synodic monthly rhythm by 180 degrees, or one-half cycle. Biologic events related to the new moon were shifted to the full moon and first quarter events were shifted to the third quarter. In using the horizontal geomagnetic vector as a spatial reference to derive the period length of its synodic monthly cycle, the organism appeared to be reading time from the geophysical environment.

A living organism can remember the geographic compass direction from which a light stimulus arrives using only information derived from its capacity to perceive weak geophysical parameters, such as the horizontal geomagnetic vector. Thus it can encode a geographically distributed pattern of light on a 360-degree spatial frame of reference constructed from information derived from directional geomagnetic field forces using a mechanism of perception that remains unknown.

The subtle pervasive geophysical parameters vary with respect to both space and time; the latter includes solar day, lunar day, synodic monthly, and annual 360-degree cyclic patterns. The natural geophysical cycles are temporally correlated with recurring patterns of numerous obvious environmental factors such as light. Thus the living biologic system senses the passage of terrestrial time in terms of the natural geophysical cycles and relates this time to the temporally varying courses of other environmental events. In terms of this view of biologic rhythmicity, the rhythmic nature of the terrestrial organism does not depend upon a capacity to measure absolute time; it depends upon its capacity to recognize the passage of time only with reference to the natural rhythmic geophysical fluctuations. Although this view of the nature of the fundamental timing system of biologic rhythms simplifies the explanation of some of the clock-like properties of the living biologic system, such as its remarkable period stability, the discovery of the fundamental mechanism is hardly much closer. Instead, the problem is now submerged in the abyss of our ignorance regarding the phenomena of perception and memory at the organizational level of molecular biology.

Explanation of the properties of certain biologic rhythms, the geophysically dependent ones, requires the existence of a continuous inflow of information regarding environmental rhythms, even under conditions traditionally considered constant for the organism. All other presently known characteristics of clock-timed biologic rhythms are compatible with either hypothesis.

Possibly the organism uses both kinds of clocks. An environmentally dependent clock component may be responsible for the remarkable resistance to temperature change or interference by chemical agents, the occasionally observed self-setting feature, the many highly accurate mean 24-hour and lunar tidal periodicities, the persistent relatively precise daily and tidal rhythms, and, particularly, the stability of the longer-period synodic monthly and annual biorhythms. An environment-independent clock component may time shorter intervals within cycles but, by itself, be insufficiently precise to account for the long-term regularity and dependability of the clock-timed biorhythms.

PERSPECTIVE

Biologic rhythms with frequencies nearly equal to those of all the environmental geophysical periodisms are common, if not universal. They are the highly specialized dependable timing system of living things. There is good reason to believe that biorhythmicity is indispensable for temporal integration of the multiplicity of physiologic processes that are essential to a healthy organism. The multiplicity of processes that must be coordinated are not innumerable isolated processes requiring separate regulation but are rather, to a large extent, components of temporally organized biologic units that must be drawn into harmonious functional relationship with each other. Temporal integration is just as important for organismic function as spatial, or structural, organization. Spatiotemporal organization is clearly evident in the response of organisms to very weak gamma radiation or to very weak horizontal magnetic fields; at any given time, the strength of the organismic response varies systematically with the geographic orientation of these fields; for any given spatial orientation of the applied field, the response varies rhythmically with time.

Man has long lived and work in artificial conditions. He is now subjecting himself to even more extreme environmental changes in submarine chambers and in spacecraft. As no species before him, he is rapidly modifying the biosphere. He is changing the atmosphere at a fearsome rate—physically with particulate matter, ionizing radiation, radar and radio waves, and other radiation and force fields used in modern technology; and chemically with nitrogen oxides, carbon monoxide, ozone, carbon dioxide, and many other pollutant substances. These artificial changes modify many environmental variables such as light and temperature, affecting fundamental geophysical periodisms. A vital concern for humanity is whether such artificial environmental changes will interfere critically with the biologic rhythms that characterize life on earth and that play essential regulatory roles in the maintenance of life itself—rhythms that we are only beginning to understand.

FALL, 1971

GLOSSARY

aperiodic variation Fluctuations in which no periodic recurrences of pattern can be distinguished.

autophasing A postulated continuing phase-shifting of a biologic clock–timed rhythm in organisms when all normally operating phase-setting factors are held constant; shifting is effected by interaction of the rhythmic organism with the environmental constancy of all Zeitgeber. The expected consequence would be the generation of overt rhythms deviating in period (frequency transformation) from the related geophysical one.

biologic clock That still unresolved means by which organisms provide the observable highly dependable timing for their overt physiologic and behavioral rhythmic patterns that, in nature, correlate with geophysical periods.

biologic clock–timed rhythm A regularly recurring pattern of variation correlated with a natural geophysical period, which persists to a significant degree under constant environmental conditions with nearly the same period and with frequency essentially independent of temperature; the rhythm exhibits a labile phase and pattern that are usually adaptively alterable in response to specific ambient environmental cycles.

biologic rhythm A periodic fluctuation or variation of a biologic process as a function of time.

celestial navigation Utilization of celestial bodies—sun, moon, or stars—as references for geographic homing or navigation during migrations. To accomplish this, solar-day, lunar-day, and sidereal-day chronometers, respectively, are required to correct for the steadily changing geographic directional relationships of these bodies that results from the earth's rotation.

chronobiology The science of biologic rhythms.

constant environmental conditions Approximate experimental constancy of known environmental variables as specified.

endogenous rhythm A persistent rhythmic variation in an organism involving physiologic or behavioral phenomena having periods of three kinds: (1) independent of all environmental periodisms, such as cardiac and respiratory rhythms; (2) clock-timed, such as circadian and tidal rhythms; and (3) dependent on fluctuations of the geophysical environment.

exogenous rhythm A rhythmic variation in an organism that is a consequence of a continuous response to rhythmically varying factors in its external environment.

external timer hypothesis The hypothesis that the biologic clock derives its regularity, precision, and high degree of resistance to physically and chemically induced disruptions from continuing utilization of and dependency upon rhythmic variations of subtle, pervasive geophysical variables.

free-running rhythm An organismic rhythm, correlated in nature with a geophysical period, that persists even when all natural phase-setting variables such as light and temperature are experimentally held constant. Under such experimental conditions a free-running rhythm commonly has a period that differs slightly from the geophysical one with which it is naturally correlated.

geophysically correlated periodism A periodic organismic fluctuation having the same, or nearly the same, period as a known fluctuation of the geophysical environment; the most obvious physical periodisms are those related to the relative movements of earth, sun, moon, and stars (the day, ocean tides, month, and year), although many other environmental periods exist.

geophysically dependent rhythm A pattern of mean physiologic variation having an exact natural geophysical period that persists under experimentally constant environmental conditions and exhibits phase and pattern stability in a fixed relationship to a mean geophysical periodism; evidence suggests that this rhythm is a response of the organism to uncontrolled periodic environmental factors.

internal timer hypothesis The hypothesis that the organism's biologic clock is a temperature-compensated, highly stable, internally regulated physicochemical system that has an inherent capacity to oscillate with nearly the same periods as those of the geophysical environment but that is fully independent of all external physical periodisms.

periodic variation Fluctuations that can be resolved into recurring patterns of change, whether clearly overt (for example, the highly regular solar-day and lunar-day variations in the acceleration caused by gravity), less regularly overt (for example, the 24-hour variations of light and temperature), or largely covert (for example, the solar and lunar tides of the earth's atmosphere).

phase-shift A resetting of the temporal pattern of events of a biologic clock–timed rhythm so that the events are induced to occur earlier (phase advance) or later (phase delay). Environmental variables that can induce such phase shifts are termed phase-setters, phasing clues, or Zeitgeber. A phase-shift occurs naturally after a rapid move to a different terrestrial longitude.

photoperiodism The phenomenon of regulation of the biologic activities of animals and plants by the cyclic annual change of the relative durations of daylight and darkness.

response curve A graphic description of the variation in character and effectiveness of an imposed phase-setting stimulus, or Zeitgeber, throughout the course of a biologic clock–timed rhythmic period.

SUGGESTED READINGS

1. Aschoff, J., editor: Circadian clocks, Amsterdam, 1965, North-Holland Publishing Co.
2. Aschoff, J.: Circadian rhythms in man, Science **148:**1427-1432, 1965.
3. Aschoff, J.: Comparative physiology: diurnal rhythms, Ann. Rev. Physiol. **25:**581-600, 1963.
4. Biological clocks, Cold Spring Harbor Symp. Quant. Biol. **25:**524, 1960.
5. Brown, F. A., Jr.: Biological chronometry, Am. Naturalist **91:**129-195, 1957.
6. Brown, F. A., Jr.: A hypothesis for timing of circadian rhythms, Can. J. Bot. **47:**287-298, 1969.
7. Brown, F. A., Jr.: Living clocks, Science **130:**1535-1544, 1959.
8. Brown, F. A., Jr., Hastings, J. W., and Palmer, J. D.: The biological clock: two views, New York, 1970, Academic Press, Inc.
9. Bünning, E.: The physiological clock, Berlin, 1967, Springer Verlag.
10. Conroy, R. T. W. L., and Mills, J. N.: Human circadian rhythms, London, 1970, J. and A. Churchill, Ltd.
11. Cloudsley-Thompson, J. L.: Rhythmic activity in animal physiology and behavior, New York, 1961, Academic Press, Inc.
12. Cumming, B. G., and Wagner, E.: Rhythmic processes in plants, Ann. Rev. Plant Physiol. **19:**381-416, 1968.
13. Ehret, C. F., and Trucco, T.: Molecular models for the circadian clock. I. The chronon concept, J. Theor. Biol. **15:**240-262, 1967.
14. Goodwin, B.: Temporal organization in cells, New York, 1963, Academic Press, Inc.
15. Halberg, F.: Chronobiology, Ann. Rev. Physiol. **31:**675-725, 1969.
16. Halberg, F., and Reinberg, A.: Rythmes circadiens et rythmes de basses fréquences en physiologie humaine, J. Physiol. (Paris) **59:**117-200, 1967.
17. Harker, J. E.: Diurnal rhythms in the animal kingdom, Biol. Rev. **33:**1-52, 1958.
18. Harker, J. E.: The physiology of diurnal rhythms, Cambridge, 1964, Cambridge University Press.
19. Kayser, C., and Heusner, A.: Le rythme nycthemeral de la dépense d'energie, J. Physiol. (Paris) **59:**3-116, 1967.
20. Luce, G.: Biological rhythms in psychiatry and medicine, Public Health Service Publication No. 2088, Washington, D. C., 1970, National Institute of Mental Health.
21. von Mayersbach, H.: The cellular aspects of biorhythms. In von Mayersbach, H., editor: Symposium on rhythmic research (sponsored by the Eighth International Congress on Anatomy, Weisbaden, 1965), New York, 1967, Springer-Verlag New York, Inc.
22. Menaker, M.: Biological clocks, BioScience **19:**681-689, 1969.
23. Mills, J. N.: Human circadian rhythms, Physiol. Rev. **46:**128-171, 1966.
24. Pittendrigh, C. S.: On temporal organization of living systems, Harvey Lect. **56:**93-125, 1961.
25. Sollberger, A.: Biological rhythm research, New York, 1965, Elsevier Publishing Co.
26. Sweeney, B. M.: Biological clocks in plants, Ann. Rev. Plant Physiol. **14:**411-440, 1963.
27. Sweeney, B. M.: Rhythmic phenomena in plants, New York, 1969, Academic Press, Inc.
28. Webb, H. M., and Brown, F. A., Jr.: Timing long-cycle physiological rhythms, Physiol. Rev. **39:**127-161, 1959.
29. Wolf, W.: Rhythmic functions in the living system, Ann. N. Y. Acad. Sci. **98:**753-1326, 1962.

3 BIOMETEOROLOGY

KONRAD J. K. BUETTNER†
N. BALFOUR SLONIM

Biometeorology is the branch of environmental physiology that deals with the effects of the geophysical and geochemical environment on living biologic systems. It is primarily concerned with the actions and interactions, direct and indirect, of atmospheric environmental elements such as temperature, barometric pressure, humidity, precipitation, visible light, infrared and ultraviolet radiation, ionization, and air flow. It deals not only with natural atmospheric effects but also with those of artificial atmospheres in shelters, buildings, and closed ecologic systems, such as aerospacecraft, satellites, and submarines. Thus the biometeorologist identifies and describes a certain set of environmental elements and measures, analyzes, and interprets their effects on living organisms. Analysis and interpretation may lead to the discovery of principles, development of concepts, and construction of models for the dynamic processes of transfer and interaction. In this chapter we discuss *human* biometeorology.

Bioclimatology, a division of biometeorology, is the branch of ecology devoted to the study of the effects on living organisms of the *natural* environmental conditions that prevail in various geographic regions. *Medical climatology* is concerned with the influence of natural climate on health; it deals with medical conditions either aggravated or caused by climatic elements. Its broad range of interest includes such topics as sunburn, skin cancer, photodermatitis, jungle rot, frostbite, and cold allergy; weather-related malaise, emotional depression, and headaches; weather-related aggravation of pain in corns, bunions, and a wide variety of arthritic, rheumatic, posttraumatic, and postsurgical conditions affecting nerves, muscles, ligaments, bones, and joints; and climatic aggravation of bronchopulmonary conditions, such as asthma and chronic bronchitis.

The terminology of the field shared by environment, physiology, and medicine has been enriched by scientists of various points of view (see Fig. 3-1). Thus the term "environmental physiology" has been used synonymously with both "quantitative biometeorology" and "autecology." Pathophysiologic effects of environmental elements are treated in medical climatology, environmental health, environmental medicine, occupational health, and industrial hygiene. The terms "hygiene," "health," and "medicine" are often used almost synonymously. Does "environmental health" mean "preventive environmental medicine"? Complicating terminology further are weather modification and environmental control. For a good discussion of the complex terminology of this field the reader is referred to Folk.[1]

The environmental elements with which biometeorology is concerned include the temperature of the atmosphere and of the liquid and solid surfaces around us, water content, ambient pressure, the partial pressures of natural and pollutant gases, and the rate and character of air flow. The intensity or quantity of any biometeorologic element may vary as a function of time and differ from place to place because of altitude, latitude, or other aspects of regional geography. Analysis of temporal variation may reveal periodicities with intervals of 1 day, 1 or 2 weeks, a month, or a year.

†Deceased.

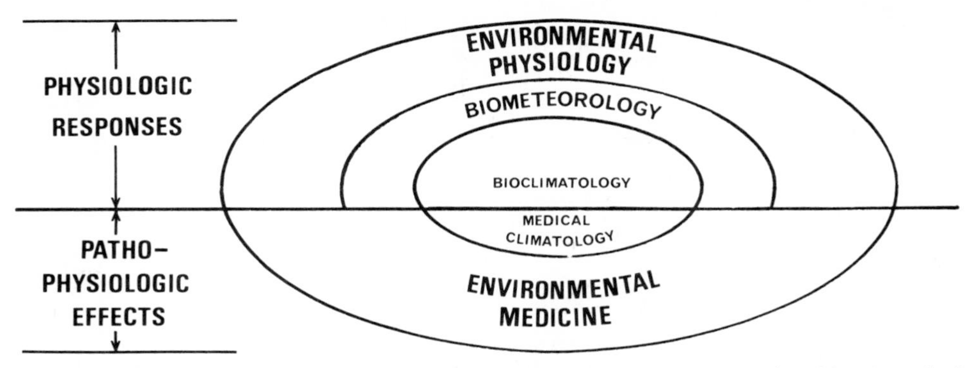

Fig. 3-1. Environment, physiology, and disease. The terminology of these related fields of study is confusing. This diagram indicates the relationship of the key terms as we use them.

Relevant physical transfers include those of heat by conduction, convection, or radiation; mass by evaporation; momentum by wind exposure; and electromagnetic radiation by absorption and emission. In the latter case we are especially concerned with those wavelengths that produce photochemical effects—sunburn and vitamin D—as well as those responsible for skin heating and cooling.

The distinction between *natural* and *artificial* meteorologic elements is not always entirely clear. Lightning has set forest fires since time immemorial; thus smoke, carbon monoxide, and radiant heat from fires may be natural. Lightning also produces minute quantities of ozone and nitrogen oxides from atmospheric oxygen and nitrogen. Volcanoes spew molten lava, ash, and sulfur oxides, constituting natural air pollution. Sunlight changes manmade nitrogen dioxide to ozone in urban smog. Outside weather almost always affects an indoor microclimate.

We live not only in a regional climate but within a personal microclimate of our own making. Our cities, conveyances, houses, furniture, and clothing produce unique environmental conditions at the surface of our skins and under our noses that often differ appreciably from the regional climatic conditions reported by the meteorologist. Protective clothing and modesty are relatively recent developments. Our ancestors lived in fire-heated caves for thousands of years. This microclimate had a rather constant temperature, artificial light, and intense air pollution. Are we reverting atavistically to a similar type of microclimate with our thermostated artificial heat, artificial light, and self-induced smoke pollution?

Research in biometeorology sometimes involves an attempt to separate a problem from the biospheric meshwork for study with standard experimental animals or well-scrubbed, male, Caucasian medical students in a clean environmental chamber. Do the results of such studies have general significance and applicability? How far may we safely extrapolate from studies on such a "standard" man?

ATMOSPHERE

Our atmosphere is a mixture of gases. In nature the ratio of partial pressures of oxygen, nitrogen, carbon dioxide, and rare gases is highly constant. Physiologically significant deviations occur only in the vicinity of strong sources or sinks, such as fires, volcanoes, mines, and mineral wells. Rebreathing exhaled air within a poorly ventilated room or within a crowded enclosed space may raise carbon dioxide concentration to 0.3% (about ten times the natural value) and reduce oxygen concentration slightly; however, such changes do not occur in open or well-ventilated spaces.

The variations of total barometric pressure and partial pressure of oxygen observed at a given meteorologic station are usually not of physiologic importance. An ambient pressure change of as much as 5% during the passage of meteorologic high or low pressure weather systems is unusual. During hurricanes and tornadoes pressure may decrease relatively rapidly by more than 10%, corresponding to a vertical ascent of about 1,000 meters (m).

Clean air has become a luxury. Volcanoes, forest fires, and duststorms are natural sources of air pollution. The air over uninhabited Canadian forests contains detectable amounts of aerosols and terpenes. Sahara dust falls frequently on Caribbean islands. To this complex natural concoction human technology adds many ingredients, especially the products of incomplete combustion of fossil fuels. Air moving across the entire Pacific Ocean contains not only sea effluents such as salt but also detectable amounts of industrial aerosol that originate in Asia.

The long-term climatic and biomedical effects of atmospheric pollution are not yet sufficiently understood; thus it is difficult to define *significant* pollution in terms of chemical specificity and concentration. Defining personal or self-induced air pollution is a difficult matter. Cigarette smoking is an undisputed factor in the causation of bronchitis, emphysema, lung cancer, and heart disease. The scientist must be aware of cigarette-generated carbon monoxide, as well as gaseous pollutants from other sources—ozone, sulfur and nitrogen oxides, hydrocarbons, and a variety of animal, plant, and mineral dusts.

Water is the most variable component of our atmosphere. Its vapor behaves like any other gas except for its condensation and sublimation. The partial pressure of water vapor (P_w) is expressed in torr.* Water vapor density (ρ_{wa}) and the maximum or saturation value ($\rho_{w, sat}$) are both expressed in gm/m^3. The gas law for water vapor is $P_w = \rho_w \times R_w \times {}^\circ K$, where the gas constant $R_w = 3.47$ ($m^3 \times$ torr $\times$ kgm$^{-1} \times {}^\circ K^{-1}$). Under ordinary room conditions a water vapor pressure of 1 torr corresponds to a water vapor density of 1 gm/m^3. The value of $\rho_{w, sat}$ increases exponentially with temperature; at 37° C, normal temperature, 1 liter of lung gas contains about 46 mgm of water vapor.

Relative humidity (rh) is 100 times the ratio $\rho_w/\rho_{w, sat}$, or $P_w/P_{w, sat}$, and is expressed in percent. In nature it varies from near zero to 100%, the latter occurring, for example, in clouds and fogs. Water vapor density varies from less than 0.1 gm/m^3 in the arctic cold to more than 30 gm/m^3 on the Red Sea and in the Indus River valley. Except near sinks and sources such as the kitchen, bath, and laundry ρ_{wa} is about the same indoors and out. Outdoor ρ_{wa} cannot exceed 0.29 gm/m^3 in a cold winter ice fog at –30° C. The same ρ_{wa} indoors at an air temperature of +25° C corresponds to a relative humidity of only 12%; at a skin temperature of 35° C the rh is only 7%, chapping the skin.

Except as modified by clothing and wind, the evaporative cooling of human skin is basically dependent on the difference between ρ_{ws} at skin surface and ρ_{wa} of the surrounding air. Thus in warm climates ρ_{wa} is very important for human well-being.

Global ρ_{wa} maps are incomplete; unfortunately, maps of precipitation do not mean the same thing. For example, some of the highest ρ_{wa} values occur in the Indus Valley, an area with very little rain. ρ_{wa} is generally low in descending air currents such as those over the hot desert belt; but even in the summer Sahara, ρ_{wa} is about 10 gm/m^3. The combined effect of air temperature and humidity (effective temperature) on man is discussed later.

Although the horny layer of nonsweating skin is a barrier, water vapor diffuses through it. Skin humidity, ρ_{ws}, is difficult to define, as is skin rh, a factor related to skin elasticity and condition. The latter may vary with skin rh from smoothness to brittleness with cracking. If the rate of sweat flow exceeds the rate of evaporation, sweat accumulates and ρ_{ws} is the saturation value at the relevant temperature. When this occurs, however, water on the skin blocks further sweat secretion!

We may describe sweating skin that is not totally wet as partially dry, where $\rho_{ws} = \rho_{wa}$, and partially wet, where $\rho_{ws} = \rho_{ws, sat}$, the saturation value at skin temperature. This fact is expressed as an areal or wetness factor in a formula.

In fog and clouds liquid and solid water float in air or fall as precipitation; in either case rh approaches 100%. Haze and mist precede fog and smog. The particles in clouds and fog usually include minute quantities of condensation nuclei, such as sodium chloride or products of combustion; such nuclei permit condensation when rh is approximately 100%. In smog, however, nuclei are so large that condensation, actually water vapor adsorption, occurs at rh values as low as 20%.

Cloud and fog particles are approximately 10 μm in diameter. The velocity at which they settle is insignificant. Their water content expressed in grams per cubic meter is usually smaller than that of the surrounding water vapor. The inhalation of fog is an interesting situation because droplets either deposit on the walls of the upper respiratory tract or evaporate on the way down; thus dissolved material may be absorbed by respiratory surfaces.

Precipitation either falls or is intercepted, the latter occurring when there is a strong horizontal motion component, such as wind-driven fog or rain. The forms of precipitation are many—drizzle, rain, undercooled rain, soft and hard hail, snow, and ice needles. Exposed surfaces, including

*1 torr = 1 mm Hg = 1.33 millibars = 133.3 newtons/m^2.
At sea level water boils at a pressure of 760 torr.

those of man, also receive atmospheric water in the processes of condensation (dew), sublimation (hoar frost), and sorption on hygroscopic surfaces such as clothes, dry sand, and soil. These processes counter evaporation, the source of atmospheric water vapor.

Liquid or solid water intercepted by skin may act as a powerful coolant because it may be colder than the skin and because the body provides the heat of evaporation. Furthermore, the insulation value of wet clothing may be only one fifth that of the same clothing when dry.

Whether indoors or out, air temperature is the chief environmental element in determining comfort or discomfort. Only a good thermometer free of radiational errors indicates this value. The extreme temperatures observed on the earth's surface range from +57° C to less than −70° C.

The temperature data of a weather bureau or climatic table refer to particular locations, frequently airport towers; the microclimatic temperature for a man on the street or in a field may be quite different. On a clear night the air temperature in a concave orographic formation may be from 5° C to 40° C colder because surfaces radiate infrared toward the sky and cosmos while cold air gravitates downward. On a clear day the air temperature near sun-heated surfaces may be from 10° C to 30° C warmer. Thus human microclimatic temperatures vary greatly. Absence of evaporating plants and soil, high heat capacity, the frictional resistance to passing winds, and home and industrial heating convert cities to heat islands, day and night, winter and summer.

The temperature variations within a heated or cooled home are even larger. In heated homes air temperature may differ significantly from the radiant temperature of walls, windows, and stoves. In a room heated by circulating warm air the radiant temperature is lower than the air temperature; in rooms heated by a stove or a water-heated radiator the opposite is true. In the latter case lower air temperature results in a higher relative humidity. We breathe air of a certain temperature and humidity, but our skin is exposed to the combined effects of air and radiant temperature. These differences affect human thermoregulation.

The concept of relative humidity can be extended to include solids and liquids near a solid-air or liquid-air interface; it is the rh of an air bubble equilibrated at constant temperature with a relatively large quantity of the solid or liquid. This concept can be applied to skin, nails, horn, hair, and lips and to some extent to the upper respiratory tract. It can also be applied to wood, paper, soil, clay, silica gel, and many other colloidal substances. All become hard and brittle at low rh but are soft, pliable, and gluey at high rh. The effects of extreme dryness are usually seen in overheated rooms during the winter when they are often wrongly attributed to cold.

HEAT TRANSFER

Before proceeding it may be well to review the phenomenon of *convection*. This process involves a fluid, liquid or gas, a heat source and sink, displacement of less dense warmer fluid by more dense cooler fluid in a gravitational field, and development of flow currents in a circulatory pattern. The fluid flow regime may or may not be assisted by a nonconvective driving pressure (forced flow versus free flow), may be open or enclosed, and may be laminar or turbulent. Convection transfers heat, mass, and momentum. These three transfers are proportional respectively to temperature gradient, concentration (for example, water vapor content), and wind velocity. The term "convection" is sometimes restricted to mean upward vertical motion; downward vertical motion is then termed *subsidence* and horizontal mass motion is termed *advection*.

Wind, or moving air, involves three transfers:

1. Heat, by convection, constituting most of the cooling effect; this transfer is proportional to $(\rho \times v)^{1/2}$, where ρ is air density and v is wind velocity
2. Mass—the amount of air transferred is $\rho \times v$
3. Momentum—wind blowing against an obstacle gives up some of its kinetic energy, creating a front-side pressure difference that is proportional to $\rho \times v^2$

If air flow is nonlaminar—which is usually the case in the open—eddies form behind the obstacle, creating a leeward pressure sump.

The processes involved in the transfer of heat between skin and air, including evaporation of sweat from the skin and radiant heat exchange, are illustrated in Fig. 3-2. On both sides of the skin-air interface are subsurfaces, or layers, that permit heat and water vapor transfer only by molecular motion. The air subsurface for heat and water vapor transfer is the same; it is about 5 mm thick in calm air but less than 1 mm in wind. The outer skin layers, lacking blood perfusion, constitute a protecting subsurface for heat transfer in the skin. A variable peripheral blood flow controls the thickness of this subsurface. In warm surroundings the insulating subsurface consists of only the epidermis; in the cold all the

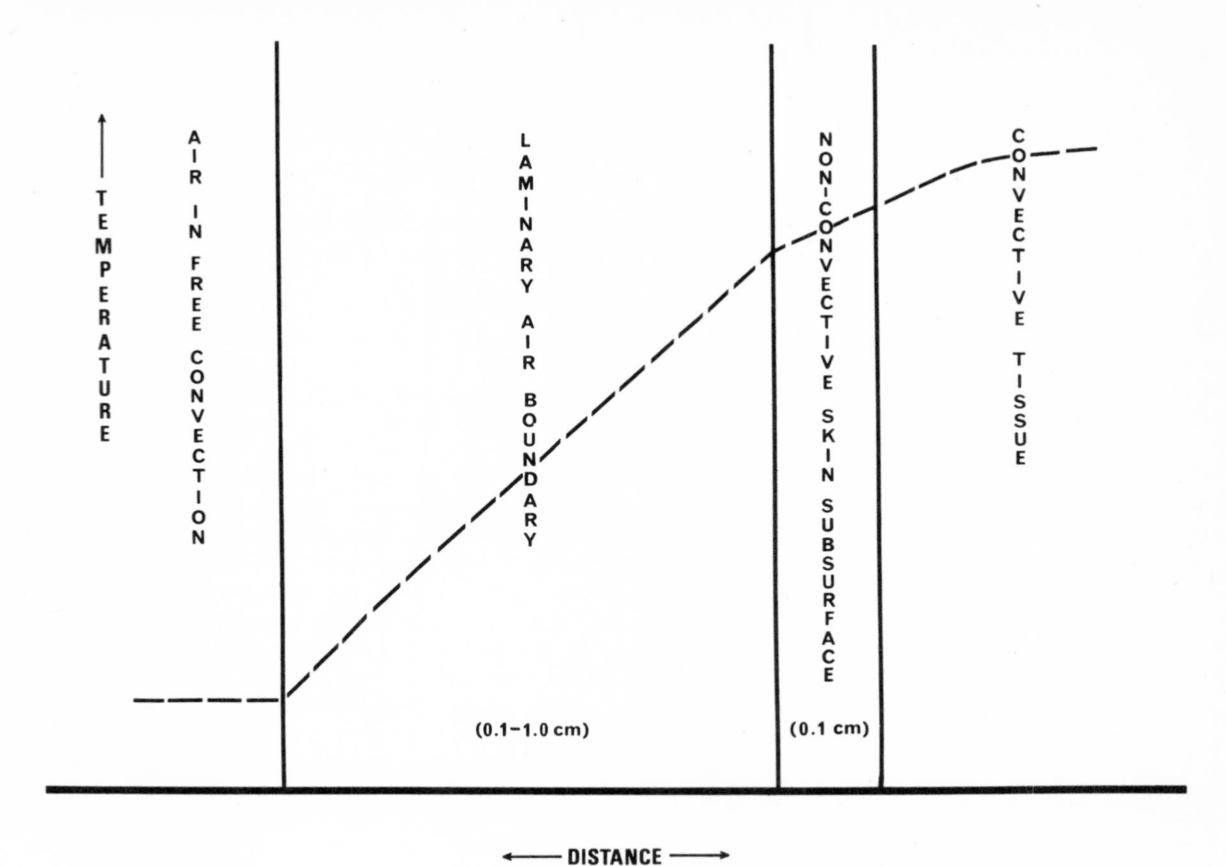

Fig. 3-2. The four thermal layers at the periphery of the unclothed body. The dashed line represents the relative temperature gradient in a cool environment.

fatty subdermal tissues may become involved. The water vapor barrier is restricted to the corneum; sweat secretion is a short circuit across this barrier.

The elements of weather have a complex effect on our heat balance.*

For the naked body:

metabolic heat *(M)* + solar heat *(H)* = convective loss *(C)* + infrared loss *(R)* + evaporative loss *(E)* + lung loss *(L)* + work *(W)* + storage *(S)* (1)

where $H = A_s \alpha_s S$; $C = A_c h_c (T_s - T_a)$ (2)

$R = A_r h_r (T_s - T_r)$; $E = A_e h_e (P_{ws} - P_{wa})$ (3)

$L = m e_a c_P (T_{ex} - T_a) + m\lambda(\rho_{wex} - \rho_{wa})$ (4)

S = body heat capacity × average body temperature change (5)

where s, c, r, and e = solar, convective, infrared radiative, and evaporative heat transfer, respectively

A = respective surface areas (may be different for each transfer, are always smaller than the geometric surface, and are also subject to climatic variation)

α_s = skin absorptivity for solar radiation (α_s = 0.6 for white skin, 0.7 for tanned skin, and more than 0.9 for black skin)

S = incident solar heat flux at noon on a sunny day (may be as great as 1,000 watts/m²)

h = combined heat transfer coefficient

*See Appendix for standard system of units and definition of symbols used in thermal physiology.

Convective loss or gain depends on exposed area, on skin-air temperature difference, and on the exchange coefficient h_c, mainly a function of wind velocity. For man lying on an insulating bed in a small quiet room h_c is approximately 4.2 watts × m^{-2} × $°C^{-1}$; and in the wind h_c = 14 × (V × watts)$^{1/2}$ × m^{-2} × $°C^{-1}$, where V is expressed in meters per second.

Note that T_s is the skin *surface* temperature, not that at some point below the surface or that measured by a gauze-covered thermocouple.

The infrared radiant heat loss or gain is:

$$R = A_r \sigma \epsilon (°K_s^4 - °K_r^4) \quad (6)$$

which can be simplified to

$$R = A_r h_r (T_s - T_r)$$

where $h_r = \epsilon 4 \sigma (°K_{sr})^3$
$°K_{sr}$ = average absolute temperature of skin and radiant environment
ϵ = infrared emissivity, $\epsilon = 0.97$

For room conditions $h_r = 5.9$ watts $\times$ $m^{-2} \times °C^{-1}$ —somewhat larger than h_c in a calm room. Although the skin temperature involved is the same for both processes (*C* and *R*), air temperature affects *C* and radiant environmental temperature —the temperature of the walls, floor, and ceiling in a room or, in the open, the temperature of the soil below and sky above—affects *R*.

In the evaporative term *E*, the air subsurface, or molecular boundary layer, controls the transfer coefficient. Substituting water vapor *pressure* for water vapor *density* and introducing the heat of vaporization, h_e is 2.2 h_c (°C/mm Hg water vapor pressure at sea level). The humidity of the external milieu is the ambient water vapor pressure (P_{wa}). The water vapor pressure at the skin can be measured directly, for example, with a hair hygrometer and a thermocouple. If the skin is totally wet, P_{ws} equals the saturation pressure $P_{ws,\ sat}$ at skin temperature; if only partly wet A_e is proportionately less. However, *E* is not zero for a nonsweating body; water vapor diffuses through even the horny layer of the skin, the largest quantity per unit area leaving the scrotum, labia majora, and plantar area.

We perceive gentle winds as a result of skin cooling; however, we cannot distinguish such wind cooling from cooling resulting from infrared loss as, for example, toward a cold wall. We sense stronger winds by their mechanical pressure effects, such as the movement of fine hair on exposed body surfaces.

Heat loss from the lung (*L*) is the heat content difference between expired gas and inspired air plus the heat of vaporization associated with the almost complete humidification of the drier inhaled air. Breathing air the rate of mass exchange ($\dot{m}$) is expressed in m^3/hr; ρ_a is air density; c_P is specific heat of air; T_{ex} is the temperature of the expired gas stream, which may be measured at the nostril; λ is the heat of vaporization of water (0.7 watt $\times$ hours $\times$ gm^{-1}); and the ρ_w values are those for both exhaled and inhaled air. The temperature of expired gas is precisely that of blood only in a hot, moist environment.

Heat conduction occurs through body contact areas, usually support surfaces such as the soles of the feet or those in contact with an insulating bed to which conduction is very slow by design.

The body can respond to temperature fluctuations in several specific ways. These include changes in metabolic rate (*M*), the skin-to-air heat-exchanging area (for example, by spread-eagling or curling up), skin temperature (T_s), ρ_{ws}, and A_e. Both cold and extreme heat increase metabolic rate. Under cold-to-comfortable conditions the rate of peripheral blood flow regulates T_s. ρ_{ws} may attain saturation values on maximally sweating skin or it may be very near ρ_{wa} during insensible perspiration. Reverse transfer may actually occur into nonsweating areas—water vapor enters the skin from a microenvironment of saturated air.

In bioclimatologic analysis of a complex set of data that define an environment, the paramount considerations are survival and safety. Both short-term and long-term health must be considered. Duration of exposure is an important variable. Limiting values define an area in which survival is possible and within this lie zones of tolerable discomfort and comfort. Fig. 3-3 gives data for man, either clothed lightly or partially or naked; environmental conditions include either a standard gentle wind or, in the experiments, an unventilated room in which $T_a = T_r$.

Circulatory failure limits exposure to moderate conditions, whereas skin failure limits exposure to extreme conditions. Some special limiting conditions are as follows. On a hot-dry desert *S*, T_a, and h_c are so high that a rate of 2 to 3 kgm/hr of sweat evaporation would be needed to produce heat balance; the human body is incapable of sweating at such a rate. In the jungle on a hot-moist desert *S*, h_c, and h_e are moderate, but T_a approaches T_s; the air is so moist that $\rho_{ws} - \rho_{wa}$ is small and evaporative cooling is ineffective.

Survival in cold requires good clothing. The following equation for h_c' is a modification of our equation for h_c that includes two clothing characteristics:

$$h_c' = \frac{1}{\frac{1}{h_c} + \frac{d}{k}} \qquad (7)$$

where d = thickness
k = heat conductivity of clothing

For heavy clothing or sleeping bags equation 7 simplifies to $h_c' = \frac{d}{k}$ and equation 1 simplifies to $M = \frac{k}{d} A_c (T_s - T_a)$ + lung loss (*L*).

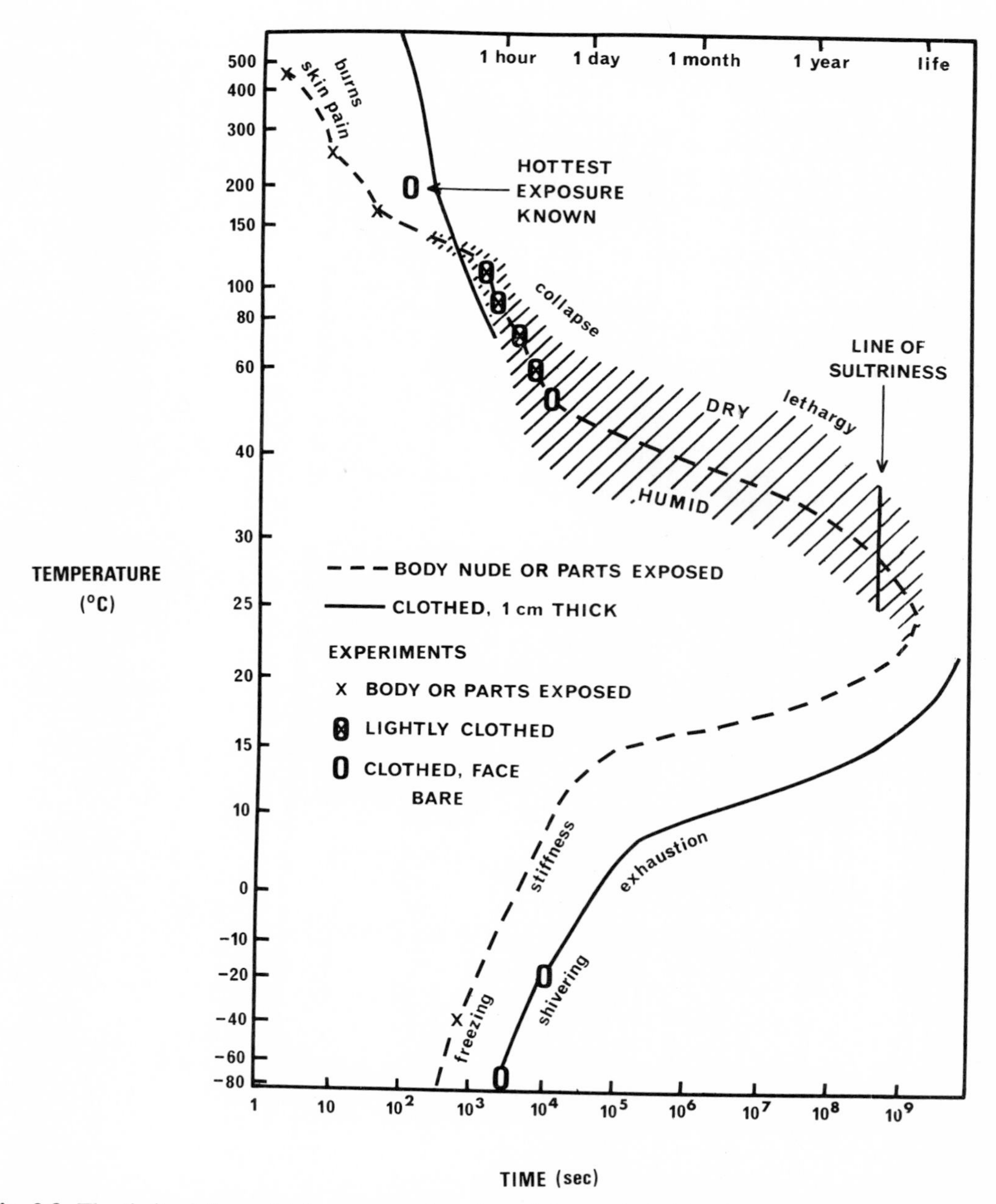

Fig. 3-3. The limit of thermal tolerance as a function of time and clothing for healthy man at rest. Hatching shows the influence of humidity. In the experiments the room is unventilated and, because air and walls are at the same temperature, free of net radiant heat transfer. (Modified from Buettner, K. J. K.: Human aspects of bioclimatological classification. In Tromp, S. W., editor: Biometeorology, Proceedings of the Second International Bioclimatological Congress, Oxford, 1962, Pergamon Press, Ltd., p. 129.)

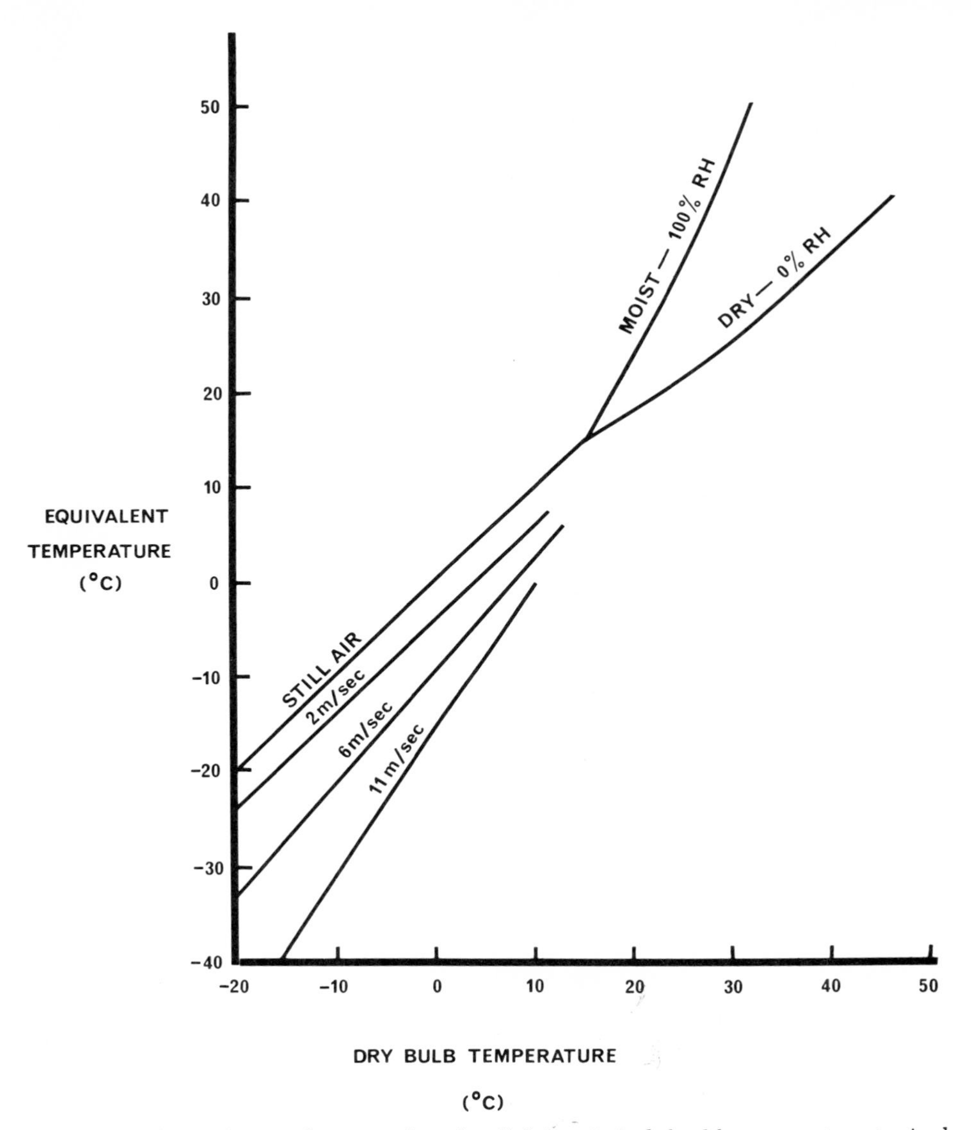

Fig. 3-4. At upper right, above 15° C, are data for lightly clothed healthy man at rest. A dry bulb (actual) temperature of 30° C at a relative humidity (*RH*) of 80% is subjectively equivalent to a desert temperature of 40° C. At lower left, below 15° C. are data for appropriately clothed healthy man at rest exposed to moving air as indicated. A strong wind at 0° C is subjectively equivalent to a temperature of –15° C.

In cool environments T_a and wind-induced h_c have the most potent effect on body temperature.

Many attempts have been made to predict human responses to temperature change. These models hold either T_s constant and measure *M* or hold *M* invariant and measure T_s. Human reactions vary widely and models must account for the highly variable effects of clothing. Precise prediction requires a set of good climatic data and thorough knowledge of the relevant physiologic regulatory systems. Models for hot environments are simpler because T_s approaches body core temperature and is thus not an important variable. The rate of metabolic heat production is low; when h is known the required sweat rate is calculable.

A widely used combination of T_a and ρ_{wa} is the *effective temperature* (T_{eff}); this is based on

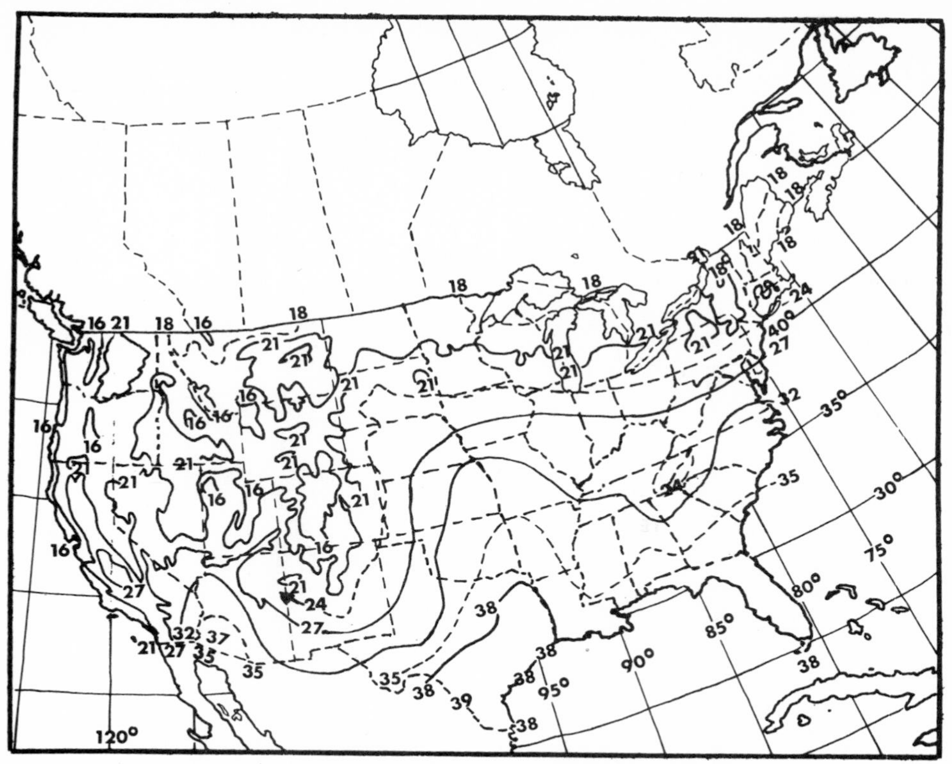

Fig. 3-5. Average desert equivalent temperatures in degrees celsius during July. The effect of humidity has been neglected for air temperatures below 21° C. All data are tentative. (Modified from Buettner, K. J. K.: Human aspects of bioclimatological classification. In Tromp, S. W., editor: Biometeorology, Proceedings of the Second International Bioclimatological Congress, Oxford, 1962, Pergamon Press, Ltd., p. 134.)

the subjective reports of volunteers rating the comfort of any combination of temperature and humidity. It is also termed *discomfort index* or *temperature-humidity index*. Effective temperature is the temperature in degrees of a saturated atmosphere that would feel like the actual one. *Desert equivalent temperature* (DET) is based on the same set of observations and seems more appropriate; it compares the real atmosphere to a fixed water vapor density of 10 gm/m^3 instead of to a saturated atmosphere (see Fig. 3-5).

Total yearly climatic hardship and the annual cost of heating and cooling can be estimated as follows: for heating, the time integral of the difference between comfortable indoor temperature (about 22° C) and outdoor temperature yields a value whose units are degrees times days. As temperature rises above comfort level physiologic strain increases about twice as fast per degree increase of effective temperature as per degree decrease. Twice the energy is required to reduce effective temperature 1° by air cooling and drying as to increase effective temperature 1°. Fig. 3-6 shows these calculations on a map of the United States. It suggests that the moderate Mediterranean-like climate of the west coast of the United States requires little heating or cooling. It also shows that total climatic strain and cost of cooling or heating is about equal in southern Texas and North Dakota.

A review of worldwide effective temperature data for summer reveals that the Red Sea and the Indus Valley are worst. In the United States the lower Rio Grande Valley and a section of the Gulf Coast are worst. These values are even worse than those for typical tropical areas, such as Devil's Island in Guiana. Note that we are comparing the Texas summer with the climate of a tropical station that has little seasonal change.

RADIANT ENERGY

Most of the radiant energy we receive comes to us from the sun. Per unit area of body surface our bodies emit about one tenth as much radiant

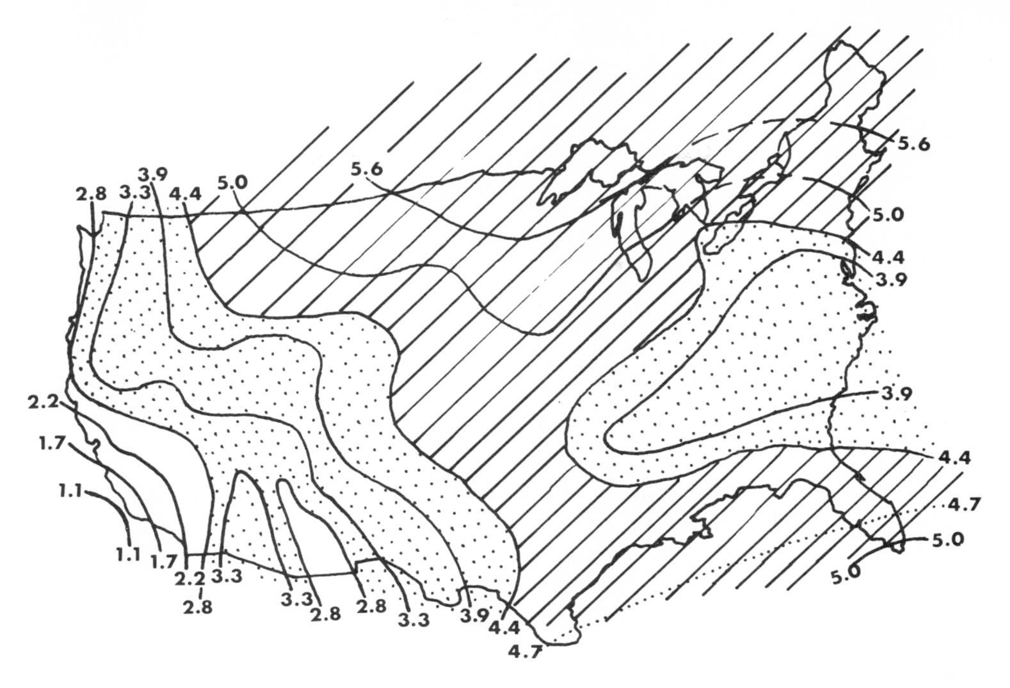

Fig. 3-6. Heating degree-days plus 2 times cooling degrees-days. Units: °C × days × 1,000. The larger this sum, the greater the average year-round outdoor discomfort. The first term is the time integral of the difference between comfortable temperature and outside air temperature. The second term uses equivalent temperature or a modified form of it instead of comfortable temperature. The numerical values of these two terms are respectively proportional to the cost of heating and cooling a well-insulated home. (Redrawn from Buettner, K. J. K.: Human aspects of bioclimatological classification. In Tromp, S. W., editor: Biometeorology, Proceedings of the Second International Bioclimatological Congress, Oxford, 1962, Pergamon Press, Ltd., p. 133.)

energy as we receive in full sunshine; compared to the latter we receive only 10^{-6} as much from the full moon and 10^{-9} as much from the stars. At the outer fringe of our atmosphere 1.38 kW/m^2 (about 1.96 cal × cm^{-2} × min^{-1}) arrive from the sun, a quantity sometimes called the solar constant. Specific bands of this electromagnetic spectrum are absorbed by atmospheric ozone, water vapor, and carbon dioxide. Remarkably, the visible band is transmitted with minimal absorption; however, pollutant gases such as sulfur dioxide may absorb certain specific colors.

A photochemically produced layer of ozone extends from an altitude of about 10 to 50 km, absorbing the shortwave end of the ultraviolet spectrum (about 300 nm*). Ozone is somewhat more plentiful at higher latitudes. More important than ozone and cloud cover in determining the amount of available ultraviolet (UV-B) is solar altitude, or elevation, a factor that varies with time of day, latitude, and season. Infrared (IR), both solar and low temperature (about 10 μm), is strongly absorbed by atmospheric water vapor.

The physical atmospheric phenomena of refraction, reflection, and diffraction also attenuate solar radiation. This scattering of sunlight by air molecules produces the blue color of skylight; the shorter the wavelength, the greater the scatter. The red band is negligibly scattered, but sunburn-producing UV-B is so effectively scattered that more UV-B comes to the earth's surface from the sky than directly from the sun. Sunlight is also scattered by clouds, aerosols, and their surfaces. For example, clouds diffusively reflect as much as 80% of the total solar radiation, including UV-B. Fresh snow, the only natural substance that substantially increases the total UV-B dose to exposed skin, may reflect as much as 90%. Light-scattering by clouds, fog, and aerosols affects visibility and is often a crude indicator of high levels

*1 m = 10^3 mm = 10^6 μm = 10^9 nm. The latter two are called micrometer and nanometer, formerly called μ and mμ.

of air or water pollution. Light scattered from surfaces provides the input for most of our visual perception. Reflectance may be high or low, gray or colored (spectrally differentiated), and specular (mirror-like) or diffuse.

Natural electromagnetic radiation of wavelength longer than the visible spectrum has been generally considered to have no specific (nonthermal) effects on man. The physiologic effects of visible and shorter wavelengths include the following: (1) heating by absorbed and cooling by emitted infrared (discussed in the section on heat transfer), and (2) specific physiologic effects that result from the physical processes of excitation, dissociation, ionization, and nuclear reactions.

The eye is a paramount example of photochemistry. We see wavelengths from the ultraviolet (about 313 nm) to the deep red (about 780 nm). The intensity of natural light that can be transduced in the retina and appreciated in the visual cortex ranges from the blinding brightness of a sunlit snowfield to the dim signals detected by the visual purple on a starry night. The ratio of these light intensities (signal strength) is about $10^{10} : 1$.

When skies are clear, blinding brightness is seen in two terrestrial zones that have highly reflective surfaces—the desert belt and the arctic-antarctic. The darkest regions on the surface of the earth are the tropical rain forests.

Both visible and ultraviolet light produce an immediate pigment in the skin of some individuals by darkening preformed melanin in the presence of oxygen. Sunlight may produce this pigment within minutes; it usually disappears rapidly.

Natural UV-B wavelengths less than 320 nm, and especially those less than 312 nm, produce a variety of photochemical reactions: (1) corneal inflammation that may result in snow blindness; (2) sunburn—skin redness, inflammation, and desquamation; (3) a characteristic pleasant odor of unknown origin; (4) delayed or normal pigmentation; (5) conversion of 7-dehydrocholesterol in the skin to the antirachitic vitamin D_3 (dihydrotachysterol); and (6) after many years of exposure, skin elastosis, keratosis, and basal cell carcinoma (a type of skin cancer). The last is more common in individuals who have a fair complexion.

The radiation intensities that produce these effects constitute about 0.1% of the total available solar energy; the spectral range is sharply limited at one end (about 312 nm) by the transmittancy of the epidermal stratum corneum and by the basic photochemical mechanisms and at the other end (about 297 nm) by ozonic absorption. Thus these effects are all produced by a narrow band of about 15 nm out of the almost 3,000 nm wide solar spectrum. In the band of highest sunburn and vitamin D–producing efficiency the maximum available UV-B is about 30 $\mu W/cm^2$ per 15-nm bandwidth.

The ultraviolet action spectra that cause corneal inflammation and sunburn and that produce vitamin D are strikingly similar at their long wavelength end because the molecules involved have a ring structure and double bonds; these include proteins, nucleic acids, porphyrins, carotenoids, and sterols. Activated proteins and nucleic acids are thought to dilate capillaries.

Susceptibility to erythema varies greatly from person to person. At high noon on a clear summer day at 50-degree latitude the time required to produce perceptible reddening of skin never exposed before varies from 4 to 40 minutes for white Caucasians. Dark races are almost insusceptible to erythema. Most prone to sunburn and eventually skin cancer are those of fair complexion—white skin, blond or red hair, and light-colored eyes. Sunbathers should note that when unadapted sun-exposed skin contrasts visibly with protected skin areas, it is time to terminate a sunbath. Although uncommon, sunburn does occur in animals less intelligent than man.

From 2 to 3 days after exposure to the sun pigment develops in the skin and may remain, albeit diminishing, for months or even years. Both sun-produced and hereditary (racial) pigmentation protect well against sunburn. Urocanic acid, a product of sweat, is another natural sun shield and may explain the reduced susceptibility to sunburn during summer. Unfortunately, all this protection from UV-B, including hair, clothing, and shelter, also prevents the production of vitamin D_3 within the skin.

It is naive to attempt to relate the amount of solar ultraviolet rays arriving at treetop level to the probability of long-term effects such as skin cancer. We absorb only a small and variable fraction of the total available UV-B. If it is too hot (Bedouin) or too cold (Eskimo) for free body exposure, we don clothing or seek shelter. The absorption of UV-B by human skin relative to the amount available (utilization ratio) is affected by climate, skin pigmentation, and custom.

Fig. 3-7 shows the global pattern of available UV-B in relation to latitude, season, and cloudiness. The latter is given only on an average basis;

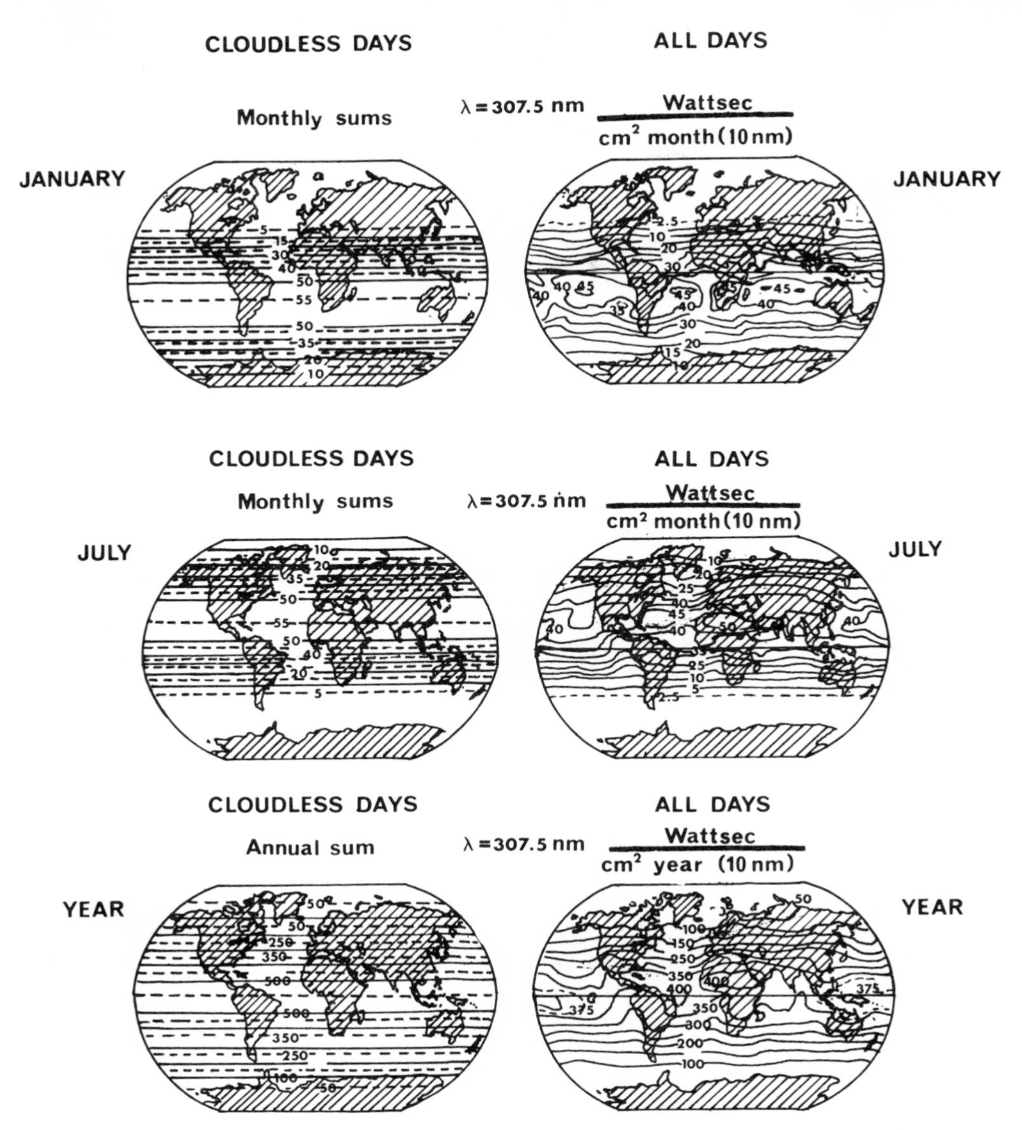

Fig. 3-7. Global geographic distribution of solar ultraviolet radiation ($\lambda = 307.5$ nm) in January for cloudless days (left) and for all days (right). Monthly sums in watt-sec $\times$ m^{-2} $\times$ $month^{-1}$ (10 nm) were computed using the measurements of Bener. (Modified from Schulze, R., and Gräfe, K.: The biological effects of ultraviolet radiation with emphasis on the skin. In Urbach, F., editor: Proceedings of the first international conference of the Skin Cancer Hospital, Temple University Sciences Center and the International Society of Biometeorologists, Oxford, 1968, Pergamon Press, Ltd., p. 366.)

UV-B is nil under a large thundercloud but may be quite intense under a typical stratus cloud or within a fog. As described earlier the atmosphere itself scatters UV-B effectively; moderate cloud cover does not add much to this. Clouds may, however, lower ultraviolet usability—when it is too cool for the faddists to toast themselves.

The evolutionary significance of racial pigmentation remains a mystery. If black man originated in equatorial rain forests, his color would have served the biologic purpose of camouflage. He would have received very little UV-B to produce vitamin D_3 in his skin. However, natural food probably contained enough vitamins D_2 (calciferol) and D_3 to compensate for this deficiency. White man probably lived in caves, obtained

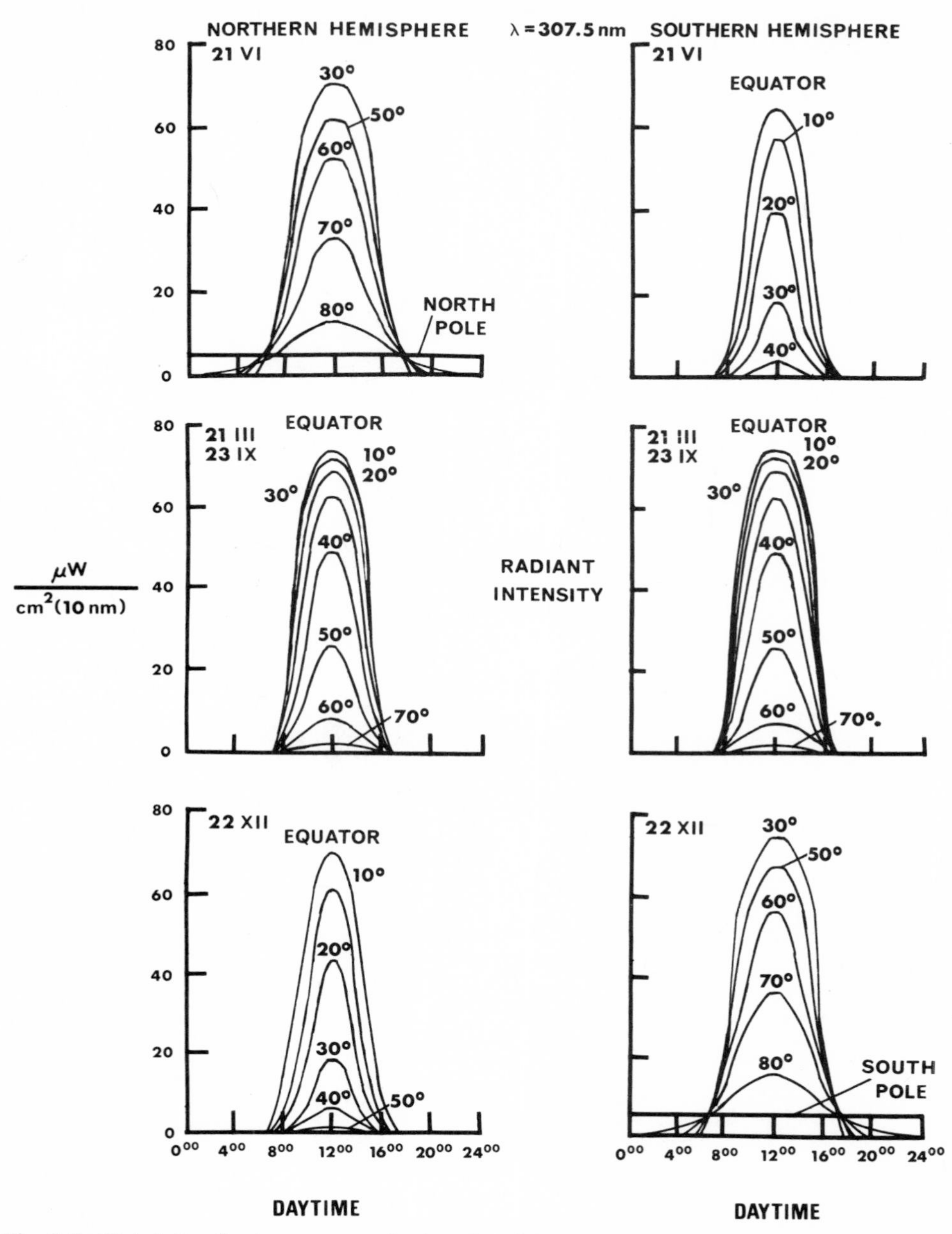

Fig. 3-8. Global distribution pattern of solar ultraviolet radiation ($\lambda = 307.5$ nm) intensity as a function of time of day and latitude computed using the measurements of Bener. (Redrawn from Schulze, R., and Gräfe, K.: The biological effects of ultraviolet radiation with emphasis on the skin. In Urbach, F., editor: Proceedings of the first international conference of the Skin Cancer Hospital, Temple University Sciences Center and the International Society of Biometeorologists, Oxford, 1968, Pergamon Press, Ltd., p. 367.)

Table 3-1. Typical bioclimates*

Climate types and examples	Average air temperature (worst months) (° C)	Average water vapor density (worst months) (gm/m³)	Worst season*	Solar UV-B	Relative magnitude of air temperature variation		Typical aerosol	Average cloudiness	Frequency of thunder-storms	Frequency of strong winds
					Daily	Seasonal or yearly				
Equatorial rain forest Microclimate: jungle Location: Para, Brazil	26	20	—	0	0	0	Microorganisms Insects	++	++	0
Hot dry desert Microclimate: sand dune Location: Bagdad, Iraq	44	7	Summer	+++	++	++	Dust storms	0	0	++
Hot moist desert Microclimate: sand dune Location: Aden	37	25	Summer	+++	++	++	Dust storms	0	0	++
Tropical savanna Microclimate: grassland Location: Saigon, RVN	28	20	Summer	++	++	++	Microorganisms Insects	++	++	+
Temperate warm			Summer	++	+	++	From combustion	+	++	++
Temperate cool			Winter	+	+	++	From combustion	+	+	++
Cool dry desert Microclimate: gravel Location: Tibet	Low	Low	Winter	+++	++	++	—	0	0	+
Cool foggy desert Microclimate: coast Location: Chile	16	11	?	+	0	+	Fog	++	0	+
Polar Microclimate: snowfield Location: Fairbanks, Alaska	–20	0	Winter	0	0	++	Ice fog	++	0	+++

*"Typical" and "worst" with respect to man.

enough vitamin D from his diet, and had an ample dose of solar ultraviolet each summer.

THE BIOCLIMATES

Bioclimates are climates that affect man in various geographic settings (see Table 3-1). In comparing bioclimates one must ask, "Bioclimate for whom?" For example, the white newcomer to the tropics responds differently to the climatic stress and feels less comfortable than either the acclimatized resident or the adapted native. During winter the thermostats in American homes are frequently set at 24° C as compared to 17° C in England. In regions of cold winter climate, homes are more effectively insulated and heating devices have greater output capacity; without these accommodations even a Mediterranean winter can be a miserable experience.

The city in temperate zones

Modern man often lives in houses and vehicles within the city rather than in a rural microclimate. The physical structure of the city offers considerable resistance to the wind, creating friction. Its streets and sewers remove rainwater and prevent surface evaporation. A city produces approximately one third of the amount of heat that is obtained from the sun. Finally, a city produces air pollution. Thus the urban microclimate differs significantly from its rural surroundings.

Urban air temperature is always higher by about 1° C, especially during winter nights; on a clear winter night it may be as much as 10° C higher. Relative humidity is about 5% lower. Clouds, fog, and precipitation are more frequent. Total solar radiation is, on the average, about 10% lower, especially during winter; ultraviolet radiation is reduced to about the same extent. The nature and intensity of air pollution vary greatly, depending on its sources: heating, industry, and internal combustion engines. During the warm season in Los Angeles, California, westerly winds push cooler air under the warmer continental air. Intense air pollution is thus held in place, the temperature inversion acting as a lid. Such conditions may produce smog at a relative humidity as low as 20%.

Equatorial tropics

Human microclimate varies widely depending on the particular location. Rain forests are perennially dark, the temperature is constant, relative humidity is high, and there is no wind. At the seashore, weather alternates between heavy rain and sunshine, and sea breezes make the afternoons bearable. Houses may consist of walls that are only insect screens and roofs to keep out sun and rain. There is plenty of UV-B in open places, but temperature and humidity are not conducive to sunbathing. Climatic conditions are ideal for the growth of microorganisms. Thus infectious diseases are prevalent in the tropics and textbooks on tropical medicine are mainly concerned with infectious diseases.

Hot-dry and hot-moist desert climates

As we move from the equator to higher latitudes, tropical rain forests give way first to periodically dry savanna, then to the intermediate steppes, and finally to the desert belt that stretches from about 15 to 35 degrees latitude if land and sea configuration does not produce a monsoon climate instead, as in India. Within the world's hot desert belts are not only the highest air and surface temperatures but also the highest summer humidities on record. The water vapor content of the Indus Valley, resulting from local evaporation of the oasis system, is more than 30 gm/m^3.

Desert is a plant-geographic term that means an area with too little precipitation to support vegetation. In summer, in terms of human comfort, there are hot-dry deserts and hot-moist deserts. During the hottest month in hot-dry Bagdad the means are $T_a = 43.5°$ C and $\rho_{wa} = 7.0$ gm/m^3; in hot-moist Aden the means are $T_a = 36.5°$ C and $\rho_{wa} = 24.6$ gm/m^3. The effective temperatures are 28° and 30° C, respectively.

A desert in summer is cloudless. Noon air temperatures of 50° C are common; the sand surface exceeds 80° C. The daily temperature variation is large. Shady areas are few. Humidity is higher than frequently assumed. Water vapor content of the summer Sahara is about 10 gm/m^3, contributing heavily to the high effective temperature, or discomfort index. The traveler faces a high radiant heat load from both direct and reflected sunlight (sand reflects about 40% of the total but only 17% of the ultraviolet) while hot sand emits infrared intensely. Appropriate clothing is white and loose enough to permit circulation of air over the sweating skin. Only fools sunbathe.

Frequent windstorms produce the typical pollutant aerosol, a mixture of sand and dust. The dust is composed of soft minerals (calcite, diatoms, and limestone) ground to a fine powder by wind-driven hard quartz sand. Residents inhale this aerosol recipe.

Mountains

For every 1,000-meter increase in altitude, air temperature decreases approximately 6° C while barometric pressure and partial pressure of oxygen (P_{O_2}) decrease by about 10%. However, air temperatures may vary from a 10° C/km decrease to an actual temporary reversal or inversion—increasing temperature with altitude within a certain atmospheric layer. The barometric pressure decrease depends on the absolute temperature and may vary by ±10% about the average decline. Pressure decreases produce hypoxia and changes are also perceived in the middle ear, whereas increases may produce pain by closing the eustachian tube while exerting pressure on the tympanic membrane.

Humidity declines sharply with altitude, hence the air in a heated mountain dwelling is extremely dry. Sweat evaporation is facilitated both by the reduced barometric pressure and by the low humidity, except in the lower jungle portions of tropical mountains. Convection of the less dense air transfers less heat (cools less) for a given wind velocity; however, mountain winds can be fierce.

As we rise above the lower atmosphere with its clouds and air pollution, the sky appears darker, although solar radiation is generally more intense; however, sky ultraviolet (UV-B) is essentially unchanged up to 4,000 m but increases sharply at solar zenith. At 3,000-m elevation in the Alps total UV-B is about twice that observed at 700 m; snow reflection may nearly double this difference again.

As a rule air pollution decreases with altitude. Exceptions occur with anticyclonic inversions in valleys that have their own sources of pollution and no connection with open plains. In valleys that run from mountain sides to open plains regular day and night drainage winds remove polluted air. The bright snowfields of mountains and polar regions blind the eye; reflected ultraviolet inflames the cornea and burns the skin. Warm air melting snow produces a special microclimate.

Cold climates

Although the climate of high latitudes at sea level supports flora that are similar to those of the high altitudes in temperate zones, the partial pressure of oxygen is an important difference between these two climates for man and other mammals. Conditions are otherwise similar in these cold, windy climates of high convective heat loss. Only good clothing and adequate shelter make them bearable; no wonder the Eskimo invented a totally protective suit of clothing.

The lowest air temperature (T_a) occurs in holes—concave areas without outflow to lower plains—on calm nights when heat is lost rapidly from the air by radiation to ground and sky. In such a hole in the Alps, air temperature was more than 25° C lower than on an adjacent plain. The cold pole in Siberia, topographically similar, is another example of such a hole.

As we move from temperate toward polar regions, the daily cycles of sunlight and temperature are replaced by pronounced yearly cycles. Reports of early arctic explorers described profound weakening of men during the long polar night. More recent expeditions have not confirmed those observations but have suggested instead that weakness may have been caused by malnutrition or inactivity. Certainly we can live without UV-B. Whether a total lack of photic eye stimulation is deleterious remains unknown.

The northern hemisphere has experienced four ice ages during the last million years. Thus cold climates have periodically extended their domain and will likely do so again in the future.

How to learn about a bioclimate

Weather bureaus, their statistics and maps, usually cater to the needs of the aviation pilot who is interested in low-lying clouds, ground level visibility, and air currents or to the farmer and hydrologist who need information about maximum and minimum temperatures and the probability of precipitation. The environmental data needed by a physiologist or bioclimatologist to study the effects of climate on man are not generally available from weather stations. For example, temperature maxima and minima may be available at a given location, such as an airport weather station, but not temperature variation as a function of time and place. Weather stations do not measure radiant environmental temperatures; solar radiant heat is measured in relatively few places, and data for solar and sky ultraviolet are even more difficult to find.

The water vapor concentration of the atmosphere cannot be estimated from rainfall. Air pollution data are just now becoming available as monitor stations are hastily improvised to avert an impending environmental catastrophe. Although air pollution measurement should be a routine aspect of weather observing and recording, it is usually the concern of other agencies. Yet other agencies may count pollens and other airborne allergens and measure UV-B, lake temperatures, river runoff, water pollution, and ionizing radiation, such as cosmic rays and gamma rays from ground sources.

The known biologic effects of hypobaria per se —whether only reduced or fluctuating ambient pressure—are limited to decompression sickness, barotitis, barosinusitis, and effects resulting from hypoxia or reduced gas density. Nevertheless innumerable correlations of ambient pressure with a wide variety of biologic functions are assumed to indicate a cause-effect relationship. When such a relationship is alleged, the data for meteorologic phenomena associated with ambient pressure changes, including wind direction (in regard to humidity and air pollution), air temperature, precipitation, sunlight intensity, visibility, lightning, and any other elements of possible relevance, should be examined. Unconscious observer bias and psychophysiologic reactions of the test subjects cannot be ignored either.

For bioclimatologic work the physiologist must grasp and use the principles of microclimatology. The student of environmental physiology should view the apparent and the obvious with a healthy, scientific skepticism. One must remember that biologic processes as well as climatic elements vary with daily and seasonal rhythms; temporal coincidence does not prove causation.

METEOROTROPISM, IONS, AND OTHER UNSOLVED PROBLEMS

In the ninth century A. D. the Lex Frisiorum set a fine (weregeld) for the crime of cutting off the hand or arm of another Frisian. This fine was to be doubled if the remaining stump were sensitive to changes in the weather. Scars, arthritis, and certain rheumatic conditions may ache as weather changes, as weather fronts pass, or during certain other synoptic weather conditions. Headaches may presage a storm. This subject, an aspect of medical climatology, has also been called "statistical biometeorology." Unfortunately the extensive literature on this subject contains little solid fact for the physiologist to analyze and interpret; in this twilight zone of science, quacks and mystics thrive.

Repeatedly, one finds descriptions of a temporal association of meteorologic phenomenon A with biologic event B, concluding that "post hoc, ergo propter hoc." However, such correlations are incapable of establishing causation and indeed may prove only that phenomenon A and event B both occur on the same rotating planet revolving about the same sun. These correlations are usually of only poor to fair degree and frequently the highly significant correlation coefficients are those near zero. Some of the medical symptoms just mentioned correlate poorly with measurable weather elements such as barometric pressure, air temperature, humidity, and sunlight intensity; however, controlled, definitive studies in the environmental chamber remain to be done. Some knowledge has emerged from this twilight zone to achieve respectability—for example, the natural history of malaria and the phenomenon of biorhythmicity.

Of the 10^{19} molecules in 1 cc of air at standard temperature and pressure, from 10^2 to 10^3 are ions, that is, clusters of air and water molecules that have a positive (+) or negative (−) charge. These smaller ions are produced by cosmic and terrestrial ionizing radiation. They tend either to recombine and disappear or to attach themselves to larger aerosol particles or droplets, forming intermediate or larger-sized ions. It is claimed that inhalation of artificial unipolar ions, either positive or negative, affects the ciliary motion of respiratory epithelium and also ameliorates the symptoms of hayfever. If indeed this were a chemical reaction, such an ion effect would be about 10^6 times more effective on a molar basis than the most powerful chemical agent yet known! The concentration of naturally occurring ions is far less than that in such environmental chamber tests and they are never unipolar. Thus we cannot explain meteorotropism as an effect of atmospheric ionization.

COMPARATIVE BIOCLIMATOLOGY

Although this chapter deals primarily with human bioclimatology, some generalizations about comparative bioclimatology may help to place the subject matter in perspective. Plants generally require water, carbon dioxide, sunlight, oxygen, and nutrients from the soil or water in which they grow. Photosynthetic green plants produce more oxygen than they themselves require, enriching the atmosphere as they remove carbon dioxide. Animals require oxygen, water, and food. All animal species are sensitive to ambient oxygen pressure; hypoxia impairs them and hyperoxia intoxicates them. As in the case of man before the advent of modern transportation, the availability of oxygen, water, and food is a basic determinant of the natural geographic distribution of animals.

Physiologically, other homeothermic animals differ from man in many ways. The most significant differences are associated with the function of organs derived from ectoderm: the central nervous system and the epidermal structures, including hair, nails, and cutaneous glands.

All living biologic systems exist within the limits of thermal constraints. Human thermoregulation involves millions of eccrine sweat glands and is facilitated by an effective epidermal stratum corneum. Certain other mammalian species also use sweat evaporation to prevent hyperthermia, for example, the chimpanzee, equidae, camel, and certain breeds of cattle. Thermoregulation in some species that do not sweat is accomplished by panting or by wetting themselves with saliva (licking).

Does natural sunlight kill microorganisms? Desiccation of microbial habitats by solar heat may be a more important antimicrobial factor.

Animals achieve protection against cold in various ways. These include high metabolism and ample responsive blood flow (birds), thick fat insulation (whales), fur insulation (polar bear), low surface/mass ratio, and effective shelter (snow cave, burrow). In terms of thermoregulation the hibernators are between the homeotherms and the poikilotherms.

Biologically, the color of skin and fur is chiefly significant for camouflage or sex attraction. This was probably also true for our hominid ancestors.

GLOSSARY

aerosol A suspension of solid particles or liquid droplets in air. Manmade aerosols are of many types, may be harmless, noxious, obnoxious, or even therapeutic, and may be a component of polluted air. Examples of natural aerosols are sea salt, windblown sand and dust, debris from plants and animals, microorganisms, insects, water droplets (rain, clouds), ice particles (snow, clouds), smoke (unburned hydrocarbons, soot) from natural fires, and volcanic ash. Examples of manmade aerosols are unburned hydrocarbons, ash, soot, coal from combustion, mineral debris from cement factories and building construction, lead compounds from motor exhausts, and rubber and oil particles from motor cars and roads.

convection The transfer of energy and transport of matter by the circulatory mass flow that occurs in an unevenly heated fluid such as the atmosphere; thermally produced and sometimes mechanically assisted upward and downward air currents involving a limited region of the atmosphere.

desert A geographic region that receives too little rain for its temperature regime and therefore having little or no plant life. Note, however, that certain rainless deserts have some of the highest water vapor densities on earth; these are called moist-hot deserts.

desert equivalent temperature (DET) The same as effective temperature except that values are for a water vapor density of 10 gm/m^3 (typical for a dry desert in summer) instead of water vapor saturation.

effective temperature (T_{eff}) A measure of the combined effects of air temperature and humidity on man based on the subjective report of comfort or discomfort by volunteers. Effective temperature is the temperature of saturated air in degrees celsius (° C) that produces the same subjective sensation as the environmental air of a given humidity and temperature.

infrared Band of electromagnetic radiation with wavelengths longer than those of visible light (about 0.78 μm). For practical purposes infrared is divided into a solar band (0.78 to 3.0 μm) and a low temperature terrestrial band (4 to 50 μm); other infrared bands of natural origin are weak at ground level.

pigmentation, solar, delayed Persistent skin pigmentation occurring several days after exposure to solar UV-B.

pigmentation, solar, immediate The brownish red pigment that appears promptly in the skin of some subjects during exposure to the sun. It is a response to visible light and ultraviolet radiation.

reflectance The ratio of reflected to incident radiation above a reflecting surface.

ultraviolet Band of electromagnetic radiation with wavelengths shorter than those of visible light. For practical purposes ultraviolet is divided into three regions:

1. *UV-A,* 320 to 400 nm. This region has only weak photochemical effects on man; it is visible and causes some sunburn and immediate pigmentation. Ultraviolet of this wavelength is naturally abundant at ground level.
2. *UV-B,* 290 to 320 nm. This invisible region produces the typical natural ultraviolet effects on skin and eye. In nature at ground level its intensity is relatively low and highly variable.
3. *UV-C,* 240 to 290 nm. This region produces most of the effects of UV-B but penetrates only superficially. At ground level it is available only from artificial sources.

REFERENCE

1. Folk, G. E., Jr.: Introduction to environmental physiology, Philadelphia, 1966, Lea & Febiger.

SUGGESTED READINGS

1. Altman, P. L., and Dittmer, D. S., editors: Environmental biology, Bethesda, Md., 1966, Federation of American Societies for Experimental Biology.
2. Becker, F., and others. In Sargent, F., and Tromp, S. W., editors: Survey of human biometerorology, No. 160 TP 78 (65), Geneva, 1964, World Meteorological Organization.
3. Buettner, K. J. K., and others: Biometeorology today and tomorrow: principal findings of study group on bioclimatology, Am. Meteor. Soc. Bull. **48**(6):378-393, June, 1967.
4. Daniels, F., Jr.: Man and radiant energy: solar radiation. In Dill, D. B., editor: Handbook of physiology, Baltimore, 1964, The Williams & Wilkins Co., pp. 969-987.
5. Landsberg, H. E.: Bioclimatic work in the weather bureau, Am. Meteor. Soc. Bull. **41**(4):184-187, 1960.
6. Licht, S., editor: Medical climatology, New Haven, Conn., 1964, Elizabeth Licht, Publisher.
7. Lowry, W. P.: Weather and life: an introduction to biometeorology, New York, 1969, Academic Press, Inc.

8. Sargent, F., and Stone, R. G., editors: Recent studies in bioclimatology, Meteor. Monograph. 2:8, 1954.
9. United States Department of Defense: Air Force: German aviation medicine, World War II, 1950.
10. United States Department of Health, Education, and Welfare: Seminar on human biometeorology, National Center for Air Pollution Control and Environmental Science Services Administration, Jan. 14-17, 1964, Cincinnati, Ohio, Washington, D. C., 1967, U. S. Government Printing Office.
11. Urbach, F., editor: The biological effects of ultraviolet radiation with emphasis on the skin, Proceedings of the First International Conference of the Skin Cancer Hospital, Temple University Sciences Center and the International Society of Biometeorologists, New York, 1968, Pergamon Press.

4 TEMPERATURE AND HUMIDITY

PART A: COLD

LOREN D. CARLSON
ARNOLD C. L. HSIEH

Maintenance of life depends on processes that lead to biosynthesis (renewal of cells, maintenance of cell structure, reproduction, growth, and production of secretions) and internal work (mechanical, chemical, electric, and osmotic). The chemical reactions involved are endergonic and cannot proceed without energy being supplied to the system. This energy is derived from exergonic processes during oxidation of food.* Many of the reactions are exothermic, but, if there is no change in the thermodynamic state of the animal, all the energy released will eventually appear as heat. Thus the rate of heat production reflects the rate of cellular activity and is often referred to as the metabolic rate. Under special conditions of rest and minimal activity the rate of heat production is called the basal metabolic rate.

Mammals are homeotherms (homoiotherm is an occasionally used synonym). This term implies that the animal has a "normal" body temperature that is kept within certain limits. Burton and Edholm[1] have pointed out that a change in temperature not only changes the rate of biochemical reactions but also changes, in complex systems, the qualitative character of the processes. The more complex the system, the more dependent it is on a constant temperature for normal function. Environmental temperature varies with geographic location (the extremes appear to be –68° and 49° C) and even in the same location there are seasonal and diurnal fluctuations. The physiologic need for some mechanism to regulate body temperature is apparent. The fact that in developing these mechanisms the animal becomes "emancipated from his thermal environment" is incidental.

This chapter describes the temperature-regulatory mechanisms of man and certain mammals and the effects of acute and prolonged exposure to temperature extremes on these mechanisms.

THERMOPHYSIOLOGY

Thermophysiology involves the physiologic measurements that reveal the mechanisms of heat exchange between the organism and its environment. The body temperature of an animal is related to its heat production by the following equation:

$$dT_b/dt = (H_{in} - H_{out})/c_p m \qquad (1)$$

where dT_b/dt = rate of change in body temperature (°C/s)
H_{in} = rate of heat input (W) (W = watt = J/s = 0.85984 kcal/hr)
H_{out} = rate of heat loss (W)
m = body weight (kgm)
c_p = thermal capacity at constant pressure ($J \times kgm^{-1} \times {}^\circ C^{-1}$) (J = joule = 0.23885 cal)

The body temperature of a 70-kgm man with a heat production of 81.4 W (70 kcal/hr) increases at a rate of 1.25° C/hr if heat loss is zero, the accumulated heat is equally distributed, and his tissue thermal capacity (Appendix, Table A-6)

*Cellular oxidation of food is extremely complex (complete oxidation of glucose to carbon dioxide and water involves more than seventy separate, sequential enzymatic reactions). The energy released is coupled to endergonic reactions by means of the formation and hydrolysis of energy-rich compounds such as adenosine triphosphate, phosphocreatine, glycerophosphate, and acetyl-coenzyme A.

is 3,350 J × kgm^{-1} × $°C^{-1}$. His body temperature will remain constant if heat production equals heat loss: $H_{in} = H_{out}$.

Analysis of the factors involved in heat exchange is made easier by the use of the following equation, which is a statement of the law of conservation of energy:

$$M + E + R + C + K + W + S = 0 \qquad (2)$$

where M = metabolic heat production (W)
E = evaporative heat exchange, negative for loss
R = radiative heat exchange, negative for loss
C = convective heat exchange, negative for loss
K = conductive heat exchange, negative for loss
W = work accomplished, negative for work against external forces
S = heat storage, negative for storage

For an animal at rest ($W = 0$), equation 2 can be rearranged:

$$M + S = -E - R - C - K \qquad (3)$$

The terms on the right side of the equation represent heat lost from the body when they have been adjusted for direction of exchange (negative for heat loss). The algebraic sum of the values gives net heat loss (H_{out}). If the body is in thermal equilibrium ($S = 0$), equation 3 states the conditions for maintaining a constant temperature.

The avenues of heat exchange between the animal and its environment are depicted schematically in Fig. 4-1. Heat produced in the body will have to traverse the tissues to the surface. Heat exchange between the body and the environment takes place at the surface of the body. It is therefore convenient to divide the discussion into skin-to-environment system and tissue-to-skin system.

Skin-to-environment system

Evaporative heat loss

A change of state from liquid to gas consumes energy. The amount of heat absorbed is related to the heat of vaporization and the amount of liquid vaporized:

$$E = 0.01667 \times \dot{m} \times \lambda \qquad (4)$$

where E = rate of evaporative heat loss (W)
$\dot{m}$ = rate of vaporization (gm/min)
λ = latent heat at constant pressure (J/gm)
0.01667 = factor to convert J/min to W

The latent heat of water varies with the temperature. Some representative values are given in

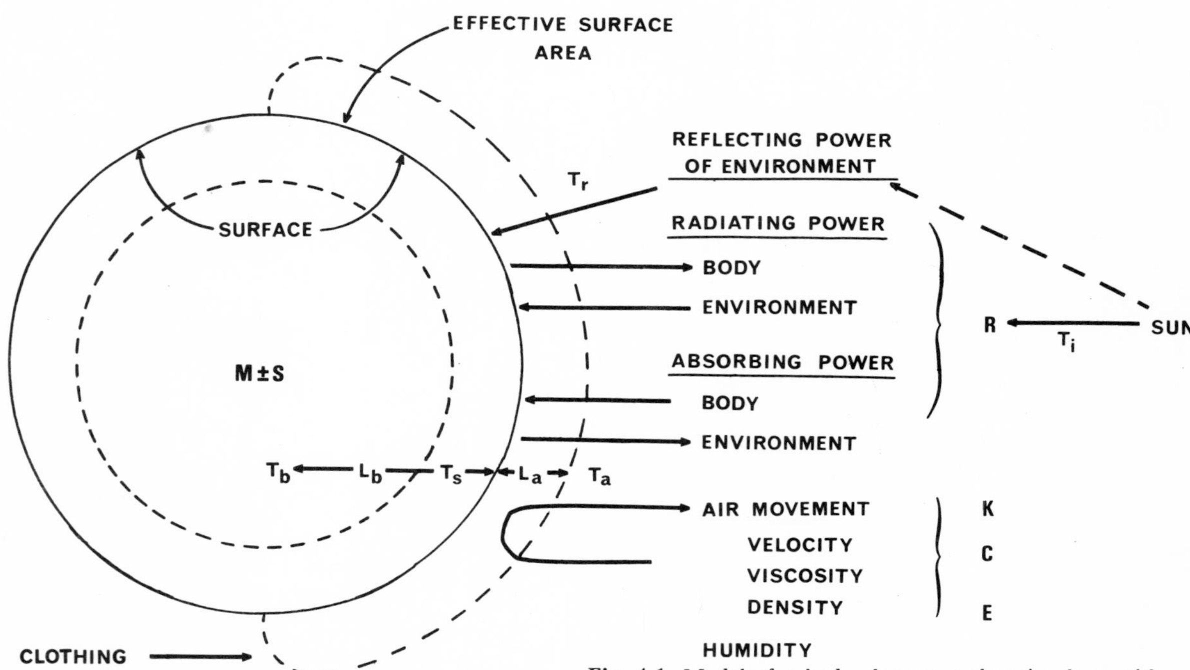

Fig. 4-1. Model of a body element exchanging heat with the environment. Details are described in the text. (Modified from Carlson, L. D., and Hsieh, A. C. L.: Control of energy exchange, New York, 1970, The Macmillan Co.)

the Appendix, Table A-7. A convenient approximation for evaporative heat loss is:

$$E = 40 \times \dot{m} \quad (5)$$

If all the sweat produced by an animal is evaporated (sweat dripping off the surface does not cool it), water loss estimated from the difference between the weight loss and the weight of carbon dioxide produced can be used to calculate E. A man at rest and comfortably warm will lose water from his respiratory tract and by insensible perspiration at a rate of about 30 gm/hr (the equivalent of 20 W or about 25% of his resting heat production). It is usually considered that one fourth of the heat loss occurs in this way regardless of the level of activity.

Heat lost via the respiratory tract can be estimated from the following equation:

$$E_{ex} = 0.01667\ \lambda\ \dot{V}\ (\rho_{ex} - \phi_a \rho_{in}) \quad (6)$$

which can be simplified to:

$$E_{ex} = 40\ \dot{V}\ (\rho_{ex} - \phi_a \rho_{in}) \quad (7)$$

where E_{ex} = heat loss via respiratory tract (W)
$\dot{V}$ = ventilation rate (L/min)
ρ_{ex} = density of saturated exhaled air (gm/L)
ρ_{in} = density of saturated inhaled air (gm/L)
ϕ_a = fractional relative humidity

The density of saturated air varies with temperature as shown in the Appendix, Table A-7. If the ventilation rate is 7.5 L/min and temperature of inhaled air is 20° C with a relative humidity of 50%, E_{ex} is 10.5 W. This calculation is based on the assumption that the exhaled air is saturated and at 37° C. During hyperventilation or in cool environments, exhaled air is not at body temperature. When the inhaled air is warm and dry the exhaled air is not saturated. At high altitude water loss by evaporation is increased because of the increased ventilation rate (a physiologic response to low oxygen tension in inhaled air) and low relative humidity.

The rate of heat loss by evaporation from the surface is:

$$E_{sw} = 40\ h_D(P_{ws} - \phi_a P_{wa})/R_w T \quad (8)$$

where E_{sw} = rate of heat loss from sweating (W)
h_D = transfer coefficient (L/min)
P_{ws} = vapor pressure of water at skin temperature (mm Hg)
P_{wa} = vapor pressure of water at ambient temperature
R_w = aqueous gas constant (3.47 L × mm Hg × gm^{-1} × °K^{-1})
T = average of skin and ambient temperatures (°K)

Equation 8 can be simplified to:

$$E_{sw} = 11.53\ h_D(P_{ws} - \phi_a P_{wa})/T \quad (9)$$

At constant temperature and pressure the rate of evaporative heat loss of a completely wet man is determined by relative humidity and the amount of air movement (in general, h_D is a function of air movement). An empirical equation, based on one derived by Nelson and co-workers[2] is:

$$\dot{E}_{sw} = 11.79\ v^{0.37}\ (P_{ws} - \phi_a P_{wa}) A_w \quad (10)$$

where E_{sw} = rate of evaporative heat loss from sweating in a nude man (W)
v = velocity of air movement (m/s)
A_w = area of wetted surface (m²)

The relationship of wetted area to operative temperature is shown in Fig. 4-2.

Radiative heat exchange

Radiative heat exchange between an animal and the walls and ceiling of a chamber is:

$$R = \sigma(\overline{T}_s^4 - \overline{T}_r^4) A_r \quad (11)$$

where R = radiative heat exchange (W)
σ = Stefan-Boltzmann constant (5.67 × 10^{-8} W × m^{-2} × °K^{-1})
$\overline{T}_s$ = average surface temperature (°K)
$\overline{T}_r$ = average radiant temperature
A_r = effective radiating surface of the animal (m²)

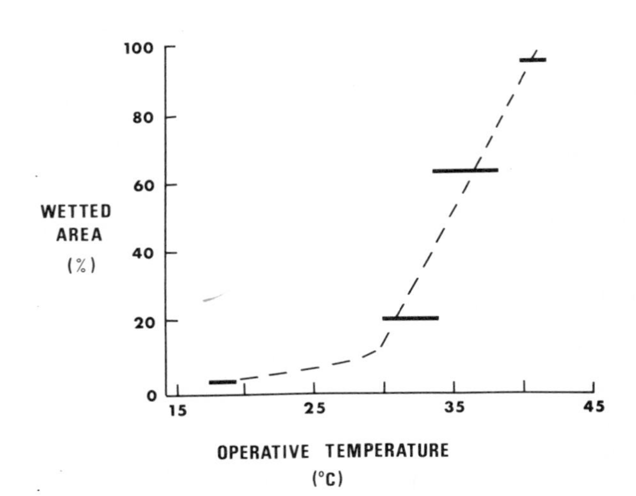

Fig. 4-2. The percent wetted area of a nude subject computed for various operative temperatures. In this experiment operative temperature was the average of wall and air temperature. The dashed line is fitted by eye. (Curve constructed from data of Winslow, C. E. A., and Herrington, L. P.: In Temperature and human life, Princeton, N. J., 1949, Princeton University Press, p. 91.)

A man with a mean skin temperature of 33° C and a surface area of 1.8 m^2 radiating to 21° C has a radiant heat loss of 132.3 W (113.8 kcal/hr). Effective radiating surface is an important determinant of the magnitude of *R*. Curling up may reduce heat loss by as much as one half.

If the difference between the temperatures of the two radiating surfaces is less than 20° K, equation 11 can be simplified to:

$$R = h_r(\overline{T}_s - \overline{T}_r)A_r \quad (12)$$

where $h_r = 4\,\sigma T^3$
T = mean of $\overline{T}_s$ and $\overline{T}_r$ in °K

The radiation heat transfer coefficient, h_r, is $W \times m^{-2} \times {}^\circ K^{-1}$.

For an unclothed man and a source radiating at a very high temperature, skin temperature may be omitted:

$$R = \sigma \epsilon_i \epsilon_s T_i^4 A_p / F_{si} \quad (13)$$

where ϵ_i and ϵ_s = emittance, dimensionless
T_i = temperature of emitting source (°K)
A_p = projected area (m^2)
F_{si} = shape factor, dimensionless

In the interior of an opaque body of uniform temperature that allows complete absorption of entering radiation, blackbody conditions are found. In such a body the rate of emission of radiant energy from its surface corresponds to the unhindered passage of energy from the interior. If the body is a nonblackbody, the rate of emission is lessened by way of reflection because of hindrance at the surface. Emittance for a body at some constant temperature is defined as the ratio of rate of emission of radiant energy by the body, in consequence of its temperature only, to the corresponding rate for a blackbody at the same temperature:

$$\epsilon_x = (R_{em})_x / (R_{em})_{blackbody}$$

The condition of the surface of the body and the condition as to opaqueness are immaterial. The emittance of a blackbody is by definition equal to 1 at all temperatures. The emittance of nonblackbodies, however, varies with temperature. At 35.6° C the emittance for skin is about 0.99. Equations 11 and 12 assume $\varepsilon = 1.0$.

Kirchoff's law relates the radiating power of a surface to its reflecting power. The law states simply that all radiation that falls on a surface is either absorbed or reflected. Consider a body receiving radiant energy, *R*. The rate of gain is proportional to the rate of absorption. The increase in heat content of the body results in a temperature rise ($\Delta T = \Delta H \times c_p^{-1} \times m^{-1}$), which in turn increases emissivity (R_{em}) of the body. The temperature of the body stabilizes when radiant heat emitted equals heat absorbed:

$$R_{em} = R_{ab}*$$

According to Kirchoff's law:

$$R_{ab} = R - R_{ref}$$

thus,

$$(R_{em})_x / (R_{em})_{blackbody} = (R - R_{ref})_x / (R_{em})_{blackbody} \quad (14)$$

Since $(R_{em})_{blackbody} = R$, equation 14 becomes:

$$\epsilon_x = 1 - \rho_x \quad (15)$$

where ρ_x = reflectance

Equation 15 holds only for conditions of constant temperature.

Reflectance of a body varies with the wavelength of the radiant energy falling on it. Skin reflects about 20% of incident radiation with a wavelength of about 1 μ, while radiations with wavelengths more than 6 μ are almost totally absorbed. Between 1 and 20 μ (the near infrared range of light) the color of skin does not have an effect on reflectance. Below 1 μ, however, darker skins reflect less energy. White skin reflects about 30% to 40% of the sun's radiation (about 0.5 μ), whereas the dark skin of a Negro reflects less than 18% of these rays.[3]

Convective heat exchange

The skin is surrounded by a laminar boundary layer several millimeters thick (in calm air). At the inner portion of this layer, heat transfer takes place by pure conduction. Convective heat loss is related to environmental temperature by the following equation.

$$C = h_c(\overline{T}_s - T_a)A_c \quad (16)$$

where h_c = surface coefficient of convective heat transfer ($W \times m^{-2} \times {}^\circ C^{-1}$)
A_c = area involved in convective heat loss (m^2)

The convective heat transfer coefficient is a function of air movement, viscosity, density, and thermal conductivity:

$$h_c d / k_a = F_c(\rho v d / \eta)^n \quad (17)$$

where $h_c d/k_a$ = Nusselt number, dimensionless

*A blackbody that absorbs all the radiant energy falling on it will attain a higher temperature than one that absorbs, for example, 50%.

d = diameter of sphere or cylinder (m)
k_a = thermal conductivity of air (2.53×10^{-2} W $\times$ m^{-1} $\times$ °C^{-1})
F_c = a constant
$\rho vd/\eta$ = Reynolds number, dimensionless
v = velocity of air movement (m/s)
ρ = density of air (1.22 kgm $\times$ m^{-3})
η = viscosity of air (dyne $\times$ s $\times$ m^{-2} = 1.79×10^{-5} kgm $\times$ m^{-1} $\times$ s^{-1})
n = a power function determined by the structure of air flow

Buettner[4] found F_c to be 1 and n to be 0.54 for nude adult males in the supine position. For practical purposes $n = 0.5$ is a suitable approximation. By rearrangement, equation 17 becomes:

$$h_c = F_{ca}(v/d)^{0.5} \qquad (18)$$

where $F_{ca} = k_a(\rho/\eta)^{0.5}$, a constant at constant temperature and pressure, 6.53 at 1 atmosphere and 15° C

Since density of air is directly related to pressure, h_c decreases with decreasing atmospheric pressure. Equation 18 indicates that h_c is inversely related to the square root of the diameter of the part involved in convective heat transfer. Thus convective heat loss per square meter from fingers is greater than from the trunk.

Assuming $d = 0.36$ for adult man:

$$h_c = 10.9\ (v)^{0.5} \qquad (19)$$

Buettner,[4] Nelson and associates,[2] and Winslow and his colleagues[5] obtained factors equal to 13.9, 8.6, and 12.1, respectively, in experiments on nude adults in the supine, standing, and sitting positions.

In the absence of forced flow, the higher surface temperature in itself produces convection. However the body is rarely surrounded by completely calm air. In rooms of normal size and without perceptible air movement, values of 3.8 and 4.5 W $\times$ m^{-2} $\times$ °C^{-1} have been found for recumbent adults and children, respectively.

Conductive heat exchange

Fourier's law describes conductive heat flow:

$$K = (k_x/d_x)(T_1 - T_2)A_k \qquad (20)$$

where k_x = thermal conductivity of a substance (W $\times$ m^{-1} $\times$ ° C^{-1})
d_x = thickness of the substance (m)
$(T_1 - T_2)$ = temperature gradient (° C)
A_k = area of the conductor (m^2)

Equation 20 is often written:

$$K = h_k(T_1 - T_2)A_k \qquad (21)$$

where $h_k = (k_x/d_x)$, conductive heat transfer coefficient (W $\times$ m^{-2} $\times$ ° C^{-1})

Thermal conductivities of different substances vary tremendously (see Appendix, Table A-8). Conductive heat loss of a nude man is negligible if he is standing on a wooden floor (most of the heat is lost through the soles of his feet) but will increase considerably if he sits on a marble bench. Heat loss from the clothed portion of a man is mainly via conduction through the clothes. Sweating increases the thermal conductivity of "wettable" clothing material. Evaporation lowers surface temperature and thus increases the temperature gradient.

General considerations and approximations

The preceding discussion illustrates the complexity of the skin-to-environment heat exchange system. For general considerations of heat loss from the body, a number of approximations are usually made. The simplest, and least accurate, of these is to assume $\overline{T}_r = T_a$ and $A_r = A_c = A_k = A$. Then by combining equations 12, 16, and 21:

$$H_{out} - E = h_{rck}(\overline{T}_s - T_a)A \qquad (22)$$

where $H_{out} - E$ = nonevaporative heat loss
h_{rck} = combined heat transfer coefficient $h_r + h_c + h_k$

The objection to equation 22 is that it gives too much weight to conductive heat loss and too little to radiative heat loss. In air, conductive heat loss is small and may be omitted.

Equations 12 and 16 give:

$$H_{out} - E = h_r(\overline{T}_s - \overline{T}_r)A_r + h_c(\overline{T}_s - T_a)A_c \qquad (23)$$

assuming $A_r = A_c = A$

$$H_{out} - E = h_{rc}A(\overline{T}_s - h_r\overline{T}_r/h_{rc} - h_cT_a/h_{rc})^{*} \qquad (24)$$

where $h_{rc} = h_r + h_c$
$h_r = 4\ \sigma T^3$ (equation 12)

h_c can be obtained from equation 19. While h_r is relatively constant, h_c varies with air movement. For a man in still air ($h_c = 3.8$) and in a room with usual wall and ambient temperatures:

$$H_{out} - E = 10\ A(\overline{T}_s - 0.62\ \overline{T}_r - 0.38\ T_a) \qquad (25)$$

is a reasonable approximation. However, if air movement is 1 m/s and equation 19 is used to calculate h_c, equation 25 becomes:

$$H_{out} - E = 17.1\ A(\overline{T}_s - 0.36\ \overline{T}_r - 0.64\ T_a) \qquad (26)$$

The effect of air movement is to increase the heat transfer coefficient and to alter the weighting given to wall and ambient temperatures.

*Winslow, Herrington, and Gagge[6] defined operative temperature as $(h_rT_r + h_cT_a)/h_{rc}$.

The skin of unclothed man is the surface exposed to the environment and $\overline{T}_{skin} = \overline{T}_s$. When clothing is worn heat loss from the body is:

$$K = h_k A_k(\overline{T}_{skin} - \overline{T}_s) \quad (27)$$

where $\overline{T}_{skin} - \overline{T}_s > 0$ to satisfy the definition of heat loss

The effect of clothing is to reduce the rate of heat loss by reducing the temperature of the surface exposed to the environment. The smaller the value of h_k, the greater the temperature gradient from skin to surface. The practical importance of the thermal conductivity of clothing material becomes apparent when it is remembered that $h_k = k_x/d_x$. If k_x is large, then thickness, d_x, will need to be increased to reduce h_k. Increasing the thickness of clothing, quite aside from being cumbersome, leads to an increase in the surface area exposed to the environment (A in equations 22 to 27).

Tissue-to-skin system

Storage

The parameters determining the rate and direction of heat exchange have been discussed on the basis of constant temperature. In fact, discrepancies between H_{in} and H_{out} lead to changes in temperature as shown in equation 1, which may be rewritten:

$$S = c_P m(dT/dt) \quad (28)$$

Since tissue cooling adds to the heat available for loss, S is given a positive sign when this takes place.

Newton's law of cooling relates the rate of fall in temperature to the temperature difference between a cooling body and the environment:

$$dT/dt = F_{cool}(\overline{T}_b - T_a) \quad (29)$$

where F_{cool} = cooling constant (s^{-1})

From equations 28 and 29:

$$S = F_{cool}(\overline{T}_b - T_a)c_P m \quad (30)$$

$$S = h_s(\overline{T}_b - T_a)m \quad (31)$$

where h_s = transfer coefficient ($W \times kgm^{-1} \times {}^\circ C^{-1}$)

Equation 27 states that heat loss via conduction is proportional to surface area, while equation 28 states that heat loss from stores is proportional to weight. Kleiber[7] pointed out an intriguing consequence of this difference. Assume $K = S$ (that is, the body is cooling by conduction), equations 27 and 28 give:

$$dT/dt = h_k A_k(\overline{T}_{skin} - \overline{T}_s)/c_P m \quad (32)$$

the rate of cooling is proportional to the ratio of surface area to mass (A_k/m). If it is further assumed that $A_k = m^{2/3}$, then the rate of cooling is inversely proportional to the cube root of body weight. In other words, larger animals cool at a slower rate per degree of temperature difference.

Conduction

Heat transfer from the interior of the body is only partially by conduction through subcutaneous tissues. Equation 20 applies.

Convection

In addition to conductive heat transfer, animals also transfer heat by convection via blood flow to the surface:

$$C_{blood} = Fc_P(\overline{T}_{ar} - \overline{T}_{ve})A \quad (33)$$

where C_{blood} = convective heat loss in the blood (W)
F = blood flow ($kgm \times m^{-2} \times s^{-1}$)
$\overline{T}_{ar}$ = average arterial temperature (° C)
$\overline{T}_{ve}$ = average venous temperature

General considerations and approximations

The amount of heat transported from the interior of the body to its surface is the amount of heat conducted through the tissues plus the amount brought to the surface by the circulating blood:

$$H_{out} = (k_{tissue}/d_{tissue})(\overline{T}_b - \overline{T}_{skin})A_k + Fc_P(\overline{T}_{ar} - \overline{T}_{ve})A \quad (34)$$

where k_{tissue} = thermal conductivity of tissue ($W \times m^{-1} \times {}^\circ C^{-1}$)
d_{tissue} = distance between T_b and T_{skin} (m)

Assuming $T_{ar} = T_b$, $T_{ve} = T_{skin}$, and $A_k = A$.

$$H_{out} = (\overline{T}_b - \overline{T}_{skin})(Fc_P + k_{tissue}/d_{tissue})A \quad (35)$$

The heat from the interior of the body is transferred to the surface via two parallel routes. While the assumptions may not be valid for all cases, equation 35 is useful as a basis for discussion of the relative importance of the two routes. Thermal capacity of blood is about 3,852 $W \times s \times kgm^{-1} \times {}^\circ C^{-1}$ and tissue thermal conductivity about 0.24 $W \times m^{-1} \times {}^\circ C^{-1}$. Assuming tissue thickness to be 0.02 m:

$$H_{out} = (\overline{T}_b - \overline{T}_{skin})(3{,}852\ F + 12)A \quad (36)$$

In order to maintain T_b and T_{skin} at 37° and 33° C respectively, a man of 1.8 m^2 size, producing heat at a rate of 90 W, would need very little blood flow to the skin (about 0.0078 $kgm \times m^{-2} \times min^{-1}$ or 0.0073 $L \times m^{-2} \times min^{-1}$; density of blood is about 1.06 kgm/L). The results of Hardy and Soderstrom[8] indicate that blood flow to the skin is zero when the gradient between rectal and skin temperatures is greater than 4.4° C. The conductive heat transfer coefficient of an obese man is about 2 $W \times m^{-2} \times {}^{\circ}C^{-1}$, in which case blood flow to the skin must be about 0.154 $L \times m^{-2} \times min^{-1}$.

There is a relationship between blood flow and $\overline{T}_{skin}$ and the conductive heat transfer coefficient that is not apparent in the preceding equations. A reduction in flow to the skin increases d_{tissue} (the heat source is now farther from the surface), decreasing the transfer coefficient. At the same time $\overline{T}_{skin}$ is reduced, but the increased temperature gradient is compensated for exactly by the decreased gradient that occurs in the skin-to-environment system ($\overline{T}_{skin}$ in equation 27 is reduced by an equal amount). Thus blood flow serves to regulate conductive as well as convective heat loss from the interior of the body. When blood flow is zero the tissue conductive heat transfer coefficient is at minimum value.

Conductive heat loss depends mainly on temperature gradients; convective heat loss depends mainly on blood flow. By increasing blood flow to the surface it is possible to increase heat loss from the interior of the body without the necessity for a large increase of temperature gradient.

COLD

Cold is not a measurable variable but is a sensation that results from heat loss. This sensation is mediated by mechanisms within the animal that monitor temperature and rate of temperature change. Changes in either the animal or its environment that lead to $H_{in} - H_{out} < 0$ also lead to a fall of tissue temperature and are interpreted as cold. Low ambient temperature is but one, albeit important, factor contributing to the "coldness" of an environment.

The responses of animals to cold are discussed in terms of (1) mechanisms available to them for maintaining body temperature, (2) thermoregulatory mechanisms, (3) the results of thermoregulation failure, and (4) changes that occur after prolonged exposure to cold.

Heat conservation mechanisms

Because they involve changes in physical parameters, mechanisms that conserve heat are usually classified as physical regulatory mechanisms. Physical regulation requires less expenditure of energy on the part of an animal; it is usually the first thermoregulatory mechanism brought into play.

Postural changes

Postural changes reflect changes in body geometry and result in a reduction of the surface area presented to the environment. Curling up or flexion of the body can appreciably reduce the rate of heat loss. Winslow and Herrington[9] pointed out that a tightly contracted posture on exposure to cold is an involuntary reflex that is a basic part of temperature regulation.

Perspiration

Sweating is inhibited when the animal feels cold. There is, however, an obligatory evaporative heat loss via the respiratory tract that is not subject to regulatory control. Fig. 4-3 gives the evaporative heat loss per liter of respired air at various ambient temperatures. Heat loss in heating the inhaled air to body temperature is added to obtain total respiratory loss.

Surface insulation

Fur-bearing animals can greatly increase the insulation of their outer coat by piloerection. This increases thickness and decreases the conductive heat transfer coefficient.

Tissue insulation

While the thermal conductivity of tissues is constant, tissue insulation (reciprocal of heat transfer coefficient) is increased by vasoconstriction. Maximum tissue insulation can be estimated from experiments in which blood flow to the surface is assumed to be zero. The results of Carlson and associates[10] indicate that maximum values are reached in men of normal adiposity when they are in turbulent water maintained at 25° C. In colder water vigorous shivering results in a reduction of the estimated tissue insulation. Fig. 4-4 shows that there is a linear relationship between maximum tissue insulation and the percent fat content of an individual.

The importance of subcutaneous fat (skinfold thickness) in preventing cooling of the body during swimming in cold water has been stressed by Pugh and co-workers,[11] who estimate that an extra 1 mm of subcutaneous fat may be equivalent to raising the water temperature by 1.5° C.* How-

*Assuming tissue thickness to be 2 cm, a 1-mm increase would lead to a 5% decrease of the rate of heat loss. A 5% reduction of temperature gradient would yield the same result. If the original gradient was 30° C (T_b at 37° C and T_w at 7° C), and if T_b remains constant, then a 5% reduction of the gradient is equivalent to raising water temperature 1.5° C.

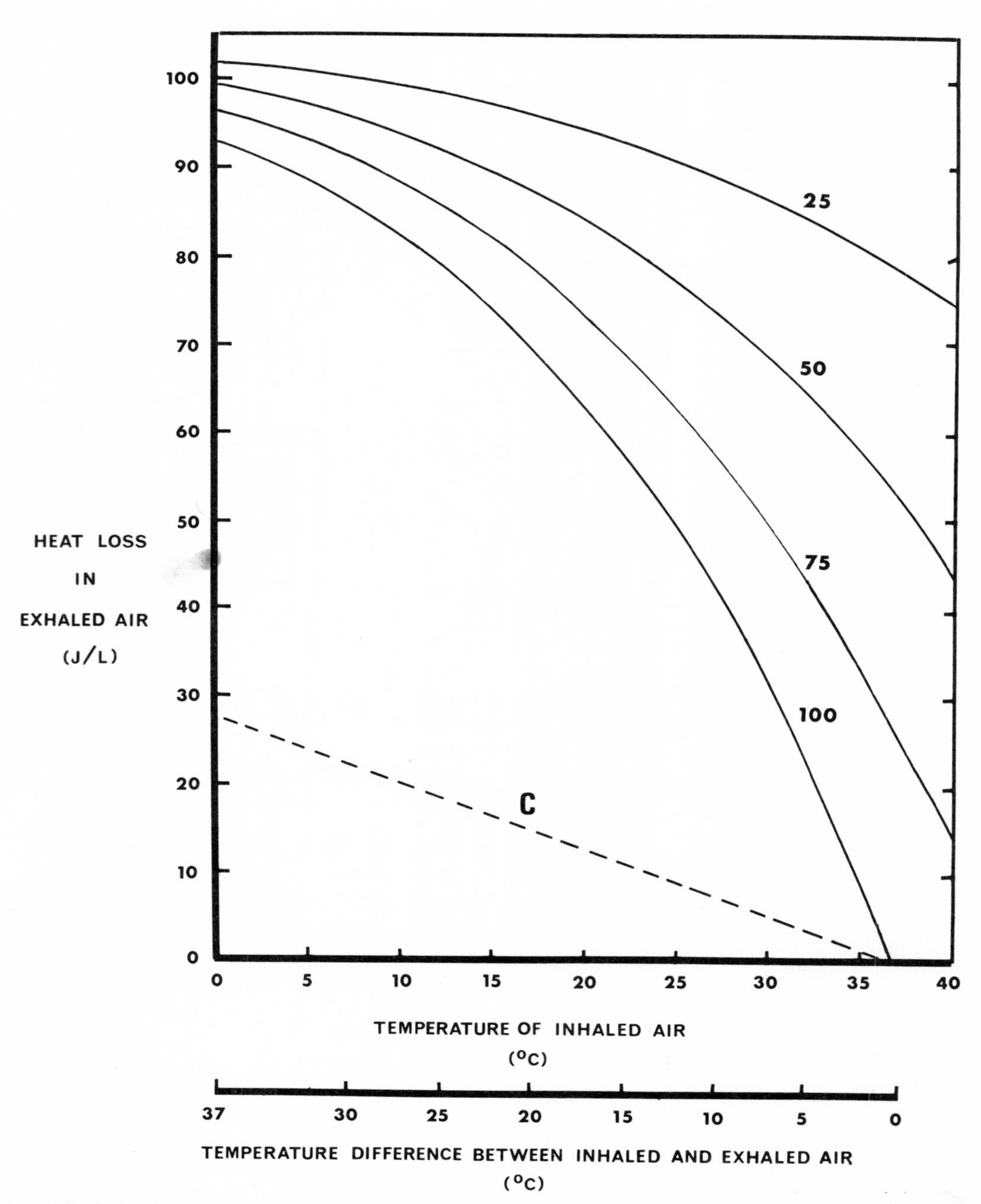

Fig. 4-3. Heat loss from the respiratory tract in joules per liter of exhaled air. The numbers near the curves indicate the relative humidity of the inhaled air. It is assumed that exhaled air is saturated with water vapor and at a temperature of 37° C. The dashed line *(C)* indicates heat loss caused by warming the air. To obtain the total heat loss add the heat loss from warming the air to the evaporative heat loss.

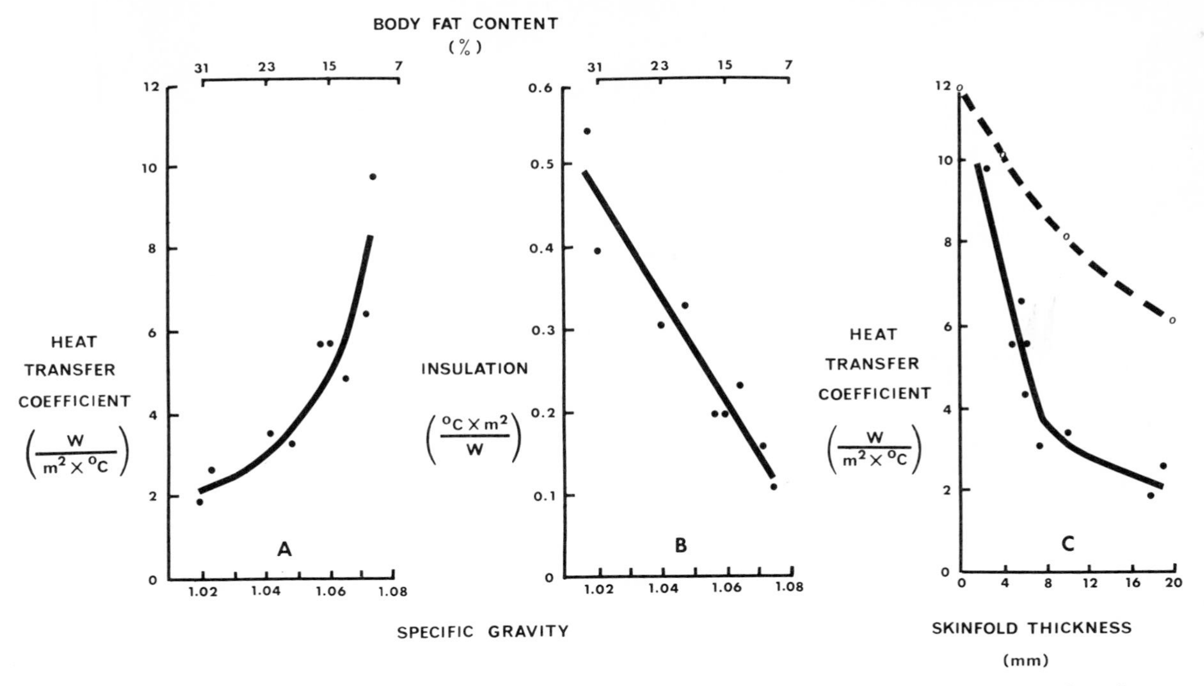

Fig. 4-4. Heat transfer coefficients (**A**) and insulation (**B**) of nine men sitting in a well-stirred bath maintained at 25° C plotted as a function of the subjects' specific gravity. The approximate body fat content of the individuals is indicated in percent at the top of the graphs. In **C** heat transfer coefficients are plotted against skinfold thickness. The dashed line in **C** represents the coefficients calculated by assuming that subcutaneous fat is added to an initial tissue thickness of 2 cm. This study indicates that initial tissue thickness is not constant for all individuals but increases with increasing body fat content. (Curves constructed from data of Carlson, L. D., Hsieh, A. C. L., Fullington, F., and Elsner, R. W.: J. Aviat. Med. **29**:145-152, 1958.)

ever, the relationship between heat transfer and skinfold thickness is not linear. Fig. 4-4, *C,* shows that the major decrease of heat transfer coefficient takes place between 0 and 8 mm of skinfold thickness. This does not mean that further increases in the amount of adiposity are of little advantage in the cold. An obese person cools at a slower rate and in a situation where time is of the essence this may mean the difference between survival and death.

The preceding calculations of heat transfer coefficients are for the whole body with the head out of water. In areas such as the hands and feet, where almost all the heat comes from circulating blood, vasoconstriction reduces heat loss to almost nothing. The effect on heat loss from the body as a whole is as if surface area were reduced.

Heat loss from the head is of special importance in the cold. Since blood flow to the brain remains constant, there is not much variation in heat transfer coefficient. Froese and Burton[12] have shown that insulation of the head does not change with environmental temperature and is independent of the thermal state of the body. Their results indicate that in still air at –4° C heat loss from the head is about 45 W (that is, about half the resting heat production).

Peripheral circulation

At rest the major source of heat is from the brain and the viscera. The approximate percentages of resting heat production are: brain, 16%; viscera, 56%; and skin and muscle, 18%. Circulation serves to cool internal organs and to conduct heat to the periphery, warming peripheral tissues and bringing heat to the surface to be dissipated. During work the muscles may provide 75% of the total heat. On exposure to cold, conservation of heat is brought about by reduction of blood flow to the surface and, in the limbs, countercurrent heat exchanging between arterial and venous blood.

Specific structure and innervation. The temperature measurements of Bazett and McGlone,[13] illustrated in Fig. 4-5, show that gradients from the skin surface to the interior form a biphasic

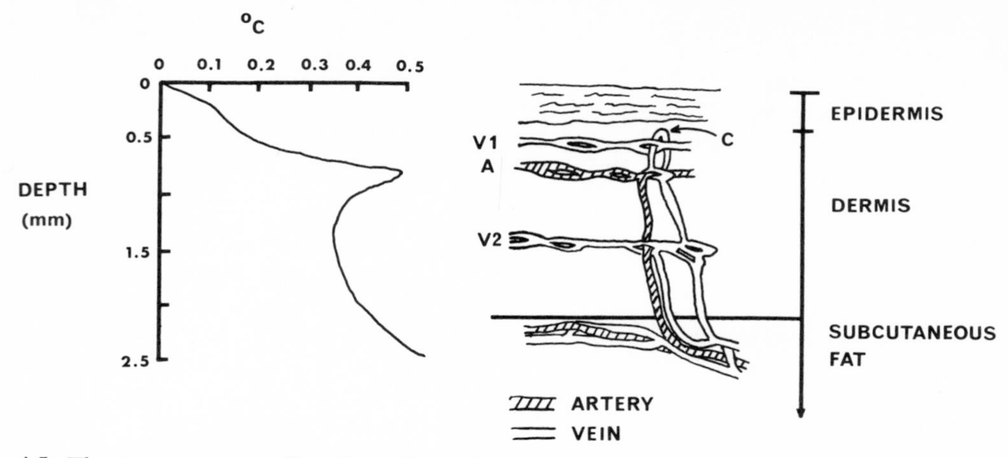

Fig. 4-5. The temperature gradient from the surface of the skin to a depth of 2.5 mm in relation to the approximate position of the superficial blood vessels. *C* represents capillaries; *V1* and *V2* are the superficial venous plexi; *A* is the superficial arteriolar plexus. In the subcutaneous layer the arteries and veins follow a similar course. (Modified from Bazett, H. C.: Temperature sense in man. From Temperature, its measurement and control in science and industry by H. C. Bazett. Copyright © 1941 by Litton Educational Publishing, Inc. by permission of Reinhold Publishing Co.)

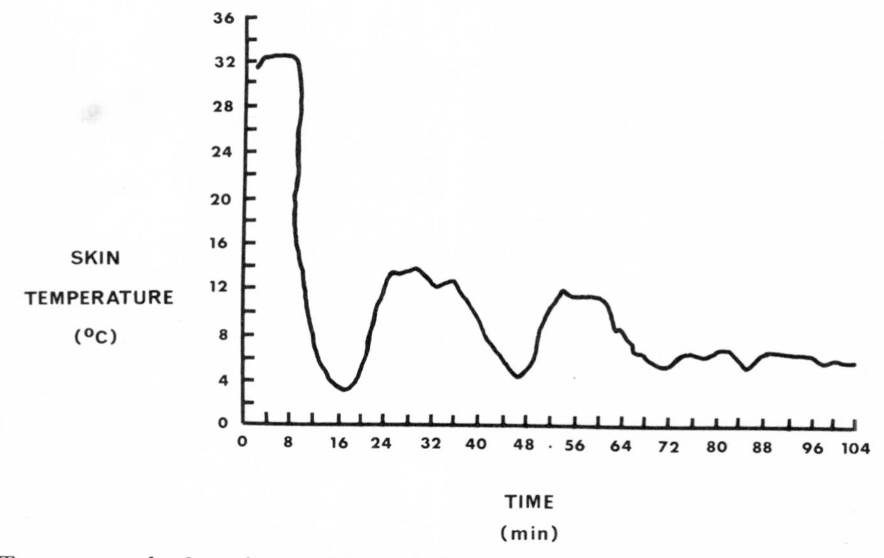

Fig. 4-6. Temperature of a finger immersed in crushed ice. Room temperature was 19° C. The water bath was unstirred. The curve shows large, prolonged temperature oscillations finally giving place to smaller, more rapid ones—Lewis' "hunting" phenomenon. (Modified from Lewis, T.: Heart **15**:177-208, 1930.)

curve over the first 2.5 mm. There is a peak at about 0.8 mm at about the level of the subepidermal arteriolar plexus. The abrupt fall on either side is ascribed to cooler blood in the venous plexi. At a depth of 2.5 mm subcutaneous arteries are accompanied by venae comitantes.

Arteriovenous anastomoses connect the arterioles with superficial plexuses. Opening of these anastomoses results in shunting of blood from arterioles to superficial veins, bringing heat close to the surface for dissipation. Grant[14] and Clark and Clark[15] made direct observations on the arteriovenous anastomoses of the rabbit ear. They noted that there were spontaneous rhythmic changes in caliber of the vessels but on heating the animal a dilation occurred. Cooling the animal led to closure of the anastomoses. Arteriovenous anastomoses in the tongue of the dog, sheep, and goat may also serve as devices for cooling the animal during panting.[16, 17]

Grant[14] noted that after the ear of the rabbit had been cooled for a while the arteriovenous anastomoses reopened. Detailed study of skin temperature changes of fingers immersed in crushed ice led Lewis[18] to describe the "hunting" phenomenon that bears his name. Upon immersion there was a steep fall of finger temperature that continued for about 5 minutes, until temperatures were about 1.5° C above bath temperature. After about 7 minutes finger temperature began to rise rapidly, increasing by 7° to 13.5° C and then decreased again (Fig. 4-6). The following quotation from Lewis is still the best description of this phenomenon, now called cold-induced vasodilation:

> The temperature curve during long immersions varies a good deal in type. It is the rule for the temperature of the finger to fall away again after the reaction has continued for some minutes. This fall may be to the original point but is more usually less than this. It is frequent for the temperature reaction to be repeated over and over again during immersion; this slow "hunting" of temperature is never quite rhythmic and the rises for the most part irregular in time and form . . . It is frequent for the amplitude of the excursions of temperature to decrease as these succeed each other, the curve taking a straighter course at an intermediate level, but "hunting" may start again after the curve has run relatively straight for a long time. The duration of the individual phases is variable; they may last as long as 30 to 36 minutes and on another occasion a curve from the same subject may show successive phases of about 10 minutes' duration. In other rarer instances little or no "hunting" is to be seen. The extent of the initial and secondary rises of temperature is greater if the mixture of ice and water remains unstirred, for the finger then becomes coated by a layer of warmer water.*

Similar responses to cold were noted in the skin of the ear, cheek, nose, chin, and toes. Lewis[18] suggested that this mechanism has an important function in cold weather in protecting exposed parts of the skin from excessive cooling and injury. Attempts have been made to relate the type of response to individual susceptibility to cold injury. However, the responses are extremely variable and are subject to many modifying factors, for example, thermal and nutritional state of the subjects, degree of training, subjective feeling of pain, and emotional response to the experimental procedure.[19]

*From Lewis, T.: Observations upon reactions of vessels of human skin to cold, Heart 15:177-208, 1930.

It is now generally accepted that the skin temperature rises result from increased blood flow through arteriovenous anastomoses. The opening of these shunt vessels without concomitant relaxation of metarterioles feeding into capillaries leads to a reduction of capillary flow. Indeed, the local rise in venous pressure could lead to stoppage or even reversal of capillary blood flow. Measurement of the ^{24}Na clearance rate from the rabbit foot suggests that capillary flow is zero during cold-induced vasodilation.[20] Since the temperature rise is at the expense of local tissue blood flow, it is difficult to understand how this is a physiologic defense mechanism. The possibility that cold-induced vasodilation is the result of failure of the vasoconstrictive mechanism should be kept in mind.

Blood flow to the periphery is regulated by alterations in the arteriolar caliber. These changes are effected by contraction or relaxation of spiral smooth muscles in the vessel walls, which are in turn controlled by intrinsic humoral and neural factors. Vasoconstriction is produced by an increase of reflex sympathetic activity (usually referred to as sympathetic tone), and dilation occurs as a result of increased local metabolic vasodilator factors (metabolites, hypoxia) and decreased sympathetic tone.

Sympathetic nerve constriction of vessels is mediated by norepinephrine stimulation of alpha receptors. Beta receptors, present in vessels supplying skeletal muscles and the heart but absent in vessels supplying skin, result in dilation when stimulated. A third receptor is postulated to mediate the effect of metabolic products. Local cooling, in addition to stimulating general reflex vasoconstrictor mechanisms, affects the intrinsic tone of vascular smooth muscle.

At least four factors are involved in the initial decrease of flow to the skin on exposure to cold: reflex excitation of vasoconstrictor fibers; direct constrictor response of smooth muscles; change of blood viscosity; and a change of local metabolism. Cold-induced vasodilation may result from axon reflexes,* release of local vasoconstriction

*Application of an irritant like mustard to the skin or subdermal histamine injection produce cutaneous vasodilation. Since the response does not occur if the sensory fibers supplying the area have degenerated following section of the peripheral nerves or the posterior spinal nerve roots, this response is thought to be caused by a local nerve reflex, the axon reflex. The similarity of this vascular response to that occurring in fingers during immersion in crushed ice led Lewis to postulate that H substance release stimulates sensory nerve endings, leading to reflex local dilation.

caused by low temperature, and progressive relaxation of the smooth muscle constrictor response.

Countercurrent heat exchange. Blood of different temperatures flowing in opposite directions through contiguous vessels exchange heat. Assuming no heat loss from the system, the rate of heat transfer from the warmer to the cooler fluid is:

$$H = (kA)c_P F\Delta\overline{T} \tag{37}$$

where H = rate of heat transfer (W or J/s)
k = the sum of all factors affecting heat flow (for example, conductivity and thickness of the tissues separating the fluids, boundary conditions within the tubes, and character of fluid flow) (m^2).
A = area available for heat transfer (m^2)
c_P = thermal capacity of blood (3,852 J $\times$ kgm^{-1} $\times$ C^{-1})
F = mass flow rate (kgm/s)
$\Delta\overline{T}$ = mean temperature difference between the artery and vein (°C)

It should be noted that $\Delta\overline{T}$ varies with F and kA and is thus not an independent variable. Since heat is lost from the arterial blood, there is a temperature gradient along the artery. The magnitude of this gradient depends upon the extent of countercurrent heat exchange (that is, an increase in H leads to a greater temperature fall along the artery).

Examination of the vascular system in many regions in most mammals and birds shows two veins accompanying each artery, not only in the limbs but also beneath the dermis and in the viscera. Bazett and co-workers[21] demonstrated the presence of longitudinal temperature gradients within the arteries of the human arm. Under warm conditions this gradient was about 0.03° C/cm, but when the arm was exposed to cold (air of 4° C for about 1 hour) the gradient was about 0.35° C/cm. Evidence was also obtained indicating that blood entering the arm was being cooled by the venous blood returning up the arm lying close to the artery. Since heat is transferred to the venous blood, this precooling allows the maintenance of blood flow to the hand without excessive loss of heat to the environment. During muscular exercise and shivering, the reverse is the case.

In many regions in which countercurrent heat exchange has been either demonstrated (by measurement of intravascular temperatures) or inferred from surface temperature distribution and vascular morphology, venous drainage provides the possibility of a superficial return path separated by a considerable thickness of tissue from the arteries and a deep path via the veins that accompany the arteries. Changes in the route of venous return appear to be part of the temperature-regulatory mechanism, but there is no information regarding the actual control.

Heat production mechanisms

Metabolic processes are the sole source of heat production. It is convenient to classify heat prodution as either *obligatory* or *regulatory*.

Obligatory heat production is the heat liberated from food (substrate) as a result of biochemical processes that supply the energy required for the maintenance of life. Obligatory heat production increases in hyperthyroidism and during muscular exercise. The efficiency of the human body in exercise is about 20%. A man working at a rate of 81.4 W (70 kcal/hr) generates heat at a rate of 325.6 W (280 kcal/hr) in addition to his resting heat production.

Regulatory heat production is the heat produced in response to a reduction of the temperature of specific thermoreceptors in the skin, hypothalamus, and perhaps the spinal cord. It is a response to cold when heat conservation mechanisms fail to maintain body temperature. Thus exposure to cold stimulates regulatory heat production only to the extent necessary to compensate for the increased rate of heat loss. This point is important in assessing the metabolic response to cold. If obligatory heat production is high, one can expect a reduced response to cold exposure. Conversely, a person with a low obligatory heat production (for example, a hypothyroid patient at rest) responds vigorously to cold exposure.

Regulatory heat production is usually caused by shivering, but in the newborn and in cold-adapted animals a nonshivering process has been described.

Metabolic pathways

Shivering is the result of activation of skeletal muscles in a characteristic phasic manner. The immediate source of energy derives from the hydrolysis of adenosine triphosphate (ATP) to adenosine diphosphate (ADP) and inorganic phosphate (P_i). The ATP content of muscle remains remarkably constant during activity. This is because ADP is rephosphorylated as soon as it is formed. If sufficient oxygen is present, rephosphorylation takes place by means of high-energy chemical groups generated during oxidation of substrates (muscle glycogen, glucose, fatty acids, and, to a lesser extent, amino acids). In the absence of oxygen, muscle can still contract for a time by utilizing the ATP generated during glycolytic conversion of glycogen and glucose to lactate. Phosphocreatine is a source of reserve high-energy phosphate groups in muscle. This compound takes part in an enzymatic reaction with ADP to yield ATP and creatine. During recovery from strenuous activity the muscle phosphocreatine and glycogen stores are replenished. Thus increased heat production during shivering can be explained entirely on the basis of ATP utilization.

Since no external work is performed during shivering, all the energy released, about 8 kcal/mole of ADP formed, appears promptly as internal heat. Because of the inefficiency of the processes regenerating ATP, additional heat is released when the energy stores are replenished. If glucose is the substrate oxidized, the efficiency of the system is about 44%:

$$\text{Glucose} + 6\ O_2 + 36\ P_i + 36\ \text{ADP} \rightarrow \rightarrow 6\ CO_2 + 44\ H_2O + 36\ \text{ATP} \quad (38)$$

Under normal circumstances the reaction is coupled tightly to the availability of ADP. In other words the reaction cannot be driven to the right by either glucose or oxygen excess. The free energy of oxidation of glucose to carbon dioxide is about –686 kcal/mole. In the reaction 288 kcal are conserved by the conversion of 36 ADP to 36 ATP (8 kcal/mole); the rest of the free energy appears as heat. The respiratory quotient for the reaction is 1 ($R.Q. = CO_2/O_2$) and the calorific value of oxygen is 5.1 kcal/L [$686/(6 \times 22.4) = 5.1$]. Since 1 mole of glucose weighs 180 gm, its caloric value is 3.8 kcal/gm, but it should be noted that its "useful" caloric value is only about 1.6 kcal/gm.

The efficiency of ATP regeneration from energy released from fatty acids is about 43%:

$$\text{Palmitic acid} + 23\ O_2 + 130\ P_i + 130\ \text{ADP} \rightarrow \rightarrow 16\ CO_2 + 16\ H_2O + 130\ \text{ATP} \quad (39)$$

The free energy of the reaction is –2,400 kcal/mole. The respiratory quotient is about 0.7 and the calorific value of oxygen is about 4.1 kcal/L. The caloric value of fatty acids is about 9.3 kcal/gm.

During vigorous shivering the oxygen consumption increases to four or five times the resting value. However, the efficiency of shivering in terms of body heat maintenance is only about 11%.[22] Therefore, there must be an increase in the body-to-environment heat transfer coefficient during shivering. The probable cause is increased circulation to skeletal muscles near the body surface and increased convective heat loss caused by body motion.

Teleologic considerations suggest that a nonshivering regulatory heat production mechanism, with tissues and organs in the body interior playing an active part, would be the method of choice. Nonshivering heat production is discussed later in more detail on p. 77. While the response appears to be mediated by catecholamines released from the sympathetic nervous system, the metabolic pathways involved remain unknown. Catecholamines promote glycogenolysis (conversion of glycogen to glucose and lactate) and lipolysis (mobilization of fatty acids from fat depots). These effects increase substrate availability but do not fully explain the resulting increased heat production. Catecholamines increase the synthesis of cyclic 3′,5′-AMP by the adenyl cyclase system located in the cell membrane of various tissues. Cyclic 3′,5′-AMP then acts on phosphorylase or lipase, inducing reactions that lead to the observed increase of substrate concentrations. The respiratory quotient of animals receiving norepinephrine infusions is about 0.74, suggesting that fatty acids are being metabolized. Conversion of acetyl-CoA to palmitic acid and back to acetyl-CoA consumes 8 ATP. Thus continuous formation and breakdown of fatty acids may utilize ATP to drive substrate oxidation. How norepinephrine increases both lipolysis and lipogenesis remains unanswered. A possible simultaneous effect of catecholamines is on an enzyme system (ATPase) regulating the reaction:

$$\text{ATP} \rightarrow \text{ADP} + P_i \quad (40)$$

The resulting increase in ATP turnover rate may explain the increased calorigenic response of cold-adapted rats to norepinephrine. In these animals the blood glucose level does not increase and fatty acid concentrations correlate negatively with norepinephrine dose.

Endocrine interaction in metabolic pathways

Insulin and glucagon. Insulin, formed in the pancreas, is a hormone that decreases liver and muscle glycogenolysis and facilitates glucose transfer into cells. Thus blood glucose is related inversely to insulin level. Glucagon, also a pancreatic hormone, increases liver glycogenolysis. The two hormones work together to maintain blood glucose concentrations within normal limits. The availability of energy from glycogen depends on those factors that influence its interconversion to or from glucose. The major stores of glycogen are in liver and muscle. Liver glycogen is rapidly depleted to maintain the blood sugar level and it is formed when the blood sugar level rises. In liver and muscle the balance depends on the insulin blood level and, in liver, glucagon is the second factor.

Epinephrine and norepinephrine. The main source of blood catecholamines is the adrenal medulla. Norepinephrine, released from sympathetic nerve endings, may also find its way into the bloodstream. Catecholamine hormones, especially epinephrine, have a calorigenic effect in mammals. They also cause a marked increase of blood glucose and fatty acids in warm-adapted animals. As mentioned before, the mechanism of the calorigenic effect is not entirely explained.

Thyroid hormones. The hormones of the thyroid gland, thyroxin and triiodothyronine, affect metabolism directly. Removal of the thyroid gland produces a gradual reduction of metabolic rate over a period of weeks to a level that is 50% to 60% of the presurgical resting level; metabolism is still increased by activity, but the maximum rate is very restricted. Thyroxin injection increases metabolic rate. The calorigenic effect of thyroxin may be caused by uncoupling of oxidation of glucose from the ADP → ATP reaction (see discussion following equation 38). Another current hypothesis postulates a general influence on protein synthesis, increasing enzyme concentration. A sodium pump is postulated to be present in tissue cell membranes and is said to be responsible for pumping sodium out of the cell against a concentration gradient. Energy for powering the pump is derived from hydrolysis of ATP. The calorigenic effect of thyroid hormone may also be caused by activation of mitochondrial ATP synthesis, resulting in stimulation of the sodium pump by local increase in ATP concentration or direct activation of the sodium pump.[23]

Adrenocortical hormones. Cortisol promotes glucose formation from protein (gluconeogenesis). The hormone alters amino acid transaminase levels, impairing protein synthesis and permitting gluconeogenesis to proceed more rapidly. Cortisol has been termed a permissive hormone because it accelerates a reaction, but does not produce it per se.

Regulation of body temperature

Body temperature

Body temperature is usually measured by placing a thermometer in the oral cavity or rectum. (The usual clinical thermometer has a stricture in the measuring column so that the *maximum* temperature is recorded.) These temperatures reflect the core, or central, temperature of the body.

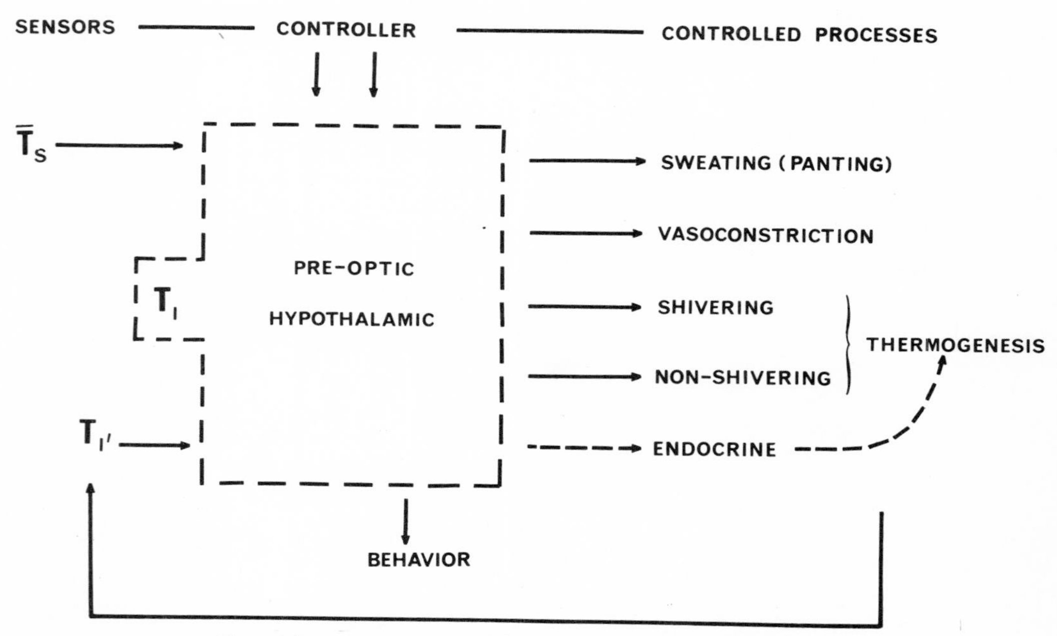

Fig. 4-7. Diagram of the temperature-regulating system.

Actually, the temperature of various body regions varies slightly. However, for clinical purposes, we rely on the fact that the body tends to regulate, or maintain, its central temperature.

In addition to temperature variations within the body core, there are interindividual temperature variations. The mean oral temperature of 276 medical students between 8:00 and 9:00 AM was 36.7° ± 0.2° C.[24] There is also a definite circadian temperature rhythm and, in the female, a temperature pattern associated with the menstrual cycle.

Temperature-regulating system

Fig. 4-7 illustrates the homeothermic mechanisms involved in responses to cold or heat. Phrased in control systems terminology, the sensor, controller, and control processes are indicated as physiologic entities. The elements involved in sensing are skin temperature ($\overline{T}_s$) and internal temperatures ($T_{I'}$ and T_I). One site for internal temperature sensing is in the hypothalamus. A second more peripheral one may be in the body core. The controlled processes are vasoconstriction, the shivering mechanism, and the nonshivering heat production mechanism; the principal controller pathways are the sympathetic nervous system controlling peripheral blood flow, the motor system that controls shivering, and cell metabolism.

Temperature regulation is primarily neural, but these responses are modified and supported by endocrine and behavioral factors. The response of the regulatory system depends on input signals reaching a central integrating area from which output signals effect changes in insulation and heat production.

Thermoreceptors. Specific cold receptors are located close to the skin surface. The number of receptors varies with the skin region; it is much higher in the skin of the face and hands than it is on the legs or chest. These cold receptors respond not only to rate of temperature change but also to absolute temperature. Abrupt temperature changes greatly increase the frequency of impulses in cold receptor fibers; the rate then decreases to a level that is characteristic of the temperature. When skin temperature increases, the cold receptors cease sending impulses along the cold receptor fibers. Individual cold receptors respond to a certain range of temperatures. The nerve fibers that originate in the cold receptors, together with pain fibers, are predominantly in the lateral spinothalamic tracts where they form synaptic connections that project to the somesthetic area of the brain. Some temperature-regulating responses may involve axon and spinal reflexes.

Areas in the hypothalamus are involved in temperature regulation as demonstrated by classic lesion procedures, electric stimulation and recording, and methods of changing hypothalamic temperature such as brain perfusion and local heating or cooling with thermodes. Such methods have established the existence of functional areas in the hypothalamus. Generally, heat loss control is in the anterior, or preoptic, hypothalamus; heat conservation mechanisms lie in the posterior hypothalamic area. The preoptic area is temperature-sensitive. There are also indications that some neurons in the spinal cord are sensitive to temperature changes.

Efferent pathways. The efferent pathways from the hypothalamus are sympathetic systems involved in peripheral blood flow regulation, the motor system that involves shivering, and the sympathetic pathways involved in nonshivering thermogenesis. The primary area for peripheral vascular control is probably in the anterior or posterior hypothalamus but has not yet been definitively localized. The primary control area for shivering is probably the dorsomedial region of the posterior hypothalamus in the wall of the third ventricle between the mamillary bodies and the tubular regions. The pathways for shivering (Fig. 4-8) proceed laterally to the red nucleus and the lateral reticular area in the pons and medulla and then through the lateral white columns of the spinal cord. The shivering mechanism is influenced by neurons in the motor and premotor cortex. The specific pathway for nonshivering thermogenesis has not yet been located.

Acclimatization to cold*

Acclimatization, acclimation, and adaptation are used synonymously in this discussion, although the terms have been distinguished with respect to the mode of acclimatization. Hart, cited by Burton and Edholm,[1] suggested that acclimatization be the term designating natural exposure in the normal habitat, acclimation the term designating artificial environments, and adaptation used in an evolutionary sense. The criteria of acclimation are also equivocal to some extent. In this discussion, we include changes in physiologic systems that occur with exposure to cold.

*The material in this section is described in greater detail by Carlson.[25]

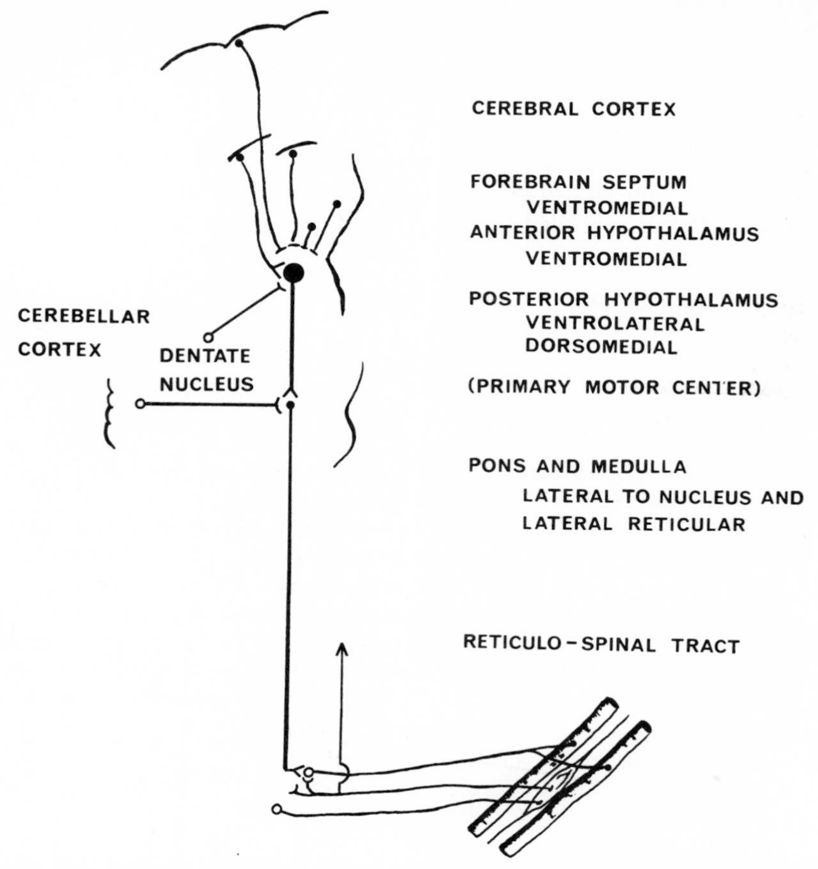

Fig. 4-8. Motor pathways for shivering. Facilitation is shown by Y type endings and inhibition by T type endings.

Metabolic changes

Overall metabolic response may be used to characterize the system's response to cold when referred to environmental temperature or indices of heat load. Mean skin temperature has been selected as a significant variable since it is the external sensing site. Continuous exposure of man to cold causes a shift in the metabolic response with reference to a given skin temperature. The data in Fig. 4-9 represent two distinctly different experiments. One represents subjects exposed to cold during a 2-week bivouac in Alaska, the second, subjects acclimated in an artificial cold room. At the end of the 2-week exposure the change in metabolism at a given skin temperature is considerably reduced.

Many studies have been made of the effects of cold exposure on man. Regression lines of heat production on average skin temperature in overnight experiments are given in Fig. 4-10. The Alacaluf and the aborigine were able to sleep during the night, but the Eskimo and white caucasian could not. Over the skin temperature range of 33° to 28° C the heat production increase in response to low skin temperature is reduced in Eskimos and absent in Alacaluf Indians and Australian aborigines. There is considerable difference between the initial metabolic rates of various groups. However, the response lines of Eskimos and Alacalufs, who appear to have similar initial metabolic rates, indicate different heat production responses at the same skin temperature. The skin temperature at which a response occurs in the Alacaluf and Australian aborigine and the nature of this response remain unknown. The undefined effect of sleep on temperature is an additional variable in these experiments, as is also the circadian temperature fluctuation. The Ama

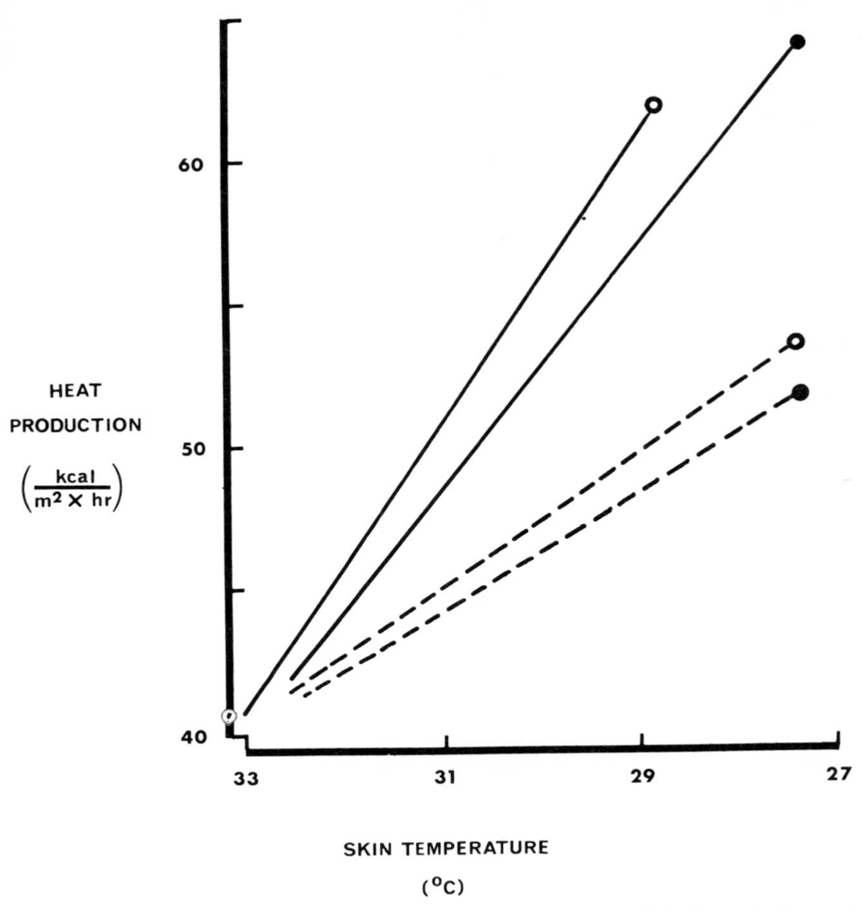

Fig. 4-9. Heat production as a function of mean skin temperature. Solid lines indicate data before acclimation, dashed lines after. Open circles indicate data from field study, closed circles from laboratory study. (Redrawn from Carlson, L. D., and Hsieh, A. C. L.: Cold. In Edholm, O. G., and Bacharach, A. L., editors: Physiology of human survival, New York, 1965, Academic Press, Inc.

Korean diving women (Fig. 4-11), whose shivering response differs greatly from that of others, are exceptional examples of acclimation.[26]

Metabolic studies on animals indicate some of the mechanisms of change in metabolic response. Sellers and associates[27] showed that rats exposed to cold gradually reduce shivering and yet are able to maintain body temperature. By curarizing rats to prevent any shivering, Cottle and Carlson[28] and Hsieh and Carlson[29, 30] discovered a metabolic response apparent only in cold-exposed animals; this mechanism, nonshivering thermogenesis, has since received considerable attention.

Keller's observations on gross shivering in dogs[31] emphasize that cold-stimulated nonshivering heat production is just as distinct an entity as shivering heat production. He found that each has a separate, distinct group of nerve cells and a unique, descending fiber tract in the hypothalamic gray matter and brainstem. DuBois and his co-workers[32] also found prevalent nonshivering thermogenesis in some women and stated that the lack of significant metabolic change in men and women with environmental change suggests a chemical regulation that increases metabolic rate in the body core to compensate for the lowering of peripheral metabolic rate.

Nonshivering thermogenesis occurs in man and other animals to a varying degree. This cold-induced nonshivering thermogenesis is mediated by the sympathetic nervous system. The adrenal medulla also participates in the mediation. Cold-exposed animals are more responsive calorigenically to injected or infused norepinephrine.

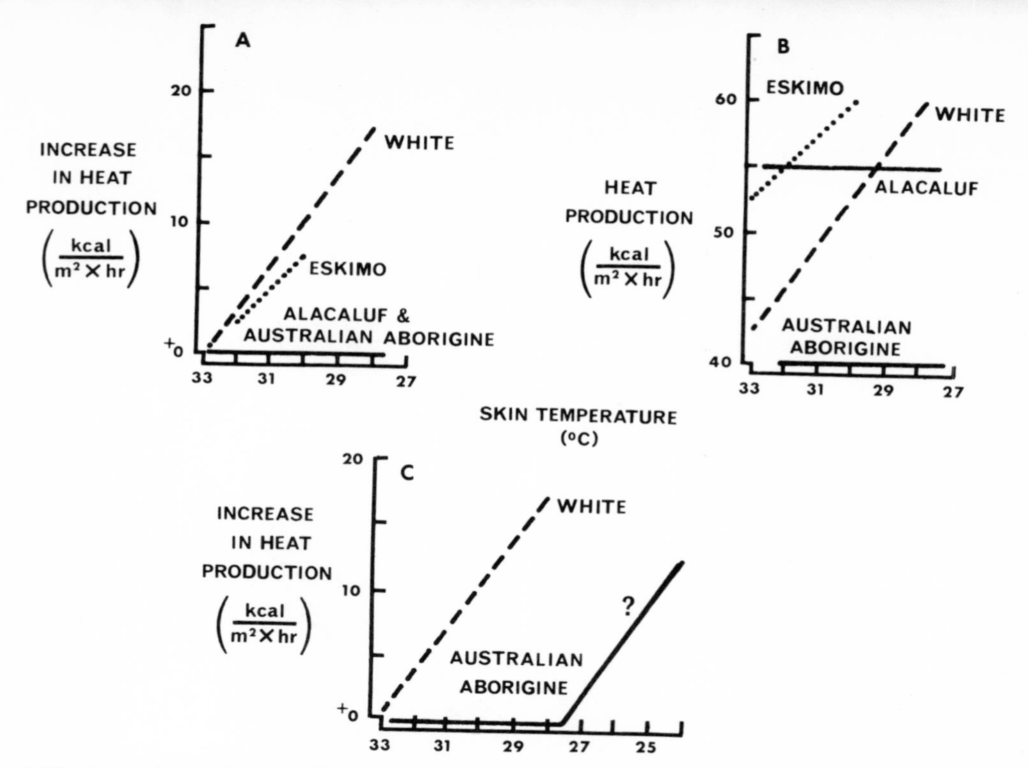

Fig. 4-10. Data from field studies showing the relationship of metabolism to skin temperature in various racial groups. **A** indicates the increase of heat production with decreasing skin temperature. **B** gives the total heat production. **A** is redrawn in **C** to illustrate the question of whether the skin temperature resulting in increased heat production is less in the aborigine and/or whether the magnitude of the response changes. (Redrawn from Carlson, L. D., and Hsieh, A. C. L.: Cold. In Edholm, O. G., and Bacharach, A. L., editors: Physiology of human survival, New York, 1965, Academic Press, Inc.)

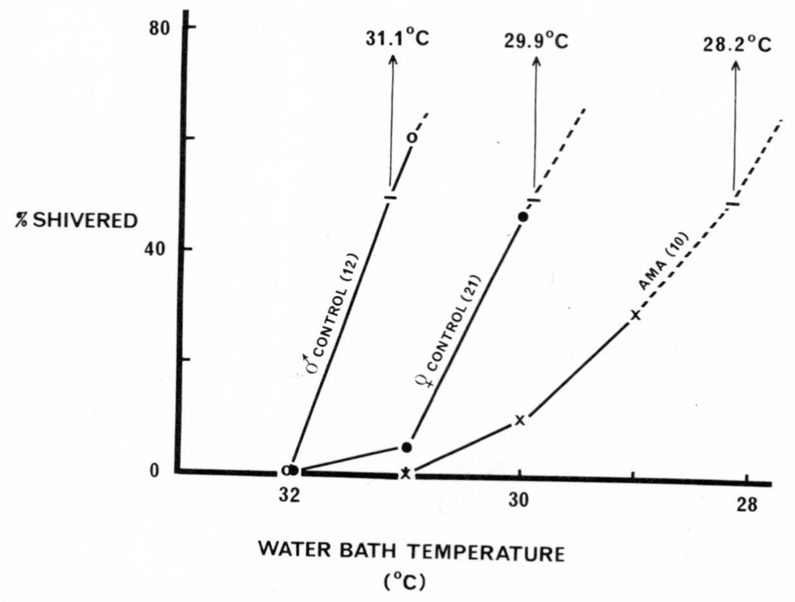

Fig. 4-11. Incidence of shivering in Ama Korean diving women as compared to that of control groups. The number of subjects in each group is indicated in parentheses. (Redrawn from Hong, S. K.: Fed. Proc. 22:831-833, 1963.)

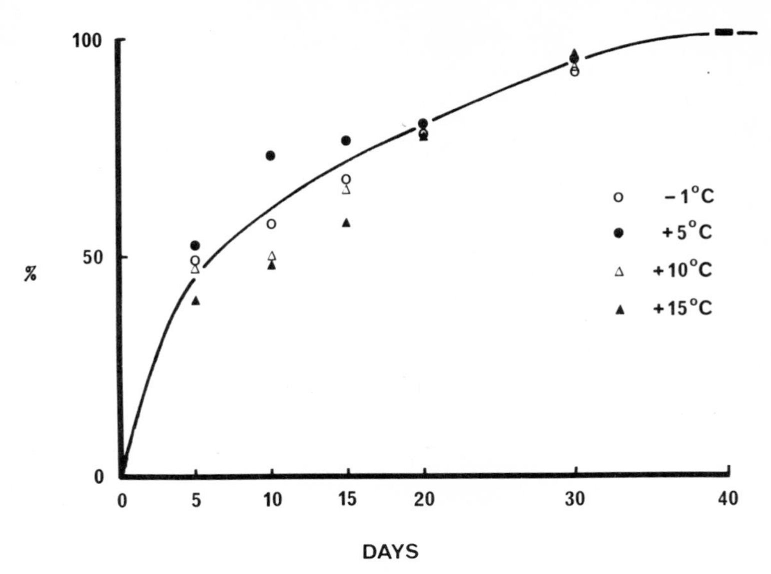

Fig. 4-12. Metabolic rate increase after norepinephrine injection (0.4 mgm/kgm) as a function of exposure time to four different temperatures. (Redrawn from Jansky, L., and others: Physiol. Bohemoslov. **16:**366-372, 1967.)

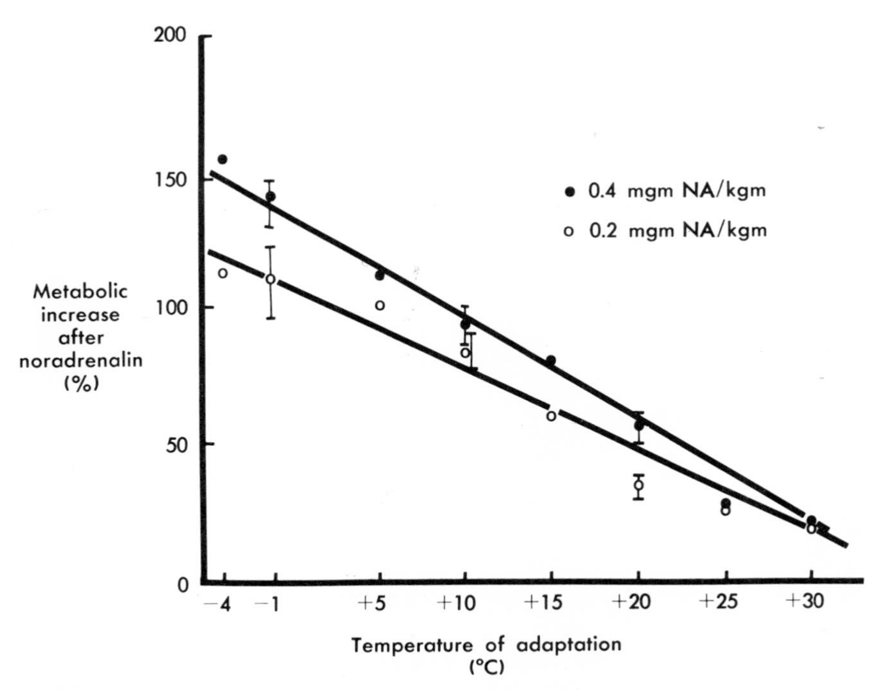

Fig. 4-13. Metabolic rate increase after noradrenalin *(NA)* injection as a function of adaptation temperature and noradrenalin dose. (Redrawn from Jansky, L., and others: Physiol. Bohemoslov. **16:** 366-372, 1967.)

Jansky and co-workers[33] elegantly described the metabolic responses to the administration of norepinephrine in the rat as a function of exposure time and exposure temperature. Fig. 4-12 shows the metabolic rate increase after 0.4 mgm/kgm plotted as a percentage of maximum rate versus exposure duration at four different temperatures. The time-course of response is very similar. The decay of this response also has a rather long time-course (days) following return of the animals to a warm environment. The relative effects of temperature in inducing the response are linear (Fig. 4-13). The maximum metabolic responses after administration of 0.2 mgm of norepinephrine per kilogram of body weight, expressed as a percentage of resting metabolism, increases as the number of hours per day of exposure at 5° C.

Striking as the norepinephrine effect is, other substances are involved in the response to cold. Adrenocortical hormones are only related to the initial cold stress and not required in augmented amounts during the course of cold exposure. Thyroxin requirements, on the other hand, increase and the turnover rate more than doubles.[34] However, thyroxin per se is not the single factor involved in acclimation since, in a thyroidectomized cold-acclimated animal, cold-induced thermogenesis still occurs.[30]

The site of heat production in nonshivering thermogenesis has been a matter of considerable debate. The liver and other viscera are possible

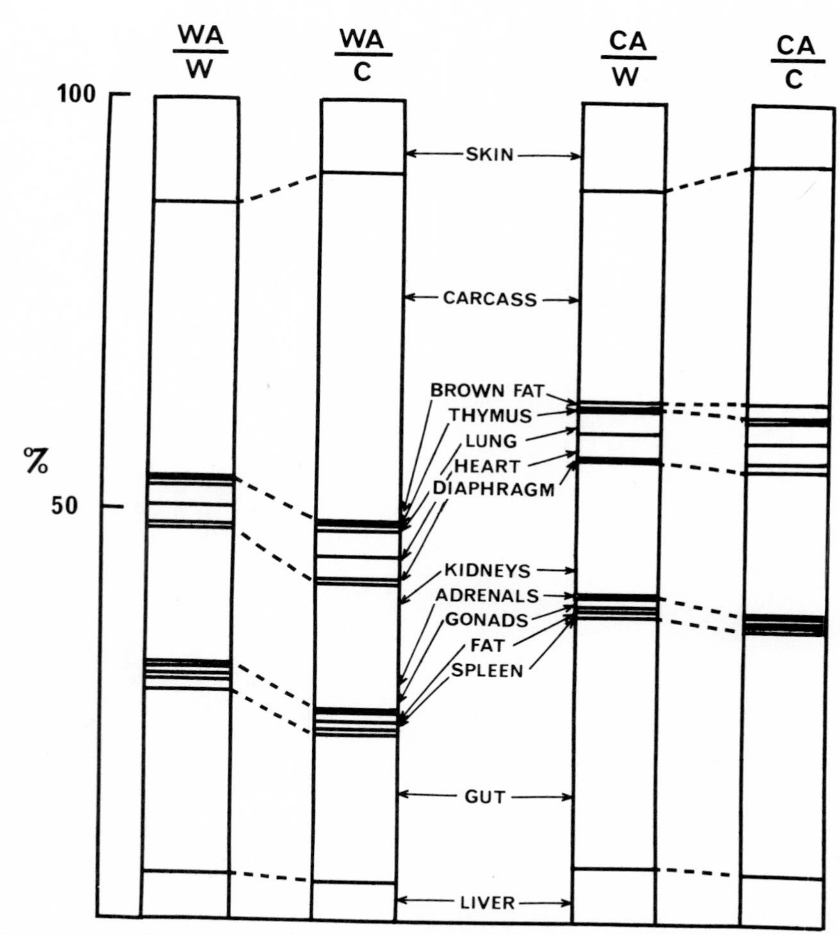

Fig. 4-14. Fractional distribution of cardiac output in warm-acclimated *(WA)* and cold-acclimated *(CA)* animals exposed to warm *(W)* and cold *(C)* environments. (Redrawn from Jansky, L., and Hart, J. S.: Canad. J. Physiol. Pharmacol. **46**:653-659, 1968. Reproduced by permission of The National Research Council of Canada.)

and teleologically attractive sites. However, functionally eviscerated animals respond to cold and infused norepinephrine. Comparative whole animal studies of circulation in warm-acclimated rats show that blood flow increases significantly in muscular organs such as the heart, diaphragm, and skeletal muscles during exposure to cold. In cold-acclimated rats blood flow to brown fat, white adipose tissue, and the splanchnic area increases significantly, accounting for 65% of the increased blood flow during cold exposure compared to only 36% in the warm-acclimated rat (Fig. 4-14). The cold-acclimated animals also increase cardiac output during cold exposure.

Before searching for the metabolic pathways that are involved, we should indicate that thermogenesis is a general phenomenon in the newborn animal. Guinea pigs exposed to cold shortly after birth increase oxygen uptake with little shivering as indicated by electromyogram. This increased oxygen consumption is blocked by alderlin, a beta-sympathetic receptor blocker.[35] The response to cold changes to predominantly shivering thermogenesis as the animal grows older. From experiments with rats and using weight as an index of age, Hagen and Hagen[36] showed that the nonshivering response in the neonate disappears within a few weeks. At tissue level Hagen and Hagen[36] showed that the norepinephrine-induced increase in metabolism of adipose tissue is high at birth and decreases with age, unless the animals are maintained in cold. Norepinephrine releases free, or nonesterified, fatty acids (FFA or NEFA) into the bloodstream. This effect changes during cold acclimation. The expected positive correlation between plasma NEFA and a dose of norepinephrine is found in warm-acclimated, but not cold-acclimated, animals. Plasma NEFA increases with increasing oxygen consumption in warm-acclimated animals, but decreases in cold-acclimated ones.[37]

It appears that adipose tissue is the primary target for norepinephrine-stimulated thermogenesis. The following remarks of George F. Cahill in his introduction of Frank L. Engel at a conference on adipose tissue are pertinent. Referring to the explosive metabolic activity of adipose tissue, Dr. Cahill remarked[38]:

> A zoologist friend reminded me that adipose tissue, particularly subcutaneous adipose tissue, does not appear in evolution until the homeotherms. Another friend, an engineer, pointed out that lipid is really not an excellent insulation material . . .
>
> Placing the knowledge of all three of them together, perhaps one should think of the recycling of triglyceride fatty acids to free fatty acids by lipolysis, activation of the acyl Co-A moiety and esterification with glycerol phosphate as a means of generating heat. Thus, we should think of the subcutaneous adipose tissue not merely as a simple insulating blanket but perhaps as an electric blanket.

The manner in which a hormone like norepinephrine acts was schematized (Fig. 4-15) by Sutherland and Robison.[39] Norepinephrine is assumed to act like epinephrine in initiating the cascade effect described by Krebs and associates[40] by way of adenyl cyclase. In fact, insofar as the effects on cyclic AMP are concerned, there appears to be no difference between epinephrine and norepinephrine. However, this fact established in vitro is not consistent with the different calorigenic effects of the two catecholamines in the warm-adapted animal nor does it offer an explanation for the changed norepinephrine sensitivity of cold-adapted animals.

The interpretation of metabolic events in adipose tissue based on data obtained with in vitro technics suggests that epinephrine and norepinephrine are rather nonspecific lipolytic hormones. To account for the generation of heat, the assumption must be made that this system cycles.

Changes in peripheral circulation

Early field studies indicated that, at any given skin temperature, acclimated subjects have a smaller increment in heat production than nonacclimated subjects. Since the avenues of heat loss from the skin were unaltered in these tests, the logical deduction was that heat is supplied from body storage and the idea of the change in ratio of core to shell was introduced. A second observation was that hand temperatures were higher and finger pulse height diminished less during the test cold exposure. These observations have been extended in a number of studies. For a summary see Carlson and Hsieh.[19]

Changes in peripheral circulation during cold exposure have been studied in the rabbit ear; the circulatory response of the ear to cold exposure and the response to norepinephrine infusion is changed in cold-acclimated animals. Honda's[41] interpretation of these observations is shown in Fig. 4-16. This model assumes that cooling blood at the surface gives up a quantity of heat, Q_2. Heat exchange occurs between artery and vein over a length, L, with a heat value of Q_1. In the transient state this exchange mechanism cools body tissues if either L or heat exchange increase. The ratio Q_1/Q_2 indicates the amount of heat exchanging. By this analysis, the cold-acclimated rabbit has a considerably greater portion of its

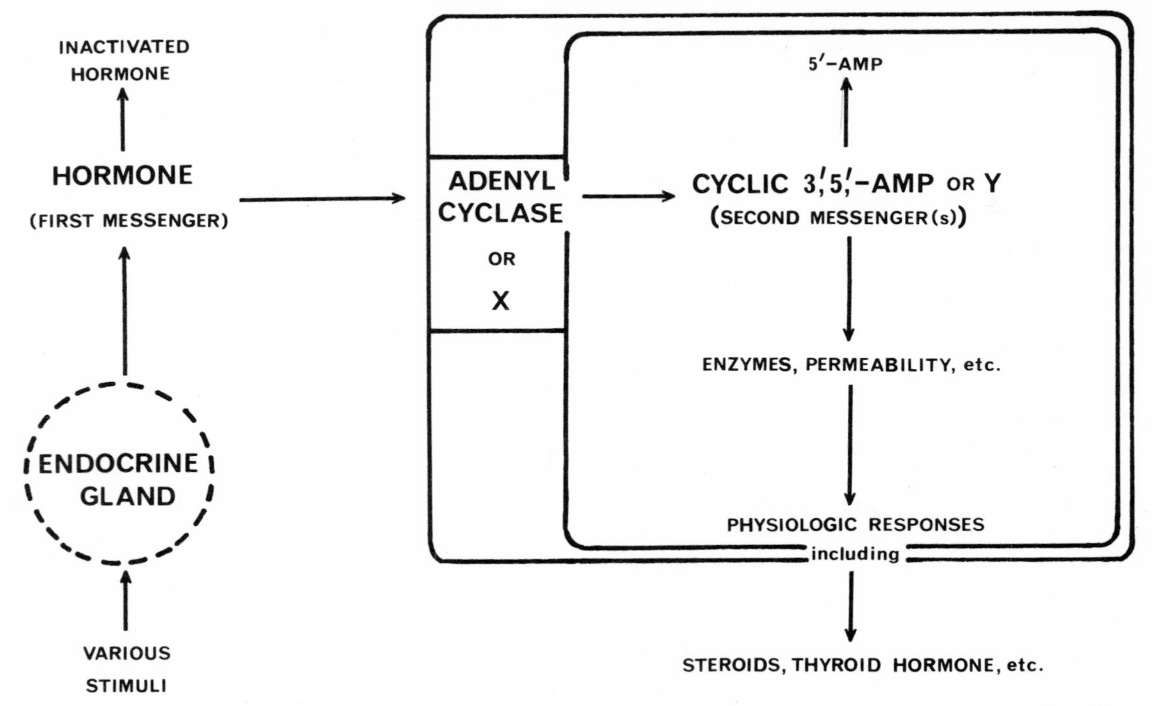

Fig. 4-15. Schematic representation of the possible mode of action of a hormone such as noradrenalin. (Redrawn from Sutherland, E. W., and Robinson, G. A.: Pharmacol. Rev. **18:**145-161, 1966.)

body participating in cooling and there appears to be increased heat exchange.

Hypothermia, hibernation, and aestivation

Failure to meet the requirements for heat production and heat conservation on exposure to cold leads to a reduction of body temperature, or hypothermia. Hibernation, as defined by the Shorter Oxford English Dictionary, is "to spend the winter in some special state suited to resist it; said especially of animals that pass the winter is a state of torpor." Aestivation is the condition of summer dormancy in certain mammals.

Hypothermia

Common usage tends to restrict the term "hypothermia" to the condition of reduced body temperature in homeotherms that do not hibernate. Any mammal exposed to sufficiently low temperature shivers and increases its metabolic rate. If the cold is too intense or of too long duration, exhaustion eventually takes place and the animal cools. Thus hypothermia results from failure of compensatory mechanisms and must be regarded as a pathologic condition. In general, mammals cannot survive body temperatures below 25° C. Death is usually caused by ventricular fibrillation or cardiac arrest. However, using special technics, Andjus[42] was able to reduce the body temperature of rats to values between 0° and −3° C and to revive about 80% of them without apparent long-term deleterious effects. In human beings hypothermia has been induced by means of icepacks and a combination of drugs that inhibits shivering and blocks sympathetic activity. Hypothermia is of interest in clinical medicine because body temperature reduction decreases tissue oxygen requirement; the circulatory system can then be stopped for many minutes. Patients undergoing brain surgery have had both common carotid and vertebral arteries occluded for 10 minutes while their body temperatures were artificially maintained at about 29° C.

Hibernation

Animals such as the hedgehog, marmot, woodchuck, hamster, dormouse, squirrel, and bat hibernate during the winter or when exposed to low environmental temperature for varying lengths of time. These animals reduce their metabolic rates and allow body temperature to fall when exposed to cold. Hibernators increase their chances of sur-

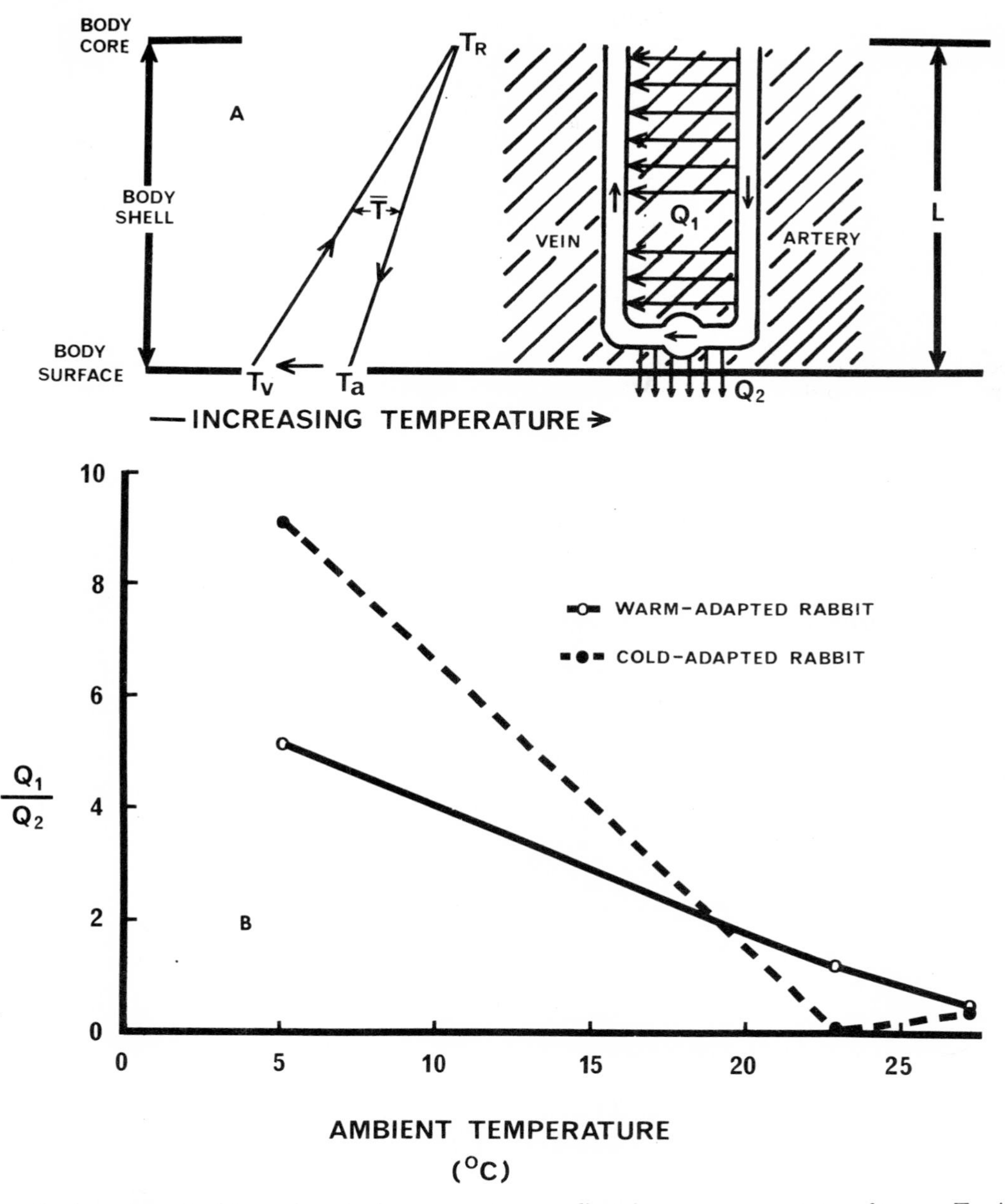

Fig. 4-16. A, Schematic illustration of the temperature gradient in a countercurrent exchanger. T_a, Arterial blood temperature; T_v, venous blood temperature; T_R, rectal temperature; $\overline{T}$, mean temperature of the body shell. **B,** Q_1/Q_2 ratio for the ear of warm-adapted and cold-adapted rabbits at 5°, 22°, and 28° C. (Redrawn from Honda, N.: J. Appl. Physiol. **20:**1133-1135, 1965.)

vival by reducing the strain on temperature-regulating systems during periods when food is scarce since the demands for homeothermia would otherwise exhaust energy reserves. However, Mayer[43] pointed out that the hibernator is in a precarious balance between maintenance of metabolism at such a level that recovery is possible and decrease to a level at which death is inevitable. The hibernation process is a precarious method of survival at best and one from which many animals do not awaken.

The classic criteria for hibernation are given by Bartholomew and Hudson[44]:

1. Body temperature within a degree of ambient temperature
2. Oxygen consumption markedly reduced
3. Prolonged periods of apnea
4. A torpor more pronounced than deep sleep
5. Arousal, either spontaneous or induced, accompanied by activation of the major heat-producing mechanisms

The mechanisms involved in hibernation are unknown. Lyman and Chatfield[45] discuss three theories. The first, which is not widely accepted, is that mammals that hibernate are primitive, have poor temperature regulation, and behave physiologically like the lower poikilothermic vertebrates. However, hibernation appears to be a specialization of the homeothermic state and not a reversion to poikilothermia. The second general theory proposes that hibernation is brought about by a decrease or increase of substances that are ordinarily present in the mammalian body. Extracts from the so-called hibernating gland have been examined with equivocal results. Most of the endocrine glands involute prior to hibernation and resume nominal functioning before winter sleep ends. Hibernators given thyroid hormone do not hibernate at all or hibernate less than controls. Ground squirrels and hamsters need adrenocortical hormones to hibernate.[46] The third concept of hibernation implicates the central nervous system. This states simply that hibernation is initiated by "turning down the thermostat" in the hypothalamus. The nature of this change is unknown and there is no evidence implicating hypothalamic centers in the process.

Aestivation

In desert areas aestivation offers an effective mechanism for survival during periods when food and water are scarce. The Mohave ground squirrel is normally active above ground only during spring and early summer. Under laboratory conditions it spontaneously enters torpor at room temperature and spends much of the summer, fall, and winter in a dormant condition.[44] It appears that aestivation and hibernation are the same physiologic phenomenon, the only difference between the two being the level of body temperature, and this depends on ambient temperature.

JULY, 1970

PART B: HEAT AND HUMIDITY

SID ROBINSON
DAVID L. WIEGMAN

Man is a tropical homeotherm who without the protection of clothing and shelter cannot tolerate cold climates as do many birds and fur-bearing mammals. On the other hand, he is capable of working at moderate rates and regulating his central body temperature within rather narrow limits in the most extreme humid heat of the wet tropics. Here water and plant growth are plentiful and the intensity of solar radiation to which he is exposed is reduced by clouds and the high water vapor content of the atmosphere. With an adequate supply of drinking water and light clothing to protect his skin from the damaging ultraviolet radiation of the sun, man can live and work even in hot dry desert environments, where his total heat gain by convection and by radiation from sun and ground may far exceed his resting metabolic heat production.

A man who is accustomed to prolonged hard work and is well acclimatized to the heat may maintain thermal equilibrium during 6 hours of treadmill work at a metabolic rate that is four times the usual resting rate at a room temperature of 50° C and low relative humidity. In this case the man must dissipate his metabolic heat (170 $kcal \times m^{-2} \times hr^{-1}$) plus about an equal amount of heat that he absorbs from the indoor environment by radiation and convection. Even with this great heat and work load he regulates his central temperature at about the same level he would have if performing the same work in a cool room. The only avenue of heat dissipation in this situation is evaporation of water, principally from his skin, supplemented by much smaller amounts from the lungs and air passages. To execute this remarkable feat our subject must drink water each hour at the rate he is expending it (1.2 kgm/hr), since his capacity for sweating and evaporative cooling is reduced by dehydration. To finish the day in good condition he must also, as he progresses, replace the electrolytes, principally sodium and chloride, lost in the sweat and urine. The task can be completed with much less fatigue if the energy he expends is also replaced with readily available nutrients during the 6-hour work period.

ENVIRONMENTAL FACTORS

Man encounters hot climates not only in tropical equatorial areas but also during summer in many of the vast land areas of the temperate zones. Dry desert areas are characterized by sparse plant growth and irregular rainfall, low relative humidity, and intense solar and ground radiation. The ground becomes so dry that its specific heat is low and there is little water to evaporate. This may result in ground surface temperatures on the desert in excess of 80° C at midday when solar radiation is high. The depth of heating in the sand is not great and with the low specific heat the ground cools rapidly at night.

In summer the diurnal variation of air temperature in the desert may range from 25° to 30° C at night to 45° or 50° C in the intense radiation of midday. The dry cloudless atmosphere of the desert has usually lost its moisture by precipitation on surrounding cold mountains. The low water vapor content of desert air, whose daytime relative humidity may be 6% to 10%, absorbs little of the sun's rays; solar radiation reaching the earth is much greater than in the moist tropics. The low water vapor pressure of desert air is physiologically advantageous to man since it provides a large vapor pressure gradient between skin and atmosphere; evaporative cooling of the skin occurs as rapidly as sweat is secreted. Desert winds are frequently high, facilitating evaporative cooling but adding heat to the body by convection in proportion to wind velocity and the temperature gradient between hot air and relatively cool skin.

A comparison by Molnar[47] of heat gain or loss of men in tropical and desert environments is given in Fig. 4-17. This illustrates the great difference in the intensity of heat stress in these two climates and the protection provided by clothing and shelter at midday in the desert. Fig. 4-18 gives the mean summer midday temperature and

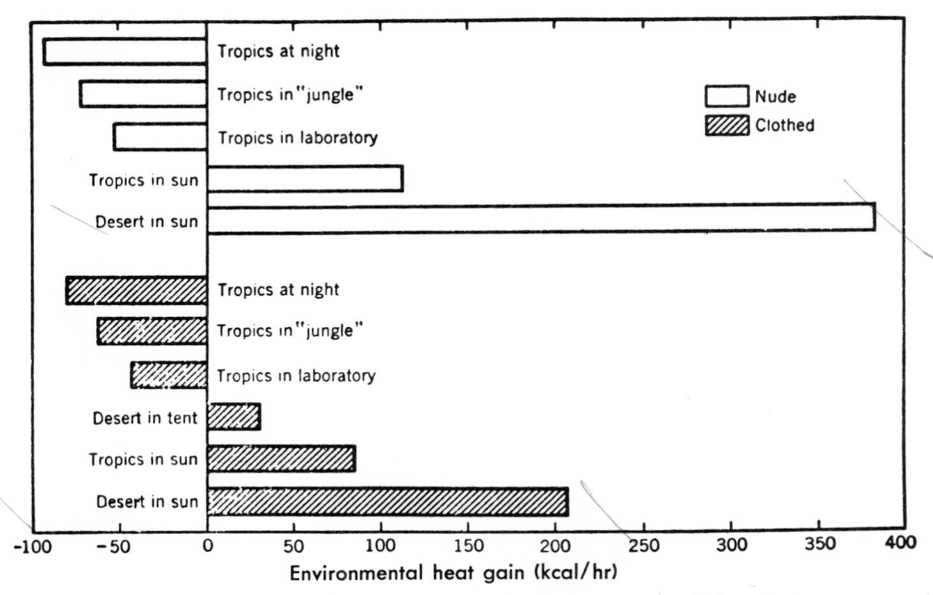

Fig. 4-17. Mean values of heat gain of men from the environment at midday during summer in various situations in the California desert and at Eglin Air Force Base, Florida. The jungle is a small wooded area with heavy undergrowth and damp ground. (Data from Molnar, G. W. In Adolph, E. F., and others: Physiology of man in the desert, New York, 1947, Interscience Publishers, Inc.)

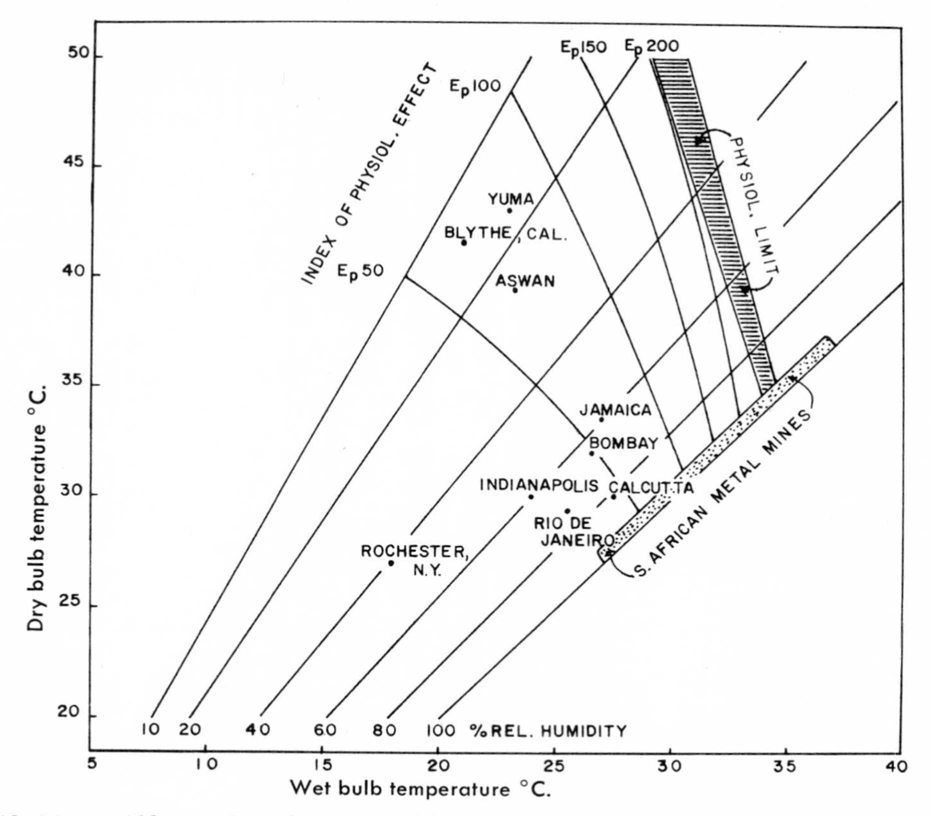

Fig. 4-18. Mean midday outdoor thermal conditions of typical desert, tropical, and temperate localities in midsummer, and the thermal conditions in deep metal mines of South Africa. Each contour line of index of physiologic effect (E_p) represents indoor conditions of temperature and humidity with air movement (55 m/min) that have the same physiologic effect on acclimatized men performing moderate work ($M = 125$ kcal $\times$ m^{-2} $\times$ hr^{-1}) and wearing 5-oz poplin suits during 2-hour exposures. The index E_p is calculated by equally weighting the men's heart rates, sweat rates, $\overline{T}_s$, and T_{re} during the second hour of exposure. The physiologic limits are the most severe conditions in which acclimatized men can maintain thermal equilibrium for 6 hours. (Data from Molnar, G. W. In Adolph, E. F., and others: Physiology of man on the desert, New York, 1947, Interscience Publishers, Inc.; and Leithead, C. S., and Lind, A. R.: Heat stress and heat disorders, Philadelphia, 1964, F. A. Davis Co. From Robinson, S., and others: Am. J. Physiol. **143**:21, 1945.)

humidity typical of tropical and desert regions.[47, 48] Also given are the hot humid conditions in the deep metal mines of South Africa. For comparison with the climatic environments in Fig. 4-18, contour lines on the psychrometric chart representing the physiologic effects (E_p) of various indoor conditions of temperature and humidity on working men ($M = 125$ kcal $\times$ m^{-2} $\times$ hr^{-1}) are given.

HEAT EXCHANGE

In part A of this chapter the authors present a comprehensive mathematical analysis of the mechanisms of heat exchange between organism and environment. In this section we emphasize the physiologic adjustments required to maintain a balance between heat gain and heat dissipation under conditions of heat loads produced by the environment and/or by increased metabolic heat production during physical work.

To maintain a constant body temperature an organism must balance its rate of heat dissipation against its rate of heat gain. Heat is added to the body by metabolism, which in work increases as the work rate increases. In warm or hot environments the body may gain heat by radiation, convection, and/or conduction, depending on the respective temperature gradients between skin and environment. Heat is dissipated from the body by evaporation of water from the skin and respiratory air passages and, in cool or cold environments, by radiation, convection, and conduction. Relatively small amounts of heat may also be gained or lost by ingestion of food and drink at temperatures different from body temperature. Deviations

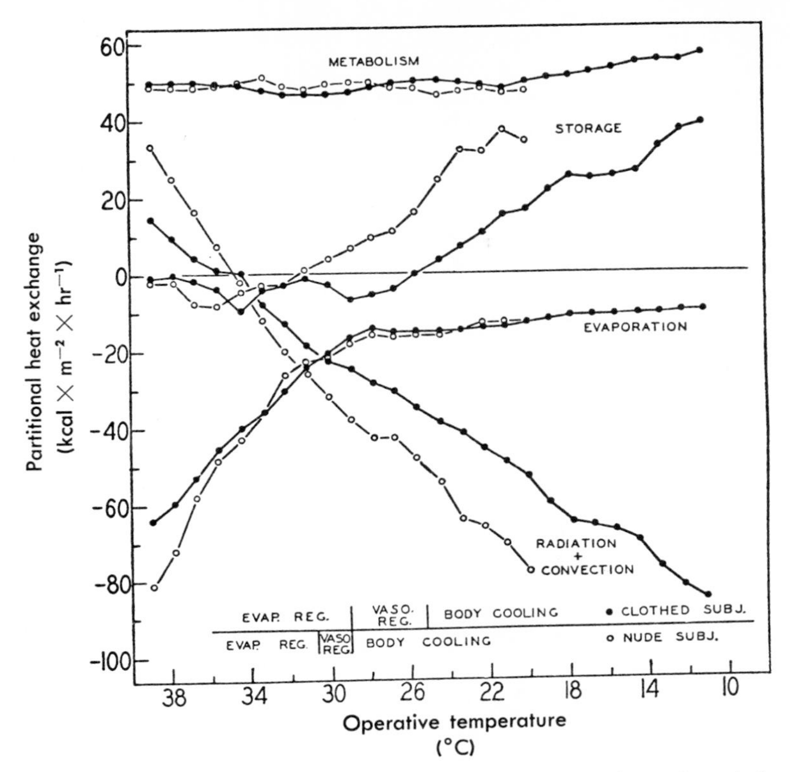

Fig. 4-19. Partitional heat exchange of clothed and nude men at rest in various indoor environmental (operative) temperatures. (Data of Gagge, A. P., Winslow, C. E. A., and Herrington, L. P.: Am. J. Physiol. **124**:30, 1938.)

from the balance of heat exchange involve alterations of the body's stored heat and thus its mean temperature. The balance of heat exchange between a man and his environment is expressed by the equation

$$M \pm R \pm C \pm K - E \pm W \pm S = 0 \qquad (41)$$

where each term is expressed in kcal $\times$ m^{-2} $\times$ hr^{-1}. *M* is the total metabolic rate of the man calculated from his respiratory exchange of oxygen and carbon dioxide. *R*, radiation, is positive when surrounding surfaces are warmer than the body surface (net radiant exchange adds heat to the body) and negative when these surfaces are cooler than mean skin surface. *C*, convection, is positive when air temperature exceeds that of the body surface (heat is added to the skin from the air) and negative when air temperature is lower than average skin temperature. The rate of convective exchange is proportional to the temperature gradient between the subject's skin and the air temperature, the rate of air movement, and the subject's movement through the atmosphere. *K*, conductive heat exchange, is positive for heat gain from objects warmer than the skin and negative for heat loss from the skin to cooler material in contact with it. *E*, evaporative heat exchange, is always negative because evaporation of water from skin and lungs absorbs heat from the body. *S*, storage, with a negative sign indicates a mean body temperature increase and a gain of stored heat; with a positive sign, it means a decrease in mean body temperature and a loss of stored heat. If thermal equilibrium is maintained, *S* is zero. *W*, work, is negative in this equation when the subject is performing measurable external work and positive when work is performed on the subject, as in walking down a grade or down stairs (negative work, equation 43). The relative magnitudes of these factors in the heat exchange of a resting man during exposure to environmental temperatures ranging from 11° to 39° C are shown in Fig. 4-19.[49]

Radiation

Exchange of heat by radiation between body and environment depends upon (1) the subject's effective radiating surface area, (2) his mean surface temperature, and (3) emissivity and reflecting power of skin and clothing, in relation to emissivity and average radiant temperature (infrared) of the environment.

The skins of both white and black men act as almost perfect blackbodies in the infrared part of the spectrum. It has been found that the white man's skin, depending on degree of pigmentation, reflects 30% to 45% of total solar radiation whereas the black man's skin reflects only 16% to 19%.[50] The principal difference between white and black skin in absorption and reflection of radiation is in the visible and ultraviolet parts of the solar spectrum, not in the infrared where the two are the same.[50]

The effective radiating surface area of a man standing with arms and legs spread is about 85% of total skin area, according to measurements made by Bohnenkamp and Ernst.[51] The remaining skin surfaces, such as the inner surfaces of the legs, arms, and corresponding sides of the trunk, radiate to other skin surfaces. Winslow and associates[5] found the effective radiating surface of a sitting man to be 70% to 75% of the body surface area.

The radiative heat exchange of a man is most complex under the outdoor conditions of sunlight with its wide spectrum of wavelengths emitted at various intensities and absorbed by the skin to varying degrees. Complicating the outdoor problem further is the fact that the spectrum and intensity of radiation from the terrain are different from those of the sun and sky.

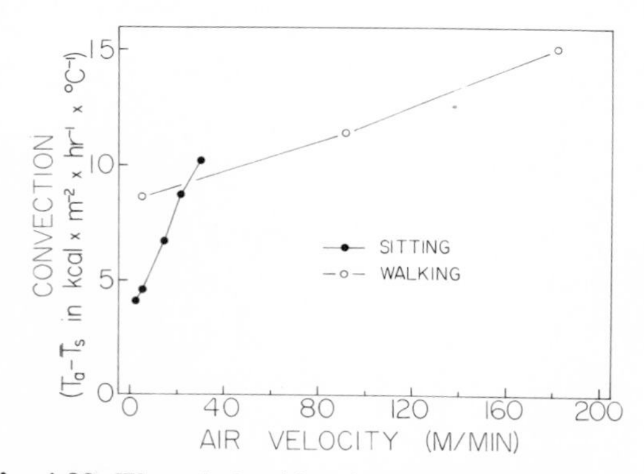

Fig. 4-20. The relationship of convective heat loss to air velocity in nude men while sitting at rest and while walking on treadmill at 5.6 km/hr. (Data for sitting from Winslow, C. E. A., Herrington, L. P., and Gagge, A. P.: Am. J. Physiol. **116**:669-684, 1936; data for walking from Robinson, S.: Physiological adjustments to heat. In Newburgh, L. H., editor: Physiology of heat regulation, Philadelphia, 1949, W. B. Saunders Co., p. 197.)

Convection

The exchange of heat between body and surrounding air depends on (1) the difference in temperature between body surface and air, which determines the heat absorbed or given up by a unit mass of air coming in contact with the skin, and (2) air movement, which determines the mass of air coming in contact with the surface. Air movement may be free convection currents resulting from different air temperature and density at the skin surface; the amount of free convection is determined by the temperature gradient between skin and air and evaporation from skin. In forced convection, as by wind, heat exchange increases with air velocity, and this factor may greatly exceed free convection in importance (Fig. 4-20). Increasing air movement above 15 km/hr has little additional cooling or heating effect on the body. Bodily movement within the atmospheric medium also increases convective exchange as a function of the rate of movement.

Conduction

Usually conduction of heat between skin and surrounding surfaces with which the subject is in contact represents a small percentage of his total heat exchange; in studies of physiologic adjustments to heat stress, conduction is not usually calculated separately but is combined with radiation and convection in the exchange. The areas of skin in contact with surrounding objects are usually small, especially in a working man, and direct contact with highly conductive materials is avoided. A man would not sit or recline directly on hot or cold metal or even on the ground at midday on the hot desert. In clothed man conduction of metabolic heat from skin to clothing occurs; it is then dissipated from the outer clothing surfaces by evaporation and/or convection and radiation, depending on air movement and temperature and vapor pressure gradients between clothing and environment.

Evaporation

Evaporation of water from the skin, lungs, and air passages is the only avenue of heat dissipation by a man in environments where air and radiant temperatures exceed those of his skin. The cooling effect of the latent heat of vaporization of water (0.58 kcal/gm at average skin temperature) is one of its most important contributions to biologic systems.

Evaporative heat loss from man is determined by the rate of evaporation of water from skin and respiratory system times the latent heat of vaporization of water. Evaporation may be determined by measuring the subject's net weight loss on a precision balance during the period of observation and subtracting from this the difference between the weight of carbon dioxide produced and the weight of oxygen consumed. Evaporative weight loss *cannot* be determined if unmeasured sweat drips from the skin as it often does during work in hot humid environments.

$$E = \frac{(g_1 - g_2) + (g_i - g_e) - (CO_2 - O_2)}{t \times A} \times (0.58) \quad (42)$$

where E = total evaporative cooling (kcal $\times$ m^{-2} $\times$ hr^{-1})
g_1 and g_2 = subject's body weight in grams at the start and end of the observation period
g_i = intake of water and food in grams
g_e = weight lost as urine or feces
CO_2 and O_2 = grams of carbon dioxide and oxygen exchange for the period
t = time (hours)
A = surface area (m^2)
0.58 = latent heat of vaporization of water from the body (kcal/gm)

Evaporation from the respiratory system is partitioned from cutaneous evaporation by calculating pulmonary evaporation as the difference between the water vapor contents of inspired and expired air based on their respective volumes, temperatures, and relative humidities.

The rate of sweat evaporation from wet skin depends upon (1) the difference between the vapor pressure of water on the skin and that in the air, (2) air movement around the subject, and (3) the percentage of the skin surface that is wet with sweat. The vapor pressure of water on wet skin is proportional to the surface temperature of the skin. In the air it is dependent upon air temperature and relative humidity. According to Gagge[52] the maximal rate of evaporation from the completely wet nude skin of a resting man is about 28.5 kcal $\times$ m^{-2} $\times$ hr^{-1} per centimeter of mercury difference in water vapor pressure between skin and air when there is turbulent air movement at 8.65 cm/sec. Evaporation is increased by increasing air movement and/or bodily movements of the subject. In humid atmospheres the percentage of wetted area increases, keeping evaporation at the rate required for thermal equilibrium. A well-regulated sweating mechanism increases the wetted area of the skin, balancing decreased moisture demand of a humid atmosphere and thus maintaining evaporative heat loss at a desirable level.

Work

If a subject is performing measurable work, it should be included in the calculation of his partitional energy exchange as expressed in equation 41. For example, in walking at a constant rate on the treadmill the rate of work ($\pm$ W) expressed in kcal $\times$ m^{-2} $\times$ hr^{-1} may be calculated as follows:

$$W = \frac{L \left[\frac{\pm \% \ Gr}{100}\right] \times kgm}{427 \ A} \quad (43)$$

where L = rate of walking (m/hr)
Gr = percent grade, (−) if walking up grade, (+) if down grade
kgm = weight of subject and load carried
A = body surface area of the subject in m^2
427 kgm-m of work = 1 kcal of heat energy

Recall that in the heat exchange equation 41 positive work performed, as in walking up a grade, is given a negative sign since this part (15% to 20%) of the metabolic energy expenditure is mechanical work and not heat; during negative work heat is added to the body in excess of the metabolic energy cost and is therefore positive in the heat balance equation.

Storage

In equation 41 changes in heat storage (S) of the body are proportional to changes of mean body temperature during the period of observation and may be calculated by the following equation:

$$S = \frac{(0.8 \times \Delta T_{re} + 0.2 \times \Delta \overline{T}_s) \ 0.84 \times kgm}{t \times A} \quad (44)$$

where S = rate of change of stored body heat (kcal $\times$ m^{-2} $\times$ hr^{-1}) during the period of observation
ΔT_{re} and $\Delta \overline{T}_s$ = subject's rectal and mean skin temperatures at the start minus those at the end of the period of observation, assuming that T_{re} and $\overline{T}_s$ represent 0.8 and 0.2 in the estimation of mean body temperature
0.84 = specific heat of the body
kgm = subject's body weight
t = time (hours)
A = body surface area (m^2)

With a rise of mean body temperature during a period of observation, a part of the heat gain is being stored and not dissipated, and thus S (storage) is given a negative sign in the heat balance equation 41. With a decrease of body temperature, a part of the stored heat is being dissipated and thus must be added in the heat exchange equation.

Water intake

Under certain circumstances values for heat loss or gain by food and water ingestion at temperatures different from body temperature may be added to equation 41. However, this avenue of heat exchange usually represents a small percentage of total heat exchange and is not calculated separately but is included with the other avenues in the partitional exchange. For example, the direct cooling effect of drinking cold water is most important during work in a hot, dry environment where both radiant and air temperatures exceed skin temperature and the subject absorbs heat from the environment. Here he must drink a large volume of water to replace evaporative losses. Thus a man working (heat production of 175 kcal $\times$ m^{-2} $\times$ hr^{-1}) in thermal equilibrium with a hot indoor environment (46° C) may drink 0.523 kgm $\times$ m^{-2} $\times$ hr^{-1} of water at mean body temperature (37° C) to replace his evaporative water loss ($523 \times 0.58 = 303$ kcal $\times$ m^{-2} $\times$ hr^{-1}). Instead he could maintain thermal and water balances by evaporating and drinking 508 ml of water $\times$ m^{-2} $\times$ hr^{-1} at 20° C since it would be warmed to mean body temperature (37° C) and absorb 8.6 kcal $\times$ m^{-2} $\times$ hr^{-1} of heat, or 3% of his evaporative heat exchange. If he were drinking cold water (3° C) its direct cooling effect would be about twice as great. If the environment were cooler than his skin and part of his heat dissipation were by radiation and convection, his water and evaporative requirements would be reduced in proportion to the difference in radiative plus convective heat exchange in the two situations.

Total environmental heat load

In many studies of the physiologic adjustment to thermal stresses and their regulation, it is not necessary to determine heat exchanges by radiation, convection, and conduction separately. For example, in the desert where air movement and the vapor pressure gradient between skin and atmosphere are high and evaporation of sweat is complete, M, S, W, and E may be measured and the channels of dry heat exchange (R, C, and K) combined to represent environmental heat load as estimated by difference from equation 41:

$$(\pm R \pm C \pm K) = -M \mp S \mp W + E \qquad (45)$$

Gagge and associates[49] in laboratory studies and Adolph and colleagues[53] in desert and tropical field studies used this technic for measuring total environmental heat gain of men during exposure to environmental heat stresses, as well as for measuring total dry heat exchange in cooler environments (Figs. 4-17 and 4-19).

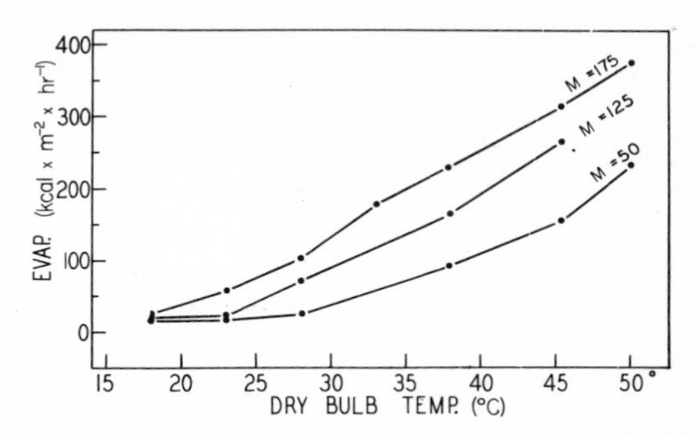

Fig. 4-21. The effects of increasing air temperature on the evaporative cooling necessary for thermal equilibrium in heat acclimatized men at three different metabolic rates. Equilibrium values were determined during 6-hour exposures of men at rest, walking at 4.5 km/hr, and walking at 5.6 km/hr on a 2.5% grade at six different room temperatures with low relative humidity. Air movement was 55 m/min and wall temperatures were within 1.3° C of air temperature in all experiments.

PHYSIOLOGIC ADJUSTMENTS TO HEAT

Adjustments of cutaneous blood flow and evaporation of sweat from skin represent the first and principal lines of the body's physiologic defense against overheating during prolonged exposure to hot environments and/or during work with increased metabolic heat production. The capacities of these mechanisms to balance heat dissipation with heat gain are remarkably great (Fig. 4-21). However, even within the range of combined environmental and metabolic heat loads that can be dissipated by these mechanisms, changes of body temperature do occur. Substantial adjustments occur in mean body temperature and in the temperature gradients and heat distribution within the body. With continued exposure to work and heat loads exceeding the body's capacity for heat dissipation, body temperature rises to intolerable levels.

The strain shown by men exposed to the stress of hot environments is indicated by increased heart rate, sweating, skin temperature, and central, or core, temperature. Circulatory strain is indicated by increased heart rate and, under conditions nearing the limits of tolerance, by a fall of systemic arterial blood pressure and syncope if the subject attempts to stand erect. The effects of increasing heat stress on men are shown in Fig. 4-22, representing average values of data from experiments in which two men, well acclimatized to work in the heat, walked on a treadmill in

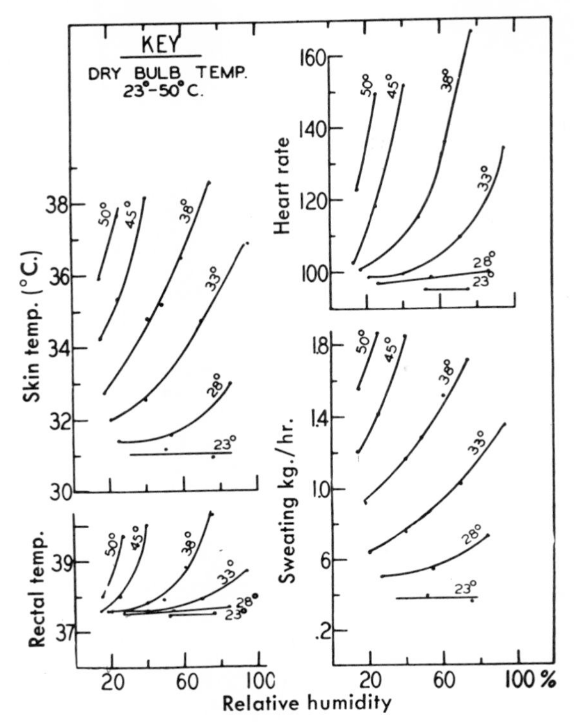

Fig. 4-22. Effects on working men ($M = 190$ kcal $\times$ $m^{-2} \times hr^{-1}$) of varying relative humidity at six different dry bulb temperatures. Each value for skin temperature (T_s) and sweat rate represents the average of data for two clothed men during the entire second hour of an exposure; each value for heart rate and rectal temperature represents the average for the two men at the end of the corresponding exposure. The men walked at 5.6 km/hr on a 2.5% grade. (Data from Robinson, S.: Physiological adjustments to heat. In Newburgh, L. H., editor: Physiology of heat regulation, Philadelphia, 1949, W. B. Saunders Co., p. 340.)

environments where air temperature and relative humidity varied. At the lowest values of relative humidity used in the study, skin temperatures and rates of sweating in these 2-hour walks were both elevated significantly with each 5° C increase in dry bulb temperature between 23° and 50° C; in the same experiments neither heart rate nor rectal temperature was seriously elevated by increasing dry bulb temperature up to 50° C with a relative humidity of 15%. At a dry bulb temperature of 23° C, raising the humidity had no measurable effect on any of these four physiologic parameters. At 28° C the working men's rates of sweating and skin temperatures increased curvilinearly with increasing humidity, but neither heart rate nor rectal temperature increased significantly above their values at 23° C. The effects of increasing relative humidity on all four functions became more and more pronounced as dry bulb temperature increased above 28° C. Significant circulatory strain in the working men, manifested by heart rates above 140, appeared only when water vapor pressure in the air became so high that heat dissipation by evaporation was limited and rectal and skin temperatures rose significantly above 38.5° and 36° C, respectively.

Body temperature changes

A man's mean skin temperature ($\overline{T}_s$) varies directly with increments in environmental temperature, while the core, or central, temperature (usually measured as rectal temperature, T_{re}) is regulated within narrow limits over a wide range of thermal stresses. This is shown in Fig. 4-22 for two clothed men working at a constant rate in 2-hour experiments in dry bulb room temperatures from 23° to 50° C, and in Fig. 4-23 for a man wearing only shorts, shoes, and socks during both rest and work experiments in effective temperatures (T_{eff}) ranging from 17° to 37° C. Effective temperature is an index of the warmth felt by the human body on exposure to various indoor temperatures, humidities, and air movements, with walls and air at the same temperature. In Fig. 4-23 the man, who was highly acclimatized to work in the heat, became overheated at T_{eff} 34° to 37° C, as the gradient $T_{re} - \overline{T}_s$ progressively decreased. The relation of skin temperature to environmental temperature is the net result of three factors: (1) direct exchange of heat by radiation, conduction, and convection between skin and environment, (2) variations of cutaneous blood flow, a factor that tends to warm the skin when increased, and (3) evaporative cooling of the skin.

The resting ($M = 46$ kcal $\times$ $m^{-2} \times hr^{-1}$) subject's central body temperature (T_{re}) at the end of the 2-hour exposures declined slightly below 37° C in the cooler environments ($T_{eff} = 17°$ to 23° C) and rose above 37° C in the higher room temperatures (Fig. 4-23). On the other hand during constant work ($M = 190$ kcal $\times$ $m^{-2} \times hr^{-1}$) rectal temperature increased to 37.6° to 37.8° C and was regulated closely at this level in all T_{eff}'s between 17° and 30° C. Rectal temperature rose further during work as resistance to heat dissipation increased with increments in effective temperature above 30° C. Burton and associates[54] found the rise of T_{re} of resting men during acute exposure to heat to be transitory and, during a few days of continuous exposure and acclimatization to heat, T_{re} gradually decreased to the usual

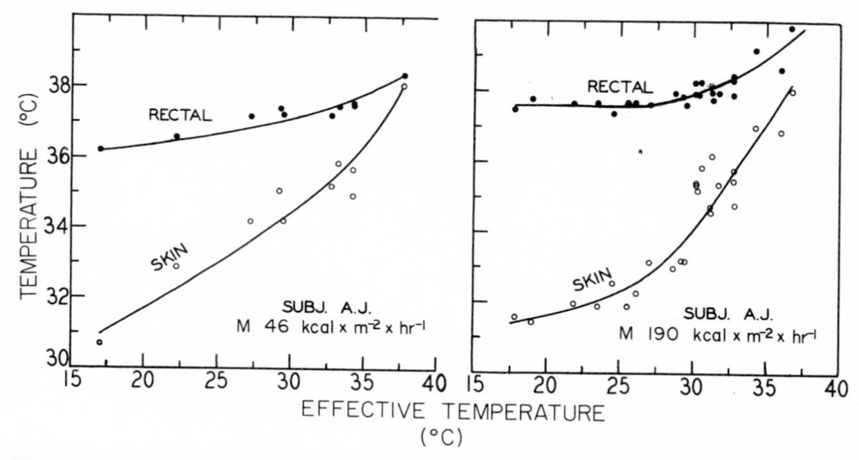

Fig. 4-23. The effects of exposing a highly acclimatized man to increasing effective temperatures (T_{eff}). In one series the subject was at rest; in the other he walked on a treadmill at 5.6 km/hr on a 2.5% grade. Each value of mean skin temperature is the mean during the second hour of a 2-hour exposure; rectal temperatures were measured at the end of the second hour. (Data from Robinson, S.: Physiological adjustments to heat. In Newburgh, L. H., editor: Physiology of heat regulation, Philadelphia, 1949, W. B. Saunders Co., p. 225.)

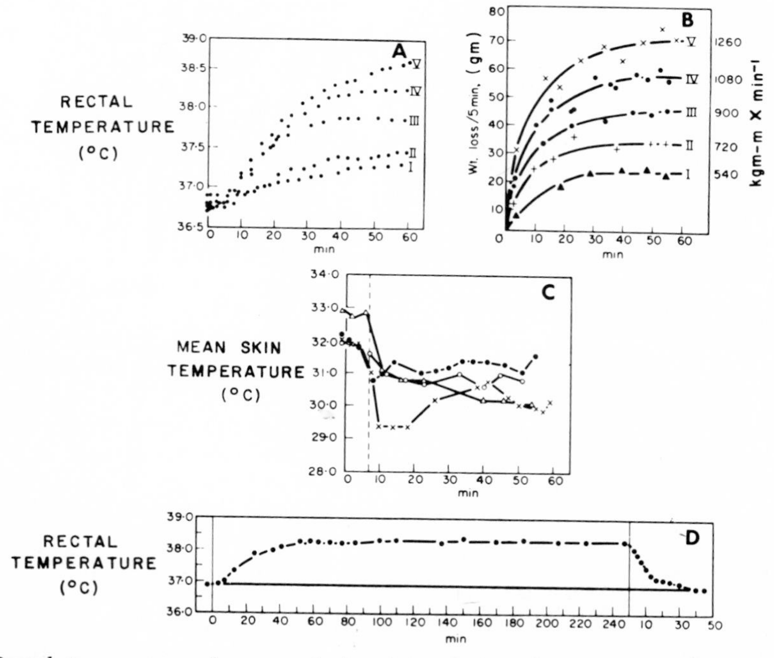

Fig. 4-24. A, Rectal temperature of a man during 1 hour of work on a bicycle ergometer at five different rates: *I,* 360 kgm-m × min⁻¹; *II,* 540 kgm-m × min⁻¹; *III,* 900 kgm-m × min⁻¹; *IV,* 1,080 kgm-m × min⁻¹; and *V,* 1,260 kgm-m × min⁻¹. Room temperature was between 22° and 23° C in all experiments. **B,** Weight loss of the subject in five work experiments described in **A.** Weight loss indicates sweat and evaporation rates of the man. **C,** Mean skin temperature of the subject during four work experiments performed in a cool room (23° C): -O-, 360 kgm-m × min⁻¹; -△-, 540 kgm-m × min⁻¹; -●-, 1,080 kgm-m × min⁻¹; -×-, 1,260 kgm-m × min⁻¹. A fan was turned on the subject beginning at 7 minutes in each experiment. **D,** Rectal temperature during 250 minutes of work at 1,080 kgm-m × min⁻¹. (From Nielsen, M.: Skand. Arch. Physiol. **79:**193, 1938.)

resting level of 37° C. It is also important to note in Fig. 4-23 that during both rest and work the $\overline{T}_s$ rise with increments of T_{eff} was significantly greater than the rise of T_{re}, with the net result that the gradient between T_{re} and $\overline{T}_s$ was progressively reduced. This means that both circulation of blood to the skin and the coefficient of tissue heat conductance increased as the gradient decreased (see p. 102 and Fig. 4-34).

The rectal temperatures of men increase with increasing intensities of work and metabolic heat production, even in a constant cool environmental temperature. Nielsen[55] found that for each level of continuous work within the limits of the steady state a man's rectal temperature rises in 40 to 60 minutes to a new level and is regulated there until the work is stopped; the temperature then returns slowly to the resting level (Fig. 4-24). The hypothalamic thermostat setting seems to increase to a level proportional to metabolic rate[55] or, more specifically, to the rate of metabolic heat production in positive or negative work.[56, 57] On the other hand, in a constant cool environment, increasing the work rate has little effect on mean skin temperature ($\overline{T}_s$) and may even lower $\overline{T}_s$ as sweating and evaporation increase (Figs. 4-24 and 4-25).

For a given level of moderate work rectal temperature is maintained in equilibrium and is the same over a wide range of environmental temperatures (Figs. 4-22, 4-23, and 4-25). However, if the combined heat and work stresses are raised beyond a certain level, a subject cannot continue to work indefinitely in thermal equilibrium and his body temperature rises to intolerable levels (Figs. 4-22, 4-23, and 4-25). Robinson and coworkers[58] have determined that the severity of indoor heat stresses under which men can maintain thermal equilibrium from the second through the sixth hours of exposure is reduced in proportion to the severity of the work and its metabolic requirement.

Fig. 4-26 compares contour lines on the psychrometric chart, denoting the most severe conditions of temperature and humidity in which heat-acclimatized men can maintain practical thermal equilibrium for 6 hours. This graph shows that in these indoor environments the severity of conditions in which subjects maintain thermal equilibrium is significantly less when clothing is worn and is also reduced with increments of work-induced metabolic rate. Under these conditions clothing interferes with evaporative skin cooling; evaporation directly from the skin surface is more

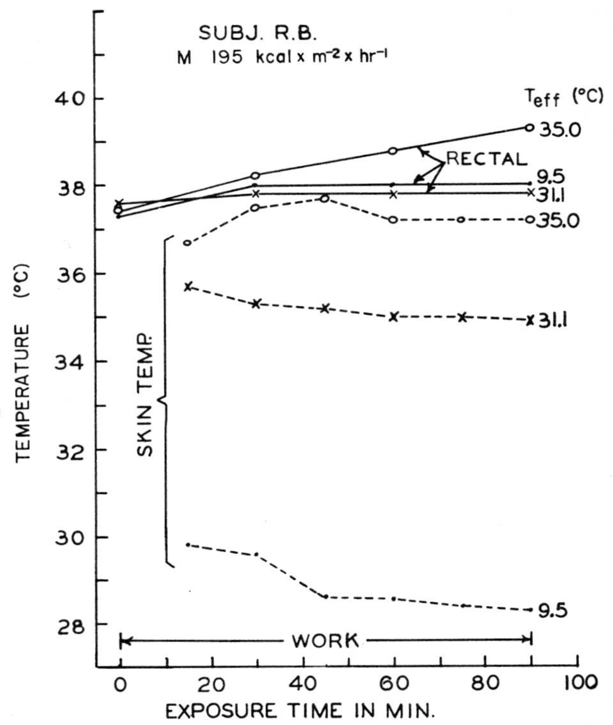

Fig. 4-25. Rectal (T_{re}) and mean skin temperatures ($\overline{T}_s$) of a highly acclimatized man during exposure to three different effective temperatures (T_{eff}). In each experiment the subject walked 90 minutes at 5.6 km/hr on a 2.5% grade. Note the steady state of rectal temperature at an effective temperature of 9.5° and 31.1° C, its continued rise throughout the exposure at T_{eff} of 35° C, and the close relationship of $\overline{T}_s$ to T_{eff}. (Data of Robinson, S. In Mountcastle, V. B., editor: Medical physiology, ed. 13, St. Louis, 1974, The C. V. Mosby Co.)

effective in cooling the body than when it takes place at the outer surface of clothing. On the other hand, Adolph and his colleagues[53] found that the protection provided by light clothing on the desert in summer against the intense radiation from sun and ground is significantly advantageous for temperature regulation (Fig. 4-17). Combinations of heat and work stresses that elevate rectal temperature to about 39° C within 2 hours result, if continued, in further rises of rectal temperature and exhaustion within the next 4 hours. Most well-acclimatized men can barely tolerate a rectal temperature of 40.5° C with a mean skin temperature of 37.5° C produced by 1 to 2 hours of work in a hot environment. This represents an increase of about 200 kcal of stored body heat in a man weighing 70 kgm. An occasional man is exhausted and heat stroke im-

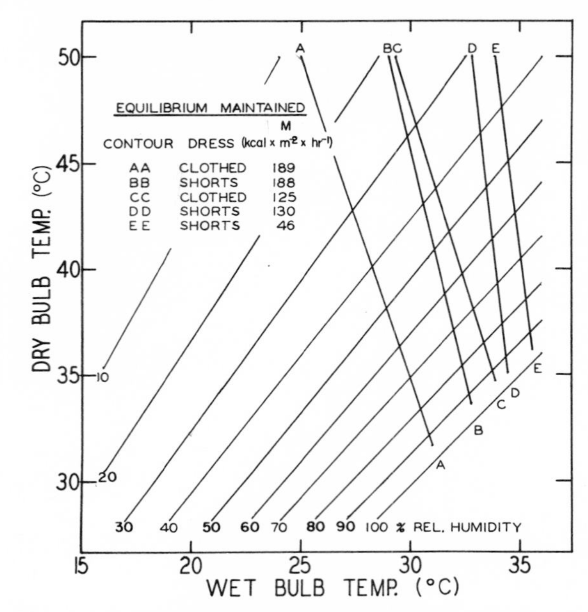

Fig. 4-26. The effects of clothing (light poplin suit or shorts) and work at the most severe indoor air temperatures and humidities in which highly acclimatized men can maintain thermal equilibrium from the second through the sixth hours of exposure. Air movement was 55 m/min during all exposures. The men rested in one experiment and walked on the treadmill at two different metabolic rates in the other experiments. The severity of temperature and humidity at which they could maintain equilibrium was significantly reduced by both clothing and increased work metabolism. (From Robinson, S., Turrell, E. S., and Gerking, S. D.: Am. J. Physiol. **143:** 21, 1945.)

minent, when rectal temperature rises to 40° C. We have observed that champion runners increase rectal temperatures to about 39.5° C in 5- to 10-km races on cool days (10° to 20° C) and to 40° to 41° C on clear summer days when air temperatures are 30° to 32° C. On the other hand, a number of runners have developed heat stroke in distance races during hot weather, and there have been numerous fatal heat strokes among football players during long practice sessions in hot weather.[59]

Evaporation and its regulation

Insensible evaporation from skin and respiratory system occurs continuously, even in cool or cold environments when sweat glands are not secreting (Fig. 4-19). The rate of insensible water loss from lungs and skin of a resting man under comfortable conditions is about 30 gm/hr, absorbing heat at about 17 kcal/hr, about a third of this taking place in the respiratory system. Evaporation from the wet surfaces of the lungs and air passages varies with the rate of pulmonary ventilation and with the temperature and water vapor content of inspired air.[60] Newburgh and associates,[61] studying subjects who live in cool environments and perform only the routines of life, found that total insensible evaporation during 24-hour periods varies directly with metabolic rate and that it dissipates about 24% of the metabolic heat. Gagge and co-workers[49] determined that in resting men total insensible evaporative heat loss varies over a narrow range from minimal values of about 10 kcal $\times$ m^{-2} $\times$ hr^{-1} in an environmental temperature of 20° C up to 18 kcal $\times$ m^{-2} $\times$ hr^{-1} at a neutral room temperature of 29° C (Fig. 4-19).

Evaporation of water from the skin without secretory activity of the sweat glands is called *insensible* perspiration and is caused by diffusion of water through the epidermis from the deeper layers of the skin. The surface of the skin is never actually wet in insensible perspiration; in fact, if the surface is wet, water diffusion through the skin stops because the gradient that brings about diffusion decreases. Burton and associates[54] and Pinson[62] found that insensible perspiration varies directly with the rate of cutaneous blood flow and with skin temperature.

When environmental temperature exceeds 29° C for resting clothed men and 31° C for nude men, heat loss by radiation, convection, and insensible evaporation becomes inadequate and evaporation from the skin is increased by sweat secretion (Fig. 4-19). Above these critical temperatures is the zone of evaporative cooling. In this temperature zone the amount of sweating and evaporation required to maintain thermal equilibrium in the body is proportional to the effective radiation, air temperature, and air movement in the environment (Fig. 4-19). The evaporative requirement for thermal equilibrium in men also increases with increased metabolism in work. These effects are illustrated in Fig. 4-21, which shows the rates of evaporative cooling required for thermal equilibrium by nude men at three metabolic rates in relation to indoor dry bulb air temperatures ranging from 18° to 50° C with the radiating temperature of the walls and objects in the room very close to air temperature.

In hot, humid atmospheres where the difference between the water vapor pressure on the skin and in the air is small and the skin is completely wet with sweat, the rate of moisture uptake by the air is the limiting factor in deter-

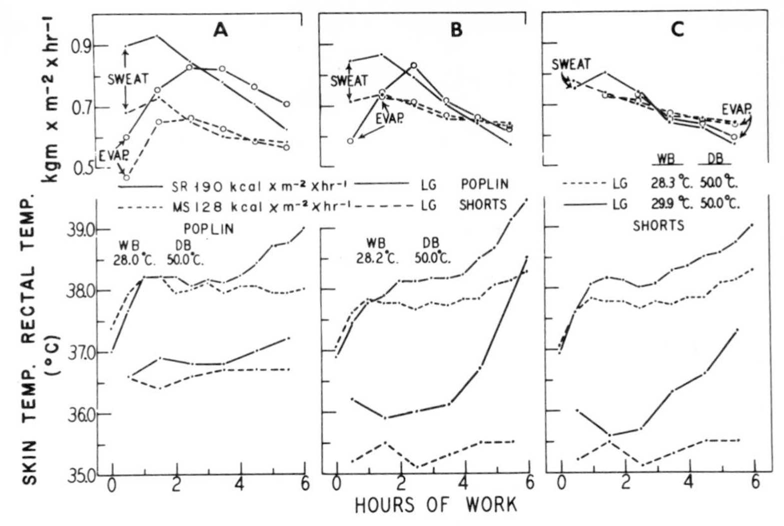

Fig. 4-27. The decline of sweat and evaporative cooling rates of highly acclimatized men during prolonged work in severe heat. In all experiments the subjects maintained water and salt balance by drinking 0.1% NaCl solution at frequent intervals. They walked at 5.6 km/hr on a 2.5% grade, except for **MS** in **A.** Failure of sweating and temperature regulation was greatest when (1) sweating was highest in the first 2 hours, (2) metabolic rate was higher, as in **A,** (3) clothing was worn, as in **B,** (4) wet bulb temperature was higher, as in **C,** and (5) decline continued in the fifth and sixth hours when both skin and rectal temperatures were rising. (From Robinson, S., and Gerking, S. D.: Am. J. Physiol. **149:**476, 1947.)

mining maximal evaporative cooling. In this case a normal man sweats profusely and much unevaporated sweat may drip from his skin. The amount of sweat wasted in this way is particularly great during work and increases as the rate of work and metabolism increase. In hot, dry atmospheres (50° C with 15% relative humidity) with moderate air movement, men walking on a treadmill with metabolic rates of 190 kcal $\times$ m^{-2} $\times$ hr^{-1} may have to evaporate sweat from the skin at 0.7 kgm $\times$ m^{-2} $\times$ hr^{-1} in order to regulate body temperature. Where the evaporative requirement is higher than this, failure of temperature regulation may result from inadequate sweating caused by fatigue of the sweating mechanism if exposure continues for 5 or 6 hours or if the subject is not well acclimatized to work in hot environments (Fig. 4-27). This decline during 4- to 6-hour work periods in severe heat occurs only at high skin temperatures (36° C and over) and high initial sweat rates (0.8 kgm $\times$ m^{-2} $\times$ hr^{-1}).[63] It is greatest in humid heat and in clothed men whose skin remains wet with sweat. It is less pronounced in acclimatized men but still occurs even when both water and salt balances are maintained and when thermal stimuli for sweating (T_{re} and $\overline{T}_s$) are rising. However, the decline in sweating increases greatly with dehydration. In less severe work-heat stresses normal men who are acclimatized to work in the heat can sweat at 0.4 kgm $\times$ m^{-2} $\times$ hr^{-1} ($\overline{T}_s$ = 35° C) for 6 hours with no decline in sweat rate. In the rare individuals who are congenitally unable to sweat, adaptation to such situations is impossible.

There are two types of sweat glands in human skin: apocrine glands, which are located chiefly in the axillae and pubic regions and probably have a secondary sexual function, and eccrine glands, distributed generally over the entire skin, that function in temperature regulation. The sweat glands of the palms and soles are always active; emotional stimuli may increase their secretion, whereas thermal stimuli do not affect them.

Normally, thermal sweating of the eccrine glands over the general body surface is activated by impulses along sympathetic motor nerve fibers that act cholinergically. (Acetylcholine is the humoral transmitter substance that activates the glands at the motor nerve endings.) The center for overall control of thermal sweating is in the anterior region of the hypothalamus. This hypothalamic reflex center is excited by (1) afferent impulses from intradermal thermoreceptors, (2) the direct effect of temperature changes within the center itself that are determined largely by local metabolism and the temperature of its arterial blood supply, and (3) impulses originating from neuromuscular activity that also contribute to the activity of the center and the regulation of sweating during exercise. Bazett[64] suggested that the principal effect of increasing temperature on the center itself may be to increase its responsiveness to afferent impulses from the cutaneous thermoreceptors or other nervous influences acting on the center. Seckendorf and Randall[65] have demonstrated spinal cord reflexes that activate sweating in paraplegic patients, but their role in normal temperature regulation is unclear. Bullard

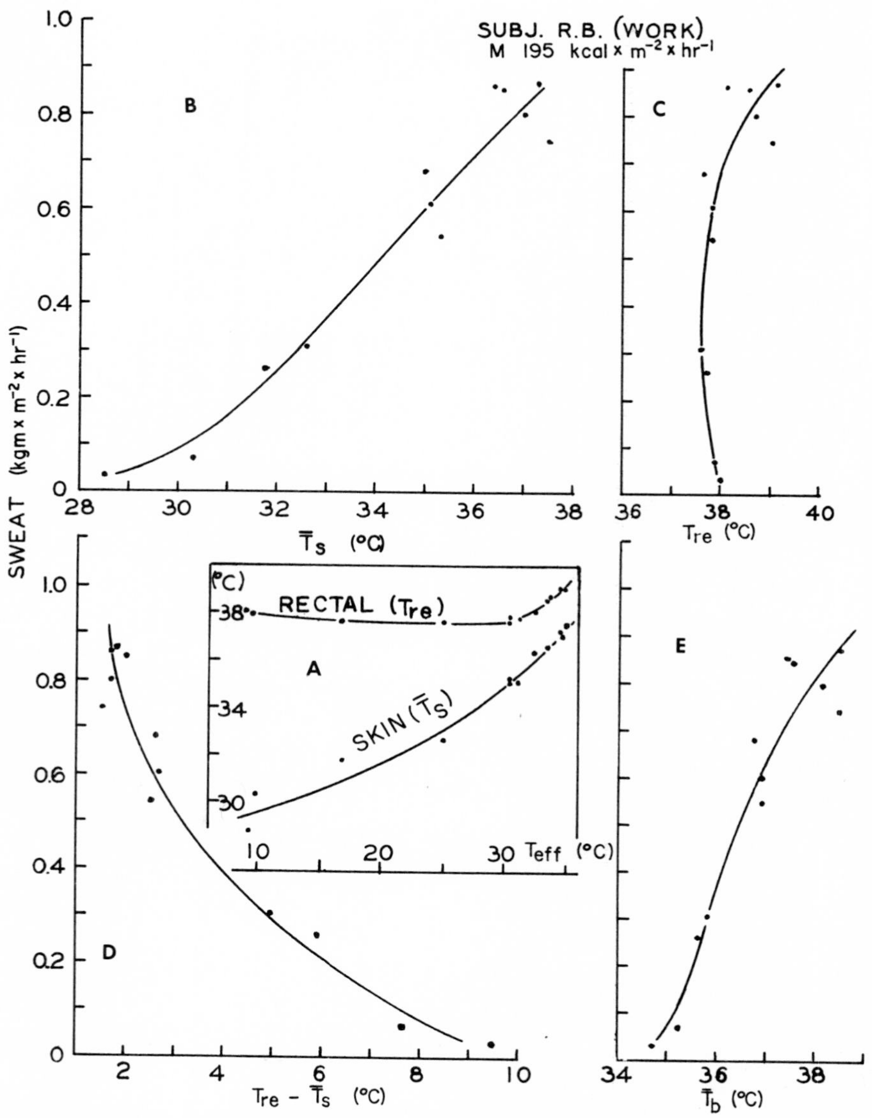

Fig. 4-28. A, Rectal (T_{re}) and mean skin temperatures ($\overline{T}_s$) of subject RB during 90-minute exposures to twelve different effective environmental temperatures (T_{eff}: 9.5° to 35° C). **B** to **E** show respectively the relationships of sweating to $\overline{T}_s$, T_{re}, gradient between rectal and mean skin temperature ($T_{re} - \overline{T}_s$), and mean body temperature ($\overline{T}_b$) in the twelve experiments. Effective room temperature was constant during each exposure as was the subject's work rate (walk at 5.6 km/hr, 2.5% grade). Each plotted value represents the average of measurements made during the last 60 minutes of a 90-minute exposure. (Data from Robinson, S.: Pediatrics 32:691, 1963.)

and associates[66] found that the temperature at the level of the sweat gland directly affects its secretory function.

Reflexes from the cutaneous thermoreceptors are the first line of defense against environmental heat; heat warms the skin, stimulating cutaneous thermoreceptors that initiate reflexes that elicit sweating and cutaneous vasodilation, thus preventing significant body temperature elevation. Numerous investigators have demonstrated a clear-cut relationship between sweat rate and skin temperature increments produced by environ-

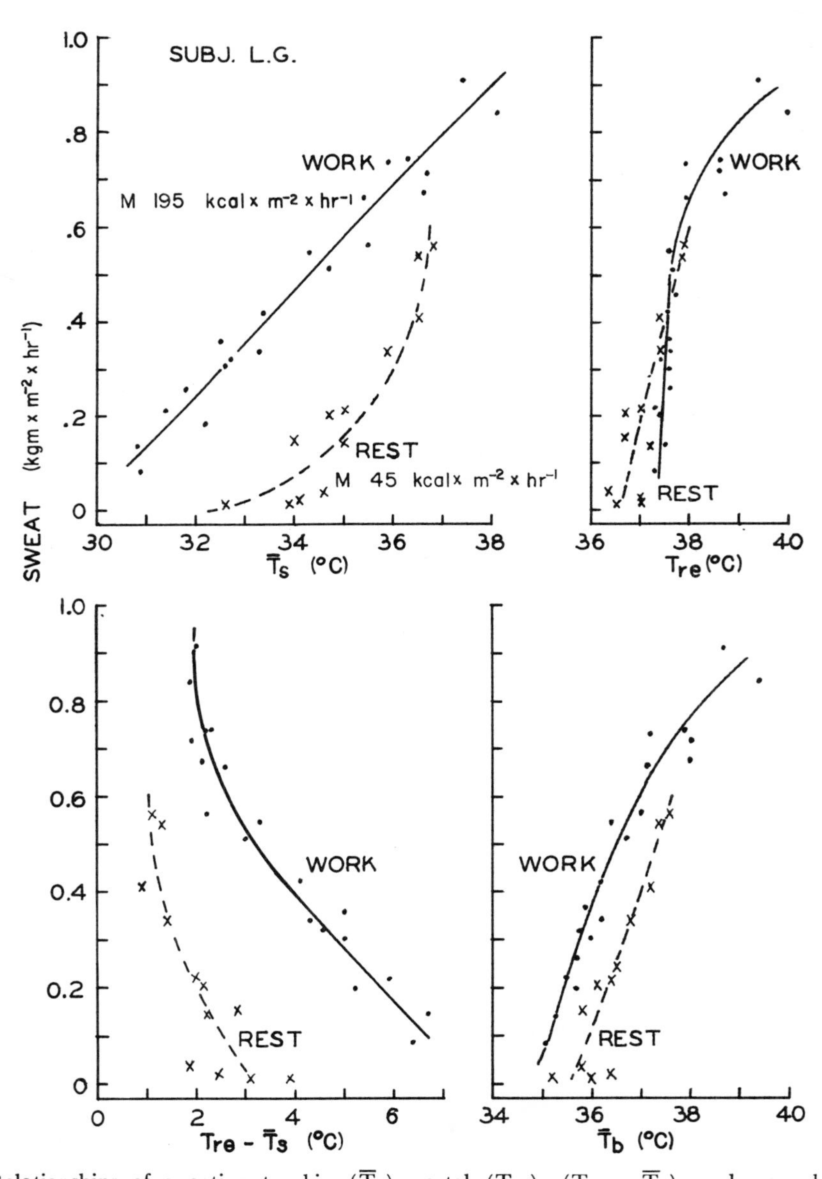

Fig. 4-29. Relationships of sweating to skin ($\overline{T}_s$), rectal (T_{re}), ($T_{re} - \overline{T}_s$), and mean body temperature ($\overline{T}_b$) of subject LG at rest and during treadmill work (5.6 km/hr on a 2.5% grade). Each plotted value represents the average of measurements made during the second hour of a 2-hour exposure to a constant environment. Environment varied (T_{eff}: 15° to 35° C) from one experiment to another to produce changes of body temperature and sweat rate. (Data from Robinson, S. In Mountcastle, V. B., editor: Medical physiology, ed. 13, St. Louis, 1974, The C. V. Mosby Co.)

mental thermal stress.[59, 67] This is true for a man working at constant rate, even under conditions in which central body temperature and metabolic rate remain constant (Figs. 4-28 and 4-29). Sweating in the average nude man at rest is initiated at $\overline{T}_s$ of 34° to 35° C, variation in the threshold $\overline{T}_s$ being inversely related to the prevailing T_{re} (Fig. 4-29). If environmental heat stress increases above that required to elevate $\overline{T}_s$ to the sweating threshold (34.5° C), skin temperature and sweat rate both increase further.

Human sweating also increases with mean body temperature ($\overline{T}_b$) elevation and with reduction of the gradient between central and mean skin temperature ($T_{re} - \overline{T}_s$) under steady state conditions of rest or work in which rectal temperature and metabolic rate are constant. Fig. 4-28 shows these relationships for a man in a series of 90-minute work experiments at constant rate. The observed variations of temperature in the subject were produced by varying environmental temperature (T_{eff} = 10° to 35° C) from one experiment to the next (Fig. 4-28, *A*). The plotted values of temperature and sweat rate represent average values observed during the last 60 minutes of the respective 90-minute exposures with the subject in a steady state of body temperature in all but the most intense heat stresses. Sweat rate during the last hour of work increased up to 0.7 kgm $\times$ m^{-2} $\times$ hr^{-1} without a rise in the steady state level of rectal temperature during the work (Fig. 4-28). The lowest sweat rates were observed in the coolest environments and were associated with a rectal temperature of 38° C; all other rates up to 0.7 kgm $\times$ m^{-2} $\times$ hr^{-1} were associated with T_{re} values of 37.6° to 37.8° C. In the more severe environments, increments of sweating above 0.7 kgm $\times$ m^{-2} $\times$ hr^{-1} were accompanied by further skin temperature increases and also by elevation of rectal temperature above control work level. Since exercise, metabolic rate, and internal temperature remained constant throughout most of this range of conditions, the corresponding sweat increments of the working man were apparently dependent upon either increasing skin temperature or the gradient between internal and surface temperatures, or both. Although rectal temperature may not have been identical with hypothalamic temperature, it seems reasonable to assume that the two probably maintained a constant relationship to each other during these steady state periods. The data thus indicate that in these steady states of work and constant internal temperature, the increases of sweating with increments of T_{eff} and $\overline{T}_s$ were probably not dependent upon increasing hypothalamic temperature but upon reflexes resulting from afferent impulses from peripheral thermoreceptors.

A man sweats much faster during muscular work than he does at rest in any comparable steady state of T_{re}, $\overline{T}_s$, $\overline{T}_b$, and $T_{re} - \overline{T}_s$; this indicates that a neuromuscular factor contributes to sweat regulation during work. This factor is shown in Fig. 4-29, which presents the sweat responses of a man (LG) exposed to a wide variety of thermal stresses (T_{eff}). Data represent average values observed during the second hour of twelve 2-hour exposures of the subject at rest (M = 45 kcal $\times$ m^{-2} $\times$ hr^{-1}) in comparison with twenty 2-hour work experiments (M = 190 kcal $\times$ m^{-2} $\times$ hr^{-1}). The effective temperature of the room varied (15° to 35° C) from experiment to experiment, increasing body temperature and sweat rate in this subject as in subject RB in Fig. 4-28, *A*. In the work experiments, the relationships of sweat rate to each of the four measures of body temperature in this subject are remarkably similar to those observed in subject RB (Fig. 4-28) over a corresponding environmental range. The responses during rest also differ from those during work in that rectal temperature declined below control resting values in cooler environments and rose above the controls in the hotter environments; in the work experiments rectal temperature was about the same in all environments except the most intense heat stresses. From these relationships at rest, the changes observed in internal temperature probably contributed to the thermal stimulus for sweating over the entire range of resting sweat rates, whereas in the work experiments performed at a constant rate, sweating increased up to 0.7 kgm $\times$ m^{-2} $\times$ hr^{-1} with practically no increase of rectal temperature above the control work level of about 37.5° C. This was also true in the work experiments by subject RB (Fig. 4-28).

Further evidence that a neuromuscular factor participates in sweat regulation during exercise is provided by data of Van Beaumont and Bullard[68] and of Gisolfi and Robinson.[69] These investigators, studying men who were warm and already sweating at low rates during rest, showed that sweat rate begins to increase within 1 to 2 seconds after a subject begins vigorous work (Fig. 4-30). If a subject is not excessively hot, sweating begins to decrease within 1 to 2 seconds after the cessation of work. Gisolfi and Robinson found that

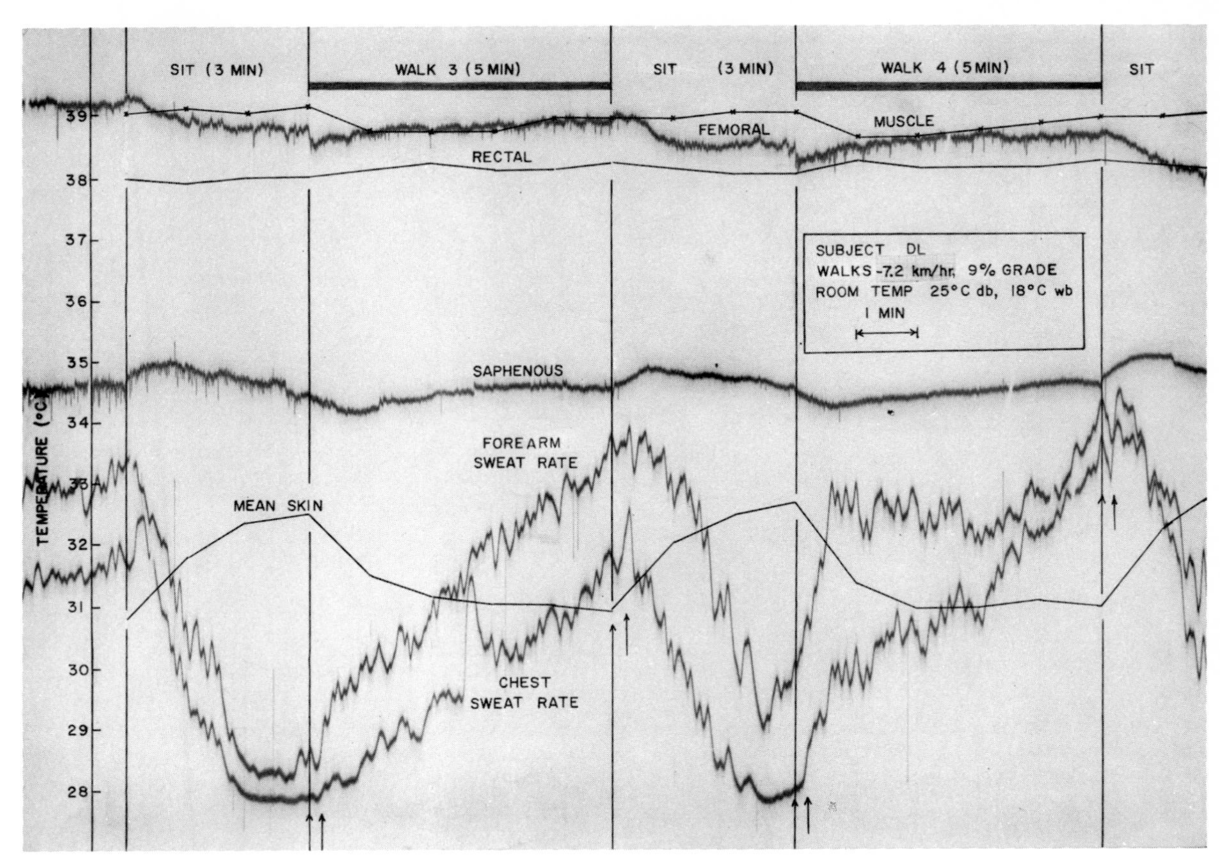

Fig. 4-30. Hygrometer records of sweat rate and corresponding body temperature changes in subject DL during alternating treadmill work ($\dot{V}_{O_2}$: 2.45 L [STPD]/min) and rest in a cool room. Note that central temperatures (rectal, gastrocnemius muscle, and femoral vein blood) were elevated by previous work periods; the subject was sweating during rest periods, and his skin was cool. The double arrows show the 12-second lag time in hygrometer recording of the sweat response. Latent intervals (1 to 2 seconds) of changes in sweat rate at the starts and stops of work periods are not clearly measurable in these records because of the condensed time scale and the cyclic nature of sweating. (Data from Gisolfi, C., and Robinson, S.: J. Appl. Physiol. **29**:761, 1970.)

these rapid sweat responses to work occur when skin is warm (35° to 36° C) and the body core cool (T_{re} = 37° C), or on a background of elevated core temperature (38° C) with cool skin (31° to 32° C), or when both $\overline{T}_s$ and T_{re} are elevated. These quick sweat gland responses probably depend on proprioceptor reflexes from muscles or joints or possibly on irradiation of impulses from motor tracts in the brainstem, acting reflexly through the hypothalamic thermoregulatory center. They occur so rapidly that they precede temperature changes even in working muscles (Fig. 4-30). Two subjects in experiments of Robinson and associates[70] and of Gisolfi and Robinson[69] showed rapid increases and decreases of femoral vein blood temperature at the start and cessation of hard treadmill work coincident with the abrupt sweat changes (Fig. 4-30). This suggests that thermoreceptors in the deep leg veins may be responsible for the rapid reflex sweat changes in these experiments. However, results on two other subjects in the studies showed abrupt sweat changes but did not show the rapid changes of femoral vein blood temperature; thus the question of thermoreceptors in the deep limb veins remains unsettled.

In a constant cool environment the sweat rate attained during work is proportional to work rate and therefore to the elevation of metabolic rate and T_{re} without a corresponding rise in $\overline{T}_s$ (Fig.

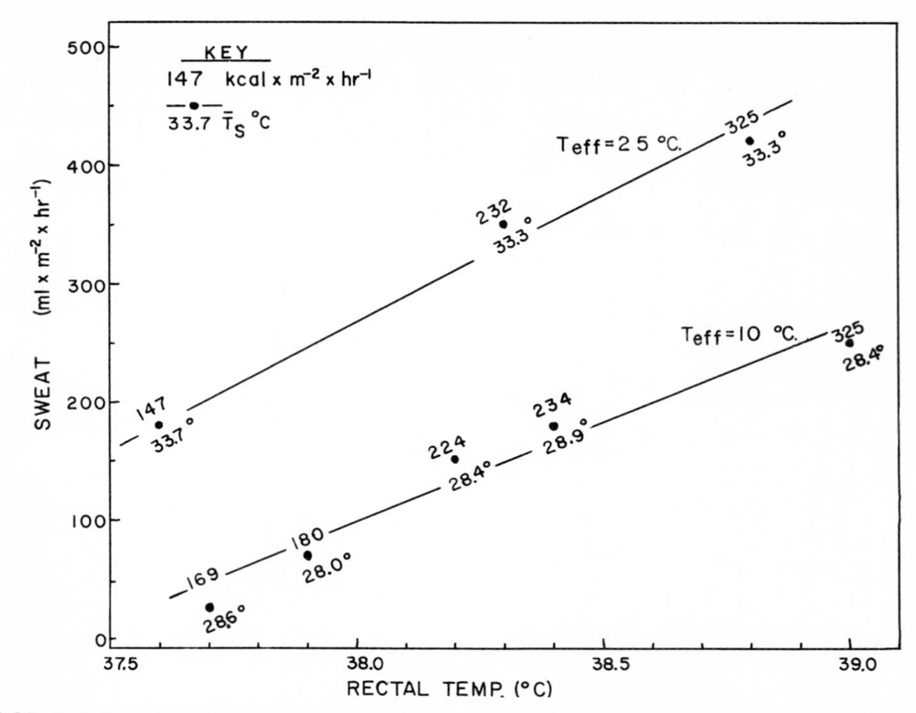

Fig. 4-31. Relationship of sweat rate to rectal temperature increments of a man (RK) during work at five different rates (M = 147 to 325 kcal × m^{-2} × hr^{-1}). The subject performed two series of 90-minute work experiments—one in a warm room (T_{eff}: 25° C), the other in a cold room (T_{eff}: 10° C). Skin temperature and sweat rate were uniformly higher in the warm room than in the cold room. Each plotted value represents the average of observations made during the last 30 minutes of a 90-minute experiment.

4-24). This is also illustrated in Fig. 4-31, representing a man's steady state responses to five different treadmill work intensities, each performed for 90 minutes at two different effective room temperatures (T_{eff} = 10° and 25° C). In these experiments metabolic rate ranged from 147 to 325 kcal × m^{-2} × hr^{-1} as indicated. Each value of sweat rate, rectal temperature, skin temperature, and metabolic rate represents the average of values observed during the last 30 minutes of a 90-minute experiment, during which environment and work rate were constant and the man was in a virtual steady state. The relationship of sweat rate to $\overline{T}_s$, T_{re}, and metabolic rate in these experiments (Fig. 4-30) are as follows: (1) $\overline{T}_s$ at each T_{eff} was about the same in all grades of work, (2) at each $\overline{T}_s$ the T_{re} and sweat rate increased linearly with metabolic rate, (3) the difference of about 5° C in the subject's $\overline{T}_s$ in the two environments produced a uniform difference in the man's sweating at all work rates, (4) at a given effective temperature there were no increases of $\overline{T}_s$ to increase cutaneous reflexes and cause the increased sweat rate. Since $\overline{T}_s$ was essentially the same at all work intensities at each of the effective temperatures, the data indicate that sweat rate during work may be increased directly by elevated internal temperatures and/or by neuromuscular reflex effects on the temperature-regulatory center itself without rise of skin temperature. Taking into account the responses in both of the series of experiments shown in Fig. 4-31, it appears that changes of both skin and internal body temperatures participate in the regulation of sweating.

The evidence just presented suggests the complexity of the mechanisms that control the temperature-regulatory responses of the sweat glands. These responses are not entirely the result of reflexes originating in the skin thermoreceptors because even in a cool or cold environment, a working man sweats profusely with a low skin temperature. Furthermore, during steady states of work, sweat rate is directly proportional to work rate and metabolic heat production and to the corresponding increase of central body temperature, even without corresponding skin temperature changes. On the other hand, when environmental temperature varies (T_{eff} = 9° to 35° C), sweat rate varies directly with mean skin temperature increments up to 36° C and inversely with the gradient between internal and skin temperatures over a wide range of conditions in which metabolic rate and internal body temperature do not vary. The stimulatory effect of neuromuscular work that combines with the effects of thermal stimuli to elicit thermoregulatory responses is probably caused by (1) neuromuscular reflexes such as those participating in the regulation of respiration, and (2) the resulting elevation of central body temperature, including the hypothalamic center, working muscles, and blood in the deep veins of the working limbs. It is significant that the steady state interrelationships among coefficient of heat conductance (Fig. 4-34), heart rate, and sweat rate to body temperature ($\overline{T}_s$, T_{re}, $\overline{T}_b$, and $T_{re} - \overline{T}_s$) during work and at rest are remarkably similar to each other (Figs. 4-28 and 4-29).

Circulatory adjustments

The circulatory system plays a major role in body temperature regulation. Since the heat conductivity and specific heat of blood are high, blood can transport large amounts of heat to the body surface and release it with only small changes of the temperature of the blood itself. Also the large vasomotor adjustments that can be made provide needed flexibility to cope with widely varying thermal environments. It is estimated that blood flow to the skin can vary from 0.16 L $\times$ m^{-2} $\times$ min^{-1} in a nude man resting at 28° C[71-73] to 2.6 L $\times$ m^{-2} $\times$ min^{-1} for men working in thermal equilibrium in a very hot environment.[74]

A heat-acclimatized man can maintain thermal equilibrium during a 6-hour bout of work in the heat (M = 190 kcal $\times$ m^{-2} $\times$ hr^{-1}, 50° C, 18% relative humidity). In this case the circulatory system is accomplishing at least three functions beyond the normal resting processes: (1) increased blood flow to the working muscles to supply nutrients and to provide for respiratory exchange and increased heat removal; (2) increased blood flow to the skin to supply water to the sweat glands and to allow heat dissipation; and (3) blood flow through the alimentary tract and liver to carry water and nutrients to the sweat glands and working muscles. To make these adjustments during work in the heat, a man must increase his cardiac output to more than twice the usual resting level and make fine vasomotor adjustments in the blood flow patterns. Furthermore, a number of investigators have found increased circulating blood volume in men during prolonged heat exposure.

Cutaneous blood flow

In general, cutaneous circulation increases with thermal stress intensity, maintaining heat transport to the skin for dissipation by evaporation, radiation, conduction, and convection. Hertzman[75] and Hertzman and Dillon[76] found the richness of blood supply to different skin areas to be arranged in descending order as follows: finger pad, ear lobe, toe pad, palm of hand, forehead, forearm, knee, and lower leg. Finger blood flow varies from 0.2 in the cold to 120 ml $\times$ min^{-1} $\times$ 100 gm^{-1} of finger tissue in the heat.[77-80] Within the hand the fingers have greater circulation than proximal parts and, in the fingers themselves, circulation is greater in distal phalanges than in proximal ones.[81] Also the greatest range of vasomotor adjustments is in the most distal parts where surface-to-volume ratio is largest. This indicates that distal skin areas (such as hands and feet) are more suited to make circulatory adjustments for temperature regulation than the skin of trunk and more proximal parts of the limbs. There is much research evidence to indicate that this is indeed the case. In animals other than man, the horns of the goat,[82] the tail of the *Ondatra zibethica* (muskrat)[83] and *Castor fiber* (beaver),[84] the ears of the rabbit,[85] and the uninsulated legs of the wood stork[86] all have important temperature-regulatory functions. In man blood flow through the hands and feet is mainly related to body temperature regulation.[87-91] However, Abramson and Ferris[87] found that the rate of blood flow through arms and legs is more closely related to the metabolic requirements of these more muscular regions.

Vasomotor regulation in hands and feet is quite different from that of forearms and legs. The hands and feet are mainly under vasoconstrictor control, so that increased flow in the heat is accomplished by graded release from vasoconstrictor tone, while cutaneous blood flow to forearms and legs in the heat is mainly under vasodilator control. There is evidence for vasoconstrictor control of the hand; sympathectomy or blocking sympathetic innervation of the hand produces maximal vasodilation in the hand, even in a cool environment.[92-95] For the degree of vasoconstriction in the hand to be useful as a thermoregulatory mechanism, it must be controlled by inputs that reflect heat stress. There are three such inputs. First, Kerslake and Cooper[96] showed that radiant heating of the skin of the legs increases flow in the hand after a latent period of only 10 to 15 seconds; this response is eliminated by lumbar sympathectomy,[97, 98] indicating neurally mediated excitation of cutaneous temperature receptors. Second, Snell[99] demonstrated central excitation of the response by adding 2.4 kcal of heat to the body with intravenous infusion of warm saline solution. Third, the blood vessels in the hands are also directly sensitive to temperature since, in the reflexly dilated hand, dilation is further increased by local heating of the hand.[100, 101]

Vasodilator nerve control of forearm blood flow is suggested by the work of Roddie and associates,[93] who found that blocking the forearm cholinergic vasodilators with atropine eliminated most of the forearm response to heat. Anesthetization of the cutaneous nerves of the forearm has the same effect.[102] Roddie, Shepherd, and Whelan[103] also showed that in the heat, the increase of blood flow in the forearm is largely cutaneous in nature, with little change of flow to resting muscles. Oxyhemoglobin saturation of venous blood from resting forearm muscles is unchanged from values found in a cool environment, whereas that of blood from the skin doubled, indicating that blood flow to the skin had increased greatly. Furthermore, iontophoresis of adrenaline into forearm skin, blocking blood flow to the skin but not affecting flow to deeper tissues, eliminates the forearm response to heat.[104-106] Thus the vasodilator response to heat involves cutaneous vessels and not deep vessels in the limbs.

Heat conductance

The concept of thermal conductance of tissues was developed by Burton[107] and other investigators. The basic measure is the coefficient of heat conductance, the rate of heat transfer from core to surface in kcal $\times$ m^{-2} $\times$ hr^{-1} $\times$ Δ°C^{-1}, expressed as difference between core and mean skin temperature. Tissue heat conductance can be used as an indicator of cutaneous blood flow,[72, 107-111] since conductance depends largely on variations of blood flow patterns and temperature gradients in the body. This is especially important in the limbs. For example, consider blood flow in an arm and the phenomenon of countercurrent heat exchange. Fig. 4-32 shows the main circulatory routes in the arm. Blood is supplied to the arms and legs by large arteries and returns via deep venae comitantes (adjacent to the large arteries) and/or via superficial veins. The apportionment of blood between these two avenues of venous return has thermoregulatory importance. In a cold environment most of the venous return from arms and legs is in venae comitantes; thus countercurrent heat exchange is high. Blood flowing out in the arteries gives up heat to cool blood returning in venae comitantes. By this mechanism external heat loss is minimized, conductance is low, and yet the total blood flow to the limbs may be high, protecting tissues of the limbs from cold injury and hypoxia.

In a warm or hot environment blood returns in superficial veins and is cooled in transit as heat is lost to the external environment—conductance is high. In a warm man Bazett and McGlone[13] found the temperature of the blood in the superficial veins of the finger to be 35.2° C and at the elbow 34.6° C. Under these conditions it appears that venous blood returning in the limb is shunted to dilated superficial veins by constriction of deep veins, including the venae comitantes.[64] Similar changes of blood temperature in superficial leg veins of men working in the heat were reported by Gisolfi and Robinson.[112] These investigators found that in men performing hard treadmill work much of the warm blood from working muscles in the lower leg drains into the superficial saphenous vein and is cooled in transit up the leg. The volume of this superficial venous blood flow directly from leg muscles is so large that during hard treadmill work in a warm room blood lactate concentration (an anaerobic metabolite) in saphenous vein below the knee rises as rapidly as in blood simultaneously collected from the deeper femoral vein, which collects blood from both deep and superficial leg veins, and far more rapidly than in simultaneously collected venous blood from the arms which are not working.

Measurements of heat conductance reveal many important variations in cutaneous blood flow.[72, 81, 107, 108, 110] Most interesting are those related to age, heat acclimatization, and metabolic rate. Hellon and associates[113, 114] found that upon exposure to heat, older men between the ages of 41 and 58 circulate 50% more blood to the forearm than men 17 to 26 years old. These results probably indicate that older men are unable to shift blood flow from deeper forearm tissues to the skin as well as younger men. In this connection Allwood[115] found that with increasing age there is a reduction of maximal cutaneous blood flow in the foot in response to excessive heat stress. Eichna and co-workers[74] found that during acclimatization of young men by repeated daily exposure to work in a constant dry heat stress there were

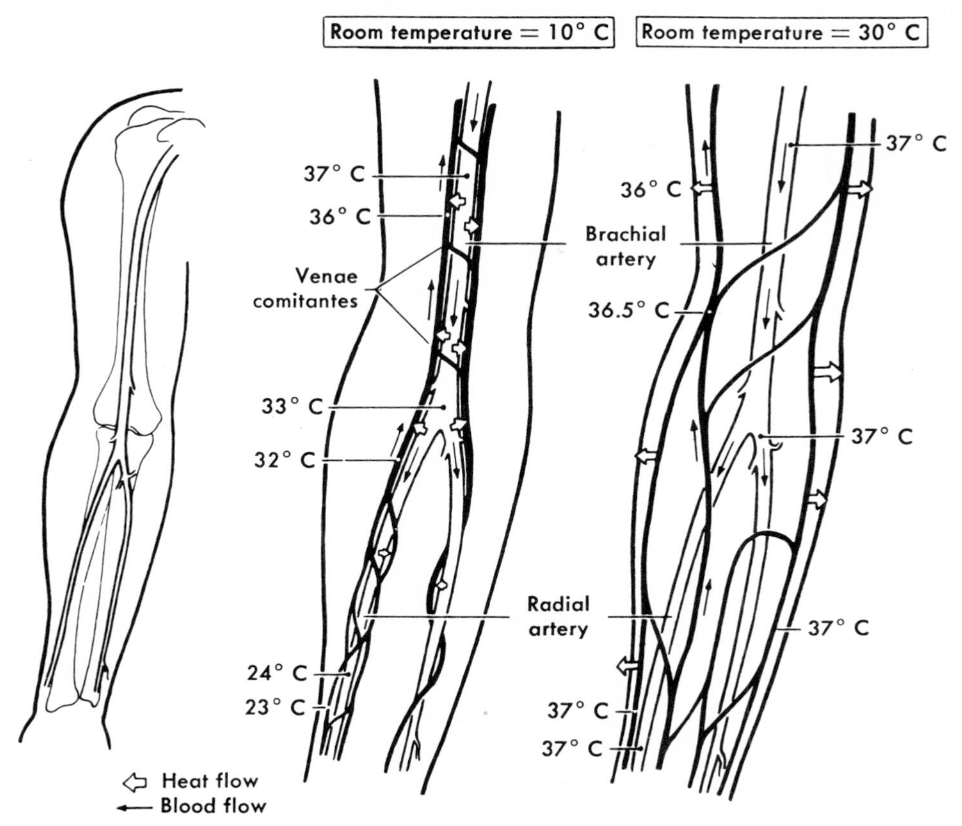

Fig. 4-32. Countercurrent heat exchange in the human arm. Intravascular temperatures show that venous blood flow patterns in the arm adjust to meet thermal environmental stress. In the cold (10° C) venous blood flow is mainly via the deep venae comitantes that receive heat from blood flowing out in the arteries and thereby minimize body heat loss. In a warm environment (30° C) venous blood flow is mainly in superficial veins which, being close to the surface, increase body heat loss. (From Bullard, R. W.: Temperature regulation. In Selkurt, E. E., editor: Physiology, Boston, 1966, Little, Brown and Co.)

gradual reductions of mean skin temperature, thermal conductance, and cutaneous blood flow as the men's sweating and evaporative cooling improved. $\overline{T}_s$ decreased and the gradient $T_{re} - \overline{T}_s$ increased, decreasing the requirement for cutaneous blood flow in heat transport (Fig. 4-33). Thus cardiovascular strain decreased as evaporative temperature-regulatory ability improved. Along with improvements of evaporative cooling and reduced cutaneous blood flow in acclimatization to a fixed submaximal heat stress, Scott and colleagues[78] found that the maximal capacity for increasing blood flow to the finger increased.

Fig. 4-34 gives conductance values for a man at rest ($M = 45$ kcal $\times$ m^{-2} $\times$ hr^{-1}) and during work ($M = 190$ kcal $\times$ m^{-2} $\times$ hr^{-1}) in a series of different 2-hour exposures to various environmental temperatures as shown in Figs. 4-22 and 4-28. The data represent average conductance values in relation to body temperature measurements observed during the second hour of exposure. Conductance during each period of observation is plotted against the subject's mean skin temperature ($\overline{T}_s$), rectal temperature (T_{re}), gradient between rectal and mean skin temperature ($T_{re} - \overline{T}_s$), and mean body temperature ($\overline{T}_b$), respectively. The insert in Fig. 4-28, taken from a similar experiment, shows the relationship between effective environmental temperature (T_{eff}) and the rectal and mean skin temperature observed during the last hour of exposure.

As expected, conductance data correlate positively with $\overline{T}_s$ and $\overline{T}_b$, negatively with the gradient $T_{re} - \overline{T}_s$, and less well with T_{re}. Thermoregulatory

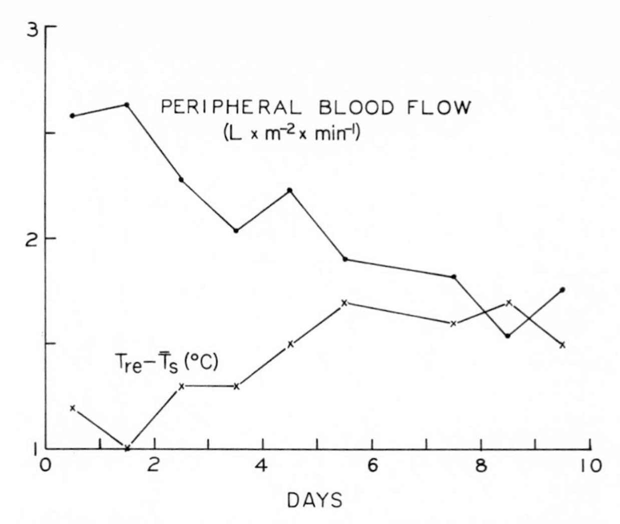

Fig. 4-33. Peripheral or cutaneous blood flow and the temperature gradient ($T_{re} - \overline{T}_s$) during acclimatization in daily exposures to work in dry heat (dry bulb temperature 50.5° C, relative humidity 15%, and air movement 450 ft/min). During acclimatization sweat rate increase resulted in a greater decrease of $\overline{T}_s$ than of T_{re}. The increased gradient ($T_{re} - \overline{T}_s$) increased heat loss per unit of cutaneous blood flow, thereby decreasing the total need for blood flow to the skin. (Data from Eichna, L. W., and others: Am. J. Physiol. **163**:585, 1950.)

mechanisms maintain relatively constant internal, or central, temperature of which T_{re} is a measure. Conductance varies with T_{re} during rest but not during work, except at the higher thermal stresses. This suggests finer control of T_{re} during work. Indeed, the insert shows that T_{re} was maintained constant except when the combined heat stresses of environment and metabolism exceeded the capacity for heat dissipation ($T_{eff} > 32°$ C). Thus conductance, cutaneous blood flow, and sweat evaporation (shown in Figs. 4-28 and 4-29) adjust to maintain T_{re} constant over the greatest possible environmental temperature range.

As shown in Fig. 4-34, both minimal and maximal tissue heat conductance rates increase during exercise. Thus the stimulus of neuromuscular work plus thermal stimuli elicit a thermoregulatory response greater than that at rest. Although the mechanism of this increased response is not fully understood, it must certainly be aided by increased cardiac output associated with increased metabolic rate and the massaging action of working muscles on veins that augments venous return. Several other possibilities bear mentioning. The effect of work on conductance is similar to its effect on pulmonary ventilation rate, sweat rate (Figs. 4-28 and 4-29), and heart rate.[116] Since we know that neuromuscular reflexes stimulate respiration during work,[117] it seems reasonable that similar reflexes may excite the thermoregulatory center.[118] Some investigators[64, 119] have suggested that the site of this stimulus might be Ruffini thermoreceptors, located in the deeper cutaneous venous plexuses that drain working muscles. This idea is supported by the work of Grant and Pearson,[89] and Cooper, Randall, and Hertzman,[120] who found a rapid increase of skin temperature (up to 3° C) over a muscle immediately after the onset of work. They also showed this rise of temperature to be dependent upon circulatory heat convection. Thompson[121] demonstrated a possible neurophysiologic pathway for this response. Another possible mechanism is excitation of the thermoregulatory center by joint mechanoreceptor reflexes and/or radiation of impulses over motor tract fibers to the hypothalamus. Evidence of neuromuscular influence on sweat regulation has already been discussed.

Compensatory adjustments

When men unacclimatized to heat are exposed to high heat stress, they develop tachycardia and often syncope. These signs of cardiovascular strain are more likely if the men attempt to stand or perform work. Associated with circulatory strain are marked elevation of skin and rectal temperatures above control values. The process of heat acclimatization increases cardiovascular stability dramatically and can be accomplished by daily work periods in the heat. This improvement, which in

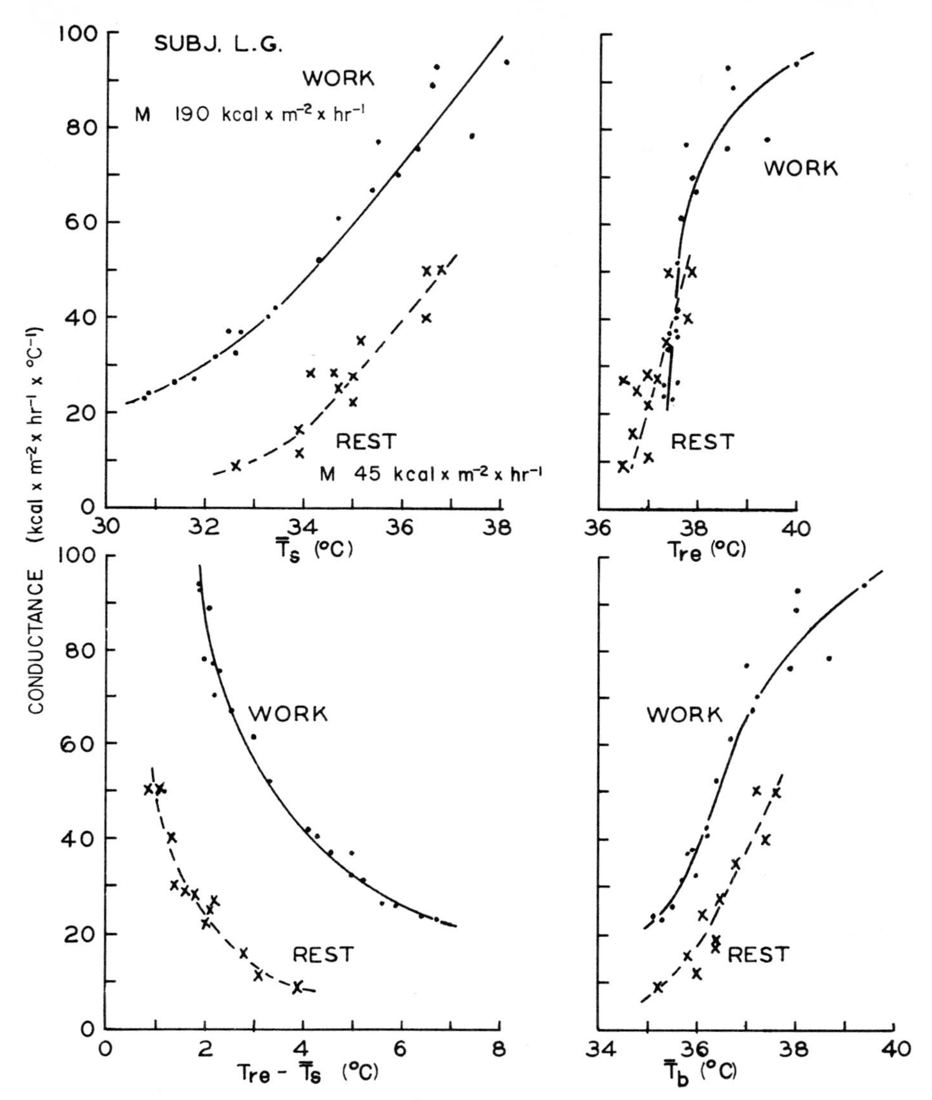

Fig. 4-34. The effect of work on the relationship of tissue heat conductance to body temperature—T_{re} (rectal temperature), $\overline{T}_s$ (mean skin temperature), $T_{re} - \overline{T}_s$, and $\overline{T}_b$ (mean body temperature). Each plotted value represents the average of measurements made on subject LG during the second hour of a 2-hour exposure. Effective room temperature was constant during each exposure but varied from 15° to 35° C in the different 2-hour exposures. (Data from Robinson, S.: Circulatory adjustments of men in hot environments. In Temperature: its measurement and control in science and industry, vol. 3, part 3, New York, 1963, Van Nostrand Reinhold Co., p. 289.)

physically fit men is virtually complete in 5 to 8 days[122-124] is observed as a reduction of both rectal and skin temperatures; if the imposed heat stress is not too great, rectal temperature returns to the control levels observed during work in a cool environment (Fig. 4-38).

With the greatly increased cutaneous blood flow needed for heat dissipation, circulatory stability could not be maintained without efficient compensatory adjustments. Two possible adjustments are: (1) compensatory vasoconstriction in other vascular beds, shunting a greater fraction of total cardiac output to the skin, and (2) an increase of total circulating blood volume, providing more blood for cutaneous flow. An increase of cardiac output would also be advan-

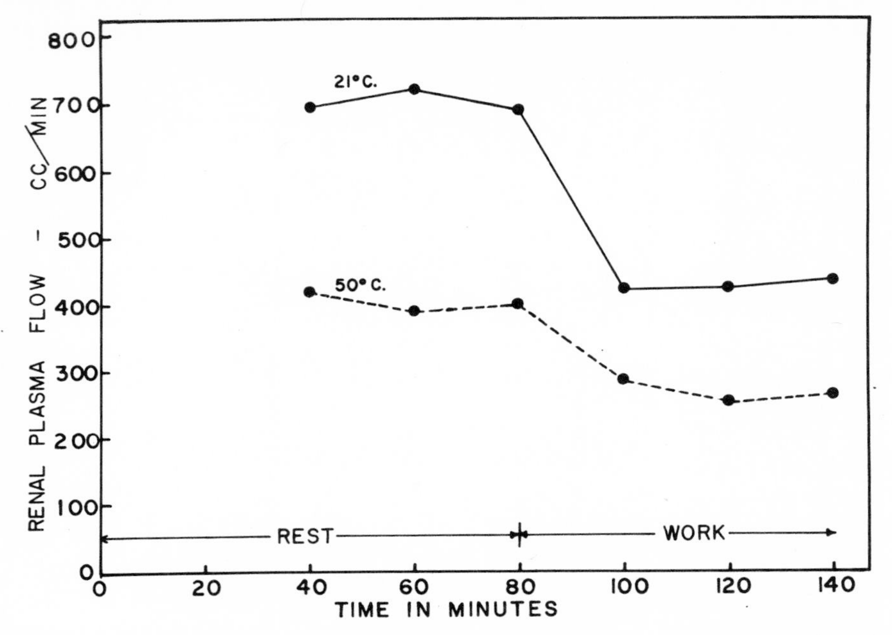

Fig. 4-35. Effects of exercise and environmental heat stress on effective renal plasma flow. Points are the mean values for five subjects. The work was walking on a treadmill at 5.6 km/hr on a 5% grade. (From Radigan, L. R., and Robinson, S.: J. Appl. Physiol. 2:185, 1949.)

tageous but does not occur consistently; in fact, the cardiac output of men standing in the heat decreases.[78, 125-128]

Renal blood flow measurements provide good evidence of compensatory vasoconstriction in the viscera of men exposed to heat; in severe heat (50° C dry bulb [db], 27° C wet bulb [wb]) renal blood flow of resting men may be reduced to 50% to 70% of the control values (Fig. 4-35) observed in a cool environment.[129-132] If the men exercise at a moderate rate (M = four times resting metabolism), renal blood flow decreases even further. In the 6-hour work experiments of Radigan and Robinson[131] and of Smith and associates,[132] heat-acclimatized subjects walked in the heat (50° C db, 27° C wb). Cutaneous blood flow was estimated from conductance values to be 1.2 L $\times$ m^{-2} $\times$ min^{-1}. Renal blood flow reductions alone accounted for more than half of the increased cutaneous blood flow. More recently Rowell and co-workers[133, 134] found that hepatic blood flow of men working in heat may decrease by 80% or more. It is probable that compensatory vasoconstriction of other vascular beds also contributes to man's circulatory adjustment to heat.

The advantage of increased circulating blood volume during heat exposure is clear; more blood is available for circulation to dilated cutaneous blood vessels. Many investigators have found that the blood volume of men increases above basal levels during heat exposure.[135-141] However, others have found no change or even reduction of blood volume during acute exposure to heat or during heat acclimatization.[142, 143] In some cases comparison of blood volume of men during summer and winter has shown inconsistent changes or no change,[144-146] whereas other studies have found significantly higher blood volume during summer than winter.[141, 147] Robinson[140] reported that men highly acclimatized to work in heat show normal basal blood volume during rest in a cool environment but, when water and electrolyte balances were maintained, show increases averaging 13% during exposure of 2 to 3 hours' duration to moderate treadmill work in severe heat.

According to Bazett and his colleagues[137] acclimatization to moderate heat by continuous exposure of resting men for 4 to 6 days increases blood volume even more than short exposures. During continuous exposure of men to more severe heat (49.0° C db, 26.6° C wb) Bass and co-workers[136] found that blood volume increases an average of 15% by the fifth day; by the fourteenth day in the heat, as evaporative cooling improved, it had returned almost to the control level previously determined in a cool environment. These blood volume increases were largely caused by plasma volume expansion associated with NaCl retention by the sweat glands and kidneys in proportion to isotonic increases in the extracellular fluid volume.

Myhre[142, 143] reported decreases of 5% to 8% of plasma volume in men during acute exposures to severe heat. The various results of different investigators probably result from differences in experimental conditions, methods, and subject variables, such as thermal state, hydration, electrolyte balance, physical fitness, and degree of heat acclimatization when blood volume was measured. Further research with improved methods and more careful standardization of relevant conditions at the time of measurement will probably produce more consistent results.

The state of body hydration greatly affects the circulatory adjustment of men to heat. Dehydration reduces heat tolerance whether it is the result directly of unreplaced water loss in sweat and urine or secondary to the NaCl deficit incurred in the same way. Dehydration of healthy men in a hot environment reduces plasma volume when it can least be tolerated. Adolph,[53] studying men dehydrated by 1% to 8% of body weight under desert conditions, found a reduction of circulating plasma volume of 2.5 times that expected if plasma water loss were proportional to total body loss. Myhre and Robinson[143] confirmed this fact in rapidly dehydrating men during 2 to 3 hours of work in severe heat. These investigators also found that if the subjects moved to a cool room and remained dehydrated for 4 hours longer, fluid gradually shifted from other compartments into the plasma, replacing 50% of the lost plasma volume. No further increase of plasma volume occurred when the men remained dehydrated for 19.5 hours after leaving the heat.

The preceding discussion of the importance of adequate circulation and cutaneous vasodilation under heat stress emphasizes that dehydration and plasma volume reduction increase circulatory strain and reduce the tolerance of man to heat stress. Pearcy and associates[148] found that men who deliberately abstain from drinking water during work ($M = 190$ kcal $\times$ m^{-2} $\times$ hr^{-1}) in the heat (44.4° C db, 26.4° C wb) reduce sweat rate 10% during the first hour and 20% during the second hour, as compared with exposure during which water balance is fully maintained by drinking. As a result of sweat rate reduction, mean skin temperature ($\overline{T}_s$) increased 1° C during the second hour at which time subjects were dehydrated by only 1% to 2% of body weight. This effect was studied by Adolph and colleagues[53] and by Pitts and associates,[149] who found that water deficits as low as 1% to 2% of body weight cause measurable evidence of increased circulatory strain indicated by heart rate and rectal temperature increases in men at rest or working in hot environments (Fig. 4-36). The strain, under otherwise constant conditions of metabolic rate and heat stress, increases linearly as water deficit increases. Accompanying the tachycardia of dehydration are parallel increments of rectal temperature, indicating definite failure of the circulation in its function of heat transfer from tissues to skin. Adolph called this failure, manifested by characteristic symptoms of decreased work capacity, drowsiness, faintness, dyspnea, dry mouth, and restlessness, *dehydration exhaustion.*

Water and electrolyte exchange

From the foregoing section on evaporation and its regulation it is clear that water exchange through kidneys, skin, and respiratory tract may range from values of 1 to 2 L/day in sedentary individuals in cool environments to 10 or 12 L/day in men working in hot environments. Water output through sweat glands increases in proportion to the need for evaporative cooling. The importance of balancing increased output by increasing water intake is also clear from the effects of dehydration on blood volume, circulation, sweating, and temperature regulation of men working in heat. The regulation of water intake to balance water loss in hot climates depends on thirst.

There are two traditional theories regarding the mechanism of thirst. Cannon[150] theorized that thirst originates locally in receptors in the mouth and pharynx stimulated by dryness resulting from decreased salivary secretion as the body becomes dehydrated. Claude Bernard[151] suggested that the sensation of thirst is of diffuse origin, dependent on general dehydration. This theory was based on his experiments with a horse and a dog that had esophageal and gastric fistulas, respectively, through which water flowed to the outside as fast as it was drunk. When dehydrated and thirsty, these animals drank until fatigued, rested, then resumed drinking again and again, despite the fact that mouth and pharynx remained wet. Thirst was unquenchable by this type of drinking, and Bernard concluded that the sensation does not arise locally but depends upon a general state of water need. Adolph[152] confirmed Bernard's finding

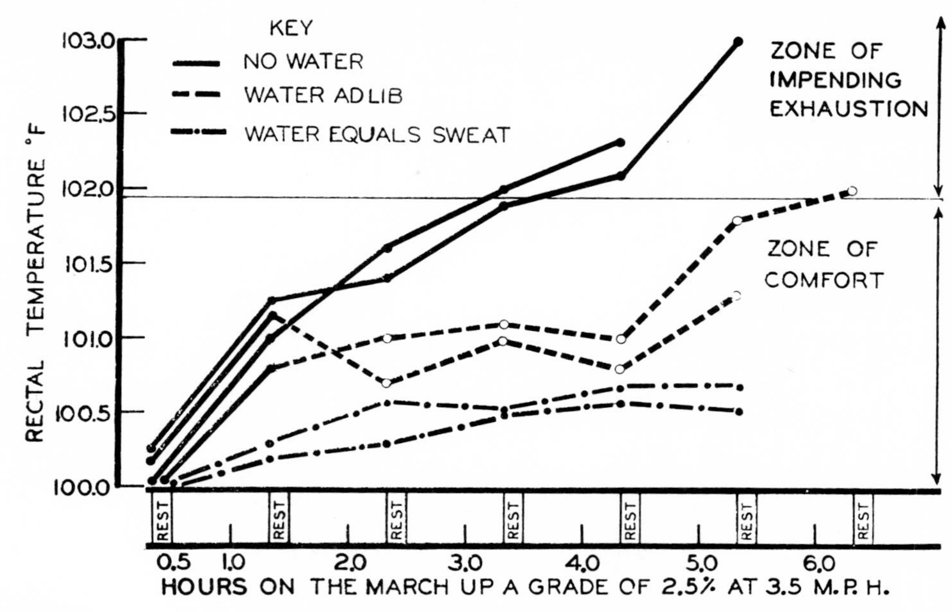

Fig. 4-36. Effect of water consumption on rectal temperature while walking on a level treadmill at 5.6 km/hr in heat (temperature 37.78° C, relative humidity 35% to 45%). Three drinking conditions were used: no water, water ad lib (subject drank at will, about two thirds of water lost), and water intake equal to sweat loss. Data show that increasing dehydration decreases thermoregulatory ability and that guided by thirst a man does not replace total water loss during work. (Data from Pitts, G. C., Johnson, R. E., and Consolazio, C. F.: Am. J. Physiol. **142**:253, 1944.)

that a dog with an esophageal fistula alternately rests and drinks far in excess of water need, the volume of the drinks being a function of the degree of dehydration. Normal thirsty dogs, dehydrated by 2% to 4% of body weight, drink within 5 minutes slightly more than the volume of water deficit and stop drinking before all the water needed is absorbed and distributed within the body.

Although the sensation of thirst may originate from a dry mouth and pharynx, resulting from decreased salivary secretion, the regulation of water intake probably also depends on other peripheral receptors. Evidence from animal experimentation suggests a drinking center in the hypothalamus, a structure itself sensitive to hypertonic NaCl solutions and to afferent impulses from peripheral receptors.[153-155] Dill[156] proposed that small increases of Na^+ concentration in extracellular fluid osmotically reduce body cell water content, including that of the receptors in the mouth and pharynx, causing intense thirst. As in dogs with esophageal fistulas, thirst may be briefly alleviated by rinsing the mouth with water or by distending the stomach with a balloon, but it can be satisfied only when water is delivered by the bloodstream to the dehydrated body cells that demand it.

A resting man acclimatized to a hot environment maintains water balance accurately by drinking enough water to satisfy thirst and, as a result, body water content is remarkably constant when measured each morning before the day's work begins. In contrast to this, Dill,[156] Adolph,[152] and Pitts and associates[149] found that thirst does not always cause an exercising man to maintain water intake equal to output rate. This is particularly true in unacclimatized men[156] who may, when working in a hot environment, secrete a large volume of sweat containing relatively high concentrations of Na^+ (40 to 65 mEq/L). In this case sweat formation involves little modification of the osmotic pressure of extracellular fluid and thus does not cause a degree of thirst proportional to the water deficit. On the other hand, an acclimatized man whose sweat is very dilute ($[Na^+]$ only 10 to 20 mEq/L) is much thirstier at a given water deficit and comes closer to maintaining water balance by voluntary drinking than does the unacclimatized subject. In this case thirst is more intense because water loss increases the osmotic pressure of extracellular fluid

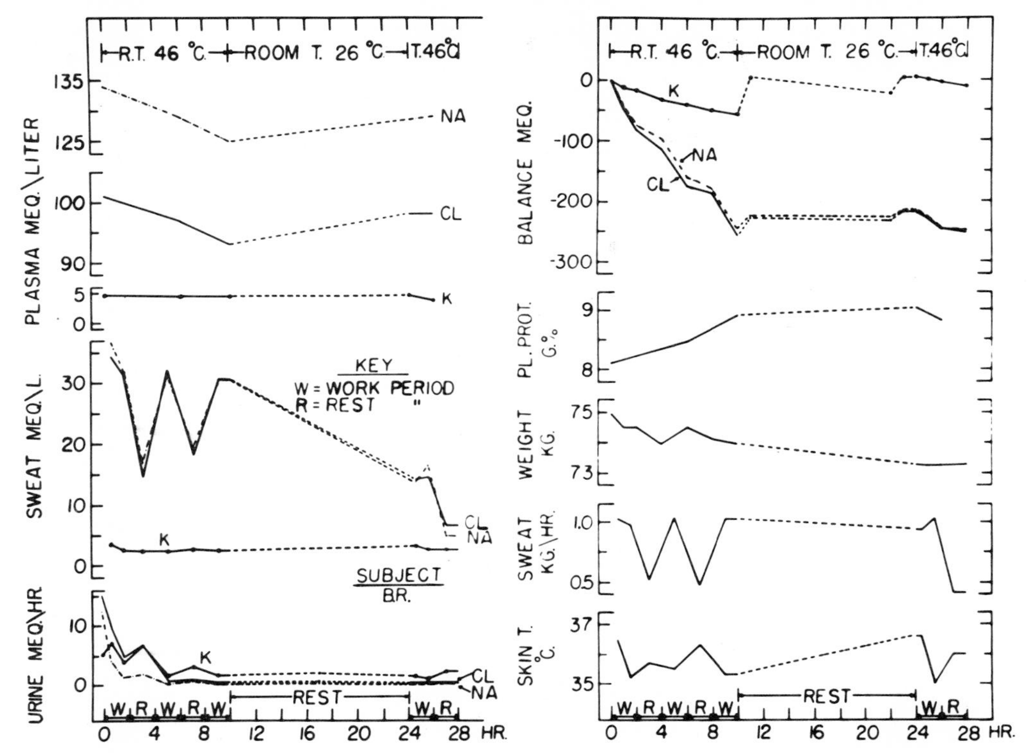

Fig. 4-37. Responses of subject BR to NaCl deficiency during a 28-hour period as follows: (1) alternating 2-hour periods of work ($M = 190$ kcal $\times$ m^{-2} $\times$ hr^{-1}) and rest in heat for 10 hours, (2) rest in a cool room during the eleventh to twenty-fourth hours, (3) work in heat during the twenty-fifth and twenty-sixth hours, and (4) rest in heat during the twenty-seventh and twenty-eighth hours. (Data from Robinson, S., and others: J. Appl. Physiol. 8:159, 1955.)

more, withdrawing fluid from cells. A number of investigators, including Dill,[156] Adolph,[152] and Pitts and co-workers,[149] found that even acclimatized men working in heat never voluntarily drink as much water as they sweat, even though this is advantageous for temperature regulation; they usually drink at a rate of about one half to two thirds of the rate of sweat water loss.

Serious dehydration, circulatory strain, and elevated T_{re} may result when working men voluntarily abstain from water despite a plentiful water supply (Fig. 4-36). Adolph terms this condition *voluntary dehydration* and finds that it varies directly with sweat and work rates. To prevent voluntary dehydration during work in heat, a subject must deliberately maintain body weight by drinking water, preferably containing salt at the same concentration as that in his sweat. If Na^+ loss is not thus replaced, thirst will demand water replacement only to restore isotonic Na^+ concentration in extracellular fluid. If water loss is fully replaced without corresponding replacement of Na^+ loss, plasma Na^+ concentration decreases, diuresis ensues, and isotonicity is restored. This dehydration is thus secondary to salt deficiency.[148]

The following experiment illustrates some of the factors involved in the handling of water and NaCl by the sweat glands and kidneys of a man during intermittent work in heat.[157] Fig. 4-37 gives the results of an experiment with subject BR alternately working 2 hours and resting 2 hours in the heat (46° C db, 26° C wb) for a total of 10 hours, resting overnight in a cool room, and then again working 2 hours and resting 2 hours in the heat the following morning. Diet during the 5 days preceding this experiment contained 420 mEq of NaCl per day. During the first 10 hours of this experiment the subject abstained from salt, ingesting only water and glucose each hour in amounts sufficient to maintain approximate water and carbohydrate balances; supper and breakfast contained little salt.

The data show gradual development of a salt deficit of 250 mEq (14.6 gm NaCl) during the first 10 hours. Attempting to maintain his weight, the subject drank water considerably in excess of that being lost through the skin but diuresed as salt concentration of extracellular fluid decreased and finished the 10-hour period with a water deficit of 1 kgm. Both sodium and chloride plasma levels decreased substantially during the 10-hour period. During the following night he did not sweat and his kidneys excreted practically no salt but did excrete enough water to partially restore plasma sodium and chloride concentrations to control levels. The next morning water deficit had increased to 1.8 kgm. Associated with the dehydration was an increase of plasma protein concentration from 8 to 9 gm%.

Both sodium and chloride urinary output had begun to decrease even during the first hour in the heat and had fallen to only 0.5 mEq/hr by the fifth hour, at which time the salt deficit of the subject was about 150 mEq. Urinary sodium excretion declined much more rapidly than chloride excretion. After the fifth hour there was no sodium and very little chloride in the urine; this low output continued through the night and during the 4-hour hot room experiment the next morning. In contrast to this rapid renal response, the sweat glands during the first 10 hours did not reduce the sodium or chloride sweat concentration in adapting to gradual salt depletion. By the time of the tests the next morning (24 to 28 hours after the start) sweat concentration had dropped about 50%. In adaptation to salt deficiency, sweat glands did not show the degree of selective retention of sodium over chloride accomplished by the kidneys. The first day the sweat contained slightly more sodium than chloride but the next morning, after sweat gland adaptation had definitely begun, two of the three sweat samples contained more chloride than sodium. During these 28-hour experiments there were no significant changes in plasma or sweat concentration of potassium although urinary output of potassium declined moderately.

One should note that the subject's sweat was uniformly much more concentrated during the work periods, when the man was secreting about 1 kgm/hr, than during the rest periods in the heat, when he was sweating only 0.5 kgm/hr. It is also important that variations of urinary sodium and chloride outputs between the rest and work periods were opposite to those observed in sweat output; sweat concentration was higher during work than at rest while urinary output was higher at rest. The difference between sweat concentrations at rest and during work probably depend on the difference in sweat rate in these two states.

Experimental evidence shows that salt output by both kidneys and sweat glands is reduced by increased mineralocorticoid activity in response to salt deficiency; however, salt outputs by kidneys and sweat glands are affected differently by at least three conditions: (1) in response to gradual salt depletion by sweating, a man's kidneys may begin to reduce salt output quickly (latent period of 1 to 2 hours) and make complete adaptation rapidly (5 to 14 hours), whereas sweat glands reduce sweat salt concentration slowly (latent period of 12 to 24 hours) and require several days to complete their adaptation to continuous deficiency; (2) in comparison with the resting state, work by a subject in heat is generally accompanied by reduced renal salt output and by increases in sweat salt concentration and increased hourly sweat output; (3) continued stress (heat and work or extreme heat alone) appears to delay (to longer than 20 hours) the salt-conserving sweat gland response to deficiency while such stress hastens the reduction of renal salt output. Intermittent relief from heat stress appears to hasten the salt-conserving response of sweat glands to deficiency.

These differences are not surprising in view of present knowledge of the functions of these two organs. Sweating functions primarily to regulate temperature and in this role the sweat glands are under nervous control; the kidneys are primarily concerned with chemical homeostasis, and regulation of renal salt output is principally governed by the adrenal cortex.

The shorter latent period of the sweat glands' salt-conserving response to salt deficiency when heat stress is intermittent than when continuous remains unexplained. It is possible that relief of the organism from stresses other than salt deficiency allows the adrenal cortex to respond more specifically to the deficiency, thus facilitating reabsorption of Na^+ by the sweat ducts, which have a much more limited capacity for reabsorption than the renal mechanism.

Pearcy and associates[148] compared the handling of water by the kidneys and sweat glands of four men in response to variations of water and salt balance during exposure to work (M = 190 kcal $\times$ m^{-2} $\times$ hr^{-1}) in heat (44° C db, 26° C wb). In one experiment the subjects maintained water and NaCl balance by drinking appropriate salt solutions throughout the experiment; in another experiment NaCl balance was maintained but they were dehydrated by 3.1% of body weight in 4 hours; in another they were dehydrated by 3.4% and depleted of an average of 157 mEq of NaCl; and in still another they attempted to maintain water balance but were depleted of NaCl by an average of 169 mEq.

When the men were deliberately dehydrated, their sweat rates during work in the heat were consistently reduced about 20% below the rates (1 to 1.2 kgm/hr) observed when they were fully hydrated, even though skin and rectal temperatures were elevated in dehydration. Sweating decreased during the first hour of dehydration; when reduced by dehydration during an experiment it could be elevated to the normal level within an hour by rapidly restoring water balance. There was a strong tendency for sweat rate to vary inversely with the variations of chloride concentration in extracellular fluid associated with changes of water or salt balance. Sweat rates were highest in salt depletion with water replacement and reduced plasma chloride.

Urine flow was lowest (23 ml/hr) when subjects were depleted of both water and salt. When they attempted to maintain water balance by drinking water during salt depletion there was a reduction of serum chloride, marked diuresis, and secondary dehydration. In working men urine flow began to increase during the second hour after

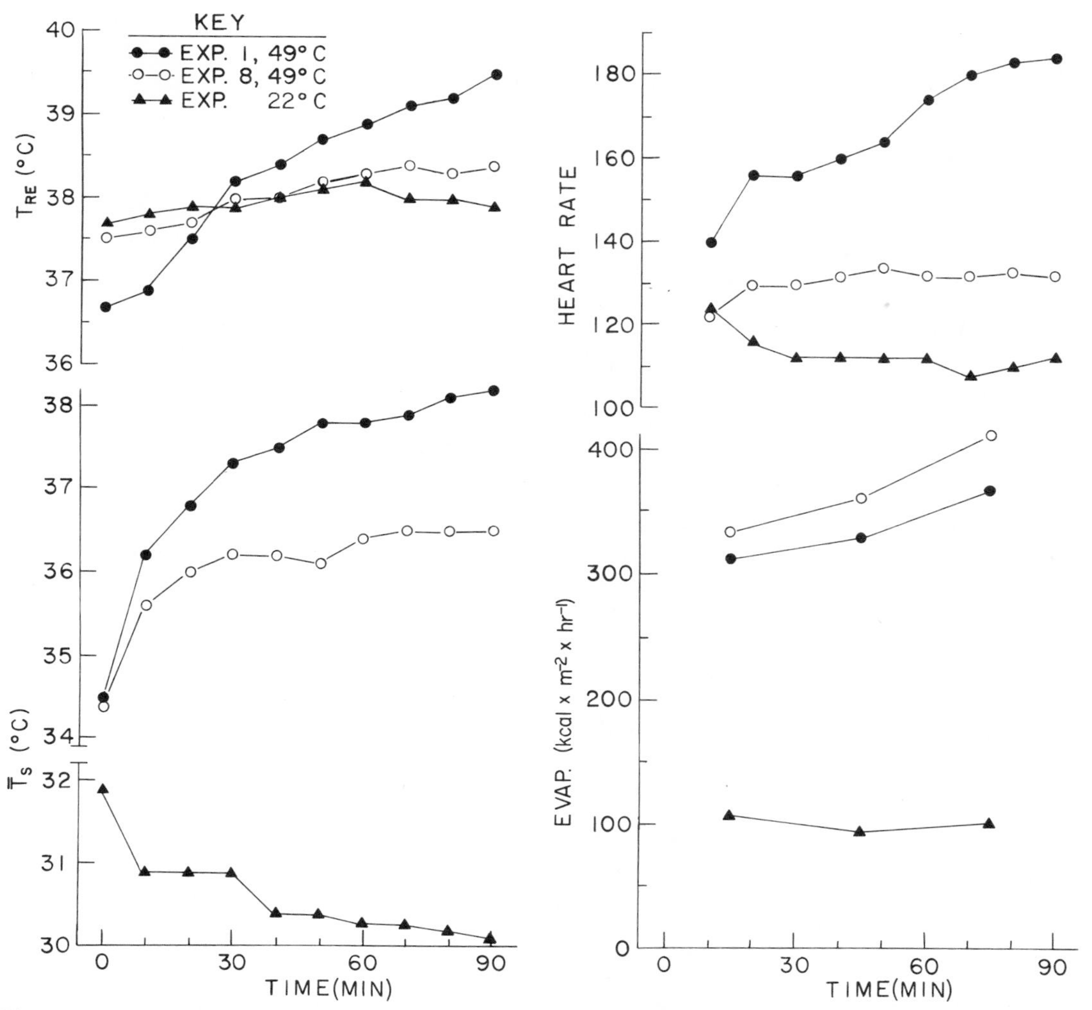

Fig. 4-38. Acclimatization of a young man during eight daily 90-minute periods of work (M = 180 kcal × m^{-2} × hr^{-1}) in heat (49° C db; 26.6° C wb). Note the marked reductions of T_{re}, $\overline{T}_s$, and heart rate and the increase of evaporative cooling in the eighth exposure to heat as compared with those on day 1. Subject regulated central temperature (T_{re}) almost as well on day 8 in heat as in the control experiment at room temperature of 22° C.

starting rapid restoration of water balance, whereas sweat rate began to increase during the first hour. The lowest urine flows and sweat rates occurred in the same experiments. ADH administered to fully hydrated men working in heat produced marked antidiuresis without altering sweat rate, thus indicating that antidiuretic hormone is probably not responsible for reduction of sweat rate in dehydration.

Acclimatization to heat

In preceding sections we referred to changes in sweating, circulation, and body temperature during heat acclimatization. We now present a summary integration of these changes. When men who are not accustomed to heat attempt to perform moderate work in a hot environment, they may overheat within 1 to 2 hours, showing markedly elevated rectal and skin temperatures and reduced gradients between internal and surface temperatures, metabolism increases in proportion to body temperature; heart rate is nearly maximal, and the subjects show signs of circulatory instability such as syncope, which is most pronounced when they attempt to stand. Exhaustion is imminent. With repeated daily exposure to the same combination of work and heat stresses physically fit men acclimatize rapidly—in four to eight daily exposures. [74, 123, 124, 136, 158] Acclimatization is characterized by (1) gradual improvement in evaporative cooling efficiency and in the sensitivity and capacity of the sweat mechanism[158]; (2) gradual improvement of temperature regulation by the eighth day of exposure, enabling men to perform work in heat (if the heat stress is not too intense) with increased rectal-to-skin temperature gradients, but with about the same elevations of core temperature and metabolism as when they perform the same task in a cool environment; (3) markedly improved circulatory stability with reduced heart rate; (4) rapid reduction of renal Na^+ output and a much slower reduction of sweat Na^+ concentration in response to Na^+ deficit[156, 159-161]; and (5) rapid compensatory reduction of renal water output if dehydration occurs. Fig. 4-38 illustrates the improvements of body temperature regulation, heart rate, and evaporative cooling of a young man during eight daily 90-minute periods of work in a hot environment.

Eichna and co-workers[74] found that improved evaporative skin cooling of acclimatized men during work in dry heat is responsible for increased temperature gradient from core to surface, and thus less cutaneous blood flow sufficed for transfer of metabolic heat to skin (Fig. 4-33). Analysis of the data in Fig. 4-38 reveals the same relationships as those found by Eichna. From Fig. 4-33 the gradient $T_{re} - \overline{T}_s$ during work in the heat increased with acclimatization from 1.2° C in the first exposure to 1.7° C in the ninth exposure. The gradient difference of 0.5° C made circulatory heat transfer more effective so that 0.6 L of blood on day 9 transferred about the same amount of heat as that transferred by 1 L of blood on day 1.

The time-course of the remarkable transformation that occurs during heat acclimatization may be affected by (1) dehydration in which sweating, evaporative cooling, and circulating plasma volume decrease, increasing circulatory strain and elevating body temperature in the heat (Fig. 4-36); (2) physical fitness of the subject, fit men adapting more quickly and completely to work in heat, and strenuous winter training programs for varsity athletes preacclimatizing them to work in heat[162]; and (3) physical activity during the acclimatization period (sedentary men do not acclimatize fully to work in heat). Lind and Bass[163] found that the most efficient way to acclimatize men for work in heat is by daily 100-minute periods of treadmill work (M = 160 kcal × m^{-2} × hr^{-1}) at room temperature of 49° C db, 26.6° C wb.

The question of whether internal body temperature of resting men is permanently elevated in hot climates was debated for many years.[164] However, it has been found that it probably is not, either in long-term residents of hot climates or in subjects during experimental studies of heat acclimatization. Burton and colleagues[108] found that when unacclimatized men are exposed continuously to a warm environment (32.4° C db, 24° C wb), rectal temperature at rest on the first day rose about 0.5° C above the usual level. This shift was gradually reversed in the next few days as adaptation to the warm environment progressed.

Knipping[165] found that basal metabolism of healthy individuals increases during the first week of exposure to a hot climate, then falls gradually until after several months' residence in the hot climate it is below ordinary basal standards. Burton[54] also found that men unacclimatized to heat undergo a moderate rise of metabolism during the first few days in a hot environment. He attributed this rise to elevated body temperature and found that after 5 days' acclimatization, rectal temperature and metabolism both returned to control levels. There is some evidence that the basal metabolism of men decreases during residence in hot climates and that this occurs in natives as well as in Europeans who have moved to the tropics. Numerous investigators[166, 167] have found evidence of reduced metabolism, ranging from 6% to 24%, in residents of hot climates. On the other hand, Eijkman[168] reported that some Europeans living in Indonesia appear to have unchanged metabolic rates.

GLOSSARY

basal metabolic rate The rate of energy exchange of human beings per unit of body surface area recorded under conditions of rest and thermal neutrality. The results are expressed as a percentage above or below a standard established for persons of the same age and sex.

blackbody A substance that completely absorbs the radiant energy falling on its surface. At thermal equilibrium all the energy absorbed is emitted.

emittance The ratio of the rate of emission of radiant energy by a body to the corresponding rate of a blackbody at the same temperature.

endergonic reaction A chemical reaction that cannot proceed without the introduction of energy into the system; the reaction absorbs energy.

exergonic reaction A chemical reaction that results in the liberation of energy from the reactants.

exothermic reaction A chemical reaction that results in the release of heat.

heat transfer coefficient The rate of heat transferred per unit of area for unit of difference in temperature; $W \times m^{-2} \times {}^{\circ}C^{-1}$ or $kcal \times m^{-2} \times hr^{-1} \times {}^{\circ}C^{-1}$. The heat transfer coefficient should not be confused with thermal conductivity; used in Part B as *tissue heat conductance* ($kcal \times m^{-2} \times hr^{-1} \times \Delta\ {}^{\circ}C^{-1}$).

homeotherm An animal that has relatively constant normal body temperature which is maintained despite environmental temperature change within limits.

insulation The reciprocal of the heat transfer coefficient.

latent heat of vaporization The quantity of heat necessary to change 1 gm of a liquid to its vapor without change of temperature; cal/gm or J/gm.

metabolic rate The rate at which an animal converts energy; expressed as $kcal \times m^{-2} \times hr^{-1}$ or W/m^2.

reflectance The ratio of the rate of radiation reflected from a body to the rate of emission of a blackbody at the same temperature.

specific heat Ratio of thermal capacity of a substance to that of water at 15° C.

thermal capacity The quantity of heat necessary to produce a unit change of temperature in a unit mass; expressed as $cal \times gm^{-1} \times {}^{\circ}C^{-1}$ or $J \times gm^{-1} \times {}^{\circ}C^{-1}$.

thermal conductivity Time rate of transfer of heat by conduction, through unit thickness, across unit area for unit difference of temperature; expressed as $cal \times cm \times (cm^{-2} \times s^{-1} \times {}^{\circ}C^{-1})$ or $W \times m \times (m^{-2} \times {}^{\circ}C^{-1})$.

thermal equilibrium The condition under which the rate of heat production of an animal is equal to the rate of heat loss, so that body temperature does not change.

ventilation rate The volume of air breathed in and out of the lungs per minute; expressed as L/min.

REFERENCES

1. Burton, A. C., and Edholm, O. G.: Man in a cold environment, London, 1955, Edward Arnold (Publishers) Ltd.
2. Nelson, N., and others: Thermal exchanges of man at high temperatures, Am. J. Physiol. **151**:626-652, 1947.
3. Hardy, J. D., and Muschenheim, C.: Radiation of heat from human body; transmission of infrared radiation through skin, J. Clin. Invest. **15**:1-9, 1936.
4. Buettner, K. J. K.: Physical aspects of human bioclimatology. In Committee on the Compendium of Meteorology: Compendium of meteorology, Boston, 1951, American Meteorological Society, pp. 1112-1125.
5. Winslow, C. E. A., Herrington, L. P., and Gagge, A. P.: The determination of radiation and convection exchanges by partitional calorimetry, Am. J. Physiol. **116**:669-684, 1936.
6. Winslow, C. E. A., Herrington, L. P., and Gagge, A. P.: Physiological reactions of the human body to varying environmental temperatures, Am. J. Physiol. **120**:1-22, 1937.
7. Kleiber, M.: Trophic responses to cold, Fed. Proc. **27**:772-774, 1963.
8. Hardy, J. D., and Soderstrom, G. F.: Heat loss from the body and peripheral blood flow at temperatures of 22° C to 35° C, J. Nutr. **16**:493-510, 1938.
9. Winslow, C. E. A., and Herrington, L. P.: Temperature and human life, Princeton, N. J., 1949, Princeton University Press, p. 66.
10. Carlson, L. D., Hsieh, A. C. L., Fullington, F., and Elsner, R. W.: Immersion in cold water and body tissue insulation, J. Aviat. Med. **29**:145-152, 1958.
11. Pugh, L. G. C. E., and others: A physiological study of channel swimming, Clin. Sci. **19**:257-273, 1960.
12. Froese, G., and Burton, A. C.: Heat losses from the human head, J. Appl. Physiol. **10**:235-241, 1957.
13. Bazett, H. C., and McGlone, B.: Temperature gradients in the tissue in man, Am. J. Physiol. **82**:415-451, 1927.
14. Grant, R. T.: Observations on direct communications between arteries and veins in the rabbit's ear, Heart **15**:281-303, 1930.
15. Clark, E. R., and Clark, E. L.: Observations on living anastomoses as seen in transparent chambers introduced into the rabbit's ear, Am. J. Anat. **54**:229-286, 1934.
16. Prichard, M. M. L., and Daniel, P. M.: Arteriovenous anastomoses in the tongue of the dog, J. Anat. (London) **87**:66-74, 1953.
17. Prichard, M. M. L., and Daniel, P. M.: Arteriovenous anastomoses in the tongue of the sheep and the goat, Am. J. Anat. **95**:203-225, 1954.
18. Lewis, T.: Observations upon reactions of vessels of human skin to cold, Heart **15**:177-208, 1930.
19. Carlson, L. D., and Hsieh, A. C. L.: Cold. In Edholm, O. G., and Bacharach, A. L., editors: The physiology of human survival, New York, 1965, Academic Press, Inc., chap. 2.
20. Edwards, M. A.: The role of arteriovenous anastomoses in cold-induced vasodilation, rewarming, and

reactive hyperemia as determined by ^{24}Na clearance, Comed. J. Physiol. Pharmacol. **43**:39-48, 1967.

21. Bazett, H. C., and others: Temperature changes in blood flowing in arteries and veins in man, J. Appl. Physiol. **1**:3-19, 1948.
22. Horvath, S. M., Spurr, B. G., Hutt, B. K., and Hamilton, L. H.: Metabolic cost of shivering, J. Appl. Physiol. **8**:595-602, 1956.
23. Ismail-Beigi, F., and Edelman, I. S.: Mechanism of thyroid calorigenesis: role of active sodium transport, Proc. Nat. Acad. Sci. **67**:1071-1078, 1970.
24. Ivy, A. C.: What is normal or normality? Quart. Bull. Northwestern Univ. Med. School **18**:22-28, 1944.
25. Carlson, L. D.: Temperature regulation and acclimation, Brody Memorial Lecture IX, Agricultural Experimental Station, Columbia, 1969, University of Missouri.
26. Hong, S. K.: Comparison of diving and non-diving women of Korea, Fed. Proc. **22**:831-833, 1963.
27. Sellers, E. A., Scott, J. W., and Thomas, N.: Electrical activity of skeletal muscle of normal and acclimatized rats on exposure to cold, Am. J. Physiol. **177**:372-376, 1954.
28. Cottle, M., and Carlson, L. D.: Turnover of thyroid hormone in cold-exposed rats determined by radioactive iodine studies, Endocrinology **59**:1-11, 1956.
29. Hsieh, A. C. L., and Carlson, L. D.: Role of the thyroid in metabolic response to low temperature, Am. J. Physiol. **188**:40-44, 1957.
30. Hsieh, A. C. L., and Carlson, L. D.: Role of adrenalin and noradrenalin in chemical regulation of heat production, Am. J. Physiol. **190**:243-246, 1957.
31. Keller, A. D.: Hypothermia in the unanesthetized poikilothermic dog. In Dripps, R. D., editor: Physiology of induced hypothermia, NAS-NRC Publication 451, Washington, D. C., 1956, The Council.
32. DuBois, E. F., Edbaugh, F. G., and Hardy, J. D.: Basal heat production and elimination of thirteen normal women at temperatures from 22° C to 35° C, J. Nutr. **48**:257-293, 1952.
33. Jansky, L., Bartunkora, R., and Zeisberger, E.: Acclimation of the white rat to cold, noradrenaline thermogenesis, Physiol. Bohemoslov. **16**:366-372, 1967.
34. Cottle, W. H., and Carlson, L. D.: Regulation of heat production in cold-adapted rats, Proc. Soc. Exp. Biol. **92**:845-849, 1956.
35. Brück, K., and Wünnenberg, B.: Influence of ambient temperature in the process of replacement of non-shivering by shivering thermogenesis during post natal development, Fed. Proc. **25**:1332-1336, 1966.
36. Hagen, J. H., and Hagen, P. B.: Actions of adrenalin and noradrenalin on metabolic systems. In Litwack, G., and Kritchevsky, D., editors: Actions of hormones on molecular process, New York, 1964, John Wiley & Sons, Inc., chap. 2.
37. Hsieh, A. C. L., Pun, C. W., Li, K. M., and Ti, K. W.: Circulatory and metabolic effects of noradrenaline in cold-adapted rats, Fed. Proc. **25**:1205-1212, 1966.
38. Cahill, G. In Kinsell, L. W., editor: Adipose tissue as an organ, part III, Springfield, Ill., 1962, Charles C Thomas, Publisher.
39. Sutherland, E. W., and Robison, G. A.: The role of cyclic 3′,5′-AMP in responses to catecholamines and other hormones, Pharmacol. Rev. **18**:145-161, 1966.
40. Krebs, E. G., DeLange, R. J., Kemp, R. G., and Riley, W. D.: Activation of skeletal muscle phosphorylase, Pharmacol. Rev. **18**:163-171, 1966.
41. Honda, N.: Pre-cooling of peripheral arterial blood in cold-adapted and warm-adapted rabbits, J. Appl. Physiol. **20**:1133-1135, 1965.
42. Andjus, R. K.: Suspended animal in cooled, supercooled, and frozen rats, J. Physiol. **128**:547-556, 1955.
43. Mayer, W. V.: Histological changes during the hibernating cycle in the arctic ground squirrel. In Lyman, C. P., and Dawe, A. R., editors: Mammalian hibernation, Bull. Museum Compar. Zool. (Harvard) **124**:131-154, 1960.
44. Bartholomew, G. A., and Hudson, J. W.: Aestivation in the Mohave ground squirrel, *Citellus mohavensis*. In Lyman, C. P., and Dawe, A. R., editors: Mammalian hibernation, Bull. Compar. Zool. (Harvard) **124**:193-208, 1960.
45. Lyman, C. P., and Chatfield, P. O.: Physiology of hibernation in mamals. In Dripps, R. D., editor: The physiology of induced hypothermia, NAS-NRC Publication 451, Washington, D. C., 1956, The Council.
46. Popovic, V.: Endocrines in hibernation. In Lyman, C. P., and Dawe, A. R., editors: Mammalian hibernation, Bull. Museum Compar. Zool. (Harvard) **124**:105-130, 1960.
47. Molnar, G. W.: Man in the tropics compared with man in the desert. In Adolph, E. F., and others: Physiology of man in the desert, New York, 1947, Interscience Publishers, Inc., chap. 19.
48. Leithead, C. S., and Lind, A. R.: Heat stress and heat disorders, Philadelphia, 1964, F. A. Davis Co., chap. 1.
49. Gagge, A. P., Winslow, C. E. A., and Herrington, L. P.: Influence of clothing on the physiological reactions of the human body to varying environmental temperatures, Am. J. Physiol. **124**:30, 1938.
50. Hardy, J. D.: Heat transfer. In Newburgh, L. H., editor: Physiology of heat regulation, Philadelphia, 1949, W. B. Saunders Co., chap. 3.
51. Bohnenkamp, H., and Ernst, H. W.: Über die Strahlungsverluste des Menschen. Das energetische Oberflächengesetz, Pflügers Arch. **228**:40, 1931.
52. Gagge, A. P.: A new physiological variable associated with sensible and insensible perspiration, Am. J. Physiol. **120**:277, 1937.
53. Adolph, E. F., and others: Physiology of man in the desert, New York, 1947, Interscience Publishers, Inc., chaps. 5, 10, 11, 16, and 19.
54. Burton, A. C., Scott, J. C., McGlone, B., and Bazett, H. C.: Slow adaptations in the heat exchanges of man to changed climatic conditions, Am. J. Physiol. **129**:84, 1940.
55. Nielsen, M.: Die Regulation der Körpertemperature bei Muskelarbeit, Skand. Arch. Physiol. **79**:193, 1938.
56. Nielsen, B.: Regulation of body temperature and heat dissipation at different levels of energy and

heat production in man, Acta Physiol. Scand. **68:** 215, 1966.
57. Smiles, K. A., and Robinson, S.: Regulation of sweat secretion during positive and negative work, J. Appl. Physiol. **30:**409, 1971.
58. Robinson, S., Turrell, E. S., and Gerking, S. D.: Physiologically equivalent conditions of air temperature and humidity, Am. J. Physiol. **143:**21, 1945.
59. Robinson, S.: Temperature regulation in exercise, Pediatrics **32:**691, 1963.
60. Cole, P.: The conditioning of respiratory air, J. Laryng. Otol. **67:**669, 1953.
61. Newburgh, L. H., and others: A respiration chamber for use with human subjects, J. Nutr. **13:**193, 1937.
62. Pinson, E. A.: Evaporation from human skin with sweat glands inactivated, Am. J. Physiol. **137:**492, 1942.
63. Robinson, S., and Gerking, S. D.: Thermal balance of men working in severe heat, Am. J. Physiol. **149:**476, 1947.
64. Bazett, H. C.: The regulation of body temperatures. In Newburgh, L. H., editor: Physiology of heat regulation, Philadelphia, 1949, W. B. Saunders Co., chap. 4.
65. Seckendorf, R., and Randall, W. C.: Thermal reflex sweating in paraplegic man, J. Appl. Physiol. **16:** 796, 1961.
66. Bullard, R. W., and others: Skin temperature and thermoregulatory sweating. In Hardy, J. D., editor: Physiological and behavioral temperature regulation, Springfield, Ill., 1968, Charles C Thomas, Publisher, chap. 40.
67. Robinson, S.: The regulation of sweating in exercise. In Montagna, W., editor: Advances in biology of skin, vol. 3, Oxford, 1962, Pergamon Press, chap. 9.
68. Van Beaumont, W., and Bullard, R. W.: Sweating: its rapid response to muscular work, Science **141:** 643, 1963.
69. Gisolfi, C., and Robinson, S.: Central and peripheral stimuli regulating sweating during intermittent work in men, J. Appl. Physiol. **29:**761, 1970.
70. Robinson, S., and others: Relations between sweating, cutaneous blood flow and body temperature in work, J. Appl. Physiol. **20:**575-582, 1965.
71. Behnke, A. R., and Willman, T. L.: Cutaneous diffusion of helium in relation to peripheral blood flow and the absorption of atmospheric nitrogen through the skin, Am. J. Physiol. **131:**627, 1941.
72. Hardy, J. D., and Soderstrom, G. F.: Heat loss from the nude body and peripheral blood flow at temperatures from 22° to 31° C, J. Nutr. **16:**493, 1938.
73. Hertzman, A.B., and Randall, W.C.: Regional differences in the basal and maximal rates of blood flow in the skin, J. Appl. Physiol. **1:**234, 1948.
74. Eichna, L. W., and others: Thermal regulation during acclimatization to hot, dry environment, Am. J. Physiol. **163:**585, 1950.
75. Hertzman, A. B.: The blood supply of various skin areas as estimated by the photoelectric plethysmograph, Am. J. Physiol. **124:**328, 1938.
76. Hertzman, A. B., and Dillon, J. B.: Selective vascular reaction patterns in the nasal septum and skin of the extremities and head, Am. J. Physiol. **127:**671, 1939.
77. Forster, R. E., II, Ferris, B. G., Jr., and Day, R.: Relationship between total heat exchange and blood flow in the hand at various ambient temperatures, Am. J. Physiol **146:600**, 1946.
78. Scott, J. C., Bazett, H. C., and Mackie, G. C.: Climate effects on cardiac output and the circulation in man, Am. J. Physiol. **139:**102, 1940.
79. Spealman, C. R.: Effect of ambient air temperature and of hand temperature on blood flow in the hands, Am. J. Physiol. **145:**218, 1945.
80. Wilkins, R. W., Doupe, J., and Newman, H. W.: The rate of blood flow in normal fingers, Clin. Sci. **3:**403, 1938.
81. Hertzman, A. B., Randall, W. C., and Jochim, K. E.: Estimation of cutaneous blood flow by the photoelectric plethysmograph, Am. J. Physiol. **145:** 716, 1946.
82. Taylor, C. R.: The vascularity and possible thermoregulatory function of the horns in goats, Physiol. Zool. **39:**127, 1966.
83. Johansen, K.: Heat exchange through the skin in the tail of the muskrat *(Ondatra zibethica),* Fed. Proc. **20:**110, 1961.
84. Steen, I., and Steen, J. B.: Thermoregulatory importance of the beaver's tail, Comp. Biochem. Physiol. **15:**267, 1965.
85. Schmidt-Nielsen, K., and others: The jack rabbit—a study in its desert survival, Hvalrådets Skrifter **48:**125-142, 1965.
86. Kahl, M. P.: Thermoregulation in the wood stork, with special reference to the role of the legs, Physiol. Zool. **36:**141, 1963.
87. Abramson, D. I., and Ferris, E. B., Jr.: Observations on reactive hyperemia in various portions of the extremities, Am. J. Physiol. **129:**297, 1940.
88. Doupe, J., Robertson, J. S. M., and Carmichael, E. A.: Vasomotor responses in the toes: effect of lesions of the cauda equina, Brain **60:**281, 1937.
89. Grant, R. T., and Pearson, R. S. B.: Further observations on the vascular responses of the human limb to body warming; evidence for sympathetic vasodilator nerves in the normal subject, Clin. Sci. **3:**119, 1938.
90. Sheard, C., Williams, M. M. D., and Horton, B. T.: Investigations on the skin temperatures of the extremities under various controlled environmental conditions, Am. J. Physiol. **123:**184, 1938.
91. Uprus, V., Gaylor, J. B., and Carmichael, E. A.: Vasodilatation and vasoconstriction in response to warming and cooling the body, Clin. Sci. **2:**301, 1936.
92. Freeman, N. E.: Effect of temperature on rate of blood flow in the normal and sympathectomized hand, Am. J. Physiol. **113:**385, 1935.
93. Roddie, I. C., Shepherd, J. T., and Whelan, R. F.: Contribution of constrictor and dilatory nerves to skin vasodilatation during body heating, J. Physiol. **136:**489, 1957.
94. Roddie, I. C., Shepherd, J. T., and Whelan, R. F.: A comparison of heat elimination from the normal and nerve-blocked finger during body heating, J. Physiol. **138:**445, 1957.
95. Warren, J. V., Walter, C. W., Romano, J., and Stead, E. A., Jr.: Blood flow in the hand and forearm after paravertebral block of the sympathetic ganglia, J. Clin. Invest. **21:**665, 1942.

96. Kerslake, D. McK., and Cooper, K. E.: Vasodilatation in the hand in response to heating the skin elsewhere, Clin. Sci. **9**:31, 1950.
97. Cooper, K. E., and Kerslake, D. McK.: Abolition of nervous reflex vasodilatation by sympathectomy of the heated area, J. Physiol. **119**:18, 1953.
98. Cooper, K. E., and Kerslake, D. McK.: Evidence for the existence of afferent pathways from legs in the lumbar sympathetic chain in man, International Physiological Congress **XIX**, Montreal, 1953, p. 279.
99. Snell, E. S.: Relationship between vasomotor response in the hand and heat changes in the body induced by intravenous infusions of hot or cold saline, J. Physiol. **125**:361, 1954.
100. Allwood, M. J., and Burg, H. S.: Effect of local temperature on blood flow in the human foot, J. Physiol. **124**:345, 1954.
101. Roddie, I. C., and Shepherd, J. T.: Blood flow through the hand during local heating, release of sympathetic vasomotor tone by indirect heating and a combination of both, J. Physiol. **131**:657, 1956.
102. Grant, R. T., and Holling, H. E.: Vascular responses of the human limb to body warming, Clin. Sci. **3**:273, 1938.
103. Roddie, I. C., Shepherd, J. T., and Whelan, R. F.: Evidence from venous oxygen saturation measurements that the increase in forearm blood flow during body heating is confined to skin, J. Physiol. **134**:444, 1956.
104. Cooper, K. E., and others: Vasodilatation in the forearm during indirect heating, J. Physiol. **125**: 56P, 1954.
105. Cooper, K. E., Edholm, O. G., and Mottram, R. F.: Blood flow in skin and muscle of the human forearm, J. Physiol. **128**:258, 1955.
106. Edholm, O. G., Fox, R. H., and MacPherson, R. K.: The effect of body heating on the circulation in skin and muscle, J. Physiol. **134**:612, 1956.
107. Burton, A. C.: The application of theory of heat flow to the study of energy metabolism, J. Nutr. **5**:497, 1934.
108. Burton, A. C., and Bazett, H. C.: A study of the average temperature of the tissues, of the exchanges of heat and vasomotor responses in man by means of a bath calorimeter, Am. J. Physiol. **117**:36, 1936.
109. Herrington, L. P., Winslow, C. E. A., and Gagge, A. P.: The relative influence of radiation and convection upon vasomotor temperature regulation, Am. J. Physiol. **120**:133, 1937.
110. LeFevre, J.: La chaleur animale et bioenergetiques, Paris, 1911, Masson et Cie.
111. Winslow, C. E. A., Gagge, A. P., and Herrington, L. P.: Heat exchange and regulation in radiant environments above and below air temperature, Am. J. Physiol. **131**:79-92, 1940.
112. Gisolfi, C., and Robinson, S.: Venous blood distribution in the legs during intermittent treadmill work in men, J. Appl. Physiol. **29**:368, 1970.
113. Hellon, R. F., and Clarke, R. S. J.: Changes in forearm blood flow with age, Clin. Sci. **18**:1, 1959.
114. Hellon, R. F., and Lind, A. R.: The influence of age on peripheral vasodilatation in a hot environment, J. Physiol. **141**:262, 1958.
115. Allwood, M. J.: Blood flow in the foot and calf in the elderly; a comparison with that in young adults, Clin. Sci. **17**:331, 1958.
116. Robinson, S.: Circulatory adjustments of men in hot environments. In Plumb, H. H., editor: Temperature—its measurement and control in science and industry, vol. 3, New York, 1963, Van Nostrand Reinhold Co., pp. 287-297.
117. Comroe, J. H., Jr., and Schmidt, C. F.: Reflexes from the limbs as a factor in the hyperpnea of muscular exercise, Am. J. Physiol. **138**:536, 1943.
118. Robinson, S.: Physiological adjustments to heat; laboratory and field studies: tropics. In Newburgh, L. H., editor: Physiology of heat regulation, Philadelphia, 1949, W. B. Saunders Co., chaps. 5 and 11.
119. Bazett, H. C.: Theory of reflex controls of body temperature in rest and exercise, J. Appl. Physiol. **4**:245, 1951.
120. Cooper, T., Randall, W. C., and Hertzman, A. B.: The vascular convection of heat from active muscle to overlying skin, WADC Technical Report 6680, part 7, Dec. 1951.
121. Thompson, F.: Central interactions of afferent projection from the femoral vein in the cat, Ph.D. dissertation, Indiana University, 1971.
122. Bean, W. B., and Eichna, L. W.: Performance in relation to environmental temperature, Fed. Proc. **2**:144, 1943.
123. Robinson, S., Turrell, E. S., Belding, H. S., and Horvath, S. M.: Rapid acclimatization to work in hot climates, Am. J. Physiol. **140**:168-176, 1943.
124. Taylor, H. L., Henschell, A. F., and Keys, A.: Cardiovascular adjustments of man in rest and work during exposure to dry heat, Am. J. Physiol. **139**:583, 1943.
125. Asmussen, E.: Cardiac output in rest and work in humid heat, Am. J. Physiol. **131**:54, 1940.
126. Burch, G. E., DePasquale, N., Hyman, A., and DeGraff, A. C.: Influence of tropical weather on cardiac output, work and power of right and left ventricles of men resting in hospital, Arch. Int. Med. **104**:553, 1959.
127. Burch, G. E., and Hyman, A.: Influence of tropical weather on output of volume, work, and power by the right and left ventricles of man at rest in bed, Am. Heart J. **57**:247, 1959.
128. Dill, D. B., Edwards, H. T., Bauer, P. S., and Levenson, E. J.: Physical performance in relation to external temperature, Arbeitsphysiol. **4**:508, 1931.
129. Kenney, R. A.: Effect of hot, humid environments on renal function of West Africans, J. Physiol. **118**: 25P, 1952.
130. Kenney, R. A.: Effect of exercise in hot, humid environments on renal function of West Africans, J. Physiol. **118**:26P, 1952.
131. Radigan, L. R., and Robinson, S.: Effects of environmental heat stress and exercise on renal blood flow and filtration rate, J. Appl. Physiol. **2**:185, 1949.
132. Smith, J. H., Robinson, S., and Pearcy, M.: Renal responses to exercise heat and dehydration, J. Appl. Physiol. **4**:659, 1952.
133. Rowell, L. B., Blackmon, J. R., and Bruce, R. A.: Indocyanine green clearance and estimated hepatic blood flow during mild to maximal exercise in upright man, J. Clin. Invest. **43**:1677-1690, 1964.

134. Rowell, L. B., and others: Hepatic clearance of indocyanine green in man under thermal and exercise stresses, J. Appl. Physiol. **20**:384-394, 1965.
135. Barcroft, J., and others: The effect of high altitude on physiological processes of the human body, Phil. Trans. Roy. Soc. (London) **B211**:351, 1922.
136. Bass, D. E., and others: Mechanisms of acclimatization to heat in man, Medicine **34**:323, 1955.
137. Bazett, H. C., Sunderman, F. W., Doupe, J., and Scott, J. C.: Climatic effects on volume and composition of the blood in man, Am. J. Physiol. **129**: 69, 1940.
138. Conley, C. L., and Nickerson, J. L.: Effects of temperature changes on the water balance of man, Am. J. Physiol. **143**:373, 1945.
139. Glickman, N., Hick, L. K., Keeton, R. W., and Montgomery, M. M.: Blood volume changes in men exposed to hot environmental conditions for a few hours, Am. J. Physiol. **134**:165, 1941.
140. Robinson, S., Kincaid, R. K., and Rhamy, R. K.: Effects of desoxycorticosterone acetate on acclimatization of men to heat, J. Appl. Physiol. **2**:399, 1950.
141. Sunderman, F .W., Scott, J. C., and Bazett, H. C.: Temperature effects on serum volume, Am. J. Physiol. **132**:199, 1938.
142. Myhre, L. G.: Shifts in blood volume during and following acute environmental and work stresses in man, Ph.D. dissertation, Indiana University, 1971.
143. Myhre, L. G., and Robinson, S.: Plasma volume during and following acute dehydration by exposure to environmental and work stresses (abstract), Proceedings of International Union Physiological Sciences XXV, International Congress, Munich, 1971.
144. Forbes, W. H., Dill, D. B., and Hall, F. G.: Effect of climate upon the volumes of blood and tissue fluid in man, Am. J. Physiol. **130**:739, 1940.
145. Gibson, J. G., Jr., and Evans, W. A., Jr.: Clinical studies of blood volume, J. Clin. Invest. **16**:301, 1937.
146. Talbott, J. H., Edwards, H. T., Dill, D. B., and Drastich, L.: Physiological responses to high environmental temperatures, Am. J. Trop. Med. **13**: 381, 1933.
147. Maxfield, M. E., Bazett, H. C., and Chambers, C. C.: Seasonal and postural changes in blood volume, Am. J. Physiol. **133**:128, 1941.
148. Pearcy, M., and others: Effects of dehydration, salt depletion and pitressin on sweat rate and urine flow, J. Appl. Physiol. **8**:621-626, 1956.
149. Pitts, G. C., Johnson, R. E., and Consolazio, C. F.: Work in the heat as affected by intake of water, salt and glucose, Am. J. Physiol. **142**:253, 1944.
150. Cannon, W. B.: The physiological basis of thirst, Proc. Roy. Soc. (London) **B90**:283, 1918.
151. Bernard, C.: Leçons de physiologie expérimentale appliquée à la médecine, vol. 2, Paris, 1856, Baillière, p. 50.
152. Adolph, E. F.: Measurements of water drinking in dogs, Am. J. Physiol. **125**:75, 1939.
153. Andersson, B., Gale, C. C., and Sundsten, J. W.: Preoptic influences on water intake. In Wayner, M. J., editor: Thirst, New York, 1964, The Macmillan Co., pp. 361-379.
154. Andersson, B., Larsson, S., and Persson, N.: Some characteristics of the hypothalamic "drinking centre" in the goat as shown by the use of permanent electrodes, Acta Physiol. Scand. **50**:140, 1960.
155. Gilbert, G. J., and Glaser, G. H.: On the nervous system integration of water and salt metabolism, Arch. Neurol. **5**:179, 1961.
156. Dill, D. B.: Life, heat and altitude, Cambridge, Mass., 1938, Harvard University Press.
157. Robinson, S., and others: Time relation of renal and sweat gland adjustments to salt deficiency in men, J. Appl. Physiol. **8**:159, 1955.
158. Wyndham, C. H.: Effect of acclimatization on the sweat rate/rectal temperature relationship, J. Appl. Physiol. **22**:27-30, 1967.
159. Conn, J. W.: Electrolyte composition of sweat, Arch. Int. Med. **83**:416-428, 1949.
160. Robinson, S., Kincaid, R. K., and Rhamy, R. K.: Effects of salt deficiency on salt concentration in the sweat, J. Appl. Physiol. **3**:55-62, 1950.
161. Weiner, J. S., and Van Heyningen, R. E.: Salt losses of men working in hot environments, Brit. J. Indust. Med. **9**:56-64, 1952.
162. Piwonka, R. W., Robinson, S., Gay, V. L., and Manalis, R .C.: Preacclimatization of men to heat by training, J. Appl. Physiol. **20**:379-383, 1965.
163. Lind, A. R., and Bass, D. E.: The optimal exposure time for acclimatization of men to heat, Fed. Proc. **22**:704, 1963.
164. Sundstrom, E. S.: The physiological effects of tropical climate, Physiol. Rev. **7**:32, 1927.
165. Knipping, H. W.: Ein Beitrag zur Tropenphysiologie, Z. Biol. **78**:259, 1931.
166. Galvão, P. E.: Human heat production in relation to body weight and body surface, J. Appl. Physiol. **3**:21, 1950.
167. Radsma, W.: The influence of a temporary sojourn in a cool environment on various vegetative functions in inhabitants of the tropics, Acta Physiol. Pharmacol. Neerland **1**:112, 1950.
168. Eijkman, C.: Le metabolisme de l'homme tropical, J. Physiol. Path. **19**:33, 1931.
169. Bedford, T.: Environmental warmth and its measurement, Medical Research Council War Memo, No. 17, London, 1946.
170. Mountcastle, V. B., editor: Medical physiology, ed. 13, St. Louis, 1974, The C. V. Mosby Co.
171. Bullard, R. W.: Temperature regulation. In Selkurt, E. E., editor: Physiology, Boston, 1966, Little, Brown and Co.

SUGGESTED READINGS

1. Adolph, E. F., and Dill, D. B.: Observations on water metabolism in the desert, Am. J. Physiol. **123**: 369, 1938.
2. Bajusz, E., editor: Physiology and pathology of adaptation mechanisms: neural, neuroendocrine, humoral, New York, 1969, Pergamon Press.
3. Carlson, L. D., and Hsieh, A. C. L.: Control of energy exchange, New York, 1970, The Macmillan Co.
4. Dill, D. B.: Life, heat, and altitude, Cambridge, Mass., 1938, Harvard University Press.
5. Dill, D. B., Hall, F. G., and Edwards, H. T.: Changes in composition of sweat during acclimatization to heat, Am. J. Physiol. **123**:412, 1938.

6. Edholm, O. G., and Bacharach, A. L., editors: The physiology of human survival, New York, 1965, Academic Press, Inc.
7. Goldsby, R. A.: Cells and energy, New York, 1967, The Macmillan Co.
8. Hannon, J. P., Evonuk, E., and Larson, A. M.: Some physiological and biochemical effects of norepinephrine in the cold-acclimatized rat, Fed. Proc. **22**:783-788, 1963.
9. Jansky, L., and Hart, J. S.: Cardiac output and organ blood flow in warm and cold-acclimated rats exposed to cold, Can. J. Physiol. Pharmacol. **46:** 653-659, 1968.
10. Kleiber, M.: The fire of life, New York, 1965, John Wiley & Sons, Inc.
11. Lehninger, A. L.: Bioenergetics, New York, 1965, W. A. Benjamin, Inc.
12. Newburgh, L. H., editor: Physiology of heat regulation and the science of clothing, Philadelphia, 1949, W. B. Saunders Co.
13. Taylor, H. L., Henschel, A., Mickelsen, O., and Keys, A.: The effect of sodium chloride intake on the work performance of man during exposure to dry heat and experimental heat exhaustion, Am. J. Physiol. **140**:439, 1943.

5 SOUND, VIBRATION, AND IMPACT

DONALD E. PARKER
HENNING E. VON GIERKE

Throughout life man is exposed to mechanical energy generated by both natural and artificial sources. The environmental physiologist seeks to specify the kinds and magnitudes of man's responses to environments that contain known levels of mechanical energy. He quantitatively describes alterations of given body structures and physiologic functions resulting from exposure to particular types of mechanical energy at specific levels. He determines the effects of mechanical energy on man's overall functional performance capability and well-being.

Purpose and goals

The major purpose of this chapter is four-fold. First, we wish to acquaint the student with the basic concepts employed in analysis of man's responses to mechanical energy. Definitions for a set of fundamental terms are given in the Glossary. Applications of these concepts will be clarified as they are discussed in subsequent sections of this chapter. Our second goal is presentation of a framework for analysis and comprehension of scientific research in the areas of sound, vibration, and impact. One aspect of this framework concerns a general theory of vibratory energy reception by organisms (particularly man) presented in The Human Body as a Mechanical Energy Receiver. Another aspect of the framework relates to levels of analysis of man's response to mechanical energy, which is developed subsequently in this introduction and forms the outline for Response to Mechanical Energy. Third, we offer a representative sample of current information concerning the consequences of exposure to mechanical energy, with particular emphasis on costs—ranging from tissue damage to annoyance and possible mental disturbance. This information is presented in Response to Mechanical Energy. Our final goal is dependent upon achievement of the preceding three goals; the fourth goal is to provide the student with a firm base for further reading and research into the biologic effects of mechanical energy exposure.

Attention will be focused upon three categories of mechanical energy normally present in our environment: (1) energy in sound fields, including airborne shock waves; (2) energy in mechanical vibration, applied to man by some solid structure; and (3) energy in the impacts between man and some fixed, solid object or in the impacts of various solid projectiles upon some part of the body surface. Our environment may contain energy in only one of these three forms at a given time; however, any two or even all three forms of energy are often present simultaneously.

Response analysis framework

Man's responses to mechanical energy exposure may be categorized in terms of (1) the level of biologic organization at which the response is described and (2) the presence or absence of interaction between the energy source and the receiver.

Levels of biologic organization. Biologic systems can be subdivided into elements of decreasing complexity, ranging from societies through organs to cells. In Response to Mechanical Energy responses are described in terms of tissue and organ changes, organ system (physiologic system) responses, and behavioral system responses. Modification of the fine sensory elements of the inner ear following intense sound exposure is an illustration of a mechanical energy–induced tissue change. The auditory system and the visual system are examples of physiologic systems; a physiologic system response to mechanical energy is the protective reflex of the middle ear elicited by intense sound. Integrated activities of the whole man are considered under the heading of behavioral systems. Speech perception or complex task performance are behavioral system responses that can be disrupted by mechanical energy. Next we proceed to consideration of possible interaction between an energy source and an energy receiver and, at the same time, illustrate some advantages of categorizing responses in terms of levels of biologic organization.

Interaction between sources and receivers. Our approach to this chapter is within the *systems* framework. Biologic systems are defined as sets of components that operate together in order to achieve some *goal:* a system is defined by what it *does.* By operating in certain ways, some systems modify the inputs that they receive. The label *closed loop (feedback) system* is used in these cases. A familiar example of a closed loop system is the pupillary reflex, which controls the amount of light entering the eye. As we consider increasingly complex levels of biologic organization, an increasing number of closed loop systems can be identified. At the physiologic systems level, reflexes are elicited that limit potential damage from loud sound exposure. At the behavioral systems level, human beings act to limit mechanical energy exposure or to eliminate mechanical energy sources. Man should *not* be considered a passive recipient of environmental energy; rather, in everyday situations, there is a continuous interaction between man and his environment.

Comment

The information presented in this chapter derives from diverse sources that cut across several academic disciplines. Physical description of environmental mechanical energy and analysis of human response biodynamics are the domain of the physicist and biophysicist. Physiologists and anatomists have been primarily concerned with description of body structures and analysis of functional action at the organ systems level. Audiologists and psychologists have compiled information concerning behavioral responses to stimulation. Data describing the various ways in which body functions can be disrupted by mechanical energy exposure have been gathered by medical scientists.

If we spoke with scientists who work in the areas of sound, vibration, and impact, few would say, "I am only interested in receptor physiology," or, "I restrict my attention to collecting audiograms." Rather, these scientists would probably express *interdisciplinary* interests. Although our focus is on the physiologic aspects of environmental science, disciplinary lines in this area are vague and an eclectic approach is required.

CHARACTERISTICS OF SOUND, VIBRATION, AND IMPACT FIELDS

A clear understanding of the biologic effects resulting from stimulation by mechanical energy is dependent upon comprehension of the basic principles of mechanical energy production. There is a symmetry between the concepts presented in this section and those developed in The Human Body as a Mechanical Energy Receiver, where biologic reception of vibratory energy is considered.

Energy transmission systems

Identification and measurement of man's responses to mechanical energy are facilitated by thinking in terms of the following system:

energy source → transmission path → energy receiver

for example,

sound source → air → human body or a specialized receptor

In this example, the sound source determines the properties of the sound field, including the energy level initially released. However, as the wave propagates (travels) from the source to the receiver, the transmission path alters the energy level of the sound field as well as some of its properties. These alterations are determined by the characteristics of the medium through which the energy travels and by several other attributes of the environment. Thus the sound in immediate contact with the surface of the receiver (the human body or one of its receptors) differs from the one generated by the source; the values of the *parameters* (frequency, intensity, and the like) of the sound field at the body surface are determined by the combined effect of source characteristics and transmission pathway properties.

For most purposes, it is sufficient to quantitatively specify the parameters of the sound field at the body surface. In practical situations, these parameters can be determined by first quantitatively describing the source sound field and then applying corrections for transmission path effects. Furthermore, in most instances where actual sound control is needed, we must know the values of the source parameters and path effects to design successful control devices or procedures. This is particularly true if people move about in the sound field or if many individuals are exposed to the field generated by one source.

If we now consider situations in which the parameters of the sound field at the surface of the body are known, a simplified system can be conceived as follows:

$$\begin{array}{c}\text{energy field}\\ \text{(at the body surface)}\end{array} \rightleftarrows \begin{array}{c}\text{energy receiver}\\ \text{(human body)}\end{array}$$

In this case, source and transmission path effects have been taken into account, and the parameters of the energy field at the body surface have been quantitatively defined. Also, an arrow from the receiver to the source has been added. This arrow indicates possible interaction between the source and the receiver, in a closed loop system.

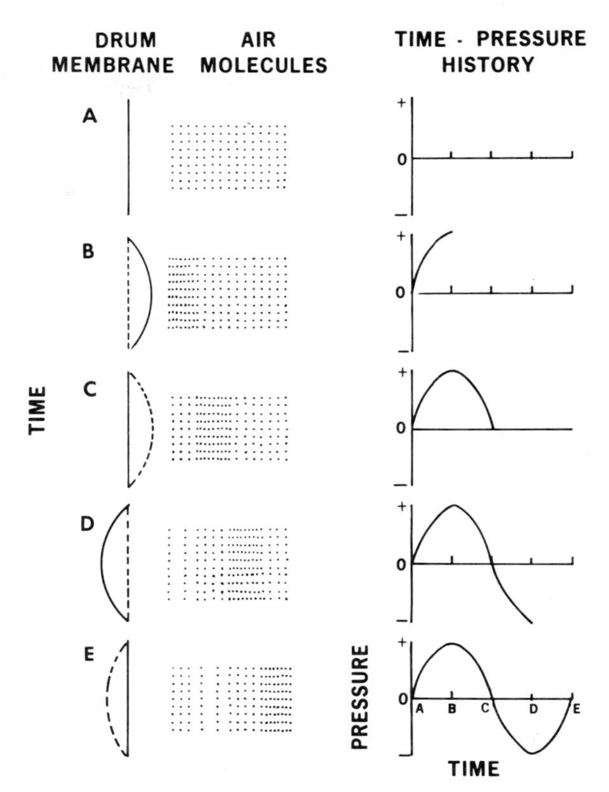

Fig. 5-1. Production of sound waves.

Fundamentals of mechanical energy generation

Generation of mechanical energy is ultimately dependent upon *displacement* of some physical structure. The displacement produced by a mechanical energy source may directly activate a biologic receptor or the displacement may be propagated through some medium (such as sound pressure waves in air) as described previously.

A mechanical energy source is a physical system that transforms some form of energy received by the source into a form of mechanical energy. Our interest is directed toward forms of mechanical energy that elicit some type of response from a man. For example, a loudspeaker transforms electric energy into mechanical energy in the form of sound waves that we can detect. Consider what happens when a musician strikes a drum. The drum membrane (head) is the mechanical (sound) energy source, and the blow by the musician is the energy input to the source. The characteristics of the mechanical energy field produced by a drum are dependent on the drum membrane's physical properties of *inertia* and *elasticity* (as well as the intensity and time course of the blow). Inertia is the property of all physical systems containing mass and refers to the tendency of a mass to resist motion change: when at rest a mass resists being set into motion and when in motion it resists changes in velocity or direction of motion. Opposing inertia is the physical property of elasticity, which describes the tendency of a body to maintain its original size and shape. Although we usually associate elasticity with objects such as springs or rubber bands, it is important to keep in mind that gases exhibit an equivalent property (compressibility) in some degree also.

The manner in which inertia and elasticity operate to produce a mechanical energy field is illustrated in Fig. 5-1. At time A the drum head is at rest and the air molecules surrounding the drum are uniformly distributed. Immediately after time A the drum is struck. At time B the drum membrane has reached its maximum rightward displacement. The magnitude of this displacement depends on the inertia and elasticity of the membrane and other factors. As it moves during the interval from time A to time B, the membrane pushes the air molecules in the path of its motion together, producing a concentration of molecules or a *conden-*

sation phase in the transmission medium (air). This condensation can be viewed as an increase in air pressure at a particular point in space near the drum, as illustrated by the graph of the *time-pressure history* on the right side of the figure. At time C the membrane's elasticity has produced a return to its original position and the concentration of the air molecules surrounding the membrane is the same as at time A. Because of the membrane's inertia at time C it does not stop in its original resting position; rather, it continues past the resting position to some maximum leftward displacement until elasticity matches inertia at time D. The air molecules are pulled apart during motion from time C to time D, producing a *rarefaction phase*. At time E the elastic forces have produced a return to the original position. The membrane continues to vibrate until frictional forces absorb all of the energy in the original blow. Pressure changes at a particular point adjacent to the drum, illustrated by the time-pressure history graphs, describe a *sinusoid* during the one *cycle* of membrane motion.

Sound is propagated through the air in all directions from a vibrating source. This propagation results from the fact that the air molecules communicate their motions to one another in a progressive fashion. Thus the condensation and rarefaction phases in the medium are propagated as compression waves (Figs. 5-1 and 5-15). Sound wave propagation is dependent on the medium's physical characteristics of inertia, elasticity or compressibility, and friction. Moreover, *density* is also a significant parameter when considering a medium, as illustrated by the fact that sound propagates approximately four times faster in water than in air, under ordinary conditions.

Parameters of mechanical energy: sound

The basic parameters used to describe acoustic energy field exposures include *frequency, intensity,* and *duration*. A fourth parameter, which is of interest in particular situations, is *phase*. In this section, we present most of the basic concepts used in mechanical energy measurement by considering sound. The additional parameters re-

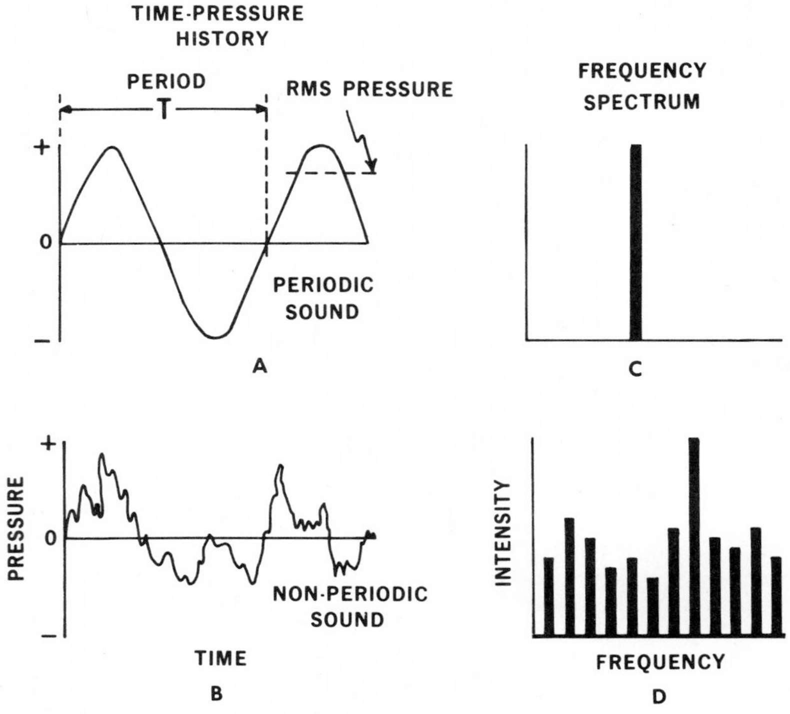

Fig. 5-2. Parameters of sounds: frequency and intensity. **A** represents the time-pressure history of a periodic sound; **B** illustrates a nonperiodic sound. Possible frequency spectra that could be associated with the time-pressure histories are given in **C** and **D.** A frequency spectrum indicates the relative amount of energy within particular bands of frequencies.

quired for specification of vibration and impact exposures are presented subsequently.

Frequency

The concept of *oscillation* at a particular frequency is readily understood, perhaps because we have all had experiences such as playing on swings or watching clock pendulums. A simple oscillation can be illustrated by a sinusoidal function, as in Fig. 5-2, *A*. The number of oscillations (cycles) per unit time defines the frequency, specified by the units *cycles per second* (cps) or the equivalent *Hertz* (Hz).

Sounds are characterized by the relationship between their major frequency components and the normal range of human hearing, usually given as 20 to 20,000 Hz. *Audiofrequency* refers to frequencies within the range of human hearing, *ultrasound* frequencies are those above 20,000 Hz, and *infrasound* contains frequencies below 20 Hz.

Period (T) is defined by the time interval required for one complete oscillation (see Fig. 5-2, *A*) and is the reciprocal of frequency. A sound is said to be *periodic* to the extent that the value of the period (and hence the frequency) remains constant throughout its duration. Most environmental sounds are *nonperiodic* since they contain many frequency components. Fig. 5-2, *B*, illustrates a nonperiodic sound.

Since many effects of interest to the environmental physiologist are frequency dependent, it is critical to specify the frequencies present in a sound field. These data are usually presented in the form of a *frequency spectrum,* which is a plot of relative intensity against frequency. Spectra that could be associated with periodic and nonperiodic sounds are illustrated in Figs. 5-2, *C* and *D*. Frequencies, on the horizontal axis, are usually specified in successive octave, half-octave, or third-octave frequency bandwidths or in single cycles. An octave refers to an interval of frequencies where the lower frequency is one half the higher frequency. For an octave band spectrum, the first line (as in Fig. 5-2, *D*) or point (as in Fig. 5-7) might represent the intensity contained in sound frequencies between 37.5 and 75 Hz, the second line illustrating intensity between 75 and 150 Hz, and so on.

Intensity

It would be desirable to measure sound intensity in terms of the displacement of air particles, their average velocity during oscillation, or the energy contained in their motion. At the magnitudes associated with ordinary sound, however, satisfactory technics are not available for performing these measurements. Fortunately, the pressure changes in the sound wave can readily be determined; pressure measurements obtained with special microphones provide the basis for assessment of sound intensity. It is interesting to note that the dynamic pressure changes associated with sound differ by several orders of magnitude from those static pressures that we normally encounter. Normal atmospheric pressure is about 10^6 dynes/cm^2, whereas the pressure fluctuations produced by sounds near the absolute threshold for hearing are in the order of 10^{-4} dynes/cm^2.

For a simple sinusoidal wave, pressure amplitude can be specified as the maximum value of the pressure increase (peak pressure, P_{max}) above the ambient (background) pressure or, alternatively, as the root-mean-square pressure (rms pressure, P_{rms}), as shown in Fig. 5-2, *A*. The relationship between these measures is:

$$P_{rms} = \frac{P_{max}}{\sqrt{2}} = 0.707\ P_{max} \qquad (1)$$

The preceding relationship refers to sinusoids. For complex waves, which contain more than one frequency, there is no unique relationship between peak pressure and rms pressure. Peak pressure varies widely in a complex sound wave; rms pressure is more stable and is therefore the preferred measure when describing environmental sounds, which usually contain many frequencies. Another reason for using the rms measure is that under specific conditions, the power present in a sound wave can be related to the rms pressure. Power (J) in microwatts/cm^2 is given by:

$$J = \frac{P^2}{10\ R} \qquad (2)$$

where P = pressure
R = specific acoustic resistance (a quantity determined by the density and sound propagation velocity of the medium)

As with pressure, the power associated with ordinary environmental sound is very small compared to power levels that we encounter in our everyday activities. Sounds that are sufficiently intense to destroy the sensory cells of the ear may produce as little as 0.002 watt.

The ear operates over an enormous range of sound intensities: the most intense sound pressures that can be received without producing damage are approximately 1 million times greater than the least intense sound pressures that can be

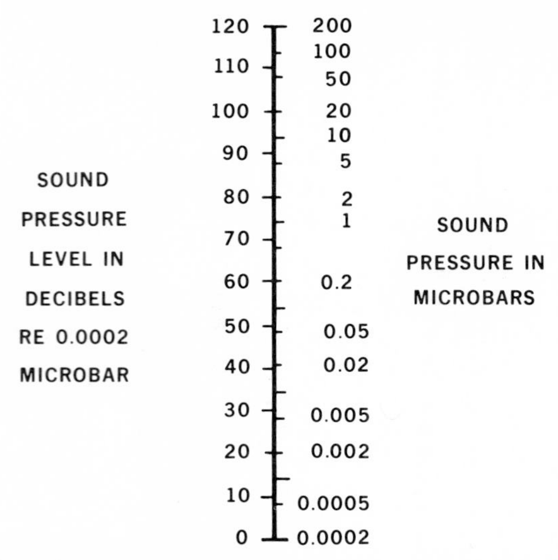

Fig. 5-3. The relationship between sound and pressure measured on linear (microbar) and logarithmic (dB SPL) scales.

detected. In order to conveniently deal with this range of physical sound intensities, a logarithmic scale has been developed. With a logarithmic scale, equal intervals are specified equal increments in the exponent of some base, for example 10^1, 10^2, 10^3, and so on. Unfortunately, a logarithmic scale has a major drawback because there is no zero point. Therefore, sound intensities must be specified with respect to some standard reference intensity. Sound intensity is commonly scaled as *sound pressure level (SPL)* in *decibels (dB)* according to the formula:

$$SPL = 20 \log_{10} \frac{P_1}{P_0} \qquad (3)$$

where P_1 = pressure of the sound in question
P_0 = reference pressure

The least pressure oscillation of sound in air that can normally be detected by the ear has a level of approximately 0.0002 dyne/cm^2 (which is equivalent to 0.0002 millionths of normal atmospheric pressure or microbar). The most widely employed reference pressure for sound measurement is P_0 = 0.0002 microbar. The relationship between SPL in dB, re 0.0002 microbar, and sound pressure in microbars is illustrated in Fig. 5-3.

For a complex (nonperiodic) sound, the pressure levels within various frequency bands can be measured, and a frequency spectrum can be constructed, as described before. The intensity of the total complex wave can also be specified with a single number as the *overall sound pressure level (OASPL)*. Equal weight is given to the frequencies between 20 and 20,000 Hz by the OASPL scale. (The OASPL scale is "flat" across this frequency range.) Several other single-value scales of complex sound intensity are in use or have been proposed. These scales apply different weights to the various frequencies. For example, the dBA scale gives more emphasis to the midrange of the audiofrequency spectrum than to the extremes. This scale approximates the ear's loudness response at relatively low sound pressure levels. The dBC scale is nearly flat across the audiofrequency range and is similar to the ear's loudness response at high sound levels. Sound level meters in common use have weighting networks that allow determination of sound intensities in dBA or dBC.

Sound power refers to the total sound energy radiation in all directions from a source per unit time. The power of sound sources is measured in watts or can also be expressed by a logarithmic scale of *sound power level (PWL)* according to the formula:

$$PWL\ (dB) = 10 \log_{10} \frac{W_1}{W_0} \qquad (4)$$

where W_1 = power of the source
W_0 = arbitrary reference power of 10^{-13} watt

As spherical sound waves diverge from a sound source, the surface area of the wave front increases as the square of the radius. Consequently, the energy per unit area decreases inversely as the square of the distance from the source. This is known as the *inverse square law* of intensity. Stated in terms of the scales for sound intensity measurement defined before, there is a 3-dB decrease in PWL (equivalent to a 6-dB SPL decrease) with each doubling of the distance from the source.

Time

Variations in sound pressure as a function of time are represented by time-pressure histories, such as those in the righthand column of Fig. 5-1. We may wish to consider the temporal pattern of the pressure change directly, as in the case of a shock wave (Fig. 5-6), or we may note the temporal change of an envelope within which a complex sound wave is contained. (Imagine a sine wave that is varied in amplitude from zero to maximum and back to zero; the sound envelope is obtained by drawing lines between adjacent peaks of the wave.)

Duration of exposure to a sound is critical when

considering possible reactions. As discussed in Response to Mechanical Energy, brief exposure to a particular sound may elicit no adverse reaction, whereas extended exposure to the same sound could result in irreversible damage. The effects of total duration also vary significantly if the sound is interrupted or if other parameters of the sound change during the exposure period.

Parameters of vibration energy

In this section on vibration and the following section on impact, we present only parameters that differ from those defined for sound. Here we consider vibratory frequencies between 0.1 Hz and several hundred Hz. This is the range of frequencies that appears to be of interest for whole body or localized vibrations.

Vibratory motion is usually quantified in terms of displacement, velocity, acceleration, or some higher derivative of body motion, with respect to a specific reference. Alternatively, motions can be specified by the stimulus force or pressure that leads to displacement of the whole body or a particular body part. The specified reference may be the body at rest or in some steady-state condition of velocity or acceleration. Two classes of difficulties are inherent in vibration measurement. First, in the real world (as opposed to the laboratory), the body is usually acted upon by a complex force spectrum, and a particular steady state condition of displacement, velocity, or acceleration is unlikely to be maintained for a long period. Consequently, determination of a reference point from which vibration can be measured may be difficult or impossible. Second, because of their differing physical properties, different parts of the body react (move) in different ways to vibratory force inputs.

Vibratory amplitude is usually defined in terms of linear or angular acceleration. One unit of linear acceleration is centimeters per second per second (cm/sec^2); linear acceleration is also frequently expressed in *G's* as multiples of the earth's gravitational constant (978 cm/sec^2). Angular acceleration is given in degrees per second per second (deg/sec^2). As with sound, rms amplitudes are generally cited, although peak-to-peak amplitudes may be used for simple oscillations.

Orientation of motion with respect to the body is a critical factor in determining the effects of vibratory energy. Vibration can have three linear and three rotational degrees of freedom, specified with reference to the human skeleton (see Fig. 5-4). Linear vibratory acceleration may be in the longitudinal (head-foot, G_z) axis, the fore-aft (chest-back, G_x) axis, or the lateral (side-to-side, G_y) axis. Rotational vibratory acceleration may be separated into pitch (rotation around the y axis), roll (rotation around the x axis), and yaw (rotation around the z axis).

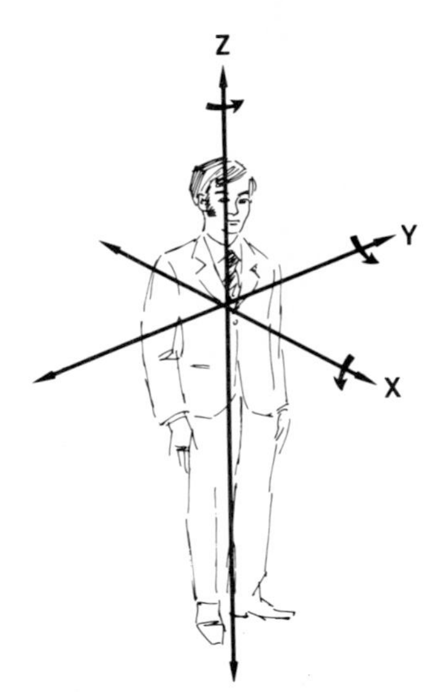

Fig. 5-4. Rectangular coordinate system for specifying orientation of motion with respect to the body. *Z* is the longitudinal body axis; *X* is the transverse axis; *Y* is the lateral axis. Vibratory motions defined with respect to these three axes may be rotational or translational.

Parameters of impact energy

In the following, we consider separately the parameters of mechanical impacts or shocks resulting from contact with solid structures and airborne impacts (impulses) from shock waves and blast.

Mechanical impact

Mechanical impacts occur when a body is rapidly accelerated or decelerated (for example, when a moving body is brought to rest by contact with a stationary structure). The most frequently encountered scale for mechanical impact is *deceleration* specified (as for vibration) in multiples of earth's gravitational acceleration field according to the formula:

$$G = \frac{V^2}{2 \times g \times S} \qquad (5)$$

where V = body's velocity just prior to impact
g = gravitational constant
S = distance traveled during the impact

A report by Kiel,[1] somewhat prosaically entitled "Hazards of Military Parachuting," illustrates this calculation:

> During one of the battalion drops, from 1,200 feet on a clear, relatively warm day, an observer noted what appeared to be an unsupported bundle falling from one of the C-119 airplanes; no chute deployed from the object. The impact looked like a mortar round exploding in the snow. When the aid men reached the spot, they found a young . . . paratrooper flat on his back at the bottom of a 3½ foot crater in the snow, which consisted of alternating layers of soft snow and frozen crust. He could talk and did not appear injured; nevertheless, he was air evacuated to a hospital. His only injuries were an incomplete fracture of the clavicle, a chip fracture of the second lumbar, and a few bruises. He was released from the hospital in time to return south with his unit.

Stapp[2] calculated the paratrooper's impact deceleration at about 141 G. This calculation is based on estimates of the paratrooper's velocity just prior to impact (54.25 m/sec) and the stopping distance during the impact (1.065 m).

Clearly, the major variables to be considered when calculating impact deceleration are the velocity change produced by the deceleration and the stopping distance. Impact duration increases with stopping distance: an impact material that increases the stopping distance also increases the duration of the impact. The energy in a deceleration is dependent upon the mass of the body and its velocity. For a given energy, the peak deceleration reached during impact depends on the duration of the impact. Fig. 5-5 presents two deceleration profiles. The same amount of energy (represented by the area under curve) is dissipated in each case. However, the long-duration impact reaches a lower peak deceleration value than the short-duration impact. These observations have clear implications for ameliorating impact injuries in accidents such as automobile collisions.

A second metric for impact is pressure P, calculated by the formula:

$$P = \frac{M \times G}{A} \qquad (6)$$

where M = body mass
G = average deceleration
A = area of impingement

Compare the impacts generated by falls when the body decelerates in a feet-first position and a prone position; to the extent that the body behaves as a rigid structure, the pressure (and consequently the damage) should be considerably greater with feet first than with body prone.

Airborne impact

Shock waves are large amplitude sound waves in which the pressure is so great that the compressibility of the air is nonlinear. Explosions, supersonic aircraft, and sudden insertion of the body into a high-velocity airstream generate shock waves. Microphones and pressure gauges are used

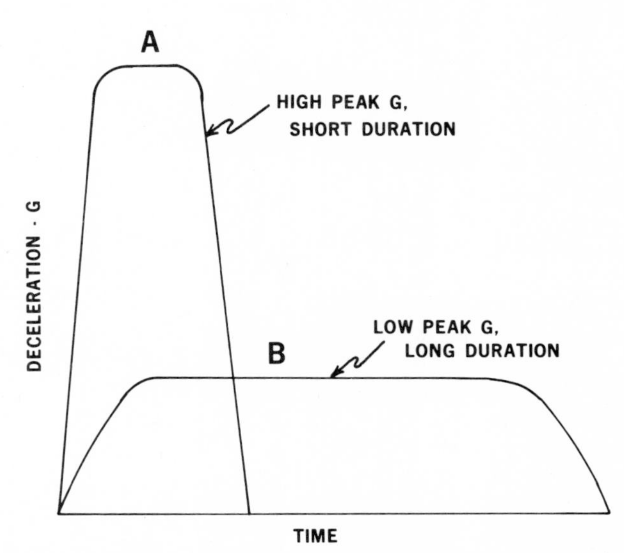

Fig. 5-5. Impact deceleration profiles. Both impacts dissipated the same amount of energy. *A* reaches a higher peak G because the energy is dissipated over a shorter period of time than in *B*.

to measure airborne impact, and these measurements can be in terms of time-pressure histories or frequency band spectra (see Figs. 5-1 and 5-2). The vertical axis for an airborne impact time-pressure history is overpressure, scaled in pounds per square foot (lbs/ft^2) or an equivalent (such as kgm/cm^2, lbs/in^2, psi, and so on). Overpressure refers to pressure changes above ambient pressure. Frequency spectra and time-pressure histories for a sonic boom are presented in Fig. 5-6.

ENVIRONMENTAL SOURCES OF MECHANICAL ENERGY

In this section, we discuss environmental sources of sound, vibration, and impact. Airborne shock waves, such as those emanating from supersonic aircraft and explosions, will be considered under Impact Sources, whereas relatively continuous (lasting at least seconds) sounds will be discussed in the following paragraphs.

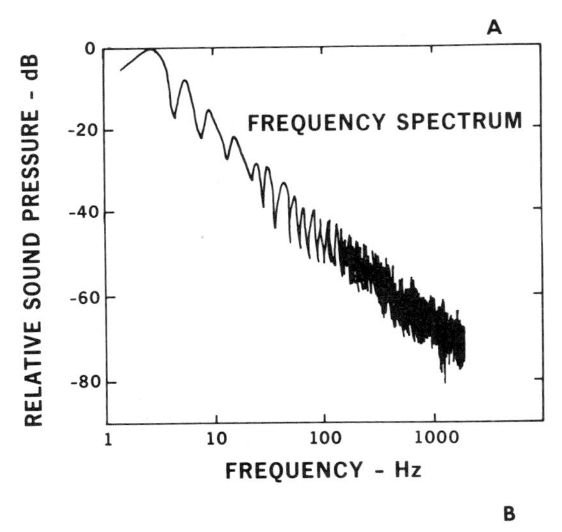

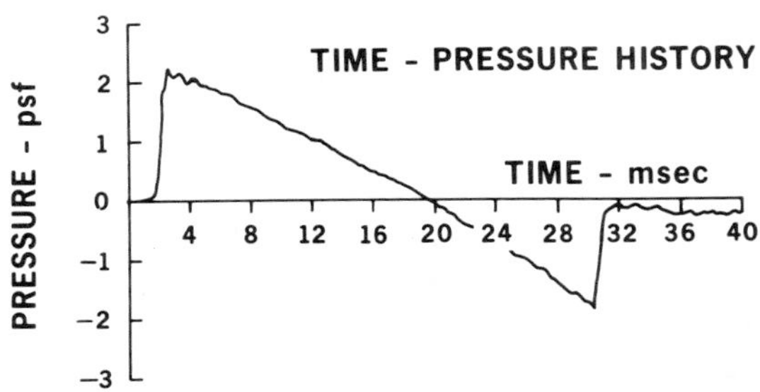

Fig. 5-6. Sonic boom measurement. **A** is the frequency spectrum of a sonic boom; **B** represents a typical time-pressure history for this event and is called an N wave.

Sound sources

Biologic consequences of sound exposure are dependent on a wide variety of factors, ranging from physical characteristics of the particular sound to the susceptibility of the individual or the goals being pursued at the time of sound exposure. For example, reactions to aircraft sound exposure differ when a person is attempting to sleep at home as opposed to when he is embarking on a transoceanic flight. In this section, we focus on physical sound parameters and reserve extra-acoustic considerations (such as attitudes toward sound sources) for Response to Mechanical Energy. Overall intensity, frequency distribution of the sound intensities, duration, and onset and termination rate are important parameters when predicting the responses that a sound is likely to elicit.

In the following, we emphasize sound produced by various manmade tools and devices. We do not discuss people as potentially annoying sound sources. However, in some situations (such as near playgrounds), sounds from human beings are more disturbing than those from other sources.

Home

Some common sounds present, at least intermittently, in the home environment are described in Fig. 5-7. The intensity levels produced by a particular type of apparatus may vary widely, depending on the model or the state of repair. Manufacturers may deliberately construct a machine, such as a vacuum cleaner, to produce high sound intensities with the expectation that the buyer will assume sound intensity is related to power and hence efficiency. Characteristics of the sound produced by an apparatus vary also with the acoustic properties of the space within which it is used; the sound at an observer's ear differs when the source is located in a highly reverberant space (a tiled bathroom) as opposed to a sound-absorbing space (a Victorian parlor).

Perhaps a word should be included about "home music centers." A revolution in music has occurred as a result of developments in sound recording and amplification technics and apparatus. Many people enjoy music played at levels far in excess of those used one or two decades ago. The sound produced in the home music center varies widely, depending on numerous factors such as the "taste" of the listener, the equipment employed, and the room in which the equipment is located. We have measured sound levels of up to 105 dBA for one set-up during the playing of a popular record.

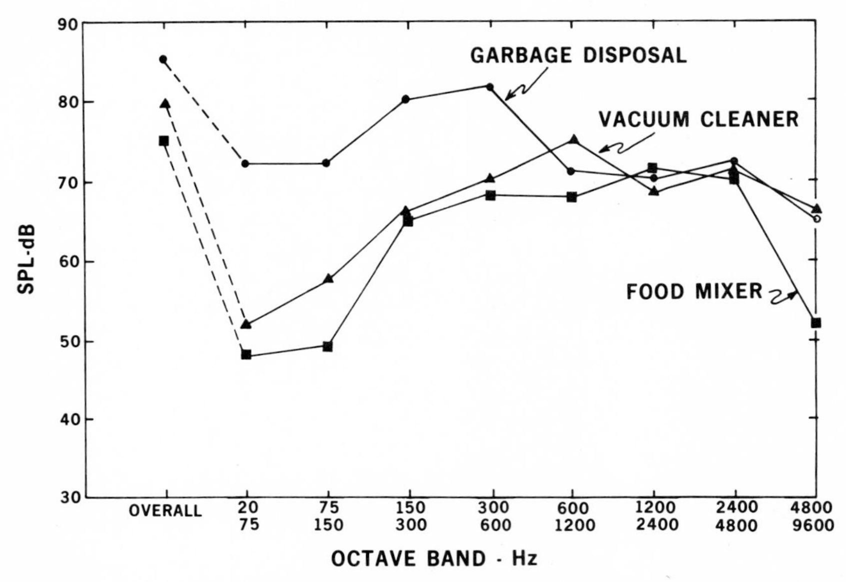

Fig. 5-7. Typical sound sources in the home environment. The curves give overall sound pressure levels and octave-band spectra. (Curves constructed from data of Mikeska, E. F.: Noise Control **4**:38, 1958.)

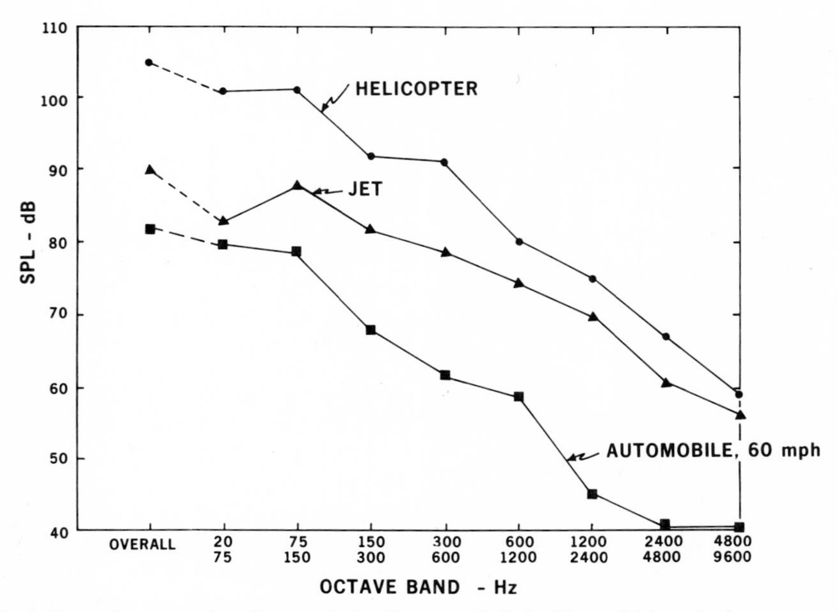

Fig. 5-8. Overall and octave-band sound levels recorded inside vehicles during operation. Aircraft recordings were from the aft cabin. Automobile sound was measured with the windows closed on a smooth road. (Based on data from Miller, L. N., and Beranek, L. L.: Noise Control **4**:19, 1958; Wiener, F. M.: Noise Control **6**:13, 1960.)

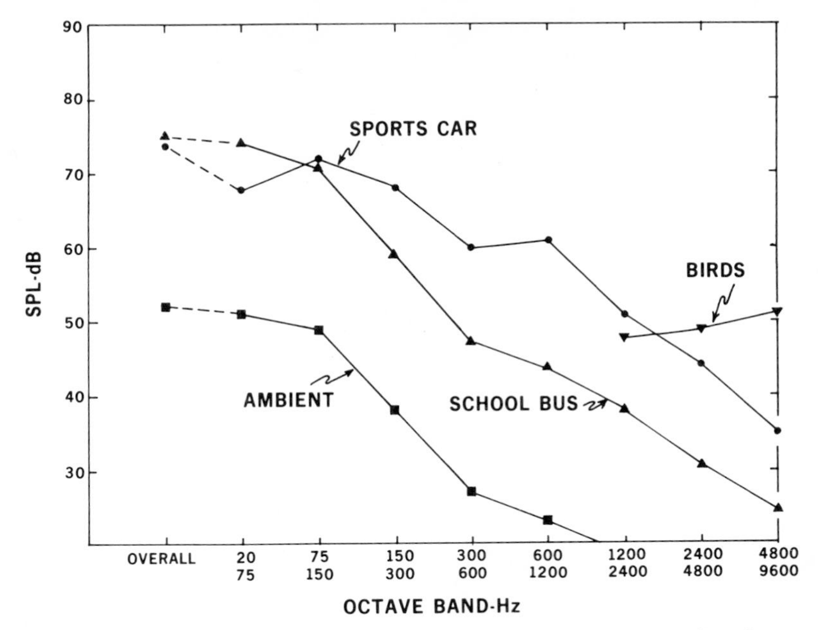

Fig. 5-9. Transportation sound levels recorded in a quiet residential area during the morning. *Ambient* refers to the background of mixed sounds. (Curves constructed from data of Veneklasen, P. S.: Noise Control 2:14, 1956.)

Casual observation suggests that sound levels in the same range can easily be found in bars and discotheques that cater to the younger members of the population.

Transportation

A large proportion of the unwanted sound in our environment derives from transportation vehicles. Let us consider two general situations: (1) exposure levels inside a vehicle and (2) sounds propagated from vehicles that come to a man who remains in a fixed place, such as a home or office. Spectra for sound inside various types of vehicles are presented in Fig. 5-8. Note the difference between aircraft and ground vehicles. Remember also that sound exposure inside vehicles during travel is relatively continuous. Potential annoyance may be ameliorated by the fact that ordinarily a person chooses a particular form of transportation to achieve some goal and that we associate travel with excitement and adventure.

However, when transportation sounds propagate to a person at his home, the situation is quite different. In this case, the sound is intermittent and probably not congruent with the immediate goals being pursued. Duration of the sound can be important in determining reactions, particularly for aircraft noise. Cohen and Ayer[3] reported that sound intensities exceeding an arbitrary speech interference level for up to 60 seconds during a flyover were not unusual in the neighborhood of an airport. Typical spectra for selected transportation sounds recorded at residences are presented in Fig. 5-9. In general, jet aircraft sound contains more energy toward the high-frequency end of the spectrum than does ground transportation sound. Also, frequency spectra produced by turbine or propeller aircraft are usually dominated by a specific frequency, representing the rotation rate of the propeller or turbine.

Use of helicopters and other types of vertical takeoff and landing (VTOL) aircraft has been proposed to alleviate congestion of ground transportation in cities. Although the frequency spectrum for a helicopter appears similar to the spectrum of a fixed-wing propeller airplane in Fig. 5-8, these curves do not show the differences between these sources at the infrasound and low-audiofrequency sound range. Annoyance resulting from these low-frequency sounds is a disadvantage of the increased use of helicopters. Another disadvantage of VTOL types of aircraft is that they remain in a particular location for extended periods during takeoff and landing; therefore, the resulting sound exposures are of greater duration than with ordinary airplanes.

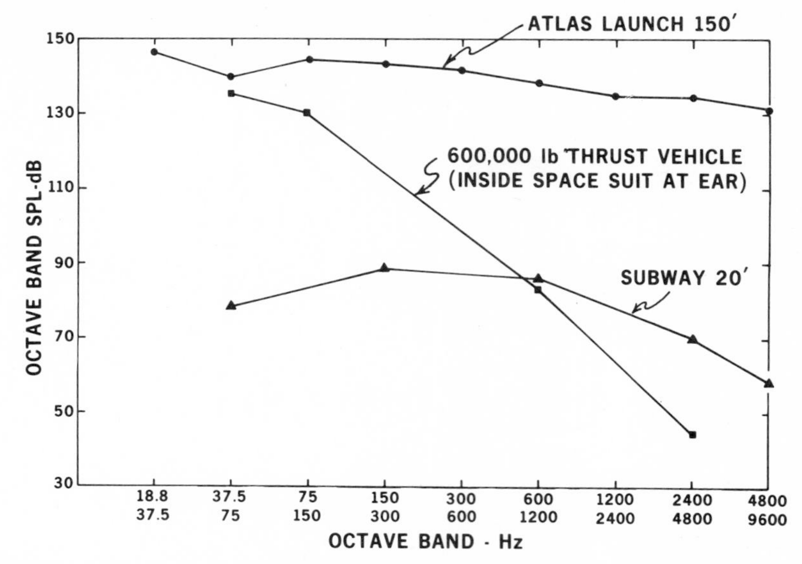

Fig. 5-10. Spectra for sound produced by space vehicles. Subway sound levels are given for comparison. Note differences between the spectra for these different types of vehicles in the infrasound and low audiofrequency sound range. (Curves constructed from data of von Gierke, H. E., and Nixon, C. W.: Noise effects and speech communication in aerospace environments. In Randel, H., editor: Aerospace medicine, Baltimore, 1971, The Williams & Wilkins Co.)

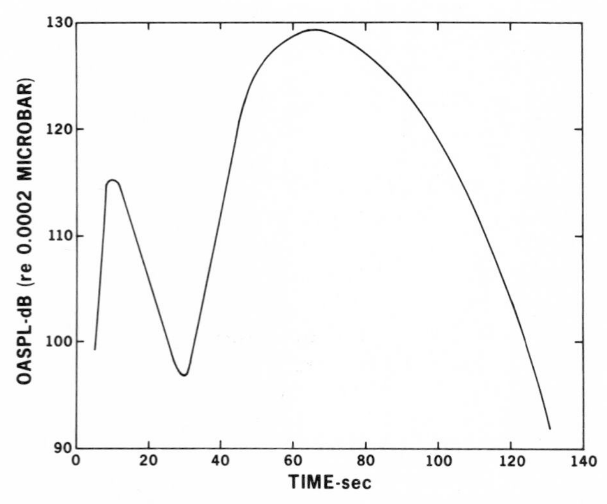

Fig. 5-11. Overall sound pressure level change as a function of time after launch ignition at crew stations inside a space vehicle. (Curves constructed from data of von Gierke, H. E., and Nixon, C. W.: Noise effects and speech communication in aerospace environments. In Randel, H., editor: Aerospace medicine, Baltimore, 1971, The Williams & Wilkins Co.)

With increasingly more powerful rocket engines constructed for use in space exploration, new sound energy environments have been developed or are anticipated. Figs. 5-10 and 5-11 illustrate the sound energies produced by representative space vehicles. Fig. 5-10 indicates that a planned 600,000-lb thrust vehicle and the Nova type boosters produce high energy levels in the infrasound and low audiofrequency sound range. Fig. 5-11 illustrates fluctuations in OASPL at crew stations in a space vehicle, such as the Apollo, during launch. The initial peak is related to the propulsion system while the space vehicle is on the ground or moving slowly, and the second broad peak is mainly a function of air turbulence, produced by pressure fluctuations in the boundary layer as air rushes over the vehicle skin at high vehicle velocity.

Offices and factories

Sounds associated with offices and industrial operations differ from those present in the average home because the machinery and office equipment sounds are relatively continuous, whereas sound from devices operated in the home (such as vacuum cleaners and dishwashers) is intermittent and infrequent. Much of the data for office equipment and industrial machinery is at least 10 years old. As a result of increasing awareness of en-

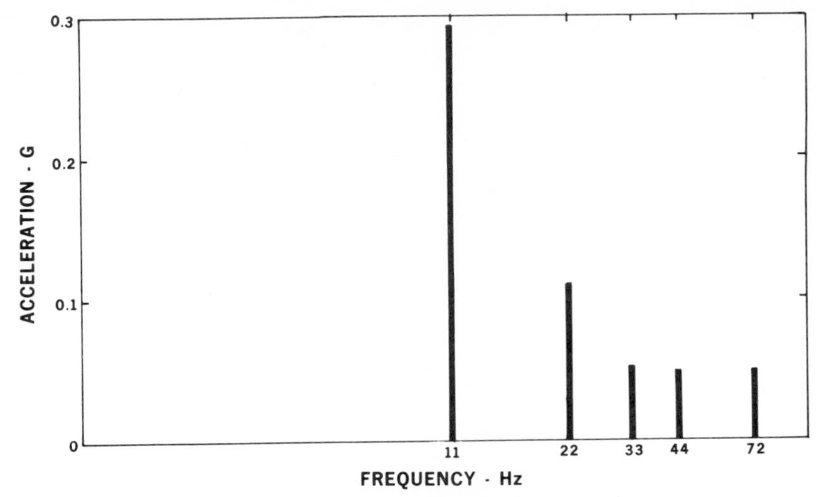

Fig. 5-12. Vibration frequency spectrum for a helicopter during operation at 120 knots. (Based on data from Rosenberg, G., and Segal, R.: The effects of vibration on manual fire control in helicopters, Technical Report 1-168, Philadelphia, 1966, Franklin Research Institute Laboratories.)

vironmental pollution in general and sound pollution in particular, and with legislation directed at limitation of industrial sound exposure, the situation concerning sound levels in offices and factories is highly variable; presentation of "typical" figures does not appear useful.

Vibration sources

Localized vibration may be produced by contact between the body and a vibrating source such as a chain saw, electric shaver, power drill, or electric toothbrush. Contact between a vibrating source and the skin produces traveling waves radiating from the contact point on the skin surface and through the deep layers, as described by von Bekesy.[4] The amplitude of vibration received by a person may vary widely, depending on where the vibration is applied, the firmness of contact between the source and the receiver, and numerous other factors. Whole-body vibration can result from exposure to intense low-frequency sound fields or exposure of the whole body to the vibrations of heavy industrial machinery. Transportation vehicles are the main source of whole-body vibration. The amplitude of whole-body vibration received by a man depends upon many of the same considerations given for localized vibration.

Surface transportation produces vibration as a result of irregularities in the path over which the vehicle travels and because of the inefficiency of the power plant. (It seems curious that we continue to produce piston engines for vehicles that have a basically inefficient design, in that they must convert linear motion into rotational motion, with considerable efficiency reduction because of power losses in the form of vibration.) Vibrations can be a problem in vehicles such as tractors, trucks, and earth-moving equipment. Helicopters are notorious as vibration sources. Fig. 5-12 illustrates the vibration frequency spectrum for a large helicopter during operation at high velocity. In fixed-wing aircraft, particularly when operated at high speeds and low altitudes, intense vibration may result from buffeting (air turbulence). Finally, very intense (+25 G) vibration or repetitive impact accelerations may occur during escape maneuvers from a military aircraft by means of an ejection seat, when after the initial ejection impact acceleration the man oscillates in the windstream. In the preceding situations, many inputs beside vibration influence the man, and it is difficult to extract pure vibration effects. However, several types of laboratory devices have been developed to allow assessment of responses attributable to vibration.[5]

Impact sources

Impact with stationary structures would seem to occur mainly as a result of accidents, such as falls or traffic mishaps. Of course, one may deliberately expose himself to impact environments by participating in contact sports. Special situations in which mechanical impacts occur are as-

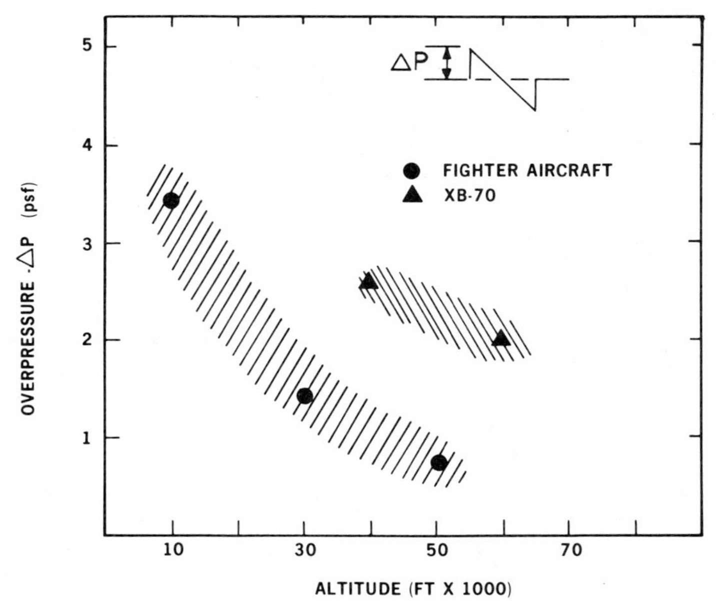

Fig. 5-13. Sonic booms: overpressure as a function of airplane altitude and size. Booms produced by large aircraft (such as XB-70) are indicated by the triangles; booms from smaller aircraft are indicated by the circles. (Modified from von Gierke, H. E., and Nixon, C. W.: Noise effects and speech communication in aerospace environments. In Randel, H., editor: Aerospace medicine, Baltimore, 1971, The Williams & Wilkins Co.)

sociated with emergency ejection from military aircraft, space vehicle operation (particularly during reentry), and parachuting.

Devices such as jackhammers, chipping hammers, and riveting tools produce localized impact exposures in industrial situations. Such repetitive impact exposures are similar to continuous vibration but are characterized by individual high amplitude peaks.

Airborne sources of impact include explosions and detonations, gunfire, and sonic booms, as well as the more prosaic hammering and pounding. A great deal of interest has centered around sonic booms as a result of debate concerning the commercial supersonic transport. Fig. 5-13 presents typical levels of overpressure during sonic boom as a function of aircraft altitude and type.

THE HUMAN BODY AS A MECHANICAL ENERGY RECEIVER

There have been many studies on the transmission of vibratory mechanical energy within and through the human body. The mechanical energy sources studied include those inside the body (such as heart sounds) as well as the numerous sources in the environment external to the body. In general, each study has had a specific and different purpose and each has been concerned with a different source of vibratory mechanical energy. Regardless of the differences in the general purpose of these studies, each has finally centered on just one question: How is vibratory mechanical energy received, propagated, changed in form, and dissipated within the human body? More generally, how can we describe the human body as a *dynamic mechanical system?* Once this question is fully answered, it will make little difference which specific case of vibratory mechanical energy transmission is under consideration; the available knowledge would then apply as well to one situation as to another.

Continuing efforts to develop a general theory of mechanical biodynamics can be justified in several ways. A major result of this effort is the development of analog models of biodynamic responses. These models offer an important advantage because they allow us to generalize our knowledge. In many cases, we can predict thresholds for damage from data obtained in sub–damage-threshold exposures. Models also allow us to construct tranformation scales by means of which responses of human beings can be compared with those of lower animals. Models allow us to predict human damage thresholds and to verify our predictions with lower animals.

In this section, we consider separately the

general principles of mechanical biodynamics as they apply to the whole body and the special principles that must be invoked to account for the behavior of specialized mechanical energy receptors (*mechanoreceptors*). Our aim is to understand the mechanisms whereby mechanical energy produces tissue displacement and distortion, whether the result is receptor stimulation or tissue damage. Some differences between mechanoreceptors, such as the ear and the semicircular canal, and nonspecialized structures, such as the heart or the brain, are as follows:

1. Specialized structures are designed to respond to particular forms of mechanical energy with a high degree of sensitivity and to reject, insofar as biologically possible, other types of energy.
2. Specialized receptors contain sensorineural elements (such as the hair cells and auditory nerve dendrites of the inner ear), interposed between structures that are designed to move differentially when excited by a particular form of mechanical energy (for example, the inner ear's basilar membrane and the tectorial membrane).

General biodynamic principles

Reception

Basic for understanding mechanical energy reception by the body or comprehending transmission of mechanical energy within the body is the concept of *impedance*. Impedance may be understood as the degree to which an input to a system is effective in producing an output from that system or, restated for our particular purpose, the degree to which environmental mechanical energy induces vibrations in body tissues. If most of the energy at the body surface is transmitted across the interface between the environment and body, the impedance would be low; if most of the energy is reflected at this interface, the impedance would be high. For example, the middle ear facilitates transfer of sound energy from air to the inner ear fluids; hence, impedance for airborne sound at the ear is low.

Within a generalized energy system, impedance Z is defined by the relationship:

$$Z = \frac{e}{f} \qquad (7)$$

where e = generalized effort
f = generalized flow

This relationship can be extended to several situations, as in the mechanical case where e is applied force at an interface and f is velocity of motion at this point, or the acoustic case where e is pressure and f is particle velocity.

For the mechanical case, impedance is determined by the three basic properties of vibrating bodies: inertia, elasticity, and friction (*damping*). The terms used to describe contributions to the total impedance caused by inertia, elasticity, and friction are *mass reactance, elastic reactance,* and *frictional resistance*. The magnitude of the reactance terms varies directly with the driving frequency, as can be understood by considering the basic characteristics of mass and elastance. For a mass, inertia, by definition, opposes every change the driving force tends to make in the velocity of motion. As the frequency of an oscillating driving force increases, the velocity of induced motion tends to increase; consequently, the mass reactance tends to increase. The converse is true for an elastance: elastic reactance decreases with increased driving frequency. The magnitude of a pure frictional resistance is independent of input frequency.

The output amplitude from a complex mechanical system (one containing mass, elastance, and friction) varies as a function of input frequency. At particular frequencies, the mass reactance and the elastic reactance tend to balance one another, and the output amplitude is greater than at other frequencies. When this occurs the system is said to *resonate,* and the frequency at which this relatively high output occurs is called the *resonance frequency*. The curve relating output amplitude to input frequency exhibits a high, narrow resonance peak if the system has relatively little damping. Conversely, the resonance curve will be low and broad if the system is highly damped. These concepts are critical for understanding the behavior of specific body components during exposure to varying frequencies of mechanical energy. In particular, the concept of resonance allows us to comprehend the relatively high susceptibility to damage of some body components when they are exposed to specified frequencies of vibrating mechanical energy.

Biologic response to impact can also be understood by considering the physical properties of body components. As with the drum membrane (Fig. 5-1), inertia and elasticity determine the initial displacement of a body component by an impulse. This initial displacement is called the *forced response*. Following the forced response, elasticity produces a return to the original position; however, because of inertia, the component moves

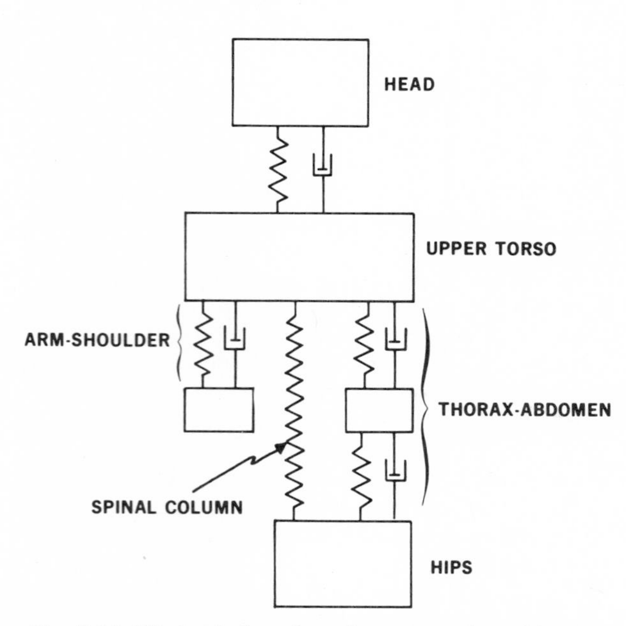

Fig. 5-14. Mechanical analog of upper portion of human body for use in z axis translational vibration research. Rectangles indicate mass, springs indicate elastance, dashpots represent viscous damping.

past its original resting position. Displacement past the original position (Fig. 5-1, *D*) is labeled *overshoot*. All of the body component's movements following the forced response are called the *free response*. A critical fact for understanding susceptibility to impact injury is that the magnitude of the free response varies as a function of input force application *rate*. Maximum free response magnitude is elicited when the rate of the input force increase corresponds to the inherent elastic deformation rate of the particular body component because elastic energy storage is most efficient when this is so.

The basic mechanical responses of body components to vibrating mechanical energy can be modeled by mechanical networks, as in Fig. 5-14, where the rectangles indicate inertial masses, the springs represent elasticity, and the dashpots represent friction. (An automobile shock absorber is an example of a dashpot.)

If the mechanical analog of Fig. 5-14 is to be of value in performing quantitative measures, information must be obtained concerning the physical properties of tissues. Table 5-1 presents typical values for selected tissue parameters.

The model of Fig. 5-14 represents a *lumped parameter system*. The arm and shoulder are represented as a single mass-spring-dashpot network, interacting with the upper torso at a particular point. In reality, the mass of the arm and shoulder is distributed through the length of

Table 5-1. Physical properties of tissues (approximate)*

	Tissue (soft)	Bone (compact)
Density (gm/cm^3)	1 to 1.2	1.93 to 1.98
Young's modulus (dyne/cm^2)	10^5 to 10^7‡	2.26×10^{11}
Volume compressibility† (dyne/cm^2)	2.6×10^{10}	1.3×10^{11} (dry)
Shear elasticity† (dyne × sec × cm^2)	2.5×10^4	7.1×10^{10} (dry)
Shear viscosity† (dyne × sec × cm^2)	1.5×10^2	—
Sound velocity (cm/sec)	1.5 to 1.6×10^5	3.36×10^5
Acoustic impedance (dyne × sec × cm^{-3})	1.7×10^5	6×10^5
Tensile strength	5×10^6 to 5×10^7	9.75×10^8
Shearing strength (dyne/cm^2)		
Parallel	—	4.9×10^8
Perpendicular	—	1.16×10^9
Breaking index		
Stretch	0.2 to 0.7	< 0.05§
Compression	—	< 0.04

*From Goldman, D. E., and von Gierke, H. E.: Effects of shock and vibration on man. In Harris, C. M., and Crede, C. E., editors: Shock and vibration handbook, New York, 1961, McGraw-Hill Book Co.; and von Gierke, H. E., and Clarke, N. P.: Effects of vibration and buffeting on man. In Randel H., editor: Aerospace medicine, Baltimore, 1971, The Williams & Wilkins Co., pp. 198-223.

†Lamé elastic moduli.

‡Depending on stretch.

§Spinal column.

the limb, and a complete model would have to take account of this fact. Lumping may provide adequate descriptions for particular, defined situations. The most obvious of these situations is when the main mass of the system is clearly physically separated from the elastic part. For example, in the case of head vibrations, the plausible assumption is made that the skull moves as a mass and that all of the displacement (elasticity) takes place in the neck area, as shown in Fig. 5-14. Also, we use lumped parameter systems because of the great difficulty in working with partial differential equations, particularly when dealing with material so patently nonhomogeneous as the human body.

Propagation

After entering the body, mechanical energy is propagated through tissues by traveling waves. These traveling waves may take several different forms (Fig. 5-15), depending on factors such as the physical properties of the tissues involved, tissue geometry, input frequency, and distance of the point being observed from the input. Numerous investigations of mechanical energy propagation in the body have been performed. One study, by von Gierke and associates,[6] showed that tissue behaves like viscoelastic material and that mechanical energy travels as shear and surface waves in the near field and compression waves in the far field. (Far field conditions can only be assumed when the distances and body dimensions under consideration are large compared with the wavelength of the traveling wave.) Tissue response, including damage, depends on the form of the traveling wave in which the energy is propagated. Energy propagation within the body can be very complex as a result of discontinuities in the physical properties and geometry of tissues and the resulting reflection, transmission, and dissipation patterns.

Dissipation

All of the energy that enters the body must eventually be dissipated in the form of heat. If the impedance between the body and the environment is low, as in the case of airborne sound with a furry animal, a high proportion of the energy in the mechanical energy field may enter the body. Small laboratory animals, such as mice, may be unable to compensate for the increase in body temperature resulting from high-intensity ultrasound exposure and die. Man, however, because his smooth skin provides a less favorable impedance match for energy transmission across the environment-body interface and because of his greater thermal capacity, does not exhibit noticeable core temperature elevation when exposed to high-intensity ultrasound.

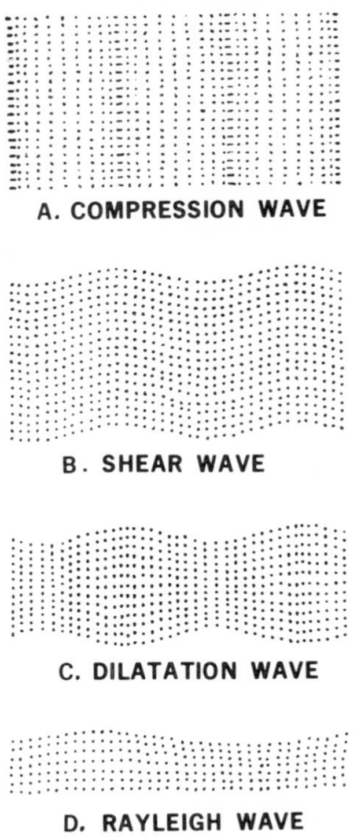

Fig. 5-15. Possible forms of traveling waves in tissues. For compression waves, **A**, tissue strain would be mainly in the direction of wave propagation. Rayleigh waves (surface waves—**D**) produce strain that is perpendicular to the wave propagation direction. Specification of strain for shear waves, **B**, and dilation waves, **C**, is complex, as can be seen from the figure. The nature and location of possible tissue damage depend on the form of the traveling wave by which mechanical energy is propagated through the body. (Redrawn from von Bekesy, G.: J. Acoust. Soc. Am. **25**:770, 1953.)

Properties of specialized receptor systems

We assume that the reader has studied the physiology of specialized receptors previously. Here we review a few basic facts and present a few illustrations for reference use. Given more space we could develop the properties of mechanoreceptors within the general framework of mechanical biodynamics presented previously.

Ears

The mechanical properties of ears have been the subject of extensive investigation by numerous

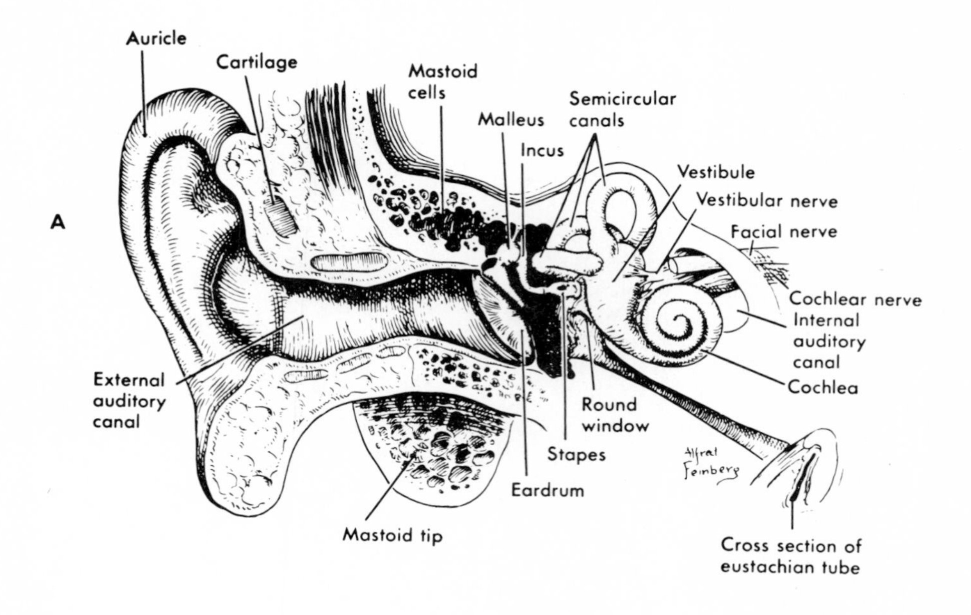

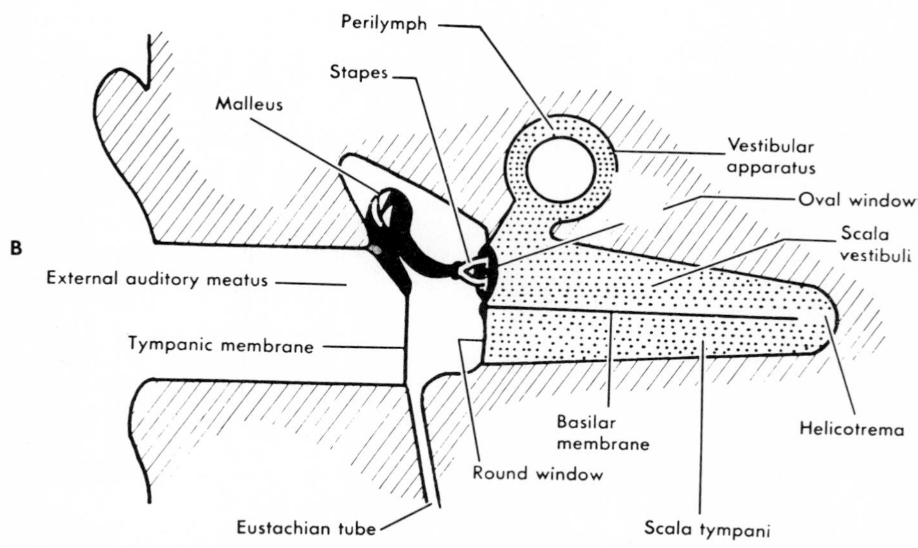

Fig. 5-16. Gross anatomy of ear. **A,** Semidiagrammatic drawing of the outer, middle, and inner parts of the ear. **B,** Schematic drawing of the ear illustrating the position of the basilar membrane in the cochlea. Sound enters the basal end of the cochlea via the oval window and is propagated to the apex of the cochlea near the helicotrema. (From Hearing and deafness, revised edition, edited by Hallowell Davis and S. Richard Silverman. Copyright 1947 © 1960 by Holt, Rinehart and Winston, Inc. Reprinted by permission of Holt, Rinehart and Winston, Inc.; and from von Bekesy, G.: Pfluger Arch. Ges. Physiol. **236:**59, 1935.)

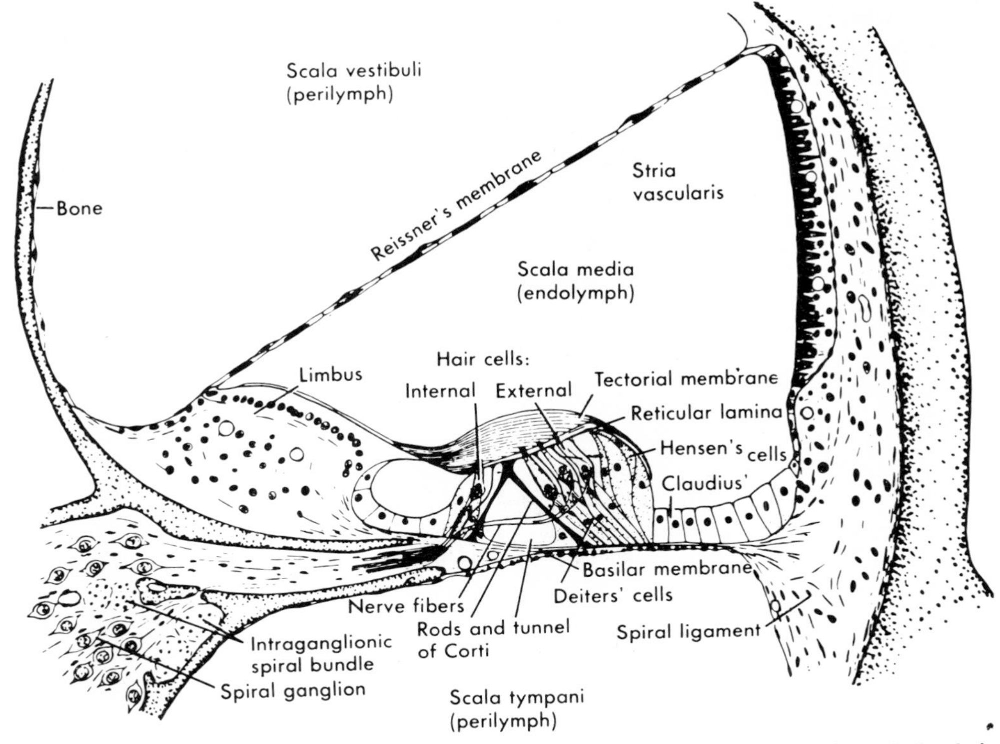

Fig. 5-17. Cross section of the guinea pig cochlear duct that illustrates the location of the hair cells on the basilar membrane. When activated by sound, a shearing force is exerted on the hair cells because of differences in the motion of the basilar membrane and the tectorial membrane. (From Davis, H., and others: J. Acoust. Soc. Am. 25:1180, 1953.)

scientists. Mammalian ears are rather odd structures because, unlike insect ears, the sensory portion is essentially designed to work under water. The solution to the problem of hearing sounds in air with a "water ear" has been magnificently solved by the evolution of the *middle ear,* which provides an excellent impedance match between airborne sound and the structures of the *inner ear.* Fig. 5-16 illustrates the gross anatomy of the ear.

Mechanical energy enters the *cochlea* via the outer and middle ears, as in the case of airborne sound (*air conduction*), or by direct vibration of the entire cochlea (*bone conduction*). Energy is propagated in the cochlea by traveling waves along the *basilar membrane* and through the inner ear fluids, as described by von Bekesy.[4] Important for understanding patterns of hearing loss are observations of differences in these traveling waves as a function of sound frequency. One of the functions performed by the ear is the transformation of sound frequency to location along the sensory receptor surface (basilar membrane). High frequencies produce maximum basilar membrane displacement at the basal end of the cochlea (near the oval window), whereas low frequencies produce waves that travel all the way to the apex of the cochlea (adjacent to the helicotrema). Also the peak of the traveling wave is sharper for high-frequency than for low-frequency sounds. Transformation from mechanical displacement to electrochemical neural energy is performed by the *hair cells,* which are located on the basilar membrane (Fig. 5-17). These sensory cells are extremely sensitive: calculations suggest that deformations of the order of 10^{-11} cm suffice to produce a functional electrochemical neural response.[7] (The diameter of a hydrogen atom is approximately 10^{-8} cm.)

Other specialized mechanoreceptors

Numerous types of mechanoreceptors, in addition to the ear, are found in the body. The skin contains several types of specialized nerve endings that are sensitive to various forms of pressure and vibratory energy. Other specialized pressure receptors are found in parts of the cardiovascular system as well as in other visceral tissues. Another type of pressure receptor is located in the joints and yields information concerning limb position. Still other mechanoreceptors are located in muscles and are important in maintaining muscle tension.

The vestibular apparatus contains receptors that are designed to respond to linear acceleration (the *utricle* and *saccule*) and angular acceleration (the *semicircular canals*). Responses from vestibular receptors can be produced by low-frequency (0.01 to 10 Hz) angular or linear vibration. Vestibular receptors provide important information for muscle control and combine with other senses to produce a basic orientation capability.

Verbal reports of pain or other behavior indicating that a particular stimulus is noxious must be related to mechanoreceptors in some way. It is clear that deformation or breakage of tissue may elicit reports of pain. However, whether it is appropriate to talk about pain receptors is a matter of debate.[8] We use reports of pain to indicate actual or potential tissue damage without a clear understanding of the mechanical-neural transformations involved in this response.

RESPONSE TO MECHANICAL ENERGY

In the preceding section our focus was on development of the general principles of mechanical biodynamics; here we emphasize responses to mechanical energy, particularly under unusual conditions of environmental stimulation. Our analysis follows the categories of organization of living material that was presented in the first paragraphs of the chapter. As we proceed through this analysis, there is a gradual reduction in the energy levels considered; high-intensity, long-duration stimulation may elicit tissue damage, whereas much smaller energy levels are associated with reports of annoyance.

Tissue and organ responses

The body is composed of tissues, organs, and physiologic systems of widely varying physical characteristics. Consequently, the body does not act as a solid homogeneous mass when it is exposed to mechanical forces; rather, it responds with various deformations and shape changes. Because of their mechanical properties, particular structures exhibit high-amplitude resonance deformations when stimulated with certain frequencies of vibratory energy or with impact impulses of particular onset and termination rates. Small deformations activate appropriate mechanoreceptors. Larger deformations can stretch tissues beyond their elastic limits, producing cell damage or membrane rupture. Intermediate amplitude deformations can produce disturbances below the level of physical damage yet result in disruption of physiologic function.

Auditory receptor structures

Sound. Several types of membrane rupture or cellular destruction can be observed in auditory receptors following overexposure to intense sound for varying durations. Stimulation with very high-intensity (140 to 170 dB SPL) audiofrequency sound can produce gross destruction of auditory structures, including rupture of the tympanic membrane, separation of the middle ear bones (particularly at the joint between the incus and the stapes), and rupture of the fine membranes or sacs of the cochlea and labyrinth.[9, 10]

Patterns of damage to the sensory cells on the basilar membrane following intense sound exposure have been extensively investigated. There has been a resurgence of interest in this area because of the development of new technics for tissue examination with a light microscope[11] and the availability of new apparatus, such as the electron microscope. It has been observed that high-frequency sounds produce damage within a limited area toward the basal end of the cochlea, whereas low-frequency sounds produce damage over a greater area toward the apical end of the cochlea (Fig. 5-18). These data are consistent with the observed patterns of traveling waves in the cochlea, discussed previously. It is interesting to note that similar patterns of cell destruction have been reported in experimental animals following extensive exposure to "rock" music. One might suggest that "hard rock" has served to reduce the "generation gap," at least insofar as hearing is concerned.

Vibration and impact. Vibration and mechanical impact effects on the auditory receptor are negligible or secondary to other types of damage. However, such is not the case with airborne impulses. Numerous cases of tympanic membrane rupture have been reported following exposure of human beings to explosions or weapons fire. Pressure levels required for tissue destruction are in the range of 6 to 8 psi.[12] Possible damage following exposure to actual or simulated sonic booms has been investigated with overpressures of up to 144 psf (1 psi); no gross tissue damage has been observed following stimulation up to these intensities.[13]

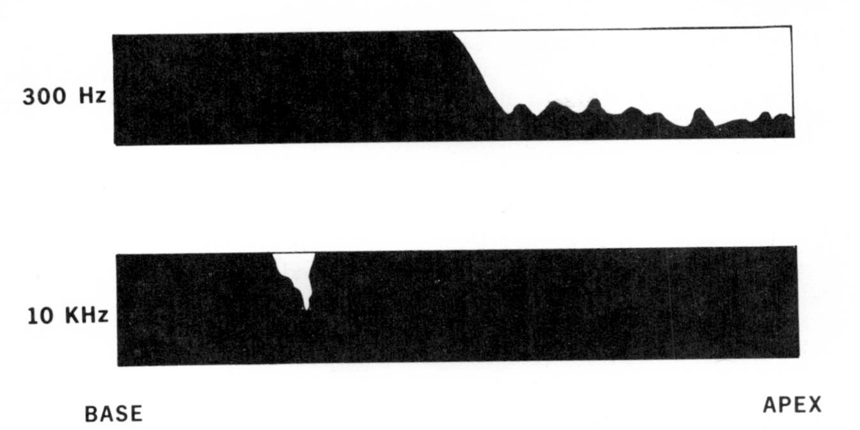

Fig. 5-18. Patterns of hair cell loss following overstimulation with intense sound at two frequencies. The degree of normality is indicated by the amount of dark filling in the rectangles. Low-frequency stimulation produces relatively more damage at the apical end of the cochlea; high-frequency sound tends to damage the basal end. (Modified from Wever, E. G., and Lawrence, M.: Physiological acoustics, Princeton, N. J., 1954, Princeton University Press.)

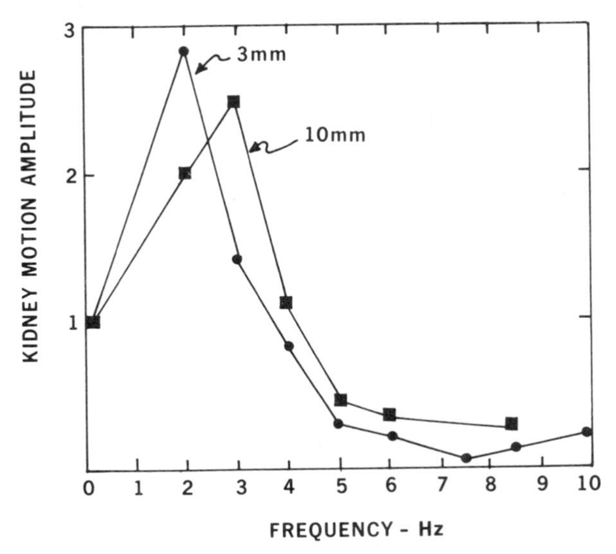

Fig. 5-19. Displacement of dog kidney during periodic translational vibration at two amplitudes. Measurements were made with an x-ray apparatus and radiopaque objects implanted in the dog. Curves indicate a damped resonance of kidney motion at stimulus frequencies of 2 to 4 Hz. (Curves constructed from data of Nickerson, J. L., and Coermann, R. R.: Internal body movements resulting from externally applied sinusoidal forces, AMRL TR 62-81, Wright-Patterson AFB, Ohio, 1962, Aerospace Medical Research Laboratories.)

Nonauditory structures

Sound. Tissue damage to structures other than the ear and the labyrinth by sound is either negligible or not demonstrated in human beings. The exception to this statement concerns structure- or liquid-borne high-intensity ultrasound. A relatively large amount of energy can be carried in ultrasound vibrations, and increased transmission across the body surface interface is provided by contact with some solid structures or liquids. The absorbed ultrasound energy is converted into heat in the body, and ultrasound is used for therapeutic purposes with good results. If the exposure is too great, tissue damage resembling burns may occur.[14] However, ultrasound fields of sufficient intensity to produce tissue damage would rarely be encountered except in special industrial or laboratory settings. In these instances, the ultrasound sources are normally well shielded or controlled.

Vibration. Response to vibratory mechanical energy is dependent on the impedance characteristics of body structures, as described in The Human Body as a Mechanical Energy Receiver. Frequency, as well as intensity, duration, and direction of motion, is critical in determining the physiologic consequences of vibration exposure. After implantation of a radiopaque object in the target organ, internal body structures can be visualized in animals during vibration using an x-ray apparatus. An example of data obtained with this technic is presented in Fig. 5-19. The frequency response curves illustrate that the dog's kidney has a damped resonance for x axis vibration exposure in the region of 2 to 4 Hz; moreover, the amplitude of kidney motion exhibits only a small increase when the vibration input amplitude is tripled. Fig. 5-19 indicates that the supporting structures for the dog's kidney are likely to be damaged by extended and/or intense x axis vibration at frequencies of 2 to 4 Hz. Newly developed technics in x-ray cinematography and image intensification hold the promise of more extensive and detailed data in the immediate future.

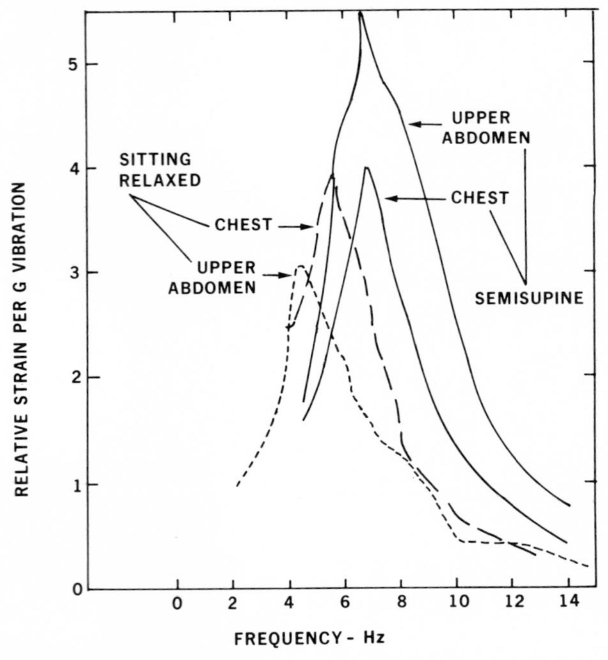

Fig. 5-20. Body strain as a function of translational vibration frequency. Strain was measured around abdominal and chest circumference. For the sitting subjects, vibration was in the z body axis; semisupine subjects received x axis stimulation. Body strain is greater for x axis than for z axis vibration. The chest and abdomen exhibit a damped resonance in the 4- to 8-Hz frequency range. (Redrawn from von Gierke, H. E., and Clarke, N. P.: Effects of vibration and buffeting on man. In Randel, H., editor: Aerospace medicine, Baltimore, 1971, The Williams & Wilkins Co.)

Strain of the chest and upper abdomen, measured with sensors on the body surface, during x axis and z axis vibration is illustrated in Fig. 5-20. These frequency response curves indicate damped resonances in the 4- to 8-Hz range. Head movements during linear vibration for standing and sitting subjects are illustrated in Fig. 5-21. These observations indicate head resonances at 1 and 2 Hz. The observations of Figs. 5-19 to 5-21 are used to account for reports of discomfort and pain during vibration exposure, as well as decrement of physiologic functioning or tissue pathology following exposure. These data allow us to set guidelines for maximum exposure levels and are particularly useful in the design of transportation vehicles and manned aerospace systems.

Vibration can destroy parts of the specialized linear acceleration sensors of the vestibular apparatus, the utricle and saccule. Loss of otoconia from the otolithic membrane has been observed in guinea pigs following vibration exposures of 1 to 2 G at 6 Hz for 4 to 6 hours.[15] Ongoing research indicates that guinea pigs may be more susceptible to damage of this structure than other species; damage thresholds in man may be higher than those noted before.

Impact. There is extensive literature concerning tissue damage following exposure to mechanical impact or airborne shock waves. Here we examine only a few fundamental principles and observations; an extended discussion of this topic can be found in Stapp.[2] Impact injuries range from minor contusions and abrasions to decapitation and evisceration. Representative of an extreme is the example of an airplane crash on land at over 500 mph. The combined weight of the identifiable remains of the three occupants was 9.5 lbs.

Among the major factors that determine the nature and extent of an impact injury are the (1) point(s) of contact, (2) area of contact, (3) onset rate of the deceleration (G/sec), and (4) peak value of the deceleration. These points can be illustrated by considering pilot injuries sustained in two light airplane crashes investigated by Swearingen.[16] In both cases, one involving a crash into hard soil and the other a much less

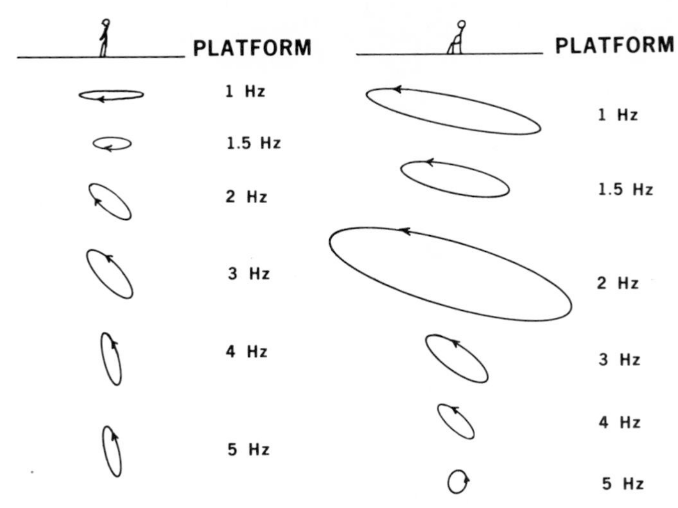

Fig. 5-21. Head movements during x axis translational vibration as a function of vibration frequency. The left column shows responses of a standing subject; the right column illustrates head movements of a sitting subject. Head movements are generally more damped for the standing than for the sitting subject. The sitting subject shows head movement resonances at 1 and 2 Hz. (Based on data of Dieckmann, D.: Intern. Z. Angew. Physiol. **17**:83, 1958.)

Table 5-2. Accelerations in simulated head impacts for two cases of light airplane crashes*

Instrument panel impact area	Velocity at impact			
	Low (15 ft/sec)		High (30 ft/sec)	
	Peak G	Rise time-sec	Peak G	Rise time-sec
Thin aluminum semicylinder	30	5	115	17
Sharp steel edge	165	3.5	295	2

*Data from Swearingen, J. J.: Tolerance of the human face to crash impact, AM 68-20, Oklahoma City, 1965, Civil Aeromedical Research Institute.

violent crash into water, the pilot's head impacted with the instrument panel. In the former case (land crash), the instrument panel had a thin aluminum semicylinder mounted on it at the head impact point. In the latter situation (water crash) the point of head impact was identified by a small dent in the sharp edge of the panel. Head damage suffered by the first pilot included minor facial lacerations and a slight fracture of the right frontal sinus. The second pilot was killed by a crushing of the skull across the bridge of the nose and the eyes.

These crashes were simulated using dummies and the results are presented in Table 5-2. Notice that the peak decelerations associated with the sharp-edge panel are much greater than those with the semicylinder. Also, note that the onset rate with the sharp-edge panel impact stayed approximately the same with the high-velocity impact as for the low-velocity impact. However, onset rate more than tripled with the increased impact velocity for the semicylinder. Finally, because of the malleability of the thin aluminum, the area of load distribution was significantly greater for the impact with the semicylinder than with the sharp-edge panel. The implications of these observations for protection of human beings from impact injuries (such as automobile crashes) are obvious.

Head injuries following impacts have been extensively investigated with particular emphasis on brain damage.[17] Brain injuries result from (1) direct tearing or shearing of tissues, (2) deformation of the brain tissue under the area of the blow, (3) differential acceleration of the skull and the brain, (4) differential motion of the brainstem and the spinal cord, and several other factors. Consequences of brain damage depend on the point of injury; a small lesion in the occiptal lobe may result in a barely noticeable sensory deficit, whereas an injury of the same size in the brainstem results in immediate death. Phenomena that are useful for illustrating the basic biodynamic principles discussed previously are "coup" and "countercoup." Brain damage beneath the point of impact results from the original coup (blow); countercoup results from rebound of the brain within the skull cavity. The brain overshoots its resting position and "slaps" against the internal skull wall on the side away from the point of input during the period of free response following impact. Localized damage on the side of the brain opposite to the impact point results from the countercoup.

Differential motion of the brainstem and the spinal cord, with the localized tension resulting in damage to the brainstem, has been investigated by Friede.[18] Additional head mass, as provided by a protective helmet, increases the likelihood of this type of injury. Some of the helmets used by football players a few years ago were so poorly designed that the rear portion of the helmet could provide a fulcrum for flexion between the head and the upper torso, resulting in an increased tension at the cervical end of the spinal cord and producing an acceleration concussion.

Impact injuries to internal organs of the body trunk have also been extensively investigated. These studies indicate that some organs, such as the spleen and liver, are less able to withstand trauma than other organs because of their limited mobility. In these instances, the forces generated by the organ's displacement produce strain in a small volume of elastic support tissue; consequently, damage to the support tissue occurs readily.[19] Research also indicates that the physiologic status of an organ at the time of impact may change its vulnerability. For example, cardiac damage appears to be directly dependent on the contractile state of the heart muscles; ventricular rupture is much more likely during ventricular systole than during the diastolic phase.[20]

Explosive blast can produce injury similar to that observed from mechanical impact. Tearing or shearing of the elastic connections between various body masses results from the very large compressive force associated with blast. Another cause of damage is the development of air emboli at various points within body tissues. Air at some locations, such as the brainstem, may produce fatal anoxia, whereas other tissues can respond by exhibiting hemorrhage or edema. The most frequent cause of death from blast exposure is tearing of lung tissue and subsequent development of air emboli. Because of the resonance of the lung system, injury depends not only on the peak pressure but also on the duration of the blast wave. Wind blast, particularly as a result of ejection from a high-speed airplane, produces many effects analogous to those seen following explosive blast.

Physiologic system responses

Here we examine responses attributable to mechanical energy exposure within specified body systems including sensory systems, such as audition, vision, and orientation, as well as the cardiovas-

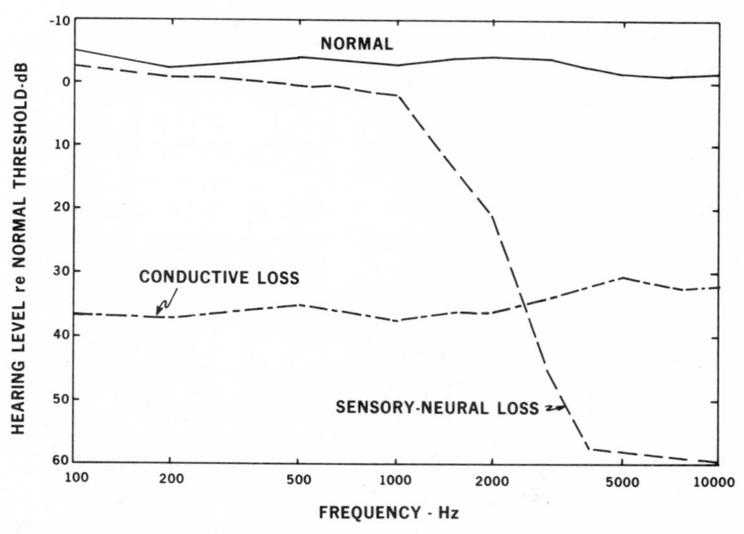

Fig. 5-22. Audiograms showing normal hearing, a conductive hearing loss that is relatively "flat" across the audiofrequency range, and a sensorineural hearing loss. Conductive hearing loss is usually associated with middle ear sound transmission structure damage or disease. The sensorineural loss audiogram exhibits a typical pattern of increased deficit with higher frequency. The sensorineural audiogram is associated with aging (presbycusis) or with daily exposure to an intense sound environment.

cular system, the respiratory system, and the endocrine system. Our focus is on energy detection and reflexive, genetically preprogrammed reactions. Influences of mechanical energy on behavior controlled or directed to a great extent by higher mental processes are considered in the section on behavioral system responses.

Auditory system

Hearing loss: PTS and TTS. Audiograms are graphic records of absolute sound intensity detection thresholds (hearing levels) at various test frequencies. *Permanent threshold shift* (PTS) refers to irreversible elevation of hearing thresholds; *temporary threshold shift* (TTS) applies to transient elevation of sound detection thresholds. Fig. 5-22 illustrates various types of PTS as described by audiograms.

Hearing losses are generally attributed to two types of deficit: conduction loss and sensorineural loss. In conductive hearing loss, there is a disturbance in the mechanism that transfers mechanical energy from the air to the cochlea (see Fig. 5-16). Conductive hearing losses are associated with such things as tympanic membrane rupture and separation of the ossicular chain. Many older people suffer from a type of conductive deafness called otosclerosis, in which the stapes becomes immobilized as a result of excessive bony growth. Sensorineural losses are attributed to destruction of the receptor hair cells, as noted on p. 138, or to lesions of the auditory neural pathway. Hearing loss caused by environmental mechanical energy exposure is nearly always of the sensory (hair cell loss) type. A gradual decline in sensitivity to high-frequency sound associated with aging is called presbycusis. It is often suggested that presbycusis is a direct result of the aging process; however, a more likely explanation is that the decline reflects the cumulative effects of acoustic trauma, infections, and other damaging circumstances that are naturally a function of age.[21]

Fig. 5-23 shows a progression of hearing loss that has been observed in factory workers. These workers were exposed to nonperiodic sound at 104 dB SPL throughout each working day. PTS is initially manifested by a dip in the audiogram at the 4- to 5-kHz region. With continued exposure, the dip broadens until PTS is exhibited across the entire frequency range.

Fig. 5-24 illustrates recovery from TTS following two intensities of sound exposure. As would be expected, recovery from high-intensity exposure requires a greater time period than re-

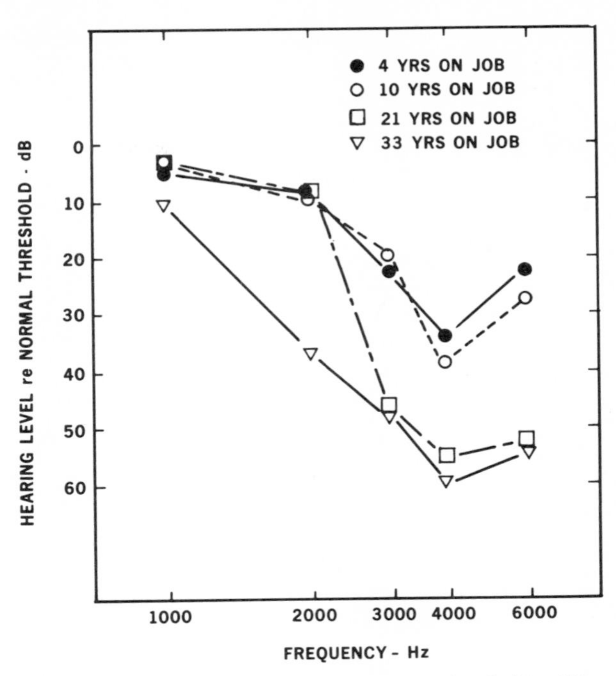

Fig. 5-23. Sound-induced permanent threshold shift. Audiograms indicate increasing hearing loss as a function of the number of years working in an intense sound environment. (Redrawn from von Gierke, H. E., and Nixon, C. W.: Noise effects and speech communication in aerospace environments. In Randel, H., editor: Aerospace medicine, Baltimore, 1971, The Williams & Wilkins Co.)

covery from low-intensity sound exposures. Periodic sound exposure exhibits a curious and unexplained phenomenon: the maximum hearing loss is found at frequencies about one-half to one octave higher than the exposure frequency.

A major reason for interest in TTS relates to the question of individual differences in susceptibility to sound-induced hearing loss. Numerous reports indicate that men exhibit wide variability in their ability to withstand prolonged sound exposure: if two workers are exposed to the same industrial sound environment on a daily basis, one may become severely deaf and the other may exhibit only moderate hearing loss. Two questions follow from these observations. First, do individuals differ in susceptibility to TTS? Second, what is the relationship between TTS and PTS susceptibility? Ward[22] has answered the first question by demonstrating that the distribution of TTS susceptibility is approximately normal (gaussian). Despite considerable research effort, agreement has not yet been reached concerning the answer to the second question. Differences in TTS are

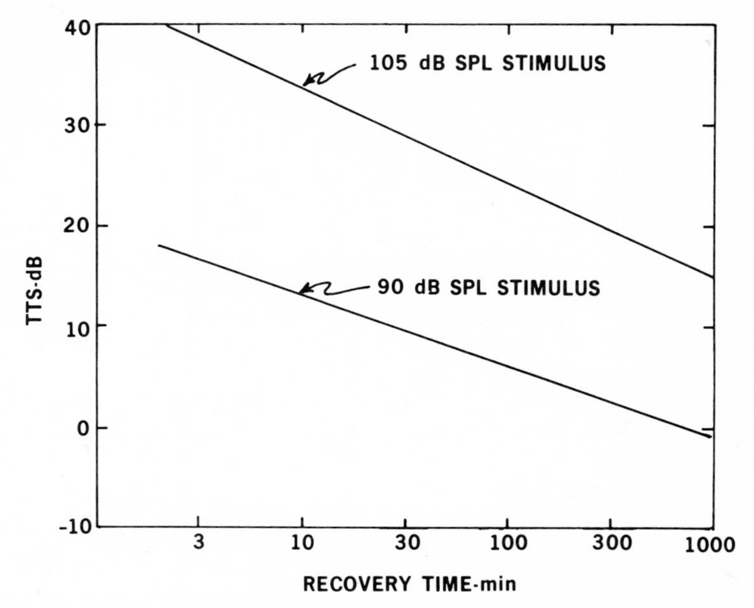

Fig. 5-24. Recovery from a temporary threshold shift (TTS) at 4,000 Hz following presentation of a band of noise between 1,200 and 2,400 Hz for a 102-minute "effective" exposure. The parameter in the figure is exposure intensity. Studies of TTS are pursued in an effort to develop measures with which individual susceptibility to permanent hearing loss can be predicted. (Curves constructed from data of Ward, W. D., Glorig, A., and Sklar, D. L.: J. Acoust. Soc. Am. **31**:522, 1959.)

associated with many of the same variables that appear important for PTS: overall sound level, sound frequency, duration of exposure, and whether the exposure is continuous or intermittent. Difficulties arise with the observation that the two ears of a particular man may differ in TTS susceptibility or even that the same ear may show different responses when tested at different times. Another reason for caution is the observation that a particular noise can produce maximum TTS at one frequency, but if the noise exposure is continued, the maximum PTS can be at quite different frequency.[23] These observations suggest that we should proceed with caution in attempts to use TTS susceptibility as a predictor of occupational hearing loss. Further discussion of this important topic can be found in Burns[24] or Ward.[25]

Sound attenuation mechanisms. The auditory system has two mechanisms for attenuation of unwanted or potentially destructive sound. The first of these is a mechanical mechanism and involves the transmission properties of the middle ear. Sudden or unexpected sounds elicit reflexive contraction of the middle ear muscles, which reduces the amount of energy conducted to the cochlea. Attenuation of up to 10 dB can be produced by this mechanism for low tones; high tone attenuation is less.[26] This *middle ear muscle reflex* cannot be effective in protecting against sudden-onset high-intensity sounds, such as explosions, because the energy enters the ear before the reflex has time to take place.

A neural mechanism that appears to attenuate particular components of a sound has been described by Rasmussen.[27] The olivocochlear bundle is the final link in an efferent auditory pathway that descends from the cortex to the ear. Nerve fibers comprising the olivocochlear bundle synapse directly on the basilar membrane hair cells. These fibers apparently function to improve *signal-to-noise ratios* by direct inhibition of cochlear hair cells (p. 137).

Visual system

Visual acuity during exposure to mechanical energy has been measured with various procedures, such as the ability to read dials or letters on Snellen test charts. Decrements of visual acuity associated with sound have seldom been reported except with exposure to very high levels (+150 dB SPL) in the infrasound or low-audiofrequency range.[28]

Visual acuity loss for whole body vibration is potentially much more severe than in sound fields.

Essentially, these losses can be attributed to difficulties in visual *tracking*. Visual tracking refers to the behavior whereby a moving target is maintained in the center of the visual field. The problem encountered during exposure to environmental vibration is more complex than that found in most visual tracking experiments; when the vibration frequency exceeds 2 Hz, the eye and the target move differentially with respect to some stable reference point. Visual acuity losses are minimized to the extent that environmental motion is regular and hence predictable; such would be the case for sinusoidal vibration. However, when the motion-time pattern is complex and unpredictable for the visual tracking control system, visual acuity losses can be severe.[29]

In general, visual acuity losses are most severe in the vibration frequency ranges where either head resonance (30 Hz) or eyeball resonance (estimates range from 30 to 100 Hz) occur. Above 100 Hz acuity decrements are negligible.[5] These visual effects are important in situations where a man is required to monitor visual displays, such as dials and gauges, during vibration exposure.

Orientation system

Information concerning where we are in space and/or in relation to objects or other organisms is provided by several sensory receptors, including the vestibular apparatus, the ear, the eye, and the skin. Little systematic research has been performed on disturbance of orientation system responses by mechanical energy exposure, although several effects, ranging from failure to detect movement in the peripheral visual field to spatial disorientation, are conceivable.

Direct vestibular apparatus activation has been observed following exposure to high-intensity sound (135 to 140 dB SPL) in the low-audio-frequency range. The vestibular responses are indicated by observations of nystagmic eye movements, dizziness, nausea, and loss of balance.[30]

Immersion in intense sound fields disrupts auditory localization and orientation cues. Although in most instances orientation cues derived from the auditory system would seem relatively unimportant because of redundant information through other sensory pathways, there are some situations or circumstances in which the loss of auditory orientation cues could be important. For example, casual observation suggests that auditory localization cues are important in table tennis, although this is primarily a visual-motor skill. It may be instructive to consider the importance of auditory cues for performance in those sports where spectators are expected to maintain silence during play.

Recent evidence supports the suggestion that the visual system operates in two basic modes: orientation and identification. The function of the orientation mode is to detect movement in the peripheral visual field and to reflexively bring the movement source into foveal vision.[31] Many observations indicate degradation of visual acuity during vibration exposure. Failure to detect moving objects in the peripheral visual field could certainly contribute to accidents with either surface or airborne transportation. Stapp[2] noted that peripheral vision is lost before central vision during intense acceleration exposure. Beyond these observations, we have been unable to find research concerning the effects of mechanical energy on visual orientation responses, and it has not yet been demonstrated that the data for visual acuity decrement can be used to predict visual orientation response disturbance.

Angular and linear vibration can activate both skin and vestibular receptors. The signals from these receptors as a result of vibratory stimulation could lead to confusions in the detection of gravitational vertical or self-induced motion. These confusions become important to the extent that maintenance of spatial orientation is required for adequate performance of particular tasks, as in the situation of a pilot moving through three-dimensional space. Finally, severe disorientation, including motion sickness, can be induced by high-amplitude vibrations in the range of 0.1 to 0.7 Hz.[32]

Visceral systems

Some of the effects of mechanical energy on visceral systems (the cardiovascular system, respiratory system, reproductive system, gastrointestinal system, and endocrine system) result directly from mechanical stimulation of a particular organ system. In other cases, however, the observed responses reflect activation of general adaptive physiologic programs.[33] Coordination of these adaptive programs is usually considered to be the province of the endocrine system and the autonomic nervous system, but the effects are observable in nearly all body components.

Direct effects. Whole-body vibration of human beings at low frequencies (2 to 12 Hz) elicits visceral responses, mainly as a direct result of mechanical action. Mean systemic arterial blood pressure, heart rate, and oxygen consumption all increase as a function of vibration amplitude and, to a lesser degree, as a function of vibration frequency.[34] The maximum response is between 4 and 8 Hz for the sitting man (z axis vibration) because this is the resonance frequency range for the thoracic-abdominal system[35] (see Fig. 5-20). The combined cardiopulmonary response to vibration is similar to that seen with exercise except for forced oscillation of respiratory air and blood flow at very low vibration frequencies. The pulsatile flow resulting from this low-frequency vibration is superimposed on the normal flow pattern. If the vibrations are synchronized and in phase with heart activity, the resulting effect on the circulatory system can probably be appreciable and deserves further study. Vibration of this type has been proposed as a therapeutic cardiac assist mechanism.

Indirect effects. Indirect activation of visceral physiologic systems by mechanical energy can be produced by the orienting reflex, startle responses, and stress reactions. Any large and unexpected change in stimulation, regardless of the direction of the change, can elicit a startle response or an orienting reflex. The orienting reflex refers to the sequence of behaviors whereby an animal shifts its attention to a novel stimulus. Peripheral (visceral systems) physiologic arousal produced by the orienting reflex is ordinarily of small magnitude. The startle response is elicited by large changes in stimulation and includes protective reflexes, such as closing the eyes, as well as quite intense physiologic arousal. Stress responses are usually evoked by stimuli that are aversive to an organism.

Intense sound exposure produces changes in endocrine function generally similar to those found with more commonly employed stressors such as immobilization, forced exercise, surgery, and cold. Activation of the pituitary-adrenal axis by sound exposure has been observed in both laboratory animals and man. Henkin and Knigge[36] traced changes in adrenal secretion of corticosterone in rats exposed to 220 Hz at 130 dB SPL for up to 48 hours. Lockett[37] observed more than a tenfold increase in urinary epinephrine levels following exposure of rats to a sound level of 100 dB at 20,000 Hz for periods as short as 2 seconds. Sound-induced increases of plasma 17-hydroxycorticosteroids and urinary epinephrine have been

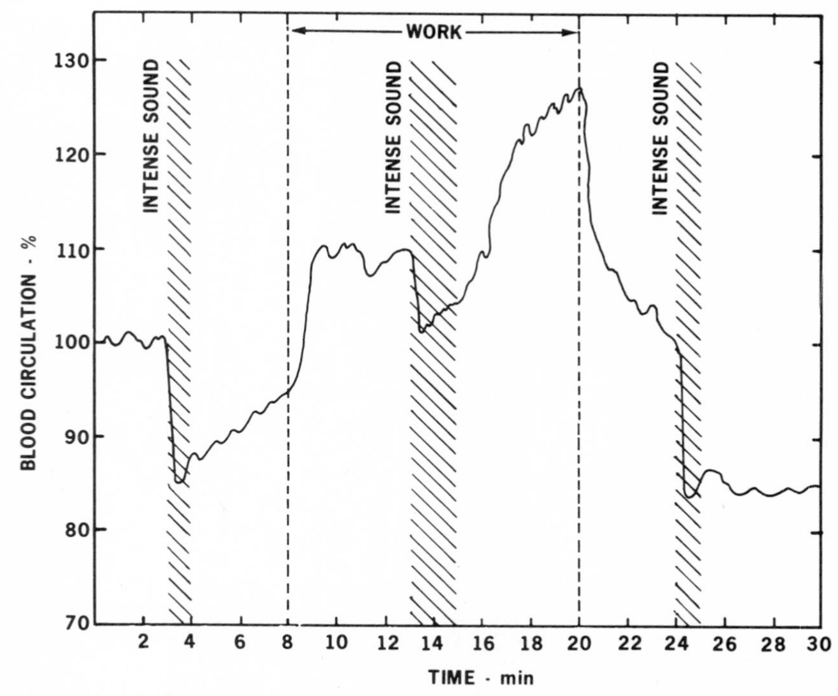

Fig. 5-25. Effects of work, rest, and intense sound on peripheral blood circulation as measured by finger pulse amplitude. Work was performed on a bicycle ergometer. Sound exposure induced marked vasoconstriction. (Curve constructed from data of Jansen, G.: Intern. Z. Angew. Physiol. **20**:233, 1964.)

observed in both control groups and in patients suffering from either cardiovascular or psychologic disorders.[38] Similar changes in endocrine system activity have been correlated with vibration and impact exposures.

Several experiments indicate that intense sound exposure can induce changes in the activity of the cardiovascular system, including increased heart rate. One widely discussed experiment demonstrates peripheral blood flow reduction resulting from vasoconstriction.[39] These findings are illustrated in Fig. 5-25 and may be related to reported increases in muscle action potentials induced by sound exposure.[40] Possible cardiovascular system damage or disease as a result of daily exposure to intense sound has been a topic of considerable concern. Jansen[41] reported a greater incidence of peripheral circulatory and heart problems for German industrial workers who were exposed to high-intensity sound environments than for workers who were exposed to lower levels of sound. Although statistically significant differences were obtained in this study, Kryter[42] has suggested several extraneous variables that could account for the observed differences. Adequate answers to questions about the effects of sound on the cardiovascular system are not available, and further research is needed.

Observations on human volunteer subjects exposed to impacts of various amplitudes in laboratory situations reveal minimal involvement of visceral systems response. In some cases, moderate clinical signs of shock were seen. X axis impacts of 15 G or more result in a significant decrease in heart rate during the postimpact period.[2]

Changes in gastrointestinal (GI) system activity have been associated with both sound and linear vibration. Stern[43] demonstrated that gastrointestinal motility can be directly related to sound level in an experimental setting: Gastrointestinal motility increased when the sound stimulus was raised from a low to a moderate level, and motility decreased when the stimulus was lowered from a high to a moderate level. Very large effects of linear vibration on the gastrointestinal system correlate with motion sickness. The symptoms of motion sickness range from epigastric discomfort through pallor and sweating to vomiting.

Tamari[44] reported several reproductive system changes following exposure of laboratory animals to intense auditory stimulation. These changes include modification of the estrus cycle and increased weight of the uterus and ovaries in rats. Rabbits exhibited responses indicative of increased gonadotropic hormone secretion. These latter observations are surprising because stress reactions are ordinarily accompanied by a decrease in anterior pituitary gonadotropic function and a resultant sex organ atrophy.

The research on visceral systems responses to mechanical energy cited in this section is but a small sample of the many reports available. During the past few years there has been a large increase in the amount of work in this area, with particular emphasis on the effects of sound. The published proceedings of a symposium on the physiologic effects of sound[45] is a source for further reading. However, from the environmental physiologist's view, many of the studies presented at the symposium are subject to criticism. For example, extension of findings from laboratory animals such as rats, rabbits, and mice to human beings is questionable because several strains of these laboratory animals are known to be highly sensitive to auditory stimulation. This sensitivity is manifested by audiogenic (sound-induced) seizures. Another important point is that few of the studies deal with long-term exposure to auditory stimulation, whereas this long-term situation is of primary interest for the prediction of physiologic responses of men exposed to high levels of environmental sound on a daily basis.

Behavioral systems responses

The focus of this section is mechanical energy responses involving integrated, goal-oriented activity of several physiologic systems or subsystems. Such activities include verbal communication and problem solving. Generally, these behaviors involve the higher mental processes: language, learning, and memory.

Concept of noise

It is at the behavioral level that the concept of noise becomes useful. Noise is usually defined as unwanted signal or stimulation. This clearly implies that *someone* is doing the "wanting." It is possible to infer that an animal, such as a monkey or rat, finds a stimulus to be unwanted or aversive on the basis of the animal's behavior. However, it is often difficult to specify precisely what aspect of the total stimulus environment is aversive to the animal. With a human being, it is somewhat easier to determine what stimulus characteristics make the stimulus unwanted. Even with human beings, as we shall see, a noise may be difficult to specify and many aspects of the environment besides the sound itself can influence the perceived noisiness.

Some investigators in this area write about the effects of "noise per se." These investigators refer to a particular class of sounds: those in which the frequency spectrum contains energy distributed across the audiofrequency range (nonperiodic sounds). In this chapter we have avoided use of the term "noise" when referring to particular types of sound. It seems to us that the issues are sufficiently complicated for the student without requiring him to remember two quite different definitions for a central term.

Even within the definition as unwanted stimulus, there are interesting implications of the concept of noise. Consider that in all goal-oriented behavior we are required to detect some type of stimulus. The stimulus may be generated internally, as in the case of blood sugar level and the feeding control system, or the stimulus source may be in the environment, as with speech communication. During goal-oriented behavior we have some idea concerning the particular stimuli that we wish to receive. More precisely, we have stored a set of templates against which we match incoming signals. We assign probabilities or expectancies to a limited set of possible signals; signals outside this set are irrelevant for our problem and could be considered noise. For example, if we are trying to receive signals that will aid us in predicting where and when a hurricane might strike the mainland, signals concerning the number of births recorded during the last decade in Vladivostok would be useless and considered to be noise.

The situation becomes more complex when we consider that, within a particular environment, people may be pursuing many different goals and that a signal for one man might constitute noise for another. The warning blast of a train air horn has quite different implications when we are listening to a baseball game or driving across a railroad crossing.

Another dimension of complexity is introduced by the consideration that behavior is guided by a multiplicity of goals that may change rapidly. In a movie theater, our goal may be understanding the film's dialogue, and the "humorous" comments of other theatergoers may be noise. Hearing the vocal signal "FIRE!" may immediately change our goals and, consequently, our behavior.

One final comment concerning the problem of defining the term "noise." As we become familiar with our environment, stimuli that were originally considered noise may become signals. Particular patterns of sound in the heating system can give us information that the system is operating properly. Departures from the familiar sound pattern may be detected and elicit appropriate behavior. Such a process could be involved in the restlessness experienced by the city dweller during the first few nights of his vacation in the countryside.

Speech communication

The presence of mechanical energy in the environment can disturb all phases of the speech communication process, including the encoding of signals at the source, modification of the signal during transmission, and the presence at the receiver of unwanted as well as wanted energies.

Audiofrequency noise can adversely affect speech communication in at least three ways. First, the noise can *mask* the speech signal. Second, the listener may suffer from a temporary threshold shift (TTS) as a result of the noise exposure and, consequently, require higher than normal signal intensities in order to correctly perceive the speech signal. Third, noise may interfere with the "efficiency" of the perceptual process.

Masking refers to the process whereby one sound is "drowned out" by another sound. Noise masking of speech signals has been extensively investigated and several procedures are available for predicting the consequences of noisy environments on speech communication. One of these procedures is called *speech interference level (SIL)* and is defined as the average sound pressure level in the three octave bands that are most important for speech communication (600 to 1,200, 1,200 to 2,400, and 2,400 to 4,800 Hz). Consequences of selected SIL values on face-to-face communication are summarized in Table 5-3.

As we noted earlier, man is not to be viewed as a passive receiver of environmental stimuli. Rather, by his actions man influences the mechanical energy field to which he is exposed. Possible actions to improve speech perception, such as tilting the head or moving closer to a speech source, are obvious. Less obvious are the mechanisms underlying the "cocktail party effect." The cocktail party effect refers to the ability to focus our attention on a particular conversation in a room where several conversations are present and to shift our attention from one conversation to another. The focusing of attention appears to involve enhancement of desired speech signals and attenuation of undesired signals; the result is an

Table 5-3. Effects of selected speech interference levels on person-to-person communication. Overall long duration rms sound levels are measured at the external ear of the listener

Speech interference level (dB)	Person-to-person communication
30-40	Communication in normal voice satisfactory, 6 to 30 ft. Telephone use satisfactory.
40-50	Communication satisfactory in normal voice 3 to 6 ft, and raised voice 6 to 12 ft. Telephone use satisfactory to slightly difficult.
50-60	Communication satisfactory in normal voice, 1 to 2 ft; raised voice, 3 to 6 ft. Telephone use slightly difficult.
60-70	Communication with raised voice satisfactory, 1 to 2 ft; slightly difficult, 3 to 6 ft. Telephone use difficult. Earplugs and/or earmuffs can be worn with no adverse effects on communication.
70-80	Communication slightly difficult with raised voice, 1 to 2 ft; slightly difficult with shouting 3 to 6 ft. Telephone use very difficult. Earplugs and/or earmuffs can be worn with no adverse effects on communication.
80-85	Communication slightly difficult with shouting, 1 to 2 ft. Telephone use unsatisfactory. Earplugs and/or earmuffs can be worn with no adverse effects on communication.

improvement in the signal-to-noise ratio. Several interesting phenomena have been observed in laboratory attempts to investigate the cocktail party effect. One of these studies indicates the advantages of listening with two ears. Pollack and Pickett[46] performed an experiment in which the same speech signal was presented to one or both ears via earphones; however, each ear received a different noise input. Speech perception was considerably better when the speech and noise stimuli were presented to both ears than in a control condition where only one ear was stimulated. As with the cocktail party effect, binaural (two ear) listening appears to result in improved signal-to-noise ratios. Possibly, both observations may be accounted for by the same neural mechanisms.

Recent experiments by Dewson[47] suggest that the efferent neural supply to the cochlea (the olivocochlear bundle) may function to improve signal-to-noise ratios. In Dewson's experiment, monkeys were trained to discriminate speech sounds with and without a background noise. Ability to perform this task was assessed before and after bilateral surgical section of the olivocochlear bundle. After the operation, the monkey's ability to discriminate speech sounds was not adversely affected if the background did not contain noise; if noise was present, speech discrimination ability was reduced.

In addition to noise intensity and signal-to-noise ratio, numerous other factors influence our ability to correctly identify speech signals. Frequency characteristics of the masking noise, rate of interruption of the speech signal, and characteristics of the vocabulary used by the speaker are among the factors that have been studied. Concerning the last factor, the language that aircraft controllers have developed to communicate with airplane pilots is more easily perceived than normal speech because this "control tower language" is more redundant than the language of everyday conversation. An extended discussion of the effects of noise on speech perception can be found in Kryter.[42]

The notion of noise-induced reductions in perceptual efficiency derives from experiments in which decrements in speech communication are observed although noise levels are not of sufficient intensity to mask the speech signals. Holloway,[48] in an overview of research on speech communication, suggested that the observed decrements may reflect limitations of information processing rate, particularly in relation to *short-term memory.* The suggestion is that noise somehow overloads a hypothetical auditory short-term storage mechanism, thus reducing the ability to retain information from the speech sounds. A particularly interesting finding in this area is that communication losses may be reduced if the noise source is in a different spatial location from the speech source, independent of the relative sound intensities from these two sources at the observer's ear. This observation suggests an interplay between hypothesized auditory localization and auditory identification systems.

Infrasound and whole-body vibration can produce disturbances in speech communication by modifying the signal at the source. A tremolo quality in the vocal signal results from vibration exposure, with the frequency of the tremolo matching the frequency of the vibration. Word recognition scores generally decrease with increased vibration amplitude and vibration frequency up to 18 to 20 Hz.[49]

Airborne impact disturbs speech communication by interruption of the speech signal. These stimuli would ordinarily be unexpected, and further disruption of communication would result from a startle response and the refocusing of attention at the time of stimulation.

Complex task performance

Speech communication must certainly be considered under the general heading of complex tasks. However, here we concentrate on those situations that require both perceptual monitoring and motor performance capability. For simple situations, deficits in complex task performance can be partially predicted on the basis of losses in sensory or motor capability. In addition, there may be a deterioration in performance as a result of disturbance of higher mental processes (for example, memory) by the mechanical energy exposure.

One of the topics most enthusiastically investigated by undergraduate experimental psychology students is the effect of noise on various types of problem solving. The results of these studies are nearly always negative: the noise does *not* result in a decline in problem-solving ability. On the contrary, many reports indicate a seemingly paradoxic improvement in test scores. However, some investigators, working with specific procedures, have been able to obtain reliable performance decrements attributable to the presence of noise in the environment. Hockey[50] has summarized the conditions under which noise can adversely affect performance: (1) the task is long (more than 30 minutes); (2) the task is continuous; (3) the places where and the times when information is presented are unpredictable (for example, where the subject is required to monitor different instruments located in different parts of the room); and (4) information is presented near the maximal assimilation rate for the subject. In other words, noise-induced performance loss can be expected in situations that demand an observer's complete attention and in which there is little margin for error. Hockey interprets these findings within the framework of optimal levels of arousal or excitation for differing tasks.

Sleep

The effects of noise on sleep can be placed in two general categories. First, the noise may arouse or awaken the sleeper. Second, the noise can produce a change in the "stage" of sleep but not awaken the sleeper. Sleep stages are classified by changes in several physiologic variables, including the electroencephalogram (EEG), eye movements, and neck muscle tension. The stages may be labeled light sleep, deep sleep, and dream sleep; however, different labels, as well as different physiologic criteria for the sleep stages, are used by various investigators.[51, 52]

Many studies indicate that the effects of noise on sleep change with repeated presentations of the noise and across nights. Comprehension of these findings requires an understanding of the basic phenomena of *habituation.* Habituation refers to the observation that, for a wide range of stimuli and responses, repeated stimulus presentations elicit gradually reduced response, as in the folk tale of the boy who cried, "Wolf!" The results of intermittent noise exposure on a sleeping person depend on (1) physical characteristics of the sound, (2) the background noise, (3) possible significance of the sound (such as, crying baby and sleeping mother), (4) the possibility of eliminating the sound (thunder versus dripping faucet), (5) the age and state of health of the person, and (6) the stage of sleep the person is in at the time of noise exposure. With regard to this last condition, it has been suggested, but not demonstrated, that noise-induced changes in stage of sleep (such as from deep to light sleep) may be physiologically harmful. Kryter[42] has reviewed the literature concerning the effects of noise on sleep. Since a considerable effort is currently being expended in several laboratories, both in the United States and around the world, we should gain a much clearer understanding of the effects of noise on sleep within the current decade.

Annoyance

Numerous technics for predicting potential annoyance as a result of noise exposure have been developed since the 1940's. A large proportion of this effort has been concentrated on annoyance evoked by aircraft noise and, more recently, the sonic boom (bang). These technics have also been employed to examine the annoyance associated with a wide range of other environmental sounds.

One annoyance measure, which is widely used in the United States and which has been recently adopted by the International Standards Organization (ISO), is *perceived noisiness.* This measure, which was originally proposed by Kryter and Pearsons,[53, 54] relates the noisiness of a particular

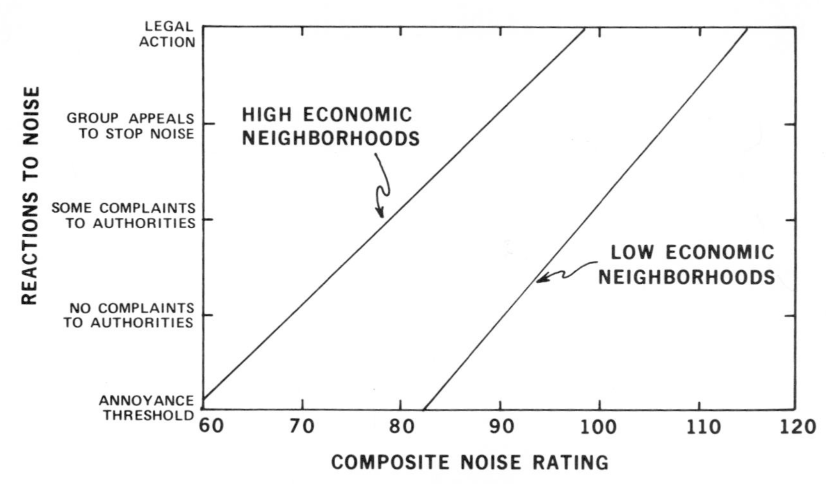

Fig. 5-26. Reactions to noise as a function of the composite noise rating (CNR). The CNR measures both perceived noisiness and frequency of occurrence of environmental noise. Intensity of response correlates with the CNR magnitude. High economic neighborhoods demonstrate less noise tolerance than low economic neighborhoods. (Modified from Kryter, K. D.: The effects of noise on man, New York, 1970, Academic Press, Inc.)

environmental sound to the noisiness of a band of noise centered around 1,000 Hz. The scale of perceived noisiness is *PNdB;* the calculation of this value is somewhat complicated and you can find the detailed procedures in Burns[24] or Kryter.[42] Studies show that several attributes of an environmental sound, in addition to its loudness, affect this measure. Perceived noisiness increases with the (1) duration of a sound (although loudness doesn't change after about 0.2 second), (2) "unwantedness" of the sound, (3) potential of the sound for speech interference, (4) degree to which the sound disrupts sleep, (5) awareness on the part of a person concerning possible damage to hearing by sound exposure, and (6) some special meaning that the sound might have for the observer. In regard to the last attribute, perceived noisiness ratings of aircraft sounds can be relatively higher for residents of homes near an airport to the extent that the observer fears personal harm and property damage as a result of an airplane crash.

One drawback to the use of the perceived noisiness scale is the fact that it does not take into account the number of occurrences of a particular sound per day. This defect has been overcome by the development of the *composite noise rating* (CNR). Use of the CNR is illustrated in Fig. 5-26, which summarizes case history findings concerning reactions to aircraft or industrial noise sources. The figure indicates that increasingly intense reactions, ranging from complaints to legal action, are correlated with increases in the value of the CNR.

Among the most interesting results of efforts to measure sound-induced annoyance are those from the sonic boom studies, which were summarized by Kryter[55] as follows:

> It is concluded that the sonic booms from the Concorde and Boeing SST operated during the daytime sometime after 1975, at frequencies now projected for long-distance supersonic transport of passengers over the United States, will result in extensive social, political, and legal action against such flights at the beginning of, during, and after years of exposure to sonic booms from the flights.

This conclusion, with its implications for cost-benefit analyses, appears to have been a significant factor in the rejection of continued funding for the development of the Boeing SST by the United States Senate in 1971. This would seem to be the most far-reaching decision that has been strongly influenced by a potential "noise pollution" problem in the history of mankind.

Psychophysiologic stress

Possible damage to health as a result of noise-induced stress is among the topics of foremost popular concern. Concerning this topic, however, there is considerable divergence of opinion among scientists. The view of many investigators is represented by the following statement[56]:

> So-called stress reactions in the human organism when continued for sufficiently long periods can be physiologically harmful. However, it appears that the psychologic and physiologic responses to noise (excluding changes in hearing) are transitory, that they adapt out with continued exposure to noise, and therefore do not constitute harmful physiologic stress.

On the other hand, in his summary of an international symposium on the physiologic effects of noise, Leake[57] stated:

> Noise is a form of pollution that isn't often recognized as such. But actually we are all aware that we are frequently annoyed by noise, and it is the overflow from this emotional disturbance that we are now recognizing as being threatening to our good health. . . . Even though we may have learned both behaviorally and physiologically to ignore noise and thus to reduce the intensity of its emotional response, some of the neurologic stimulus may spill into the autonomic nervous system, producing cardiovascular-renal disturbance, together with endocrine, metabolic, and reproductive abnormalities.

The apparent conflict in the above statements leads us to conclude that the scientific community has *not* yet agreed upon a conceptual framework within which the topic of noise-induced damage to health can be discussed, much less what kinds of experimental evidence can provide the base for answers to our questions. Perhaps the ideas of Selye[33] concerning diseases of adaptation, integrated with concepts from the emerging discipline of psychophysiology and developments in the field of endocrinology, will lead to formulation of the needed conceptual framework.

The foundations are currently being laid for investigation of psychophysiologic stress, and it appears that included among the basic principles will be (1) a systems approach, (2) recognition and quantification of individual differences, and (3) analysis of influences of learning on visceral response systems.

The systems approach will emphasize integration of visceral responses by broad, but as yet unspecified, adaptive control mechanisms. During the past three decades the locus of control for visceral responses has gradually moved from the periphery to the brain. We have seen that the autonomic nervous system and the endocrine system are less autonomous than we previously thought and that some "higher" mechanism integrates the action of these systems. We recognize that the adaptive control mechanisms are able to maintain physiologic equilibrium in the face of stressful stimulation, but the physiologic "costs" associated with equilibrium maintenance, particularly over periods of months or years, have not been determined. Another aspect of the systems approach is the recognition of various possible reactions with negative feedback networks. Important in this regard is the realization that both the magnitude and the direction of a response are dependent on the status of the system at the time a particular stimulus is received.

We are gradually coming to recognize the wide differences that exist between individuals with respect to physiologic reactions following annoying or potentially damaging stimulation. Dimensions on which individual differences in reaction to noise exposure can be specified include (1) rate and degree of habituation with repeated stimulus presentations, (2) the degree to which "stereotyped" responses are elicited by different types of stimulation, (3) the relative precision of visceral response control, and (4) the relative balance between the reactions of the adrenergic and cholinergic branches of the autonomic nervous system. The concept of habituation is considered in the previous section on sleep, and here we indicate that this same process can account for observed changes in visceral responses. Response stereotypy has been extensively investigated, particularly by the Laceys.[58] Some individuals exhibit the same stereotyped pattern of physiologic reactions, such as changes in heart rate or stomach motility, regardless of the type of stimulation, whereas other people show widely different responses to the same stimulus presented at different times. Precision of control is reflected in the amount of rebound or overshoot in a particular response, such as heart rate, following stimulation. Although the idea of rebound has been postulated to account for various experimental results, this phenomenon has not been extensively studied. Finally, an apparently reliable measure of the relative activity of the two autonomic nervous system branches (autonomic balance, $\bar{A}$) has been developed by Wenger.[59] While useful at present, the foregoing dimensions are not completely adequate and new ways of scaling physiologic responses are currently being sought.[60]

We are in the midst of a major revision in our thinking about the ways in which learning might influence physiologic responses. Recent experiments indicate that *instrumental conditioning technics* can induce change in response systems that we thought previously were inaccessible to behavioral technics. In one experiment, rats were conditioned to increase or decrease their heart rate by reinforcement with electric stimulation of the brain "pleasure center."[61] In another investigation, rats were trained to increase the blood flow to one outer ear but not change the blood flow to the opposite ear.[62] These observations indicate that highly specific responses can be conditioned. Clearly, these new insights must be taken into account as we attempt to build a coherent discipline of psychophysiologic stress.

SOCIAL IMPLICATIONS

As a society, we strive to eliminate or control unwanted mechanical energy, and we seek to protect individuals from hazardous levels of mechanical energy exposure. *Complaint behavior,* ranging from appeals to authorities to legal action, is one of the ways in which the first goal is achieved. If dissatisfaction is widespread, *legislation* may be adopted. Examples of such legislation include *antinoise ordinances* and *building codes.* The second goal is realized by development of *damage risk exposure criteria* and adoption of programs to ensure that individuals are not exposed to energy levels in excess of these criteria. Procedures for implementation of sound damage risk criteria include *hearing conservation programs.*

Complaint behavior

The mechanisms underlying complaint behavior are complex and not completely understood.[63] An adequate discussion of this topic would take us far afield into the domain of psychology. Superficially, we can discuss complaint behavior within the framework of *value judgements.* Mechanical energy sources (for example, manufacturing plants and transportation systems) have been designed and built by us to make products or accomplish functions that satisfy our needs. We assign *positive values* to the outputs or products of these mechanical energy sources. Also, we tend to value the source itself, be it a plant or a transportation vehicle. However, if the mechanical energy generated by a particular facility or device is a byproduct that injures us, impairs our communication with others, reduces our performance capability, or otherwise disturbs us, these energy byproducts are assigned *negative values.* Complaint behavior results when negative values outweigh positive values. The intensity of the complaint behavior must somehow be related to the magnitude of the difference between the positive and negative values. This last point can be illustrated by considering Fig. 5-26. It is likely that the negative values assigned to noise sources are positively correlated with the CNR. The figure illustrates that the intensity of complaint behavior is also positively correlated with the CNR. The observation that high economic neighborhoods exhibit complaint behavior at lower CNR values than low economic neighborhoods suggests that esthetic factors influence this behavior.

An anecdote concerning difficulties encountered by the government of the United Kingdom during a research program to assess the effects of sonic booms illustrates some of the problems in interpretation of complaint behavior. During these tests the Concorde SST produced sonic booms in a rural area of Wales. Officials of a Welsh farmers' organization obtained a government complaint form and sent copies to all of their members with instructions on how to complete the form and where to send it. In some instances, the forms were even partially completed before they were mailed to the farmers. It was hardly surprising that, for some counties, the incidence of complaint was nearly 100%.[64]

Legislation and governmental regulations

Legislators at the local, state, and federal levels have moved into the area of mechanical energy control in various ways. For example, through the Walsh-Healey Public Contracts Act (1968) the United States government set standards for industrial noise exposure. At that time, the health and safety standards were revised to include the provision that sound levels in the working environment not exceed 85 dBA. This legislation applies to all firms that enter into contracts with the United States government to supply goods or services for amounts exceeding $10,000.

Table 5-4. Maximum sound levels in residential area*

Octave band-Hz	SPL-dB
37 to 75	80
75 to 150	68
150 to 300	61
300 to 600	55
600 to 1,200	51
1,200 to 2,400	48
2,400 to 4,800	45
4,800 to 9,600	43
dBA	56

*From zoning ordinance of Dallas, Texas. Corrections for nighttime: −7dB; pure tone or impulsive noise: −7 dB; and intermittent: +10 dB for 1:12 duty cycle.

Most of the large cities in the United States have adopted antinoise ordinances during the past 10 years. Typical residential limits are given in Table 5-4. We selected ten ordinances for cities ranging in size from Chicago, Illinois, to Hemet, California, that specify noise limits at residential boundary lines and calculated the maximum permissable nighttime dBA levels. The average from these ordinances was 50.2 dBA.

Noise control requirements have been included in European building codes since 1938. These codes quantitatively specify the sound attenuation characteristics of walls and floors in apartment buildings, schools, and hospitals. On the basis of personal experience, most people are probably aware that cities in the United States have been slow to adopt building codes that contain noise control requirements.

United States government agencies and departments have published numerous regulations to protect citizens from hazardous or annoying mechanical energy exposure. For example, automobile manufacturers have been required to incorporate several devices and design features into their products to protect occupants during mechanical impacts. Another example is a recent circular from the U. S. Department of Housing and Urban Development (HUD) which prohibits HUD support to new construction on sites having unacceptable noise exposures. Finally, the U. S. Department of Defense has published damage risk exposure criteria to protect military personnel, and it is to this topic that we turn next.

Damage risk exposure criteria

Since the 1930's, a great deal of effort has been directed toward establishment of criteria for maximum sound exposure. This effort would be considerably simplified if there were a sharp line of demarcation between "safe" and damaging sound exposures across the population. Unfortunately, as we have noted repeatedly, this is not the situation; rather, individuals differ widely in their susceptibility to sound-induced hearing loss. Those who propose exposure criteria are faced with a dilemma: if criteria are set low enough to protect highly susceptible persons, the majority of the population would find the restrictions unnecessary and the "credibility" of the criteria would be undermined. On the other hand, if the criteria are set too high, susceptible individuals may suffer unnecessary hearing loss.

The most detailed and elaborate sound damage risk exposure criteria published to date are those proposed by the Committee on Hearing, Bioacoustics and Biomechanics (CHBAB) of the U. S. National Academy of Sciences–National Research Council.[65] Damage risk contours proposed by the CHBAB group for periodic and nonperiodic sound exposures are presented in Figs. 5-27 and

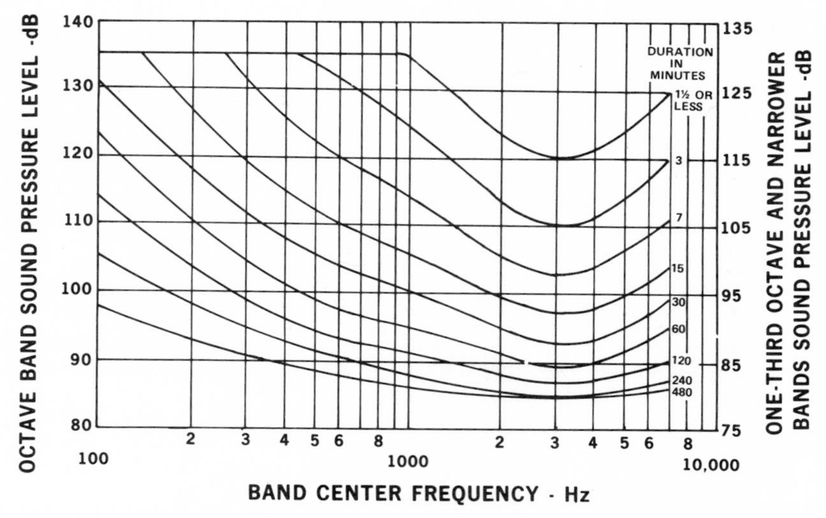

Fig. 5-27. Damage risk contours for noise exposure. The curves indicate daily noise exposures that would probably produce damage in terms of noise level, frequency, and duration. If noise exposure approaches the levels indicated by the contours, hearing conservation procedures should be implemented. (Redrawn from Kryter, K. D., Ward, W. D., Miller, J. D., and Eldredge, D. H.: J. Acoust. Soc. Am. **39**:451, 1966.)

5-28. The curves of Fig. 5-27 indicate the noise levels at which one exposure per day is likely to produce damage. Octave band sound levels are given on the vertical axis at left and one-third octave band levels are given on the right. The parameter in the figure is exposure duration, ranging from 1.5 minutes to 8 hours. The curves can be applied to particular band levels that may be present in a broad band noise. For example, the bottom curve indicates that exposure to a one-third octave band of noise centered at 3,000 Hz with a level of 80 dB for 8 hours daily constitutes a potential threat to hearing. The curves of Fig. 5-28 for pure tone exposures are interpreted in the same manner as the curves of Fig. 5-27. Further details concerning the development and interpretation of these damage risk contours can be found in Burns[24] and Kryter.[42]

Another set of criteria have been proposed by the American Academy of Ophthalmology and Otolaryngology (see Table 5-5). These criteria cover a narrower frequency range and are slightly more conservative than the contours of the CHBAB group.

Damage risk exposure criteria for impulse noise, resulting from weapons fire and other airborne impacts, are not so complete as the criteria for continuous sound. In part, this results from the fact that impulse noise measurement is not so convenient and reliable as continuous sound measurement. Tentative guidelines for impulse noise exposure limitation suggest that ear protection should be used in the following situations: (1) the impulse noise is subjectively judged as uncomfortable, (2) the level of the impulse noise exceeds 135 dB, and (3) the impulse noise is presented repeatedly.[12]

Proposed vibration exposure criteria are presented in Fig. 5-29. Acceleration in G is given on the vertical axis. The horizontal axis is vibration frequency, and the parameter in the figure is vi-

Table 5-5. Recommended maximum daily sound exposure*

Average level of 300 to 600, 600 to 1,200, and 1,200 to 2,400 Hz bands (dB)	On-time per day (minutes)
85	Less than 300
90	Less than 120
95	Less than 50
100	Less than 25
105	Less than 16
110	Less than 12
115	Less than 8
120	Less than 5

*From the Guide for conservation of hearing in noise, American Academy of Ophthalmology and Otolaryngology, Los Angeles, 1965.

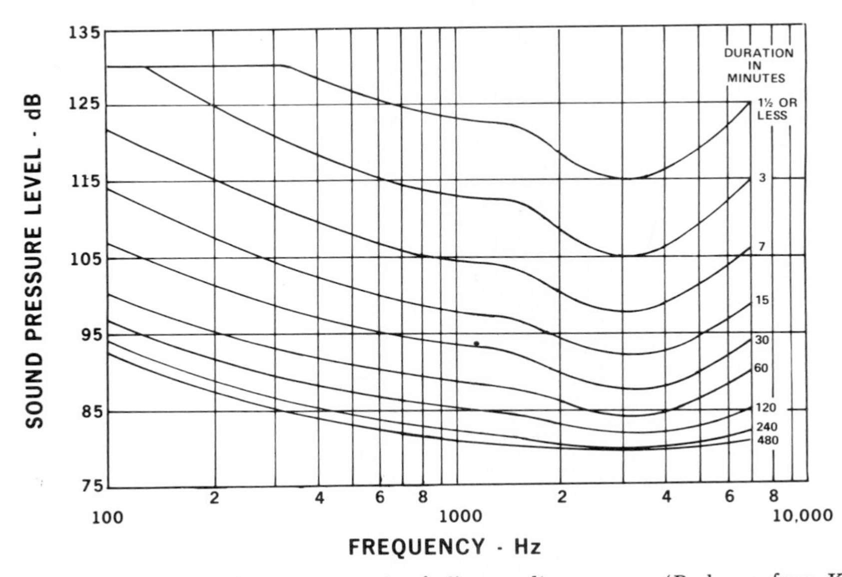

Fig. 5-28. Damage risk contours for pure tone (periodic sound) exposure. (Redrawn from Kryter, K. D., Ward, W. D., Miller, J. D., and Eldredge, D. H.: J. Acoust. Soc. Am. 39:451, 1966.)

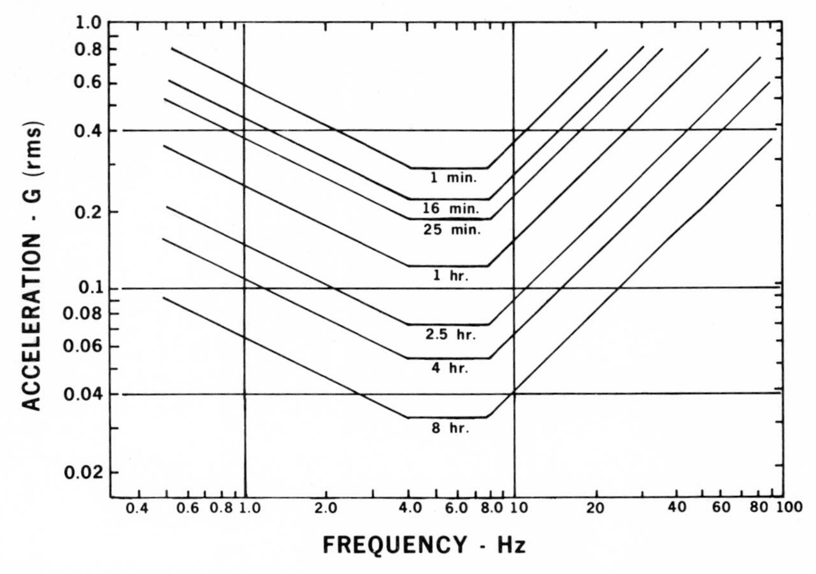

Fig. 5-29. "Fatigue-decreased proficiency" boundaries for vibration exposure in terms of vibration amplitude (acceleration), frequency, and duration. Multiply acceleration values by 2 for damage risk criteria. Divide acceleration values by 3.15 for "reduced comfort" boundaries. (Modified from von Gierke, H. E., and Clarke, N. P.: Effects of vibration and buffeting on man. In Randel, H., editor: Aerospace medicine, Baltimore, 1971, The Williams & Wilkins Co.)

bration exposure duration. The curves indicate "fatigue-decreased proficiency" boundaries. Damage risk criteria are obtained by doubling the acceleration values, and "reduced comfort" boundaries can be determined by dividing the acceleration values by 3.15. For example, the curves indicate that 16 to 25 minutes of vibration exposure to 0.2 G at 4 to 8 Hz is likely to result in reduced performance proficiency. Vibration exposure of the same duration and frequency at 0.4 G is likely to result in damage.[5]

Mechanical impact tolerance criteria, which are similar to the damage risk exposure criteria for continuous sound and vibration, have been published.[66] Different exposure criteria curves are required for different orientation of the impact deceleration with respect to the body. For example, men show less tolerance for z axis than for x axis deceleration impacts. Tolerance curves are plotted in terms of peak deceleration and deceleration duration. As impact duration increases, the peak deceleration likely to produce damage is reduced.

Hearing conservation programs

Practical programs for hearing preservation have three basic components: (1) the sound environment must be quantitatively specified, (2) sound exposure must be limited according to a set of criteria specifying potential hearing damage, and (3) hearing abilities must be monitored before and after sound exposure to ascertain that the exposure criteria are correct and to attempt identification of individuals who are highly susceptible to damage.

Overall sound intensities are ordinarily measured with the dBA and dBC scales of sound level meters. One-third octave band analyses of the sound spectrum are also required. If it is suspected that a great deal of the sound energy is concentrated within a narrow frequency range, a more selective spectrum analysis is indicated. Also, information concerning sound level fluctuations and total daily exposure time is necessary.

Damage risk exposure criteria are presented in the preceding section. In practical situations, individual workers may move unpredictably from one sound environment to another. Consequently, it can be difficult to determine whether or not a person is receiving potentially damaging sound exposure. Sound dosimeters, analogous to those worn by workers in radiation facilities, are currently being developed and tested as one approach to the solution of this problem.

If the sound exposures approach the damage risk criteria, some form of protection for exposed individuals is indicated. Ideally, sound is controlled at the energy source, or attenuation is introduced into the transmission pathway. Alternatively, work procedures are altered so that human beings are not exposed to high-intensity sound environments for long periods. If the sound cannot be controlled and the human being's presence is required, then the man must be protected in some way. Earplugs and earmuffs have been developed for these situations. Some types of earmuffs provide over 40-dB attenuation at moderately high frequencies (2,000 to 6,000 Hz). Earplugs are somewhat less effective. As can be predicted from studying the physiologic activity of the ear, speech communication can be improved by wearing earmuffs or earplugs when noise levels approach 85 dB or more.[42]

DIRECTIONS FOR FUTURE RESEARCH

Research on the biologic effects of mechanical energy exposure is currently being pursued in all of the areas discussed on p. 138. Throughout the chapter particularly glaring deficiencies in our knowledge have been pointed out. Here our focus is not on particular research problems; rather, we are concerned with broad methodologic issues and strategies. First, we consider some advantages to formulation of research problems within the framework of *analysis of variance*. Second, we discuss some of the problems encountered with traditional experimental procedures and suggest advantages of *unobtrusive measures*.

Analysis of variance

Analysis of variance experimental designs allow an experimenter to investigate not only the effects of particular independent variables considered separately but also to examine possible interactions in the effects of two or more independent variables. Fig. 5-30 illustrates the results of a hypothetical experiment in which sound intensity and duration are independently varied. Examination of the results reveals that long-duration sounds elicit a larger response than short-duration sounds and that high intensities have a greater effect than low intensities. In other words, there are *significant main effects* associated with each of the independent variables (intensity and duration). As can be seen from the figure, however, that is not the whole story. In addition, the data reveal a significant interaction: the combination of high intensity and long duration produces a greater response than would be expected from the additive results of duration and intensity examined separately. In some instances, observation of interactions may be especially valuable in determining the biologic processes underlying a response. Several cases of interaction between parameters of a given type of mechanical energy have been investigated (for example, the relationship between sound level and exposure duration in producing TTS). However, we know relatively little about possible interactions across mechanical energy types (such as sound and vibration) or between mechanical energy and other types of stimulation (such as sound and heat). It is important that we obtain more information about possible interactions between types of stimulation because the environments to which people are exposed in the course of their daily activities are complex; more than one form of stimulation is ordinarily present.

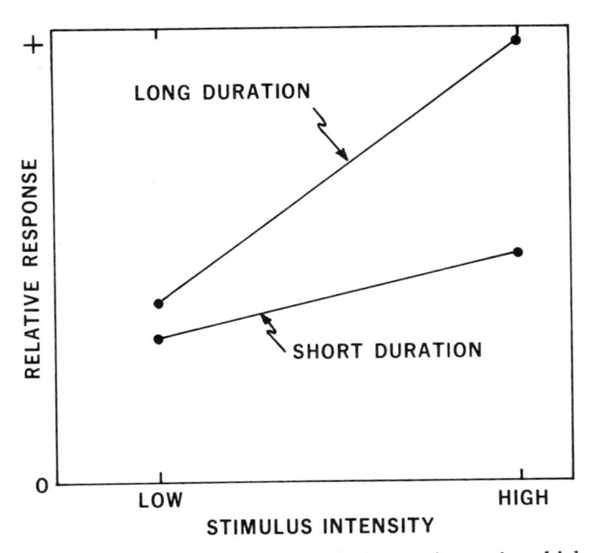

Fig. 5-30. Results of a hypothetical experiment in which stimulus intensity and stimulus duration are independently varied. These data illustrate a significant interaction between the two independent variables. Advantages of experimental design within the framework of analysis of variance are considered in the text.

Unobtrusive measures

Psychologists have become increasingly concerned with possible artifacts in behavioral research as a result of experimenter or subject bias. For example, in experiments on the physiologic effects of intense sound exposure, the subject's responses can be biased by the ways in which the experimenter describes the purpose of his research to the subject. Whether or not a subject is told the purpose of an experiment, he is likely to formulate hypotheses about what the experimenter expects to observe. The dominant attitude among experimental subjects seems to be cooperation and compliance: the subjects generally want the experiment to succeed. In conjunction with recent observations on voluntary control of visceral systems responses,[62] possible subject bias is a cause for some concern with regard to the validity of sound-induced changes that have been reported. Further discussion of subject and experimenter bias effects can be found in Rosenthal and Rasnow.[67]

Some scientists have suggested that the problems with traditional experimental technics are sufficiently great to warrant adoption of a new set of procedures for research. Proposed new procedures include unobtrusive measures. Popularity of museum exhibits can be determined with the unobtrusive measure of relative rates of floor tile replacement in various parts of the museum. Interest in a lecture can be unobtrusively measured by the number of handouts left in the lecture hall by the audience after the lecture is over. Noise interference with speech communication can be measured by an obtrusive participant-observer who records the durations of telephone conversations in different parts of an airport where different levels of background noise are present. Unobtrusive experimental technics may require a greater effort than the more traditional questionnaire or interviewer procedures, but quite different and probably more valid results are frequently obtained when unobtrusive measures are employed.[68]

OCTOBER, 1971

GLOSSARY

A scale sound level (dBA) In sound level measurements with a sound level meter, a measure of sound intensity. The A scale employs a frequency weighting network to discriminate against low and high frequencies, with the result that the sensitivity (loudness response) of the human ear at moderate sound levels is approximated.

acceleration Rate of change of velocity with respect to time. Acceleration may be either linear (centimeters per second per second, cm/sec^2) or angular (degrees per second per second, deg/sec^2). Linear acceleration may also be expressed in G's, or multiples of the earth's gravitational acceleration, 978 cm/sec^2.

acoustic Pertaining to sound, its generation, propagation, or effects (primarily audible sound). Acoustics is the science of sound.

acoustic trauma Damage to the auditory receptor resulting from sound exposure.

air conduction Sound transmitted through air to the ear.

ambient (sound or vibration) Measurable energy in the background; energy sources not clearly identifiable.

amplitude Amount of stimulus or response measured on some scale. For periodic oscillation amplitude may be peak to peak (difference between maximum and minimum), half wave (difference between average and maximum), or root mean square (rms) (difference between average and maximum multiplied by 0.707).

analog; analog model A working model or simulation of some aspect of a system's behavior by mechanical, electric, or mathematic means; for example, computer simulation of space vehicle motion.

anatomic axes A system of three orthogonal axes used to specify body motion or direction of energy application with respect to the body (see Fig. 5-20).

angular vibration Vibration around an axis.

annoyance The feeling of being annoyed; verbal report of disturbance and the expression of dislike, usually as a result of stimulation.

attenuation Reduction in signal strength resulting from some element (attenuator) in the transmission pathway.

audiofrequency range Interval of sound or vibration frequency spectrum from 20 to 20,000 Hz, which is the extent of the audible range for human beings.

audiogram A chart, table, or graph showing hearing level for pure tones as a function of frequency.

audiometer An instrument for measuring hearing acuity.

autonomic balance Relative predominance in function of the adrenergic and cholinergic branches of the autonomic nervous system. Seven weighted variables are combined to yield a single score, $\overline{A}$.

band pressure level The sound pressure level of the acoustic energy within a specified frequency band.

bandwidth The difference between upper and lower cutoff frequencies, expressed either in units of frequency or as a percentage of the center frequency of the pass-band.

biodynamics The science of the dynamic mechanical properties of the body.

breaking index Percentage increase in length required for breakage; equal to breaking strength/Young's modulus.

broad-band noise (vibration) Noise (or vibration) for which the energy is distributed over a wide frequency range; nonperiodic sound (or vibration).

buffeting Intense vibration of an aircraft, particularly as a result of turbulence.

C scale sound level (dBC) In sound level measurements using a sound level meter, a measure of sound intensity. The C scale employs a frequency weighting network that is nearly flat across the audiofrequency range and approximates the sensitivity (loudness response) of the human ear to high-intensity sound.

center frequency The geometric mean of the nominal cutoff frequencies of a pass-band.

circadian rhythm Regular fluctuation of body chemistry, physiologic activity, or behavior, related to the 24-hour or day-night cycle.

closed loop system A system where the energy or information pathways are arranged so the output from the system influences the input to the system.

composite noise rating (CNR) For exposures during the daytime, CNR = average peak PNdB $- 12 + 10 \times \log_{10} N$, where N is the number of occurrences of the noise.

compression wave (longitudinal wave) Propagation of a wave as a result of compressive stress in an elastic medium (such as sound in air). Particle motion is in the direction of wave propagation.

conductive deafness Hearing loss as a result of a lesion in the sound-conducting mechanism of the ear.

Corti, organ of Portion of inner ear consisting of sensorineural elements that transduce displacement into electrochemical neural energy.

damage risk contours Sound levels, defined in terms of intensity, frequency, and exposure duration, producing, for the average ear, a permanent hearing loss after 10 years of more than 10 dB at 1,000 Hz or 15 dB at 2,000 Hz or 20 dB at 3,000 Hz and above. Damage risk contours for vibration are currently being developed.

damage risk criteria (DRC) Sound levels, defined in terms of frequency and intensity, to which exposure for 8 hours per working day over a person's lifetime is likely to produce hearing loss. DRC for impulse sound and vibration are under development.

damping Diminution of an oscillation caused by energy dissipation.

decibel (dB) One tenth of a bel; scale for measurement of level (intensity) of pressure or power. Decibel is the ratio between two quantities, one of which is an arbitrary reference level. For sound pressure, 0 dB = 0.0002 microbar.

dissipation The process by which energy is lost in the form of heat.

dosimeter; noise dosimeter An instrument that measures the intensity of sound above a set level and relates this to the exposure time.

edema Effusion of serous fluid between cells in tissue spaces or into body cavities.

elasticity (elastance) Property of a substance that allows it to change its length, shape, or volume in response to a force and to recover its original form when the force is removed.

elastic reactance Contribution to the total impedance by the elastic properties of an energy receiver; inversely proportional to the frequency of the input energy.

embolus (plural, **emboli**) Undissolved material impacted in some vascular tissue space such as tissue fragments, gas bubbles, and the like.

frequency (of a periodic quantity) The reciprocal of the period; the number of complete cycles in a unit of time (units: the hertz [Hz] = 1 cps; also for convenience in dealing with high-frequency vibrations in acoustics, ultrasonics, and electronics, the kilohertz, 1 kHz = 1,000 Hz; and the megahertz = 1,000,000 Hz).

frequency response The magnitude of output as a function of the frequency of the input signal.

friction Resistance to relative motion at an interface between components of a system; results in thermodynamically irreversible energy dissipation.

G The acceleration of gravity; 978 cm/sec^2.

habituation Reduction of a response following repeated presentation of a stimulus, which is assumed to be mediated by changes in neural signal processing; as opposed to adaptation, which refers either to receptor changes (visual dark adaptation) or central nervous system changes (adaptation to sensory distortion).

hair cell Cells of mechanoreceptors that transduce mechanical displacement into electrochemical energy; found in the inner ear and the vestibular receptors.

hearing level A threshold of hearing, determined by audiometry, related to a specified standard of normal hearing (units: dB).

impact Collision between two objects (mechanical); or collision of object with energy contained in a pressure front (airborne).

impedance (Z) Ratio of generalized effort to generalized flow; has been defined for electric, mechanical, and acoustic cases. Impedance indicates the degree to which a stimulus (input) produces a response (output).

infrasound Vibration at a frequency below the audible range.

inner ear (cochlea) Portion of the ear that contains the sensorineural elements that transduce displacement into electrochemical energy.

instrumental conditioning Behavior modification technic wherein presentation of reinforcement (reward or punishment) is contingent upon performance of a specified response by the organism.

intelligibility; speech intelligibility An expression of the ease or efficiency with which a listener understands speech in the presence of sound or vibration; assessed in various ways, often as the percentage of correct scores made out of a spoken or recorded message or word list in prescribed conditions.

level In acoustics, the logarithm of the ratio of the value of some quantity (such as sound pressure) to a reference quantity of the same kind.

lumped parameter system System of components with differing physical characteristics but specified with a single set of parameters for the purpose of analysis.

masking The amount (expressed in decibels) that the threshold of audibility for one sound is raised by the presence of another sound.

mechanoreceptor Sensorineural system that transduces mechanical energy into electrochemical energy, thereby producing neural activity.

middle ear reflex Contraction of the muscles of the middle ear elicited by intense sound; attenuates sound transmission from the outer ear to the cochlea.

narrow-band The pass-band of a filter or the spectral distribution of a noise or vibration. The term commonly means a bandwidth of one-third octave or less.

near field The part of the field relatively close to a source of sound radiating in free-field conditions, in which the sound pressure and particle velocity are not in phase. The particle velocity does not necessarily decay in linear proportion to the distance from the source.

noise (1) (acoustics) Disturbing or unwanted sound. (2) (instrumentation; communications) A spurious, unwanted, or irrelevant signal. (3) Any nonperiodic waveform.

nonperiodic Vibrations containing more than one and usually many frequencies.

octave The interval or band between two frequencies, one of which is twice the other.

overshoot A transient response exceeding the forced response of a dynamic system to a sudden change of input.

particle velocity The vibratory velocity of the components of the medium conveying the sound wave; independent of the velocity of sound in the medium.

peak; peak value The maximum value, irrespective of sign, of a variable such as the magnitude of vibration.

peak-to-peak amplitude The algebraic difference between extreme values of an oscillating quantity.

perceived noise level (PNL) The level assigned to a noise judged by normal observers to be equally as noisy as a reference noise consisting of a band of random noise between one-third and one octave wide, centered on 1,000 Hz (units: PNdB—perceived noise decibels).

period The reciprocal of frequency.

permanent threshold shift (PTS) Hearing level shift that shows no progressive reduction with the passage of time when the apparent cause has been removed.

phase The fractional part of the cycle of a sinusoidal oscillation through which the motion has advanced as measured from a specific point in the cycle; or the difference in time, expressed as a fraction of the cycle, between the motion and an equivalent point in the cycle of a harmonically related oscillation.

presbycusis; presbyacusia Decline in hearing acuity with age, usually manifested by increasing losses at the high frequency end of the audible range.

resonance The condition in which any change in the frequency of excitation causes a decrease in the response of a vibrating system.

resonant frequency; resonance frequency A frequency at which resonance occurs.

root mean square (rms) value The square root of the average of the squared values of any set of numbers.

semicircular canals Angular acceleration receptors located in the labyrinth adjacent to the cochlea.

sensorineural hearing loss Deafness resulting from a lesion of the cochlear end-organ or its nerve supply.

shear The lateral deformation produced in a body by an external force; expressed as the ratio of the lateral distance between two points lying in parallel planes to the vertical distance between the points.

shearing strength Shear magnitude required for breakage.

shock (1) (engineering) A sudden force or displacement causing transient vibration or shock motion of a mechanical system. (2) (medicine) Acute physiologic disturbance resulting from severe stress or injury.

short-term memory Hypothesized initial storage stage in the encoding of sensory information.

signal-to-noise ratio (S/N ratio) In hearing, the level of the desired auditory stimulus (signal) relative to the level of unwanted stimuli (noise).

sonic boom; sonic bang The abrupt sound (characteristically a double bang) heard following the passage of an aircraft flying faster than the speed of sound.

sound Vibration of the material of a medium; also the sensation produced by such vibration when transmitted to the ear.

sound level A weighted sound pressure level obtained by the use of a sound level meter incorporating a prescribed frequency weighting network; loosely, sound intensity.

sound power The total acoustic energy radiated by a source in unit time (units: watts).

sound pressure level (SPL) A level of sound pressure that is twenty times the logarithm to the base 10 of the ratio of the pressure of this sound to a prescribed reference pressure (units: dB).

spectrum A description (commonly graphic or tabular) of a quantity in terms of frequency or as a function of frequency; also used loosely to refer to a frequency range, as in "audiofrequency spectrum."

speech interference level (SIL) Of a noise, the average of the sound pressure levels of the noise in the three octave bands centered at 500, 1,000, and 2,000 Hz (units: dB).

startle The physiologic response to a sudden stimulus, typically a loud unexpected noise.

stress (1) (physiology) Reversible or irreversible changes in the body, measurable by physiologic or biochemical methods or by performance, caused by external factors, such as severe environmental conditions, or demanding psychologic influences; also used loosely to refer to any stressful environmental condition such as vibration or sound capable of producing a physiologic response or deterioration of performance. (2) (engineering; biomechanics) The ratio of force to effective area in a load-bearing member or structure.

temporary threshold shift (TTS) The component of threshold shift that shows progressive reduction with the passage of time when the apparent cause has been removed.

tensile strength Minimum longitudinal stress required for breakage.

third-octave; ⅓ octave The interval between two frequencies having a frequency ratio of $2^{1/3}$ (1.2599).

threshold For a specified sensory modality (vision, hearing, vibration), the lowest level (absolute threshold) or smallest difference (difference threshold, difference limen) of intensity of the stimulus discernible by normal subjects in prescribed conditions of stimulation.

translation; translational vibration Linear motion (in a particular axis) as opposed to rotatory.

transient; transient vibration A vibration that is maintained for a relatively short time or that decays to negligible magnitude in a relatively short time. NOTE: Transient vibration may be free or forced.

ultrasonic Having a frequency above the audible frequency range.

utricle Linear acceleration receptor located in labyrinth adjacent to cochlea. The saccule has a similar structure but its function in man is uncertain.

viscous damping Damping in which the dissipative force is proportional and oppositely directed to the velocity of the damped motion.

weighting; frequency-weighting The selective modification of the values of a complex signal or function for the purposes of analysis or assessment, according to prescribed rules or formulas. This may be done by computation or by the use of electronic frequency-weighting networks to modify input signals.

whole-body vibration Vibration of man or animals applied through one or more principal supporting surfaces (a seat or platform).

x axis vibration Vibration acting along the anteroposterior (chest-to-back; fore-to-aft) axis of the human body.

y axis vibration Vibration acting sideways upon the human body; mutually perpendicular to x axis and z axis vibration.

Young's modulus Ratio of stress to strain; tensile coefficient of elasticity.

z axis vibration Vibration acting along the longitudinal (head-to-foot; cephalocaudal) axis of the human body. It is commonly called vertical vibration and, sometimes, longitudinal vibration.

REFERENCES

1. Kiel, F. W.: Hazards of military parachuting, Milit. Med. **130:**512, 1965.

2. Stapp, J. P.: Biodynamics of deceleration impact and blast. In Randel, H., editor: Aerospace medicine, Baltimore, 1971, The Williams & Wilkins Co., pp. 118-166.
3. Cohen, A., and Ayer, H. E.: Some observations of noise at airports and in the surrounding community, Am. Ind. Hyg. Assoc. J. **25**:139, 1964.
4. von Bekesy, G.: Experiments in hearing, New York, 1960, McGraw-Hill Book Co.
5. von Gierke, H. E., and Clarke, N. P.: Effects of vibration and buffeting on man. In Randel, H., editor: Aerospace medicine, Baltimore, 1971, The Williams & Wilkins Co., pp. 198-223.
6. von Gierke, H. E., and others: Physics of vibration in living tissues, J. Appl. Physiol. **4**(12):886-900, 1952.
7. von Bekesy, G., and Rosenblith, W. A.: The mechanical properties of the ear. In Stevens, S. S., editor: Handbook of experimental psychology, New York, 1951, John Wiley & Sons, Inc., pp. 1075-1115.
8. Melzack, R., and Wall, P.: Pain mechanisms: a new theory, Science **150**:971, 1965.
9. McCabe, B. F., and Lawrence, M.: Effects of intense sound in the non-auditory labyrinth, Acta Otolaryngol. **49**:147, 1958.
10. Albernaz, P. L., Covell, W. P., and Eldredge, D. H.: Changes in the vestibular labyrinth with intense sound, Laryngoscope **69**:1478, 1959.
11. Engstrom, H., Ades, H. W., and Hawkins, J. E.: Cellular pattern, nerve structures, and fluid spaces of the organ of Corti. In Neff, W. D., editor: Contributions to sensory physiology, vol. 1, New York, 1965, Academic Press, Inc.
12. von Gierke, H. E., and Nixon, C. W.: Noise effects and speech communication in aerospace environments. In Randel, H., editor: Aerospace medicine, Baltimore, 1971, The Williams & Wilkins Co., pp. 224-253.
13. Nixon, C. W., Hille, H. K., Sommer, H. C., and Guild, E.: Sonic booms resulting from extremely low-altitude supersonic flights: measurements and observations on structures, livestock and people, AMRL TR 68, Wright-Patterson AFB, Ohio, 1968, Aerospace Medical Research Laboratories.
14. Parrack, H. O.: Ultrasound and industrial medicine. Noise—causes, effects, measurement, costs, control, Ann Arbor, 1952, University of Michigan.
15. Parker, D. E., Covell, W. P., and von Gierke, H. E.: Exploration of vestibular damage in guinea pigs following mechanical stimulation, Acta Otolaryngol., Suppl. 239, 1968.
16. Swearingen, J. J.: Tolerance of the human face to crash impact, AM 68-20, Oklahoma City, 1965, Civil Aeromedical Research Institute.
17. Gurdjian, E. S., Roberts, V. L., and Thomas, L. M.: Tolerance curves of acceleration and intracranial pressure and protective index in experimental head injury, J. Trauma **6**:600, 1966.
18. Friede, R. L.: Experimental concussion from acceleration, Arch. Neurol. **4**:449, 1961.
19. Widman, W. D.: Blunt trauma and the normal spleen, peacetime experiences at a military hospital in Europe, Milit. Med. **134**:25, 1969.
20. Life, J. S., and Pince, B. W.: Response of the canine heart to thoracic impact during ventricular diastole and systole, J. Biomechanics **1**:169, 1968.
21. Wever, E. G.: Theory of hearing, New York, 1949, John Wiley & Sons, Inc.
22. Ward, W. D.: Auditory fatigue and masking. In Jerger, J., editor: Modern developments in audiology, New York, 1963, Academic Press, Inc., chap. 7.
23. Miller, J. D., Watson, C. S., and Covell, W. P.: Deafening effects of noise on the cat, Acta Otolaryngol., Suppl. 176, 1963.
24. Burns, W.: Noise and man, Philadelphia, 1968, J. B. Lippincott Co.
25. Ward, W. D.: Susceptibility to auditory fatigue. In Neff, W. D., editor: Contributions to sensory physiology, vol. 3, New York, 1968, Academic Press, Inc.
26. Wever, E. G., and Lawrence, M.: Physiological acoustics, Princeton, N. J., 1954, Princeton University Press.
27. Rasmussen, G. L.: Efferent fibers of the cochlear nerve and cochlear nucleus. In Rasmussen, G., and Windel, W.: Neural mechanisms of the auditory and vestibular systems, Springfield, Ill., 1960, Charles C Thomas, Publisher.
28. Mohr, G. C., Cole, J. N., Guild, E., and von Gierke, H. E.: Effects of low frequency and infrasonic noise on man, Aerospace Med. **36**:817, 1965.
29. Milsum, J. H.: Biological control systems analysis, New York, 1966, McGraw-Hill Book Co.
30. Parker, D. E., von Gierke, H. E., and Reschke, M. F.: Studies of acoustical stimulation of the vestibular system, Aerospace Med. **39**:1321, 1968.
31. Trevarthen, C. B.: Two mechanisms of vision in primates, Psychol. Forschung **31**:299, 1968.
32. Sjoberg, A.: Experimental studies of the eliciting mechanism of motion sickness. In Fourth symposium on the role of the vestibular organs in space exploration, NASA SP-187, Washington, D. C., 1968, National Aeronautics and Space Administration.
33. Selye, H.: The stress of life, New York, 1956, McGraw-Hill Book Co.
34. Hood, W. B., and others: Cardiopulmonary effects of whole body vibration in man, J. Appl. Physiol. **21**:1725, 1966.
35. Goldman, D. E., and von Gierke, H. E.: Effects of shock and vibration on man. In Harris, C. M., and Crede, C. E., editors: Shock and vibration handbook, New York, 1961, McGraw-Hill Book Co., chap. 44.
36. Henkin, R. I., and Knigge, K. M.: Effect of sound on the pituitary adrenal axis, Am. J. Physiol. **204**:710, 1963.
37. Lockett, M. F.: Effects of sound on endocrine function and electrolyte excretion. In Welch, B. L., and Welch, A. S., editors: Physiological effects of noise, New York, 1970, Plenum Press, pp. 21-42.
38. Arguelles, A. E., Martinez, M. A., Pucciarelli, E., and Disisto, M. V.: Endocrine and metabolic effects of noise in normal, hypertensive and psychotic subjects. In Welch, B. L., and Welch, A. S., editors: Physiological effects of noise, New York, 1970, Plenum Press, pp. 43-56.
39. Jansen, G.: Noise effect during physical work, Intern. Z. Angew. Physiol. **20**:233, 1964.
40. Davis, R. C., Buchwald, A. M., and Frankman, R. W.: Autonomic and muscular responses and their

relation to simple stimuli, Psychol. Monographs, vol. 69(2), No. 405, 1955.
41. Jansen, G.: Adverse effects of noise on iron and steel workers, Stahl Eisen **81**:217, 1961.
42. Kryter, K. D.: The effects of noise on man, New York, 1970, Academic Press, Inc.
43. Stern, R. M.: Effects of variation in visual and auditory stimulation on gastrointestinal motility, Psychol. Rep. **14**:799, 1964.
44. Tamari, I.: Audiogenic stimulation and reproductive function. In Welch, B. L., and Welch, A. S., editors: Physiological effects of noise, New York, 1970, Plenum Press, pp. 117-130.
45. Welch, B. L., and Welch, A. S., editors: Physiological effects of noise, New York, 1970, Plenum Press.
46. Pollack, I., and Pickett, J. M.: Stereophonic listening and speech intelligibility against voice babble, J. Acoust. Soc. Am. **30**:131, 1958.
47. Dewson, J. H.: Efferent olivo-cochlear bundle: some relationships to stimulus discrimination in noise, J. Neurophysiol. **31**:109, 1968.
48. Holloway, C.: Noise and efficiency: the spoken word, New Scientist **42**:247-248, 1969.
49. Nixon, C. W., and Sommer, H. C.: Influence of selected vibrations upon speech. III. Range of 6 cps to 20 cps for semi-supine talkers, Aerospace Med. **34**:1012, 1963.
50. Hockey, R.: Noise and efficiency: the visual task, New Scientist **42**:244-246, 1969.
51. Jouvet, M.: Biogenic amines and the states of sleep, Science **163**:32, 1969.
52. Williams, H. L., and others: Responses to auditory stimulation, sleep loss, and the EEG stages of sleep, Electroencephal. Clin. Neurophysiol. **16**:269, 1964.
53. Kryter, K. D., and Pearsons, K. S.: Some effects of spectral content and duration on perceived noise level, J. Acoust. Soc. Am. **35**:866, 1963.
54. Kryter, K. D., and Pearsons, K .S.: Modifications of the noy table, J. Acoust. Soc. Am. **36**:394, 1964.
55. Kryter, K. D.: Sonic booms from supersonic transport, Science **163**:359, 1969.
56. Kryter, K. D., and others: Nonauditory effects of noise, NAS-NRC Committee on Hearing Bioacoustics and Biomechanics, Washington, D. C., 1971, National Academy of Sciences.
57. Leake, C. D.: Summary of the symposium. In Welch, B. L., and Welch, A. S., editors: Physiological effects of noise, New York, 1970, Plenum Press, pp. 337-340.
58. Lacey, J. I., and Lacey, B. C.: Verification and extension of the principle of autonomic response-stereotypy, Am. J. Psychol. **71**:50, 1958.
59. Wenger, M. A.: Studies of autonomic balance in army air forces personnel, Comp. Psychol. Monogr., vol. 19(4), serial 101, 1948.
60. Sternbach, R. A.: Principles of psychophysiology, New York, 1966, Academic Press, Inc.
61. Miller, N. E., and Banuazizi, A.: Instrumental learning by curarized rats of a specific visceral response, intestinal, or cardiac, J. Comp. Physiol. Psychol. **65**: 1, 1968.
62. Miller, N. E.: Learning of visceral and glandular responses, Science **163**:434, 1969.
63. Parrack, H. O.: Community reactions to noise. In Harris, C. M.: Handbook of noise control, New York, 1957, McGraw-Hill Book Co., chap. 36.
64. Whiteside, T. D. C.: Personal communication, April, 1971.
65. Kryter, K. D., Ward, W. D., Miller, J. D., and Eldredge, D. H.: Hazardous exposure to intermittent and steady-state noise, J. Acoust. Soc. Am. **39**: 451, 1966.
66. von Gierke, H. E.: Biodynamic response of the human body, Appl. Mech. Rev. **17**:951, 1964.
67. Rosenthal, R., and Rasnow, R. L., editors: Artifact in behavioral research, New York, 1969, Academic Press, Inc.
68. Webb, E. J., Campbell, D. T., Schwartz, R. D., and Sechrest, L.: Unobtrusive measures: nonreactive research in the social sciences, Chicago, 1966, Rand McNally and Co.

SUGGESTED READINGS

1. von Bekesy, G.: Ueber akustische reizung des vestibularapparates, Pfluger Arch. Ges. Physiol. **236**: 59, 1935.
2. von Bekesy, G.: Description of some mechanical properties of the organ of Corti, J. Acoust. Soc. Am. **25**:770, 1953.
3. Broadbent, D. E.: Effects of noise on behavior. In Harris, C. M.: Handbook of noise control, New York, 1957, McGraw-Hill Book Co., chap. 10.
4. Davis, H., and Silverman, R. S., editors: Hearing and deafness, New York, 1960, Holt, Rinehart and Winston, Inc.
5. Davis, H., and others: Acoustic trauma in the guinea pig, J. Acoust. Soc. Am. **25**:1180, 1953.
6. Dieckmann, D.: Einfluss horizontaler mechanischer schwingungen auf den Menschen, Intern. Z. Angew. Physiol. **17**:83, 1958.
7. Leukel, F.: Introduction to physiological psychology, ed. 2, St. Louis, 1972, The C. V. Mosby Co.
8. Mikeska, E. F.: Noise in the modern home, Noise Control **4**:38, 1958.
9. Miller, L. N., and Beranek, L. L.: Noise levels in the Caravelle during flight, Noise Control **4**:19, 1958.
10. Nickerson, J. L., and Coermann, R. R.: Internal body movements resulting from externally applied sinusoidal forces, AMRL TR 62-81, Wright-Patterson AFB, Ohio, 1962, Aerospace Medical Research Laboratories.
11. Randel, H., editor: Aerospace medicine, Baltimore, 1971, The Williams & Wilkins Co. (especially chaps. 8, 11, and 12).
12. Rosenberg, G., and Segal, R.: The effects of vibration on manual fire control in helicopters, Technical Report 1-168, Philadelphia, 1966, Franklin Research Institute Laboratories.
13. Veneklasen, P. S.: City noise—Los Angeles, Noise Control **2**:14, 1956.
14. Ward, W. D., Glorig, A., and Sklar, D. L.: Temporary threshold shift from octave-band noise: applications to damage-risk criteria, J. Acoust. Soc. Am. **31**:522, 1959.
15. Wiener, F. M.: Experimental study of airborne noise generated by passenger automobile tires, Noise Control **6**:13, 1960.

6 ACCELERATION, GRAVITY, AND WEIGHTLESSNESS

JOHN L. PATTERSON, Jr.
ASHTON GRAYBIEL

Acceleration and accelerative force are in some degree an integral part of the lives of all human beings on earth. The force of gravitational attraction and accelerative forces imposed by various machines in or on which we ride are familiar components of our environment, even though most of us do not stop to analyze these in detail. Increasing interest in many aspects of the environment of humans, and of animals and plants as well, invites a review of the subject of acceleration, for both its intellectual and its practical value. As will become evident, differences of concept and terminology have arisen in this field and are substantial enough to require clarification, particularly for those unfamiliar with the subject.

Acceleration is defined as a change in velocity, a vector quantity that possesses both speed and direction. Acceleration therefore results when either of these components is changed. Thus if speed is changed in rectilinear (straight-line) motion, there is associated acceleration; or if motion takes place at constant speed along a circular path, there is also acceleration, owing to the change of direction.

Accelerations do not take place without the action of forces on the body being accelerated. Force can be applied by direct mechanical contact with the body or through action-at-a-distance forces, the most familiar of which is gravity. Except in what may be termed the "ideal" weightless state, the bodies of man and animals are always subject to certain contact* forces. Such forces differ in some respects from gravitational forces in their effects on man; contact forces primarily affect the body through the areas of immediate contact, from which forces are distributed throughout the rest of the body, whereas gravitational forces act on every atom in the body, whether or not contact forces are applied.

All bodies possess mass and density, and in the case of living organisms, the total masses not only vary but the distribution of mass is not homogeneous. All masses possess the property of inertia, that is, resistance to acceleration. The nonhomogeneous organs and tissues of humans and animals and the physical means by which these diverse structures are held together comprise bodies with the capability of major internal displacements during acceleration. As will be seen, the dynamics of rigid bodies that can be treated as discrete (as opposed to distributed) mass must be modified in considering the living body, in which only the bones possess any considerable degree of rigidity. The varying degrees of displacement of the different components of the body relative to one another during acceleration can result in a wide range of physiologic changes. When these are extreme, they are incompatible with life, but in this chapter the major focus will be on physiologic effects over tolerable magnitudes of the different forms of acceleration.

The accelerative force with which we live throughout our lives, except for the fortunate few who become astronauts or cosmonauts, is that of the earth's gravitational attraction. This force must be continuously opposed if we are not to fall toward the center of mass of the earth. This op-

*Contact forces from clothing alone are commonly omitted from force analyses.

posing force is provided by structures such as floors, chairs, beds, the ground, aircraft in level flight, and the like. Since, as we have said, these contact forces are more narrowly applied on the body than the pervasive attraction of gravity, which acts on every element of mass within the body, physiologic changes take place throughout our daily lives, sometimes to the point of symptom production. This is a fact with which we all have some familiarity.

Much confusion has arisen from differences in methods of physical analysis, differences in viewpoint, differences in reference frames, and differences in terminology. As a simple example, the physiologic effects of quiet standing on the floor under the influence of the force of gravity are the same as if gravity were not present but one's body were being accelerated away from the earth at the same rate as in a free fall toward it. Either of these viewpoints can be adopted in the physical analysis, but not both at the same time. Thus it is imperative that rigorous and internally consistent methods of physical analysis be employed, or confusion will inevitably result in attempting to "explain" a physiologic event in terms of physical forces and their associated accelerations. It has therefore been considered essential to include in this chapter a section on how to deal with common force and acceleration situations, although space precludes exhaustive analysis. The references should enable the reader to overcome this limitation. It will be shown that the matter of frame of reference is of crucial importance and that from the standpoint of an "observer" within the body, the forces that appear to produce physiologic changes are not the same as those apparent to an observer external to the body undergoing acceleration. The important distinction between these viewpoints is discussed in the next section of this chapter.

It is a peculiar and rather mysterious fact that apparent inertial forces, generated as a result of applied accelerative force, are indistinguishable from gravitational forces in their effects. These gravitoinertial forces can therefore be treated in the same vectorial analysis. As will be seen, the resultant of these forces is of great service in explaining many of the physiologic changes during acceleration.

The special problems of angular acceleration, that is, change in the rate of rotation, are treated in some detail. Any reader familiar with physics will immediately recognize that angular acceleration does not occur without centripetal acceleration, except along the exact axis of rotation. Nevertheless, the sensory organs of the semicircular canals of the inner ear are stimulated primarily by angular acceleration, or a direct physical correlate, and must be afforded special treatment from this viewpoint. It is important to emphasize, however, that in the daily activities of most individuals, the angular accelerations are such as to be below the threshold of awareness and symptom production and that special conditions are required to elicit the various specific responses to angular acceleration.

The phenomenon of weightlessness (Fig. 6-1), with which mankind has had no prolonged experience since the dawn of time until recently, represents a special case of acceleration physiology and one clearly deserving of some treatment in this chapter. An important and neglected subject that should receive greater attention concerns internal or immanent forces in the living body, forces that must be present even in the weightless state. It is apparent that there are many such forces, ranging from the musculoskeletal to those of electric attraction or repulsion in membranes and electromagnetic forces of chemical bonding. For the most part, acceleration physiology has been concerned with the effects of forces on the body greater than that of gravity, and this subject will comprise much of what is presented. One may allude to the hypergravitational state that will be experienced when man lands on planets with heavier mass than earth—Jupiter, for example. The hypogravitational state (one sixth of the earth's gravity) on a satellite of lesser mass, the moon, has already been experienced by man. There is little doubt that it is comparable in physiologic effects to the artificial gravity of the same magnitude induced by rotation of spacecraft producing the same accelerative force directed in the same relation to the body axes.*

Since our knowledge of acceleration physiology precedes World War II, although greatly augmented during and since that conflict, some historic perspective is appropriate. In this chapter we have, however, attempted to present acceleration physiology from a unified, current, and interpretive viewpoint. We have emphasized those aspects of the subject that seem to us of greatest present interest, or that we feel have been insufficiently emphasized in the past. A carefully

*This is true at least when the human subject is not moving so as to alter significantly his distance from the axis of rotation.

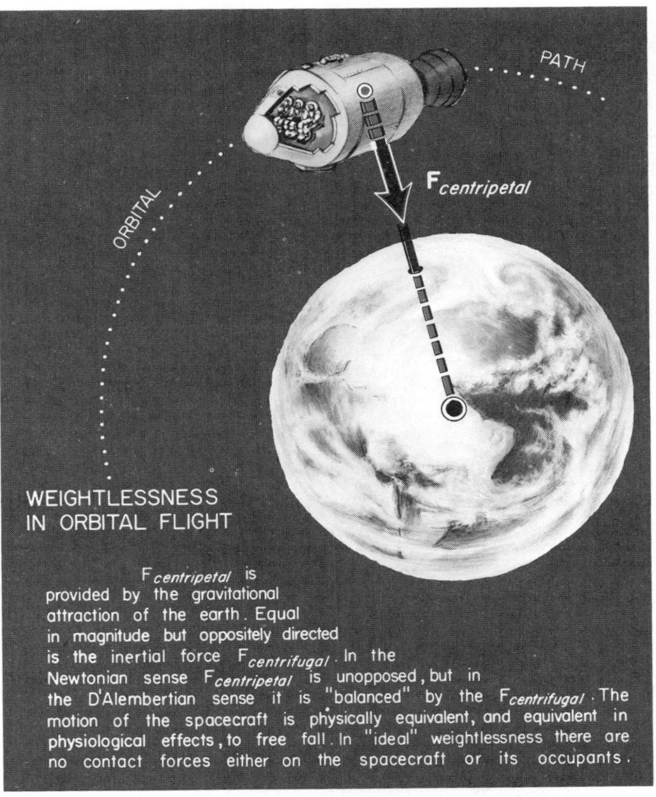

Fig. 6-1. Manmade satellite in orbit about the earth and the physical principles involved.

selected bibliography has been provided for those who need or wish to probe the subject more deeply.

BACKGROUND PHYSICS

A working knowledge of certain fundamentals of acceleration physics is essential to an understanding of acceleration physiology. The problems confronting the physicist and engineer are sufficiently different from those analyzed by the physiologist as to require an appreciation of the differences in physical situation. In this section, some of the basic physics that has been found of greatest use to the physiologist, physician, and biologist working in this field is presented. Emphasis is placed on the type of analysis that most readily explains the physiologic alterations produced by accelerations to which man, animals, or plants may be subjected. It has been found that most physiologists think more readily of the relative displacements of structures within the living body in terms of the accelerative *forces* that produce them rather than the *accelerations* alone, and the material in this section is presented primarily from this viewpoint. The student beginning the study of this field is at once confronted by differences in concept, usage, terminology, and symbols in the literature of acceleration. An attempt has therefore been made to clarify and simplify this aspect of the subject. Texts and monographs that have been found particularly useful are given in the annotated list of Suggested Readings.

Most of the physics that the physiologist requires in dealing with acceleration and its effects on the living body lies in the branch known as *mechanics,* which is concerned with the action of forces on bodies. One of its divisions, *statics,* deals with bodies in static equilibrium, whereas the other, *dynamics,* is concerned with bodies in motion. *Kinematics* is that portion of dynamics that is concerned with the analysis of the motion of bodies apart from the causes of this motion. The other subdivision of dynamics is *kinetics,* which is concerned with forces acting on bodies and the changes in motion produced by these forces. Some authors employ this as a definition for the whole of dynamics, since the term "kinetics" is in decreasing usage. A study of the effects of accelerative forces on man thus falls properly under the subject of dynamics or kinetics.

Newtonian mechanics as it has developed provides the physiologist with most of the tools required for physical analysis of acceleration situations. It is perhaps best to begin by reviewing Newton's three laws of motion, with which most readers already have some familiarity.

Laws of motion; frames of reference

Newton's first law of motion, in his words, states: "Every body persists in its state of rest or of uniform motion in a straight line unless it is compelled to change that state by forces impressed on it." This first law may also be termed the "law of inertia" and requires inertial reference frames for its application. The important question of reference frames is discussed later.

Newton's second law is commonly defined by the equations:

$$\mathbf{F} = m\mathbf{a} \quad (1A)$$

or

$$\Sigma\mathbf{F} = m\mathbf{a} \quad (1B)$$

where $\mathbf{F}$ = force
$\Sigma\mathbf{F}$ = sum or resultant of external forces acting on the body
m = mass
$\mathbf{a}$ = acceleration

It should be noted that **F** and **a** are in boldface type to denote that they are vectors, having magnitude and direction, whereas mass *m*, in italics, is a scalar quantity possessing magnitude only. In words, Newton's second law simply states that the acceleration (rate of change of velocity) of a particle or body is equal to the resultant of the external forces acting on the body divided by its mass. This second law can also be considered a definition of *force*. It is also apparent that Newton's first law represents a special case of the second law, since if $\mathbf{F} = 0$, $\mathbf{a} = 0$.

Newton's third law of motion, in his words, states: "To every action there is always opposed an equal reaction; or, the mutual actions of two bodies upon each other are always equal, and directed to contrary parts." It should be emphasized that in newtonian mechanics the forces of action and reaction act on *different* bodies. The laws of motion cannot be understood without an appreciation of the importance of the frames of reference.

The question of *frames of reference* is of fundamental importance in mechanics. Newton's laws of motion are valid only if the observer is positioned in an *inertial reference frame*. It is common to define an inertial or unaccelerated reference system as one that shows no acceleration in relation to the so-called fixed stars—those at great distances from us in which we can detect no acceleration. It is important to emphasize that observers in different frames of reference moving at constant velocity relative to each other will measure the same acceleration for the same moving body. This fact is one of the consequences of Einstein's special theory of relativity. For all situations that will be considered in this chapter, an inertial system is defined with precision for an observer who is motionless on the face of the earth by taking into account the acceleration of the earth about its axis and the acceleration involved in its orbit about the sun. For most purposes the acceleration physiologist can utilize a reference frame that is simply fixed in relation to the earth. Thus in the human centrifuge work, it would rarely be necessary to take into account the motion of the earth. On the other hand, in the case of Coriolis acceleration, considered in more detail later, the rotation of the earth has a major effect on certain phenomena, such as currents in the sea and in the atmosphere, and this rotation would have to be accounted for in the physical analysis. A *noninertial reference frame* is one that is accelerating in relation to the fixed stars. To apply classic mechanics in a noninertial reference frame, additional forces must be introduced into the analysis. These forces are termed by some *pseudoforces.*

The term "vector" is defined as a quantity possessing both magnitude and direction, as distinguished from a "scalar" which possesses magnitude only. Vectors are commonly denoted by an arrow, the head of which denotes the "sense" of the vector, since the shaft alone would be bidirectional, just as a highway is bidirectional; the length of the arrow in arbitrary units denotes the magnitude of the vector quantity. Vector diagrams can be constructed in different ways, but the magnitudes and directions of the vectors must not be altered. Vectors can be added in any sequence provided that the tail of one vector is placed at the head of the preceding one. The vector required to join the tail of the first vector to the head of the last vector represents the vector *sum* or *resultant* of the individual vectors. Where only two vectors are involved, the above method of addition of vectors can be seen to amount to completion of the parallelogram, the diagonal of the parallelogram representing the resultant (vec-

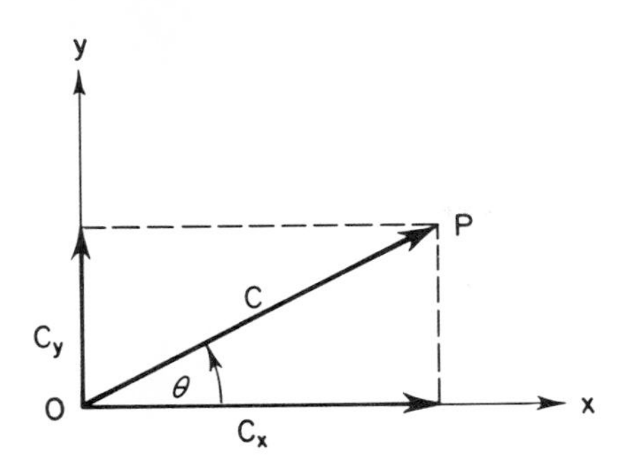

Fig. 6-2. Rectangular coordinate system in two dimensions.

tor **C** in Fig. 6-2). Vectors can also be subtracted. If a vector **M** is to be subtracted from another vector **K**, simply establish a vector –**M**, equal in magnitude but oppositely directed to **M**, and proceed as in vector addition.

When equation 1A is rewritten in the form $\mathbf{a} = \frac{\mathbf{F}}{m}$, it is apparent that the acceleration **a** can be represented as a vector having the same direction as the causative force, since the only difference in the force and acceleration diagrams will be the scalar factor m. It should be noted that force and acceleration are always in the *same direction*, whereas this relationship is not necessarily true of force and velocity. This becomes apparent from analyzing the situation of a moving body that is being decelerated by an applied force. In this case, the force and the deceleration (negative acceleration) are in the same direction, whereas velocity is in the opposite direction.

Coordinate systems

In dealing with motions of bodies and their positions at different instants in time, spatial reference systems are required. The *Cartesian* or *rectangular coordinate* system has been found to be the most useful to the acceleration physiologist. The simplicity of this system is particularly useful in vectorial analysis. Fig. 6-2 depicts a Cartesian coordinate system in one plane, with mutually perpendicular x and y axes. These axes are marked off in units of distance from the point of origin O. The positive side of each axis is designated by the arrowhead; the negative side is the opposite. The position of a point P is specified by its perpendicular projections on the x and y axes. If the projection of P on the y axis is designated as 1 unit in length, and the projection of P on the x axis 2 units, the conventional notation for the Cartesian coordinates of point P would be (2, 1). If we draw a vector **C** to point P, we can resolve this vector into its x and y components, namely C_x and C_y, respectively. Multiple vectors can be represented in such a Cartesian coordinate system and each resolved into its x and y components. Addition of these vectors is then possible by addition of their components. By this means the x and y components of the resultant single vector are obtained. It should be emphasized that when vectors are added in this manner, the sign of each vector, + or –, must be taken into account.

A *polar coordinate* system in one plane provides an alternative means of describing the position of point P. The essentials of this system are also shown in Fig. 6-2. In this system the reference point is again the origin O and a single polar axis or initial ray is drawn from it, corresponding to the x axis in the diagram. The radius vector or position vector, **C** in Fig. 6-2, is usually designated by **R** or **r** in polar coordinates, and the angle θ, measured counterclockwise from the polar axis to the radius vector, provides the other coordinate. The position of the point P thus is identified as (r, θ), or in Fig. 6-2 as (C, θ). The interrelations between Cartesian and polar coordinates are useful:

$$C_x = C \cos \theta \quad (2A)$$

$$C_y = C \sin \theta \quad (2B)$$

$$C = \sqrt{C_x^2 + C_y^2} \quad (2C)$$

$$\theta = \arctan \frac{C_y}{C_x} \quad (2D)$$

Each of these coordinate systems has its own merits. In cases where the emphasis in the analysis is on vector rather than scalar properties of mechanical quantities, the polar coordinate system can be the more useful, since the magnitude and direction of the position vector of a point is being specified.

The Cartesian coordinate system in three dimensions is of the greatest value in acceleration physics and physiology. A vector in space is most usefully represented in this coordinate system. Fig. 6-3 represents such a system with the x, y, and z axes orthogonal, that is, mutually perpendicular. The concept of unit vectors is essential

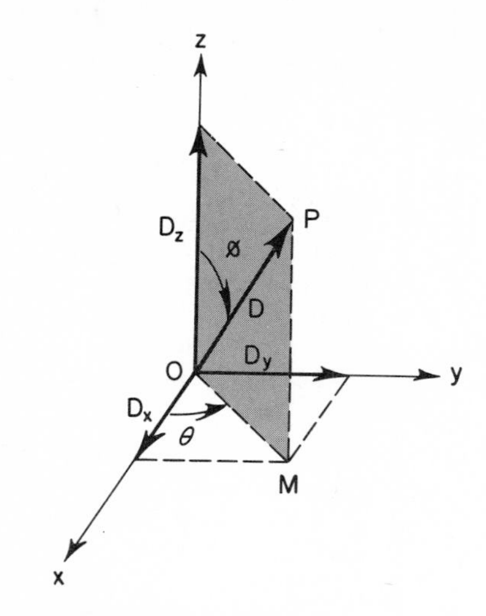

Fig. 6-3. Rectangular coordinate system in three dimensions.

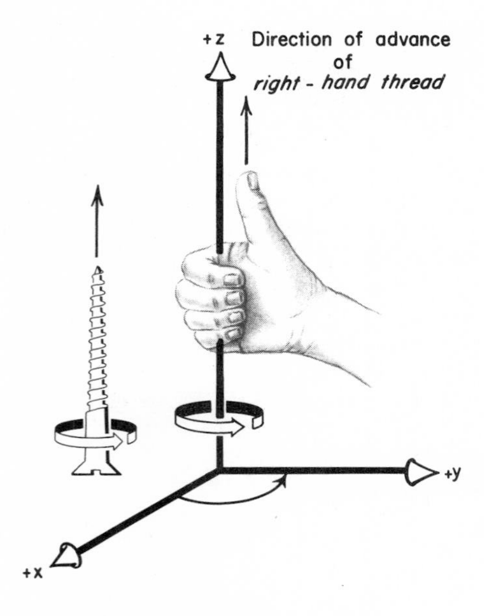

Fig. 6-4. Righthand thread rule for mathematic parity.

to vectorial analyses in such a system. The angles ϕ and θ in the figure may be omitted from consideration for present purposes. A *unit vector* is simply a vector of unit length; for the Cartesian coordinate system of Fig. 6-3 the unit vectors of the axes are $\hat{\mathbf{x}}$, $\hat{\mathbf{y}}$, and $\hat{\mathbf{z}}$. The notation $\hat{\mathbf{i}}$, $\hat{\mathbf{j}}$, $\hat{\mathbf{k}}$, respectively, is also commonly used.

The *sense* or direction of the z axis relative to the other two axes is determined by what is known as the *righthand thread rule* (Fig. 6-4). The sense of z is such that if the x axis is rotated about the z axis so as to sweep through the shortest angular distance between x and y, the positive end of z is in the direction of advance of a righthanded screw rotated in the same way. In Fig. 6-4 are shown both the screw and a hand with the fingers curled in the direction of rotation. The thumb points in the positive z direction, just as the direction of motion of the righthanded screw when rotated is toward the positive end of the z axis. In terms of the unit vectors, the meaning of the righthand rule is such that:

$$\hat{\mathbf{z}} = \hat{\mathbf{x}} \times \hat{\mathbf{y}} \quad (3)^*$$

The vector **D** in Fig. 6-3 can be written in terms of the unit vectors and of the perpendicular projections of **D** on each of the axes:

$$\mathbf{D} = \hat{\mathbf{x}}D_x + \hat{\mathbf{y}}D_y + \hat{\mathbf{z}}D_z \quad (4)$$

The length D is given by repeated application of the Pythagorean relationship between hypotenuse and sides of a right triangle:

$$D = D_x^2 + D_y^2 + D_z^2 \quad (5)$$

The *spherical polar coordinate* system is also used in acceleration physics and physiology. Fig. 6-3 represents such a system. **D** is the position vector or radius vector of the point P. The position of point P in three-dimensional space is fixed by the length of D and by the angles ϕ and θ. The angles ϕ and θ are sometimes termed the *colatitude* and *longitude,* respectively. The spherical polar and rectangular coordinates of point P have the following interrelationships:

$$D_x = D \sin\phi \cos\theta \quad (6A)$$

$$D_y = D \sin\phi \sin\theta \quad (6B)$$

$$D_z = D \cos\phi \quad (6C)$$

$$D = x^2 + y^2 + z^2 \quad (7A)$$

$$\phi = \arctan \frac{x^2 + y^2}{z} \quad (7B)$$

$$\theta = \arctan \frac{y}{x} \quad (7C)$$

Before presenting the next set of equations, which will be in the G system of units, it is appropriate first to consider systems of physical units.

*The indicated vector "cross product" is explained later in this chapter, p. 179.

Table 6-1. Fundamental physical standards

Quantity	British unit	Metric unit	Conversion factor
Length	Yard	Meter	1 yard = 0.9144018 meter
Mass	Pound	Kilogram	1 pound = 0.4535924277 kilogram
Time	Second	Second	1 second = 1 second

Table 6-2. Mechanical units of the British gravitational system

Quantity	Symbol	Units	Abbreviation
Length or distance*	s	Feet	ft
Force*	F	Pounds	lb
Time	t	Seconds	sec
Mass	m	Slugs	—
Velocity*	v	Feet/second	ft/sec
Acceleration*	a	Feet/second squared	ft/sec²

*Vector quantities.

Systems of physical units

Among the various quantities with which physics deals, three are fundamental—*length, mass,* and *time.* If we set up arbitrary standards of unit length, unit mass, and unit time, all other unit mechanical quantities can be defined in terms of these standards.

Two groups of arbitrary standards are in common use today: the *British units* and the *metric units.* Table 6-1 below gives the names of, and relationships between, the fundamental units of the two groups.

Having defined suitable standards of length, mass, and time for universal reference, we can employ any multiples or fractions of these standard quantities as *practical units.* Moreover, in the so-called gravitational systems of physical units set up for engineering convenience, a unit of force rather than a unit of mass is selected as one of three basic standards. (This force unit, however, is ultimately defined in terms of the fundamental quantities of Table 6-1.) There are two principal practical systems for the purposes of aerospace physiology and medicine: the British gravitational system and the metric system. The British system has as its basic units the foot, the pound-force, and the second. The basic units in the metric system are the meter, kilogram-mass, and the second. Although the metric system may ultimately be adopted in the United States, at present the British gravitational system is so commonly used that it has been adopted for the majority of this chapter. The magnitude of the basic units of the British gravitational system are:

1 foot	one third of the standard yard
1 pound	the force that accelerates a standard pound-mass at the standard rate of 32.174 ft/sec every second
1 second	1/86,400 of a mean solar day

Secondary units are derived from these three basic units by means of the various equations of physics. For example, unit mass in the British gravitational system is obtained by writing the Newtonian force equation (1A) in the form:

$$m = \frac{F}{a} \qquad (8)$$

and then substituting unit force (the pound) and unit acceleration (the foot divided by the second squared) in the righthand side. The unit mass so defined is called the *slug,* and by equation 8:

$$1 \text{ slug} = 1 \text{ pound} \times \text{ft}^{-1} \times \text{sec}^{-2}$$

That is, a slug is that mass that will be accelerated by a force of 1 pound at a rate of 1 ft/sec every second; it is 32.174 times as great as the pound-mass.

The above systems of physical units have been established in conjunction with the equations of physics in such a way that the equations will balance only when all quantities involved are expressed in units from the same system. It is, therefore, essential always to use *consistent units* in solving physical problems. Table 6-2 lists the mechanical units for the British gravitational system, together with their symbols.

Dimensional analysis

Dimensional analysis is the method of studying the relations among physical quantities apart from the units in which these quantities may be expressed. Since the concepts of *length, mass,* and *time* are sometimes termed *dimensions* (not to be confused with reference to the physical size of an object), these basic concepts can be used to build the concepts of other physical quantities. Thus *area* represents the product of two lengths and its dimensions are L^2. Similarly, *velocity* represents distance divided by time and therefore has the dimension L/T. *Acceleration* is velocity divided by time, or L/T^2. *Force,* representing the product of mass and acceleration, has the dimensions ML/T^2.

Dimensional analysis is of the greatest use in verifying the correctness of equations. Both sides of an equation must reduce to the same dimensions for the equation to be complete and correct. As an example we will consider two formulas for distance s traversed by a body that is accelerated from rest at a rate a for a period of time t, one of these equations being incorrect. Neglecting the constant ½ (dimensionless) we have:

$$\text{Dimensions of } at/2 = \frac{L}{T^2} = (T) = \frac{L}{T} \text{ (incorrect)}$$

$$\text{Dimensions of } at^2/2 = \frac{L}{T^2} = (T^2) = L \text{ (correct)}$$

We see that the first equation is incorrect since it reduces to the dimension $\frac{L}{T}$, which does not represent distance alone, whereas the second equation reduces to L, which is correct since L represents distance. A cautionary note may be added: dimensional validity is but one requirement for the validity of an equation. In addition, the *units* used must balance. If one system of physical units is consistently utilized, the balance of the units is automatically obtained.

G system of units

In aviation medicine it is convenient to measure any acceleration a as a multiple of the standard acceleration (g_o) and any force (F) as a multiple of the standard weight (W_o) of the body upon which F is acting. In so doing, however, only the *magnitudes* of the vectors $\mathbf{g}_o$ and $\mathbf{W}_o$ are employed; the *directions* of the vectors $\mathbf{a}$ and $\mathbf{F}$ are quite independent of the fixed direction of the gravitational quantities $\mathbf{g}_o$ and $\mathbf{W}_o$.

It is evident that the preceding procedure is equivalent to setting up an auxiliary system of practical units, in which:

Unit acceleration $1\ g_o$ in magnitude $= 32.2$ ft/sec^2

Unit force $1\ W_o$ in magnitude $= mg = 32.2\ m$ lb

It should be noted that in the above system, which we term the *G system,* the unit of acceleration is a true constant but the unit of force differs for bodies of differing masses. For this reason, the unit of force is somewhat artificial; however, in dealing with a single body and the forces acting upon it, the difficulty mentioned does not appear.

The process of dividing a variable quantity by some special reference value that the variable may assume is known as *normalization* of the quantity. Accordingly, when specifying forces and accelerations in terms of the G system of units, it may be said that the two physical variables are being normalized with respect to gravitational values. If, then, an acceleration is expressed in multiples of the standard gravitational acceleration, it would be logical to describe the result as a *gravitationally normalized acceleration,* which might be conveniently abbreviated to GNA. Thus we would write the defining equation:

$$\text{GNA} = \frac{\mathbf{a}}{g_o} \qquad (9A)$$

Similarly, a force expressed in multiples of the standard weight of the body upon which it acts could be called a *gravitationally normalized force,* abbreviated to GNF and mathematically defined by:

$$\text{GNF} = \frac{\mathbf{F}}{W_o} \qquad (9B)$$

Upon substituting $\mathbf{F} = m\mathbf{a}$ and $W_o = mg_o$ into the above, there results:

$$\text{GNF} = \frac{m\mathbf{a}}{mg_o} = \frac{\mathbf{a}}{g_o} \qquad (9C)$$

from which, by comparison with equation 9A, it is seen that the gravitationally normalized force acting on a body and the gravitationally normalized acceleration that is produced are numerically equal. Furthermore, since the original force and acceleration have the same direction, the GNF and GNA as vector quantities are identical. It would be desirable to employ a single symbol to represent the common ratio formed by normalizing $\mathbf{F}$ and $\mathbf{a}$ with respect to standard gravitational values. Accordingly, we define:

$$\mathbf{G} = \frac{\mathbf{F}}{W_o} = \frac{\mathbf{a}}{g_o} \qquad (10A)$$

The quantity **G** is dimensionless, since it is the ratio of two accelerations or of two forces. However, by virtue of its definition, it is capable of telling how much force is acting and what acceleration is taking place in a given dynamic situation. Thus when it is stated that a body is experiencing a certain amount of G, we can determine the actual amount of force in ordinary units if we multiply the given value of G by the standard weight W_o of the body. That is:

$$F = GW_o \qquad (10B)$$

In a similar way, G defines the acceleration also:

$$a = Gg_o \qquad (10C)$$

Finally, although dimensionless, **G** possesses the same vector property of direction as do the force and acceleration that it connotes. Hence, it is legitimate to construct **G** vector diagrams in place of force or acceleration diagrams when analyzing dynamic problems in aviation medicine.

We have said that any force can be expressed as a **G** when it is divided by the standard weight of the body upon which it acts. The actual force on a body caused by the earth's gravity—that is, the weight **W** of the body—can therefore be so expressed, and we introduce the symbol $\mathbf{G}_w$ for this purpose:

$$\mathbf{G}_w = \frac{\mathbf{W}}{W_o} = \frac{mg}{mg_o} = \frac{\mathbf{g}}{g_o} \qquad (11A)$$

As shown in the above, the *gravitationally normalized weight* ($\mathbf{G}_w$) reduces by virtue of $\mathbf{W} = m\mathbf{g}$, or $\mathbf{W} = mg$, and $W_o = mg_o$ to the ratio of actual to standard values of gravitational acceleration. It is clear, therefore, that the magnitude of $\mathbf{G}_w$ will normally be approximately unity and will deviate from unity in the same way that g deviates from g_o. Thus from $g = g_o \left(\frac{r_o}{r}\right)^2$ we have:

$$G_w = \left(\frac{r_o}{r}\right)^2 \qquad (11B)$$

Similarly, for flights at altitudes that are considerably less than the radius of the earth, we can utilize the approximate formula $g \cong g_o \left(1 - \frac{h}{2{,}000}\right)$:

$$G_w \cong 1 - \frac{h}{2{,}000} \qquad (11C)$$

where h is expressed in *miles*.

In the case of bodies located on the earth's surface or flying at relatively low altitudes, $\mathbf{G}_w$ can safely be regarded as a vector of unit magnitude pointing vertically downward. However, in the case of space vehicles traveling at large distances from the earth, we find significant reductions in the earth's gravitational pull and corresponding decreases in the value of G_w from approximately unity to approximately zero.

Modified usages of G

A statement often made by workers in the field of aviation medicine and physiology is that "an individual was subjected to *a force of 3 G*." This has the same significance as "an individual was subjected to *a force of 3 W*," that is, a force three times the weight of this particular individual. If the standard values for the magnitudes of W and g are considered to be the unit quantities in a G system of physical units, they are properly spoken of as the G unit of force and the G unit of acceleration, respectively. After a valid expression has been established and agreed upon, contractions are permissible to avoid repetition or awkwardness. Thus we would not speak of "a force of 3 G units of force" but rather "a force of 3 G units." In so doing, we assume that the audience is acquainted with the G system of units and knows that the G unit of force is 1 W_o. By the same token, it is correct to speak of "an acceleration of 3 G units"; here one has in mind 3 G units of acceleration, or 3 g_o's. To illustrate, we write the following set of equivalent statements:

A force of 3 G units is acting on the body: $F = 3\ W_o$
The acceleration of the body is 3 G units: $a = 3\ g_o$
A GNF of 3 is acting on the body (or, the body has a GNA of 3): $G = 3$

It has become common practice to contract the sense of "a force of 3 G units" to "a force of 3 G." This has in fact become such standard usage in the acceleration literature that it is doubtful that it could be changed. Nevertheless, it should be recognized as questionable practice and interpreted to mean "a force of 3 G units" or "a force of 3 W_o."

It should be emphasized that the small letter g has been widely established by physicists as a symbol for a specific physical quantity (32.2 ft/sec^2, normally). Any other usage in a similar connection constitutes a distortion of meaning that creates unnecessary confusion.

Clark, Hardy, and Crosbie[1] have summarized the various symbols and usages over the years. Their Table I lists the various means of describing what these authors term "an acceleration of

2 G_z." The z axis is parallel to the long axis of the trunk of the body and is discussed later under body axes. These authors utilize the G as the unit for physiologic acceleration and state that it "includes both displacement and resisted gravitational acceleration effects." This corresponds to the gravitoinertial resultant force in G units that we prefer to designate G★ ("G star") to avoid confusion with the resultant external or resultant contact forces on the body. The question of G★ is discussed more fully in the section on relative accelerative force in semirigid (deformable) systems.

As stated earlier, the gravitationally normalized force (GNF) or acceleration (GNA) of a dynamic situation can be obtained by dividing the expression for ordinary acceleration a by the mechanical constant g_o. We thus have the following modified formulas for G magnitudes derived from the preceding theory.

Instantaneous and average linear G

$$G = \frac{\dot{v}}{g_o} = \frac{\ddot{s}}{g_o} \tag{12A}$$

$$\overline{G} = \frac{1}{g_o} \cdot \frac{\Delta v}{\Delta t} \tag{12B}$$

$$\overline{G} = \frac{v_2^2 - v_1^2}{2g_o\ \Delta s} \tag{12C}$$

$$\overline{G} = \frac{2}{g_o}\left[\frac{\Delta s}{(\Delta t)^2} - \frac{v_1}{t}\right] \tag{12D}$$

When Δs and Δt are measured from zero and, hence, are equal to s and t, respectively, the last two formulas become:

$$\overline{G} = \frac{v^2 - v_0^2}{2g_0 s} \tag{13A}$$

$$\overline{G} = \frac{2}{g_0}\left[\frac{s}{t^2} - \frac{v_0}{t}\right] \tag{13B}$$

where v_o = speed at $t = 0$
v = speed at any other time (t) for which the distance traversed is s

Centripetal G

From equations $a_c = \dfrac{v^2}{R}$ and $a_c = R\omega^2$ we obtain:

$$G_c = \frac{v^2}{g_o R} \tag{14A}$$

$$G_c = \frac{R\omega^2}{g_o} \tag{14B}$$

Angular acceleration

Just as we have linear acceleration when there is a change in linear speed of a body, so we can have *angular acceleration* (α), defined as the time rate of change of angular speed. If a body moving in a circular path has its angular speed (ω) changed by an amount $(\Delta\omega)$ during an interval of time (Δt), then the body is said to have undergone an *average* angular acceleration given by:

$$\overline{\alpha} = \frac{\Delta\omega}{\Delta t} \tag{15A}$$

Instantaneous angular acceleration would be written either as the first time derivative of an expression for instantaneous angular speed or as the second time derivative of an expression for instantaneous angular position (θ):

$$\alpha = \dot{\omega} = \ddot{\theta} \tag{15B}$$

Since angular velocity or speed is expressed in radians per second (rad/sec), angular acceleration will have units of *radians per second per second* (rad/sec²).

As shown by $v = R\omega$, an angular speed ω implies a linear speed v at a radius R. With R constant, therefore, $\Delta\omega = \Delta v/R$. By this substitution, equation 15A becomes:

$$\overline{\alpha} = \frac{1}{R} \cdot \frac{\Delta v}{\Delta t} \tag{16A}$$

Now utilizing $\overline{a} = \dfrac{\Delta v}{\Delta t}$ and rearranging equation 16A, we have:

$$\overline{a} = R\overline{\alpha} \tag{16B}$$

or, the corresponding "instantaneous" formula can be written:

$$a = R\alpha \tag{16C}$$

Equations 16A to 16C are not true vector equations and, again, apply validly only to circular motion in a plane. However, they serve to indicate that, in general, linear acceleration occurs whenever angular acceleration occurs. Furthermore, since at every instant during angular acceleration there exists some value of angular velocity (ω), even though this quantity is changing, $a_c = R\omega^2$ shows that centripetal acceleration also must be present. Thus *angular acceleration implies the existence of linear and centripetal accelerations simultaneously.*

As with all angular quantities, the direction of an angular acceleration vector is along the axis of rotation. Linear and centripetal vectors, on the other hand, lie tangential to and perpendicular to the path of motion, respectively. While one can add linear and centripetal accelerations together vectorially (in the manner described previously) to obtain a resultant acceleration, angular acceleration is never included in such an addition because angular quantities do not have the same units as corresponding linear quantities. An angular acceleration vector should be regarded as an *alternative* representation of a dynamic situation, the reasons for its use being similar to those discussed later in the chapter.

Angular G

Because G must be a dimensionless ratio of two forces or accelerations, we cannot divide angular acceleration α by g_o and obtain a quantity that fits the definition of G; the term "angular G," as such, is therefore meaningless. However, a G does exist during angular acceleration and is determined by taking the resultant of the linear and centripetal G's. Since the latter vectors are at right angles to each other, the magnitude of their resultant is simply the square root of the sum of their squares.

Total force, acceleration, and G

In the most general dynamic situation possible, a body moving in a rectangular coordinate system will have components of acceleration (and also of force and G) along each of the three axes. If these components are denoted by a_x, a_y, and a_z, we may determine the magnitude and direction of the resultant or *total acceleration* vector **a** through use of equations 7A to 7C:

$$a = \sqrt{a_x^2 + a_y^2 + a_z^2} \qquad (17A)$$

$$\phi = \arctan \frac{\sqrt{a_x^2 + a_y^2}}{a_z} \qquad (17B)$$

$$\theta = \arctan \frac{a_y}{a_x} \qquad (17C)$$

From equation 1A, the corresponding *total force* vector must have a magnitude given by:

$$F = ma = m\sqrt{a_x^2 + a_y^2 + a_z^2} \qquad (18)$$

Alternatively, we can express the total force as the resultant of three component forces F_x, F_y, and F_z directed along the three axes:

$$F = \sqrt{F_x^2 + F_y^2 + F_z^2} \qquad (19A)$$

$$\phi = \arctan \frac{\sqrt{F_x^2 + F_y^2}}{F_z^2} \qquad (19B)$$

$$\theta = \arctan \frac{F_y}{F_x} \qquad (19C)$$

Finally, we can divide the acceleration (equation 17A) by g_o to obtain the magnitude of the *total G* (gravitationally normalized force and acceleration) acting:

$$G = \frac{a}{g_o} = \frac{1}{g_o}\sqrt{a_x^2 + a_y^2 + a_z^2} \qquad (20A)$$

Or, similarly, division of the force (equation 19A) by W_o yields an expression for this quantity:

$$G = \frac{F}{W_o} = \frac{1}{W_o}\sqrt{F_x^2 + F_y^2 + F_z^2} \qquad (20B)$$

Again, however, by first expressing the total G in terms of components G_x, G_y, and G_z along the three axes, we can obtain the somewhat simpler alternative forms:

$$G = \sqrt{G_x^2 + G_y^2 + G_z^2} \qquad (21A)$$

$$\phi = \arctan \frac{\sqrt{G_x^2 + G_y^2}}{G_z} \qquad (21B)$$

$$\theta = \arctan \frac{G_y}{G_x} \qquad (21C)$$

This last set of equations will be found of greatest value in the solution of problems in aviation physiology, since it is usual to express all forces immediately in G units.

Relative accelerative forces in semirigid (deformable) systems—concept of the gravitoinertial resultant force

The physical approach utilized to derive expressions for the magnitude and direction of a resultant G vector existing in various dynamic situations carries with it an important tacit assumption: that the position, velocity, and acceleration of the body involved are taken with respect to a coordinate system fixed at some point on the earth. As mentioned previously, the earth is not a perfect inertial reference frame but can be treated as one for a number of practical purposes. Where great accuracy is required, corrections for the rotation of the earth can be inserted. The concern of the biologist, physiologist, and physician interested in physiologic effects is not primarily with how or why a body as such moves in three-dimensional earth space but rather on the relative displacements that take place among various organs and tissues while the body as a whole is undergoing acceleration. Thus the physiologist must know the *relative internal dynamics* of the body with which he deals.

This section is concerned with the mechanics of bodies, which are not perfectly rigid, and with the concept of a resultant force, which to the stationary external observer is fictitious or apparent but to an observer moving with the body is real and explains the disturbances of function within the living subject.

External forces that act *on* a body may be divided into two distinct types: *contact* and *action-at-a-distance* forces. An example of a contact force would be the force exerted by the floor on an individual's shoes when he is standing without other support. The force of gravitational attraction represents, for our purposes, the most important action-at-a-distance force.

The *total force* acting *on* any body near the earth can always be expressed as the vector sum of its weight and the various contact forces that may be applied to it. In equation form:

$$\mathbf{F} = \mathbf{W} + \Sigma\mathbf{F}^{C} \qquad (22)$$

where the summation symbol Σ indicates that there may be more than one contact force $\mathbf{F}^{C}$ (the reason for the C as superscript rather than subscript becomes apparent in the later section on body axes).

Since equation 22 is a vector equation, it is of course equivalent to three scalar equations involving the force components along coordinate axes—say x, y, and z:

$$F_x = W_x + \Sigma(F^C)_x \qquad (23A)$$

$$F_y = W_y + \Sigma(F^C)_y \qquad (23B)$$

$$F_z = W_z + \Sigma(F^C)_z \qquad (23C)$$

The same situation holds for the total **G** vector, which is given in compact vector form by division of equation 22 by W_o.

$$\mathbf{G} = \mathbf{G}_w + \Sigma\mathbf{G}^{C} \qquad (24)$$

There is a significant difference between the anatomicophysiologic effects of contact and of action-at-a-distance forces. The gravitational force acts on every element of a body simultaneously and in direct proportion to the mass of that element, with the result that each molecule of the body tends to be accelerated at the same rate. *Of itself,* therefore, gravitational action does not distort a body. On the other hand, contact forces, no matter how widely distributed on the surface of a body, can be applied to a fraction only of the total number of mass elements and must be transmitted to the remaining elements through tensions and compressions developed within the body by deformation.

Biologic semirigid (deformable) organism

In discussing biologic organisms such as man, it is helpful to conceive of the body as a *relatively* rigid framework (the skeleton) supporting a variety of nonrigid mechanical systems. The subsidiary systems may, by and large, be regarded as smaller mass elements flexibly connected to the more massive rigid structure. The flexible connections themselves can, for some purposes, be treated as having negligible mass.

It is very important to recognize that *unbalanced mechanical forces*—by definition, as it were—are capable of one and only one thing: the *acceleration of mass.* We must therefore look at internal tissue displacements (which accompany relative internal accelerations) as essential intermediate steps in the production of physiologic disturbances caused by externally applied accelerative forces.

The human body, or the body of an experimental animal, represents such a complex structure from a mechanical standpoint that it is useful to abstract essential mechanical properties of simplified subunits and study the characteristics of such elementary "models" in combination. Such a useful simple model of a semirigid system would be represented by a cubical box with rigid walls, with certain moveable internal elements. Imagine that from the center of each wall of the box is attached a spring of uniform compliance, the other ends of these springs being attached to a mass A. The box is completely filled with a damping fluid. For simplicity of analysis we assume that the springs possess compliance only (no mass or friction) and the damping fluid is purely resistive, that is, it has no mass or compliance. The fluid presents viscous resistance to any motion of the mass and we therefore have in this model a system that possesses *mass, elasticity,* and *friction.* More elaborate systems contain many subunits with the same properties of mass, elasticity, and friction, but here we are concerned with what can be learned from this simplest model. It is readily seen that in the absence of any external forces, mass A will occupy the geometric center of box B.

It is assumed for simplicity that the equilibrium position of A is coincident with the center of gravity of B. If accelerative forces are applied to the outside surface of B, a displacement s will take place, which is given by the vector difference between straight line distances to the center of gravity of A and of B, as measured from an arbitrary observation point on the earth. These distances are denoted by s_A and s_B, respectively, and we write the vector equation:

$$\mathbf{s} = \mathbf{s}_A - \mathbf{s}_B \tag{25A}$$

Two differentiations with respect to the time of equation 25A yield the relationship:

$$\ddot{\mathbf{s}} = \mathbf{a}_A - \mathbf{a}_B \tag{25B}$$

From this simple starting point an equation can be derived (the full derivation is given in Dixon and Patterson[2]), which is a mathematical representation of the dynamic condition of mass A relative to box B at any instant:

$$\frac{\ddot{\mathbf{s}}}{(1 + q_m)} + \frac{\mu\dot{\mathbf{s}}}{m_A} + \frac{\lambda\mathbf{s}}{m_A} = g_0\mathbf{G}\star \tag{26}$$

where $\mathbf{s}$ = displacement
$\dot{\mathbf{s}}$ = velocity
$\ddot{\mathbf{s}}$ = acceleration of the movable mass A relative to the rigid framework B

q_m = W_{A_0}/W_{B_0} (standard weight of A/standard weight of B)

μ = damping constant

λ = coefficient of stiffness of the spring—or simply the spring constant

m_A = mass of A

g_0 = standard acceleration of gravity

$$\mathbf{G}\star = \frac{-\Sigma\mathbf{F}^C}{W_0} \quad \text{(the reason for this definition becomes apparent later)} \tag{27}$$

where $\Sigma\mathbf{F}^C$ = totality of *externally* applied *contact* forces

The meaning of equation 27 is that the dynamics of A relative to B can be considered to be determined solely and completely by the *contact-force* pattern. The force of gravity on B, as such, does not appear, and we may conclude that the same internal response would obtain (for the same $\mathbf{G}\star$) whether or not the semirigid system were in the presence of a gravitational field.*

*The quantity q_m, although defined above as a weight ratio (that is, as involving the pull of gravity on B), is just as surely a mass ratio, hence a constant independent of the effects of gravity. The gravitational acceleration g_0, which also appears in equation 26, is likewise a constant, being the standard value 32.2 ft/sec^2, the G unit of acceleration.

Biologic organisms characteristically involve continua of tissues with varying mechanical properties, a fact that makes the physiologic analysis so difficult. We are dealing with distributed mass rather than discrete mass, and it is difficult to discuss the action of forces on such systems. Nevertheless, the intrinsic dynamic behavior of systems of "distributed properties" is reasonably well illustrated by systems of "lumped properties." If we recognize that we are discussing principle rather than detail, the mathematics of the semirigid model that we have been discussing provides insight into the nature of internal body dynamics.

For simplicity we will assume that the displaceable mass in our semirigid biologic organism is very small compared with the mass of the system as a whole. Then the weight or mass ratio q_m, given by the above, is a number that is very small compared with unity. With this assumption, the denominator of the first term in equation 26 may be simplified by approximation. In addition, we let:

$$b = \mu/m_A \tag{28A}$$

$$c = \lambda/m_A \tag{28B}$$

Then equation 26 becomes:

$$\ddot{\mathbf{s}} + b\dot{\mathbf{s}} + c\mathbf{s} = g_0\mathbf{G}\star \tag{29}$$

Because of its fundamental importance, equation 29 may be referred to as the *physiologic acceleration equation.* This second-order linear differential equation provides an admittedly idealized but usefully representative picture of the position of a small particle at every instant relative to a rigid framework, while this framework is being subjected to contact forces. This equation is thus a *prototype* from which we may draw useful inferences on the probable behavior of more complex systems, built up of many interconnected particles ultimately attached to the framework.

The important distinction between the internal and external force viewpoints can be illustrated and clarified in its simplest form by the physical model shown in Fig. 6-5. This figure depicts a horizontally rotatable tube as viewed from above. This tube is closed at its outer end and contains a spring, on the inner side of which is attached a ball of sufficient mass to demonstrate readily its inertial properties. When the system is at rest, the spring is extended and the ball is stationary, with its weight balanced by an equal but upward force exerted by the tube. When the central shaft

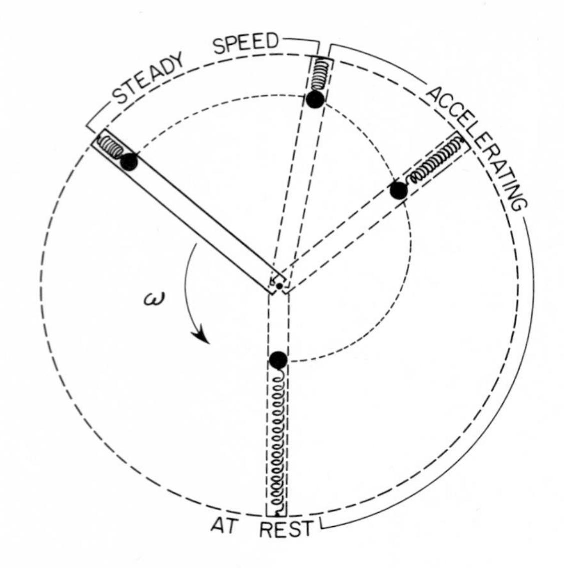

BALL AND SPRING IN ROTATING TUBE

Fig. 6-5. Tube with contained ball and spring rotating in horizontal plane illustrating internal and external force viewpoints.

and tube are subjected to angular acceleration up to a steady angular speed, the horizontal trajectory of the ball will show a gradual approach to a true circular path of increased radius. If the tube is transparent, a stationary external observer notes the pathway of the ball and accounts for its behavior by reference to Newton's laws, according to which the ball *tends* to follow a rectilinear path at constant speed. In so doing, the ball compresses the spring, since the latter is continually being pulled centripetally into the natural path of the ball by the rotating tube. As a result of this compression, the spring exerts an inwardly directed (centripetal) force on the ball, which causes it to deviate from a straight-line path and finally, at steady angular speed, keeps the ball at a fixed radial distance.

Quite different is the picture presented to an internal observer located at the center of the system in such a way that he turns with and sights down the tube. He observes that the ball travels a one-dimensional path away from him at a gradually decreasing rate. Here the **F** or **G** calculated by the external observer does not help the internal observer account for the behavior of the ball. The internal observer infers that there is an outwardly directed force accounting for the movement of the ball and the compression of the spring and from his standpoint this outwardly directed force is a "real" force.

A more complete physical model of a semirigid deformable system is shown in Fig. 6-6. *A* of this figure shows a beam that rotates horizontally about a vertical shaft. On the end of this beam, attached by a universal joint, is a hinged rod down which a ball with a hollow cylindrical core is free to slide. One end of the spring is attached to the upper end of the movable arm and the other is attached to the ball itself. The position of the ball relative to the end of the beam requires for its complete description (1) the angle from the vertical of the pivoted arm, (2) the backward-forward angle of the pivoted arm, and (3) the radial distance of the ball along the pivoted arm. With the system stationary the arm hangs straight down, the spring being slightly extended by the weight of the ball. With the drive shaft rotating at constant angular speed, the spring is more extended and the beam is swung outward from the vertical. The resultant external forces acting on the swinging arm-ball-spring system are diagrammed in *B*. The force exerted by the beam and by gravity on the system can be resolved into a single horizontal force, and it is this centripetal force that at steady angular speed accounts for the constant horizontal acceleration of the ball toward the center of the circle.

In *C* the "contact" force exerted by the beam at the universal joint on the deformable part of the system is illustrated. This G_{beam} can be resolved into a vertical component G_z, which balances the weight of the swinging arm and a component $G_{\text{centripetal}}$, shown also in *B*.

Gravitoinertial resultant force

From a number of standpoints, the most useful force that we can measure or calculate to explain the behavior of semirigid deformable systems is that force we now term the *gravitational resultant force*. This force, in G units, was designated in an earlier publication by the symbol **G★** and was termed "physiologic G."[2] More recently one of us (A. G.) has devised the more satisfactory term "gravitoinertial resultant force," and we employ this throughout the remainder of this chapter. A mathematical definition of this gravitoinertial resultant force follows.

We begin the derivation with the general vector equation 22 for the force acting on a body and divide each term of this equation by the standard weight of the body. The result is a vector equation expressing the gravitationally normalized force

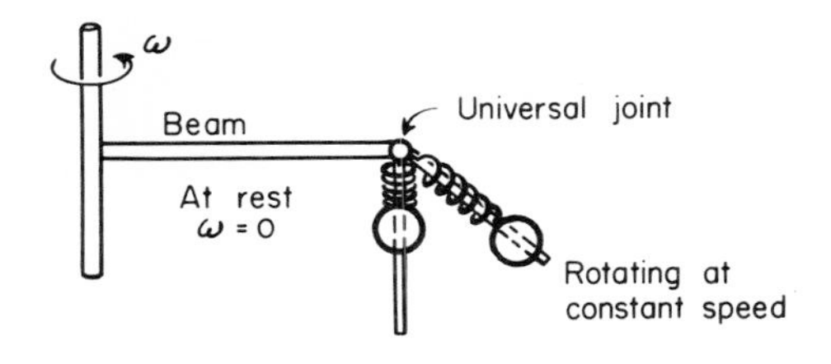

(A) SEMIRIGID (DEFORMABLE) SYSTEM

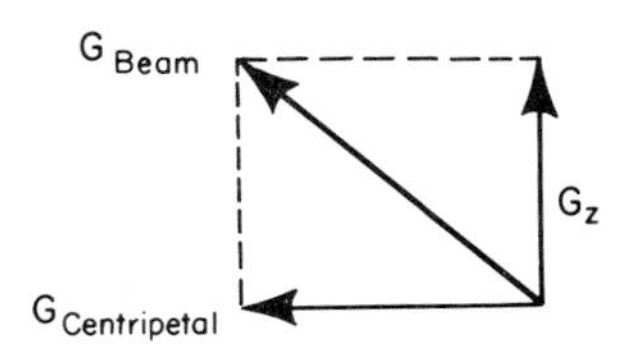

(C) "CONTACT" FORCES EXERTED BY THE BEAM AT UNIVERSAL JOINT

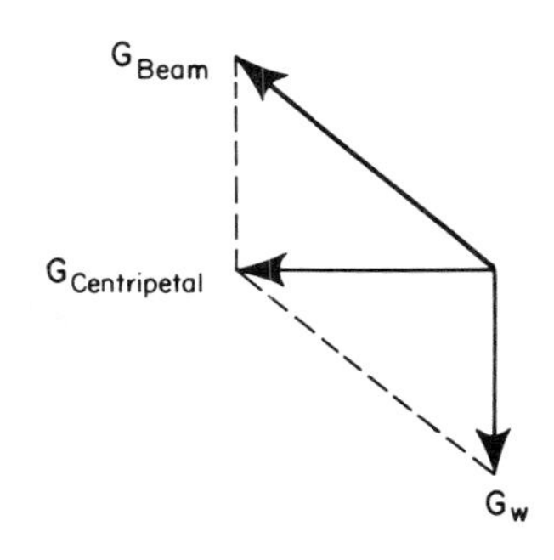

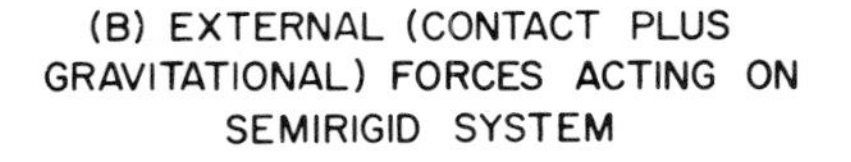
(B) EXTERNAL (CONTACT PLUS GRAVITATIONAL) FORCES ACTING ON SEMIRIGID SYSTEM

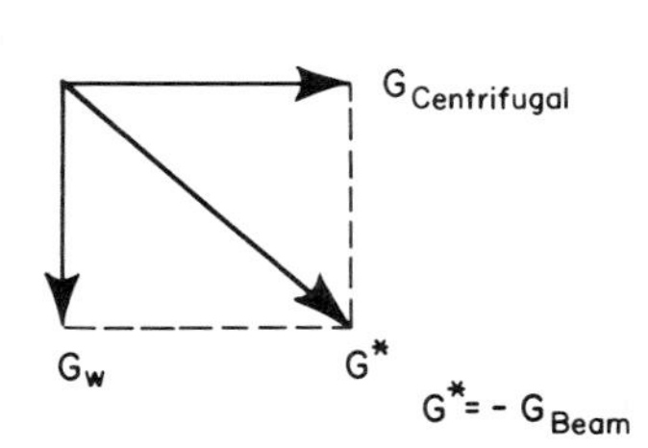

(D) GRAVITOINERTIAL FORCES IN SEMIRIGID SYSTEM

BALL AND SPRING SLIDING DOWN HINGED ROD

Fig. 6-6. A semirigid (deformable) system undergoing centrifugation, with vector diagrams of the forces involved.

acting:

$$\mathbf{G} = \frac{\mathbf{F}}{W_o} = \frac{\mathbf{W}}{W_o} + \frac{\Sigma \mathbf{F}^C}{W_o} \qquad (30A)$$

where $\mathbf{F}^C$ = each of the individual contact forces applied by *external* agents to the body
$\mathbf{W}$ = actual gravitational force due to the earth
W_o = standard acceleration of gravity at the surface of the earth

The first term $\mathbf{W}/W_o$ on the righthand side of equation 30A is the gravitationally normalized force (GNF) vector $\mathbf{G}_W$, defined and discussed previously. The second term involving the sum of the contact forces is the negative of $\mathbf{G}\star$, by the definition in equation 27. Hence equation 30A—or its equivalent, equation 24—may be written:

$$\mathbf{G}\star = \mathbf{G}_W - \mathbf{G} \qquad (30B)$$

This equation states that in order to determine the vector $\mathbf{G}\star$, we need merely find the vector difference between the gravitational quantity $\mathbf{G}_W$ and the resultant external $\mathbf{G}$, the total GNF acting on a body.

The physical picture may be clarified somewhat by writing a similar vector equation for forces in *ordinary dimensional units.* If we define:

$$\mathbf{F}\star = -\Sigma \mathbf{F}^C \qquad (31A)$$

then equation 22 may be placed in the form:

$$\mathbf{F}\star = \mathbf{W} - \mathbf{F} \qquad (31B)$$

By comparison of equation 31A with equation 27, we have also:

$$\mathbf{G}\star = \mathbf{F}\star / W_o \qquad (31C)$$

Thus we have defined a new resultant force, $\mathbf{F}\star$ (F-star), equal to the negative of the sum of the contact forces acting on a body; the $\mathbf{G}\star$ is just this $\mathbf{F}\star$ expressed in G units, as seen from equation 31C. According to equation 31B, the vector $\mathbf{F}\star$ can be found by taking the vector difference between the pull of gravity on a body and the total *external* force that is responsible for the visible acceleration of the body relative to the earth.

The physiologist approaches an *intuitive definition* for G★ when he recognizes that, in order to arrive at a resultant force that can be correlated with the displacement of body tissues accompanying acceleration, he must take the sum of the gravitational action and what may be termed the "net inertial reaction" (the fictitious force of d'Alembert). In so doing he arrives at a vector that can be exhibited as follows:

resultant force = vector sum of weight and inertial force
resultant force $= \mathbf{W} + (-ma)$ (32A)

Now, by Newton's laws, any body of mass (m) being accelerated at a rate (a) must have acting on it an unbalanced force whose magnitude is $F = ma$. Hence equation 32A becomes:

$$\text{resultant force} = \mathbf{W} + (-\mathbf{F}) = \mathbf{W} - \mathbf{F} \qquad (32B)$$

which is the same expression we have already obtained for F★ in equation 31B.

The vector diagram in Fig. 6-6, *D*, shows the two components G_W and $G_{\text{centrifugal}}$ of the gravitoinertial force vector and their summation to yield G★ for the situation depicted in Fig. 6-6, *A*. The general relationship between the gravitoinertial resultant force vector and contact forces on earth, in earth orbit, or in interstellar space is such that G★ is always equal to the negative of G_{beam}, the latter being "contact" force applied at the universal joint by the rotating beam. It will be intuitively appreciated that the longer the hinged arm, the greater will be the effective radius of rotation and therefore the greater the centrifugal force $m\omega^2R$. Thus the longer the swinging arm, the greater will be the angle that it will make with the vertical.*

*The G★ of interest is, strictly speaking, equal and opposite to G_{beam} only if the mass of the hinged arm can be considered negligible compared to that of the "load" (test subject, enclosure, and so on) located at its free end. Any mass distributed along the arm (and thus acting at differing radii of rotation) requires an additional component of supporting force on the part of the beam. The added effect tends to become more significant as the arm is made longer and therefore usually sturdier, but the effect on the angle of the swing is not great. To take an extreme example, if the arm were half as long as the beam and if the mass of the arm (assumed uniformly distributed) were as much as half the mass of the load, then at a rotational speed that caused the system to swing out to an angle of 60 degrees with the vertical, the direction of the G★ acting on the load center itself would be approximately 60.5 degrees, a deviation of only 0.5 degree from the direction of G_{beam}. The discrepancy becomes even smaller if either the mass ratio or length ratio or rotational speed is made less, and hence we disregard it for all practical purposes.

Vector products

A familiarity with the mathematical concepts of some forms of *vector product* is requisite to an understanding of the physics underlying acceleration physiology. The vector product is particularly useful for describing certain physical situations involving angular velocity and angular acceleration of a rotating human or animal body.

First it must be emphasized that quantities that possess both magnitude and direction are not, in every case, vector quantities. For a readily verified example, consider a human subject seated in a chair that we rotate in a clockwise direction (as viewed from above) by 90 degrees about a vertical axis fixed in space. If the chair is subsequently rotated by 90 degrees about a fixed horizontal axis, we have performed two 90-degree rotations each about a specified axis in space. If now the sequence of these rotations is reversed, it can be seen that the final position of the individual will not be the same, even though the rotations have occurred about the same spatial axes. Thus the *commutative* law of addition is not satisfied. (This is the law of addition that states that for all values of a and b the following holds true: $a + b = b + a$). Since it is not true for the case described, we conclude that finite rotations must not be represented as vectors. On the other hand, mathematicians have shown that if the angles of rotation can be considered infinitesimally small, the vector law of addition can be applied to them.

The above considerations serve to introduce the complex subject of *vector product,* which will here be treated in an elementary manner.

The most useful vector products, for our purposes, satisfy the *distributive* law of multiplication. This law states that:

$$a(b + c) = ab + ac \qquad (33)$$

If we have two vectors, **C** and **D**, their scalar product is calculated by taking the magnitude of **C** times the magnitude of **D** times the cosine of the angle between them when the vectors are given a common origin. Thus:

$$\mathbf{C} \cdot \mathbf{D} \equiv CD \cos (\mathbf{C}, \mathbf{D}) \qquad (34)$$

In this equation "≡" is to be read as "is identical with." This scalar "*dot* product" obeys the *commutative* law of multiplication, so that $\mathbf{C} \cdot \mathbf{D} = \mathbf{D} \cdot \mathbf{C}$.

If we take the unit vectors of **C** and **D**, that is, $\hat{\mathbf{C}}$ and $\hat{\mathbf{D}}$, their scalar product is simply the cosine of the angle between them.

If the head of vector **D** is used as the initiating point for the tail of another vector **E** and the head of **E** coincides with the head of vector **C**, we have a graphic representation of the relation **E** = **C** − **D**. If one takes the dot product of each side by itself we obtain:

$$C^2 + D^2 - 2\mathbf{C} \cdot \mathbf{D} = E^2 \quad (35A)$$

which is equivalent to the important expression known as the *law of cosines:*

$$C^2 + D^2 - 2CD \cos(\mathbf{C}, \mathbf{D}) = E^2 \quad (35B)$$

Now consider the vector **D** in the Cartesian coordinate system (Fig. 6-3). As shown here this vector may be written as:

$$\mathbf{D} = \hat{\mathbf{x}}(\mathbf{D} \cdot \hat{\mathbf{x}}) + \hat{\mathbf{y}}(\mathbf{D} \cdot \hat{\mathbf{y}}) + \hat{\mathbf{z}}(\mathbf{D} \cdot \hat{\mathbf{z}}) \quad (36)$$

If $\hat{\mathbf{D}}$ is taken as the unit vector in the direction of **D** in the figure then:

$$\hat{\mathbf{D}} = \hat{\mathbf{x}} \cos(\hat{\mathbf{D}}, \hat{\mathbf{x}}) + \hat{\mathbf{y}} \cos(\hat{\mathbf{D}}, \hat{\mathbf{y}}) + \hat{\mathbf{z}} \cos(\hat{\mathbf{D}}, \hat{\mathbf{z}}) \quad (37)$$

The cosines in equation 37 are known as the *direction cosines* of the vector **D**. If the scalar product of each side of this equation with itself is taken, we see that the sum of the squares of the direction cosines equals unity:

$$1 = \cos^2(\hat{\mathbf{D}}, \hat{\mathbf{x}}) + \cos^2(\hat{\mathbf{D}}, \hat{\mathbf{y}}) + \cos^2(\hat{\mathbf{D}}, \hat{\mathbf{z}}) \quad (38)$$

Consider a vector product **C** × **D** (which we read as "**C** *cross* **D**"). This "cross product" is defined as a vector perpendicular to the plane of **C** and **D**, with magnitude $CD\,|\sin(\mathbf{C}, \mathbf{D})|$. In equation form we have:

$$\mathbf{E} = \mathbf{C} \times \mathbf{D} = \mathbf{E}\, CD\,|\sin(\mathbf{C}, \mathbf{D})| \quad (39)$$

The righthand thread rule previously discussed gives the sense of **E**. If in Fig. 6-3 the vector **C** is substituted for the x axis and the vector **D** for the y axis, the vector product **E** will be in the +z direction of the figure.

It is important to note that this vector product does *not* obey the *commutative* law, since by the righthand thread rule the sign changes:

$$\mathbf{D} \times \mathbf{C} = -\mathbf{C} \times \mathbf{D} \quad (40A)$$

The vector product does, however, obey the *distributive* law, so that:

$$\mathbf{C} \times (\mathbf{D} + \mathbf{E}) = \mathbf{C} \times \mathbf{D} + \mathbf{C} \times \mathbf{E} \quad (40B)$$

Relative motion

Some portions of this large subject are important to the acceleration physiologist, and a few aspects have been selected and presented in this section. The texts described in the literature cited should be consulted for more extensive treatment.

Motion as a concept is always *relative,* and its description requires that the frame of reference be defined in every case. If two observers of the motion of a given body are using different frames of reference for the description of this motion, the relation of one frame of reference to the other must also be defined. The problem of description of the motion of two objects by a single observer using a single frame of reference also commonly arises. In this physical situation it is often important to know the *relative* velocity of the two bodies in motion. If these bodies are denoted by B_1 and B_2, the velocity of B_2 relative to B_1 is equal in magnitude but opposite in direction (sense) to the velocity of B_1 relative to that of B_2.

If two human observers, H_1 and H_2, are in motion relative to each other and are observing a third body, as for example an aircraft in flight, an important simplification is possible. If the observers are in *uniform relative translational motion,* both H_1 and H_2 will measure the *same acceleration* of the body (or more properly *particle*). Uniform relative translational motion would obtain when, for example, one observer was stationary and the other moving at uniform speed in a straight line, or both observers were moving along straight line paths at uniform speed. In situations where H_1 and H_2 are not in uniform relative translational motion, the mathematical treatment can be extremely complex and will not be considered here.

The motion of a particle in relation to two coordinate systems, such as those of observers H_1 and H_2, can be described graphically by an x-t diagram, x being the ordinate and t the abscissa. When the velocities of the coordinate systems relative to each other are constant, it is possible to depict in one diagram the t and x coordinates of the motion of the particle for a given time interval in relation to each of the coordinate systems. These useful graphs are known as Brehme diagrams and are well described in Sears and Zemansky.[3]

The subject of uniform relative rotational motion will lead us to the important topic of Coriolis acceleration. We begin by considering our two human observers, H_1 and H_2, fixed in relation to two orthogonal coordinate systems such as that of Fig. 6-3. We further assume that the origins of these frames of reference coincide. In this situation H_1 in reference frame $x_1y_1z_1$ observes that H_2 in his reference frame $x_2y_2z_2$ is rotating with an angular velocity ω. H_2 in turn observes H_1 in his frame of reference to be rotating with angular velocity $-\omega$. We adopt the position vector D of point *P* in Fig. 6-3 as our notation for present purposes. Let $\mathbf{V}'$ represent the velocity of particle P measured by H_1 while the velocity of particle P measured by H_2 is $\mathbf{V}''$.

Without giving the derivations we present the equation relating these velocities:

$$\mathbf{V}' = \mathbf{V}'' + \omega \times \mathbf{D} \tag{41}$$

The acceleration of P measured by H_1 is:

$$\mathbf{a}' = \frac{d\mathbf{V}'}{dt} = \mathbf{x}' \frac{dV'_{x'}}{dt} + \mathbf{y}' \frac{dV'_{y'}}{dt} + \mathbf{z}' \frac{dV'_{z'}}{dt} \tag{42}$$

The acceleration of P measured by H_2 is:

$$\mathbf{a}'' = \frac{d\mathbf{V}''}{dt} = \mathbf{x}'' \frac{dV''_{x''}}{dt} + \mathbf{y}'' \frac{dV''_{y''}}{dt} + \mathbf{z}'' \frac{dV''_{z''}}{dt} \tag{43}$$

By differentiation of these expressions with respect to time and by substitution and simplification we obtain the following equation:

$$\mathbf{a}' = \mathbf{a}'' + 2\omega \times \mathbf{V}'' + \omega \times (\omega \times \mathbf{D}) \tag{44}$$

Equation 44 is of interest and importance to the acceleration physiologist. The term $2\omega \times \mathbf{V}''$ represents what is known as *Coriolis* acceleration, and the term $\omega \times (\omega \times D)$ represents *centripetal* acceleration. Both of these acceleration components derive from the relative rotation of observers H_1 and H_2. The Coriolis term is of particular interest and is discussed more fully in the next section. In situations characterized by changing angular velocity, an additional term is required. In equation 44 we assume a constant angular velocity for the earth, although a small correction for its actual variability should be made where extreme precision is important.

Coriolis acceleration and the associated "fictitious" Coriolis force result from relative rotational motion and the inertial properties of matter. These physical phenomena derive their names from the work of the French engineer and mathematician G. G. de Coriolis in the beginning of the last century. Most commonly, Coriolis acceleration is applied to the physical situation of a body undergoing translation in a rotating reference frame. By Newton's first law a body in inertial space tends to remain in straight line motion unless acted on by an external force. Such motion relative to an inertial frame will appear to the observer in a rotating frame to be following a curvilinear pathway, and this observer therefore postulates a force to account for the observed acceleration; hence the terms *Coriolis acceleration* and *Coriolis force*. From a rigorous standpoint, the earth does not provide an inertial frame of reference for a coordinate system that is fixed in relation to earth coordinates. A frame of reference fixed in relation to the earth represents, rather, a rotating frame relative to a true inertial frame. The first term $\mathbf{a}'$ of equation 44 can be considered to represent acceleration in an inertial frame and to be equal to the second term $\mathbf{a}''$ (acceleration in a rotating frame) plus the Coriolis and centripetal accelerations in the rotating frame. In this equation the angular velocity ω is assumed to be constant. For the rotating earth the magnitude of ω is 7.292 $\times 10^{-5}$ rad/sec directed along the axis of rotation.*

Coriolis acceleration, the direction of which is at right angles both to $\mathbf{V}''$ and ω, accounts for major deviations of airflow in the atmosphere and fluid flow in the sea, as well as certain properties of the motion of projectiles and of vehicles on the surface of the earth.

When a body is moving along the surface of the earth or above the surface in a horizontal plane parallel to a plane tangent to the earth, the Coriolis acceleration has a horizontal component that is maximal at the poles and zero at the equator, together with a small and usually neglected vertical component. The horizontal component causes a deviation of the path of the moving body to the right in the northern and to the left in the southern hemispheres. This rightward deviation can readily be observed in the repetitive swings of a Foucault pendulum (the original pendulum demonstrated in Paris in 1851 was 67 m or approximately 220 ft in length).

In the case of a body undergoing free fall the $\mathbf{V}''$ of equation 44 is pointing directly downward. The coriolis acceleration given by $-2\omega\mathbf{V}''$ points east and the falling body therefore deviates toward the east. The centrifugal acceleration $-\omega \times (\omega \times R)$ has, in the northern hemisphere, a component along the north-south line that points south. The combination of these effects on the falling body is such that in the northern hemisphere the body will fall southeast of the point of intersection with the earth's surface of the radial line from the earth's center to the body at the onset of its fall.

*One consequence of this rotation is a reduction in the acceleration of gravity by the amount of the centrifugal acceleration, given by $-\omega \times (\omega \times R)$, where R is the radius of the earth (6.37×10^6 m or 2.09×10^7 ft). At the equator, where the centrifugal acceleration is maximal, its magnitude is only 0.3% of g_0, the standard acceleration of gravity; $g_{centrifugal}$ progressively diminishes toward the poles, concomitantly with a small progressive increase in g, the effective acceleration of gravity.

Expansion of the original meaning of the term "Coriolis acceleration" has created some confusion in the literature, a question that is discussed on p. 218.

Physical basis of weightlessness

Fig. 6-1, showing a satellite in orbit about the earth, provides the starting point for our discussion. For simplicity we assume a circular orbital path. Orbital flight is possible by virtue of the universal gravitational attraction between objects in the universe. This action-at-a-distance force is proportional to the products of the masses of the bodies and inversely proportional to the square of the distance between their centers of gravity. Thus for the earth and a satellite in circular orbit about it, we have a gravitational attraction capable of exerting the centripetal force required to maintain the circular flight path of the satellite, given by the equation:

$$\text{Force}_{\text{gravity}} = G\frac{m_E m_S}{R^2} \tag{45}$$

where G = universal constant of gravitation*
m_E = mass of the earth
m_S = mass of the satellite
R = distance between the centers of gravity of the earth and the satellite

Since, for circular motion:

$$F_{\text{centripetal}} = \frac{mv^2}{R} \tag{46}$$

we can equate 45 and 46:

$$G\frac{m_E m_S}{R^2} = \frac{m_S v^2}{R} \tag{47}$$

We obtain by simplifying:

$$v = \sqrt{G\frac{m_E}{R}} \tag{48}$$

where R = orbital radius
v = linear speed of the satellite in orbit

The mass of the satellite is not a factor in equation 47 and therefore any object, including the human body either within the satellite or engaged in extravehicular activity, with linear speed v and with an orbital radius R, can be considered to be in a stable orbit. The same can be said of every portion of the human body, and indeed every atom within the human body, each of which has its own small "gravitational string" to the earth.

The strength of this gravitational string, of course, diminishes with altitude. A simple aid to remembering the magnitude of this diminution is the fact that the fractional change in this force is twice the fractional change in R.

Since the spacecraft and its occupants, including all components of both, are in stable orbit, there is no physical basis for the existence of *contact* forces on the human travelers. This can perhaps best be visualized by grasping the fact that the spacecraft and its occupants are undergoing the physical equivalent of a *free fall.* Free fall can be visualized in familiar terms by imagining an elevator at the top of a tall building on earth starting its downward travel at an acceleration of 32 ft/sec.^{2} In this situation the elevator and its human occupants are both traveling downward accelerated by gravity and the floor of the elevator would exert no contact force on the feet of the occupants. This is in fact analogous to the situation in a spacecraft in stable orbit, in which the centripetal acceleration at any instant is equal to the gravitational acceleration at that altitude. This condition obtains in what we have termed *ideal weightlessness* (Fig. 6-1). Owing to the fact that the earth is not a homogenous mass and to the fact that the spacecraft in its total mass and center of mass undergoes slight changes, there are slight departures from a geometrically perfect orbit that can be sensed by the occupants of the vehicle. In ideal weightlessness, with its absence of contact forces, there is no gravitoinertial resultant G★, since G★ is the negative of the sum of all contact forces acting.

PHYSIOLOGY OF ACCELERATION ALONG RECTILINEAR AND CURVILINEAR PATHWAYS

With the preceding section as background, we now begin a brief survey of the major physiologic effects of acceleration on the human or animal subject moving along rectilinear or curvilinear paths. Motion along a curvilinear path can be, but is not necessarily, associated with angular acceleration, but it is always associated with centripetal acceleration directed along the radius. We prefer the term "centripetal" to "radial" as being more accurate, although "radial" is in common usage. We defer consideration of the special effects of angular acceleration on the labyrinthine apparatus of the inner ear until later. The effects of acceleration on plant life, particularly the gravi-

*The actual value of G is 6.673×10^{-11} nt-m^2/kgm^2, or in the British system 3.436×10^{-8} lb-ft^2/slug^2. It will be recalled that the *newton* is the unit of force in the MKS system of physical units.

tational effects, are of great interest, but these are entirely omitted here for lack of space.

In the study of physiologic effects, the importance of determining the magnitude and direction of the gravitoinertial resultant force vector and of its history, in relation to a chosen coordinate system, must be emphasized repeatedly. The concept of the gravitoinertial resultant is in common though not universal use today, but troublesome problems of terminology and symbol usage remain. These problems, together with proposals for their solutions, are next considered in order to provide a framework for the discussion of physiologic changes that follows.

Force and acceleration vectors

There are three major accelerative force vectors of physiologic interest:

1. *The gravitoinertial resultant force vector.* This is the single most useful force vector in the analysis of physiologic events, and it has been considered in detail from the standpoint of physics earlier.

2. *The total external force vector.* This vector is the resultant of the forces acting on the body through actual contact and through the action-at-a-distance property of gravity.

3. *The resultant contact force vector.* This force vector is the resultant of all of the forces making physical contact with the body. Thus in the standing position there is a contact force on the feet acting upward equal to 1 G unit of magnitude and equal to the oppositely directed force of gravity. Contact forces and their distribution are of particular importance during so-called impact, where large forces are rapidly applied to the surface of the body, and in the situation of prolonged application of lesser forces, as for example in a prolonged flight mission.

In addition to the above three force vectors, analysis of acceleration from the kinematic standpoint requires determination of the *acceleration vector* solely from the *motion* of the body itself. The kinematicist will treat the acceleration of gravity as the inertial reaction caused by a headward acceleration in space. This type of analysis is confusing to some but is entirely legitimate so long as internal consistency in analysis of an acceleration situation is maintained.

These vectors as a group, plotted in relation to the axes of the body, provide the essential information for the analysis of the various physical situations of acceleration, except for the special effects of angular acceleration involving human

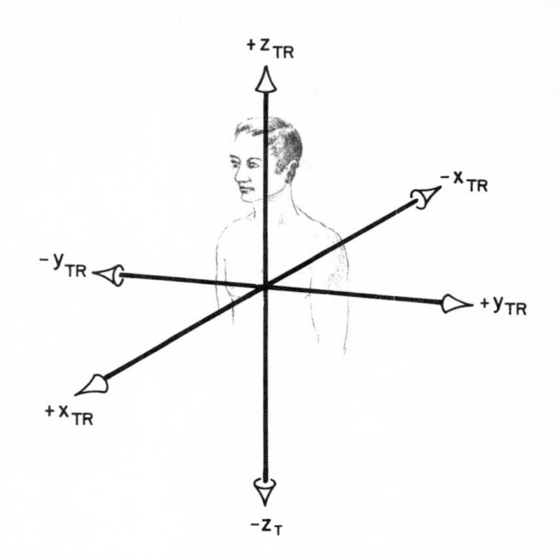

Fig. 6-7. Anatomicomathematic axes for rectilinear and centripetal acceleration. As with the "true" axis system of Fig. 6-8, the subscripts identify the system being used in any given analysis. To maintain mathematic parity throughout without transformation of axes, the +y axis for kinematic analysis and the +y axis for the gravitoinertial vector analysis have been made to coincide. In this pair of axis systems the action and reaction force vectors along the y axis are therefore in the same direction.

or animal bodies. It would be highly desirable to employ a single coordinate system in relation to axes of the body in which all of these vectors may be plotted, either in respect to their instantaneous values or preferably their "history." A variety of axis and terminology systems have been employed or suggested over the years,[1, 2, 4-6] with none completely satisfactory for every purpose. Axis systems must be suitable for calculation of the acceleration vectors in the kinematic and also the gravitoinertial type of analysis, but in addition, as mentioned, a system is needed into which all of these vectors can be transposed for simultaneous display. Fig. 6-7 portrays two of the axis systems that have been proposed[6] and utilized. These may be termed the action (kinematic) and reaction (gravitoinertial) systems. We have previously indicated (Fig. 6-4) the necessity for adherence to the righthand thread rule for preservation of mathematical parity. This being the case, it is not possible to have an action and a reaction system in which every reaction axis is opposite in direction and sense to the corresponding action axis. This results in the ambiguous situation in which we are forced to adopt a common axis for

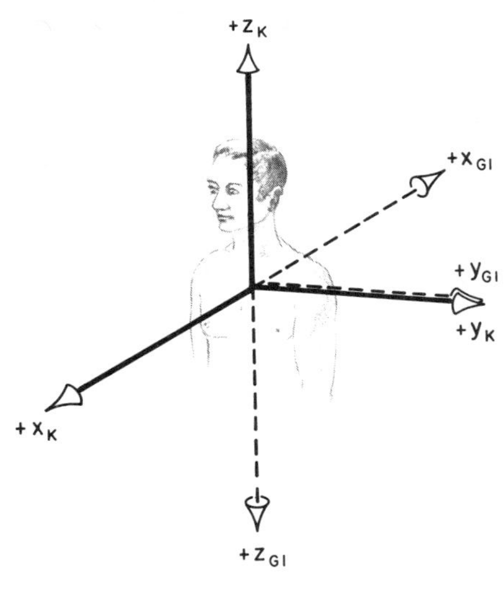

Fig. 6-8. Anatomicomathematic axis system for the human body. In this "true" axis system, the positive and negative sides of each axis are on opposite sides of the origin. The x, y, z axes are mutually perpendicular. The subscript T denotes the axis system being used without further identification.

the positive side of one axis in each of these systems, shown in Fig. 6-7 for the +y axis. It is clearly impossible for acceleration to be in the +y direction and the inertial reaction to this acceleration to be in the same direction. Insofar as a coordinate system is simply used for calculation of the magnitude and direction of the vector, no great problem is presented; the correct analysis can be carried out and mathematical rigor maintained in either the kinematic or gravitoinertial axis system. It is, however, incorrect to plot gravitoinertial vectors in the kinematic axis system or acceleration vectors in the gravitoinertial system, even though these vectors are expressed in G units. It is therefore essential for display of these multiple vectors expressed in G units that some type of true reference system be adopted. What we are proposing is such a true reference system, taking the standard three-dimensional Cartesian coordinate system in common use in physics today, in which the +x axis is toward the reader, the +y axis directed toward the right side of the page, and the +z axis directed vertically upward, with each of these axes perpendicular to the other two. Into this system we place the individual in the standard anatomic position facing the reader (+x direction) with his left shoulder toward the +y direction and the vertex of his head toward the +z direction, as shown in Fig. 6-8.

Table 6-3 presents a set of symbols for use in the kinematic, the gravitoinertial, and the true reference systems. Ideally a vector should be identified with subscripts and superscripts to denote both the type of vector and its direction (relative to human or animal body axes) without ambiguity. There are two ways that appear to us to be suitable for this purpose:

1. Denote the *axis* system being used with a *subscript.* This we have done for the kinematic with the subscript K, for the gravitoinertial with the subscript GI, and for the true reference system with the subscript TR. The type of vector is therefore identified by the subscript denoting the axis system. For example, only gravitoinertial vectors would be shown in a gravitoinertial axis system.

2. In relation to a true reference system in which several vectors of different types expressed in G units are to be plotted, it is proposed the *superscript* identify the *vector* as follows:

Total external force vector—superscript E
Resultant contact force vector—superscript C
Gravitoinertial resultant force vector—superscript*

*This is not to suggest that accelerative motion and accelerative force are dissociated. The equation $\mathbf{F} = ma$ shows that these are inseparable.

Table 6-3. Symbols for earth gravity acting on individual at rest in different body positions—three anatomicomathematic axis systems

Anatomico-mathematic axis system	Standing on feet	Standing on head	Horizontal, right lateral decubitus	Horizontal, left lateral decubitus	Horizontal, dorsal decubitus (supine)	Horizontal, ventral decubitus (prone)
Kinematic	$+1G_{z_K}$	$-1G_{z_K}$	$+1G_{y_K}$	$-1G_{y_K}$	$+1G_{x_K}$	$-1G_{x_K}$
Gravitoinertial	$+1G_{z_{GI}}$	$-1G_{z_{GI}}$	$-1G_{y_{GI}}$	$+1G_{y_{GI}}$	$+1G_{x_{GI}}$	$-1G_{x_{GI}}$
True reference	$-1G^{\bigstar}_{z_{TR}}$	$+1G^{\bigstar}_{z_{TR}}$	$-1G^{\bigstar}_{y_{TR}}$	$+1G^{\bigstar}_{y_{TR}}$	$-1G^{\bigstar}_{x_{TR}}$	$+1G^{\bigstar}_{x_{TR}}$

These symbols are obvious except for the asterisk. The last was proposed some years ago[2] as a simple means of identifying what was then termed "physiologic G." We now believe "gravitoinertial resultant vector" or "gravitoinertial resultant G" to be better terms. The choice of an asterisk as a superscript commended itself to us as being simple, readily identified, and available on all typewriters. Alternatively, the superscript GI could be used, but since this is also proposed as a subscript for denoting the gravitoinertial axis system, the double use would be confusing. It would therefore seem preferable to continue the use of the asterisk (star) for this purpose. Finally, if the resultant acceleration vector derived from calculation of the motion of a body (kinematic analysis), expressed in G units, is to be displayed in the true reference system, it is proposed that the superscript A be employed for this purpose. Thus we have three accelerative force* vectors and one accelerative motion* vector that can be readily displayed in the true reference system and identified without ambiguity from the proposed symbols alone. In a complex acceleration situation the vector often will not correspond in direction with one of the cardinal axes of the trunk of the body (Figs. 6-7 and 6-8). In this situation there are two possible presentations: (1) the magnitude of the vector can be specified as well as two orientation angles[1]; or (2) the magnitude of the vector's projection on the x, y, and z axes can be given (Fig. 6-3).

In Table 6-3 we give the proposed symbols in three reference systems for earth gravity acting on an individual at rest in six primary body positions. As stated previously, from the standpoint of the kinematicist an individual standing motionless is actually being accelerated headward at 32 ft/sec^2 (981 cm/sec^2). The physiologic effects of being accelerated in the headward direction in a gravity-free environment in interstellar space will be precisely the same as the effects of motionless standing on earth. *In the actual use of these symbols an author, for simplicity and brevity, may wish to state at the outset of his manuscript that a given axis system is employed and that all vectors presented in the analysis are of a specified type. In this case, after initially showing a vector with the complete symbol notation, an author may choose to omit both the superscripts and subscripts denoting the type of vector and the axis system in the remainder of his manuscript.* This short cut has the disadvantage demonstrated by the current situation in the literature, namely that the vector being presented is not always evident with certainty. With the system of symbols that we are proposing there should be no confusion of meaning.

In Table 6-4 we show the symbols for the three force vectors in G units that are of major interest in the notation of the true reference system for six body positions. Since the total external (net) force on an individual at rest is zero regardless of body position, the first horizontal row in the table consists of zeros. In order to show the notation for the total external force vector, we have in the footnote indicated the proper symbol for an individual standing in an elevator that is accelerating upward at 32.2 ft/sec^2. The reader will be able without difficulty to fill in the notations for this same situation with the individual in the other body positions.

One disadvantage of the almost universal use in the literature of plus or minus G_z, G_x, G_y (as in the gravitoinertial axis system) is the fact that this usage implies a more precise relationship between the anatomic axes of the body and the gravitoinertial resultant force vector than often is the case. There are common situations in modern combat maneuvers, in experiments on the human centrifuge, in blastoff and reentry of spacecraft, and in daily life in which the force vector will deviate to a physiologically significant degree from the cardinal axes as listed. Precision in analysis of physiologic events in relation to the accelerative forces causing them will require in the future a more careful designation of the relation of the force vector to the body axes and of the history of this relation in the physiologically significant immediate past.

The *magnitude* of a force vector is almost universally expressed in G units, although one can visualize types of precise analysis of internal body dynamics that would require forces to be expressed in absolute units. Where spherical polar coordinates are used, the magnitude of the vector itself is given, as well as the two orientation angles θ and ϕ, which may be termed longitude and colatitude respectively (discussed in more detail later). Alternatively, the magnitudes of the vector's projection on the x, y, and z axes can be given, a presentation that still provides information on both the magnitude and the spatial orientation of the force or the acceleration vector, albeit less explicitly. The *orientation angles* that we employ are measured as follows: ϕ is measured from the

*This is not to suggest that accelerative motion and accelerative force are dissociated. The equation $\mathbf{F} = ma$ shows that these are inseparable.

Table 6-4. Symbols for total external, contact, and gravitoinertial force vectors in G units acting on individual at rest in different body positions—true reference system

Force vector	Standing on feet	Standing on head	Horizontal, right lateral decubitus	Horizontal, left lateral decubitus	Horizontal, dorsal decubitus (supine)	Horizontal, ventral decubitus (prone)
Total external*	0	0	0	0	0	0
Contact	$+1G^{C}_{Z_{TR}}$	$-1G^{C}_{Z_{TR}}$	$+1G^{C}_{Y_{TR}}$	$-1G^{C}_{Y_{TR}}$	$+1G^{C}_{X_{TR}}$	$-1G^{C}_{X_{TR}}$
Gravitoinertial	$-1G\bigstar_{z_{TR}}$	$+1G\bigstar_{z_{TR}}$	$-1G\bigstar_{y_{TR}}$	$+1G\bigstar_{y_{TR}}$	$-1G\bigstar_{x_{TR}}$	$+1G\bigstar_{x_{TR}}$

*In order to show the usage of the proper symbol, imagine the individual standing on feet in an elevator accelerating upward at 32.2 ft/sec². In this case, the total external force vector is represented as follows: $+1G^{E}_{Z_{TR}}$

positive side of the z axis to the vector itself. Thus ϕ can vary from 0 to 180 degrees. Theta is measured from the +x axis in the direction indicated by the righthand thread rule to the projection of the vector on the x-y plane and can vary from 0 to 360 degrees. In this way the magnitude of the vector in the true reference or kinematic systems is completely specified. For the gravitoinertial system, since the positive direction of z is downward, the angle ϕ is measured upward from the polar axis, which is inverted in this case. Similarly, θ is measured from the +x axis in accordance with the righthand thread rule as before. It is evident that angles other than the longitude and colatitude can be used to specify a vector's direction in space[1] one simply has to adopt a given system and use it with rigorous consistency.

It is common practice to speak of the acceleration or accelerative force vector in terms of one of the cardinal axes of the body or of the vehicle. Thus, in the literature we often see an expression such as "an acceleration of 3 G_z" ($-3G\bigstar_{z_{TR}}$)—that is, downward from the head in the direction of the long axis of the body. It is evident from studying the accelerative force pattern in even such an elementary maneuver as the loop (shown in Fig. 6-15 as a vertical circle for simplicity) that both the magnitude and the direction of the gravitoinertial resultant force vector (G★) in relation to the anatomic axes of the pilot and of the aircraft are undergoing major changes throughout the maneuver. Modern combat maneuvers comprise such complex flight patterns as the barrel-roll attack, the high- and low-speed yoyo attack, and the scissors defense. Analysis of these maneuvers indicates that extreme changes are occurring in the magnitude and direction of G★ relative to the axes of the aircrew and aircraft as the pilot flies through the eggshell-shaped envelope of possible flight patterns.

To describe accurately such situations of maximal stress, the time history of the acceleration or accelerative force in relation to body or vehicle axes needs to be specified. A method must be adopted for readily describing the instantaneous force or acceleration vector and its direction in space, as well as a time sequence of such vectors. In the following material, we are proposing what appears to us to be a reasonable approach to the solution of this problem. Any of the accelerative force vectors previously discussed regarding total external force, resultant contact force, or gravitoinertial resultant force can be placed in a three-dimensional Cartesian coordinate system (Fig. 6-3). The vector can be displayed in a human (or animal) anatomicomathematic axis system (Fig. 6-8) or in the axis system of the aircraft (x-y-z in Fig. 6-11). As we have indicated, two options are available, each with advantages for certain applications: (1) the projections of the vector on each of the three axes can be given or (2) the magnitude of the vector and two angles can be given.

The essential elements in both options can be seen in Figs. 6-3 and 6-8 considered in combination. For the vector **D** in Fig. 6-3 we substitute the symbol for the vector that applies to any given situation being analyzed. In the following examples we will employ the gravitoinertial resultant vector G★, but any other of the vectors of interest could be similarly used. The true reference system is chosen for this example, in which both the sense and direction of each axis, shown in Fig. 6-8, correspond to those of Fig. 6-3. We wish to describe the behavior of G★ throughout a certain flight pattern in relation to the pilot. Let us assume that 10 seconds after an arbitrary zero time point in the flight maneuver, the magnitude of G★ is 4.0 G units but that its direction does not correspond to any of the cardinal axes of the pilot's body.

The magnitude of G★ will have been obtained from the recorded readings of three linear accelerometers with their axes mounted orthogonally. For simplicity we assume that these accelerometer axes were fixed at the time of mounting to correspond with the cardinal axes of the pilot's body as shown in Fig. 6-8. For our example, the reading of the three accelerometers is as follows: $2.449G\star_x$, $1.414G\star_y$, $2.828G\star_z$. The same figures could theoretically be obtained if we knew the precise flight pattern of the aircraft in three-dimensional space and could calculate the instantaneous acceleration for the point of interest. The three-accelerometer method is far more practical. The resolution of the three components of G★ into a single vector is given by equation 21A. A continuous solution by computer is the ideal method, but the calculations can be made from recordings of each accelerometer. In the proposed system of usage the representation of the above situation at the 10-second point in the flight pattern is the following:

$$^{10}(4.0G\star,\ 2.449G\star_x,\ 1.414G\star_y,\ 2.828G\star_z)_{TR}$$

where superscript 10 = time of the measurements (10 seconds into the maneuver)
subscript TR = axis system being used

A sequence of such groupings of vector information at appropriate time intervals throughout a flight maneuver would enable the presentation of a large amount of essential data with accuracy. The number of figures carried after the decimal can, of course, be altered as required. Continuous recording of computer calculations of the four items of information between the parentheses would provide more thorough analysis of the maneuver, particularly if the aim is prediction of, or correlation with, physiologic effects.

The alternative method of presenting the magnitude and direction of the gravitoinertial resultant force vector in relation to the pilot's body axes is to calculate the angles ϕ and θ from equations 21B and 21C, respectively. These data in the symbols of our proposed system are as follows:

$$^{10}(4.0G\star,\ 30°\ \theta,\ 45°\ \phi)_{TR}$$

It can be seen that this presentation makes for easier visualization of the direction of G★ in relation to the pilot's body. On the other hand, the x, y, and z components of G★, or of other vectors, provide information not quickly appreciated from the magnitude of the vector and its two orientation angles.

The choice of the method of presenting the data thus reduces to the question of the type of application. The use of both presentations would, of course, be the most informative. As is shown later, the use of spherical polar coordinates has major applications in the calculation of angular accelerations for analysis of labyrinthine physiology. Analogous representations of angular acceleration vectors, their magnitude, components, and direction can be employed. Fig. 6-9 shows the anatomicomathematic axis systems for displaying such vectors ($\boldsymbol{\alpha}$ or $\dot{\boldsymbol{\omega}}$) together with the subscripts for identifying the system being used. Angular accelerometers with their axes mutually perpendicular, as with the linear accelerometers, provide the raw data for the components of angular acceleration of the resultant angular accelerative force vector.

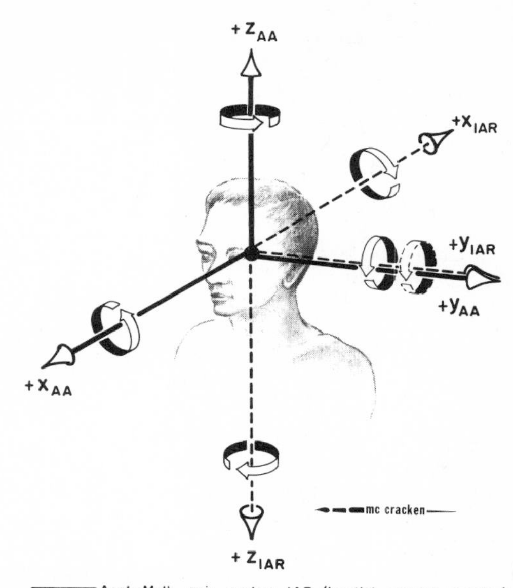

Fig. 6-9. Anatomicomathematic axes for angular acceleration and angular reaction. As with Fig. 6-8, the acceleration axis system and the reaction axis system coincide in their +y axes.

Physiologic effects of acceleration

The genesis of the major physiologic changes produced by acceleration along rectilinear or curvilinear pathways lies in the relative displacements produced within the body. It is not possible to distinguish between the displacements produced by gravitational forces and those pro-

INSTRUCTIONS FOR THE DETERMINATION OF THE ANGLE ϕ AND THE MAGNITUDE OF THE GRAVITOINERTIAL RESULTANT FORCE: *SELECT HORIZONTAL COMPONENT OF ACCELERATIVE FORCE ON ABSCISSA SCALE, GO UP TO INTERSECT GRAVITY LINE FOR PLANET OF INTEREST, THEN HORIZONTALLY TO THE LEFT TO READ ϕ, OR HORIZONTALLY TO RIGHT TO INTERSECT RESULTANT FORCE LINE AND BACK DOWN TO ABSCISSA SCALE TO OBTAIN MAGNITUDE.*

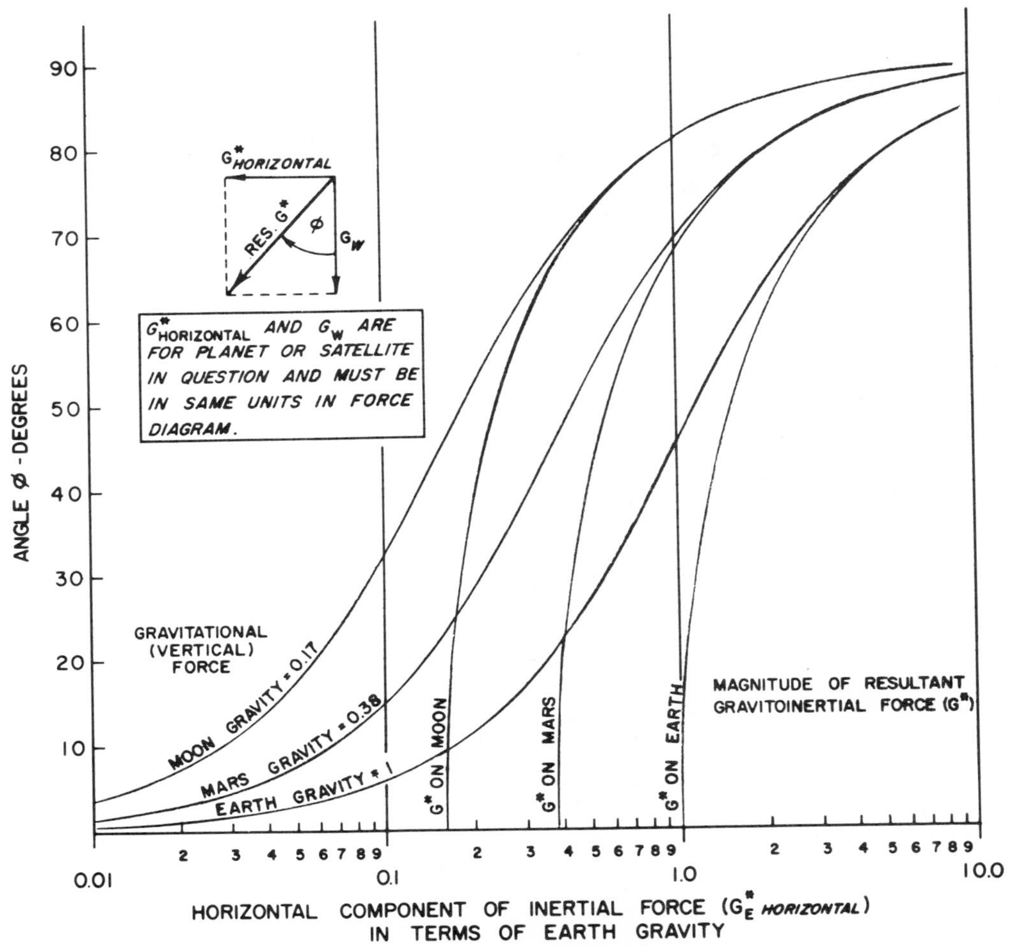

Fig. 6-10. A graph for the determination of the magnitude and deviation from the vertical (angle ϕ) of the gravitoinertial resultant force on earth, moon, and Mars. The required instructions for use of the graph are contained in the two enclosed rectangles.

duced by inertial forces, although from the standpoint of theoretic physics there are certain differences.[7] Some investigators have felt that it is not necessary to introduce the gravitoinertial resultant force in explaining physiologic phenomena. We strongly disagree. The equivalence of gravitational and inertial forces in the production of physiologic effects makes it mandatory to include the gravitational component of G★, particularly where gravity makes a significant contribution to the magnitude and direction of the gravitoinertial resultant. Gravity is, of course, all-pervasive and constant (except for slight variations on the earth's surface) and thus differs in these respects from the induced "impulse" accelerations of manmade machines. Except in situations of extremely high accelerations in relation to that of gravity, the gravitational contribution will be found to be physiologically significant; such significance will be the case in most of the acceleration situations to be considered in this section.

It is appropriate to comment that we are here considering the effects of earth gravity in relation to inertial forces. On the moon, where gravitational force is one sixth that on earth, inertial forces produced by acceleration or deceleration of a vehicle such as the lunar rover, or by an astronaut's own kangaroo-like locomotion, could have effects on the direction of **G★** far greater than on earth. These comparative effects of earth gravity and moon gravity on the gravitoinertial resultant (**G★**) are depicted in Fig. 6-10. On a planet with greater mass and greater gravitational force than that of earth, gravity would make a correspondingly larger contribution to the gravitoinertial resultant force vector.

Much of what is considered in the present section will, of necessity, be concerned with acceleration directed approximately parallel to the cardinal axes of the body. Most acceleration experiments have been done on human centrifuges, which have a variety of structural and performance characteristics and radii of rotation. Human centrifuges were developed surprisingly early, and by 1814 a 13-ft diameter machine with 5 G unit capability was in use in treating psychiatric patients at La Charité Hospital, Berlin. Jongbloed[8] in 1934 reported extensive studies on an animal centrifuge, and by the beginning of World War II most of the major countries of the world had constructed human centrifuges and had performed important experimental work on the physiologic effects of acceleration. Particularly outstanding was the early work of von Diringshofen,[9] Gauer,[10] and Ruff[11] (Germany), Armstrong and Heim[12] (United States), Franks[13] (Canada), and their colleagues. The references cited are only representative of their many publications. An early civilian centrifuge, that at the Mayo Clinic, was and remains one of the most productive. The U. S. Navy's acceleration laboratory at Pensacola, Florida, became a center for angular acceleration studies early in World War II and has maintained that role. The history of these and other centrifuges has been well reviewed by Ham,[14] White,[15] and Leverett.[16] At present there appears to be a tendency toward decline in activity in a number of these laboratories.

It is clear that with a radius of rotation of, for example, 50 ft, there would only be slight differences in the magnitude of accelerative force experienced at the head and at the feet of a subject in the seated position. White[17] carried out studies on very short radius centrifuges in which the accelerative force at the feet was quite different from that at the head. Such a large force gradient along the body produces unusual differential effects. On the other hand, even a 50-ft radius for a human centrifuge fails to an important degree to mimic the accelerative force environment in flight maneuvers. One cannot obtain the same combinations of angular and centripetal accelerations in a human centrifuge that are present in an aircraft with its much longer radius of turn. Following the early stages of their development, most of the centrifuges carried the subjects in a cab mounted on a freely swinging arm.[2] In these machines, except for periods of increasing or decreasing angular velocity with the associated tangential component of acceleration, the direction of **G★** usually roughly corresponded to the long axis of the trunk of the body (z axis) or to one of the transverse axes (x or y). Too little has been done either in thorough study of the patterns of magnitudinal and directional change in **G★** that exist during the complex flight maneuvers of combat or acrobatics or in the attempt to approximate on the centrifuge, insofar as possible, the acceleration profiles of these complex patterns. Centrifuges with their planes of rotation either vertical or between vertical and horizontal would be required for more exact simulation of certain types of maneuvers, but none is known to have been constructed, except for amusement park devices.

Centrifuges have found application in the space programs of the United States and also that of the Soviet Union. For example, the Navy's centrifuge at Aerospace Medical Research Laboratory, Johnsville, Pennsylvania, was extensively used in astronaut training for the simulation of the acceleration profile of reentry.[18] It is not possible with a centrifuge to simulate weightlessness, which would require a vertical parabolic profile. A centrifuge can play a useful role, however, in training a subject to walk under hypogravitational conditions. In this application, developed at NASA's Langley Research Center[19] and also employed at Manned Spacecraft Center, Houston, a subject is supported by a special harness that holds him at a precalculated angle to the vertical. The harness is in turn attached to the arm of a slowly turning centrifuge while the subject's feet are in contact with an inclined surface resembling a banked racetrack. By this technic the subject can practice walking around the inclined surface with his feet bearing the same weight as they would under the one-sixth earth's gravity conditions existing on the moon.

A potentially important future use of centrifuges on deep space missions has been proposed by White[15]; his suggestion is to utilize a small centrifuge in a spacecraft to simulate partially or wholly the gravitational attraction on earth and thus maintain the astronaut's physiologic integrity for the many months of the mission and to ensure his safe return to earth. The question as to whether the same result might be obtained by specially designed exercises against resistance has been raised[20] but remains to be answered in the future.

The variable *time* is all-important, not only in the physical generation of acceleration but also in the production of physiologic alterations. Thus acceleration is the second derivative of distance with respect to time and the first derivative of velocity with respect to time. The inertial properties of deformable structures in the body, including blood, require significant time even under constant gravity for completion of their deformations of shape and change of position. If acceleration is sufficiently high and prolonged, a new dynamic anatomic and physiologic equilibrium cannot be reached, the normal negative feedback defensive responses become overwhelmed, and positive feedback takes over. In positive feedback, any further deterioration in time of a vital function, such as systemic arterial pressure, leads to additional deterioration of the same function, and the condition of the organism rapidly becomes critical.

The activation and response times of various physiologic defensive mechanisms can be important influences on the sequence of physiologic events. Continuous recordings of systemic arterial pressure on human centrifuges suggest that effective activation of baroreceptor mechanisms leading to arteriolar constriction and improvement in recorded systemic arterial blood pressure requires about 8 seconds.[21] Interestingly, this is the same length of time that complete cessation of blood flow requires for the production of unconsciousness.[22] The more rapidly an acceleration curve reaches its peak, the less is the time available for physiologic defense mechanisms to become usefully operative. The tilt table can be used to demonstrate the importance of the time rate of change of the gravity vector in relation to the long axis of the body, a demonstration with implications for situations of higher acceleration. An individual with inadequate postural responses during head-up tilting to +70 degrees will show a far greater fall in systemic arterial pressure at heart or head level if the tilting is accomplished in a matter of a few seconds as compared with tilting over a period of, for example, 20 seconds. In the latter situation the effective component of gravitational force—that is, its component along the long axis of the subject's body—is changing direction sufficiently slowly so that physiologic compensation can be normal or nearly so even though quite inadequate with rapid tilt. Another important aspect of the time-acceleration relations is the effect of the physiologic changes produced by acceleration on the reaction times of individuals. This reaction time of 0.2 to 0.3 second (in youth and in health) to a simple stimulus and its components (sensing, perceiving, decision, and response times)[23] can be crucial in an individual's ability to command a vehicle undergoing accelerations of large magnitude at high speed. Reaction time can be unfavorably altered by significant reduction in a vital function, such as brain blood flow.

It has been thought that *jolt* (the first derivative of acceleration with respect to time) would be useful in the analysis of situations of rapid change in the acceleration profile although to date it has seen limited application except in the analysis of *impact*. This term is used for high but brief accelerations, usually lasting 1.0 second or less, such as occur during ejection from aircraft, parachute opening shock, or crash situations either in military or civilian life. One can intuitively predict that analysis of the rate of change of acceleration could be fruitfully applied to accelerations of longer duration. If, for example, an accelerative force reaches a peak magnitude of 5 G units within 2 or 3 seconds, that is, before compensatory responses become operative, it will have more severe physiologic effects than an acceleration reaching the same peak value under a slow rate of change, even though the peak acceleration is maintained for the same duration in both instances.

The existence of a *latent period* of 0.2 second may be required for the development of hydrostatic effects,[24] yet inertial properties are always present and theoretically can become operative instantly. In a semirigid deformable system such as the human or animal body, the more deformable structures can cushion the effects of sudden acceleration for a brief period, so that there exists a deformation time or latent period for the ef-

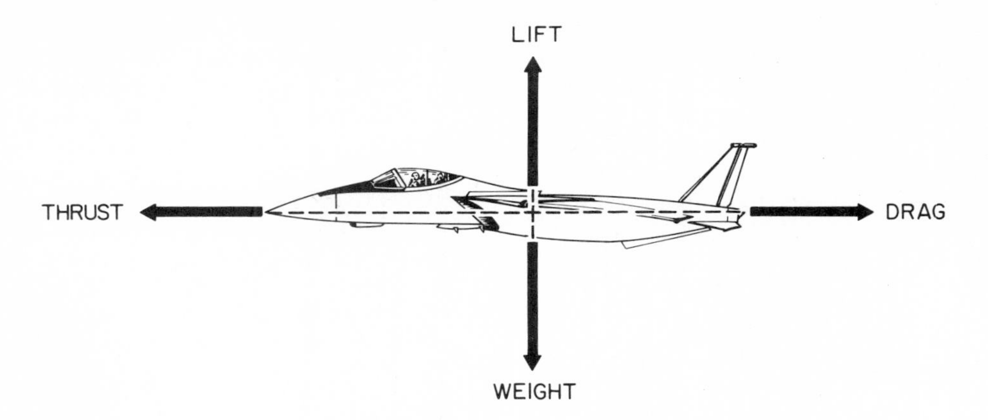

Fig. 6-11. Forces on an aircraft in level flight.

fects of tissue displacement. Fraser[24] has suggested that the 0.2 second duration be used as an arbitrary division between *abrupt acceleration* and *brief acceleration,* the latter defined as lasting 0.2 to 10 seconds. He would term *prolonged acceleration* that which lasts longer than 10 seconds.

Positive acceleration

Classically, the term "positive acceleration" has been applied to situations in which the accelerative force drives the blood away from the head and toward the seat and feet. The term is ingrained in the literature and would be difficult to alter. The kinematicist will also call this "headward acceleration." Vernacular terms have been applied to acceleration situations; "eyeballs down" is one of such terms used to denote positive acceleration. A table of these terms, which are not utilized in the present chapter, is given in Allen[25] and Randall.[26] The symbols for earth gravity acting on an individual standing or seated with the long axis of the body vertical and head up (Table 6-3, p. 183) are: $+1G_{z_K}$, $+1G_{z_{GI}}$, $-1G\star_{z_{TR}}$; these are the symbols in the kinematic, gravitoinertial, and true reference systems, respectively. It will be remembered that the kinematic and gravitoinertial systems have the disadvantage of requiring that one axis of the coordinate systems for action and reaction vectors be shared in common (Fig. 6-7, +y axis). In the true reference system (Fig. 6-8) more than one vector of interest can be displayed simultaneously, after the calculations have been made using any appropriate axis system.

The symbol in common usage in the current literature of aviation medicine and physiology for positive acceleration is $+G_z$. This is the same as the symbol in the kinematic and gravitoinertial axis systems without the additional subscripts. We have indicated our feeling that if the abbreviated symbology (G_x, G_y, G_z) is utilized in the future, authors of every publication should specify what axis system is being used and what vector has been calculated. There is one additional use of the symbol G that deserves comment, namely its substitution for the term "acceleration." This vernacular usage is so common—"G tolerance" for "acceleration tolerance"—that we accept it and utilize G without boldface or italics for this purpose.

Physical situations. The simplest positive acceleration situation *in flight* is represented by Fig. 6-11. The aircraft is in straight and level flight and the many forces on it are shown as reduced to four major resultant forces: lift, weight, thrust, and drag. At constant speed in perfectly smooth air, thrust balances drag and lift balances weight; the personnel in the aircraft are subjected by gravity to positive acceleration equal to 1 G unit in magnitude, assuming that the long axis of the trunk of each of the crew is vertical or nearly so. The synonymous symbols for earth gravity acting on the crew are given above. Fig. 6-12 also shows the aircraft in straight and level flight, but with additional important details. *P* is the *angle of pitch* between the longitudinal or x axis of the aircraft and its direction of motion (that is, flight path) and equals in magnitude the angle between the anatomic and functional z axes (z and z′). This is to be distin-

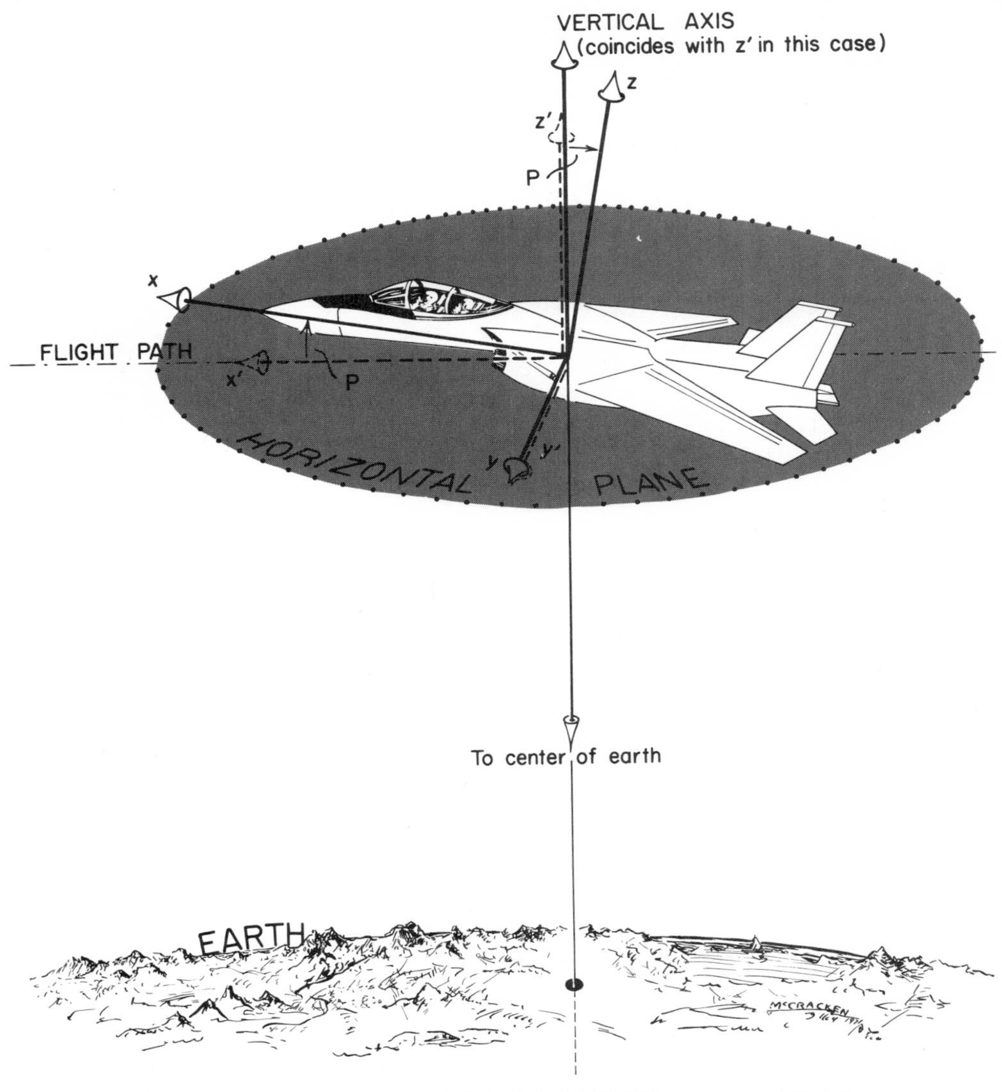

Fig. 6-12. Aircraft in level flight showing anatomic and functional axes in relation to the horizontal plane in which the aircraft is flying.

guished from the *angle of attack (A),* which is *P* plus the angle of incidence *(I)* of the wing (the angle between the chord line of the wing and the x axis of the aircraft). Since *I* is a constant for a fixed wing aircraft, *P* and *A* vary together. The physiologically significant point is that the angle of attack can change appreciably and thus change the direction of the gravity vector or of the gravitoinertial resultant relative to the long axis of the individual's body. This change in angle of attack can be sufficiently large to be significant with respect to otolith organ function and possibly cardiovascular function as well. For example, the angle of attack for zero lift in contemporary high performance fighter aircraft is not far from 0 degrees, but the stalling angle of attack in some aircraft will run to 25 degrees or greater; thus, the possible range in *A* is wider than is often appreciated.

The *production of positive accelerations* of higher magnitude than that of gravity is accomplished by a coordinated turn in an aircraft, shown

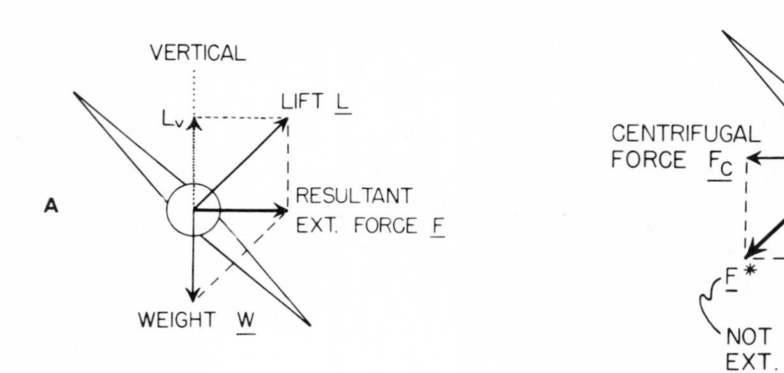

Fig. 6-13. Forces on aircraft in coordinated horizontal turn. The vertical component of lift (L_v) balances the weight (**W**) of the aircraft. The resultant of the lift and weight (gravitational) forces represents the unbalanced force that causes the turn. Note that the gravitoinertial resultant force vector is at right angles to the y axis of the aircraft.

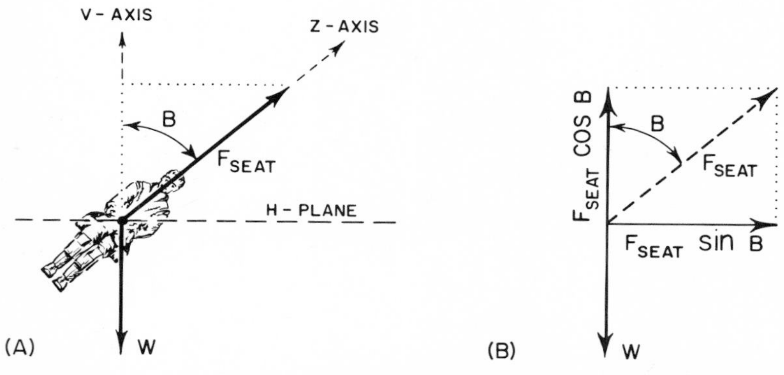

Fig. 6-14. Forces on pilot seated in the conventional position in an aircraft in a coordinated horizontal turn. The gravitoinertial resultant vector (not shown) is equal in magnitude and opposite in direction to the force (F_{seat}) exerted by the seat on the pilot.

diagrammatically in Fig. 6-13. Assuming that the z axis of the pilot's body coincides with the functional z axis (z′) of the aircraft and that the tangential speed of the aircraft (that is, its airspeed) is constant, a true correspondence will exist between the z axis of the pilot's body and the resultant accelerative force vector. Thus, we will have an accurate $+G_{z_K}$, or $-G\star_{z_{TR}}$, the magnitude of which will be determined by the angle of bank. It should be mentioned that the steeper the angle of bank in a coordinated turn, the greater the angle of attack. This is true because the total lift vector must be greater for its vertical component to equal the weight of the airplane. In Fig. 6-13, *A,* the forces on the aircraft are the *lift* (**L**) and *weight* (**W**), leaving the unbalanced resultant external force (**F**). In accurate symbology this would be identified as $\mathbf{G}^E$ (with subscripts denoting aircraft axes if desired—for example, AAC for anatomic, aircraft) followed by the value of its x, y, and z components in the reference system chosen or followed by the magnitudes of the orientation angles ϕ and θ. Since our primary interest is in the gravitoinertial resultant force rather than the resultant external force, we would commonly use the analysis of forces shown in Fig. 6-13, *B.* Note that *weight* appears both as an external force and an inertial force. The inertial force in this diagram is the centrifugal force F_C, which combined with the weight **W** yields the resultant **F★** that "explains" the physiologic phenomena in the crew. If **F★** is expressed in G units, we have the familiar gravitoinertial resultant **G★**. The quantitative relationships of Fig. 6-12 are as follows:

$$\text{lift} = \text{weight} \times \text{secant of the angle of bank} \quad (49)$$

$$\text{centripetal force} = \text{lift} \times \text{sine of the angle of bank} \quad (50)$$

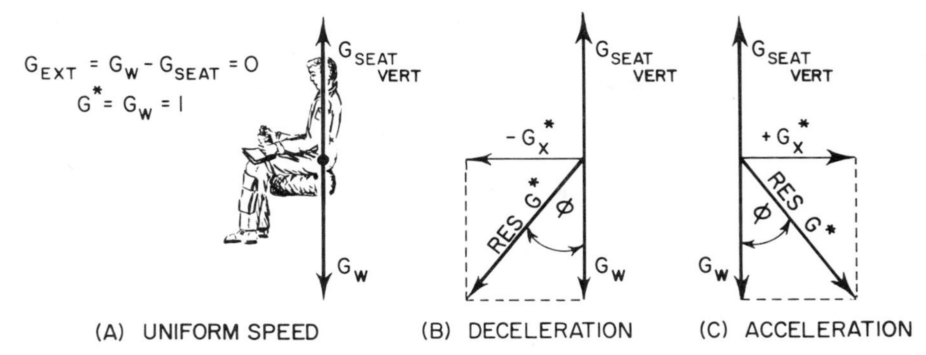

Fig. 6-15. Forces on pilot in horizontal flight. **A,** Uniform speed, **B,** during deceleration, and **C,** during acceleration of the aircraft. The gravitoinertial resultant vector is indicated by the symbol G★. To simplify the diagram only the vertical component ($G_{seat_{vert}}$) of the external force exerted by the seat on the pilot is shown; this force balances the pilot's weight G_w. The horizontal force exerted by the seat and harness in **B** and **C** is equal and opposite to $G\bigstar_x$. The small changes in the relation of the pilot's body axes to these forces, owing to changes in angle of attack of the aircraft, are omitted in these diagrams.

$$\text{centripetal force} = \text{weight} \times \text{tangent of the angle of bank} \quad (51)$$

Equation 49 can be used to obtain the total force of the seat acting on the pilot, depicted in Fig. 6-14 to be on his seatpack and in a coordinated turn comparable to that diagrammed in Fig. 6-13. In Fig. 6-14, *A*, $F_{seat} = W \sec B$, the vertical component of which balances the pilot's weight, leaving the unbalanced (except in d'Alembertian analysis) horizontal component, $F_{seat} \sin D$, which causes the pilot to accelerate centripetally at the same rate as the aircraft. An inertial force equal and opposite to $F_{seat} \sin B$ can be vectorially combined with weight to give the gravitoinertial resultant force (G★), equal in magnitude but opposite in direction to F_{seat}.

When the pilot of Fig. 6-14 rolls out of his turn and returns to straight and level flight in quiet air he of course returns to the 1 G positive acceleration situation previously described for this type of flight (Fig. 6-15, *A*). If, however, he reduces power, the deceleration of the aircraft will cause significant change in the direction and some change in the magnitude of the gravitoinertial resultant force (Fig. 6-15, *B*). Here $-G\bigstar_x$ is directed forward ($-x_{GI}, +x_K, +x_{TR}$) and the resultant of this force and weight causes the gravitoinertial resultant to swing forward of the vertical by the angle ϕ. We have now departed from "pure" positive acceleration. In Fig. 6-15 the upward force of the seat balances the pilot's weight, and a backward contact force exerted on him by his restraints (not shown) balances the inertial force $-G\bigstar_x$. Similarly, during acceleration (Fig. 6-15, *C*) the gravitoinertial resultant swings backward. It should be noted that the ϕ as shown in Fig. 6-6 would be in the GI axis system in which the positive side of the z axis is directed downward. Deviations of ϕ up to 25 degrees or so can readily occur in fighter aircraft. Such deviations can have profound effects on the functioning of the otolith organs, particularly in flying at night or in an overcast sky when visual cues are markedly reduced. They can also have certain circulatory effects, which are discussed below.

The *human centrifuge* has been the standard laboratory method for producing positive acceleration (as well as negative and transverse accelerations, discussed later). The original U. S. Army Air Force centrifuge at Wright-Patterson Air Force Base, Dayton, Ohio,[27] and others had no swinging arm and the subject lay horizontal with the long axis of his body parallel to the beam, and with his head directed toward the center of rotation for positive G studies. One disadvantage of this arrangement was that the gravitoinertial resultant force at no time was directed exactly along the long axis of the subject's body. The centrifuge with the passively pivoted arm, mentioned earlier on p. 188, became in one form or the other the standard laboratory machine for acceleration studies and still remains in extensive use today despite the development of actively pivoted cabs or gondolas. The passively pivoted centrifuge is shown diagrammatically in Fig. 6-16. The subject rides in the cab at the end of the swinging arm.

In theory, with constant angular velocity of the drive shaft the passively pivoted arm and its structurally integral cab swing outward so that the gravitoinertial resultant force is parallel to the

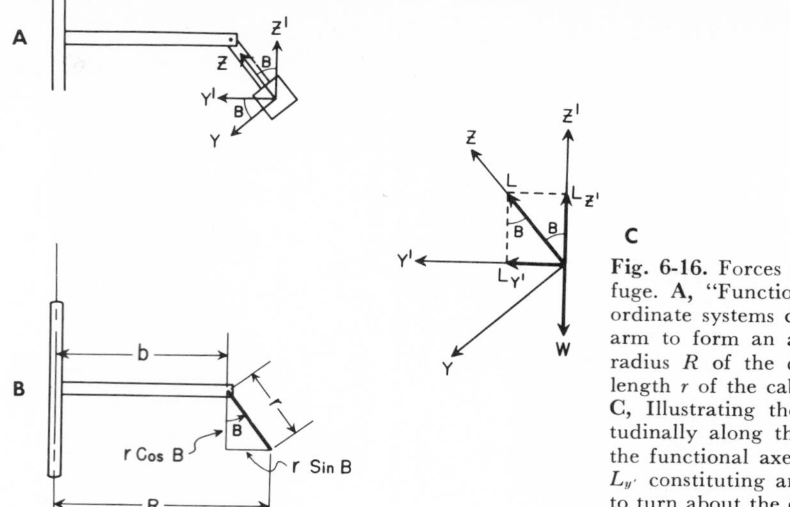

Fig. 6-16. Forces on passively pivoted (1-degree) centrifuge. **A,** "Functional" x′-y′-z′ and "anatomic" x-y-z coordinate systems differ as a result of pivoting of the cab arm to form an angle *B* with the vertical. **B,** Effective radius *R* of the cab is determined by the angle *B* and length *r* of the cab arm, as well as by the beam length *b*. **C,** Illustrating the resolution of "lift" *L* (acting longitudinally along the cab arm) into its components along the functional axes, $L_{z'}$ opposing the cab weight *W*, and $L_{y'}$ constituting an unbalanced force that causes the cab to turn about the drive shaft.

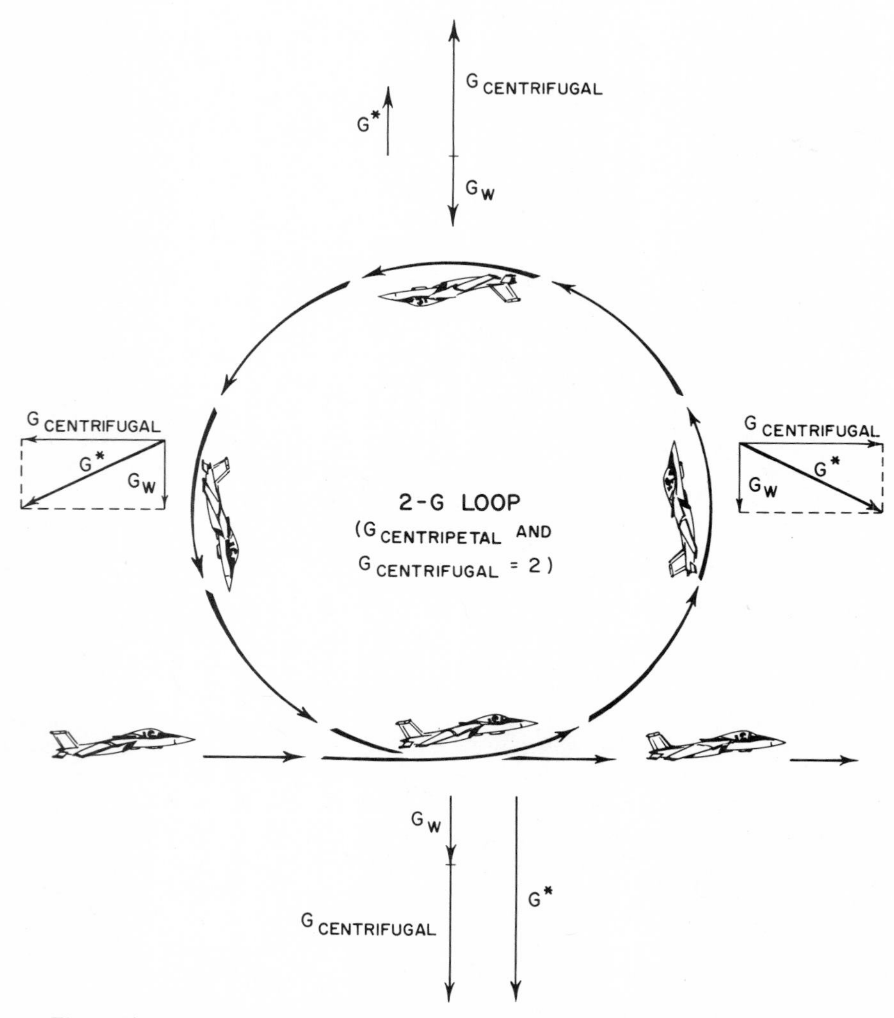

Fig. 6-17. Gravitoinertial forces on the aircraft and pilot during a 2-G loop.

anatomic z axis of the cab and thus differs from the true vertical by *B,* the angle of bank. The angle *B* will be affected to some extent by the uniformity, or the lack of it, of distribution of mass on either side of the line joining the pivot and the anatomic center of the cab. As a result G★ may not be perfectly parallel with the anatomic z axis of the cab. Furthermore, the longer the swinging arm, the farther outward it swings (larger angle *B*) for a given angular velocity. The gravitoinertial resultant G★ at the pivot and at the cab will differ, since the radius of turn will differ, although the angular speed will be the same in the two locations. The magnitude of *G*★ will also differ at the head and seat and feet in the seated positive-G situation, particularly if the centrifuge radius is short. With an effective radius of 50 feet, the magnitude gradient of G★ is negligible for most purposes, but where unusual precision is required, G★ should be determined as to both magnitude and direction at the major point of interest in the subject's body or at more than one location.

In situations of changing angular velocity of the drive shaft, there will be an acceleration component tangential to the spatial pathway of the subject. For example, with increasing angular velocity, with the subject facing the direction of motion, there will exist tangential components of acceleration that we can represent in symbols as $+G_{x_K}$, $+ G_{\bar{x}_{GI}}$, $-G\star_{x_{TR}}$. These components must be included in the calculation of the total G★ vector.

The behavior of the gravitoinertial resultant with regard to its magnitude and direction relative to the subject, under both transient and

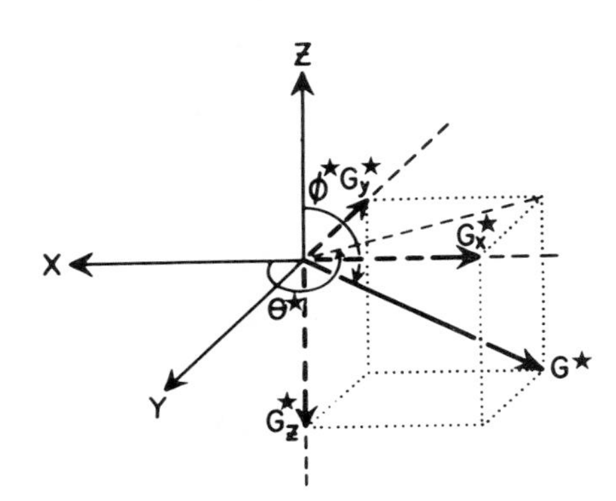

Fig. 6-18. General vector descriptions of G★ in rectangular x-y-z coordinate system. The figure shows the components $G\star_x$, $G\star_y$, and $G\star_z$ of the resultant G★. An alternative description consists in specifying the magnitude *G*★ and the direction (orientation) angles $\phi\star$ and $\theta\star$.

steady state conditions in a given centrifuge experiment, should be specified. In Fig. 6-18 we indicate the presentation of G★ in a three-dimensional rectangular coordinate system, a treatment comparable to that of Fig. 6-2 with respect to the orientation angles and the components of the vector along the x, y, and z axes. The presentation of these vectors and orientation angles for practical use is discussed on p. 185.

Physiologic effects. In considering these physiologic effects, certain *background facts,* in addition to these previously presented, should be kept in mind. These will also be useful in the interpretation of negative and of transverse acceleration.

Three organs of the body are characterized by an interface between epithelium and gas at or near atmospheric pressure: the skin, the lungs, and to a lesser extent the gastrointestinal tract. In these organs it is necessary for the vertical hydrostatic pressure gradient in the bloodstream resulting from gravity to be opposed by forces built up within the vascular walls and surrounding tissues rather than by a gradient of pressure in the gaseous environment. This hydrostatic gradient—that is, rate of change of pressure with elevation (height)—is given by:

$$\frac{dp}{dh} = -\rho g \tag{52}$$

where ρ = density of blood
g = acceleration of gravity

The specific gravity of blood (its density relative to that of water) is 1.057 ± 0.005.[28] The hydrostatic difference in pressure between two levels in the circulation is given by:

$$\Delta p = \rho g \Delta h \tag{53}$$

where h = elevation
g = acceleration of gravity

On earth, Δp can be conveniently calculated in millimeters of mercury by:

$$\frac{\Delta h \times 1.057}{1.36} \tag{54}$$

where Δh is in cm

(NOTE: 13.6 is the specific gravity of mercury; the decimal is shifted one point to the left, since Δh is more commonly expressed in centimeters than in millimeters.)

The gradient of pressure in the atmosphere over the very small distances involved is negligible. A uniform gaseous counterforce (barometric pressure) is therefore opposing intravascular pressure, with its large vertical pressure gradient, and tissue pressure, with its less well-known vertical pressure gradients. Since gravitational and inertial forces are indistinguishable in their effects on the body, the vertical hydrostatic gradients will be multiplied by the gravitationally normalized magnitude of the vertical component of the gravitoinertial resultant force vector.

It is well known that many human beings, even normally healthy ones, at times show symptoms of postural circulatory inadequacy under the force of gravity alone.[29] The development of major disturbances in the circulation when multiples of the accelerative force of gravity are applied is therefore to be expected. The most immediately vulnerable circulations to fall in perfusion pressure are (1) the circulation of the brain and its extension into the eye, (2) the pulmonary circulation, and (3) the coronary circulation.

Since blood is the most readily displaceable "tissue" in the body, much of the symptomatology from positive acceleration ($+G_{z_K}$, $-G^{\star}_{z_{TR}}$), or of acceleration in any direction relative to the body axes, is based on shift of blood volume and its effects on circulatory dynamics. In positive acceleration the shift of blood is toward the seat and feet. With the human trunk upright, the base of the brain is about 34 cm (plus or minus a few centimeters depending on height and body habitus) above the heart; thus the brain normally receives blood at an arterial pressure that is about 24 mm Hg lower than at heart level. Approximately one third of the arterial pressure drop at brain level on changing from the horizontal to the upright position on earth is offset by the 7 mm Hg fall in internal jugular bulb pressure.[30] The internal jugular system is not a perfect siphon since it is capable of partial collapse, which increases the resistance to blood flow and reduces the pressure gradient from the jugular bulb to the right atrium. Where there is higher positive acceleration of, for example, 3 to 4.5 G units, the internal jugular venous pressure becomes markedly subatmospheric and the arterial pressure at brain level falls to, or nearly to, zero.[31] Since symptoms of cerebrovascular insufficiency begin when brain blood flow has fallen by approximately 40%,[32] it is clear why mild symptoms can be readily produced early in positive acceleration, even of low magnitude (2 or 3 G units).

A highly pertinent group of observations was made by Rossen, Kabat, and Anderson,[22] during World War II. These investigators throttled most of the brain blood flow of volunteer subjects with a pneumatic collar, although in their experiments some vertebral arterial flow almost certainly continued. Consciousness was lost within a mean time of approximately 8 seconds. The reason becomes apparent when we consider the minimal oxygen reserves available to the ischemic brain: the brain consumes about 45 ml of oxygen per minute but has a dissolved pool of only about 2 ml of oxygen and is unable to enter rapidly into significant anaerobic metabolism. Thus the brain is highly vulnerable to the severe hypoxia that accompanies marked reduction or cessation of its blood flow.

The lungs, by virtue of their mechanical properties[33, 34] and their dual system of gas-filled airways and liquid-filled blood vessels, are peculiarly susceptible to the effects of accelerative force. This vulnerability has been emphasized by research in recent years. Even under earth gravity there is a definite vertical gradient of intrapleural pressure amounting to 0.21 cm H_2O/cm of vertical distance.[35] Accompanying this is a gradient of lung density, which, in the anesthetized dog held in the vertical head-up position, amounts to a threefold increase in density from apex to base.[36] Since the mean density of the human lung and of the dog lung are approximately the same, one can anticipate that in upright man under earth gravity the lung density at the apex will be approximately 0.1 gm/cm^3 and at the base 0.4 gm/cm^3. When multiples of earth's gravitational acceleration are applied, it is evident why positive acceleration can be associated with severe congestion of the lung bases.

Other aspects of respiratory function as influenced by gravity have been studied in recent years. The well-known work of West[37] has shown that a ventilation/perfusion ratio gradient exists from apex to base, a low $\dot{V}_A/\dot{Q}$ ratio being present at the base and a high ratio at the apex. As a result, the alveolar carbon dioxide tension (P_{CO_2}) is lower at the apex than at the base, while the converse is true for the alveolar oxygen tension ($P_{A_{O_2}}$). These relationships hold true for the normal individual upright under the 1 G unit of earth gravity. West also showed that the inequality of the ventilation/perfusion ratio produces a gradient of alveolar nitrogen tension (P_{N_2}): alveoli with a low $\dot{V}_A/\dot{Q}$ have a high P_{N_2} and vice versa. The P_{N_2} at the apex is approximately 555 mm Hg and at the base 582 mm Hg. Acceleration in multiples of that of gravity would be expected to exaggerate these differences. Quite recently Michaelson and colleagues[38] contributed an important study on vertical differences in pulmonary diffusing capacity and capillary blood flow in human subjects. Diffusing capacity per liter of lung volume (D_L/V_a) and pulmonary capillary blood flow per liter of lung volume ($\dot{Q}_c/V_A$) were calculated from the rates of carbon monoxide and acetylene disappearance relative to that of neon. When the bolus of neon, carbon monoxide, and acetylene was inspired from residual volume, the D_L/V_A was 5.26 ml/min × mm Hg/L. When the bolus was inspired from the functional residual capacity position, D_L/V_A was 6.54 ml/min × mm Hg/L. At RV and FRC $\dot{Q}_c/V_A$ was 0.537 L × min^{-1} × L^{-1} and 0.992 L × min^{-1} × L^{-1}, respectively. When similar maneuvers were performed using ^{133}Xe, the results confirmed the authors' interpretation that during inspiration more of the bolus was restricted to the upper zone of the lung if introduced at residual volume and more was distributed to the lower if introduced at functional residual capacity. The authors' lung model, constructed to explain their data, shows a steep gradient $\dot{Q}_c/V_A$, increasing from apex to base, and a directionally similar although less steep gradient of D_L/V_A. The final conclusion of Michaelson and colleagues is that a more generalized unevenness of the ratios D_L/V_A and $\dot{Q}_c/V_A$ must exist throughout the lung, independent of gravity. These authors compared their measurements with those of West and colleagues[39] and found agreement that the vertical increase in blood flow from apex to base is steeper than the vertical increase of diffusing capacity (D_{Lco}), a situation that creates an uneven distribution of $D_L/\dot{Q}_c$ and red cell transit times from apex to base.

The literature on the effects of positive acceleration contains several excellent reviews of the subject.[16, 24, 40-42] We will briefly review the sequence.

The clinical and physiologic events with positive acceleration of brief duration usually take the following course. As the gravitoinertial resultant force increases by multiples of gravitational G, every atom and every tissue in the body have been effectively increased in weight by the same multiple. (The G levels in relation to specific events in the descriptions that follow represent an average, with a range of about ± 0.5 G.) An individual feels this increased weight immediately and at 2.5 G he or she can only arise from the seated position with difficulty and at 4 G strenuous effort must be made even to lift the arms. If the rate of change of acceleration (jolt) is rapid—for example, 1.5 G/sec—the initial symptoms of inadequate cerebral blood flow will appear after approximately 3 seconds. With an intraocular pressure of 20 mm Hg opposing retinal blood flow the peripheral vision begins to dim at about 3.5 G and is lost at about 4 G; complete loss of vision (blackout) follows at about 4.5 G. Blood flow in the brain is continuing at this point sufficiently to maintain consciousness and hearing, but about 0.75 G above the blackout level consciousness is lost if the acceleration is continued. This loss of consciousness occurs after approximately 6 seconds, an interval that agrees well with the work previously cited on duration of consciousness,[22] since decrease in brain blood flow has obviously begun during the early period of increasing G. After loss of consciousness there is a relaxation of the skeletal muscles and, even though deceleration is started immediately after loss of consciousness, half of the subjects may show some sort of convulsive movement.

The physiologic defense mechanisms against positive G are similar to those acting against change in posture from horizontal to upright with the 1 G (earth gravity) alone acting. Primarily these mechanisms comprise arterial and venous constriction under carotid sinus and aortic baroreceptor stimulus. It is also probable that the intrinsic myogenic response originally described by Bayliss has some effect. In this response, which resides in the blood vessel itself, increased tension in the vascular wall from increased intraluminal pressure is opposed by increased contraction of vascular smooth muscle. Physiologic defenses are commonly adequate to support the circulation of healthy young individuals up to positive accelerations of 3 G or somewhat higher, but beyond this these defenses become increasingly inadequate, with the development of right atrial pressure too low to fill the right heart and consequent fall in cardiac output and systemic arterial pressure. The ischemia of the heart during positive acceleration with a thin cardiac shadow was demonstrated many years ago cineradiographically by German investigators.[10]

The remarkable protection to the circulation of the brain afforded by the venous column from jugular bulb to right atrium in the positive-G situation has been commented on. Usually consciousness is lost when mean arterial pressure at brain level falls to 25 mm Hg; yet during high accelerations to the point of visual blackout, sufficient brain blood flow for reasonable alertness may be maintaned by the development of markedly subatmospheric jugular bulb pressure, which can range from 20 to 60 mm Hg below ambient

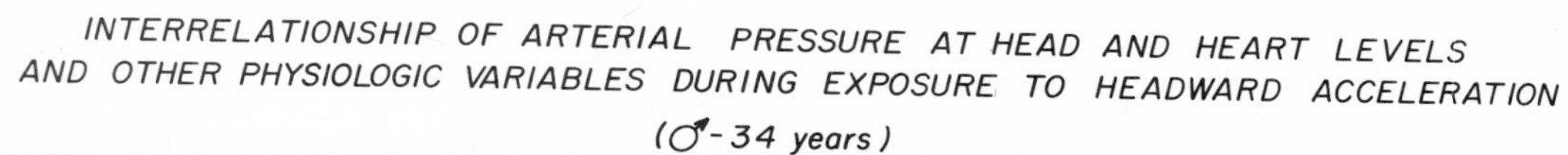

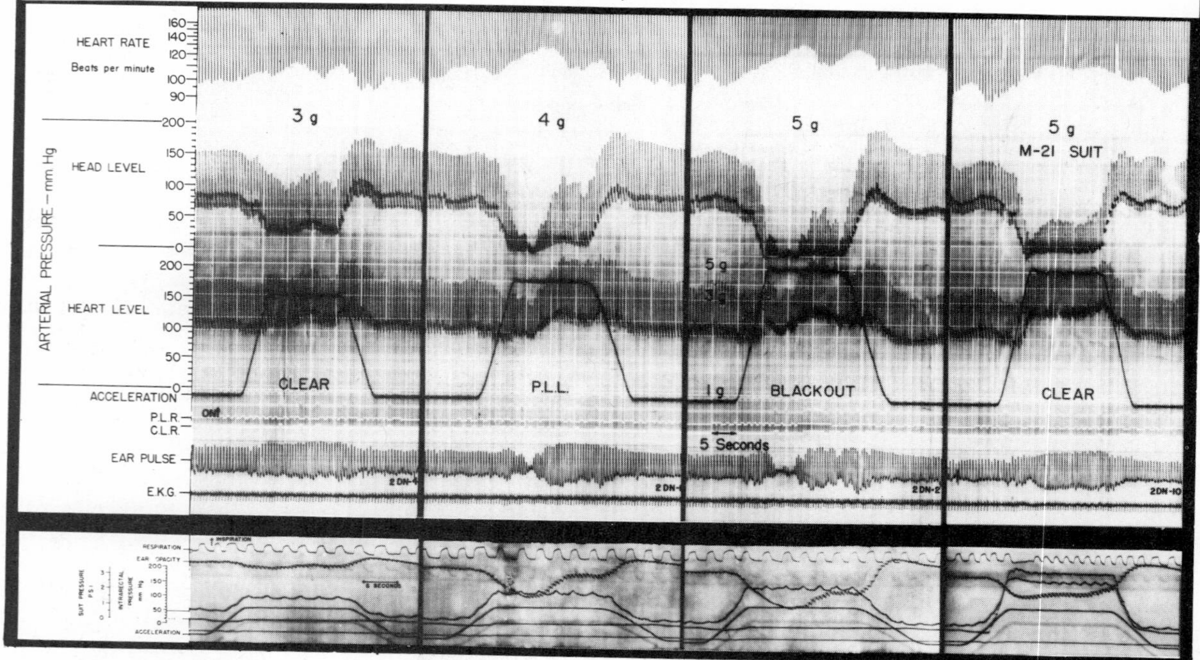

Fig. 6-19. Relationships of arterial pressure at head and heart levels and other physiologic variables during exposure to headward ($-G\bigstar_{zTR}$) acceleration. Subject: human male, age 34 years. In the figure, "Clear," "P.L.L." (peripheral lights lost), and "Blackout" refer to state of vision. (From Wood, E. H., and others: Wadd Technical Report 63-105, Wright-Patterson Air Force Base, Fig. 14, Dec., 1963.)

pressure.[31] We have measured cerebrospinal fluid (CSF), internal jugular bulb, and arterial pressures at brain level with head-up and head-down tilting.[43] It was found that the changes in CSF and jugular venous pressure were virtually identical, with tilting in either direction. This being the case, it appears that the fall in intracranial pressure in the head-up position, or its increase in the head-down position, effectively balances the changes in venous pressure and protects the caliber of intracranial veins. This may well hold true for higher accelerations than 1 G, but without actual measurement, possibly hazardous at high gravity, the question cannot be given a final answer for man. The role of autoregulatory mechanisms in the circulation, for example in the circulation of the brain, is probably not great during accelerations of brief duration but may be of some importance in prolonged acceleration.

The sequence of a number of physiologic variables during positive acceleration ("headward" acceleration to the kinematicist) is shown in Fig. 6-19.[44] This is representative of the now classic recordings obtained over many years by E. H. Wood and associates on the Mayo Clinic centrifuge. The effects of 3, 4, and 5 G accelerations without protection and of 5 G with the protection of the M-21 antiblackout suit are shown. The minimal effect of 3 G, maintained for several seconds, on arterial pressure at heart level is evident as well as the moderate fall in this pressure at 4 G. Significant fall in coronary perfusion pressure might have occurred in the 4 to 5 G range, but this is conjectural. We can be sure, however, that at this level of acceleration the effect of the hydrostatic column between heart and brain becomes important, and we see that the subject's peripheral vision suffers. At 4 G the arterial pressure at brain level is only about 10 mm Hg, but despite this the subject of Fig. 6-19 is conscious, owing to the action of the venous siphon. The slight notch in the arterial pressure tracing at 4 G

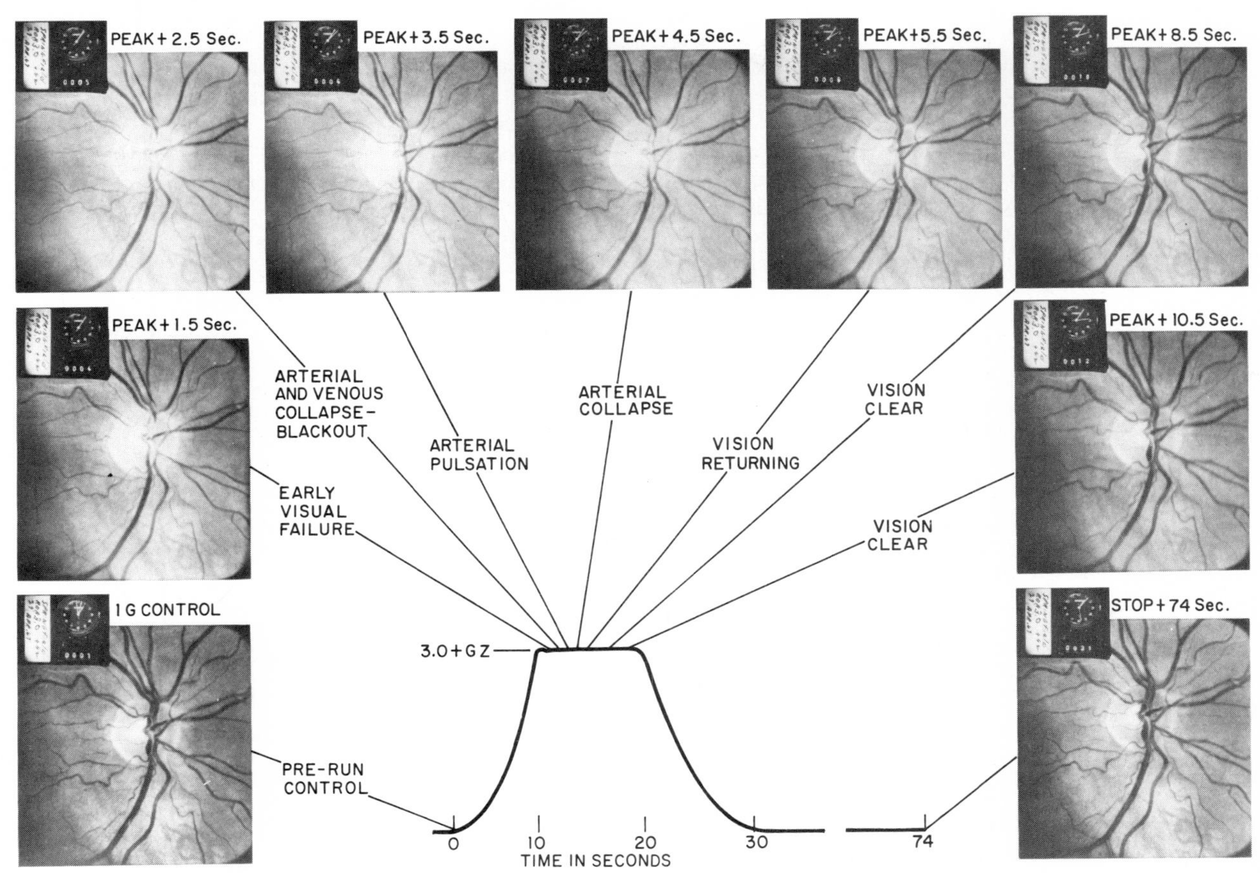

Fig. 6-20. The ocular fundus pre- and postacceleration and after different periods of exposure to -3.1 $G\bigstar_{zTR}$ acceleration. The arterial pressure during acceleration stabilized and began to rise slightly toward the end of the acceleration plateau. Note the marked retinal reactive hyperemia following the acceleration. (Courtesy Dr. Sidney D. Leverett, Jr., Chief, Biodynamics Branch, USAF, Brooks Air Force Base, Tex.)

soon after the plateau in the accelerometer recording is reached can be seen best in the head level recording. This may represent the onset of the effective vasoconstriction. Circulatory compensation continues, since blood pressure begins to recover while acceleration is maintained. The overshoot after cessation of acceleration is probably based both on vasoconstriction and on the effects of return of cardiac filling. At 5 G the arterial pressure at head level remains at or near zero for a considerable period, the restoration of arterial pressure through compensatory mechanisms is inadequate, and loss of vision (blackout) occurs. In the last panel the effects of an antiblackout suit that applies pressure to the calves, thighs, and abdomen is seen to be sufficient to maintain clarity of vision at 5 G. (A Valsalva maneuver alone would probably have afforded brief protection.) Other features of Fig. 6-19 are worthy of study: heart rate increases as arterial pressure at head level falls and returns to approximately control value after the 3 and 4 G acceleration runs are discontinued. At 5 G the heart rate remains somewhat elevated for several seconds after return to 1 G. The ear pulse tracing is seen to become markedly attenuated only when arterial pressure falls nearly to zero at brain level, but the blood content of the ear ("ear opacity" tracing) falls earlier at 3 and 4 G and remains reduced for a longer period. Respiratory rate shows some tendency to increase at 4 and 5 G but the effect is not great; there is, however, tachypnea at 5 G with the antiblackout suit, but some tachypnea is evident in this record in the control state.

Fig. 6-20 shows clearly the events in the circulation of the retina during positive acceleration of 3.1 G units, sustained for approximately 10 seconds. The increasing pallor of the retina and diminished filling of both the arterial and venous vessels, particularly the venous, is evident, as well as some improvement that occurs in the vascularity of the retina toward the end of the acceleration plateau. Finally, a marked retinal reactive hyperemia develops. In our own laboratory we have seen increase in internal jugular bulb hemoglobin-oxygen (HbO_2) saturation to values well above control level following return to the horizontal position after human vasodepressor syncope, an observation suggesting that some reactive hyperemia can occur in the brain as well as in the retina.[43]

A tabular summary has been prepared by E. H. Wood and colleagues showing the changes in major cardiovascular variables during positive acceleration of 2, 3, and 4 G units.[45] The values in this table were obtained by careful and sophisticated technic on seven human subjects and can be considered the best available. Rosenhamer[46] has, however, raised a question about the validity of the indicator-dilution curves under acceleration and concludes that no thoroughly evaluated method is available.

The effects of *sustained acceleration* has been studied by a number of investigators, among them Bondurant.[47] This investigator found that 3 G could be tolerated for 60 minutes and 4 G for 20 minutes. On the other hand, E. H. Wood and colleagues[45] reported that syncopal attacks occurred in some of their subjects exposed to 3.5 G for 10 minutes.

Table 6-5. Changes in cardiovascular quantities during positive acceleration, as compared with control values*

Quantity	Percentage increase (+) or decrease (−)		
	$+2\ G_z$	$+3\ G_z$	$+4\ G_z$
Cardiac output	−7	−18	−22
Stroke volume	−24	−37	−49
Heart rate	+14	+35	+56
Mean aortic pressure	+9	+21	+27
Systemic vascular resistance	+17	+41	+59

*After Wood, E. H., and others: Effective headward and forward accelerations on the cardiovascular system, NASA SP-103, Washington, D. C., 1966, U. S. Government Printing Office.

The importance of *changes in respiratory function* has received emphasis in recent years. Abnormalities of gas exchange have been found with positive acceleration. Although this change has been less well studied than with transverse acceleration, HbO_2 saturation with positive acceleration[48] has been found to fall from the normal 97% at 1 G to a 93% at 4 G and to decrease further to 89% after the antiblackout suit was inflated. Only the first 15 seconds before injection of the indicator dye could be measured in these studies, since the dye precluded further measurement of percent HbO_2 saturation.

Prior to World War II the German workers von Diringshofen[49] and Gauer[50] demonstrated increase in tidal volume and respiratory rate during positive accelerations up to 5 G. Later American work[47] showed that decreases occur in pulmonary compliance with accelerations of 3 and 3.5 G, with concomitant increase in functional residual capacity amounting to between 100 and 600 ml. More recently Glaister[51] reported that the stiffness of the lungs and the airway resistance changed little up to 3 G but that there was a definite increase in the work of breathing owing to increased stiffness of the chest wall and increased intra-abdominal pressure.

Increased respiratory minute volume was found by Barr[52] in studies on positive acceleration in nine human subjects exposed to 5 G for 2 minutes. Minute volume increased from 8.6 to 20.8 L/min and alveolar ventilation from 4.9 to 9.6 L/min. Oxygen consumption rose from a control value of 269 ml/min to 410 ml/min and the respiratory exchange ratio from 0.80 to 0.96. This investigator attributed the increased arterial-end tidal P_{CO_2} of 8.0 mm Hg to ventilation of unperfused areas of the lung.

Barr and colleagues[53] also studied effects of positive acceleration (head-to-tail) on respiratory function in anesthetized dogs. These workers found that moderate G forces produced severe hypoxemia during a period of several minutes even though 100% oxygen was breathed and the animals were hyperventilating.

West and colleagues,[54] using an isolated lung preparation in the dog divided the lung into three zones on the basis of *blood flow patterns* as determined by alveolar pressure (P_A), pulmonary arterial pressure (P_a), and pulmonary venous pressure (P_v). In *zone 1* at the apex $P_A > P_a$ and therefore there is no blood flow; in *zone 2* $P_a > P_A$

$> P_v$. Since the driving pressure $P_a - P_A$ increased down the lung, flow did likewise. In *zone 3* the most dependent portion of the lung, $P_a > P_v > P_A$ and the driving pressure is $P_a - P_v$. Despite relative constancy of driving pressure in zone 3, both P_a and P_v continued to increase downward, and it was postulated that this increase in intravascular pressures had distended the resistance vessels, resulting in the increase in flow that was observed down zone 3. The flow increase found in zone 3 was, however, of lesser magnitude than in zone 2.

The distribution of blood flow at different levels in the lung has been studied with the ^{133}Xe technic.[55] In the uppermost 2.9 cm of the human lung (zone 1) blood flow appeared to be absent, then increased rapidly down zone 2 of the lung for 15 to 20 cm, where the rate of increase in flow became less (zone 3). Increased accelerative force would be expected to increase the zone of nonperfusion or poor perfusion.

Adaptation of sophisticated technics such as these to study of the basic respiratory variables under multiples of the gravitational accelerative force should make a major contribution to our understanding of the abnormalities of respiratory physiology associated with acceleration.

Flight surgeons have commonly believed that the emotional state can diminish tolerance to acceleration stresses. This is a difficult question to prove, yet at times there are identifiable emotionally traumatic life situations that are followed by decreased acceleration tolerance. One also sees such cases in the nonflying situation characterized by orthostatic hypotension on change of body position from the horizontal to the upright. Occasionally attempts to prove, even by intra-arterial pressure recording, that momentary hypotension is the cause of blackout of vision in such patients may not be successful unless the postural change is effected quite suddenly, rather than gradually as on a motor-driven tilt table. In sudden change to the upright position the blood pressure may fall to grossly inadequate levels, only to rebound rapidly to levels higher than control through the delayed overreaction of the negative feedback baroreceptor-autonomic nervous system-effector mechanism.

Anxiety-induced hyperventilation is not uncommon in flight, particularly if motion sickness of any degree is also present.[56] The cerebral vasoconstrictor response to the resulting decrease in arterial carbon dioxide tension,[57, 58] together with the hypotension that at times accompanies hyperventilation, will render the individual more likely to develop insufficiency of the cerebral circulation during acceleration-induced hypotension at brain level. Theoretically, an exceptionally tall individual will have a greater fall in arterial pressure at brain level than a short individual, given equal mean arterial pressures at heart level. Conversely, for the same arterial pressure at brain level the tall individual undergoing positive acceleration will require somewhat higher mean blood pressure at heart level than the short individual. Furthermore, the hydrostatic column from heart to foot will be longer and its effect therefore greater, especially the effect of the venous column.

In markedly exaggerated form one can see this situation progressively as one goes from man, to ox, to giraffe, each of which has adequate arterial pressure at brain level but, in the sequence named, correspondingly higher blood pressure at heart level.[59] A systematic study of the effect of body height on G tolerance has not, as far as we are aware, been made, perhaps because the heart-to-brain distance of flying personnel does not have a wide range. One cause of lowering of blackout threshold that is insufficiently thought of is severe sunburn, with the associated redistribution of blood volume toward superficial areas.[60] G tolerance returns as the sunburn recedes. The development of severe fatigue, by limiting the muscular counterforce and tissue pressure against blood vessels during acceleration, as well as by making the temporary Valsalva maneuver or the M-1 maneuver[61] less effective, can be an important factor in reducing acceleration tolerance. Some decrease in tolerance to positive acceleration after 2 to 4 weeks of bed rest has been reported,[62] with higher mean heart rates and, in one subject of a gradual onset run (rate of acceleration 0.07 G/sec), bursts of premature atrial contraction. Obviously a number of clinical conditions, among them myocardial disease, can contribute to diminution in acceleration tolerance, but these will not be considered here in any detail.

Measures to *improve tolerance* to acceleration obviously should be designed to remove or ameliorate the causes wherever possible. Temporary shift from the positive acceleration to the transverse acceleration position, in which the gravitoinertial resultant force vector is directed from front to back ($-G\star_{x_{TR}}$, $+ G_{x_K}$), will greatly increase acceleration tolerance, and tilting seats for aircraft have been devised for this purpose. Such seats have not proved practicable as yet, since there is an attendant cross-coupled angular acceleration that can produce severe disorientation through stimulation of multiple semicircular canals. The couch used during blastoff and reentry of spacecraft makes use of the higher acceleration tolerance associated with transverse orientation to the gravitoinertial vector.

Theoretically, the ideal protection would be for the individual in the positive acceleration situation to be surrounded by fluid of the same density as the blood. Even immersion to the clavicles in fluid will not protect the lungs, since there are no liquid-filled airways to provide an effective hydrostatic gradient to oppose the gradient in the pulmonary and bronchial vessels. Fluid-filled antiblackout suits have the inherent disadvantages of weight and cumbersomeness but were developed by Canadian and German flight surgeons in World War II. (Experiments have shown that an individual or an animal immersed in a tank does have much greater acceleration tolerance.) The breathing of liquids under high oxygen tension has been accomplished in experimental animals[63] but appears to be an impractical approach for man without a great deal more research into the problem.

Protective suits with gas inflation of calf, thigh, and lower abdominal bladders, either with the same pressure or with differential pressures, have proved to be the most practical. Inflation of G suits augments venous return and produces a transient increase in cardiac output[64] or a slight fall in output.[65] Tolerance to positive acceleration is raised about 2 G, and fatigue in pilots is diminished.[61] Prolonged inflation is disadvantageous, since the resulting reactive hyperemia may lower the peripheral vascular resistance excessively on deflation of the suit.

Improvement in physical fitness probably raises acceleration tolerance in the physically unfit, although superior athletic capability does not appear necessary. Navy pilots with excellent performance records do not always possess a high degree of athletic fitness.[66] A favorable effect of leg exercise on the tolerance of younger men to positive acceleration of 3 G units has been reported.[67] The arterial mean and pulse pressures at heart level actually rose in these studies more rapidly in response to acceleration when the subjects were exercising than when they were at rest. The authors felt that their findings emphasized the importance of the leg muscle pump as a booster of the cardiac pump.

Negative acceleration

Negative acceleration exists when the gravitational or gravitoinertial force vector is directed from seat to head. Thus a child hanging by his heels is exposed to $-1G_{z_K}$ or $+1G\star_{z_{TR}}$. Negative G is experienced at the bottom of an outside loop, a maneuver first performed by the great aviator James Doolittle. When a high-speed aircraft is rapidly dived from climbing or level flight in a pushover maneuver, negative acceleration can readily be produced since the aircraft's vertical (downward) component of velocity exceeds the acceleration of gravity. Exposure to negative acceleration is usually brief and not as common as exposure to positive acceleration, so that investigations as extensive as those on positive acceleration have not been performed.

A number of profound physiologic changes are associated with the reversal of the normal relationship of the individual's body axes to the gravity vector. The apex-to-base density gradient is eliminated or, in a minority of the animals studied, actually reversed.[36] The carotid sinus and brain are now on the high end of the hydrostatic gradient in the circulation, with the heart at a lower pressure level. Man does not have a rete mirabile caroticum at the base of his brain; this structure in the giraffe is thought to be an effective system for equalizing the pressure changes in the arterial and venous systems, such as when the animal lowers its head to drink and produces negative G at brain level. If negative acceleration continues for sufficient time in man or animals, a large shift of blood volume not only headward but into the thorax can occur, and pulmonary blood volume is redistributed. The weight of the abdominal viscera moves the diaphragm headward, even above the position it normally occupies in recumbency on earth. Accessory respiratory muscles, in particular the scalenes and also the less important sternomastoids, no longer serve their antigravity role and cannot assist significantly in the work of breathing. Although the implications of all of these physiologic changes have not been explored fully, it is clear that both from the standpoint of the circulation and respiration the organism faces serious physiologic disadvantages in the experience of negative acceleration.

The experience of negative G can be disturbing and uncomfortable. We have observed transitory mental confusion in a healthy 43-year-old subject exposed to −1 G acceleration in an inverted restraining chair for 10 minutes, and younger individuals in the same experiment reported that the experience was a decidedly uncomfortable one. Even at −1 G there is a marked sense of fullness about the head and neck that includes the tissues around the eyes and the face. After a time the discomfort is such that the experimental subject becomes aware that he is not quite as alert as he was 8 or 10 minutes earlier. Apparently some type of tolerance can be developed, since those who are practiced in the disciplines of Yoga can tolerate the negative G of a headstand for hours.

The question of human tolerance to negative G is not fully settled. It would appear that few symptoms are experienced if the acceleration lasts for less than 2 seconds, a period of time that does not permit the development of the maximal extracranial venous pressures, so that 10 or more G of this duration have been tolerated. It has been reported by Ryan and colleagues[68] that disturbing symptoms develop after 5 seconds' exposure to –2 G. On the other hand, Maher[69] found from his studies in aircraft that –3.6 G were well tolerated for about 7 seconds. The experience of British investigators on the RAF centrifuge was that –2.5 G could be tolerated for 2 minutes and –2 G for 5 minutes without undue discomfort.[40] British stoicism to discomfort may have been a factor here. This emphasizes the fact that endpoints for use in human tolerance measurements are difficult to establish. Visual changes in the form of "red-out" (a reddish, misty, visual experience) appear in some subjects and later vision is lost and finally unconsciousness sometimes occurs. Petechial hemorrhages in the conjunctivae and about the face may develop owing to an increase in capillary pressure of about 30 mm Hg per negative G unit.

The earliest major experimental study in animals was performed by Jongbloed and Noyons in 1933.[8] These workers showed that –2.5 G produced a fall in heart rate that could be abolished by carotid sinus denervation. Some years later Gamble and colleagues[70] investigated the effects of –3 G on six human subjects during 15-second exposures and –7 G on dogs and rabbits for 2-minute exposures. In the human subjects, pressures were measured in the frontal vein and in the radial artery held at the same level and in the dogs pressures were recorded from the carotid artery and external jugular vein. In the human subjects the initial response was a rise of pressures in the radial artery and frontal vein of 70 to 90 mm Hg, but as the run continued arterial pressure began to fall and venous pressure began to rise still further, thus producing a decreased cerebral perfusion pressure. In the animals at the higher G levels blood pressure in the carotid sinus rose to as much as 260 mm Hg and that in the jugular vein to 100 mm Hg. A variety of arrhythmias were seen on the electrocardiogram of the types seen with strong vagal stimulation. In the animals a marked bradycardia began within 3 seconds following the onset of acceleration together with varying degrees of heart block and nodal and ventricular escape beats. Asystole occurred for 10 to 20 seconds in some animals and at times idioventricular rhythm was seen. These changes were largely abolished by vagal nerve section. Electroencephalograms after –7 G for 1 minute showed delta rhythm, interpreted as indicating brain injury.

In 1950, Ryan, Kerr, and Franks[68] published their wartime studies on over 100 subjects who had been subjected to –1 G on a tilt table and up to –3 G on a centrifuge. The human subjects were seated on the centrifuge with the long axis of the trunk 30 degrees below the horizontal but with legs and thighs extending vertically upward. The authors calculated that the acceleration gradient along the subject's body varied from –0.175 to –3.28 G. The symptoms were carefully described and were similar to those listed earlier. Bradycardia developed comparable to that in studies by other investigators. The initial (and near maximal) fall in heart rate occurred within 3 seconds, although the full bradycardia response required 10 to 15 seconds and then rapidly disappeared after cessation of the negative G, to be replaced at times by slight tachycardia. Some degree of tolerance apparently was acquired, since the subjective discomfort became less with repeated studies. No anginal symptoms were experienced.

The behavior of the cerebrospinal fluid (CSF) pressure in relation to venous pressure represents one of the defenses against negative G. This aspect of physiology was investigated by Rushmer, Beckman, and Lee in the cat.[71] Venous and CSF pressures at the vertex and at neck level were recorded during positive acceleration of 6 G as well as negative acceleration of the same magnitude. The duration of the negative G was 15 seconds at –2 to –6 G. CSF pressures as high as 80 cm above control were recorded during positive G and 190 cm below control during negative G. Both CSF and venous pressures at the same level in the circulation changed by very nearly the same amount, but both of these pressures near heart level changed very little. The authors' interpretation was that the CSF and the venous blood columns responsible for these pressure changes must either have been of approximately the same length or that the CSF pressure represented a reflection of local venous pressure. They pointed out that intracranial veins, including very small vessels, must be almost perfectly protected from the effects of rapid changes in intravascular pressure by comparable changes in CSF pressure. In most of the animals arterial pressure at neck level increased by greater amounts than the venous pressure during negative acceleration. A similar correlation between CSF and venous pressures was found by Beckman[72] in goats subjected to negative accelerations of –1.2 to –8.9 G. He demonstrated that the change in systolic pressure per unit of negative G decreased as the acceleration becomes greater and appeared to approach an asymptote. Autopsies on his animals showed no cerebral hemorrhages.

Transverse acceleration

Transverse accelerations are those in which the gravitoinertial resultant vector acts at right angles to the long axis of the body. It is therefore possible for the vector to vary through 360 degrees in a plane at right angles to the long axis of the body and still be considered to represent a situation of transverse acceleration. In practice, most discussions and research into the problem are concerned with conditions in which the gravitoinertial resultant vector is directed from front to back through the body ($+G_{x_K}$, $-G\bigstar_{x_{TR}}$, termed by some "forward" acceleration), or from back to front ($-G_{x_K}$, $+G\bigstar_{x_{TR}}$, termed by some "backward" acceleration); under the normal conditions of earth gravity these conditions would be represented by the supine and prone positions, respectively. Higher transverse accelerations are generated by catapult takeoff and carrier landing with arresting gear; in these the transverse component of the gravitoinertial resultant (G★) is directed front-to-back and back-to-front, respectively. The vector can also be directed from the right shoulder to left shoulder ($-G_{y_K}$, $+G\bigstar_{y_{TR}}$) or from the left shoulder to right shoulder ($+G_{y_K}$, $-G\bigstar_{y_{TR}}$). We have previously discussed the fact that the gravitoinertial resultant vector will not always be directed in such simple anatomic relations to the axes of the human body or human head. Where this is true the magnitude of the projection of the resultant vector on the x, y, and z axes of the orthogonal coordinate system should be given or the magnitude of the vector and two orientation angles.

Transverse acceleration is assuming more interest and research attention than formerly, primarily as a consequence of the accelerations produced between blastoff of spacecraft and attainment of orbital velocity and particularly during the deceleration (negative acceleration) required for reentry of the spacecraft into the earth's atmosphere. It was fortunate that the U. S. Navy and Air Force had carried out work on the tilting or supine seat for aircraft, since this experience was available to assist in the development of the contour couch utilized by The National Aeronautics and Space Administration (NASA) for space vehicles.

In transverse acceleration the gradients of hydrostatic pressure that can be developed are much smaller than for either positive or negative acceleration. This is certainly true where the gravitoinertial vector is directed from front-to-back or vice versa but is somewhat less true for lateral acceleration, in which G★ is directed from the right to left shoulder or vice versa. Except in barrel-chested individuals the distances available for the development of hydrostatic effects in lateral acceleration are somewhat greater. The forward acceleration or supine situation ($+G_{x_K}$, $-G\bigstar_{x_{TR}}$) is the best tolerated, and magnitudes as high as 20 G may not be associated with symptoms of cerebral or retinal circulatory insufficiency. Petechial hemorrhages may develop, however, on the "dependent" side of the skin, particularly if contact forces are not uniformly distributed. If the head has to be raised, as for example, to make adequate forward vision possible in prone or supine position flying, a significant heart-to-brain distance develops and decreases the transverse acceleration along the x axis that can be tolerated from 20 to about 12 G before blurring or darkening of vision occurs. Human and animal radiographic studies have shown that a definite posterior displacement of the cardiac shadow occurs.[73, 74] X-ray studies in experimental animals (dogs) by Soviet investigators[75] showed similar findings. Heart rate increases up to about 100 beats per minute may be seen, but usually this is the maximum rate. Other anatomic changes that occur during forward acceleration are a decrease in the dorsoventral diameter of the thorax and upward elevation of the dorsal portion of the diaphragm together with a simultaneous depression of the ventral portion of the diaphragm. Ventral lung fields become radiolucent and spatial distribution of pulmonary blood flow probably shifts in the direction of the gravitoinertial resultant G vector. This conclusion is based on the assumption that the directional trends shown by scintiscan on lateral acceleration also apply to forward and backward acceleration[76] applies.

The major effects of forward acceleration are

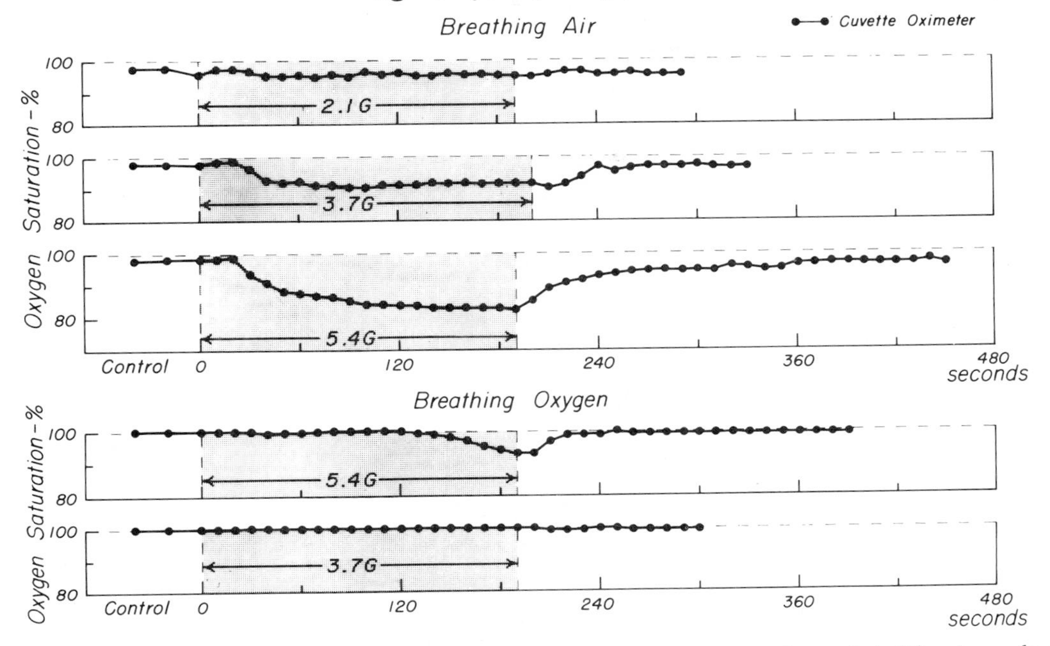

Fig. 6-21. Changes in arterial oxyhemoglobin saturation during exposure to forward ($-G\bigstar_{xTR}$) acceleration. Subject: human male, age 37 years, weight 81 kgm. (From Nolan, A. C., and others: Aerospace Med. **34:**798-813, 1963.)

on respiratory functions. Tidal and respiratory minute volume show early increase in $+G_{x_K}$ ($-G\bigstar_{x_{TR}}$) acceleration up to the 6- or 8-G level, followed by later decreases.[77] These falling tidal and minute volumes are associated with a sense of increasing difficulty in breathing by the subject. That this difficulty is restrictive rather than obstructive in nature is suggested by studies on the maximum breathing capacity, vital capacity, and 0.5-second timed vital capacity.[78]

Gas exchange can be markedly impaired during forward acceleration. Fig. 6-21 is taken from the work of Nolan and associates[79] on the Mayo Clinic centrifuge. It can be seen that 2.1 G produced almost no change in arterial HbO_2 saturation during approximately 3 minutes of exposure, while 3.7 G produced moderate and 5.4 G very marked oxygen unsaturation. Oxygen breathing abolished or greatly reduced the hypoxia, except toward the end of the 5.4-G run. Pulmonary atelectasis was demonstrable on x-ray after the 5.4-G runs. These investigators attributed the oxygen desaturation to perfusion of the atelectatic alveoli, associated with increased segmental blood volume and blood pressure. In another paper on this and similar work, Wood[80] discusses possible mechanisms for the hypoxemia, atelectasis, and disruption of pulmonary parenchyma. Lung disruption was emphasized dramatically on a +6 G_x (in our symbols $+G_{x_K}$, $-6\ G\bigstar_{x_{TR}}$) centrifuge run. A healthy male subject suddenly experienced severe chest pain and was found to have developed physical signs of acute mediastinal emphysema. Wood calculates that in a 5-G run in the same acceleration mode the pulmonary pressure in the dorsal aspects of the lungs would be increased to 70 cm H_2O and pulmonary venous pressure in the same regions increased to 60 cm H_2O. If, as seems probable, the capillary pressure would be in excess of 60 cm H_2O—far higher than the colloidal osmotic pressure of plasma—rapid development of pulmonary edema would be expected in the dependent lung areas. On the other hand, opposite effects would probably be produced in the superior portions of the lungs, where perfusion would be abolished. Wood also postulates that the increase in weight of the pulmonary parenchyma during acceleration would have disruptive effects. He describes ex-

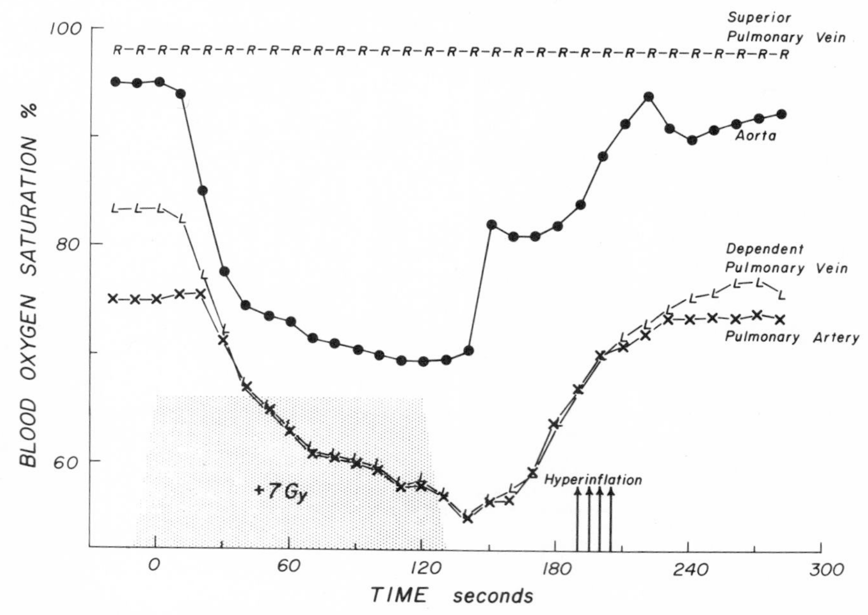

Fig. 6-22. Blood oxyhemoglobin saturation in different vascular sites before, during, and after 7G acceleration. Animal was in the left decubitus position $(+G\bigstar_{y_{TR}})$. Subject: dog weighing 16 kgm, morphine pentobarbital anesthesia. (From Vandenburg, R. A., Nolan, A. C., Reed, J. H., Jr., and Wood, E. H.: J. Appl. Physiol. 25:516-527, 1968.)

periments on intrapleural pressures at various sites in the thorax of the dog, as well as on esophageal pressures, that convinced him of the development of large transalveolar pressure gradients in the superior portions of the lungs. Wood views these pressure gradients as potentially dangerous, particularly since the human thoracic topography suggests that pressure differences associated with acceleration might be greater than in the dog; on the other hand he points out that the human heart has better anatomic support than that of the dog. Wood's conclusion, perhaps a slight overstatement, is that the lungs represent man's most vunerable organ with respect to his capability to withstand high levels of acceleration.

Lateral acceleration, that is, acceleration with the resultant vector directed from side to side in the body along its y axis, is the least studied of the transverse acceleration modes. There has been recent increase in interest in this aspect of the subject, however, and the results of one animal experiment are presented in Fig. 6-22.[81] These studies of blood oxygen saturation, again taken from the work of the Mayo Clinic group, were done on a dog in the left decubitus position $(-G_{y_K}, +G\bigstar_{y_{TR}})$. It should be noted that one of the advantages of the system of symbols that we are using is seen in this position, since the action (kinematic) and reaction (gravitoinertial) accelerations are of opposite sign, which is proper. The rather high acceleration of 7 G, maintained for 2 minutes, was associated with the rapid development of marked systemic arterial hypoxemia with a fall of the percent oxyhemoglobin saturation to approximately 70%. The explanation is seen in the oxygen saturation of blood from a dependent pulmonary vein as compared with that in a superior pulmonary vein; in the former the oxygen saturation fell to extremely low levels, whereas in the latter it remained at its normal value of nearly 100% oxyhemoglobin. The estimated percentage pulmonary arteriovenous shunting increased progressively as acceleration continued. The figure shows that the degree of atelectasis and other causes of severe disturbance of the ventilation/perfusion ratios in the dependent areas was not completely cleared up by hyperinflation.

A later paper from the same group[76] outlines a

highly sophisticated electronic data processing and computer analysis system for multiple continuously recorded cardiovascular and respiratory variables. In this study correlation was demonstrated between the vertical distance separating the pulmonary venous catheter tips and the difference in oxygen saturation of blood withdrawn from the two catheters with the animals (dogs) in the supine and left decubitus positions. The P_{CO_2} of dependent pulmonary venous blood was uniformly higher than that from a superior vein and from the aorta, an observation indicating the presence of regional differences in ventilation/perfusion ratios. In this same work are presented three-dimensional views of regional isotopic distributions recorded during scintiscan of the ventral surface of the chimpanzee thorax. The authors believe their data to be indicative of regional pulmonary blood flow distribution. Their three-dimensional maps clearly show a shift of isotope and presumably blood flow in the direction of the gravitoinertial resultant acceleration vector. Other chimpanzees in this study showed redistribution of blood flows toward the midthoracic rather than the more dependent regions.

It is clear that far more work is required on the changes in cardiorespiratory physiology in the different transverse acceleration modes. One may raise the question as to whether or not the common view that the cardiovascular receptors are not strongly invoked in lateral acceleration is accurate. In lateral acceleration where the left side of the thorax is dependent, the question can be raised as to whether or not increased pressures in the left atrium and pulmonary venous bed might invoke the respiratory depressor response described by Hardie.[82] This investigator found that sudden increase in pressure in the left atrial and pulmonary veins was frequently associated with marked depression of respiratory rate and respiratory minute volume in dogs. The complexity of the cardiovascular system and its receptors is the subject of a recent excellent review.[83] One may doubt that the pressure changes and deformations in the vascular system, even in the transverse acceleration modes, would be left untouched.

Impact

The subject of acceleration or deceleration of large magnitude and short duration, usually less than 1 second, is highly important to a number of practical situations: escape from aircraft or spacecraft, parachute opening shock, the rocket accelerations of space vehicles, reentry decelerations of these vehicles, crash of aircraft, and accidental human falls. In such situations the forces of inertia can be enormous and the distribution of the contact forces over the human body are of critical importance in survival. The sudden changes of kinetic energy of the body, which varies as the square of the velocity, are of large magnitude and again emphasize the importance of the distribution of contact forces on the surface of the body. More than three decades ago DeHaven[84] became interested in human survival after falls from high buildings and instituted a classic series of studies under Cornell University sponsorship. He found remarkable instances of survival after impact with the earth. From measurements of distance traveled after impact and height of fall he calculated probable acceleration values as high as 162 G. DeHaven's suggestions for improving chances of survival from aircraft and passenger car accidents stimulated an entire field of endeavor. Stapp[85] carried out over many years his pioneering studies in rapid deceleration on various linear accelerators of his design. His review of the subject should be consulted by all those interested in this field.

More recently Ewing and Thomas have published an elaborate and detailed study of the human head and neck responses to impact acceleration.[86] They have kindly permitted reproduction of Fig. 6-23 from their motion pictures. Comparison of the prints of the subject at rest with his head tipped forward in a voluntary nod and at the time of acceleration reveals some of the effects of the impact. Acceleration of the sled was to the left, and inertial forces were therefore directed to the right. Although the skin markings on the neck were for reference only, since more precise measurements were done utilizing target mounts at the mouth, top of the head, neck, and back of the chair, the widening of the distance between the markings is evident as well as the effect of the wide restraining straps on the subject's body. It is this type of study, done jointly by the Army-Navy-Wayne State University research team, that is expected to make major contributions to the field of impact acceleration. The monograph by Ewing and Thomas should be consulted as essential to an understanding of present approaches to this field.

Physiology of weightlessness

One of the most significant developments of this or any age has been man's venture into space and his experience of the weightless state. Man as

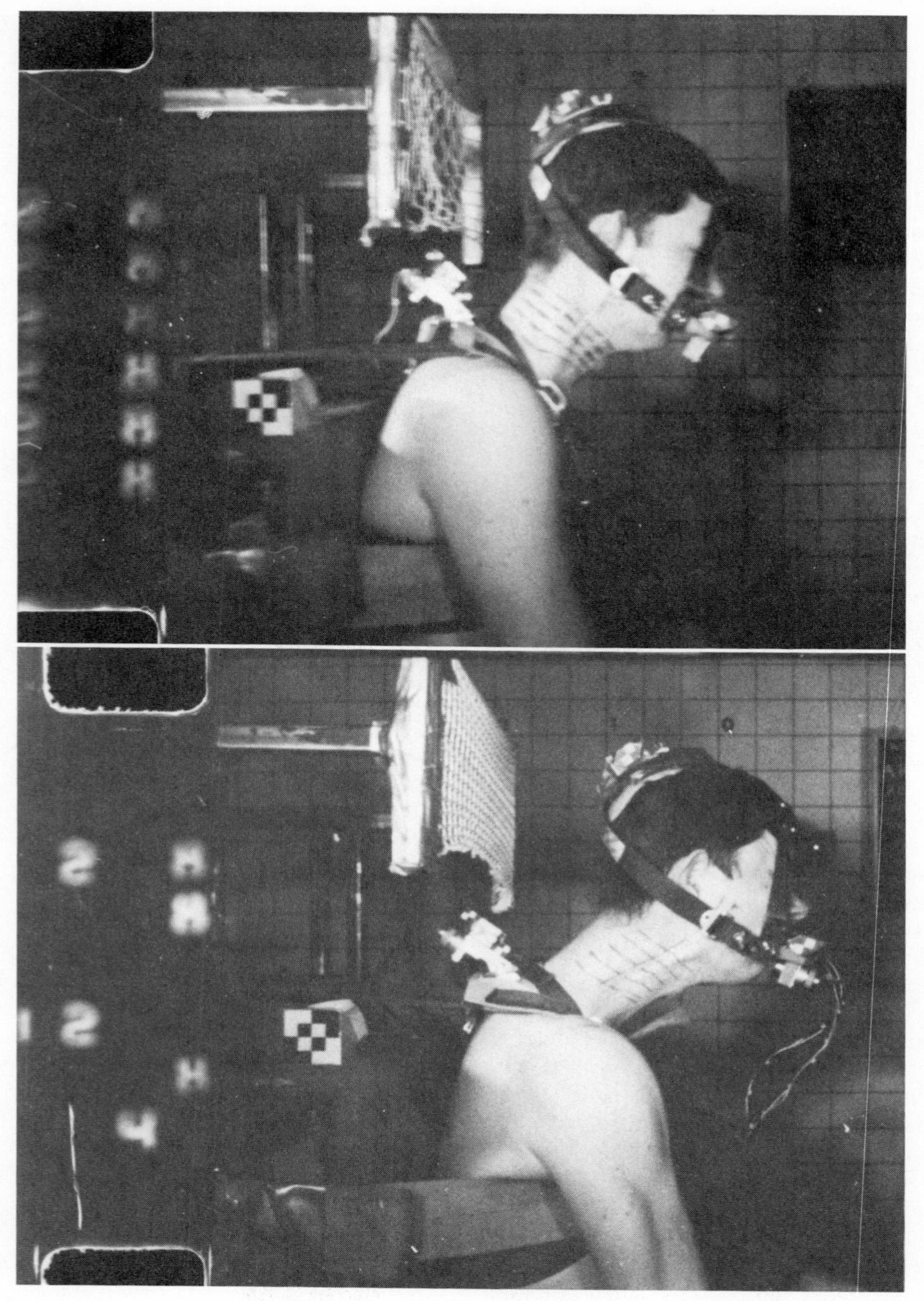

Fig. 6-23. Frames taken from 16-mm motion pictures (500 frames/sec) of human volunteer undergoing impact acceleration; experiments done by a joint Army-Navy-WSU research team on the WHAM II horizontal accelerator at Wayne State University, Detroit, Michigan. The upper photograph is for purposes of comparison and illustrates displacement of the head and neck during a voluntary head "nod" that created the same angular relationship between head and neck as in the lower photograph. The sled is at rest. The lower photograph illustrates the motion of the head and neck at the time of maximum displacement of these structures from the sled chair. The acceleration of the sled was to the left ($-G_{x_K}$) and the acceleration was 9.6 G units. The markings on the neck were for reference only, since the target mounts at the mouth, top of the head, neck, and back of the chair were used to record the motion photographically for precise measurement. An instrumentation system in the same mounts measured linear acceleration and angular velocity. (Courtesy Capt. Channing L. Ewing, MC, USN, and Dr. Daniel J. Thomas, Naval Aerospace Medical Research Laboratory Detachment, Michoud, La.)

a species has evolved through processes acting over a great span of time, during which all of the evolutionary mechanisms have taken place under the influence of the earth's gravitational field. It would indeed be surprising if profound physiologic changes did not accompany the new historic experience of extended periods of weightlessness. One may wonder, perhaps, that man has done so well during his exposures to weightlessness thus far.

Fig. 6-1 summarizes the physical basis of the weightless state; the legend should be read before continuing with the present section.

In "ideal" weightlessness, with its absence of contact forces, the organs based on the balance of internal forces and structures of the body are free to assume their positions of rest uninfluenced by external forces. It is physically an accurate concept to think of the weightless state as comparable to that of a freely falling object, since the individual in an orbiting spacecraft does have an instantaneous acceleration toward the center of mass of the earth equal to the acceleration of gravity at the orbiting altitude. Just how far reaching and how profound will be the effects of prolonged exposure to the weightless state is not known at present, although the American and Soviet space programs have provided a very considerable amount of significant data on the effects of weightlessness of moderate duration. The American Skylab program, with orbital flights of 4 weeks' and 8 weeks' duration, is adding another extremely important body of data to this field.

Dr. Charles A. Berry, NASA Director for Life Sciences, presented a succinct review in 1971 at the Fourth International Symposium on Basic Environmental Problems of Man in Space, held at Yerevan, Armenia, U.S.S.R.[87] This summary should be consulted by those wishing a brief authoritative review of the major findings on the adaptation of American astronauts to weightlessness; some of the findings on Russian cosmonauts are also included. It is of interest to compare these findings with the report of Miller, Johnson, and Lamb[88] on the effects of 4 weeks of absolute bed rest on the circulatory responses of human subjects. An annotated bibliography of the subject of weightlessness and subgravity up to 1963 is available.[89] Berry's review covered data on fifty-four astronauts exposed to the environment of space, some of them for periods as long as 22 days. No lasting ill effects have been observed, but there were many signs of adaptive changes. The subject of man's adaptation to weightlessness is too large and complex for full summary in the present chapter; selected aspects only are discussed.

The physiologic disturbance caused by irregular work/sleep cycles initially was a source of considerable difficulty, but by the Apollo 11 lunar landing flight, sleep periods were improved. The problem of cardiovascular adaptation illustrates two important aspects of the physiology of weightlessness: (1) the inflight adaptations and (2) the effect of changes taking place during weightlessness on the ease or difficulty of returning to gravity. Inflight heart rate and blood pressures were within normal limits and abnormalities of rhythm were rarely noted. On the other hand, orthostatic intolerance was consistently observed, both in American astronauts and in Russian cosmonauts. Berry states that despite tilt table tests, lower body negative pressure (LBNP) tests, and static standing tests, the most sensitive index of reduced orthostatic tolerance appears to be heart rate. The more even distribution of blood volume throughout the body that occurred during weightlessness was followed by increase in leg volumes from 12% to 82% after return to gravity. One puzzling finding in the Apollo series, in which LBNP was used as a test of orthostatic tolerance, was that the changes in leg volumes seen on tilt table tests were not observed; Berry suggests that this may be indicative of a basic difference in the two tests.

Weight losses were largely regained within 1 day after reentry, a fact strongly suggesting that the weight loss was based primarily on fluid and electrolyte loss. Average weight loss immediately postflight ranged from a low figure of 1.3 pounds for Gemini 11 to 10.5 pounds for Apollo 14.

Electrolyte changes were variable. Potassium excretion was depressed both inflight and postflight for 24 hours. The decrease in both serum and urinary potassium levels in Apollo crewmen was believed to have been associated with decrease in total potassium. Gamma spectrometry utilizing ^{40}K showed that total body potassium for Apollo 14 crewmen was significantly below preflight values. Decreases in total body potassium were found 17 days postflight in Apollo 14 as compared with increased total body potassium in control subjects. It was postulated that the potassium loss was associated with intracellular fluid cation loss. Total exchangable potassium in Apollo 15 crewmen was studied postflight with ^{42}K and found decreased. Reduced sodium and chloride excretion has also been consistently observed immediately after flight.

From the teleologic standpoint, these diminished

excretions would promote fluid retention to compensate for the loss of fluid volume in the weightless state.

Some endocrine studies related to fluid regulation were possible. The levels of antidiuretic hormone, aldosterone, and plasma angiotensin were increased postflight—changes consistent with the cardiovascular and fluid-electrolyte findings—but Berry cautions against overinterpretation of these data.

Attempts to determine the compartments in which the fluid loss occurred showed that the mean loss of body weight was several times larger than the diminution in plasma volume. More elaborate studies were possible on the Apollo 15 crew and showed that intracellular fluid loss accounted for most of the deficit in total body water; on the other hand extracellular fluid volume remained virtually unchanged.

The elimination of the weight-bearing and postural maintenance roles by bone and muscle in the weightless state can lead to calcium loss, but evidence was not found to suggest that depletion of bone minerals increased with increasing duration of flight. Both American and Russian space crews had a routine of daily exercise, but nevertheless there was loss of bone mass demonstrable by x-ray densitometry. An exception was the Apollo 14 studies, which showed no significant mineral losses on this 10-day mission. Muscular atrophy was not seen in American astronauts, although slightly negative nitrogen balances were found in some men despite increased nitrogen intake. Impairment in coordination was not found in American astronauts, but Soyuz 9 cosmonauts stated that their limbs felt unusually heavy postflight and that walking and lifting of objects was difficult. The Russians reported muscle soreness and diminished muscle strength, and measurable reduction in circumference of the lower extremities was observed.

Assessment of work capacity was done after all of the Gemini and Apollo missions, utilizing an electronic bicycle ergometer that could produce a fixed work load. One problem with such ergometers is that the fixed work load can be done at different pedaling rates, and a subject may not select the optimum rate for his own maximum muscular efficiency. The oxygen consumption on these tests immediately after the Apollo 14 flight was about 75% of the mean of preflight measurements. This was interpreted as representing a significant reduction in work capacity. The two crew members of the Apollo 14 team who carried out moderate work on the moon's surface under one-sixth earth gravity did not show the reduction in work capacity found in the postflight studies on the command module pilot. The reverse situation obtained for Apollo 15, and the possibility was considered that excessive work had been required on the moon's surface in this mission.

The course of adaptation to weightlessness has been formulated into a hypothesis by Leach, Alexander, and Fischer.[90] These authors divide the adaptation into (1) stress stage with redistribution of blood volume, (2) adaptation stage with major fluid and electrolyte shifts, and (3) adapted stage with establishment of a new fluid and electrolyte balance.

New data from the preliminary mission report of Apollo 16 have been kindly provided by Dr. Berry. During launch the heart rates of the crew were in the range 77 to 125 beats per minute, with the Commander, who had flown on three previous space missions, showing lower heart rates than the other two crewmen without previous space flight experience. The graph of heart rates during his 5½ hours of extravehicular activity on the moon's surface showed a tendency for the rate to fall gradually from an initial value of approximately 98 beats per minute down to between 70 and 80 beats per minute, followed by a tendency to rise after the 3½-hour point. This same individual's heart rate curves during his second period of extravehicular activity tended to be somewhat lower and showed smaller oscillations. Wider fluctuations occurred in the heart rate of the lunar module pilot. Calculated metabolic rates and heart rates tended to parallel each other. Since both Apollo 15 lunar surface crewmen had shown cardiac arrhythmias following extravehicular activities, a high-potassium diet was instituted for Apollo 16, beginning 72 hours before launch and continuing 72 hours beyond the mission. No medically significant arrhythmias occurred except for occasional isolated premature beats. It was believed that the high potassium intake, better fluid and electrolyte balance, more adequate sleep, and less fatigue produced the more favorable picture of heart rhythm on Apollo 16.

Studies of the body density changes in Apollo 16 crewmen, compared with their known caloric deficits, suggested that losses of fatty tissue predominated in the lunar surface crewmen, in contrast to the proportionately greater loss of body water on the part of the command module pilot.

Bone mineral measurements from the central os calcis on the Apollo 16 mission did not show a significant loss of bone mineral in any of the crewmen of the mission, yet during the first 3 postflight days the os calcis of the command module pilot showed progressive loss of mineral that later reversed spontaneously.

The effects of weightlessness on the vestibular system are largely unknown; two types of tests were therefore performed after Apollo 16. Postural stability, studied with and without visual cues, showed decreased performance in two of the crewmen 3 days following recovery when they were deprived of all visual sensory cues. Four days later their performance was comparable to the preflight control. Irrigation of the right and left auditory canals with water at 34° and 35.5° C, with simultaneous recording of the resulting nystagmus, showed in two of the crewmen an apparent hypersensitivity of the semicircular canals (increased frequency of nystagmus and increase in the velocity of the slow phase). One week after recovery the studies were normal.

An almost unlimited number of questions can be raised regarding the complex processes and mechanisms of adaptation to weightlessness and return to gravity conditions. One can postulate, for example, that the motion sickness seen in some crewmen during space flight might produce the hyperventilation observed during induced motion sickness in the laboratory.[56] Such reduction in arterial P_{CO_2} would produce cerebral vasoconstriction and might be a factor in prolonging the motion sickness.[91] The troublesome motion sickness itself that developed in some men is not fully explained. One interesting possibility is that the deprivation of the stimulus of the otolith organs during weightlessness, with preservation of semicircular canal stimulation, created an imbalance of vestibular traffic into central processing centers.

The possibility must be considered that prolonged space missions, of many months' duration, might introduce subtle and profound changes in human physiology not seen on the shorter missions thus far. We do not know, for example, whether the intrinsic myogenic response of Bayliss makes an essential contribution to defense of the circulation against gravity. If it does, prolonged absence of the normal exercise of this response by change in posture could lead to deterioration of an essential cardiovascular mechanism, with resulting serious consequences on return to gravity, either that of earth or of a planet possessing greater mass and gravity.

The Skylab experiments, with orbiting periods of 28 to 56 days, may furnish a variety of important clues as to the effects of prolonged weightlessness.

PHYSIOLOGY OF ANGULAR ACCELERATION

The gravitoinertial resultant force has been shown to explain and correlate well with a variety of phenomena during acceleration in semirigid, deformable biologic organisms including man. An exception to this fact is provided by the semicircular canals of the labyrinthine system. In these ducts the effects of centripetal acceleration and of gravity are largely nullified by the special anatomic features and fluid content of these canals, and their mechanoreceptors appear to be stimulated primarily by angular acceleration, or forces correlating with it.

As stated previously, angular acceleration represents a change in magnitude or direction of the angular velocity vector. The behavior of the semicircular canal receptors is in contrast with that of the mechanoreceptors in the otolith organs of the saccule and utricle, which are also portions of the labyrinthine system. These otolith organs respond primarily to the direction of the gravity vector or to the direction of the gravitoinertial resultant force where gravity and linear acceleration coexist. For this to be true, however, there must be sufficient time to permit stabilization of response to change in the force vector.

This section provides an introduction to the subject of angular acceleration from the standpoint of the semicircular canal system, discussed in detail later. In the attempt to correlate physical stimuli and physiologic responses, no subject has perhaps given more difficulty than the analysis of angular velocity and acceleration, particularly where more than one axis of rotation is involved. The semicircular canal system is complex and involves three pairs of roughly circular tubes, the planes of which in a given ear are approximately at right angles to each other. These organs are bilateral and represent mirror images of each other. One cannot approach the study of such a system from the standpoint of the effective physical stimuli without developing some understanding of a simpler model. It will become apparent that many questions regarding the biophysics and physiology of the semicircular canal system and its performance under a variety of normal and abnormal physical situations remain unanswered. This subject represents one of the most challenging areas in aerospace and terrestrial physiology.

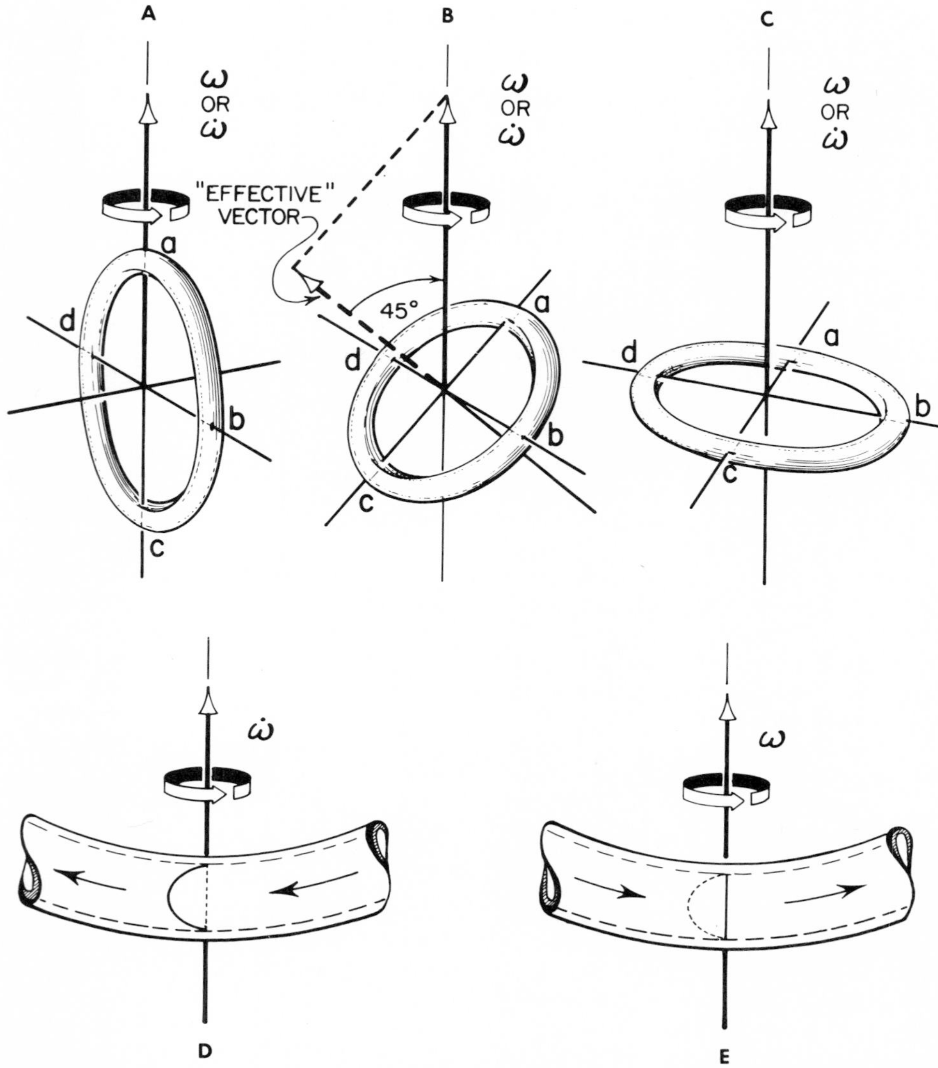

Fig. 6-24. Simplest model of a semicircular canal. The diagram shows a rigid transparent fluid-filled circular tube. Orthogonal axes in the plane of the tube intersect it at points a, b, c, and d, separated in sequence by 90-degree intervals. **A,** Uniform rotation or angular acceleration about an axis through the plane of the tube. There is no tendency for the rotation to produce relative motion between tube and contained fluid. **B,** Rotation about an axis 45 degrees from the plane of the tube. The "effective" angular velocity or angular acceleration is obtained by projection of the actual $\boldsymbol{\omega}$ or $\dot{\boldsymbol{\omega}}$ onto an axis perpendicular to the plane of the tube. **C,** Rotation about an axis perpendicular to the plane of the tube. Angular acceleration about the axis as shown has maximal effect in producing relative motion between tube and contained fluid. **D,** Effect of angular acceleration of the tube in position shown in **C,** but with an elastic partition across the tube. Angular acceleration as shown by the ribbon arrow produces displacement of the elastic partition (except at its attachments). **E,** Tube now subjected to constant angular velocity. The elastic partition, in returning to its original rest position, produces relative displacement of the fluid in a direction opposite to that in **D.**

Simplest model

A useful starting point is the study of a fluid-filled circular tube with a rigid wall, lying in a single plane, or more accurately lying between two parallel planes. We are interested in the question of relative motion between the wall of the tube and the contained fluid. Such a simple system is shown in Fig. 6-24, *C*. This figure shows the tube being rotated about an axis perpendicular to the plane of the tube. It is perhaps surprising that the effects are the same whether the axis of rotation is central or eccentric as long as the direction is the same. If such a tube is at rest and then is brought to a constant angular velocity, ω, the interval between rest and constant velocity is of necessity associated with angular acceleration, denoted by $\dot{\omega}$ or α. If the fluid were frictionless but possessed inertia, no motion relative to inertial space would take place in the fluid and the relative motion between tube and fluid would simply be the motion of the tube itself in inertial space. The actual physical situation is, however, that frictional forces between the last layer of fluid molecules and the wall of the tube cause this layer of fluid to move with the wall. The motion of progressively more central concentric "shells" of fluid is produced by viscosity, that is, the resistance to shear of the fluid. The innermost fluid molecules, at the geometric center of the tube, are therefore accelerated very slightly later than the layer of fluid immediately adjacent to the wall of the rigid tube. If the angular velocity becomes stabilized at a constant value, damping factors finally result in a steady state with no further relative motion between tube and fluid. If an angular deceleration now occurs, relative motion between fluid and tube again takes place but in the opposite direction. This motion continues beyond the moment when the rotation of the tube ceases, following which the relative motion is finally damped out. It is shown on p. 224 that it is this relative motion between the wall of the semicircular canal and the contained fluid that stimulates the mechanoreceptors of a given canal.

The behavior of the simplest model as described can be readily verified qualitatively by simple experiments. If we take a straight glass tube and fill the central portion with a colored fluid (such as water containing dye) and place this on a flat surface, no motion between fluid and tube will occur. If we then manually make a sudden rectilinear (straight line) acceleration of the tube in one direction, it will be seen that there is relative motion of the fluid within the tube in the opposite direction. This relative motion is based on the inertial property of the fluid itself. The circular tube undergoing angular acceleration can be conceptually reduced to a series of infinitely short straight tubes, with each infinitesimal element undergoing tangential acceleration in one direction and each element of fluid relative acceleration in the opposite direction. With respect to the simple circular model, if colored particles are suspended in the fluid and angular acceleration of the tube is produced either manually or on a human centrifuge during visual observation, the effects of angular acceleration on relative motion between fluid and tube become readily apparent.

Consider now the circular tube filled with fluid as before and being rotated about an axis parallel to the plane of the tube. This can be readily realized by the analogy of a coin spinning on its edge. Angular acceleration in this situation produces no relative motion between fluid and tube. We can state as a general rule that angular acceleration is without effect when the axis of rotation is parallel to the plane of the tube and is maximally effective when the axis of the acceleration vector is perpendicular to the plane of the tube. We now consider an example of the situation in which the axis of rotation is intermediate between these two positions, as for example 45 degrees from the plane of the circular tube. The "effective" angular acceleration vector is calculated by dropping perpendiculars from the actual angular acceleration vector to a line perpendicular to the plane of the tube. This new vector has the length of the cosine of the angle between the two, in this case, cos 45 degrees or 0.707. Therefore, theoretically, an acceleration of 1 rad/sec^2 about an axis 45 degrees from the perpendicular has the same effect as 0.707 rad/sec^2 acting through an axis directly perpendicular to the plane of the tube. Unfortunately, quantitative experiments have not (to our knowledge) been performed on a simple model in which the exact history of angular velocity and acceleration is known in relation to simultaneous precise measurements of motion of the fluid.

The sequence of positions shown in Fig. 6-24 from *A* to *B* to *C* is relevant to motions experienced by the semicircular canals. We may ask: What then is the effect on the fluid within the tube? The answer is not simple, as might first appear. The reader's attention is directed to positions in the tube 90 degrees apart, indicated as *a, b, c,* and *d*. In section *A* of the diagram, the particles at *b* and *d* are seen to be describing circles in space, while those at *a* and *c* are in effect stationary along the axis of rotation. As the tube is gradually turned in its orientation to the axis of rotation from tube position *A* through *B* to *C*, the particles *b* and *d* are now describing helixes in space. They will, however, have possessed some of the kinetic energy in situation *A* that they must have in situation *C* if fluid and tube are to move together. This would be true, to a lesser degree, for particles near *b* and *d* but not for particles at *a* and *c*. Suppose, by contrast, that the tube begins in the flat position, as in *C*, but is at rest, and it is then accelerated in the same period of time to the same angular velocity as required for the shift from *A* through *B* to *C*. In this situation there is no initial kinetic energy in the fluid and inertial effects in the fluid would be maximal. Thus the fluid displacement may well be different in the two situations. If this is true, these physical principles would have important bearing on the behavior of the semicircular canal system in life situations. Another question is raised for the reader's consideration. The effect on fluid displacement of the rotating tube shifting from *A* through *B* to *C* may well be different depending on which "side" of the tube is turned down, despite the fact that the angular velocity remains constant. The reader can readily verify this for himself with a coin on which the positions *a, b, c,* and *d* are marked on the "heads" side. This coin is then rotated in the manner described. It will be seen that when the coin is turned heads-up in the final position *C*, the sequence of points in rotation as shown in the figure will be *a-b-c-d*. On the other hand, when the coin is turned heads-down the sequence will be *a-d-c-b*. This suggests that the direction of fluid displacement may be opposite in the two situations depending on which "side" of the rotating circular tube is turned downward in the movement from the "coin-on-edge" position to the flat position. These puzzling and important problems, not solved to our knowledge as yet, should readily demonstrate to the reader that there are wide and challenging gaps in our knowledge of the physics of the semicircular canals.

Simplest model with elastic partition across the tube

Before proceeding to methods for calculating angular accelerations involving rotations about more than one axis, it is important to emphasize that the simplest model we have been discussing must be modified to approximate more closely the behavior of a semicircular canal. It is necessary to add an elastic partition across the tube, as shown in the enlarged cutaway drawings *D* and *E* in Fig. 6-24. As is discussed later, the cupula in the ampulla of a semicircular canal provides such an elastic partition, although, unfortunately, too little is known of its physical properties. It is apparent that the behavior of the fluid within the tube will be altered by the presence of the elastic partition. In Fig. 6-24, *D*, the simple tube with this partition is shown undergoing angular acceleration. In this situation the fluid tends to move in the same direction as in the simple tube without a partition, but the degree of movement is less. If now the angular acceleration is diminished in magnitude or is replaced by constant angular velocity, the elastic restoring force, the importance of which has been strongly emphasized by G. Melvill Jones,[92] comes into play and produces movement of fluid in the opposite direction. This opposite movement would not have occurred without the partition.

It is shown later that the direction of fluid movement in a semicircular canal is of critical importance in terms of the physiologic phenomena produced. Thus considerations of factors affecting direction and amount of fluid movement in the simple tube model are relevant to the biologic situation. In Fig. 6-24, *E*, it is evident that with constant angular velocity the restoring force will finally be spent and the motion of the fluid under its impetus will cease. One may raise the question of the particular damping properties of the system and whether or not the elastic partition will "overshoot" and oscillate before finally returning to its original position of rest. There is not as yet a reliable and precise physiologic indicator of the physical behavior of the semicircular canal system. Nystagmus, often used as an indicator of labyrinthine events, is subject to alteration by influences in the central nervous system not directly related to physical events in the canal system. Experiments on animals[93, 94] in which a portion of a semicircular canal has been photographed have demonstrated that motion of fluid (endolymph) does in fact occur in the canal during angular acceleration. Beyond this, too little is known regarding precise dynamic physical characteristics of the system.

Mechanics of the semicircular canals

The simple model of a semicircular canal with an elastic partition across the rigid fluid-filled tube, discussed on p. 214, requires modification to take into account other potentially important aspects of the anatomy and physical behavior of the system. The actual anatomic situation consists of a tube within a tube, in which a membranous tube containing endolymph, as well as the sensing organ the cupula, lies within a slightly larger tube within the bony labyrinth of the skull. The inner aspect of the bony labyrinth is lined with adherent epithelium; between this layer and the inner membranous tube is another fluid, the perilymph. The inner membranous tube is not tightly attached to the outer except in the dilated portion of the canal (ampulla) that houses the cupula. Significant motion of the inner tube relative to the outer is at least theoretically possible and has been postulated (by R. W. Steer, quoted by Young[95]) but not generally accepted.

The most popular mechanical model for a single semicircular canal has been the torsion pendulum, a model utilized by several of the major workers in this field.[93, 95, 96] If a semicircular canal is idealized and considered to be perfectly circular, the differential equation of the torsion pendulum, presented by Groen,[96] is applicable:

$$\theta\ddot{\xi} + \Pi\dot{\xi} + \Delta\xi = 0 \tag{55}$$

where ξ, $\dot{\xi}$ and $\ddot{\xi}$ = the angular deviation from a reference position, the angular velocity, and the angular acceleration of the fluid ring, respectively

θ = moment of inertia of the fluid ring (endolymph)

Π = dynamic friction couple (moment of friction) at unit angular velocity

Δ = stiffness couple at unit angular displacement (of the cupula)

All of these are taken with respect to the center of the canal. Groen compares the system to the balance wheel of a watch, in which the wheel itself is the analog of the fluid ring and the spiral spring represents the cupula. Friction, however, is far more powerful in the case of the fluid ring than in that of the balance wheel. Recently, Young,[97] using the original notation of Groen and colleagues, has altered equation 55 to a more useful form:

$$\theta\ddot{\xi} + \Pi\dot{\xi} + \Delta\xi = \alpha\theta \tag{56}$$

where α = component of angular acceleration of the skull with respect to inertial space, normal (perpendicular) to the plane of the semicircular canal under consideration

Using this model, Young believes that the angular deviation of the endolymph in relation to the skull, and therefore the angular deviation of the cupula, can be predicted for a variety of forcing functions of the head: constant acceleration, sinusoidal oscillation, and step change in angular velocity, that is, an impulse of acceleration. "Appropriate" values for the parameters Π/θ and Δ/θ must be determined or assumed in order to make the calculations. Young presents other forms of the torsion pendulum equation, which will not be given here, and concludes that this approach can account for over 95% of the cupula response. He presents Steer's physical model for the cupula, in which the motion of the fluid in the canal is opposed by (1) elastic restraining forces that are proportional to the displacement of the cupula and (2) a viscous force that is proportional to the rate of change of the cupula angle. Young feels that Groen erred in attributing the drag in the canal solely to viscosity and that a major share of this drag is the force required to slide the cupula in the ampulla. He emphasizes, however, that detailed canal dynamics, inferred from nystagmus recordings, are only known with confidence for the horizontal canals. The dangers of using the nystagmus "output" to determine the "input" of this system are commented on later.

The reader should not be left with the feeling that our knowledge of semicircular canal mechanics is solely theoretic. In his classic studies on the shark, Steinhausen[94] demonstrated, by India ink injection into the exposed membranous ampulla, that the cupula actually deviates during angular acceleration. Some years later Groen and colleagues[93] made oscillographic recordings of potentials from the nerve supplying the horizontal canal of an elasmobranch, the ray. Their observations with this preparation as a torsion swing and turntable convinced these workers that the cupula-endolymph system behaves like a true pendulum.

Calculation of angular accelerations where more than one axis of rotation is involved

The human body, and in particular the human head, is commonly subjected to angular velocities and angular accelerations about more than one axis. This is true in everyday living; an example would be the act of turning the head while rounding a curve in an automobile. Situations in flight in which there is rotation about two axes of the aircraft are readily visualized, such as entry into a climbing turn, which involves rotation about both the pitch and roll axes of the aircraft. If during this maneuver the head is turned, it would be subjected to the complex situation of rotation about three axes simultaneously. In a spacecraft, if there were a basic rotation about the longitudinal axis to produce some degree of "artificial gravity," a superimposed head motion would provide the situation of rotation of the head about at least two axes. It is a somewhat surprising fact that constant angular velocity about two nonparallel axes can generate the physical equivalent of an angular acceleration. This question of the physics of rotation about multiple axes and its physiologic effects has been the subject of a number of classic and more recent papers.[98-101]

Stone and Letko[100] have developed the equations for the effects of angular acceleration and angular velocity about two axes. The equations given in their original publication have a number of typographic errors that have been kindly corrected and checked for us by Stone. The following corrected material is taken from their paper, with some rearrangement and additions:

DEFINITIONS AND SYMBOLS

nodding A rotation of the head about the y axis (Fig. 6-9).

turning A rotation of the head about the z axis (Fig. 6-9).

rolling A rotation of the head about the x axis (Fig. 6-9).

α_{G_θ} Cross-coupled nodding acceleration (component of cross-coupled acceleration about the y axis)

α_{G_ψ} Cross-coupled turning acceleration (component of cross-coupled acceleration about the z axis)

α_{G_ϕ} Cross-coupled rolling acceleration (component of cross-coupled acceleration about the x axis)

$\omega_{G_\theta} = \alpha_{G_\theta} \int dt$

$\omega_{G_\psi} = \alpha_{G_\psi} \int dt$

$\omega_{G_\phi} = \alpha_{G_\phi} \int dt$

ω_{h_θ} **Nodding velocity** A fore-and-aft motion of the head at the neck or from the whole body*

ω_{h_ψ} **Turning velocity** A motion about the neck or long-body axis*

ω_{h_ϕ} **Rolling velocity** A sideways motion of the head or from the body*

ω_V Vehicle rotational velocity (relative to inertial space)

ω_{h_x} Total angular velocity of head about rolling axis (x axis) (relative to inertial axes)

ω_{h_y} Total angular velocity of head about nodding axis (y axis) (relative to inertial axes)

ω_{h_z} Total angular velocity of head about turning axis (z axis) (relative to inertial axes)

t Time

ϕ_e, θ_e, ψ_e Euler angular displacement* using this order of rotation. (Euler angles represent instantaneous values that relate one set of moving axes to another set [normally inertial and therefore nonmoving] and are used in Euler's classic method for this transformation of axes.)

θ_{sc} backward tilt angle of semicircular canals from $x_b y_b$ plane

ψ_{sc} angle of rotation of semicircular canals from $x_b z_b$ plane

x, y, z inertial space axes

x_b, y_b, z_b body axes

Subscripts:

lr, ll right and left lateral canals, respectively
pr, pl right and left posterior canals, respectively
ar, al right and left anterior canals, respectively

Head axes used by Stone and Letko:

x is (+) facing forward out of the nose
y is (+) extending out of the right ear
z is (+) extending downward (caudad) from the head

Note: These head axes are part of a rectangular coordinate system. The negative of each axis has the opposite sense and direction from the positive.

Semicircular canal axes:

Equations 63A to 63F and 64A to 64F relate to the axes of the semicircular canals themselves. Each acceleration vector in these equations represents an angular acceleration vector that is perpendicular to the plane of the canal in question. The angles θ_{sc} and ψ_{sc} in these equations represent the transformation of the head axes into the semicircular canal axes for a given individual.

Calculation of angular velocities and angular accelerations

These equations would apply to the calculation of these velocities as experienced by any structure within the head, such as the semicircular canals, if the axes of the structure and the axes of the head maintain a constant relationship. The same equations could also be used for calculating the angular velocities experienced by any organ

*These are angular head motions and may be from motions at the neck and shoulders, from body bending, or from rotations of the entire body within the vehicle. These angular velocities are relative to the axes of the *rotating vehicle.*

*Whittaker, E. T.: Analytical dynamics, New York, 1944, Dover Publications, Inc. (Original edition published by the Cambridge University Press, 1904.)

for which the essential data can be obtained. The total angular velocities represent the sum of the various angular velocities acting:

$$\omega_{h_x} = \omega_{h_\phi} + \omega_V \cos\theta_e \cos\psi_e \quad (57A)$$

$$\omega_{h_y} = \omega_{h_\theta} - \omega_V \cos\theta_e \sin\psi_e \quad (57B)$$

$$\omega_{h_z} = \omega_{h_\psi} + \omega_V \sin\theta_e \quad (57C)$$

where ω_V = angular or rotational velocity of the vehicle (assumed to be constant)

The differentiation of equations 57A to 57C with respect to time gives these angular accelerations, where $\dot{\omega}_{h_x}$, $\dot{\omega}_{h_y}$, and $\dot{\omega}_{h_z}$ are the angular accelerations of the head in inertial space (the accelerations that will stimulate the semicircular canals) and $\dot{\omega}_{h_\phi}$, $\dot{\omega}_{h_\theta}$, and $\dot{\omega}_{h_\psi}$ are the angular accelerations of the head in the rotating frame of reference (the vehicle).*

$$\dot{\omega}_{h_x} = \dot{\omega}_{h_\phi} - \omega_V (\sin\theta_e \cos\psi_e \, \dot{\theta}_e + \cos\theta_e \sin\psi_e \, \dot{\psi}_e) \quad (58A)$$

$$\dot{\omega}_{h_y} = \dot{\omega}_{h_\theta} - \omega_V (\cos\theta_e \cos\psi_e \, \dot{\psi}_e - \sin\theta_e \sin\psi_e \, \dot{\theta}_e) \quad (58B)$$

$$\dot{\omega}_{h_z} = \dot{\omega}_{h_\psi} + \omega_V \cos\theta_e \, \dot{\theta}_e \quad (58C)$$

Applying the principles of classic mechanics, these rates of change of the Euler angles in equations 58A to 58C can be calculated from the following:

$$\dot{\phi}_e = (\omega_{h_\phi} \cos\psi_e - \omega_{h_\theta} \sin\psi_e) \frac{1}{\cos\theta_e} \quad (59A)$$

$$\dot{\theta}_e = (\omega_{h_\phi} \sin\psi_e + \omega_{h_\theta} \cos\psi_e) \quad (59B)$$

$$\dot{\psi}_e = \omega_{h_\psi} - \tan\theta_e (\omega_{h_\phi} \cos\psi_e - \omega_{h_\theta} \sin\psi_e) \quad (59C)$$

These equations will be found to be more convenient for most purposes than their equivalents (equations 58A to 58C). A substitution of equations 59A to 59C into equations 58A to 58C results in these general expressions for total angular accelerations:

$$\dot{\omega}_{h_x} = \dot{\omega}_{h_\phi} - \omega_V (\omega_{h_\theta} \sin\theta_e + \omega_{h_\psi} \cos\theta_e \sin\psi_e) \quad (60A)$$

*In accord with previous usage, a dot over a symbol indicates its first derivative (instantaneous rate of change) with respect to time. Thus, if ω represents angular velocity, $\dot{\omega}$ represents angular acceleration.

$$\dot{\omega}_{h_y} = \dot{\omega}_{h_\theta} - \omega_V (\omega_{h_\psi} \cos\theta_e \cos\psi_e - \omega_{h_\phi} \sin\theta_e) \quad (60B)$$

$$\dot{\omega}_{h_z} = \dot{\omega}_{h_\psi} + \omega_V (\omega_{h_\theta} \cos\theta_e \cos\psi_e + \omega_{h_\phi} \cos\theta_e \sin\psi_e) \quad (60C)$$

When the vehicle is not rotating ($\omega_V = 0$) the equations simplify to:

$$\dot{\omega}_{h_x} = \dot{\omega}_{h_\phi} \quad (61A)$$

$$\dot{\omega}_{h_y} = \dot{\omega}_{h_\theta} \quad (61B)$$

$$\dot{\omega}_{h_z} = \dot{\omega}_{h_\psi} \quad (61C)$$

On earth, these equations imply that the influence of the rotation of the earth is negligible. The rate of the rotation of the earth could be substituted for ω_V in equations 60A to 60C to eliminate this slight error.

When equations 61A to 61C are subtracted from equations 60A to 60C, there results a set of expressions caused by the rotation of the vehicle. Stone and Letko[100] have adopted the long-used engineering term "cross-coupled" for these accelerations, which are given by the equations:

$$\alpha_{G_\phi} = -\omega_V (\omega_{h_\theta} \sin\theta_e + \omega_{h_\psi} \cos\theta_e \sin\psi_e) \quad (62A)$$

$$\alpha_{G_\theta} = \omega_V (\omega_{h_\phi} \sin\theta_e - \omega_{h_\psi} \cos\theta_e \cos\psi_e) \quad (62B)$$

$$\alpha_{G_\psi} = \omega_V (\omega_{h_\theta} \cos\theta_e \cos\psi_e + \omega_{h_\phi} \cos\theta_e \sin\psi_e) \quad (62C)$$

It has been believed by a number of investigators concerned with the physics of the stimulation of the semicircular canals that the "cross-coupled" angular accelerations are those components of the total accelerations experienced by the canals that primarily cause the disturbing symptoms and signs during rotation of a vehicle. It should be noted that the instantaneous angular velocities (ω) in these equations are not necessarily associated with a condition of angular acceleration but can be a part of a constant angular velocity profile. Thus *angular accelerations* (α) *can result from the "cross-coupling" effects of constant angular velocities about more than one axis*—an extremely important principle.

The subject of Coriolis acceleration was introduced earlier. It was pointed out that a body moving in a rectilinear path relative to a true inertial frame will appear to be following a curvilinear path to an observer in a rotating frame.

Such a situation is found when a mass is moving on the surface of a rotating carousel or when a projectile is fired from a gun on the surface of the rotating earth. It should be noted that the apparent acceleration measured by the observer in the rotating environment is equal to the acceleration that would be required to cause the object to move in a straight path relative to this rotating environment.

The point should be emphasized that there is increasing use in physics, engineering, and physiology of the term "Coriolis force" or acceleration where there is rotation of a body about more than one axis simultaneously. Thus the cross-coupled accelerations that we have been discussing are now commonly referred to as Coriolis accelerations. Strictly speaking, this represents a misusage, although perhaps not a serious one, but an increasingly common one. We believe, however, that cross-coupled acceleration is the preferable term.

The effective component of the cross-coupled angular acceleration that applies to each of the six semicircular canals can be derived from equations 62A to 62C. These derived equations have been kindly provided us by Stone and are as follows:

$$\Delta\dot{\omega}_{sc_{lr}} = \mathbf{a}_{G_\phi} \sin \theta_{sc} + \mathbf{a}_{G_\psi} \cos \theta_{sc} \quad (63A)$$

$$\Delta\dot{\omega}_{sc_{ll}} = \mathbf{a}_{G_\phi} \sin \theta_{sc} + \mathbf{a}_{G_\psi} \cos \theta_{sc} \quad (63B)$$

$$\Delta\dot{\omega}_{sc_{ar}} = \mathbf{a}_{G_\theta} \cos \psi_{sc} - \mathbf{a}_{G_\phi} \cos \theta_{sc} \sin \psi_{sc} + \mathbf{a}_{G_\psi} \sin \theta_{sc} \sin \psi_{sc} \quad (63C)$$

$$\Delta\dot{\omega}_{sc_{al}} = \mathbf{a}_{G_\theta} \cos \psi_{sc} - \mathbf{a}_{G_\phi} \cos \theta_{sc} \sin \psi_{sc} + \mathbf{a}_{G_\psi} \sin \theta_{sc} \sin \psi_{sc} \quad (63D)$$

$$\Delta\dot{\omega}_{sc_{pr}} = \mathbf{a}_{G_\phi} \cos \theta_{sc} \cos \psi_{sc} + \mathbf{a}_{G_\theta} \sin \psi_{sc} - \mathbf{a}_{G_\psi} \sin \theta_{sc} \cos \psi_{sc} \quad (63E)$$

$$\Delta\dot{\omega}_{sc_{pl}} = \mathbf{a}_{G_\phi} \cos \theta_{sc} \cos \psi_{sc} - \mathbf{a}_{G_\theta} \sin \psi_{sc} - \mathbf{a}_{G_\psi} \sin \theta_{sc} \cos \psi_{sc} \quad (63F)$$

The total angular acceleration experienced by each of the six semicircular canals are given by the following equations. These include the cross-coupled components of angular acceleration. It will assist visualization of these relationships if the reader consults Fig. 6-32, which shows the orientation of the semicircular canals within the cranium.

$$\dot{\omega}_{sc_{lr}} = \dot{\omega}_{h_x} \sin \theta_{sc} + \dot{\omega}_{h_z} \cos \theta_{sc} \quad (64A)$$

$$\dot{\omega}_{sc_{ll}} = \dot{\omega}_{h_x} \sin \theta_{sc} + \dot{\omega}_{h_z} \cos \theta_{sc} \quad (64B)$$

$$\dot{\omega}_{sc_{ar}} = \dot{\omega}_{h_y} \cos \psi_{sc} - \dot{\omega}_{h_x} \cos \theta_{sc} \sin \psi_{sc} + \dot{\omega}_{h_z} \sin \theta_{sc} \sin \psi_{sc} \quad (64C)$$

$$\dot{\omega}_{sc_{al}} = \dot{\omega}_{h_y} \cos \psi_{sc} - \dot{\omega}_{h_x} \cos \theta_{sc} \sin \psi_{sc} + \dot{\omega}_{h_z} \sin \theta_{sc} \sin \psi_{sc} \quad (64D)$$

$$\dot{\omega}_{sc_{pr}} = \dot{\omega}_{h_x} \cos \theta_{sc} \cos \psi_{sc} + \dot{\omega}_{h_y} \sin \psi_{sc} - \dot{\omega}_{h_z} \sin \theta_{sc} \cos \psi_{sc} \quad (64E)$$

$$\dot{\omega}_{sc_{pl}} = \dot{\omega}_{h_x} \cos \theta_{sc} \cos \psi_{sc} - \dot{\omega}_{h_y} \sin \psi_{sc} - \dot{\omega}_{h_z} \sin \theta_{sc} \cos \psi_{sc} \quad (64F)$$

The equations that have been presented enable the calculation of instantaneous angular accelerations of the osseous labyrinth (or of any rigid structure, the axes of which remain constant relative to the axes of the head). Since the inertial, viscous, and other damping properties provide the semicircular canal system with various delays, which can be expressed as time constants, the canals do not represent transducers whose output of nerve impulses at any moment is likely to be directly proportional to the instantaneous angular acceleration experienced by each canal. There may well be, and probably are, moments when the afferent nerve traffic from the cupula of a given canal is proportional to the instantaneous angular acceleration, or some cases where the momentary physical conditions combine to produce a response proportional to the instantaneous angular velocity. One would expect these moments to be exceptional.

It would appear, therefore, that what is actually needed for the solution of a number of physiologic problems is a continuous readout of the solutions to these equations. Presently available accelerometers and computing circuits make it technically feasible to provide continuous recording of angular velocity and angular acceleration in the axes of the instruments. If the accelerometers were aligned with the x, y, and z head axes, equations 64A to 64F above could be programmed to transform the measured data into the values applicable to the plane of each semicircular canal. In this way, the history of the major variables in the physiologically significant immediate past could be studied in relation to the subjective and objective physiologic phenomena. It is true that anatomic differences among different subjects will introduce some, usually small, error in the assumed positions of the semicircular canals within a given subject's head; yet it is probable that such on-line computation of these functions would make a major contribution toward our understanding of the responses both of physical models and of biologic systems, including man.

A final point regarding the use of these equations is worthy of emphasis. Equations presented in this section have been concerned with the calculation of *angular acceleration* but *not* of *inertial angular reaction*. It will have become apparent to the reader that in the angular acceleration field the angular reaction has not been accorded the importance attached to the gravitoinertial resultant force treated earlier, yet the two concepts are comparable and are both based on the property of inertia. The calculations can be made quite simply from the equations in the present section, since the inertial angular reaction is equal in magnitude but opposite in direction to the angular acceleration. That is, the axis of rotation is the same but the sense (direction) is reversed. This is also true in regard to the cross-coupled angular acceleration and its angular reaction, since these represent a real acceleration and an equally real inertial angular reaction. It is the inertial angular reaction of the endolymph of the semicircular canals that explains the directionally similar movement of the cupula.

SPECIAL PHYSIOLOGY OF THE VESTIBULAR SYSTEM

In primitive teleosts the "ear" is comprised of two highly developed organs of equilibrium but is only a rudimentary organ of hearing. The evolutionary development of the organ of Corti was associated with the expansion of its forebrain representation in the neocortex in mammals, and auditory reception gained the status of a "special sense." In striking contrast, the basic components of the vestibular system were virtually complete with the appearance of the cerebellum, an outgrowth of the hindbrain, and its cortical representation is meager. As we view the phylogenetic scale from nonmammalian vertebrates to man, the basic structure and the cardinal purpose of the organs of equilibrium remain essentially unchanged.

The primitive character of the vestibular system in man accounts, in large part at least, for the almost incredible delay in man's discovering that sensory receptors in the inner ear serve functions other than hearing. This curious history divides into two with the publication of a report in 1870 by Goltz,[102] who drew his important inference from Flourens' studies[103] showing that loss of equilibrium in pigeons followed sectioning of the semicircular canals. Goltz reasoned that if disequilibrium is caused by labyrinthine lesions, the same site must be involved with equilibratory function.

Before Goltz, what we now term vestibular side effects were of course well known; familiarity with seasickness and dizziness dates back to antiquity. Some of the earliest systematic observations were made by Erasmus Darwin,[104] who described in his *Zoonomia* (1794) the eye motions (nystagmus) and vertigo associated with rotation. Purkinje[105] described a form of vestibular vertigo that he thought had its origin in the cerebellum. Meniere,[106] who was familiar with Flourens' work, described the syndrome (deafness, vertigo, nausea) that bears his name and correctly ascribed the cause to labyrinthine disease; formerly, this syndrome had been included under the term "apoplectiform cerebral congestion."

After Goltz, within a period of 5 years Mach,[107] Crum-Brown,[108] and Breuer[109] developed the theoretic basis underlying stimulation of the mechanoreceptors in the semicircular canals and otolith organs. The theory that the canals are stimulated by impulse angular accelerations was elaborated by Ewald[110] and confirmed by Steinhausen[111] and Dohlman.[112] Stimulation of the otolithic receptors by gravity and by impulse linear accelerations was demonstrated by Kreidl,[113] Versteegh,[114] and Tait and McNally,[115] among others. Bárány deserves the credit for introducing tests of vestibular function into the clinic.[116] Magnus and his co-workers devised tests to define the role of the vestibular system in postural mechanisms, summarized in his classic monograph, *Körperstellung*.[117]

With regard to the etiology of vestibular side effects, it was the philosopher William James who in 1882[118] reasoned that if Goltz's deduction was correct, deaf mutes (with loss of labyrinthine function) should not experience dizziness. Among 519 such subjects tested, 186 were classified "not dizzy," 134 "slightly dizzy," and 199 "dizzy." James interviewed "many deaf mutes" about their susceptibility to seasickness because of the "high probability" that it was of labyrinthine origin. Fifteen in other than the "dizzy" category had been exposed to rough seas and none had experienced seasickness. Sjöberg's[119] classic studies on motion sickness demonstrated that labyrinthectomized dogs no longer were susceptible to motion sickness and, incidentally, proved the correctness of James' presumptive evidence.

Thus it was established that the vestibular system plays a dual role in the present-day lives of

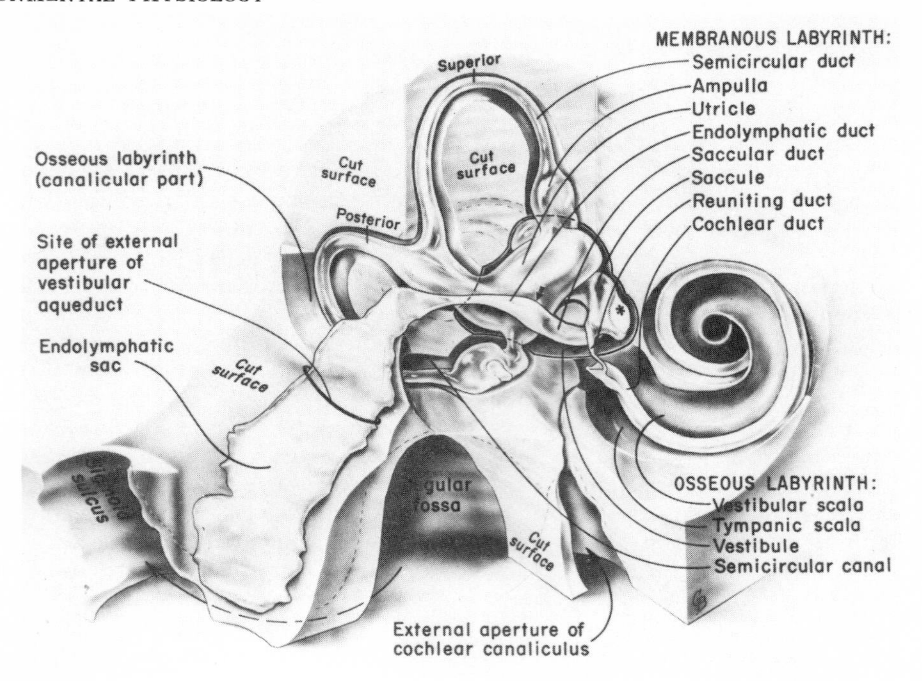

Fig. 6-25. Reconstruction of the membranous labyrinth and related anatomy. (From Anson, B. J., Harper, D. B., and Winch, T. G.: The vestibular and cochlear aqueducts: developmental and adult anatomy of their contents and parities. In Graybiel, A., editor: Third symposium on the role of the vestibular organs in space exploration, NASA SP-152, Washington, D. C., 1968, U. S. Government Printing Office.)

even typically normal women and men. One role is represented by the elegant manner in which the vestibular system functions under natural terrestrial stimulus conditions and the other, by the ease with which this system either provides unwanted information or is rendered unstable under unnatural stimulus conditions. This dual role will be seen to pervade, overtly or covertly, nearly every aspect of the presentation to follow.

End organs

The labyrinth of the human inner ear comprises the cochlea (the organ of hearing) and the otolith organs and semicircular canals (collectively termed the vestibular organs). These are paired end organs with a common blood supply and a shared secondary lymph circulation; their mechanoreceptors have similar histologic features and are innervated by the vestibular division of the eighth nerve. These sensory organs are situated in hollowed-out channels in the petrous portion of the temporal bone (Fig. 6-25),[120] and, within the bony channels, the enclosed membranous labyrinth is surrounded by perilymph and filled with endolymph. Thus the sensory receptor mechanisms are protected from the effects of superimposed body weight by the bony labyrinth and, by virtue of the contained fluids, receive additional protection from impact accelerations.

Otolith organs

The four otolith organs appear as thickened areas on the inner walls of the paired utricle and saccule (Fig. 6-26) that are termed "maculae." The sketch in Fig. 6-27,[121] based on electron micrograph studies, shows the general structure of a perpendicular section of the utricular macula in the *Saimiri sciureus* (squirrel monkey). Superimposed on the sensory epithelium (*e*) is the otolith membrane subdivided into four zones (*a* to *d*). Sharp demarcations between the zones are not seen in electron micrographs, but the zonal contents are distinguishable. The stimulus from gravity and the stimuli from impulse linear or Coriolis accelerations, generating shearing forces within the bony labyrinth, result presumably in movement of the otolithic zone (Fig. 6-27, *a*) relative to the cupular (*b* and *c*) and subcupular (*d*) zones. In the squirrel monkey there is not a direct connection between the otoliths (specific gravity = 2.71) and the cilia of the sensory hair cells (seen to project upward into the subcupular zone). The rod-shaped struc-

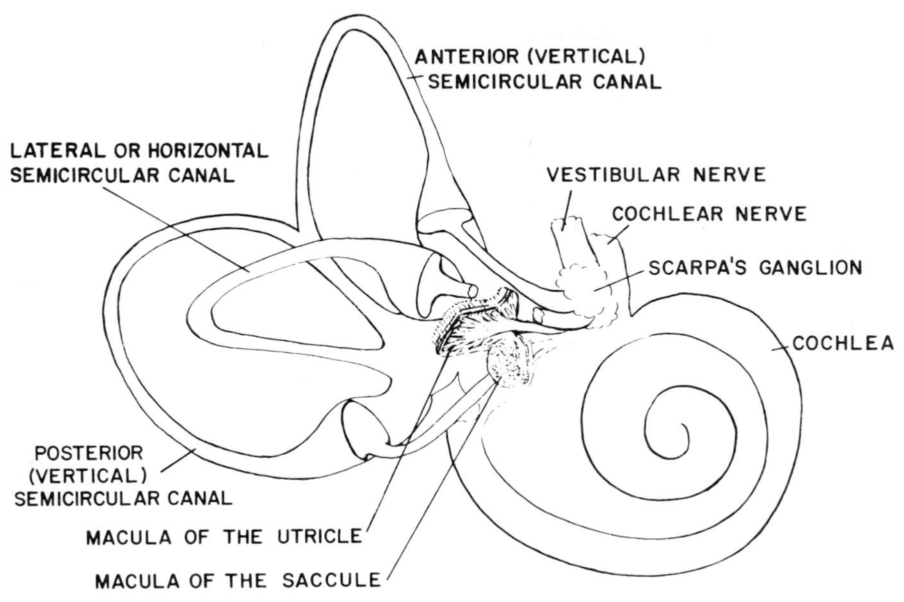

Fig. 6-26. Labyrinth of the right ear as viewed from the lateral aspect.

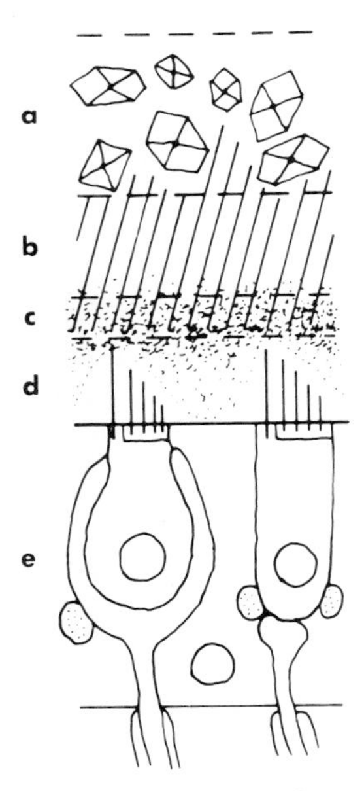

Fig. 6-27. Schematic drawing based on electron micrograph findings in perpendicular sections of the utricular macula in the squirrel monkey. *a,* Otolithic zone with hexagonal-shaped otoliths; *b,* cupular zone 1 shows rod-shaped structures that reach upward to the otoliths (specific weight about 2.71) and downward into cupular zone 2 *(c)*; *d,* subcupular zone showing upward projections of cilia from hair cells in the neuroepithelium *(e)*. (From Igarishi, M., and Kanda, T.: Acta Otolaryng. **68:** 43-52, 1969.)

tures in the cupular zones, however, extend upward, reaching the hexagonal-shaped otoconia.

The sketch in Fig. 6-28 was drawn from electron micrographs of the sensory epithelium of the utricular macula of the squirrel monkey.[122] Two types of hair cells, each with two types of cilia, are depicted. Each cell has sixty to seventy stereocilia and one kinocilium laid out in strict geometric arrangement. In different regions of the macula the kinocilia (which play the major role in the energy transfer) are polarized in different directions; hence, a shearing force in one plane will result in kinocilia moving in different directions with reference to the kinociliar pole. The result is mechanical deformation of the cilia, which in turn causes chemical changes resulting in the generation of bioelectric potentials (neural impulses). The apparatus just described may be termed the cilia-otolith mechanism.

Under natural terrestrial conditions the adequate stimuli in the otolith organs are gravity caused by a central field factor and impulse linear accelerations generated by body movements. Note that adequate stimuli can penetrate the bony labyrinth but that their derivatives, gravitoinertial forces, constitute the effective stimulus causing relative movement between otolithic and cupular membranes. On the one hand the bony labyrinth is a sanctuary preventing entry of both gravitoinertial and nongravitoinertial (mechan-

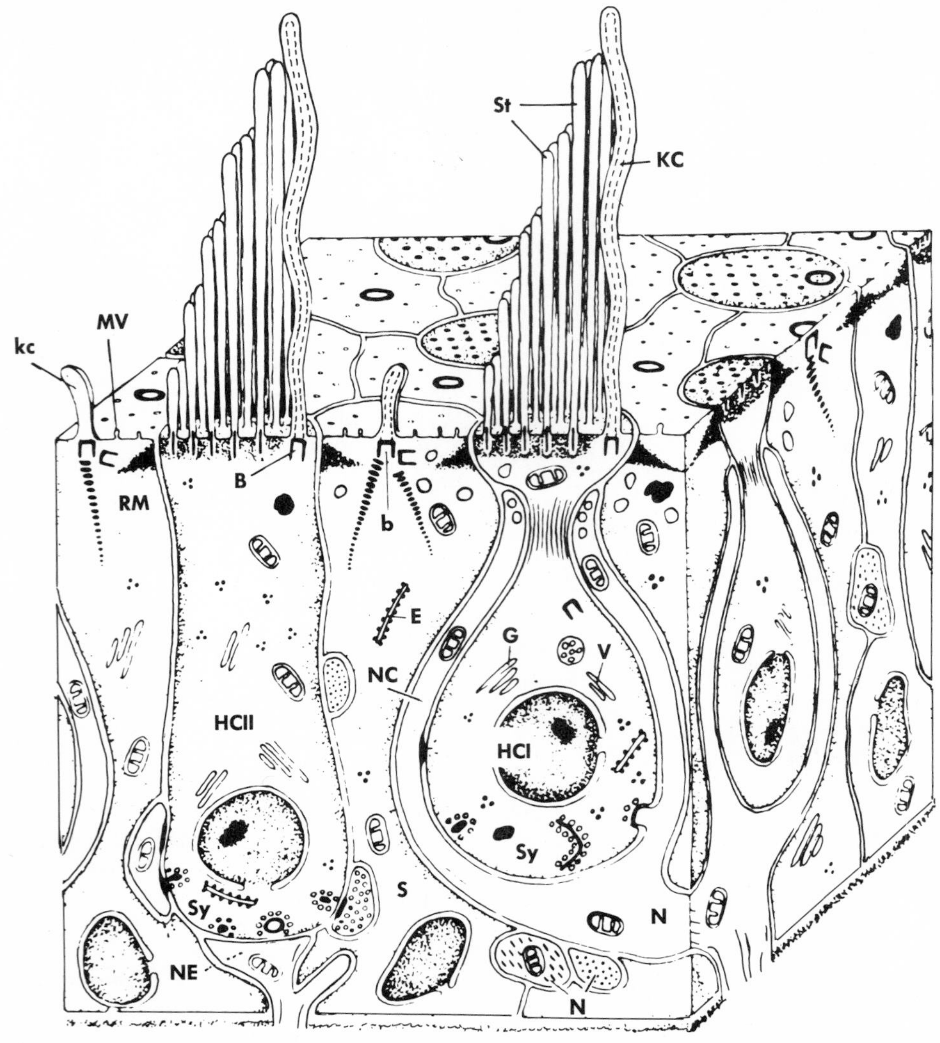

Fig. 6-28. Sketch drawn from electron micrographs of the fine structure of the utricular macula of *Saimiri sciureus* (squirrel monkey). Two types of hair cells (*HC I* and *HC II*) and two types of cilia (*KC,* kinocilia; *St,* stereocilia) are seen. (From Spoendlen, H.: Schweiz. Arch. Neurol. Psychiat. **96**[2]:219-230, 1965.)

ical) forces, but on the other hand the otolithic organs must respond slavishly within their response characteristics to gravity and to linear accelerations. Mechanoreceptors serving touch, pressure, and kinesthesis (TPK) are stimulated directly or indirectly not only by gravitoinertial forces but also by strictly mechanical (nongravitoinertial) forces that cannot penetrate the bony labyrinth. It is important to distinguish among gravitoinertial forces within and without the bony labyrinth and strictly mechanical forces outside the nonacoustic labyrinth. In Fig. 6-29, *A,* TPK receptor systems are differentially affected by gravitoinertial forces from head to foot but otolithic receptors are not influenced by extralabyrinthine weight even if the subject stood on his head. In Fig. 6-29, *B,* the subject is exposed to a vertical acceleration generating an inertial force of 1 G; both otolithic and TPK receptor systems are affected the same, as if the subject were on a planet with twice the earth's gravitational attraction. In Fig. 6-29, *C,* otolithic receptors are not influenced by the object supported on the head but TPK systems are affected along the lines of support.

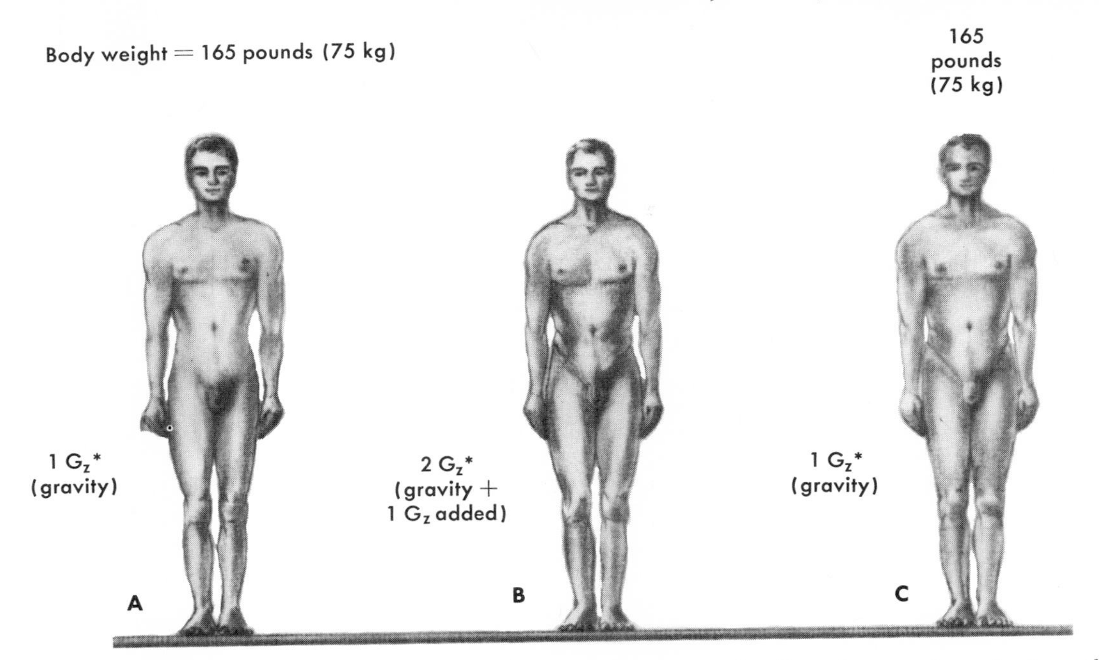

Fig. 6-29. Differential effects of doubling the earth's standard of gravity, **B,** and supporting a load equal to body weight, **C.** Otolithic and touch, pressure, and kinesthetic receptor systems affected under conditions in **B** but in **C** otolithic receptors stimulated the same as in **A** while TPK systems affected along the lines of support for the added weight.

The nomogram in Fig. 6-10, *A,* makes it possible to compare the changes in the direction (angle ø) and magnitude of the gravitoinertial resultant force (**G★**) during exposure to linear accelerations in the horizontal plane under lunar, martian, and earth standards of gravity. The performance of American astronauts in carrying out their tasks under lunar gravity, for example, was accomplished with remarkably little difficulty in view of the novel stimulus conditions. Under dynamic conditions a linear acceleration of 0.18 G unit on earth causes a change of 10 degrees in the direction of the angle ø; under lunar conditions only 0.03 earth G units would be required to produce the same change in ø (Fig. 6-10). The force environment on the moon did not directly affect the visual and canalicular systems of the astronauts but had differential effects on their otolithic and TPK systems. Only the TPK system was affected by their backpacks, approximately equal to body weight. Assuming the sensory inputs from the otolith organs played a role in the lunar walk, it would seem that a brief period of adaptation was required before optimal functioning by the astronaut on the lunar surface was possible.

Semicircular canals

The mechanoreceptors in the two vestibular organs are similar, but the gross structure of the semicircular canals bears little resemblance to that of the otolith organs. The three canals in each human labyrinth are mutually perpendicular, and each so-called semicircular canal actually forms a complete circuit by virtue of its connections with the utricle, near which the canal expands into what is called the ampulla. A section through the ampulla of an exceptionally well-preserved human specimen* is shown in Fig. 6-30. The crista is a transverse ridge of tissue covered by a sensory epithelium containing hair cells (similar to those in the maculae) whose cilia extend into a gelatinous body called the cupula. The kinocilia of the hair cells are uniformly polarized; in the horizontal canals they are directed toward the utricle (utricular pole) and in the vertical canals toward the opposite pole. The cupula, a meshwork of fibers, presumably collagen, extends to the roof of

*Kindly provided by Professor Makoto Igarashi, Department of Otolaryngology, Baylor University College of Medicine, Texas Medical Center, Houston, Tex.

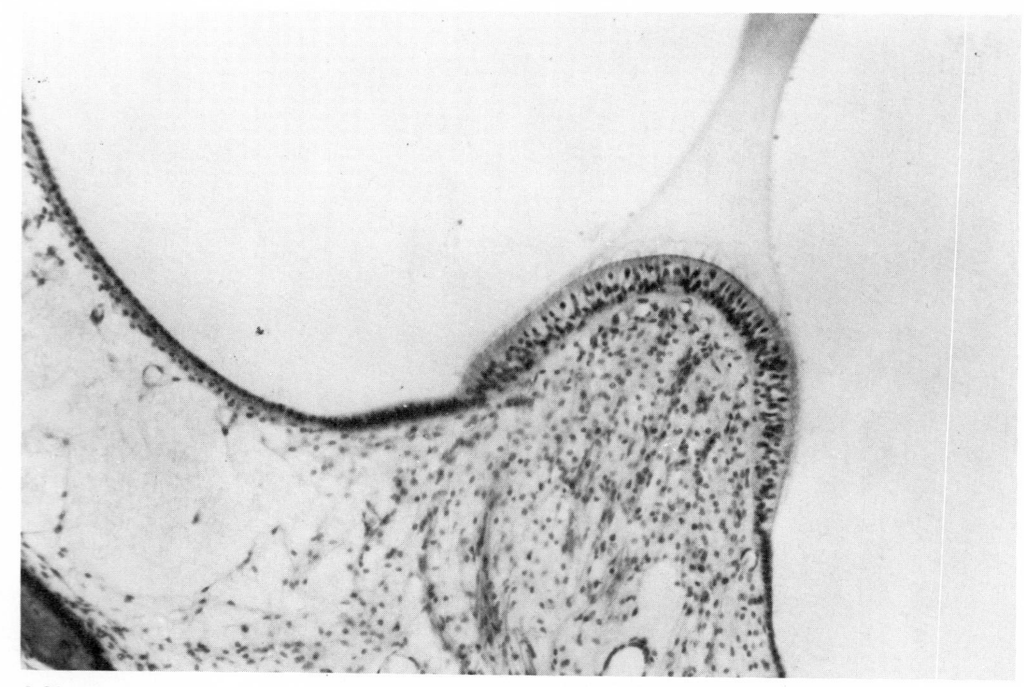

Fig. 6-30. Cross section of well-preserved human ampulla (posterior semicircular canal) showing the crista with its sensory epithelium surmounted by the cupula. (Courtesy Professor M. Igarashi, Houston, Tex.)

the ampulla (not shown in Fig. 6-30), completing a fluid-tight gate across the ampulla, hinged at the crista and free to move back and forth in response to movements of the endolymph. This apparatus constitutes the cupula-endolymph mechanism.

The cupula-endolymph mechanism is similar in man and fish, and its operation will be described with the aid of Fig. 6-31 showing drawings made from photographs of the ampulla of the living pike. A drop of opaque oil was injected into a single canal and served to indicate any movement of the endolymph. Fig. 6-31, *A,* shows the position of the oil droplet and cupula prior to stimulation. In response to angular acceleration, the endolymph lags behind the movement of the bony and membranous canal, thus displacing the droplet and the cupula in a direction counter to the rotary motion. The cupula-endolymph system, responding only to impulse angular accelerations in the plane of the canal, has been likened to a fluid-filled torus, with the cupula responding to movements of the endolymph in the manner of a spring-mass system with viscous damping. Cupula-endolymph dynamics have been discussed in detail previously.

Under natural stimulus conditions (for man) angular deceleration quickly follows angular acceleration and, inasmuch as the areas under the curves depicting acceleration and deceleration are the same, the cupular deflection is immediately restored to, or almost to, its natural resting position. Under unnatural stimulus conditions such as rotation at constant angular velocity following an initial angular acceleration, the cupula will slowly return to its natural position by virtue of its inherent elasticity but is then poised to be deflected on deceleration, again necessitating elastic restoration of the cupula. Such abnormal stimulus conditions in man may give rise to the oculogyral illusion and other nystagmus phenomena that are described later.

The orientation of the six semicircular canals with reference to the head (see Fig. 6-32 and p. 216 for discussion of reference system) is shown in Fig. 6-32. It will be noted that, although the three canals on one side lie approximately in mutually perpendicular planes, only the plane of the horizontal canals even approaches being parallel to one of the coordinate planes of the skull, and even here the divergence is about 25 degrees. The superior and posterior canals by contrast de-

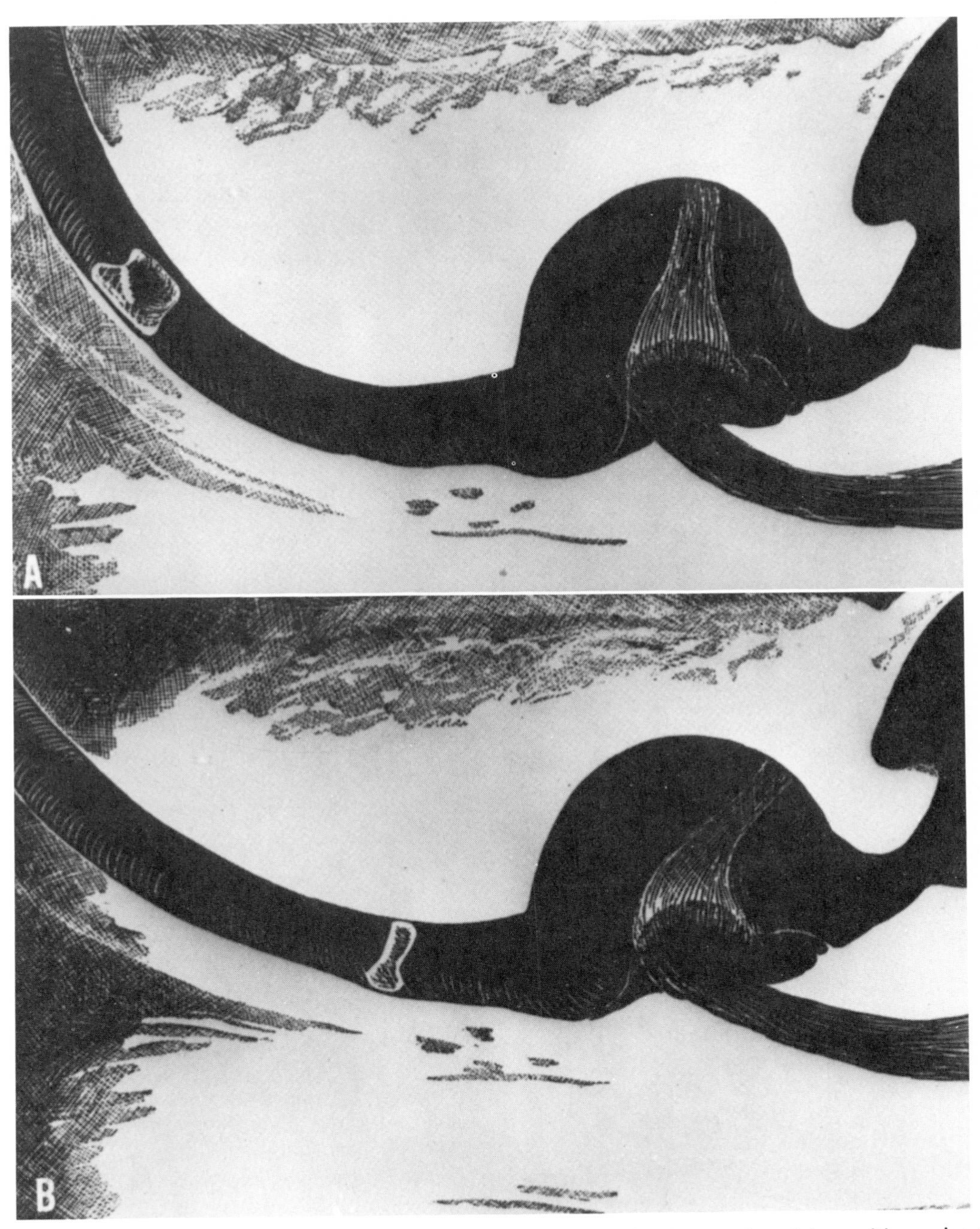

Fig. 6-31. A drop of opaque oil has been injected into the semicircular canal and its position prior to, **A**, and after an accelerative stimulus, **B**, indicate changing relationships, that is, endolymph flow with consequent displacement of the cupula. (Adapted from photographs taken in situ by Dohlman; from Research in the service of medicine, vol. 28, Chicago, G. D. Searle & Co.)

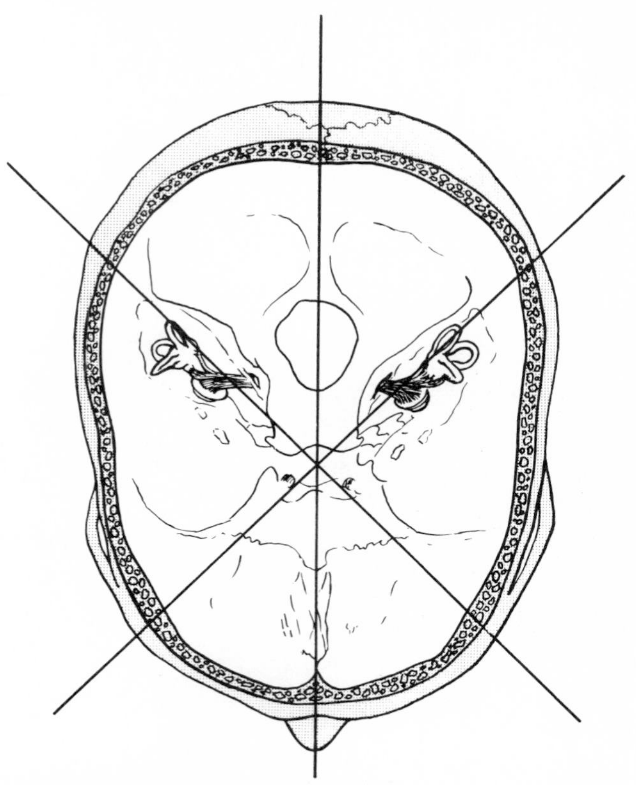

Fig. 6-32. Orientation of semicircular canals (enlarged) as viewed in the skull from above.

viate by 45 degrees from the sagittal and frontal planes. Thus rotary motions in the horizontal plane generating impulse angular accelerations in the horizontal plane would stimulate the horizontal pair of canals (although not maximally) with the subject's head upright; but, with head tilted forward about 25 degrees, near maximal stimulation would result. Rotation in the sagittal and frontal planes would generate angular accelerations in planes almost 45 degrees from the planes of the vertical (superior and posterior) canals. It is interesting that only small differences have been demonstrated for perception of rotation about the x, y, and z body axes,[123] even though only in the z axis (rotation in the horizontal plane) are right and left canals coplanar.

Functional neurology

The reflex character of the vestibular system differs markedly from the predominantly sensory character of the auditory system; auditory pathways to the cortex are believed to have been almost fully delineated anatomically, whereas a corresponding vestibular pathway still remains to be demonstrated. The vestibular system functions automatically, mainly through motor effector mechanisms, and this very automaticity undoubtedly accounts for the fact that it is not placed in the same category as somatic sensory systems and that the vestibular organs are termed special sense organs rather than organs of special sense. The great differences in structure and modes of stimulation of the two vestibular end organs alone would indicate that these organs serve different functions by providing different information to the central nervous system; yet, when we leave the periphery, their individual identity is usually ignored by the use of the combining term "vestibular." Added to this vagueness of terminology is the frequent neglect of the major differences between vestibular influences under natural and unnatural stimulus conditions.

What follows is a brief review of vestibular system neurology based on anatomic, physiologic, and behavioral studies. The reader is referred to reports covering many important but exceedingly complex details in this area.

Anatomic aspects

Morphologic studies using classic technics have provided definitive information with important functional implications. These studies possess the great advantage of high relevance for man of investigations conducted on animals. Important contributions prior to the fourth (1952) edition of Rasmussen's *The Principal Nervous Pathways*[124] were made by Cajal, Lorente de Nó, Retzius, Burlet, Camis, and Vilstrup among others. During the past 20 years extensive morphologic studies have been carried out to which the reader is referred for details.[125-129] Here it is possible to present only a few highlights.

Fig. 6-33 summarizes present knowledge of the main pathways, revealed by classic anatomic technics, that comprise the reflex vestibular system. The vestibular nerve carries primary afferent fibers to the vestibular nuclear complex, cerebellum, and reticular formation and returns efferent fibers to the mechanoreceptors in the cristae and maculae. The vestibular nuclear complex, a term introduced by Brodal and his associates,[125] comprises not only subdivisions within the confines of the four "classic" nuclei (superior, medial, lateral, and descending) but also small cell groups (known as f, g, i, x, y, z, Sv) and the interstitial nucleus of the vestibular nerve. Cerebellar connections extend beyond the archicerebellum or classic vestibulocerebellum and include much of the vermis but

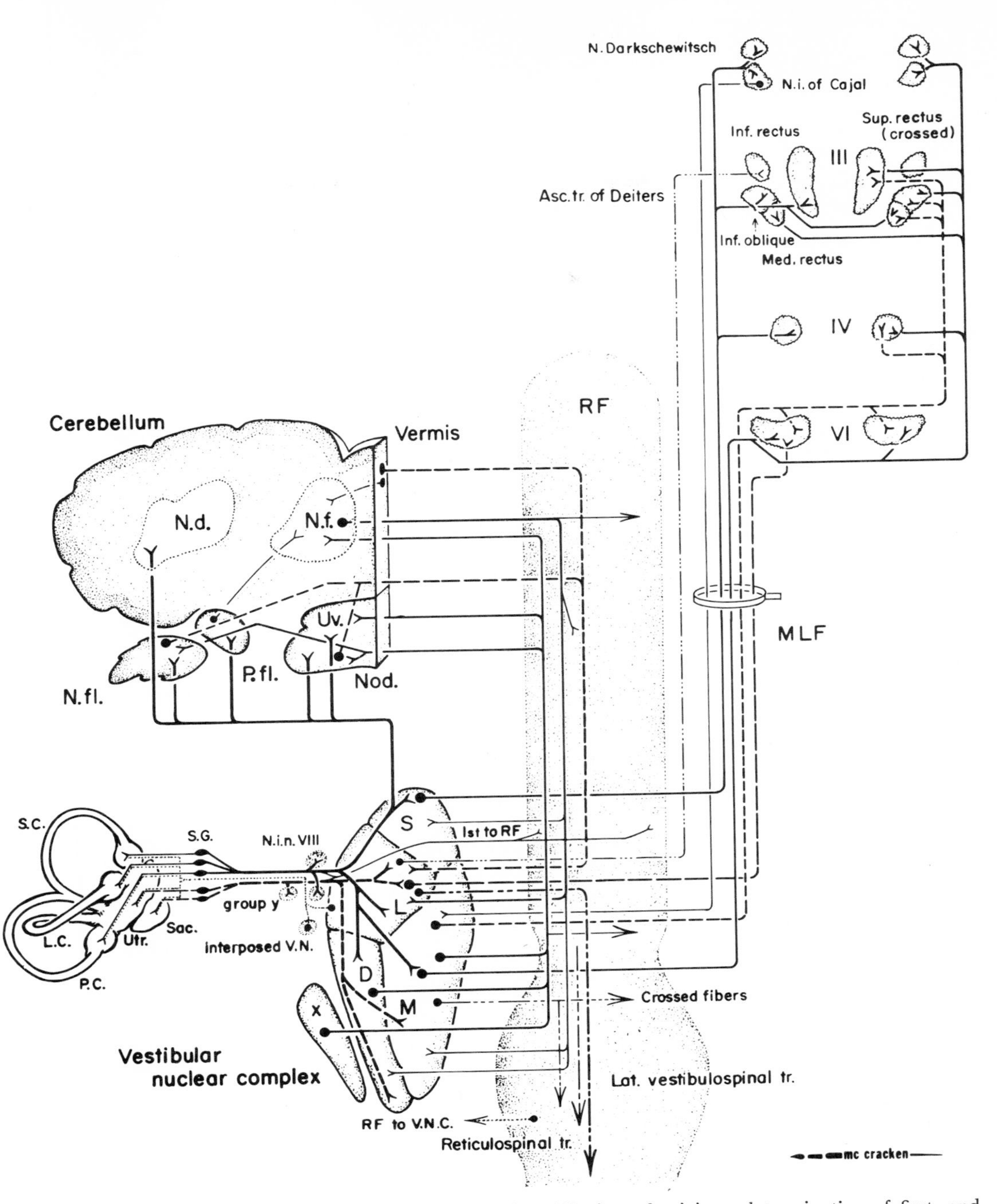

Fig. 6-33. Schema of the reflex vestibular system showing (1) sites of origin and termination of first- and second-order neurons, (2) efferent fibers of the vestibular nerve, and (3) third- (or higher) order fastigial fibers. Note absence of pathways to the cerebral cortex. *III, IV,* and *VI,* Cranial motor nerve nuclei; *D,* descending vestibular nucleus; *group y,* small cell group; *interposed V.N.,* interposed nucleus of the vestibular nerve; *L,* lateral vestibular nucleus (Deiters); *L.C.,* lateral semicircular canal; *M,* medial vestibular nucleus; *MLF,* medial longitudinal fasciculus (ascending); *N.d.,* dentate nucleus; *N.f.,* fastigial nucleus; *Nod.,* nodulus; *Nin. VIII,* interstitial nuclei (above and below afferents from cristae) of the vestibular nerve; *N.i.,* interstitial nucleus of Cajal; *P.C.,* posterior semicircular canal; *P.fl.,* paraflocculus; *RF,* reticular formation; *S,* superior vestibular nucleus; *Sac.,* saccule; *S.C.,* superior semicircular canal; *S.G.,* Scorpa's (vestibular) ganglion; *Utr.,* utricle; *U.V.,* uvula; *V.N.C.,* vestibular nuclear complex; *X,* small-celled group x.

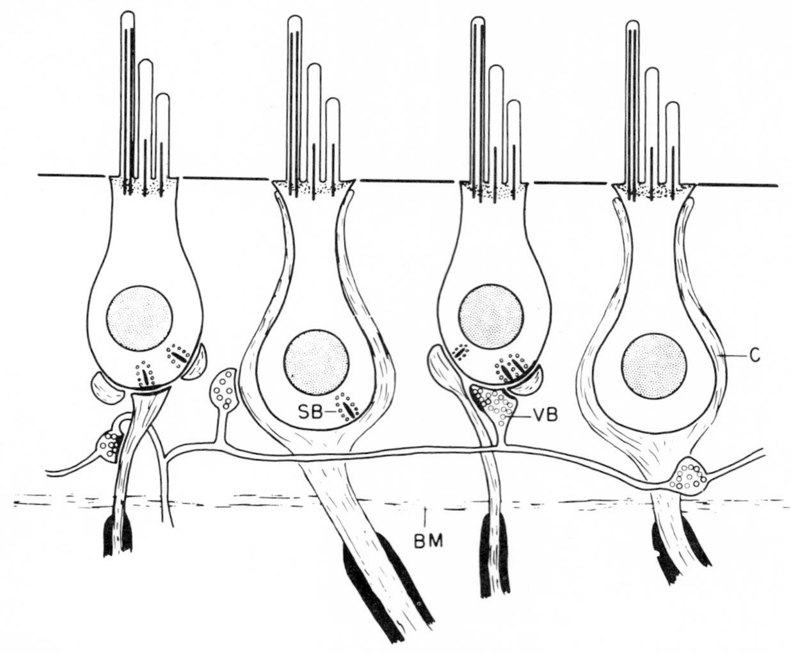

Fig. 6-34. Diagrammatic drawing showing four hair cells and their nerve endings and the relationships of vesiculated boutons *(VB)* to hair cells, chalice terminals *(C)*, other boutons, and nerve fibers in the chinchilla maculae. *BM,* Basement membrane; *SB,* synaptic bar. It is believed the efferent nerves form a sort of horizontal plexus as drawn.

not the hemispheres. Only a few first- or second-order fibers have been traced to the pontine reticular formation, but the absence of discrete nuclei may account for part of the sparsity. In general, sites of termination of primary fibers are sites of origin of secondary fibers that not only consolidate interrelations among the three major recipients of primary fibers but also ascend, descend, and cross the neuraxis.

The vestibular nerve is the smaller division of the eighth nerve coursing from the internal auditory meatus to the cerebellopontine angle, where it enters the dorsolateral aspect of the brainstem, medial and somewhat ventral to the cochlear division of the eighth nerve.

Rasmussen, whose 1952 schema[124] had not shown efferent fibers to the end organs, led the way in their subsequent discovery[127] and participated in defining their origin, course, and termination.[130] Efferent fibers apparently arise in the lateral vestibular nucleus (Fig. 6-33) and, according to Rossi and Cortesina,[131] in the nearby "interposed vestibular nucleus." They leave the brain in company with the cochlear efferent fibers but reach the end organs in company with the distal ends of primary sensory fibers and terminate as second-type vesiculated boutons (Fig. 6-34) associated with all hair cells in cristae and maculae.[130] Anatomically, these efferent fibers complete a feedback loop, a fact holding out the possibility that central influences of an inhibitory or regulatory nature can be brought to bear on the end organs.

Primary vestibular fibers have been intensively studied by many investigators, but the recent findings of Gacek,[132] who traced their course from specific end organs to specific central terminations in the vestibular nuclei of the cat, deserve special mention here.

In Fig. 6-33 it can be seen that primary canalicular neurons, after giving off short collaterals to the interstitial nucleus of the vestibular nerve, enter the brainstem, where each axon divides into an ascending and a descending branch. The ascending branches terminate in the superior vestibular nucleus and also the cerebellum. The descending branches give off collaterals to the lateral, medial, and descending vestibular nuclei. Gacek was able to trace large and small fibers from the posterior canalicular cristae to large and small cells in the superior nucleus. One should remember that in higher vertebrates the sensory epithelium of the horizontal canal was split off from the superior (vertical) canal; hence, the fibers from the two are intermixed and impossible to trace as individual bundles. Primary utricular fibers divide into ascending and descending branches, the former terminating in the lateral and medial nuclei, the latter in the medial and descending nuclei. Primary saccular fibers terminate mainly in the small-group y nucleus, with some fibers terminating in the lateral and descending nuclei. In summary, afferents from cristae and maculae are differentially distributed to sites in the vestibular nuclear complex and only canalicular fibers terminate in the interstitial nucleus of the vestibular nerve and in the superior vestibular nucleus.

Primary afferents have been traced to the flocculonodular lobe and ventral part of the uvula, which together comprise the archicerebellum, and, in addition, to the ventral and dorsal paraflocculus and to the dentate nucleus.[133] It is noteworthy that first-order neurons apparently do not reach the fastigial nucleus, as was formerly thought. Primary vestibulocerebellar fibers end as a particular type of mossy fiber not only in the flocculonodular lobe, comprising what Brodal[133] termed the classic vestibulocerebellum, but also in the ventral and dorsal paraflocculus.

As seen in Fig. 6-33 only a few primary fibers have been traced to the reticular formation.

While in general it is difficult, using classic anatomic procedures, to trace pathways in the vestibular system beyond second-order neurons, an important exception is found in the fastigial nucleus of the cerebellum, which is not a receiving site for primary vestibular fibers. In addition to its role as a relay station for cerebellovestibular fibers, the fastigial nucleus would appear to serve as a major center in the vestibular system. The fastigial nucleus receives fibers from the vermis, paramedian lobule (both sides), dorsal paraflocculus, crus II, and vestibular nuclei. The fastigial nucleus sends fibers to the reticular formation, all vestibular nuclei, mainly ipsilaterally, and especially to the lateral vestibular nucleus.

Secondary sensory vestibulocerebellar fibers, mainly from the descending nucleus but also from the medial nucleus and group x, project, chiefly ipsilaterally, to end as mossy fibers in the flocculonodular lobe, the ventral part of the uvula, and the fastigial nucleus.

With regard to cerebellovestibular fibers a distinction is made between second-order fibers and fibers from the fastigial nucleus. Second-order fibers have their origin in the archicerebellum and in vermal cortices (mainly anterior lobe) and project to the reticular formation and to all vestibular nuclei. Those fibers projecting to the lateral nucleus (which show a somatotopic arrangement) have been intensively studied by Brodal and his group.[125] Fig. 6-35 shows the preservation of an orderly arrangement from vermal cortices via the fastigial nucleus to ipsilateral and contralateral Deiters' nuclei.[133] The ipsilateral system, which involves forelimb and hindlimb, projects to rostral and caudal parts of the lateral nucleus. These fibers have their origin, respectively, in rostral and caudal parts of the anterior vermis with their relay stations in the rostral part of the fastigial nucleus. The contralateral system is analogous except that it projects to the ventral half of the lateral vestibular nucleus and crosses the neuraxis in the Hook bundle via a relay station in the caudal part of the fastigial nucleus.

Gacek[134] and Tarlov[135] have recently traced the vestibulo-ocular pathways in the cat. Five major pathways described by Gacek are shown in Fig. 6-33; all except the ascending tract of Deiters comprise the (ascending) medial longitudinal fasciculus (MLF). Two pathways in the MLF may be activated by the ascending or descending primary canalicular afferents. The pathway arising in the superior vestibular nucleus and continuing the ascending canalicular branch ascends ipsilaterally, gives off fibers to cranial nerves IV and III (some crossing to the subnucleus of the medial rectus on the contralateral side), and terminates in the interstitial nucleus of Cajal and the nucleus of Darkschewitsch. The continuation of the descending branch of the primary canalicular fibers arises in the medial vestibular nucleus and, after giving off fibers to cranial nerve VI bilaterally, ascends contralaterally, giving off fibers to nerves IV and III (subnuclei for the superior rectus and inferior oblique ocular muscles), and terminates in the interstitial nucleus of Cajal and the nucleus of Darkschewitsch.

Three pathways may be activated by the ascending and descending branches of the primary macular afferents. The pathway continuing the ascending branch itself splits in the lateral vestibular nucleus: one component courses outside the MLF as the ascending tract of Deiters to terminate in the subnucleus of the inferior rectus in cranial nerve III, the other part terminates ipsilaterally in nerve VI. The continuation of the descending branch (of the primary canalicular fibers) arises chiefly in the medial vestibular

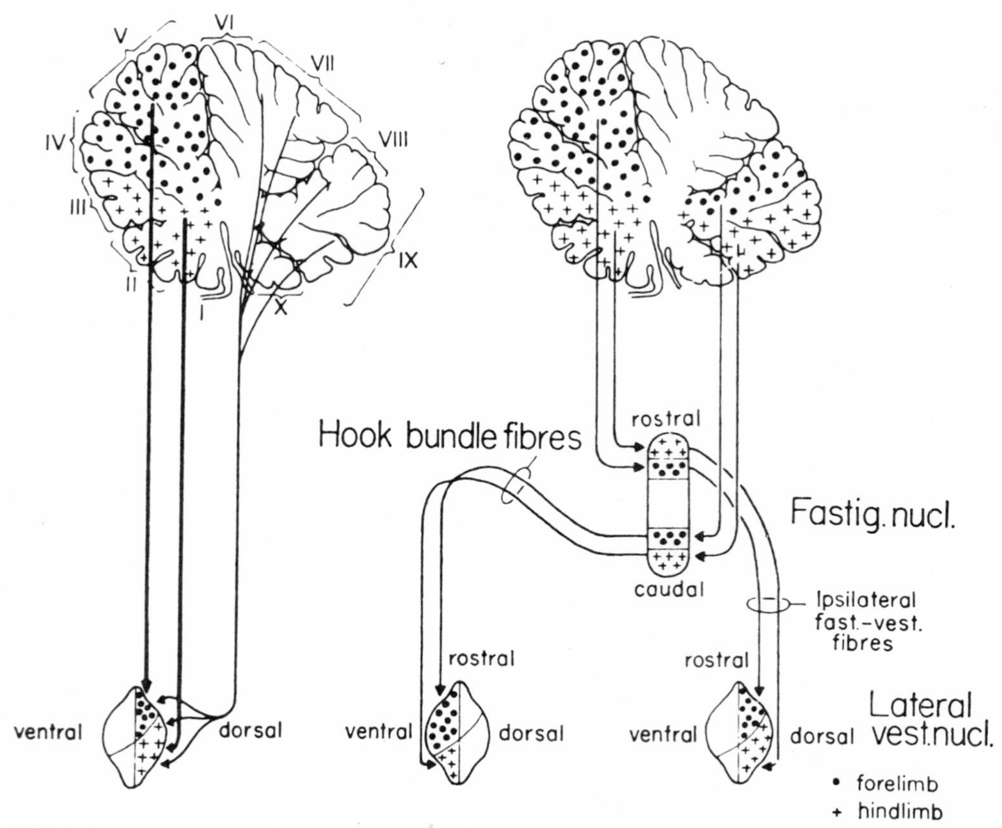

Fig. 6-35. Diagram illustrating major features in the projections from the cerebellar cortex onto the nucleus of Deiters (to the left) and (to the right) in the projections from the cerebellar cortex onto the fastigial nucleus and from this to the lateral vestibular nuclei. Note that the direct cerebellovestibular fibers and the projection from the rostral part of the fastigial nucleus end in the dorsal half of the ipsilateral lateral vestibular nucleus, while the fibers from the caudal part of the fastigial nucleus via the hook bundle supply the ventral half of the contralateral lateral vestibular nucleus. Within each of these projections there is a somatotopic localization. (From Brodal, A., Pompeiano, O., and Walberg, F.: The vestibular nuclei and their connections: anatomy and functional correlations, Ramsay Henderson Trust Lectures.)

nucleus, gives off fibers ipsilaterally to nerve VI, then crosses the midline and gives off fibers to nerves VI and IV and the subnuclei for the inferior oblique, medial rectus, and superior rectus in the oculomotor nuclear complex.

Not shown in Fig. 6-33 is a direct cerebellar (contralateral) projection to cranial nerves IV and III and ipsilateral fibers to the reticular formation that might comprise the independent vestibulo-ocular pathway readily demonstrated in physiologic studies.

The three major descending pathways, arising in the vestibular complex, comprise the lateral and medial vestibulospinal tracts and the reticulospinal tract (Fig. 6-33). The lateral vestibulospinal tract arises in the somatotopically arranged part of the lateral vestibular nucleus and projects to the length of the spinal cord, preserving its somatotopic arrangement. Nyberg-Hansen[136] has described the terminations in great detail, based on Rexed's subdivisions of the spinal gray matter, and points out that fibers in the vestibulospinal tract influence the entire cord by modulating stretch reflexes and muscular tone. The medial tract (Fig. 6-33) arises chiefly in the medial vestibular nucleus and descends bilaterally in the descending MLF, terminating in the upper half of the cord without evidence of a somatotopic arrangement.

In summary, the anatomic organization of the reflex vestibular system, while not extensive (few third-order pathways have been traced outside the cerebellum) is complex. The complexities are seen in the high degree of differentiation involving hair cells of cristae and maculae, sensory pathways, and their interconnections both within the vestibular system and with nonvestibular systems. As Brodal[133] stated: "The anatomical data available at present indicate functional differentiation between cell groups and parts of nuclei which go beyond what has so far been clarified in physiological studies."

Physiologic aspects

Until recently, only a few experiments using electric stimulation had been conducted in normal unanesthetized animals. In addition to anesthesia or decerebration, the use of abnormal stimulation raises the question as to whether the vestibular response is normal or whether a vestibular side effect is involved. Moreover, depending on the animal used, there is the question of relevance for man; for instance, physiologic consequences of labyrinthectomy may be strikingly different in animal and human subjects. All of these considerations are being taken more and more seriously—witness the use of human subjects and of technics that minimize the departure from physiologic conditions and the emphasis upon organizational levels characterized by similarity among species. Some examples of vestibular activity involving pathways that have not yet been identified by classic anatomic procedures have resulted. These involve (1) vestibular projections beyond the reflex system, (2) interactions between vestibular and nonvestibular systems, and (3) the intrinsic organization of the vestibular system.

Razumeyev and Shipov[137] have recently reviewed evidence for a vestibulocortical pathway. The reader is referred for details to this excellent summary of these authors' studies and of representative reports by other investigators.[138-143]

Early on it was established that stimulation of the vestibular nerve in the lightly anesthetized cat elicited short-latency responses (around 0.6 msec) in parts of contralateral anterior suprasylvian and ectosylvian gyri that depended on the functional integrity of the nonacoustic labyrinth. It is interesting that stimulation of the flocculonodular lobe elicits short-latency responses from the labyrinth. Changes in the activity of single neurons in cortical and subcortical regions as the result of linear accelerative stimuli have been placed into four classes by Razumeyev and Shipov,[137] based on impulse frequency: (1) increase, (2) decrease, (3) phasic (with acceleration), and (4) no change. Convergence of vestibular and nonvestibular afferent "signals" was reviewed, categorized, and summarized by these authors as follows:

> Electrophysiological experiments which have been performed to date show that the so-called specific cortical convergence of visual, vestibular auditory, and also, in all probability, somatic afferentation takes place almost exclusively in the anterior portions of the ecto- and suprasylvian gyri; i.e., in the portions of the cortex defined as *cortical projection fields* of the vestibular analyzer. Therefore, the assumption of Gorgiladze and Smirnov (1967), which states that "the vestibular cortical field is the coordination center which integrates afferent impulses from various sense organs and creates images of spatial relationships between the individual and surrounding objects of the visible world," appears to be correct.

It should be emphasized that while these physiologic studies indicate the presence of a cortical projection for the vestibular system, neuroanatomic studies have so far failed to identify a vestibular lemniscus.

Vestibular connections with the visceral nervous system have been described on the basis of physiologic experiments, but some of the early findings reported by Akert and Gernandt[139] may have to be amended insofar as they infer vestibulovagal connections. In a recent report by Tang and Gernandt[143] the vestibular "influences" were demonstrated above and not below the point at which the recurrent laryngeal nerve parts company with the vagus. Tang and Gernandt report that vestibular stimulation in cats elicited responses in recordings from the phrenic and recurrent laryngeal nerves. These responses were associated with an increase in rate and depth of respiration and an increase in blood pressure. A study to be reported by Tang[144] raises the possibility of artifacts vitiating many experiments that involve electric stimulation of the vestibular nerve.

Electrophysiologic studies have done much to demonstrate connections between vestibular and nonvestibular systems. The reader is referred to some recent studies that contribute to our knowledge of this aspect of vestibular neurology. Pompeiano[145] pointed out that in deep sleep, activity of second-order neurons in the vestibular nuclei increases phasically as a result of extralabyrinthine inputs and that this results in rapid eye movement (REM) sleep. Wilson[146] has demonstrated that impulses from peripheral nerves ascending the spinal cord facilitate cells in the lateral vestibular nucleus that are sites of origin of the lateral vestibulospinal tract. Fredrickson and Schwarz[147] investigated cells in the vestibular nuclei by means of single-unit analysis in unanesthetized cats and found that among units responsive to labyrinthine stimulation (99%) no less than 80% were also sensitive to movement of joints. There were no responses to muscle pressure and optic or acoustic stimuli, and cerebellectomy did not "grossly alter" the joint influence. Brandt, Wist, and Dichgans[148] have reported responses of cells (rabbit) in the vestibular nuclei to optokinetic stimulation.

Just as physiologic studies indicate more widespread extrinsic ramifications of vestibular influence than anatomic methods have yet demonstrated, so also the reverse is true—namely, that stimulation of peripheral receptors elicits responses that imply a degree of central intrinsic organization that has not yet been defined anatomically. Two among many possible examples must suffice here for illustrative purposes.

The vestibulo-ocular reflex arc has long been an object of high interest. Fluur[149] found that selective stimulation of the nerve from individual semicircular canals in cats yielded two types of responses. Type I responses are characterized by spontaneous activity in the extraocular muscles and conjugate deviation of the eyes, and Type II responses are characterized by the absence of spontaneous activity and of nonconjugate movements during stimulation. The differences in the types of responses were considered in the light of differences in end-organ receptors, and the functional state either of the extraocular muscles and their proprioceptive mechanisms or of the brainstem. Type I responses to selective stimulation of the nerve to the left lateral canal resulted in conjugate eye movement to the right, with activation and reciprocal inhibition of the appropriate muscles. Similar stimulation in the left anterior canal caused upward deviation of the eyes; of the left inferior canal, downward deviation; and of both left vertical canals, counterclockwise rotary deviation. It is worth noting that while stimulation of the horizontal canal caused deviation in that plane, stimulation of either vertical canal caused movement in the sagittal plane. In brief, impulses from a single canal must carry messages not only to the extraocular nuclei but also to their functional subdivisions as well. Gacek[134] raised the question whether one of the two canalicular pathways might be inhibitory and the other facilitory, thereby making unidirectional eye movements possible.

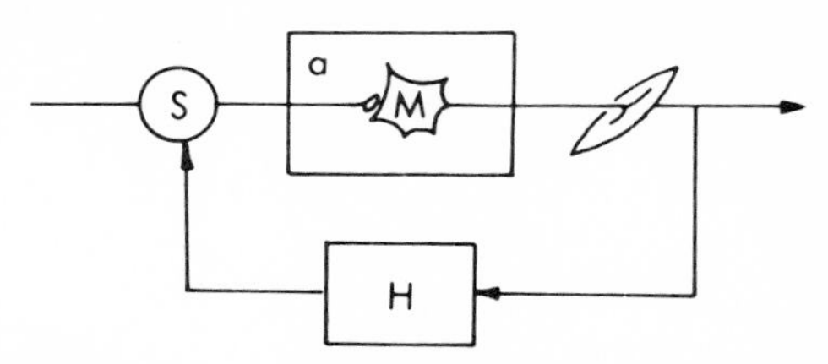

Fig. 6-36. Block diagram illustrating the development of the motor control system. *M,* Motoneurons; *S, S', S'',* sensory part of the system; *H, H', H'',* feedback loop. (From Ito, M.: The cerebellovestibular interaction in cat's vestibular nuclei neurons. In Graybiel, A, editor: Fourth symposium on the role of the vestibular organs in space exploration, NASA SP-187, Washington, D. C., 1970, U. S. Government Printing Office.)

Cortical, cerebellar, and reticular influences on vestibular activity have generally been regarded as inhibitory, but this broad generalization must be modified in the light of recent studies. For instance, Ito[150] has proposed a model (Fig. 6-36) based on motoneurons in combination with certain receptors and muscles that would form a simple control system with a negative feedback loop. With the insertion of the cerebellum into this control system, a more complex performance is possible (Fig. 6-37, *A*). Ito reports that insertion of the cerebellar nuclei between the cerebellar cortex and the brainstem may modify the ability of the cerebellum-brainstem unit in two respects:

1. The integration of extracerebellar excitatory inputs with the inhibitory Purkinje cell signals is performed at the level of the cerebellar nuclei (see Fig. 6-37, *B*).
2. A reverberating circuit may be formed between neurons of the cerebellar nuclei and those originating in certain cerebellar afferents (see Fig. 6-37, *C*).

Ito also pointed out that there is anatomic evidence to suggest a reverberating connection between the descending vestibular nucleus and the fastigial nucleus, between the paramedian reticular formation and the fastigial nucleus, and between the pontine and the lateral cerebellar nuclei. These connections, according to Ito, would favor the maintenance of a certain standard of activity in the cerebellum-brainstem system, or the bias around which the dynamic characteristics of the system may be rendered optimum.

Behavioral aspects

Vestibular responses elicited in healthy persons have important neurologic implications. Indeed, our point of departure might well have been reversed by discussing behavioral rather than

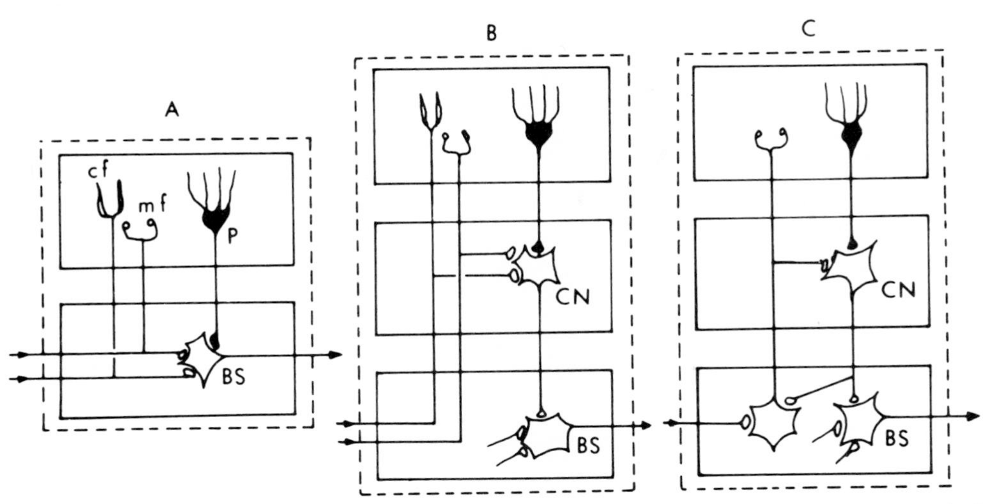

Fig. 6-37. Diagrammatic illustration of variation in cerebellar corticosubcortical connections. *BS,* Brainstem; *CN,* cerebellar nuclei.

anatomic aspects first, because behavioral phenomena indicate functional connections that demand anatomic explanation. Moreover, man's nervous system may contain elements that are redundant or vestigial in terms of natural living conditions today; conversely, under unnatural conditions (strong motion environments) we must seek explanations for phenomena that only recently have become parts of our lives.

In addition to the distinction between behavioral phenomena elicited under natural and unnatural stimulus conditions, vestibular side effects may be categorized according to the immediacy and nature of the response. Immediate reflex responses may be divided into those that represent near normal responses—such as perception of small magnitudes of the oculogravic illusion (see below)—and those that are frankly abnormal—such as nystagmus (see p. 240). Delayed responses comprise epiphenomena best known under the term "motion sickness." Inasmuch as vestibular side effects will be discussed in later sections, here it will suffice to point out a few examples that illustrate neurologic aspects not mentioned in the preceding sections.

Under favorable conditions, thresholds for the perception of rotation by subjects passively exposed for 10 seconds to angular acceleration about the vertical (z) axis are in the range of about 0.17 deg/sec^2.[123] On sudden stop, after constant rotation at 1 rpm for 60 seconds, the perception is immediate in behavioral terms. Persons with bilateral loss of vestibular (canalicular) function do not perceive rotation during a sudden-stop deceleration from 30 rpm to zero velocity.

When a person is exposed to a change in direction of a gravitoinertial force relative to himself, he experiences an apparent reorientation of his body, more or less in accord with the direction of this force, and objects in the visual field likewise appear to assume a new position in space. Important distinctions must be made between the postural and visual effects. Mechanoreceptors in otolith organs and the somatosensory receptors serving touch, pressure, and kinesthesis contribute to the feeling that one is being tilted with reference to his support. These sensory systems are functioning in an expected manner that differs from physiologic functioning mainly in the discrepancy between the change in intensity (magnitude) and the change in direction of the gravitoinertial resultant vector. In other words, these sensory systems are performing in a near normal manner in signaling the alignment of the body to the change in the gravitoinertial force upright. Inasmuch as the subject is constrained in his seat, he feels tilted with respect to his support, with all portions of his body retaining their same relative positions, and, given the opportunity to set a rod that can swivel about its center in the plane of the changing direction of force, he will set the rod approximately in accord with the gravitoinertial upright or horizontal (Fig. 6-38, *A*).

A corollary type of observation also indicates that, within certain functional limits, the otolith organs respond with precision to the gravito-

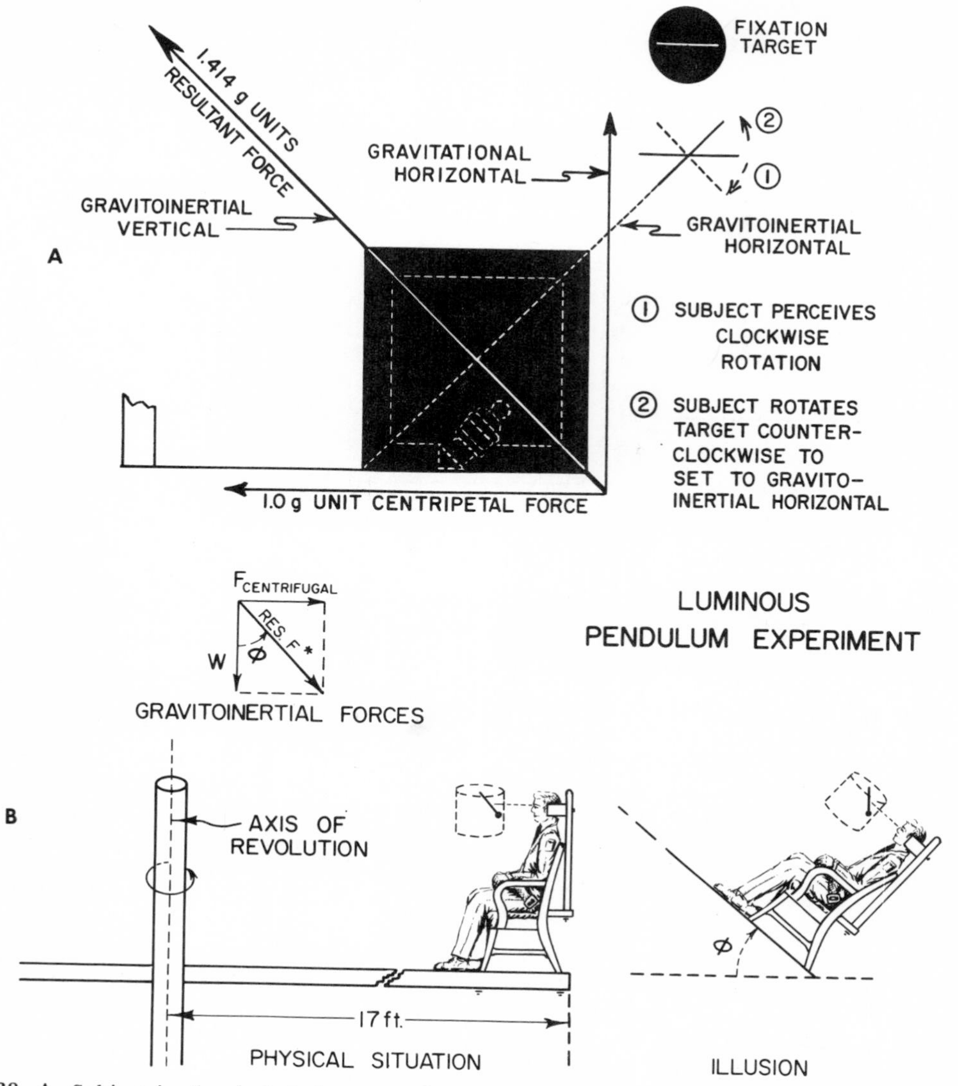

Fig. 6-38. A, Subject in the dark facing away from direction of rotation fixating a luminous line regards himself as tilted in an upright room. **B,** A luminous pendulum experiment demonstrating the accuracy of the otolith organs in their response to the gravitoinertial resultant force.

inertial resultant force. In Fig. 6-38, *B*, is shown the physical situation of the experiments on the original Naval Air Station, Pensacola, centrifuge. The subject, seated in a fixed position in the dark, observed a luminous string that carried a small steel ball at the free end to make the system an effective pendulum. The pendulum was protected from air currents by an airtight transparent cylinder, to the top of which the upper end of the pendulum was attached. At all times during constant rates of revolution, up to an angle ø of 45 degrees (deviation of pendulum from the true vertical), the luminous pendulum appeared to point *straight downward*. The upper limits of the phenomenon were, however, not explored. In other words, the otolith organs were responding to the direction of the gravitoinertial resultant force (**F★** or, in **G** units, **G★**). At the same time the subject's entire body felt as though tilted in space by the same number of degrees. This was a demonstration of the power of otolithic influences, although TPK receptor systems may also have contributed to the effect.

This feeling of being tilted is just as compelling

as orientation to the gravitational upright, hence is illusory only in the sense that it does not correspond to earth reference.

The visual component of the illusion, termed the "oculogravic illusion," has its genesis mainly in the otolith organs. The small contribution from somatosensory inputs can be demonstrated when subjects are exposed to a change in direction of the gravitoinertial vertical during head-out water immersion.[151] The apparent change in the visual world is properly termed an illusion, although it represents an attempt to make the visual upright conform to the gravitoinertial force upright.

The postural and especially the visual effects have been studied mainly by exposing subjects in darkness to centripetal force on a human centrifuge. If the change in direction of the gravitoinertial vector with reference to the subject occurs within a few seconds, the behavioral effects (both postural and visual) lag behind. This so-called lag phenomenon[105, 106] indicates sluggish responses under the circumstances, but they may be caused by an artifact, namely concomitant exposure to angular and linear accelerations in the initial (dynamic) phase of exposure on the centrifuge. There is some direct and indirect evidence that the lag phenomenon is an artifact and not a result of a sluggish otolithic response.

Recent studies by Brandt, Wist, and Dichgans[148] demonstrate what have been termed pseudo-Coriolis effects, namely, a visually induced perception of self-rotation and a pseudo-Coriolis illusion. The Coriolis (or oculogyral) illusion is readily perceived under favorable circumstances in a room rotating at constant velocity if a person rotates his head out of the plane of the room's rotation. The pseudo-oculogyral illusion can be elicited by substituting rotation of the visual environment (a striped drum, for example) for rotation of the "room." Rotation of the head in the stationary room is essential, thus implicating the vestibular organs, although the head movements generate normal accelerative stimuli. After abolition of the visual stimulus, abnormal effects can be elicited for as long as 30 seconds. That the probable sites of interaction between the normal vestibular inputs and abnormal visual inputs are in the medial and lateral vestibular complex and adjacent reticular formation was demonstrated by single fiber recordings in rabbits. Some fibers responded not only to accelerative but also to optokinetic stimuli. In this connection it is important to recall, however, that subjects who have never perceived light nevertheless may be highly susceptible to motion sickness when exposed to angular (Coriolis) accelerations,[152] proving that symptoms may be elicited in the absence of any visual memory and possibly in persons with defective visual pathways.

Motion sickness (discussed in the next section) represents a constellation of delayed epiphenomena precipitated by repetitive vestibular sensory inputs that are either abnormal or, if normal, encounter an abnormal central neural integrative pattern. Cardinal symptoms have their immediate origin in nonvestibular systems; hence, first-order responses, at least, must reach cell groups via pathways (presumably in the brainstem reticular formation) not used under natural stimulus conditions.

In summary, the vestibular system under artificial stimulus conditions readily evokes responses that range from near normal (the oculogravic illusion) to the absurd (motion sickness). Preferential pathways and unusual interactions with nonvestibular systems deserve study for scientific reasons and for potential practical benefits.

Input-output relations

There is no vestibular counterpart to the elegant studies on vision and hearing that are based on fine control of stimulus and precise measurement of response in human subjects. It is virtually impossible to conduct systematic investigations using natural canalicular and otolithic stimulus patterns. The reason is that natural activities greatly limit the investigator in terms of stimulus manipulation and measurement and in the use of specific meaningful responses that are available for measurement. Thus the investigator must resort to unnatural stimulation of canals or otoliths, or both, which usually elicits abnormal responses that can be measured. Advantages gained by these means are offset in greater or lesser degree by individual differences with respect not only to susceptibility to these side effects but also to change in susceptibility as a consequence of adaptation.

Unnatural stimulus conditions also prevail under many conditions in modern life, especially in connection with the use of conveyances in traveling. This has generated a second important category of laboratory studies. Under these stimulus conditions great individual differences are manifested not only in susceptibility to reflex vestibular disturbances and motion sickness but also in the rate of acquisition and decay of adaptation effects. Although these studies are designed to obtain information useful in predicting responses under specific operational conditions, these probes may also reveal important functional characteristics of the vestibular system. Thus procedures play

g = Acceleration Of Gravity
A cen = Centripetal Acceleration
A cor = Coriolis Acceleration
Vt = Tangential Velocity
Vr = Radial Velocity
ω = Angular Velocity Of Rotating Room
ϕ = Angle Between Gravitoinertial And Gravitational Upright

Fig. 6-39. Responses to the force environment in a rotating room. Crewmen 1 and 2, in articulated molds supported by air-bearing devices, are "walking on the wall," simulating the orientation in a rotating spacecraft. Crewman 2, walking in the direction of rotation, becomes somewhat heavier because his angular velocity, hence, centripetal acceleration, is increased and sums with the Coriolis accelerations generated. Crewman 1, walking opposite to the direction of rotation, becomes somewhat lighter because his centripetal accelerations are decreased and the Coriolis accelerations must be subtracted. Crewman 3, walking toward the periphery of the room, is exposed to increasing levels of centripetal acceleration and constant levels of Coriolis accelerations. Crewman 4, standing, is demonstrating two phenomena: (1) as he moves his arm or leg sideways, a tendency to veer backward, the so-called giant-hand effect; (2) as he makes his head move (rotates) in the plane of the room's rotation, cross-coupled angular accelerations and illusions are not generated, a so-called free movement; (3) the angle phi (ϕ) indicates the change in direction of the gravitational upright caused by centripetal force. Crewman 5 is making a head movement out of the plane of rotation, which does generate cross-coupled angular accelerations, producing characteristic illusions described. (Courtesy Dr. D. B. Cramer.)

an important role and the brief discussion to follow will center around (1) manipulating the motion environment, (2) measuring responses, and (3) selecting and indoctrinating subjects.

Procedures

Among the purposes served by different procedures are (1) determining the functional integrity of the canalicular and otolithic system, (2) measuring the response characteristics of the two end organs, (3) measuring susceptibility to side effects, (4) measuring the rate of acquisition and decay of adaptation effects, (5) identifying the mechanisms underlying vestibular responses, and (6) evaluating the usefulness of countermeasures for the prevention of side effects.

It is helpful to keep in mind that unnatural stimuli may be arbitrarily placed in three categories:

1. A "near normal" stimulus that does not depend on disturbing the vestibular system to elicit responses
2. Stimuli that elicit immediate reflex phenomena by disturbing the vestibular system
3. Stimuli that elicit delayed epiphenomena (motion

sickness) that are superimposed on whatever immediate reflex phenomena are present

The semicircular canals are so structured that they can be selectively stimulated (that is, almost to the exclusion of otolithic and TPK receptors) using rotary devices of different kinds. A variety of rotating chairs and larger devices are available; some permit programming of the stimulus profile and the use of physiologic sensors. More elaborate devices permit programming about the x, y, and z axes.[123]

The generation of cross-coupled angular accelerations (discussed on p. 217) usually involves the active rotation of the head (and body) in combination with rotation at constant velocity in a chair, cubicle, or rotating room. The head movements must be out of the plane of rotation of the device. These head movements may be standardized by means of a tape-recording furnishing the cadence, while the angular velocity of the room is manipulated to vary the level of stress. The otolith apparatus is always implicated, but its level of activation is far less than the canalicular component and often is not mentioned. Rotating chairs, cubicles, and rooms have been used.

A slow rotation room (SRR) in a laboratory setting (Fig. 6-39) provides an excellent facility for the study of vestibular side effects. Its versatility as a human centrifuge and as a device for generating rotary motions depends on its performance characteristics and the opportunity for measurements to be made with sophisticated laboratory equipment. The cardinal advantages include (1) its habitability, (2) the experimenter's control over accelerative stimuli, (3) easy manipulation of the visual environment, and (4) the fact that observers (or experimenters) can be with the subjects. Factors contributing to habitability are housekeeping and entertainment facilities, the fact that the room if fully enclosed does not appear to rotate, and the capacity to avoid stressful accelerations. Near the center of the room, with head fixed, one is scarcely aware that the room is rotating, and, by limiting head movements to those in the plane of rotation, the experimenter can cautiously move about. It is apparent that the stressful stimuli to the vestibular organs can be avoided in the SRR, which differs from stimulus conditions such as those that may be experienced on turbulent seas or in turbulent air that cannot be avoided. A pitch-and-roll device has been mounted in an SRR for passive exposure of the subject (Fig. 6-40) to cross-coupled angular accelerations.[153]

Stimulation of the otolith organs (selective in the sense that the canals are not stimulated) is possible using devices that generate rectilinear accelerations[154] or by exposing the subject to a rotating linear acceleration vector in a counter-rotating cubicle or room.[155] Selective stimulation of otolithic receptors can also be achieved by the generation of centripetal accelerative forces at constant angular velocity; periods of change in angular velocity that stimulate the semicircular canals must, however, be taken into account. Concomitant stimulation of TPK receptor systems cannot be avoided in these devices. Using a heavy-duty human centrifuge, subjects can be exposed to head-out water immersion, thereby greatly reducing stimulation of TPK receptor systems.[151] Linear accelerations add vectorially with gravity, producing a change in direction and magnitude of what has been termed the gravitoinertial vertical.[156] See p. 211 for a more complete discussion.

Exposing a person to rotation about a body axis other than in the earth-vertical stimulates the otolithic receptors; at constant angular velocity the canals are, in all likelihood, not stimulated.[157-159] An off-vertical rotating (OVR) chair (Fig. 6-41) can be used in either the tilted or vertical mode, and its usefulness in either mode is determined in large part by its performance characteristics and by its control instrumentation. In the vertical mode, it can be used to test the functional integrity of the semicircular canals and for conducting provocative tests for susceptibility to reflex vestibular disturbances and motion sickness.

In the off-vertical mode the canals are not stimulated at constant velocity, but the changing position of the otolith organs with respect to gravity results in a continually changing stimulus to the otolith organs. Provocative tests for susceptibility to reflex vestibular disturbances and motion sickness are easily conducted, and there is evidence that certain dynamic response characteristics of the otolith organs can be studied.[159]

Among the devices for manipulating the gravitoinertial force environment, none equals in elegance the transition into weightlessness (parabolic or orbital flight), which removes the gravitational load on the receptors in the macular plates of both the utricle and saccule. It has no corresponding effect on the semicircular canals; indeed, the stimulus to these organs is affected little if at all. The effect of weight on the TPK mechanoreceptors is lifted, although in sharp contrast to the otolith organs these receptors can still be stimu-

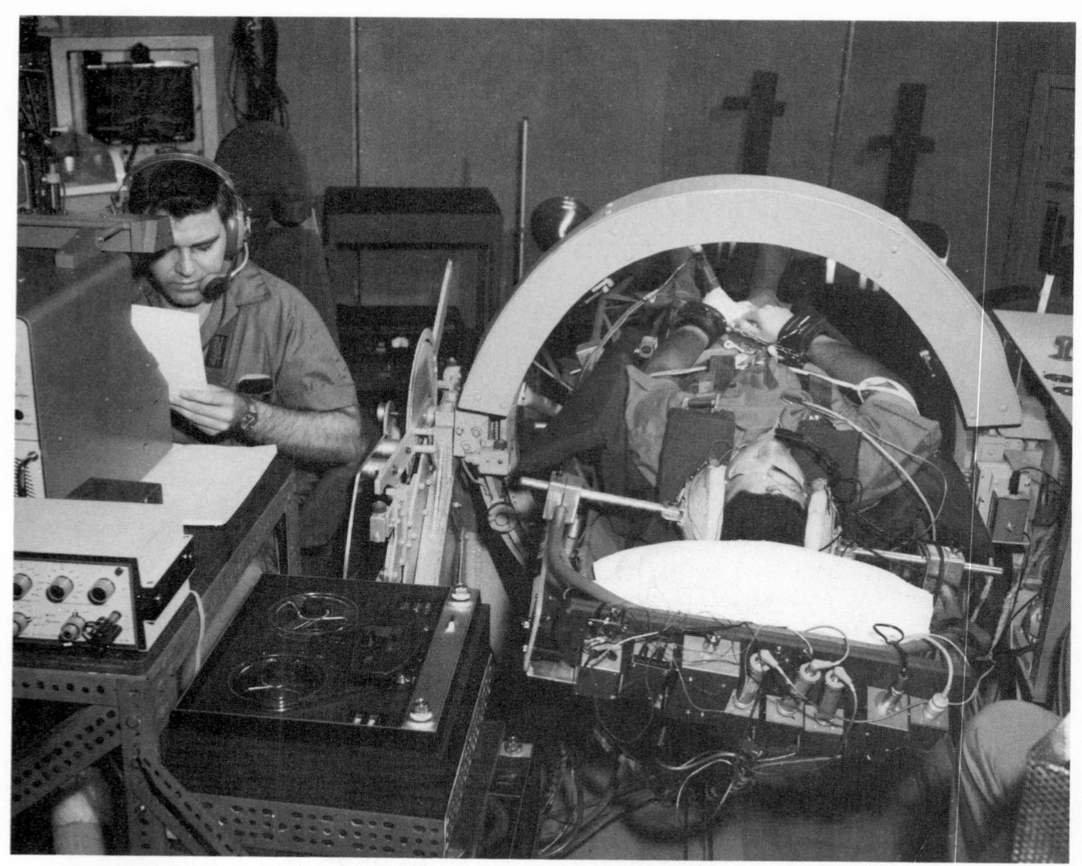

Fig. 6-40. Pitch-and-roll device mounted in a slow rotation room that permits passive exposure to cross-coupled angular accelerations. The advantage gained in conducting experiments that require that the subject remain passive (such as the effect of sleep on susceptibility to motion sickness) is offset in part by technical difficulties.

Fig. 6-41. For legend see opposite page.

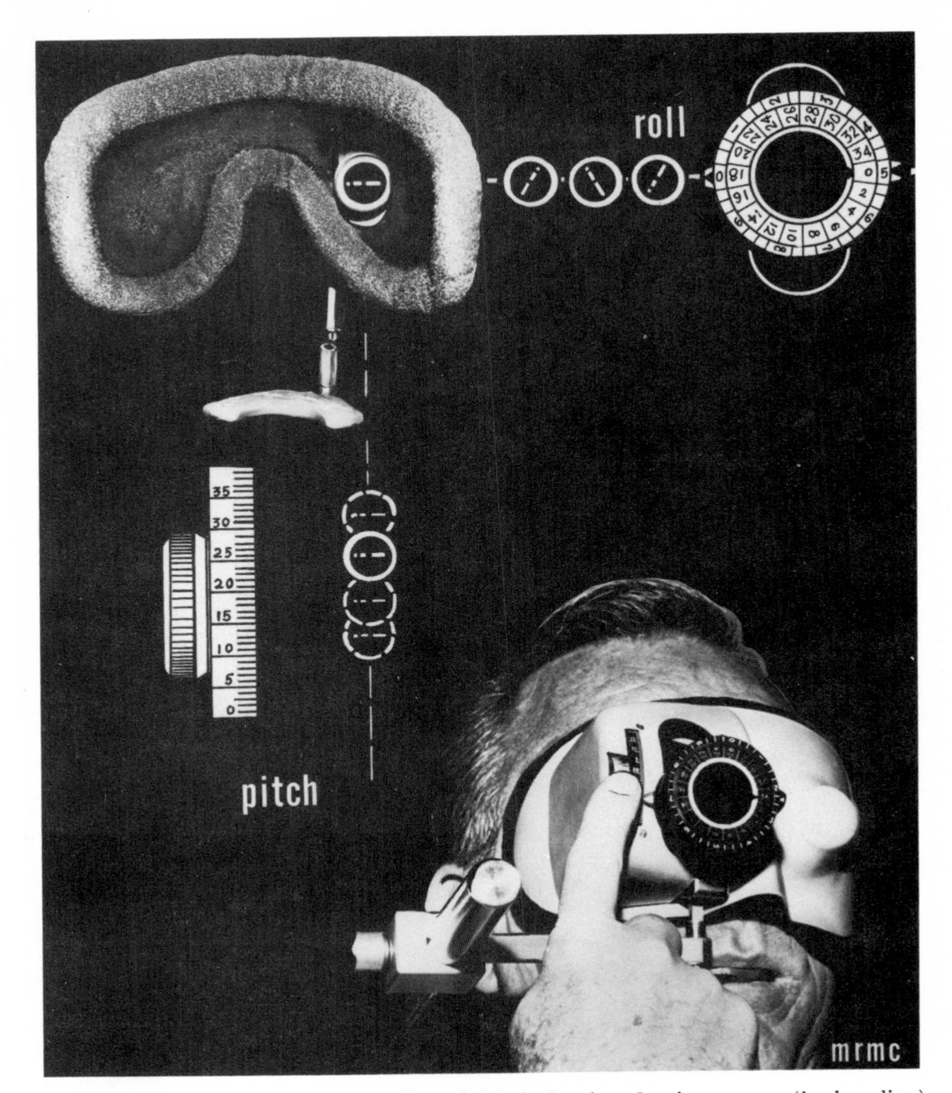

Fig. 6-42. Goggle device. Upper left, View from behind showing fixation target (broken line) and custom-fitted bite-board (below). Upper right, Changes in "roll" position of fixation target and readout. Lower right, Goggle device in use, finger on knurled knob for adjustment of target in pitch plane. Lower left, Changes in target position in the pitch plane.

Fig. 6-41. Off-vertical rotating chair device with recording equipment. The chair may be used in the upright mode for stimulating the semicircular canals and in the off-vertical mode (at constant angular velocity) for stimulating the otolith organs. Instrumentation permits independent programming of angular velocity and angle of tilt. (From Graybiel, A., and Miller, E. F., II: Aerospace Med. **41**:407-410, 1970.)

lated by external or internal mechanical forces. A distinction between the gravitoinertial and agravitoinertial (nongravitoinertial) force environments is obligatory (see p. 217) if one is to deal with man's biologically effective force environment in the weightless or near-weightless spacecraft. Studies aloft might open the way toward making a similar distinction under terrestrial conditions.

A large body of information has been obtained by having subjects set a visual target or rod either to internal or external spatial coordinates. Many investigators since Aubert have explored the effects of tilting their subjects in the gravitational field, and with the introduction of the human centrifuge it has become possible to manipulate the direction of the gravitoinertial force vector with respect to the subject. There are interesting and important performance differences between the responses obtained with tilt and with centrifugation, not all of which have been satisfactorily explained.

Devices for studying the visually perceived direction of space in the absence of patterned visual surroundings have long been in use, but the principle underlying such devices is so simple that its elegance is seldom appreciated. The basic element is a visual target, usually a lighted line pattern on a dark background that in itself affords no clue to spatial orientation, yet it can be manipulated to indicate a direction in space. The apparatus typically designed for this purpose incorporates a relatively large target system that is remote from the subject and moved by means of mechanical, electric, or hydraulic linkages. These structural disadvantages were overcome by the development of a miniature device that can be worn as a goggle.

The goggle device (Fig. 6-42), described elsewhere in detail,[160] consists essentially of a collimated line of light in an otherwise dark field. This line can be rotated about its center by means of a knurled knob. A digital readout of line position is easily seen and is accurate within $\pm$ 0.25 degrees. With the aid of the goggle device the oculogyral illusion can be used either to determine response thresholds to stimulation of the horizontal pair of canals or to obtain cupulograms and to measure the oculogravic illusion. The goggle device was used by astronauts prior to, during, and following Gemini flights V and VII[161] to compare measurements of "horizontality" under gravitoinertial and agravitoinertial stimulus conditions.

A rod-and-sphere device has been fabricated (Fig. 6-43) to aid in measuring the direction of personal and extrapersonal space, using nonvisual cues.[162] The rod, held to the sphere by magnetic attraction, slides over the sphere as the subject attempts to indicate the "upright." The absence of any constraint limiting the setting of the rod in a particular plane constitutes its novel feature.

Among the many responses that may be elicited under abnormal stimulus conditions, sensations, illusion, and eye motions have received the most attention, although other responses such as past pointing, postural disequilibrium, and electromyography have been employed. Response thresholds for perception of canalicular responses can be determined more accurately than the thresholds for perception of otolithic responses because man does not have a good substitute for the canals, but TPK systems can substitute in part for the otolith organs.

The word "nystagmus" (Gr. *nystagmos,* drowsiness; Fr. *nystazein,* to nod in sleep) refers to involuntary rapid oscillations of the eyeball. Many prefer the more convenient definition, namely, eye movements characterized by a slow deviation in one direction followed by a quick return. Spontaneous nystagmus may result from disease, defect, or functional disorder but rarely occurs in normal persons and then under certain special conditions. By contrast nystagmus can readily be triggered by optic or vestibular stimulation in normal persons. Vestibular nystagmus may be elicited by exposure to unusual motion environments in conveyances or in the laboratory. Mechanical, galvanic, and especially thermal stimulation are also used to stimulate the vestibular system. The resulting nystagmus may be observed or recorded and, with the advent of electronystagmography,[163, 164] it became possible to easily distinguish between nystagmus and other eye movements and to establish criteria for systematic measurement and classification.[165-169] Nystagmus is an important, reliable clinical sign, and nystagmography has now reached the status of a subspecialty in the area of neuro-otology.

Electronystagmography is based on the demonstration that the eyeball functions as a dipole in which the cornea is positive and the retina (when the biopotential is generated) negative. When the gaze is directed straight ahead, a single coordinate system may represent the eyeball and its electric field within the framework of the coordinates of the skull. When the eyeball is rotated, the coordinates representing the electric field change

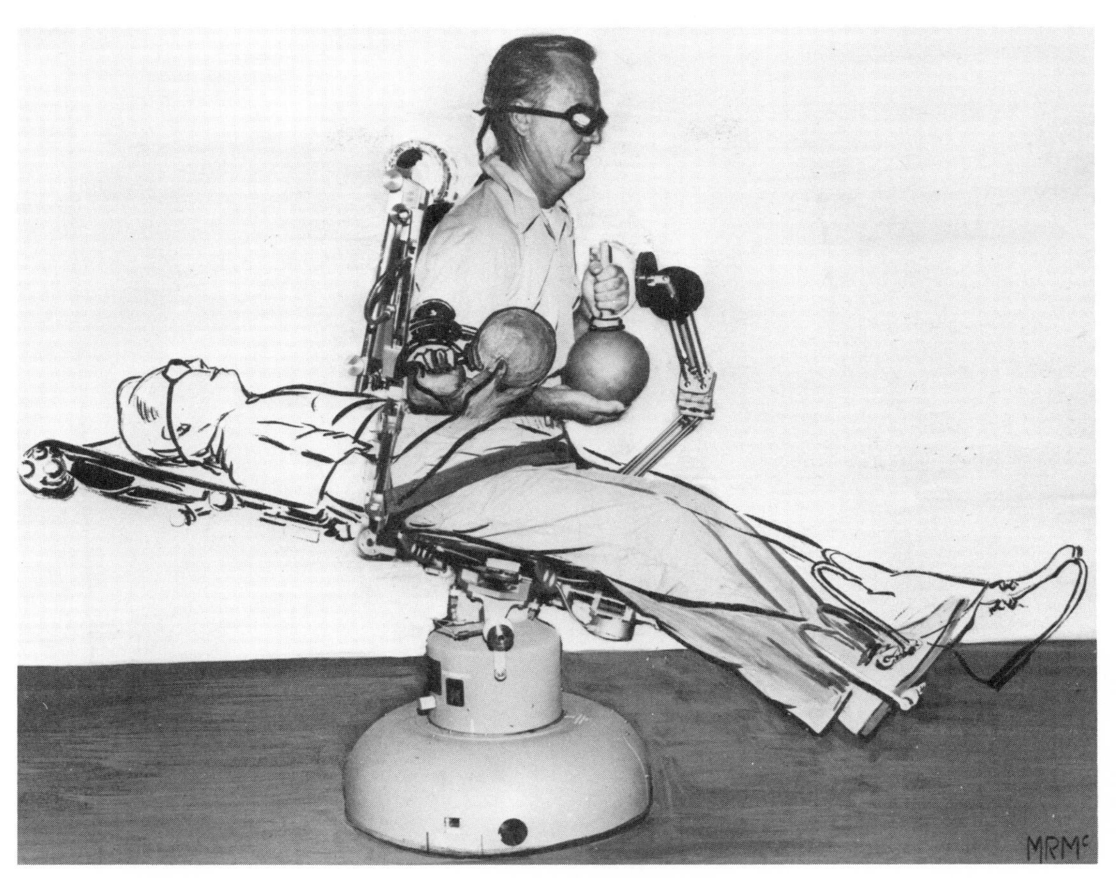

Fig. 6-43. Rod-and-sphere device for measuring nonvisually perceived personal and extrapersonal space. The device consists of a pointer held by magnetic attraction to a hollow steel ball. With eyes covered, the subject slides the rod over the ball until spatial localization is achieved; note absence of constraints that are usually provided to maintain the indicator in one plane. The readout is semiautomatic.

with respect to the fixed coordinates of the head. If electrodes are attached above and below (z axis) and on the right and left (y axis) of the eye, with leads connected to a suitable preamplifier and recording unit, deviations of the eye other than in the frontal plane can be detected and recorded. The changes in potential registered are a function of the tangent of the angle of rotations of the eye about the z and y axes.

Vestibular nystagmus. A large body of medical literature and much of the experimental work dealing with induced vestibular nystagmus involves the horizontal (lateral) pair of semicircular canals. There are important differences between a nystagmus elicited by stimuli applied to the semicircular canals and that induced by otolithic stimulation, however, so that separate descriptions with illustrative nystagmograms are necessary. Canalicular nystagmus (Fig. 6-44) shows some important stimulus-response relations when the horizontal semicircular canals are stimulated by exposing a subject to counterclockwise rotary acceleration about the vertical axis. Movement of the cupula is opposite to the direction of turn. This movement initiates the slow phase of the nystagmic beat that represents the compensatory movement that contributes to the maintenance of visual fixation. The nystagmus is usually identified, however, by the fast phase, which represents a correcting movement of central origin; visual perception is usually lacking during the quick phase. Note that the oculogyral illusion and perception of rotation are in the direction of turn but that the subject "past points" in the opposite direction.

Fig. 6-45 shows on idealized right-beating horizontal nystagmus and the conditions under which nystagmograms are commonly recorded. Single

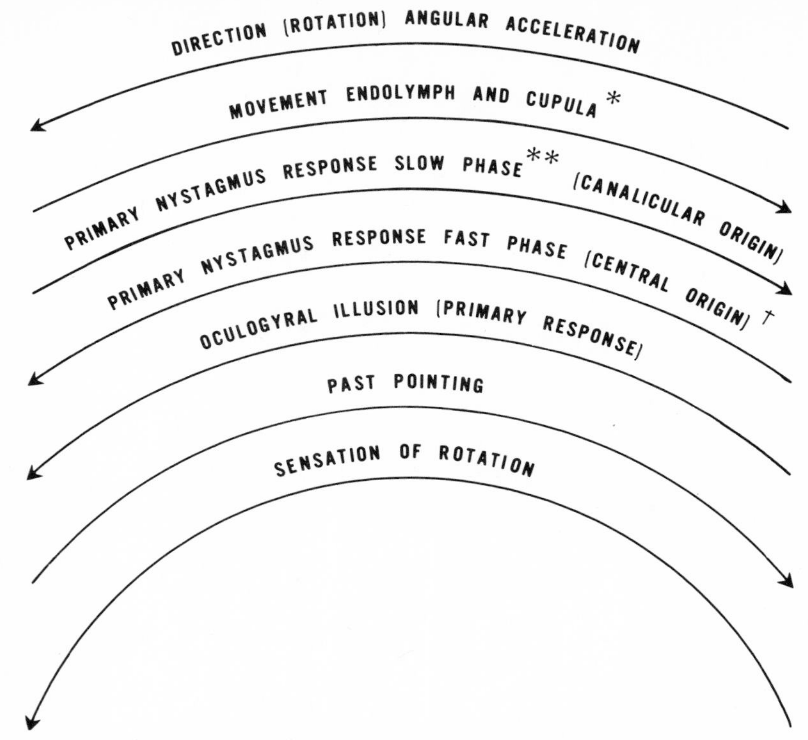

* LEFT SEMICIRCULAR CANAL AMPULLOPETAL FLOW
RIGHT SEMICIRCULAR CANAL AMPULLOFUGAL FLOW
SAME FLOW RESPONSES INDUCED BY THERMAL STIMULATION; L EAR HOT, R EAR COLD

** THRESHOLDS FOR RESPONSE DIFFERENT FOR HORIZONTAL AND VERTICAL (H AND V) SEMICIRCULAR CANALS. AIDS IN MAINTAINING SAME DIRECTION OF GAZE WHILE HEAD TURNS

† FAST PHASE HAS ITS OWN THRESHOLD FOR RESPONDING BUT SAME FOR H AND V SEMICIRCULAR CANALS

Fig. 6-44. Horizontal semicircular canals stimulus-response relations.

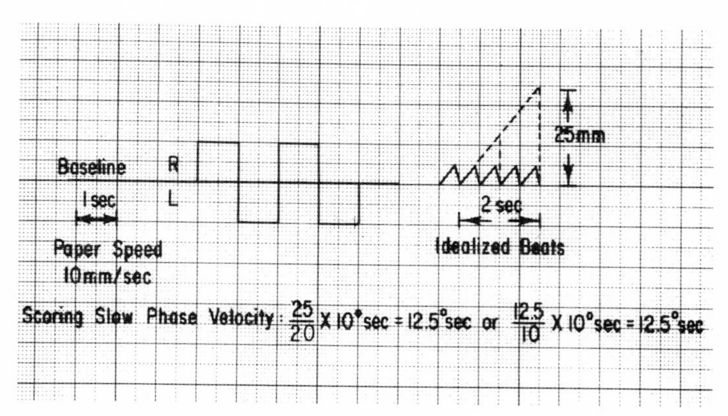

Fig. 6-45. Calibration of eye movements to right and left of straight (10 degrees of arc = 10 mm) and method for measuring and calculating slow-phase velocity.

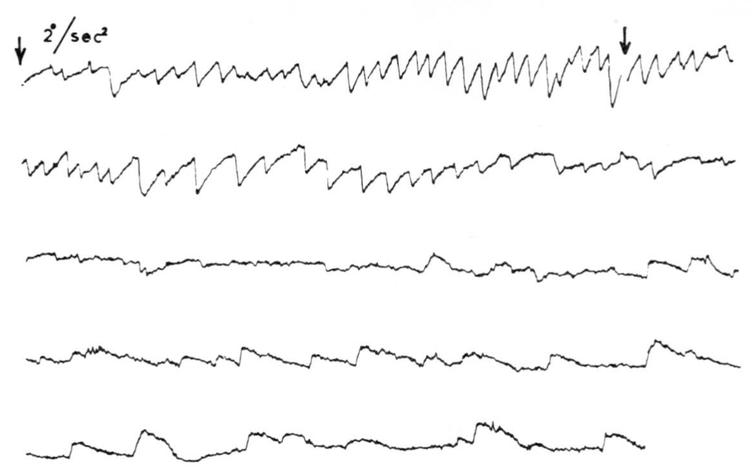

Fig. 6-46. Electronystagmogram showing primary left-beating nystagmus in response to angular acceleration for 25 seconds (between arrows) followed successively by a gradual decline (strip 2), disappearance (strip 3), and a secondary right-beating nystagmus (onset, end of strip 2). (From Aschan, G., and Bergstedt, M.: Acta Soc. Med. Upsaliensis **60**[3-4]:113-122, 1955.)

beats reveal different forms and direction; measurements may include amplitude and duration of slow and fast phase; calculations may include displacement, velocity, and even acceleration of both phases. Slow phase velocity reflects cupulo-endolymph mechanics and is regarded as the single best measure of the intensity of the response. A means of calculating the velocity of the slow phase is also shown in Fig. 6-45; when horizontal and vertical beats are recorded simultaneously, the plane and magnitude of the actual deviation of the eye can be calculated because these represent resultant values that can be calculated from measurements of the horizontal and vertical components.

The entire recording may be analyzed in terms of rhythm, duration, and number of beats as well as the measurements that can be made on single beats and can be plotted as a function of elapsed time or of arbitrary time intervals. The duration is considered less important an indicator than frequency, which reflects mechanisms of central origin, and far less important than slow phase velocity. Henriksson in 1955[170] introduced a method for obtaining a continual readout of slow-wave velocity. More recently, computer-assisted methods[171] have greatly reduced the time required in analysis of the recordings, hence in the usefulness of electronystagmography.

The nystagmogram in Fig. 6-46 illustrates a common finding, namely, the spontaneous appearance of a secondary nystagmus following the elicitation of a primary nystagmus.[172] In this instance the primary left-beating nystagmus was elicited by exposing the subject for 25 seconds to an angular acceleration of 2 deg/sec² (interval between the arrows in strip 1). Following the angular acceleration the subject continued to rotate at constant angular velocity throughout the remainder of the test. With the abolition of the stimulus the cupula returned to its neutral position by virtue of its elasticity, and this was associated with a decline and finally the disappearance of left-beating nystagmus. After an interval of about 25 seconds a right-beating nystagmus appeared. This secondary nystagmus is, in all likelihood, of central origin and in the nature of an adaptive influence opposing the primary nystagmus. Fig. 6-47, *A*, shows the change from left- to right-beating nystagmus when a normal subject was exposed to a sinusoidal acceleration with a frequency of 0.10 cps and a peak acceleration of 60 deg/sec².[173] Note the characteristic decline in frequency and amplitude of left-beating nystagmus. This is followed by a pause marking the transition that occurs shortly after the change in stimulus acceleration from counterclockwise to clockwise direction. Fig. 6-47, *B*, shows a transition from a right-beating (thermally induced) primary nystagmus to a left-beating secondary nystagmus after transition into the weightless phase of parabolic flight.[174] At least three factors were involved: (1) cessation of movement of endolymph, (2) abolition of the tonic influence of the otolithic

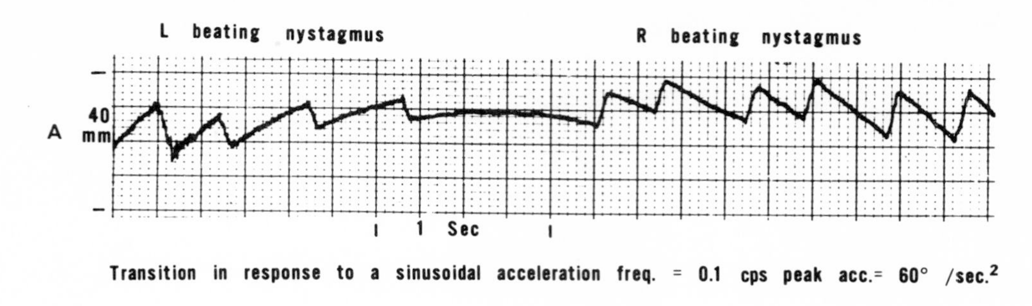

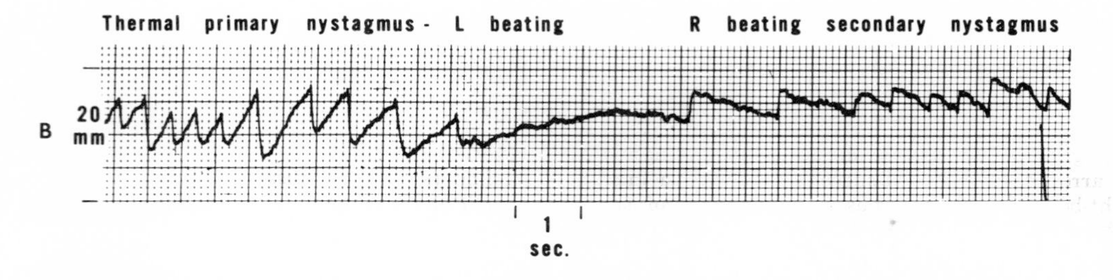

Fig. 6-47. Electronystagmogram showing the similarity in transition from right-beating to left-beating nystagmus in response to sinusoidal acceleration and the release of a latent secondary nystagmus on transition into weightlessness.

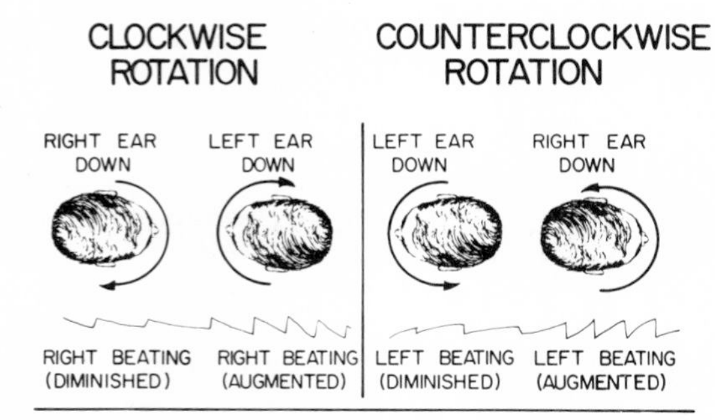

Fig. 6-48. Normal nystagmic responses (drawings) to rotation about an earth-horizontal axis. Note directional bias and cyclic modulation. (From Graybiel, A., Stockwell, C. W., and Guedrý, F. E., Jr.: Acta Otolaryng. **73:** 1-3, 1972.)

system (as distinct from the influence of the resting discharge), that (3) favored the appearance of a latent secondary nystagmus. The last presumably represented a competing influence in the nature of an adaptation effect of opposite sign.

Otolithic nystagmus. Fig. 6-48 shows the type of nystagmus elicited when a normal subject is rotated at constant velocity about an earth-horizontal axis.[175] During rotation in a clockwise direction, the subject displays a right-beating nystagmus that is diminished as he rotates through the right-ear-down position and is augmented as he rotates through the left-ear-down position. During counterclockwise rotation, he displays a left-beating nystagmus that is diminished as he rotates through the left-ear-down position and enhanced as he rotates through the right-ear-down position. Both a directional bias and a cyclic modulation about the bias level are manifested, indicating that two etiologic factors are operant.[176]

Until motion sickness has been defined there are, understandably, important differences in opinion with regard to scoring the severity of the manifestations. Both clinical appraisal and objective measurements have limitations. Table 6-6 shows clinical diagnostic criteria for different levels of severity of acute motion sickness[177] that have been validated in testing the effectiveness of anti-motion sickness drugs.[178]

Other simpler rating scales rely heavily on evaluating the severity of the nausea syndrome. Shortcomings of clinical appraisal, in addition to the obvious difficulty in measurement, are revealed when the exposures are prolonged and when individual differences in susceptibility to particular responses are great. The use of physiologic sensors has the advantage of objectivity but poses prob-

Table 6-6. Diagnostic categorization of different levels of severity of acute motion sickness

Category	Pathognomonic (16 points)	Major (8 points)	Minor (4 points)	Minimal (2 points)	AQS* (1 point)
Nausea syndrome	Nausea III,† retching, or vomiting	Nausea II	Nausea I	Epigastric discomfort	Epigastric awareness
Skin		Pallor III	Pallor II	Pallor I	Flushing/subjective warmth ≥ II
Cold sweating		III	II	I	
Increased salivation		III	II	I	
Drowsiness		III	II	I	
Pain					Persistent headache ≥ II
Central nervous system					Persistent dizziness: Eyes closed ≥ II; Eyes open III

Levels of severity identified by total points scored

Frank sickness (FS)	Severe malaise (M III)	Moderate malaise A (M IIA)	Moderate malaise B (M IIB)	Slight malaise (M I)
≥ 16 points	8-15 points	5-7 points	3-4 points	1-2 points

*Additional qualifying symptoms.
†III, Severe or marked; II, moderate; I, slight.

lems that are both qualitative (the nausea syndrome) and quantitative (changes in skin color). A combination of subjective and objective appraisal using semiautomatic recording may be one answer.

Many vestibular side effects are elicited under conditions that require the attention and cooperation of the subject and may be influenced by nonvestibular etiologic factors over which the subject has little or no control. Hence, in addition to a comprehensive medical examination and assessment of the sensory organs of the inner ear and of vision, special attention should be given to such factors as maturity, mental stability, and freedom from psychobiologic defects. Even in typically normal persons with normal canalicular and otolithic functions, considerable individual differences in susceptibility to reflex vestibular disturbances (RVD) and great individual differences in susceptibility to motion sickness may be manifested.

A questionnaire is useful in the appraisal of the sensory organs of the inner ear, both from the clinical standpoint and with regard to susceptibility to vestibular side effects in different motion environments. Because of the structural proximity and commonality of certain features of the acoustic and nonacoustic labyrinths, loss of hearing other than that explained by acoustic trauma raises the possibility that the nonacoustic labyrinth was affected by the same disease process; hence, audiometric findings provide an important clue to vestibular function. Tests of canalicular function[167, 179-183] are usually limited to evaluation of the horizontal pair of canals, and "normality" is often interpreted as "normality of vestibular function." Just as in the case of audiometry, the implication is usually but not always correct. Ocular counterrolling is probably the most reliable test[184] of otolithic function but is costly in terms of time and equipment. A postural equilibrium test battery, described elsewhere in detail,[185] is useful, although the procedure has, of course, limitations in that many systems other than the canalicular and otolithic systems are involved. If the findings are typically normal or abnormal, they indicate, respectively, the probability of normal and the possibility of abnormal vestibular function.

Tests dealing with susceptibility to vestibular side effects and with the rate of acquisition and decay of adaptation effects are especially important in motion sickness studies and will be described in connection with ground-based studies.

Central nervous system mechanisms and conceptual schema

The schema in Fig. 6-49 represents an attempt to fit important elements concerned with vestibular input-output relations into a conceptual framework.

In Block I are shown the types and combinations of natural and artificial accelerative stimuli

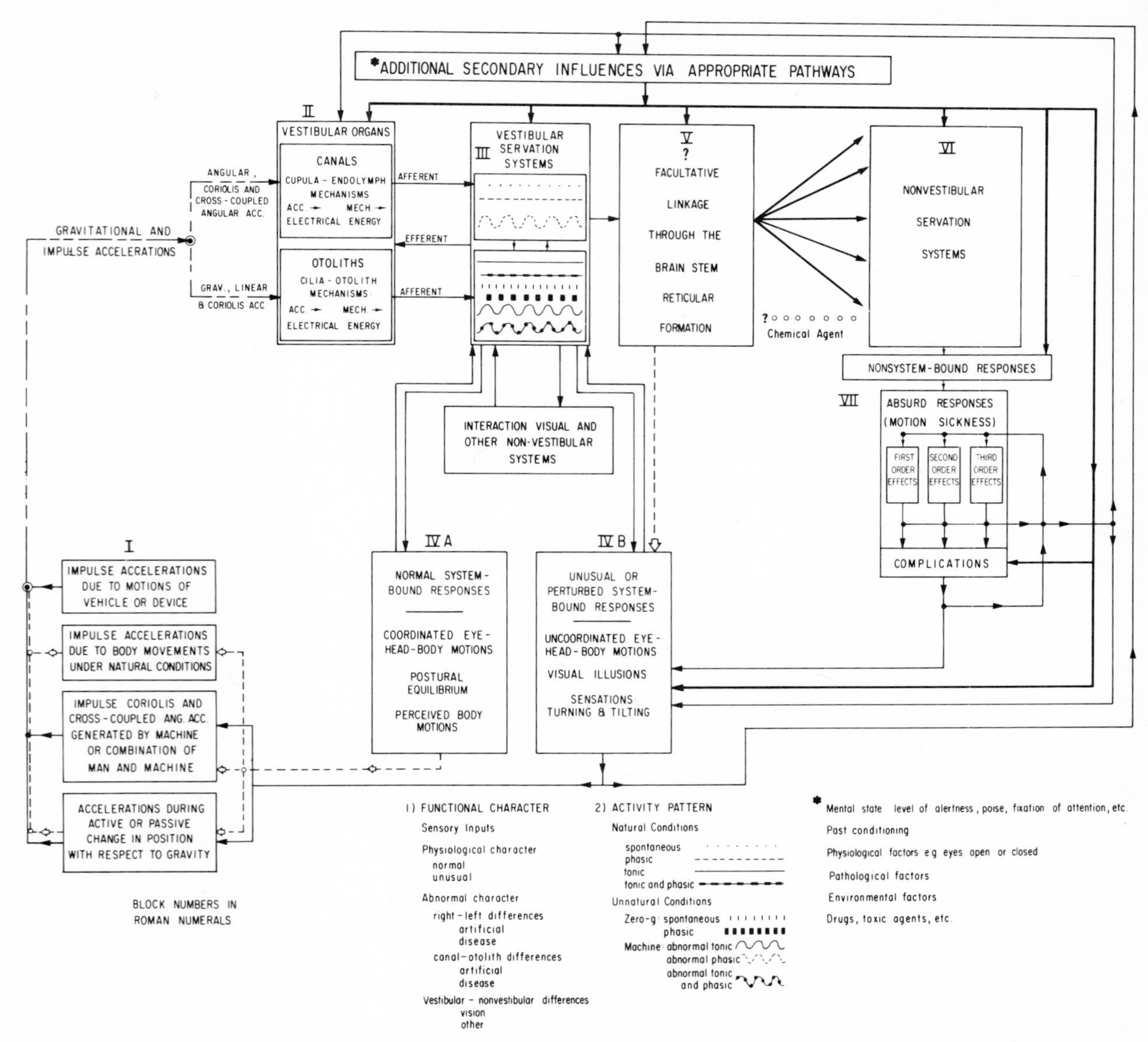

Fig. 6-49. Schema depicting input-output relations that make distinctions among normal vestibular responses, reflex phenomena, and motion sickness.

that reach the semicircular canals and otolith organs. The very important contribution to artificial stimulus patterns made by man's motions, especially those involving rotation of the head, deserves emphasis. In the adjacent footnotes are shown (1) categories of activation of the vestibular system, some of which are not accelerative in nature, such as disease process, and (2) some typical activity patterns in different motion environments.

Block II symbolizes the transducer functions of the semicircular canals and otolith organs, thereby altering the temporal and spatial pattern that constitutes the sensory input.

Block III shows the vestibular servation system and its two components (canalicular and otolithic) that have reciprocal modulating influences. An attempt has been made to indicate typical normal and abnormal canalicular and otolith activity patterns. Also depicted are the major acting or interacting influences, including vestibular efferent fibers ensuring a return flow of impulses to the end organs, thus closing one loop. This efferent vestibular activity is under intensive study; the morphologic evidence is on a firm basis[127, 128, 130, 131] and the functional role in man, while uncertain, appears likely to be inhibitory.

Astonishingly little is known concerning the normal function of the vestibular system in man under natural conditions. Block IV A of Fig. 6-49 illustrates typical responses to which the vestibular organs are known to contribute under natural terrestrial stimulus conditions and the entire chain of events involves Blocks I through IV A.

A classic experimental approach to understanding normal vestibular functions involves the use of human or animal subjects with bilateral loss of canalicular and otolithic functions. Experiments on animals alone, however, will never suffice because the findings are not directly applicable to man. The identification of human subjects with bilateral loss of vestibular function has been accomplished by screening groups of deaf persons, but experimentation on subjects identified in this way is complicated by the variety of differences, some of these great, between persons who hear and those who do not. Moreover, in all such subjects not only is there the need to make sure that the pathologic changes are quiescent and that adaptive changes are complete following any loss of function but also the need to take into account the unmeasurable factor of compensatory adjustment.

Despite these limitations, the best information that we have has been derived from a comparison in performance of persons with and without vestibular defects. Under ordinary present-day living conditions severe losses of vestibular function have commonly gone undetected. This is dramatically illustrated by the rare cases in which there has been loss of vestibular function early in life with retention of hearing.[186] Two such persons, discovered by chance, revealed that neither they, their families, nor their physicians were aware of the loss. Despite the fact that loss of function was readily demonstrated, this observation takes little away from the fact that they not only met the ordinary demands of present-day living but also were proficient above the average in a variety of sports. When apprised of their loss, they stated that they had experienced difficulties under circumstances in which visual cues were inadequate and possibly also in situations requiring complex eye-head-body coordination when visual cues were adequate.

A number of investigations comparing subjects with bilateral labyrinthine defects (who were also deaf) with normal subjects under near-natural stimulus conditions may be found in the proceedings of the International Symposium on Otophysiology held at Ann Arbor, Michigan, in May, 1971.[187]

The vestibular responses under abnormal stimulus conditions fall mainly into two categories, system-bound and nonsystem-bound responses. The main chain of events leading to system-bound responses involves Blocks I, II, III, IV B, and V of Fig. 6-49.

Some but not all system-bound responses reflect instability of the vestibular system; these will be referred to as reflex vestibular disturbances (RVD). Typical manifestations in normal persons include nystagmus, the oculogyral illusion, past pointing, and postural disequilibrium. Systematic studies of reflex manifestations reveal characteristics of the various responses that may be observed or inferred and, in general, they have the following in common: (1) short latencies, (2) maximal response to the initial stimulus, (3) no persistence of responses unless explicable by continuation of stimulation (including after-effects), and (4) response decline with acquisition of adaptation effects.

Nonsystem-bound responses (Blocks I to III and VI to VII) constitute epiphenomena elicited by certain repetitive accelerative stimuli that not only disturb the vestibular system but also allow vestibular influences to stimulate, by means of a facultative or temporary linkage, cells or cell groups outside the vestibular system. These responses include the symptoms of motion sickness, superimposed on any reflex manifestations also present. Inasmuch as they are not elicited in response to physiologic stimuli and serve no useful purpose, they may be properly characterized as illogical responses or absurd manifestations.

Little is known concerning the facultative linkage (Block V). The fact that irradiating vestibular activity is demonstrably open to modulating influences points to the use of common pathways in the brainstem reticular formation; mild symptoms of motion sickness have disappeared under the influence of experimenter-directed tasks that may have preempted neural pathways used by irradiating vestibular activity. What makes the vestibular facultative linkage unusual but not unique is the readiness with which vestibular activity may get out of bounds and elicit the widespread responses including the typical symptoms of motion sickness. The sometimes long delay between the onset of stimulation and appearance of motion sickness suggests that a chemical linkage may also be involved.

Certain secondary etiologic influences are categorized in Fig. 6-49 (right lower corner). Some of these influences, such as eyes open or closed, are always present, tending to increase or decrease susceptibility to motion sickness. Also, it may be assumed that any factor tending either to evoke or to inhibit a response characteristic of motion sickness will affect susceptibility accordingly.

Although the typical symptoms of motion sickness are well known,[188, 189] a list of first-order responses (let alone the precise sites of origin) has not been compiled (Block VII). At least some first-order responses also act as stimuli that elicit second and higher-order responses until the disturbances involve the organism as a whole.

In a typically normal person the clinical course of motion sickness is similar during repeated exposures to stress under controlled laboratory conditions if the stress profile and previous exposure history are properly taken into account. The cardinal symptoms useful in making a clinical diagnosis include cold sweating, pallor, drowsiness, increased salivation, and the nausea syndrome. Release of antidiuretic hormone[190] and urinary excretion of 17-hydroxycorticosteroids and catecholamines[191] are among the many biochemical changes that can be measured.

Recovery during continued exposure to stress is complicated. At some point in time adaptation is achieved and nonvestibular systems are freed from vestibular influences (and, possibly, from certain secondary etiologic factors) allowing restoration to take place spontaneously through homeostatic mechanisms. The time course of adaptation in the vestibular system, the disappearance of vestibular influences, and the restoration in the nonvestibular systems tend to overlap and have not been clearly defined.

From an analysis of the symptomatology in a typical case, the following characteristics of non-system-bound responses emerge: (1) delay in appearance of symptoms, (2) variation in order of appearance of responses as a function of stimulus intensity, (3) temporal summation, (4) perseveration after cessation of stressful stimuli, and (5) acquisition (and decay) of adaptation effects. An analysis of data obtained on groups of persons emphasizes individual differences in (1) susceptibility to motion sickness, (2) order of appearance and intensity of responses, (3) acquisition and decay of adaptation effects, (4) amount of performance decrement for comparable levels of severity of symptoms, and (5) the manifestation of conditioned responses.

Diagnosis is easy when typically normal symptoms are manifested in close temporal relation to exposure in motion environments regarded as stressful, but it becomes increasingly difficult both with increasing delay between exposure to motion and elicitation of symptoms and with increasing departure from typical symptomatology. General agreement has not been reached either with regard to what constitutes the symptomatology of motion sickness or the criteria for diagnosing different levels of severity. This points up at once areas where an effort should be made to reach agreement and emphasizes the difficulties experienced by the nonprofessional in self-diagnosis, especially in the absence of nausea and vomiting.

Prevention of motion sickness is the key to good therapy. Countermeasures include selection (where feasible or applicable), adaptation prior to exposure in motion environments, minimization of stressful accelerations aloft, attention to secondary etiologic factors both before and during flight, and the use of drugs.

Ground-based studies in preparation for space missions

This section is divided into discussions of provocative tests, adaptive capacity tests, and simulation studies in parabolic flight and rotating environments. Functional tests were discussed previously. Only simulation studies will be discussed in detail here, partly because some of this material is not readily accessible elsewhere but mainly because they comprise the most important studies performed in preparation for space missions.

Provocative tests

Provocative tests of many types are widely used[119, 155, 159, 192, 193] and serve the important purpose of evaluating a person's susceptibility to reflex vestibular disturbances and to motion sickness. In addition, they may measure his ability to cope with such disturbances either with or without the aid of countermeasures, including the use of drugs. The distinction between provocative and simulation tests involves both their duration and their specificity in terms of the global exposure conditions.

Factors of etiologic significance in addition to

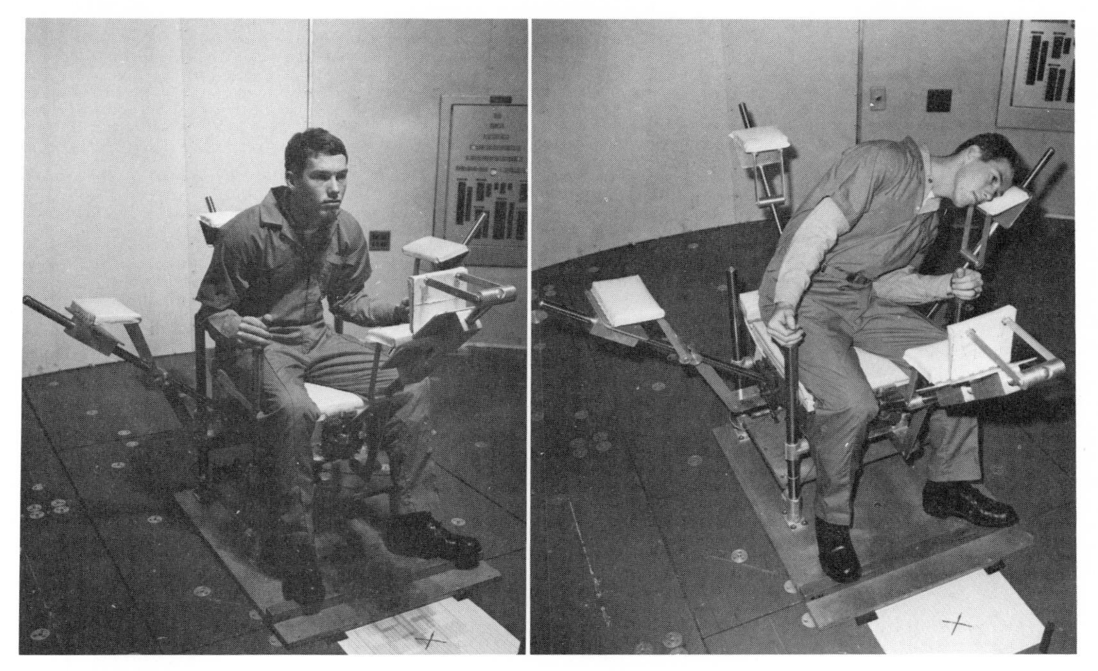

Fig. 6-50. Subject seated on a specially designed chair in a slow rotation room. The handholds facilitate the execution of head movements and the adjustable stops ensure control over the arc of rotation.

the motion environment may be introduced to simulate more fully the anticipated operational conditions or to explore their role in affecting an individual's susceptibility to novel circumstances; thus, the predictive value of provocative tests for specific operational conditions is less than that of simulation tests.

Both in conducting provocative tests and in interpreting the results, the following factors must be taken into account: (1) the individual differences in susceptibility with regard to a given test, (2) intraindividual differences in susceptibility when exposed in different gravitoinertial force environments, (3) preternaturally high susceptibility if insufficient time has not elapsed between exposures, (4) the fact that adaptation occurs as an inevitable consequence of every test, with much individual variation in the rate of acquisition and of loss of adaptation, and (5) the difficulty in expressing the results in absolute values. Great advantage occurs from the use of normalized scores and standardization of technics.

Standardized tests have been devised for determining susceptibility to vestibular side effects (usually motion sickness) both in an SRR with eyes open[193] and in a rotating chair device with eyes closed.[192] In the SRR the experimenter can exercise control over the stressful accelerations by standardizing the head motions executed by the subject and varying the angular velocity of the room. The subject, seated on a specially designed chair (Fig. 6-50), actively rotates his head (and body) about an axis other than that of the room's rotation. The head movements (front, back, left, and right) are limited by "stops," usually placed at 90 degrees of arc. The head movements, "over" and "return," are randomized in the four quadrants, and a taped recording sets the cadence.

Fig. 6-51 shows the stress profile (provocative incremental test schedule) used in comparing susceptibility to motion sickness with eyes open and eyes closed in an SRR.[193] The endpoint was a motion sickness score of approximately 12 points. The advantage of using the terminal rpm reached as the "normalized" score has the advantage of allowing comparison of susceptibility within and between subjects.

The Coriolis sickness susceptibility index. This test, a modification of the procedure just described, uses a rotating chair instead of a room. The subject is rotated with his eyes closed.[192] A noteworthy feature of this test is the method of

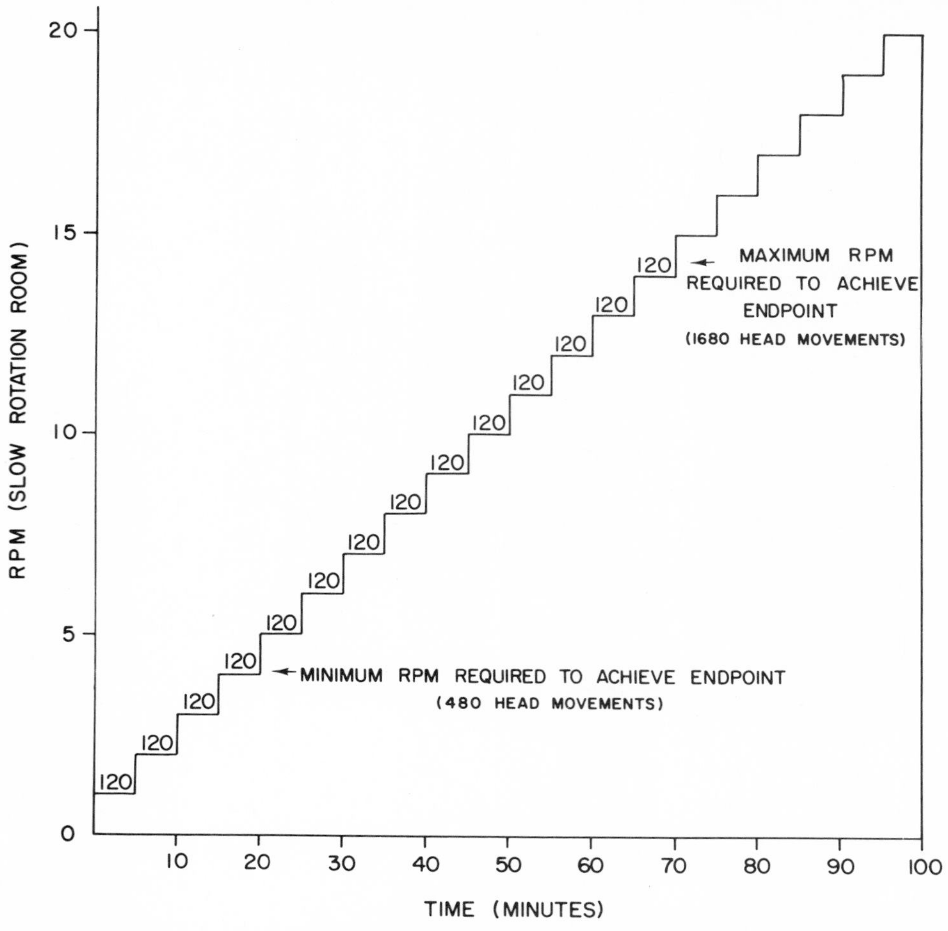

Fig. 6-51. Stress profile used for testing motion sickness susceptibility in twenty-four subjects. There were 120 head movements made in four quadrants at each step increase in velocity of the room. Endpoint was 12 units on a scale used in grading severity of motion sickness. (From Oosterveld, W. J., Graybiel, A., and Cramer, D. B.: Aerospace Med. **43**:1005-1007, 1972.)

scoring, which yields a single value, the index enabling the investigator to make comparisons within and between subjects.

Off-vertical rotation test. In contrast to the two tests just described, which initially disturb the canalicular system, the off-vertical rotation test (during constant angular velocity) involves rotating linear acceleration vectors that initially disturb the otolithic system. In provocative testing both the angle of tilt and the rpm are manipulated in different ways. At a predetermined angle of tilt the rotation, programmed on a time axis, involves periods of acceleration at 0.5 deg/sec^2 for 30 seconds followed by periods of constant velocity for 6 minutes, until either the predetermined motion sickness endpoint (point score) is reached or 6 minutes completed at 25 rpm, the cutoff point. In effect, this program represents unit increases of 2.5 rpm every 6.5 minutes after the initial step. The endpoint can be expressed in terms of elapsed time at terminal velocity, as total elapsed time at terminal velocity, or as total elapsed time, which serves as an index of susceptibility to motion sickness. The findings in a group of healthy men, the great majority attached to a naval air station, are shown in Fig. 6-52. All but twelve men reached the predetermined endpoint (M IIA) at a 10-degree tilt; all but five of the remainder reached it when the angle of tilt was increased to 20 degrees. Thus the scores rank ninety-five subjects in terms of their susceptibility to this unusual gravitoinertial force en-

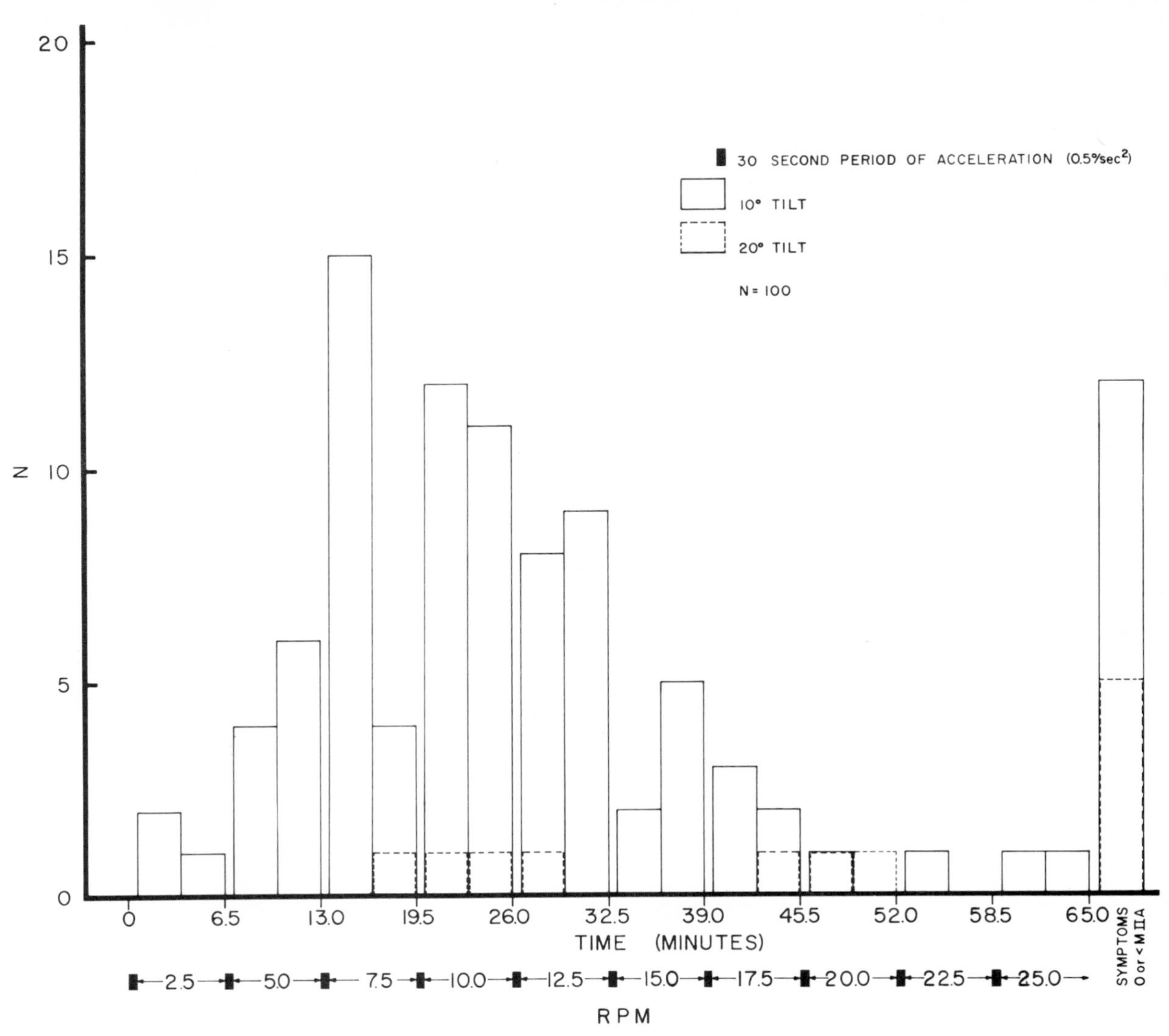

Fig. 6-52. Motion sickness susceptibility index in subjects exposed to off-vertical rotation according to programmed stress indicated on horizontal axis. (From Graybiel, A., and Miller, E. F., II: Aerospace Med. **41**:407-410, 1970.)

vironment and demonstrated that five of them were highly resistant.

Fig. 6-53 shows plots comparing susceptibility to motion sickness with scores obtained in testing the function of the semicircular canals and the otolith organs.[192] Although it appears that significant relationships were not found between functional test scores and susceptibility to motion sickness, it is worth adding that when extreme values are compared, susceptibility was lower in subjects with high rather than low values for the counterrolling index.

Adaptive capacity tests

On a given occasion a person's susceptibility to motion sickness is determined by (1) the ease with which the vestibular system is disturbed in a particular motion environment so that vestibular activity can escape beyond its normal bounds and (2) the thresholds above which the escape of this neuronal activity can occur over certain preferential pathways (presumably pathways in the brainstem reticular formation) not normally used toward sites where first-order symptoms of motion sickness have their origin. If a person is exposed

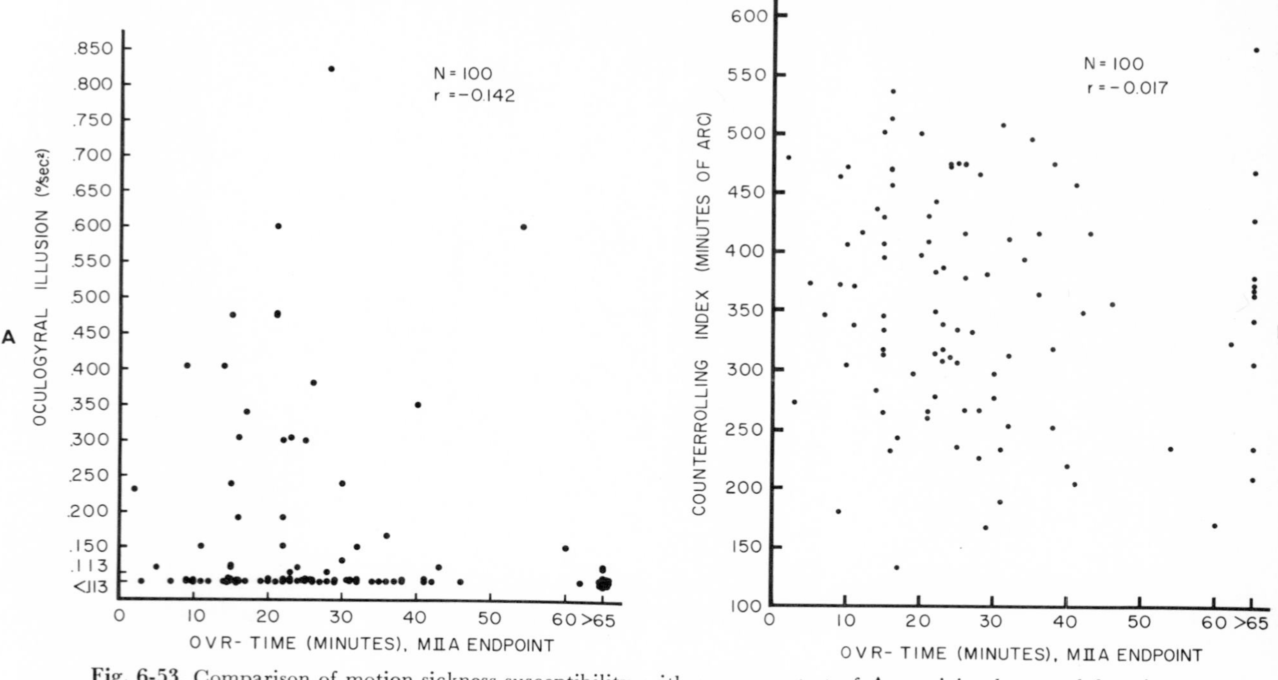

Fig. 6-53. Comparison of motion sickness susceptibility with scores on test of **A,** semicircular canal function (the oculogyral illusion), and **B,** otolith function (counterrolling index). (From Miller, E. F., II, and Graybiel, A.: Acta Otolaryng., Suppl. 274, 1970.)

to an incremental adaptation schedule, additional information is gained regarding adaptation to the motion environment, but the level of susceptibility measured now comprises the two factors determining susceptibility (just mentioned) minus the amount of adaptation acquired. To some extent, these factors can be separated using a modification of the stress profile mentioned above, termed an incremental adaptation schedule (IAS).[194]

Two standard IAS stress profiles are illustrated in Fig. 6-54. One requires the execution of 120 head movements at each 1-rpm increase in rotation (clockwise or counterclockwise) between 0 and 6 rpm and, after a single-step gradual return to zero velocity, the execution of 120 head movements either immediately after the return ("no delay"; Fig. 6-54, *A*) or after delay periods varying from 1 to 24 hours (Fig. 6-54, *B*). The delay period is included for the purpose of measuring the decay of adaptation effects described in the next section. The other standard stress profile differs from the first by the addition of a second IAS in which the direction of rotation is reversed either immediately after return to zero velocity (Fig. 6-54, *C*) or after delay periods measured in hours (Fig. 6-54, *D*). The terms "initial IAS" and "reverse IAS" are used because the initial direction of rotation (clockwise or counterclockwise) may be semirandom. During short delay periods the subject remains in the SRR with head fixed. With longer delay periods the subject remains in an adjacent ward under constant attendance to ensure that his head is always fixed relative to the thorax and that his needs are met with minimal moving about.

Fig. 6-55 summarizes findings in six normal young Navy ensigns exposed to a typical IAS and, after a single-step return to zero velocity, to a (weak) challenging stimulus in the stationary SRR. The motion sickness scores are shown in the force profile. Testing was aborted in the most susceptible subject (Fig. 6-55, *A*) after the execution of twenty-four head movements at 4 rpm, while overt symptoms of motion sickness were not elicited in the least susceptible subject (Fig. 6-55, *F*). The absence of symptoms during the challenge at zero velocity indicates that little or no adaptation occurred during the initial IAS. By reversing the direction of rotation (a strong chal-

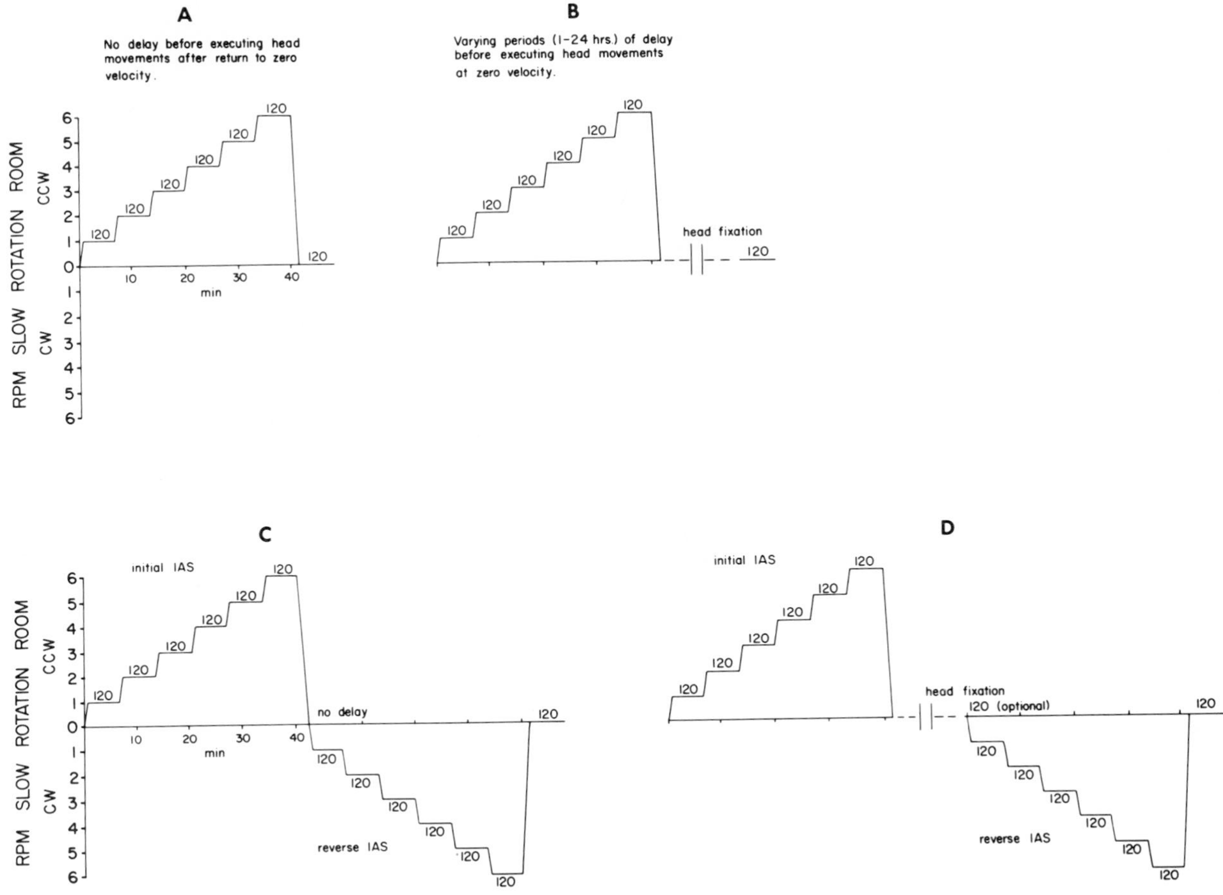

Fig. 6-54. Variations in a typical stress profile. The initial adaptation schedule (IAS) is followed either by a one-step return to zero velocity, **A**, or by a reverse IAS that follows immediately, **C**, or after a delay, **D**. A delay also may folow the initial IAS, **B**. The number of head movements executed at each "step" in this profile was 120 unless testing was aborted.

lenge) a further distinction can be made between subjects who manifest overt symptoms (evidence of adaptation) and those who do not (no evidence of adaptation). Among the three remaining subjects who experienced symptoms during the initial IAS (Fig. 6-55, *B* to *D*) the subject (Fig. 6-55, *C*) who experienced the most severe symptoms during the initial IAS (13 points) manifested the least severe symptoms during the challenge at zero velocity. In summary, a single test reveals, during the IAS, individual differences in susceptibility to motion sickness and, during the execution of head movements on return to zero velocity, the amount of direction-specific adaptation acquired as revealed by symptoms of motion sickness.

Although in general the lower the susceptibility to motion sickness, the faster the acquisition of adaptation effects, there are many exceptions. Similarly, although the more rapid the acquisition of adaptation effects, the slower their decay, there are many exceptions. The findings in Fig. 6-56 indicate that both subjects 5 and 7 manifested similar responses during their first IAS test. Comparing their initial IAS scores in three successive tests, subject 5 became symptom-free after the second test while subject 7 demonstrated little change. The scores made during the reverse are not comparable because the conditions were different.

These findings are only suggestive of what might be accomplished by exposing subjects every

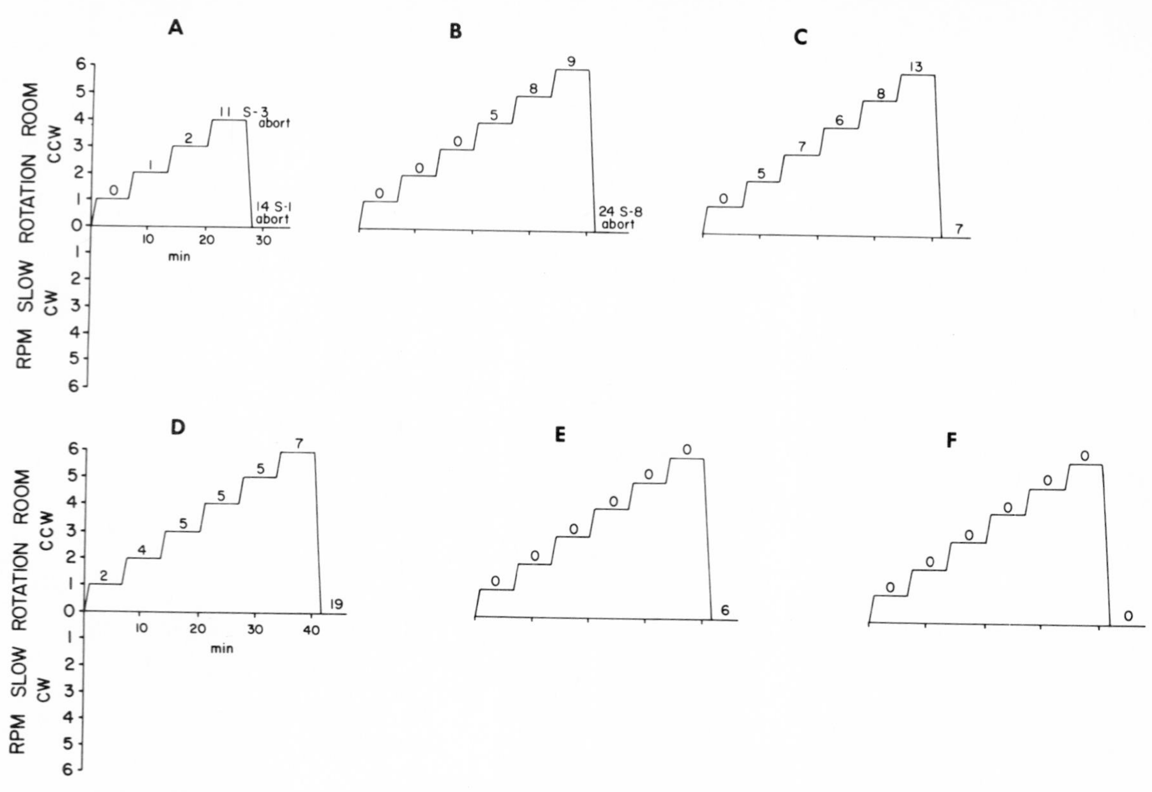

Fig. 6-55. Stress profiles and motion sickness scores in six healthy young subjects exposed to the same initial adaptation schedule and, after a gradual return to zero velocity, the added requirement to execute 120 head movements. Note the individual differences in susceptibility to motion sickness during the six 1-rpm increases in rpm and the variations in scores on return to the stationary environment when head movements generated normal vestibular inputs.

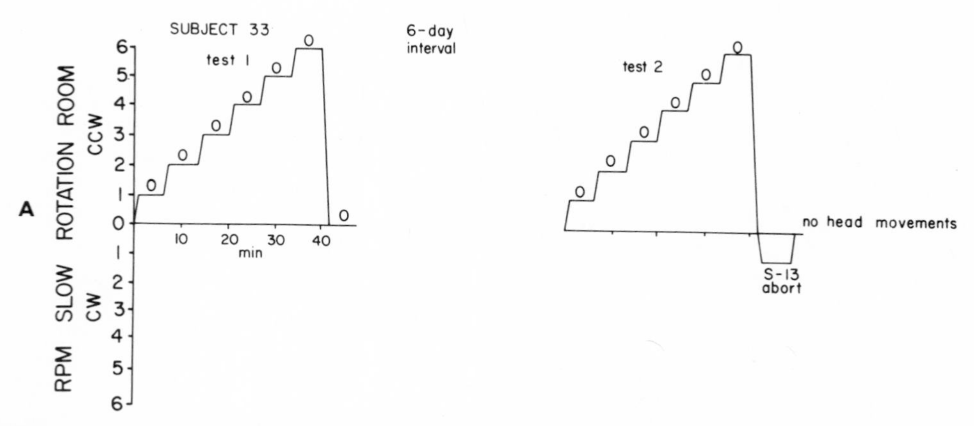

Fig. 6-56. Stress profiles and motion sickness scores in two subjects. Both were symptom-free in test 1 during the incremental adaptation schedule and after executing head movements on return to zero velocity. In test 2 the direction of rotation was reversed immediately after return to zero velocity; testing was aborted in one subject (33) (indicating the previous acquisition of direction-specific adaptation effects) while the other subject was symptom-free, implying a high degree of insusceptibility.

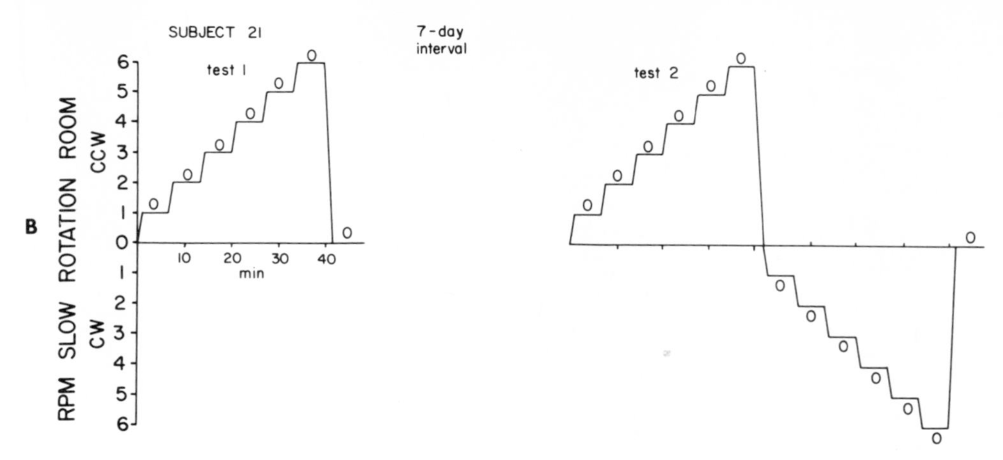

Fig. 6-56, cont'd. For legend see opposite page.

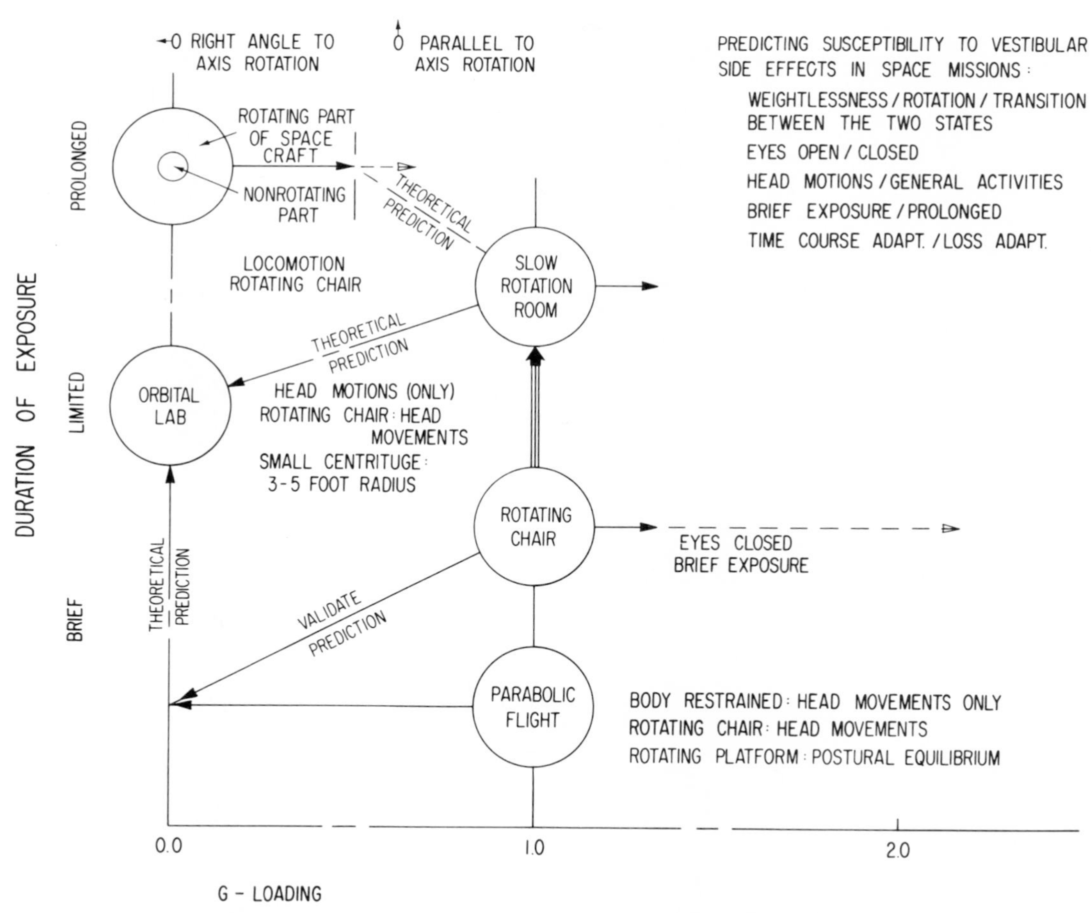

Fig. 6-57. Problems in predicting vestibular side effects in rotating space base.

day or every other day to a suitable IAS schedule.

Simulation studies

Some of the problems posed in attempting to predict susceptibility to vestibular side effects under the novel conditions that would exist in a rotating space base are pointed out in Fig. 6-57. An enclosed SRR can be used to simulate the angular velocity of a rotating space base and can provide prolonged exposures and sudden transitions between the rotating and nonrotating states. The SRR fails to simulate space-base conditions in such notable aspects as (1) weightlessness, (2) subgravity levels, (3) man's orientation with respect to the axis of rotation (parallel rather than at right angles), and (4) the Coriolis forces while walking and handling objects. Stated differently, the SRR is a very useful simulation device for the important study of the effects of cross-coupled angular accelerations. The SRR is useful in demonstrating the qualitative aspects of the role of the vestibular organs in postural equilibrium and in walking, but here nonvestibular factors also play an important role. The necessary use of small rotating devices poses limitations in terms of visual reference, length of exposure, and postural equilibrium.

Parabolic flights offer the opportunity of studying the effects of weightlessness and fractional subgravity levels for brief periods. Orbital flights prior to the establishment of a space base offer the opportunity to use small or even fairly large rotating devices for validation of ground-based experimental findings and the advantages of prolonged exposure to study adaptation effects.

Insofar as they have used similar methods, the results of studies involving parabolic flight con-

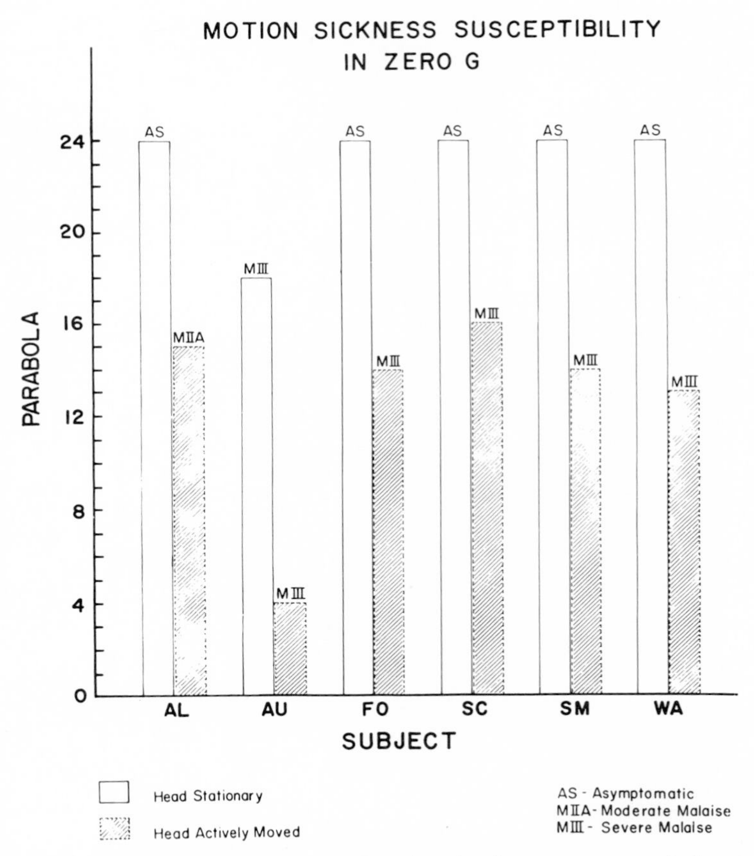

Fig. 6-58. Effect among six susceptible subjects of active head movements relative to the restrained condition upon sickness susceptibility measured in terms of the number of parabolas required to provoke severe malaise. (From Miller, E. F., II, Graybiel, A., Kellogg, R. S., and O'Donnell, R. D.: Aerospace Med. **40**:862-868, 1969.)

ducted in the United States and Soviet Union[195-201] not only are concordant but also agree with findings on astronauts[202] and cosmonauts[203, 204] in orbital flight.

Studies dealing with susceptibility to motion sickness in the weightless phase of parabolic flight have been mainly of two types. In one type subjects were restrained in their seats and required to make standardized head motions during the weightless phase only. The findings of one such experiment[198] are summarized in Fig. 6-58 and demonstrate that, among the twelve subjects tested in this manner, six were asymptomatic. Five of the remaining six experienced symptoms only when making head motions; the last subject demonstrated increased susceptibility when making head motions as compared to the head restraint (control) condition.

The second kind of experiment involved the use of a rotating chair device described previously. Subjects were required to make standardized head motions similar to those used in the SRR but with eyes blindfolded. Each subject served as his own control, and comparisons were made between susceptibility under terrestrial conditions and during parabolic flight, using similar periods of rotation and nonrotation. The findings on seventy-four subjects are shown in Fig. 6-59.[199] Susceptibility in weightlessness compared with ground-based conditions is ranked on the y axis, the topmost subject experiencing the greatest increase in susceptibility in weightlessness compared with terrestrial conditions. This ranking was made possible by the use of equivalent head movements (EHM), a universal scoring procedure described elsewhere in detail.[199] Scores on subjects tested on more than one occasion are given in chronologic order, and the open circles indicate that the moderate malaise II A (Table 6-7) endpoint was not reached.[199] The data indicate that more subjects have decreased than increased susceptibility in weightlessness and that in these subjects the endpoint frequently was not reached. Susceptibility under ground-based conditions proved to be a poor indicator of susceptibility aloft.

The effects of preadaptation to the stressful ac-

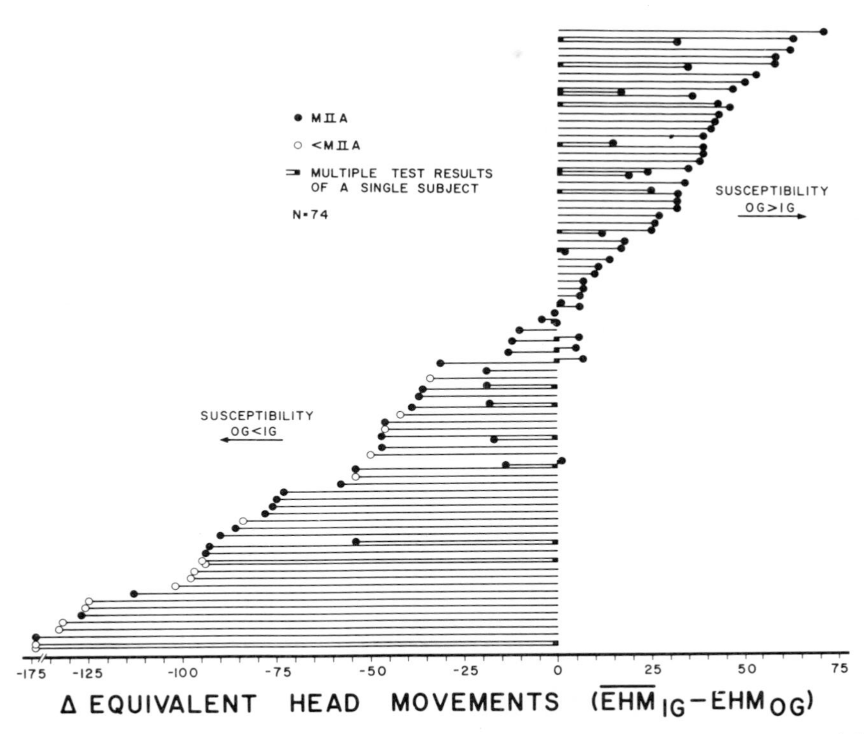

Fig. 6-59. Range of altered susceptibility among 74 subjects tested in zero G as generated by parabolic flight. Their susceptibility change is expressed as the difference between the number of equivalent head movements executed under zero- and earth-gravity conditions, reaching the endpoint of malaise IIA, or a limit of head movements imposed by the test conditions. (From Miller, E. F., II, and Graybiel, A.: Space Life Sci. **4**:295-306, 1973.)

Table 6-7. Change in motion sickness susceptibility score in eight normal male subjects*

	A					B				
Subject	Experimental trial number	Initial IAS scores		Score on return to zero velocity	Experimental trial number	Initial IAS scores		Reverse IAS scores	
		1-5 rpm†	6 rpm			1-5 rpm	6 rpm	1-5 rpm	6 rpm
2	1	0	0	6	2	2	3	a‡ 3 rpm	
3	1	0	0	0	2	0	0	9	4
18	1	1	1	1	2	1	1	a 2 rpm	
21	1	0	0	0	2	0	0	0	0
22	1	0	0	4	2	1	3	a 1 rpm	
31	1	0	0	0	2	2	0	a 2 rpm	
33	1	0	0	0	2	0	0	a 1 rpm	
39	1	0	0	5	2	0	0	a 3 rpm	

*Head movements were executed either **A**, immediately after a one-step return from a six-step incremental adaptation schedule (IAS) or **B**, during exposure without delay to a reverse IAS.
†Maximum score.
‡Abort.

celerations generated by standardized head movements with eyes open during rotation have been evaluated in ten subjects. Preadaptation to terminal velocities of either 7 or 10 rpm was accomplished by the use of IAS in an SRR. Preadaptation was always beneficial, often to a striking degree. In one subject, prior to adaptation, susceptibility was far greater in weightlessness than under terrestrial conditions, but after adaptation the subject was symptom-free in weightlessness. Moreover, whereas prior to adaptation the subject was susceptible to motion sickness in parabolic flight even when not rotating, after preadaptation he was symptom-free. In varying degrees this transfer of adaptation from rotating room to nonrotating conditions in weightlessness has been demonstrated in other subjects.

A large number of experiments have been carried out in which normal subjects have been exposed to continual rotation in the SRR at varying angular velocities for periods ranging up to 25 days. Many of these investigations were concerned with overall response patterns[205-210] and other experiments were directed toward more specific goals.

A series of experiments was carried out to determine whether there were differences in susceptibility to vestibular side effects dependent upon man's orientation to the axis of rotation of the SRR and whether the adaptation effects acquired in one orientation mode were transferred to the other. A unique feature of this experiment was the use of air-bearing supports and custom-fitted articulated Fiberglas molds that made it possible for subjects to walk on the "wall" of the circular SRR and carry out their tasks while horizontal with respect to the earth-vertical.[211] Four subjects participated in two different experiments involving adaptation to these stimulus conditions with the room rotating at 4 rpm for a period of either 4 or 5 days. One pair of subjects, initially exposed in the horizontal mode, were changed to the vertical mode near the middle of the perrotation period when symptoms of motion sickness had disappeared; in the second experiment they began in the vertical mode. The order was reversed for the second pair. When in the horizontal mode, the subjects spent approximately 6 hours a day in the airbearing device, 6 to 10 minutes upright, and the remainder of the time recumbent on a bunk. The findings, summarized in Fig. 6-60, indicate that there was no significant difference in susceptibility in the two modes and that transfer of adaptation was excellent. On cessation of rotation only mild symptoms of motion sickness were manifested. A by-product of the experiment was the demonstration of important differences between motion sickness and postural disequilibrium during adaptation to the rotating environment and subsequent return to the stationary one. In the start-horizontal mode, adaptation that ensured freedom from symptoms of motion sickness on change to the vertical mode failed to prevent ataxia. In the start-vertical mode the adaptation resulted in a great decrease in ataxia; this adaptation persisted throughout the finish-horizontal mode and as long as 36 hours afterward. This pattern implies that the dynamic processes underlying postural homeostasis involved muscular activities largely rendered static when subjects were in the horizontal mode.

With regard to adaptation schedules, three at-

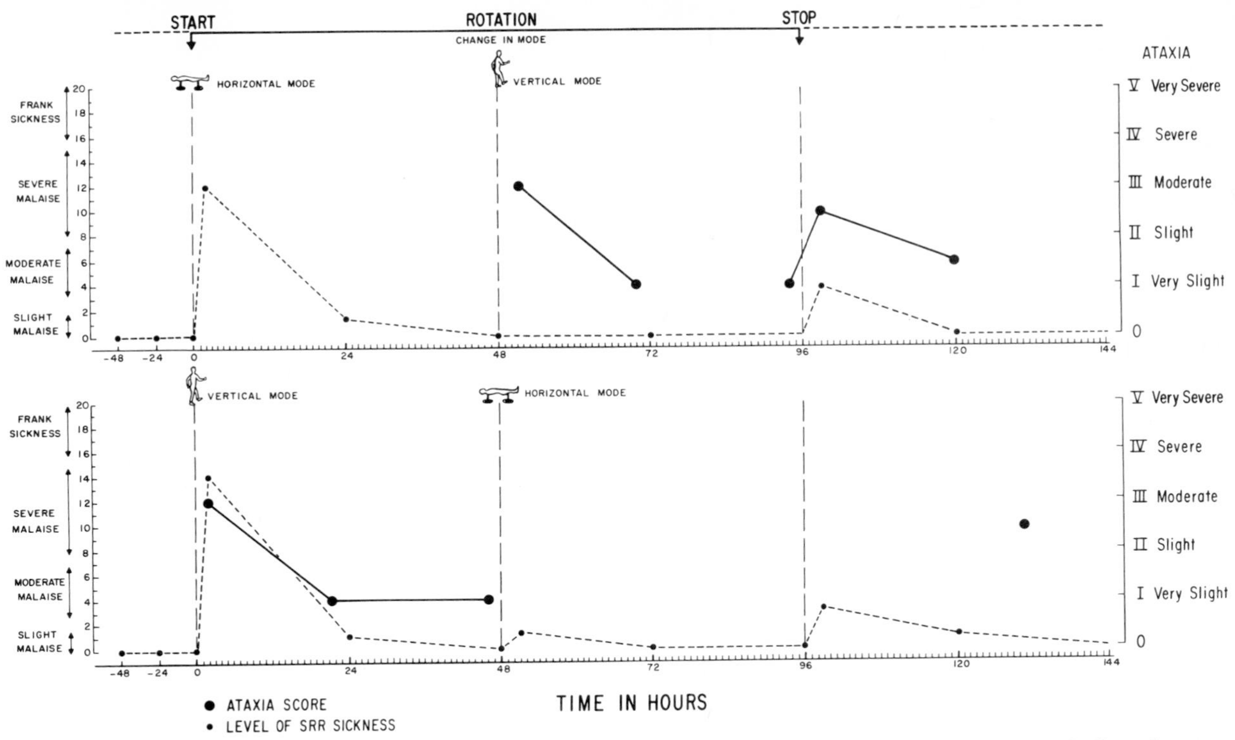

Fig. 6-60. Approximate mean changes in level of symptoms of motion sickness and in postural disequilibrium in four young healthy subjects exposed to sudden changes in body orientation during continuous rotation at 4 rpm.

tempts to prevent motion sickness by step increases to a terminal velocity of 10 rpm at Pensacola were unsuccessful[212]; two involved three incremental steps over a period of approximately 3 days and the third a series of forty incremental steps over a period of 40 hours. In the next attempt[213] overt symptoms (with the probable exception of drowsiness) of motion sickness on exposure to otherwise intolerable stressful accelerations were prevented solely by means of nine stepwise increases in rotational speeds over a period of 25 days to a terminal velocity of 10 rpm. This experiment demonstrated the possibility of achieving virtually symptom-free adaptation to a rotation velocity of 10 rpm, but the long time required was too "costly" for operational use even for a terminal velocity of 4 rpm.

An attempt was therefore made to effect asymptomatic incremental adaptation in an experiment in which three subjects executed experimenter-paced head-body movements.[214] The actual time spent making 1,000 head movements was a little over half an hour. Fig. 6-61 shows the stress profile, the number of head movements made at each step (each up-down counting as one movement), and the level of symptoms experienced by the subjects. One subject, T.A., was quite susceptible, becoming very drowsy at 2 rpm, experiencing epigastric discomfort at 5 rpm, and minimizing or refraining from making head motions at the higher rpm. The two remaining subjects experienced mild symptoms at terminal velocity, which became more severe on cessation of rotation. T.A. resorted to the use of an antimotion sickness drug. Noteworthy features of the experiment were (1) the inability of T.A. to keep up with the schedule, (2) the appearance of symptoms resulting from inadequate adaptation in the remaining two subjects, and (3) the increase in symptoms experienced by all subjects on cessation of rotation.

Fig. 6-62 shows the findings of a similar test in which more head movements were made at the higher angular velocities.[214] Symptoms of motion sickness were trivial except in subject R.O., who experienced very mild symptoms at 8 and 9 rpm

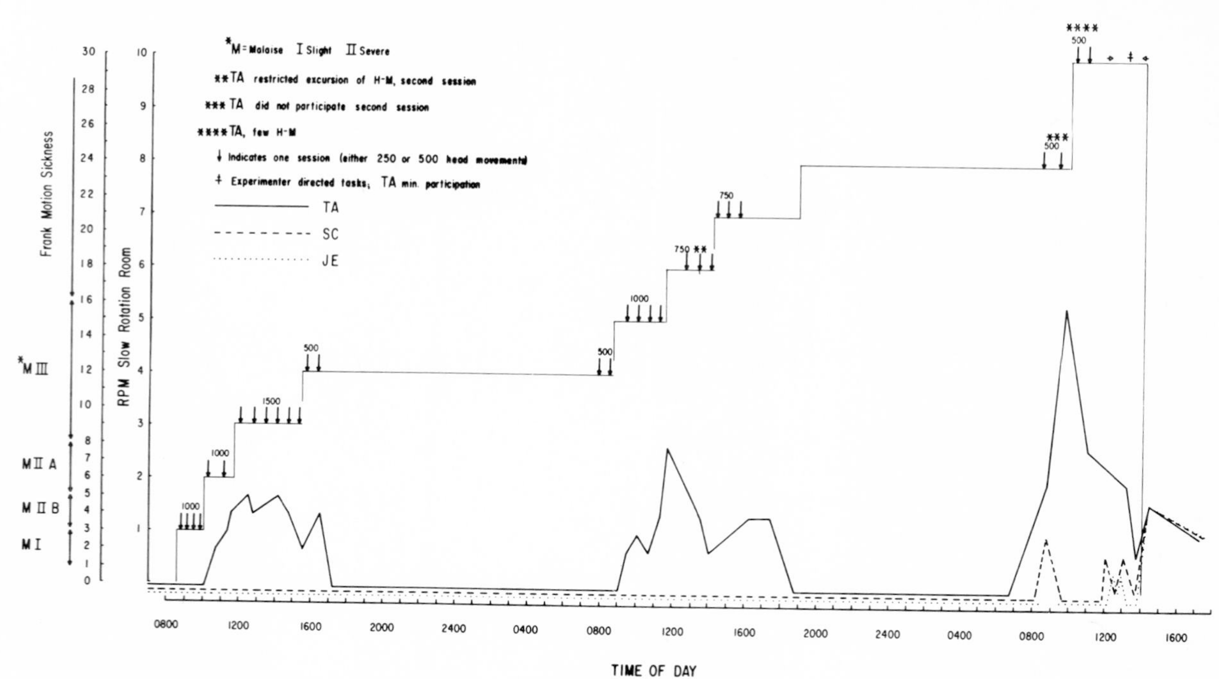

Fig. 6-61. Stress profile in the SRR and manifestations of motion sickness in three healthy subjects exposed to rotation for over 2 days. (From Graybiel, A., and Wood, C. D.: Aerospace Med. **40**:638-643, 1969.)

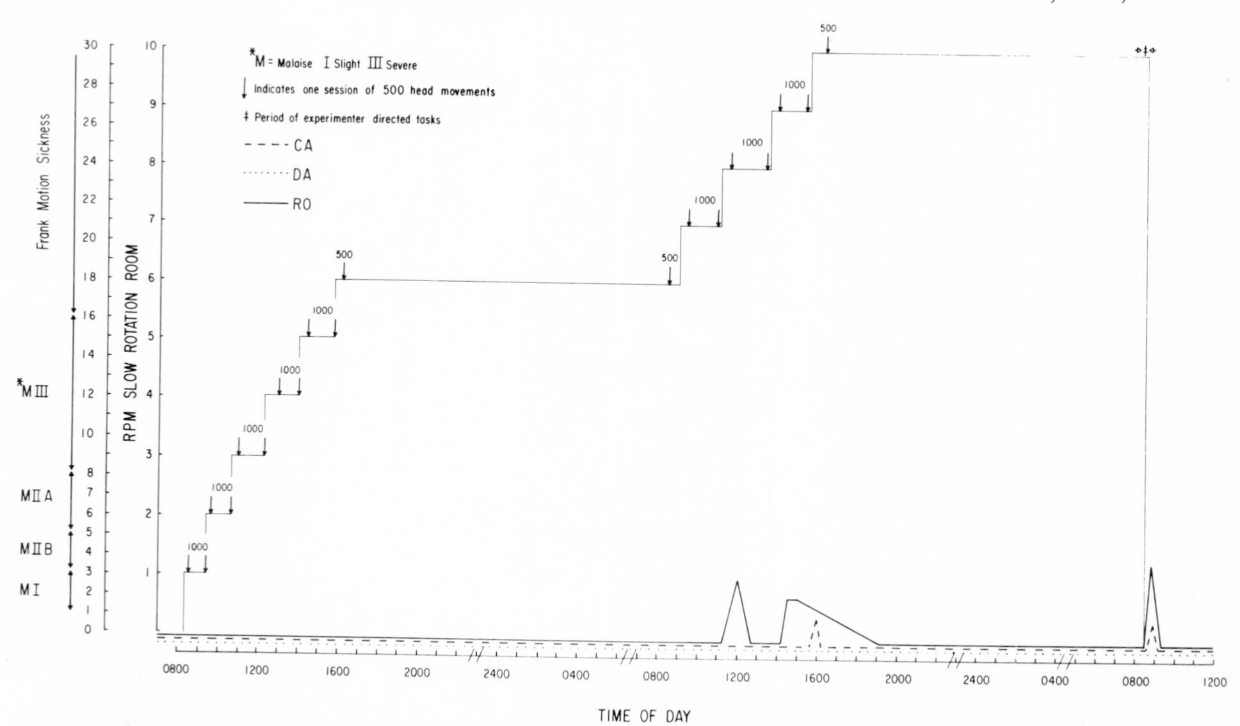

Fig. 6-62. Stress profile in the SRR and manifestations of motion sickness in three healthy subjects exposed to rotation for about 2 days. The large number of head motions accounted for the rapid adaptation. (From Graybiel, A., and Wood, C. D.: Aerospace Med. **40**:638-643, 1969.)

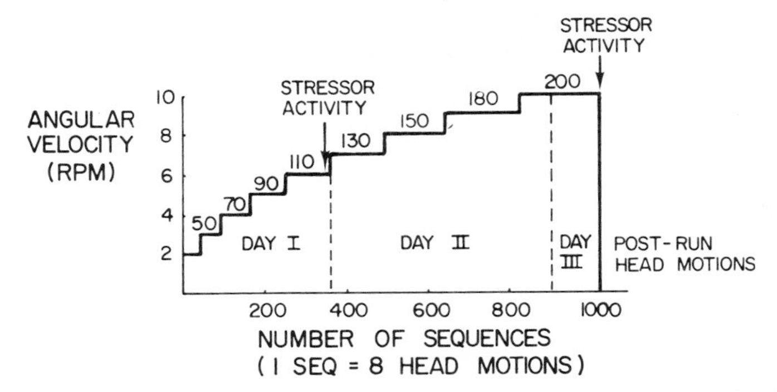

Fig. 6-63. Stimulus profile for a 3-day adaptation schedule on the SRR.

and on cessation of rotation. Except for ataxia, which was aggravated by head movements, complaints were minimal on cessation of rotation.

These findings confirmed the inferences drawn from the earlier studies that the time required to effect adaptation can be greatly shortened through control of head movements and the use of IAS. It should be mentioned that the problems encountered were greater at relatively high velocities compared with relatively low ones but that except in one instance, problems were not experienced if the unit increase was 1 rpm.

It has long been known that normal persons with mild symptoms resulting from exposure to stressful accelerations in a SRR might experience an aggravation of motion sickness on cessation of rotation. In consequence, it appeared that sudden transitions between the weightless (nonrotating) and rotating parts of a space station posed the most serious aspect of generating artificial gravity. A series of studies pertaining to this question demonstrated that (1) persons remaining symptom-free during exposure to an IAS (counterclockwise rotation) experience motion sickness when the direction of rotation is reversed,[215] (2) head movements executed on return to zero velocity, after achieving symptom-free adaptation in an incremental fashion, will elicit symptoms of motion sickness,[216] and (3) adaptation to rotation in one direction transfers to rotation in the other direction.[217]

The findings of an experiment to be reported[217] can be briefly summarized with the aid of Fig. 6-63. Three subjects participated and the adaptation schedule was identical for the three. On day 1, while rotating counterclockwise, each subject executed 40 head movement sequences at 2 rpm, 50 at 3 rpm, 70 at 4 rpm, 90 at 5 rpm, and 110 at 6 rpm. The subjects, while still being rotated, were then required to carry out highly stressful generalized activities in an attempt to evoke motion sickness. Their performance indicated that the paced head motions had produced a substantial degree of protection with respect both to reflex vestibular disturbances and to motion sickness. On day 2 the subjects executed 130 head movement sequences at 7 rpm, 150 at 8 rpm, 180 at 9 rpm, and 80 at 10 rpm, and they were again transferred to generalized activities. Their performance was similar to that on day 1. On the morning of day 3 after 120 head movement sequences at 10 rpm the room was brought to a stop, and the subjects executed the same head motions as during rotation. No symptoms of motion sickness appeared, and all reflex effects quickly disappeared.

In Fig. 6-64 are shown findings obtained on the same three subjects when they executed an incremental adaptation test before and after participating in the 3-day experiment just described. This test is also identical with the incremental adaptation test described in connection with Fig. 6-63. The noteworthy findings are (1) the small number of affirmative responses 6 hours after the 3-day experiment ended, (2) increasingly better performance after weekly exposures, and (3) transfer of adaptation effects acquired during counterclockwise rotation to rotation in the opposite direction. These observations support the conclusion that sudden transfers between the rotating and nonrotating environments are not only feasible in the SRR but also that the adaptation effects may be retained and improved.

In a recent series of experiments[194] it was demonstrated that during exposure to an IAS normal persons simultaneously acquire both short-term (direction-specific) adaptation effects and long-term (nondirection-specific) effects. The best evi-

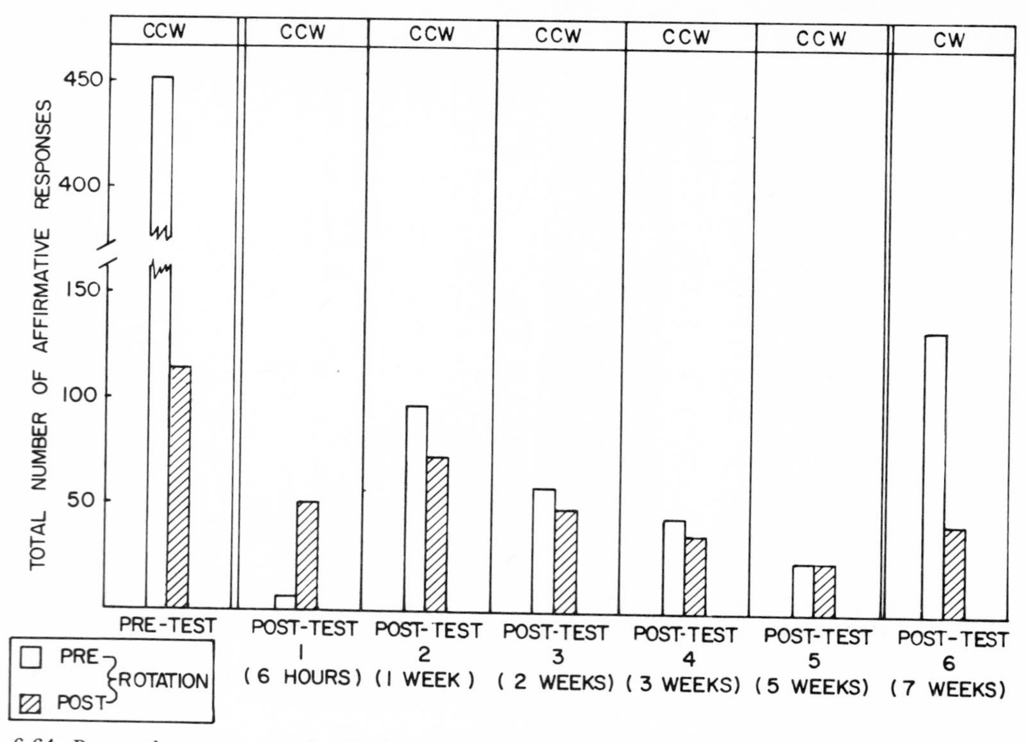

Fig. 6-64. Pre- and posttests associated with a 3-day adaptation schedule for three subjects on the SRR.

dence for demonstrating the acquisition of direction-specific adaptation effects is provided by subjects who, exposed to an IAS, nevertheless remain symptom-free or nearly symptom-free; for when symptoms (especially severe symptoms) of motion sickness are present, the time course for complete restoration of preexposure conditions is longer than for the disappearance of all overt responses. The subjects in Table 6-7 were selected on the basis that, with one exception, they did not manifest motion sickness during the initial IAS. The table compares the maximum scores elicited when head movements were executed immediately after return to zero velocity (column 5) with those elicited upon reversing the direction of rotation (last two columns). The data demonstrate that the motion sickness experienced under these conditions may be used as an indicator of previous acquisition of direction-specific adaptation built up in symptom-free subjects during an initial incremental exposure in the SRR. Note that all of the findings are compared in the same subject. It is presumed that this adaptation occurs in the vestibular system proper and not at sites where symptoms of motion sickness have their immediate origin.

A series of experimental tests was conducted with subjects serving as their own controls to plot the decay of direction-specific adaptation as a function of delay at zero velocity. The chief procedural problem here involved long-term (intertest) adaptation effects. Subjects were selected from among the original thirty-eight on the basis that, in repeated tests, long-term adaptation effects were either not evident or demonstrably small. In some instances the interval between the initial IAS and the challenge was progressively lengthened until symptoms were no longer elicited; in some instances the intervals were progressively shortened. The results, summarized in tabular form elsewhere,[94] are plotted in Fig. 6-65. The first two points on each curve represent motion sickness susceptibility scores in the first experimental tests, that is, the score at the 6-rpm step of the initial IAS and the score after return to zero velocity, respectively. When an abort occurred at zero velocity, 2 points were added to the motion sickness score for every uncompleted sequence; hence, these scores reflect levels of hypersusceptibility rather than actual motion sickness score. Each subsequent point in the plot was obtained in a subsequent test conducted after a

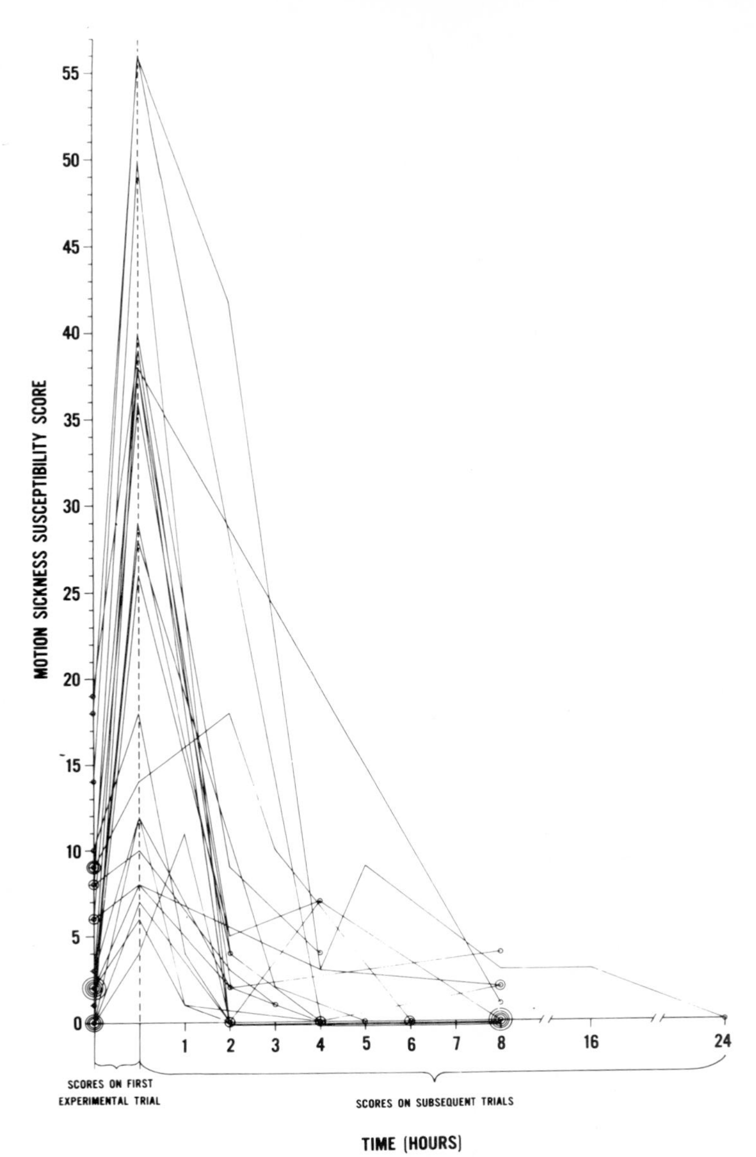

Fig. 6-65. Decay in direction-specific adaptation effects as a function of the time elapsed between completion of a standard incremental adaptation schedule (and return to zero velocity) and execution of head movements at zero velocity. The first two points on the graph represent susceptibility scores obtained in test 1 of the series, respectively, at 6 rpm and after executing head movements of zero velocity. Thereafter, each point on the graph (or circles in lieu of points) represents susceptibility scores obtained in subsequent tests. An exponential curve characterizes the decay trend. (From Graybiel, A., and Knepton, J.: Aerospace Med. **43:**1179-1189, 1972.)

delay. It can be seen at a glance that the decay in direction-specific effects for most of the twenty-two subjects took place chiefly or completely within 4 hours. The general configuration of the individual curves suggests that the decay in direction-specific effects occurs exponentially.

With the spontaneous disappearance of direction-specific (short-term) adaptation effects, long-term adaptation effects are revealed that are nondirection-specific. This nondirection-specific adaptation was demonstrated incidentally in the study just described when subjects were exposed a number of times either in a clockwise or counterclockwise direction before reversal. It was studied in a systematic manner in a recent experiment[217] to be reported.

The practical significance of these findings is two-fold. On making transitions from a rotating to a nonrotating portion of a space station, it is necessary to ensure that short-term direction-specific adaptation effects are not present and that long-term adaptation to weightlessness has not been lost. On the transition from the weightless to the rotating part of the space station, it is only necessary to ensure that adaptation to rotation has not been lost.

It is interesting to speculate on the problems of ataxia and past pointing that may occur in making transitions between rotating and non-rotating parts of a space station. Based on the incidental findings[211] that adaptation to rotation acquired by walking in the SRR was preserved during exposure to walking under simulated fractional G loads and during a subsequent period over 24 hours, it would seem that exposure to weightlessness would not result in loss of adaptation to walking in the rotating environment. With respect to past pointing, visual cues would serve to minimize any tendency to past point in making transitions between the weightless and rotating environments.

Space missions

Prevention of vestibular side effects in weightlessness

The prevention of motion sickness in the weightless spacecraft is the most important problem of immediate concern, although tumbling sensations and illusory phenomena have been reported.[218, 219] The data based on studies in the weightless phase of parabolic flight suggest not only that most persons experience either an increase or a decrease in susceptibility to motion sickness on transition into weightlessness but also that ground-based tests are not very reliable. Until validation studies have been carried out, the first step in prevention would involve the determination of susceptibility to motion sickness in parabolic flight.

The astronaut (or candidate) demonstrably susceptible to motion sickness in parabolic flight would benefit by incremental adaptation to the weightless phase in parabolic flight and, it would appear,[199] by exposures to IAS's in a rotating environment. In addition, the most effective anti-motion sickness drug (or drug combination), determined by individual bioassay procedures, should be identified and made available as a preventive measure (capsule form) or treatment (injectable form).

The susceptible astronaut on transition into weightlessness experiences maximum stressful effects on executing *any* head movement that stimulates the semicircular canals; the analogous stressful situation in a rotating environment would be a sudden transition from zero velocity to terminal velocity and the execution of head movements out of the plane of rotation. Once symptoms are experienced, especially the nausea syndrome, some time must elapse before dynamic changes in the vestibular system disappear and before restoration through homeostatic mechanisms has taken place in nonvestibular systems. If it is feasible for the astronaut to restrict head movements for a period well beyond the time overt symptoms have disappeared, this would be highly desirable even though the restriction would compromise performance and prevent adaptation. Antimotion sickness drugs taken by mouth may be slowly absorbed and administration by injection could be optional. In actual space flight, adaptation to weightlessness in the case of a susceptible space flyer has been a slow process because of restriction of head movements. The direction of ground-based experimental efforts would be determined in large part by the findings obtained under space flight conditions. Until that time all guidelines must perforce be tentative in nature.

Prevention of vestibular side effects in making transition between weightless and rotating environments

Long before rotating space stations are a reality the design characteristics will be known, including the terminal angular velocity and fractional G loadings in occupied quarters. Moreover, extensive studies will have been conducted in parabolic flight and validation studies in orbital flight. Although the motion sickness problem looms larger in a rotating environment than in a weightless environment aloft, it is nevertheless more easily manageable. In the event that motion sickness is experienced by an astronaut, antimotion sickness drugs could be administered and soon thereafter an IAS could be used. The latter could be accomplished in a Bárány-type chair rotating in the direction opposite to that of the space station. The angular velocity (ranging between zero velocity and the terminal velocity of the space station) could be manifested.

The use of antimotion sickness drugs in this situation can be predicted on ground-based studies such as those summarized in Fig. 6-66. The bioassay experiments revealed that only those drugs with a central parasympatholytic or sympathomi-

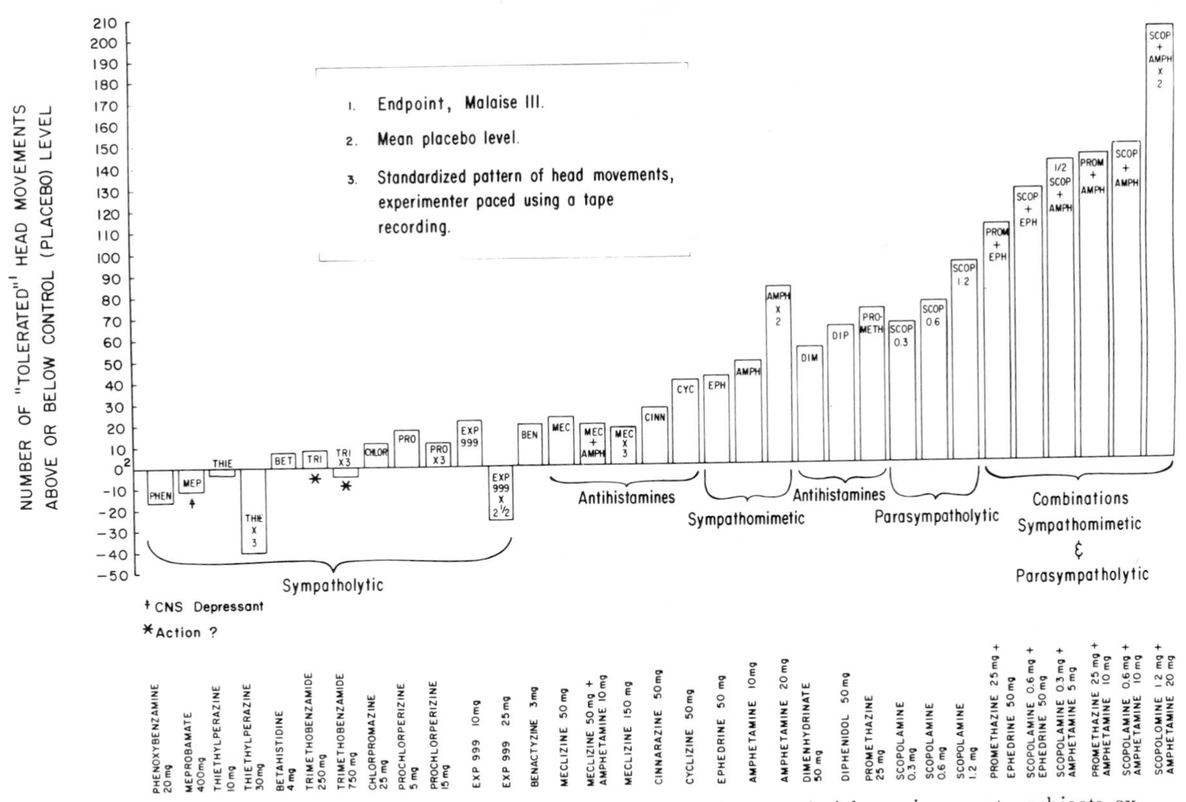

Fig. 6-66. Effectiveness of antimotion sickness drugs in preventing canal sickness in seventy subjects exposed in a rotating environment with the use of the "dial test." The antimotion sickness drugs are arranged according to their relative effectiveness in preventing motion sickness. It should be noted that drugs fall into groups according to their mechanisms of action. Different doses are placed adjacent to the usually recommended dose for each drug. (From Wood, C. D., and Graybiel, A.: Aerospace Med. **38:** 1099-1103, 1967.)

metic action were notably effective in preventing motion sickness in the SRR.[220] Recently, it was demonstrated[221] that a combination of promethazine (25 mgm) with *d*-amphetamine (10 mgm) had the same range of effectiveness as that found for scopolamine (0.6 mgm) plus *d*-amphetamine (10 mgm) and that the substitution of ephedrine (50 mgm) for the amphetamine in this combination, while slightly less effective, was superior in terms of freedom from side effects. These findings are not without theoretic interest and may have far-reaching significance in the etiology of vestibular side effects.

In summary, vestibular side effects in space exploration pose important but short-term problems. The likelihood of long-term health hazards resulting from loss of the stimulus of gravity to otolithic receptors seems highly unlikely in view of the clinical studies indicating man's tolerance for complete loss of vestibular function.

Vestibular investigations in space explorations: operational needs and scientific opportunities

Operational requirements

The major operant need is to be able to predict, on the basis of ground-based tests, the responses of space flyers (that are mainly or partly of vestibular origin) in the novel environments of the weightless space vehicle and the rotating space station. The limitations in simulating such novel stimulus conditions force an extension of ground-based studies to include validating experiments conducted aloft. The space flyer necessarily plays the key role in this integration. In the role of subject, he can serve as his own control in validating studies; in the role of onboard experimenter, he is essential in conducting experiments and making observations aloft. Among his duties are responsibilities in connection with the prevention of vestibular side effects during the mission. As indicated earlier, prevention involves taking charge rather than responding to events, and this will require close cooperation between the astronaut in the space base and the biomedical representatives in the ground-based control center during the period of adjustment.

The vestibular problem is one item to be considered, along with others, in reaching a compromise that determines vehicle design criteria. Among the interested parties are (1) specialists knowledgeable not only in the vestibular area but also in regard to the other human element problems, (2) engineers who must ensure inertial stability of the vehicle, and (3) space flyers who must assume responsibilities for making measurements and conducting experiments aloft.

The character of prolonged space missions will inevitably change with advances in spacecraft technology and the changing emphasis from exploration to the exploitation of extraterrestrial opportunities. Today the fact that travelers must be able to withstand the severe stresses incidental to launch and reentry demands a state of fitness that seems to dwarf the subtle effects of weightlessness. Even in the case of travelers in superb health, the maintenance of fitness during prolonged exposure in a weightless spacecraft will make large and continuous demands on the astronaut's or astroscientist's time, and illness aloft, making exercise impossible, could pose a hazard. Tomorrow, with the advent of a space shuttle or its equivalent, stresses incidental to launch and reentry will be greatly reduced and the limiting factor with regard to fitness for travel will be conditions aloft, of which the most important is weightlessness. The generation of artificial gravity by rotating the spacecraft represents a "technologic fix," alleviating an otherwise potentially continuous hazard, weightlessness. Vestibular side effects are of a temporary character; alleviation of the hazard is permanent.

Scientific opportunities

Transition into weightlessness constitutes not only an elegant procedure for abolishing the stimulus to the otolith organs resulting from gravity but a procedure that could not be carried out even in animal subjects under terrestrial conditions. Within limits, the TPK receptor systems also affected can be stimulated by agravitoinertial mechanical forces, thus allowing unique investigations to be conducted aloft. Highest priority should be given to testing ground-based theories under new conditions. With the lifting of the gravitational load to the otolith organs, it should be possible to measure what changes in vestibular function occur in addition to any temporary side effects. The study of resting discharges[222] and the contribution of tonic otolithic sensory inputs modulating stretch reflexes and muscle tone in man can be investigated.

GLOSSARY

acceleration The rate of change of velocity of a body in motion. A body is accelerated when either its speed or direction of motion changes.

accelerometer An instrument for measuring the acceleration of a body to which it is attached.

acoustic labyrinth The primary receptor organ of hearing, consisting of a spiral tube containing the organ of Corti. At its base the organ is in apposition with the lateral end of the internal acoustic meatus. The apex of the organ is directed anterolaterally.

adaptation Broadly defined as any alteration in a living organism that favors survival in a changed environment; applicable to the process of adapting or the state of adaptation. Meanings are so numerous that the reader must depend on text for specific use of the term. "Habituation" is favored by many psychologists; see **habituation.**

ampulla In vertebrate anatomy, the dilated portion of the semicircular canals of the ear.

analogous Similar in certain essential respects, such as appearance or function. The similarity does not extend to origin or development.

angular velocity A physical quantity related to rotational motion. Average angular velocity is the angular displacement (Θ) divided by the time required for the displacement to take place. Instantaneous angular velocity (ω) is defined by: $\omega = \lim_{\Delta t \to 0} \frac{\Delta\Theta}{\Delta t} = \frac{d\Theta}{dt}$.

Angular velocity is a vector in that it possesses both magnitude and direction. The magnitude is the angular speed, the direction is given by the direction of the axis of rotation, and the "sense" of the angular velocity vector is given by the righthand thread rule.

angular acceleration The time rate of change of angular velocity, either as to angular speed or direction or both.

bony labyrinth The bony portion of the inner ear apparatus.

Cartesian coordinate system Mathematical coordinate system of mutually perpendicular axes, either in two or three dimensions. The name is derived from that of the french mathematician and philosopher René Descartes.

centrifugal force The force exerted by a body undergoing centripetal acceleration on the physical mechanism (rod, beam, spring, and so on) producing that acceleration. Centrifugal ("fleeing the center") can be considered as a fictitious force that "balances" the centripetal force and produces zero acceleration of a mass in a rotating noninertial reference frame.

centripetal acceleration Acceleration directed toward the center of curvature of the path of a body moving along a curvilinear path. It is produced by the application of centripetal "seeking the center") force.

colatitude In the spherical polar coordinate system as used in this text, the angular distance measured from the positive side of the z axis to the vector. The tail of the vector must be at the origin of the coordinate system. This angle is denoted by ϕ.

contact force Force directly applied to the *surface* of a body.

Coriolis acceleration The acceleration or tendency toward acceleration relative to the surface of the rotating earth of an object moving above or on the earth with constant space velocity. This acceleration (deflection) was originally discussed by the French scientist G. G. de Coriolis toward the middle of the nineteenth century. The deflection is toward the right in the northern hemiphere and toward the left in the southern.

Coriolis force The apparent force responsible for Coriolis acceleration and based on the inertial property of the mass in question.

cross-coupled angular acceleration The angular acceleration resulting from the combined effects of simultaneous rotation about two or more axes.

cupula A mass of viscid, gelatinous fluid covering the crista of the ampulla of a semicircular canal.

damping The checking of a motion by friction, for example, as in the progressive diminution of the amplitude of an oscillating motion.

deceleration Negative acceleration

deformation The change in the shape of a body under stress. It can be measured as "strain," which is the deformation per unit of length in a given direction.

density The ratio of mass to volume.

elasticity The property that enables a body, when deformed by external forces, to recover without assistance its normal configuration as the forces of deformation are removed.

endolymph The fluid within the membranous labyrinth of the inner ear.

epithelium Tissue covering outer surfaces of the body and inner surfaces of hollow organs and tissues. Epithelial tissues are characterized by close apposition of cells to form sheets or layers, usually with one free surface.

friction Resistance to motion of one body across the surface of or through another body.

gravitationally normalized acceleration Acceleration expressed in multiples of the standard acceleration of gravity.

gravitationally normalized force Force expressed in multiples of the force of gravity on a given body.

gravitoinertial resultant force The resultant of the inertial (centrifugal) and gravitational force vectors.

habituation Diminution or cessation of response to a repetitive irrelevant stimulus. Habituation has numerous meanings but is not a synonym for adaptation, the more general term. Habituation of nystagmus is a common phrase.

hypogravitational Less than earth gravity. For example, the moon's gravity is one sixth of the earth's gravity.

inertia A property manifested by all matter characterized by resistance to any change of motion as to either speed or direction. Inertial force acts through the center of mass of the body.

kinematics The branch of mechanics that deals with the possible motions of a material body.

kinetics The branch of dynamics that pertains to the turnover, or rate of change of a specific factor, commonly expressed as units of amount per unit time.

labyrinth The system of intercommunicating cavities and canals that constitute the inner ear.

longitude In the spherical polar coordinate system as used in this text, the angular distance from the +x direction measured in the direction indicated by the righthand thread rule to the projection of the vector on the x-y plane. The tail of the vector must be at the origin of the coordinate system. Longitude is denoted by Θ.

magnitude A number applied to a quantity to enable its comparison with other quantities of the same class.

mass A measure of the quantity of matter in a body. It can be determined by comparing the changes in velocities when the body impinges upon a standard body or by measuring the acceleration produced by a known force.

mechanics That branch of science that deals with the effects of forces upon bodies at rest or in motion.

mechanoreceptor A receptor stimulated by differences of pressure, such as the baroreceptors in specialized vascular areas or the receptors of touch and hearing.

nausea Sickness referred to "stomach" or epigastric area. Stomach awareness is a just-noticeable change from the symptom-free state; stomach distress is more than stomach awareness but less than slight nausea.

nonacoustic labyrinth The vestibule and semicircular canals.

normalization The process of bringing or referring to the normal standard.

normal subject Difficult to define but often used term. *Normal range* may be applied to intelligence, mental stability, and fitness to perform tasks involving dexterity and whole-body effort. Much systematic data have been collected for females and males of all ages, and, excepting borderline values, classification within or without the normal range does not pose a problem. In addition to mean, median, and modal values, subjects may be selected on the basis of special tests that render them ideally suited for a particular experiment. In conducting experiments, for example, involving the semicircular canals and otolith organs, the functional integrity not only of these systems but also the visual and TPK systems must be determined. Moreover, if the experimenter is interested in determining the role of the vestibular system in the elicitation of motion sickness, then the attempt must be made to minimize unwanted etiologic factors through primary and secondary selection. In primary selection particular attention must be given to psychologic factors and ruling out psychobiologic defects. The secondary selection should be made on the day of the test (with the aid of a questionnaire) to ensure that the subject qualifies on that occasion.

nystagmus An involuntary rapid motion of the eyeball, either from side to side, up and down, circular, or in a combined motion.

oscillation A repetitive backward and forward or up and down motion.

otolith organs Organs of the nonacoustic labyrinth that contain minute crystals of calcium carbonate in a gelatinous mass that covers the maculae acusticae of the saccule and utricle of the inner ear.

perception Usually defined as an awareness based not only on sensory inputs but also inputs from higher centers; now preferred by many to sensation; see **sensation.**

perilymph The fluid within the space separating the membranous from the osseous labyrinth. There is no continuity with the endolymph within the membranous labyrinth.

pitch, roll, and yaw axes The y, x, and z axes, respectively, of the aircraft.

polar coordinate system System for fixing the position of a point. A point in a plane may be determined in position by two numbers called coordinates. One simple method for fixing the position is by means of polar coordinates. In this system, the coordinates are the polar axis, which is a line from the origin to the point, and the angle that the polar axis makes with the horizontal (x) axis.

rectangular coordinate system The simplest and most useful coordinate system. In the rectangular Cartesian system the axes of a two- or a three-dimensional system are orthogonal (mutually perpendicular). The position of the point is specified by its perpendicular projection on each of the axes.

righthand thread rule Mathemathical rule stating that the positive direction ("sense") of a rotational axis is the direction of forward motion of a screw with a righthand thread, when rotated clockwise as viewed from the head of the screw.

saccule The smaller of the two divisions of the membranous labyrinth of the vestibule containing one of the otolith organs. The saccule communicates with the cochlear duct via the ductus reuniens.

scalar A physical quantity described by a single, numerical value at each point in space. It is distinguished from a vector quantity by the fact that the scalar possesses magnitude only.

semicircular canals Three canals of the bony labyrinth of the ear, roughly circular and communicating with the vestibule by five openings. These canals contain the semicircular ducts or canals of the membranous labyrinth.

sensation Awareness based on a specific sensory stimulus without the aid of higher centers.

spherical polar coordinate system The polar coordinate in three dimensions, utilizing the polar axis and two angles to specify the position of a point.

statics The branch of mechanics that treats the action of forces, or systems of forces, on bodies that are at rest.

torsion pendulum A pendulum made up of a body suspended by an elastic fiber or wire that performs rotational oscillations by the winding and unwinding of the suspension.

trajectory The curve that a body such as a projectile fired from a gun describes in space.

utricle The larger of the two divisions of the membranous labyrinth. It is located in the posterosuperior region of the vestibule and contains an otolith organ that appears to be more important in man than the corresponding organ in the saccule. It provides information on the position and movements of the head and of the gravitoinertial resultant vector in relation to the head.

vector A physical quantity that has both magnitude and direction at each point in space. It is to be distinguished from a scalar, which has magnitude only.

velocity A vector quantity expressing both the speed and the direction of motion of a body at any instant.

vertical (1) Perpendicular to the plane of the horizontal; top, upright vertex. (2) Gravitational vertical (gravitoinertial vertical). (3) Behavioral vertical; the postural upright standing erect in a natural position.

vertigo (medical) A person wrongly perceiving the outer world as turning with respect to his body (objective vertigo) or himself turning with respect to the outer world (subjective vertigo). (aviation) *Aviator's vertigo* is a frequently used collective term to include nearly all forms of disorientation.

viscosity The resistance of fluids to flowing motion, in effect the resistance to shear of adjacent layers of the fluid (both liquids and gases).

weightlessness A condition of zero acceleration as measured by an observer within the system. Any object undergoing free fall in a vacuum is weightless, as is an unaccelerated satellite in orbit about the earth.

REFERENCES

1. Clark, C. C., Hardy, J. D., and Crosbie, R. J.: Human acceleration studies. Panel on acceleration stress for the armed forces, National Research Council, Committee on Bio-Astronautics, NAS-NRC Publ. 913, Washington, D. C., 1961, The Council.
2. Dixon, F., and Patterson, J. L., Jr.: Determination of accelerative forces acting on man in flight and in the human centrifuge, Pensacola, Fla., 1953, Naval School of Aviation Medicine, Report No. NSAM-515.
3. Sears, F. W., Zemansky, M. W.: University physics, ed. 3, Reading, Mass., 1964, Addison-Wesley Publishing Co.
4. Gell, C. F.: Table of equivalents for acceleration terminology, recommended for general international use by the acceleration committee of the Aerospace Medical Panel, Advisory Group for Aeronautical Research and Development, Aerospace Med. **32:**1109-1111, 1961.
5. Wunder, C. C.: Life into space, an introduction to space biology, Philadelphia, 1966, F. A. Davis Co., p. 121-183.
6. Jones, G. M.: The vestibular contribution to stabilization of the retinal image. In Graybiel, A., editor: First symposium on the role of the vestibular organs in the exploration of space, NASA SP-77, Washington, D. C., 1965, U. S. Government Printing Office, pp. 163-172.
7. Bergmann, P. G.: The riddle of gravitation, New York, 1968, Charles Scribner's Sons.
8. Jongbloed, J., and Noyons, A. K.: Der Einfluss von Beschleunigungen auf den Kreislaufapparat, Pflüg. Arch. Ges. Physiol. **233:**67-97, 1934.
9. von Diringshofen, H.: Neue Ergebnisse über Erträglichkeitsgrenzen gegenüber Zentrifugalkräften, Luftfahrtmed. Abh. **1:**314-316, 1937.
10. Gauer, O.: The physiological effects of prolonged acceleration. In German aviation medicine World War II, vol. I, Washington, D. C., 1953, Department of the Air Force.

11. Ruff, S.: Brief acceleration: less than one second. In German aviation medicine World War II, vol. I, Washington, D. C., 1953, Department of the Air Force.
12. Armstrong, H. G., and Heim, J. W.: The effect of acceleration on the living organism, J. Aviat. Med. **9**:199, 1938.
13. Franks, W. R., Kerr, W. K., and Rose, B.: Description of a centrifuge and its use for studying the effects of centrifugal force on man, J. Physiol. **104**:8-9, 1945.
14. Ham, G. C.: Effects of centrifugal acceleration on living organisms, War Med. **3**:30-56, 1943.
15. White, W. J.: A history of the centrifuge in aerospace medicine, Missile and Space Systems Division, Douglas Aircraft Company, Inc., 1964.
16. Leverett, S. D., Jr.: Aerospace physiology and medicine, Chicago, 1971, Year Book Medical Publishers.
17. White, W. J.: Space-based centrifuges. In Graybiel, A., editor: First symposium on the role of the vestibular organs in the exploration of space, NASA SP-77, Washington, D. C., 1965, U. S. Government Printing Office, pp. 209-213.
18. Woodling, C. H., and Clark, C. C.: Studies of pilot control during launching and re-entry of space vehicles utilizing the human centrifuge, Institute of Aeronautical Sciences Report 59-39, January, 1959.
19. Letko, W., Spady, A. A., Jr., and Hewes, D. E.: Problems of man's adaptation to the lunar environment. In Huertas, J., editor: Second symposium on the role of the vestibular organs in space exploration, NASA SP-115, Ames Research Center, Moffett Field, Calif., January 25-27, 1966, pp. 25-32.
20. White, W. J., and others: Biomedical potential of a centrifuge in an orbiting laboratory, Technical Documentary Report No. SSD-TDR-64-209—Supplement, July, 1965.
21. Leverett, S. D., and Clark, N. P.: A technique for determining changes in force of cardiac contraction during acceleration, Aerospace Med. **30**:832, 1959.
22. Rossen, R., Kabat, H., and Anderson, J. P.: Acute arrest of cerebral circulation in man, Arch. Neurol. Psychiat. **50**:510-528, 1943.
23. McCormick, E. J.: Human engineering, New York, 1957, McGraw-Hill Book Co.
24. Fraser, T. M.: Human response to sustained acceleration, NASA SP-103, Washington, D. C., 1966, U. S. Government Printing Office, p. 2.
25. Allen, W. H., editor: A dictionary of technical terms for aerospace use, NASA SP-7, Washington, D. C., 1965.
26. Randall, H. W., editor: Aerospace medicine, ed. 2, Baltimore, 1971, The Williams & Wilkins Co., pp. 169-171.
27. Armstrong, H. G.: Principles and practice of aviation medicine, ed. 3, Baltimore, 1952, The Williams & Wilkins Co.
28. Hamilton, W. F.: Blood: specific gravity and solids. In Glasser, O., editor: Medical physics, Chicago, 1944, Year Book Medical Publishers, Inc.
29. Engel, G. L.: Fainting, physiological and psychological considerations, ed. 2, Springfield, Ill., 1962, Charles C Thomas, Publisher.
30. Patterson, J. L., Jr., and Warren, J. V.: Mechanisms of adjustment in the cerebral circulation upon assumption of the upright position, J. Clin. Invest. **31**:653, 1952.
31. Henry, J. P., Gauer, O. H., Kety, S. S., and Kramer, K.: Factors maintaining cerebral circulation during gravitational stress, J. Clin. Invest. **30**:292, 1951.
32. Finnerty, F. A., Jr., Witkin, L., and Fazekas, J. F.: Cerebral hemodynamics during cerebral ischemia induced by acute hypotension, J. Clin. Invest. **33**:1227, 1954.
33. Fry, D. L., and Hyatt, R. E.: Pulmonary mechanics, a unified analysis of the relationship between pressure, volume and gasflow in the lungs of normal and diseased human subjects, Am. J. Med. **29**:672-689, 1960.
34. Mead, J.: Mechanical properties of lungs, Physiol. Rev. **41**(2):281-330, 1961.
35. Krueger, J. J., Bain, T., and Patterson, J. L., Jr.: Elevation gradient of intrathoracic pressure, J. Appl. Physiol. **16**:465-468, 1961.
36. Patterson, J. L., Jr., and others: The elevation gradients of lung density and of its blood, gas and tissue components. In press.
37. West, J. B.: Ventilation/blood flow and gas exchange, ed. 2, London, 1972, Blackwell Scientific Publications, Ltd.
38. Michaelson, E. D., Sackner, M. A., and Johnson, R. L., Jr.: Vertical distributions of pulmonary diffusing capacity and capillary blood flow in man, J. Clin. Invest. **52**:359-369, 1973.
39. West, J. B., Holland, R. A., Dollery, C. T., and Matthews, C. M.: Interpretation of radioactive gas clearance rates in the lung, J. Appl. Physiol. **17**:14, 1962.
40. Gillies, J. A., editor: A textbook of aviation physiology, Elmsford, N. Y., 1965, Pergamon Press, Inc.
41. Gauer, O. H., and Zuidema, G. D.: Gravitational stress in aerospace medicine, Boston, 1961, Little, Brown and Co.
42. Lindberg, E. F., and Wood, E. H.: Acceleration. In Brown, J. H., editor: Physiology of man in space, New York, 1963, Academic Press, Inc.
43. Patterson, J. L., Jr.: Unpublished data.
44. Wood, E. H., and others: WADD-TR-63-105, Wright-Patterson Air Force Base, Fig. 14, December, 1963.
45. Wood, E. H., and others: Effective headward and forward accelerations on the cardiovascular system, NASA SP-103, Washington, D. C., 1966, U. S. Government Printing Office, p. 14.
46. Rosenhamer, G.: Influence of increased gravitational stress on adaptation of cardiovascular and pulmonary function to exercise, Acta Physiol. Scand. **68**(Suppl. 276):5, 1967.
47. Bondurant, S.: Effective acceleration on pulmonary compliance, Fed. Proc. **17**(Suppl. 2):18, 1958.
48. Wood, E. H., and others: Effect of headward and forward accelerations on the cardiovascular system, WADD-TR-60-634, Wright-Patterson Air Force Base, 1961.
49. von Diringshofen, H.: Die Wirkung von grad-

linigen Beschleunigungen und von Zentrifugalkräften auf den Menschen: Experimentelle Unter suchungen über den Einfluss höher Beschleunigungen auf Blutdruck, Herzschlag und Atmung des Menschen im Motorflug, Z. Biol. **95:**551-566, 1934.

50. Gauer, O. H.: Die Atemmechanik unter Beschleunigung, Luftfahrtmed. **2:**291-294, 1938.
51. Glaister, D. H.: Breathing symposium, Nature **192:** 106, 1961.
52. Barr, P. O.: Pulmonary gas exchange in man as affected by prolonged gravitational stress, Acta Physiol. Scand. **58** (Suppl. 207):1-46, 1963.
53. Barr, P. O., Brismar, J., and Rosenhamer, G.: Pulmonary function and G-stress during inhalation of 100% oxygen, Acta Physiol. Scand. **77:** 7-16, 1969.
54. West, J. B., Dollery, C. T., and Naimark, A.: Distribution of blood flow in isolated lung; relation to vascular and alveolar pressure, J. Appl. Physiol. **19:**713-724, 1964.
55. Anthonisen, N. R., and Milic-Emili, J.: Distribution of pulmonary perfusion in erect man, J. Appl. Physiol. **21**(3):760-766, 1966.
56. Baker, J. B., Propert, D., and Patterson, J. L., Jr.: Circulation and respiration during acute motion sickness. In press.
57. Shapiro, W., Wasserman, A. J., and Patterson, J. L., Jr.: Mechanism and pattern of human cerebrovascular regulation after rapid changes in blood CO_2 tension, J. Clin. Invest. **45:**913, 1966.
58. Raper, A. J., Kontos, H. A., and Patterson, J. L., Jr.: Response of pial precapillary vessels to changes in arterial CO_2 tension, Circ. Res. **28:**518-523, 1971.
59. Patterson, J. L., Jr., and others: Cardio-respiratory dynamics in the ox and giraffe, with comparative observations on man and other mammals, Ann. N. Y. Acad. Sci. **127:**393, 1965.
60. Graybiel, A., Patterson, J. L., and Packard, J. M.: Sunburn as a cause of temporary lowering of blackout threshold in fliers, J. Aviat. Med. **19**(4):270-275A, 1948.
61. Christy, R. L.: Effects of radial, angular and transverse acceleration. In Randel, H. W., editor: Aerospace medicine, ed. 2, Baltimore, 1971, The Williams & Wilkins Co., p. 189.
62. Miller, P. B., and Leverett, S. D., Jr.: Tolerance to transverse ($+G_x$) and headward ($+G_z$) acceleration after prolonged bed rest, Aerospace Med. **36:** 13-15, 1965.
63. Sass, D. J., and others: Effects of $+G_y$ acceleration on blood oxygen saturation and pleural pressure relationships in dogs breathing first air, then liquid fluorocarbon in a whole body water immersion respirator. In Advisory Group for Aerospace Research and Development, North Atlantic Treaty Organization, 1970, p. 7-7-5.
64. Gray, S., III, Shaver, J. A., Kroetz, F. W., and Leonard, J. J.: Acute and prolonged effects of G-suit inflation on cardiovascular dynamics, Aerospace Med. **40:**40-43, 1969.
65. Eich, R. H., Smulyan, H., and Chaffee, W. R.: Hemodynamic response to G-suit inflation with and without ganglionic blockade, Aerospace Med. **37:**247-250, 1966.
66. Patterson, J. L., Jr., Graybiel, A., Lenhardt, H. F., and Madsen, M. J.: Evaluation and prediction of physical fitness, utilizing modified apparatus of the Harvard Step Test, Am. J. Cardiol. **14:**811, 1964.
67. Linnarsson, D., and Rosenhamer, G.: Exercise and arterial pressure during simulated increase of gravity, Acta Physiol. Scand. **74:**50-57, 1968.
68. Ryan, G. E. A., Kerr, W. K., and Franks, W. R.: Some physiological findings on normal men subjected to negative G, J. Aviat. Med. **21:**173, 1950.
69. Maher, P. J.: Memorandum MCREXD-695-69B, U. S. Air Force, 1948.
70. Gamble, J. L., Shaw, R. S., Henry, J. P., and Gauer, O. H.: Cerebral dysfunction during negative acceleration, J. Appl. Physiol. **2:**133, 1949.
71. Rushmer, R. F., Beckman, E. L., and Lee, D.: Protection of the cerebral circulation by the cerebrospinal fluid under the influence of radial acceleration, J. Physiol. **15:**355, 1947.
72. Beckman, E. L.: Protection afforded the CVS by the CSF under the stress of negative G, J. Aviat. Med. **20:**430, 1949.
73. Hershgold, E. J.: Roentgenographic study of human subjects during transverse acceleration, Aerospace Med. **31:**213, 1960.
74. Sandler, H.: Cineradiographic observations on human subjects during transverse acceleration of $+5G_x$ and $+10G_x$, J. Aerospace Med. **37:**445, 1966.
75. Agadzhanya, N., and Mansurov, A. R.: The effect of oxygen deficiency in prolonged radial acceleration on an animal organism, Bull. Exp. Biol. Med. **53:**42, 1962.
76. Woods, D. H.: Studies of effects of variation in the direction and magnitude of the gravitational inertial force environment on cardiovascular and respiratory systems. In Buzby, D. E., editor: Recent advances in aerospace medicine, Dordrecht, Holland, 1970, D. Reidal Publishing Co.
77. Barer, A. S., Galov, G. A., and Sorokina, Y. I.: Physiological reactions of the human body under the influence of acceleration of critical duration and magnitude directed along the back-chest axis, Bull. Exp. Biol. Med. **56**(8):33, 1963.
78. Cherniack, N. S., Hyde, A. S., Watson, J. F., and Zechman, F. W.: Some aspects of respiratory physiology during forward acceleration, Aerospace Med. **32:**113, 1961.
79. Nolan, A. C., and others: Decreases in arterial oxygen saturation and associated changes in pressures and roentgenographic appearance of the thorax during forward ($+G_x$) acceleration, Aerospace Med. **34:**798-813, 1963.
80. Wood, E. H.: Some effects of gravitational and inertial forces on the cardiopulmonary system, The Harry G. Armstrong Lecture, Aerospace Med. **38:**225, 1967.
81. Vandenburg, R. A., Nolan, A. C., Reed, J. H., Jr., and Wood, E. H.: Regional pulmonary arterial-venous shunting caused by gravitational and inertial forces, J. Appl. Physiol. **25:**516-527, 1968.
82. Hardie, E. L.: Investigation of the bradypnea response in dogs to left atrial distention, doctoral thesis, Medical College of Virginia, Virginia Commonwealth University, Richmond, 1969.
83. Paintal, A. S.: Cardiovascular receptors. In Hunt,

C. C., editor: Handbook of sensory physiology, vol. 3, Berlin, 1972, Springer-Verlag.

84. DeHaven, H.: Accident survival—airplane and passenger car, Society of Automotive Engineers Annual Meeting, Detroit, January 14-18, 1952.
85. Stapp, J. P.: Biodynamics of deceleration, impact, and blast. In Randel, H. W., editor: Aerospace medicine, Baltimore, 1971, The Williams & Wilkins Co.
86. Ewing, C. L., and Thomas, D. J.: Human head and neck responses to impact acceleration, Army-Navy Joint Report, NAMRL Monograph 21, USAARL 73-1, August 1972.
87. Berry, C. A.: Man's adaptation to weightlessness. In Biomedical findings on American astronauts participating in space missions, The Fourth International Symposium on Basic Environmental Problems of Man in Space, Yerevan, Armenia, U.S.S.R., October 1-5, 1971.
88. Miller, P. B., Johnson, R. L., and Lamb, L. E.: Effects of 4 weeks of absolute bed rest on circulatory functions in man, Aerospace Med. **35:** 1194-1200, 1964.
89. Price, J. F.: Physiological and psychological effects of space flight: a bibliography, vol. 2: Weightlessness and sub-gravity, Research Bibliography No. 44, Space Technology Laboratories, Inc., Redondo Beach Calif., 1963.
90. Leach, C. S., Alexander, W. C., and Fischer, C. L.: Compensatory changes during adaptation to the weightless environment, Physiologist **13:**246, 1970.
91. Patterson, J. L., Jr.: Theoretical aspects of cerebral circulation and prolonged weightlessness. In Moosey, J., and Janeway, R., editors: Cerebral vascular diseases, New York, 1971, Grune & Stratton, Inc.
92. Jones, G. M.: Origin, significance, and amelioration of Coriolis illusions from the semicircular canals: a non-mathematical appraisal, Aerospace Med. **41:**483-490, 1970.
93. Groen, J. J., Lowenstein, O., and Vendrick, A. J. H.: The mechanical analysis of the responses from the end-organs of the horizontal semicircular canal in the isolated elasmobranch labyrinth, J. Physiol. **117:**329-346, 1952.
94. Steinhausen, W.: Über den Nachweis der Bewegung der cupula in der intakten Begengangsampulle des Labyrinthes bei der natürlichen rotatorischen und calorischen Reizung, Pflügers Arch. Ges. Physiol. **228:**322-328, 1931.
95. Young, L. R.: On biocybernetics of the vestibular system. In Proctor, L., editor: Biocybernetics of the central nervous system, Boston, 1969, Little, Brown and Co.
96. van Egmond, A. A. J., Groen, J. J., and Jonkees, L. B. W.: The mechanics of the semicircular canal, J. Physiol. **110:**1-17, 1949.
97. Young, L. R.: The current status of vestibular models, Automatica **5:**369-389, 1969.
98. Newberry, P. D., and Bryan, A. C.: Effect on venous compliance and peripheral vascular resistance of headward (+G_z) acceleration, J. Appl. Physiol. **23:**150-156, 1967.
99. Meda, E.: A research on the threshold for the coriolis and purkinje phenomena of excitation of the semicircular canals, Arch. Fisiol. **52:**116-134, 1952. (Translated by E. R. Hope, 1954.)
100. Stone, R. W., Jr., and Letko, W.: Some observations on the stimulation of the vestibular system of man in a rotating environment, NASA SP-77, Pensacola, Fla., Washington, D. C., 1965, pp. 263-278.
101. Guedry, F. E., and Montague, E. K.: Quantitative evaluation of the vestibular coriolis reaction, Aerospace Med. **32:**487-500, 1961.
102. Goltz, F.: Über die physiologische Bedeutung: ung der Bogengange des ohrlabyrinths, Arch. Ges. Physiol. **3:**192, 1870.
103. Flourens, P.: Récherches expérimentales sur les propriétés et les fonctions du systeme nerveux vertebrés, ed. 2, Paris, 1842, J. B. Bailliere, pp. 438-501.
104. Darwin, E.: Zoonomia: on the laws of organic life, vol. 1, London, 1794, J. Johnson, p. 227.
105. Purkinje, J.: Beiträge zur näheren Kenntnis des Schwindels aers Heautognostischen, Datn. Med. Jb. (Wien) **6:**79-125, 1820.
106. Ménière, P.: Menaire sur les lesions de l'oreille, Paris interne donnant lieu in des symptomes de congéstion cérébrale apoplectiforme, Gaz. Med. Paris **16:**597-601, 1861.
107. Mach, E.: Grundlinien der Lehre von den Bewegungsempfindunger, Leipzig, 1875, Wilhelm Engelmann, p. 127.
108. Crum-Brown, A.: On the sense of rotation and the anatomy and physiology of the semicircular canals of the internal ear, J. Anat. (London) **8:** 327-331, 1874.
109. Breuer, J.: Über die Funktion der otolithenapparate, Pflüger. Arch. Ges. Physiol. **48:**195-306, 1891.
110. Ewald, J. R.: Physiologische Untersuchungen über das Endorgan des Nervus Octavus, Wiesbaden, 1892, Bergmann.
111. Steinhausen, W.: Über Sichtbarmachung und Funktionsprüfung der Cupula terminalis in den Bogengangsampullen des Labyrinthes, Pflüger Arch. Ges. Physiol. **217:**747-755, 1927.
112. Dohlman, G.: Some practical and theoretical points in labyrinthology, Proc. Roy. Soc. Med. **50:**779-790, 1935.
113. Kreidl, A.: Weitere Beiträge zur Physiologie des ohrlabyrinthes, Sitzungsb. Akad. Wissench. Math.-Naturw. Cl. **102:**149-179, 1893.
114. Versteegh, C.: Ergebnisse partieller Labyrinthexstirpation bei kaninchen, Acta Otolaryng. **11:**393-408, 1927.
115. Tait, J., and McNally, W. J.: Some features of the action of the utricular maculae (and of the associated action of the semicircular canals) of the frog, Phil. Trans. **224:**241-286, 1934.
116. Bárány, R.: Die Untersuchung der reflektorischen vestibularen und optischen Augenbewegungen und ihre Bedeutung für die topische Diagnostik der Augenmuskellohmungen, Münchn. Med. Wschr. No. 22-23, 1907.
117. Magnus, R.: Körperstellung. Experimentelle physiologische Untersuchungen über die Einzelnen bei der Körperstellung in Tätigkeit tretenden Reflexe über ihr Zusammenwirken und ihre Störungen, Berlin, 1924, Julius Springer.

118. James, W.: The sense of dizziness in deaf mutes, Am. J. Otol. **4**:239-254, 1882.
119. Sjöberg, A.: Experimentelle Studien über den Auslösungmechanismus der Seekrankheit, Acta Otolaryng. (Stockholm), Suppl. 14, pp. 1-136, 1931.
120. Anson, B. J., Harper, D. B., and Winch, T. G.: The vestibular and cochlear aqueducts: developmental and adult anatomy of their contents and parietes. In Graybiel, A., editor: Third symposium on the role of the vestibular organs in space exploration, NASA SP-152, Washington, D. C., 1968, U. S. Government Printing Office, pp. 125-146.
121. Igarashi, M.: Dimensional study of the vestibular end organ apparatus. In Huertas, J., editor: Second symposium on the role of the vestibular organs in space exploration, NASA SP-115, Washington, D. C., 1966, U. S. Government Printing Office, pp. 47-53.
122. Spoendlin, H.: Structurelle Eigenschaften der Vestibulären Rezeptoren, Schweiz. Arch. Neurol. Psychiat. **96**(2):219-230, 1965.
123. Clark, B.: Thresholds for the perception of angular acceleration in man, Aerospace Med. **38**:443-450, 1967.
124. Rasmussen, A. T.: The principal nervous pathways, ed. 4, New York, 1952, The Macmillan Co.
125. Brodal, A., Pompeiano, O., and Walberg, F.: The vestibular nuclei and their connections: anatomy and functional correlations, Ramsay Henderson Trust Lectures, Springfield, Ill., 1962, Charles C Thomas, Publisher.
126. Engström, H., Lindeman, H., and Engström, B.: Form and organization of the vestibular sensory cells. In Stahle, J., editor: Vestibular function on earth and in space, Oxford, 1970, Pergamon Press, Ltd., pp. 87-96.
127. Rasmussen, G. L., and Gacek, R.: Concerning the question of an efferent fiber component of the vestibular nerve of the cat (abstract), Anat. Rec. **130**:361-362, 1958.
128. Rasmussen, G. L., and Windle, W. F., editors: Conference on neural mechanisms of the auditory and vestibular systems, Springfield, Ill., 1960, Charles C Thomas, Publisher.
129. Wersäll, J.: Studies on the structure and innervation of the sensory epithelium of the cristae ampullares in the guinea pig, Acta Otolaryng., Suppl. 126. pp. 1-85, 1956.
130. Smith, C., and Rasmussen, G. L.: Nerve endings in the maculae and cristae of the chinchilla vestibule, with a special reference to the efferents. In Graybiel, A., editor: Third symposium on the role of the vestibular organs in space exploration, NASA SP-152, Washington, D. C., 1968, U. S. Government Printing Office, pp. 183-200.
131. Rossi, R., and Cortesina, G.: The "efferent cochlear and vestibular system" in Lepus Cuniculus L., Acta Anat. **60**:362-381, 1965.
132. Gacek, R. R.: The course and central termination of first order neurons supplying vestibular end organs in the cat, Acta Otolaryng., Suppl. 254, pp. 1-66, 1969.
133. Brodal, A.: Anatomical aspects on functional organization of the vestibular nuclei. In Huertas, J., editor: Second symposium on the role of the vestibular organs in space exploration, NASA SP-115, Washington, D. C., 1966, U. S. Government Printing Office, pp. 119-141.
134. Gacek, R. R.: Anatomical demonstration of the vestibulo-ocular projections in the cat, Acta Otolaryng., Suppl. 293, pp. 1-63, 1971.
135. Tarlov, E.: Organization of vestibulo-oculomotor projections in the cat, Brain Res. **20**:159-179, 1970.
136. Nyberg-Hansen, R.: Anatomical aspects on the functional organization of the vestibulospinal projection, with special reference to the sites of termination. In Graybiel, A., editor: Fourth symposium on the role of the vestibular organs in space exploration, NASA SP-187, Washington, D. C., 1970, U. S. Government Printing Office, pp. 167-180.
137. Razumeyev, A. N., and Shipov, A. A.: Problems of space biology. In Parin, V. V., editor: Nerve mechanisms of vestibular reactions, vol. 10, NASA TT F-605, Washington, D. C., 1970, U. S. Government Printing Office.
138. Walzl, E. M., and Mountcastle, V.: Projection of vestibular nerve to cerebral cortex of the cat, Am. J. Physiol. **159**:595, 1949.
139. Akert, K., and Gernandt, B. E.: Neurophysiological study of vestibular and limbic influences upon vagal outflow, Electroenceph. Clin. Neurophysiol. **14**:383-398, 1962.
140. Megirian, D., and Manning, J. W.: Input-output relations of the vestibular system, Arch. Ital. Biol. **105**:15-30, 1967.
141. Mickle, W. A., and Ades, H. W.: Rostral projection pathway of the vestibular system, Am. J. Physiol. **176**:243-246, 1954.
142. Spiegel, E. A., Szekely, E. G., Moffett, H., and Egyed, J.: Cortical projection of labyrinthine impulses: study of averaged evoked responses. In Graybiel, A., editor: Fourth symposium on the role of the vestibular organs in space exploration, NASA SP-187, Washington, D. C., 1970, U. S. Government Printing Office, pp. 259-268.
143. Tang, P. C., and Gernandt, B. E.: Autonomic responses to vestibular stimulation, Exp. Neurol. **24**:558-578, 1969.
144. Tang, P. C.: Artifacts produced during electrical stimulation of vestibular nerve in cat. In Graybiel, A., editor: Fifth symposium on the role of the vestibular organs in space exploration, NASA SP-314, Washington, D. C., 1973, U. S. Government Printing Office, pp. 115-123.
145. Pompeiano, O.: Interaction between vestibular and nonvestibular sensory inputs. In Graybiel, A., editor: Fourth symposium on the role of the vestibular organs in space exploration, NASA SP-187, Washington, D. C., 1970, U. S. Government Printing Office, pp. 209-235.
146. Wilson, V. J.: Vestibular and somatic inputs to cells of the lateral and medial vestibular nuclei of the cat. In Graybiel, A., editor: Fourth symposium on the role of the vestibular organs in space exploration, NASA SP-187, Washington, D. C., 1970, U. S. Government Printing Office, pp. 145-156.
147. Fredrickson, J. M., and Schwarz, D.: Multisensory influence upon single units in the vestibular nucleus. In Graybiel, A., editor: Fourth symposium

on the role of the vestibular organs in space exploration, NASA SP-187, Washington, D. C., 1970, U. S. Government Printing Office, pp. 203-208.

148. Brandt, T., Wist, E., and Dichgans, J.: Visually induced pseudocoriolis-effects and circularvection: a contribution to opto-vestibular interaction, Arch. Psychiat. Nerv. **214**:365-389, 1971.
149. Fluur, E.: Influences of semicircular ducts on extraocular muscles, Acta Otolaryng., Suppl. 149, 1959.
150. Ito, M.: The cerebellovestibular interaction in cat's vestibular nuclei neurons. In Graybiel, A., editor: Fourth symposium on the role of the vestibular organs in space exploration, NASA SP-187, Washington, D. C., 1970, U. S. Government Printing Office, pp. 183-199.
151. Graybiel, A., Miller, E. F., II, Newsom, B. D., and Kennedy, R. S.: The effect of water immersion on perception of the oculogravic illusion in normal and labyrinthine-defective subjects, Acta Otolaryng. **65**:599-610, 1968.
152. Graybiel, A.: Susceptibility to acute motion sickness in blind persons, Aerospace Med. **41**:650-653, 1970.
153. Oosterveld, W. J., Graybiel, A., and Cramer, D. B.: Susceptibility to reflex vestibular disturbances and motion sickness as a function of mental states of alertness and sleep, presented at the meeting of the Bárány Society, Toronto, Ontario, Canada, August, 1971.
154. Jongkees, L. B. W., and Philipszoon, A. J.: Nystagmus provoked by linear accelerations, Acta Physiol. Pharmacol. Neerl. **10**:239-247, 1962.
155. Graybiel, A., and Johnson, W. H.: A comparison of the symptomatology experienced by healthy persons and subjects with loss of labyrinthine function when exposed to unusual patterns of centripetal force in a counter-rotating room, Ann. Otol. **72**:357-373, 1963.
156. Graybiel, A.: Is there a need for a manned space laboratory? In Fleisig, R., Hine, E. A., and Clark, G. J., editors: Lunar exploration and spacecraft systems, Proceedings of Symposium on Lunar Flight, American Astronautical Society, New York, December 27, 1960, New York, 1962, Plenum Press, pp. 177-187.
157. Benson, A. J., and Bodin, M. A.: Interaction of linear and angular accelerations on vestibular receptors in man, Aerospace Med. **37**:144-154, 1966.
158. Correia, M. J., and Guedry, F. E.: Modification of vestibular responses as a function of rate of rotation about an earth-horizontal axis, Acta Otolaryng. **62**:297-308, 1966.
159. Graybiel, A., and Miller, E. F., II: Off-vertical rotation: a convenient precise means of exposing the passive human subject to a rotating linear acceleration vector, Aerospace Med. **41**:407-410, 1970.
160. Miller, E. F., II, and Graybiel, A.: Goggle device for measuring the visually perceived direction of space, Minerva Otorhinolaryng. **22**:177-180, 1972.
161. Graybiel, A., and others: Vestibular experiments in Gemini flights V and VII, Aerospace Med. **38**: 360-370, 1967.
162. Miller, E. F., II, and Graybiel, A.: Experiment M-131—human vestibular function, Aerospace Med. **44**:593-608, 1973.
163. Aschan, G., Bergstedt, M., and Stahle, J.: Nystagmography: recording of nystagmus in clinical neuro-otological examination, Acta Otolaryng., Suppl. 129, 1956.
164. Henriksson, N. G.: Speed of slow component and duration in caloric nystagmus, Acta Otolaryng., Suppl. 125, 1956.
165. Janeke, J. B.: On nystagmus and otoliths: a vestibular study of responses as provoked by a cephalocaudal horizontal axial rotation, Amsterdam, 1968, Drukkerij Cloeck En Modeigh N. V.
166. Claussen, C.: Das frequenzmaximum des kalorisch ausgelösten nystagmus i als kennlinien funktion des geprüften vestibulärorganes, Acta Otolaryng. **67**: 639-645, 1969.
167. McNally, W. J., and Stuart, E. A.: Physiology of the labyrinth, American Academy of Ophthalmology and Otolaryngology, Rochester, Minn., 1967, pp. 1-495.
168. Hamersma, H.: The caloric test: a nystagmographical study, Bergen Op Zoom, 1957, N. V. Drukkerij Van Gebr. Juten.
169. Henriksson, N., Rubin, W., Janeke, J., and Claussen, C.: A synopsis of the vestibular system, Basel, Switzerland, 1970, Sandoz, A. G.
170. Henriksson, N. G.: An electrical method for registration and analysis of movements of the eyes in nystagmus, Acta Otolaryng. **45**:25-41, 1955.
171. Honrubia, V., and others: I. Computer analysis of induced vestibular nystagmus: Rotatory stimulation of normal cats; II. Computer analysis of optokinetic nystagmus in normal and pathological cats; III. Modifications in the optokinetic nystagmus resulting from monocular stimulation and lateral tilting in normal and pathological cats, Ann. Otol. **80**(6):7-42, Suppl. 3, 1971.
172. Aschan, G., and Bergstedt, M.: The genesis of secondary nystagmus induced by vestibular stimuli, Acta Soc. Med. Upsaliensis **60**(3-4):113-122, 1955.
173. Niven, J. I., Hixson, W. C., and Correia, M. J.: An experimental approach to the dynamics of the vestibular mechanisms. In Graybiel, A., editor: The first symposium on the role of the vestibular organs in the exploration of space, NASA SP-77, Washington, D. C., 1965, U. S. Government Printing Office, pp. 43-56.
174. Graybiel, A., O'Donnell, R. D., Nagaba, M., and Smith, M. J.: Caloric nystagmus in parabolic flight: changes in primary and secondary nystagmic responses in sequential parabolas. In press.
175. Graybiel, A., Stockwell, C. W., and Guedry, F. E., Jr.: Evidence for a test of dynamic otolith function considered in relation to responses from a patient with idiopathic progressive vestibular degeneration, Acta Otolaryng. **73**:1-3, 1972.
176. Stockwell, C. W., Turnipseed, G. T., and Guedry, F. E., Jr.: Nystagmus responses during rotation about a tilted axis, NAMRL-1129, USAARL 71-15, Pensacola, Fla., 1971, Naval Aerospace Medical Research Laboratory and U. S. Army Aeromedical Research Laboratory.
177. Graybiel, A., Wood, C. D., Miller, E. F., II, and

Cramer, D. B.: Diagnostic criteria for grading the severity of acute motion sickness, Aerospace Med. **39**:453-455, 1968.
178. Wood, C. D., and Graybiel, A.: Evaluation of anti-motion sickness drugs: a new effective remedy revealed, Aerospace Med. **41**:932-933, 1970.
179. McLeod, M. E., and Meek, J. C.: A threshold caloric test: results in normal subjects, NSAM-834, Pensacola, Fla., 1962, Naval School of Aviation Medicine.
180. Fitzgerald, G., and Hallpike, C. S.: Studies in human vestibular function. I. Observations on directional preponderance ("Nystagmusbereitschaft") of caloric nystagmus resulting from cerebral lesions, Brain **65**:115-137, 1942.
181. Collins, W. E.: Repeated caloric stimulation of the human labyrinth and the question of vestibular habituation. In Graybiel, A., editor: The first symposium on the role of the vestibular organs in the exploration of space, NASA SP-77, Washington, D. C., 1965, U. S. Government Printing Office, pp. 141-150.
182. Fluur, E., and Mendel, L.: Relation between strength of stimulus and duration of latency time in vestibular rotatory nystagmus, Acta Otolaryng. **61**:463-474, 1966.
183. Preber, L.: Vegetative reactions in caloric and rotatory tests: a clinical study with reference to motion sickness, Acta Otolaryng., Suppl. 144, pp. 1-119, 1958.
184. Miller, E. F., II: Evaluation of otolith organ function by means of ocular counterrolling measurements. In Stahle, J., editor: Vestibular function on earth and in space, New York, 1971, Pergamon Press, Inc.
185. Fregly, A. R., Graybiel, A., and Smith, M. J.: Walk on floor eyes closed (WOFEC): a new addition to an ataxia test battery, Aerospace Med. **43**:395-399, 1972.
186. Graybiel, A., and others: Idiopathic progressive vestibular degeneration, Ann. Otol. **81**:165-178, 1972.
187. Graybiel, A.: Otolith function and human performance. In Hawkins, J. E., Jr., Lawrence, M., and Work, W. P., editors: Advances in oto-rhino-laryngology: Otophysiology, Basel, 1973, S. Karger, pp. 485-519.
188. Money, K. E.: Motion sickness, Physiol. Rev. **50**: 1-39, 1970.
189. Reason, J. T.: Motion sickness: some theoretical considerations, Int. J. Man-Machine Studies **1**: 21-38, 1969.
190. Taylor, N. B. G., Hunter, J., and Johnson, W. H.: Antidiuresis as a measurement of laboratory induced motion sickness, Canad. J. Biochem. Physiol. **35**:1017-1027, 1957.
191. Colehour, J. K., and Graybiel, A.: Biochemical changes occurring with adaptation to accelerative forces during rotation, Aerospace Med. **37**:1205-1207, 1966.
192. Miller, E. F., II, and Graybiel, A.: A provocative test for grading susceptibility to motion sickness yielding a single numerical score, Acta Otolaryng., Suppl. 274, 1970.
193. Oosterveld, W. J., Graybiel, A., and Cramer, D. B.: The influence of vision on susceptibility to acute motion sickness studied under quantifiable stimulus-response conditions, Aerospace Med. **43**: 1005-1007, 1972.
194. Graybiel, A., and Knepton, J.: Direction-specific adaptation effects acquired in a slow rotation room, Aerospace Med. **43**:1179-1189, 1972.
195. Loftus, J. P.: Symposium on motion sickness, with special reference to weightlessness, AMRL-TDR-63-25, Aerospace Medical Research Laboratories, Wright-Patterson Air Force Base, Ohio, 1963.
196. Kellogg, R. S., Kennedy, R. S., and Graybiel, A.: Motion sickness symptomatology of labyrinthine defective and normal subjects during zero gravity maneuvers, Aerospace Med. **36**:315-318, 1965.
197. Kolosov, I. A., Labedev, V. I., Khlebnikov, G. F., and Chekirda, I. F.: On the problem of the importance of parabolic flight to reproduce brief periods of weightlessness in vestibular evaluation of cosmonauts. In Parin, V. V., and Yemel'yanov, M. D., editors: Physiology of the vestibular analyzer, NASA TT F-616, Washington, D. C., 1970, U. S. Government Printing Office, pp. 225-229.
198. Miller, E. F., II, Graybiel, A., Kellogg, R. S., and O'Donnell, R. D.: Motion sickness susceptibility under weightless and hypergravity conditions generated by parabolic flight, Aerospace Med. **40**:862-868, 1969.
199. Miller, E. F., II, and Graybiel, A.: Altered susceptibility to motion sickness as a function of subgravity level, NAMRL-1150, Pensacola, Fla., 1971, Naval Aerospace Medical Research Laboratory.
200. Miller, E. F., II, and Graybiel, A.: The effect of gravitoinertial force upon ocular counterrolling, J. Appl. Physiol. **31**:697-700, 1971.
201. Yuganov, Y. M., Sidel'nikov, I. A., Gorshkov, A. I., and Kas'yan, I. I.: Sensitivity of vestibular analyzers and sensory reactions in man during brief weightlessness. In Sisakyan, N. M., editor: Biological studies under conditions of space flight and weightlessness, Izv. Akda. Nauk SSSR Ser. Biol. (Moscow) No. 3, pp. 369-375, 1964.
202. Berry, C. A., and Homick, G. L.: Findings on American astronauts bearing on the issue of artificial gravity for future manned space vehicles, Aerospace Med. **44**:163-168, 1973.
203. Mandrovsky, B. N.: Soyuz-9 flight, a manned biomedical mission, Aerospace Med. **42**:172-177, 1971.
204. Gazenko, O. G.: Medical problems of manned space flight, Space Sci. Rev. (Dordrecht) **1**:369-398, 1963.
205. Graybiel, A., Clark, B., and Zarriello, J. J.: Observations on human subjects living in a "slow rotation room" for periods of two days, Arch. Neurol. **3**:55-73, 1960.
206. Clark, B., and Graybiel, A.: Human performance during adaptation to stress in the Pensacola Slow Rotation Room, Aerospace Med. **32**:93-106, 1961.
207. Guedry, F. E., Jr., Kennedy, R. S., Harris, C. S., and Graybiel, A.: Human performance during two weeks in a room rotating at three rpm, Aerospace Med. **35**:1071-1082, 1964.
208. Newsom, B. D., and Brady, J. F.: Observations

on subjects exposed to prolonged rotation in a space station simulator. In Graybiel, A., editor: The first symposium on the role of the vestibular organs in the exploration of space, NASA SP-77, Washington, D. C., 1965, U. S. Government Printing Office, pp. 279-292.

209. Stone, R. W., Jr., and Letko, W.: Some observations on the stimulation of the vestibular system of man in a rotating environment. In Graybiel, A., editor: The first symposium on the role of the vestibular organs in the exploration of space, NASA SP-77, Washington, D. C., 1965, U. S. Government Printing Office, pp. 263-278.
210. Graybiel, A., and others: The effects of exposure to a rotating environment (10 rpm) on four aviators for a period of twelve days, Aerospace Med. **36**:733-754, 1965.
211. Graybiel, A., and others: Transfer of habituation of motion sickness on change in body position between upright nad horizontal in a rotating environment, Aerospace Med. **39**:950-962, 1968.
212. Bergstedt, M.: Stepwise adaptation to a velocity of 10 rpm in the Pensacola Slow Rotation Room. In Graybiel, A., editor: The first symposium on the role of the vestibular organs in the exploration of space, NASA SP-77, Washington, D. C., 1965, U. S. Government Printing Office, pp. 339-344.
213. Graybiel, A., Deane, F. R., and Colehour, J. K.: Prevention of overt motion sickness by incremental exposure to otherwise highly stressful Coriolis accelerations, Aerospace Med. **40**:142-148, 1969.
214. Graybiel, A., and Wood, C. D.: Rapid vestibular adaptation in a rotating environment by means of controlled head movements, Aerospace Med. **40**:638-643, 1969.
215. Reason, J. T., and Graybiel, A.: The effect of varying the time interval between equal and opposite Coriolis accelerations, Brit. J. Psych. **62**: 165-173, 1971.
216. Reason, J. T., and Graybiel, A.: Adaptation to Coriolis accelerations: its transfer to the opposite direction of rotation as a function of intervening activity at zero velocity, NAMI 1086, Pensacola, Fla., 1969, Naval Aerospace Medical Institute.
217. Reason, J. T., and Graybiel, A.: Progressive adaptation to Coriolis accelerations associated with 1-rpm increments in the velocity of the slow rotation room, Aerospace Med. **41**:73-79, 1970.
218. Vasil'yev, P. V., and Volynkin, Y. K.: Some results of medical investigations carried out during the flight of Voskhod, NASA TT F-9423, Washington, D. C., 1965, U. S. Government Printing Office.
219. Graybiel, A., and Kellogg, R. S.: The inversion illusion in parabolic flight: its probable dependence on otolith function, Aerospace Med. **38**:1099-1103, 1967.
220. Wood, C. D., and Graybiel, A.: Theory of antimotion sickness drug mechanisms, Aerospace Med. **43**:249-252, 1972.
221. Wood, C. D., and Graybiel, A.: The effectiveness of benactyzine hydrochloride and other antimotion sickness drugs in new combinations, NAMRL-1152, Pensacola, Fla, 1971, Naval Aerospace Medical Research Laboratory.
222. Gualtierotti, T., Bracchi, F., and Rocca, E.: Orbitport on the data reduction and control experimentation, prepared by the University of Milan, ing frog otolith experiment (OFO-A), final re-Second Department of Human Physiology, Milan, Italy, for NASA, Contract NASW-2211, January, 1972.

SUGGESTED READINGS

Physics

1. Alonso, M., and Finn, E. J.: Fundamental university physics, Reading, Mass., 1967, Addison-Wesley Publishing Co.: vol. 1, Mechanics.
2. Bergmann, P. G.: The riddle of gravitation, New York, 1968, Charles Scribner's Sons.
3. Borowitz, S., and Bornstein, L. A.: A contemporary view of elementary physics, New York, 1968, McGraw-Hill Book Co.
4. Condon, E. U., and Odishaw, H., editors: Handbook of physics, ed. 2, New York, 1967, McGraw-Hill Book Co.
5. Kittel, C., Knight, W. D., and Ruderman, M. A.: Mechanics, New York, 1965, McGraw-Hill Book Co., vol. 1.
6. Krauskopf, K. B., and Beiser, A.: Fundamentals of physical science, ed. 6, New York, 1971, McGraw-Hill Book Co.
7. Resnick, R., and Halliday, D.: Physics, part 1, New York, 1967, John Wiley & Sons, Inc.

Mathematics

1. Kleppner, D., and Ramsey, N.: Quick calculus, New York, 1965, John Wiley & Sons, Inc.
2. Martin, W. T., and Reissner, E.: Elementary differential equations, ed. 2, Reading, Mass., 1961, Addison-Wesley Publishing Co.
3. Selby, S. M.: Standard mathematical tables, ed. 19, Cleveland, Ohio, 1971, The Chemical Rubber Company.
4. Thomas, G. B., Jr.: Calculus and analytic geometry, ed. 4, Reading, Mass., 1968, Addison-Wesley Publishing Co.

Aerodynamics

1. Dommasch, D. O., Sherby, S. S., and Connolly, R. F.: Airplane aerodynamics, ed. 4, New York, London, 1967, Pitman Publishing Corp.

7 RADIANT ENERGY

PART A: SOLAR RADIATION

FARRINGTON DANIELS, Jr.

Solar radiation is the major source of energy for the terrestrial biosphere. As such, it has implications for many of the topics discussed in this book. The key role of our sun to life on earth has long been recognized; in many religions the sun is god. In the Hebrew bible, Book of Genesis, Chapter I, the third verse reads, "and God said, let there be light and there was light." In the King James version, the fourth verse reads, "and God saw the light, and it was good: and God divided the light from the darkness." Modern cosmologists attribute no less importance to light than did the ancient Hebrews.

The sun is about 93 million miles from the earth. As stars go, our sun is quite ordinary in size and brightness. It is an average yellow star about 100 times as bright as an average red star and about one hundredth as bright as an average blue star. In its interior thermonuclear reactions transmute about 564 million tons of hydrogen into 560 million tons of helium every second. The matter destroyed in this process is released from the sun's surface as radiant energy. It radiates outward in every direction at the speed of light and arrives at the outer limit of the earth's atmosphere at an intensity of 135.30 mW/cm^2, or slightly less than 2 gm-cal $\times$ cm^{-2} $\times$ min^{-1}. This value is called the *solar constant*. The solar spectrum ranges from short-wave high-energy x-rays to long-wave low-energy radiowaves, with ultraviolet, visible, and infrared between. About one half of the solar energy reaching the surface of the earth is infrared radiation and the remainder is largely visible and ultraviolet light.

Fortunately for us, our present atmosphere screens out x-rays, much of the ultraviolet, and certain other radiations discussed later in this chapter.

The student of biology may have difficulty visualizing elementary particles such as protons and electrons. Books on optics or physics sometimes state that these particles cannot be understood by analogy with the familiar perceptible world around us, but only in terms of the mathematics of quantum mechanics. It is said that our own body mass is about halfway between that of an electron and the universe. Newton's laws of mechanics do not apply to objects in our universe that are either very large or very small. We also have difficulty visualizing the universe as infinite, bounded, or curved. Physicists are still struggling to understand magnetic, electric, gravitational, and other forces.

It was long argued whether electromagnetic radiation consists of particles or waves. This argument was settled with the dualistic recognition that it has the characteristics of both and that it occurs in units called quanta, or photons. The simplest definition is that "a photon is the smallest quantity of radiation."

Electromagnetic radiation travels through a vacuum at a speed of 2.9978×10^{10} cm $\times$ sec^{-1} and is composed of oscillating magnetic and electric fields that are perpendicular to each other and perpendicular to the direction of propagation. Frequency, wavelength, wave number, and the electron-volt are units used to characterize electromagnetic radiation. Frequency is the number of cycles per second, a cycle being one complete oscillation into the magnetic and into the electric vectors. Wavelength is the length of the same oscillations and is expressed in meters (m), millimeters (mm), micrometers (10^{-6} m, formerly called micron) (μm), nanometers (10^{-9} m, formerly called millimicron) (nm), and a unit now less commonly used, the Angstrom, or 0.1 nm (Å).

The wave number is the number of cycles per centimeter or now, more correctly, the number in 10 mm, where wavelength is given in the same units. Wave number and frequency are proportional to the energy of a photon. Energy and frequency are related in the following famous equation:

$$E = h\nu \quad (1)$$

where E = energy
h = Planck's constant (6.624×10^{-27} erg × sec)
ν = frequency

Solar electromagnetic radiation is a long continuum; the physical, chemical, and biologic effects of different spectral regions depend upon the energy per photon.

An electron-volt (ev) is the energy acquired by an electron accelerated by a potential of 1 volt. It is a useful unit for describing the energy of photons of different wavelengths. Another useful unit is the Einstein, or the energy of the number of photons equal to Avogadro's number (6.022×10^{23}). The properties of different regions of solar electromagnetic radiation are given in Fig. 7-1. It is important to distinguish between radiation that produces thermal effects, such as infrared (IR) radiation and most visible radiation, and radiation in the ultraviolet (UVR) and visible that produce photochemical effects.[2]

Fig. 7-1 shows that the energies of infrared photons are much less than those of the ultraviolet; upon absorption they produce only rotation of the molecule about some axis and molecular vibrations that change the bonds associated with displacements of the atomic nuclei relative to each other within a molecule. The thermal effect may increase reaction rates. Below 200 nm (6.25 ev) radiant energies are greater than most chemical bond energies, and thus bond breakage and ionization occur. Energy levels that produce

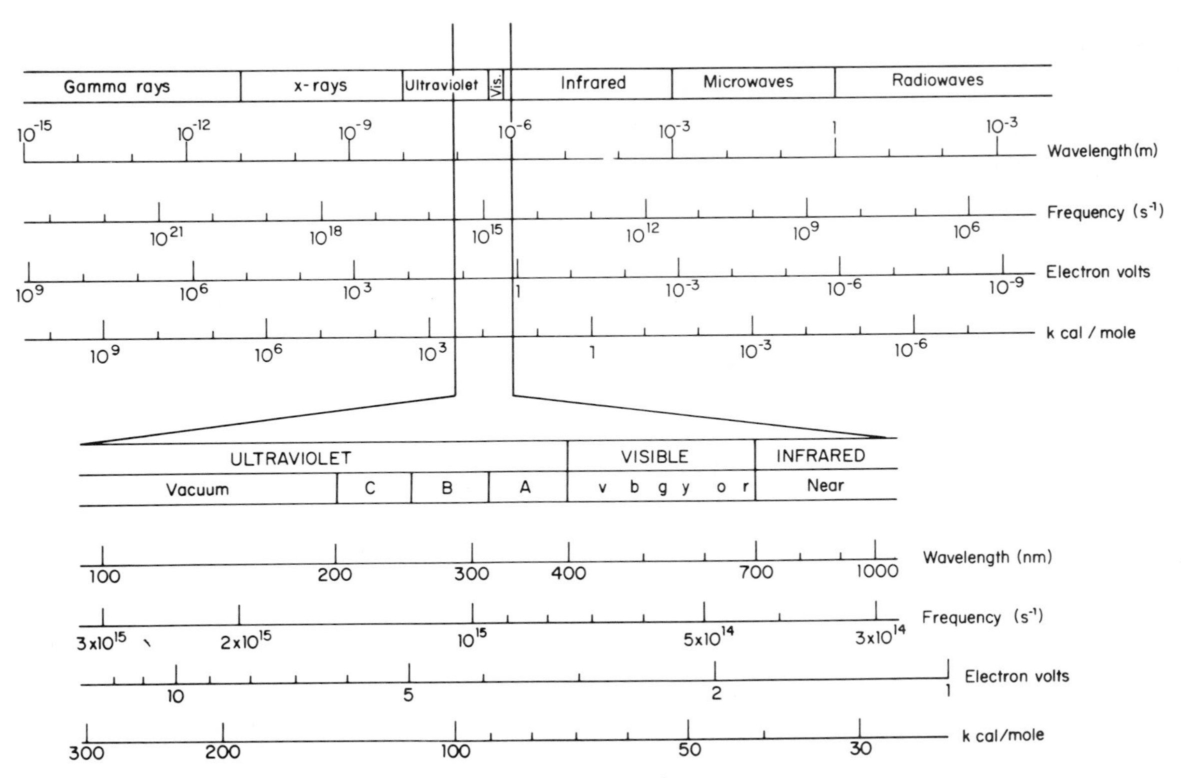

Fig. 7-1. Electromagnetic spectrum.

electronic excitation are significant for photochemistry and photobiology and thus also for environmental physiology in considering the photobiologic aspects of solar radiation. It is important to distinguish thermal effects from photochemical effects with radiation in the ultraviolet and visible regions that produces both.

Spectroscopic analysis of chemical structure depends upon the fact that a given wavelength of radiation is absorbed only when its energy matches that of a particular "allowed" excited state. The biologic compounds in man that absorb in the ultraviolet between 290 and 400 nm (those absorbing the ultraviolet found in sunlight) are molecules with conjugated double bonds such as those containing aromatic ring structures. Proteins and nucleoproteins absorb most strongly at wavelengths shorter than 290 nm, but absorption does occur even beyond 315 nm.

Heating increases the average energy of atoms and molecules and thus their chemical reaction rates. Photochemical and photobiologic effects depend upon the absorption of a photon by a single atom or molecule, which is then raised to a more reactive excited state. By exciting single molecules, photons affect biologic systems, producing reactions that could not be caused by temperature increases compatible with life.

Absorption of ultraviolet radiation, and in certain molecules visible radiation, raises electrons to higher energy orbitals, producing excited states. From their spectroscopic effects excited states are referred to as singlets or triplets. Many different reactions occur when an excited singlet loses some of its energy to become a triplet, which is longer-lived and referred to as metastable.

RADIATION, EVOLUTION, AND THE ATMOSPHERE

As solar ultraviolet, visible, and infrared radiation pass through the atmosphere, they are absorbed, scattered (change of direction without loss of energy), and reflected in every direction, depending on their wavelength and on atmospheric turbidity and cloudiness. Shorter wavelengths are scattered much more than longer wavelengths; thus, the sky appears blue because blue light is scattered more than red. Sunsets are colorful because the atmosphere transmits the longer visible wavelengths while blue is scattered. Infrared photography produces clear photographs despite haze. The wavelengths of incoming solar radiation range from approximately 290 nm to nearly 8 μm. Radiation from the earth, which acts as a radiator at about 10° C, extends from 4 μm to wavelengths longer than 30 μm.[3] Some visible radiation is reflected back to space; the percent of reflectance is called the *albedo.* Atmospheric carbon dioxide absorbs in bands near 2 and 4 μm and broadly from 13 to 17 μm; carbon dioxide is thus a barrier to outgoing radiation, and the "greenhouse effect" helps to maintain atmospheric temperature. Water vapor absorbs in bands between 1 and 2 μm and broadly between 5 and 8 and beyond 17 μm. Other considerations of atmospheric transmission of ultraviolet radiation will be discussed in connection with sunburn.

Oxygen is basic to physiology, but the student should not take it for granted. Because animal life depends on oxygen and because the atmospheric oxygen concentration appears quite constant during the relatively brief time it has been measured, one might assume that atmospheric oxygen is a geophysical law; this is far from true. The photosynthetic activity of green plants has been adding oxygen to the earth's atmosphere as it removed carbon dioxide for approximately a billion years. The total atmospheric oxygen is replaced by the photosynthesis of chlorophyll-containing plants in approximately 40,000 years, which is a short time in geologic perspective.

Prior to organic evolution there were eons (an eon is 10^9 years) of chemical evolution during which the action of ionizing radiation, ultraviolet radiation, lightning, and sound pressure waves created important organic compounds from hydrogen, ammonia, and methane in the primitive reducing atmosphere.[4] In the laboratory, purines, pyrimidines, amino acids, and some of their precursors and simple polymers can be made using electric discharges, ultraviolet radiation, and the other modalities indicated. Another theory of the origin of the early atmosphere is that it was composed of water, carbon dioxide, cyanogen, sulfur dioxide, and hydrochloric acid and other trace gases rather than ammonia, hydrogen, and methane.[5] In either case, carbon, nitrogen, and oxygen were present for chemical evolution.

It is visualized that in this primeval "soup" of organic compounds macromolecules aggregated and ultimately became self-replicating, and chemical evolution became organic evolution. Ultraviolet radiation shorter than 310 nm is extremely destructive to unpigmented microorganisms, and the habitat for organisms during the time of the reducing atmosphere must have been restricted. However, after chlorophyll evolved, oxygen release to the atmosphere produced important changes.[6] Atmospheric oxygen is converted to ozone (O_3) by wavelengths of radiation shorter than 240 nm. Ozone absorbs ultraviolet radiation from about 200 to 320 nm and blocks radiation below 290 nm. The ozone layer, which would be only 3 mm thick if reduced to standard temperature and pressure at sea level, protects the biosphere from the biocidal radiation shorter than 290 nm. The small band between 290 and 310 nm is essential for life because it produces vitamin D and is also responsible for killing microorganisms in the atmosphere; it also has several harmful effects on man.

Plants appear green because green light is not used in photosynthesis or plant regulatory functions as are blue and red light; 500 to 600 nm radiation is used little by plants. By coincidence or by evolutionary adaptation this light not used by the plant is the wavelength of maximum sensitivity of our eyes. The near infrared is also reflected from plants, protecting them from the thermal effects of radiation they do not use.

The direct effects of solar radiation are much better known for plants and animals than they are for man. Solar radiation has the following biologic effects on animals and plants:

1. It provides direct and indirect heating.
2. It permits vision.
3. In insects and a variety of other animals there are light-sensitive structures that do not form images. In some animals the pineal gland is light sensitive. In arthropods ocelli are light-sensitive structures without image resolution.
4. It facilitates synthesis of vitamin D and possibly other biochemical substances.
5. It directs hormonal and neural changes of pigmentation in certain species.
6. It regulates photosynthesis and growth in plants.
7. It causes photoperiod effects and is a time trigger (Zeitgeber) for many rhythmic biologic processes and biologic clocks.
8. Many animals, especially birds, navigate by direct vision or by detecting solar position by light polarization; the position is compensated by a biologic clock mechanism.
9. Ultraviolet radiation damages unprotected cells.

STRUCTURE OF SKIN

Although thin, the human skin is remarkably complex, consisting of many cells and tissues with a variety of functions. The dermis is composed of collagen, elastic, and other finer fibers; in the deep dermis the collagen and elastic fibers are large and course. The fibers are embedded in a homogeneous ground material that provides mechanical viscosity and is also a diffusion barrier. This ground substance is largely composed of acid mucopolysaccharide (glucosaminoglycans) and has a very high electric charge that is responsible for some of its properties relating to ionic movement and to water storage in the tissue. The outer dermis is composed of fine fibers and many elastic fibers have fine filaments that connect with the dermal-epidermal junction. The dermis also contains blood vessels, nerves, lymphatics, sweat glands, hair follicles, and sebaceous glands attached to hair follicles. The basketwork of nerve endings around the hair follicles is particularly concerned with the sense of touch, whereas undifferentiated nerve endings near the dermal-epidermal junction sense heat, cold, pain, and touch; specialized nerve endings are not required to transduce these stimuli.

Human skin is supplied with much more blood than necessary for its own nutrition, an adaptation for variable heat loss involving skin temperature change and sweat formation. The dermal-epidermal interface is irregular, with capillaries that loop into the papillae between deeper epidermal ridges called rete ridges. Fig. 7-2 is a simplified diagram of the skin and its circulation. Compare the diagram to the histologic section shown in Fig. 7-3. Organized arteriovenous shunts are located particularly in the tips of the extremities and in the ears; these permit large circulatory changes that subserve thermoregulation rather than local nutritional needs.

The dermis and epidermis are connected by a basement membrane. This may be a poly-

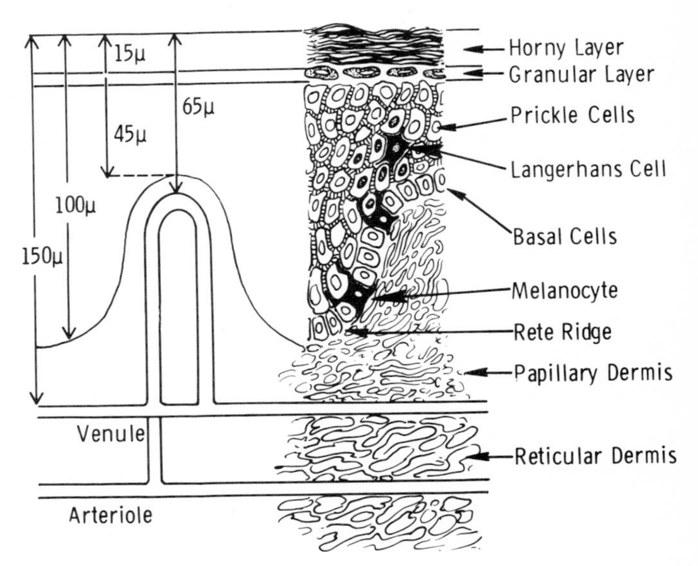

Fig. 7-2. Diagram of human skin.

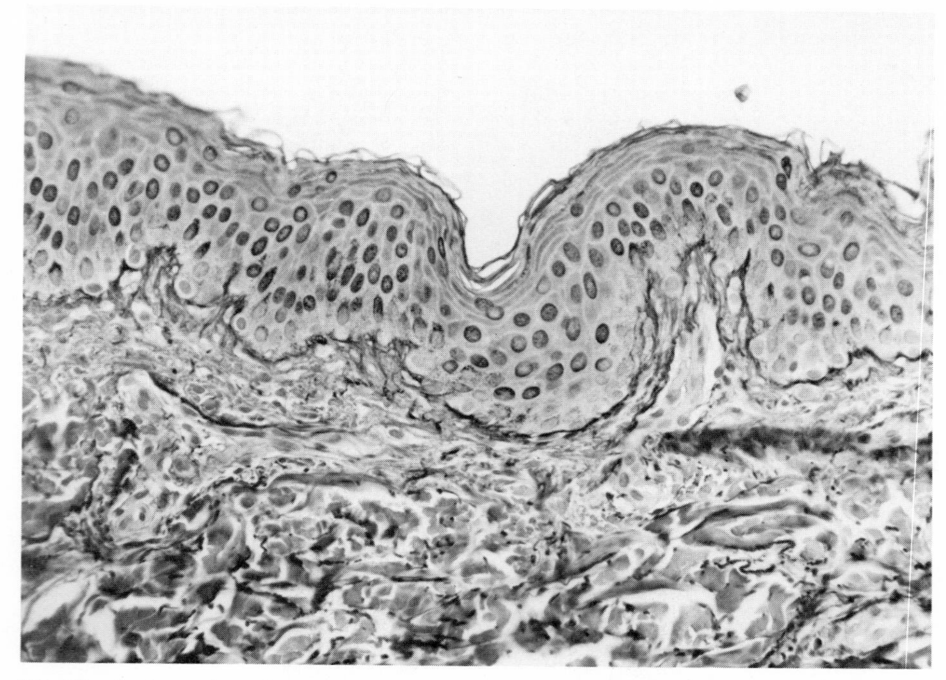

Fig. 7-3. Histologic section of human skin excised from the back. (Silver-orcein–aniline blue stain.)

saccharide glue into which various projections and filaments from the epidermis and dermis insert. The epidermis may be removed from the dermis by suction pressure; temperature greatly affects the time to produce a blister at a given suction pressure. Even though the epidermis is only about 60 to 100 μm thick, or roughly a twentieth to a tenth of a millimeter, it is a complex structure with many interacting cells types. An essential product of the epidermis is the stratum corneum, or horny layer, on the outer surface. This is approximately 15 μm thick and is an extremely tough, chemically resistant, almost impermeable layer, resembling a plastic film. It is composed of many layers of flat cells that have lost most of their cytoplasm and nucleus but that contain filaments of keratin in a matrix. The cells are held together by an extremely tough cement substance of unknown composition. Much of the barrier function of the stratum corneum involves lipids that can be extracted and then replaced, restoring the barrier function. The thickness of this dead layer is maintained with remarkable constancy, posing an intriguing physiologic question: How is a dead layer physiologically regulated?

The horny layer is produced by a cell line called *keratinocytes*. The basal cell is the keratinocyte adjacent to the basement membrane. It replaces itself by division and also divides to provide the cells that migrate. Attachments between these migrating cells suggested the name *prickle* cells for this spinous layer. The spaces between prickles are filled with an intercellular cement. Keratinocytes migrate outward and as they approach the horny layer they accumulate granules, becoming the *granular* layer. In healthy skin there is a sharp delineation of these cells with nuclei from the flat stratum corneum cells.

Interspersed among the basal cells are *melanocytes,* formerly called melanoblasts. During embryonic development these cells migrate into the skin from the neural crest and retain the appearance of dendritic nerve cells. Melanocytes synthesize the pigment melanin, starting with the amino acid tyrosine, which is acted upon by the enzyme tyrosinase. Melanin is a large random polymer that is deposited with protein in specific subcellular organelles called *melanosomes.* Melanosomes are transferred through the cell processes, or dendrites, into the basal cells and lower level keratinocytes of the prickle cell layer. After endocytosis by the keratinocyte they are found in vacuoles associated with lysosomes. Lysosomes are membrane-bound intracellular packages of enzymes that digest materials within them and, upon rupturing, digest materials around them. The mature melanosome, or melanin granule, is dispersed into smaller melanin particles as the cells move outward.

Melanins are brown to black pigments found widely in animals and in some plants; they are chemically very resistant. They have a relatively flat absorption spectrum throughout the ultraviolet, visible, and near infrared radiation. Melanin is not only a good absorber but the melanin granule, which is about 1 μm in size, also scatters well. Fig. 7-3 shows that melanin granules (silver stain) tend to localize as a cap over the nucleus of the keratinocyte. This localization may protect nuclear DNA, which contains the genetic information for the maintenance of the epithelium throughout life. The degree of melanization may be the largest factor determining the tolerance of the human skin for ultraviolet radiation. Dark races differ from light races in the amount of melanin formed, although the number of melanocytes is approximately the same. Negroes form larger melanin granules than Caucasians and upon transfer into the keratinocyte the melanin granules are packaged individually.[7] In the caucasoid, two or more are found in each lysosome. Negroes presumably achieve greater dispersion of pigmentation by this means, and the granule remains intact into the stratum corneum, rather than being showered out in smaller fragments as it is in Caucasian skin. Oriental skin appears to behave as the Caucasian, whereas that of the Australian aborigine behaves more as the African Negro although, in terms of other criteria, they are apparently not closely related. The alleged difference of stratum corneum thickness between black and white persons disappeared under careful study[8]; however, it was found that the Negro stratum corneum is quite opaque because of the persistance of melanin granules.

The significance of racial differences is not entirely understood, and their role in adaptation to different environments is the subject of much speculation. The standard explanation is that the dark-skinned races are protected from the aging, carcinogenic, and other damaging effects of ultraviolet radiation by their high melanin content. The light skin of the European, Caucasian, Chinese, and Japanese is considered an adaptation to low levels of ultraviolet radiation.

SOLAR RADIATION AND ENVIRONMENTAL HEAT LOAD

The environmental heat load is the sum of the heat transferred to the body by (1) *conduction* via feet or other body areas in direct contact with a warmer surface such as the ground; (2) *convection* from moving air warmer than the skin surface; and (3) *radiation*. The radiation component includes direct solar radiation, radiation scattered from the sky, and both reflected and scattered radiation from the ground and surroundings. The last also emit long-wave infrared radiation as a function of their absolute temperature.

Desert environments feature extensive heat exchange by infrared radiation between man and surroundings. This produces an oven-like feeling during the day and a sharp chill at night. Surprisingly, the cool part of the sky is a source of cooling; in the northern hemisphere the northern sky is often the coolest part of the surroundings and the skin may lose heat to it by infrared radiation.

Heat exchange between man and his environment is complex. During daylight hours he receives infrared, visible, and ultraviolet radiation directly from the sun, scattered radiation from the sky, and reflected radiation from the ground. The surfaces of his skin and clothing exchange energy in the long infrared region, depending upon the respective temperatures of the radiating environmental, clothing, and skin surfaces. Heat exchange under these conditions is a function of the difference of the fourth powers of the absolute (°K) temperatures, the Stefan-Boltzmann relationship. Thus even when exposed to solar radiation, the body radiates heat to the cool part of the sky. Nocturnal radiation to the sky is an important feature of cooling in many environments, particularly the desert. Even though the ground is dry, radiation exchange with a cool sky may decrease the temperature of the ground surface to the dew point. Heat exchange between man and air occurs by conduction and by transfer of heat to moving air, or convection. The film of air, or boundary layer, at the skin surface is either relatively static or rises as a chimney effect of body heat.

Stoll and Hardy,[9] using a panradiometer and thermistor radiometer simultaneously, studied direct solar radiation, scattered and reflected solar radiation, average radiant temperature of the total environment, radiant sky temperature, radiant ground temperature, air temperature, and wind velocity. Their data for New York City, Fairbanks, Alaska, and Death Valley, California, during the months of July and August are given in Fig. 7-4. The findings shown in the diagram are based upon a solar radiation of 660 kcal $\times$ m^{-2} $\times$ hr^{-1} and a wind velocity of 5 mph. The sky temperature at

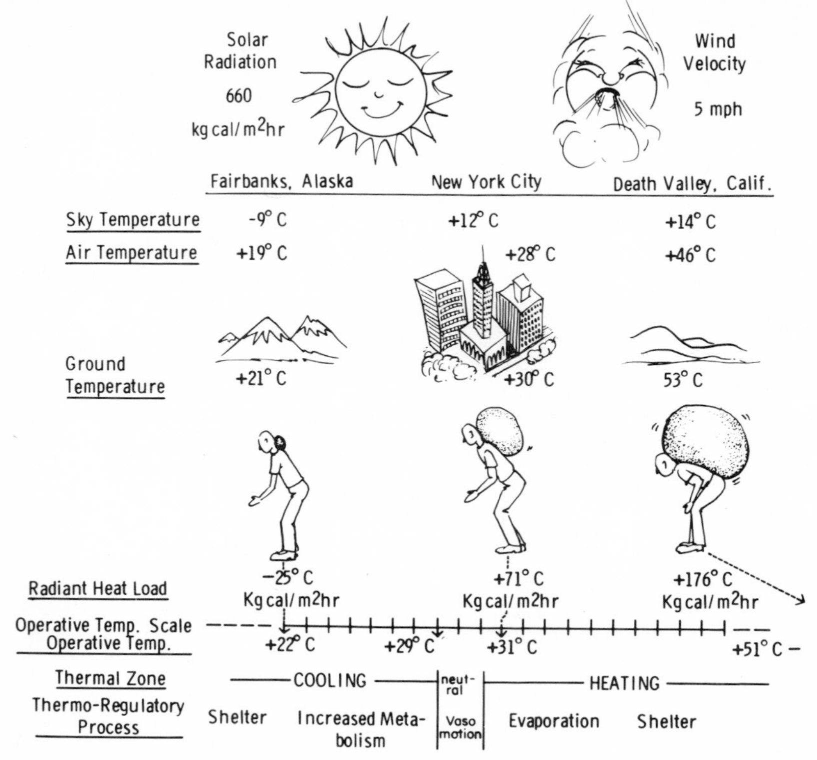

Fig. 7-4. Interaction of bioclimatologic components in three different environments during July and August. (Modified from Stoll, A. M.: The measurement of bioclimatological heat exchange. In Hardy, J. D., editor: Temperature: its measurement and control in science and industry, part 3, New York, 1963, Van Nostrand Reinhold Co.)

Fairbanks was –9° C; at New York, 12° C; and at Death Valley, 14° C. Air temperature was 19° C in Fairbanks, 28° C in New York, and 46° C in Death Valley. Ground temperature was 21° C in Fairbanks, 30° C in New York, and 63° C in Death Valley. These thermal conditions produced a radiant heat load on the man in Fairbanks of –25 kcal $\times$ m^{-2} $\times$ hr^{-1}, in New York of +71 kcal $\times$ m^{-2} $\times$ hr^{-1}, and in Death Valley of +176 kcal $\times$ m^{-2} $\times$ hr^{-1}. Compare these to the heat production of 40 kcal $\times$ m^{-2} $\times$ hr^{-1} for a resting man.

Man's erect posture gives him an advantage over four-legged animals in using and in avoiding solar radiation. Vertical posture increases the area available for warming when the sun is low in morning or evening but minimizes heat load and sunburn radiation during the middle of the day. This advantage was mathematically expressed by Underwood and Ward[10] but has received little attention from students of human evolution. The effect can be visualized in Fig. 7-5, which is redrawn from Underwood and Ward. When a man faces the sun rising on the horizon, 24.7% of his body surface receives direct radiation. When the sun is directly overhead a standing man receives direct radiation on only 4.4% of his body surface. The amount of indirect radiation he receives depends upon scatter from the sky and reflectance from the ground and surrounding objects.

The role of skin pigment in solar radiation uptake was studied in the southwest desert by Baker.[11] Black and white soldiers matched for height and weight were exercised in the desert. When black soldiers were protected from the sun, the body temperature rise was the same or less than that of the white soldiers. When the two groups exercised essentially unclothed in the sun, the black's rectal temperature rose substantially more than the white's. The average rise for white subjects walking almost nude in the sun was 0.5° C, whereas for black subjects it was 0.7° C. Lee,[12] studying men in the Arizona desert, found that the radiation load was equivalent to a 6° to 9° C increase of air temperature.

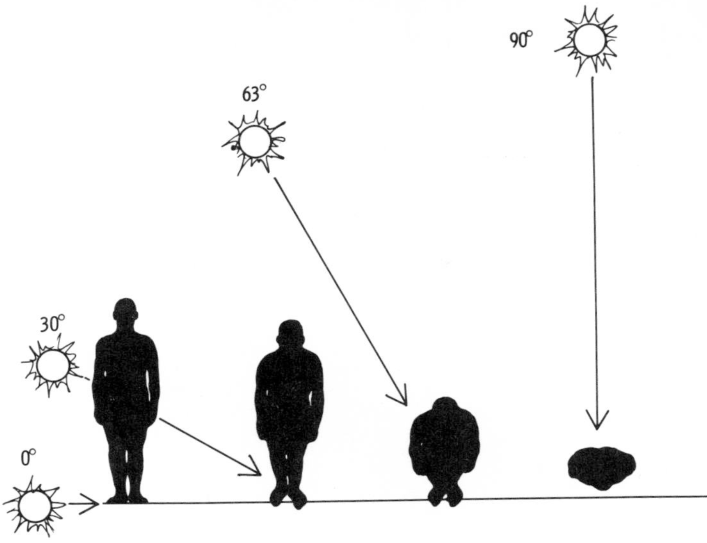

Fig. 7-5. Surface area of direct sunlight exposure on a standing man facing the sun at different solar elevations. (Redrawn from Underwood, C. R., and Ward, E. J.: Ergonomics 9:155-168, 1966.)

The amount of solar radiation varies from moment to moment, hour to hour, and with both latitude and season. Thus it has many indirect effects on weather and climate and may also have indirect effects on man, other animals, and plants; these are in addition to direct effects of intensity, duration, and wavelength composition of solar radiation.

When Europeans moved to the tropics during the fifteenth through nineteenth centuries, understanding of tropical diseases was limited. The impression developed that sunshine itself produced sunstroke. Indeed, it was believed that solar radiation penetrated the brain and spinal cord, causing high fever and death.[13] Actually, the only solar rays that penetrate more than a fraction of a millimeter are the long visible red and near infrared; ultraviolet radiation is absorbed in a fraction of a millimeter. There is no solar radiation that penetrates the human brain, although in certain animals light appears to stimulate receptors in the brain directly. As in many other animals the effects on man of daylength, light period, and dark period are apparently mediated through the eye, the sympathetic nervous system, and the pineal gland.[14]

The injurious effects of sunlight on cells and tissues are produced by ultraviolet radiation that causes sunburn and, with chronic exposure, skin cancer. The direct solar radiation load is one factor increasing body heat; many animals bask or otherwise orient themselves to minimize or maximize the solar contribution to body temperature. If the solar load on the human body exceeds the capacity of the body to cool by radiation, convection, and sweat evaporation, cardiovascular inadequacy, dehydration, electrolyte deficiency, mental confusion, pallor, hypotension, headache, tachycardia, vomiting, low grade fever, and fainting occur; this condition is *heat exhaustion*. Rest, fluid replacement, and protection from further heat exposure may be all that is necessary. If thermoregulation fails to meet the thermal stress in a subject exposed to a high environmental temperature, a high radiant heat load, or other heating factors, a condition characterized by high body temperature (hyperpyrexia), irritability, prostration, delirium, hot dry flushed skin, diminished to absent sweat, hyperventilation, and coma may occur, causing temporary or permanent damage to the thermoregulatory system and even death. This condition is much more serious than heat exhaustion and is the sense in which *heat stroke* is used.

The unacclimatized individual is more vulnerable to heat stress, whether this includes direct solar radiation or not. While melanin pigment protects the skin from local injury, it increases the uptake of direct solar radiation. The theory that black pigmentation is adaptive to solar radiation in the tropics is simplistic and has been challenged on various grounds.

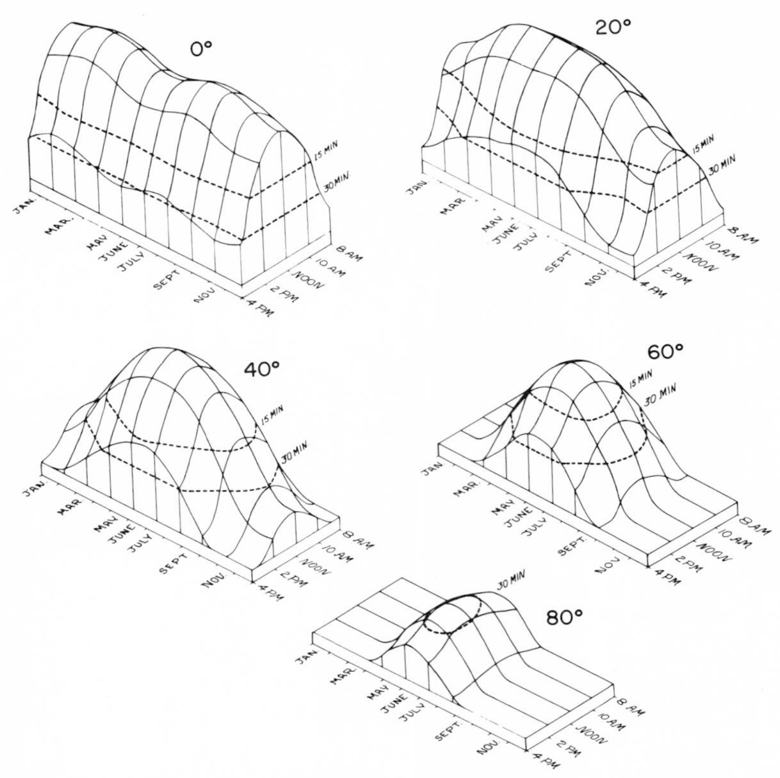

Fig. 7-6. Predicted minimal erythema time calculated from sunburn radiation intensity as a function of time of day, day of year, and latitude. Calculations based on measurements of direct sunlight and skylight incident to a horizontal surface. (Modified from Daniels, F., Jr.: Arch. Derm. **85**:358-361, 1962.)

EFFECTS OF ULTRAVIOLET RADIATION ON HUMAN SKIN

The amount of sunburn ultraviolet radiation reaching the earth's surface is a function of the amount of atmosphere through which it passes; it is also attenuated by absorption in the ozone layer and scattered by atmospheric turbidity. Latitude, season, and time of day all affect the intensity of sunburn-producing radiation. A series of isometric graphs prepared by Wallman[15, 16] display the effects of latitude, day of the year, and time of day on predicted minimal erythema time (Fig. 7-6). These show the sunburn ultraviolet intensity at different times of year and at different times of day for 80, 60, 45, and 30 degrees north latitude and for the equator, based on the radiation incident to a horizontal surface. These graphs were drawn by extrapolating from measurements made in clear air and should be considered pictorial rather than safe predictors of sun tolerance times under different circumstances. As would be expected, these graphs show a rise and fall to a lower intensity as one approaches the poles. Sunburn radiation is also the same wavelength that produces vitamin D, skin cancer, and cataracts of the lens of the eye and that kills atmospheric microorganisms. Thus the relationships shown in Fig. 7-6 have general biologic significance as well as importance for the white-skinned man going to the beach.

Sunburn is the complex reaction of skin to injury by 290- to 320-nm ultraviolet radiation. Ultraviolet radiation produces photochemical effects as well as thermal effects; it is inappropriate to consider sunburn as an example of a first-degree thermal burn. Photochemical reactions have the characteristic of reciprocity; a given quantity produces the same effect whether received in a short time or distributed over a longer interval. Some artificial sources of ultraviolet radiation produce reactions in a few seconds or minutes equivalent to those produced by hours of natural

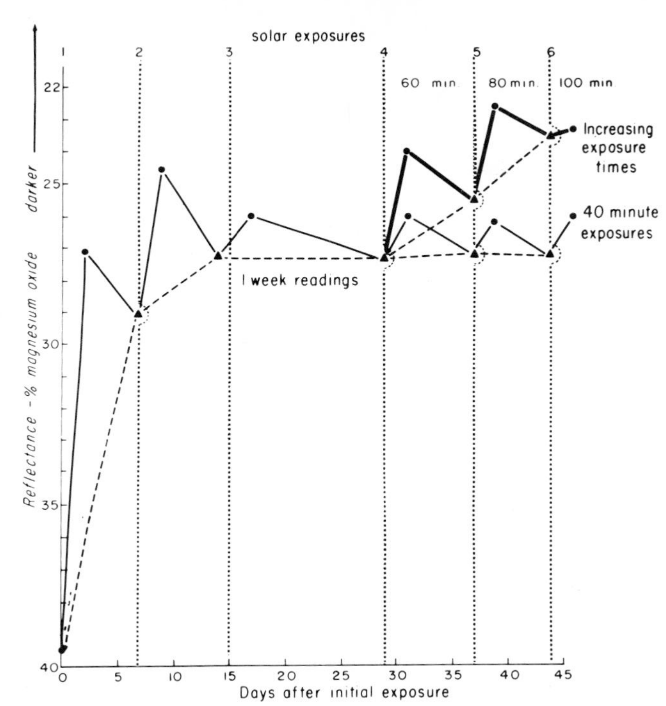

Fig. 7-7. Skin acclimatization to ultraviolet radiation as shown by melanin pigmentation. Repeated exposures of equal duration compared to exposures of increasing duration. (Redrawn from Daniels, F., Jr.: Ultraviolet radiation and dermatology. In Licht, S., editor: Therapeutic electricity and ultraviolet radiation, ed. 2, New Haven, Conn., 1967, E. Licht, Publisher.)

sunlight. However, there are limits since we do not accumulate from one day to the next.

Sunburn is an experience familiar to most white-skinned and to many dark-skinned people.[17] Redness may not be apparent during the sun exposure, there being a latent period of several hours. Blood vessels in the exposed areas dilate, usually reaching a maximum between 8 and 24 hours. Sunburn erythema is localized sharply to exposed areas. Sunburn is associated with discomfort, lowering of the pain threshold, severe blistering, sweat suppression, and changes in the pharmacologic and mechanical responses of blood vessels that persist for many months. Repeated sunburn leads to degenerative changes in both dermis and epidermis; skin cancer is one of the consequences of repeated exposure.

Adaptation of human skin to repeated ultraviolet exposure is so familiar that it is frequently overlooked as a simple model of acclimatization in general. The amount of melanin pigmentation in human skin determines the tolerance to ultraviolet radiation. This is true at the racial level, at the individual genetic level, and in terms of personal adaptation. In Fig. 7-7, reflectance readings are shown for a group of men exposed to summer sunlight in Idaho. A series of 40-minute exposures are shown in the lower dashed line. After reaching a plateau at the third exposure, which was given 2 weeks after the first exposure, melanin pigmentation had reached a plateau and with continued 40-minute exposures the skin remained acclimatized to the 40-minute exposure with the same degree of pigmentation. In contrast, in the right upper portion of the graph it is seen that with increasing exposures of 60, 80, and 100 minutes pigmentation increases progressively. This is somewhat analogous to a man in poor physical condition who runs a quarter mile daily for a period of time, developing an improved capacity to run a quarter mile that reaches a plateau. If he then begins to run a mile daily, he starts with the improved acclimatization gained while running the quarter mile, but he has to develop additional increments of adaptation for the more strenuous exertion. This is characteristic of many kinds of adaptation to the environment. The reflectance meter following repeated sun exposure is a convenient method of demonstrating this phenomenon.

Almost everyone enjoys going to the beach on a sunny day unless he gets a severe sunburn. However, the good feeling associated with sun exposure is not explained in metabolic or biochemical terms. The only known beneficial effect of solar irradiation of normal human skin is that of vitamin D production. In contrast to the ignorance just expressed, much is known about the beneficial role of ultraviolet radiation in providing energy for vitamin D synthesis in the skin. Until the 1920's rickets was a serious medical problem in the United States and in Europe. Rickets involves a defect of calcium absorption and as a result bones form poorly during growth; the child grows up with bowed legs and other skeletal malformations. Women with rickets have deformed pelves that may prevent normal childbirth. Thus an environmental factor can have the same effect as a lethal mutation. Anthropologists believe that the short stature and curved bones of Neanderthal man resulted from rickets. Since rachitic deformity complicates childbirth and since dark-skinned children are much more susceptible to rickets than light-skinned children, some scientists believe that natural selection for light skin was extremely rapid.

The fascinating story of vitamin D [18, 19] includes the early knowledge in some regions that cod liver oil prevents rickets. It was later found that exposure to sunlight also prevents or cures rickets. Then it was found that feeding rachitic animals certain ultraviolet-irradiated foods prevents or cures rickets. Biochemical and photochemical studies showed that 7-dehydrocholesterol in animal skin is converted to vitamin D_3 by the same wavelengths of ultraviolet radiation that cause sunburn and that ergosterol in yeast and other plant sources is converted to vitamin D_2. The D vitamins have been added to milk and other foods, resulting in the virtual disappearance of rickets. The body has no effective mechanism for excreting excess vitamin D and, paradoxically, today vitamin D intoxication is more of a problem than rickets. Predictably, the light-skinned child is more prone to hypervitaminosis D than the dark-skinned child, presumably because the extra vitamin D synthesized in his lighter skin is in addition to that ingested.

In the liver vitamin D is converted to 25-hydroxycholecalciferol, which enters the circulation and is subsequently bound to the nuclei of the intestinal epithelial cells. Here it presumably serves as a messenger, telling DNA to transcribe the information to RNA, which in turn transcribes the message to form the enzyme needed for Ca^{++} transport from gut lumen to the circulation. Thus ultraviolet radiation of skin, fur, or feathers produces vitamin D, which, through a chain of biochemical events, regulates calcium absorption, making an internal calcified skeleton possible.

PHOTOSENSITIZING CHEMICALS

For thousands of years in Egypt and India plants containing psoralens (furanocoumarins) have been used in conjunction with sunlight to stimulate melanin formation in the skin of patients with vitiligo, an acquired white spotting of the skin. Psoralens are found in the plant around the seeds and are thought to play a role in the regulation of germination. St. John's wort (*Hypericum* sp.) and buckwheat (*Fagopyrum sagittatum*) contain chemical substances that photosensitize the domestic animals that eat them; the effects are observed in exposed light-colored skin areas. Animals with liver disease may be photosensitized by the chlorophyll breakdown product phylloerythrin. The photosensitizing effects of plants on animals are discussed in Blum's classic monograph[20] and in the review by Pathak and associates.[21]

Photosensitization is a process in which the energy of molecules excited by photon absorption is transferred to an attached or nearby molecule, producing an effect or starting a biochemical chain reaction. Many medications, drugs, and chemical substances in our increasingly artificial environment are photosensitizers. Those in soaps, cosmetics, and perfumes are especially troublesome. Tars, such as those used in roofing and road construction, also photosensitize, causing hyperpigmentation. Some drugs and hormones cause the liver to produce an excess of normal porphyrins, compounds in the hemoglobin synthesis pathway. Some porphyrins photosensitize skin. Werewolves were people with claw-like scarred hands, hairy and scarred faces, and red teeth who feared light and went out only at night; they probably had a rare form of congenital porphyria.

Photosensitization is not confined to biologic systems; for example, various dyes are used to sensitize the silver grains in photographic films to yellow, red, and even infrared radiation whereas silver grains alone react only to ultraviolet and blue light. Fluorescence indicates that a chemical substance is excited by absorbed radiation; almost all photosensitizers fluoresce under long-wave ultraviolet radiation, or "black light." Although many photosensitizing chemical substances also produce cancer, not every substance that fluoresces is carcinogenic.

Skin photobiologists suspect that there are many

unknown effects of solar radiation because visible and ultraviolet radiation penetrate the skin and reach the blood circulating in the capillaries. Red light passes readily through a fingertip and sufficiently through the cheeks and forehead to permit detection of fluid in the sinuses thus transilluminated. About 16% of the 300-nm ultraviolet radiation, or sunburn radiation, penetrates the epidermis of white skin. More than 50% of 400-nm blue light penetrates untanned white epidermis, and about 1% reaches the subcutaneous tissue. More than 70% of 550 nm green light reaches the dermis and 5% penetrates even farther.

The vitamin riboflavin photosensitizes some biologic systems. Erythrocyte protoporphyrin photosensitizes before the iron atom is inserted to form heme. I once put some of my own plasma into a spectrophotofluorimeter and found eleven bands of fluorescence. The suspicion that light has many unknown effects is reasonable.

SUMMARY

Solar radiation played a vital role in the chemical and then organic evolution of early life and then provided, via photosynthesis, the oxygen that made metazoan animals possible. It provides the biosphere with the energy required by most living things and continues to provide the oxygen that we recombine with food from the chain that starts with photosynthesis. It provides the light for vision, which gives us about 70% of our information input. Life on earth is strongly regulated by the direct and indirect effects of solar radiation. Electromagnetic radiation gives us what information we have about the universe and its evolution.

AUGUST, 1972

PART B: IONIZING AND NONIONIZING RADIATION

FRANCES L. ESTES

Soon after the discovery of x-rays in 1895, evidence of the harmful effects of indiscriminate exposure began to accumulate. Observations included hair loss, burns, chronic skin ulcers, and cancers. In the early 1900's workers using radium in the luminous paint industry developed bone cancer and aplastic anemia from radium ingestion.[22] The high rate of lung cancer in miners in Joachimsthal, Silesia, had been attributed to the high concentration of radon and radioactive daughter products in the mine atmosphere.[23] Observations on workers in the United States uranium mining and milling industry confirmed this.[24]

With the ongoing development of numerous electronic devices for use in the home, as well as in industry and hospitals, artificial sources of radiation have become a part of daily life. Of these, the major source may be the 214,000 x-ray units used in the health professions. In 1967 more than 1 million patients were medically exposed to x-radiation. Furthermore, the use of radioactive by-product materials and accelerator-produced materials is increasing rapidly. Every year twenty-five to thirty nuclear reactor installations begin operation. These reactors as well as fuel processing are a source of ^{131}I, tritium, ^{41}Ar, ^{133}Xe, ^{135}Xe, and ^{85}Kr. These isotopes may contribute to environmental radionuclide contamination, as do nuclear explosions, industrial and research use of radioactive materials, and uranium mining and milling.

X-rays are emitted from many consumer products, the most ubiquitous of which is the color television receiver. There are estimated to be almost 20 million color sets in use. Electronic cooking ovens that are coming into use in the home are another source of environmental radiation.

With the development of the cyclotron and interest in exploration beyond the earth's atmosphere, other types of ionizing radiation must be considered. There are three sources of radiation in space: the Van Allen radiation belt, solar cosmic rays (ionizing particles emitted by the sun during solar flares), and galactic cosmic rays (heavy particles outside the solar system). The effects of heavy particles, electrons, alpha particles, and protons must be considered in relation to space travel. Because of their energy and potential abundance, protons are the greatest biologic threat. This fact has prompted studies of heavy particle and proton radiation on whole animals as well as on cells and tissues.

Standards are being developed for x-ray emission of color television receivers, medical and dental x-ray equipment, and such nonmedical equipment as accelerators and x-ray diffraction units. The standard proposed by the U. S. Environmental Control Administration for microwave emission from electronic cooking ovens is 1 mW/cm^2 at the time of sale of the new product. Taking deterioration during use by the consumer into account, the standard goes up to 5 mW. Of more interest to us here is the elucidation of the mechanism of radiation damage and the development of indices of radiation damage to human biologic systems.

In the summary of a Conference on Radiobiological Factors in Manned Space Flights,[25] it was noted that there is considerable quantitative data available for a wide range of phenomena at the atomic, molecular, and cell levels relating the quality and quantity of radiation to the specific magnitude of the effects. For multicellular organisms and man, despite many investigations, there is little reliable quantitative data to serve as guidelines for the protection of astronauts during prolonged spaceflight, to say nothing of the limits of exposure for the stay-at-home watching the flight on color television. This lack of solid information persists, even though in 1966 there were 783,615 rats and 111,084 mice used in research programs of the Atomic Energy Commission alone.[26]

To understand the problems involved in evaluating the effects of radiation on man and the difficulties in setting standards for radiation control, it is necessary to have some concept of the nature of radiation and also of the various types of biologic response evoked. Thus we must understand the physics of radiation, the chemistry of molecules, and the biochemical interactions of the component parts, each of which contributes to the physiologic responses to radiation.

In considering the physiologic effects of radiation, be it ionizing or nonionizing, it must be realized that the total response of the organism is to the sum of a number of factors including amount and duration of exposure and other environmental factors, as well as rate of repair. Thus, as in the consideration of the responses of an organism to any other environmental stressor, here also the nature of the challenging system as well as the component parts and function of the receptor system must be considered. One of the means by which we try to understand the response of a whole organism is by observing the responses of individual cells and tissues.

To that end the electromagnetic spectrum will be examined along with the responses of simple molecules. This should permit understanding of the magnitudes and types of forces that elicit physiologic responses and, finally, the examination of the initiation and mechanisms of the physiologic responses.

RADIATION AND MOLECULAR RESPONSE

Electromagnetic spectrum

Before considering the impact of electromagnetic waves on man and his reactions to them, a look at the physical characteristics of the electromagnetic spectrum is in order.[27] Since the beginning man has been aware of light, color, and color differences and has developed numerous theories to explain his observations. The realization that the human eye perceives as light but a small fraction of the total electromagnetic wave spectrum has come from a number of observations during the nineteenth and early twentieth centuries. Thus the ultraviolet region, not visible to the eye, consists of light having wavelengths somewhat shorter than violet light. Ordinary x-rays have a wavelength of approximately 1 Å, whereas even shorter wavelengths of 0.1, 0.01, and 0.001 Å occur in the rays produced by nuclear decomposition and the action of cosmic rays. Conversely, beyond the red region at wavelengths longer than the eye can see are the infrared, microwaves, and radiowaves.

Table 7-1 is a schematic diagram of the electromagnetic spectrum showing the common names of its various regions. Spectroscopic responses are also shown. Note that with decreasing wavelength and the concomitant increase of frequency, absorption produces electronic and nuclear changes. Conversely, at longer wavelengths and lower frequencies molecular events occur. For simplicity, this diagram does not include all of the Hertzian (10^7 to 10^{14} Å) and longer wavelength region in which the unit of measurement is frequency in cycles per second rather than wavelength. For convenience the units of wavelength are not identical throughout the electromagnetic spectrum.

Regardless of what region of the electromagnetic spectrum is to be considered, its radiation can be characterized in terms of two properties—frequency, which is directly related to the wavelength, and velocity, as follows:

$$\nu = c/\lambda \tag{2}$$

where ν = frequency (cps)
c = a constant, the velocity of electromagnetic radiation (3×10^{10} cm/sec)
λ = wavelength (cm)

Table 7-1. Schematic diagram of the electromagnetic spectrum*

Wavelength	Spectral region			Structural response
1 Å			Gamma rays	Nuclear transitions
		X-rays		
10 Å			Soft x-rays	Inner shell translations
2,000 Å		Ultraviolet rays	Far ultraviolet	Valence electronic transitions
4,000 Å			Near ultraviolet	
	Visible			
8,000 Å				
0.8 μm				
2.5 μm		Infrared rays		Molecular vibrations
25 μm				
0.04 cm			Far infrared	Molecular rotation
400 μm				
		Microwaves		
25 cm				
		Radiowaves		Spin orientation (in magnetic field) Electron spin resonance Nuclear magnetic resonance

*Note that the scale is nonlinear and that the boundaries between adjacent regions are in general arbitrary.

The amount of light energy of wavelength λ that is absorbed or emitted in a single act was found by Planck to be proportional to the frequency:

$$E = h\nu \tag{3}$$

where E = energy of the light of frequency ν
h = proportionality constant

$$h = 6.624 \times 10^{-27} \text{ ergs} \times \text{seconds} \tag{4}$$

Note that Planck's constant h has the dimension of energy and of time. With time constant it is apparent that the shorter the wavelength, the larger the bundle of energy involved. This is shown in Table 7-2 in terms of energy of radiation. As seen in the table, energy in electron-volts (ev) is approximately 12,420 Å.

The 4,000- to 5,000-Å region may be regarded as representative of the energies of light in the visible region that produce such responses as photosynthesis and the events of vision and color perception. The ultraviolet region of the spectrum includes the so-called near (2,000 to 4,000 Å) and the far (2,000 to 1,000 Å); the ultraviolet region below 3,000 Å is removed from the solar spectrum by ozone absorption in the upper atmosphere.[28] The significance of this absorption and its importance for life on earth is well illustrated by the bactericidal action of ultraviolet lamps.

Table 7-2. Wavelength and energy of electromagnetic radiation

Energy relationships		Chemical energy (kilocalories)
Angstroms	Electron-volts*	
10,000	1.24	28.6
5,000	2.48	57.2
4,000	3.10	71.5
3,000	4.14	95
2,000	6.21	143
1,000	12.42	286
100	124.2	2,864
10	1,242	28,641
1	12,420	286,405

*The amount of energy acquired by an electron falling through a potential difference of 1 volt. 1 ev = 1.6×10^{-12} erg. kev = 10^3 ev and mev = 10^6 ev are frequently used multiples.

As we progress to still shorter wavelengths, the increasing energies anticipate the more complicated reactions and responses that ionizing radiation evokes. Conversely, the low energies found even at 10,000 Å in the near infrared region suggest that the responses to the microwave and infrared wavelengths involve a different phenomenon.

Just as increasing frequency is associated with increasing energy, so too is the decreasing frequency of the longer wavelengths (infrared and radiowaves) associated with less energy. Thus emphasis is on magnetic rather than on electric effects; the responses are no longer electronic transitions per se. Molecules, unlike atoms, do not respond as if they were rigid spheres but rather as amebas whose shape and orientation depend on their surroundings. The response to long wavelengths is one of molecular orientation in which the component parts respond in harmonic fashion. The quantum energies of infrared radiation are of the order of magnitude of the energies of vibrations of the atoms of the molecule. Hence, with infrared spectroscopy one observes the vibrations of the molecule and thus its fingerprints.

The atoms are considered harmonic oscillators, moving in simple harmonic motions at frequencies determined by their mass and by the strength of their binding forces. Models for demonstrations of the various frequencies and directions of motion of such a system have been made using steel balls and spiral springs. It becomes evident immediately that a molecule of *n* atoms, most simply thought of as a mass point, has a spatial configuration that may be represented as 3 n. Three of these may be used to represent the position of the center of the mass of the molecule, and three more are required to represent the rotational position of the molecular axis. Thus with respect to a fixed center of mass and fixed rotational position, the configuration is represented by $3n - 6$ internal coordinates. These coordinates also represent the vibrations of the molecule. If the atoms of the molecule move harmonically without displacing the molecule's center of mass or its rotational position, they must move in a precise manner, all with the same frequency and with definite relative amplitudes. Thus, the internal motions do not require $3n - 6$ coordinates for each of the many positions of the atoms, because these positions are not independent of each other.

In terms of the wave theory of radiation, the oscillating electric field of the radiation produces electric vibrations in the molecule at the same frequency as the radiation. This molecular electric vibration may be represented as a changing charge distribution, or dipole moment. Since the forces binding the atoms together and the forces available in the radiation are commensurate with the latter, the primary effect of such radiation is to vibrate the atoms so that the varying dipole moment produced by the absorption of the radiation is primarily a periodically varying separation of the charged centers. For radiation frequencies corresponding to the normal modes of vibration of the molecule, the resulting molecular vibrations are large, whereas at other frequencies the vibrations are negligible. Note that for symmetric molecules some of the normal vibrations involve such symmetric motion of the charged centers, or nuclei, that there is no net change of dipole moment. Absorption of radiation occurs only when it produces a periodic change of molecular dipole moment.

To this simple classic picture must be added the unfortunate complication that when one atom vibrates, it modifies the potential field in the region of another atom. Interaction of the different vibrations produces "overtones." Because atomic motions are harmonic, radiation of the frequency of these overtones or combined frequencies interacts with the molecule, causing it to vibrate. Although these overtones complicate the assignment of fundamental modes from the observed absorption spectrum, they are useful in analytic work.

Table 7-3 shows that ionizing radiation includes both electromagnetic waves, which have neither mass nor charge but produce ions on interaction with gases, liquids, or solids, and particles of diverse mass and charge. These diverse types of ionizing radiation vary in the extent to which they penetrate tissue as well as in the number of ions left in their tracks as they move through the tissue. One of the important variables with respect to biologic effect is ion density, or the number of ions produced per unit length of track[29]; this variable is termed *linear energy transfer* (LET).

Ionizing radiation results from naturally occuring radioactive decay processes or are produced by high-energy accelerators. A radioactive nucleus is one that spontaneously changes to a lower energy state and in so doing emits gamma rays and particles; the commonly emitted particles

Table 7-3. Charge and relative mass of ionizing radiation and elementary particles

Ionizing radiation	Mass	Charge
Electron	1	−
Proton	1,836	+
Neutron	1,839	0
Deuteron	3,671	+
Alpha particle	7,296	++
X-ray or gamma ray	0	0

are the alpha and beta. In addition to these particles, protons and x-rays occur in cosmic radiation or are produced in high-energy accelerators.

Elementary aspects of atomic and molecular excitation

As the name implies, an electromagnetic wave involves both an electric and a magnetic field. In free space electromagnetic oscillations travel at the speed of light and close locally when the source is turned off or removed. Any process that occurs in an atom or molecule associated with the absorption or emmission of radiation is called a *transition*. In general, energy absorption occurs only if the incident photons possess precisely the amount of energy for the transition to occur in the atom or molecule. Energies that are too high are no more effective than energies that are too low.[30]

When a molecule is exposed to electromagnetic radiation of suitable wavelength and polarization, the electronic interractions induce changes in the internal motion of the molecule. Since changes in electronic distribution are rapid compared to nuclear motions, the immediate effect of light absorption is to produce an unstable molecule with the electron charge distribution of the excited state but retaining the overall atomic distribution of the ground state.

The assumption of Planck, and also that of Einstein, that light (visible electromagnetic radiation) of frequency ν is emitted or absorbed only in quanta of energy $h\nu$ led to several problems that were resolved by Bohr in what is known as the Bohr frequency rule:

$$h\nu = E'' - E' \qquad (5)$$

where the frequency of the light emitted is related to the energy difference of an atom in the excited electronic state E'' and the same atom in the lower energy state E', usually referred to as the electronic ground state. The equation is applicable to the emission or absorption of light by molecules and even more complex systems.

The photochemical effects induced by monochromatic visible and ultraviolet light (wavelength $> 1{,}000$ Å) involve the quanta of energy absorbed in toto and the excitation of only the valence electrons of the receptor molecule. Thus the effects of ultraviolet radiation are uncomplicated by the secondary effects of ionizing radiation of shorter wavelength.

When radiation enters matter, any of several effects occur. Although light passes through a liquid with little absorbance and little energy loss, its velocity decreases and refraction is observed; however, if the radiation is absorbed entirely or in part, this absorption transfers energy to the medium. The response of the medium depends on chemical structure, environmental conditions, and the specific radiation. In the simplest case, a molecule exposed to electromagnetic radiation of suitable wavelength and polarization, electric interaction induces change in the internal motion of the molecule.

All high-energy radiation, such as x-rays and gamma rays, alpha and beta particles, protons, and neutrons interact with matter, producing ionization.

$$M \rightsquigarrow M^{+} + e \qquad (6)$$

The symbol $\rightsquigarrow$ is used to indicate *under the influence of high-energy radiation*. The resulting electron may be captured by either the parent positive ion to give an excited molecule or by a neutral molecule giving a negatively charged species.

$$M^{+} + e \rightarrow M^{*} \qquad (7)$$

$$M + e \rightarrow M^{-} \qquad (8)$$

The excited molecule may decompose to give either free radicals or a stable product.

$$M^{*} \rightarrow R_1 + R_2 \qquad (9)$$

$$M^{*} \rightarrow P \qquad (10)$$

In addition the negative ion may undergo decomposition to give a free radical and a more stable ionic species.

$$M^{-} \rightarrow R_3 = C^{-} \qquad (11)$$

Thus even for a simple irradiation system, ionic, excited, and free radical species are present. In the absence of a suitable electron trap, the lifetime of the ions may not be very long and the reactions of free radicals and excited species is of greater importance. These observations indicate that the absorption of ionizing radiation by matter is nonselective.

The high-energy quantum of fast particles does not give up its energy in a single encounter but degrades progressively to lower energies. Thus ionizing radiation is a broad term that refers to electromagnetic waves and to alpha and beta particles capable of producing ions directly or indirectly during passage through matter. In general, ionization is followed by molecular disruption. Secondary electronic interactions with neighboring matter produce many excited molecules as well as more ions. From a chemical point of view surprisingly selective effects may occur. Thus, x-rays penetrate to the inner shell electrons while gamma rays can produce nuclear transitions.

The mass and charge associated with various types of ionizing radiation are shown in Table 7-3. The wavelengths associated with these particles may be calculated using the expression of deBroglie:

$$\lambda = \frac{h}{mv} \qquad (12)$$

where λ = wavelength
h = Planck's constant
m = mass (gm)
v = velocity (cm/sec)

Ionizing radiation is often called high-energy radiation (Table 7-2) because its units of radiation (mev) are vastly greater than the binding energies of the valence electrons in the molecules (10 ev) and atoms (5 ev). The most conspicuous property of ionizing radiations is the ability to eject electrons from atoms, producing ions—negatively charged electrons and electron-deficient positively charged atoms and molecules. This classic picture comes from the study of the gaseous state in which the most useful investigative tools, such as cloud chambers and ion counting devices, detect ions. As a result, more emphasis has been placed on ionization and fragmentation phenomena than on energy dispersion by simple excitation processes; in fact, about twice as many molecules are excited as are ionized.

Let us first consider an energy range in which the effect may be regarded as independent of the chemical bond present. If a pulse of gamma or x-radiation passes an electron at some distance, the perturbation induces a transition. The electron is excited either to a bound state (true excitation) or to an unbound state (ionization producing a low-energy electron). In a closer encounter the radiation may give up almost all of its energy to the electron; this corresponds to a collision in heavy particle radiation. If the incident energy is high, a fast electron is ejected, which, in turn, may ionize other atoms.

If one neglects the phenomenon of *pair production* (a gamma ray producing an electron and a positron) which may be important at energies above 1 mev, the effects of radiation are of two types, known as the photoelectric effect and the Compton effect.

Photoelectric and Compton effects

The energy of the ejected electron is the difference between the energy of the radiation quantum and the electron binding energy. For x-rays up to about 50 kev, photoelectrons are the major cause of ionization. Above 50 kev there are already more Compton electrons than photoelectrons; however, they are still of much lower energies and contribute little to ionization.

The Compton effect is best thought of as a mechanical collision, the photon displaying a "particle-like" character. The electron recoils under the impact of the photon. The recoil energy of the electron depends upon the shift of photon wavelength, which in turn is related to both its deflection and the initial energy of the radiation quantum.

When the wavelength shift is small compared to the initial wavelength, the ratio of recoil to incident energy is low. For example, for 10-kev electrons (λ = 1.24 Å) the recoil energy is about 4% of the initial energy. As the incident wavelength becomes comparable to or smaller than the wavelength shift, the Compton energy rises rapidly. At 1 mev, an increase of incident energy by a factor of 100, essentially all electrons may be considered to be of the Compton recoil type.

If the incident radiation consists of protons or other particles, atomic electrons are still the primary target. With increased mass, wavelength has decreased and frequency has increased. If the collision is head-on and the incident particles have an energy of 50 kev or more, they move so rapidly that the orbital electrons may be regarded as stationary during the collision; the resulting effect is essentially independent of the forces binding the electrons in the irradiated material.

The situation is complicated by the fact that the incident particles may pass many atoms before a head-on collision occurs. In such encounters, sometimes referred to as glancing collisions, the particles must be treated as a fast pulse of perturbing electromagnetic radiation. Even at a distance of several angstroms, when the electrons are affected for 10^{-16} to 10^{-17} second, the electrons may absorb energy from the radiation field and make transitions. These transitions may be atomic, molecular, intermolecular (electron transfers), or to unbound states (ionizations). Furthermore, neighboring electrons in the same molecule, or in a colliding molecule, may be polarized in the radiation field and may affect the subsequent path of the particles.

Thus it is evident that with increasing frequency and energy the simple selective transition of valence electrons observed by the photochemical effects on wavelength in the visible and ultraviolet regions is replaced by a variety of photoelectric and mechanical effects that produce a diverse multiplicity of transitions.

EFFECTS OF RADIATION AT WAVELENGTHS SHORTER THAN THE VISIBLE RANGE

Ultraviolet radiation

Ultraviolet light has been known to exist since 1801; it has also been called "chemical" or "actinic" rays. Although the bactericidal action of sunlight was discovered in 1877, it was not until 1904 that Bie demonstrated that carbon arc radiation in the 2,000 to 2,905 Å region is ten to twelve times more effective bactericidally than the more intense radiation above 2,950 Å.[31]

Of the early studies of ultraviolet light, perhaps the most significant was that of Thiele and Wolf,[32] who discovered that sunlight filtered through window glass does not kill bacteria, whereas that filtered through quartz glass does. It was subsequently observed that the bactericidal effect, erythrocyte hemolysis, and the inactivation of the enzymes invertase and peroxidase by ultraviolet light proceed with or without oxygen. A little later Roffo[33] observed that rats exposed to a mercury arc light develop skin tumors; however, if the light is filtered through ordinary window glass none develops a tumor. The most reasonable explanation for the loss of effect of ultraviolet light filtered through window glass is that some specific substance or substances absorb the radiation.

Of relevance to our subsequent discussion is the fact that light in this region is specifically absorbed by nucleic acids (2,600 Å) and by proteins (2,800 Å). The analytic use of this absorption is well known to every student of biochemistry and physiology. This suggests a mechanism for the bactericidal action of ultraviolet radiation. It is of interest that ultraviolet light at 2,800 Å has minimal sunburn effect in the human being, a fact that correlates with the minimal epidermal transmittance of 2,800 Å radiation. In general, the longer wavelength ultraviolet, visible, and near infrared radiations are selectively absorbed by skin pigment systems, producing changes in the electron population at electronic and vibrational levels. The ultraviolet region of the spectrum is of interest because the energy involved excites only valance electrons of the receptor molecule and the effects are uncomplicated by the secondary absorption characteristic of ionizing radiation. Thus ultraviolet radiation contributes to our understanding of the primary effects of ionizing radiation.

Seliger and McElroy[28] estimate that at the wavelength of peak nucleic acid absorption bacteria are killed (LD) by 3×10^6 ultraviolet quanta, or 2.2×10^{-5} erg. By comparison, this is 100 times more energy and approximately 2×10^6 times as many quanta as are required for x-irradiation having the same killing effect.

This observation is one of several differences between the effects of ultraviolet and ionizing radiation. Thus with respect to the bactericidal effect of x-rays, the removal of free oxygen reduces the killing effect to 30% of that observed in the presence of oxygen; with respect to the bactericidal effect of ultraviolet radiation, removal of oxygen does not decrease the lethal effect. Furthermore, reactivation after radiation damage and the dependence on temperature and chemical treatment are quite different for ionizing radiation from that for ultraviolet light.

For a variety of organisms, studies of ultraviolet-induced mutation have been paralleled by studies of the photochemistry of pyrimidines, purines, and polynucleotides. Ultraviolet radiation is nonionizing and causes relatively few chromosome breaks; its absorption is limited essentially to conjugated molecules. Because of the present interest in the mutagenic action of ultraviolet and ionizing radiation, that of chemical compounds such as nitrogen mustards, and the potentially mutagenic action of drugs such as LSD, and because of the possible similarity of targets, mutagenic activity will be discussed separately.

Photodynamic action

By the turn of the century there were several reports of the lethal effect of fluorescent dyes on bacteria in the presence of light, and the term "photodynamic action" was developed. A correlation between the photodynamic action of dyes and the light-mediated carcinogenic effect of these same dyes was also observed. By 1937, Bungeler[34] demonstrated specific correlation between photodynamic materials such as eosin and hematoporphyrin and their photosensitization to carcinogenesis.

Early observations indicated that ingestion of plants belonging to the genus *Hypericum* produce a disease in grazing animals characterized by sensory stimulation and surface lesions. Subsequent studies indicated that photosensitive material is formed during digestion of the plant material.

Since that time numerous materials have been found to have a photodynamic action. Most photodynamic substances are fluorescent, either in solution or when absorbed on surfaces. However, not all fluorescent dyes exhibit photodynamic action, and substances such as sodium benzoate, which is nonfluorescent, can sensitize photo-oxidation.

The term "photodynamic action" is usually reserved for the killing effect on the entire organism or for enzyme inactivation. Because the basic mechanism of action is the photosensitized oxidation by molecular oxygen, a more descriptive term is *sensitized photo-oxidation*.

The characteristics of photodynamic action are: (1) oxygen is specifically required and used stoichiometrically with a quantum yield close to unity; (2) the reaction is essentially independent of temperature, corresponding to a true primary photochemical reaction; (3) the reaction rate is zero order and the sensitizing molecule acts as a catalyst.

There is no obvious relationship between tumor production in vertebrates and the death of protozoa or enzyme inactivation. There is much evidence linking photodynamic activity with photosensitized carcinogenesis and with photosensitized mutagenesis, both in bacterial and in chromosomal DNA. Although ultraviolet damage and certain x-ray damage occur in the absence of oxygen, photodynamic action is oxygen dependent. This fact suggests differences in the mechanism of damage; however, the common factor in these correlations is apparently damage to DNA. As we will discuss, it appears that the common denominator may be free radical formation and the involvement of peroxides.

Photoreactivation

Photoreactivation is the light-mediated recovery of a biologic system from radiation damage. The photoreactivating light is of longer wavelength than that of the damaging radiation. There are extensive observations of the reversal of ultraviolet-induced inactivation of a variety of bacteria, as well as the photoreversibility of ultraviolet-induced mutagenesis. During the past 20 years, a wide distribution of photoreactivations have been demonstrated in both the plant and the animal kingdoms. Apparently, whatever ultraviolet light does to a cell—loss of viability, growth inhibition, enzyme synthesis, or mutation—is photoreversible. However, the time at which the photoreactivating light is applied relative to the ultraviolet irradiation is quite important. In general, under normal metabolic conditions photoreactivation falls off rapidly with time and is not necessarily related to cell division.

It appears that photoreactivation involves the repair or replacement of ultraviolet-produced damage to DNA. In addition to the evidence for photoreactivation of DNA synthesis, it is apparent that ultraviolet-induced damage to self-duplicating structures, whether nuclear or cytoplasmic and not necessarily DNA, are also photoreactivated.

The different responses of lethality and mutagenesis to photoreactivation, together with diverse observations of photoprotection, are interpreted as indicating that not all ultraviolet damage occurs by the same pathway. More specifically, different receptor sites are involved that in turn relate to the organism and to other environmental factors. Thus the effect may be mediated by different initial absorbing materials that may be activated by different wavelengths.

As a result of the widespread use of ultraviolet light in a variety of industrial processes, occupational exposures may be hazardous. Ultraviolet light generated during welding can cause keratoconjunctivitis as well as skin burns. Some of the newer welding processes, such as metal arc welding with consumable electrodes, produce high-intensity ultraviolet light, which greatly increases the risk of eye and skin burns.

As previously noted, many photosensitizing agents have action spectra in the ultraviolet range. Many plants such as figs, limes, parsnips, and pinkrot celery contain photosensitizing materials thought to be furocoumarins and psoralens, producing exaggerated sunburn and frequently blisters upon contact. The most important industrial photosensitizers are coal tars, which have action spectra in the visible light range. The increased incidence of skin cancer in coal tar workers re-

Table 7-4. Early effects of acute whole-body exposure to ionizing radiation in man*

Acute exposure (roentgen units)	Probable effect
0 to 25	No obvious injury
25 to 50	Possible blood changes but no serious injury
50 to 100	Blood cell changes, some injury, no disability
100 to 200	Injury, possible disability
200 to 400	Injury and disability certain, death possible
400	Fatal to 50%
600 or more	Fatal

*From Atomic Energy Commission: Effects of atomic weapons, Washington, D. C., 1950. U. S. Government Printing Office.

Table 7-5. Summary of effects resulting from whole-body exposure to ionizing radiation in man*

Mild dose		Moderate dose			
0 to 25 r	**50 r**	**100 r**	**200 r**	**Median lethal dose (400 r)**	**Lethal dose (600 r)**
No detectable clinical effects Probably no delayed effects	Slight transient reductions in lymphocytes and neutrophils No other clinically detectable effects Delayed effects possible, but serious effect in average individual very improbable	Nausea and fatigue, with possible vomiting above 125 r Reduction in lymphocytes and neutrophils with delayed recovery Delayed effects may shorten life expectancy as much as 1%	Nausea and vomiting within 24 hours Latent period of about 1 week, perhaps longer Following latent period, epilation, loss of appetite, and general malaise Sore throat, pallor, petechiae, diarrhea Moderate emaciation Possible death in 2 to 6 weeks in a small percentage of individuals Recovery likely unless complicated by previous poor health, superimposed injuries, or infections	Nausea and vomiting in 1 to 2 hours Latent period, perhaps as long as 1 week Beginning epilation, loss of appetite, and general malaise accompanied by fever and severe inflammation of mouth and throat in the third week Pallor, petechiae, diarrhea, nosebleeds, and rapid emaciation about the fourth week Some deaths in 2 to 6 weeks; possible eventual death of 50% of the exposed individuals	Nausea and vomiting in 1 to 2 hours Short latent period following initial nausea Diarrhea, vomiting, inflammation of mouth and throat toward end of first week Fever, rapid emaciation, and death as early as the second week with possible eventual death of 100% of exposed individuals

*From Atomic Energy Commission: Effects of atomic weapons, Washington, D. C., 1950, U. S. Government Printing Office.

sults not only from coal tar carcinogens but also from the repeated occurrence of photosensitization.

Effects of ionizing radiation

Man has always been exposed to ionizing radiation from natural sources. However, with increasing availability and use of ionizing radiation, the magnitude of the hazard has grown. Early in this century, it was observed that a sufficient dose of ionizing radiation causes sterility and changes in the composition of the blood. Since that time, numerous experiments have been performed to determine the dose-time relationships for high-intensity, brief exposures that produce detectable effects on individual organisms and, at the other extreme, to determine the dose-time relationships for low-intensity chronic exposures that produce effects too faint to be recognized.

Acute exposure to intense ionizing radiation produces a complex set of signs and symptoms known as *acute radiation syndrome*. A summary of the acute effects of varying levels of radiation intensity is shown in Tables 7-4 and 7-5; radiation is expressed in terms of roentgen units (r). The roentgen is defined as the amount of gamma or x-radiation required to produce 1 electrostatic unit (esu) of ions per cubic centimeter of dry air under standard conditions. Since the roentgen is a measure of interaction with air, the dose absorbed by different materials (in rads) varies for a given exposure in roentgens. The rad, a basic unit of absorbed dose of ionizing radiation, is defined as the energy imparted by the ionizing radiation (in ergs) per unit mass of irradiated material; 1 rad equals 100 ergs/gm. Thus with moderate-energy gamma rays of 0.2 to 3 mev, an exposure of 1 r produces an absorbed dose in muscle of about 0.97 rad.

Because all radiation does not produce identical biologic effects for a given amount of energy, another unit is necessary to express human biologic dose. To compare the effective dose of different types of radiation, the relative biologic efficiency factor (RBE) is used. It is the ratio of the dose of gamma or x-rays to the dose of the radiation in question that produces the same biologic effects (Table 7-6). Hence the roentgen equivalent for man (rem) is the product of RBE and the absorbed dose in rads:

$$\text{RBE} \times \text{rad} = \text{rem} \qquad (13)$$

Referring back to Table 7-5, it is now apparent that when exposure is expressed in roentgen units, the administered dose does not directly define the amount of radiation absorbed by a human target but gives only the composite effect of the relative amount of radiation.

Table 7-6. Comparison of types of ionizing radiation

Type of radiation	Relative biologic effectiveness
X-rays or gamma rays	1
Beta particles	1
Fast neutrons	10
Thermal neutrons	4 to 5
Alpha particles	10 to 20

In any consideration of the biologic effects of ionizing radiation, numerous variables must be kept in mind: (1) type of radiation; (2) amount of radiation; (3) structure and density of the target; (4) condition of the target at the time of irradiation; and (5) capacity of the target to repair itself. Furthermore, it should be noted that radionuclides do not emit a single type of radiation to the exclusion of all others.

The most important somatic effects of ionizing radiation are leukemia and other types of cancer, cataracts, and reduced life expectancy. These observations suggest that tissues differ with respect to their sensitivity and reaction to radiation. Reams of data on the somatic effects of radiation have been obtained from animal experiments, observation of patients undergoing radiotherapy, studies of occupationally or accidentally exposed workers, and observation of the Japanese survivors of Hiroshima.[25, 29, 36]

When an animal receives a dose of ionizing radiation at or near the critical level, the most sensitive tissues are the intestinal mucosa, the bone marrow, and lymphoid tissue. Intestinal damage results in a loss of water and electrolytes. If death does not occur from intestinal damage, the effects of bone marrow damage become apparent. It was observed some time ago that mice whose hematopoietic system was destroyed by a lethal dose of radiation can be kept alive by implanting or injecting normal blood-forming tissue from healthy animals; this approach is used in the clinical treatment of total body irradiation. Subsequently, it was found that radiation also interferes with the immune response.

The various tests used for early detection of the effects of radiation involve such hematologic findings as leukopenia, relative lymphocytosis, appearance of abnormal monocytes,[37, 38] increased incidence of bilobed lymphocytes, and increased lymphocyte DNA. Obviously, these tests are nonspecific and do not of themselves constitute evidence of exposure to ionizing radiation.

The increased incidence of leukemia from exposure to ionizing radiation has been known from animal experiments for a long time and is being substantiated by observation of exposed human beings. According to the reports of the Atomic Bomb Casualty Commission,[39, 40] the surviving persons from Hiroshima and Nagasaki who were exposed to radiation from the atomic bombs exploded in 1945 now have a high incidence of leukemia, which began a few years after exposure. There is a correlation between leukemia incidence and the distance of the individual from the site of the bomb. The distance is presumably correlated with the amount of radiation that each individual received. Each individual received a single substantial dose of radiation at very high intensity.

Although prior to 1950 there was an increased incidence of leukemia in radiologists,[41] this is no longer the case. This decrease in leukemia incidence in radiologists is attributed to the improvement of protective facilities and technics so that the amount of radiation received is less than the current recommended maximum permissible dose. These maximum permissible levels allow for a total lifetime dose of 250 rems.

Radiation has produced skin cancer, thyroid cancer in children irradiated in the neck region, lung cancer in miners and millers exposed to radon and its daughters, and bone cancer after ingestion of radium. Bone cancer (osteogenic sarcoma) can result from irradiation from radioelements similar to calcium, such as radium, radioactive strontium, radioactive thorium, and radioactive lead. These elements are ingested and deposited in the bone. Cases of bone cancer from external radiation have been reported, although a high dose (3,000 to 4,000 rads) is required.[25]

Peculiar to the radionuclides is their deposit in tissues. In experiments at the University of Utah, it was found that beagles injected with ^{224}Ra, ^{226}Ra, ^{228}Ra, ^{239}Pu, ^{241}Am, ^{228}Th, ^{210}Pb, and ^{90}Sr deposited these radionuclides in the skeleton. Plutonium and americium were also deposited in other tissues. These animals developed bone and other cancers as well as liver tumors. The experiments indicated that injected alpha emitters are more toxic and, conversely, a beta emitter such as ^{90}Sr is less toxic than was initially believed.

Commercial animals are an important link in the food chain and thus a means by which radioactive contamination finds its way to man. Grazing animals receive most of their radiation by ingestion of airborne nuclides deposited on forage. Since the ingested radionuclides are poorly absorbed, the tolerance of these animals is high. However, they concentrate ^{90}Sr, ^{131}I, and ^{137}Cs in their muscle tissue and secrete them in their milk. The maximum tolerable level for man for contamination by this route remains unknown.

Some of the ^{90}Sr released into the atmosphere by nuclear weapons tests finds its way into the human body by way of the food cycle. Measurements on young children whose rapidly growing bones incorporate ^{90}Sr at a rapid rate have shown burdens up to a few thousandths of a microcurie. This corresponds to a dose rate of about 0.01 rad per year.

Lung cancer as a result of the inhalation of radioactive dusts or gases containing radon and its daughter products has been observed in miners in Europe as well as in North America. Wagoner and co-workers[24] reported excessive occurrence of respiratory cancer among uranium miners in the United States, indicating a dose-response relationship. However, the number of individual studies and the periods of observation at low levels of exposure are not yet adequate for a quantitative statement of the dose-response relationship.

The risk of inhaling radioactive aerosols is greater than that of inhaling radioactive gas because the lung retains particulates. The dose of radioactivity received by different regions of the lung from inhaled radioactive dust depends on the concentration of the radionuclide in the inhaled air, the physical properties of the radionuclide, the rate of dust inhalation, the region in which the dust is deposited, and the rate at which it is removed. Although lung models have been developed for calculating the dust deposition in and clearance from the bronchopulmonary system as a basis for determining lung dosimetry and setting exposure limits,[42] the dose required to produce lung cancer in man is not known.

Most studies of the effects of irradiation of the thymus gland of adults and of children indicate increased risk of thyroid cancer. Data from Hiroshima and Nagasaki indicate that the child's thyroid gland[43] is more sensitive to irradiation than that of the adult. Because the thyroid gland concentrates iodine and because irradiation of the thyroid gland increases the risk of thyroid cancer, there is good reason to minimize the amount of radioactive iodine released into the atmosphere.

Cataracts were observed among the survivors of Hiroshima and Nagasaki and have also been reported as a result of exposure of the lens of the eye to x-rays, gamma rays, beta particles, and neutrons. Although lens changes have been reported for doses as low as 200 rads, the minimal dose of x-ray and gamma ray capable of producing clinically significant cataracts is 550 to 950 rads in adults and perhaps less in children. Exposure to an average dose of about 800 rads delivered over a 2-week to 3-month period produces opacity in 70% of those exposed; for 30% the opacities were progressive and eventually impaired vision. Cataracts are much more likely to result from exposure to neutrons than to x-rays or gamma rays.[25]

Irradiation of nerve tissue produces a delayed response. Doses of x-rays in the 2,000- to 4,000-rad range to the spinal cord of rats or monkeys produce spinal cord lesions. Carsten and Innes[44] found that these lesions are associated with hind-leg paralysis and loss of sphincter control of the urinary bladder and rectum. These signs may be caused by demyelinating lesions that proceed to malacia and liquefaction in the ventral and lateral columns of spinal white matter. Monkeys receiving a 3,500-rad x-ray exposure to the brain showed electroencephalographic changes after 6 to 8 weeks; pathologic examination revealed necrobiotic lesions, demyelination, and vascular changes.

Prior to 1935 there was evidence of decreased lifespan of radiologists that could not be attributed to radiation-induced disease such as leukemia but that appeared to involve acceleration of the aging process. With the use of more rigid radiation protection technics, the evidence for decreased lifespan of radiologists has disappeared. The initial observations are consistent with numerous observations that whole-body irradiation of experimental animals shortens life. The absence of definitive data regarding the dose-response relationship to shortening of life in man has led to the development of theoretic models for extrapolating radiation exposure and lifespan shortening in experimental animals to man. This extrapolation predicted a theoretic lifespan shortening of 17 days per rad. However, this relationship is not evident in the data from the Hiroshima survivors, and the value is currently regarded as being too high.

Generalized exposure to penetrating ionizing radiation producing acute radiation syndrome is usually the result of an accident. Conversely, occupational exposure is usually localized and produces acute or chronic radiodermatitis. Radiotherapy is an acute exposure; chronic exposure is usually industrial. As with ultraviolet radiation, the cutaneous effects of ionizing radiation are cumulative. Acute radiodermatitis is classified into three degrees of damage.[45] The first degree is characterized by erythema after several hours or days followed by hyperpigmentation and temporary alopecia. The second degree is characterized by erythema and edema followed by superficial ulceration, permanent alopecia, permanent loss of nails and glands (sweat glands may regenerate), and atrophic telangiectatic scarring. The third degree is characterized by erythema and deep edema, followed by necrosis and sloughing; healing is slow and difficult or never occurs, leaving ulcers; and underlying tissue such as cartilage and bone may be involved.

Dose-response relationships

In any consideration of an organism's responses to a potential environmental stressor, specific knowledge of the dose-response relationship is of interest. Such knowledge provides the means for establishing the limits of exposure to environmental hazards. These relationships have not been established for low doses of ionizing radiation. As a result there is disagreement among scientists working in this field and criticism of them by spectators and an aroused citizenry.

Let us sort out the various aspects of this problem. Most of the correlations of leukemia incidence with ionizing radiation involve substantial doses given at high dose rates. Above 100 rads the development of leukemia and other forms of cancer is a linear function of dose. When these data were extrapolated to lower doses, the conclusion was reached that the incidence of leukemia was about one in a million per year per roentgen received. This conclusion, of course, assumes a linear relationship between leukemia incidence and the integrated dose of radiation irrespective of dose rate. The results of experiments on mice exposed to various levels of radiation are not in agreement and, in fact, raise serious questions concerning the validity of extrapolation from existing human data on the relationship between dose and leukemia incidence to low intensity exposures. Conversely, it has been proposed that there is a threshold intensity for radiation-induced leukemia somewhere below 100 rems and that lower intensities of radiation are ineffective. Other scientists regard the number of cases of leukemia in-

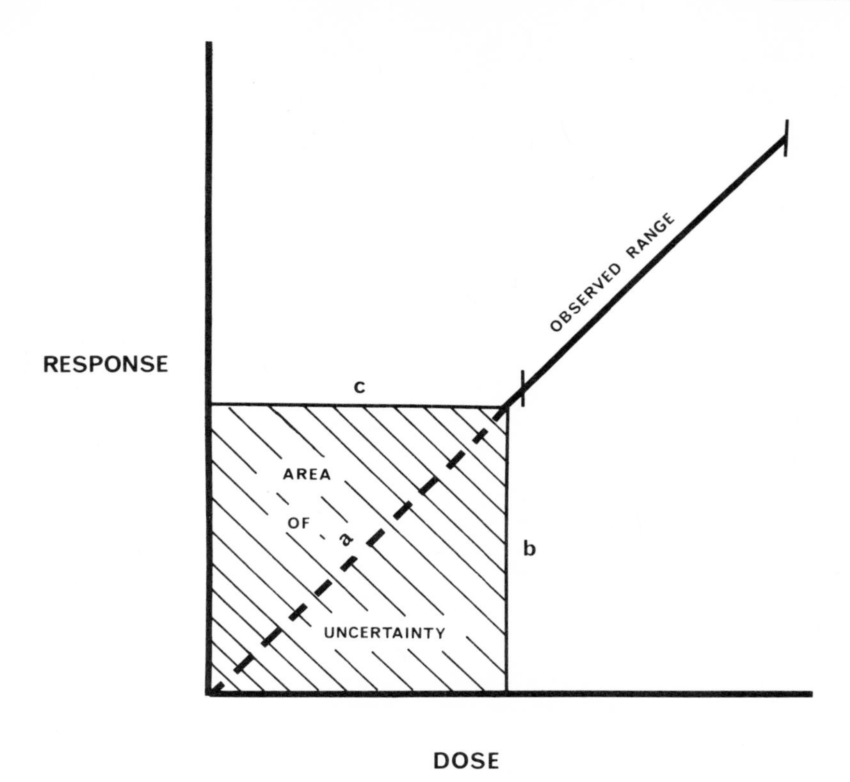

Fig. 7-8. Possible observations in the area of uncertainty. *a,* Linear dose-response relationship; *b,* threshold dose must be exceeded to observe a response; and *c,* observed response requires a specific dose intensity.

duced by low doses of ionizing radiation as too few to permit firm conclusions.

The variety of opinions on this subject does not result from simple statistics nor from the relationship of mice and men but must be considered together with the other variables involved. An idealized representation of this problem as it exists for other environmental stressors as well as radiation is shown in Fig. 7-8.

In the range of high doses a dose-response relationship is linear. As lower and lower dose intensities are used, an area of uncertainty is reached. The limit of uncertainty is defined on the vertical axis by the sensitivity of the method for determining the response as well as by the response system itself. Similarly, on the horizontal axis the intensity of the administered dose and the intensity of the dose reaching the response system may or may not have a constant relationship. For radiation this could involve the orientation of the animal to the source of radiation or limits in the selection of the area irradiated. If the response system consists of cell or tissue components, an exposure is more uniform and this variable is more easily controlled.

Thus the observation of a threshold dose that is real in terms of raw data may result from (1) determination of response that is within the limits of error of the method; or (2) the capacity of the response system for self-repair has not been exceeded by the dose (the rate of repair is greater than the rate at which damage is inflicted); or (3) protective effects or materials are present, for example, free radical scavengers.

Conversely, if no threshold intensity is observed, several factors may be involved, such as (1) potentiation by specific absorbing compounds, or (2) the response system is sufficiently limited in size or concentration to give maximum response at all intensities used. In either situation, for whole-animal experiments the observations become a function of the condition of the target animal at the time of exposure and reflect other stressors. Although such other stress factors as well as other variables can be controlled or defined in a laboratory experiment, this is not true for man. The question of the existence of threshold limits, particularly for genetic effects and leukemia, has led to the concept of maximum permissible dose as the basis of radiation standards.

In developing quantitative modeling of the physiologic factors in radiation lethality, Iberall[46] found that deterministic descriptions of phenomena are often regarded as actuarial or statistical. Thus his evaluation of the available data led to a spectrum of effects. Work of this type could be useful in designing discriminating experiments.

The permissible dose is defined as the amount of ionizing radiation that, in the light of present knowledge, is not expected to cause appreciable damage to a person during his lifetime.[47] The practice recommended by the Federal Radiation Council[48, 49] would limit the maximum whole-body dose for an individual from nonmedical sources to 0.5 rem/yr and suggests that the average dose to the population during a 30-year period should not exceed 5 rem to the male gonads. The dose from natural background radiation is not taken into account.

Genetic effects of ionizing radiation

Ionizing radiation produces mutations in human gametes that appear in future generations.[50] The contribution of natural radiation to the estimated 1% of liveborn infants who suffer from the effects of spontaneous chromosomal abnormalities is not known.[51] It is estimated that exposure to a total of 10 to 20 rads during an individual's reproductive lifetime would double the mutation rate.

As might be expected, most of the data regarding the genetic effects of ionizing radiation have been obtained from animal experiments. After the initial observations that x-rays produce mutations in *Drosophila,* experiments with mice showed an even greater radiation-induced mutation frequency. Such observations have been interpreted as indicating that the genetic hazard to man is greater than initially assumed. Extrapolations from the results of irradiation experiments on mice suggest that the genetic hazards to man are as follows[25, 51, 52]:

1. The more mature male germ cells (spermatozoa) are more sensitive to genetic damage than are the stem cells (spermatogonia). The risk of transmitting genetic damage can be reduced by postponing procreation after irradiation of the male gonads until the irradiated spermatozoa are no longer present. The time interval is about 5 weeks for the mouse and about 10 weeks for man.
2. The interval between irradiation and conception has a definite effect on mutation frequency in the female irradiated with neutrons. The hazard decreases as the time between irradiation and conception increases.
3. Mutation frequency is less per unit dose of irradiation when exposure is distributed over a long period of time. Low dose-rate exposures do not produce as many mutations as high dose-rate exposures. The dose-rate effect is considerably more pronounced in females than in males.
4. There does not appear to be a threshold dose-rate, or dose-rate below which no mutations occur.

Contradictory evidence includes the long-term observation of a herd of Hereford cattle located about 15 miles from the detonation site of the first atomic bomb in 1945; the cattle were exposed to high levels of radiation from radioactive fallout particles. Except for superficial damage to the skin of the animals from direct contact with radioactive particles, their general condition, productivity, and death rate were comparable to those of control cattle. One cow from the herd lived for 20 years, producing sixteen healthy calves.[26]

The current disagreement over the allowable levels of radiation is based in part on the unresolved problem of threshold dose-rates. Perhaps the strongest evidence for a nonthreshold, linear relationship between dose and effect comes from genetic studies. Several experiments have shown that when animals mate soon after irradiation, mutations occur in their offspring; these mutations are not correlated with specific mutations in the chromosomes. Since the dose did not alter mutation rate, the effect was assumed to be linear. Conversely, the possibility of finding a correlation between damage and exposure to chronic low-level radiation is being explored in areas where background levels are high (1.5 to 5 rads) as a result of a high content of thorium in the soil. This situation occurs in Guarapary, Brazil, a coastal town north of Rio de Janeiro, and in the Kerala area of India. A limited population survey indicated a statistically significant increase of chromosomal aberrations; however, a later report contradicted this conclusion. Obviously, a complete analysis of a large population sample is required before conclusions can be drawn. Holcomb[53] concluded that, at least in terms of chromosomal aberrations resulting from this high radiation background, nothing disastrous is happening. Thus, in establishing standards for exposure of human populations, we face unknown factors such as the importance of background radiation and the unresolved problem of threshold dose-rates. Furthermore, there is a growing list of chemical compounds that are mutagenic for bacteria and that are regarded as possibly mutagenic for man. In addition to LSD and alkylating agents, the list includes various anticancer agents, tranquilizers, antidepressants, antibiotics, and antinausea agents.

Since biochemical alterations must precede the development of pathologic changes, let us look at cellular and subcellular events. Various changes may occur in the chromosomes as a result of ionizing radiation.[54] Some of these changes are related to the condition of the cell at the time of exposure.[55] The types of chromosomal change observed at metaphase depend upon the mitotic stage at which the cell was exposed. In general, if the chromosomes have not yet duplicated at the time of irradiation, the aberrations involve both of the chromatids at identical loci and are termed *chromosome-type* aberrations. Prior to detectable DNA synthesis, throughout the synthesis phase, and during much of the postsynthesis phase, chromosomes behave as double-stranded structures with respect to radiation, and the aberrations involve individual chromatids. A cell in late prophase or metaphase produces no detectable aberrations at that division, but structural changes appear at subsequent divisions.

Samples of peripheral blood were taken from individuals occupationally exposed to alpha particles (total dose: 7.7 to 36 rem during a period of 7 to 15 years) and from individuals exposed to fast neutrons and gamma rays (total dose: 4.3 to 15.9 rem). Examination of the leukocytes gave no evidence of increased aneuploidy. The blood samples from these exposed individuals and samples from normal controls both showed an incidence of chromosomal aberration of 1%, which precluded evaluation of a possible relationship between aberration yield and dose kinetics.[56]

The involvement of DNA in the mediation of genetic effects requires a look at molecular genetics. First it should be noted that recent advances in microbiology and the biochemistry of the genetic process has made the insistence on heritability, in the usual sense of phenotypic changes, an arbitrary one. In their famous papers on the structure of DNA[57] and its genetic consequences,[58] Watson and Crick outlined the way in which replication of the genetic molecule occurs. In most organisms the genetic information is carried by DNA and mutation is regarded as a process that leads to changes in the DNA base sequence. Therefore, rather than considering chromosomal structure and its disorganization by radiation or radiomimetic agents, let us look at what happens when nucleotides are irradiated.

For a variety of organisms, studies of ultraviolet-induced mutation have been paralleled by studies of the photochemistry of pyrimidines, purines, and polynucleotides. Ultraviolet radiation does not ionize and causes relatively few chromosome breaks; it is absorbed almost entirely by conjugated molecules.

The main target of ultraviolet radiation is undoubtedly the pyrimidine bases. The two most conspicuous reactions are dimerization and hydration. A typical dimerization reaction is that of thymine[59]:

Note that several stereoisomers of the dimer are possible. Irradiation of thymine dinucleotides and polynucleotides produces intramolecular dimerization.[60] Irradiation of DNA causes dimerization of the adjacent thymidylic acid residues.

Hydration of pyrimidines results in saturation, or hydrogenation, of the 4-5 double bond, as shown for uracil:

Such hydration reactions are partially, if not completely, reversed by heat or by treatment with acid.

The dimers are photodissociated by ultraviolet light of relatively short wavelength and are also dissociated enzymatically, a biologic reactivation; whether or not other photochemical reactions of nucleotide bases are reversed enzymatically is not known. Study of the effects of ultraviolet irradiation on the coding properties of the polyuridylic acids indicate that both dimerization and hydration occur. As assessed by the Nirenburg system, it is the hydration reaction that changes the coding properties of the polynucleotide. Hydration of some of the uracil residues in polyuridylic acid leads to incorporation of serine as well as phenylalanine into the polypeptide. In the presence of

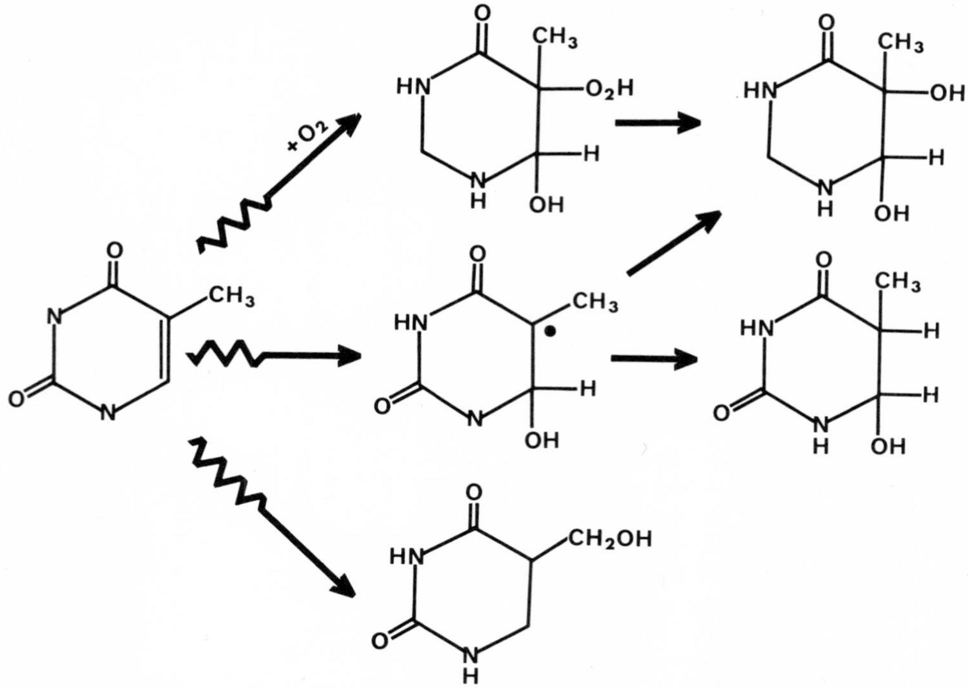

Fig. 7-9. Irradiation of thymine in the presence of molecular oxygen yields a peroxide that decomposes to a mixture of *cis* and *trans* glycols. However, irradiation of a *deaerated* thymine solution yields glycols, including the hydrated form of thymine, by a free radical mechanism.

irradiated polyuridylic acid, RNA polymerase incorporates some guanine instead of adenine.[61] These and similar observations suggest that in base-pairing reactions a hydrated uridylic acid residue may behave as a cytidylic acid residue and vice versa.

Recent observations indicate that cytosine is extensively deaminated by ultraviolet irradiation to uracil and suggest a similar reaction for adenine.[62] Note that deamination of adenine, cytosine, and guanine produces hypoxanthine, uracil, and xanthine, respectively.

In general, most of the high-energy radiation damage to DNA in solution results from the secondary reactions of the radicals produced by the decomposition of water. In the presence of oxygen, the O_2H radical may also be important.

Purines are somewhat more resistant than pyrimidines to high-energy, as well as to ultraviolet, radiation. The effect of high-energy radiation on pyrimidine bases has been extensively studied. The chemistry is complex and beyond the scope of this chapter; however, the following observations are indicative.

When thymine is irradiated in the presence of oxygen, a peroxide of thymine forms (80% cis and 20% trans isomers), which decomposes to give a mixture of the cis and trans glycols. If a deaerated solution of thymine is irradiated, another set of reactions occurs,[63] producing glycols from different intermediates. Thus glycols are formed in the presence of air from a peroxide and in the absence of air by a free radical mechanism. These reactions are summarized in Fig. 7-9. Of more importance is the fact that the hydrated form of thymine, implicated as an intermediate in radiation-induced mutation, results from the free radical. The dominant free radical observed by electron spin resonance after ultraviolet or gamma irradiation of thymine or DNA at low temperatures[64] has the following structure:

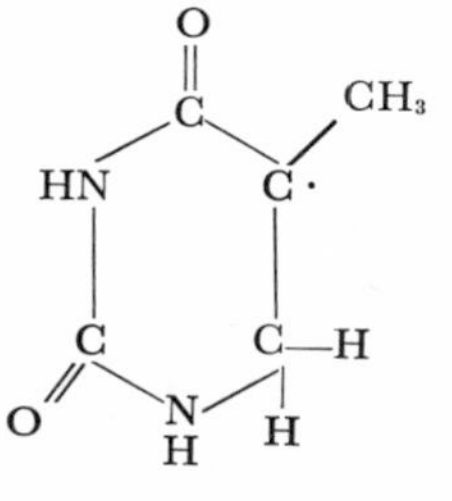

When a solution of cytosine is irradiated in the presence of oxygen, a peroxide forms that decomposes by way of the cis glycol to isobarbituric acid. Uracil, formed by deamination of cytosine, constitutes about 10% of the products of gamma irradiation in the absence of oxygen. This reaction could certainly produce mutations. Another major product is the glycol formed by hydroxylation and subsequent deamination. Thus although ultraviolet radiation of the purine bases produces dimerization that may be reversed, ionizing radiation leads to breaks in the chromosomes and interconversions of the pyrimidine bases so that some replication may be misdirected.

In their examination of the genetic effects of ionizing radiation on yeasts, Nakai and Mortimer observed that densely ionizing particles are more effective than sparsely ionizing radiation in producing lethality, mitotic segregation, allelic recombination, and reverse mutations. Differences in haploid inactivation and mutation as compared with diploid inactivation, mitotic crossing over, and recombination suggest that there are differences in the molecular mechanism of indirection of these effects. Very different effects of heavy, as compared to light, ions were not observed with respect to the indirection of mitotic recombination in different intergenic or intragenic regions, even though the dimensions differed by two orders of magnitude. These observations and many others were explained by assuming that the nature and size of the targets for effects at different levels within the chromosome are similar or that the effects are mediated by a common diffusible intermediate.[65]

Theory of action of ionizing radiation

The primary effect of radiation is to start a chain of events that proceeds through a number of intermediate steps and is finally expressed in the observed pathobiologic change. This was briefly noted in our discussion of genetic effects and will now be examined in more detail.

From observations of the effects of ionizing radiation on the structure and function of living biologic systems, two different explanatory concepts have developed. The so-called target theory of direct action[66] was proposed as an approach to the problems involving specific macromolecules, such as gene mutation, chromosomal breakage, and virus inactivation. The basic assumption of this theory is that ionization remote from a specific structure has little effect on the structure itself even though other events of biologic significance may be induced. Thus the biologic event is a direct result of a single direct hit of ionizing radiation occurring in or in the immediate vicinity of a vital part of the cell, or target.

The target theory originated with the observation of "survival" curves of microorganisms given a lethal dose of radiation. The fraction of surviving organisms appeared to be a direct function of dose and target cross-section. This is a frequently observed empiric relationship and raised the question of the significance of the cross-sectional area of the target for the subsequent biologic events.[67]

Confirmation that target area is related to the effects comes from observations of two types: (1) the effectiveness of various ionic particles on subcellular structures as opposed to whole cells, and (2) the relationship between target size and inactivation dose. Thus for trypsin inactivation, the target theory estimates the molecular weight to be from 34,000 to 39,000; 36,000 is the accepted value as determined by other methods. With respect to larger molecules, the theory predicts the diameter of phage particles within 50% of recognized values. When this technic is applied to larger unicellular organisms in which multihits are required, predictions deviate even more.[30]

A second theory proposes an indirect attack by way of free radicals originating from the splitting of water molecules in the cell. This concept recalls our previous general discussion of the molecular events produced by ionizing radiation. Applied to water, the reactive species include hydrogen, hydroxyl, and O_2H radicals as well as molecular hydrogen peroxide.

The concept that many normal biochemical interactions in intact tissues proceed by way of a one-electron transfer mechanism is not new. The extension of measurements of energy absorption to the microwave region (electron spin resonance) has provided evidence of extensive free radical formation in fundamental biochemical structures, including proteins, nucleic acids, and enzymes. This emphasizes the importance of all types of free radicals in the biochemical reactions to ionizing radiation. The identity of radical species and the significance of radiation-produced radicals has been extensively studied.

These concepts of direct and indirect action developed independently and were initially regarded as contradictory; however, as with many such inconsistencies, they are now thought to be complementary. The extent to which either type of action predominates depends not only on local conditions but also on the general environment of the target;

this is illustrated by the difference of the products when thymine is irradiated in the presence or absence of oxygen.

In general, the evidence indicates that for dried materials the chemical changes result from direct action, whereas in very dilute solution indirect reactions predominate. Conditions in a living cell are between the two extremes.[68, 69] Observations of changes in intracellular DNA molecules and DNA phages under various experimental conditions are helpful.

1. With oxygen absent but with free radical scavengers present, larger doses of x-ray are required to produce a given effect, and breaks in the double-stranded DNA account for most of the lethal events.
2. In an unprotected aqueous system containing extracted DNA, there were single-strand DNA breaks in addition to breaks in double-strand DNA. Single-strand breaks do not require oxygen and are suppressed by free radical scavengers, such as histidine.
3. With the addition of histidine to a well-oxygenated system, the double-strand DNA breaks account for less than half of the lethality. The predominant reaction is pyrimidine modification.

The foregoing observations are consistent with both theories—direct and indirect attack. In the first set of experimental conditions suppression of free radical formation requires a larger dose of x-rays to reach the endpoint. Such observations emphasize the importance of the indirect attack. The more common environment of intracellular DNA in well-oxygenated cells suggests that peroxide formation plans a prominent role, whereas in anoxic cells free radicals per se become important.

Pullman and Pullman[70] pointed out that the free radicals formed from water by ionizing radiation can secondarily produce organic free radicals:

$$RH + H^{\bullet} \text{ or } HO^{\bullet} \rightarrow R^{\bullet} + H_2 + H_2O \qquad (14)$$

$R^{\bullet}$ would be less reactive than $H^{\bullet}$ or $HO^{\bullet}$ but would have a longer lifetime and a higher degree of selectivity in its chemical transformation. This secondary $R^{\bullet}$ would be identical with the $R^{\bullet}$ from the direct effect of ionizing radiation on the intact molecule, suggesting that there is really no fundamental chemical difference between the direct and indirect effects.

The response of DNA to ionizing radiation differs importantly from that to radiomimetic cytotoxic agents, such as nitrogen mustards and ethylenimines, which are electrophilic agents. Whereas ionizing radiation attacks pyrimidine, radiomimetic agents attack the purine of the pyrimidine-purine base pair.

Combination with associated protein may appreciably alter the radiosensitivity of nucleic acids; irradiation of a nucleoprotein solution with doses as high as 10^5 rad left the nucleic acid almost undamaged[71]; the active species produced by the radiation reacted with the protein but did not penetrate the protein sheath.

The reaction of protein to ionizing radiation varies with the particular protein involved. An important chemical change involves the sulfhydryl groups. It is suggested that energy migrates along oriented stacks or chains of molecules and funnels into sulfhydryl groups. Thiol enzymes are much more sensitive to ionizing radiation than nonsulfhydryl enzymes. Glutathione, cysteine, and other sulfhydryl compounds added before exposure to ionizing radiation protect appreciably and added after exposure facilitate reactivation. Such observations suggest the importance of sulfhydryl compounds in the reaction of cells and tissues to ionizing radiation.

Both ionizing and nonionizing radiation can change the charge-carrier concentration of inorganic semiconductor solids.[72] Radiation may similarly affect the particulate enzymes of living organisms and thus participate in normal photochemical processes or in pathologic events such as radiation sickness. Such a mechanism of radiation effects should be influenced synergistically or antagonistically by absorbants that change protein charge-carrier concentrations, especially by small organic molecules that are easily oxidized or reduced. Indeed, such compounds have been found to alleviate or intensify radiation sickness.[73]

From the preceding discussion it is evident that there are still many unanswered questions about the biologic effects of ionizing radiation. An enormous gap separates what is known of the initial physical events in the interaction of a living biologic system with ionizing radiation, such as energy absorption, excited states, energy transfer, and active radicals, and what is known of the biologic manifestations that finally result from these events.

EFFECTS OF RADIATION AT WAVELENGTHS LONGER THAN THE VISIBLE RANGE

At wavelengths above the visible range, responses result from magnetic rather than electronic effects. Radiation of this spectral region orients molecules instead of producing electron transition. Absorption of radiation occurs only when it produces periodic changes in molecular dipole moment. Thus the response to radiation of longer wavelengths is one of molecular orientation in which the component parts of the molecule react harmonically. Radiation of this region of the spectrum quantizes energy states, rotates molecules, and vibrates atomic nuclei.

With the development of equipment for producing and detecting these electromagnetic waves, spectroscopic technics became available for numerous analytic applications, including the study of biologic materials. These technics include infrared, microwave, nuclear magnetic resonance, and electron spin resonance. Some of these methods involve magnetic fields as well as higher frequency electromagnetic waves. Examination of a variety of biologic materials is revealing the importance of free radicals in biologic systems as well as their response to irradiation.[74-76]

Observations such as the synergistic effect on plant development of infrared radiation when combined with x-ray as a pre- or posttreatment[36] have increased interest in the biologic responses to this spectral region. The strong absorption of water vapor in this region may be relevant to this observation, as well as to the thermal burn that results from human exposure to infrared radiation.

Low-intensity exposure of human beings to infrared radiation causes fatigue and headache. Likewise, low-intensity infrared radiation that does not even burn the skin may produce eye damage. The classic eye lesion after years of exposure is posterior cataract, sometimes called glass blowers' cataract. Since protein molecules, as well as water, absorb in the infrared region, the possible mechanism of such a reaction includes direct orientation of protein molecules as well as water molecules and raises the question of a direct effect of thermal agitation per se.

In some industrial environments both infrared and ultraviolet radiation are present; the processes involved include welding, open-hearth, electric furnace, and foundry pouring. The possibility of synergistic or potentiating reactions to these radiations merits investigation.

In general, microwave radiation is similar to infrared radiation; it heats skin locally but penetrates deeper. Observations of laboratory animals exposed to stationary pulsed beams shows that microwave radiation can cook underlying muscle, produce cataracts, and cause death by hyperthermia. There are a few reports of harmful effects on human beings, such as transient superficial skin heating, heating of steel fracture plates, and cataracts.

Microwave radiation includes electromagnetic frequencies ranging from 300 to 30,000 megacycles/sec. These waves are propagated into the atmosphere by the rotating antennas of search radar, the stationary type of tracking radar, radio relay links, and television transmitters; these waves are also utilized in radiofrequency ovens and medical diathermy devices. Waves are propagated in two discrete modes known as continuous wave (CW) and pulsed. The CW mode is used in communication transmitting devices, whereas the pulsed mode is associated with radar, industrial, and medical equipment. Because of the high power intensities and energy distribution, the pulsed mode is considered of more biologic significance.

Although there have been no reports of serious human injury or death from radar exposure per se, serious injury has occurred when men working on radar generators were exposed to x-rays. In regard to the Lockport incident[77] it is pointed out that the dose absorbed by various body regions varied; the differing conditions that would acutely injure masses of people were compared with those that cause occasional casualties in occupational exposures. The increased aminoaciduria of the exposed individuals and, in particular, the excretion of cystine, proline, and tryptophan[78] indicate structural protein breakdown that is an effect of x-irradiation.

Such incidents emphasize the frequent coexistence of various types of radiation as well as the possibility of combined effects. It is speculated that under certain conditions microwave ovens could cause malfunction of implanted cardiac pacemakers. A fluctuating microwave field modulated by 60-Hz current induces an EMF in the pacemaker circuitry, which after demodulation and rectification could produce 60- or 120-Hz surges or affect detection circuits; a current of 20 μamp at a frequency of 60 Hz can produce a fatal arrhythmia in the human heart. The potential incompatibility of these two electronic devices prompted a warning to patients who have such pacemakers. This is only one of many examples of incompatibility, potential or manifest, in our modern technology.

Although whole-animal observations thus far suggest that the hazards of microwave radiation are specifically thermal, the absence of conclusive information as to the existence or nonexistence of nonthermal effects may be caused by a lack of appropriate detection technics. Indirect evidence suggests that nonthermal effects may be significant.

Although ouabain increases the cardiac output of intact anesthetized rats, this effect is reversed by exposure of the rats to microwaves after administration of the drug. Pinakatt and Richardson[79] found that nonirradiated ouabain in a dose of 0.09 mgm/kgm increases cardiac output 16%, whereas the same dose of irradiated ouabain from the same vial decreases cardiac output 9%. When rats that received irradiated ouabain were exposed to total-body microwave irradiation to a body temperature of 40.5° C, the cardiac output increase was comparable to that of undigitalized rats. These experiments suggest that microwave irradiation inactivates ouabain by a nonthermal molecular effect.

The picture is complicated by the observation that microwave pretreatment of mice, dogs, and rats decreases the lethal effect of x-irradiation; the mean survival time of microwave pretreated animals after a lethal dose of x-rays is longer than that of animals not protected by microwave pretreatment.[80]

Microwave radiation interacts with macromolecule dipole moments to cause internal heating and denaturation by altering effective charge equilibria. Bach and colleagues[81] exposed human gamma globulin to radiofrequency radiation in vitro. At certain frequencies the electrophoretic pattern changed and antigenic reactivity increased. The particular frequencies depend on solution temperature and were of the order predicted by Debye's equation for the relaxation time of polar particles in a viscous medium—2.4%/° C for water at a temperature of 30° to 40° C.

Neither the mass heating of the medium nor the average power absorbed was related to the changes of electrophoretic pattern and antigenic reactivity. In phosphate-buffered physiologic saline at 37.5° C and pH 7.6, the effective frequencies for producing the changes in human gamma globulin were 13.1, 13.2, 13.3, 13.5, 13.6, and 14.4 megacycles. The authors concluded that these frequencies may be the second harmonics of a series of harmonics that are also effective.

Even broader inferences may be drawn from the observations of Piccardi,[82] who found that reaction rates in several chemical systems and sedimentation rates in colloidal systems are influenced by natural electromagnetic radiation. Laboratory examination of these phenomena revealed that the sedimentation rate of a colloidal system is decreased by a moderately strong electric as well as moderately strong static or alternating magnetic fields.[83] Conversely, a low or moderately strong alternating electric field increases the sedimentation rate of a colloidal system. These environmental conditions also increased electrophoretic mobility, viscosity, and shock-freezing temperature while decreasing the conductivity of water. These effects may result from changes in water structure induced by the radiation fields, such changes occurring either directly or by way of a modification of the electric double layer in a heterogeneous system. In colloidal systems the change modifying the electric double layer may involve either electric charge reduction or reduction of the effect of the charge because of local increase of dielectric constant.

These interesting effects of microwave radiation on inanimate physicochemical systems suggest not only the possibility of nonthermal physiologic responses but also the possibility that the interaction of natural and artificial electromagnetic waves with organisms is significant for biology and medicine.

LASER ENERGY

The development of laser energy broadened the scope of the study of the effects of radiant energy on biologic systems. The rapid development of laser technology and the many applications in science and industry suggest that laser technology may affect the lives of many people.

When an atom drops from an excited state to a lower energy level, it emits a photon; this event occurs in ordinary fluorescence. If the emitted photon hits another atom that is in the same excited state, the second atom also drops to a lower energy level, emitting a second photon of the same wavelength as the first. The first photon is not absorbed. The second photon travels in the same direction as the first and thus doubles the light flux. This is the basic principle of light amplification by stimulated emission of radiation, for which *laser* is an acronym. In ordinary fluorescence there is little stimulated emission because so few atoms are in the excited state. Furthermore, what stimulated emission does occur is quickly absorbed by the relatively large number of atoms in the ground state.

If more atoms are in an excited state than in the ground state, the photon initiating the cascade travels in a random direction, as do the cascades of photons produced in the chain reaction, and the coherency and intensity of laser energy are not achieved. To enhance stimulated emission in one direction, laser materials are made in long rods with a mirror at each end. Photons travel the length of the rod, bouncing from end to end, hitting more excited atoms, and building up a phase-coherent light beam that is usually monochromatic.

How the majority of atoms in a laser material are raised to an excited state depends upon the material. Single crystal and glass lasers, such as ruby and neodymium-doped glass, are pumped with a high-energy lamp; a xenon flash lamp may be used for *pulsed* operation, whereas a high-pressure mercury lamp may be used for *continuous* operation. Gas lasers, which are usually continuous, are generally pumped by direct current discharge.

Most early lasers had emission wavelengths of 8,400 to 31,000 Å with power outputs of 1 to 1.5 watts.[84] Exceptions were the well-known ruby laser (6,943 Å) and the helium-neon laser (6,328 Å). The argon ion laser emits at 4,880 and 5,145 Å, whereas nitrogen lasers, which are now on the market, emit at various wavelengths between 3,300 and 4,000 Å. A helium-cadmium laser is reported to be a reliable, low-threshold source of continuous coherent light in the ultraviolet (3,250 Å) or in the deep blue (4,416 Å) range, depending on the adjustment of the instrument. The increased power output of laser instruments is most dramatically evident in the development of a 1,000-watt, continuous-wave, carbon dioxide laser by Sylvania Electric Products.

Even the early less powerful instruments emitted light that could be focused to an intensity of 10^{15} W/cm^2—10^8 times that on the surface of the sun.[28] The associated optic frequency electric field is 10^6 volts/cm, an energy in excess of that binding the outer electron in most atoms and thus sufficient to disrupt severely even transparent materials. It is calculated that four to forty photons per nanosecond are available per molecule in a 1- to 10-megawatt experiment. It is estimated from experiments with dyes that a temperature rise of 400° C results from the absorption of one quantum. As an example of a reactive species with appropriate absorption characteristics for ruby laser studies, the half-life of a solvated electron is about 16 μsec, and within this timespan thousands of quanta per molecule can be delivered.[85] High-power, pulsed lasers develop photon intensities of about 10^{30} photons $\times$ cm^{-2} $\times$ sec^{-1}.[86] When matter is subjected to photon densities of this order of magnitude, processes which are ordinarily highly improbable are observed. Multiple photon absorption[87] is an example of such a process. Thus a molecule that usually does not exhibit transition at the energy of a laser photon is induced by simultaneous or successive absorption of two or more laser photons to undergo transition to an excited electronic state. An example is the absorption of two ruby laser photons (6,943 Å) by anthracene to produce an excited state that initially contains twice the energy (6,943 Å/3,471.5 Å) of a single red photon. This phenomenon also occurs with other materials.

Of the various possible physiologic effects of laser radiation, considerable attention has been given to effects on the human eye. There is particular interest in the pulsed-ruby injection type and in the neodymium-doped lasers because of the spectral regions involved as well as the high energy. It is quite likely that eye damage can occur before the victim is aware of the hazard.[88] Note that the commercially available protective eyewear is designed to protect against specific wavelengths; no single device protects against all wavelengths.

The power density required to reach the threshold of retinal injury increases with reduction of pulse duration.[89] The retinal damage threshold, defined as the energy or power density required to produce a lesion barely observable after a lapse of 5 minutes or more postexposure, is several tenths of a joule per square centimeter for pulsed white light (xenon) and normal-mode pulsed ruby systems, and an order of magnitude less for Q-switched ruby. Wilkening[89] points to the need for data on other laser systems, including continuous wave (CW) systems.

Some scientists believe that retinal injury by nanosecond and shorter pulses results from absorption of thermal energy.[90, 91] Although a thermal model describes the observed histologic damage for laser pulses of microsecond duration reasonably well, Wilkening[89] suggests that pulses of the order of 10^{-9} second may involve nonlinear mechanisms, such as Raman and Brillouin scattering, ultrasonic resonance, and acoustic shock waves.

The use of laser radiation as a research or clinical tool has already resulted in many other interesting observations. It is obviously a useful tool for many biologic research problems. As expected from availability of the instruments, the reported observations of laser effects on biologic systems involve the ruby (6,940 Å), neodymium (6,328 Å), and argon ion (5,145 to 4,880 Å) lasers.

Peroxidase is inactivated by ruby laser emission, although trypsin, alcohol dehydrogenase, and amylase are not.[92] Rounds[93] found that cells in tissue culture from pigmented skin, retinal epithelium, and a malignant melanoma are killed by exposure to ruby laser emission, whereas similar nonpigmented cells are unaffected unless pretreated with a photosensitizing dye. Exposure to argon ion laser emission after photosensitization with a dye produces chromosomal and nuclear lesions. Conversely, lesions are produced in the large mitochondria of myocardial cells without the use of photosensitizing agents.[94] This effect is consistent with the absorption of respiratory enzymes and cytochromes in the blue-green region of the spectrum.

Although laser radiation differs importantly from ionizing radiation, the chromosomal anomalies produced by exposure to laser emissions are of the same type as those that result from ionizing radiation.[93] The persistence of abnormalities in the mitotic apparatus through three subcultures suggests that laser irradiation is mutagenic. This is consistent with the report of Litwin and Glew[95] that a laser-induced lesion on the forearm of a human volunteer was classified histologically 4 days after irradiation as carcinoma in situ.

The reported inhibition of the contractility of cardiac, smooth, and skeletal muscle by ruby laser emission[93] may be related to absorption by myoglobin. Similarly, the induction of intravascular microthrombosis and interstitial hemorrhage[96] may result from absorption by hemoglobin. An obvious factor such as heme-absorbance is probably involved but is certainly not the only one to be considered.

In selecting material for this chapter emphasis was placed on current trends of scientific thought in the various areas of the radiation field. We have attempted to portray the current state of our knowledge, the turbulence at the advancing front, and the pressing unsolved problems. The active areas of research are contributing bits and pieces to our understanding of the complex of events that are the physiologic response of a living biologic system to an ambient radiant energy stressor.

GLOSSARY

alpha (α) particle A particle consisting of two protons and two neutrons and thus having a double positive charge. An alpha particle is the nucleus of a helium atom; it usually has energies of 4 to 10 mev. Alpha particles interact readily with matter, producing ions; they travel a few centimeters in air and as far as 60 μm in tissue. Their high energy and short path means that they produce a dense track of ionization along their path and can thus cause serious damage to the tissues with which they interact. Since alpha particles do not penetrate the cornified layer of the skin, they are not a serious external hazard. However, a serious exposure problem results if alpha-emitting elements are retained within the body; radium, radiothorium, and polonium are examples of such elements.

angstrom (Å) Unit of length equal to 10^{-8} cm, 10^{-4} μm, and 0.1 mμ.

beta (β) particle High-speed, negatively charged electron or a positively charged positron created by nuclear processes in which a neutron changes to a proton and an electron, or a proton changes to a neutron and a positron. Beta particles from radioactive decay have energies ranging from essential zero to a definite maximum characteristic of each element. One of the more energetic of such naturally produced beta particles has an energy of 3.1 mev. By contrast, beta particles produced by the betatron may have energies up to 100 mev.

calorie The quantity of heat required to increase the temperature of 1 gm of water at 15° C by 1° C.

Compton recoil effect The elastic impact of an incident quantum, or photon, and scattering by essentially free electrons. The impact is characterized by conservation of energy and momentum; it produces a recoil electron while decreasing the frequency and increasing the wavelength of the scattered x-rays and gamma rays. It is also termed incoherent scattering.

curie A unit of radioactivity. One curie is that quantity of radioactive nuclide, or isotope, that disintegrates at the rate of 3.700×10^{10} atoms per second; thus a microcurie (one millionth of a curie) is that quantity that disintegrates at 3.7×10^{4} atoms per second. A curie is approximately the rate of decay of 1 gm of radium. The relationship between the rate of disintegration of radioactive material in curies and the radiation dose rate in rads per second depends on (1) the energy of the radiation emitted, (2) the type of radiation emitted, (3) the geometric pattern between the radioactive material and the receptor, and (4) the amount of absorbing material between the radioactive material and the receptor.

daughter product A product of radioactive decay; a nuclide resulting from the radioactive disintegration of a radionuclide, formed either directly or as the result of successive transformations in a radioactive series; a decay product that may be radioactive or stable; an atomic species that is the immediate product of the radioactive decay of a given element.

dipole moment A pair of equal and opposite electric charges or magnetic poles of opposite sign separated by a small distance.

electron A small particle having unit negative electric charge, a small mass, and small diameter. Its charge is $(4.80294 \pm 0.00008) \times 10^{-10}$ absolute electrostatic units, its mass 1/1,837 that of a proton, and its diameter about 10^{-12} cm. It displays wave properties when in motion. Every atom consists of one nucleus and one or more electrons. Cathode rays and beta rays are electrons.

electron-volt (ev) The energy acquired by an electron moving through a potential difference of 1 volt.

electrostatic unit (esu) That quantity of electric charge that, in a vacuum 1 cm distant from an equal and like charge, repels it with a force of 1 dyne.

fluorescence Radiation-stimulated luminescence that is the result of absorption of radiation, either electrified particles or waves, from some other source. Photon absorption excites an electron, which in turn emits a photon as the excited electron returns to its ground state. If emission does not continue longer than about 10^{-6} second after the stimulating radiation ceases, the phenomenon is called *fluorescence*. Fluorescent radiation usually has a longer wavelength than that of the absorbed radiation. *Resonance radiation* and *reemission* are types of fluorescence in which a gas or vapor emits radiation of a certain wavelength when absorbing incident radiation of that same wavelength. Resonance radiation is exhibited by matter that has a line absorption spectrum, whereas reemission occurs with matter that has a continuous or unresolved band absorption spectrum. *Selective reflection* is closely related to fluorescence. A vapor or gas at high pressure reflects radiation of wavelength nearly equal to that of its resonance radiation but does not reflect radiation of other wavelengths. *Phosphorescence* is a radiation-stimulated luminescence that persists for an appreciable time after the stimulation process ceases. An electronic state intermediate between the first excited state and the ground state reemits radiation after a time interval of 10^{-6} second or longer. The time for decay to a certain fraction of the initial intensity depends upon the nature of the phosphorescing matter and on the temperature. The decay time for phosphorescence may be from 10^{-3} second to many days. Fluorescence differs from phosphorescence primarily with respect to decay time. This is the time required, after stimulation ceases, for luminescent intensity to decay to a value of $1/e$ times the intensity at time zero.

gamma (γ) **rays and x-rays** Short wavelength electromagnetic radiation emitted from the nucleus, ranging in wavelength from about 10^{-8} to 10^{-11} cm. Although x-rays and gamma rays have similar properties, x-rays in general have lower energies. Gamma rays are produced by nuclear processes, whereas x-rays result from the interaction of high-speed electrons with atoms. Since x-rays constitute a broad region of the spectrum, they include the low energies of those adjacent to the ultraviolet region as well as the energies of high-speed electrons that produce x-rays, often above 100 kev. The same is true for gamma rays; those naturally produced have energies ranging from kev to mev, whereas high-energy accelerators produce gamma rays of several hundred mev. Gamma rays and highly penetrating x-rays produce a low ion density in the substance with which they interact. Their biologic effects are better known than those of other ionizing radiations.

heat A form of energy associated with the random and chaotic motion of the individual atoms or molecules of which matter is composed. It is therefore fundamentally measured in mechanical energy units, such as ergs or joules. As with other forms of energy, thermal energy involves an intensive, or potential, factor (temperature) and an extensive, or capacity, factor (entropy). Infrared radiation is sometimes called radiant heat, but heat results from absorption of radiation at any wavelength. Thus the *solar heat load* on a man is about half visible radiation and half infrared radiation.

Hertzian An electromagnetic wave produced by the oscillation of electricity in a conductor such as a radio antenna, ranging in wavelength from a few centimeters to many kilometers.

ionizing radiation Any electromagnetic or particulate radiation capable of producing ions, directly or indirectly, in passing through matter.

kiloelectron-volt (**kev**) One thousand electron-volts, or 10^3 ev.

micron A unit of length equal to one thousandth of a millimeter.

million-electron-volt (**mev**) One million electron-volts, or 10^6 ev.

nanometer A unit of length equal to 10^{-9} m; formerly called millimicron.

neutron An uncharged elementary particle of mass number approximately 1; the rest mass of the neutron is 1.00894 atomic mass units. It is a constituent particle of the nuclei of all elements of mass number greater than 1. It is unstable with respect to beta decay and has a half-life of about 12 minutes. It produces no primary ionization in its passage through matter but interacts predominantly by collision and, to a lesser extent, magnetically. Neutrons lose energy by (1) direct collision with nuclei or (2) entering a nucleus and initiating a nuclear reaction. They are produced from sources such as radium-beryllium mixtures or by nuclear reactors. Reactors produce neutrons with energies ranging up to several mev. The relative biologic effectiveness of neutrons is dependent on their energies. The biologic effects result from the charged particles and secondary gamma rays produced by collisions and interactions.

nuclide A species of atom characterized by the constitution of its nucleus. Nuclear constitution is specified by the number of protons (Z), the number of neutrons (N), and the energy content or, alternatively, by the atomic number (Z), the mass number ($A = [N + Z]$), and the atomic mass. To be a distinct nuclide, an atom must be capable of existing for a measurable time; thus, nuclear isomers are separate nuclides, whereas promptly decaying excited nuclear states and unstable intermediates in nuclear reactions are not.

pair production The simultaneous and complete transformation of a quantum of radiant energy into an electron and a positron when the quantum interacts with the intense electric field near a nucleus.

photon The smallest and indivisible unit of electromagnetic radiation. The photon has no rest mass and has properties of both waves and particles. In a collision between a photon and an electron, the electron gains the energy and momentum that the photon loses. A quantity of electromagnetic energy whose value in ergs (E) is the product of its frequency in cycles per sec (ν) and Planck's constant (h): $E = h\nu$. A synonym is quantum or light quantum.

positron A positively charged particle of the same mass and magnitude of charge as the electron.

proton An elementary particle having a single positive charge equivalent to the negative charge of the electron but a mass approximately 1,837 times as great. The positive nucleus of the hydrogen atom is a proton. The protons produced by high-energy accelerators usually have energies of a few mev. They produce significant tissue ionization and, as might be expected from their smaller mass, their path length is longer than that of an alpha particle of equivalent energy.

rad Acronym for radiation absorbed dose; a unit of measurement of the absorbed dose of ionizing radiation; an energy transfer of 100 ergs/gm of mass of the irradiated absorbing material.

radiomimetic substance A chemical substance that produces biologic effects similar to those of ionizing radiation.

radon A heavy radioactive gaseous element formed by disintegration of radium. Radon belongs to a group of chemically inert gases.

relative biologic effectiveness (rbe) Ratio of an absorbed dose of x-rays or gamma rays to the absorbed dose of a particulate radiation required to produce an identical biologic effect in a particular experimental tissue or organism.

roentgen A unit of measurement of ionizing radiation; that amount of ionizing radiation that produces 1 electrostatic unit of ions per cubic centimeter of volume.

roentgen equivalence for man (rem) The product of relative biologic effectiveness and the absorbed dose in rads.

scattering A change in the direction of electromagnetic radiation resulting from interaction with matter without a change of energy. Rayleigh scattering is produced by molecules that are very small in relation to the wavelength of light. Molecular scattering produces the blue sky as seen from earth and the blue appearance of the earth as seen from outer space. The orange and red of sunrise and sunset are the result of light from which blue has been removed by scattering. Scattering from particles comparable to or larger than a given wavelength is termed *Mie scattering.* Because scatter occurs from multiple points on a particle surface, variable interference is observed from different directions. One consequence of this phenomenon is that there is more scatter in a forward than in a sideward direction.

solar elevation angle The angle of the sun above the horizon.

Van Allen radiation belt Either of two broad zones, inner and outer, of intense natural ionizing radiation encircling the earth at varying levels in the upper atmosphere.

REFERENCES

1. Thekaekara, M. P., and Drummond, A. J.: Standard values for the solar constant and its spectral components, Nature Phys. Sci. **229**:6-9, 1971.
2. Smith, K. C., and Hanawalt, P. C.: Molecular photobiology, inactivation and recovery, Molecular Biology Series, vol. 18, New York, 1969, Academic Press, Inc.
3. Newell, R. E.: The circulation of the upper atmosphere, Sci. Am. **210**:62-74, 1964.
4. Calvin, M.: Chemical evolution, Eugene, 1961, University of Oregon Press.
5. Engel, A. E. J.: Time and the earth, Am. Sci. **57**: 458-483, 1969.
6. Cloud, P., and Gibor, A.: The oxygen cycle, Sci. Am. **223**:110-123, 1970.
7. Szabo, G., and others: Ultrastructure of racial color differences in man. In Riley, V., editor: Pigmentation: its genesis and biologic control, New York, 1972, Appleton-Century-Crofts, pp. 23-41.
8. Thomson, M. L.: Relative efficiency of pigment and horny layer thickness in protecting the skin of Europeans and Africans against solar ultraviolet radiation, J. Physiol. (London) **127**:236-246, 1955.
9. Stoll, A. M.: The measurement of bioclimatological heat exchange. In Hardy, J. D., editor: Temperature: its measurement and control in science and industry, part 3, New York, 1963, Holt, Reinhart & Winston, Inc., pp. 73-82.
10. Underwood, C. R., and Ward, E. J.: The solar radiation area of man, Ergonomics **9**:155-168, 1966.
11. Baker, P. T.: Racial differences in heat tolerance, Am. J. Phys. Anthrop. **16**:287-305, 1958.
12. Lee, D. H. K.: Terrestrial animals in dry heat: man in the desert. In Dill, D. B., editor: Handbook of physiology, sec. 4, American Physiological Society, Baltimore, 1964, The Williams & Wilkins Co., pp. 551-582.
13. Renbourn, E. T.: Life and death of the solar topi: a chapter in the history of sunstroke, J. Trop. Med. Hyg. **65**:203-218, 1962.
14. Axelrod, J., and Wurtman, R. J.: The pineal gland, Sci. Am. **213**:50-60, 1965.
15. Daniels, F., Jr.: Physical factors in sun exposure, Arch. Derm. **85**:358-361, 1962.
16. Daniels, F., Jr.: Man and radiant energy: solar radiation. In Dill, D. B., editor: Handbook of physiology, sec. 4, chap. 62, American Physiological Society, Baltimore, 1964, The Williams & Wilkins Co.
17. Daniels, F., Jr., van der Leun, J. C., and Johnson, B. E.: Sunburn, Sci. Am. **219**:38-46, 1968.
18. DeLuca, H. F.: Medical intelligence, current concepts: vitamin D, New Eng. J. Med. **281**:1103-1104, 1969.
19. DeLuca, H. F.: Vitamin D: a new look at an old vitamin, Nutr. Rev. **29**:179-181, 1971.
20. Blum, H. F.: Photodynamic action and diseases caused by light, New York, 1941, Van Nostrand Reinhold Publishing Co.
21. Pathak, M. A., Daniels, F., Jr., and Fitzpatrick, T. B.: The presently known distribution of furocoumarins (psoralens) in plants, J. Invest. Derm. **39**:225-239, 1962.
22. Evans, C. J.: Protection of radium dial workers and radiologists from injury by radium, J. Ind. Hyg. Toxicol. **25**:253, 1943.
23. Loranz, E.: Radioactivity and lung cancer: critical review of lung cancer of miners of Schneeberg and Joachimsthal, J. Nat. Cancer Inst. **5**:1, 1944.
24. Wagoner, J. K., and others: Radiation as the cause of lung cancer among uranium workers, New Eng. J. Med. **273**:181, 1965.
25. Langham, J. L.: Radiobiological factors in manned space flight, National Academy of Science–National Research Council Publ. 1487, Washington, D. C., 1967.
26. Riccioti, E. R.: Animals in atomic research, U. S. Atomic Energy Commission Technical Information Division, Oak Ridge, Tenn., August, 1967.
27. Bauman, R. P.: Absorption spectroscopy, New York, 1962, John Wiley & Sons, Inc.
28. Seliger, H. H., and McElroy, W. D.: Light: physical and biological action, New York, 1965, Academic Press, Inc.
29. Whipple, H. E., editor: Physical factors and modification of radiation injury, Ann. N. Y. Acad. Sci. **114**:1-717, 1964.
30. Reid, C.: Excited states in chemistry and biology, New York, 1957, Academic Press, Inc.

31. Thiele, H., and Wolf, K.: Über die Abtötung von Bakterien durch Licht, Arch. Hyg. **57**:29-55, 1906.
32. Thiele, H., and Wolf, K.: Über die Abtötung von Bakterien durch Licht, Arch. Hyg. **60**:29, 1906.
33. Roffo, A. H.: Cáncer y sol; carcinomas y sarcomas producidos por la acción del sol total, Bol. Inst. Med. Exper. Estud. Trat. Cáncer **11**:353-469, 1934.
34. Bungeler, W. Z.: Über den Einfluss photosensibilisierender Substanzen auf die Entstehung von Hautgeschwülsten, Z. Krebsforsch. **46**:130, 1937.
35. Atomic Energy Commission: Effects of atomic weapons, Washington, D. C., 1950, U. S. Government Printing Office.
36. Swanson, C. P.: Contrasts of initial biological effects resultant from different types of radiation. In Burton, M., Kirby-Smith, J. S., and Magee, J. L., editors: Comparative effects of radiation, New York, 1960, John Wiley & Sons, Inc.
37. Cronkite, E. P.: Evidence for radiation and chemicals as leukemogenic agents, Environ. Health **3**: 297, 1961.
38. National Research Council, Committee on Pathologic Effects of Atomic Radiation: Effects of ionizing radiation on human hemopoietic system, National Academy of Science–National Research Council Publ. 875, 1961.
39. Bizzozero, O. J., Jr., Johnson, K. G., and Ciocco, A.: Radiation-related leukemia in Hiroshima and Nagasaki, 1946-1964. 1. Distribution, incidence and appearance time, New Eng. J. Med. **274**:1095, 1966.
40. Bizzozero, O. J., Jr., and others: Radiation-related leukemia in Hiroshima and Nagasaki, 1946-1964. 2. Observation on type-specific leukemia, survivorship, and clinical behavior, Ann. Intern. Med. **66**:522, 1967.
41. Henshaw, P. S., and Hawkins, J. W.: Incidence of leukemia in physicians, J. Nat. Cancer Inst. **4**: 339-346, 1944.
42. Marrow, P. E., chairman, Task Group: Lung dynamics, deposition and retention models for internal dosimetry of the human respiratory tract, Health Phys. **12**:173, 1966.
43. Lewis, E. B.: Thyroid radiation doses from fallout, Proc. Nat. Acad. Sci. **45**:894, 1959.
44. Carsten, A., and Innes, J.: Physical factors and modification of radiation injury, Ann. N. Y. Acad. Sci. **114**:316, 1964.
45. Gafafer, W. M., editor: Guide to the recognition of occupational diseases, U. S. Department of Health, Education, and Welfare, Public Health Service Publ. No. 1097, 1964.
46. Iberall, A. S.: Quantitative modeling of the physiological factors in radiation lethality, Ann. N. Y. Acad. Sci. **147**:1-81, 1967.
47. Western, F.: Developing radiation protection standards, Arch. Environ. Health **9**:654, 1964.
48. Background material for the development of radiation protection standards, Federal Radiation Council Report No. 1, Washington, D. C., 1960, U. S. Government Printing Office.
49. Background material for the development of radiation protection standards, Federal Radiation Council Report No. 2, Washington, D. C., 1961, U. S. Government Printing Office.
50. Effects of radiation on human heredity, World Health Organ. Report, Ser. 106, 1957.
51. General Assembly Official Record, 19th Session: Report of the United Nations Scientific Committee on the Effects of Atomic Radiation, Suppl. 14, 1964.
52. Russell, W. L.: Studies in mammalian radiation genetics, Nucleonics **23**: 53, 1965.
53. Holcomb, R. W.: Radiation risk: a scientific problem? Science **167**:853, 1970.
54. Wolff, S., editor: Radiation induced chromosome aberrations, New York, 1963, Columbia University Press.
55. Savage, J. R. K.: Sites of radiation induced chromosome exchanges. In Ebert, M., and Howard, A, editors: Current topics in radiation research VI, New York, 1970, John Wiley & Sons, Inc., pp. 131-194.
56. Brown, J. K., and McNeill, J. R.: Aberrations in leukocyte chromosomes of personnel occupationally exposed to low levels of radiation, Radiat. Res. **40**: 534, 1969.
57. Watson, J. D., and Crick, F. H. C.: Molecular structure of nucleic acids; a structure for deoxyribose nucleic acid, Nature **171**:737-738, 1953.
58. Watson, J. D., and Crick, F. H. C.: Genetical implications of the structure of deoxyribonucleic acid, Nature **171**:964-967, 1953.
59. Beukers, R., and Berends, W.: Isolation and identification of the irradiation product of thymine, Biochim. Biophys. Acta **41**:550-551, 1960.
60. Johns, H. E., Rapaport, S. A., and Delbrueck, M. J.: Photochemistry of thymine dimers, Mol. Biol. **4**:104-114, 1962.
61. Grossman, L.: The effects of ultraviolet-irradiated polyuridylic acid in cell-free protein synthesis in Escherichia coli. II. The influence of specific photoproducts, Proc. Nat. Acad. Sci. U. S. A. **50**:657-664, 1963.
62. Daniels, M., and Grimison, A.: Photochemical deamination of cytosine at 2,537 Å, Biochem. Biophys. Res. Commun. **16**:428, 1964.
63. Latarjet, R., Ekert, B., Apelgot, S., and Reboyrotte, N.: Radiobiochemical studies of desoxyribonucleic acid (DNA), J. Chim. Phys. **58**:1046, 1961.
64. Eisinger, J., and Shulman, R. G.: UV-induced electron spin resonances in DNA and thymine, Proc. Nat. Acad. Sci. U. S. A. **50**:694-696, 1963.
65. Nakai, S., and Mortimer, R.: Induction of different classes of genetic effects in yeast using heavy ions, Radiat. Res. **7** (Suppl.):172-181, 1967.
66. Lea, D. E.: Action of radiations on living cells, Cambridge, 1955, Cambridge University Press.
67. Powers, E. L.: Contributions of electron paramagnetic resonance techniques to the understanding of radiation biology. In Snipes, W., editor: Electron spin resonance and the effects of radiation on biological systems, National Academy of Science–National Research Council, Washington, D. C., 1966, p. 137.
68. Szybalski, W.: Molecular events resulting in radiation injury, repair and sensitization of DNA, Radiat. Res. **7** (Suppl.):147-159, 1967.
69. Ebert, M.: Direct and indirect initial effects on biological systems. In Burton, M., Kirby-Smith, J. S.,

and Magee, J. L., editors: Comparative effects of radiation, New York, 1960, John Wiley & Sons, Inc., p. 214.
70. Pullman, B., and Pullman, A.: Quantum biochemistry, New York, 1963, John Wiley & Sons, Inc.
71. Emmerson, P., and others: Radiation chemistry: chemical effects of ionizing radiations on nucleic acids and nucleoproteins, Nature **187:**319, 1960.
72. Bube, R. H.: Photoconductivity of solids, New York, 1960, John Wiley & Sons, Inc.
73. Bacq, Z. M., and Alexander, P.: Fundamentals of radiobiology, rev. ed., Oxford, 1961, Pergamon Press.
74. Kowalsky, A., and Cohn, M.: Applications of nuclear magnetic resonance in biochemistry, Advances Biochem. **33:**481, 1964.
75. Phillips, G. O.: Energy transfer in radiation processes, New York, 1966, Elsevier Publishing Co.
76. Blois, M. S., Jr., and others, editors: Free radicals in biological systems, New York, 1961, Academic Press, Inc.
77. Ingram, M., Howland, J. W., and Hansen, C. H., Jr.: Sequential manifestations of acute radiation injury vs. "acute radiation syndrome" stereotype, Ann. N. Y. Acad. Sci. **114:**356-367, 1964.
78. Ganis, F. M., Hendrickson, M. W., and Howland, J. W.: Amino acid excretion in human patients accidentally exposed to large doses of partial-body ionizing radiation (the Lockport incident), Radiat. Res. **24:**278, 1965.
79. Pinakatt, T., and Richardson, A. W.: Effect of microwave irradiation on cardiac activity of ouabain, Fed. Proc. **24:**702, 1965.
80. Thomson, R. A. E., Michaelson, S. M., and Howland, J. W.: Modification of X-irradiation lethality in mice by microwaves (radar), Radiat. Res. **24:** 631, 1965.
81. Bach, S. A., Luzzio, A. J., and Brownell, A. S.: Effects of R-F energy on human gamma globulin, J. Med. Electron., September-November, pp. 9-14, 1961.
82. Piccardi, G.: The chemical basis of medical climatology, Springfield, Ill., 1962, Charles C Thomas, Publisher.
83. Fischer, W. H., Sturdy, G. E., Ryan, M. E., and Pugh, R. A.: Laboratory studies on fluctuating phenomena, Int. J. Biometeor. **12:**15-19, 1968.
84. Townes, C. H.: Optical masers and their possible application to biology, Biophys. J. **2:**325, 1962.
85. Wiley, R. H.: Laser organic chemistry, Ann. N. Y. Acad. Sci. **122:**685, 1965.
86. Wilson, M. W.: Prospects in laser chemistry, Ann. N. Y. Acad. Sci. **168:**615, 1970.
87. Petocolas, W. L.: Multiphoton spectroscopy. In Eyring, H., Christensen, C. J., and Johnson, H. J., editors: Annual reviews in physical chemistry, vol. 18, Palo Alto, Calif., 1967, Annual Reviews Inc.
88. Staub, H. W.: Protection of the eye from laser radiation, Ann. N. Y. Acad. Sci. **122:**773, 1965.
89. Wilkening, G. M.: A commentary on laser-induced biological effects and protective measures, Ann. N. Y. Acad. Sci. **168:**621, 1970.
90. Hayes, J., and Wolbarsht, M.: Thermal model for retinal damage induced by pulsed lasers, Aerospace Med. **39:**5, 1968.
91. Kohtiao, A., Newton, J., and Schwell, H.: Hazards and physiological effects of laser radiation, Ann. N. Y. Acad. Sci. **122:**777, 1965.
92. Igelman, J. M., Rotte, T. C., Schecter, E., and Blaney, D. H.: Exposure of enzymes to laser radiation, Ann. N. Y. Acad. Sci. **122:**790, 1965.
93. Rounds, D. E., Chamberlain, E. C., and Okigaki, T.: Laser radiation in tissue culture, Ann. N. Y. Acad. Sci. **122:**713, 1965.
94. Berns, M. W., and Rounds, D. E.: Laser microbeam studies on tissue culture cells, Ann. N. Y. Acad. Sci. **168:** 550, 1970.
95. Litwin, M. S., and Glew, D. H.: The biological effects of laser radiation, J.A.M.A. **187:**842, 1964.
96. Kochen, J. A., and Balz, S.: Laser-induced microvascular thrombosis, embolization and recanalization in the rat, Ann. N. Y. Acad. Sci. **122:**728, 1965.

SUGGESTIONS FOR FURTHER READING

1. Alvarez, L. W.: Recent developments in particle physics, Science **165:**1071-1090, 1969.
2. Cook, J. S.: Photoreactivation in mammalian cells. In Giese, A. C., editor: Photophysiology, vol. 5, chap. 7, New York, 1970, Academic Press, Inc.
3. Epstein, J. H.: Ultraviolet carcinogenesis. In Giese, A. C., editor: Photophysiology, vol. 5, chap. 8, New York, 1970, Academic Press, Inc.
4. Giese, A. C., editor: Photophysiology, vol. 1, New York, 1964, Academic Press, Inc.
5. Giese, A. C.: Studies on ultraviolet radiation action upon animal cells. In Giese, A. C., editor: Photophysiology, vol. 2, chap. 12, New York, 1964, Academic Press, Inc.
6. Goodman, J. M.: Light, Sci. Am., vol. 219, 1968.
7. Johnson, B. E., Daniels, F., Jr., and Magnus, I. A.: Response of human skin to ultraviolet light. In Giese, A. C., editor: Photophysiology, vol. 4, chap. 11, New York, 1970, Academic Press, Inc.
8. McElroy, W. D., and Glass, B., editors: Light and life, Baltimore, 1961, The Johns Hopkins Press.
9. Painter, R. B.: The action of ultraviolet light on mammalian cells. In Giese, A. C., editor: Photophysiology, vol. 5, chap. 6, New York, 1970, Academic Press, Inc.
10. Ponnampheruma, C.: Ultraviolet radiation and the origin of life. In Giese, A. C., editor: Photophysiology, vol. 3, chap. 8, New York, 1968, Academic Press, Inc.
11. Ripps, M., and Weale, R. A.: The photophysiology of vertebrate color vision. In Giese, A. C., editor: Photophysiology, vol. 5, chap. 5, New York, 1970, Academic Press, Inc.
12. Urbach, F., editor: The biological effects of ultraviolet radiation, New York, 1969, Pergamon Press.
13. Weisskopf, V. F.: Physics in the twentieth century, Science **168:**923-930, 1970.
14. Wolfson, A.: Animal photoperiodism. In Giese, A. C., editor: Photophysiology, vol. 2, chap. 12, New York, 1964, Academic Press, Inc.

8 MAGNETOBIOLOGY

MADELEINE F. BARNOTHY
JENO M. BARNOTHY

The starting point of any discussion of magnetic fields is generally a description and rationalization of terms. This approach is appropriate in view of the many sets of units in use and the different ways in which magnetic effects have historically been defined. Throughout this chapter we will use the cgs unit of field intensity, the *oersted* (Oe), defined as that field that exerts a force of 1 dyne on unit magnetic pole. Although customary, it is incorrect to express field *intensity* in gauss units; the latter is a measure of magnetic *induction*. However, in a vacuum, and very nearly also in air, a magnetic induction of 1 gauss is numerically equal to a field intensity of 1 oersted. The range of measurable magnetic fields is very wide; it extends from only a few gammas ($1 \gamma = 10^{-5}$ Oe) through the geomagnetic fields (approximately 0.5 Oe) to those produced by spectacular implosion technics (about 10^6 Oe).

Study of the biologic effects of magnetic fields is both a very old and a very recent area of investigation. Since the dawn of human culture a connection between health and the mysterious force of the lodestone (magnetite) has been suspected. During the sixteenth century it was believed that the human body had magnetic properties. Such ideas were expressed in the writings of Paracelsus and subsequently in the medical literature of the eighteenth and nineteenth centuries. Messmer used the expression "animal magnetism" to describe the phenomenon of hypnotic suggestion. It was also believed that magnetic fields could cure some manifestations of hysteria. In the process of rejecting these early pseudoscientific ideas, biomagnetism and magnetobiology fell into disrepute, where they remained for a long time. Cast out from among respectable scientific disciplines, they have emerged only during recent decades from the "dark ages" by producing reliable evidence of the biologic effects of magnetic fields.

During the 1930's seed germination, orientation of roots, growth of bacteria and yeast, and growth of tumors were investigated in magnetic fields with varying results. Most of these early studies suffered from improper documentation and particularly from lack of control specimens in dummy magnets. Invariably, when an organism is placed in a magnetic field, other environmental factors are also changed. The most obvious such changes involve confinement, temperature, illumination, and air circulation. Unless all other environmental changes are also carefully duplicated in the environment of a control specimen, no serious consideration can be given to the experimental results.

Biologic processes are mainly complex chemical reactions. The chemical properties of molecules result from the arrangement and motion of electrons and nuclei, as determined by the interactions of the magnetic and electric fields of these particles. Thus the principles of chemistry and biology are understood in terms of electrodynamics and quantum mechanics. These two theoretic tools enable us to analyze and interpret chemical reactions. Consequently, electric and magnetic fields would seem to be the natural experimental devices to be used in obtaining information regarding the basic phenomena themselves.

The biologic medium in which these chemical reactions take place is not an insulator; rather, it is a conductor, although not a very good one.

An external electric field polarizes this conductor and the body acts as a Faraday cage, shielding all organs, cells, and molecules within from the external electric field. This effect does not occur with a magnetic field because the magnetic permeability of most body constituents is nearly equal to that of air; hence, the body does not act as a shield and the field within an organism is practically the same as that outside it. Thus the magnetic field seems to be an ideal means for investigating biologic function.

In his general theory of relativity Einstein showed that the mysterious gravitational force acting between masses can be interpreted as a peculiar state of space, characterized in the neighborhood of masses by a deviation from the Euclidean geometry of flat space. Masses cause flat Euclidean space to become curved, to bulge out. A test particle possessing mass "senses" the varying curvature of this bulge as a real physical force, which causes the particle to deviate from its straight-line path. We may even say that mass is no more than a local curvature of generally flat space, the bulge representing all the properties of the mass.

Physicists still have a long way to go before a unified gravitational-electromagnetic field theory is developed. Nevertheless, we believe that we are not very far from the truth when we surmise that a magnetic field can be interpreted similarly as a peculiar state of space, but with the difference that the deviation from Euclidean geometry is now sensed by test particles endowed with an electric charge or magnetic moment.

Faraday succeeded in presenting a vivid description of the magnetic field and its properties by introducing the concept of lines of force. In the case of a bar magnet, with the north pole at one end and the south pole at the other, the lines of force start from the north pole and, following a shorter or longer curved path, return to the magnet at the south pole. In the case of a moving charge or a conductor in which an electric current is flowing, the lines of force are circles with the moving charges or the current-carrying conductor as the center.

Contrary to electricity whose positive and negative charges can exist independently, it is impossible to separate a north pole from a south pole; magnetic poles always exist in pairs. Thus Faraday introduced the concept of the magnetic dipole. He called the product of pole strength and distance between north and south pole the magnetic moment of the dipole. Whereas the electric field around an electric charge decreases as the square of the distance from the charge, the magnetic field around a magnetic dipole decreases as the cube of the distance from the dipole.

A magnetic field is always associated with an electric current. The simplest way to produce a magnetic field is to take a conductor, such as a wire, and wind it around a cylinder, forming a solenoid. Inside a solenoid lines of force are parallel to the axis of the solenoid and the field is relatively homogeneous; that is, the gradient is very small. Outside the solenoid lines of force spread out and converge again at the other end of the solenoid. In a distance large compared to the length of the solenoid, the field is the same as that produced by a bar magnet. This similarity is a consequence of the fact that a permanent bar magnet produces an external field with the help of a large number of aligned ring currents. The lines of force continue inside the bar magnet and inside the molecular ring currents, just as they continue inside the loops of the solenoid.

Lines of force and magnetic poles are extremely useful concepts for describing the properties of a magnetic field, but actually neither exists. It is a mistake to imagine lines of force as some kind of radiation emanating from a north pole that is absorbed by a south pole.

From a strictly quantitative viewpoint it is difficult to understand how magnetic fields affect a biologic system; the energy that a fairly strong magnetic field (for example, 10,000 Oe) can transfer to a molecule with an unpaired electron, and thus a magnetic moment, is less by two orders of magnitude than the thermal energy of molecules at body temperature. Thus we would expect any ordering, orienting, or rotating effect of a magnetic field to be randomized by thermal motion of the molecules. In the discussion of theoretic models some possible mechanisms of the effects of magnetic fields on biologic systems are mentioned.

During the last 20 years the ready availability of strong magnets and the advent of space travel have enhanced interest in the biologic action of magnetic fields. Many carefully planned and executed experiments have been conducted in laboratories throughout the world. In the following sections of this chapter the results of these experiments are presented and discussed.

MAGNETIC ENVIRONMENTS

Living beings on earth, or when exploring surrounding space, must live and function in the natural or artificial magnetic fields to which they are exposed. Although not so readily sensed as temperature, pressure, atmosphere, light, and gravity, the magnetic field is an environmental factor of importance.

Geomagnetic field

The earth is a large magnet with its south pole close to the geographic north pole and its north pole close to the geographic south pole. The magnetic field at the earth's surface has a mean average strength of about 0.5 Oe. The geomagnetic field is thus an environmental factor common to all organisms on the earth's surface. It is logical to ask whether, during the evolution of life on earth, physiologic processes have become adjusted to the geomagnetic field as an ever-present environmental factor. The question also arises whether living organisms can be removed from this geomagnetic environment without effect.

Over most of the inhabited surface of the earth the horizontal component of the geomagnetic field points northward. It has a downward component through the northern and an upward component throughout the southern hemisphere. It is presumed that this field arises from electric current flowing inside the earth, rather than from a permanently magnetized core.

To study the physiologic and psychologic effects of the geomagnetic field, one can either study the effect of removal of the field or one can study the effects of variations of the intensity and direction of the field. Most research has been done with plants and lower animals; very little research has been done with human beings.

To "remove" an organism from the omnipresent geomagnetic field is not simple. A specimen must either be moved as far from the earth as about 10 earth radii from the surface of the earth or the varying geomagnetic field must be automatically canceled by an artificial field of equal intensity and opposite direction. Beischer[1] exposed two healthy male volunteers to such a "zero field" environment for 12 days. Subsequently the number of exposed subjects was increased to twenty-five. No abnormal variation of body weight, temperature, respiratory rate, blood pressure, electrocardiogram, electroencephalogram, or number and distribution of blood cells was found. Psychophysiologic and psychologic performance were also assessed by means of tests of space perception, hand-eye coordination, visual spatial memory, body image, visual fields, visual auditory conflict, and conceptual reasoning. No unusual physiologic effects beyond the effects of confinement during exposure were observed. The only significant change observed was a decrease of peripheral visual critical flicker-fusion frequency; however, even this effect could be a consequence of confinement.

In a magnetic field of about one thousandth the strength of the geomagnetic field the response of young male mice to induced foreign biopolymers as measured by the effect on enzyme activity was reduced on the average by about 30%.[2]

Swiss/Webster mice and their progeny were kept in mu-metal cages within which the field was one five-hundredth the strength of the geomagnetic field.[3] Many startling effects were observed. The mice became less active, more docile, often assumed a dorsal position (unusual for mice), developed alopecia, showed signs of premature aging, and died early. These observations, if true, would suggest that chronic exposure to low-intensity magnetic fields, such as the lunar field, may have a deleterious effect on mammals. However a repetition of the experiments by Beischer in a 50 γ field did not show either alopecia or any of the other abnormalities even after 1 year's exposure.

A second way that inferences regarding the biologic, physiologic, and psychologic effects of the geomagnetic field can be evaluated is to determine whether field strength changes are correlatable with observed effects. The geomagnetic field intensity on the surface of the inhabited areas of the earth varies from 0.58 to 0.25 Oe and the field strength decreases by about 0.04% per kilometer elevation.

The geomagnetic field varies in both direction and intensity. It has a diurnal (24-hour) and lunar monthly (27-day) periodicity, varying with an amplitude of less than 1%. The fact that living organisms exhibit cyclic phenomena, with periods closely approximating the major circadian and lunar month cycles, in the apparent absence of any other environmental cues such as light, temperature, and barometric pressure suggests the possibility that the timing of biologic rhythms may depend on subtle rhythmic changes of geomagnetic field intensity. Much larger variations occur during sporadic magnetic storms, usually associated with intense solar activity, or solar flares.

Experiments involving exposure of lower animals to magnetic fields ranging in strength from 1 to 10 Oe demonstrated that altering the vectorial relationship between photic stimulus and magnetic field changes the cyclic pattern of planaria activity.[4] These results suggest the possibility that planaria sense some component of the earth's fluctuating magnetic field and that this geophysical circadian fluctuation may be the driving force for biologic circadian rhythms. This idea is further suggested by reports of behavioral alterations in human psychopathologic population groups by parameters associated with naturally occuring variations in the elements of the geomagnetic field.[5] However, an attempt to substantiate this finding by applying steady artificial, low-strength magnetic fields to the head in a bitemporal direction and measuring psychomotor reaction time of normal and schizophrenic individuals was unsuccessful.[6] On the other hand, 5- to 11-Oe fields modulated at rates of 0.1 to 0.2 Hz produced significant changes in reaction time.

If the cyclic behavior of living organisms is viewed as a rhythmic variation in the level of irritability, do naturally occurring magnetic storms produce a demonstrable variation of irritability level? A statistical study of the time of onset of pathologic states and suicide in Denmark and Switzerland and the occurrence of magnetic storms revealed a strong correlation.[7] The peak of the magnetic intensity change preceded the peak of the incidence of morbid conditions by 2 to 3 days. A relationship was reported between the rate of human blood coagulation and solar activity. Total leukocyte count was found to be reduced during sun spots. Statistical correlations were established between the incidence of epidemics, such as influenza, diphtheria, cholera, and plague, and periods of enhanced solar activity.[8]

The orientation of birds during their period of migratory unrest was shown to be altered by experimental manipulation of the environmental magnetic field direction. A mechanism was proposed to explain the fact that migrating birds can find their way when flying at great altitude or in clouds when no geographic landmarks are visible.[9] According to this theory an electromotive force is generated in the wings by polarization currents, the direction of which change with the wing-flapping cycle; the magnitude of these currents depends on the flight direction relative to the geomagnetic field. The potential differences that could arise in this manner are very small (of the order of microvolts); however, the African river fish *Gymnarchus niloticus* can detect currents caused by application of a potential difference of only 1 μv to the body between head and tail.

Orientation effects of the geomagnetic field on the motion of planaria and diptera imagoes, and on early seed germination has been reported. Snails can distinguish geographic direction and orientation of an artificial magnetic field; the curves of emergence of snails from a narrow opening in an artificial field were mirror images of emergences of the same snails in the geomagnetic field.[10] Artificial fields of strength close to that of the natural background field produced the greatest changes in emergence direction. It was also shown that planaria can associate the direction of an experimental light source with the direction of the horizontal vector of a magnetic field, whether natural or experimentally imposed, and retain this information for a period of many minutes. The orientation of the bee dance on a vertical honeycomb is affected by very small, experimentally induced alterations of the ambient magnetic field.[11] Either increased or decreased field strength can induce a change in the bee's orientation, suggesting that bees have made an evolutionary adjustment to geomagnetism. Usually about 2 hours is needed before bees can adjust to a change in field environment. In "zero field" the direction of the bee dance is isotropic.

An even more surprising finding is that human beings can sense muscle tone alterations caused by geomagnetic field gradient variations of as little as 0.3 to 0.5 mOe/m and, in this way, may sense underground cavities and water currents.[12] On the other hand, Foulkes',[13] careful studies did not corroborate the sensitivity of humans to magnetic gradients. The question still remains controversial.

We may infer from the foregoing phenomena that organisms have developed a kind of sense organ during the millenia of biologic evolution; thus, living organisms may have a mechanism that acts as a biologic compass and/or biologic clock by which they orient themselves relative to the fluctuating geomagnetic field. If such a mechanism truly exists, it is probably "tuned" to weak magnetic fields. In laboratory investigations of the biomagnetic effects of strong fields, such a mechanism may be entirely inhibited or suppressed. It would have to be adaptive in relation to environmental changes and function within the organism linked to the system that coordinates the vital processes of organisms as they interact with the environment.

The geomagnetic field serves the biologically important function of shielding against the influx of high-energy charged particles arriving from the sun (solar wind) and from all directions within and without the galaxy (cosmic rays). The geomagnetic field deflects or traps these particles except at the magnetic poles of the earth. High-energy charged particles produce biologic and genetic effects as do x-rays and gamma rays.

Thus far we have discussed the intensity and variations of the geomagnetic field and their biologic implications. However, the direction of the field is not constant; it undergoes secular variations and is known to have reversed direction several times, the south pole having become the north pole and vice versa. During the process of polarity reversal, there exists a certain time interval when the intensity of the geomagnetic field is zero; thus, the magnetic shield does not exist and only absorption in the atmosphere reduces the intensity. Hence, more charged particles will bombard the earth's surface.

Several investigators[14] have shown a close correlation between the extinction of planktonic organisms and geomagnetic field reversals. On the other hand, only a few organisms have become extinct near the time of such reversal while some have survived many reversals before extinction. This may mean that field reversal per se, despite the associated drastic change of the geomagnetic environment and removal of the protective shield, is probably not sufficient to result in the extinction of a species. Reversal would affect only certain species especially vulnerable to slight environmental change.

A challenging speculation has recently been proposed; it links the appearance of *Homo sapiens* with a mutation caused by cosmic ray particles during a period of the magnetic field polarity reversal about 2 million years ago. This theory has been questioned on the basis of whether cosmic ray intensity could have increased sufficiently during reversal to produce such a mutation.

Questions relating to space travel

Technologic sophistication has made space travel a reality; astronauts and space travelers will be exposed to new intensities and patterns of magnetic environment. The particle flux from the sun interacting with our geomagnetic field produces three regions of extraterrestrial space: the interplanetary region, where the influence of the earth and its magnetic field is not felt; the magnetosheath, where the interaction of solar wind and geomagnetic field occurs; and the innermost region, the magnetotosphere, where the geomagnetic field prevails. Between magnetosheath and magnetosphere there is a 100- to 1,000-km thick region, called the magnetopause. The magnetopause extends from 8 to 14 earth radii on the earth's sunlit side and as much as 31 earth radii on its dark side; the intensity in this region is about 40 γ, one hundredth of the intensity of the geomagnetic field at the earth's surface. In the interplanetary region around the earth, field strength is usually about 5 γ, but it may vary in other parts of the solar system between 1 and 100 γ.

The magnetic field at the sun's surface is approximately 1 Oe, although during sunspot activity intensities up to several thousand oersteds may occur. At the moon's surface the field is about 40 γ, on Venus, presumably 500 γ, whereas it is estimated that the field strength on Jupiter is about 1,000 Oe.

Interstellar magnetic fields are probably very weak, a fraction of 1 γ. During lunar and interplanetary missions to Mars and Venus man will be exposed to magnetic environments generally much lower than those existing on earth. Not only will field intensities be lower but they will vary with a different time-intensity pattern.

There are plans to use artificial high-intensity magnetic fields in and around future spacecraft to shield astronauts from solar wind and cosmic radiation. Depending on the technic suggested, the astronauts' cabin would be exposed to fields of 100- to 1,000-Oe strength, but probably only for the brief duration of solar flares. Should magnetohydrodynamic propulsion systems be used in spacecraft, the strong fields needed to contain the plasma flow, even if shielded from the living quarters, may expose astronauts to fields of several thousand oersteds' intensity for brief periods while servicing the engine.

Most biomagnetic effects have a cumulative nature, leading to detrimental effects after long-term exposure. It is also true that most biomagnetic effects are vector effects. Barnothy[15] therefore suggested that cumulative effects could be avoided during prolonged aerospace flight if either manned compartments or the entire spacecraft or space station were rotated around an axis perpendicular to the direction of the external field. He inferred from animal experiments that very slow rotation around an axis at a right angle to the magnetic field direction would suffice. Such slow rotation would not induce polarization currents or cause dizziness.

Weightlessness and its physiologic consequences may pose a serious problem during prolonged space voyages, despite the fact that thus far manned orbital flights suggest that adaptation is possible. During weightlessness unaccustomed sensations in the balance organs of the inner ear might produce a feeling of discomfort. The gravity sensing organ consists of a gelatinous substance penetrated by hair tufts that are nerve endings. Embedded in this jelly are microscopic crystalline bodies, the otoliths, consisting of calcium carbonate, proteins, and magnesium salts. The force of these otoliths on nerve endings, caused either by the gravitational field or by forces of acceleration, are important for the sensations of weight and balance. The otoliths are paramagnetic. Placing a small permanent magnet, an artificial inhomogeneous magnetic field, close to the inner ear may produce an accelerative force on the otoliths, stimulating the nerve endings and creating the sensation of exposure to a gravitational field.[15]

HEMATOLOGIC AND CARDIOVASCULAR EFFECTS

Hematologic effects

The blood circulatory system transports many essential substances to and from cells, tissues, and organs. An almost constant cell environment is necessary for normal biologic function. The circulation facilitates maintenance of a constant internal environment for cells throughout the body. In this section we discuss the effects of magnetic environments on the concentration and distribution of cell types in circulating blood; this is another way in which magnetic fields affect homeostasis. The effects of magnetic fields on number and distribution of circulating blood cell types has been investigated most extensively in mice exposed to medium-strong magnetic fields.[16]

Erythrocytes

One would expect magnetic fields to affect erythropoiesis for two reasons. First, as we shall see later (p. 335), magnetic fields produce atrophy of the zona fasciculata of the adrenal cortex, extirpation of which causes anemia. Second, iron and the paramagnetic molecule oxygen that has two unpaired electrons are important in erythrocyte metabolism. Erythrocytes carry the pigment hemoglobin which combines with molecular oxygen, thus transporting large amounts of this vital gas from lungs to tissues.

Hemoglobin consists of a protein, globin, bound to an iron-containing prosthetic group called heme. Oxygen combines with Fe^{++} in heme to form oxyhemoglobin. At rest arterial blood hemoglobin is more saturated with oxygen (97%) than venous blood (75%). The simple change of ferrous iron (Fe^{++}) in a hemoglobin molecule to ferric iron (Fe^{+++}) produces the nonfunctioning brown pigment, methemoglobin. Heme is also part of myoglobin, an oxygen-binding muscle pigment, and is found in the respiratory enzyme, cytochrome c. Acid hydrolysis of hemoglobin liberates the prosthetic group as a complex ferric salt called hemin.

Diamagnetism of whole blood was known to Faraday. Hemoglobin, methemoglobin, myoglobin, heme, and hematin are paramagnetic; oxyhemoglobin and carboxyhemoglobin are diamagnetic. Thus the magnetic properties of arterial blood differ from the magnetic properties of venous blood. The affinity of hemoglobin for oxygen is a function of both temperature and pH; temperature dependence may be a consequence of the Curie effect, paramagnetic susceptibility changing with temperature, whereas pH dependence may result from the fact that in solution different acid groups having different magnetic properties interact with heme at different pH values.

A large number of mice placed in vertical homogeneous static magnetic fields did not show a change in the number of circulating erythrocytes.[16] During and after this magnetic exposure red blood cell count remained constant within $\pm$ 0.8%. One explanation for this may be the relatively long lifespan (about 120 days) of erythrocytes; perhaps longer exposures than those used in these experiments would be necessary to produce an effect. On the other hand, inhomogeneous magnetic fields produced a 60% increase in red blood cell count within 13 days of exposure.

Blood hemoglobin content was also determined by measuring optic density with a spectrophotometer at 5,400 Å of sample diluted with 0.04% ammonia. A significant difference was found from the fourth to the sixteenth day of exposure between mice kept in magnets and those kept in dummy magnets. Blood hemoglobin content of the magnetic group mice was 5.4% $\pm$ 1.0% lower than that of the dummy magnet mice.

In rabbits with an inflammatory process produced by intra-abdominal turpentine injection, blood hemoglobin concentration in the animals exposed to magnetic fields was lower than that of unexposed control rabbits.[17] Likhachev proposed that a magnetic field stresses the fragile membrane of the oldest erythrocytes, producing hemolysis and releasing hemoglobin. These experiments also showed that the elevated erythrocyte sedimentation rate associated with the inflammation returned to normal faster in rabbits exposed to medium-strong magnetic fields for 15 minutes daily.

Magnetic fields have a very striking effect on sickled erythrocytes. Sickled red cells orient themselves with their long axes perpendicular to the magnetic lines of force.[18] It has been proposed that sickled erythrocytes contain deoxygenated S-hemoglobin molecules whose paramagnetic heme groups are oriented by the magnetic field.

Red blood cells tend to form cylindric aggregates, or rouleaux. The rate at which erythrocytes gravitate to the bottom of a container (erythrocyte sedimentation rate) is largely determined by rouleaux formation, because when two or more erythrocytes form a single assembly, the terminal velocity of this assembly falling through a viscous liquid is greater than that of a single cell.

Carbohydrates on the erythrocyte surface are antigens. When red blood cells having a certain genetically determined antigenic group are suspended in plasma containing reactive antibodies, the cells tend to aggregate and agglutination occurs.

The effect of magnetic fields on agglutination reactions was studied in vitro.[19] Blood samples were obtained from human volunteers; the cells were centrifuged, washed, and made up into a 2% saline suspension. Antisera were added in various concentrations to the cell suspensions and incubated at 37.5° C. Half of each sample was incubated in magnetic fields of various strengths. After incubation the mixture of cells and antiserum was withdrawn with a pipet and deposited on a slide for visual examination or placed into a beaker of saline for counting with a Coulter particle counter. A magnification of 100× was used for visual scoring and the degree of agglutination assessed on a point scale ranging from 0 (no agglutination and cells evenly distributed) to 10 (agglutination clearly visible to the naked eye). With anti-D (anti-Rh⁰) and D-positive cells, enhanced agglutination in a magnetic field of 2,000-Oe strength was observed, amounting on the average for different serum dilutions to 32%, the difference being greater at intermediate and low concentrations. At field strength less than 50 Oe no difference was found between samples incubated in magnetic fields and controls; at higher field strengths the difference increased. D-negative cells incubated with anti-D serum did not agglutinate, indicating that a magnetic field does not produce nonspecific agglutination. Anti-C (anti-rh′) and anti-E (anti-rh″) sera against C-positive and E-positive cells, respectively, yielded essentially the same pattern. Magnetic fields did not enhance agglutination in tests of antibodies of the ABO and MN systems.

No effect on agglutination was observed in strongly inhomogeneous fields when the particle counting method was used. The authors suggested that this difference is a possible consequence of the method. With visual observation erythrocytes lying flat with cell rims touching were designated aggregates, while those lying flat but above one another could not be distinguished from a single erythrocyte. The latter would pass through the orifice of the particle counter as multiplets, whereas those lying side by side in the same plane might have been destroyed in the streaming of the liquid through the orifice. This explanation is supported by the fact that the anti-D reaction produces weak bonds that are easily disrupted by mechanical action.

Platelets

Blood platelets derive from bone marrow megakaryocytes and normally circulate for 8 to 9 days. Platelets and platelet factors are necessary for blood coagulation; platelet thromboplastic factor is one of the most important of the many factors thus far implicated in the coagulation process. Platelets also play a major role in clot retraction, probably by contraction of filamentous pseudopodia of intact platelets caught within the fibrin network.

One-hundred-day-old DBA/J2 female virgin mice were exposed to a 9,000-Oe vertical homogeneous magnetic field and an equal number of mice of the same batch were housed in dummy magnets.[20] Platelets in tail vein blood were counted in a hemocytometer with a phase contrast microscope (600×). Blood was taken at the same time of day to avoid circadian fluctuations of platelet count. Within a few days after the start of exposure the magnetic field produced a 25% ± 3% increase in platelet count compared to the dummy magnet group. The count gradually diminished to zero by the tenth day of residence in the field. After 20 days of exposure the mice were removed from the magnet and the platelet count again increased initially, reaching a maximum around the third day; this was followed by a slight decrease, returning to normal by the twentieth day after exposure. This pattern is strikingly similar to that observed in various other stress conditions. A 30% platelet count decrease was observed after exposing a human subject's head to a 5,000-Oe field for 15 minutes.[17]

Leukocytes

Granulocytes (neutrophils, eosinophils, and basophils) form from stem cells in bone marrow. They mature rapidly and enter the circulation where they survive for 2 weeks or less. Although some lymphocytes form in the bone marrow, most originate in the thymus, spleen, and lymph nodes. Granulocytes are phagocytes; lymphocytes are involved in immunity, hypersensitivity, and antibody production. Glucocorticosteroids secreted by the adrenal cortex decrease the number of circulating eosinophils and lymphocytes but affect the number of circulating granulocytes and erythrocytes much less.

Barnothy[16] investigated the effect of a vertical homogeneous 4,000-Oe magnetic field on leukocyte count. One hundred seventy mice were investigated, half having been housed concurrently in control dummy magnets. At regular intervals blood was taken from the tail vein and white blood cell (WBC) counts were done with a Coulter particle counter. In mice, unlike human subjects, lymphocytes constitute by far the largest percentage of all blood leukocytes; thus the changes in WBC count were primarily changes in the number of circulating lymphocytes.

Confinement of mice to experimental cages decreased total WBC count in both magnet and dummy magnet group mice during the first few days (Fig. 8-1). Thereafter, however, there were definite differences between the two groups. The most significant feature seen on the graph is an experimental decrease of 20% to 40% in leukocyte count within the first 2 weeks of exposure. This minimum is followed by a temporary increase in leukocyte count, which may reach base line; thereafter, a second more prolonged, more pronounced decrease occurs with a minimum around the thirtieth day of residence. If mice are removed from the field at the time of the first or second minimum, the leukocyte count increases drastically and overshoots the base line by 20% in about 2 weeks. This increase does not occur if the mice are left in the field for a prolonged period after the second minimum. The transitory maximum in leukocyte count early in exposure indicates that the initial drop is probably caused

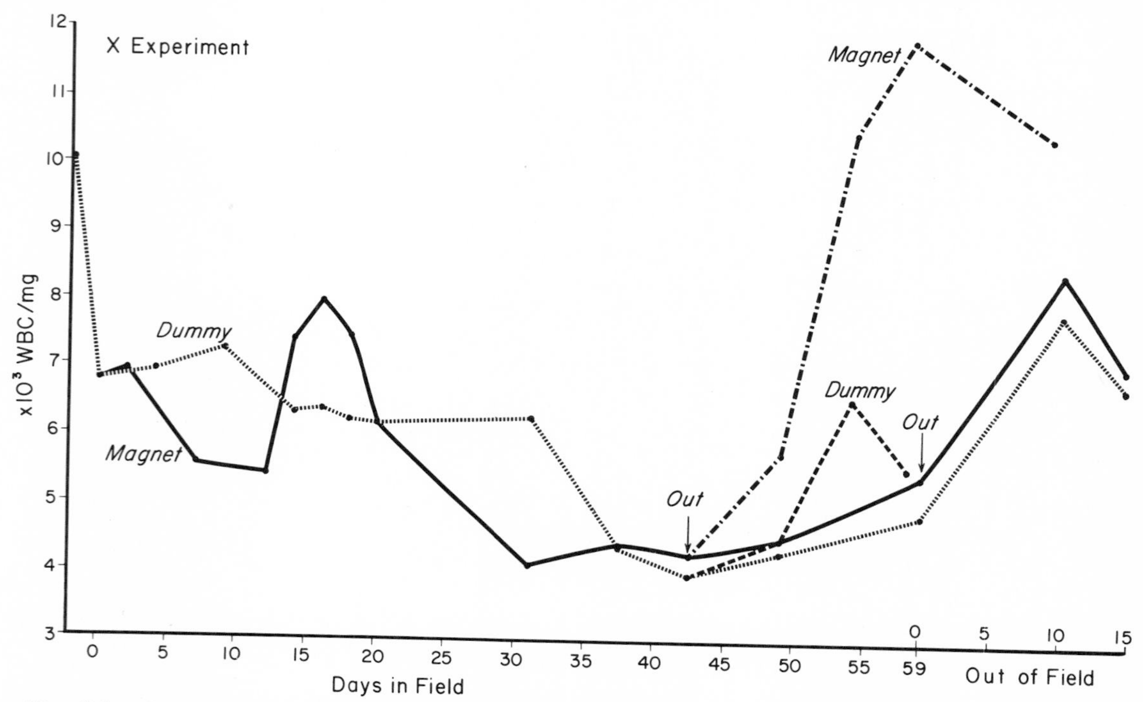

Fig. 8-1. Change in mouse leukocyte counts. The solid line is the average leukocyte count of mice exposed to a 4,200-Oe field; the dotted line is the average leukocyte count of mice kept in "dummy magnet" cages. The arrows marked "out" indicate when the mice were removed from their cages. The dash-dotted and the dashed lines are, respectively, leukocyte counts of magnet and "dummy magnet" mice after transfer to standard cages. About one third of each group was removed from their cages on the forty-second day. The magnet group relative to the "dummy magnet" group shows an initial drop, followed by a transient increase, and a subsequent second drop while in the field; they also show an increase above base line after termination of magnetic exposure. This rise is absent if mice are removed after prolonged (60-day) residence in the field.

by decreased leukocyte lifespan. The decreased lifespan stimulates a more abundant replacement by cells from extravascular compartments. However, after prolonged exposure, inhibition of hematopoiesis is manifest, leading to the second protracted minimum on the graph.

The changes of blood composition just described are similar to those in the reaction of an organism to diverse harmful stimuli. According to Selye[21] such changes are mediated by the hypophysis, acting either directly on the adrenal gland or via release of adrenalin from the adrenal medulla. The three phases of nonspecific stress reaction—alarm, adaptation, and exhaustion—are associated with an initial decrease in leukocyte count, followed by a temporary increase, and eventually a further decrease. During the phase of adaptation blood platelets increase but then they decrease during the third stage of the syndrome. We thus infer that a static magnetic field may constitute an environmental stressor.

In the course of these experiments stained blood films were examined and more than a quarter million leukocytes were classified. During the phase of transient leukocyte increase, granulocytes increased relative to lymphocytes, suggesting that the decreased leukocyte count resulted mostly from a lymphocyte decrease. Together with decreased eosinophils, this finding suggests that magnetic fields influence corticosteroid production.

Cardiovascular effects

Young[22] observed arrhythmic contractions of frog vagal heart preparations in 17,000-Oe magnetic fields, whereas 4,000-Oe fields did not have such an effect. The vagus nerve was stimulated by acetylcholine administration (see also p. 331). Arrhythmic contraction usually occurred 20 minutes after initiation or after cessation of exposure. This suggests that a high-intensity field may affect the cardiac pacemaker as well as the conduction system. Some preparations reverted spontaneously to normal cardiac rhythm within 1 to 3 days; in others, irregularity persisted for as long as 4 days after cessation of exposure. Thus it appears that the heart was permanently affected.

The mechanism of this magnetically induced arrhythmia was further investigated using organic compounds effective in treating certain clinical arrhythmias; however, glucose, digitalis, and oxygen were ineffective against this arrhythmia. On the other hand, acetylcholinesterase injection corrected the arrhythmia, irregularities reappearing when acetylcholinesterase was washed out of the heart. This suggests that a magnetic field impairs the activity of this enzyme; the hydrolysis rate, which normally matches the rate of presynaptic acetylcholine release, is decreased, resulting in the arrhythmia (see also p. 322).

In experiments on rabbits in a 10,000-Oe field the electromotive force (emf) of blood flow was

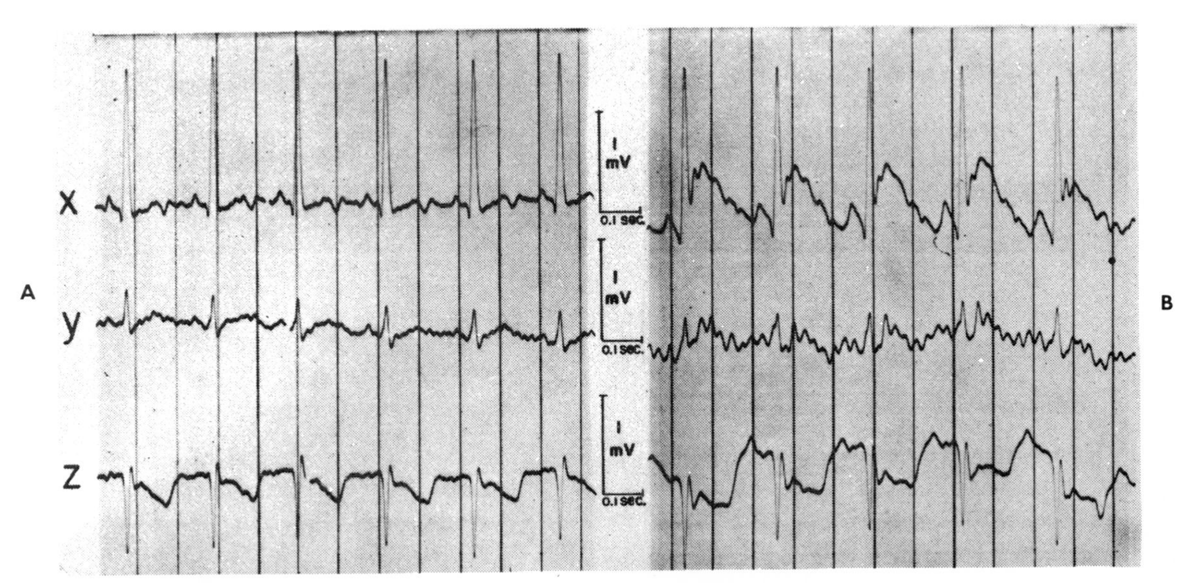

Fig. 8-2. Electrocardiogram (Frank lead system) of a squirrel monkey. **A,** Control outside the magnetic field; **B,** in a field of 100,000 Oe. Note absence of the T wave in **B,** which is almost completely overshadowed by the strong signal generated by aortic blood flow.

superimposed on the electrocardiogram, the peaks occurring between the S and T waves.[23] Beischer[24] observed major electrocardiographic changes in squirrel monkeys exposed for 15 minutes to very strong fields (up to 100,000 Oe) produced by superconductive magnets. The T wave was completely overshadowed by the strong signal generated by aortic blood flow (Fig. 8-2). The flow signal had the fast ascending and gradually descending limb characteristic of aortic blood velocity graphs recorded with a flowmeter. The maximum scalar value of the aortic blood flow potential was a linear function of magnetic field strength. Potentials strongly reflected respiratory changes. However, breathing frequency remained constant at about 50 to 60 breaths per minute during exposure. Well-defined peaks and notches that recurred at regular intervals cycle after cycle were superimposed on the ascending and descending limbs of the flow potential.

In the strong field, P wave amplitude increased, and the P wave also showed respiratory variations of both amplitude and duration that were superimposed on the stronger flow signal. The P waves of animals in magnetic fields began earlier and lasted longer than control P waves. Neither axis, duration, nor amplitude of the QRS complex changed in the field. A slowly decreasing heart rate and increasing sinus arrhythmia previously observed in 1-hour exposures of squirrel monkeys to a 70,000-Oe field were not seen during 15-minute exposures of the same species to the stronger field. Possibly, during longer exposure to a magnetic field, the flow potential stimulates vagal branches of the parasympathetic nervous system via carotid sinus and aortic pressoreceptors, eventually decreasing heart rate.

These experiments on squirrel monkeys in strong fields encourage us to believe that human beings may safely be exposed to high-intensity fields for short periods; longer exposures may possibly decrease heart rate.

METABOLISM

Metabolism is the general sum of all processes by which chemical bonding energy is made available to and utilized by the body. The chemical reactions by which energy is made available to the body are termed catabolism; as a rule these are oxidative or hydrolytic, converting either ingested foods or body materials into carbon dioxide and water. Changes of the normal metabolism of an organism exposed to a static magnetic field can be studied by observing respiratory and growth processes.

Respiratory processes

Oxygen consumption of mice during exposure to a 4,200-Oe field was compared to that of controls kept in dummy magnets.[16] The carbon dioxide produced was absorbed by a potassium hydroxide solution and measured by titration. Oxygen consumption increased about 10% above that of controls during the first 2 weeks of residence in the field but decreased about the same amount below that of controls during the fourth and fifth weeks in the field. After observing the same animals in activity cages, it was concluded that the difference in oxygen consumption is not the result of different activity and/or motility. Lower oxygen consumption coincided with decreased blood hemoglobin concentration and lower blood oxygen capacity (see p. 318).

Breathing frequency did not change in squirrel monkeys during rather short exposure to very strong fields up to 100,000 Oe.[24]

Amer[25] studied the combined effects of several environmental factors on the development of *Tribolium confusum* from the pupal to the adult stage. He found that the effect of 6,000- to 10,000-Oe magnetic fields on development depends on ambient oxygen pressure; a magnetic field increases oxygen toxicity. At higher incubation temperatures a magnetic field reduced the incidence of heat-induced molting failure, suggesting that a magnetic field exerts a cooling effect. Indeed, mouse rectal temperature decreased by 0.7° C during and immediately after exposure to 4,200-Oe magnetic fields.[26]

That magnetic fields reduce the rate of oxidative processes was indirectly inferred from decreased rate of restoration of normal body temperature after acute hypothermia.[27] Albino mice were chilled at −10° C until rectal temperature fell from about 37° to 19° C. After removing the mice from the refrigerator, the rate of temperature restoration of the magnet group mice was definitely slower than that of the controls.

Cook and co-workers[28] studied cell respiration extensively, using a constant pressure differential respirometer designed for small samples (1 to 4 mgm) of tissue or cell suspensions. Tissue samples were exposed to a 7,300-Oe homogeneous field for 3 to 4 hours. Oxygen uptake of the ascites form of Sarcoma-37 was depressed by 29% ± 6% compared to control suspensions, whereas yeast respiration was stimulated by the field. To decrease error resulting from variations between individual samples in both magnet and control groups, they turned the field alternately on and then off for

10 minutes; thus the samples served as their own controls. These measurements corroborated earlier findings and showed that the magnetic effect on cell respiration has a minimum threshold field strength (about 80 Oe) but that an increase to 10,000 Oe does not increase the magnitude of the effect.

Membrane permeability

Histochemical investigations were conducted to determine whether there is a change in the activity of oxidative enzymes in the livers of albino mice exposed to inhomogeneous magnetic fields of 5,000-Oe strength.[27] The greatest change was observed in the activity of mitochondrial enzymes (succinate-dehydrogenase, glutamate-dehydrogenase, and malate-dehydrogenase), whereas no change was found in the activity of enzymes located in hepatic cell hyaloplasm. These data are consistent with the mechanism of mitochondrial membrane permeability disturbances in general; it was speculated that, by disturbing mitochondrial membrane permeability, magnetic fields may produce changes in the coupling of oxidation and phosphorylation.

Changes of mitochondrial membrane permeability produced by magnetic fields should alter active transmembrane transport. It is generally assumed that two different ion transport systems operate in parallel: the "pump mechanism" transporting Na^+ and K^+ against, and the "spike mechanism" transporting these ions along, concentration gradients. These two mechanisms are quite separate. Magnetic fields may produce changes in either mechanism—decreasing the permeability of the membrane to K^+ or increasing the permeability to Na^+.

In vitro studies of frog skin revealed that magnetic fields of strength above a certain threshold value produce a sudden change in the free skin potential.[29] This change of skin polarization was perfectly reproducible for 4 to 5 hours and was always accompanied by decreased influx of Na^+. The magnetic effect disappeared when active sodium transport was totally blocked by either potassium cyanide or ouabain. Decreasing active sodium transport by substituting Li^+ for Na^+ in the bath solution changed the magnetic effect in proportion to the extent of decreased active sodium transport. One may infer, barring a possible artifact, that a magnetic field influences the Na^+ pump mechanism. If, as assumed by some, the conductance of Na^+ results mainly from the orientation of molecules in the membrane, medium gradé molecular paramagnetism could produce observable permeability changes in moderately strong magnetic fields. Others have independently observed increased urinary sodium and potassium excretion in magnetic fields.[30]

Exposure of the myocardial fibrils of a common snail to a field of 15,000 Oe reduced diastolic activity and heart rate; when the field was removed, both increased above the initial level. At higher temperatures the magnitude of the magnetic effect increased.[31] This could also be caused by a change of membrane permeability. Electrocardiograms of isolated turtle hearts in constant magnetic fields up to 15,000 Oe showed changes in electric potential indicative of depolarization and varying degrees of tetanus.[32] These results suggest a change in the transmembrane ion transport mechanism.

A link between these ion transport changes and those found with body temperature variation is suggested by studies that indicate that the body temperature of many mammals is regulated by a mechanism involving a constant inherent balance between sodium and calcium ions within the posterior hypothalamus.[33] An excess of Na^+ in extracellular fluids in the absence of Ca^{++} change increases body temperature, whereas an excess of Ca^{++} in the absence of Na^+ change decreases body temperature.

The properties of liquid crystalline systems may explain changes in ionic and electronic charge transport in magnetic fields.[34] When heated, most complex lipids (such as those found in adrenal cortex, ovaries, and myelin) change from crystalline to isotropic structure, through an intermediate liquid crystalline phase. During the intermediate phase energies of the order of body thermal energy could produce twists in the crystals. Nonspherical or rod-like molecules might orient themselves in a magnetic field no stronger than 1,000 Oe. It has been found, for example, that diffusion of m-nitrophenol in a liquid crystal system increases when a magnetic field is applied parallel to the direction of flow and decreases when the applied field is perpendicular.

Growth

Normal growth pattern is the result of several factors, the most important of which are the genetically determined biochemical constitution and the intake, digestion, absorption, and assimilation of food; hormonal influences interact with these factors. Growth hormone increases the size and weight of all body tissues, but optimal growth requires the presence of thyroid hormone and insulin.

A direct effect of magnetic fields on growth hormone has not been established; hence we discuss the growth process in terms of descriptive indices —size, number of cells, body weight, and so on.

Embryonic development

Because organisms are generally more susceptible to injury during development, it is not surprising that magnetic fields have been found to affect embryonic development in several ways. Strong inhomogeneous fields delayed development of sea urchin eggs and caused excess abnormal cleavage.[35] Fertilized sea urchin eggs exposed to homogeneous and inhomogeneous fields of varying strength show significant retardation of early cleavage of ova; the magnitude of this effect is strongly gradient dependent, particularly at lower field strength.[36]

Neurath[37] placed freshly fertilized eggs of the leopard frog, *Rana pipiens,* in strong inhomogeneous fields with the animal-vegetal axis parallel to field and gradient directions. Fifty-three percent of the control eggs developed through the hatching stage, whereas only 31.6% of the eggs exposed to the magnetic field did so. In both groups, most eggs that stopped developing did so after gastrulation. This investigation was to ascertain whether a cell component of large volume and relatively high paramagnetic susceptibility mediates the observed biomagnetic effects. He postulated that such a cell component would be acted upon by great enough force in an inhomogeneous magnetic field to redistribute within the cell, establishing a concentration gradient. Ferritin was thought to be such a cell component because of its high iron content and because it is used during early embryonic development for synthesis of the first hemoglobin supply; despite observation of a significant magnetic effect on embryonic development, he failed to substantiate the ferritin hypothesis. He could not locate such particles with light microscopy, electron microscopy, or electron microprobes, nor could he by chemical means demonstrate modified iron distribution in the embryo.

Adverse effect on embryonic development of salamander and frog larvae subjected to strong inhomogeneous fields has also been observed.[38] Abnormalities included microcephaly, abnormal development, and edematous growth.

Profound developmental changes were observed in artificially inseminated eggs of the fresh water fish *Rhodeus ocellatus.*[39] When the animal poles were placed parallel to the direction of a homogeneous magnetic field of 12,000 Oe, the surviving fish at 6 months of age were 29% shorter and 40% narrower than controls (Fig. 8-3). If eggs were placed with animal poles perpendicular to the field, fish were 7.5% longer and 20% wider than the controls. Unfortunately, neither of the last two mentioned experiments involved sufficiently large control groups to assure valid samples of the population, nor were dummy magnets used. Thus one cannot exclude the possibility that the proximity of the relatively large magnetic probe changed the environment in other ways—air circulation, temperature, sound, and vibration.

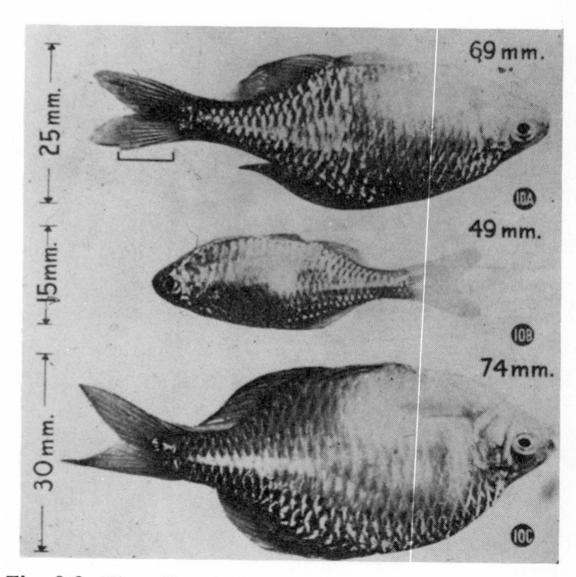

Fig. 8-3. The effects of magnetic fields on the growth of *Rhodeus ocellatus* hatched from artificially inseminated eggs. Top, Control fish; middle, fish hatched from egg with animal pole directed parallel to the magnetic lines of force in a 12,000-Oe field; bottom, fish hatched from egg with animal pole at right angle to the magnetic lines of force.

Cytogenic variations were observed when a high field gradient probe was applied to the ovaries and testes of *Drosophila.*[40] Embryonic and post-embryonic development was retarded through the thirtieth generation of descendants of the treated flies. With successive inbred crosses the initially decreased rate of development gradually approached that of the control cultures. Furthermore, the yield of progeny increased as the rate of development increased.

In another investigation mice were placed in homogeneous magnetic fields on the fifteenth day of their pregnancy.[41] Newborn mice were removed from both magnet and dummy cages within 6 hours. After maturation successive generations were produced by brother-sister matings during which process magnet-born pregnant mice were again placed in the field and dummy magnet-born pregnant mice were again placed in dummy magnets. The 52 magnet-born litters were 14.4% ± 4.2% smaller than the 32 dummy magnet-born litters. At 30 days of age magnet-born mice were an average of 3.5 gm lighter than their dummy magnet-born counterparts. Magnet-born mice remained stunted at 100 days of age.

Lengthened gestation period was also observed in pregnant mice in a 2,000-Oe field.[42] In mice placed in stronger fields before the sixteenth gestational day, pregnancy ceased; inspection of the uterus indicated fetal resorption. Some uteri were filled with a clear fluid; others were small and contained embryos of an earlier gestational stage. In both cases the resorbed tissue was redeposited as thick fat layers in the abdomen. Pregnancy was not arrested if the mice were placed in the field during the last few days of gestation. Rats placed in a 9,000-Oe field between the second and fifteenth day of pregnancy and killed before parturition showed preferential nidation that was always on the same side of the uterus; surprisingly, this asymmetry proved to be independent of field direction.

Body growth and aging

The best indication that magnetic fields affect metabolism would be a change of basal metabolic rate; however, because so many factors affect metabolic rate, because it fluctuates spontaneously, and particularly because it is subject to large interindividual differences, the effects of magnetic fields on basal metabolism per se have not been studied. Thus one must infer metabolic changes from other effects.

Barnothy[43] observed the weight of approximately 1,000 mice in magnetic fields. Weighing them always during the early afternoon, when mice usually sleep, he determined that the day-to-day fluctuation of mouse weight does not exceed ± 0.5 gm and that the average weight of a group of mice in a magnetic field can be established within 0.20 gm for the purpose of comparison with an equal number of mice kept in dummy magnets. He found that during the first 11 days of residence in fields of several thousand oersted strength the weight of 30-day old Swiss female mice was 2.7 ± 0.7 gm less than that of mice kept in dummy magnets. The weight difference between magnet and dummy magnet mice was greater in young mice (20 to 40 days of age) than in older mice (60 days or more). However, the relative decrease of growth rate (difference in growth rate divided by the growth rate of controls) was greater for older mice. Young mice continue to grow in the field, whereas older mice sometimes even lose weight. He also found that homogeneous fields retard growth more than inhomogeneous fields.

In addition to this growth-retarding effect, an initial weight decrease of about 1 gm was observed during the first 48 hours of exposure to the field.[44] However, this weight loss was only temporary; after 96 hours of exposure the mice had returned to their original weight (Fig. 8-4). The weight loss was interpreted as a "shock" effect of the magnetic field. Repeated exposures of the same animal at 14-day intervals revealed that mice do not adapt to this "shock" effect.

Developmental rate and the rate of aging are generally linked; the slower the rate of development, the slower the rate of aging. The results of experiments on mice seem to fit this pattern. Mice kept in magnetic fields during youth are stunted for at least 100 days (Fig. 8-5). At the age of 400 days (quite old for this strain), mice kept during youth in magnetic fields appeared younger and were physically more active than littermates raised in the natural geomagnetic environment.[45] An explanation of this age-retarding effect of magnetic fields is discussed on p. 339.

Morphologic examination of plant material showed that a magnetic environment of 500 to 5,000 Oe interferes with normal growth and development, promoting early maturation and senescence of cells, tissues, and organisms.[46] Fig. 8-6 illustrates the growth of *Coleus* cuttings in 3,000-Oe fields. In the exposed sample (left) the roots were, on the average, one fourth to one fifth as long as those of the matching control cuttings.

Growth of cells in vitro

That a magnetic field affects cell growth in vitro was first reported by Julia Lengyel,[47] who found growth retardation of chick heart cells and giant abnormal cells in cultures exposed to a magnetic field. During the following three decades many investigations of cell cultures were undertaken; in most instances cell growth retardation

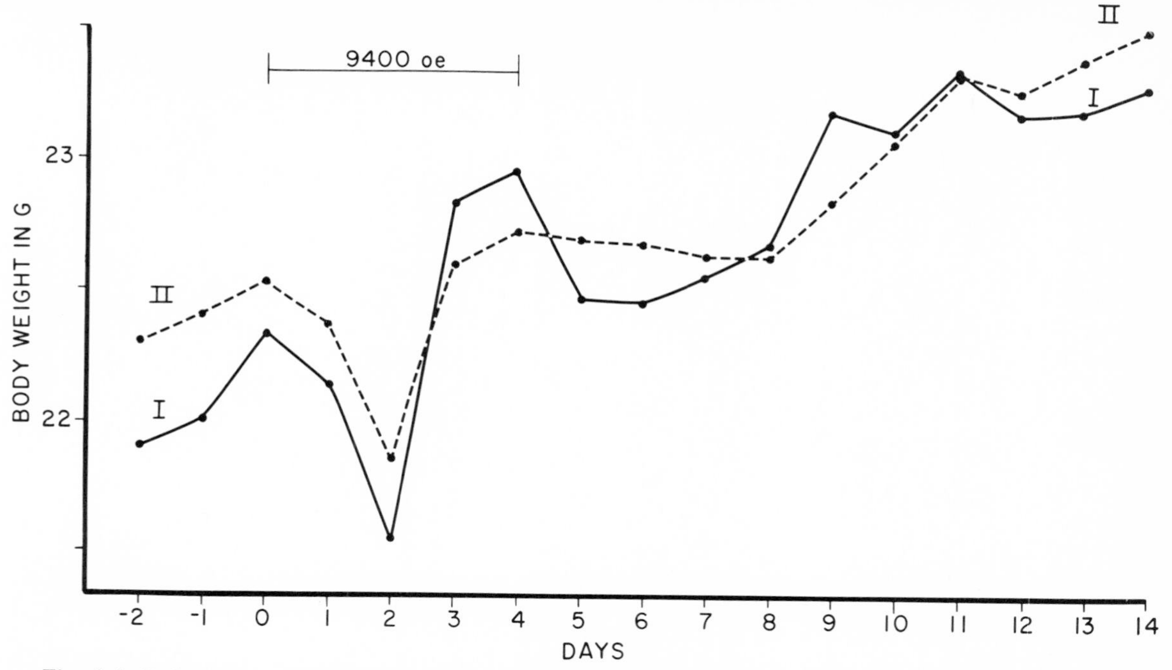

Fig. 8-4. Body weight change of mice in 9,400-Oe homogeneous field. Group I and Group II mice were alternately placed every 14 days for a total of 4 days in the magnetic field; the other group was kept during the same time in a "dummy magnet" cage. Thus the two groups served as their own controls. The lines represent averages of five cycles each. There was a sharp drop in weight during the second day of residence. Repetition of exposure did not decrease the minimum, implying that mice do not become accustomed to the stress of the magnetic field.

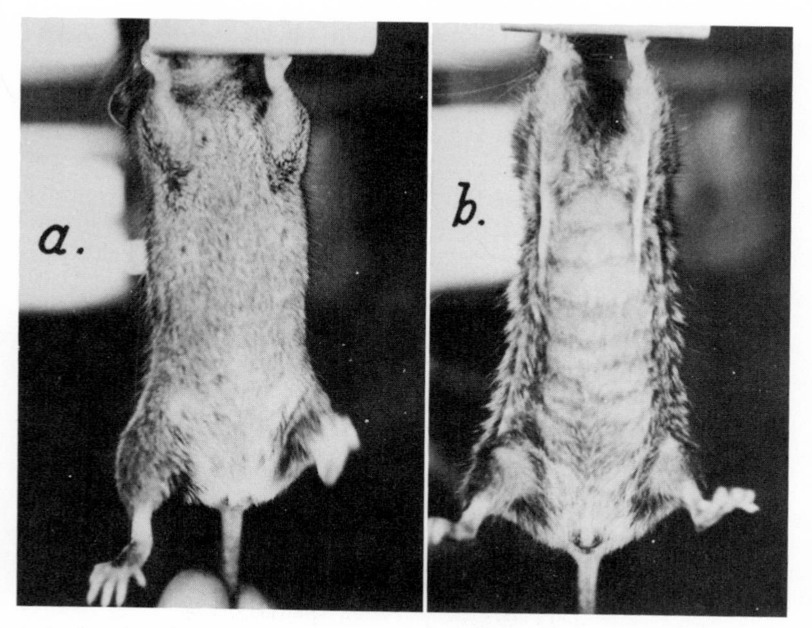

Fig. 8-5. C3H strain female mice at an age of 400 days (average life span of this strain is 308 days). **a,** Mouse was treated for 4 weeks in a 4,200-Oe field 11 months earlier; **b,** control mouse of same age, kept 4 weeks in a "dummy magnet" cage 11 months earlier. Note the smooth fur of a compared to the wrinkled fur of **b,** showing the effects of age.

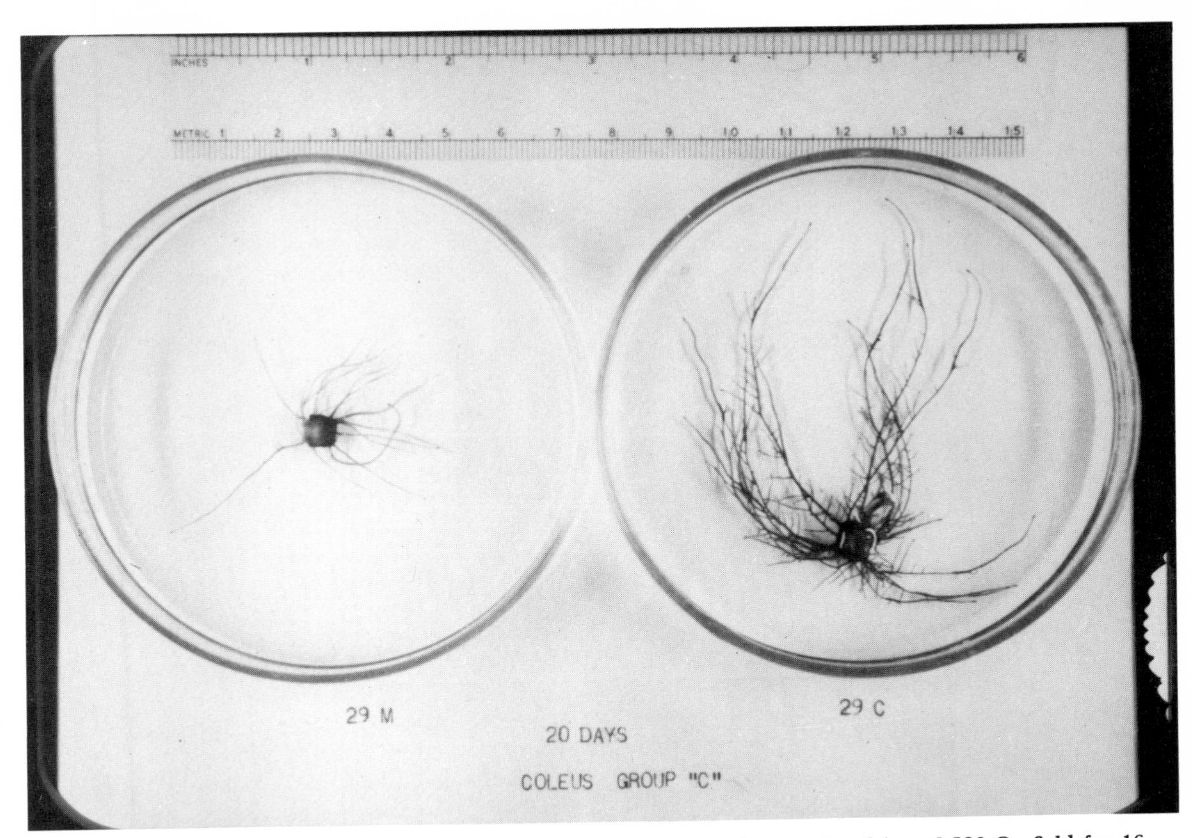

Fig. 8-6. *Coleus* cuttings with adventitious roots. The cutting at left was placed in a 3,500-Oe field for 16 days; the right paired cutting is the control. The first adventitious roots of the exposed sample emerged 1 to 2 days later than those of the control. The exposed cuttings had 10% fewer roots on the average and total root length was 42% less than that of control cuttings.

was found, but there were some experiments in which enhancement occurred, and some in which no effects were observed. The only morphologic change to be established was increased degeneration of ascites Sarcoma-37 cells.[48] Tumor production by *Bacterium tumefaciens* in *Pelargonium zonale* was inhibited by a magnetic field, but growth and morphology of the organism were unaffected.

Gerencser and Barnothy[49] found that highly inhomogeneous fields inhibit bacterial growth during the fifth to eighth hours of incubation. They suggested that magnetic fields increase the death rate of the original nonresistant bacterial population, this increased mortality clearing the biologic space for multiplication and emergence of a more resistant strain of cells. They emphasize, however, that the term "resistant" does not necessarily mean "mutant." In their experiments both exposed and control bacterial cultures had a lag phase of about 3 hours. The lag phase can be called a phase of rejuvenescence, because during this time the protoplasm of the old but still viable bacteria in the inoculum acquires the characteristics of young protoplasm; the cells increase in size but do not divide.

The fact that the lag phase of the exposed culture did not lengthen suggests that a magnetic field does not inhibit rejuvenation. Inhibition did occur during the logarithmic growth phase, but only after the fifth to eighth generations; during the logarithmic phase cell division and hence DNA replication are vigorous. They thus speculate that a magnetic field does not affect cell metabolism and growth but rather the process of cell division and, specifically, DNA replication. That the presumed inhibition of DNA replication is not immediately observable suggests the possibility that a magnetic field may alter the genetic code; such an alteration is magnified by a factor of 100 to 1,000 at the fifth to eighth generations. This inference is supported by direct observations of DNA synthesis in magnetic fields.[50] When a freshly harvested ascites Sarcoma-37 tumor cell suspension was placed in a magnetic field of several thousand oersted strength and tritiated thymidine was added during and after removal from the field, decreased thymidine uptake was found in the exposed suspensions, suggesting a 20% decrease of DNA synthesis.

Gerencser and Barnothy[49] found that at lower field gradient values rod-shaped bacteria, or bacilli, show a greater degree of temporary inhibition than spherically shaped bacteria, or cocci. This suggests an orientation effect of the magnetic field on the bacterium as a whole, which is para- or diamagnetic relative to the medium. A nonspherically symmetric cell would be more easily oriented than a spherically symmetric one. This inference is corroborated by the finding that a homogeneous field, which does not inhibit the logarithmic growth period does inhibit the stationary phase reached after 16 hours of exposure.[51] With interrupted exposures the inhibition disappeared, probably because during the interruptions magnetically oriented cells resumed natural orientation. It was also noted that the normal grape-cluster configuration of the cells was somewhat disarranged and there was a slight change in hydrogen gas production during dextrose fermentation by *Escherichia coli* after 40 hours of exposure to a magnetic field.

Strong magnetic fields affect the growth of serum-free cell cultures of rabbit myocardium and of mouse lung fibroblast isolates.[52] Both cell lines underwent several hundred in vitro passages. Fields of 20,000 Oe with a 15,000 Oe/cm vertical gradient were used; the temperature difference between exposed and control cultures was less than 0.05° C. After 3 to 7 days' growth in the field, the harvested cells were counted with a Coulter particle counter. In each experiment growth was enhanced, but the difference in cell number between magnet and dummy magnet cultures decreased with increasing final cell count. This supports the finding on bacterial cultures that the effect of an inhomogeneous magnetic field is observable during the logarithmic growth phase and vanishes after the biologic space becomes saturated; both suggest a factor affecting rapid cell division. Using other cell lines they found that sensitivity to a magnetic field varies greatly with cell type and even with the age of a serial culture. Some results indicate that the effect is greater if a culture grows at temperatures lower than the optimal 37.5° C. This temperature dependence could result from an action on intracellular enzymes while they are incorporated in a multimolecular structure that is near a transition point. Forces that cause molecular reorientation are most apparent near transition temperatures as, for example, the temperature dependence of heat-induced molting failure of *Tribolium confusum*.[25]

Tumor growth

The literature regarding the influence of magnetic fields on tumor growth is abundant; this is partly because of the specific medical importance of the subject and partly because of the early recognition that magnetic field effects are enhanced in rapidly developing tissues. The uncontrolled proliferation of cancer cells involves an imbalance between cell production and cell destruction. Szent-Gyorgyi believes that two enzymes, one a growth promoter, the other a growth inhibitor, are important; thus excessive proliferation from either too much growth-promoting enzyme or subnormal destruction from an insufficient amount of growth-inhibiting enzyme would favor tumor growth.

Barnothy[53] found that exposing Swiss mice to magnetic fields 5 days after transplantation of adenocarcinoma T 2146 arrested tumor growth, after which the tumors were rejected. This is illustrated by the upper curve in Fig. 8-7. The middle curve shows tumor arrest, despite which the mouse succumbed to the tumor. The lower curve is interesting in that the tumor was rejected 12 days after transplantation and the mouse was subsequently removed from the field. Ten days later a new tumor formed on the wound circumference, the growth of which was also arrested with subsequent rejection. All control mice died 14 to 24 days after tumor transplantation. However, using tumors grown in control mice for transplantation, the rate of tumor rejection decreased with each successive homotransplant. He concluded that a magnetic field does not act directly on a tumor but rather increases host homograft reaction. No isotransplant tumors were rejected. However, magnetic fields did affect tumor-bearing mice with isotransplants, lengthening survival time by as much as 35%. At the time of death the average tumor weight of the magnet group mice was 200% greater than the average tumor weight of mice kept in dummy magnets, indicating that the magnetic field did not arrest isotransplanted tumor growth. Locally

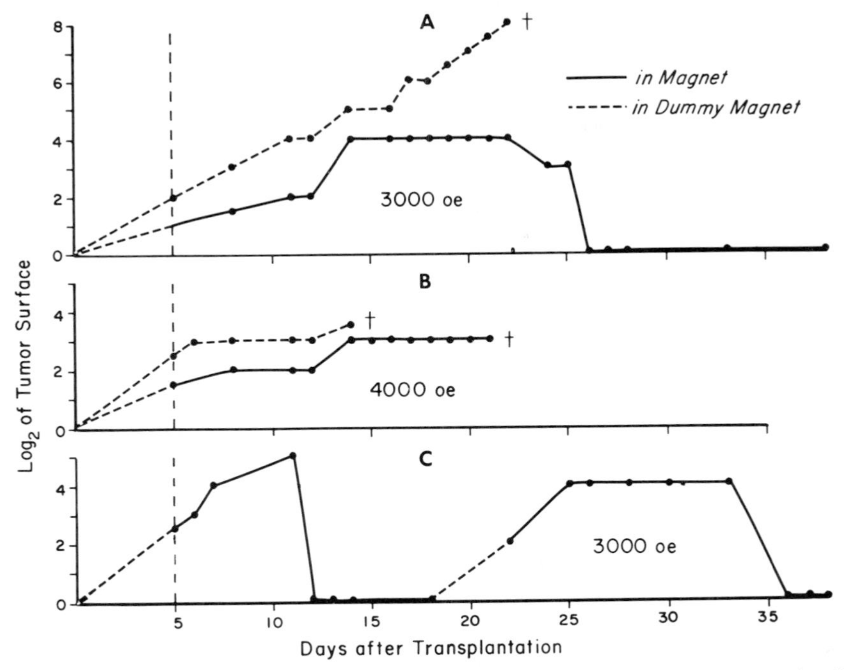

Fig. 8-7. Growth of transplanted tumor in mice. Dashed curves represent mice kept in "dummy magnets." Tumor size increased continuously in the controls and they died between the fourteenth and twenty-fourth day after transplantation. Solid lines represent tumor growth in mice placed in magnetic fields 5 days after transplantation. **A,** Tumor increased in size up to the fourteenth day, at which time growth was arrested; the tumor was rejected on the twenty-sixth day. **B** is a case in which tumor growth was arrested but the mouse died. **C** is a case in which the tumor was rejected as early as the twelfth day of residence in the field. The mouse was removed from the field and 10 days later a new tumor was noted on the circumference of the wound. The second tumor was also treated in the field and finally rejected.

applied highly inhomogeneous fields arrested growth of the spontaneous mammary gland cancer of the C3H strain mice for a period of 2 months; after removal from the field, the mice died of the malignant tumor.

The average lifespan of tumor-bearing mice also increased if, prior to tumor transplantation, they were kept in magnetic fields.[54] This effect is probably attributable to the lymphocytosis that occurs after removal of animals from the magnetic field; a variety of agents that produce lymphocytosis also extend survival time of tumor-bearing animals. Increased survival time of mice was also observed with a variety of ascites-producing and solid transplanted tumors when they were exposed to slowly varying magnetic fields.

In one experiment leukemic cells were injected into 57 mice.[55] Three days after injection 25 mice were placed in an 18,000-Oe homogeneous magnetic field and kept there for 1 week; the rest were used as controls. All 32 control mice died within 4 weeks. Seven of the magnetic group mice died during that time, but two of these died from causes other than leukemia. The 18 survivors of the magnet group were sacrificed 4 months after injection, at which time none showed evidence of leukemia. The possibility of affecting the course of leukemia deserves further study.

Although magnetic fields affect cell and embryonic growth, they do not seem to influence tumor growth per se. Löwdin[56] saw a parallelism between mutation and cancer, both involving errors in the genetic code of DNA molecules. The genetic code is essentially a linear message in terms of nucleotide bases that is read by messenger RNA in a sequence of triplets, each triplet corresponding to a certain amino acid in the synthesis of a protein. Changing a single letter of this code may lead to a mutation. If such a change occurs in a sex cell of a multicellular organism, it may lead to a hereditary mutation. If it occurs in another type of cell, it may cause a somatic mutation, leading to the development of a tumor. However, mutagenic chemical compounds rarely show carcinogenic activity. Although mutagenesis and carcinogenesis both influence the fundamental reaction—DNA → messenger RNA → protein—and the associated replication and transcription process, it seems likely that mutagens attack the genetic template and lead to rearrangement of proton-electron pair patterns, whereas carcinogenesis is related to an error in the reading mechanism. Such transcription errors may permit differentiated cells to read "forbidden" parts of the genetic code, thereby synthesizing false proteins without necessarily activating the immunologic mechanism of the organism. Such an error would lead to correct proteins situated in the wrong cell. The still inconclusive evidence that homogeneous magnetic fields do not arrest tumor growth suggests that magnetic fields do not influence the transcription process of the genetic code.

Effect on enzymes

The processes of digestion, absorption, and assimilation of nutrients involve catalysis by a variety of enzymes that hydrolyze peptide bonds and degrade carbohydrates to monosaccharides. It is thus reasonable to ask whether the metabolic effects of a magnetic field are mediated by an effect on enzymes.

Unfortunately, studies of enzyme activity in magnetic fields are inconclusive. In low magnetic fields of the order of one thousandth of the geomagnetic field the total acid phosphatase activity of serosal macrophages was significantly decreased compared to control groups kept in the natural geomagnetic environment.[2]

Young[22] found that 17,000-Oe fields produce arrhythmic contractions of frog vagal-heart preparations, although no effect on cardiac rhythm was observed at 4,000 Oe. Injection of microquantities of acetylcholinesterase (AChE) corrected the magnetically induced arrhythmia, which immediately reappeared when AChE was washed out of the heart. This suggests that the magnetic field inactivated AChE in exposed postsynaptic regions. Thus the rate of hydrolysis, which normally matches the rate of presynaptic acetylcholine release, decreases and arrhythmia results. Acetylcholinesterase activity was measured as the time necessary for acetylcholine hydrolysis. The force of cardiac contraction was recorded with an integrating digital voltmeter. Injection of acetylcholine simulates vagal inhibition of the heart; the duration of this inhibition is a measure of acetylcholinesterase activity. This inhibition period was much shorter in the exposed preparations, suggesting that acetylcholinesterase hydrolyzed acetylcholine more rapidly (Fig. 8-8). He also investigated whether a magnetic field affects the rates of hydrolysis of common and deuterated acetyl-

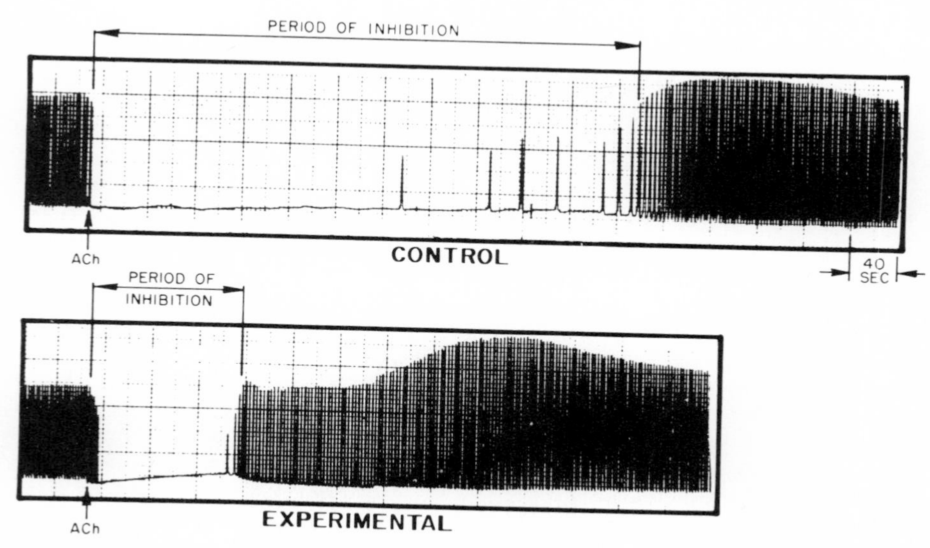

Fig. 8-8. Cardiac contractions in a vagal heart preparation of *Rana pipiens*. Top, control; bottom, exposed to 15,000-Oe field. Arrows mark acetylcholine injection. Cardiac contraction is inhibited during hydrolysis of acetylcholine. Hydrolysis lasted 450 seconds in the control and only 120 seconds in the exposed preparation. On the average the rate of hydrolysis was 250% faster in the exposed samples than in the control samples.

choline differently. With or without magnetic field the hydrolysis rate for common acetylcholine is faster than deuterated acetylcholine, and a magnetic field increases the hydrolysis rate of both. The magnetic field probably affects the substrate molecule, altering the interaction of its ligand field with the active site of the enzyme molecule.

A 2-hour exposure to a 13,000-Oe field increased trypsin activity 15% to 25% relative to control values.[57, 58] Care was taken to eliminate differences between exposed and control samples with regard to concentration, age of solution, autolysis, denaturation, and temperature. The change in trypsin activity was found to depend on pH; at pH 3 activity increased 24%, whereas at pH 8 activity decreased 10%. Since the decreased activity at high pH cannot be attributed to autolysis or denaturation, we must conclude that the magnetic field exerts a greater stress on the stable form of the enzyme at pH 3, where the critical amino acid residues are not hydrogen bonded, providing an ordered array or more favorable orientation of the electronic dipoles. Carboxydismutase activity increased (or denatured more slowly) in a magnetic field.[59]

Under different experimental conditions and using fields in the kilo-oersted (kOe) range, the activity of L-glutamic dehydrogenase decreased from 5% to 12%[60]; in an inhomogeneous field the decrease was 90%. Concentration gradients were proposed as the cause of the activity change. Whether this decreased enzyme activity, compared to the increased activity observed by others, must be attributed to differences in field characteristics or to the enzyme systems used is not yet known. Enzyme activity is extremely temperature dependent, hence minute differences in the temperature of samples exposed to magnetic fields relative to that of control samples may simulate an effect in either direction.

A mechanism by which magnetic fields may affect enzyme activity involves the fact that macromolecules, including enzymes, exhibit the property of superconduction. The transition temperature at which the state of superconduction ceases depends strongly on the strength of the ambient magnetic field. Thus in the presence of a magnetic field greater than the critical value, any enzyme that utilized superconduction for achieving high catalytic rates would be catastrophically denatured and reaction rate would decrease markedly. This idea was tested by Maling and associates[61] on various enzymes in 50 kOe fields with negative results. However, it is likely that enzymes are not ideal macromolecules, as required by the theoretic computation of the strength of the field at which a macromolecule will lose its superconducting properties. Thus an actual enzyme might undergo this transition at a much higher field intensity than those used in the experiments just described. Failure to observe an effect might also have been caused by the fact that experiments were conducted at temperatures at which the thermal coefficient of enzyme reaction was low. Molecular reorientation is facilitated near phase transition temperatures. A third reason why experiments in very high-intensity fields yielded no results, while others at lower field strength produced enzyme activity changes, may be the result of experimental differences. Maling used only a few minutes of exposure, whereas the other investigators used exposures of several hours before the enzymes reacted with their respective substrates. Thus a comparison between positive and negative results at high field strength is not meaningful.

The positive effects of magnetic fields on enzyme activity are undoubtedly attributable to the orienting effects of magnetic torque and force on paramagnetic centers within macromolecular complexes. That magnetic fields are known to alter the optic properties of polymers also points in this direction. For example, in a 3,000-Oe field spinach ferredoxin protein shows rotary transitions in addition to those exhibited in the absence of a field.[62]

CENTRAL NERVOUS SYSTEM

The central nervous system is a collection of cells that specialize in the conveyance of signals in rapid tempo and with great fidelity, carrying information of the immediate external environment and dispatching common signals for corrective action. It thus plays a major role in homeostatic regulation. Among other things it automatically adjusts body temperature and blood pressure and regulates all metabolic and endocrine functions of the organism. Sensory information is received, analyzed, and stored in the cerebral cortex, which interacts with virtually all other regions of the central nervous system.

Under ordinary conditions a magnetic field is not readily felt by man, although in the state of hypnosis or in the state of mescaline intoxication a magnetic field is said to modify visual images.

We will discuss the effects of static magnetic fields on the behavior of animals, unconditioned reactions to such fields, the changes produced in previously elaborated conditioned reflexes, and effects on the electric activity of the brain.

Behavioral changes

Changes in motor activity have been observed in birds, fish, and mammals. A magnetic field increased the total motor activity of birds and also altered the normal distribution of activity within the day. The activity of birds in natural illumination gradually decreases from morning to evening; in a weak magnetic field a second maximum was noted at about 3:00 PM, suggesting that the field had disturbed the natural circadian activity rhythm.[63]

Brown and co-workers[4, 10] made extensive studies on the orientation of planaria that were allowed to emerge from a narrow channel in the geomagnetic field and in artificial fields produced by rod magnets. The effect of the field depends on the rhythmic behavior of the organism. They suggest that the system that perceives magnetism is not functionally isolated from the rest of the living organism.

Cyclic behavior of organisms may be viewed as a rhythmic variation of sensitivity level; thus magnetic fields may act by varying irritability. In an open field experiment albino rats exposed continuously during prenatal development to a 3- to 30-Oe magnetic field that rotated 360° every 2 seconds traversed significantly fewer squares than controls but defecated significantly more in that situation.[64]

The controversial behavior of dowsers was interpreted by Rocard[12] as perception of magnetic anomalies. Using a nuclear resonance magnetometer he established that at locations where a dowser can detect water, it is never still water in a pond, or running water in a river, but rather water filtering through porous media. In this situation water produces electric currents (electrofiltration) that generate magnetic anomalies of the order of a few millioersteds per meter at the soil surface. Rocard also used an artificial magnet (a frame with a coil), ahead of which the dowser walked without knowing whether the current and field were on or off. He found that a gradient greater than 0.1 mOe/m can be sensed by almost everyone; at 50 mOe/m saturation occurs. He proved that a dowser really detects a magnetic anomaly and not water; thus the dowser reaction can be triggered by iron or an unexploded shell buried in the ground or by certain rocks such as basalt that have been magnetized by lightning. He suggested that in the geomagnetic field the protons in body substances precess at a rate of about 2,000 rps and that in a nonuniform field not all protons would have this Larmor frequency, some having 2,001 rps (a 0.25-mOe field in addition to the 0.5-Oe geofield). Thus a beat frequency could exist between those protons that are mobile (in blood) and those fixed in other tissues. If the number of protons involved is large, a measurable effect may result, which could change muscle tone and thus flip the "divining rod."

Conditioned reflexes

A magnetic field can be used as a conditioning stimulus for fish[65]; a 10- to 20-second exposure to a very weak field of 10- to 30-Oe strength can produce a food-seeking conditioned reflex. On the other hand, strong fields appear to extinguish established conditioned reflexes; monkeys stopped punching a lever for food in 60,000-Oe fields.

Lesions in different cerebral regions such as the forebrain, tectum opticum, and cerebellum did not modify the reaction of fish to magnetically conditioned reflexes, although such lesions derange conditioned reflexes established by light and sound stimulation. Only diencephalic lesions disturbed reflexes conditioned by a magnetic field. This finding suggests that a magnetic field exerts direct influence on the diencephalon.

Combining electric and magnetic stimulation, it was found that the strength of an electric current necessary to provoke a slight quiver in fish had to be increased in a magnetic field of 100-Oe strength by an average of 45%; the magnetic field reduced the sensitivity of the fish to electric stimulation.

Much stronger fields of 1,000 Oe are needed to produce electrodefensive conditioned reflexes in rabbits. Comparing the establishment of an electrodefensive reflex by a magnetic field and by acoustic stimulation, magnetic stimulation required about twice as many trials for the first occurrence and twice as many trials again to establish the reflex. Furthermore, the latent period of the effect was almost twice as long as that of acoustic stimulation, suggesting that a magnetic stimulus is weaker than an acoustic stimulus. Investigation of inhibitory reflexes conditioned to a magnetic stimulus revealed that the inhibitory action persists for several minutes after the magnetic field was switched off. From the wide variety of experiments with the field as a stimulus, it appears that a magnetic field is a nonspecific relatively weak stimulus that produces predominantly inhibitory effects that persist after the field has been turned off.

Effects similar to those evoked by a static magnetic field can be produced by variable electromagnetic fields within a wide range of frequencies, providing that the electromagnetic waves penetrate the brain. Waves in the centimeter and millimeter range are absorbed within the skin and produce either no change or much less change in conditioned reflex activity.[8] Alternating magnetic fields generate induced currents in conductors; thus they initiate induced currents within a biologic system. These electric currents exert biologic effects that may overshadow the effect of the field alone. Of the effects produced by alternating magnetic fields, that discovered in 1898 by d'Arsonval is of particular interest—magnetophosphenes. Magnetophosphenes are diffuse flashes of light seen when the temporal areas of the head are exposed to pulsing or alternating magnetic fields.[66, 67] Phosphene intensity is greatest at about 30 Hz, and they can be evoked by fields as low as 200 Oe. The light patterns of magnetic phosphenes appear identical to those produced by electric stimulation with electrodes attached to the area surrounding the eye. Probably, induced electromotive force is responsible for the phenomenon, induced currents being generated within each retinal cell, or at different points along the visual pathway. However, other photomagnetic effects may also be involved.

Effects on electric activity of the brain

Oscillations of electric potential occur almost continuously between any two electrodes placed on the surface of the head or on the cerebral cortex itself; records of such potential differences as a function of time constitute the electroencephalogram (EEG). From fish to man, brains all exhibit basically similar EEG's. In all species the dominant sinusoidal rhythm, or alpha rhythm, has a frequency less than 50/sec. In awake healthy individuals waves slower than the alpha rhythm occur only rarely; when they do they usually indicate disease or brain injury. During sleep alpha rhythm frequency decreases and "sleep spindles" appear; the stage of deep sleep is characterized by the appearance of very slow (1/sec) delta waves. Sensory stimulation usually blocks the alpha pattern.

The most extensive study of EEG changes in magnetic fields was carried out by Kholodov and associates,[68] who placed the whole body, head, and isolated encephalon of various animals in fields of different strength. The observed EEG changes were compared with those produced by acoustic and optic stimulation and those occurring after injection of stimulants. These studies showed an increased number of high-amplitude slow waves and sleep spindles. A latent period of the order of minutes elapses before the magnetic field effect is observed and the effect persists after discontinuation of the field.

The effect of a magnetic field on the EEG increases after administration of stimulants such as caffeine or adrenaline and decreases after pentobarbital (Nembutal) or chlorpromazine (Aminasine). Prolonged increase of slow wave amplitude was not seen after brief exposures of a few minutes' duration repeated every 5 to 20 minutes; however, a kind of summation occurred when exposures lasted several hours and were repeated daily.

The nature of EEG changes as recorded from the cerebral cortex is not altered by destruction of the visual, auditory, or olfactory analyzers, nor by injury to the hypothalamus, thalamus, or midbrain reticular formation. The reaction of the intact brain is the same as that of an isolated brain transected at midbrain level or that of a neuronally isolated strip of cerebral cortex, with the sole difference that the reaction of these isolated preparations is more pronounced; the strength of the reaction is greater and the latent periods are shorter. These results suggest that a magnetic field does not act on specific brain structures but rather affects all parts of the brain, the reaction of the whole brain being an integration of local responses from various regions.

Small permanent magnets fastened to the skull above the sensorimotor area of the cerebral cortex for periods of time up to 40 days produced a predominance of slow high-amplitude biopotentials; the reaction to continuous local exposure is thus the same as that produced by briefer exposure of the entire brain.

Kholodov also investigated individual neurons by a microelectrode method during exposure of a rabbit's head to a magnetic field. Neurons usually discharge more frequently during exposure to adequate sound or light stimuli; conversely, a magnetic field, having a predominantly inhibitory effect on the spontaneous electric activity of neurons, also inhibits light- or sound-stimulated neuronal activity.

The increased number of EEG spikes and spindles after a few minutes' exposure to a 700- to 1,000-Oe field resembles the changes usually observed during sleep and anesthesia. Kholodov's observations that mice kept in magnetic fields are easier to handle, quieter, and less excitable are of interest in this regard.

Using tranquilized amphibians Becker[69] demonstrated that the application of a 2,500-Oe magnetic field perpendicularly to the brainstem (midbrain) reduced consciousness and altered the EEG pattern to one resembling moderate to deep anesthesia.

At variance with these observations Beischer and Knepton[70] noted an increase in the frequency (from 8 to 10 Hz to 14 to 60 Hz) and amplitude (from 25 to 50 μv to 50 to 400 μv) of the EEG tracings of squirrel monkeys exposed to 92,000-Oe fields. Dependence on field polarity or gradient strength was not observed. He proposed that, unless the observed effects were artifacts, these high frequencies represent a sequence of rapid discharges that disrupt the coordinated function of the central nervous system.

The latent periods of reaction for other agents of adequate strength are about 100 times as long as that for magnetic stimulation. This long latent period suggests that in addition to the electrically active elements, electrically inactive elements, such as glia or vessels, may also play a role. To investigate this possibility histologically, astrocytes, oligodendrocytes, microglia, and neurons were stained and counted.[68] All glial cells were found to be increased in proportion to the magnetic exposure. For example, after a 3-minute exposure of the head of a rabbit, the number of silver-stained astrocytes increased by 85% ± 7%; after a 1-hour exposure they increased by as much as 156%. This change is of particular interest because it has been suggested variously that this neuroglial cell may play a role in neuronal impulse conduction and synaptic transmission, serve as conduits between vessel and neuron, function as reservoirs of metabolites for neurons, or play an active symbiotic role in neuronal metabolism. They may release glycogen from their stores during prolonged and intense activity in response to slow glial depolarization evoked by K^+. Glial cells contain respiratory enzymes, fat, and glycogen. There is preliminary experimental evidence that learning involves the transfer of specific nucleotides from glia to neurons. This controversial suggestion is based on the finding that during both sleep and learning RNA and the ratio of purine to pyrimidine bases of neuronal RNA increase in certain brain areas. Furthermore, the production of RNA molecules with these highly specific base ratios is paralleled by a decrease of the same specific RNA in adjacent glial cells.

In addition to the increase of glial cells after magnetic exposure, cell wall hypertrophy and hyperplasia were noted, changes suggestive of gliosis. Kholodov, however, states that the magnetic field does not change the total number of astrocytes but rather enhances the number of cells that stain with silver, a property depending on the oxidation-reduction potential. Hence the finding may reflect a change of metabolic activity, particularly a change in the process of biologic oxidation. On the other hand, should the increased number of glial cells represent gliosis, this may place a serious limitation on any future therapeutic use of magnetic fields. Before drawing final conclusions, it should be noted that experimental rabbit brains often contain histopathologic evidence of enzootic disease, rendering interpretation of the lesions attributed to the magnetic field rather difficult.*

Neurons are electrically and magnetically insulated from other cell elements and extracellular fluid; however, local changes of ionic concentration occur as ions exchange between neuron and environment during each action potential. Local ion concentrations—particularly Ca^{++}, Mg^{++}, Na^+, and K^+—may be affected by magnetic fields and thereby influence synaptic transmission. It has also been proposed that Na^+ transport across membranes may depolarize axons.

Becker[71] believes that the interaction of an applied magnetic field with the central nervous system is related to the direct current potentials of nerve tissue. All animals, even those with a rudimentary central nervous system, have a direct current system, displaying in each case a field pattern that expresses the general anatomic brain structure. Possibly, this direct current system has semiconductor ionic properties. If current flow were of semiconductor nature, the interaction between charge carrier and an applied magnetic field would be many orders of magnitude greater than that of ionic or metallic conduction systems.

Some isolated observations are worth mentioning. Young,[22] studying the acetylcholinesterase activity in frog vagal heart preparations, noted that the application of a magnetic field always decreases the duration of vagal inhibition. In other experiments the frog gastrocnemius muscle preparation was first exhausted by rhythmic electric stimulation at a rate of 2 pulses/sec and thereafter exposed to a magnetic field in the 500-Oe range.[72] Total exhaustion and myoneural shock occurred later in the exposed preparation than in

*Friedman conjectures that the observed gliosis could have been caused by parasites that become active in a magnetic field.

controls. Cohen demonstrated weak alternating magnetic fields (about 10^{-9} Oe) outside the human scalp produced by alpha rhythm currents; the magnetic signals coincided with the electric signals of the EEG. Thus the magnetoencephalogram may be a sensitive, noninvasive, remote investigatory tool.

ENDOCRINE AND RETICULOENDOTHELIAL SYSTEMS

Endocrine system

The system of ductless endocrine glands regulates various cell processes through hormonal secretion, facilitating maintenance of a constant internal environment. Directly or indirectly, the central nervous system controls several endocrine glands and is concerned with rapid adjustments: the endocrine system is concerned with slower metabolic adjustments. The integrated neuroendocrine system is a homeostatic negative feedback control system, the nervous system stimulating the endocrine system to elaborate hormones that in turn affect the nervous system. This reciprocal interplay is particularly significant in maintaining constancy of the internal environment during exposure to environmental stressors. It is thus important to determine whether a magnetic field, as an environmental factor, produces effects that are mediated by the endocrine system.

Stress is monitored by the central nervous system and affects reproductive, adrenal, and thyroid activity. The hypothalamus together with the hypophysis, or pituitary, constitutes the neurohumoral regulatory subsystem. In health hormonal balance is maintained by the negative feedback control system: changing levels of certain circulating hormones increase or decrease release of specific pituitary tropic hormones, which in turn stimulate a target gland. Psychologic stressors, such as noise and hostile environments, release adrenocorticotropic hormone (ACTH), acting on the posterior lobe of the pituitary gland, whereas systemic stressors, such as hemorrhage and trauma, affect the anterior lobe. Possibly the persistent growth retardation of mice exposed to a magnetic field at an early age is the result of changes in the pituitary gland, although this has not been demonstrated.

Adrenal gland

The adrenal cortex is the only endocrine structure in which histopathologic changes induced by exposure to magnetic fields have been found.[73] Histologic study of the adrenal glands of exposed mice

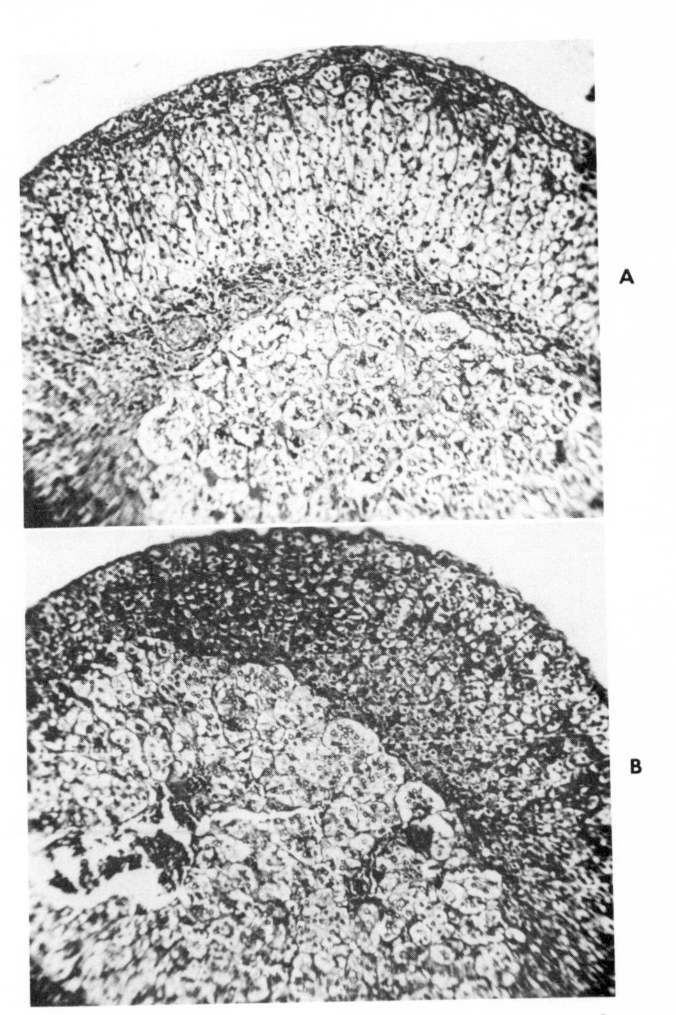

Fig. 8-9. Adrenal sections of mice. **A,** Adrenal gland of a mouse kept in a "dummy magnet." **B,** Adrenal gland of a mouse kept in a 4,200-Oe field. In **A** the three zones of the adrenal cortex and the usual cord-like arrangement of the cells in the zona fasciculata are clearly visible; in **B** zone demarcations are hardly visible, the zona fasciculata is unusually narrow, and the cord-like cell arrangement is blurred. (Hematoxylin-eosin, ×320.)

show a narrowed zona fasciculata, unchanged zona glomerulosa, and normal to slightly widened zona reticularis; cells of the zona reticularis appear to infiltrate the zona fasciculata. The normal cord-like arrangement of cells in the zona fasciculata is somewhat disorganized (Fig. 8-9) and the normally sharp zone demarcation is less well defined. In 67% to 78% of all exposed mouse adrenals more than half of the zona fasciculata is disorganized; in only 17% of the mice is adrenal cell arrangement normal. However, even in these normally arranged fasciculatae cytoplasmic vacuolization suggested subnormal lipid content. Since the studies were made on sections embedded in paraffin, the finding of reduced lipid content does not prove abnormal hormonal content. Between 2,000 and 9,000 Oe the extent of the lesion is not a function of field strength, nor had it changed much 6 months after the termination of exposure. The adrenals of mice kept in dummy magnets were normal.

Indistinctness of adrenocortical zone demarcations is associated with disturbance of normal synthesis and secretion of adrenocortical hormones. Hypothalamic neurohumoral stimulation releases ACTH from the pituitary gland, which in turn stimulates adrenocorticosteroid secretion. Cells of the zona fasciculata are most strongly affected by ACTH. Thus it is not surprising that static magnetic fields disorganize and narrow this zone. However, under ordinary conditions of stress the zona fasciculata is widened, not narrowed. Certain chemical substances, such as acetaldehyde, narrow the zona fasciculata; rat tumor implants first widen and later narrow the zona fasciculata while widening the zona reticularis. Nevertheless, the observed narrowing of the zona fasciculata and especially the persistence of this feature long after termination of exposure suggest that a magnetic field does not act merely as a nonspecific stressor producing a general adaptation syndrome. The adrenal medulla was unchanged. The most direct way of assessing the functional status of an endocrine gland is to determine the concentration of circulating hormone. Rats exposed to magnetic fields for periods as long as 14 days showed no significant change relative to dummy magnet rats in either adrenal weight (9% decrease) or in gland or plasma corticosterone content.

If a magnetic field were to increase ACTH production, certain effects would be expected that have actually been observed. Glucocorticosteroids favor carbohydrate formation at the expense of protein synthesis (gluconeogenesis), thus retarding growth as actually observed. Increased corticosteroid production causes splenic lymph follicle involution, decreasing spleen size. The spleens of mice born in magnetic fields are 22% smaller than those of dummy magnet controls. Corticosteroids increase the number of circulating lymphocytes; this is followed by marked lymphopenia and eosinopenia, a pattern observed in 75% of exposed mice.[16] Even small doses of ACTH usually increase the thrombocyte count.[20] ACTH and glucocorticosteroids inhibit wound healing by suppressing fibroblast formation and delaying appearance of collagen fibers. Retarded wound healing associated with marked suppression of fibrosis was observed in mice exposed to magnetic fields.[74] Increased ACTH secretion shifts the balance toward suppression of inflammation. The inflammation produced by intra-abdominal turpentine injection in rabbits subsided more rapidly in animals exposed to a magnetic field as evidenced by the much more rapid return to normal of erythrocyte sedimentation rate.[17] Magnetic fields reduce peak antibody titer,[75] an effect that may be related to changed corticosteroid secretion. The changes in weight and leukocyte count resembling ACTH effects were observed in different magnetic fields by different investigators at different times; however, in no case was a direct hormonal cause established. Even if it were confirmed that all these manifestations are related to a general adaptation syndrome caused by a magnetic field, they may have been observed at different stages of the syndrome.

Reproduction

In addition to the effect on embryonic development discussed on p. 324, some data are available on the process of reproduction. Normal reproductive function is particularly dependent on a delicate and complex balance of neuroendocrine processes, involving hypothalamus, hypophysis, and gonads. Function is affected by thyroid, adrenal cortex, emotions, and nutrition.

In the female reproductive system three steroid hormones—estrogens, progestins, and androgens—are important. Estrogen cornifies vaginal mucosa and produces other genital changes, evoking the state of estrus in lower animals. The estrus cycle of mice is deranged during residence in strong magnetic fields.

Twenty female mice were kept for 21 days in a 4,200-Oe field, twenty in dummy magnets, and twenty in standard cages. During the first 14 days in the field vaginal smears were taken daily at 3:00 PM, always from the same eight mice in the field and from the same four mice in dummy magnets. Seven of the eight magnet mice showed evidence of estrogen secretion but no cycling. Of the four dummy magnet mice two showed a 5-day cycle, one a 6-day cycle, and one no cyclicity. On the twenty-first day smears were taken from all sixty mice; the ratio of magnet to dummy mice in the state of proestrus or diestrus was fourteen to six, respectively, whereas in estrus or metestrus, it was six to fourteen, respectively. There was no difference between standard cage mice and dummy magnet mice. Assuming the difference in estrus cycle phase to indicate cyclicity, and using the Chi square test, the probability is 1,000 to 1 that magnetic fields derange the estrus cycle of mice.

Mutations

Whether magnetic fields produce mutations was investigated in *Drosophila melanogaster*. Exposure of *Drosophila* to a field of 3,000- to 4,000-Oe produced many more corneous deformations in the first generation, but this characteristic was not hereditarily transmitted.

The hereditary changes produced in the first and second generations of *Drosophila* by magnetic fields are increased yield of pupae and flies; exposure of inbred flies increases fecundity and viability of the second generation by 15%.[76] Interstrain hybridization reduced fecundity.

When flies mated in the field and eggs, larvae, pupae, and emerging flies were constantly exposed during the entire life cycle, repetition through successive generations produced a significantly higher percentage of deformities: wrinkled or rudimentary wings, three wings, or missing legs.[77] Some deformities were probably of genetic nature since they were transmitted through several generations. Mutations were induced by fields both higher and lower than the geomagnetic field. All mutations appeared autosomal, the majority of mutagenic effects occurring in gonadal tissue. This was interpreted as a mutagenic effect on the DNA molecule. Since mutations are nearly always biologically disadvantageous, sex-linked lethal characteristics in the exposed female parent tend to produce fewer male offspring. Male flies in the F_2 and F_3 generations increased significantly.

Levengood[78] used a special microphobe to treat selected regions of egg, larva, pupa, and adult, having a field strength of 21,800 Oe with 9,000 Oe/mm gradient in the 0.125-mm diameter pole tip. The duration of exposures varied from a few minutes to a half hour, a short time compared to the average development time (13.6 ± 1.12 days) of the experimental *Drosophila*. One-hour-old pupae usually died of magnetic exposure before hatching into flies; the head region proved most vulnerable. A 10-minute exposure produced 50% mortality; of those that matured into imagoes, 5% to 10% had severe wing anomalies and body deformities and as a rule did not survive for more than 1 hour after emergence. Furthermore, exposing 4.5-day-old pupae for 30 minutes almost doubled development time of the second generation; however, development time then decreased in subsequent generations, remaining detectably increased even at the thirtieth generation. This pattern suggests adaptation to the magnetic environment. When fertilized female adult flies were treated, development time of their progeny did not increase whereas males emerging from pupae treated magnetically for 10 minutes at the age of 21 hours exhibited greatly increased development time. When crossed with untreated females, development time of the offspring of these males was increased up to the tenth generation.

Reticuloendothelial system

The reticuloendothelial system is comprised of specialized cells in liver, spleen, lymph nodes, and bone marrow that have a unique ability to remove materials from the bloodstream by active phagocytosis and subsequent biochemical degradation. The reticuloendothelial system plays a role in the general defense of the organism and in its reaction to a variety of stimuli.

Liver

The liver is an organ that performs a variety of essential biochemical and metabolic functions, including hormone inactivation. Factors that affect liver function may thus change the rate of hormone inactivation. Steroid hormones are transformed, inactivated, or destroyed during passage through the hepatic portal circulation; the liver can change blood corticosteroid content even while the adrenal cortex releases these hormones at constant rate. Its wide variety of complex functions renders the liver susceptible to many noxious agents that usually damage hepatic parenchymal and reticuloendothelial cells. Noxious chemical agents and hepatic diseases generally produce pathologic changes differing more in degree than in nature and distribution.

Barnothy and Sümegi[73] made histopathologic examinations of the livers of mice killed immediately after exposure to magnetic fields. They found some edema at the periphery and center of hepatic lobules. Capillaries around the portal vein were dilated and Disse's spaces were much wider than usual. Cells at the center of lobules stained poorly or, conversely, were smaller with small dark nuclei (pyknosis). Glycogen was absent from the central portion of hepatic lobules. Fatty deposits were rare. Cells in the peripheral zone and midzone were generally larger than normal; their cytoplasm was vacuolated and contained some glycogen. Many enlarged cells in peripheral zones and midzones contained mitotic figures, suggesting reaction to injury—replacement of destroyed and/or necrobiotic cells. The mitotic index, or number of mitoses per 400 liver cells, is commonly used to measure the intensity of a regenerative process. A 4-day exposure of mice to a 9,000-Oe field increased the mitotic index 74%; after 13 days in the field, the increase was 129%, suggesting that the intensity of the process initiated by a magnetic field increases with the length of exposure. In addition to parenchymal injury, reticuloendothelial irritation is suggested by the increased number of Kupffer cells and their increased glycogen phagocytic activity. Connective tissue proliferation was not seen. Mice killed 200 days after the termination of magnetic exposure showed much less pronounced central parenchymal injury and only a few pyknotic liver cells; nevertheless, the number of enlarged liver cells and the mitotic index remained above normal.

The abnormalities observed in the livers of mice exposed to magnetic fields differ only quantitatively from those seen after injuries of various kinds. The livers of stressed animals have been reported to show cloudy swelling, nuclear pyknosis, distention of Disse's spaces, and edema.[79]

The similarities between the livers of mice exposed to magnetic fields and those subjected to other stressors, and the fact that magnetically induced changes decrease in intensity but nevertheless persist after cessation of exposure, suggest that a magnetic field, acting as an environmental stressor, produces hormonal imbalance. However, it is also possible that the observed hepatic changes result from general stimulation of the reticuloendothelial system and are thus a manifestation of a defense reaction evoked by the magnetic field.

Spleen

A variety of environmental and chemical stressors affect lymphatic structures, including the spleen. As described earlier, the splenic weight of magnetic field–exposed mice decreases. Histologic study of the spleens of exposed mice revealed that increased number of megakaryocytes is the most characteristic abnormality.[73] Healthy rodent spleens always contain a certain number of multinucleated giant cells that are difficult to differentiate morphologically from typical bone marrow megakaryocytes. These giant cells usually occur in smaller groups scattered throughout the splenic pulp, but never in follicles. Groups of megakaryocytes are always separated by large areas without any such giant cells. Conversely, the spleens of mice kept for a least 4 days in fields stronger than 4,000 Oe contained an average of 40% more megakaryocytes than controls. The megakaryocytes of exposed mice were rather diffusely scattered or, if in small groups, such groups were close together and were also seen at the periphery of the follicles. Another splenic abnormality was the indistinctness of the boundaries of megakaryocytes that were connected through extended processes to proliferating reticular cells, the latter being definitely increased in number. Appearance suggested that the giant cells resulted from fusion of reticular cells.

Although the mechanism of the splenic abnormalities of mice exposed to a magnetic field is unknown, various interpretations are possible:

1. The spleens of animals stressed with tumor implants contain an increased number of megakaryocytes, but such proliferation is also characteristic of the reaction to neoplastic tissue; it is not usually seen with nonspecific stress, such as restraint, cold, and trauma. Furthermore, such proliferation is usually associated with erythropoiesis and myelopoiesis in the splenic red pulp, a change not seen in the spleens of mice exposed to a magnetic field.

2. After forming in bone marrow, some megakaryocytes enter the bloodstream and migrate through the capillaries into the lung, where they produce thrombocytes.[80] During periods of increased thrombocytogenesis more megakaryocytes enter the pulmonary circulation, the smaller and more immature of which probably enter the systemic circulation and thus branches of the celiac artery, reaching the spleen and, to some extent, also the liver. This explanation is supported by the facts that mice exposed to a magnetic field show increased blood platelet count, decreased bone marrow megakaryocytes, and megakaryocytes in the liver.

3. According to another theory megakaryocytes arise locally whenever endomitosis of the reticular cells is followed by cytokinesis and formation of many large multinucleated cells[81]; if true, megakaryocyte formation would be reactive reticulosis and an increase in the number of splenic megakaryocytes would be a manifestation of a general organismic reaction to stimulation of the reticuloendothelial system. Such stimulation is a part of the general organismic defense reaction to acute hepatitis and subcutaneous casein injection, in which cases it is always associated with an increased number of splenic megakaryocytes. Thus a magnetic field would act by stimulating the

reticuloendothelial system. Supporting this interpretation is the finding that spleens of field-exposed mice reveal a close anatomic relationship between megakaryocytes and surrounding reticular cells.

Bone marrow

Erythrocytes, granulocytes, and thrombocytes, as well as some lymphocytes and monocytes, form in bone marrow. Platelets arise from fragmentation of the cytoplasm of the multinucleated megakaryocytes. The bone marrow of mice exposed to magnetic fields for extended periods contained significantly fewer megakaryocytes immediately after termination of exposure, and this decrease persisted somewhat for 200 days.[73] Between 4,000 and 10,000 Oe, the magnitude of this decrease does not depend on field strength, but the megakaryocyte decrease doubled when exposure duration increased from 4 to 35 days.

THEORETIC MODELS

A variety of magnetobiologic phenomena have been produced with homogeneous and inhomogeneous magnetic fields ranging in strength from 1/1,000 to 100,000 times that of the natural geomagnetic field. Despite these observations we still have neither a uniform characterization of magnetobiologic phenomena nor an explanation of the mechanism by which a magnetic field produces biologic effects. The probable existence of an amplification mechanism in living biologic systems interposed between the basic phenomenon and the observed biomagnetic manifestation complicates the problem of identification of fundamental mechanisms.

Some common aspects of magnetobiologic effects are as follows:

1. A threshold field strength exists below which a magnetic field is biologically ineffective. The value of this limiting field strength is different for each parameter investigated. Furthermore, above this threshold value the effect does not increase proportionally with field strength.
2. Some effects persist for long periods of time after termination of magnetic exposure, the duration ranging from hours to 200 days.
3. Often, magnetobiologic effects are not manifest immediately after the beginning of exposure, but only after a latent period. Such a latent period is associated with the effects of many stressors, such as confinement, cold, starvation, and hemorrhage; the manifest response of the organism to the noxious agent is not immediate.
4. Several observations suggest that magnetic fields produce organismic responses characteristic of regenerative processes.
5. Some findings suggest that a magnetic field affects oxidation-reduction processes. This may be related to the paramagnetism of the oxygen molecule.
6. The activity of enzyme-substrate systems is affected by a wide range of magnetic field strengths from "zero" field to extremely high intensity.
7. Embryogenesis is very susceptible to the action of a magnetic field, suggesting particular sensitivity of rapidly growing tissues.
8. It is the opinion of these authors that many, but not all, magnetobiologic effects follow the pattern of Selye's general (or local) adaptation syndrome. This syndrome is considered to be the sum of all nonspecific systemic responses to long-term exposure to a stressor.

The stress of exposure to a magnetic field produces effects similar to those of other stressors. The effect is observable only after a latent period of continuous exposure to a field and is attenuated if the exposed specimen changes its position frequently relative to the direction of the field and gradient vectors. Many effects observed in mice exposed to vertical magnetic fields are not seen in horizontal fields. Unrestrained mice never assume the dorsal position; hence in a vertical field they are constantly exposed to the same field direction, whereas in a horizontal field the field direction changes continuously as the subject moves.

Magnetobiologic effects that appear to belong in the "stress-effect" group are as follows: mouse growth retardation, rejection of transplanted tumors, hematologic changes, retardation of wound healing and tissue regeneration, central nervous system effects, body temperature decrease, estrus cycle derangement, in utero resorption of embryos, tissue respiration decrease, inhibition of bacterial cultures during the maximum stationary phase, and pathologic changes in the liver and adrenal glands of mice.

The time-course of some of these effects is consistent with the three stages of the general adaptation syndrome. Fig. 8-10 shows the general pattern and the changes observed in mouse body weight and leukocyte count during exposure to a magnetic field. The first stage, termed alarm reaction, is the response of the organism to sudden exposure to a stressor; it consists of two phases—shock and countershock. The shock phase is characterized by nervous and vascular depression fol-

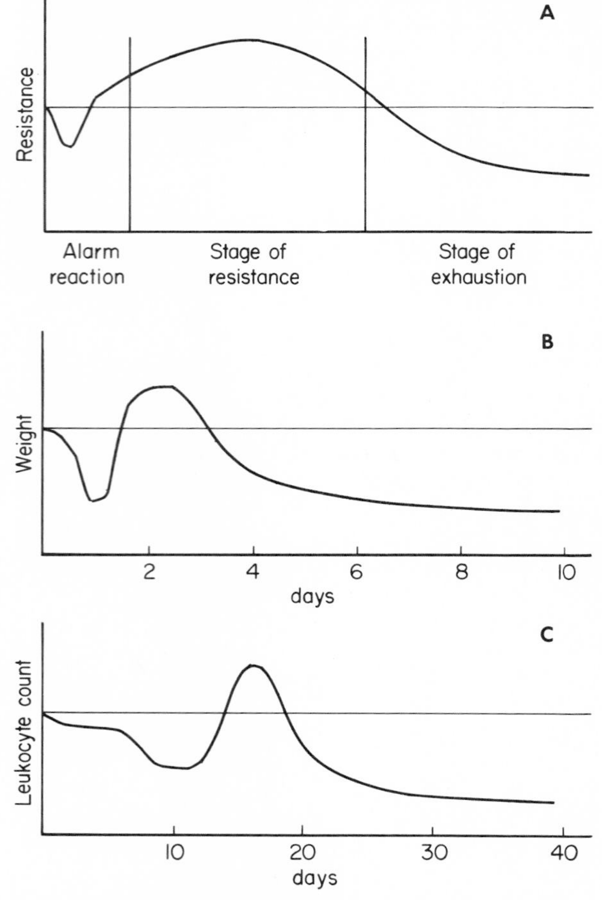

Fig. 8-10. A, Schematic diagram of the general adaptation syndrome consisting of a biphasic alarm reaction, a stage of resistance, and a stage of exhaustion. **B,** Change in body weight and **C,** the leukocyte count of mice exposed to a homogeneous magnetic field. In **B** all three stages of the syndrome are present, even the shock and countershock phases of the alarm reaction are manifest. In **C** the two phases of the alarm reaction are not discernible, but the transient increase of leukocyte count caused by adaptation and the stage of exhaustion are definitely visible.

lowed, in nonfatal cases, by a countershock phase. Increased ACTH release stimulating corticosteroid release increases resistance to the stressor. During stage 2, termed the stage of resistance, the adjustment initiated during the countershock phase predominates; adaptation has occurred. The organism now has increased resistance to the initiating stressor but decreased resistance to other stressors. In stage 3, termed the stage of exhaustion, the state of adaptation can no longer be maintained and the shock syndrome reappears.

If, during the general adaptation syndrome an excess or imbalance of hormones whose secretion is evoked in the process of adaptation produces a naturally occurring pathologic condition, this is termed a "disease of adaptation."

We propose the following theoretic models for magnetobiologic phenomena:

1. A homogeneous magnetic field exerts a torque on magnetic dipoles. Paramagnetic molecules, those with unpaired electrons and those with nonspherically symmetric susceptibility distribution, have magnetic moments and are thus also subject to torque. To produce a magnetic orientation, a rotational energy change must be appreciably larger than the thermal rotational energy at body temperature. Assuming reasonable values for size and paramagnetic susceptibility differences, magnetic fields of 10^5 Oe, or even higher intensity, would be needed. If bonds in a molecule break in such a way that each fragment keeps one electron, free radicals form; in some molecules this occurs spontaneously at room temperature. Free radicals commonly result from photolysis of molecules by light or x-rays, but they can also form from other free radicals and in oxidation-reduction processes. Radicals associated with mitochondrial particles and macromolecules that have transition metals complexed within their structures (for example, catalase, myoglobin, and cytochrome c) should be particularly susceptible to orientation in a magnetic field.

It has been suggested that by distorting the bond angles of paramagnetic molecules, magnetic fields may impair the closeness of fit of enzymes and substrates; this would reduce the rate of synthesis of large molecules because thermal randomization by agitation would be comparatively less in larger molecular aggregates.[82]

Orientation may also act in a different manner. It is commonly assumed that a reaction between two molecules occurs when the interacting molecules are momentarily fixed at constant distance but can still rotate. Interaction occurs when the specific structural sites of two molecules face each other; they rotate until their respective reactive sites are in apposition. If paramagnetic free radical intermediates are involved, magnetic fields may slow this rotational diffusion.[83] In this way the biochemical reaction rates of enzymes or ribosomes may be decreased.

Liquid crystals are strongly anisotropic; hence fields of a few thousand oersteds should suffice to orient them in such a manner that the optic axis becomes parallel to the field direction.[84] The

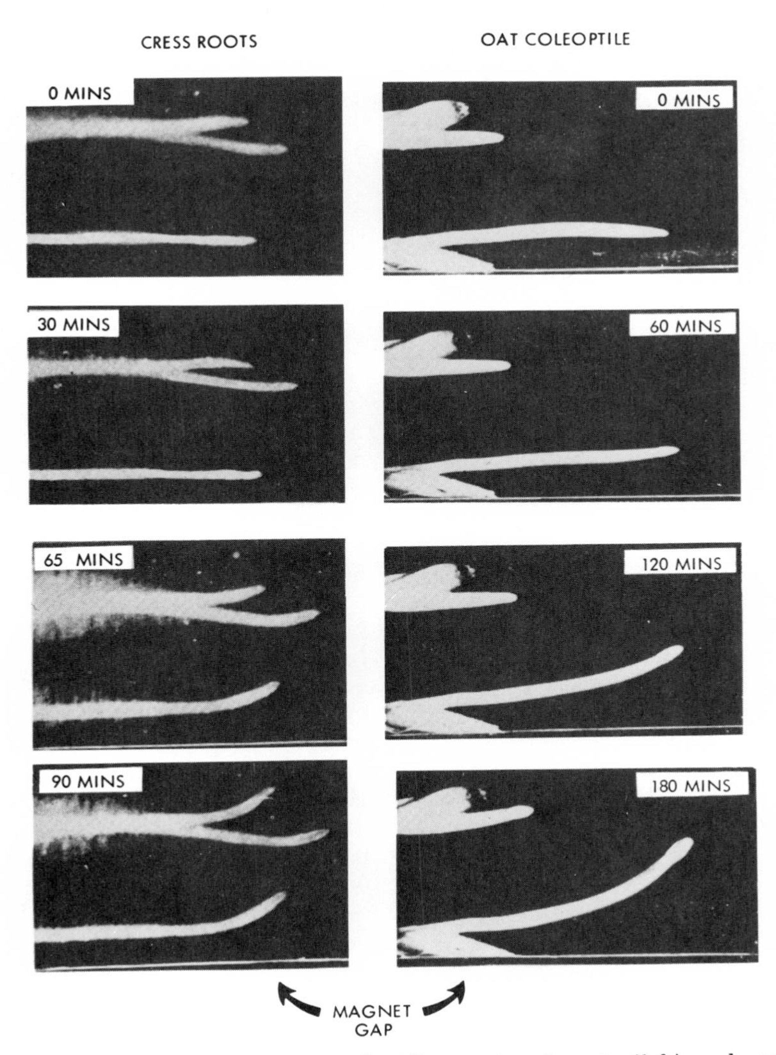

Fig. 8-11. Typical growth curvature of cress *(Lepidium sativum)* roots (left) and oat *(Avena sative)* coleoptiles (right) in an inhomogeneous magnetic field of 4,000 Oe with a 5,000 Oe/cm gradient. Rotation of the magnet around a horizontal axis eliminated the growth curvature caused by gravity. In geotropism roots curve downward and coleoptiles upward; thus they have oppositely oriented curvatures. In magnetotropism both organs curve in the same direction.

structure of liquid crystals is very sensitive to all kinds of stresses—thermal, mechanical, and electric. Many enzymes have several allosteric forms; transition between these forms involves small amounts of energy that could be supplied by magnetic fields. Similarly, the melting curve of DNA and its relaxation behavior indicate that it has liquid crystal properties. Membranes within and between cells also show such properties. Some biomagnetic effects may result from such reorientation near transition temperatures.[85]

2. An inhomogeneous field exerts an accelerating force in the direction of the gradient on molecules that have an intrinsic magnetic moment or those whose susceptibility differs from that of their environment. In general, this acceleration is much smaller than that caused by the gravitational field of the earth, but when the product of field strength and gradient amounts to 10^9 Oe^2/cm and the difference in susceptibility amounts to 2×10^{-6}, the acceleration may be twice that of gravity. In such fields the paramagnetic oxygen molecule may migrate in the direction of the gradient, while the diamagnetic nitrogen molecule

migrates in the opposite direction, producing intracellular concentration gradients.[36] In a macromolecule of 2.5×10^5 molecular weight, a concentration gradient of 0.5% may result.

Such an accelerative force acting on some cell components may explain the phenomenon of magnetotropism discovered by Audus and Whish.[86] They eliminated geotropism in their experiments by rotation of the sample; indeed, roots and coleoptiles curved in the direction of the magnetic gradient (Fig. 8-11).

3. A static magnetic field generates an electromotive force in a moving conductor and exerts a Lorentz force on moving charge carriers, redistributing electric charges and creating polarization currents. In inhomogeneous fields, where different parts of the conductor (experimental specimen) are exposed to different field strengths, conduction currents arise. In general, polarization currents and conduction currents are distributed over the entire specimen, the precise distribution depending on conductivity, capacity, speed, and acceleration. Such currents can generate heat, produce electrolytic dissociation, and affect the central nervous system.

It was speculated that the direction-sensing mechanism of migrating birds might be caused by their ability to sense potential differences of a few microvolts arising from in-flight polarization currents and current changes that result from flapping their wings in different directions relative to the geomagnetic field.[9] If this were true each wing might contribute a different input, creating a guidance pattern similar to the signals transmitted to pilots to guide an airplane along a gliding path as it approaches a landing strip.

4. An electric current is a stream of electrons. When a magnetic field is applied at right angles to the direction of current flow, electrons moving in the conductor are deflected. Thus an electric field, or potential difference, is set up in the direction transverse to both magnetic field and electric current (Hall effect). In the magnetic fields that are commonly used, the magnitude of the Hall effect in metallic or ionic conductors is rather small.

Becker[71] postulated that the current flow in nerves is of semiconductive nature. If this is true the interaction between charge carriers and a magnetic field would be many orders of magnitude greater. A magnetic field applied perpendicularly to the neuron axis of a salamander creates a potential difference.[69]

5. The cell growth process is related to the diffusion of dissociated salts across the plasma membrane, from the extracellular domain into and out of the cytoplasm, or through the nuclear membrane between nucleus and cytoplasm. A magnetic field might affect a cell by interfering with the diffusion of ions.[87] It has been estimated that fields of the order of 100,000 Oe are needed to perturb significantly the dynamics of charged cytoplasm. Although forces other than magnetic determine the physiologic orientation of intracellular particles, nuclei, and organelles, we may speculate that inhomogeneous fields orient these, enhancing diffusion effects, or that a magnetic field affects mitochondria that are fortuitously in optimum orientation for diffusion effects to alter reaction kinetics. According to the electro-osmotic theory of protoplasmic movement, it should be possible to affect protoplasmic movement, active transport, and mitosis with strong magnetic fields, because such fields would distort the ionic currents associated with such cell activity.[88]

6. A special mechanism involving hydrogen bonds and, in particular, the DNA molecule was suggested to explain mutations and other genetic changes produced by magnetic fields.[45] This mechanism is based on Löwdin's hypothesis that quantum mechanical effects may influence the DNA code.[89] In terms of quantum mechanics the proton represents a wave packet that may have nonzero amplitude even in regions classically forbidden. The hydrogen bond between complementary bases in DNA consists of a proton shared by two lone electron pairs situated on the nitrogen or oxygen atoms of the nucleotide bases (thymine, cytosine, adenine, and guanine).

The great stability of the genetic code suggests that the double-well potential, the minima of which represent the two possible positions of the proton, is highly asymmetric, the proton being comparatively stationary in the lower-lying energy level of the deeper well; tunneling through the potential barrier would thus occur rarely. The tunneling of one proton and the reverse tunneling of another proton for reasons of charge conservation form tautomeric bases. If breakage of the hydrogen bond and DNA replication occur while these tautomeric bases exist, they cannot combine with their normal complementary bases; this would lead to coding errors and, in some instances, to code deletion. Such an error would be multiplied by a factor of 2 in each subsequent DNA replication. A change of base pairs is the essence of mutation. Whenever accumulation of errors exceeds a certain limit, it may upset the balance between enhancing and controlling enzymes in the growth cycle and may possibly lead to the occurrence of tumors.

Because of the multiplication of any one error by a factor of 2^n after n replications, this imbalance will more likely manifest itself at an advanced age after many replications, when n is large. Thus accumulation of code errors may also be a cause of aging. An external magnetic field would change the energy levels of the nucleotide bases, changing the potential difference between the minima of the two wells and the height of the potential barrier between them. This would change the probability that a proton in the deeper well is above the lowest tunneling level as well as the transmission coefficient of the proton through the potential barrier. We should therefore expect a magnetic field to decrease or increase the number of code errors. An estimation of the effect of a magnetic field on proton tunneling is rather complicated because the form of the double-well potential depends not only on the attraction between the proton and the lone electron pairs but also on the polarizing effect of the proton on the σ and the π electron clouds. Barnothy made a rough estimation that in a 10,000-Oe field a bacterial culture having about 0.4% mutants per generation would have a 40% mutation rate after 100 replications. Mice of the C3H inbred strain kept during their youth in magnetic fields had a 36.3% ± 4.5% higher activity and a more youthful appearance (Fig. 8-5) at the age of 400 days than control mice of the same age. The spontaneous mammary gland carcinoma incidence in the magnet group mice was 40% compared to 73% for the controls. These results were interpreted as indicating that a magnetic field lowers the tunneling probability and the number of DNA code errors.

The polypeptide chains of proteins are also linked by hydrogen bonds in which proton tunneling may occur. Although a change produced by proton tunneling would not be multiplied by replication (as in the DNA molecule) in this case, if an external magnetic field simultaneously causes protons in a large number of protein molecules to prefer a position different from normal, and if these protein molecules are enzymes, then enzyme balance could change enough to produce observable effects. Moreover, a change of the shape of the double-well potential in the hydrogen bond is equivalent to a change of bond energy.

Acetylcholine hydrolysis in frog vagal heart preparations is markedly increased by partial replacement of potassium ions by cesium ions.[22] However, when a 15,000-Oe field was maintained for 30 minutes, the already elevated rate induced by cesium ions increased further by a factor of more than 3. Urea and guanidine, typical hydrogen bond–breaking agents, act on vagal heart preparations in a manner similar to that of a magnetic field, suggesting that the magnetic field affects hydrogen bonding energy levels.

7. The fact that all the effects one can theoretically expect from known physical interactions are of extremely small magnitude requires the assumption that in most instances some amplification mechanism in the organism is involved in the production of observable biomagnetic manifestations. In biologic processes involving a positive feedback mechanism, such as growth and reproduction, an extremely slight change in the feedback factor may, in time, produce a large change in the final value.

The concept of an amplification mechanism is supported by the finding that it is much easier to produce a magnetobiologic effect in vivo than in vitro. While theoretically one would expect a magnetic field to alter chemical reaction rates, with the field retarding reaction velocity when a more paramagnetic substance changes into a less paramagnetic one and accelerating the reaction from a diamagnetic to a paramagnetic substance, an extensive search for such reaction rate changes proved futile.[90] Single-electron tunneling between superconductive microregions (for example, purine and pyrimidine rings in DNA and RNA) may rate-limit various nerve and growth processes. Magnetic fields could impair the biologic process by destroying the superconductivity.[91]

POSSIBLE APPLICATIONS FOR THERAPY AND DIAGNOSIS

Magnetobiology is rapidly becoming an important tool for the study of fundamental physiologic processes. In addition to this purely scientific use, there may be therapeutic and/or diagnostic applications.

Because magnets large enough to accommodate human beings and produce high-intensity fields are heavy, bulky, and expensive,[92] knowledge regarding the effects of magnetic fields on human subjects has accumulated slowly. Beischer surveyed nuclear physics laboratory personnel regarding their experiences in high-intensity magnetic fields. Their answers suggest that brief partial or total body exposure to fields up to 20,000 Oe is tolerated without sensation. Furthermore, exposure to fields of 5,000 Oe for a cumulative total of 3 days within 1 year produced no detectable effects.

Therapeutic possibilities

Despite the many magnetobiologic effects observed in plants and animals, it is still not possible to decide whether magnetic fields are harmful or beneficial. The observation that mice reared in magnetic fields look and behave younger[45] suggests that the field has a salutary effect. A method for retardation of aging without substantially increasing lifespan would be of great interest. Since embryos were resorbed in the uteri of pregnant mice,[42] we may also speculate about a "magnetic abortion" that would obviate the surgical procedure.

The arrest of spontaneous tumors in mice for at least 2 months and the failure of metastases to occur during residence in the field suggest the possibility that magnetic fields may be useful in cancer therapy, even before we know the precise mechanism of this remarkable effect. The arrest of tumor growth per se has therapeutic significance, because it would extend the period during which other anticancer agents could be used; possibly it would reduce the required dose of these agents, resulting in fewer side effects. It is promising that a high percentage of leukemic mice survived when exposed to a strong magnetic field.[55]

It is speculated that the tensile strength of cancer cells is less than that of normal cells.[93] A strong external magnetic field might generate forces sufficiently strong to tear malignant cells apart, leaving normal cells intact.

A change in the number of circulating leukocytes and thrombocytes prevents some phases of the radiation syndrome that occurs after whole body irradiation with x-rays or gamma rays. Magnetic field pretreatment decreased radiation mortality by about 27% at a dose that killed 80% of the controls.[94] The death rate decrease was greatest between the seventh and fourteenth days after irradiation, a time when death usually occurs as a result of radiation-induced leukopenia. Possibly, the magnetic field effect on blood platelet count would be of value in controlling certain hemorrhagic disorders.

Electroencephalographic changes in animals exposed to static magnetic fields were similar to those seen during deep anesthesia.[95] This observation suggests the possibility of a new method of anesthetization, avoiding the side effects of chemically induced anesthesia.

Diagnostic possibilities

The electromotive force generated by blood flow was superimposed on the electrocardiograms of monkeys exposed to 70,000- to 100,000-Oe fields.[24] Aortic blood flow velocity could be determined from the ECG tracing. Thus in a single record the magnetic field provides information on both electric and mechanical cardiovascular events; this suggests a new, noninvasive method for study of the cardiovascular system. The flow potential generated by cardiac and large vessel blood flow in the magnetic field did not affect the cardiac activity of the monkey during a brief exposure, although heart rate decreased after a 1-hour exposure. However, it would be advisable to monitor

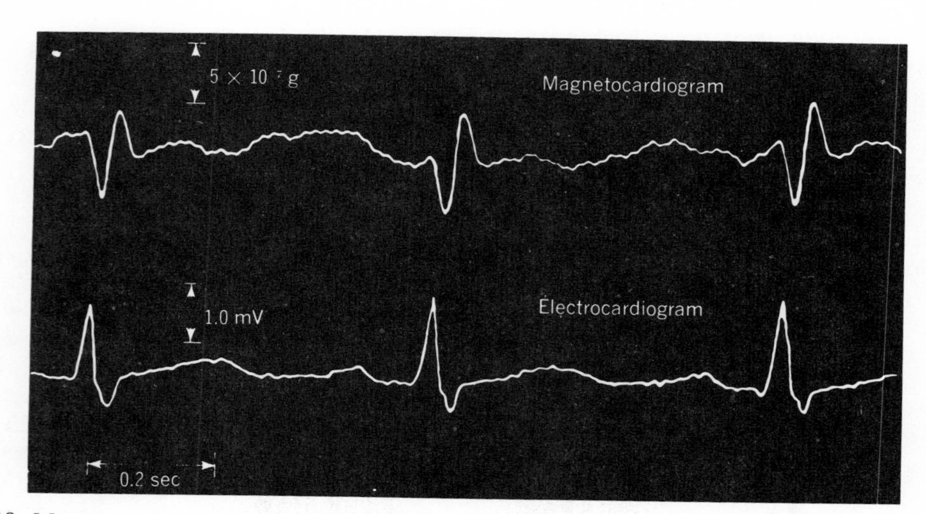

Fig. 8-12. Magnetocardiogram and electrocardiogram of a human subject. The magnetocardiogram was obtained with a magnetic probe placed 10 cm from the chest, showing signals of 10^{-7} to 10^{-8} Oe amplitude that are clearly related to the electrocardiogram.

cardiovascular function during any human exposure to strong magnetic fields. Electromagnetic blood flow transducers have been designed that permit the diagnosis of circulatory disorders without resort to surgical intervention.[96]

The complex pattern of rhythmic contraction and relaxation of cardiac muscle produces a time-dependent spatial distribution of electric currents in the chest. Detected with electrodes and amplified, the regular pattern of potential variations recorded or displayed as a function of time constitutes the diagnostically valuable electrocardiogram. The electric impulses generated by the beating heart produce magnetic fields that extend beyond the body. Cohen[97] detected regular magnetic field fluctuations near the human chest that were related to the electrocardiogram. Fig. 8-12 compares a human magnetocardiogram with the same subject's electrocardiogram.* The strength of the recorded magnetic fields was about 10^{-7} that of the geomagnetic field and about 10^{-4} that of the magnetic field fluctuations produced by electric currents in a building. Hence a magnetocardiogram must be recorded in a special room that is very well shielded against magnetic fields.

The weak alternating magnetic fields outside the human scalp produced by alpha-rhythm currents may also be useful as a magnetoencephalogram.[99] The three-fold greater sensitivity of the magnetoencephalograph permits detection of more subtle brain events than those detectable by electroencephalography, yielding new and different information.[100] However, the clinical value of magnetocardiograms and magnetoencephalograms as compared to their electric analogs remains to be evaluated.

GLOSSARY

cosmic rays Rays of extremely high penetrating power of extraterrestrial origin, arising both within and outside of our galaxy. They are responsible for some ionization in the earth's atmosphere and are probably capable of producing biologic effects such as mutations. The geomagnetic field shields us from the cosmic ray flux to some extent, except in the polar regions.

Curie point All ferromagnetic substances have a definite transition temperature at which the phenomena of ferromagnetism disappear and the substances become merely paramagnetic. This temperature, the Curie point, is usually lower than the melting point of the substance.

diamagnetic substance A substance in which the intensity of magnetization induced by an applied field is less than that produced in a vacuum by the same field. Almost all organic and inorganic compounds, with the exception of free radicals and compounds of the transition elements, are diamagnetic. Diamagnetism is a consequence of the interaction of the applied field with the field induced in the electronic orbits of the atoms. The magnetic force on the orbital electrons causes a change in their angular frequency; they precess around the field direction with an angular frequency called Larmor frequency. Since the Larmor frequency is independent of the electron energy, all randomly oriented atoms precess as a whole about the applied field at a rate independent of temperature. The permeability of a diamagnetic substance is less than unity and its susceptibility is negative.

dummy magnet To ensure identical environmental conditions in biomagnetic experiments for exposed and control samples, the latter must be placed in a set-up that is an exact replica of the set-up used with the magnet, including pole caps, except that the magnetic field is absent.

Ettinghausen's effect When an electric current flows at right angles to the lines of force of a magnetic field, a temperature gradient develops that is opposite in direction to the Hall electromotive force.

ferromagnetism Property of a material in which atomic and molecular moments are aligned even in the absence of an applied external field. These alignments extend to domains over 10^{22} atoms. Application of an external magnetic field rotates the domains as a whole in the field direction and/or the total volume of the domains oriented in the external field direction grows at the expense of those unfavorably oriented; the relative contribution of these two factors depends on the applied field strength and the material involved. Some ferromagnetic materials remain permanently magnetized even after removal of an external magnetic field and thus become permanent magnets. In ferromagnetic materials susceptibility and permeability have large values; intensity of magnetization does not change linearly with field intensity but also depends on the past history of the material (hysteresis). Ferromagnetism is not a property of biologic materials.

field intensity (H) Measured by the force acting on a unit pole. The unit of field intensity is the oersted (Oe), or that field that exerts a force of 1 dyne on a unit magnetic pole. A smaller unit is the gamma: $1\ \gamma = 10^{-5}$ oersted.

flux density Number of lines of force per unit area. Its unit is the gauss, corresponding to 1 line/cm^2. In a vacuum a field intensity of 1 Oe produces a flux density of 1 gauss; in other materials flux density B = permeability $\mu \times$ field intensity H. If permeability has a value close to 1 (as in most biologic materials), flux density and field intensity are numerically equal, $B \approx H$.

force in a homogeneous magnetic field The force of a magnetic field on an isolated pole is proportional to pole strength and field intensity. A homogeneous field does not exert a force on a magnetic dipole; it exerts a torque.

force in an inhomogeneous magnetic field A force on a magnetic dipole that is proportional to the gradient of the field and to the magnetic moment of the dipole.

force exerted on a paramagnetic substance Force propor-

*Magnetocardiograms can be used to detect myocardial infarctions.[98]

tional to field strength, gradient, and volume susceptibility of the paramagnetic material.

force between two magnetic poles Force proportional to the product of the two pole strengths and inversely proportional to the square of the distance between the two poles and the permeability of the medium.

gradient of a magnetic field ($G = dH/dx$) A measure of the change of field intensity within a field. Unit gradient is a field intensity change of 1 Oe over a distance of 1 cm.

Hall effect When a steady current flows in a steady magnetic field, electromotive forces develop that are at right angles to both magnetic force and current and proportional to the product of current intensity, magnetic force, and the sine of the angle between the direction of these two quantities.

homogeneous (or uniform) magnetic field Magnetic field in which the flux density is the same over the entire volume. The lines of force are parallel.

inhomogeneous magnetic field A magnetic field in which field intensity (flux density) changes. The lines of force diverge. Their number is greater in regions where field intensity is higher.

intensity of magnetization The magnetic moment of a magnet divided by its volume. Unit intensity of magnetization is the intensity of a magnet that has unit magnetic moment per cubic centimeter.

Lenz's law When a conductor moves relative to a magnetic field, the electromotive force thus induced produces a current whose magnetic field in turn opposes the movement that produced it.

line of force A line of direction such that every point is the same as the direction of the force that would act on a small positive charge or magnetic pole placed at that point. This term is used to describe an electric or a magnetic field.

magnetic field caused by a current The intensity of the magnetic field at the center of a circular conductor of radius r in which a current of 1 ampere is flowing is $H = 2\pi I/10 r$, in oersteds.

magnetic field caused by a magnet At a point on the extended magnetic axis r cm from the center of the magnet whose magnetic moment is M, the field strength $H = 2 M/r^3$ oersteds; at a point on a line bisecting the magnet at right angles $H = M/r^3$.

magnetic induction The magnetic flux per unit area (flux density) in a substance subjected to a magnetic field. If a substance of permeability μ is placed in a magnetic field H, the magnetic induction in the substance is $B = \mu H$. As a consequence of this magnetic induction, the substance possesses an intensity of magnetization I, whereby $I = (B - H)4\pi$.

magnetic moment of a magnet Measured by the torque when it is at right angles to a homogeneous field of unit intensity. The value of the magnetic moment μ is given by the product of magnetic pole strength and the distance between the poles. The elementary unit of magnetic moment is the Bohr magneton, which is the magnetic moment of an electron $-eh/4\pi mc = 0.917 \times 10^{-20}$ erg/Oe.

magnetic permeability (μ) A property of materials modifying magnetic flux through the material. It equals the ratio of flux density B to field intensity H. A vacuum has a permeability of 1. The permeability of diamagnetic substances is slightly less than 1, of paramagnetic substances slightly greater than 1; ferromagnetic substances have high permeability, the value depending also on field intensity, temperature, and previous magnetization.

magnetic susceptibility Measured by the ratio of the intensity of magnetization I produced in a substance to the intensity of the field H to which it is subjected. $\kappa = I/H$. Diamagnetic materials have negative susceptibilities, paramagnetic materials positive susceptibilities.

magnetic saturation The permeability of ferromagnetic materials first increases then decreases with field strength, reaching unity above 60,000 gauss for most high-permeability alloys.

magnetometer An instrument to measure magnetic field strength. The most sensitive magnetometers operate on the principle of nuclear magnetic resonance.

mu-metal A ferromagnetic alloy of very high permeability at very low field strength. Boxes made of mu-metal sheets are excellent shields against weak external magnetic fields.

Nernst effect When heat flows in a body exposed to a magnetic field, an electromotive force is observed in the mutually perpendicular direction. The resulting potential difference is proportional to magnetic field intensity, temperature gradient, and the sine of the angle between the directions of these two quantities.

paramagnetic bodies Those bodies that tend to move toward a stronger magnetic field. The permeability of a paramagnetic substance is greater than unity and its magnetic susceptibility is positive. A molecule that has at least one unpaired electron has a magnetic moment. The uncompensated spin angular momentum of the electron tends to align itself in the direction, or opposite to the direction, of the applied field. If the molecule aligns itself with its magnetic moment parallel to the applied field, its energy is lower than when it aligns itself opposite to the applied field. Consequently, there will be statistically more electrons aligned with the field than opposite to the field. This positive contribution to magnetization more than offsets the diamagnetic contribution of the paired electrons, and the system as a whole behaves paramagnetically; that is, it has a positive susceptibility. At higher temperature the more intense thermal motion tends to randomize the orientation of the magnetic moments and accordingly paramagnetic susceptibility is temperature dependent, decreasing proportionally to absolute temperature.

semiconductors Crystals, including the metallic oxides and sulfides, cupric oxide, zinc oxide, and lead sulfide and the elements germanium, silicon, and selenium. Semiconductors have feeble electronic conductivity that increases extremely rapidly with increasing temperature. As temperature increases, more electrons "boil off" from their parent atoms, becoming free to wander about in the material and to conduct electricity.

sunspots Dark spots that appear from time to time on the sun's surface, usually visible only with a telescope. Their appearance is frequently accompanied by magnetic storms on the earth.

torque acting on a magnet Torque proportional to magnetic moment, field intensity, and sine of the angle between field direction and magnetic moment direction.

vector magnetic effects Physical and biologic effects that change direction or sign when the magnetic field direction is reversed relative to the coordinate system of the biologic sample. Most physical effects are vector effects, and some cumulative biologic effects seem also to have vector character.

REFERENCES

1. Beischer, D. E., and Miller, E. F.: Exposure of man to low intensity magnetic fields, NAV-SAM-MR-OOS. 13-9010-1-5, 1962, U. S. Naval School of Aviation Medicine.
2. Conley, C. C.: Effects of near-zero magnetic fields upon biological systems. In Barnothy, M. F., editor: Biological effects of magnetic fields, vol. 2, New York, 1969, Plenum Publishing Corp., p. 29.
3. Halpern, M. H., and van Dyke, J. H.: Very low magnetic fields: biological effects and their implications, Aerospace Med. **37**:281, 1966.
4. Brown, F. A., Jr., and Park, Y. H.: Association formation, photic and subtle geophysical stimulus patterns—a new biological concept, Biol. Bull. **132**:311, 1967.
5. Friedman, H., Becker, R. O., and Bachman, C. H.: Psychiatric ward behavior and geophysical parameters, Nature **205**:1050, 1965.
6. Friedman, H., Becker, R. O., and Bachman, C. H.: Effect of magnetic fields on reaction time performance, Nature **213**:949, 1967.
7. Düll, T., and Düll, B.: Connection between disturbances of geomagnetism and occurrence of death, Deutsch. Med. Wschr. **61**:95, 1935.
8. Presman, A. S.: Electromagnetic fields and life, New York, 1970, Plenum Publishing Corp.
9. Barnothy, J. M.: Proposed mechanism for the navigation of migrating birds. In Barnothy, M. F., editor: Biological effects of magnetic fields, vol. 1, New York, 1964, Plenum Publishing Corp., p. 287.
10. Barnwell, F. H., and Brown, F. A., Jr.: Responses of planarians and snails. In Barnothy, M. F., editor: Biological effects of magnetic fields, vol. 1, New York, 1964, Plenum Publishing Corp., p. 263.
11. Lindauer, M., and Martin, H.: Earth's magnetic field affects orientation of honeybees in gravity field, Z. Verg. Physiol. **60**:219, 1968.
12. Rocard, Y.: Actions of a very weak magnetic gradient. The reflex of the dowser. In Barnothy, M. F., editor: Biological effects of magnetic fields, vol. 1, New York, 1964, Plenum Publishing Corp., p. 279.
13. Foulkes, R. A.: Dowsing experiments, Nature **229**: 163, 1971.
14. Hays, J. D., Saito, T. O., Opdyke, N. O., and Burcke, L.: Pliocene-pleistocene sediments of the equatorial Pacific, Geol. Soc. Am. Bull. **80**:1481, 1969.
15. Barnothy, J. M.: The vector character of field and gradient and its possible implications for biomagnetic experiments and space travel. In Barnothy, M. F., editor: Biological effects of magnetic fields, vol. 1, New York, 1964, Plenum Publishing Corp., p. 60.
16. Barnothy, M. F.: Hematological changes in mice. In Barnothy, M. F., editor: Biological effects of magnetic fields, vol. 1, New York, 1964, Plenum Publishing Corp., p. 109.
17. Likhachev, A. I.: Changes in the erythrocyte sedimentation rate of rabbits due to exposure of the central nervous system to a constant magnetic field. In Barnothy, M. F., editor: Biological effects of magnetic fields, vol. 2, New York, 1969, Plenum Publishing Corp., p. 137.
18. Murayama, M.: Magnetic orientation of sickled erythrocytes, Third International Biomagnetic Symposium, Chicago, 1966, University of Illinois, p. 40.
19. Hackel, E., Smith, A. E., and Montgomery, D. J.: Agglutination of human erythrocytes. In Barnothy, M. F., editor: Biological effects of magnetic fields, vol. 1, New York, 1964, Plenum Publishing Corp., p. 218.
20. Barnothy, J. M., and Barnothy, M. F.: Magnetic fields and the number of blood platelets, Nature **225**:1146, 1970.
21. Seyle, H.: The diseases of adaptation: introductory remarks, Recent Progr. Hormone Res. **8**:117, 1953.
22. Young, W.: Magnetic field and *in situ* acetylcholinesterase in the vagal heart system. In Barnothy, M. F.: Biological effects of magnetic fields, vol. 2, New York, 1969, Plenum Publishing Corp., p. 79.
23. Togawa, T., Okai, O., and Okina, M.: Observation of blood flow in squirrel monkeys in strong superconductive electromagnets, Med. Bio. Eng. **5**:169, 1967.
24. Beischer, D. E.: Vectorcardiogram and aortic blood flow of squirrel monkeys in a strong superconductive electromagnet. In Barnothy, M. F., editor: Biological effects of magnetic fields, vol. 2, New York, 1969, Plenum Publishing Corp., p. 241.
25. Amer, N. M.: Effects of homogeneous magnetic field, ambient gas composition and temperature on the metamorphosis of *Tribolium confusum,* thesis, University of California at Berkeley, 1966.
26. Barnothy, J. M.: Biologic effects of magnetic fields. In Glasser, O., editor: Medical physics, vol. 3, Chicago, 1960, Year Book Medical Publishers, Inc., p. 61.
27. Shyshlo, A. A., and Shimkevich, L. L.: The effect of static magnetic fields on the oxidative processes in albino mice, Third International Biomagnetic Symposium, Chicago, 1966, University of Illinois, p. 16.
28. Cook, E. S., Fardon, J. C., and Nutini, L. G.: Effects of magnetic fields on cellular respiration. In Barnothy, M. F., editor: Biological effects of magnetic fields, vol. 2, New York, 1969, Plenum Publishing Corp., p. 67.
29. Gualtierotti, T.: Decrease of the sodium pump activity in the frog skin in a steady magnetic field, Physiologist **7**:150, 1964.
30. Hanneman, G. D.: Changes in sodium and potassium content of urine from mice subjected to intense magnetic fields. In Barnothy, M. F., editor: Biological effects of magnetic fields, vol. 2, New York, 1969, Plenum Publishing Corp., p. 127.
31. Chalzonits, N., and Arvanitaki, A.: Effects of constant magnetic fields on autoactivity of *Helix pomatia* myocardial fibres, C. R. Soc. Biol. (Paris) **158**:1902, 1964.

32. Beischer, D. E., and Reno, V. R.: Cardiac excitability in high magnetic fields, Aerospace Med. **37:** 1229, 1966.
33. Myers, R. D., and Veale, W. L.: Body temperature: possible ionic mechanism in the hypothalamus controlling the set point, Science **170:** 95, 1970.
34. Labes, M. M.: A possible explanation for the effect of magnetic fields in biological systems, Nature **211**:968, 1966.
35. Perakis, N.: Cell division in a magnetic field, Compt. Rend. Acad. Sci. **208**:1686, 1939.
36. Reno, V. R.: Sea urchin mitosis in high magnetic fields, NASA R-39 Rept. No. NAMI-954, 1966.
37. Neurath, P. W.: The effect of high-gradient, high-strength magnetic fields on the early embryonic development of frogs. In Barnothy, M. F., editor: Biological effects of magnetic fields, vol. 2, New York, 1969, Plenum Publishing Corp., p. 177.
38. Levengood, W. C.: A new teratogenic agent applied to amphibian embryos, J. Embryo. Exp. Morph. **21**:23, 1969.
39. Minnaar, P. C.: The influence of external magnetic fields on fertilized fish eggs, S. Afr. Cancer Bull. **10**:164, 1966.
40. Levengood, W. C.: Cytogenetic variations induced with a magnetic probe, Nature **209**:1009, 1966.
41. Barnothy, J. M., and Barnothy, M. F.: Observations on mice born for several generations in magnetic fields, Third International Biomagnetic Symposium, Chicago, 1966, University of Illinois, p. 31.
42. Barnothy, M. F.: Biological effects of magnetic fields on small mammals, Biomed. Sci. Instrum. **1**:127, 1963.
43. Barnothy, J. M.: Development of young mice. In Barnothy, M. F., editor: Biological effects of magnetic fields, vol. 1, New York, 1964, Plenum Publishing Corp., p. 93.
44. Barnothy, J. M., and Barnothy, M. F.: Second-day minimum in the growth curve of mice subjected to magnetic fields, Nature **200**:189, 1963.
45. Barnothy, M. F.: A possible effect of the magnetic field upon the genetic code. In Barnothy, M. F., editor: Biological effects of magnetic fields, vol. 1, New York, 1964, Plenum Publishing Corp., p. 80.
46. Dunlop, D. W., and Schmidt, B. L.: Sensitivity of some plant material to magnetic fields. In Barnothy, M. F., editor: Biological effects of magnetic fields, vol. 2, New York, 1969, Plenum Publishing Corp., p. 147.
47. Lengyel, J.: The effect of the magnetic field on fiber formation and tissue growth, Arch. Exp. Zellforschung **14**:255, 1933.
48. Mulay, I. L., and Mulay, L. N.: Effect on *Drosophila melanogaster* and S-37 tumor cells; postulates for magnetic field interactions. In Barnothy, M. F., editor: Biological effects of magnetic fields, vol. 1, New York, 1964, Plenum Publishing Corp., p. 146.
49. Gerencser, V. F., and Barnothy, M. F.: Inhibition of bacterial growth in fields of high paramagnetic strength. In Barnothy, M. F., editor: Biological effects of magnetic fields, vol. 1, New York, 1964, Plenum Publishing Corp., p. 229.
50. D'Souza, L., Reno, V. R., Nutini, L. G., and Cook, E. S.: The effect of a magnetic field on DNA synthesis by ascites sarcoma 37 cells. In Barnothy, M. F., editor: Biological effects of magnetic fields, vol. 2, New York, 1969, Plenum Publishing Corp., p. 53.
51. Hedrick, H. G.: Inhibition of bacterial growth in homogeneous fields. In Barnothy, M. F., editor: Biological effects of magnetic fields, vol. 1, New York, 1964, Plenum Publishing Corp., p. 240.
52. Pumper, R. W., and Barnothy, J. M.: The effect of strong inhomogeneous magnetic fields on serum-free cell cultures. In Barnothy, M. F., editor: Biological effects of magnetic fields, vol. 2, New York, 1969, Plenum Publishing Corp., p. 61.
53. Barnothy, J. M.: Rejection of transplanted tumors in mice. In Barnothy, M. F., editor: Biological effects of magnetic fields, vol. 1, New York, 1964, Plenum Publishing Corp., p. 100.
54. Gross, L.: Lifespan increase of tumor-bearing mice through pretreatment. In Barnothy, M. F., editor: Biological effects of magnetic fields, vol. 1, New York, 1964, Plenum Publishing Corp., p. 132.
55. Kirsten, W.: Unpublished results.
56. Löwdin, P. O.: Some aspects of the possible importance of the reading mechanism of DNA in carcinogenesis, International Symposium on Carcinogenesis, Jerusalem, 1968.
57. Cook, E. S., and Smith, Sister M. J.: Increase of trypsin activity. In Barnothy, M. F., editor: Biological effects of magnetic fields, vol. 1, New York, 1964, Plenum Publishing Corp., p. 246.
58. Smith, Sister M. J.: Effect of a homogeneous magnetic field on enzymes, Third International Biomagnetic Symposium, Chicago, 1966, University of Illinois, p. 22.
59. Akoyunoglou, G.: Effect of a magnetic field on carboxydismutase, Nature **202**:452, 1964.
60. Haberditzl, W.: Enzyme activity in high magnetic fields, Nature **213**:72, 1967.
61. Maling, J. E., Weissbluth, M., and Jacobs, E. E.: Enzyme substrate reactions in high magnetic fields, Biophys. J. **5**:767, 1965.
62. Marlborough, D. I., Hall, D. O., and Cammach, R.: Magneto-optical rotary dispersion studies on spinach ferredoxin, Biochem. Biophys. **35:** 410, 1969.
63. El'darov, A. L., and Kholodov, Y. A.: The effect of constant magnetic fields upon the motor activity of birds, Zhurn. Obshchei Biol. **25**:224, 1964.
64. Persinger, M. A.: Open-field behavior in rats exposed prenatally to a low intensity-low frequency rotating magnetic field, Develop. Psychobiol. **2:** 168, 1970.
65. Kholodov, Y. A.: Effects on the central nervous system. In Barnothy, M. F., editor: Biological effects of magnetic fields, vol. 1, New York, 1964, Plenum Publishing Corp., p. 196.
66. Valentinuzzi, M.: Phosphenes provoked by magnetic fields. In Magnetobiology, Downey, Calif., 1961, North American Aviation Inc., p. 77.
67. Oster, G.: Phosphenes, Sci. Am. **222**:82, 1970.
68. Kholodov, Y. A., Alexandrovskaya, M. M., Lukjanova, S. N., and Udarova, N. S.: Investigations of the reactions of mammalian brain to static magnetic fields. In Barnothy, M. F., editor: Biological effects of magnetic fields, vol. 2, New York, 1969, Plenum Publishing Corp., p. 215.
69. Becker, R. O.: Search for evidence of axial current flow in the peripheral nerves of salamanders, Science **134**:101, 1961.

70. Beischer, D. E., and Knepton, J. C., Jr.: The electroencephalogram of squirrel monkeys in a very high magnetic field, NAMI-972, NASA No. R-39, 1966.
71. Becker, R. O.: The effect of magnetic fields upon the central nervous system. In Barnothy, M. F., editor: Biological effects of magnetic fields, vol. 2, New York, 1969, Plenum Publishing Corp., p. 207.
72. Amineev, G. A., and Khasanove, R. I.: The effect of a constant magnetic field on the myoneural synapse, Proceedings of the Conference on the Effect of Magnetic Fields on Biological Objects, Moscow, 1960, p. 7.
73. Barnothy, M. F., and Sümegi, I.: Effects of the magnetic field on internal organs and the endocrine system of mice. In Barnothy, M. F., editor: Biological effects of magnetic fields, vol. 2, New York, 1969, Plenum Publishing Corp., p. 103.
74. Gross, L.: Wound healing and tissue regeneration. In Barnothy, M. F., editor: Biological effects of magnetic fields, vol. 1, New York, 1964, Plenum Publishing Corp., p. 140.
75. Gross, L.: The influence of magnetic fields on the production of antibody, Biomed. Sci. Instrum. **1:** 137, 1963.
76. Akhmerov, U. S., and others: The effect of a constant magnetic field on the development phases of *Drosophila,* Proceedings of the Conference on the Effects of Magnetic Fields in Biology, Moscow, 1966, p. 6.
77. Tegenkamp, T. R.: Mutagenic effect of magnetic fields on *Drosophila melanogaster.* In Barnothy, M. F., editor: Biological effects of magnetic fields, vol. 2, New York, 1969, Plenum Publishing Corp., p. 189.
78. Levengood, W. C.: Morphogenesis as influenced by locally administered magnetic fields, Biophys. J. **7:**297, 1967.
79. Selye, H., Vega, J., and Richter, C. L.: Role of the adrenal in the production of ectopic hemopoietic tissue by walker tumors, Acta Haemat. **21:** 378, 1959.
80. Kaufman, R. U., Airo, R., Pollack, S., and Crosby, W. H.: Circulating megakaryocyte and platelet release in the lung, Blood **26:**720, 1965.
81. Goldberg, E. D., Goldberg, D. I., and Gobulev, I. V.: Disturbance of the mitotic activity of bone marrow hemocytoblasts in acute radiation sickness and the action of chemical mutagenic agents, Arkh. Pat. **29:**25, 1967.
82. Gross, L.: Distortion of the bond angle in a magnetic field and its possible magnetobiological implications. In Barnothy, M. F., editor: Biological effects of magnetic fields, vol. 1, New York, 1964, Plenum Publishing Corp., p. 74.
83. Valentinuzzi, M.: Rotational diffusion in a magnetic field and its possible magnetobiological implications. In Barnothy, M. F., editor: Biological effects of magnetic fields, vol. 1, New York, 1964, Plenum Publishing Corp., p. 63.
84. Go, Y., Ejiri, S., and Fukada, E.: Magnetic orientation of poly-γ-benzyl-L-glutamate, Biochem. Biophys. Acta **175:**454, 1969.
85. Silver, I. L., and Tobias, C. A.: Magnetic fields and their biological effects. In Tobias, C. A., and Todd, P., editors: Space radiation biology, AEC and NASA. In press.
86. Audus, L. J., and Whish, J. C.: Magnetotropism. In Barnothy, M. F., editor: Biological effects of magnetic fields, vol. 1, New York, 1964, Plenum Publishing Corp., p. 170.
87. Liboff, R. L.: A biomagnetic hypothesis, Biophys. J. **5:**845, 1965.
88. Ambrose, E. J., Shepley, K., and Bhisey, A. N.: Effects of magnetic fields on cell growth *in vitro,* British Empire Cancer Campaign Research, London, 1963.
89. Löwdin, P. O.: Quantum genetics and the aperiodic solid, Rev. Mod. Phys. **35:**724, 1963.
90. Bhatnagar, S. S., and Mathur, K. N.: Physical principles and applications of magnetochemistry, London, 1935, The Macmillan Company.
91. Cope, F. W.: Evidence from activation energies for superconductive tunneling in biological systems at physiological temperatures, Physiol. Chem. Physics **3:**403, 1971.
92. Abler, R. A.: Magnets in biological research. In Barnothy, M. F., editor: Biological effects of magnetic fields, vol. 2, New York, 1969, Plenum Publishing Corp., p. 1.
93. Winterberg, F.: Some theoretical considerations on inhibition of tumor growth by ultrastrong magnetic fields, Arch. Biochem. Biophys. **122:**594, 1967.
94. Barnothy, M. F.: Reduction of irradiation mortality through pretreatment. In Barnothy, M. F., editor: Biological effects of magnetic fields, vol. 1, New York, 1964, Plenum Publishing Corp., p. 127.
95. Becker, R. O.: Relationship of geomagnetic environment to human biology, N. Y. J. Med. **63:** 2215, 1963.
96. Kolin, A.: Magnetic fields in biology, Phys. Today **21:**39, 1968.
97. Cohen, D.: Magnetic fields around the torso; production by electrical activity of the human heart, Science **156:**652, 1967.
98. Cohen, D., Norman, J. C., Molokhia, F., and Hood, W., Jr.: Magnetocardiography of direct currents: S-T segment and baseline shifts during experimental myocardial infarction, Science **172:**1334, 1971.
99. Cohen, D.: Magnetoencephalography: evidence of magnetic fields produced by α-rhythm currents, Science **161:**784, 1968.
100. Cohen, D.: Magnetoencephalography: detection of the brain's electrical activity with a superconducting magnetometer, Science **175:**664, 1972.

SUGGESTED READINGS

1. Barnothy, M. F., editor: Biological effects of magnetic fields, vols. 1 and 2, New York, 1964, 1969, Plenum Publishing Corp.
2. Becker, R. O.: The biological effects of magnetic fields, Med. Elect. **1:**293, 1963.
3. Beischer, D. E.: Human tolerance to magnetic fields, Astronautics **7:**24, 1967.
4. Busby, D. E.: Biomagnetics, NASA Contractor Report, NASA CR-889, 1967.
5. Kholodov, Y. A.: Formation of conditioned reflexes in fish to a magnetic field, Proceedings of the Conference on Physiology of Fish, Izv. Akad. Nauk. S.S.S.R., 1958.
6. Presman, A. S.: Electromagnetic fields and life, New York, 1970, Plenum Publishing Corp.

9 ALTITUDE

A. T. MILLER, Jr.

GENERAL FEATURES OF HIGH-ALTITUDE ENVIRONMENTS

The term "altitude" refers to vertical distance above sea level, usually measured in meters or feet. "Simulated" altitude is an experimental situation in which the biologically most important feature of high altitude, the low oxygen pressure, is reproduced at sea level by reduction of either oxygen concentration or barometric pressure of the air breathed. It is often convenient to conduct physiologic studies at simulated altitude, and this is permissible provided it is remembered that real high altitude has features other than oxygen deficiency (for example, cold, wind, radiation) that may not be duplicated by simulated altitude. On the other hand, the simulated condition is often more suitable for the separate study of individual features of altitude.

In order to understand the physiologic effects of altitude, it is necessary to consider briefly certain features of the earth's atmosphere. The *composition* of the atmosphere* (that is, the relative proportions of the constituent gases) is remarkably constant, not only in various regions of the earth but also at various vertical distances above sea level. This is true despite the fact that some of the gases (oxygen and carbon dioxide in particular) are being produced and consumed at different rates in different geographic localities, because of the turbulence of the air that results from temperature gradients between different regions. There is evidence, however, that the composition of the atmosphere is gradually changing because of the ever-increasing use of fossil fuels. The only constituent of the atmosphere that does fluctuate widely is water vapor. Strictly speaking, the constancy of the composition of the atmosphere refers only to the atmosphere from which all the water vapor has been removed (dry gas). Water vapor, when present, simply dilutes all the other gases.

The *pressure* of the atmosphere (and therefore of each of its constituent gases) decreases with increasing vertical distance above sea level, since the pressure at any elevation results from the weight of the column of air extending upward from that point to the limit of the atmosphere. The relation between altitude and barometric pressure is not linear, since the density of the atmosphere diminishes with altitude. As a convenient approximation, the atmospheric pressure diminishes by half for each 18,000-ft increase in altitude. This means that the partial pressure of each of the atmospheric gases (with the exception of water vapor) also diminishes by the same proportion. Water vapor, as we have seen, may be considered to be a diluent rather than a constituent of the atmosphere.

It is customary to subdivide the atmosphere into several concentric layers. The layer nearest to the earth's surface, the troposphere, extends from sea level upward to an average distance of 10 km, and it thus includes the terrestrial altitudes. The temperature falls progressively in the troposphere, reaching an average temperature of –60° C at an altitude of 10 km. It is the region of greatest air turbulence, which not only keeps the composition constant but also distributes both heat and water over the surface of the earth.

*See Table A-1 in the Appendix.

The distribution of water over the earth's surface depends largely on convection currents in the troposphere. Large quantities of water are evaporated by the heat of the sun and the water vapor thus formed rises with the warm air. As a result of the fall in barometric pressure, rising air expands. Since this expansion is adiabatic (without gain or loss of heat), the air is cooled and the relative humidity increases. The average air temperature decrease is about 6° C/km altitude increase. As the humidity of the air approaches saturation, water vapor condenses about aerosol particles to form clouds. The clouds are borne by the winds, and so water is distributed over the surface of the earth.

The atmosphere also acts as a selective filter of the radiant energy reaching it from the sun. There is no absorption of the visible spectrum or of the near ultraviolet, but wavelengths shorter than this, down to and including x-rays, are almost completely absorbed by oxygen, ozone, and nitrogen. Clouds reflect and diffuse light but do not absorb it; as a result, much of the sunlight falling on the cloud cover is reflected back into space, the remainder reaching the earth as diffused light.

Ozone is formed in the stratosphere (the layer of the atmosphere between the altitudes of 10 and 20 km) by the photochemical action of ultraviolet light on oxygen. Although the total amount of ozone in the atmosphere is small, it is extremely important because of its action in absorbing ultraviolet radiation. From the upper stratosphere, where its concentration is greatest, ozone diffuses downward and is destroyed at lower levels by the oxidation of aerosol particles in the atmosphere. Its sea-level pressure is only about one tenth that in the region of its highest concentration. There is thus a constant diffusion of ozone toward the earth, balanced by a constant rate of destruction.

The most remarkable feature of the earth's atmosphere, compared to that of the other planets, is the presence of an abundance of free oxygen. It is generally thought that all the oxygen in the earth's atmosphere has been produced by photosynthesis in green plants, which implies that the earliest forms of life must have obtained their energy from anaerobic metabolism. It is possible, however, that prephotosynthetic organisms may have obtained oxygen derived from the photochemical dissociation of water, associated with the loss of hydrogen from the atmosphere by diffusion into space.

The contributions of the atmosphere to the "fitness of the environment" (L. J. Henderson) can be summarized as provision of optimal concentrations of the gases essential for oxidative and photosynthetic metabolism (oxygen and carbon dioxide), application of suitable pressure on the external surface of living organisms, distribution of heat and water over the earth's surface, and differential filtration of radiant energy reaching the upper atmosphere, allowing passage of biologically essential forms of energy while absorbing much of the energy having harmful effects.

Life at high altitudes is conditioned not only by the properties of the atmosphere but also by those of the terrain. Except when motorized vehicles are used, ascent to and travel at high altitudes require physical effort under difficult conditions (ice, snow, steep grades). Physical fatigue, resulting from the interaction of terrain and atmospheric factors (especially reduced oxygen pressure), is an ever-present deterrent to strenuous activity by newcomers to high altitude, though natives often show an amazing capacity for hard work.

More than 10 million people live permanently at altitudes above 12,000 ft, most of them in the Andes and the remainder in Tibet and surrounding areas. The altitude limits for human habitation in these regions is determined as much by climate as by barometric pressure. Above 13,000 ft agriculture is seriously hampered by cold, and life is largely pastoral. The communities at the highest elevations in the Andes are engaged in mining and depend for their subsistence on food brought up from lower elevations. The highest permanently occupied settlement in the world is a mining camp at an elevation of 17,500 ft in the Peruvian Andes. Daily the miners climb to work at even higher elevations but return to the lower elevation of their homes at night. They refuse to live permanently at the higher altitude, giving as their reasons loss of appetite and weight and inability to sleep. On this evidence, 17,500 ft has been generally accepted as the limit of altitude to which man can become permanently adjusted. It is possible, of course, to ascend to much higher altitudes for short periods of time, but there is a general impression of an underlying process of physical deterioration in persons who stay at altitudes above 20,000 ft for any length of time.

The occurrence of amphibians and reptiles at high altitudes appears to be restricted by cold and the scarcity of food rather than by oxygen deficiency. Mammals of all sizes, from the very small deer mouse to large members of the camel family (llama and vicuna), are native to high altitudes.

Human populations native to high altitudes show remarkable capacity for performing hard work despite the rarified atmosphere. Some authorities are convinced that these natives have achieved a level of fitness that is impossible for sojourners, even after a period of years at high altitude, implying a genetic superiority resulting from selection over the course of generations.

Plant life at high altitudes has many points of resemblance to plant life in arctic regions. The same climatic conditions appear to be limiting in both cases (cold, high winds), though the two environments differ in many other respects (oxygen pressure, humidity, annual distribution of hours of daylight).

HISTORIC ACCOUNT OF HIGH-ALTITUDE RESEARCH

The scientific study of high altitudes dates from the invention of the balloon in 1783, although the Spanish conquistadors of the sixteenth century had observed a greatly reduced birth rate in themselves and their animals when they attempted to live in the high Andes of South America. In the year 1783, successful flights were made, first in hot air balloons and later in hydrogen-filled balloons.* These flights aroused great enthusiasm, and many others were made during the next few years. The first balloon ascension made in the interest of science occurred in 1803, when two French physicists reached an altitude of 7,170 m and made a number of observations of electric and magnetic phenomena. Two birds were carried in the gondola, and one of them died. The first physiologic observations on human subjects were made in 1804 by the French scientists Gay-Lussac and Biot. In their first flight they reached an altitude of 3,960 m and reported an increase in pulse rate but no other physiologic changes. The second flight was made by Gay-Lussac alone; he reached an altitude of 7,020 m and reported difficulty in breathing, accelerated pulse and respiration rates, and difficulty in swallowing that he attributed to dryness of the mouth and throat resulting from breathing the "dry attenuated air."

Scientific interest in balloon flights diminished after this, until the famous ascent of Glaisher and Coxwell in 1862. Hitchcock quotes extensively from their own account of the flight, in which they claimed to have reached an altitude of 29,000 ft. Both men were incapacitated and Glaisher lost consciousness for about 7 minutes. They observed increases in pulse and respiration rates and cyanosis of the hands and lips. Glaisher made repeated flights after this and believed that he became somewhat adapted to high altitude, since he could then breathe without discomfort at an altitude of 4 miles (about 21,000 ft). He stated that he reached his limit of ability to breathe at an elevation of 6 or 7 miles (32,000 to 37,000 ft). It is difficult to accept these altitude measurements as accurate in the light of modern research on the acute effects of hypoxia, even in altitude-adapted subjects. Another famous balloon flight was made in 1875 by three scientists (Croce-Spinelli, Sivel, and Tissandier); two of the three died after reaching an altitude of 8,000 m, but they were presumably not altitude-adapted.

The founder of the science of high-altitude physiology was Paul Bert, a student of Claude Bernard, whom he succeeded at the Sorbonne. Bert built the first low-pressure chamber for purely scientific studies on human subjects. His extensive experiments are summarized in his classic book, *La Pression Barometrique*.[2] Bert established that the harmful effects of decompression could be prevented by breathing oxygen. He also defined natural altitude adaptation in the following statement:

> The organisms at present existing in a natural state on the surface of the earth are acclimated to the degree of oxygen tension in which they live; any decrease, any increase, seems to be harmful to them when they are in a state of health.

Bert also observed that repeated exposure to simulated altitude is associated with an increased blood erythrocyte count.

The next important phase of high-altitude research was a series of expeditions to high mountains, including the Alps, the Andes, and the American Rocky Mountains. Viault, in 1890, "made the first careful counts of red cells during his trip through the Andes of Peru and Bolivia, and found that his red cell count and that of his companion had risen from 5 million per mm^3 at sea level to 7.5 to 8 million per mm^3 at Morococha (4,540 meters) and that this increase had occurred during a short time."[3]

The first organized mountain expeditions for the purpose of biologic studies of the effects of altitude were made in the Alps, beginning in 1894. The first Pike's Peak expedition was made in 1907, and then for some years attention turned to the Peruvian Andes, where a railroad was available from Lima (sea level) up to altitudes of 4,200 m (Cerro de Pasco) and 4,540 m (Moro-

*The following discussion of balloon flights is based largely on a review by F. A. Hitchcock.[1]

cocha). In recent years a number of scientific expeditions to the Himalayas have permitted physiologic observations at altitudes higher than those attainable in other regions, culminating in the conquest of Mt. Everest (29,000 ft) in 1953.

These early expeditions to the mountains were hampered by logistic problems, including the difficulty in providing adequate laboratory facilities at the high altitudes. This led to the establishment of permanent laboratories at high altitudes, many of them open during the entire year, in which the most sophisticated studies can be carried out. Another advance in the technology of high-altitude research has been the construction of improved altitude chambers in which human and animal subjects can live for weeks or months at reduced barometric pressure (produced by partial evacuation of the chamber with a pump). Most of the climatic features of high altitudes (oxygen pressure, temperature, humidity, wind) can be simulated in these chambers, and they have provided a large amount of useful information.

METHODS FOR STUDYING THE EFFECTS OF HIGH ALTITUDE

Biologically, the most important feature of high altitude is the reduced barometric pressure; this is important not because of the mechanical effects on the body of a reduction of the pressure acting on it but because of reduced oxygen pressure of the air breathed. It is natural, then, that the central theme of altitude physiology is hypoxia, or a deficiency of oxygen in the body.

Two technics are commonly used in altitude studies: field or laboratory studies in a natural high-altitude environment and low-pressure (decompression) chambers. Natural altitude is preferred when the objective is information about the biologic effects of the *total* altitude environment, including diurnal and seasonal variations in temperature, humidity, and radiation. However, when the primary purpose of a study is to learn about the effects of hypoxia alone, or about hypoxia as influenced by controlled values of the other climatic variables, the low-pressure chamber has many advantages. A third technic, sometimes used in studies on laboratory animals, is exposure to gas mixtures low in oxygen concentration but at sea-level barometric pressure. In general, this method is less convenient than the low-pressure chamber.

A description of the more important permanent high-altitude laboratories may be found in the review by Luft.[4] Many of these stations have facilities for guest investigators in addition to their permanent staffs. The limited number of such laboratories and their relatively great distance from the majority of scientists who may wish to conduct high-altitude research have led to increasing reliance on low-pressure chambers.

Soon after the invention of the vacuum pump, Boyle (1670) placed animals in a transparent jar to observe their behavior at reduced pressures, but, as mentioned earlier, the pioneer studies on human subjects in low-pressure chambers were performed by Paul Bert. Problems of military aviation in World Wars I and II provided a great stimulus for chamber studies, and this interest has continued through the postwar periods. In its simplest form a low-pressure chamber is a single compartment large enough to hold one or more subjects (men or laboratory animals) and strong enough to withstand a pressure difference across its walls of several atmospheres (to allow a margin of safety). A pump with a large free air flow capacity evacuates the chamber to the desired pressure, which is then maintained by adjusting the inflow valve so that the rates of inflow and outflow of air are the same. Since the chamber may, at least in animal experiments, be left unattended for hours at a time, it is essential that it have safety devices that protect the subjects against power failure, occlusion of the air intake valve, and so on. Chambers containing human subjects are never left unattended and are designed to be opened quickly in case of emergency. They contain oxygen equipment, which is controlled from the chamber operating station, and pressure suits for use at very high simulated altitudes. Human chambers also usually contain an air lock so that observers or subjects may enter or leave the main chamber without disturbing its pressure. The free air flow through the chamber should be adequate for ventilation, including the removal of carbon dioxide and excess moisture. Many types of low-pressure chambers for small laboratory animals have been described, including glass or plastic vacuum desiccators, discarded pressure sterilizers, and plastic boxes. Any laboratory can easily and cheaply assemble the necessary equipment for simulated altitude studies on small animals.

BIOLOGIC EFFECTS OF HIGH ALTITUDE

Effects of acute altitude exposure

General systemic effects

The unique physiologic result of exposure to high altitude is hypoxia. Hypoxia is a state of the organism in which the rate of oxygen utilization by the cells is inadequate to supply all their energy requirements. This may result from an insufficient delivery of oxygen to the cells or from an inability of the cells to utilize oxygen at a normal rate. Textbooks of physiology present classifications of varying degrees of complexity of the "types" of hypoxia, based on the oxygen tension and content of the arterial blood and the adequacy of the tissue blood flow. These schemes actually classify the causes of hypoxia rather than hypoxia itself.

Hypoxia results in an energy deficit that may be compensated, within limits, by an increased energy production from anaerobic metabolism. This is much less efficient than the aerobic type that predominates when oxygen is abundant; it not only yields far less energy per gram of foodstuff utilized but has the further disadvantages of being dependent on the availability of carbohydrate (glycogen or glucose) and producing a fixed acid (lactic acid) that may disturb the acid-base balance.

The distinction between acute and chronic hypoxia is not a sharp one. We think of acute hypoxia as the immediate effect of a reduction of oxygen availability lasting a few minutes, hours, or perhaps several days. Chronic hypoxia is a continuation of the condition for longer periods, sometimes for years or life, but the transition from the acute to the chronic state is difficult to define and is probably arbitrary. The significance of the distinction lies partly in the differences in the organism's compensatory adjustments to acute and chronic hypoxia.

There are certain almost immediate responses to oxygen deficiency. Minute volume of breathing increases because of stimulation of the carotid and aortic body chemoreceptors by decreased arterial P_{O_2}. The maximal increase in breathing is never very great because (1) arterial hypoxemia has a direct depressant effect on the medullary respiratory center, and (2) hypoxia-induced hyperventilation results in decreased arterial P_{CO_2} (hypocapnia) which reduces the effectiveness of the normal stimulation of the medullary respiratory center (actually the lateral medullary chemoreceptors) by arterial carbon dioxide.

The cardiovascular system is also affected by acute oxygen deficiency. Heart rate increases at altitudes above 10,000 ft, but the mechanism of this increase is not clear. Some attribute it to stimulation of the carotid and aortic chemoreceptors, whereas others believe that arterial hypocapnia and vagal afferent impulses from pulmonary stretch receptors are more important. Because cardiac stroke volume is unchanged or slightly increased, the increase of cardiac output is roughly proportional to the increase of heart rate. Increased heart rate produces a moderate elevation of both systolic and diastolic blood pressures. If hypoxia becomes severe (equivalent to an altitude of 20,000 ft or higher) a circulatory crisis may occur, characterized by decreased systolic and diastolic blood pressures and by decreased heart rate. If cardiac insufficiency progresses, cardiac output becomes inadequate to supply the tissues with oxygen and death follows.

Moderate hypoxia not only increases cardiac output but dilates cardiac and cerebral blood vessels, selectively redistributing blood flow in their favor. At the same time, blood flow to less vital regions is reduced by vasoconstriction of central origin. However, there is a complicating factor: reduction of arterial P_{CO_2} that results from increased pulmonary ventilation causes cerebral vasoconstriction, which in itself reduces cerebral blood flow. If this effect predominates over hypoxic dilation of cerebral vessels, the net result is a reduction of cerebral blood flow. Myocardial blood flow usually increases in hypoxia since the coronary vessels are more responsive to decreased arterial P_{O_2} than are cerebral vessels.

One of the most effective adjustments to acute hypoxia is the result of the unique properties of hemoglobin, as reflected in the oxyhemoglobin dissociation curve. With a decrease of arterial P_{O_2}, oxygen unloading in the tissue capillaries is now represented by the steeper portion of the curve. This means that a smaller decrease of blood P_{O_2} is required for unloading a given volume of oxygen, and the oxygen pressure head responsible for driving oxygen from capillary blood to tissue cells is maintained at a higher level. By the same token, oxygen loading in the lung capillaries is less complete, but this effect is minimal, as shown by the flat upper portion of the dissociation curve.

The principal adjustments to acute hypoxia may be summarized as:

1. Increased pulmonary ventilation, which increases alveolar P_{O_2} and improves the oxygenation of blood flowing through the pulmonary capillaries
2. Increased cardiac output, caused mainly by increased heart rate
3. Selective redistribution of blood flow favoring the heart and brain
4. Increased ease of oxygen unloading in the tissue capillaries from operation in the steep portion of the oxyhemoglobin dissociation curve

The net result of these adjustments is delivery

of a larger amount of oxygen to the tissues than would otherwise occur. The primary responses to a reduction of inspired oxygen tension result directly from decreased arterial blood P_{O_2}. There are also second- and third-order responses. For example, increased pulmonary ventilation, by increasing the elimination of carbon dioxide, lowers arterial P_{CO_2} and produces respiratory alkalosis, which has secondary effects on cell metabolism. The acute mountain sickness experienced by most persons shortly after their arrival at high altitude is considered by some to be caused by this transient disturbance in acid-base balance. The kidneys respond to the disturbance by increasing the rate of bicarbonate ion excretion in the urine; within a few days the normal ratio of carbonic acid to bicarbonate has been restored and the blood pH returns to normal. The subject is no longer in a state of alkalosis, but his supply of blood buffer base (bicarbonate) has been reduced, and he is less able to neutralize large amounts of fixed acid (for example, the lactic acid formed during strenuous muscular exercise).

The effects of acute hypoxia on the organism are many and varied. The functions of the different organs are impaired by the intracellular oxygen deficiency, which results in an energy deficit and an increase in anaerobic metabolism. All tissues and organs are affected, but the most serious functional impairments are those involving the nervous system.

The effects of hypoxia on the nervous system are conveniently described in terms of the sequence of alterations observed during progressive reduction of the oxygen pressure of the inspired air. These effects may be manifested as impaired psychologic (intellectual and behavioral), sensory, and motor functions of the brain; much severer hypoxia is required to impair the functions of the spinal cord and peripheral nerves.

Psychologic impairment

Barcroft[5] suggested that "acute hypoxia resembles alcoholic intoxication: the symptoms are headache, mental confusion, drowsiness, muscular weakness, and incoordination." There is often an "initial state of euphoria, accompanied by a feeling of self-satisfaction and a sense of power. The subject may become hilarious and sing or shout and manifest other emotional disturbances." This is probably a release phenomenon, that is, the normal inhibition of lower brain centers by the cerebral cortex is diminished by the hypoxic depression of the cortex. This false sense of well-being is one of the most insidious effects of hypoxia, since it may prevent the subject from making the appropriate behavioral responses to save his life. Loss of critical judgment and initiative is perhaps the most characteristic cerebral manifestation of hypoxia. Familiar operations (for example, simple mathematical calculations and manipulation of scientific instruments) may be performed with a high degree of efficiency, but performance of unfamiliar tasks or difficult calculations is often considerably impaired. In general, the ability to learn a new procedure is affected more severely than is the performance of a previously learned procedure. This may represent defective consolidation of short-term memory into a permanent memory trace. It is interesting to speculate that selective impairment of the memory of recent events in older persons who have reduced cerebral blood flow may have a similar basis.

Effects of hypoxia on sensory function

Of the various sensory modalities, vision is the most sensitive to hypoxia. A diminution in sensitivity to light was reported by McFarland and Evans[6] at a simulated altitude of only 7,400 ft and at an altitude of 15,000 ft an intensity 2.5 times greater than normal was required for the light to be seen. Light sensitivity appears to be one of the first physiologic functions to show a decrement at altitude. Visual acuity is also reduced by hypoxia, especially at low light intensities; the effect of hypoxia is slight at high light intensities. Other visual disturbances produced by hypoxia involve judgment of distance, extent of peripheral visual fields, accommodation, and convergence. There is some uncertainty about the effect of hypoxia on color vision. It is probable that hypoxia does not produce color blindness, but it may intensify any color blindness that is present at sea level. The site of action of hypoxia on the visual system might be the photochemical processes in the retina, the retinal synapses, or the central projections of the visual pathway in the brain. Evidence suggests that the retinal synapses are the most vulnerable to hypoxia and the photochemical reactions the least sensitive.

In progressive hypoxia the sense of hearing is the last to disappear. There is a moderate decrease in auditory acuity at altitudes of about 14,000 to 15,000 ft. The site of action of hypoxia on hearing is not certain; experiments on laboratory animals indicate that cochlear potentials are reduced only with very severe grades of hypoxia (equivalent to altitudes of 20,000 to 40,000 ft). The relative resistance of the sense of hearing and the sensitivity of vision to hypoxic impairment should be borne in mind in the design of experiments on the behavioral effects of hypoxia.

Vestibular sensibility seems to be little affected by tolerable severities of hypoxia, but more studies are needed to determine whether activities requiring highly accurate equilibrium function are, in fact, unaffected. The little information there is on other sensory modalities (smell, taste, touch, pain) indicates that they are not significantly affected by moderate hypoxia.

Effects of hypoxia on motor function

Tissandier in describing the 1875 balloon flight reported that all three subjects experienced great muscular weakness at about 24,000 ft; this may have been the early stage of ascending paralysis. Muscular weakness and incoordination have been mentioned as features of acute hypoxia by numerous workers since that time; the muscular weakness may persist for some time following an acute episode of hypoxia. The great feats of strength and endurance attributed to Peruvian Indians working at altitudes above 14,000 ft indicate that muscular weakness may disappear with adaptation. It is not clear whether the muscular weakness associated with acute hypoxia represents a diminished functional capacity of the muscle itself or a lack of excitation of the muscle (caused by depression of the central nervous system or failure in the transmission of impulses at the neuromuscular junction).

Neuromuscular control is definitely impaired in acute hypoxia. There may be mild tremors and increased reflex irritability at elevations above 14,000 ft, but this may be the result of increased neuromuscular excitability associated with hyperventilation-induced alkalosis. The severity of neuromuscular impairment increases with increasing altitude, and above 20,000 ft handwriting may become unintelligible. Small laboratory animals often exhibit tonic-clonic convulsions at simulated altitudes of 40,000 ft. The cause of hypoxic convulsions is not clear; perhaps the hypoxic brain energy deficit leads to depolarization of neuronal membranes and massive neuronal discharge, comparable to the convulsions induced by electroconvulsive shock or drugs such as pentylenetetrazol (Metrazol).

Severe hypoxia is associated with an ascending paralysis, beginning with the legs. This was dramatically illustrated by Coxwell in his famous balloon ascent. He became completely paralyzed except for his head and neck and was able to save his life only by seizing the rope to the escape valve with his teeth, thus allowing the balloon to descend.

Effects of hypoxia on the autonomic nervous system

Cannon suggested many years ago that the sympatheticoadrenal system plays an important role in the adaptation of an organism to hypoxia. He believed that hypoxia induces increased secretion of epinephrine and that when this is prevented by removal of the adrenal glands the animal's resistance to hypoxia is greatly reduced. Woods and Richardson[7] observed a marked increase in heart rate and blood pressure in vagotomized dogs breathing 100% nitrogen. This response was abolished by bilateral adrenalectomy and preganglionic sympathetic blockade, indicating that the response to acute hypoxia results mainly from sympathetic discharge. Several other workers have confirmed the earlier collapse of sympathectomized animals made acutely hypoxic. Results on the effect of hypoxia on the parasympathetic nervous system are conflicting; both stimulation and inhibition have been reported. Monge, quoted by Van Liere and Stickney,[8] reported increased vagal tone in high-altitude residents. Keys and associates[9] believed that hypoxia disturbs the balance between the sympathetic and parasympathetic divisions of the autonomic nervous system, with an initial tendency for sympathetic dominance that gives way to a marked parasympathetic dominance.

It should be pointed out that many of the effects of hypoxia on the gastrointestinal tract, the kidneys, and perhaps other organs can be explained at least in part by the altered function of the nerve supply to these organs. The effects of acute hypoxia on various organs and systems have been reviewed by Van Liere and Stickney.[8]

Pathologic responses to acute altitude exposure

We have seen that acute exposure of an organism to high altitude evokes a variety of physiologic responses. Some of these appear to be beneficial in reducing the severity of oxygen lack in the cells. Others are less obviously beneficial, perhaps reflecting overreaction to hypoxia or the secondary results of primary responses that are themselves beneficial. Finally, acute altitude exposure may result in biologic alterations that range from unpleasant to life-threatening (sometimes actually fatal). The most important of these conditions are acute mountain sickness and high-altitude pulmonary edema.

Acute mountain sickness

Acute mountain sickness occurs during the first few days of exposure of normal human subjects to high terrestrial altitudes. One of the most characteristic features of this condition is its great variability with respect to the lowest altitude at which it appears, the time lag of development, the duration of the illness, and the relative prominence of the different symptoms. This variability complicates the explanation of the pathogenesis of acute mountain sickness.

The symptoms reported most often include headache, anorexia, nausea and vomiting, marked exertional dyspnea, and difficulty in sleeping. There may be increased sensitivity to cold, dizziness, weakness, palpitation, inability to concentrate, impairment of judgment, transient leg, back, and chest pains, and a variety of neurologic disturbances. Nasal symptoms (congestion and rhinorrhea) are often prominent.

The rapidity of onset of symptoms varies with the rate of ascent, the altitude reached, and other less well-defined factors, including individual variability. Onset may be sudden when ascent is very rapid, but more often it develops gradually over a period of hours, with maximal severity reached after 24 to 48 hours. The symptoms then subside even more slowly and usually disappear within 5 or 6 days.

The lowest altitude at which symptoms appear varies greatly in different subjects; it may be as low as 6,000 ft for some, while occasional subjects are unaffected at 15,000 ft. At 20,000 ft all non-adapted subjects are affected. At any altitude the onset of mountain sickness is probably hastened by physical exertion. It has been reported that women and children are less severely affected than are men, but the basis for this is unknown. Consumption of alcohol is said to increase the rapidity of onset and the severity of the symptoms.

The pathogenesis of acute mountain sickness is obscure. The factors most often implicated are hypoxia caused directly by high altitude and hypocapnic alkalosis resulting from hypoxia-induced hyperventilation. Metabolic acidosis secondary to increased anaerobic metabolism, cold, fatigue, and potassium deficiency resulting from renal compensation for respiratory alkalosis are thought by some to contribute to the conditions.

Most evidence indicates that hypoxia is probably the primary cause of acute mountain sickness. Many of the symptoms can be produced experimentally by inducing hypoxia and relieved by the administration of oxygen. In experiments of this type it is important, of course, to prevent alterations in the arterial P_{CO_2} while lowering or raising the arterial P_{O_2} if hypocapnia and respiratory alkalosis are to be eliminated as causal factors. If hypoxia is accepted as the basic cause of acute mountain sickness, several important questions remain:

1. How does hypoxia produce the symptoms?
2. Why is the condition so variable in different individuals?
3. Why don't the symptoms appear immediately?
4. What causes the symptoms to subside after a few days at high altitude?

It may be helpful to sort the symptoms of acute mountain sickness into categories corresponding to organ systems or biologic functions affected. When this is done, it is apparent that, aside from dyspnea on exertion, the important symptoms are referable, directly or indirectly, to the nervous system. This is consistent with the well-known sensitivity of the brain to oxygen lack, but it does not explain how the specific symptoms occur or the time course of their appearance and subsidence. Evidence is accumulating that gives some insight into possible mechanisms of pathogenesis, but it must be admitted that many of the conclusions are based on argument from analogy.

One of the most striking changes that has been observed in persons going to high altitude is a shift of water from the extracellular compartment into the cells, together with a slight increase of total body water.[10] The result is an intracellular edema that, in the case of the brain, may be responsible for some of the neurologic symptoms of acute mountain sickness. It should be noted, however, that the increase in intracellular water outlasts the symptoms of mountain sickness by some days. Acute exposure to high altitude is also associated with a redistribution of blood within the vascular system, resulting in a consistent increase of pulmonary blood volume. While this may contribute to the respiratory symptoms of acute mountain sickness, it is probably more important in the pathogenesis of high-altitude pulmonary edema (described later).

The headache that is such a prominent feature of acute mountain sickness is very similar to migraine and may also be of vascular origin. Both types of headache are alleviated by ergotamine preparations. It is significant that cerebral vasodilation and increased cerebral blood flow are usually observed in experimentally induced hypoxia.

The maximal severity of the symptoms in acute mountain sickness coincides with the maximal *rate* of shifts in blood and fluid distribution rather than with the maximal *degree* of the shifts. This suggests that compensatory adjustments may occur to the redistributions of fluid and blood, but the nature of these compensations, if they occur, is unknown.

While major emphasis has been placed on hypoxia as the basic cause of acute mountain sickness, hypocapnia and respiratory alkalosis resulting from hyperventilation may also be involved. The evidence on this point is not conclusive, partly because experiments are seldom designed to separate clearly the effects of hypoxia and hyperventilation. Wayne[11] compared the frequency of symptoms in a large group of normal subjects when they were exposed to a simulated altitude of 25,000 ft and when they voluntarily hyperventilated at sea level. In each case the experiment continued until the subjects were unable to write their names legibly. Following the test each subject filled out a checklist of symptoms. The frequency of the various symptoms corresponded fairly well between the two groups, and none of the symptoms was found in only one of the groups. The most prominent symptoms (dizziness, tingling, numbness, lightheadedness, visual disturbances, inability to think clearly, muscular incoordination, and the like) occurred with much greater frequency in these experiments than in cases of acute mountain sickness. On the other hand, headache rarely occurred in the acute experiments and no mention was made of nausea or other gastrointestinal symptoms. Unfortunately, experiments of this type may have little relevance to the pathogenesis of acute mountain sickness both because of their failure to distinguish between the contributions of hypoxia and hypocapnia and because of the great differences in the time course of symptom development in the two conditions. Critical experiments would simulate the time course of symptom development in acute mountain sickness and would compare (1) hyperventilation with air and with low oxygen mixtures and (2) hyperventilation with low oxygen mixtures with and without prevention of hypocapnia by the addition of carbon dioxide to the inspired gas.

Prevention and treatment of acute mountain sickness are necessarily empiric since the cause of the symptoms is poorly understood. General principles include gradual ascent to high altitude, avoidance of exertion and alcohol, restraint in eating, and, perhaps, a high-carbohydrate diet. The use of drugs, hormones, and other chemical agents has been reported to reduce the incidence and severity of symptoms. In general, the suggested treatments have been designed to correct a condition produced by high altitude or to hasten the physiologic adjustment to the altitude. One of the earliest attempts was the administration of ammonium chloride, an acidifying agent, with the objective of hastening the correction of the respiratory alkalosis; the results were not promising. More recently, carbonic anhydrase inhibitors such as acetazolamide have been used; inhibition of renal carbonic anhydrase increases renal excretion of bicarbonate ions and thus hastens correction of the respiratory alkalosis. Several such studies have indicated a reduction of both severity and duration of symptoms.[12, 13] The mechanism of this beneficial effect of the drug is not entirely certain; not only does it hasten renal compensation of respiratory alkalosis, but it also elevates arterial P_{O_2}, presumably by stimulating breathing since arterial P_{CO_2} is reduced.

To summarize, acute mountain sickness is a transient condition that affects all newcomers to high altitude. It begins after a latent period of some hours, reaches peak severity after 24 to 48 hours, and usually subsides by the fifth or sixth day. The most prominent symptoms (headache, anorexia, nausea, dizziness, weakness, impairment of judgment, and sleep disturbances) are indicative of disturbed cerebral function. Aside from oxygen administration, the only treatment that has been shown to be beneficial is administration of a carbonic anhydrase inhibitor beginning several days before exposure to high altitude.

High-altitude pulmonary edema

After the turn of the century reports of a condition resembling pneumonia occurring shortly after acute exposure to high altitude appeared occasionally. Hurtado,[14] suggested in 1937 that the clinical picture is that of acute pulmonary edema. He described the case of a high-altitude resident who, after a visit to sea level, returned to the mountains and developed acute pulmonary edema that subsided after he descended to sea level. This sequence of events has been observed in many cases and must be considered in any analysis of the pathogenesis of this condition. Many cases.

however, have been reported in sea-level residents going to high altitude for the first time. This was a serious medical problem for the Indian Army when sea-level soldiers were rushed to the mountainous frontier between India and China during the border skirmishes in 1965.

In a typical case there is a lag of 12 to 36 hours between high-altitude exposure and the onset of symptoms. Cough and dyspnea are usually observed; headache, hemoptysis, and vomiting are less frequent. Chest films show bilateral patchy areas of density and enlarged central pulmonary blood vessels. Improvement is usually rapid with bed rest and oxygen administration, but some deaths have been reported, caused mainly by delayed or inadequate treatment. Corticosteroids have recently been used with good results. It is sometimes necessary to transport patients to lower altitude; this should always be done when adequate medical facilities are not available at the high altitude. The most important preventive measure appears to be avoidance of physical exertion for a few days after going to high altitude for the first time or after returning to high altitude following a sojourn at sea level.

The pathogenesis of high-altitude pulmonary edema is uncertain, but recent clinical and experimental studies have explained some of the features. The factors involved in common instances of pulmonary edema are well known; they involve a disturbance in the Starling forces that normally prevent the net transudation of fluid from pulmonary capillaries (that is, increased pulmonary capillary pressure or increased pulmonary capillary permeability permitting leakage of plasma proteins).

The difficulty in explaining high-altitude pulmonary edema has been the absence of obvious reasons for disturbance in either of the Starling forces. Visscher[15] showed that pulmonary capillary permeability does not increase in experimental animals breathing 5% oxygen in nitrogen. This suggests that increased capillary pressure is the likely cause of pulmonary edema and, indeed, an increase of pulmonary arterial pressure has long been known to occur at high altitude. However, since pulmonary venous pressure is usually normal, elevated pulmonary arterial pressure has been regarded as resulting from precapillary vasoconstriction, which would not raise capillary filtration pressure. In autopsy reports on patients dying from acute high-altitude pulmonary edema, one of the outstanding features is the extensive plugging of alveolar capillaries with sludged red blood cells; these plugs are also seen in some of the thin-walled veins. It has been suggested[16] that "at high altitude pulmonary vascular obstruction occurs unevenly throughout the lung, resulting in areas of the vascular bed where essentially no blood flow occurs. Other areas may not be obstructed and in these areas high flow may be present and the pulmonary capillaries in these areas will 'see' the high pulmonary artery pressure." This would explain the patchy nature of the edema observed in chest roentgenograms of patients with high-altitude pulmonary edema. In animal experiments[16] occlusion of 75% to 83% of the total pulmonary vascular bed while maintaining normal cardiac output is associated with increased pressure and flow in the unoccluded portion of lung. This occlusion produced pulmonary edema in the presence of a normal left atrial pressure and a normal pulmonary artery wedge pressure.

The sequence of events may be as follows: Hypoxia is associated with increased pulmonary blood volume and pulmonary vasoconstriction. If vasoconstriction is more severe in some areas of the lung, because of less favorable ventilation-perfusion relations, there might be reduction of blood flow and sludging of platelets and erythrocytes in these areas. The result of vasoconstriction and sludging would be to divert blood flow to non-occluded areas. The combined result of restricted vascular capacity and increased pulmonary blood volume would be an increased filtration pressure in open capillaries and a patchy distribution of edema.

Adaptation to high altitude

There is much confusion in the use of terms describing changes in organisms resulting from environmental stressors. With respect to high altitude it has been suggested that *acclimatization* be used to describe the physiologic responses to the totality of the high-altitude environment, including decreased barometric pressure, cold, dryness, radiation, and the like, and that *acclimation* be used to describe adjustments to any single feature of the environment (for example, decreased oxygen pressure). *Adaptation* is used by some to describe the genetic changes occurring in successive generations of organisms exposed to a different environment, beneficial changes being perpetuated by natural selection. For example, it has been claimed that the superior altitude tolerance of the Indians living in the Andes is the result of selection over many generations and that it is unattainable during the lifetime of newcomers to high altitude.

This terminology seems unnecessarily complicated. I suggest that *adaptation* be used to describe all the changes occurring in response to alteration of the environment that enable organisms to function more efficiently in the new environment. The exact implications of the term "adaptation" can be made clear by suitable modifying terms: immediate adaptation, delayed adaptation, genetic adaptation, adaptation by natural selection, adaptation to a total environment (altitude, desert, tropics), adaptation to a particular feature of an environment (barometric pressure, temperature, humidity). The proper use of such precise terminology would remove much of the ambiguity now encountered in the literature.

There is often a tacit assumption that all the changes that occur at high altitude are beneficial, that is, that they increase altitude tolerance. While there is no doubt that altitude tolerance is increased by exposure to high altitude, the relative contributions of individual changes can be evaluated only by testing them both separately and in relation to each other. For example, hyperventilation increases alveolar and arterial oxygen pressures and arterial blood oxygen content. It also results in hypocapnic alkalosis, which decreases oxygen unloading in systemic capillaries and reduces cerebral blood flow. Only by direct experimentation can it be determined whether the benefits of hyperventilation outweigh the disadvantages. Another example is the polycythemia that develops on exposure to high altitude. Blood oxygen content is thereby increased, but so also is blood viscosity, and the latter change, by increasing resistance to blood flow, might actually reduce oxygen delivery to the tissues. These and similar problems will be analyzed in greater depth in relation to the individual components of the adaptive process.

The immediate responses to acute hypoxia described in an earlier section include increased pulmonary ventilation and cardiac output and greater ease of oxygen unloading in systemic capillaries as shown by a shift of the unloading region to the steep portion of the oxyhemoglobin dissociation curve. The direct effect of each of these responses is to elevate mean capillary P_{O_2} and thus to increase the rate of oxygen diffusion from blood to tissue cells. That these adjustments are not completely adequate is shown by the fact that newcomers to high altitude suffer persistently reduced exercise capacity, disturbed sleep, and other indications of hypoxia.

Following the initial period of high-altitude exposure, which is usually characterized by the onset and subsidence of the symptoms of acute mountain sickness, more gradual changes occur, some of which require many weeks or months for completion. Cardiac output, which increases at first, usually returns to normal, while the minute volume of breathing remains elevated and may actually show further increase. There is prompt stimulation of erythropoiesis (indicated by an increased blood level of erythropoietin derived from the hypoxic kidneys), but the maximal increase of the total number of circulating erythrocytes is not reached for a month or longer, and a further increase may continue very slowly for many months. Another very important gradual change is increased capillary density of various organs, especially skeletal muscle, heart muscle, and the brain. This is probably beneficial in two ways: first, by decreasing the average distance over which oxygen must diffuse from capillary blood to cells, and second, by providing additional channels for flow of the more viscous blood.

Much has been written about the role of cell changes in adaptation to high altitude, but there is no general agreement concerning the nature of the changes or their relative contribution to the total adaptive process. There is an element of wishful thinking in many of the discussions, implying that there must be cell changes if only we had the wit to discover them. This is acceptable as a stimulus for further study but obviously not as a factual contribution to the description of altitude adaptation.

Most of our information about adaptation to high altitude has been derived from longitudinal studies of changes that occur in sea-level natives (men and laboratory animals) transported to high altitude or made chronically hypoxic in a low-pressure chamber. The general similarity of results obtained with these two procedures indicates that low oxygen pressure is the most important physiologic feature of the high-altitude climate and supports the validity of chamber studies. Additional information of great importance has been gained from studies of men and animals native to high altitudes; these studies suggest that the high-altitude adaptation of newcomers may be incomplete even after a year and that some changes observed in newcomers may actually represent overreactions to a hostile environment rather than adaptation.

In the discussion that follows, altitude-adapted newcomers and high-altitude natives will be compared and the relative contributions of various systemic and cell changes to the adaptive process will be evaluated. The limitations of altitude adaptation and efficacy of various treatments designed to hasten the course or extend the range of altitude adaptation will be considered.

When sea-level natives are first exposed to real or simulated high altitude, there is an immediate drop of alveolar and arterial oxygen tensions. This results in increased pulmonary ventilation, cardiac output, and increased ease of oxygen unloading in systemic capillaries. Increased pulmonary ventilation lowers arterial P_{CO_2}, decreasing the ease of oxygen unloading in systemic capillaries (Bohr effect) and reducing cerebral blood flow (caused by cerebral vasoconstriction). The net result of these changes is believed to be increased oxygen transport to the tissues, despite the adverse influence of reduced arterial P_{CO_2}. Respiratory alkalosis, resulting from hyperventilation, has complex effects on cell metabolism including an increased rate of aerobic glycolysis and an elevated blood lactic acid concentration that may be mistakenly interpreted as being the result of hypoxia.

It was pointed out earlier that there is no sharp dividing line between these immediate responses to acute hypoxia and the more slowly developing changes that characterize chronic hypoxia. Many of the delayed adjustments begin in the first few hours at high altitude even though the results are first manifested days or weeks later. For this reason restriction of the term "adaptation" to delayed responses only is considered to be arbitrary.

Within a few hours after arrival at altitudes of 14,000 ft or higher, nearly all nonadapted subjects begin to experience the discomforts of acute mountain sickness. The symptoms reach peak severity in 24 to 48 hours and then subside gradually over the next few days. The condition is described in detail elsewhere in this chapter, along with speculations concerning its pathogenesis. One of the most promising of these is that the condition represents a mild form of intracellular edema resulting from the shift of fluid into cells, including those of the brain. The cause of the fluid shift is not clear, but the brief duration of acute mountain sickness suggests that it is associated with one of the transient changes observed during the early phase of acute altitude exposure, for example, the uncompensated respiratory alkalosis.

The various systemic and cell responses, which together constitute the state of altitude adaptation, develop at different rates; some increase progressively for many months, others reach an early peak and then subside, while still others may be present in some subjects and absent in others. In the following discussion individual responses will be analyzed with respect to rate of development, mechanism (if known), and contribution to the state of altitude adaptation.

Systemic adjustments to high altitude

Increased pulmonary ventilation. Increased pulmonary ventilation is the most obvious immediate systemic response. Decreased arterial blood P_{O_2} stimulates chemoreceptors in the carotid and aortic bodies. These receptors send a few impulses to the respiratory center even when arterial P_{O_2} is normal; tension must fall to about half the normal value before impulses from these chemoreceptors increase the rate and depth of breathing appreciably. Even when arterial P_{O_2} falls to quite low levels, the increase of breathing is not very striking in resting subjects, for reasons already given. The full magnitude of the increase of breathing is not achieved immediately, but only after a few days. The reason for this fact was obscure until the discovery that the chemical regulation of breathing involves a second set of chemoreceptors located in the lateral walls of the medulla, which respond primarily to hydrogen ions in the cerebrospinal fluid (CSF). Since hydrogen ions as such diffuse slowly from blood to CSF, while carbon dioxide diffuses rapidly, CSF pH is determined largely by arterial blood P_{CO_2}. If this rises, carbon dioxide diffuses into the CSF where it hydrates to form carbonic acid, which in turn increases hydrogen ion concentration. Since CSF is a poorly buffered fluid, a small rise of carbonic acid concentration produces a relatively large change of pH and considerable stimulation of the respiratory center. The versatility of this mechanism is further increased by the fact that bicarbonate ions are actively transported from CSF to blood when the P_{CO_2} of CSF decreases, thus restoring CSF pH to its normal value of 7.32. This appears to be important in the breathing adjustments that occur during the initial stage of altitude exposure. The prompt breathing increase mediated by reflexes from the peripheral (carotid and aortic body) chemoreceptors reduces arterial P_{CO_2}. This decreases the P_{CO_2} and central chemoreceptor drive of CSF to the respiratory center. If this situation persisted, the respiratory response to high altitude would be damped. What actually happens is that bicarbonate ion is actively transported from CSF to blood, restoring normal CSF pH and normal carbon dioxide sensitivity of the central chemoreceptors. This accounts for the fact that the hyperventilation that begins immediately on exposure to high altitude not only persists but usually increases somewhat for a few days. Blood pH is restored to normal more slowly by increased renal bicarbonate ion excretion.

One difference between the control of breathing in newcomers to high altitude and in those who are altitude-adapted that remains unexplained is the tendency for the increased breathing to become automatic in adapted subjects. For newcomers, on the other hand, every breath may require conscious effort. If the nonadapted person makes this effort oxygen uptake improves, but it is virtually impossible to maintain this effort indefinitely.

Newcomers who remain at high altitude long enough to become well adapted retain their respiratory sensitivity to hypoxia, that is, the carotid and aortic chemoreceptors continue to send impulses to the respiratory center in response to the decreased arterial blood P_{O_2}. Surprisingly, high-altitude natives appear to have lost most of their respiratory sensitivity to hypoxia—their breathing is stimulated only by a very large decrease of alveolar P_{O_2}. It seems strange that one of the most important compensatory adjustments of the newcomer to high altitude is lacking in those who should be best adapted to altitude. This does not appear to be a genetically determined trait, since Peruvians of high-altitude descent who are born and raised at sea level have a normal respiratory response to hypoxia. The most widely accepted explanation is that the respiratory response to hypoxia is permanently depressed or eliminated by chronic hypoxia dating from birth. This view is supported by the fact that patients who have congenital cyanotic heart disease with hypoxia from birth resemble high-altitude natives in their blunted respiratory sensitivity to hypoxia; furthermore, correction of their heart defects does not restore normal sensitivity.

We must now examine the experimental evidence concerning the importance of increased breathing in adaptation to high altitude. Hyperventilation improves blood oxygenation in the lungs in two ways. First, the concentration of oxygen in the alveolar gas is increased in proportion to the decrease of alveolar carbon dioxide concentration, nitrogen concentration being unaffected. For example, doubling the alveolar ventilation rate decreases alveolar P_{CO_2} from 40 to 20 mm Hg and raises alveolar P_{O_2} by an equal amount. The second effect of hyperventilation on blood oxygenation results from the fact that some alveoli that are poorly ventilated during normal breathing are adequately ventilated during hyperventilation.

A possible adverse effect of hyperventilation is reduction of cerebral blood flow resulting from hypocapnic constriction of cerebral blood vessels. It has long been debated whether hyperventilation, even with oxygen, may produce cerebral hypoxia in this way. Miller, Curtin, Shen, and Suiter[17] investigated this problem by hyperventilating rats with a respirator to arterial P_{CO_2} levels below 10 mm Hg (a very severe grade of hyperventilation) and observing the effects on brain metabolism. When the rats breathed air, there was a slight tendency toward brain hypoxia, but when they breathed low oxygen mixtures, hyperventilation improved brain metabolism. There was also a striking improvement of cardiovascular function (maintenance of arterial blood pressure). If we may extrapolate from rats to men, hyperventilation at altitude should be beneficial despite reduced cerebral blood flow.

Transport of oxygen from atmosphere to tissues involves a succession of steps, each decreasing the oxygen pressure. It is of obvious advantage to the organism to minimize the magnitude of the fall in oxygen pressure at each step at high altitude so that oxygen may be delivered to the tissues at a sufficiently high pressure. As already pointed out, the P_{O_2} drop in two of these steps in oxygen transport is so reduced: the P_{O_2} difference between atmosphere and alveolar gas (caused by elevation of alveolar P_{O_2} by hyperventilation) and between arterial blood and mixed venous blood (resulting from operation in the steep portion of the oxyhemoglobin dissociation curve). We shall see later that the oxygen pressure fall between systemic capillary blood and tissue cells is reduced, at least in some organs, by increase of the numbers of capillaries, which reduces the average distance through which oxygen must diffuse to reach the cells.

While the fall of oxygen pressure between alveolar gas and arterial blood is not great in resting subjects at sea level, it has been suggested[18] that adaptation to altitude may reduce this pressure difference, referred to as the A-a D_{O_2} so that the arterial blood P_{O_2} corresponding to a particular alveolar P_{O_2} would be raised. In a more recent study[19] A-a D_{O_2} was compared in normal sea-level residents and fully altitude-adapted Andean Indians. The A-a D_{O_2} was actually greater in the high-altitude natives than in the sea-level residents; this was attributed to greater inequality of ventilation/perfusion ratios in different regions of the lung and to increased venoarterial shunting. This conclusion is supported by the observation[20] that the oxygen diffusing capacity of the lungs (the volume of oxygen diffusing from alveolar gas to pulmonary capillary blood per minute per millimeter of mercury oxygen pressure difference between alveolar gas and pulmonary capillary blood) does not increase in sea-level natives after 7 to 10 days at high altitude (4,560 m). It might be objected that the altitude was not very great and that the time spent at high altitude may have been too brief to permit adaptive changes to occur.

At any rate, the available evidence suggests that altitude adaptation does not reduce the oxygen pressure difference between alveolar gas and arterial blood. It seems that nature has neglected an opportunity for yet another adaptation to altitude.

In summary, breathing increases immediately on exposure to high altitude because of stimulation of the respiratory center by impulses from the carotid and aortic chemoreceptors. The magnitude of the response is limited by hypoxic depression of the respiratory center itself and by hyperventilation-induced hypocapnia. The initial fall of the P_{CO_2} of CSF caused by arterial hypocapnia reduces central chemoreceptor drive to the respiratory center. Active transport of bicarbonate ions from CSF to blood then restores normal CSF pH and normal central chemoreceptor sensitivity to carbon dioxide, and breathing shows a further increase. Adaptation to high altitude does not appear to facilitate oxygen diffusion from alveolar gas to blood or to diminish the P_{O_2} difference between alveolar gas and blood. The major feature of respiratory adaptation to high altitude is thus the elevation of alveolar P_{O_2} at the expense of the reduced alveolar P_{CO_2}, with resulting increase of arterial blood P_{O_2} and content.

Cardiovascular adjustments at high altitude. There is considerable difference of opinion about cardiac output changes at high altitude. The controversy concerns the stroke volume; it is generally agreed that heart rate increases.

The mechanism responsible for tachycardia at high altitude is not fully understood, and there seem to be important species differences. In man, central nervous system hypoxia increases activity of the sympathetic cardioaccelerator nerves that stimulate beta-adrenergic receptors in the heart; hypoxia-induced tachycardia is reduced or abolished by pronethalol, a beta-adrenergic blocking agent.[21] Neither hyperventilation nor hypocapnia is essential for hypoxia-induced tachycardia. In the anesthetized dog, on the other hand, the response to hypoxia is altered when hyperventilation is prevented by muscular paralysis and pump-ventilation; there is now a slowing of the heart.[22] This confirms an earlier report[23] that hypoxia slows the heart of dogs with denervated lungs. Both groups of workers suggested that hypoxic tachycardia in the dog depends on hyperventilation, which stimulates stretch receptors in the lungs and thus initiates a reflex acceleration of the heart in which the afferent pathway is the vagus nerves and the efferent pathway is the cardioaccelerator nerves. Although demonstrable in the anesthetized dog, this reflex is less important in man, as is also the respiratory reflex arising from human pulmonary stretch receptors.

Tachycardia appears in man immediately on his arrival at high altitude and persists even though the sojourn lasts many months. The heart rate of high-altitude natives is also increased by hypoxia, since the administration of oxygen or descent to sea level results in slowing of the heart rate. Heart rate is faster during exercise at altitude than at sea level for any given work load. In addition, maximal heart rate is reduced in many individuals at altitudes above 10,000 ft. Both factors reduce maximal exercise capacity at high altitude.

There is a continuing controversy concerning the effect of high altitude on cardiac stroke volume. At first glance, it would appear that stroke volume increases according to one school of thought[24, 25] and decreases according to another opinion.[26] The difference between these two views may be more apparent than real and may be largely a matter of timing, including rate of ascent to high altitude. In some studies, it was reported that stroke volume remained unchanged during exercise initiated immediately on exposure to decreased atmospheric pressure; since heart rate increased, cardiac output also increased. After 10 days at high altitude, cardiac output for a given work load was less than at sea level, entirely as a result of reduced stroke volume attributed to hypoxic impairment of myocardial function. According to the authors, reduced cardiac output compared to sea-level values persists for the duration of the sojourn at high altitude and requires about 10 days at sea level to return to normal.

According to the "opposing" view, cardiac output at rest and during exercise increases over sea-level values during the first day or two at high altitude and returns gradually to sea-level values, or slightly below, by the end of a week or two. Since the heart rate increase persists, this means, of course, a stroke volume decrease. Furthermore, the degree of initial cardiac output increase appears to be related to the rate of ascent; the less rapid the ascent, the less pronounced is the initial increase of cardiac output. The proponents of this second view suggest that the decline of cardiac output after the initial rise results not from myocardial depression but from return of the arterial oxygen content to normal.

Despite some unanswered questions, it appears likely that cardiac output during exercise increases slightly above sea-level values during the first week or so at high altitude and that it then returns to values the same as, or slightly less than, those observed at sea level. This means, of course, that a relative increase of cardiac output is not a permanent feature of altitude adaptation, as, for example, is the persisting hyperventilation.

Other features of the cardiovascular response to high altitude should be mentioned. Coronary blood flow increases in acute hypoxia. Little is known about coronary flow in chronic hypoxia, but it has been suggested that it may decrease.[26] However, chronic hypoxia is a potent stimulus to development of coronary interarterial anastomoses in sea-level patients, and increased myocardial capillary density has been reported in chronically hypoxic experimental animals.

Systemic arterial blood pressure of Andean natives is lower than that of sea-level natives, and hypertension is reported to be rare. It is not clear to what extent this is caused by cultural factors.[27] On the other hand, it is generally agreed that pulmonary arterial blood pressure is elevated at high altitude, both in sojourners and in natives, and that this is usually associated with some degree of right ventricular hypertrophy. Since cardiac output is not increased in high-altitude natives and their pulmonary artery wedge pressure (a reflection of pulmonary venous pressure) is normal, the elevated pulmonary arterial pressure must result from increased arteriolar or precapillary resistance. This also explains the absence of pulmonary edema that might result from elevated pulmonary capillary pressure. The increased vascular resistance is partly functional, representing sustained vasoconstriction caused by decreased alveolar P_{O_2}, since pulmonary arterial pressure decreases to normal on return to sea level. There is also an anatomic component of the increased pulmonary arterial pressure consisting of a thickening of the walls of small pulmonary arteries. Blood viscosity does not appear to affect pulmonary arterial pressure to an important degree. Perhaps elevated pulmonary arterial pressure at altitude is beneficial in improving blood flow through lung regions that are normally less adequately perfused (for example, the upper lobes). If this were true, it should increase pulmonary diffusing capacity, which appears not to happen in newcomers to high altitude, although it may be a characteristic of high-altitude natives. At present, there seems to be no teleologic advantage to the development of pulmonary arterial hypertension at high altitude. This problem is analyzed in depth by Hultgren and Grover.[26]

The cardiovascular effects of altitude-induced alterations of blood volume and blood viscosity will be considered in the next section.

Alterations in characteristics of blood at high altitude. The principal function of the blood affected by high altitude is that of oxygen transport to the cells of the body. The significance of the various changes in the blood at high altitude must be judged in terms of their effect on this function, and this is sometimes difficult.

The first, and most important, of the adjustments in the function of the blood at high altitude is inherent in the characteristics of the oxyhemoglobin dissociation curve. With decreased arterial blood P_{O_2}, oxygen unloading in the systemic capillaries occurs on the steep portion of the dissociation curve where large volumes of oxygen are released with a small decrement of arterial blood P_{O_2}. An adequate oxygen diffusion gradient from blood to cells is thus maintained without excessive reduction of cell P_{O_2}. This effect is potentiated by a shift to the right of the oxyhemoglobin dissociation curve. This was long an unexplained phenomenon, since a shift to the left caused by respiratory alkalosis was anticipated. It has been suggested[28] that the decreased affinity of hemoglobin for oxygen at high altitude results from increased concentration of erythrocyte 2,3-diphosphoglycerate (2,3-DPG). It had previously been shown[29] that this compound plays an important role in determining the affinity of hemoglobin for oxygen.

During the first few days spent at high altitude by sea-level natives, blood hemoglobin concentration increases. This results from the movement of water from the blood, with consequent decrease of plasma volume and no change of red cell volume. This initial decrease of plasma volume may be the result of a hypoxia-induced increase in venous vasomotor tone, reducing the capacity of the venous blood reservoir. At any rate, the net result is an early increase of hematocrit (the proportion of total volume occupied by erythrocytes in a blood sample) and a reduction of blood volume. This combination (increased hematocrit and reduced blood volume) has been reported[30] to cause a reduction of cardiac output in anesthetized dogs and could be a factor in the return of cardiac output to normal after the initial rise.

At altitudes high enough to reduce arterial blood oxygen tension below 60 mm Hg, renal erythropoietin production increases, stimulating the bone marrow to produce more erythrocytes. There follows a progressive increase of the total number of circulating erythrocytes and a further hematocrit increase. Plasma volume may remain unchanged or decrease still further, but the net result is usually a modest increase of total blood volume. Polycythemia and increased hematocrit increase blood oxygen content and capacity, and this should increase oxygen delivery to the tissues. However, increased hematocrit increases blood viscosity and hence flow resistance in small vessels. The result might be either reduced flow and oxygen delivery to the tissues or increased cardiac work. Smith and Crowell[31] concluded, from acute experiments on dogs, that oxygen transport (cardiac output times arterial blood oxygen content) is optimal with the normal hematocrit of about 40% and that oxygen flow is reduced when the hematocrit is higher. Whether this is a significant factor at high altitude is not certain. It is possible that the increased vascularity of organs observed at high altitude (see following section) might so increase the capacities of peripheral vascular beds as to compensate for the effect of increased blood viscosity. It is possible that excessive hematocrit increase, such as occurs in chronic mountain sickness, makes blood so viscous that flow through small vessels is impeded and tissues are poorly supplied with oxygen and other essentials. It is of interest that the hematocrit is generally lower in high-altitude natives than in altitude-adapted newcomers and that many high-altitude natives show no increase of hematocrit.

Increased vascularity of organs at high altitude. It has been known for many years that exposure to real or simulated high altitude results in a gradual increase of capillary density in a number of organs, including brain, heart, and skeletal muscles.[8] The beneficial effect of this vascular response is obvious, since it not only provides extra channels for the more viscous blood to flow in but also reduces the distances through which oxygen in the capillaries must diffuse to reach tissue cells. There are some (myself included) who believe that this may well be the most important of all the delayed adaptive changes that occur with exposure to high altitude. In light of this statement, it is surprising that very little effort has been made to identify the mechanism responsible for the growth of new blood vessels or even to establish definitely that the process involves the growth of new vessels rather than the opening of previously closed vessels. In a preliminary study of this problem[32] it was demonstrated that increased vascularity of rat organs is produced not only by chronic hypoxia in an altitude chamber but also by injections of cobalt chloride and by repeated transfusions of erythrocyte suspensions. This suggests that hemodynamic factors (increased blood volume or blood viscosity) may be as important as tissue hypoxia in stimulating blood vessel proliferation at high altitude. The possible role of increased tissue acidity is also being investigated, since it has been reported[33] that the growth of blood vessels in the kitten retina is stimulated by local injection of lactic (but not acetic) acid.

Cellular adaptive changes at high altitude

There are strong differences of opinion concerning the importance of cellular adaptive changes at high altitude. Some of the uncertainty results from the difficulty of extrapolating from in vitro data to the intact organism and some from lack of uniformity in the methods used to detect in vitro changes in tissues removed from normal and from altitude-exposed animals.

The argument for cellular adaptive changes is occasionally based on the conviction that the systemic adjustments to high altitude are inadequate to explain the increased tolerance observed. In other cases, the evidence consists of demonstrations of increased activities of various enzymes, which may or may not be rate-limiting, in hypoxic cells. Finally, in vitro changes that are demonstrable only under such severe conditions of oxygen deficiency as to be incompatible with life are sometimes assumed to represent useful adaptation to high altitude.

One of the most vigorous advocates of the importance of cellular altitude adaptation is Barbashova.[34] According to her, adaptation to hypoxia is not explained completely by respiratory adjustments, cardiac output, and blood oxygen capacity. Furthermore, she states that these adjustments do not always occur and that when they do they may represent overreactions to hypoxia; thus they may not necessarily be beneficial. Barbashova divides adaptive mechanisms into two categories: (1) the "struggle for oxygen"—adaptive changes that tend to increase oxygen delivery to cells—and (2) "adaptation to hypoxia"—

changes at the tissue level that increase the ability of cells to function despite oxygen deficiency. By "struggle for oxygen" she means all systemic adjustments plus certain cell adjustments (increased myoglobin concentration and oxidative enzyme activities). "Adaptation to hypoxia" includes decreased cell oxygen consumption, increased energy production by anaerobic glycolysis, and a nonspecific "increase in resistance of tissues."

In the following discussion, the evidence for cell adaptation to hypoxia is presented under three headings: myoglobin, aerobic metabolism, and anaerobic metabolism.

Myoglobin. Myoglobin is a heme protein with a molecular weight of about 18,400 that is found mainly in heart and skeletal muscles. It has one heme group per molecule; oxygen reacts with the iron atom at the center of the heme group. The iron is in the ferrous state in both oxymyoglobin and reduced myoglobin. The oxygen dissociation curve of myoglobin differs from that of hemoglobin; the myoglobin curve has the form of a rectangular hyperbola and lies well above and to the left of the hemoglobin curve. This means that myoglobin has a much greater affinity for oxygen than does hemoglobin. At a venous P_{O_2} of 40 mm Hg, at which hemoglobin is only 66% saturated, myoglobin is still 94% saturated. At a P_{O_2} of 10 mm Hg, hemoglobin is only 10% saturated, whereas myoglobin is 80% saturated. Since mitochondrial cytochrome oxidase can operate at full capacity at a P_{O_2} well below 5 mm Hg, it follows that myoglobin can accept oxygen from hemoglobin and store it in the cells for release to cytochrome oxidase as needed. In addition to this function of myoglobin as an oxygen storage-release mechanism, Wittenberg[35] and Scholander[36] independently demonstrated that myoglobin facilitates the diffusion of oxygen in aqueous media. By this mechanism, the rate of oxygen diffusion from cell membrane to the mitochondria is increased, and this may be especially beneficial when delivery of oxygen to cells is reduced. In the light of this protective role of myoglobin in maintaining optimal rates of intracellular oxygen diffusion, one wonders that the brain, the most vulnerable of all the organs to hypoxia, is not provided with myoglobin.

Because of the stoichiometric relationship between myoglobin and oxygen, an increase of myoglobin concentration in cardiac and skeletal muscle cells should be a beneficial adjustment to life at high altitude. The experimental evidence, with some exceptions, indicates that myoglobin concentration increases in heart muscle and in some (if not all) skeletal muscles as a result of residence at high altitude, provided the duration and severity of hypoxia are great enough. Neither the quantitative improvement in altitude tolerance resulting from the increased myoglobin concentration nor the mechanism responsible for this increase has been determined.

Capacity of cells for aerobic energy production. When oxygen is abundant, the rate of cell oxygen consumption is determined primarily by the mitochondrial concentration of the breakdown products of ATP, that is, inorganic phosphate and especially ADP. The demand for energy is met by ATP breakdown, and the stimulation of oxygen consumption by the ADP liberated serves to restore the ATP level by oxidative phosphorylation of ADP to ATP. In this manner, the rate of oxygen consumption is regulated automatically in accordance with the need for energy production. So long as the cell oxygen tension is above the critical level, oxygen plays no part in regulating the energy metabolism of the cells, and increased oxygen supply serves no useful purpose. The critical level of oxygen is defined as the oxygen tension below which cell oxygen consumption begins to decrease. It is of the order of 1 to 5 mm Hg and is probably less than 1 mm Hg in the interior of the mitochondria.

Oxygen functions as the final electron acceptor in the mitochondrial respiratory chain. The electron carriers nearest the electron-donating substrate are *relatively* reduced and those nearest oxygen are *relatively* oxidized. The intermediate carriers exist in a gradient that is increasingly more oxidized in passing from NADH to cytochrome oxidase. This steady state condition is maintained rather constant during the flow of electrons to oxygen. When ADP concentration is high, oxygen is consumed at a rapid rate, and the members of the respiratory chain become more oxidized. When ADP concentration is low (that is, ATP is not being broken down rapidly to supply energy), the rate of oxygen consumption is low and the respiratory chain carriers become more reduced. If the supply of oxygen is decreased, the terminal member of the chain, cytochrome oxidase, becomes less oxidized, and its capacity to accept electrons is thereby diminished. This in turn slows the flow of electrons down the chain and reduces the rate at which ATP is formed by oxidative phosphorylation. The result is an energy deficit.

The studies of possible adaptive increases in the capacity for aerobic energy production in the cells of organisms in response to chronic hypoxia have been largely empiric. They have, for the most part, involved measurements of the activities or concentrations of individual enzymes, cofactors, and the like, with little regard for the usefulness of such changes if they did occur. The initial question that every investigator of these problems should ask is whether an increase in the activity or concentration of the factor that he proposes to study would compensate in any way for reduced oxygen availability.

After careful examination of the system for aerobic energy generation in cells, it should be possible to predict the changes that would increase the tolerance of cells to hypoxia. We shall make such an examination before considering the experimental evidence. For convenience, only the mitochondrial system will be considered at this point; the interrelations of this system with that of the cytoplasmic glycolytic system will be discussed when adaptive changes in anaerobic metabolism are reviewed.

The function of the Krebs (citric acid) cycle is, by successive substrate dehydrogenations, to provide reducing equivalents for the respiratory electron transfer chain; the passage of electrons down the chain to oxygen is associated with the production of ATP by oxidative phosphorylation. The rate of operation of the Krebs cycle during resting metabolism is determined by the availability of oxidized mitochondrial NAD (NAD^+) to which the reducing equivalents removed from the substrates can be passed. This is dependent on the rate of oxidation of NADH via the respiratory chain, which in turn is regulated by the concentration of ADP. During maximal rates of oxidative phosphorylation, mitochondrial NAD is about 50% oxidized, and the rate of operation of the Krebs cycle may then be limited either by the availability of substrates or by the capacity of the enzymes of the cycle itself.

The rate of mitochondrial oxygen uptake depends on both availability of oxygen and the amount of reduced cytochrome oxidase with which oxygen can react. When oxygen is abundant, the amount of reduced cytochrome oxidase is very low, and the rate of oxygen consumption is sensitive to the capacity of the substrate dehydrogenation system (Krebs cycle). When the oxygen supply is restricted, cytochrome oxidase becomes progressively more reduced, but the rate of oxygen consumption falls relatively slowly until nearly all cytochrome oxidase is in the reduced form. At extremely low intracellular P_{O_2} (less than 0.05 mm Hg), cytochrome oxidase should be almost completely reduced, and the rate of oxygen consumption would then depend mainly on P_{O_2}. At intermediate intracellular P_{O_2}, the rate of oxygen consumption depends on the product of the concentrations of oxygen and reduced cytochrome oxidase. In the intermediate range of intracellular P_{O_2} an increase in the amount of cytochrome oxidase might be of adaptive value by increasing the rate of uptake of the oxygen that is available, thus reducing cell P_{O_2} and increasing the oxygen diffusion gradient from capillary blood to cells. There is no identifiable rate-limiting step in the respiratory chain; accordingly, oxygen consumption might be increased by an increase in the amount of any of the components of the chain, since this would increase the relative degree of reduction of the cytochrome oxidase and hence the rate of interaction of reduced cytochrome oxidase and oxygen.

In summary, increases of any member of the respiratory chain might be beneficial when cell P_{O_2} is reduced. Likewise, an increase of the concentration of the individual substrate dehydrogenases of the Krebs cycle might be beneficial by increasing the rate of delivery of reducing equivalents to the respiratory chain, provided (1) that the enzymes under consideration are normally rate-limiting and (2) that substrate dehydrogenation is limiting for the respiratory chain phosphorylation system. With respect to the first provision, the control points in the Krebs cycle are probably determined more by such factors as feedback inhibition, enzyme activation by precursors, and the like, than by the actual amounts of individual enzymes, so that it is difficult to predict the effect on the rate of operation of the cycle of increases in the various member enzymes. With respect to the second provision, it appears that the rate of substrate oxidation is limited by the phosphorylation system (even when this is maximally active) with certain substrates, including succinate, so that no benefit should result from increased amounts of the corresponding dehydrogenases. However, with other substrates, including glutamate, alpha-ketoglutarate, and isocitrate, dehydrogenation appears to be the limiting factor, and increased amounts of their dehydrogenases might result in more rapid passage of reducing equivalents down the respiratory chain. In the present state of our knowledge, it would seem premature to attribute adaptive value to increases at altitude of individual enzymes of the Krebs cycle.

Let us now examine the experimental evidence for enzyme changes at high altitude. The cytochrome oxidase activity of various tissues has been reported to be unchanged by exposure to high altitude.[37, 38] Barbashova[34] reported no change in the cytochrome oxidase activity of tissues from altitude-exposed animals when the enzyme test was performed in a high oxygen environment (usual procedure) but a considerable increase when the test was performed in a low oxygen environment. The significance of this result is obscure. The cytochrome c concentration of tissues from altitude-exposed animals has been reported to be unchanged[34, 39] or lowered.[40] An increase in the NADH-cytochrome c reductase activity has been reported in skeletal muscle from high-altitude natives[41] and in the livers of rats exposed to a simulated altitude of 18,000 ft.[42]

Of the substrate dehydrogenases of the Krebs cycle, only succinic dehydrogenase has been examined carefully. The activity of this enzyme was reported to be increased at altitude in liver[43] and in diaphragm, liver, kidney, and red rectus femoris muscle, but not in other skeletal muscles, red or pale.[37] In view of the analysis presented above, it is uncertain whether this represents a significant adaptive change.

In summary, the evidence that changes in the aerobic enzyme systems play a significant role in altitude adaptation is not convincing. In addition to the studies on specific enzymes mentioned above, the numerous investigations of the overall rate of oxygen consumption and oxidative phosphorylation of tissue slices and tissue homogenates have uniformly failed to reveal any change caused by altitude exposure.

Capacity of cells for anaerobic energy production. Glycolysis consists of a chain of reactions that convert glucose (or glycogen) to pyruvic acid, with net production of two molecules of ATP per molecule of glucose utilized. The reactions occur in the cytoplasm and are thus separated from the mitochondrial energy-generating system (Krebs cycle plus respiratory chain). When oxygen is abundant, pyruvic acid is converted to acetic acid, which enters the Krebs cycle for oxidation, resulting in the production of large additional amounts of ATP. In the glycolytic sequence there are two oxidative steps in which substrate oxidation is made possible by its coupling to the reduction of NAD^+ to NADH. The reoxidation of the cytoplasmic NADH is essential for continued operation of the glycolytic process. Under aerobic conditions, this is accomplished indirectly by transfer of reducing equivalents, via the so-called shuttle systems, from cytoplasm to the mitochondrial respiratory chain (since NADH does not readily penetrate the mitochondrial membrane). Of course, under anaerobic conditions this is not possible, and cytoplasmic NADH is oxidized to NAD^+ by being coupled to the reduction of pyruvic acid to lactic acid. Thus the end product of anaerobic glycolysis is lactic acid, a metabolic dead end that must be disposed of when oxygen is available again. The formation of lactic acid from pyruvic acid is not associated with energy production; it is solely a device for restoring the cytoplasmic NAD^+ that is necessary for glycolysis.

Thus the essential features of anaerobic metabolism are a low energy yield and the formation of a fixed acid, lactic acid, that may disturb the acid-base balance. Its redeeming feature is that it can provide energy when oxygen is deficient. This occurs, for example, in skeletal muscle during the initial stage of exercise (before the rate of aerobic metabolism has caught up with the increased demand) and during very strenuous exercise that exceeds the aerobic capacity of the muscles.

Studies on anaerobic metabolism at high altitude have been of several types, including measurement of the overall rate of tissue metabolism in the complete absence of oxygen, estimation of the rate and degree of recovery following a period of oxygen deprivation, and measurement of the maximal capacity of the individual enzymes that participate in glycolysis.

Barbashova[34] believes that "the intensification of the anaerobic release of energy is an essential link in cellular adaptation to the prolonged influence of an oxygen deficiency." In support of this view is work[44–46] that demonstrates increased capacity for anaerobic energy production by heart muscle strips and phrenic nerve–diaphragm preparations from altitude-adapted animals. According to Anderson and his co-workers[47] some of the increased tolerance to anoxia in tissues from altitude-adapted animals may result from increased tissue buffering capacity, since they observed a protective effect of metabolic alkalosis on myocardial muscle function during acute anoxia.

There are certain control points in the glycolytic sequence, including reactions catalyzed by the enzymes hexokinase, phosphofructokinase, aldolase, and lactic dehydrogenase. Lenti and Grillo[48] observed an increase of about 50% in the aldolase activity of heart and skeletal muscles of rats exposed to high altitude for only 3 weeks. Increases of hexokinase and lactic dehydrogenase activity were reported in the skeletal muscles of rats in which muscle hypoxia was the result of muscular exercise training.[34] However, neither activity nor isoenzyme pattern of lactic dehydrogenase was altered in the brain, heart, liver, and skeletal muscles of rats exposed to a simulated altitude of 18,000 ft for many weeks.[49] Alterations of phosphofructokinase activity at high altitude have not been reported; alterations in the activity of this enzyme would probably not be of adaptive value anyhow, since it controls the rate of glycolysis primarily through the allosteric effects of ATP, ADP, inorganic phosphate, and citrate rather than by its actual activity level.

To summarize, it has been claimed that exposure to chronic oxygen deficiency may increase (1) capacity for anaerobic glycolysis of tissue homogenates, (2) anaerobic functional capacity of heart muscle strips and phrenic nerve–diaphragm preparations, and (3) activity levels of some of the individual enzymes involved in glycolysis. These alterations are usually demonstrated in tissues removed from the animals and tested under extreme conditions (complete absence of oxygen). It remains to be shown that increased capacity for energy production by anaerobic metabolism is an important feature of altitude adaptation, with the possible exception that the capacity for brief spurts of strenuous exercise might be increased somewhat.

In none of the studies of the capacity for aerobic and anaerobic energy production in altitude-adjusted animals has a mechanism been proposed to account for the observed changes. In some way, oxygen deficiency must be assumed to act as a stimulus to the production of increased amounts of enzymes, cofactors, and so on. The simplest explanation is one based on the well-known phenomenon of substrate induction. The supposition applied to hypoxia would be that oxygen deficiency results in the accumulation of the reduced forms of substrates (or carriers in the case of the respiratory chain), which then induce the formation of increased amounts of the appropriate enzymes.

Barbashova[34] believes that an important feature of altitude adaptation is development of the ability of tissues to function more efficiently by reducing metabolic rate. As evidence, she refers to studies in which tissues removed from altitude-exposed animals were reported to have a lower rate of oxygen consumption, even in an atmosphere of oxygen, than tissues from control animals. It is not certain whether these results indicate a reduced energy requirement of the tissues (which might be beneficial when the oxygen supply is limited) or of tissue ability to generate energy. Duckworth,[40] commenting on her observation of decreased oxygen consumption by diaphragms removed from altitude-adapted rats, suggested that this was the result of reduced thyroid activity at high altitude, which in turn led to decreased formation of oxidative enzymes.

In conclusion, while cell changes may play a role in high-altitude adjustments, this has not been proved. The best documented cell response to high altitude is increased myoglobin concentration in cardiac and skeletal muscle, which probably aids rapid oxygen transport from cell membrane to mitochondria. Evidence of increased capacity for aerobic energy formation as a result of exposure to high altitude is not convincing; it consists of scattered reports of increased cytochrome oxidase and succinic dehydrogenase activities unsupported by demonstration of increased aerobic capacity of intact cells and tissues. Evidence from several sources indicates that tissue preparations from altitude-exposed animals have increased capacity for obtaining energy from anaerobic sources in the complete absence of oxygen. It remains to be determined whether this is a significant advantage to organisms subjected to partial, rather than absolute, oxygen deprivation. In my opinion, adaptation to high-altitude hypoxia is much more likely to result from adjustments that increase oxygen delivery to cells than from cell changes improving the efficiency of oxygen utilization by cells.

Attempts to hasten or improve adaptation to high altitude

Attempts to accelerate the changes occurring during the first few days at high altitude and to lessen the severity of acute mountain sickness were described previously. Here we deal with various suggested treatments for hastening the attainment, or improving the level, of the ultimate degree of altitude adaptation achieved.

Physical conditioning produces some of the same changes as exposure to high altitude,[50] and the most frequently used criterion of altitude adaptation is improved exercise tolerance at high altitude. It is natural, then, that attempts should be made to hasten the course of altitude adaptation by physical conditioning before or during exposure to high altitude. Interpretation of results is complicated by the difficulty in distinguishing between increased hypoxia tolerance and increased capacity for muscular exertion. In the study of Balke and

Wells[51] this complication was avoided, since the effect of physical training was evaluated in terms of increased tolerance for acute hypoxia under resting conditions. These authors observed an increase of tolerance to altitude of about 3,000 ft and an extension of the "time of useful consciousness" at 24,000 ft from 2 minutes to 6 to 8 minutes, as the result of a physical conditioning program in young men. Additional improvement occurred in both trained and nontrained subjects after exposure to high altitude.

When the capacity for strenuous exercise at high altitude is used as an indication of the effect of physical conditioning on altitude tolerance, there seems to be a general impression that physical conditioning improves work capacity but that it has no *specific* effect on altitude tolerance per se.[52–54] The question of a specific effect of physical conditioning on altitude tolerance is largely semantic. The outstanding feature of high-altitude impairment is reduced exercise capacity, and if this is improved it can be said that tolerance is improved, whether the improvement results from physical conditioning, altitude exposure, or both.

A matter of considerable practical importance, as all climbers of very high mountains know, is that adaptation to intermediate altitudes improves tolerance for much higher altitudes and shortens the time required for satisfactory adjustment to higher altitude.[55] If exposure to intermediate altitude is combined with exercise, the degree of adaptation is further increased.

The similarity between the effects of altitude exposure and physical conditioning has led to the suggestion that intermittent periods of hypoxia may increase exercise capacity at sea level. Evidence supporting this idea is not very convincing and, at any rate, it is simpler to exercise subjects at sea level than it is to make them hypoxic.

Many attempts have been made to influence the course of altitude adaptation with drugs, hormones, and chemical agents. Aside from the possible beneficial effects of drugs such as acetazolamide on the adjustments occurring during the first few days at high altitude, the results have been disappointing. This is not surprising when one considers the nature of the gradual adjustments to altitude discussed in an earlier section. Time is required for such changes as the increased total number of erythrocytes, growth of new blood vessels, and increase of myoglobin concentration in cardiac and skeletal muscles. A decrease of cell metabolic requirement would be beneficial, and there is evidence that thyroid function may decrease in response to altitude exposure.[56] This would not, however, warrant the use of antithyroid drugs, with their attendant risks, for the purpose of increasing altitude tolerance. It has been claimed that administration of a mixture of amino acids, vitamins, and purine bases to rats undergoing muscular training at a simulated altitude of 2,500 m resulted in greater increases of hemoglobin, myoglobin, and creatine phosphate concentrations and in the activities of succinic dehydrogenase, cytochrome oxidase, and NADH–cytochrome c reductase in liver and skeletal muscle. Albrecht and Albrecht[57, 58] reported improved work capacity, but no relief of subjective complaints, from administration of anabolic steroids to sea-level subjects who climbed to a height of 20,200 ft (Mt. Aconcagua in Argentina). When other subjects were given intravenous injections of an antigen-free, nonprotein extract of calf blood, marked improvements were reported in both working capacity and subjective feeling of well-being. The composition of the blood extract is unknown, but it is said to stimulate tissue respiration (as measured by the Warburg technic) and to increase cardiac stroke volume in human subjects. As might be expected, all sorts of drugs, hormones, vitamins, and other agents were used by trainers preparing athletes for competition in the Mexico City Olympics (1968), but so far as is known, there were no proved benefits.

In summary, aside from the possible beneficial effects of acetazolamide in relieving the discomforts of acute mountain sickness, it has not been shown conclusively that the course of altitude adaptation is favorably influenced by any drug, hormone, vitamin, or other chemical agent. On the other hand, it has been shown that physical conditioning, either before or during altitude exposure, increases exercise capacity at high altitude (which is commonly regarded as an important result of altitude adaptation). There is much to recommend exposure to intermediate altitudes, together with physical conditioning, as a means of facilitating the adaptive process at still higher altitudes.

Altitude adaptation compared in high-altitude natives and sojourners

It is often claimed that high-altitude natives have a degree of adaptation that is not achieved by sea-level natives even after they have spent months or years at high altitude. This is attributed to natural selection acting over many generations. The amazing capacity of the Andean Indians to perform hard physical labor at high altitude is often quoted as evidence of their superior adaptation. We have seen, however, that physical conditioning and altitude exposure produce results that are similar in many respects, so that a valid comparison of altitude-adapted sea-level natives and high-altitude natives requires careful matching of the two groups with respect to their state of physical conditioning. The assumption is commonly made that the Andean Indians are trained for hard physical labor and therefore should be compared to a similar group (for example, athletes) that has spent some weeks or months at altitude. This may or may not be a fair requirement.

Aside from anecdotal statements, objective evidence is derived mainly from measurements of exercise capacity and the tolerance of resting subjects to very severe, acute hypoxia. Additional useful information is provided by the general cultural and anthropologic studies made during recent years.

The most reproducible index of aerobic metabolic capacity, and hence of capacity for strenuous exercise, is the maximal oxygen uptake per minute (Max $\dot{V}_{O_2}$). When sea-level natives go to high altitude, Max $\dot{V}_{O_2}$ decreases, the decrease being greater in physically conditioned than in sedentary subjects. Residence at high altitude results in only a slight increase in Max $\dot{V}_{O_2}$ of sea-level natives. Max $\dot{V}_{O_2}$ was not appreciably greater in naturally adapted Peruvian Indians than in newcomers to high altitude who were trained athletes.[59] Moreover, when exercising on bicycle ergometers, the Indians had to utilize a greater fraction of their aerobic capacity and required a larger recovery oxygen uptake (how much of this resulted from unfamiliarity of the Indians with bicycles is uncertain). While the evidence is far from complete, it seems that the aerobic work capacity is not greatly different in altitude-adapted sea-level natives and high-altitude natives when the level of physical conditioning is comparable in both groups. This argues against the importance of natural selection over many generations in accounting for the reputed great work capacity of Indians native to high altitude. More direct comparisons are needed of sojourners and natives who have been carefully matched for body size, physical condition, nutritional status, and similar factors and who are tested by methods appropriate to their capabilities and experiences.

The tolerance of high-altitude natives to sudden exposure to a much higher (simulated) altitude has been found to be considerably greater than that of sea-level residents.[60] This test has the advantage over measurements of exercise capacity that physical condition is not a factor, since the test is made on resting subjects. If high-altitude natives and altitude-adapted newcomers were compared, the question of superior adaptation of the high-altitude natives might be nearer to an answer.

Aside from any question of selection over generations as a factor in the adaptation of high-altitude natives, there is evidence that hypoxia extending from birth onward may produce results that differ from those of even long periods of hypoxia beginning in adulthood. An example is the enlarged chest characteristic of Andean Indians. This may prove to be an important adaptation that does not occur in sea-level natives who become adapted to high altitude; by increasing the functional residual capacity and improving the uniformity of ventilation of alveoli in normally poorly ventilated regions of the lung (such as, the apex), oxygen exchange in the lungs may be significantly improved.

There are other differences between high-altitude natives and altitude-adapted newcomers whose physiologic significance is uncertain. For example, the natives tend to have decreased respiratory sensitivity to hypoxia compared to newcomers who have adapted. The mechanism responsible for this difference is unknown, and we can only speculate that it reduces the dangers of hyperventilation hypocapnia, which are ever present at high altitude. There is also a general impression that high-altitude natives usually have lower hematocrits (and hence lower blood viscosity) than do adapted newcomers. It may be that slowly developing adjustments of other types (tissue vascularity, cell adaptations, and the like) have decreased the importance of respiratory and hematologic adjustments to high altitude in the natives.

Great socioeconomic and cultural differences between sea-level and high-altitude native populations make direct comparisons of their altitude adaptation very difficult. There are, however, about a dozen species and subspecies of rodents that form altitudinally stable populations in the Andes. Morrison[61] reported that the exercise tolerance (in terms of the critical inspired P_{O_2} that permitted a metabolic rate twice basal) increased progressively with the altitude at which the mice lived; the critical P_{O_2} of lowland mice was about 125 mm Hg, while that of some of the highland races was only 50 to 60 mm Hg. Morrison comments, "Highland men and highland mice exhibit a similar advantage in physiologic performance over their less capable lowland relatives." A surprising observation was that the hematocrit is about the same in highland and lowland mice, although, at all altitudes, the more vigorous mice had higher hematocrits than did their more sedentary relatives. Thus increased hemoglobin concentration is not an adaptive feature in these animals; however, the affinity of the hemoglobin for oxygen is greater in the highland animals. Myoglobin concentration in the diaphragm and leg muscles is about twice as great in highland as in lowland forms. This high concentration persisted when they were transferred to sea level, so that it appears to be a genetic trait not dependent on a hypoxic stimulus. Morrison believes that these observations indicate a real difference between adaptation of the individual and adaptation of the race through natural selection. Studies of this type are instructive in indicating the types of adaptive changes available to different organisms.

Failure of altitude adaptation

The condition known as chronic mountain sickness serves as a constant reminder that adaptation to high altitude is achieved at a cost and that it cannot be regarded as necessarily a permanent state of affairs, even in natives.

Chronic mountain sickness was first described as a distinct clinical entity by Monge in 1928. It is basically an intolerance to high altitude occurring in acclimatized residents, including natives born and reared at high altitude. Although it may be diagnosed for the first time following an acute attack, giving the appearance of a disease of sudden onset, it is probably an insidious disease whose only manifestations for years may be diminished mental and physical capacity. It is cured by descent of the patient to sea level, but it may recur if he returns to high altitude.

Nervous system symptoms usually dominate the clinical picture; respiratory symptoms are often present but circulatory manifestations are rare, except for shortness of breath in advanced cases. The pathophysiologic basis of the condition appears to be an exaggeration of the hypoxia of high altitude caused by a decrease of the usual degree of hyperventilation. The result is a fall of arterial blood P_{O_2} (just as would occur if the patient had ascended to a still higher altitude), with a secondary increase in the degree of polycythemia. It has been proposed that the reduction of pulmonary ventilation is caused by a decrease in the sensitivity of the respiratory center to carbon dioxide, and this is supported by the experimental demonstration of a subnormal response to the administration of carbon dioxide in the inspired air. The reason for the altered sensitivity of the respiratory center to carbon dioxide is unknown, but the condition may have features in common with idiopathic cases of hypoventilation occurring at sea level.

Monge and Monge[14] give a detailed clinical description of this disease. The following brief discussion is based chiefly on their exhaustive studies.

> The mental symptoms are usually mild and easier to detect in people of higher education (i.e., the engineer or business man who suddenly finds it difficult to solve elementary arithmetic problems). Changes of character and tendency to depression are common. In advanced cases there are pronounced mental alterations, such as irritability, loss of volitional control, a belligerent attitude, maniac behavior, hallucinations, paranoid reactions, and tendency to suicide.

Nervous system symptoms are frequent and varied; they include headache, dizziness, and paresthesias (numbness, tingling, aches and pains, hot and cold sensations) commonly felt in the hands and feet. Visual and auditory disturbances may occur. Some advanced cases are predisposed to cerebral crises of sudden onset and great intensity, resembling very severe migraine attacks. The crises may last for minutes or hours and may be followed by unconsciousness, sometimes associated with convulsions. Spontaneous remission is common even without treatment.

Respiratory symptoms are usually mild, but there may be copious hemoptysis, probably related to pulmonary thrombosis secondary to the excessive polycythemia characteristic of chronic mountain sickness (hematocrit values as high as 85% have been reported in some of these patients). Circulatory symptoms are rare except in advanced cases when shortness of breath is common and mild anginal symptoms may occur; there may be right heart failure, with liver congestion and marked edema, probably representing an exaggeration of the usual tendency to pulmonary arterial hypertension and right heart hypertrophy in high-altitude dwellers.

The outstanding features of the disease revealed by physical examination and laboratory studies are marked cyanosis, vascular engorgement, and a very high hematocrit. Arterial blood oxygen saturation is lower than that predicted for the altitude, and arterial P_{CO_2} is elevated; both findings reflect the characteristic hypoventilation.

It is important to emphasize that the clinical picture of chronic mountain sickness can be produced by any respiratory disease aggravating a hypoxic condition (such as silicosis, tuberculosis, pulmonary fibrosis), and this must be distinguished from true chronic mountain sickness for purposes of treatment.

The basis of the symptoms of chronic mountain sickness is probably either (1) intensification of hypoxia caused by relative hypoventilation, (2) interference with tissue perfusion (and oxygen transport) resulting from the high viscosity of very polycythemic blood, or (3) a combination of these two factors.

Adaptive disease

The low atmospheric oxygen pressure at high altitude acts as a physiologic stressor, and the resulting state of hypoxia is a stress condition. Monge and Monge[14] suggested that this may be regarded as a special case of the general adaptation syndrome of Selye. The immediate responses to high altitude, including acute mountain sickness, would then be considered analogous to the alarm reaction; the succeeding state of adaptation would be comparable to Selye's state of resistance (a successful adjustment of the organism to the stressor), and chronic mountain sickness would correspond to the state of exhaustion, or failure of adjustment. This analogy must not be carried too far, however, since Selye's general adaptation syndrome involves the hypothalamic pituitary-adrenal axis as the central theme of the stress reaction, and this may not be true of the responses to high altitude.

Influence of high altitude on some common diseases

Monge and Monge[14] have reviewed the special problems encountered in the medical and surgical management of a variety of clinical conditions and their monograph should be referred to for details.

FUTURE OF RESEARCH IN HIGH-ALTITUDE PHYSIOLOGY

There are many reasons why research into the physiologic effects of high altitude should be continued and expanded. Some of them are practical; people must live at high altitudes for various reasons, and the time may come when the inhospitable regions of the earth, such as deserts and mountains, may be called upon to accommodate some of the expanding world population. The study of high-altitude adaptive processes may contribute information of medical importance, since many disease states are characterized by an inadequate delivery of oxygen to the tissues. From a less immediately practical standpoint, the study of the physiologic adjustments occurring at high altitude is important for the insight it gives into the capacities of biologic systems to adapt to changing conditions.

There are several areas of high-altitude research that appear to be especially promising. One of these is the continued effort to determine the limits of adaptation, both of individuals and of races, together with attempts to extend these limits by various means. This will require a better understanding of the relative importance of the biologic alterations that occur as a result of exposure to altitude, as well as identification of the stimulus-response mechanisms that bring about these changes. Some of these studies will be possible only in the laboratory, where the various responses to hypoxia can be studied both separately and together.

The new ecology emphasizes that man must learn to live in harmony with his environment, rather than attempting to alter it to fit his desires. Nowhere is this goal better exemplified than in the study of man in relation to what are usually referred to as the hostile environments—mountains, deserts, and oceans. This is the ultimate reason for the study of environmental physiology.

JUNE, 1970

GLOSSARY

acclimation Adaptation to a single feature of an altered environment (for example, low oxygen pressure at high altitude).

acclimatization Adaptation to all the features of an altered environment (for example, low oxygen pressure, cold, radiation, and wind at high altitude).

adaptation Any change in an organism resulting from exposure to an altered environment that enables the organism to function more efficiently in the new environment.

adjustment Any change in an organism produced by alteration of the environment, without regard to whether the change is beneficial to the organism.

altitude Vertical distance above sea level; usually measured in feet or meters. The term is often used to indicate high altitude.

atmosphere The mantle of gases surrounding the earth (and other planets).

barometric pressure The pressure exerted by the column of air extending vertically from the reference point to the upper limit of the atmosphere; measured in millimeters of mercury or in torr (1 torr = 1/760 atmosphere = 1 mm Hg).

decompression chamber A chamber in which the barometric pressure may be reduced by means of a vacuum pump.

simulated altitude An artificial environment in which the oxygen pressure has been reduced by lowering the barometric pressure in a chamber. Less often, oxygen pressure is reduced by decreasing oxygen concentration of a gas mixture at atmospheric pressure.

stratosphere The layer of the atmosphere between the limits 10 and 20 km above the surface of the earth.

troposphere The layer of the atmosphere extending from the surface of the earth upward to a distance of 10 km.

REFERENCES

1. Hitchcock, F. A.: Animals in high altitudes: early balloon flights. In Dill, D. B., editor: Handbook of physiology, sect. 4, chap. 52, Washington, D. C., 1964, American Physiological Society.
2. Bert, P.: Barometric pressure—researches in experimental physiology (translated by M. A. and F. A. Hitchcock), Columbus, Ohio, 1943, College Book Company.
3. von Muralt, A.: Where are we? A short review of high altitude physiology. In Weihe, W. H., editor: The physiological effects of high altitude, Oxford, 1964, Pergamon Press, p. 15.
4. Luft, U. C.: Laboratory facilities for adaptation research: low pressures. In Dill, D. B., editor: Handbook of physiology, sect. 4, chap. 19, Washington, D. C., 1964, American Physiological Society.
5. Barcroft, J.: Presidential address on anoxaemia (abridged), Lancet **199**:485, 1920.
6. McFarland, R. A., and Evans, J. N.: Alterations in dark adaptation under reduced oxygen tensions, Am. J. Physiol. **127**:37, 1939.
7. Woods, E. F., and Richardson, J. A.: Effects of acute anoxia on cardiac contractility, Am. J. Physiol. **196**:203, 1959.
8. Van Liere, E. J., and Stickney, J. C.: Hypoxia, Chicago, 1963, University of Chicago Press.
9. Keys, A., Stapp, J. P., and Violante, A.: Responses in size, output and efficiency of the human heart to acute alteration in the composition of inspired air, Am. J. Physiol. **138**:763, 1943.
10. Hannon, J. P., and Chinn, K. S. K.: Effects of high altitude on body fluid volumes, Fed. Proc. **26**:719, 1967.
11. Wayne, H. H.: Clinical differentiation between hypoxia and hyperventilation, J. Aviat. Med. **29**:307, 1958.
12. Cain, S. M., and Kronenberg, R. S.: Effects of carbonic anhydrase inhibition on the responses of men to 14,000 feet simulated altitude: changes caused by acetazolamide after 24 hours at altitude. In Hegnauer, A. H., editor: Biomedicine problems of high terrestrial elevations, Natick, Mass., 1969, U. S. Army Research Institute of Environmental Medicine, p. 501.
13. Landowne, M., Forward, S. A., and Hansen, J. E.: Evaluation of acute mountain sickness at 12,000 ft. altitude and the effect of acetazolamide. In Hegnauer, A. H., editor: Biomedicine problems of high terrestrial elevations, Natick, Mass., 1969, U. S. Army Research Institute of Environmental Medicine, p. 64.
14. Monge, C. M., and Monge, C. C.: High altitude diseases, Springfield, Ill., 1966, Charles C Thomas, Publisher.
15. Visscher, M. B.: Basic factors in the genesis of pulmonary edema, and a direct study of the effect of hypoxia upon edemogenesis. In Hegnauer, A. H., editor: Biomedicine problems of high terrestrial elevations, Natick, Mass., 1969, U. S. Army Research Institute of Environmental Medicine, p. 90.
16. Hultgren, H. N.: High altitude pulmonary edema. In Hegnauer, A. H., editor: Biomedicine problems of high terrestrial elevations, Natick, Mass., 1969, U. S. Army Research Institute of Environmental Medicine, p. 131.
17. Miller, A. T., Jr., Curtin, K. E., Shen, A. L., and Suiter, C. K.: Brain oxygenation in the rat during hyperventilation with air and with low O_2 mixtures, Am. J. Physiol. **219**:798, 1970.
18. Houston, C. S., and Riley, R. L.: Respiratory and circulatory changes during acclimatization to high altitude, Am. J. Physiol. **149**:565, 1947.
19. Kreuzer, F., Tenney, S. M., Mithoefer, J. C., and Remmers, J.: Alveolar-arterial oxygen gradient in Andean natives at high altitude, J. Appl. Physiol. **19**:13, 1964.
20. Kreuzer, F., and Campagne, P. V. L.: Resting pulmonary diffusing capacity for CO and O_2 at high altitude, J. Appl. Physiol. **20**:519, 1965.
21. Richardson, D. W., Kontos, H. A., Raper, A. J., and Patterson, J. L., Jr.: Modification by beta-adrenergic blockade of the circulatory responses to acute hypoxia in man, J. Clin. Invest. **46**:77, 1967.
22. Kontos, H. A., Mauck, H. P., Jr., Richardson, D. W., and Patterson, J. L., Jr.: Mechanisms of circulatory responses to systemic hypoxia in the anesthetized dog, Am. J. Physiol. **209**:397, 1965.
23. Daly, M. de B., and Scott, M. J.: The cardiovascular responses to stimulation of the carotid body chemoreceptors in the dog, J. Physiol. **165**:179, 1963.
24. Klausen, K.: Cardiac output in man in rest and work during and after acclimatization to 3,800 m, J. Appl. Physiol. **21**:609, 1966.
25. Vogel, J. A., Hansen, J. E., and Harris, C. W.: Cardiovascular responses in man during exhaustive work at sea level and high altitude, J. Appl. Physiol. **23**:531, 1967.
26. Hultgren, H. N., and Grover, R. F.: Circulatory adaptation to high altitude, Ann. Rev. Med. **19**:119, 1968.
27. Baker, P. T.: Human adaptation to high altitude (biocultural mechanisms of adaptation are explored in a population native to the high Andes), Science **163**:1149, 1969.
28. Eaton, J. W., Brewer, G. J., and Grove, R. F.: Role of red cell 2,3-diphosphoglycerate in the adaptation of man to altitude, J. Lab. Clin. Med. **73**:603, 1969.
29. Benesch, R., and Benesch, R. E.: The effect of organic phosphates from the human erythrocyte on the allosteric properties of hemoglobin, Biochem. Biophys. Res. Commun. **26**:162, 1967.
30. Murray, J. F., Gold, P., and Johnson, B. L., Jr.: The circulatory effects of hematocrit variations in normovolemic and hypervolemic dogs, J. Clin. Invest. **42**:1150, 1963.

31. Smith, E. E., and Crowell, J. W.: Influence of hematocrit ratio on survival of unacclimatized dogs at simulated high altitude, Am. J. Physiol. **205:** 1172, 1963.
32. Miller, A. T., Jr., and Hale, D. M.: Increased vascularity of brain, heart and skeletal muscle of polycythemic rats, Am. J. Physiol. **219:**702, 1970.
33. Imre, G.: Studies on the mechanism of retinal neovascularization: role of lactic acid, Brit. J. Ophthal. **48:**75, 1964.
34. Barbashova, Z. I.: Cellular level of adaptation. In Dill, D. B., editor: Handbook of physiology, sect. 4, chap. 4, Washington, D. C., 1964, American Physiological Society.
35. Wittenberg, J. B.: Oxygen transport—a new function proposed for myoglobin, Biol. Bull. **117:**402, 1959.
36. Scholander, P. F.: Oxygen transport through hemoglobin solutions (how does the presence of hemoglobin in a wet membrane mediate an eightfold increase in oxygen passage?), Science **131:**585, 1960.
37. Tappan, D. V., Reynafarje, B., Potter, V. R., and Hurtado, A.: Alterations in enzymes and metabolites resulting from adaptation to low oxygen tensions, Am. J. Physiol. **190:**93, 1957.
38. Albaum, H. G., and Chinn, H. I.: Brain metabolism during acclimatization to high altitude, Am. J. Physiol. **174:**141, 1953.
39. Tappan, D. V., and Reynafarje, B.: Tissue pigment manifestations of adaptation to high altitudes, Am. J. Physiol. **190:**99, 1957.
40. Duckworth, M. W.: Tissue changes accompanying acclimatization to low atmospheric oxygen in the rat, J. Physiol. **156:**603, 1961.
41. Reynafarje, B.: Myoglobin content and enzymatic activity of muscle and altitude adaptation, J. Appl. Physiol. **17:**301, 1962.
42. Reynafarje, B., and Green, J.: Pyridine nucleotide-cytochrome c reductases in rats exposed to low oxygen tensions, Proc. Soc. Exp. Biol. Med. **103:**224, 1960.
43. Mefferd, R. B., Jr., Nyman, M. A., and Webster, W. W.: Whole body lipid metabolism of rats after chronic exposure to adverse environments, Am. J. Physiol. **195:**744, 1958.
44. Poupa, O., Krofta, K., Procházka, J., and Chvapil, M.: The resistance of the myocardium to anoxia in animals acclimated to simulated altitude, Physiol. Bohemoslov. **14:**233, 1965.
45. Komives, G. K., and Bullard, R. W.: Function of the phrenic nerve-diaphragm preparation in acclimation to hypoxia, Am. J. Physiol. **212:**788, 1967.
46. McGrath, J. J., and Bullard, R. W.: Altered myocardial performance in response to anoxia after high-altitude exposure, J. Appl. Physiol. **25:**761, 1968.
47. Anderson, G., Souhrada, J., Bullard, R. W., and McGrath, J. J.: The "protective" effect of metabolic alkalosis on myocardial muscle during acute anoxia, Physiologist **12:**158, 1969.
48. Lenti, C., and Grillo, M. A.: Über die Wirkung des Hochgebirges auf die Aldolase des Skelett-und des Herzmuskels, Naturwissenschaften **45:**68, 1958.
49. Miller, A. T., Jr., and Hale, D. M.: Comparisons of lactic dehydrogenase in rat and turtle organs, Comp. Biochem. Physiol. **27:**597, 1968.
50. McFarland, R. A.: Experimental evidence of the relationship between aging and oxygen want: in search of a theory of aging, Ergonomics **6:**339, 1963.
51. Balke, B., and Wells, J. G.: Ceiling altitude tolerance following physical training and acclimatization, J. Aviat. Med. **29:**40, 1958.
52. Kollias, J., and others: Work capacity of long-time residents and newcomers to altitude, J. Appl. Physiol. **24:**792, 1968.
53. Turner, H. S., Hoffler, G. W., Billings, C. E., and Bason, R.: An attempt to produce acclimatization to hypoxia by intermittent altitude exposure with vigorous exercise, Aerospace Med. **40:**971, 1969.
54. Klein, K. E., Wegmann, H. M., Brüner, H., and Vogt, L.: Physical fitness and tolerances to environmental extremes, Aerospace Med. **40:**998, 1969.
55. Pugh, L. G. C. E.: Muscular exercise on Mount Everest, J. Physiol. **141:**233, 1958.
56. Mulvey, P. F., Jr., and Macaione, J. M. R.: Thyroidal dysfunction during simulated altitude conditions, Fed. Proc. **28:**1243, 1969.
57. Albrecht, H., and Albrecht, E.: Ergometric, rheographic, reflexographic, and electrocardiographic tests at altitude and effect of drugs on human physical performance, Fed. Proc. **28:**1262, 1969.
58. Albrecht, E., and Albrecht, H.: Metabolism and hematology at high altitude and the effect of drugs on acclimatization, Fed. Proc. **28:**1118, 1969.
59. Buskirk, E. R.: Decrease in physical working capacity at high altitude. In Hegnauer, A. H., editor: Biomedicine problems of high terrestrial elevations, Natick, Mass., 1969, U. S. Army Research Institute of Environmental Medicine, p. 204.
60. Velásquez, T.: Tolerance to acute anoxia in high altitude natives, J. Appl. Physiol. **14:**357, 1959.
61. Morrison, P.: Adaptation and acclimatization of mammals to high altitude, Navy Res. Rev. **17:**4, 1964.

SUGGESTED READINGS

1. Barcroft, J.: The respiratory function of the blood, Cambridge, 1925, Cambridge University Press.
2. Cannon, W. B.: Organization for physiological homeostasis, Physiol. Rev. **9:**399, 1929.
3. Dill, D. B.: Life, heat, and altitude, Cambridge, 1938, Harvard University Press.
4. Dill, D. B., editor: Handbook of physiology, sect. 4, Washington, D. C., 1964, American Physiological Society.
5. Edholm, O. G., and Bacharach, A. L., editors: The physiology of human survival, New York, 1965, Academic Press, Inc.
6. Jokl, E., and Jokl, P.: Exercise and altitude, Basel, 1968, S. Karger.
7. Licht, S., editor: Medical climatology, New Haven, 1964, E. Licht.
8. Margaria, R., editor: Exercise at altitude, New York, 1967, Excerpta Medica Foundation.
9. Proceedings of the International Symposium on Altitude and Cold, Fed. Proc. **28:** No. 3, 1969.
10. Weihe, W. H., editor: The physiological effects of high altitude, New York, 1964, The Macmillan Co.
11. Wulff, L. Y., Braden, I. A., Shillito, F. H., and Tomashefski, J. F.: Physiological factors relating to terrestrial altitudes: a bibliography, Columbus, Ohio, 1968, Ohio State University Press.

10 AEROSPACE ENVIRONMENTS

ULRICH C. LUFT

The early aspirations of man to free himself from the confines of the earth and to explore the reaches of the stars, once expressed so dramatically in the myth of Daedalus and Icarus, are being fulfilled in our time. Never before has a greater challenge been posed to science, technology, and medicine than the task of preparing man to survive and master the hazards of a hostile and unexplored environment toward which his own unbounded intellectual curiosity inevitably drives him.

The environmental problems of aeronautics and space travel are so universal that a comprehensive treatment would involve the subject matter of practically every chapter of this book. The present chapter will therefore be limited to the most fundamental environmental factor peculiar to aerospace operations, namely the attenuation and complete absence of the terrestrial atmosphere, and will include the physiologic requirements for a self-sustaining artificial environment for orbital flight and exploration of the solar system by man. Other important factors such as weightlessness, radiation, thermal stress, and chronobiology are dealt with authoritatively elsewhere in this volume.

THE ATMOSPHERE

The earth is surrounded by a tenuous envelope of gas and vapor; the kinetic energy of the molecules, imparted by solar radiation, tends to move them out into space. Opposing these forces is the gravitational attraction of the earth's mass. But since the influence of gravity decreases with the square of the distance from the center of the earth, the density of the atmosphere and the resulting barometric pressure decline with altitude following the exponential curve shown in Fig. 10-1. At approximately 5.5 km the pressure is only half that at sea level, at 11 km one quarter, and at 32 km only one hundredth of 1 atmosphere.

In spite of this rapid attenuation of the atmosphere with altitude, there is evidence from thermal and electromagnetic observations that it extends as a continuous medium for more than 1,000 km above the earth's surface.[1] Based on certain physical criteria, three major subdivisions of the inner atmosphere can be distinguished with boundaries that are as yet not precisely defined.

The *troposphere* extends approximately 10 km above the earth with variations in depth depending on season and latitude. It is characterized by decreasing temperature with increasing altitude (–1° C/100 m), variable humidity, and cloud formation associated with marked vertical and horizontal air movement, so that the gaseous constituents are well mixed and maintain uniform composition (Table 10-1).

The lowest layer of the *stratosphere* is essentially isothermal at –55° C from 11 to 30 km, but the temperature rises again up to 50 km because of the interaction of ultraviolet solar radiation with ozone, most of which occurs between 25 and 45 km. The ozone layer is a highly effective shield that protects all living things on earth from harmful ultraviolet rays below 3,000 Å. Above this region, from 50 to 80 km, temperature again drops to about –35° C, increasing turbulence and mixing.

Above 80 km the remaining gas particles, bombarded by intensive shortwave ultraviolet radiation, are more or less completely ionized. The *ionosphere,* ranging from 80 to 700 km, is a strong reflector of certain electromagnetic waves and is utilized for long-range earth-bound com-

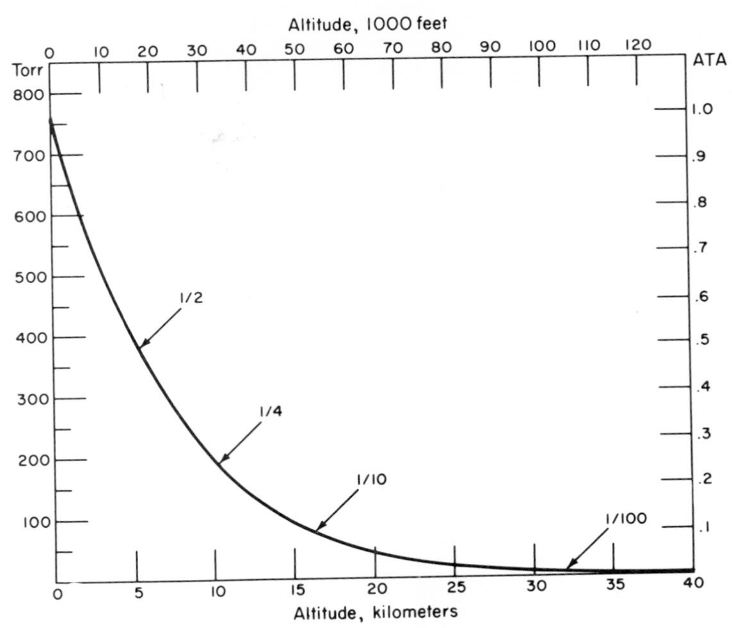

Fig. 10-1. The relationship between barometric pressure and altitude. (From U. S. standard atmosphere, Washington, D. C., 1962, U. S. Government Printing Office.)

Table 10-1. Composition of dry atmospheric air

Gas	Concentration (vol/vol [%])
Nitrogen	78.08
Oxygen	20.95
Argon	0.93
Carbon dioxide	0.03
Trace elements	0.01

munications. Fortunately, the ionosphere is not impermeable to all radio wavelengths but acts as a sieve permitting waves from a few millimeters to 10 m to penetrate. Bands in this range can be employed for communications and telemetry with space vehicles in orbit or on excursions to other celestial bodies as well as for relaying audio and video signals from earth to and from communication satellites. The temperature within the ionosphere rises progressively, reaching more than 1,000° C at 700 km. This temperature, however, relates to the kinetic energy of the gas particles and, in view of their extremely low density at this point, does not correspond to heat in the conventional sense.

The *exosphere* or outer atmosphere extends several thousand kilometers beyond, and it is believed to contain mainly helium and hydrogen. These solitary particles with high kinetic energy rarely collide with each other and are lost in space when they exceed the escape velocity from earth's gravity (11.2 km/sec).

For aeronautic purposes the borderline between atmosphere and space is much closer to the earth, about 80 km; above this altitude aerodynamic lift and guidance by controlled surfaces of an airplane are no longer effective, and propulsion and directional control by rockets are mandatory.

FLIGHT WITHIN THE ATMOSPHERE

The first successful flights with hot air and hydrogen balloons in the latter part of the eighteenth century were events that stirred the imagination of scientists and the public at large as much as the exploits of the astronauts of our time. These intrepid balloonists had very little knowledge of the nature of the environment to which they were committing themselves and were much less prepared and equipped against the hazards of high altitude than the astronauts on their voyages to the moon. It was not until Glaisher and Coxwell narrowly escaped death in the clouds in 1862 and the fatal outcome of the record-breaking ascent of the "Zenith" in 1875 in which Tissandier lost his two companions and barely survived himself that the grave dangers of high altitude were taken seriously. The French physiologist Paul Bert had predicted these effects from ingenious experiments he had conducted on animals and on himself in a low-pressure chamber in Paris.[2] He proved unequivocally that the deleterious effects of altitude were caused by lack of oxygen and could be overcome by addition of sufficient oxygen to the inspired air.

Aerohypoxia

Physiology

All animal life is sustained by energy derived from biologic oxidation of organic fuels. This requires a continuous supply of molecular oxygen to the site of energy production in the cell mitochondria. Since the oxygen reserves of the organism are negligible, sudden interruption of the oxygen stream anywhere from the atmosphere to the brain cells, for instance, causes loss of function within seconds and irreversible damage within a few minutes. Oxygen lack at cell level, or *histohypoxia,* is the cause of death in all forms of disease where oxygen transport to the vital organs is impaired by respiratory or circulatory insufficiency. At high altitude the air we breathe is deficient in oxygen and the appropriate term is *aerohypoxia.*

According to Dalton's law, the total barometric pressure is equal to the sum of the partial pressures of all constituent gases and water vapor. Therefore the partial pressure of each component gas is the product of the barometric pressure and the fractional concentration of the individual gas. Thus for oxygen:

$$P_{O_2} = P_B \times F_{O_2} \qquad (1)$$

where P = partial pressure
P_B = barometric pressure
F = fractional concentration

For example, at sea level the partial pressure of oxygen in dry air is:

$$P_{O_2} = 760 \times 0.2095 = 159 \text{ torr} \qquad (2)$$

Equation 2 was used to calculate P_{O_2} as a function of barometric pressure with altitude in Fig. 10-2, line *A*. It must be taken into consideration, however, that as the inspired air enters the airways it is warmed to approximately body temperature and saturated with water vapor, which exerts

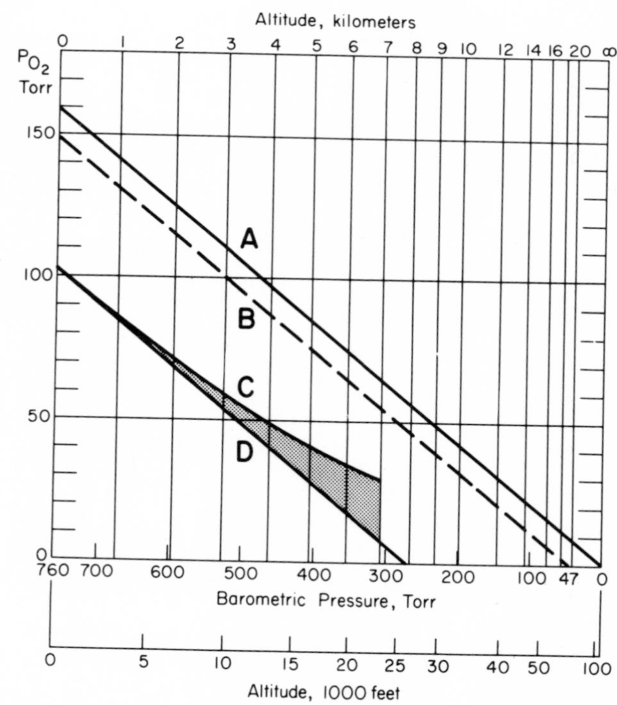

Fig. 10-2. Partial pressure of oxygen on ascent to altitude. *A,* Dry air; *B,* inspired air saturated with water vapor at body temperature; *C,* alveolar gas from actual samples[3]; *D,* alveolar gas calculated for constant ventilation.

a vapor pressure of 47 torr, regardless of barometric pressure. The P_{O_2} as it enters the lungs is:

$$P_{O_2} = (P_B - 47) \times 0.2095 = 149 \text{ torr} \qquad (3)$$

and is represented in line *B* of Fig. 10-2. This is referred to as tracheal P_{O_2}. Curve *C* in the same figure represents mean values for alveolar P_{O_2} based on more than 1,000 determinations obtained in a low-pressure chamber at various simulated altitudes.[3] By way of comparison, line *D* indicates the course of alveolar P_{O_2} predicted for a hypothetical individual who maintains constant pulmonary ventilation and consequently the same alveolar P_{CO_2} (40 torr) during ascent to altitude, as calculated by the alveolar gas equation:

$$P_{A_{O_2}} = P_{I_{O_2}} - P_{A_{CO_2}} \left(F_{I_{O_2}} + \frac{1 - F_{I_{O_2}}}{R} \right) \qquad (4)$$

where $P_{A_{O_2}}$ = alveolar P_{O_2}
$P_{I_{O_2}}$ = inspired P_{O_2}
$P_{A_{CO_2}}$ = alveolar P_{CO_2}
$F_{I_{O_2}}$ = inspired O_2 fraction
R = respiratory exchange ratio (carbon dioxide output/oxygen uptake)

The shaded area between lines *C* and *D* signifies the improvement in alveolar P_{O_2} gained by the spontaneous increase in ventilation stimulated by hypoxia via the carotid and aortic chemoreceptors.[4] A person exposed acutely to high altitude is usually on the verge of unconsciousness when his alveolar P_{O_2} is reduced to 30 torr. On the average this occurs at about 7 km (23,000 ft), whereas without any increase in ventilation (line *D*) this point would be reached at 4.8 km (17,000 ft), a net gain of 2.2 km or 6,000 ft in altitude tolerance.

The hyperventilation resulting from hypoxia is actually not nearly so great as it would be if it were not inevitably associated with a loss of carbon dioxide from the blood (hypocapnia). The action of carbon dioxide, ordinarily the predominant stimulus for respiration, is reduced on acute exposure to high altitude. The loss of carbon dioxide also tends to reduce blood flow to the heart and the brain, while severe hypoxia increases it. This conflict calls for physiologic compromise. Under these circumstances one can understand why attempts to improve altitude tolerance by voluntary additional ventilation have not been successful, because they interfere with physiologic autoregulation. On the other hand, stimulation of breathing by controlled admixture of carbon dioxide to the inspired gas or by partial rebreathing does alleviate the effects of altitude, but it is more practical to add oxygen.

The most effective physiologic safeguard protecting the oxygen supply to the cells under the adverse conditions of high altitude is inherent in the manner in which hemoglobin combines with oxygen in the lungs and releases it to the tissues (Fig. 10-3). Near sea level the blood coursing through the lungs is exposed to any oxygen tension of 100 torr and hemoglobin is nearly saturated *(A)*. On its way through the body it releases about 30% of its capacity and the P_{O_2} in the mixed venous blood is 64 torr less *($\overline{V}$)*. At an altitude of 6.7 km (22,000 ft) alveolar P_{O_2} is close to 30 torr and arterial blood oxyhemoglobin saturation is less than 60% *(A')*. Assuming the same arteriovenous oxyhemoglobin saturation difference here as at sea level (30%), the P_{O_2} only drops 12 torr because of the steep slope of the oxyhemoglobin dissociation curve *($\overline{V}'$)*. By this means the P_{O_2} in the tissues is maintained above the critical level.

The loss of oxygen tension between arterial and venous blood is still further minimized in acute hypoxia by increased blood flow, which reduces oxygen extraction per unit volume of blood at the same oxygen consumption rate. Elevated heart rate is a consistent response to acute hypoxia of any kind. Probably even more important than increased cardiac output are regional vascular adjustments serving the preferential distribution of blood supply to the brain and heart at the cost of other organs that are not so susceptible to hypoxia and whose optimal function may be dispensed with temporarily in the interest of survival, such as skeletal muscle, intestines, and skin.

Symptoms

The manifestations of aerohypoxia vary widely, depending upon rate of ascent, final altitude, and duration of exposure. A particularly insidious feature is the lack of any special discomfort or embarrassment that might serve as a warning signal. Loss of critical judgment and self-assessment preclude all subjective manifestations, so that the victim does not recognize his predicament or attempt to remedy it. Indeed, a sense of well-being and euphoria may suppress any foreboding of danger, as described in the classic account of altitude sickness by Tissandier, the sole survivor of the tragic flight of the balloon "Zenith" in 1875, as cited by Bert[2]:

> Toward 7.5 km the body and mind weaken little by little. One does not suffer at all. On the

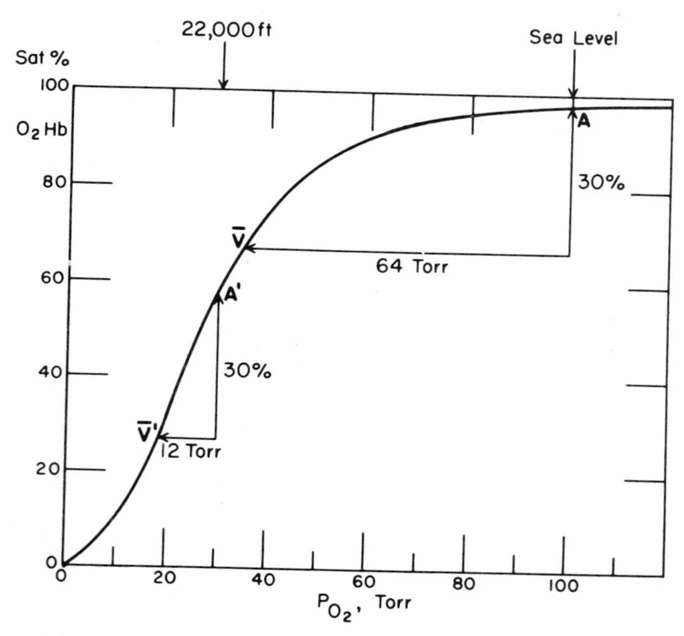

Fig. 10-3. Oxyhemoglobin dissociation curve for whole blood with arterial and mixed venous points $(A - \bar{V})$ at sea level and at 22,000 ft $(A' - \bar{V}')$, assuming that the arteriovenous oxyhemoglobin saturation difference remains unchanged. Oxygen transport is accomplished at high altitude despite minimal blood oxygen pressure decrement.

contrary, one experiences inner joy as if it were an effect of the inundating light. One becomes indifferent, one no longer thinks of the perilous situation, one rises and is happy to rise. The rapture of high altitude is no vain phrase. But as far as I can judge, this sensation appears at the last moment. It immediately precedes annihilation, sudden, unexpected and irresistible.

In the following paragraphs the various manifestations of aerohypoxia are summarized as they are observed during gradual ascent from near sea level at a rate of 300 m (1,000 ft) per minute, as it might occur in a balloon or an airplane without oxygen equipment or pressure cabin. Most of the observations were actually made under controlled conditions in low-pressure chambers in the presence of experienced investigators who used supplemental oxygen at high altitude.

Up to 3 km (10,000 ft) there are no remarkable respiratory or cardiovascular responses at rest, but a slight decrement in learning capacity[5] and a delay in dark adaptation[6] have been demonstrated beginning at 2 km (6,500 ft).

Between 3 and 5 km (10,000 and 16,500 ft) increased heart rate and respiration signify beginning physiologic compensation. Physical work capacity is reduced and tests of choice reaction time,

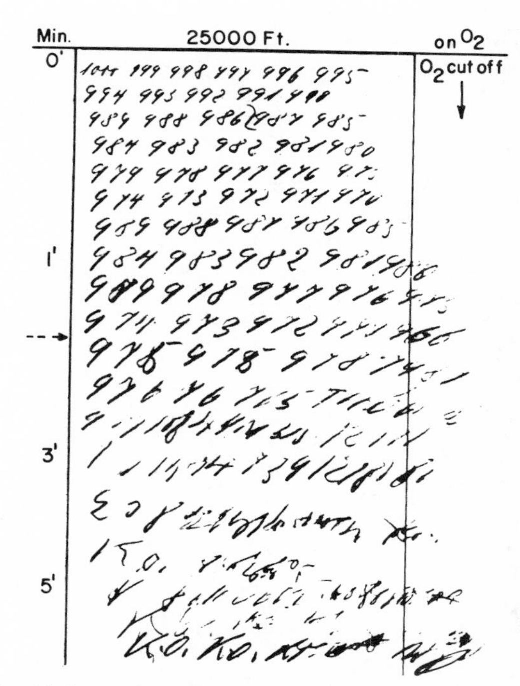

Fig. 10-4. Handwriting test while breathing air at 7.5 km (25,000 ft) after ascent with oxygen equipment, Oxygen supply was interrupted at 0 minute.

Table 10-2. Distribution of 100 young men according to the altitude equivalent at which they were incapacitated by aerohypoxia during gradual decompression in a low-pressure chamber

Altitude	Number of subjects
9,000 m (29,500 ft)	4
8,000 m (26,200 ft)	20
7,000 m (23,000 ft)	55
6,000 m (19,700 ft)	18
5,000 m (16,400 ft)	3

which require judgment before responding to a stimulus, show a decrement in mental performance while simple reaction time is not affected.[5]

From 5 to 7 km (16,500 to 23,000 ft) respiration and heart rate increase markedly without subjective sensation of dyspnea or palpitation. Mild headache and dizziness are common but usually dulled by progressive lassitude and indifference. Emotional behavior varies between individuals from euphoria and hilarity to depression and aggressive attitudes reminiscent of alcohol intoxication. Neuromuscular coordination declines rapidly, as demonstrated by deterioration in performing a simple writing test (Fig. 10-4).

Finally, comprehension and responsiveness wane rapidly, merging into unconsciousness. However, postural tone is maintained at this point, occasionally with mild convulsive movements of the extremities, and consciousness is regained within 10 to 15 seconds if oxygen is administered immediately. The altitude and exposure time at which unconsciousness sets in varies widely between individuals. Table 10-2 gives the distribution in altitude "ceiling" for 100 healthy men 20 to 30 years of age in graded ascent in a low-pressure chamber at 300 m/min with 5-minute pauses every 1,000 m, using a handwriting test to determine critical threshold. The majority reached their limit at 7 km (23,000 ft). The altitude "ceiling" on acute exposure is considerably higher in individuals acclimatized to high terrestrial elevations. This has been found in participants in a Himalayan expedition who had lived for a month at or above 6,000 m and were tested before the expedition and after return to the lowlands. Even a short sojourn of 2 to 3 weeks at an elevation of 3.5 km (11,500 ft) markedly improves acute altitude tolerance.[7] Similar observations have been made on Peruvian natives who live at 4.5 km (15,000 ft).[8]

Protection

Oxygen. The most effective remedy against aerohypoxia is to increase the P_{O_2} in the inspired gas so that it approaches the level we are accustomed to breathe at or near sea level. This can be accomplished at high altitude either by increasing the fractional concentration of inspired oxygen by controlled admixture or by boosting the total pressure in cockpit and passenger compartment of the aircraft by compressing ambient air to the appropriate level. A combination of both principles may be expedient under certain circumstances.

For reasons of economy, it is desirable to limit the supply of additional oxygen to the minimum necessary to prevent hypoxia at the altitude at which the aircraft is traveling. For example, to obtain the oxygen fraction (F_{O_2}') that produces the same partial pressure at a lower barometric pressure (P_B') as at sea level, equation 3 is modified as follows:

$$F_{O_2}'(P_B' - 47) = 0.2095\ (760 - 47) \qquad (5)$$

$$F_{O_2}' = \frac{149}{P_B' - 47}$$

This expression is plotted as a function of altitude in Fig. 10-5 (sea-level equivalent). Since no functional impairment is to be expected in healthy persons up to 3 km (10,000 ft), the equivalent for this altitude is permissible for practical purposes and is the minimal requirement in military aircraft (lower curve in Fig. 10-5).

The earliest type of oxygen equipment with a continuous flow and a simple pipestem mouthpiece was superseded by a mask of rubber or, later, plastic material covering nose and mouth. Oxygen was introduced through a reservoir bag of about 500 ml capacity attached to the mask to store oxygen during expiration. Several small openings or a single orifice with a foam rubber throttle in the mask served for expiration and also to admit air during inspiration at lower altitudes where less oxygen is needed. The oxygen supply was governed by a regulator controlled manually or by an aneroid, calibrated to provide the appropriate flow for the flight altitude. This type of equipment with disposable masks is available in most current commercial airlines for emergency use in the event of loss of cabin pressure and, if properly used, provides adequate protection for emergency descent from altitudes up to 12,000 m (40,000 ft), but it is not recommended for extended use above 7,000 m (23,000 ft).

Requirements for flight at much higher alti-

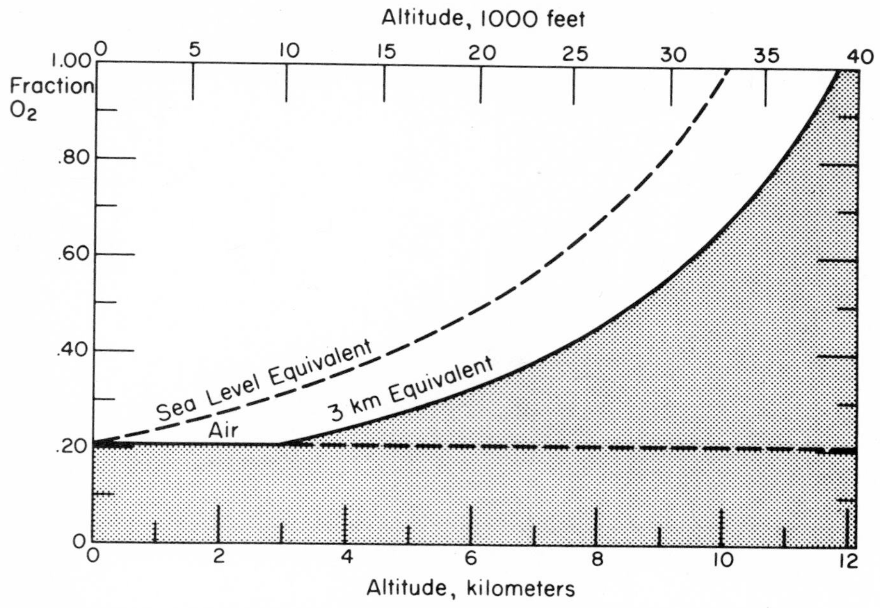

Fig. 10-5. Oxygen fraction required to maintain an oxygen pressure in the lungs equal to that at sea level or at 3 km (10,000 ft). Shaded area is hypoxic zone.

Table 10-3. Comparison of physiologically equivalent altitudes of aerohypoxia while breathing air at lower and oxygen at higher altitudes based on equal tracheal oxygen pressure

Altitude on air		Tracheal	Altitude on 100% oxygen		Hypoxic
km	ft	P_{O_2}	km	ft	symptoms
0	0	149	10.0	32,800	None
1	3,300	132	10.7	35,000	None
2	6,600	115	11.3	37,000	None
3	9,800	100	11.9	39,000	None
4	13,100	87	12.5	41,000	Mild
5	16,400	75	13.1	43,000	Mild
6	19,700	64	13.7	45,000	Severe
7	23,000	55	14.2	46,600	Critical

tudes in unpressurized aircraft in the late 1930's led to the development of a much more efficient system using a tightly fitted mask and a regulator that delivers oxygen only during inspiration triggered by mask pressure. The proper admixture of air is precisely controlled by the "diluter-demand" regulator. Its main advantages over the continuous flow principle are greater oxygen economy and, at the same time, instant response to changes in ventilation according to the physical activity of the user.

The limit of usefulness for ambient pressure oxygen equipment is reached at about 12 km (39,000 ft). Above this point the barometric pressure, even using 100% oxygen, is too low to maintain oxygen pressure in the lungs above the minimum for safety. At higher altitudes hypoxic symptoms develop just like those at lower altitudes breathing air. This is apparent in Table 10-3, where certain altitudes are compared that are physiologically equivalent, based on their tracheal P_{O_2} (saturated with water vapor at 37° C), either breathing air at lower altitudes or pure oxygen further aloft. It is also evident that the safe operational ceiling for aviation in ambient pressure aircraft is raised from 3 km (10,000 ft) to nearly 12 km (39,000 ft) by the use of oxygen equipment.

Pressure breathing. At a time when high-performance aircraft were already capable of operating above 12 km (40,000 ft) but reliable pressurized cockpits were not yet available, attempts were made to "supercharge" the lungs by pressure breathing in order to increase altitude tolerance. This is accomplished by supplying a controlled constant positive pressure from the oxygen regulator, which increases the total pressure in the lungs, airways, and face mask, which is fitted with a counterbalanced exhalation valve. By raising the mean intrapulmonic oxygen pressure by 25 to 30 torr above ambient, altitudes up to 13.7 km (45,000 ft) can be tolerated without hypoxic embarrassment. The equipment has been used successfully in emergencies up to 15.2 km (50,000 ft). However, constant positive pressure breathing (CPPB) imposes certain physiologic penalties that partly offset its advantages at altitude. Not only is breathing laborious because of the necessary expiratory resistance, but the mechanics of breathing are also altered. The active work of inspiration is supplied by pressure from the regulator and the lungs are expanded passively. On the other hand, considerable effort is required for expiration, which is normally passive. More important is the circulatory embarrassment caused by the elevation of intrathoracic pressure, which tends to impede the return of venous blood into the chest and right heart unless sufficient backpressure is built up in the extrathoracic veins to overcome the hindrance. In response to this, the arterial blood pressure rises immediately by about the same amount as the intrathoracic pressure and circulation is maintained; however, cardiac output is reduced and intrathoracic blood volume is depleted. Furthermore, urine output is consistently depressed during CPPB. This observation has initiated a number of revealing experimental studies that have brought new insight into the neuroendocrine control of blood volume mediated by intracardiac mechanoreceptors.[9] The unbalanced type of pressure breathing is not well tolerated for longer periods because it pools blood in peripheral veins and also causes loss of plasma fluid into the tissues. This can lead to collapse from loss of circulating blood volume.

Much higher pressures can be sustained by the lungs without respiratory or circulatory embarrassment if counterpressure is applied to the trunk with pneumatic bladders and rubber capstans along the limbs, which, when inflated, apply pressure to a tight-fitting garment called a partial pressure suit. Equipped with a full helmet, this garment provides pressures up to 150 torr in the lungs. Using 100% oxygen, this gives adequate protection for escape to lower altitudes in emergencies up to 30 km (100,000 ft). The study of the physiology of pressure breathing has focused attention on important interactions between respiratory, cardiovascular, and renal function, with implications for clinical medicine as well as for aerospace and underwater physiology. For example, the therapeutic application of intermittent positive pressure breathing (IPPB) for patients suffering from obstructive pulmonary disease is the outcome of developments in aviation.

Hypobaric decompression sickness

It was not until a number of years after the introduction of reliable oxygen equipment had enabled aviators to fly at altitudes well above 7 km (23,000 ft) that they occasionally became aware of certain symptoms that were not alleviated by oxygen but were similar to those experienced by divers and caisson workers on decompression to 1 atmosphere from a hyperbaric environment. Indeed, it was after an ascent to a simulated altitude of 12.8 km (42,000 ft) for the purpose of testing oxygen equipment that Jongbloed[10] first described acute joint pain and a peculiar itching of the skin from personal experience and identified them as manifestations of decompression sickness. It did not become a matter of concern in flying operations until World War II, when unpressurized aircraft were frequently operating for hours at altitudes above 10 km (33,000 ft). Subsequent experience gained in the indoctrination of large numbers of airmen in low-pressure chambers showed that at an altitude of 11.5 km (38,000 ft) above 50% developed decompression symptoms in less than 1 hour and that 30% were incapacitated within 2 to 3 hours. The incidence and severity are inversely related to barometric pressure, and only a few cases have been reported at less than 7 km (23,000 ft). Even on rapid decompression to quite high altitudes symptoms rarely develop in less than 5 minutes. The incidence increases beyond 1 to 2 hours, then tapers off. Individual susceptibility varies a great deal; in general, obese individuals and older people are more prone to develop decompression symptoms.

Symptoms

In contrast to aerohypoxia, hypobaric decompression sickness is characterized by the unpredictability and diversity of its manifestations. The most common of these are dull gnawing pains around one or more joints, known as "bends." The knees, shoulders, and elbows are most frequently, but not exclusively, afflicted. Physical activity is known to elicit pain more readily in the joints in motion, and cold exposure increases susceptibility. Recompression to lower altitude almost invariably gives complete relief, but returning to high altitude, even after a period of many hours, is likely to produce pain in exactly the same location within a shorter period of time. Much less frequent but more alarming is a feeling of discomfort in the chest, accentuated by deep inspiration that induces a rapid shallow breathing pattern interrupted by paroxysmal coughing bouts known as "chokes." This is a serious condition that may lead to a state of asphyxia and collapse unless recompression is initiated promptly. Other manifestations include itching or burning of the skin with occasional local discoloration in the form of a rash or "marbling." Various neural manifestations such as blurring of vision, defects in the visual field, regional anesthesia, and selective paralyses of the extremities have been described that disappear rapidly on recompression. Unlike the effects of hypoxia, decompression sickness does not usually interfere with mental function, and full consciousness and critical judgment are maintained. However, syncope and collapse are seen in about 10% of those stricken with severe "bends" and more frequently in those with "chokes." As a rule such persons recover at lower altitude without any after-effects. However, in a few rare instances a state of intractable shock may develop with marked hemoconcentration, coma, and bizarre neurologic symptoms.[11] The only effective therapy for postdecompression shock appears to be exposure to high pressures (3 ATA) in a hyperbaric chamber with oxygen.[12]

Etiology in aviators and divers

Current concepts on the etiology of hyperbaric decompression sickness, as experienced during or after decompression from a high-pressure environment under water or in caissons, are presented in Chapter 11. Basically, the same considerations apply to the hypobaric variety with which we are dealing here. But there are certain differences that deserve emphasis because they are important in understanding the manifestations and the course of the disorder.

If the pressure of the environment is constant long enough, all body fluids become saturated with nitrogen (or other inert gases), depending upon its partial pressure in the lungs. This is not true for oxygen and carbon dioxide that are being removed or produced in the tissues. When environmental pressure decreases, a state of supersaturation with nitrogen exists in the body until a new equilibrium is reached by elimination of excess nitrogen through the lungs. The rate of removal, being determined by regional blood flow and relative nitrogen content, is not equal for all parts of the body. It takes many hours to reach a new level of saturation. If the pressure of nitrogen tending to escape from the tissues greatly exceeds the containing force of the surrounding tissues, a critical condition may occur, leading to the formation of bubbles originating from minute nuclei created by cavitation, surface tension, or turbulence in blood vessels. Once a bubble has evolved, water vapor and other gases diffuse into it and participate in its growth, which will be directly proportional to the expansion pressure of the gas it contains and inversely related to the forces containing it, namely the external pressure plus the pressure of the tissue or fluid surrounding it. If the tissue is compliant enough to accommodate expansion of the bubble, an equilibrium may be reached whereby most of the expansion pressure is dissipated. If, on the other hand, the surrounding medium is nonelastic, a high pressure persists and may cause pain or tissue damage or obstruct blood flow.

A highly simplified model in Fig. 10-6 (soap bubble) serves to illustrate the difference between an ascent to 11.75 km (38,500 ft) from sea level (hypobaria) and surfacing from a depth of 40 m (132 ft) of water (hyperbaria). Conditions are chosen so that the potential for relative gas expansion following Boyle's law is the same for both instances, based on the ratio of initial to final pressure (5:1). The expansion pressure (ΔP), disregarding surface tension and other factors, is much greater in the hyperbaric model than in the hypobaric case. Under the same conditions a diver allowed to stay at a depth of 40 m (132 ft) until fully saturated with nitrogen would have five times more of the gas in his body as a potential source of bubbles than his counterpart ascending to 11.75 km (38,500 ft) from sea level. Thus decompression sickness is much more dangerous for divers than for aviators, as practical experience has shown; not only are the symptoms usually more severe and the mortality higher, but there is also a greater predilection for lesions of the central nervous system and bone damage in

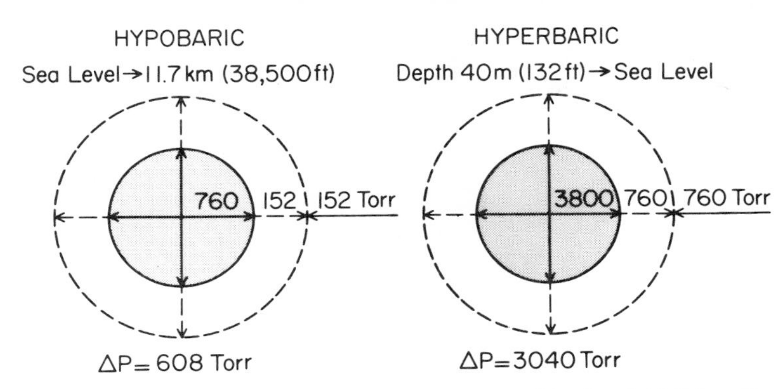

Fig. 10-6. Decompression of a soap bubble from sea level to an altitude of 11.7 km (hypobaria) and surfacing from a depth of 40 m under water (hyperbaria). The relative expansion, if unopposed, is the same in both cases, but the pressure head in the bubble (ΔP) is five times greater in the hyperbaric decompression. As the bubble expands, ΔP approaches zero.

caisson disease. Another difference is that the diver is exposed to high pressures for relatively short periods of time and his tissues are not usually saturated with nitrogen corresponding to his depth before he returns to his habitat at lower pressure. Therefore, those organs that equilibrate faster are more prone to bubble formation because they contain more nitrogen. The opposite is true for the aviator, who lives at the higher pressure of his range so that all tissues are saturated for 1 atmosphere. On decompression to high altitude, those tissues with slowest nitrogen elimination are generally more likely to cause trouble. Furthermore, the best remedy for decompression symptoms, namely recompression, is implicit on return to lower altitudes by the aviator, whereas recompression for the diver requires special hyperbaric equipment for treatment, followed by gradual return to a normal atmosphere.

Practical experience in diving has shown that the incidence of decompression sickness is related to the ratio of initial to final pressures on emerging; Haldane[13] introduced a successful schedule for stage decompression, based on the observation that no difficulties were encountered if the pressure ratio is not greater than 2:1. Thus it is permissible to surface without stopping after a dive of long duration to less than 10 m (33 ft) without any danger. If one applies this concept to altitude, no indication of decompression sickness should ever occur on ascent to altitudes where the barometric pressure is more than 380 torr, that is, 5.5 km (18,000 ft). In fact, no cases have been reported below this altitude and they are exceptional below 6 km (20,000 ft). Apparently, the rule applies under hypobaric as well as hyperbaric conditions. An interesting corollary to this can be seen in several incidents of serious bends while flying in commercial aircraft with a cabin pressurized to 2.4 km (8,000 ft) after scuba diving to depths of 10 m (33 ft) a few hours before takeoff. It has since been recommended that an interval of at least 12 hours on the ground elapse after diving before taking to the air.[14] Similar reasoning applies to diving operations in lakes or caisson operations at high terrestrial elevations, where decompression symptoms may occur on surfacing rapidly from depths much less than the 10-m (33-ft) limit considered safe at sea level.

Prevention

The stage decompression principle used so successfully in diving and caisson activities for protection against decompression ailments is, unfortunately, not practicable for aerospace operations, for it would take a stage ascent of about 3 hours' duration to provide protection for flight at an altitude of 10 km (33,000 ft). The most obvious alternative is to pressurize the compartments for crew and passengers to a level not less than 380 torr corresponding to 5.5 km (18,000 ft), and this is now universally employed in civil aviation.

Before pressurized aircraft were introduced, it was found that the incidence of decompression sickness could be substantially reduced by breathing pure oxygen for a certain length of time prior to flight to minimize the nitrogen content of the body (preoxygenation, or denitrogenation). This

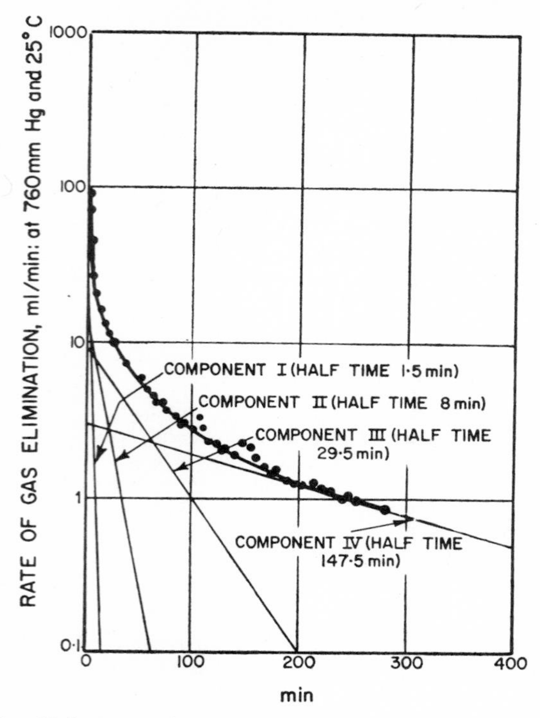

Fig. 10-7. Rate of nitrogen elimination from the body while breathing oxygen. The experimental curve is broken down into its fast and slow components attributed to tissues with different removal rates. (From Fryer, D. I., and Roxburgh, H. L.; after Jones, H. B.: Preoxygenation and nitrogen elimination. In Decompression sickness, National Research Council Report, Committee on Aviation Medicine, Subcommittee on Decompression Sickness, Philadelphia, 1951, W. B. Saunders Co., chap. 9.)

procedure is still in use for certain military flight profiles and in aerospace operations where the cabin pressure is kept at a lower level for technical reasons.

When pure oxygen is breathed, the partial pressure of nitrogen in the lungs drops to a minimum within a few minutes, creating a pressure gradient that tends to draw nitrogen out of the tissues toward the lungs, where it is vented. Since the pressure head becomes ever smaller as nitrogen is removed, the rate of nitrogen elimination when plotted against time (Fig. 10-7) follows a decay curve with multiple components. More than two thirds of the entire nitrogen store in the body (about 1 L at 1 atmosphere pressure) is removed during the first hour of oxygen breathing, mainly from blood, muscles, and viscera that have high blood flow. The slowest components of the curve (Fig. 10-7) are attributed to bone and fatty tissues because of their minimal blood supply. Besides, nitrogen is five times more soluble in fat than in muscle. Complete removal of this less accessible nitrogen fraction takes more than 15 hours, and it is not surprising that preoxygenation even for 4 to 5 hours does not give absolute protection against bends. However, the incidence and severity of decompression symptoms are appreciably reduced after only 1 hour of oxygen breathing preflight, which is more realistic in terms of flight operations. In this context it is noteworthy that people living at moderate elevations, even less than 1.5 km (5,000 ft), are unusually resistant to decompression. This is surprising, because at that altitude the nitrogen content of the body is only 17% less than at sea level, whereas more than 60% is eliminated by 1 hour of oxygen breathing. Possibly decompression sickness is caused primarily by the slowly removable nitrogen in poorly perfused tissues. Even a slight reduction of this fraction may be more important than a much greater loss of total body nitrogen.[15]

Barotrauma

Considerable discomfort can arise in flight as a result of the mechanical effects of changes in barometric pressure on undissolved gas trapped inside body cavities without immediate access to the environment. Most frequently involved are the middle ear, the paranasal sinuses, the teeth, and the gastrointestinal tract. Although rarely dangerous in a lethal sense, this type of affliction can cause severe pain and embarrassment and may distract from or interfere with activities essential for the operation of an aircraft. In severe cases damage may occur to the tissue with hemorrhage, effusion, or other sequelae that may persist after landing.

The middle ear lies in a small cavity of the temporal bone. It is separated from the external auditory canal by the eardrum and communicates with the nose and mouth via the eustachian tube, a narrow canal surrounded by bone and cartilage and lined with mucosal and adenoid tissue. The orifice of the tube to the oronasal cavity is normally closed but opens momentarily during swallowing or yawning. During ascent the air trapped in the middle ear expands but escapes readily through the eustachian tube whenever the pressure gradient exceeds 10 to 15 torr unless the passage is obstructed by inflammatory congestion. Thus the middle ear vents automatically during ascent and creates no subjective sensations beyond audible clicking at intervals. During descent the air in the middle ear contracts, producing a negative gradient against ambient pressure. Since the orifice of the eustachian tube acts as a flutter valve, the passage remains closed unless opened by swallowing or yawning. If this is unsuccessful, more forceful efforts are required; the nose and mouth are held closed and the cheeks blown out by forced expiration (Valsalva maneuver). Even better is the Frenzel maneuver in which the glottis, nose, and mouth are closed and the muscles of the floor of the mouth and cheeks contracted to build up pressure.

If descent continues without venting the middle ear, sufficient negative pressure may build up to make it impossible to open it by any means. The eardrum is forced inward, which can be very painful, and hearing is impaired (barotitis, or aerotitis media). This can be immediately relieved by reascending to high altitude and the suggested maneuvers may be repeated with more success. There may be bleeding or serous effusion into the middle ear that clears up in 4 to 8 days. Flying should be discontinued during recovery. A similar condition occurs much less frequently but also predominantly during descent in any of the paranasal sinuses (barosinusitis). Normally, these communicate freely with the nasal passage through small rigid apertures, so that no pressure gradients arise during flight. Barosinusitis is almost always caused by preexisting sinusitis with obstructing discharges or deformities of the nasal septum or turbinates occluding the sinus openings. Active measures to clear the sinuses are much less effective than those described for venting the middle ear.

Toothache during ascent to high altitude has been known since the early days of aviation, but it is a rare occurrence. It was first believed to be caused by gas trapped in or around faulty teeth, but this is not always the case. Circulatory changes in the dental pulp caused by reduced barometric pressure in the presence of dental pathology are now believed to be the most frequent cause. Barodontalgia subsides immediately on descent, and this makes it easy to differentiate from maxillary barosinusitis if a tooth in the upper jaw is afflicted.

The gastrointestinal tract always contains small amounts of gas that have either been swallowed or have evolved from the digestive process. Its quantity varies from 150 to 500 ml, depending upon the digestive state and the diet, and it also tends to expand with altitude. Owing to the flexibility and distensibility of the gut, uncomfortable sensations are not experienced unless there is kinking around the trapped gas pocket. Abdominal pain from barometeorism is rarely severe below 9 km (30,000 ft), but it may cause respiratory embarrassment with splinting of the diaphragm, and it has been known to be followed by reflex vasomotor reactions with cold sweat, pallor, and syncope. Dietary precautions including avoiding the ingestion of gas-forming foods such as cabbage, legumes, rye bread, and beer prior to flight to high altitudes reduce the probability of barometeorism.

As far as civilian air travel is concerned, the incidence of barotrauma in its various forms has been minimized by the use of pressurized cabins, although occasionally discomfort may be felt in the ears when cabin pressure increases too rapidly before landing.

Pressure cabins

The preceding sections of this chapter have dealt with the physiologic effects of acute exposure to high altitude as experienced by the aviator, particularly with the biologic limitations to human endeavors in aerospace imposed by the nature of the earth's atmosphere. Nearly 50 years ago the British physiologist J. S. Haldane came to the conclusion that an aviator enclosed in an airtight garment or compartment, similar to a diver's suit, providing him with the necessary pressure would have no limit to the height he could attain.[13] Indeed, the implementation of this principle has not only opened up a new era of aerial travel to the public but has also made possible the exploration of outer space and man's landing on the moon.

The primary purpose of pressure compartments is to prevent aerohypoxia and decompression sickness by maintaining a more or less constant pressure inside, regardless of how low the pressure may be outside the aircraft. This can be accomplished by either of two methods. In the lower strata of the atmosphere it is feasible to impel air from outside the aircraft into the cabin continuously by compressors and to bleed it out by loaded outlet valves balancing the flow of air so as to maintain the necessary pressure differential and also to provide adequate ventilation. Pressure cabins of this type are currently in use in most commercial airliners and military aircraft. For operations in the upper reaches of the atmosphere and in space there is no alternative to a sealed cabin with a completely self-contained internal environment.

The use of pressure cabins became mandatory

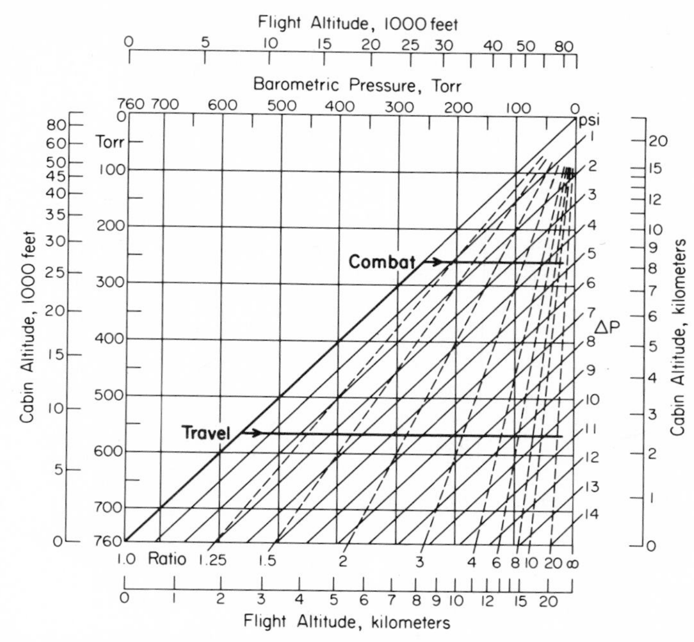

Fig. 10-8. Pressure profiles in aircraft cabins. Below, Civil transport with high differential profile and isobaric cabin pressure at or below 2.4 km (8,000 ft). Above, Combat aircraft with low differential profile using oxygen equipment above 3 km (10,000 ft) and isobaric cabin above 8.2 km (27,000 ft).

with the advent of jet-propelled aircraft that operate more efficiently and economically at much higher altitudes than reciprocating engines. With sufficient cabin pressure the use of cumbersome individual oxygen equipment is not necessary and decompression symptoms and barotrauma caused by rapid changes in pressure are avoided. Well-protected from the cold thin air outside, crew and passengers enjoy freedom of movement within the cabin and the ability to take meals and beverages without special protective equipment or garments.

Since the additional weight and power requirements for pressurization impose an appreciable penalty on the performance and economy of an aircraft, the type of pressure system employed and its control must be determined with close consideration given to the type of aircraft and its mission. Thus aircraft traveling at supersonic speeds generate high ram pressures by the impact of their speed, and this can be utilized with proper controls to pressurize the cabin. Traveling at an altitude of 15 km (50,000 ft) at a speed of mach 2 (multiple of the speed of sound) an aircraft creates sufficient ram pressure to maintain a cabin at the equivalent of 3 km (10,000 ft). However, since ram pressure varies with air speed and density, auxiliary compressors would always be required for slower phases of flight.

From a physiologic point of view the ideal cabin pressure would be as close to that at sea level as possible. In practice concessions can be made without forfeiting health or comfort of the occupants in view of the considerable latitude of human tolerance to altitude. For any given cabin "altitude" the force of the pressure differential that must be supported by the walls of the cabin and the ratio between cabin pressure and outside pressure, which reflects the power required of the compressors, increase with altitude. The relationships between these variables are described graphically in Fig. 10-8, where flight altitude is plotted on the horizontal axis against cabin "altitude" on the vertical axis. The parallel diagonal lines indicate the pressure differential (ΔP) and the rays originating from the top right corner show how many times greater the pressure is inside the cabin than outside (R = ratio). It can be seen that in an aircraft traveling at 12 km (40,000 ft) with the cabin pressurized equivalent to sea level, the walls would have to withstand a force of 12 psi. But if the cabin pressure were set to the equivalent of 3 km (10,000 ft)—still providing adequate protection—the differential pressure would be only 7 psi. At the same time the power requirements, corresponding to the pressure ratio, would be reduced from 5 to 3.6, a saving of 28%.

In commercial air travel the cabin "altitude" is generally controlled to maintain the equivalent of 2.4 km (8,000 ft) or less to accommodate a wide range of age and state of health in the passenger population. The pressure profile of the cabin would follow the course indicated in Fig. 10-8 (travel), becoming isobaric at 2.4 km (8,000 ft) as the aircraft ascends to whatever its cruising altitude may be. For military aircraft, particularly in fighter types, much lower cabin pressures are employed in combat because of greater risk of sudden loss of cabin pressure from enemy action. In the upper example in Fig. 10-8 (combat) cabin pressure is not engaged before reaching 8.2 km (27,000 ft) followed by an isobaric phase. In this case the differential pressure is only 3 psi at 15 km (50,000 ft) as compared to 9.4 psi in the transport plane at the same altitude. The crew of the combat plane, however, must use supplementary oxygen equipment from 3 km (10,000 ft), if not from takeoff, and continue to do so throughout the mission.

At altitudes above 20 km (66,000 ft) the task of pressurizing the cabin from the atmosphere becomes increasingly difficult because it is so rare. For example, to maintain adequate cabin pressure in flight at 21 km (70,000 ft) would take a compression ratio of 17, at 24.3 km (80,000 ft) a ratio of 28, and at an altitude of 30.5 km (100,000 ft), of 68. Obviously, the power requirements become excessively high. An additional problem is the intense heat created by the rapid, high compression. The air coming out of the compressors into the cabin at 21 km (70,000 ft) would have a temperature of more than 300° C, requiring more machinery and power to cool it to a comfortable level. The same would apply if the cabin were supplied by ram pressure during supersonic flight. There is also concern about possible toxic effects of ozone, which is more abundant at these altitudes (ozone layer) and would be concentrated by the compression process. For these reasons the operation of aircraft that rely on outside air for their cabin environment is limited to less than 25 km (82,000 ft) altitude. Beyond this, only a completely closed, or sealed, system with an artificial atmosphere of its own offers the means to penetrate the outermost fringes of the earth's atmosphere into space.

FLIGHT BEYOND THE ATMOSPHERE

Space cabins

The design and development of a sealed compartment with a life-sustaining environment that meets all bodily needs for a number of occupants who are isolated in the midst of a lifeless, hostile environment on a journey of days, weeks, or longer are indeed a gigantic undertaking. To perform their arduous tasks the crew must have the best possible conditions, not only with regard to the pressure and composition of the gas they breathe but also with respect to cabin temperature and humidity, food and water supply, hygienic facilities, and accommodations for sleep and recreation. In many respects the requirements for space travel are very similar to those for long submarine voyages. However, prolonged space missions require even more reliable life-support systems and confront man with physical and psychologic stresses that cannot be simulated on earth. The absence of gravity, a force to which all life on earth is conditioned, is unique to the space environment. How prolonged weightlessness may affect physiologic processes is a matter of grave concern and much speculation. Naturally, the investigation of this unknown feature of the space environment continues to be the primary objective of physiologic observations on animals in biosatellites and on the astronauts themselves. These in-flight studies are fraught with difficulties mainly because of the interaction of other unusual and unphysiologic conditions associated with space flight, such as prolonged limitation of physical activity and certain essential features of the artificial gaseous cabin environment.

Pressure and composition

In choosing the total pressure and gaseous composition of the artificial atmosphere for a manned space vehicle, the first consideration is the maintenance of mental and physical integrity of the occupants with adequate safety margins for all levels of activity including necessary modifications of the gas environment for extravehicular operations in docking maneuvers with other spacecraft or landings on other celestial bodies. At the same time the nature of space flight demands that the requirements be met with a minimum penalty in terms of weight, power, and operational complexity.

With this in mind let us first consider certain special circumstances over and above the routine flight activities that must be taken into account in the choice and control of a respiratory gas mixture for the astronaut. During launch the several boosting stages necessary to achieve escape velocity (11.2 km/sec, or 7 miles/sec) subjects him to strong forces of acceleration with peaks as high as 6 G for a period of several minutes. The tremendous friction of the earth's atmosphere during reentry of the spacecraft at enormous speed slows it at such a rate that even greater G forces act on his body. Acceleration and deceleration are best tolerated if their direction is in the transverse, rather than longitudinal, axis of the body. Therefore, during launch and reentry astronauts are strapped into contoured couches in recumbent position. Under high G forces breathing is very labored, and the distribution of pulmonary blood flow is so disrupted that gas exchange suffers and arterial hypoxemia results despite normal or even higher oxygen tension in the alveolar gas.[16] Thus inspired gas oxygen concentration should be as high as possible during these relatively short periods.

Extravehicular activities (EVA) require a spacesuit with its own self-contained life support system. For mechanical reasons it has not yet been possible to design a spacesuit that permits satisfactory mobility of the joints of the body if it is pressurized to more than 3.8 psi (196 torr). This is equivalent to an altitude of 10 km (33,000 ft) when used in the vacuum of space and is just sufficient to provide a normal oxygen pressure if 100% oxygen is breathed (Fig. 10-5). Finally, possible emergency situations in which the cabin pressure is reduced either deliberately or by accident, such as perforation by a meteorite, are contingencies that must be considered in selection of a cabin environment affording optimal conditions for rescue and survival.

From the technologic point of view, the choice of total pressure and composition of the cabin environment largely determines the structural rigidity and weight of the walls, the minimal leak rate and storage capacity for gases, thermal and humidity control, as well as the necessary monitoring and supply systems required on board. Another highly important factor is the potential fire hazard of a cabin atmosphere.

Reconciliation of the physiologic necessities with the technical constraints and overall safety precautions is best understood by scrutinizing the merits and disadvantages of several possible cabin atmospheres that have either already been employed or are being seriously considered for future space missions (Fig. 10-9).

As in the pressure cabin discussed above, the first option of the physiologist for prolonged space flights is whether to use a sea-level equivalent (14.7 psi, 760 torr) atmosphere. Unfortunately, there are several disadvantages, both physiologic and tech-

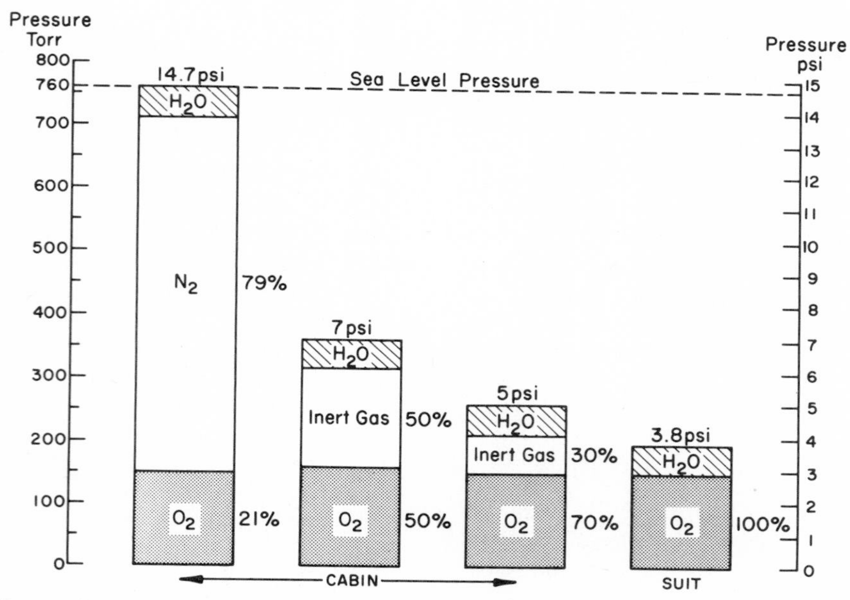

Fig. 10-9. Proposed space cabin atmospheres with different cabin pressures and composition showing the oxygen concentrations necessary to maintain a partial pressure equivalent to that at sea level in each case.

nical, to this choice. In the event of scheduled or unscheduled decompression to spacesuit pressure (3.8 psi, 196 torr), there would be a high probability of developing decompression sickness unless the transition occurred in slow stages, as in diving. A second disadvantage is the danger of mechanical trauma in the unlikely event of accidental rapid decompression, and the reduced chances of recovery from fulminating aerohypoxia because of the low oxygen concentration in the inspired air. From the technical aspect a cabin with a pressure of 1 atmosphere requires the strongest, and thus the heaviest, construction to withstand a pressure head of 14.7 psi. Another consideration is the high leak rate of gas through minimal pores and cracks in sealing materials (diffusion leakage) that can never be totally eliminated. More gas would have to be carried along adding to the liftoff weight of the vehicle.

The lowest cabin pressure compatible with a normal oxygen tension in the inspired gas using 100% oxygen is 3.8 psi (196 torr), as shown on the far right of Fig. 10-9, and this is the environment in the spacesuit. If the same pressure were employed in the cabin throughout the mission there would be a high incidence of decompression symptoms during the early phase of flight, a problem that might jeopardize the mission. In case of emergency with loss of cabin pressure, there would be less time to don the spacesuit as an alternative than in a cabin with higher pressure. Breathing pure oxygen for many hours in experiments at simulated altitude in a low-pressure chamber has produced pulmonary atelectasis[17]; alveoli that have poor or no communication with the air passages collapse because the oxygen they contain is absorbed and removed by the blood. A similar process can occur in the middle ear, leading to retraction of the eardrum and impaired hearing. Incidentally, animal experiments have shown that the rate of collapse of a lung filled with oxygen and then occluded is 3.5 times faster at a pressure of 256 torr than at sea level because of the much lower density of the gas[18] and the fact that a large part of the alveolar space is occupied by carbon dioxide and water vapor. Both conditions are avoided by the presence of an inert gas that is not so readily absorbed.

Technically, the 3.8 psi (196 torr) 100% oxygen atmosphere has the advantage of a low leak rate and the simplicity of a single gas atmosphere with the same pressure for the cabin and the suit. On the negative side of the balance is the greater weight and power requirement for temperature and humidity control in the cabin by convective heat and water transfer at these low pressures. On earth some body heat is dissipated by convection; air, warmed and moistened at the skin, being less dense, gives rise to air currents in the 1-G field. In the weightless state such thermal convection is absent and its effect must be replaced by an air conditioning system that also serves to remove water vapor and carbon dioxide as well as to maintain a comfortable temperature. As pressure decreases, the heat-convective properties of a gas decrease exponentially and a greater volume must move to control the temperature, thus increasing the weight and power requirements of the life-support system. When all technical factors are taken into account, calculations show that the weight penalty attributable to the choice of a cabin atmosphere increases at high and low pressures and that the most effective and economic system would operate between 5 and 7 psi (258 to 362 torr).[19] There are also several physiologic arguments in favor of a cabin pressure in this range. The risk of decompression sickness, either coming from the earth's atmosphere on launch or transferring to the low-pressure spacesuit for EVA after an appropriate intervening period (at least 5 hours) in the cabin atmosphere at 5 to 7 psi would be minimal since neither step exceeds the critical ratio of 2:1 according to Haldane's diving regimen. Furthermore, more time is available for corrective measures or for donning the spacesuit in emergencies involving loss of cabin pressure.

A cabin atmosphere of 5 psi (258 torr) with pure oxygen was used in all flights of the Mercury, Gemini, and Apollo space programs. This pressure would have permitted the use of a two-gas system such as 30% nitrogen and 70% oxygen and a normal oxygen pressure (Fig. 10-9). However, the complexity of monitoring and regulating two gases appeared prohibitive in the planning stages of these programs.

The most convincing reason for adopting a two-gas atmosphere with an inert diluent gas and just enough oxygen to meet physiologic needs is the serious fire hazard in a pure oxygen atmosphere. Although the rate of flame propagation once a fire has started is less in the weightless state of space because of the lack of convection, the tendency of electric circuits to arc is greatly enhanced at low barometric pressure. The overall fire hazard in a cabin at 5 psi (256 torr) with 100% oxygen is estimated to be three to four times greater than in a commercial transport plane with a cabin pressure equivalent to 2.4 km (8,000 ft).[19]

There is also concern as to whether prolonged exposure to oxygen at a pressure of 256 torr is completely innocuous to an astronaut. Experiences in hyperbaric environments such as diving and caisson work clearly indicate the toxic effects of oxygen on the central nervous system from exposure to partial pressures of oxygen of more than 2,280 torr (3 ATA). Dizziness, nausea, and convulsions are common after a symptom-free latent period.[20] A reduction of red cell mass during hyperoxia, considered an adaptive response of the hematopoietic system, is also well established. The deleterious effects of breathing pure oxygen at 1 atmosphere for extended periods of time have been recognized clinically as the cause of retrolental fibroplasia of the newborn. Pulmonary manifestations of hyperoxia are observed frequently if pure oxygen is breathed at or near sea level for periods of more than 24 hours; these include substernal irritation and cough attributed to toxic effects on the epithelial cells of airways and alveoli and damage to the pulmonary surfactant elements that predisposes to atelectasis. Pulmonary symptoms have been reported at oxygen pressures much lower than 1 atmosphere and the threshold value below which oxygen is innocuous for indefinite periods remains unknown. According to experiments simulating the 5 psi (256 torr) 100% oxygen environment with human subjects, exposure for periods up to 30 days is tolerated without significant deleterious effects.[18] Another reason for keeping oxygen pressure as low as possible is the greater susceptibility of hyperoxic tissues to the effects of ionizing radiation that astronauts may encounter. As mentioned previously, the presence of an inert gas decreases the risk of pulmonary atelectasis.

It is generally agreed, from the biomedical as well as from the technical point of view, that the cabin atmosphere should contain two gases—oxygen, in concentration sufficient to provide a normal partial pressure of 150 torr, and an inert gas diluent, constituting the remainder. Selection of the latter presents interesting problems. Nitrogen, the major component of air, is naturally the first choice. However, other inert gases have also been closely scrutinized for suitability. Argon, krypton, and xenon were eliminated because their potential for bubble formation in the body under conditions of supersaturation is clearly greater than that of nitrogen. Helium has been used extensively in diving operations in place of nitrogen to prevent hyperoxia at high pressures. However, this is not because it is less likely to produce decompression sickness than nitrogen but rather to prevent nitrogen narcosis at great depths. There is little apparent difference in the incidence and severity of decompression sickness between nitrogen and helium, although the onset may be more rapid with helium. On theoretic grounds the only gas that might be less conducive to bubble formation is neon,[19] but practical experience with it is minimal thus far. Helium has certain physical properties that give it an advantage over nitrogen with regard to heat transfer in air conditioning a space cabin. Whereas the specific heat capacity of helium is less than that of nitrogen, the thermal conductivity of the former is six times greater than that of the latter. This means that excess heat and water vapor can be eliminated at a lower turnover rate and with less power expenditure. Even in a cabin at 5 psi (256 torr) in which the thermal conductivity of all gases is reduced and the atmosphere is 70% oxygen and only 30% inert gas, a helium system would require 20% less power for environmental control than a nitrogen system.

Regeneration of the atmosphere

At an average metabolic rate for routine activities of 2,900 kcal/day an astronaut consumes 600 L (STPD) of oxygen weighing 0.6 kgm in 24 hours. During the same time he produces 500 L (STPD) of carbon dioxide weighing 1.0 kgm. The oxygen is replaced from converters, where it is stored in liquid form, and the carbon dioxide is removed by passing the gas through a scrubber containing lithium hydroxide or through a molecular sieve. Incomplete removal of carbon dioxide and its subsequent accumulation within the cabin increases pulmonary ventilation because carbon dioxide is a physiologic respiratory stimulant normally supplied by the metabolic process. The presence of appreciable amounts of carbon dioxide in the inspired gas raises its tension in the blood and increases the H^+ concentration in body fluids (respiratory acidosis). The immediate response of the body is to augment pulmonary ventilation, diluting the carbon dioxide in the lungs. Another, but much slower, response is a reduction of the rate of renal bicarbonate ion secretion, tending to restore pH to normal. These changes affect the electrolyte balance between intracellular and extracellular fluids and may have severe physiologic consequences. It is generally assumed that a P_{CO_2} in the inspired air of 8 torr (equivalent to about 1% at sea level) is of no physiologic consequence, and this is the permissible level for manned spacecraft. There is a measurable increase in respiration at this low level, but it does not affect metabolic rate or performance. At an inspired partial pressure of 15 torr (2% sea-level equivalent) respiration increases noticeably, particularly during exertion, and the subject fatigues more readily. If P_{CO_2} is more than 21 torr (3% sea-level equivalent) headaches are frequent and even moderate exertion becomes uncomfortable because of excessive dyspnea. Experience in submarines indicates that it is possible to adapt oneself to 3% carbon dioxide, but it is definitely associated with a performance decrement.

The fact that carbon monoxide is produced in the body from the breakdown of porphyrin, a derivative of hemoglobin, at a rate of approximately 10 ml (STPD)/day has led to some speculation on the possible accumulation of this highly toxic gas in a sealed cabin on a space journey of 100 days or more. Fortunately (in this case), space cabins have an irreducible constant leak rate of about 0.5 kgm of gas per day, a rate that would probably be sufficient to reduce the accumulation of endogenous carbon monoxide to an innocuous level (less than 25 ppm). Otherwise it would have to be oxidized catalytically to carbon dioxide and removed as such. Several hundred other trace contaminants have been identified in the atmospheres of spacecraft and submarines; these derive either from the occupants or the equipment. Ways and means to monitor and eliminate contaminants that have toxic or nuisance potential are being developed.[21]

The behavior of particulate matter and aerosols in the space environment at zero gravity has stimulated some speculation, but so far no in-flight measurements of particle size or deposition characteristics have been reported. Among the aerosols most likely to occur are oils, fibers, chemical dust from paint and insulating material, food, skin, and dandruff. Two significant differences are expected in the behavior of an aerosol in space as compared to that on earth. Gravitational sedimentation being absent, a considerably greater range of particle sizes, even as large as 1 mm, can stabilize permanently and, if inhaled by mouth, penetrate as far as the bronchioles. Second, aerosol mobility is enhanced at low pressure, favoring diffusion deposition, which is the principle mechanism of deposition of submicronic particles in the lungs. On the other hand, deposition by inertial impaction is unaffected in space, and many of the particles that pass through the upper airways and are ordinarily deposited in the bronchial tree remain airborne and are expired.[22]

Simple arithmetic leads to the conclusion that the regeneration of space cabin atmospheres by continuously adding oxygen and absorbing carbon dioxide with expendable scrubbing material is feasible only for space flights of a few months at most. When it comes to interplanetary expeditions and temporary or even permanent lunar or planetary stations, it will be necessary to resort to some means of recycling respiratory gases and probably also water and nutrients. The most successful and effective ecologic regenerating system we know of is the cycle involving the animals and plants in the biosphere of the earth, in which carbon dioxide combines with water by photosynthesis in plants using solar energy to produce oxygen and organic matter. Biologic cycling of this kind now appears the method of choice to maintain extended space operations, and intensive research is under way to find the most suitable plant organisms, possibly algae or plankton, for this purpose.

RAPID DECOMPRESSION AND SURVIVAL IN SPACE

While development of aircraft with pressurized cabins has greatly advanced air travel within the atmosphere at high speeds and altitudes and the use of sealed cabins has made the vastness of the universe accessible to manned space flight, the implications of sudden loss of cabin pressure from structural failure, hostile action, or meteorite impact have become increasingly formidable. However remote the probability of such an event may be, an assessment of the effects of rapid decompression on body structure and function is necessary to develop the means of survival and rescue.

In analyzing the sequence of events following rapid decompression at altitudes above 20 km (60,000 ft), one must distinguish between the immediate mechanical effects of a sudden drastic ambient pressure decrease and the subsequent exposure to almost total anoxia with vapor formation in body fluids.

Mechanical effects

Since the liquid and solid body components are not subject to appreciable deformation by ambient pressure change, only those organs that contain significant amounts of gas are immediately affected by dynamic expansion during rapid decompression. Because the lungs contain a large amount of gas and their anatomic structure is delicate, they are much more susceptible to injury than any other organ. Lung damage and collapse (pneumothorax), gas embolism, and even death have occurred when men accidentally held their breath during relatively slow decompression from depth during submarine escape maneuvers. In animal experiments the mammalian lung ruptures when passively overdistended by a transthoracic pressure greater than 80 torr.[23] Fortunately, the probability of the airways being closed at the instant of unexpected rapid decompression during flight is small unless one is caught in the act of swallowing or coughing. However, even with open airways, serious damage can occur under certain circumstances if decompression is extremely rapid and profound.

The severity of the mechanical effects of decompression on the body is directly proportional to three factors: (1) the difference between initial and final pressures (ΔP); (2) the ratio of initial to final pressure (R); and (3) the rate of pressure change. Items 1 and 2 can be found for any given combination of flight altitude and cabin pressure ("altitude") at their intercept in Fig. 10-8. The time factor is a complex function of the pressure ratio,[24] but for any given ratio it depends upon cabin volume relative to the effective surface area of the opening through which the gas escapes; this is the key to understanding how rapid decompression causes lung damage. Consider the pressure changes that occur at the moment of decompression in the lungs of a person in a pressurized cabin in terms of two bottles, one inside the other (Fig. 10-10); the larger bottle corresponds to the cabin and the smaller one inside it to the chest and lungs with open airways. In this simplified model the inner bottle has a smaller opening relative to its volume than the outer one representing the cabin. Initially, the pressure inside both bottles is the same but considerably higher than that outside. The instant the large bottle opens, the pressure within it rapidly equalizes with the pressure outside and the pressure head across its walls (transmural pressure) disappears, as shown in curve I. Pressure curve II for the small bottle lags far behind curve I because of its different geometry, creating a transient pressure difference across the walls of the inner bottle as shown in curve III. Obviously, the lungs and chest are not a rigid container and actually expand within their anatomic limits, reducing the pressure gradient. However, the lungs can rupture at this point if the residual pressure at the peak exceeds 80 torr. It is also clear that the transthoracic pressure will be much higher if the airways are closed.

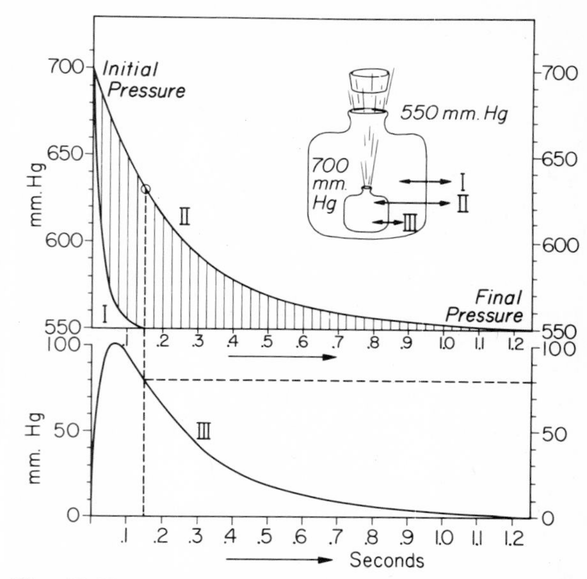

Fig. 10-10. Rapid decompression of a bottle within a bottle representing the chest in a cabin. Pressure curves *I* and *II* indicate the decay in the cabin and chest respectively. Curve *III* shows the resultant pressure difference across the lungs and chest wall if they were a rigid container. (From Luft, U. C., and Bancroft, R. W.: J. Aviat. Med. **27**:208, 1956.)

Pulmonary lesions are not likely to occur with open airways unless decompression is extremely fast and the pressure difference is high. In a large transport plane decompression after blow-out of a window or door takes several minutes because of the large volume. Conditions in a small pressurized cockpit with loss of the canopy are more critical, and low differential pressures are recommended for this type of aircraft (Fig. 10-8).

In space flight outside the atmosphere the pressure ratio, R, becomes infinite, because the outside pressure is zero, and the pressure difference, ΔP, is identical with the absolute pressure in the cabin or the spacesuit. In view of the relatively fragile construction and material of the suit as compared to the spaceship, it is under a pressure of not more than 196 torr (3.8 psi). This has the additional advantage of less tubular rigidity at the joints, providing greater mobility for the astronaut. The danger of trauma from rapid decompression per se is negligible under these circumstances.

In a space cabin with a pressure of at least 256 torr (5 psi) the possibility of sudden loss of cabin pressure from a meteoroid penetrating the shell of the vehicle cannot be entirely excluded, although this hazard can be reduced by suitable impact-absorbing layers in the shell. Even if the cabin wall were perforated there would be no chance of lung damage unless the ratio between the area of the hole and the volume of the cabin were greater than 1:200.[24] Thus, in the Gemini cabin, whose free volume when fully equipped is 2.83 m^2 (100 ft^3), a circular hole of this proportion would be 13.4 cm (5¼ in) in diameter.

Aerohypoxia

While the mechanical effects of sudden decompression should not be underestimated, the probability of fatal injury from such an event is extremely small. In comparison, the effects of fulminating aerohypoxia on sudden exposure to extremely low pressure within the atmosphere or in the hard vacuum of space are absolutely inevitable and their time course is precisely predictable.

Fig. 10-10 shows that during decompression the total gas pressure within the lungs follows the cabin pressure transient within a fraction of a second, reducing the partial pressure of the gases with the total pressure in proportion to their dry gas fraction. At the same time mixed venous blood entering the lungs contains the same amount of gas as it did immediately before decompression, at least for the period of one circulation time. The lungs, which ordinarily act as a buffer vessel between external environment and blood, are instantaneously deprived of this important function and the blood is suddenly exposed to drastically altered conditions for gas exchange. Some insight as to what occurs in the lungs under these circumstances[25] is gained by analysis of end-expiratory alveolar gas samples between 3 and 5 seconds after rapid (2-second) decompression from ground level to various altitudes while breathing air (Table 10-4). Since decompression per se does not alter lung gas composition but only rarifies it, changes after decompression must result from gas transfer into or out of the blood and concomitant mixing with inspired air. In the last column of Table 10-4 the nitrogen fraction is consistently lower with each additional increase of altitude until, at 15 km (49,200 ft), it constitutes only 32% of the total. Evidently nitrogen is diluted by a profuse flow of carbon dioxide and oxygen out of the blood during and immediately after decompression. It is not surprising that carbon dioxide is the largest fraction, up to 42%, at the highest altitudes. Its concentration in mixed venous blood is fifty times greater than that of nitrogen and three to four times greater than that of oxygen. Thus it appears that both nitrogen and oxygen are displaced by the profuse outflow of carbon dioxide. Nevertheless, the oxygen concentration of alveolar gas is consistently higher than that of inspired air (20.95%) at altitudes above 10 km (32,800 ft). This proves unequivocally that under these unusual circumstances oxygen transfer is temporarily reversed and it diffuses out of the blood and into the alveoli. This is understandable because the samples taken at 10 km (32,800 ft) show a P_{O_2} less than 30 torr, well below the usual level in mixed venous blood during rest (35 to 40 torr). Therefore, the oxygen diffusion gradient is reversed and the blood is deprived of oxygen as well as carbon dioxide. Obviously, the reverse oxygen flow cannot persist longer than the time required for blood to recirculate that has already lost most of its oxygen to the lungs. The brain, which is most susceptible to hypoxia, ceases to function shortly after this extremely hypoxemic blood reaches it from the lungs; circulation time from lungs to brain is about 5 seconds.

Fig. 10-11 shows the time elapsed to unconsciousness after rapid (within a few tenths of a second) decompression from ground-level pressure to various altitudes while breathing air (lower curve) and after rapid decompression from

Table 10-4. Composition and partial pressure of alveolar gas samples obtained 3 to 5 seconds after rapid decompression to the indicated altitudes while breathing air

Number of runs	Altitude	Oxygen		Carbon dioxide		Nitrogen	
	km (1,000 ft)	Percent	Torr	Percent	Torr	Percent	Torr
20	Ground	14.3	102	5.6	39.9	80.1	571
3	7 (23.0)	15.0	39	10.8	28.2	74.2	193.6
6	10 (32.8)	19.8	29.9	22.0	33.3	58.0	87.7
6	12 (39.4)	23.7	23.0	30.7	30.0	45.6	44.6
2	14 (46.0)	26.0	15.3	42.0	24.7	32.0	18.8
3	15 (49.2)	27.0	11.7	40.3	17.4	32.7	14.2

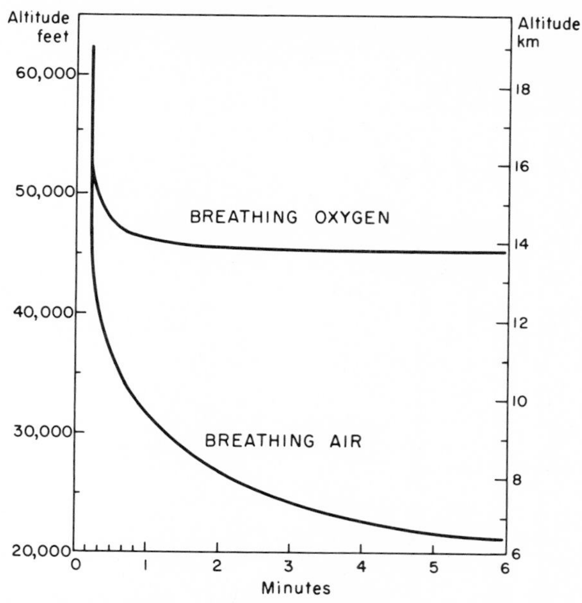

Fig. 10-11. Time to unconsciousness after sudden exposure to various altitudes by rapid decompression while breathing oxygen (above) and air (below) during decompression.

10 km (33,800 ft) to altitudes up to 18 km (60,000 ft) while breathing pure oxygen (upper curve). Note that at 10 km oxygenation of the blood while breathing pure oxygen is equal to that while breathing air at sea level (Table 10-3). Both curves converge at about 16 km (52,500 ft), indicating that it makes no difference whether one is breathing air or oxygen during decompression to this or higher altitudes, because the blood contains too little oxygen in either case. However, the chances for survival and recovery on recompression or rapid descent are immensely improved if one is breathing oxygen throughout. It is interesting that the minimal time to unconsciousness of 13 to 15 seconds remains remarkably constant at altitudes much higher than 16 km (52,500 ft). Experiments have been performed to learn the minimum duration of exposure to a given altitude necessary to produce unconsciousness, despite recompression of the subject before loss of consciousness.[26] When recompression to a safe altitude is performed within 6 seconds, cerebral function is unaffected. However, exposures of 8, 10, and 12 seconds invariably produce brief loss of consciousness, but not before 13 seconds elapse. The practical conclusion is that an aviator flying above 16 km (52,500 ft) or an astronaut in space who suffers rapid decompression has only 5 to 6 seconds available for corrective action of any kind.

Ebullism and survival time

A liquid boils when its vapor pressure equals the ambient pressure. At 1 atmosphere (760 torr) water boils at 100° C; if the barometric pressure decreases, the boiling point also decreases. The vapor pressure of water at the normal body temperature of 37° C is 47 torr, which is also the barometric pressure at an altitude of 19.2 km (63,000 ft). One might thus expect the body fluids of a homeothermic animal to boil at or above this altitude. Although boiling under these circumstances is also associated with vapor bubble formation, hypobaric boiling, also known as ebullism,[27] differs in certain respects from the usual process induced by applying heat to the bottom of a vessel. In conventional boiling bubbles start to appear at the bottom of the fluid because it is hotter and the vapor pressure exceeds the barometric pressure plus the additional hydrostatic pressure of the liquid above. The bubbles expand as they rise toward the surface because hydrostatic pressure decreases and also because dissolved gases diffuse into them. If one boils a glass of water at standard pressure to free it of dissolved gases and then, after cooling, exposes it to a constant temperature of 37° C at a pressure of 47 torr, it evaporates very rapidly without necessarily forming bubbles at the bottom.

Another important feature of hypobaric boiling of body fluids is that there is no protein coagulation or other chemical changes from excessive heat as in conventional boiling; in contrast, the rapid evaporative process at altitude has a marked cooling effect. In an open container bubbles form in blood at slightly less than 47 torr because of the presence of nonaqueous components. Vaporization within the body is neither instantaneous nor uniform because temperature varies from place to place and because hydrostatic and tensile forces in tissues and blood vessels tend to increase the boiling point. One might expect bubbles to occur first in the pleural cavity where pressure is usually subambient. In experimental animals, bubbles are observed first in the large veins entering the heart where blood pressure is lowest, and their formation progresses retrograde toward the tissue capillaries, creating a "vapor lock" that completely occludes venous return. As blood pressure falls, bubbles are also found in the heart and in systemic arteries. Complete circulatory arrest occurs regularly 10 to 15 seconds after decompression to 30 to 40 torr. Circulatory failure caused by mechanical factors is much more rapid and drastic at altitudes where the boiling point is reached than at altitudes where fulminating hypoxia is not complicated by bubble formation. For example, decompression to 68 torr, corresponding to an altitude of 16.7 km (55,000 ft) also leads to circulatory failure, but the decrease of blood pressure and heart rate extends over several minutes and recovery on recompression is much faster. As for survival time after exposure to the hard vacuum of space after failure of cabin or suit pressure, observations on several different animal species, including trained primates,[28] indicate that survival without permanent cerebral damage is possible if recompression to 196 torr (3.8 psi) with oxygen is accomplished within 3 minutes of the accident. Whether the same is true for man remains unknown.

GLOSSARY

aeroembolism See **decompression sickness.**
aerohypoxia Reduced pressure of oxygen in the inspired gas.
aerotitis media See **barotitis.**
anoxia Complete absence of oxygen.
ATA Unit of absolute pressure in atmospheres.
barodontalgia Toothache experienced on ascent to altitude.
barometeorism Expansion of intestinal gas on ascent to altitude.
barosinusitis Painful condition of the paranasal sinuses on compression or recompression.
barotitis Painful condition of the middle ear on compression or recompression. Synonym is aerotitis media.
barotrauma Any affliction caused by mechanical effects of change in pressure.
bends Painful sensation at or near the joints experienced in decompression.
caisson Water-tight box filled with air under pressure for construction work under water.
caisson disease Disorder experienced by divers and caisson workers on transition from high to normal pressure. Synonyms are decompression sickness, dysbarism, aeroembolism.
chokes Substernal distress with cough; symptom of decompression sickness.
chronobiology Science of the temporal relationships of biologic phenomena, particularly daily, monthly, and yearly cycles.
compression Transition to an environment of higher pressure, particularly in diving or in a hyperbaric chamber.
decompression Transition to an environment of lower pressure, either from depth to sea level (hyperbaric decompression) or from sea level to altitude (hypobaric decompression).
decompression sickness Disorder associated with transition from higher to lower pressure in the environment. Synonyms are dysbarism, aeroembolism, caisson disease.
denitrogenation Elimination of gaseous nitrogen physically dissolved in body fluids by breathing pure oxygen in order to reduce incidence of decompression sickness. A synonym is preoxygenation.
dyspnea Sensation of shortness of breath.
ebullism The effervescent evaporation of body fluids at barometric pressures equal to or below the saturated vapor pressure at body temperature. Synonym is hypobaric boiling.
emphysema The abnormal presence of gas in tissues, for example, mediastinal, subcutaneous; also a chronic diffuse obstructive lung disease.
EVA Abbreviation for extravehicular activities in space.
exosphere Outermost layer of the atmosphere.
Frenzel maneuver Procedure used to vent middle ears during compression or descent from altitude—consists of contracting muscles of mouth and cheeks with closed glottis, nose, and mouth.
histohypoxia Reduced pressure of oxygen in the tissues of the body.
hyperbaria Atmospheric pressure higher than that at sea level.
hypercapia Excess of carbon dioxide in the body.
hyperoxia A condition with higher partial pressure of oxygen than in air at sea level.
hyperventilation Pulmonary alveolar ventilation greater than that required to match the carbon dioxide production rate.
hypobaria Atmospheric pressure less than that at sea level.
hypoventilation Pulmonary ventilation less than that required to match carbon dioxide production rate.
hypoxemia Reduced oxygen pressure in the blood.
hypoxia Reduced partial pressure of oxygen.
isobaria Barometric pressure constancy.
ionosphere Layer of earth's atmosphere above the stratosphere where particles are in ionized form.
mach number A dimensionless number giving the ratio of the speed of an object traveling through the air to the speed of sound for the existing conditions of temperature and density; named for the Austrian physicist Ernst Mach (1838-1916).
ozone layer Layer of the stratosphere with increased ozone concentration.
partial pressure Pressure exerted by an individual gas in a gas mixture according to Dalton's law.
pneumothorax The presence of gas in the intrapleural space.
positive pressure breathing
  constant Technic used to increase oxygen pressure in the lungs above ambient at altitudes where 100% oxygen is not sufficient to protect against hypoxia.
  intermittent Used to treat bronchopulmonary diseases; pressure applied during inspiration only.
preoxygenation See **denitrogenation.**
pressure cabin Aircraft compartment pressurized by compressors from ambient air or by ram pressure.
ppm Parts per million.
psi Pounds per square inch, a unit of pressure equal to 51.7 torr.
ram pressure Pressure created by impact of air on fuselage of an aircraft traveling at high speed.
recompression Return to a condition of higher pressure after decompression; applicable to pressures both above and below that at sea level.
scrubber Device for cleansing gas of impurities, specifically of carbon dioxide.
sealed cabin Compartment with a completely self-contained gaseous environment.
stratosphere Layer of the atmosphere above the troposphere.

STPD Standard temperature, pressure, and dry; an abbreviation of the physical conditions used to define the mass of a gas contained in a stated volume. Standard physical conditions are 0° C, 760 torr, and dry.

torr Unit of pressure equal to 1/760 of a standard atmosphere; named for the Italian mathematician E. Torricelli (1608-1647).

troposphere Lowest layer of the atmosphere.

Valsalva maneuver Forced exhalation against a closed glottis; also a procedure used to vent the middle ears during compression or descent from altitude by forced exhalation against the closed nose and mouth.

REFERENCES

1. Strughold, H.: The earth's environment and aviation. In Randel, H. W., editor: Aerospace medicine, ed. 2, Baltimore, 1971, The Williams & Wilkins Co.
2. Bert, P.: Barometric pressure (translated by M. A. Hitchcock and F. A. Hitchcock), Columbus, Ohio, 1943, College Book Company.
3. Boothby, W. M., Lovelace, W. R., II, Benson, O. O., and Strehler, A. F.: Volume and partial pressure of respiratory gases at altitude. In Handbook of respiratory physiology in aviation, Randolph Field, Texas, 1954, U. S. A. F. School of Aviation Medicine.
4. Comroe, J. H., Jr., and Schmidt, C. F.: The part played by reflexes from the carotid body in the chemical regulation of respiration in the dog, Am. J. Physiol. **121**:75, 1938.
5. Denison, D., and Ledwith, F.: Complex reaction times at a simulated altitude of 8000 ft, R. A. F. Institute of Aviation Medicine, Report No. 284, 1964, Ministry of Defense.
6. Whiteside, T. C. D.: The problems of vision in flight at high altitudes, London, 1957, Butterworth & Co. (Publishers) Ltd.
7. Luft, U. C.: Die Höhenahpassung, Ergeb. Physiol. **44**:256, 1942.
8. Velasquez, T.: Tolerance to acute anoxia in high altitude natives, J. Appl. Physiol. **14**:357, 1959.
9. Gauer, O. H., and Henry, J. P.: Circulatory basis of fluid volume control, Physiol. Rev. **43**:423, 1963.
10. Jongbloed, J.: The composition of the alveolar air in man up to 14000 m partly without oxygen: the mechanical effects of very low pressure, dissertation, University of Utrecht, 1929.
11. Adler, H. F.: Dysbarism, Aeromedical Reviews, Rev. No. 1-64, Brooks, A. F. B., Texas, 1964, U. S. A. F., School of Aviation Medicine.
12. McIver, R. G., and Kronenberg, R. S.: Treatment of altitude dysbarism with oxygen under high pressure, Aerospace Med. **37**:1266, 1966.
13. Haldane, J. S., and Priestley, J. G.: Respiration, ed. 2, New Haven, 1935, Yale University Press, p. 344.
14. Clamann, H. J.: Decompression sickness. In Randel, H. W., editor: Aerospace medicine, ed. 2, Baltimore, 1971, The Williams & Wilkins Co.
15. Fryer, D. I., and Roxburgh, H. L.: Decompression sickness. In Gillies, J. A., editor: A textbook of aviation physiology, Oxford, 1965, Pergamon Press.
16. Banchero, N., and others: Effects of transverse acceleration on blood oxygen saturation, J. Appl. Physiol. **22**:731, 1967.
17. DuBois, A. B., Turaids, T., Mammen, R. E., and Nobrega, F. T.: Pulmonary atelectasis in subjects breathing oxygen at sea level or simulated altitude, J. Appl. Physiol. **21**:828, 1966.
18. Carstens, A. I., and Welch, B. E.: Space craft atmospheres. In Randel, H. W., editor: Aerospace medicine, ed. 2, Baltimore, 1971, The Williams & Wilkins Co.
19. Roth, E. M.: Gas physiology in space operations, New Eng. J. Med. **275**:144, 196, 255, 1966.
20. Lambertsen, C. J.: Effects of oxygen at high partial pressure. In Fenn, W. O., and Rahn, H., editors: Handbook of physiology, sect. 3, chap. 39, Washington, D. C., 1965, American Physiological Society.
21. Nelson, N.: Atmospheric contaminants in space flight, Report of Space Science Board, N. A. S., October, 1968.
22. Beekmans, J. M.: Alveolar deposition of aerosols on the moon and in outer space, Nature **211**:209, 1966.
23. Adams, B. H., and Polak, I. B.: Traumatic lesions produced in dogs by simulating submarine escape, U. S. Nav. Med. Bull. **31**:18, 1933.
24. Luft, U. C., and Bancroft, R. W.: Transthoracic pressure in man during rapid decompression, J. Aviat. Med. **27**:208, 1956.
25. Luft, U. C., Clamann, H. J., and Adler, H. F.: Alveolar gases in rapid decompression to high altitudes, J. Appl. Physiol. **2**:37, 1949.
26. Luft, U. C.: Aviation physiology—the effects of altitude. In Fenn, W. O., and Rahn, H., editors: Handbook of physiology, sect. 3, chap. 44, Washington, D. C., 1965, American Physiological Society.
27. Ward, J. E.: The true nature of the boiling of body fluids in space, J. Aviat. Med. **27**:429, 1956.
28. Koestler, A. G.: Exposure limits for chimpanzees at near vacuum following rapid decompression, 6571st Aeromedical Research Laboratory Technical Report, ARL-TR-62-2, 1969.
29. U. S. Standard Atmosphere, Washington, D. C., 1962, U. S. Government Printing Office.
30. Jones, H. B.: Preoxygenation and nitrogen elimination. In Decompression sickness, National Research Council Report, Committee on Aviation Medicine, Subcommittee on Decompression Sickness, Philadelphia, 1951, W. B. Saunders Co., chap. 9.

SUGGESTED READINGS

1. Gillies, J. A., editor: A textbook of aviation physiology, Oxford, 1965, Pergamon Press.
2. Henry, J. P.: Biomedical aspects of space flight, New York, 1966, Holt, Rinehart, and Winston, Inc.
3. McFarland, R. A.: The psychological effects of oxygen deprivation (anoxemia) on human behavior, Arch. Psychol. **145**:135, 1932.
4. Randel, H. W., editor: Aerospace medicine, ed. 2, Baltimore, 1971, The Williams & Wilkins Co.

11 MARINE AND OTHER HYPERBARIC ENVIRONMENTS

PART A: HYPERBARIC ENVIRONMENTS

ALBERT R. BEHNKE, Jr.

Since 1958 man has made numerous orbital flights around the earth, landed on the moon, explored the ocean floor to the deepest depths of the Continental Shelf, and extended practical diving to depths of about 650 ft. He has demonstrated a capacity in dry and dry-wet chambers for productive activity down to a depth of 1,500 ft. The littoral undersea domain of the Continental Shelf embraces an area about the size of Africa; it contains a wealth of biologic and mineral resources and is now amenable to habitation. Unfortunately, it is also the repository of chemical and biologic wastes and generally subject to unabashed defilement.

Beyond conventional diving, it has been necessary to fabricate various types of undersea habitats and to devise new methods of locomotion for transfer of divers from submerged habitats to surface decompression chambers. Extensive commercial installations include underwater welding chambers for pipeline construction. Diving with self-contained underwater breathing apparatus (scuba) engages millions of enthusiasts interested in recreation, exploration, and marine biology.

Pressurized tunneling in compressed air (up to 50 psi) involves another type of environment required for construction of facilities for rapid transit or sewage disposal. Renewed interest in hyperbaric therapy and the use of dry-wet chambers in hyperbaric research require meticulous environmental control of temperature, humidity, and the partial pressures of component gases.

These exciting developments challenge both physiologic capability and engineering ingenuity. Primary physiologic problems involve the effects of temperature, humidity, and compression; nitrogen narcosis; oxygen tolerance and toxicity; inadequate pulmonary ventilation; and the complicated problem of safe decompression to normal atmospheric pressure. The sea, always wet and chemically corrosive, may appear warm, serene, and clear, but for the professional diver it is often cold, turbulent, and murky. In addition to physiologic stresses there are the hazards of predatory and toxic marine organisms. In warm waters, fungous growth is difficult to control, especially in abraded external auditory canals. Injury from coral and associated wound contamination tend to cause indolent skin ulceration. In performing physical tasks the slightly negative bouyancy deprives a diver of stability as he attempts to work with conventional tools in the same way as he does on land. Coping with the problems and stresses of a hyperbaric environment requires courage, assiduous training, acquisition of new skills, and an extended period of adaptation. The special equipment thus far developed for respiratory needs and cold protection does not meet the challenge of this formidable environment.

NORMOBARIC CAPSULE ENVIRONMENTS

Capsule space of the submarine

Nuclear-powered submarines circumnavigate the globe underwater, and various miniature submarines operate for brief periods at depths of several thousand feet. The capsule climate of these vehicles, lacking seasons and weather, can be stabilized for months with monotonous uniformity at about 21° C and 50% relative humidity. This control has been achieved as a result of 60 years of engineering experience, chiefly in air cooling, humidity regulation, oxygen replenishment, and removal of carbon dioxide and noxious trace substances.[1] With respect to atmosphere in this motile type of "metabolic chamber," submarine personnel may even be unaware of passage from the tropics to the arctic. Although bacterial and viral infections have not been brought entirely under control, the engineer has provided a submarine atmosphere generally conducive to health and efficient performance. Nevertheless, recirculated air within a submarine becomes burdened with gaseous and aerosol contributions from all volatile substances, with sulfuric acid vapor from storage batteries, with impurities such as arsenic and antimony (which in trivalent form combine with hydrogen to produce arsine and stibine, respectively), with acrolein from hot fat in the galley ranges, and at times with vapors from sanitary tanks that literally "breathe into the boat."

Man himself is an offender with his contributions of tobacco smoke and paint solvents; more than 400 contaminants have been identified by gas chromatography and mass spectrometry.[1] Aerosols, finely divided solid and liquid particles suspended in air, deserve particular attention. Tobacco smoke is a pernicious agent that bypasses upper respiratory filtration, introducing minute aerosols (0.3 μ or less) into the pulmonary alveoli. Airway resistance increases and this restriction of pulmonary ventilation is further intensified in a carbon dioxide–enriched atmosphere.

Carbon dioxide accumulation

During World War II studies were conducted in manned sealed compartments and submarines within which carbon dioxide rose to 5% while oxygen fell to 13% during a period of 35 hours.[2] During additional periods of 20 to 27 hours these gas concentrations were maintained by absorption of carbon dioxide above the 5% level and replenishment of oxygen to maintain the 13% level. Except for minor symptoms such as headache, subjects and submarine crews remained in good condition. The chief nuisance was mouth-breathing associated with the three-fold tidal volume increase, interference with sleep, and increased heat loss from the body surface and lungs in the carbon dioxide–enriched atmosphere. Although the ambient carbon dioxide was about 5% (35 torr), the alveolar P_{CO_2} increased by only about 7 torr because of increased alveolar ventilation rate.[2] As ambient carbon dioxide exceeded 5%, the alveolar P_{CO_2} increased proportionately because greater alveolar ventilation was not maintained.

The long-term effects of increased ambient carbon dioxide concentrations above 1.5% have not been systematically examined. However, an operational ambient carbon dioxide limit of 1.5% does not produce gross deterioration as evaluated by crew performance in submarines.

FEATURES OF EXPOSURE IN PRESSURIZED ENVIRONMENTS

Psychrometric considerations

Several types of pressurized environment merit comment. One is the relatively innocuous *dry* hyperbaric environment of compression chambers and medical hyperbaric facilities. A second is the potentially nocuous pressurized environment of tunnel workers. A third type, which poses formidable obstacles to temperature control, is the *wet* pressurized medium of the diver.[3]

The dry hyperbaric facility

The ventilation engineer provides comfortable conditions using standard principles of heat transfer and air conditioning. For example, the surgeon repairs congenital cardiovascular defects of infants and children in chambers pressurized to 4 atmospheres (Table 11-1). At higher pressures helium replaces atmospheric nitrogen to avoid nitrogen narcosis. In the dry hyperbaric medium a recirculated atmosphere of helium at

Table 11-1. Pressure expressed in equivalent units* for hyperbaric environments

Atmospheres absolute (ATA)	Pounds per square inch (psi, gage)	Sea water		Millimeters of mercury (mm Hg)
		ft	m	
1.0	0	0	0	760
2.0	14.7	33	10.06	1,520
2.5	22.1	49.5	15.1	1,900
3.0	29.4	66.0	20.1	2,280
3.5	36.8	82.5	25.2	2,660
4.0	44.1	99	30.2	3,040
5.0	58.8	132	40.2	3,800
6.0	73.5	165	50.3	4,560
7.0	88.2	198	60.4	5,320
8.0	102.9	231	70.4	6,080
9.0	117.6	264	80.5	6,840
10.0	133.3	297	90.5	7,600
31.0	444.0	990	304.5	23,560

*Conversion factors:
1 psi = 2.245 ft = 0.6842 m = 51.7 mm Hg
1 ft = 0.4823 psi = 0.3048 m = 23.03 mm Hg
100 mm Hg = 4.34 ft = 1.324 m = 1.934 psi
760 mm Hg = 1.0133 bars = 1.0332 kgm (force)/cm²

a total pressure of 31.3 ATA (1,000 ft), containing 0.3 ATA oxygen, is compatible with normal psychomotor performance of moderate work; this was also true in one test at a total pressure as high as 46.5 ATA (1,500 ft).

Because of rapid convective heat loss in a helium atmosphere, ambient temperature must be maintained in the narrow range between 29° and 32° C. Relative humidity is maintained between 50% and 70%. The P_{O_2} averages 0.3 ATA and is confined to the range between 0.2 and 0.4 ATA. The P_{CO_2} is maintained less than 2 torr (0.0025 ATA). Accurate instrumentation is mandatory for monitoring in hyperbaric environments.

Formerly, without air conditioning, thermal stress in compression chambers was severe. Aboard ship, in chambers exposed to the sun, the heat of compression produced temperatures as high as 49° C that, with relative humidity approaching 100%, rendered personnel highly uncomfortable and inefficient. In winter, the cold unlagged chamber surfaces facilitated radiant cooling that, combined with water vapor condensation during decompression, created a miserably cold atmosphere.

Environment of the tunnel worker

To some extent pressurized tunneling involves the same adverse conditions mentioned previously. Workers in pressures as high as 5 ATA, to prevent ingress of water during subsurface and subaqueous operations, may labor in a cold damp atmosphere distal to the excavation shield at the tunnel face. Here, with modern automated machinery, heat radiation from surfaces at 55° C, combined with high humidity, causes profuse sweating. Despite enormous volumes of air forced into a tunnel to keep out water, mechanical air removal may be inadequate. Not only does temperature tend to rise, but welding fumes and other contaminants accumulate in air pockets. Certain soils, such as those containing limestone, may be oxygen deficient so that when tunnel pressure decreases, an influx of hypoxic air may cause collapse. The tunnel environment is far from optimal; the objective is to provide a reasonable degree of safety, not comfort.

COLD STRESS AND PROTECTIVE MEASURES

Problem of accelerated heat loss in water

It is difficult to maintain the thermal balance of a diver during long periods of submersion. Skin divers cool in water at 26.6° C and suited divers cannot be protected by insulating garments alone at water temperatures below 10° C. Hydrostatic pressurization compounds the thermal problem because (1) insulating materials that utilize a dead gas space to impede heat transfer are compressed and lose some insulating capacity, and (2) filling the spaces with helium greatly increases thermal conductivity (Table 11-2). Thus, at a depth of 600 ft, where water temperature is 4° C, cold intensified by a helium atmosphere limits diving operations.

In 1925, during salvage operations in the cold water off the New England coast, divers were first restricted to a limited number of decompression stops. *Surface decompression,* or *decanting,* to complete decompression in a relatively warm chamber aboard ship was imperative for divers who ascended rapidly to emerge from cold water after a minimum number of stops. Between 1937 and 1939 the U.S. Navy Experimental Diving

Table 11.2. Physical properties of gases relevant to narcosis and heat loss in deep sea diving relative to nitrogen*

Gas	Relative solubility		Van der Waals' constants a/b†	Relative narcotic potency		Relative thermal conductivity
	Water	Oil		Loss of righting reflex[37]	Criteria of Brauer and Way[18]	
Nitrogen	1.00	1.00	1.00	1.00	1.00	1.00
Helium	0.61	0.25	0.04	0.18	0.045	5.83
Neon	0.69	0.31	0.35	0.32	—	1.87
Hydrogen	1.13	0.82	0.26	0.41	0.26	7.19
Argon	2.00	2.30	1.18	1.46	—	0.69
Krypton	3.46	7.1	1.64	8.9	—	0.37
Xenon	6.1	25.4	2.31	31.8	—	0.27
Nitrous oxide	33.3	23.9	2.41	28.3	25.3	0.65

*Nitrogen is taken as 1.00. Although not used in diving, krypton, xenon, and nitrous oxide are included for comparison.
†Essentially, intermolecular forces of attraction divided by molecular volume.

Unit made intensive efforts to utilize helium; it was then apparent that measures are necessary to protect a diver from the high convective heat loss in a helium atmosphere. Divers wearing electrically heated underwear with power supplied from surface batteries were comfortable in cold water. However, dives to depths of 240 ft were not longer than 20 minutes' duration, and the full significance of convective heat loss in a helium atmosphere was not appreciated.

With the advent of saturation diving it became very evident that efficient diving could not be achieved without thermal equilibrium. Furthermore, the mouthpiece-mask type of scuba diving gear deprives a diver of the head and neck protection afforded by a conventional helmet. Breathing through a mouthpiece with head immersed in water predisposes to drowning in case of mishap. The following outline of progress in solving the important thermal problem of diving and, in particular, supplying heat to divers wearing modified scuba gear has been compiled largely from U. S. Navy sources.

A modern saturation diving complex to support prolonged residence underwater requires a surface ship, or platform, a personnel transfer capsule (PTC), and an *undersea habitat*. In SEALAB II (1965) three teams of divers comprising ten men each lived and worked for 15 days per team at a bottom depth of 205 ft. The pressurized atmosphere consisted of 3.5% to 5% oxygen (188 to 265 torr), about 18% nitrogen, and 77% to 79% helium (5,400 torr). An open hatch afforded free access to the ocean. Within the habitat temperature was maintained in the narrow range between 30° and 32° C and relative humidity was about 76%. Outside the habitat the ocean temperature was about 10° C and visibility was poor. Although it was possible after an initial adaptation period of several days for aquanauts to sleep normally in the helium atmosphere, there were some unexpected observations. For example, when a thoroughly chilled aquanaut returned to his bunk and covered himself with an electric blanket, he shivered violently and sweated profusely. Although shivering after exposure in cold water and subsequent removal to a warm atmosphere is known to result from the circulation of blood through cold subcutaneous tissues (Fig. 11-1), sweating was a novel phenomenon.

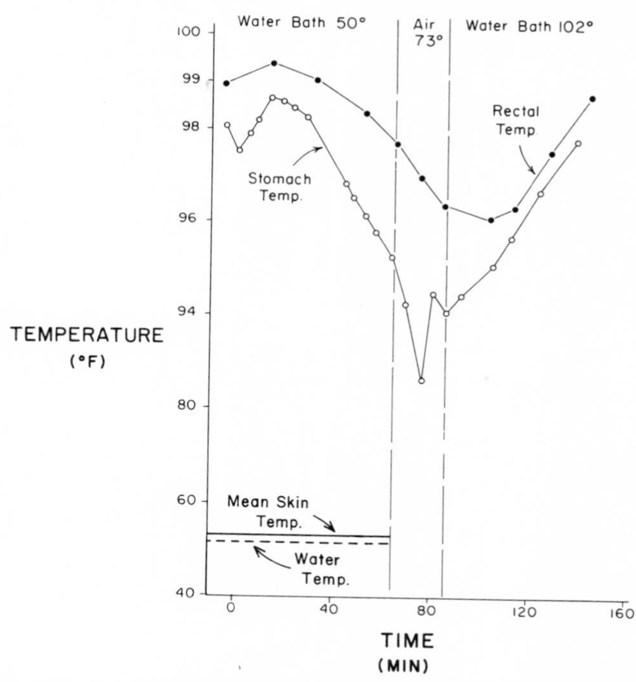

Fig. 11-1. Changes of body core temperature in a subject immersed in cold water and subsequently rewarmed first in air and then in warm water. Note that body core temperature continues to fall while the subject is in relatively warm air. Body fat was 22% of total body weight and amounted to 20 kgm. A second subject with 12.2% body fat (10 kgm) experienced a much more abrupt temperature decrease.

During the initial phases of SEA-LAB III (1969) the cold stress during dives to 600 ft, and especially during transfer in the PTC, was much more severe than anticipated. An electrically heated suit supplying 500 watt-hours (430 kcal equivalent) did not provide thermal comfort, even in conjunction with two wet suits.

Heat requirement of the diver in cold water

Preliminary tests suggest the adequacy of a tubing suit worn over two neoprene garments through which hot water (45° C) circulates. Spillover of outlet water into gloves, boots, and around the neck greatly enhances comfort in this water-circulation suit. The following data and calculations are based on U. S. Navy test data.

CONDITION I. At surface. Analysis of heat balance at 1 ATA in a *wet tank*. Water tubing suit worn over two neoprene wet suits. Water temperature: 4° C.

Heat in (from water supply + metabolic heat) = Heat out (in outflow water, from flushing [b] and respiratory heat loss [c])

For a resting diver in thermal balance metabolic gain cancels b and c. Thus heat replacement equals suit heat loss.

Calculations using test data:

Body surface temperature (34°C) – water temperature (4° C) = 30°C

Body surface area of diver = 2 m^2

Q (overall heat transfer) = 414 kcal (474 watt-hours)

K (hr), heat transfer coefficient (kcal × m^{-2} × $°C^{-1}$) = 6.90.

CONDITION II. At 600 ft. Analysis of heat balance at 19.2 ATA, diver breathing a helium-oxygen mixture. Water tubing suit worn over two neoprene wet suits. Water temperature: 4°C. Under these conditions respiratory heat loss is balanced by metabolic heat production for any level of physical activity. Thus heat replacement equals suit heat loss.

Q = 1,308 kcal, or 1,521 watt-hours, and K = 21.8 kcal

Required heat/min = 1,308/60 = 21.8 kcal

Heat supply/kgm of water = 45°C inlet temperature – 34°C outlet temperature = 11 kcal

Required water flow rate ≅ 2 kgm/min, or 2 L/min.

Heat loss of channel swimmers

It is of interest to compare the heat loss of English channel swimmers with that just calculated for insulated divers breathing helium-oxygen mixtures at great depth. I have made the following calculations:

The maximal heat equivalent for sustained physical work *in air* for a period of 8 hours is 3,600 kcal, or 450 kcal/hr. This is equivalent to an oxygen consumption rate of 1.5 L (STPD)/min. Assume that the channel swimmer maintains an increased oxygen uptake rate not for 8 but for the 15 hours usually necessary for the channel swim. Then,

450 kcal/hr – 90 kcal/hr (estimated mechanical equivalent of work) – 90 kcal/hr (estimated respiratory heat loss) = 270 kcal/hr, or Q (total body heat transfer)

K (hr) = 6.75 kcal × m^{-2} × $°C^{-1}$ (based on body surface area of 2 m^2 and body core temperature [35°C] – water temperature [15°C] = 20°C)

I (insulation) = 1/K = 0.148 m^2 × °C × $kcal^{-1}$

Pugh's data[4], *based on measurements,* suggest a representative value for K (hour) of 13.5 kcal × m^{-2} × $°C^{-1}$, and I (insulation) = 0.074 m^2 × °C × $kcal^{-1}$. From these data the total caloric expenditure during a 15-hour English channel swim is 10,310, or 687 kcal/hr; this is equivalent to an oxygen consumption rate of 2.29 L (STPD) /min.

The discrepancy between my calculations and Pugh's quantified estimates emphasizes the need for additional study of an intriguing problem—the analysis of energy expenditure and heat production during cold water swimming. One phase of the problem, the remarkable insulative value of fat, has been systematically investigated.

EFFECTS OF COMPRESSION

Arthralgia, tremor, vertigo, nausea, and tachycardia

Arthralgia is reported frequently during rapid descent to depths greater than 300 ft. In an earlier era of diving hundreds of brief helium-oxygen dives were made to depths of 300 to 500 ft at a descent rate of 100 ft/min without compression pain. However, all divers who participated in a U. S. Navy experimental program involving descent to simulated depths of 600 ft or more reported symptoms of joint stiffness and crepitation. Quadriceps myalgia was common. Both weight-bearing and nonweight-bearing joints, including metacarpal, metatarsal, and spine, were involved. This pattern contrasts with decompression sickness bends that spares the spine, hands, and joints of the feet.

Slow compression at a rate equivalent to 40 ft/hr eliminates pain and discomfort. The etiologic basis of this discontinuous, or exposure duration–related, phenomenon is unknown. One plausible hypothesis relates the pain to fluid shifts produced by osmotic gradients created during equilibration of inert gas in blood and other tissues, such as joint cartilage and synovial membranes.[5]

Adverse reactions other than arthralgia, such

as tremor, vertigo, nausea, and tachycardia, are associated with compression to 31 ATA within 20 to 30 minutes.[6] The same investigators reported that one of two subjects in a 4-hour simulated dive to 1,000 ft (31.3 ATA) developed a tremor shortly after completing compression in a total time of 20 to 30 minutes; this was followed 10 minutes later by marked vertigo and tachycardia. One subject's blood pressure rose transiently to 160/110 mm Hg; the second subject was nauseated. However, after 1.5 hours at 31.3 ATA both subjects felt well.

Effects of compression on gas-filled body spaces

Middle ear spaces

During compression it is necessary to resort to such procedures as the Valsalva maneuver, swallowing, yawning, and thrusting the jaw forward to equalize pressure on both sides of the tympanic membrane. In some clinics routine myringotomies are performed in connection with hyperbaric therapy; given more time some patients are able to "clear" their ears with decongestants. During the past 35 years U. S. submarine personnel have undergone about 150,000 compressions to 50 psi in connection with escape training. From 5% to 25% or more of trainees have been unable to accommodate readily to excess pressure, chiefly because of subclinical infection of the upper respiratory tract. In one study 158 men (36%) suffered some degree of barotrauma to the middle ear that was readily diagnosed and classified by otoscopic examination of the tympanic membrane. Postcompression audiograms showed no significant loss of acuity between 3,000 and 8,000 cps, a range that includes the frequencies of speech. All examinees regained baseline auditory acuity within 2 to 3 weeks.

It is remarkable that the delicate mechanisms of hearing, specifically the attachment of the stapes to the cochlear oval window, are not disrupted by barotrauma. The rarity or even absence of proved deafness from barotrauma contrasts sharply with the immediate deafness caused by explosive noise. However, some degree of residual acoustic impairment cannot be excluded with certainty. As a group divers sustain progressive hearing loss, as do industrial workers, in the high-frequency range; however, it is also possible that excessive noise and aging may account for this impairment.

The mechanism of injury is compression, or cupping action, on the membranous lining of the middle ear spaces. Comparatively small "negative" pressures (1 to 2 psi, or 52 to 103 mm Hg) cause hyperemia and elicit pain. If pressure equalization in the ear does not occur during sleep, even exposure to an oxygen atmosphere may cause a mild cupping action by oxygen absorption, resulting in middle ear congestion and fluid. However, only rarely does barotrauma require medical intervention.

Sinal and dental odontalgia

Lancinating pain is felt over the affected sinus or sinuses if the ostia are occluded. The mechanism of injury is presumed to be the same as that of "ear block"; likewise, medical treatment is conservative. Occasionally, a gas pocket, often minute and not observable in a dental x-ray, may be present in the dental pulp; as pressure changes, pain may be felt in the involved tooth. In such a case dental treatment is mandatory.

Cardiopulmonary effects

Breath-hold swimming

Apneic underwater swimming is a strenuous exercise that can cause serious cardiac arrhythmias. In the United States approximately 5,500 deaths (about 5% of total accidental deaths) occur yearly from drowning; this figure excludes deaths from water transportation accidents. In a certain sense some of these deaths are self-inflicted in that hyperventilation prior to breath-holding delays the urge to breath.[7] With removal of the usual carbon dioxide stimulus to break apnea, hypoxia may produce unconsciousness. Exercise further raises the underwater swimmer's tolerance for hypercapnia while a breath-hold Valsalva maneuver of some degree impedes venous return, decreasing cardiac output; hypoxic collapse may thus supervene with little or no warning. Oxygen inhalation prior to breath-holding would circumvent hypoxia, but pure oxygen may render the diver more susceptible to syncope by producing cerebral vasoconstriction.

Breath-hold diving

During descent hydrostatic forces diminish chest volume so that at 99 ft (4 ATA) the lung volume is only one fourth of what it was at the surface. Alveolar P_{O_2} increases but the alveolocapillary carbon dioxide gradient is reversed and carbon dioxide accumulates in the circulating blood. Man underwater is subject to a degree of atavistic protection; however, this falls far short of the extraordinary tolerance for apnea exhibited by such diving mammals as the beaver, seal, and whale. Nevertheless, native skin divers develop a bradycardia that may amount to a 50% reduction of the control heart rate at the surface. The majority of divers observed by Scholander and associates[8] developed cardiac arrhythmias during and after breath-hold dives.[9]

It was always assumed that the depth threshold for this type of diving is limited by thoracic compression to residual lung volume; thus, if at the surface residual volume is 1,200 ml and total lung volume is 7,200 ml, then maximal diving depth is 7,200/1,200 = 5 ATA, or 132 ft. However, a breath-hold dive was recently made to a depth of 240 ft, well below the depth of 197 ft predicted from the ratio of lung volumes. Schaefer and coworkers,[10] using impedance plethysmography, showed that at breath-hold depths of 90 to 130 ft blood is forced into the thorax, compensating for the chest volume decrement produced by hydrostatic pressure; the volume of such blood shifts into the thorax may amount to as much as 1 L.

Immersion diuresis

Another hydrostatically induced blood volume shift into the thorax results from diuresis. This response, whose underlying mechanism remains unknown, is associated with breath-holding, head-out flotation, scuba, and "hard-hat," or conventional helmet, diving. McCally,[11] analyzing "head-out" immersion diuresis, concluded that antidiuretic hormone (ADH) release is inhibited by atrial distention and stimulation of stretch receptors associated with increased intrathoracic blood volume resulting, in turn, from hydrostatically produced pressure gradients. The physiologic impact of immersion diuresis is apparent from the work of Graveline and Jackson.[12] A subject was immersed in water up to the neck for 7 days. During the first 3 days urinary output doubled in each 24-hour period and urinary nitrogen excretion increased. This striking diuresis was accompanied by thirst, hemoconcentration, and weight loss.

In our helium-oxygen tests in a wet tank (1938) helmet divers in conventional dress, subject to regulated fluid intake, secreted on the average 466 ml of urine during a 2-hour period that included bottom time of 20 minutes, followed by 100 minutes of decompression. Per hour, urine secretion in the wet tank was 236 ml compared with 131 ml for the first hour postdive and 89 ml for the second hour postdive. The relevance of these data compared with the results of "head-out" immersion is the absence of a "negative," transthoracic pressure gradient; the ambient pressure on the diver's chest is about the same as his intrapulmonary pressure. Hydrostatic pressure, however, is uncompensated up to the waistline. The determination of diuretic response to graded hydrostatic levels of water up to the xiphoid in the standing subject would be relevant to this problem.

Resistance to breathing

Resistance to the turbulent flow of air or other gas mixtures within the bronchopulmonary system is proportional to the square root of the gas density. For more than a century caisson workers have performed heavy physical work in air pressurized to 4.5 atmospheres. It is thus not surprising that at a simulated depth of 1,000 ft (31.3 ATA) the pattern of respiratory response to exercise in a helium (99.1%)-oxygen (0.9%) atmosphere that is 4.4 times as dense as air at 1 ATA is similar to that at 1 atmosphere. At 19.2 ATA (600 ft) divers breathing helium-oxygen mixtures are able to sustain pulmonary ventilation rates of more than 70 L (BTPS)/min. Resistance to breathing unencumbered by apparatus is not a limiting factor in residence at simulated depths down to 1,500 ft.

At lower pressures (4 ATA) divers equipped with scuba gear and breathing nitrogen-oxygen mixtures are handicapped during exercise by a reduced ventilatory response. Lanphier[13] first described this impairment, reporting an average alveolar P_{CO_2} of 55 torr and individual values as high as 70 torr in the carbon dioxide–retaining divers. Appreciation of the magnitude of these values is enhanced by comparison with the previously reported alveolar P_{CO_2} level of 47 torr in response to inhalation of the carbon dioxide–enriched (5%) submarine atmosphere. Divers whose pulmonary ventilation in response to exercise is inadequate for carbon dioxide elimination are more susceptible to carbon dioxide intoxication and to oxygen poisoning. The tendency of some individuals to retain carbon dioxide during exertion in a pressurized atmosphere emphasizes the importance of physical fitness for divers. Specifically, in healthy trained men the rise of respiratory quotient (RQ) above 0.75 during heavy exercise of 5 minutes' duration is not more than 0.4; in untrained men the increase may be double this value. Physical training restricts the rise of blood lactate level and hence the production of "excess" carbon dioxide. It would appear that with tests to detect inadequate ventilatory response to exercise, physical training, and use of low-resistance diving equipment this critical problem can be solved.

Impaired pulmonary oxygen uptake

Unless the partial pressure of oxygen is increased above normal, hypoxia may occur at extremely high pressures. Chouteau,[14] using goats, found it necessary to increase inspired P_{O_2} from 0.21 to 0.28 ATA in the range of 70 to 80 atmospheres. In the range of 80 to 100 ATA P_{O_2} had to be further increased to 0.48 ATA. Hypoxic manifestations, paresis of the hind legs, and complete paralysis were reversed by increasing the P_{O_2}. However, about 101 ATA (1,000 m) was the upper limit compatible with recovery. Above 101 ATA hypoxia persisted despite additional increase of oxygen pressure.

Other manifestations of compression dysbarism

If care is not taken to equalize pressure within the face mask, conjunctival bleeding may occur and, in some cases, retrobulbar hemorrhage may cause tension on the optic nerve. If vision is impaired the patient is referred to an ophthalmologist. Lung compression may cause intrapulmonary hemorrhage in a breath-hold diver if the irreducible lung volume is further compressed. The circulatory adjustments previously discussed compensate to some extent for overcompression.

EFFECTS OF GASES AT INCREASED PRESSURE

Nitrogen narcosis

In both marine and dry hyperbaric environments we are concerned with the narcotic effect of nitrogen and with similar or related impairment that may attend the inhalation of helium-oxygen, hydrogen-oxygen, and other possible gas mixtures at pressures of 10 to 100 ATA. *Narcosis* is an apt term to denote the untoward psychic responses, including stupefaction, and the neuromuscular impairment associated with inhalation of inert gases that may not induce loss of consciousness. *Anesthesia* is a special instance of the general phenomenon characterized by loss of pain sensation and other signs familiar to the anesthetist. On a comparative scale, it may be inappropriate to include gases having narcotic potential for divers with those gases used to induce anesthesia in the operating room. Xenon, for example, is a potent anesthetic at pressures at which argon is merely narcotic. Both gases are relatively unreactive in chemical terms, and differing physical properties do not appear to account for their different physiologic effects (Table 11-2).

Mood changes have long been associated with work in compressed air. Thus in a well-lighted chamber at 4 ATA, individuals experience euphoria and are subject to fixation of ideas, impaired judgment, and neuromuscular coordination. Behnke and his colleagues[15, 16] attributed these aberrations to the narcotizing effect of nitrogen and invited attention to the similarity of the molar concentration of nitrogen at 14 ATA to that of diethyl ether under conditions of light anesthesia. Quantification subsequently revealed that nitrogen-induced impairment may occur at pressures as low as 3 ATA. A surgeon, for example, may have some doubt as to his clinical judgment if it becomes necessary to depart from usual procedure. On the other hand, routine tasks may be executed efficiently by highly motivated divers who are accustomed to pressurization. The depressant effects of nitrogen are similar to those of ethanol ingestion. Electroencephalography, flicker-fusion frequency, and auditory-evoked cortical potentials have been used to quantify the degree of impairment associated with inhalation of various hyperbaric gas mixtures.

Narcotic potencies of gases used in diving correlate with the interrelated physical properties of lipid solubility, surface tension decrease of monomolecular membranes, thermodynamic activity, solubility in proteins, formation of gas hydrates, and ionic shifts of Na^+ and K^+ involved in neural transmission. In their comprehensive analysis Featherstone and Muehlbaecher[17] point out that many of the physical properties correlated with anesthetic potency are merely reflections of the intermolecular van der Waals' forces of attraction and hence of the size and polarizability (ease with which the negatively charged electrons can be shifted about a positive nucleus) of the agents. The ratio of the van der Waals' constants a and b (Table 11-2) approximates the ratio of polarizability to volume. Either ratio correlates well with the loss of righting reflex in mice induced by gases used in diving, but not by xenon or nitrous oxide. In the limited series given in Table 11-2 solubility of the gases in oil correlates better with narcotic potency or loss of righting reflex.

A definitive study of the limitations of helium, hydrogen, nitrogen, and their mixtures as hyperbaric atmospheres was made by Brauer and Way.[18] They evaluated the potential usefulness of hydrogen-oxygen mixtures at depths between 365 and 800 ft. Furthermore, these investigators suggest that hydrogen does not remain inert in animals respiring this gas in mixture with oxygen for ex-

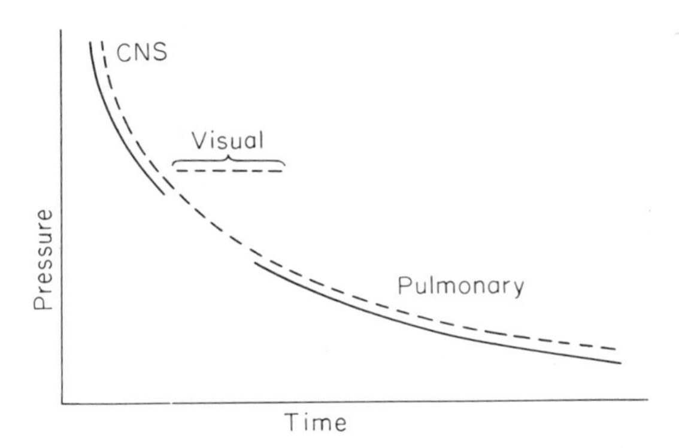

Fig. 11-2. Oxygen tolerance curve showing the separation of primary bronchopulmonary from central nervous system effects. The break occurs at about 2.5 ATA. Adverse visual changes, however, extend as a continuum with *exposure time* versus *pressure*. The shape of the curve has not been determined for long-term daily exposure to hyperbaric oxygen breathing.

tended periods of time. It is an arresting fact that, of all gases, only helium[14] is suitable for inhalation at pressures of 30 to 46 ATA (man) and to 100 ATA (goats). Except for its high heat conductivity, helium approaches the ideal gas for use in diving. With regard to the problem of narcosis, the usefulness of helium greatly exceeds predictions based on the physical properties given in Table 11-2.

Hyperoxia and hyperbaric oxygenation

Oxygen tolerance and limiting reactions

The duration and intensity of exposure to hyperoxia correlate directly with toxicity. The extensive use of oxygen in diving and in hyperbaric therapy is predicated on the concept of a relatively safe latent period of inhalation. Many of the reactions associated with inhalation of oxygen at increased partial pressure are paradoxic. There is the variable latent period of apparent well-being terminated precipitously at times by severe central nervous system disturbances resembling the convulsive seizures of idiopathic epilepsy. There is also bronchopulmonary irritation attending prolonged exposure to 2 ATA. Despite thousands of exposures involving seizures, man has escaped the chronic paralysis and pneumonitis observed in test animals. Hyperbaric oxygen inhalation (OHP) is feasible clinically and in diving because the early adverse symptoms of toxicity are readily reversible and the heart, in contrast to the sensitive respiratory centers, is resistant to injury.

Although there is a dichotomy on the time-pressure curve between central nervous system (CNS) and bronchopulmonary responses (Fig. 11-2), it is possible to study the adverse effects of OHP on the visual system of rabbits throughout the entire range of oxygen pressure (Table 11-3).

Table 11-3. Oxygen toxicity as a function of oxygen pressure and exposure duration

Oxygen pressure (ATA)	Exposure duration	Symptoms and signs of toxicity
> 0.4	Many hours	Retrolental fibroplasia of premature human infants
0.4	Days	Well tolerated by human adults
0.5	12 days	No evidence of visual cell death in rabbits[19]
0.6	7 days	Visual cell death in 50% of rabbits[19]
2.0	> 6 hours	Pulmonary irritation and decreased vital capacity
2.7	4 hours	Limit in *recompression therapy*; intermittent oxygen breathing alternating with brief periods of air breathing
3.0	2 hours	Usual limit for one or two daily sessions of hyperbaric oxygen therapy, as, for example, in treatment of gas gangrene
3.0	6 hours	Onset of visual cell death in rabbits[19]
4.0	0.5 hour	Used occasionally as adjunct to radiation therapy; high incidence of convulsions

Specific tolerance data

Oxygen at 0.33 ATA has been inhaled for more than 30 days without apparent adverse reaction. However, signs and symptoms of respiratory irritation have been reported after exposure of about 2 weeks' duration to 0.6 ATA (428 torr), which is equivalent to the P_{O_2} in air at 26.7 psi (60 ft). After about 6 hours of exposure, inhalation of oxygen at 2 ATA causes substernal distress and progressive decrease of vital capacity. Sharp limits cannot be set because of wide individual variation. At a pressure of 3 ATA (29.4 psi) oxygen can usually be inhaled by healthy men *at rest* in a dry chamber for 2 to 3 hours; however, occasionally seizures occur at this level. Striking bradycardia and intense peripheral vasoconstriction are characteristic. Termination of tolerance at 3 ATA is attended by abrupt rise of systolic and diastolic systemic arterial blood pressure, reversal of bradycardia, greatly diminished visual acuity, contraction of visual fields, extreme pallor, and periodic waves of nausea.

At a pressure of 4 ATA (44.1 psi) seizures grossly similar to those of idiopathic epilepsy, and rarely syncope, terminate oxygen inhalation. Toxicity is enhanced by exercise while breathing oxygen at 3 ATA. Divers breathing oxygen in closed-circuit systems may have convulsions while swimming underwater at depths no greater than 30 ft. In the wet tank symptoms are more prevalent than in the dry chamber. In a dry chamber at 2 ATA an athlete can engage in 5-miniute periods of maximal exercise with improved tolerance for heavy physical exertion compared with control tests in air. Between 2 and 3 ATA (Fig. 11-2) many more exercise tests are required to determine the tolerance break relative to exposure time.

Interstitial edema—a prime pathologic reaction

Bennett and Smith[20] found that rats develop perivascular edema and, subsequently, thickening and hyalinization of pulmonary artery walls after about 3 days' exposure in compressed air at 4 ATA, or a P_{O_2} of 635 torr. These changes precede pulmonary arterial hypertension, which occurs after 24 to 31 days.

In rats, injury to capillary endothelial cells occurs after 48 hours in an atmosphere of pure oxygen at 1 ATA. The initial lesion and key pathogenetic factor is accumulation of fluid in the interstitial spaces around large blood vessels and bronchioles (Fig. 11-3).

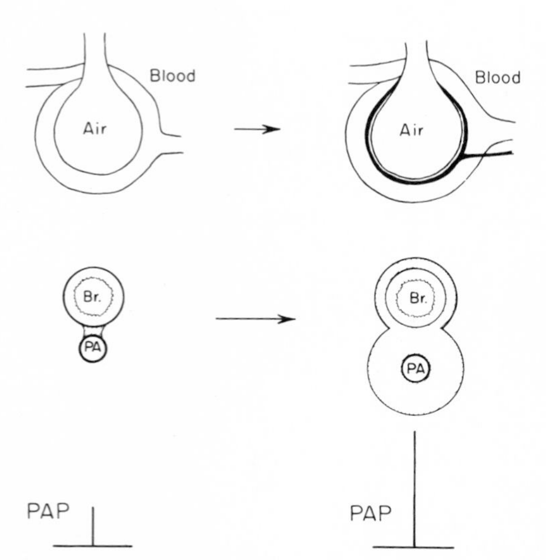

Fig. 11-3. Acute bronchopulmonary oxygen toxicity. Vascular endothelium is injured first, permitting edematous enlargement of perialveolar, peribronchiolar, and vascular interstitial spaces. Early changes are reversible and may also result from dysbaric embolism, which increases pulmonary arterial pressure (PAP). Chronic bronchopulmonary oxygen poisoning also increases PAP.

In man substernal distress on deep inspiration with paroxysmal coughing after 8 to 12 hours of oxygen inhalation at 2 ATA may well indicate an incipient stage of interstitial edema. Decreased vital capacity and pulmonary compliance are not associated with airway obstruction, atelectasis, or fluid in the lungs. Likewise in the pulmonary form of decompression sickness discussed later in this chapter bubbles in pulmonary arteries cause hypertension and fluid extrusion into the periadventitial spaces. We have long associated nascent bubbles in the pulmonary blood vessels of man with shallow respiration and *substernal distress* on deep inhalation, which elicits paroxysms of nonproductive coughing. Possibly the pulmonary manifestations of decompression sickness and of oxygen poisoning involve the same pathologic mechanism (Fig. 11-3).

Hormonal influence

Specific etiologic factors regarding central nervous system aberrations remain obscure despite information that certain enzyme systems and substrates are eventually impaired or altered. It is certain that adrenal, thyroid, and pituitary hormones shorten the latent period of apparent well-being. Oxygen inhalation produces a subjectively detectable adrenergic response. Furthermore, the apprehensive subject, in contrast to the phlegmatic one, is more susceptible to certain of the untoward reactions to hyperoxia. Experimentally, hyperbaric oxygenation (OHP) stimulates the sympathetic nervous system. In the rat, the adrenal

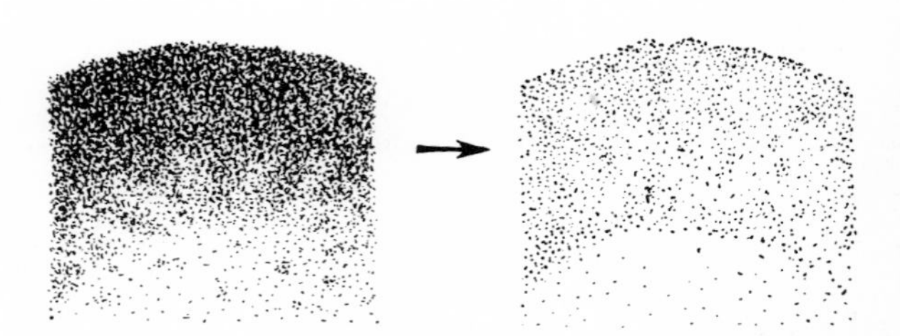

Fig. 11-4. Effect of hyperbaric oxygen on the rat adrenal gland. The zona fasciculata of a normal rat shows a large amount of lipid (black), which is depleted during exposure to oxygen pressures as low as 3 ATA. The adrenal gland is a "morphologic mirror" that reflects hyperbaric oxygen toxicity. (Redrawn and modified from Bean, J. W.: Hormonal aspects of oxygen toxicity. In Goff, L. G., editor: Underwater physiology symposium, Publ. 377, Washington, D. C., 1955, National Academy of Science–National Research Council.)

cortex hypertrophies and is depleted of lipid.[20] Recovery after OHP exposure is reflected by restoration of the normal histologic appearance of this chameleon gland (Fig. 11-4).

Role of carbon dioxide in oxygen intoxication

The addition of carbon dioxide to the breathing mixture potentiates the adverse effects of oxygen and provides a substantial reason in addition to the shift of blood from periphery to core for the greater oxygen toxicity in the wet, compared to the dry, chamber. The apparent augmentation of oxygen pressure by the addition of carbon dioxide may also be a major factor in the reduction of oxygen tolerance by exercise. Lanphier[13] noted the higher incidence of oxygen poisoning during underwater work in carbon dioxide–tolerant divers who were unaware of excess retention at carbon dioxide levels as high as 70 torr.

If alveolar P_{CO_2} is decreased to a level of 22 to 27 torr with a Drinker respirator, convulsions do not occur at 4 ATA in anesthetized dogs during an exposure of 2½ to 3 hours' duration. On the other hand, if alveolar P_{CO_2} is increased to a level of 60 to 68 torr, all animals convulse within 80 minutes.

Lambertsen[22] observed respiration, blood-gas transport, and cerebral circulation in male subjects who inhaled oxygen for less than 1 hour at 3.5 ATA. Their data for man confirm previous findings for the dog that, despite relatively constant alveolar P_{O_2}, hypercapnia produces wide P_{O_2} fluctuations in cerebral and right ventricular venous blood. By vasodilation carbon dioxide may increase cerebral blood flow and cerebral venous P_{O_2} by a factor of 10 or more. Carbon dioxide–induced cerebral blood flow increase may also potentiate the narcotic action of nitrogen and other gases at high pressure.

The unpredictable wide individual variation in response to oxygen inhalation, especially CNS effects at 2.5 ATA or more, is attributable in large measure to the factors just enumerated—fluctuating levels of P_{CO_2} in the lungs, blood, and fixed tissues; metabolic changes during exercise; blood flow changes induced by hydrostatic forces; and adrenergic hormones.

PROBLEMS OF DECOMPRESSION

Decompression blockage of aural and sinal spaces

Occasionally during decompression, if the auditory tubes (eustachian canals) are not patent, the distending internal middle ear pressure tends to evert the tympanic membrane, producing excruciating pain. A similar obstruction of a sinal ostium, usually the frontal, also causes pain. This complication is both prevented and treated by slow decompression and use of topical and systemic decongestants; the objective is to maintain or restore patency and drainage.

Complications of lung overinflation

Air embolism, interstitial emphysema, and pneumothorax

Lung overinflation may result from abrupt ascent after a dive, failure to vent the lungs during ascent (breath-holding), or excessively rapid decompression of tunnel workers to the first stage, producing physical collapse and unconsciousness. In tunnel workers such complications during or immediately after decompression have been attributed to air embolism secondary to pathologic lung conditions such as pulmonary cysts. Localized pulmonary overinflation may occur during decompression as a result of "ball-valve" or other types of airway blockage in a lung segment containing cysts, scar tissue vesicles, or emphysematous bullae. Despite underwater supervision of submarine escape training exercises from a diving bell, there have been twenty-five casualties, including several fatalities, in 150,000 ascents from depths ranging from 18 to 90 ft. Indeed, death has occurred during escape training after ascent from a diving bell at a depth of only 15 ft. Virtually all such accidents are attributable to air embolism or other complications of lung overinflation. On the other hand, trained personnel can make safe ascents from depths of 500 ft if care is taken to vent the lungs continuously during rapid ascent to the surface.

In generalized pulmonary overdistention two distinct types of injury may occur, either independently or simultaneously. In the more serious type alveolar tears and rupture of small pulmonary blood vessels permit aspiration of gas emboli, which are then carried to the left chambers of the heart and, from there, into the systemic arterial circulation. In the other type of injury excessive pressure forces gas along fascial planes of the pulmonary vascular tree. Pockets of gas forced into the mediastinum may produce subcutaneous emphysema and compressive pneumothorax. Paradoxically, such pneumothorax may splint the collapsed lung, preventing aspiration of gas emboli.

Cerebral embolization is indicated by a wide spectrum of symptoms and signs, including headache, dizziness, visual disturbance, chest pain, paresthesias, paralysis, loss of vital signs, and focal convulsive seizures. By contrast, the central nervous manifestations of decompression sickness (DCS) are delayed in onset and usually associated with prodromal symptoms related to the release of bubbles of inert gas from body tissues. Air embolism is characterized by sudden collapse after ascent from a dive or during chamber decompression and is unrelated to the duration of previous exposure. A prompt favorable response to recompression is usual; without recompression, mortality is high.

Interstitial emphysema per se is not an emergency, since it resolves itself spontaneously. However, pneumothorax during decompression may progressively embarrass respiration and circulation; surgical intervention, thoracocentesis, is then imperative.

Decompression sickness

Etiology and salient features

Experimental evidence suggests that intravascular bubbles are the primary etiologic agent of decompression sickness. Although fat emboli have been observed in pulmonary and cerebral blood vessels, they are usually not pathogenic. The presence of circulatory bubbles may produce complications that mask primary etiology and confuse the investigator. Secondary changes involve products of cell disintegration (lipases, K^+, peptides, histamine, serotonin, and bradykinins), altered physical properties of blood such as clumping of cells and platelets, decreased coagulation time, and a striking degree of hemoconcentration. Obstruction of peripheral veins by bubbles predisposes to an indurate type of edema. Bubbles accumulating in branches of the pulmonary arteries cause hypertension and interstitial pulmonary edema (Fig. 11-3).

Signs and symptoms of decompression sickness are manifest singly or in combination, subjectively as pain (bends), and objectively as respiratory distress that may be asphyxial (chokes), paralysis involving primarily the spinal cord, and chronic bone lesions. The manifestations of decompression sickness in tunnel worker, diver, aviator, and hospital personnel working in hyperbaric chambers are generally similar; however, the aviator is rarely afflicted by paralysis but prone to develop "chokes." Depending on the time-pressure pattern of decompression, symptoms develop within several hours after return to normal pressure and decrease in severity as an exponential function of time during the ensuing 12 hours. Late cases often reflect resistance of subjects, especially tunnel workers, to recompression therapy.

In Rivera's analysis of 935 cases[23] the following rounded percentages apply to the relative incidence of signs and symptoms: localized pain, 92%; numbness (paresthesia), 21%; muscular weakness, 21%; rash and pruritus, 15%; temperature change, nausea, vomiting, 8%; vertigo, 9%; visual impairment, 7%; paralysis, 6%; headaches, 4%; loss of consciousness, 3%; urinary dysfunction, 3%; dyspnea, 2%; fatigue, 1%; convulsions, 1%; and edema, 1%. We have observed fatigue, edema, and pulmonary distress elicited by deep inspiration much more frequently.

Classification

The British classification of decompression sickness consists of two types: type I, simple pain in the region of a joint (bends); and type II, any manifestation of a serious nature. However, joint pain may precede or follow type II symptoms. Thus visual scotomata may precede joint pain or, if bends are untreated, respiratory distress may occur, complicated by shock and, occasionally, paralysis. A better working definition of type I is mild decompression sickness with good response to recompression and no residuals. Type II decompression sickness is *serious;* favorable response to recompression is delayed and uncertain, and there may be residual impairment despite prolonged recompression, fluid administration, and intensive care.

Cardiopulmonary manifestations

Perhaps the most obvious manifestations of decompression sickness and those of special interest to the environmental physiologist are those involving the cardiopulmonary system. A medical officer exposed for several hours 3 days a week to a pressure of 4 ATA reported the following:

> After decompression, but preceded usually by a period of several hours of well-being, there was burning substernal discomfort, initially only on deep inspiration, which also produced paroxysmal coughing. Concomitant with this substernal distress, which tended to cause rapid shallow breathing, was debilitating malaise. Either recovery was spontaneous over a period of hours or incipient asphyxia and unremitting cough (chokes) required recompression therapy.

The pathophysiology of chokes and complicating shock has been studied in detail in anesthetized dogs decompressed within 15 seconds after exposure of about 2 hours' duration at 65 psi. After an interval of 15 to 20 minutes using a glass cannula, bubbles are observed in arteries and veins and in blood withdrawn from the right ventricle. The appearance of bubbles visible to the unaided eye is associated with the following signs: (1) bradycardia; (2) rapid shallow breathing, or tachypnea; (3) increased pulmonary arterial pressure; and (4) decreased systemic arterial pressure. Without recompression there was decreased circulating

Decompression Embolism
(Pulmonary Circulation)

Lungs	Blood	Circulatory
normal	Catecholamines	Distortion of vessels
A.D. Alv.	Histamine	Disruption of pulsatile flow
	Serotonin	Pulm. Art. Hypertension
micro-lung	Bradykinin	
Alv. A.D.	Platelet Aggregration	Bradycardia
	Cell Clumping	Fall Art. Pressure
Dyspnea		
Hypoxia	Hemoconcentration	

Circulatory Shock
Cardiopulmonary Failure

Fig. 11-5. Sequential changes associated with decompression embolization of pulmonary arteries. In addition to edematous injury (Fig. 11-3), the previously normal lung is reduced to a severely impaired microlung with constricted alveolar ducts *(A. D.)* and shrunken alveoli. The aggregation of formed blood elements and the circulatory disruption, if untreated by recompression, produce circulatory shock and cardiopulmonary failure.

blood volume, striking hemoconcentration, and a general condition of circulatory shock. Pulmonary arterial hypertension and tachypnea are signs of pulmonary embolization and occur during intravenous infusion of particulate matter, such as starch, barium sulfate, or glass beads. Despite increased breathing frequency the lungs are largely nonfunctional and arterial hypoxemia is extreme. A normal lung is reduced to a "microlung." Histamine, bradykinins, and serotonin are not only powerful smooth muscle constrictors but also produce striking aggregations of blood cells and platelets. The manifestations of decompression sickness are outlined in Fig. 11-5. Pulmonary interstitial edema is a complication of pulmonary arterial hypertension induced by the nascent bubbles that result from excessively rapid decompression (Fig. 11-3).

Bone lesions

Ischemic osteonecrosis, also termed avascular, or aseptic, bone necrosis, is prevalent in older tunnel workers who have been exposed to pressure. Similar bone changes occur in divers who receive little or no decompression and in divers engaged in deep helium-oxygen saturation tests. McCallum and others[24] reported potentially disabling juxta-articular lesions in 10% or more of workers engaged in tunnel construction under the river Clyde. The impairment was related to the number of times a worker was decompressed, to the pressure during work, and to the number of attacks of bends.

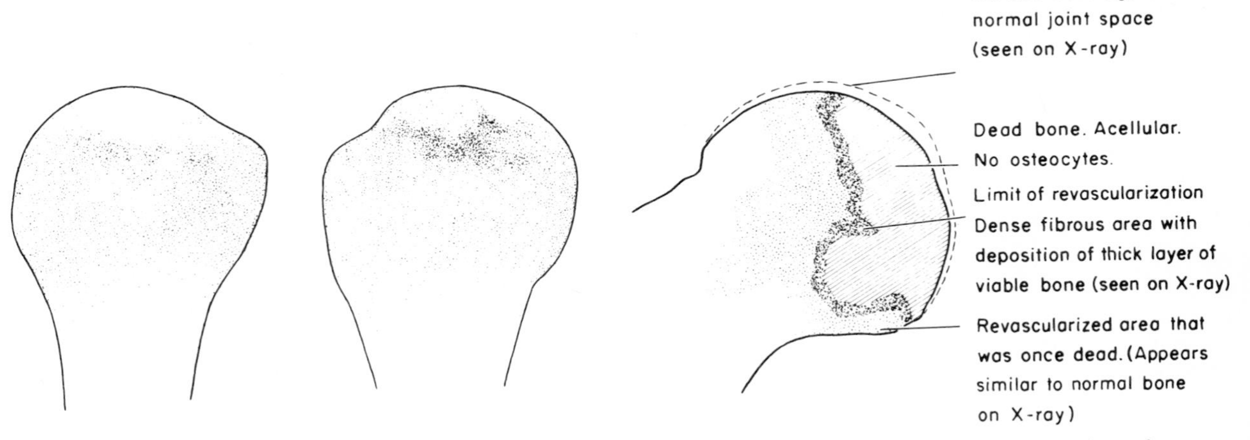

Fig. 11-6. Aseptic bone necrosis. Schematic drawings show early (left) and advanced (middle) lesions observed in roentgenograms of the head of the humerus. The sketch (right) shows details of histopathology in the head of the femur. (Courtesy Drs. Walder and Griffiths, Subcommittee of Decompression Sickness, Medical Research Council, Great Britain.)

Ischemic osteonecrosis appears in roentgenograms as (1) serpentine strands of increased density in the medullary cavities of long bones, and (2) wedge-shaped zones of infarction and small irregular dense areas adjacent to and underlying articular cartilage in the heads of the femur and humerus (Fig. 11-6). Complicating diagnosis is the latent period of a year or more before roentgenographic changes occur. Functional impairment resulting from irregularity or breakdown of the cartilaginous joint surface may not manifest itself for many years. Complaints of pain in the hip and shoulder joints of tunnel workers or divers deserve assiduous attention, including roentgenographic survey. Medical treatment of joint involvement involves absolute nonweight-bearing and judicious surgical intervention.

Bone necrosis occurs not only as a complication of decompression sickness but also in patients with hemoglobinopathies, alcoholism, hypercortisonism, arteriosclerosis, and other diseases. Considerable evidence suggests that, in addition to gas emboli, fat embolism of bone is involved in the pathogenesis of bone necrosis. Clinical and experimental studies have both shown that a fatty liver can spontaneously release fat globules into the circulation, which may then reach the vulnerable heads of the humerus and femur. This fact is important for understanding the case of the older worker.

Factors predisposing to decompression sickness

Factors influencing susceptibility to decompression sickness can be classified in two categories. The first includes conditions that increase the gas content of tissues, such as amount of fat and carbon dioxide accumulation as a result of exercise, pressure level, and duration of exposure. The second group includes factors that affect tissue perfusion. With regard to age, 18-year-old subjects are less susceptible to bends than 25 year olds. Anxiety or cold stress may cause vasoconstriction, impairing gas transport. Fat, which absorbs about five times as much nitrogen as blood, may prolong decompression many hours after saturation exposure of obese subjects in compressed air atmospheres. Serum surface tension, lipid and platelet concentration, and tendency of the formed blood elements to aggregate may further explain the variation of individual susceptibility to decompression sickness. Even extraneous circumstances merit consideration. For example, in decompression of tunnel workers immobilization and flexion of lower extremities during intensive card playing may hinder gas transport from these members. It is puzzling that among tunnel workers with similar physical characteristics and living habits some are unduly susceptible to decompression sickness. Decompression tables can be formulated that reduce the overall incidence of disability to 1% of man-shifts, but there is wide variation in the decompression time necessary to accomplish this objective.

Acclimatization

The fact that men who work daily in compressed air and divers who breathe helium-oxygen mixtures become less susceptible to decompression sickness remains unexplained. Early in this century, during construction of the East River tunnels in New York City, 330 men were employed for 36 days on a two-shift daily schedule. Men worked 3 hours per shift at 40 to 42 psi with a rest interval of 3 hours at normal pressure between shifts. Total decompression time was 48 minutes per shift or less than one half (one fourth on some schedules) of current stipulations. In 8,510 decompressions the incidence of bends was only 1.6% and there were no serious cases. The statement of the medical director of the project[25] that "only seasoned men were employed" reflects recognition of the phenomenon of acclimatization that has been systematically investigated by Walder.[26] The relationship of repeated daily exposures during work in compressed air expressed as cumulative number of compressions to the mean daily bends rate (percent bends) in the same group of ninety men is as follows:

Cumulative number of compressions	Bends (%)
1 to 5	7.3
6 to 10	3.6
11 to 15	1.3
16 to 20	0.44
46 to 50	0.67

A paradox of decompression of both practical and theoretic interest is the lack of a good inverse correlation between the incidence of bends and extension of decompression time beyond a minimum requirement. The work periods during construction of the Hudson tunnels, for which 48 minutes' decompression for a single shift appeared adequate at the time, would currently require 98 minutes (British tables), 162 minutes (U. S. Navy Schedule),[27] and 199 minutes (Washington State Code recently adopted by several states). The reported incidence of decompression sickness is similar for the three schedules despite wide differences in decompression time. However, it is likely that ischemic osteonecrosis can be eliminated by prolonged decompression.

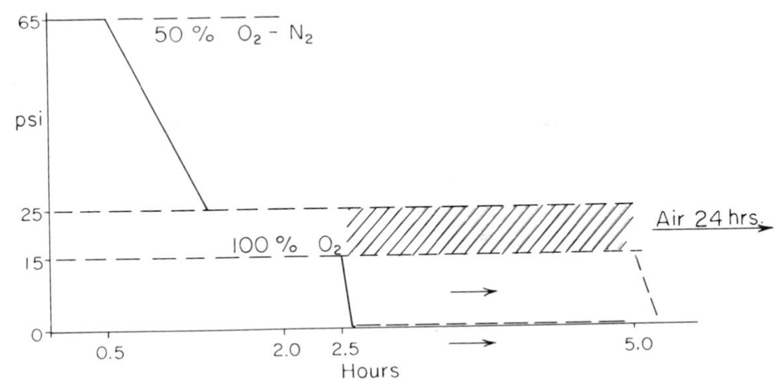

Fig. 11-7. Outline of oxygen recompression therapy. In U. S. Navy Tables 5 and 6 (1968) oxygen inhalation is limited to 27.8 psi (60 ft) and has proved highly effective. The 50% oxygen-nitrogen (or oxygen-helium) mixture at higher pressures has not been systematically evaluated. The abrupt drop in pressure from 15 psi to normal is the author's practice and has no official sanction. The "overnight soak" (air for 24 hours or longer) is the innovation of Yarbrough and Behnke.

Treatment of decompression sickness and gas embolism

Oxygen recompression therapy

For a number of years the U. S. Navy has used oxygen inhalation at pressures of 30 psi or lower after an initial sojourn in air at higher pressures (to 44 psi in type I cases and to 73.4 psi in type II cases). A schematic outline of the principles of this type of therapy is presented in Fig. 11-7. The experience of the experimental diving unit recorded by Goodman[28]* supports the value of oxygen recompression as exclusive therapy; the procedure is outlined in U. S. Navy Treatment Tables 5 and 6[27] for decompression sickness.

Of a total of 123 cases of decompression sickness, 105 divers completely relieved after one treatment at pressures not exceeding 26.7 psi (60 ft) with oxygen inhalation begun at the start of compression. Nine patients required additional recompression, and one patient was relieved only after a third treatment. The remaining eight patients, civilians who reported for treatment after some delay with type II decompression sickness, predominantly spinal cord injury, improved substantially.

Of 135 tunnel workers recompressed according to the new U. S. Navy Tables, despite an average delay of about 5 hours before initiation of treatment, only two required a second recompression.[30] The worker's ability to resume his regular work shift, often within 6 hours after therapy, was solid proof of the effectiveness of oxygen recompression therapy.

*See also reference 29.

Treatment of traumatic gas embolism

The treatment tables used for treatment of gas embolism are essentially U. S. Navy Tables 5 and 6 modified to include recompression to 73.4 psi (6 ATA) for 15 to 30 minutes, followed by reduction of pressure to 26.7 psi (60 ft) within a 5-minute period.[29] Recompression then follows the basic tables. The duration of recompression at maximal pressure depends upon the rapidity of recovery. The objective of therapy is to reduce the size of cerebral gas emboli and to restore occluded circulation without unduly saturating the tissues with inert gas. Thus recourse may be had to inhalation of high oxygen-gas mixtures (Fig. 11-7).

Recompression therapy at high pressures

In experimental diving to depths deeper than 500 ft utilizing helium-oxygen mixtures, symptoms and signs of decompression sickness may be observed during early stages of decompression. Surgeon Cdr. Barnard, R.N., pioneered treatment at high pressures previously considered infeasible. The following selected case data show the pressure required for relief of symptoms in relation to depth at onset of complaints.

Onset of complaints (ft)	Relief (ft)
40*	366
125*	165
210†	450
230†	450
340†	400

Using Barnard's method for calculating the time

*Neurologic involvement.
†Vertigo.

required for decompression, but not strictly his constants,

(depth of relief + 33) × 0.05 = rate of initial ascent (ft/hr)

At 5-hour intervals, the rate of ascent can be decreased 25%. Obviously, the patient's condition is the prime consideration in therapy, not rigid adherence to a schedule.

Measures ancillary to recompression

The life-saving importance of other modalities such as inhalation of nebulized drugs to relieve bronchiolar spasm and administration of dextran[31] has been demonstrated in severely decompressed animals. In man, infusion of plasma expanders has been effective without recompression in the treatment of shock complicating altitude decompression sickness and in the treatment of a limited series of cases of divers who did not respond well to recompression. The broad spectrum of pathologic manifestations may require sedative drugs, anticoagulants such as heparin, and isoproterenol, aminophylline, and potassium chloride for relief of "microlung" and maintenance of cardiac function.

Extensive use of ancillary measures has usually occurred when recompression facilities were not available. Despite recompression and other measures, an occasional patient develops neurologic involvement that responds initially to recompression but relapses, possibly because of cerebral edema, as pressure decreases; repressurization is ineffective, at least to 75 psi, and the patient dies. Hypothermia may be tried but assumes heroic proportions as an emergency procedure. Removal of blood foam from the right ventricle by catheter has not been established as a life-saving procedure.

PREVENTION OF DECOMPRESSION SICKNESS

Safe decompression—physical and physiologic principles

Exponential nature of gas transport

The transport of an inert gas such as nitrogen from a tissue (Fig. 11-8, lower curve) is an exponential function of time:

$$Y = A(1 - e^{-kt}) \qquad (1)$$

where Y = quantity of nitrogen recovered at time (t) during oxygen inhalation
A = total quantity of nitrogen in the tissue at time zero
e = base of natural logarithms
k = constant indicating the rate of change of the slope of the exponential curve (Fig. 11-8, lower curve)

$$K = \frac{1}{t} \times \log_e\left(\frac{A}{A - Y}\right) \qquad (2)$$

The value of A in equation 1 can be computed from Y_1 at time t_1 and from Y_2 at time t_2, if the time interval t_2 is twice t_1 as follows:

$$A = \frac{Y_1^2}{2Y_1 - Y_2} \qquad (3)$$

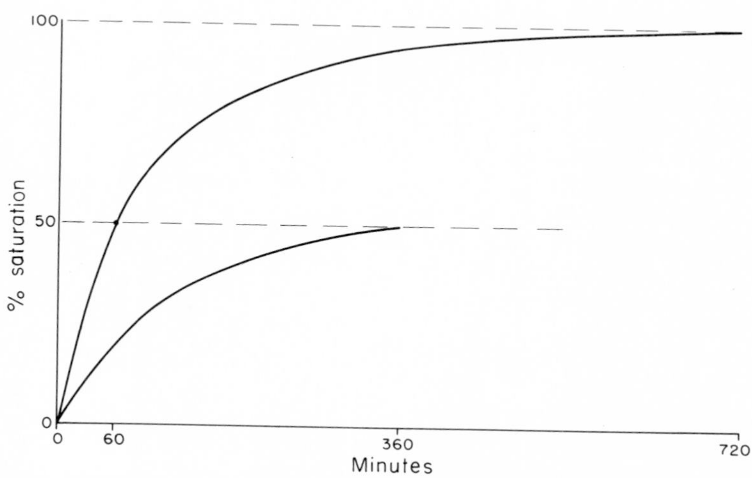

Fig. 11-8. Exponential elimination of excess inert gas (nitrogen). Lower curve shows washout from a tissue that *saturates* at a half-time rate of 60 minutes. *Desaturation* half-time is greater. Upper curve depicts desaturation and saturation of the body as a whole. Note the difference in elimination rate for the same quantity of nitrogen (50% of total). Lower curve can be represented by a single exponential expression, the upper curve, by a series of tissues.

If the rate constant (k) for any tissue is known, then the half-time $T_{1/2}$ can be computed from the following equation:

$$\log_e 2 = 0.693 \text{ and } \frac{0.693}{k} = T_{1/2} \qquad (4)$$

Multiples of 0.693 are *time units* (TU) derived as follows:

$e^{-kt} = 0.5$, when $kt = 1$ (0.693) and $(1 - e^{-kt}) = 50\%$ desaturation
$e^{-kt} = 0.25$, when $kt = 2$ (0.693) and $(1 - e^{-kt}) = 75\%$ desaturation
$e^{-kt} = 0.016$, when $kt = 6$ (0.693) and $(1 - e^{-kt}) = 98.4\%$ desaturation

For whole body nitrogen elimination (Fig. 11-8, upper curve) may be represented by a series of functions similar to equation 1:

$$Y \text{ (whole body)} = A_1(1 - e^{-k_1t}) + A_2(1 - e^{-k_2t}) + A_3(1 - e^{-k_3t}) + \ldots + A_n(1 - e^{-k_nt}) + \ldots \qquad (5)$$

Typical $T_{1/2}$ values for man (lean diver) are 1.5, 15, and 120 minutes and corresponding k values, respectively, are 0.44, 0.044, and 0.0054.

As implied by equation 1, an exponential curve for a tissue is linear on a log-linear plot; thus, log nitrogen recovered plotted as a function of time is a straight line for *fluids* and for tissues of variable fat content in *lean, medium,* and *obese* divers (Fig. 11-9).

Importance of body fat

For dives of short duration the amount of body fat per se is not an important consideration, if the fat reservoir is only partially saturated; it may possibly even be of some advantage besides its unquestionable value as insulation against cold. If 10% (1 unit) of body weight is lipid (adipose tissue triglycerides) then a 70-kgm man would have about 490 ml of nitrogen *per atmosphere* (760 torr nitrogen) dissolved in fat. If blood flow to adipose tissue is approximately 0.4 L/min, half-time for nitrogen elimination from *1 unit* of fat would require about 2 hours, and 98.4% elimination, 12 hours (Fig. 11-9, *lean*). As units of fat increase, blood perfusion of adipose tissue does not keep pace and diffusion assumes a more important role. A diver with 3 units of fat would require more than 6 hours for 50% nitrogen clearance (Fig. 11-9, *obese*). The following data show the amounts of nitrogen and helium in a lean (1-unit), a medium (2-unit), and an obese (3-unit) diver for several different partial pressures of nitrogen and helium.

	Nitrogen (ml in adipose tissue lipid)	
	1 ATA	3.7 ATA
Lean	490	1,813
Medium	980	3,626
Obese	1,470	5,439

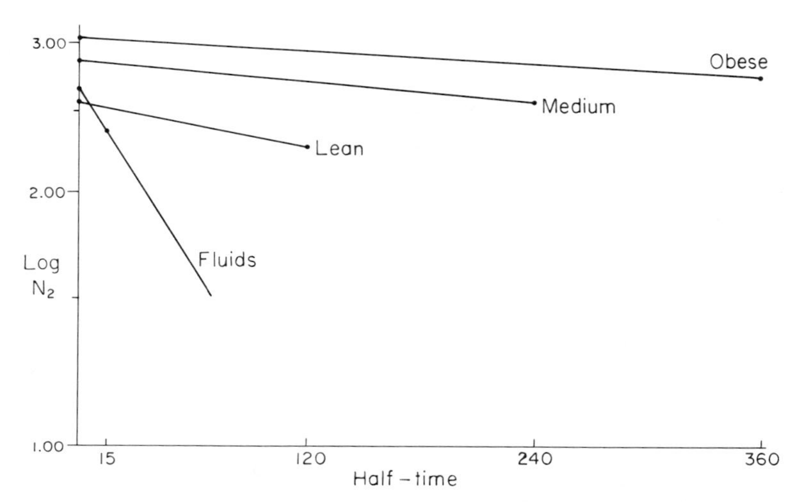

Fig. 11-9. Log-linear plot of nitrogen elimination from *lean* ($T_{1/2}$ = 120 minutes), *medium* ($T_{1/2} \doteq$ 240 minutes), and *obese* ($T_{1/2}$ > 360 minutes) divers. In each diver the nitrogen content of body *fluids* is the same, and half-time ($T_{1/2}$) is 15 minutes.

	Helium (ml in adipose tissue lipid)		
	1 ATA	15 ATA	30 ATA
Lean	105	1,575	3,150
Medium	210	3,150	6,300
Obese	315	4,725	9,540

Inert gas transport

At 1 ATA (P_{N_2} = 573 torr) about 9 ml of nitrogen is dissolved in 1 kgm of body fluid and 53 ml in 1 kgm of fat. In the *lean* diver, about 400 ml of nitrogen dissolves in body fluids, 100 ml in bone marrow and spinal cord, and 370 ml in adipose tissue lipid. About 5.2 L/min of blood perfuse the largely nonlipid cell mass and only 0.4 and 0.2 L circulate through adipose tissue and bone marrow–spinal cord, respectively. This discrepancy between perfusion and dissolved nitrogen in tissue is reflected in the nitrogen washout data for the dog from Groom and associates,[32] which are similar to those for man.

Nitrogen compartments (percent of total nitrogen)	Half-time desaturation (minutes)	Cardiac output (percent of total)
69	117	9
24	13.5	25
7.4	1.5	66

After brief exposures or dives to depths at which air must be breathed, decompression can be accomplished either nonstop or in short stages (Fig. 11-10). It is postulated that bubbles form in circulating blood but not in sufficient number to produce symptoms. After saturation exposures, however, decompression must be conducted essentially in accord with the dashed line in Fig. 11-10. It is postulated that under these conditions supersaturation or phase separation of inert gas in any appreciable quantity will cause symptoms at a later stage in decompression.

Helium as compared to nitrogen elimination

After saturation with helium, which is much less soluble in fat than nitrogen, a *lean* diver requires about 40% less time for decompression. Bühlmann and co-workers[33] conducted a series of graded exposures sufficient to establish gas pressure equilibrium between tissue and air and a comparable series with a helium-oxygen atmosphere (4 ATA). Only 3 hours of oxygen inhalation was necessary for decompression in the helium tests compared with 6 to 7 hours of oxygen breathing necessary to ensure adequate decompression after tissue saturation in air.

Diffusion of inert gases through skin

Until recently little attention was accorded a factor that may play a significant role in the process of decompression after saturation dives. Earlier tests (Fig. 11-11, subject reclining in a separated helmet-body enclosure) indicated that at room temperature helium diffuses through skin at a rate of about 55 ml/hr. This amounts to about 10% of the total body store and can be

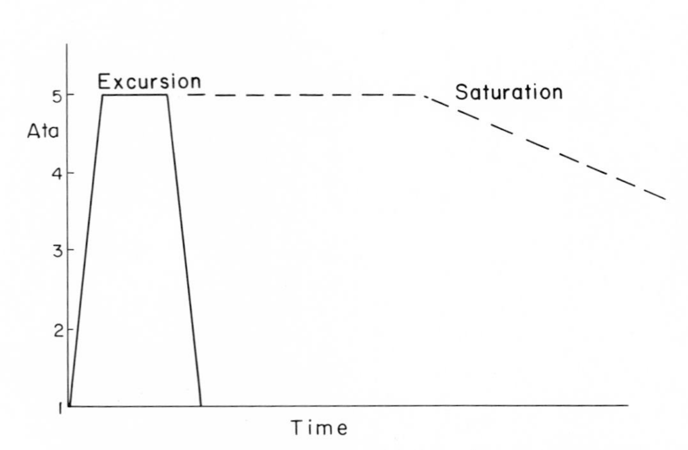

Fig. 11-10. Two different principles of decompression. For excursion dives (solid lines) a nonstop ascent is safe within tested exposure limits. After saturation, however, ascent from deep depths (>500 ft) may require from 10 to 20 min/ft (dashed line) in contrast to a rate of 1 sec/ft that is routine for ascent after excursion dives.

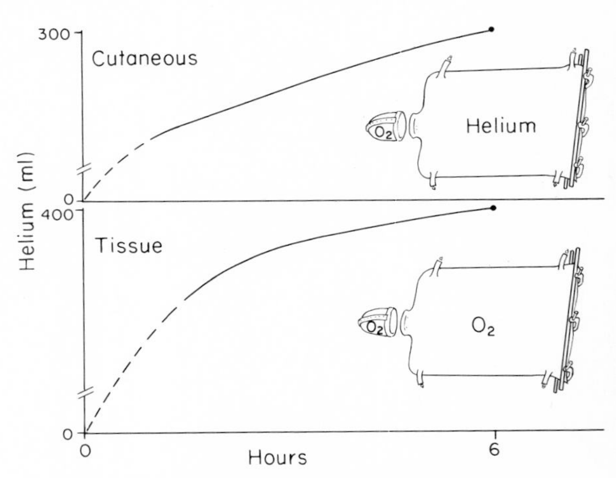

Fig. 11-11. Diadermic diffusion of inert gas. Lower curve shows helium washout after helium saturation with the body (excluding head) surrounded by oxygen. If helium subsequently replaces ambient oxygen in the bag, it diffuses through the skin and is transported to the lungs for elimination from the helmet (upper figure).

Apparent ratio (excursion depth/ saturation depth)	Saturation depth (ft)	Excursion (ft)	Duration (min) Helium-oxygen	Air
4 to 1	Surface	100	35	25
1.96 to 1	35	100	—	60
1.43 to 1	200	100	60	—
1.26 to 1	350	100	100	—

increased several-fold by raising ambient temperature to the level of 35° to 37° C. Groom and Farhi[34] found that percutaneous nitrogen diffusion in the dog per 24 hours equals the entire body store of nitrogen. The half-time of the slowest washout compartment of a dog immersed in air is 332 minutes, in contrast to 117 minutes in an oxygen atmosphere.

Safe decompression—calculations and test data

Excursion nonstop decompression

Two different methods of decompression are safe; one pertains to the excursion dive, the other to the saturation dive (Fig. 11-10). Thousands of excursion dives from surface to depth and return have been completed safely. In such dives ascent to the surface at a rate of 60 ft/min is continuous. With reference to gas transport either (1) gas has separated from solution and there is a stable gas phase comprising silent bubbles or (2) limited supersaturation persists during the restricted period of nonstop ascent. The arresting empiric finding, for which Hills[35] derived a theoretic basis, is the extension of work time for a given excursion relative to the basic saturation depth (Fig. 11-12). Over 1,000 test dives were completed without incidence of decompression sickness in the U. S. Navy program supervised by Bornmann.[36] The above data are representative.

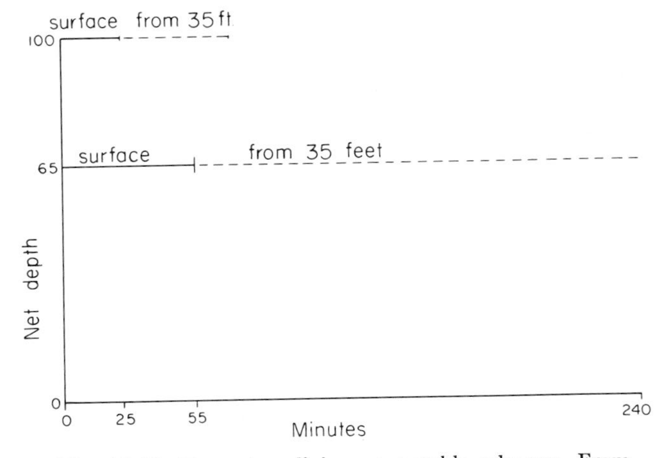

Fig. 11-12. Excursion diving—a notable advance. From the surface a nonstop excursion can be made to a depth of 65 ft for a period of 55 minutes, or to 100 ft for 25 minutes (solid lines). From a saturation depth of 35 ft the duration of the excursions for the same net depths can be greatly extended (dashed lines).

Saturation diving

There are no quantitative data for gas transport during decompression. In developing decompression tables reliance has been placed on the ingenious format of maximal saturation (M) values compiled by Workman.[37] An *M* value is defined as the maximal pressure (M) of inert gas in a specific tissue that does not permit bubbles to evolve in that tissue at a given absolute pressure. In accord with the Workman format, it is assumed (1) that limited supersaturation, scaled to depth, holds during decompression, and (2) that inert gas (nitrogen or helium) transport is one half to one fourth as fast as earlier data indicated. Supporting this second assumption are realistic half-time values of 480 minutes (nitrogen) and 180 to 240 minutes (80%-20% helium-oxygen mixture) for saturation exposures (4 ATA). In these tests[33] the inert gas was not eliminated under isobaric conditions but rather on the basis of an assumed 1.6:1 oversaturation ratio. Conditions may not have been conducive to normal gas transport. The tests, however, demonstrate unequivocally that there are unidentified deterrents to inert gas transport.

Isobaric principle

In contrast to limited supersaturation that may exist for short periods of time is the principle that a dissolved gas is transported from tissue to lungs during an extended period of time at a partial pressure that cannot exceed ambient pressure. The mechanism of gas transport depends upon the gas pressure difference (ΔP), or "oxygen window," illustrated in the example on p. 418.

	Partial pressures of gases (torr)	
	Venous blood	Lungs
Helium	2,210	2,000
Oxygen	40	257
Carbon dioxide	47	40
Water vapor	47	47
Total	2,344	2,344

P_{He} = 210 torr (~ 10 ft)
"oxygen window" = 217 − 7 = 210 torr

The pressure difference (ΔP_{He}) for helium transport from tissues to lungs is 210 torr (~ 10 ft). If the half-time (helium) of the slowest tissue is approximately 70 minutes, then if ΔP is 10 ft, rate of *isobaric* ascent is 10 min/ft. On the other hand, if we assume as in current calculations that the *M* value for the slowest tissue permits an oversaturation of 20 ft, the assumed ΔP is 20 + 10, or 30 ft. If the ascent rate is also computed as 10 min/ft, then the half-time of the slowest tissue is 3 × 70 minutes, or 210 minutes.

In the U. S. Navy test program the rate of ascent after a saturation dive is 10 to 15 min/ft for acclimatized divers.[36] To ascend from a simulated depth of 600 ft, from 100 to 150 hours were utilized for decompression in contrast to 88 hours used in the tests of Bühlmann and associates[6] to ascend from a depth of 1,000 ft. However, even the extended U. S. Navy schedule was inadequate for unacclimatized divers recruited from the fleet for training. In revising the format for postdive decompression, time was extended and consideration was also given to circadian rhythms. Thus divers are decompressed only during two 8-hour periods each day (between 0600 and 1400 and between 1600 and 2400). Rate of ascent is depth dependent in accord with the following data:

Depth (ft)	Rate (min/ft)	Rate (ft/hr)
600 to 200	10	6
200 to 100	12	5
100 to 50	15	4
50 to surface	20	3

Toward safe decompression of caisson workers

Two procedures would eliminate present difficulties. One is saturation tunneling, which entails a pressurized holding facility analogous to a surface SEA-LAB; the other is oxygen inhalation *during decompression*. The first idea may be put aside temporarily because rugged individualists would resist residence in a work environment. Thus oxygen inhalation, which has been highly effective in recompression therapy, is probably preferable. Tests with inhalation suggest that it not only prevents decompression sickness but possibly also avascular bone necrosis. The pressure range is entirely feasible—25 to 15 psi— and would decrease the time now required for decompression, a cost to industry, by one half.

The following conservative table, tested by the author,[30] outlines the procedure. Oxygen inhalation may be intermittent or cumulative. *The prime objection is the fire hazard.* For each hour of work at the indicated ambient pressure (psi), oxygen is breathed for the indicated time interval (minutes) at the indicated pressure (psi):

At 20 psi, O_2 for 10 minutes at 15 psi	Work shift: 6 hours
At 30 psi, O_2 for 20 minutes at 20 psi	Work shift: 6 hours
At 40 psi, O_2 for 30 minutes at 20 to 25 psi	Work shift: 4 hours

FITNESS AND SELECTION FOR THE HYPERBARIC ENVIRONMENT

Using a standardized test or series of tests in the altitude chamber, it is feasible to eliminate individuals who are unduly susceptible to dysbaric symptoms, both minor and serious. Body fat content can be estimated accurately from whole body density and total body water determinations. A battery of anthropometric measurements should be routinely used for definitive classification of body build, or somatotype. It is essential to obtain a careful health history with particular attention to such symptoms as syncope and head injury. Examinees who have a history of migraine are susceptible to a similar syndrome as a result of decompression. Subjects who have anatomic arteriovenous (right-to-left) shunts are at risk of disability and death from arterial gas embolism when bubbles bypass the pulmonary shield. Pulmonary abnormalities, such as adhesions at the costophrenic angle or cysts, invite precipitous collapse if the lungs are overinflated.

With regard to fitness maintenance, divers of an earlier era were carefully observed during softball games that replaced the noon lunch. Such exercise mobilized fatty acids from adipose tissue, diminishing hunger, controlling weight, and most important, permitting observation of neuromuscular coordination. The chief requirement for diving is a program of regular physical training and assessment that assures fitness.

Physiologic tests and criteria are now available to examine several facets of fitness that have heretofore not been systematically followed. Divers can be screened for adequacy of pulmonary ventilation during underwater work; thus the carbon dioxide–retainers can be eliminated or their underlying impairment corrected. Determination of the respiratory quotient (RQ) during heavy exercise was mentioned previously. Tests of vestibular and auditory nerve function are especially important. Tests of neuromuscular coordination, periodic electroencephalographic survey, and annual roentgenograms of the humerus and femur should be routine.

Fluid intake and urine secretion

We have already discussed the dramatic increase of urinary output during a dive as compared with predive and postdive control values. Rigid monitoring of fluid intake and urine volume can be achieved during diving operations. This involves restriction of fluid intake to measured quantities during stated time periods in the day, initially in accord with individual preference. The range of variation of 24-hour urine collections in divers so stabilized can be maintained within ± 300 ml during a 10-day period.

Diet

Dietary regulation has not been routine in diving; indeed, the tendency has been to provide *luxus* meals, particularly under conditions of saturation exposure. In small animals, alimentary lipemia predisposes to decompression sickness. Likewise, blood lipid levels are related to the tendency of platelets to aggregate; small gas bubbles may denature lipoproteins to the extent of disrupting weak lipid-protein linkages. Faulty carbohydrate metabolism and high blood insulin levels may be associated with hyperlipidemia.

We believe that the dietary regimen during periods of intensive diving should consist of simple items tailored to individual requirements and regulated by biochemical analysis of blood and urine. The same limited number of food items of known composition (*idem-diem* diet) are ingested daily during diving operations. Usually, the nutritional problem during the first decade of maturity is overeating, a practice that initiates subtle, but ominous, biochemical deterioration. The physical fitness program and dietary regimen are similar to those advised for the astronaut.

PART B: MARINE ENVIRONMENTS

RALPH W. BRAUER

The term "high-pressure environment" requires definition. In this chapter a high-pressure environment refers to one whose pressure* lies between 10 ATA and the upper pressure limit to which man may reasonably expect to penetrate in the foreseeable future. Ten atmospheres is approximately the limit to which man can dive breathing air; even at this pressure the narcotic properties of nitrogen are a hazard and the oxygen concentrations are too high to be tolerated for any appreciable length of time. The lower pressure limit we have chosen here implies the use of exotic artificial atmospheres designed to avoid the deleterious pharmacologic effects of the components of air. This section deals with the physiology of man at pressures prevailing in the ocean at depths between 300 and 3,000 ft (10 to approximately 100 ATA) and in atmospheres composed largely of light gases such as helium or hydrogen. To avoid unnecessary duplication we assume throughout that P_{O_2} is maintained within the physiologically tolerable limits of 0.2 to 0.4 ATA.

It is useful to place this pressure range in the broader perspective of pressures that exist within the terrestrial biosphere. Fig. 11-13 shows this pressure spectrum, extending from little more than 0.1 to almost 1,000 ATA. Within this range of values the environmental pressures presently accessible to man constitute a limited, but growing, segment. The growth of this segment is largely the result of advancing technology and knowledge of the physiology of high-pressure environments. Fig. 11-13 indicates where a particular pressure might prevail and the physiologic effects most characteristically associated with various pressures.

*See reference 39 for the units of pressure.

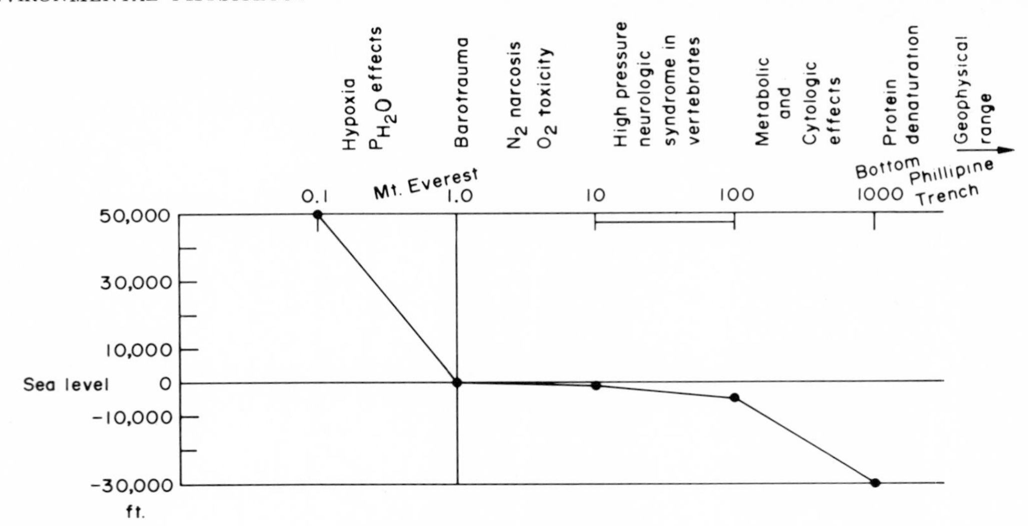

Fig. 11-13. Pressure spectrum.

High-pressure environments, as we define them, are likely to be encountered by man only under special circumstances—either when performing work as a diver at depths greater than about 300 ft or when exposed to such pressures in the controlled environment of a hyperbaric chamber. In either situation, men are working under stressful conditions near the limits of human tolerance. It is worth reiterating that the study of human physiology under such conditions serves a dual purpose. It results in more efficient, safer working conditions for men who must penetrate hostile environments, and it extends the range of environments accessible to man on his own planet. Furthermore, the experimental situations afforded by men who voluntarily place themselves in such extreme environments afford unequaled opportunities for exploring the full range of biologic adaptability of the human organism. Such study can be expected to yield a rich harvest of new insight into processes relevant not only for the scientist-explorer but also for the rest of mankind living under the supposedly normal conditions of our everyday life.

PHYSICAL FACTORS DETERMINING THE PHYSIOLOGY OF MAN IN HIGH-PRESSURE ENVIRONMENTS

The study of high-pressure environments involves four major considerations:

1. Subjects must pass from the 1 ATA sea-level environment to a high-pressure environment. This poses problems because of the mechanical effects of the differential compressibility of various body compartments and because of changes in the amounts of gaseous constituents dissolved at various partial pressures.

2. Once on the bottom, man is exposed to relatively high hydrostatic pressures determined by the conditions of each particular dive; when pressure becomes high enough it affects thermodynamic properties, reaction kinetics, and molecular arrangements of virtually every constituent of living tissues and produces a range of effects that we are only now beginning to explore.

3. In man, as in all air-breathing animals, any increase of external hydrostatic pressure must be balanced by increased total pressure of the gas within the lungs. High-pressure atmospheres most notably affect the mechanics of breathing as a result of increased gas density and increased respiratory resistance.

4. The high "inert" gas pressures required under hyperbaric conditions imply relatively high concentrations of these gases dissolved in blood and tissue fluids, creating the possibility of pharmacologic effects even from gaseous constituents normally considered quite innocuous. Nitrogen narcosis is an example. Each individual case must be scrutinized anew for possible pharmacologic effects. Fig. 11-14 illustrates one type of study required. In this case, the question concerned the relative narcotic potency of particular inert gas mixtures, information required to determine the suitability of various synthetic atmospheres for use at some predetermined high pressure. The two parts of Fig. 11-14 illustrate the two elements to be considered: (1) experimental proof that rel-

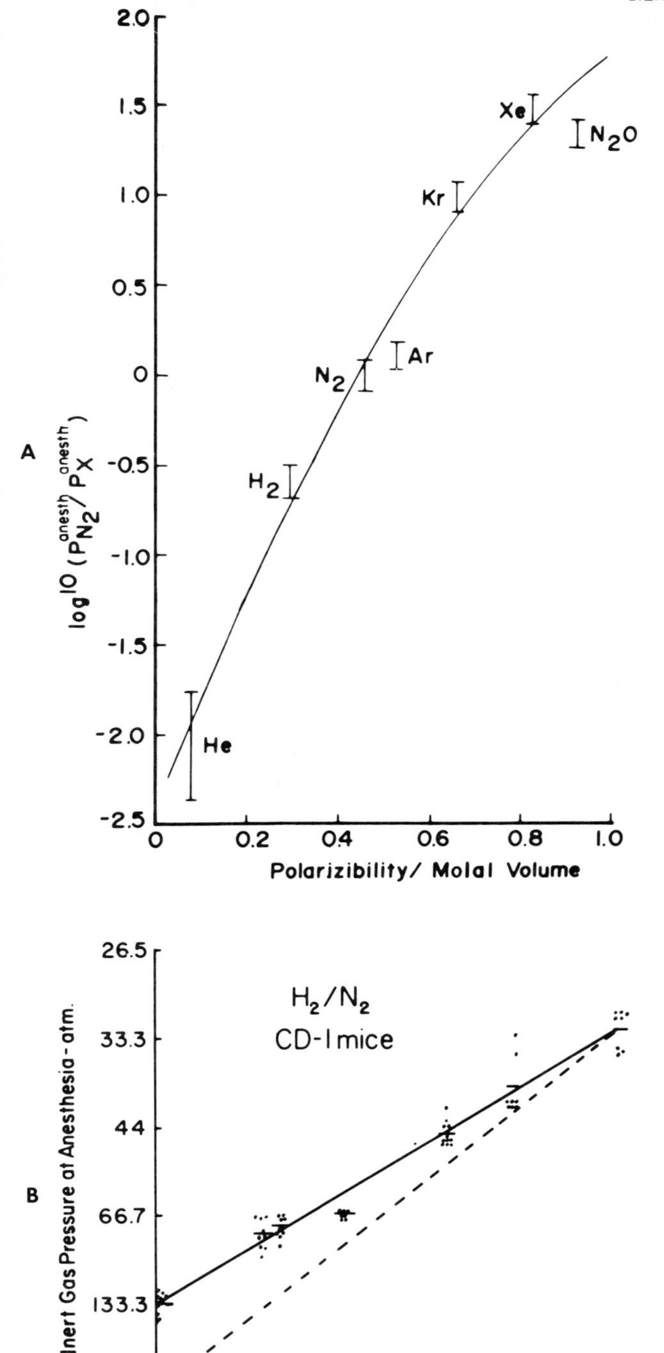

Fig. 11-14. Correlation of relative narcotic potency (measured by the reciprocal of the partial pressure of the nitrogen = 1.0) with various physical properties.
inert gas at the level where mice lose righting reflexes,

ative narcotic potency correlates sufficiently well with some physical property of the gas under study (in this case the physical property is a factor related to molecular interaction energy), and (2) the way in which the narcotic potency of a mixture of two inert gases varies with composition. In this case we assume for the purpose of calculation that the narcotic potencies of the two component gases are additive.

Review of the literature on the physiology of man in hyperbaric conditions reveals a bewildering multiplicity of effects on human performance. However, close examination of the experimental evidence suggests that the many physiologic effects are the result of a limited number of physical factors. Once these physical factors are understood, one can make qualitative and, in some cases, quantitative predictions concerning the nature and extent of the anticipated effects. Table 11-4 is an arrangement of the most important of these physical factors in relation to the physiologic effects they produce, revealing the dependence of these effects upon three key factors—the chemical nature of the gas mixture employed, the absolute pressure, and the time rate of pressure change. For those interested in further study, Fig. 11-14 presents the data of experimental studies correlating the physical factors that describe each particular high-pressure environment with the intensity of specific physiologic effects.

Two general relationships are implicit in all the effects listed in Table 11-4: the dependence of gas volume on pressure and temperature and the dependence of the amount of gas dissolved in body fluids on temperature and the partial pressure of each gas constituent. A convenient expression of the basic relationship of gas density to the physical parameters is as follows:

$$\rho = \frac{P}{RT} \sum V_i MW_i \qquad (6)$$

where ρ = gas density
V_i = mole fraction of the i-th constituent
MW_i = molecular weight of the same constituent
R = ideal gas constant
T = ambient absolute temperature

Equation 6 assumes that the ideal gas law applies. Within the pressure range under consideration here, deviations from ideality, while perceptible, are in general small enough to be ignored in this discussion; however, a discussion of the Joule-Thomson effect is given later. Regarding the solubility of metabolically inert gases in various

Table 11-4. Physiologic responses to physical factors in hyperbaric environments

	Physical factors	Related to: MW and species	P	ΔP/ΔT	Effect produced	Reference
Effective viscosity	Laminar flow	+	(+)	0	Respiratory resistance and pulmonary ventilation	38
	Turbulent flow	+	+	0	Respiratory resistance; maximal voluntary ventilation	39
Heat	Adiabatic heating or cooling	+	0	+	Acute thermal stress	40
	Heat capacity per unit volume	(+)	+	0	Heat regulation, especially convective and respiratory heat loss	41
	Heat conductance (nonconvective)	+	(+)	0	Heat regulation—surface temperature and insulating value of garments	42
	Insulating value of neoprene (foam)	(+)	+	(+)	A special case of important base material for diving suits	
	$[H_2O]/P_{H_2O}$	(+)	+	0	Psychrometric relationships and evaporative losses	43
Solubility	Phase separation	+	0	+	Decompression sickness	44
	Osmotic pressure	+	0	+	"No joint juice" syndrome; fluid shifts	5
	Narcosis	+	+	(+)	Inert gas narcosis—selection of deep diving gas mixtures	45
Sound	Velocity of sound	+	0	0	Pitch of speech	46
	Resonant chamber effect	+	+	0	Intelligibility of speech; hearing	46
Hydrostatic pressure	Central nervous system	0*	+	(+)	High-pressure neurologic syndrome	47
	Metabolism	0	+	0 or (+)	Multiple biochemical and cytophysiologic effects	48

*Some interaction with narcosis.

body constituents, the general relationship

$$M_i = KP_i \quad (7)$$

where M_i = measure of the molecular concentration of dissolved gas
P_i = the partial pressure of the i-th constituent of the gas phase
K = constant incorporating the solubility constant for the gas constituent in the liquid phase as a function of temperature and a factor to reconcile the particular units employed

As in the case of gas density, equation 7, termed *Henry's law,* assumes ideal gas behavior, this time in solution; as with other ideal gas predictions, those based on equation 7 are reasonably accurate. Physiologically significant deviations from predictions based on this equation involve especially bimolecular or high molecular weight gases; indications are mounting that stoichiometric binding of such gases to proteins and other body constituents may be of great importance in producing specific biologic effects. Among these effects, narcosis is included in Table 11-4 because it is of such importance as to influence greatly the considerations of the present chapter (Fig. 11-14).

The basic relationships among pressure, density, mean molecular weight, and solubility determine the values of other parameters that, in various combinations, enter into the somewhat more complex relationships that govern the magnitude of most of the effects listed in Table 11-4. Thus, for example, effective resistance to laminar flow is independent of gas *density*; it is wholly determined by gas *viscosity,* which is a function only of molecular size and configuration. On the other hand, the relationship of resistance to turbulent flow is strongly dependent upon gas density and hence total pressure. Similarly, while the heat capacity of a given volume of gas is dependent upon the molar concentration of the gas and secondarily upon molecular complexity, nonconvective heat conductance is strongly influenced by molecular weight, a distinction of great physiologic significance that has been overlooked frequently in the past in discussing the thermal effects of high-pressure environments.

Confusion may also arise in relation to the acoustic properties of high-pressure environments; sound velocity depends solely on the molecular properties of the gas and is independent of pressure, whereas interaction with the walls of a resonant chamber, and consequently the degree of speech distortion in a given environment, involves both sound velocity in the gas phase and a coupling factor that is strongly influenced by gas density. The behavior of water vapor in a given gas phase involves even more complex relationships. Evidence suggests that this behavior is dominated by the tendency of water molecules to form aggregates and, specifically, by the tendency to form an increasing proportion of relatively high molecular weight aggregates in a vapor phase coexisting with a relatively concentrated population of other gas molecules. This tendency is related to the marked increase of water vapor content in hyperbaric gas atmospheres that are in equilibrium with a liquid water phase at a given temperature.

In addition to these effects, which reflect the thermodynamic properties of high-pressure gas atmospheres, several other factors are rather closely linked to the rate of compression. For example, the marked change of chamber temperature as ambient pressure varies at such a rate that heat exchange between chamber atmosphere, chamber walls, and chamber environment lags behind the changes of gas temperature produced by compression or expansion. The magnitude of such adiabatic temperature changes is determined not only by the absolute rate of pressure change but also by the chemical nature of the gas being compressed, which determines the magnitude of the Joule-Thomson coefficient, by the precise configuration of the conduiting and gas reservoirs supplying the chamber, by chamber geometry, and by the extent of mixing of the gas within the chamber. We will discuss this in more detail later.

Phase separation is a second phenomenon directly linked to the rate of pressure change; this is the formation of a discrete gas phase as a result of ambient pressure reduction in the presence of a liquid phase saturated with unmetabolized gases at a total pressure sufficiently in excess of the ambient one. Under these conditions decompression sickness results. The precise mathematical functions describing the tolerable pressure excess and the time course of phase separation under various conditions of supersaturation and nucleation remain unknown. The best we can do at present is to incorporate all empiric knowledge into one of several models that predict with reasonable accuracy the probability of occurence of clinical decompression sickness as a result of a given decompression time pattern. To date all such models, however useful they may be for interpolation, have proved of limited value for extrapolation; for each new depth achieved and for each increment of exposure time beyond tested limits, prior formulations have had to be revised, almost invariably in the direction of more conservative decompression schedules.

A third effect that may depend upon the rate of compression involves the fact that gases dissolved in an aqueous phase exert osmotic pressure. If the compression rate is sufficiently rapid so that gas equilibrium between various body fluid compartments falls behind the rate of gas accretion in the most rapidly equilibrated phases, then the osmotic pressure of the dissolved gas may cause transient fluid shifts. Such osmotic effects have been invoked to explain transient aspects of the symptomatology of men being compressed to relatively great depths and, in particular, a type of joint pain facetiously called "no joint juice," or "NJJ." Whether this entity does indeed represent osmotic pressure effects or whether it represents the result of differential compressibility of discrete joint components remains unknown.

Table 11-4 contains an item regarding the insulating value of foams. This is another example of a transient effect and has been a source of annoyance during saturation dives. The problem is that the most widely used insulating garments for thermal protection of divers contain elastomer foams enclosing air-filled spaces. As pressure increases, the volume of the gas phase within the garment decreases more slowly than would the volume of a free gas phase, but enough to reduce the insulating value of the material. Second, helium, which commonly constitutes the bulk of high-pressure atmospheres, diffuses through the elastomer, reexpanding the foam to its initial value. Since the pores of the foam are now filled with helium or other highly conductant gas, the insulating value of the garment is seriously impaired. The various attempts to solve this problem have included storage of garments in atmospheres other than helium, such as argon or carbon dioxide, so that when reexpansion occurs the desirable thermoprotective properties are restored.

Hydrostatic effects per se, although listed in Table 11-4, probably constitute a separate entity. Whereas all the remaining effects summarized in Table 11-4 are related directly or indirectly to a change of density of the gas phase surrounding the subject and to providing a respiratory phase, hydrostatic pressure effects are largely independent of the nature of the gas phase; this is true except insofar as the gas phase exerts specific pharmacologic effects that facilitate or antagonize hydrostatic pressure effects and can be demonstrated experimentally equally well in animals breathing a liquid phase as in animals breathing heliox or other suitably inert gas phase.

This review of the wide variety of physiologic effects that assail men penetrating high-pressure environments raises the question of the relative significance of the various factors. In this matter every investigator has his own assessment. However, to provide the reader with a basis for making his own judgment, we present a brief discussion of our own point of view.

The acoustic effects appear most amenable to resolution by purely technical means, either by construction of increasingly effective voice "unscramblers" or by providing alternative means of communication. Although of considerable practical importance, acoustic effects are relatively unimportant in limiting the depth of penetration of the oceans by man. Similarly, there is no evidence that osmotic effects, demonstrated or hypothesized, constitute a serious limitation. Narcosis, although a serious problem indeed to personnel using air as a diving medium, is unlikely to be a serious obstacle to the operations of men using synthetic atmospheres containing helium, at least not until depths far below the deepest now in sight become practicable. This statement refers only to the effects of helium as an inert gas narcotic. The two other contenders for a major place as inert gas constituents of deep diving atmospheres, hydrogen and neon, are not devoid of narcotic effects. The extent to which narcotic effects may limit the use of these gases or, conversely, the extent to which narcotic properties may prove valuable in extremely deep diving operations where they might counteract the high-pressure neurologic syndrome remains to be determined by future work.

Surely the importance of decompression sickness as a factor controlling the pattern of diving operations cannot be exaggerated. The seriousness of decompression sickness is underscored by evidence concerning the possibility of latent adverse effects, in particular aseptic bone necrosis, even in personnel who never manifested decompression sickness. However, ischemic osteonecrosis, discussed earlier in this chapter, is not unique to deep diving and high-pressure exposures.

In the past great importance has been attached to the effects of high gas density on respiratory resistance and pulmonary ventilation. Certainly, theoretic considerations lead to the conclusion that at some level of hyperbaria these factors will become so severe as to be limiting. However, review of the history of this subject shows that it has been characterized thus far by a series of retreats from more or less dire predictions for each greater depth and subsequent admission that for this particular depth respiratory factors have not yet become limiting. The latest such development involved a series of simulated dives to 1,000 ft or slightly below in which limitation to normal operations were not evident and in which, even at maximum pulmonary ventilation rates, the effects were so tenuous as to be equivocal. In another experiment subjects were compressed to a simulated depth of 1,200 ft breathing heliox atmospheres which were then replaced by other gas mixtures of increasing density to the point where these were as dense as heliox atmospheres at a depth of 5,000 ft of sea water. Even under these conditions, pulmonary ventilation and blood-gas measurements revealed only marginal impairment. Obviously, this subject requires further study. It appears, however, that at the present time and at all depths attainable in the foreseeable future, pulmonary ventilation is unlikely to be a limiting factor of major importance in the high-pressure environment.

In contrast to the preceding factors, which are either not limiting or are readily amenable to correction by purely technical means, are two remaining factors. Thermal stress in high-pressure environments is the factor that has most seriously limited human comfort, performance, and, indeed, survival itself in hyperbaric chambers or at depth in the open ocean. Mild hydrostatic pressure effects have been encountered even in diving operations conducted thus far. We now have enough evidence to predict that, given our present technical and physiologic capability, man will reach depths within the next decade at which catastrophic changes in central nervous system excitability as a result of extremely high hydrostatic pressures will be the factor limiting further progress. Therefore, the remainder of this chapter will concentrate first on the problem of thermal stress in the hyperbaric environment and second on the problem of the effects of high hydrostatic pressure on the central nervous system.

TEMPERATURE REGULATION AND THERMAL COMFORT IN HIGH-PRESSURE ENVIRONMENTS

Scientific study of the effects of thermal stress on man and the development of reasonably comfortable thermal environments under hyperbaric conditions have been delayed to some extent by the peculiar pride of professional divers in their ability to withstand stressful conditions in general and by the willingness of observers to give credence to the "Oh, it's fine, Doc" response to a query about comfort. As scientifically trained observers experience hyperbaric exposures, it becomes increasingly apparent that such situations are highly complex and that the problem of heat loss is only one of the factors disturbing the thermal balance of man in high-pressure helium-oxygen environments. To elucidate the problems of hyperbaria it may be useful to describe the sequence of events as they are experienced by an aquanaut going through a complete cycle of descent (compression), bottom sojourn, sortie, work, and ascent (decompression).

The first step in the diving sequence is the passage from sea level to the working pressure, commonly simulated by raising the ambient pressure in some type of hyperbaric chamber. This process is often noisy and may also be associated with substantial discomfort if a diver has difficulty maintaining pressure equilibrium between the air-filled cranial sinuses, the middle ear, and the external chamber atmosphere. During this period the only fully reliable communication system is direct line-of-sight observation between the diver in the chamber and the chamber operator on the outside. Design of hyperbaric facilities and installation of control panels must take account of this fact; failure to do so will lead to discomfort and create a potential hazard. With regard to thermal comfort, the problem during the compression period is the fact that pressurizing a container with gas increases the enthalpy of the container atmosphere. The temperature of the chamber may rise to intolerable levels unless the excess heat thus generated is dissipated. Adiabatic pressurization of a container that is initially at temperature T_1 and pressure P_1 ($P_1 = \frac{n_1 RT_1}{V}$) with a gas at temperature T_{01} and reaching a final pressure P_2 ($P_2 = \frac{n_2 RT_2}{V}$) results in the final temperature T_2:

$$T_2 = KT_{01} - \frac{n_1}{n_2}(KT_{01} - T_1) \qquad (8)$$

where $K = \frac{c_P}{c_V}$, or 1.67 for helium

Equation 8 assumes ideal gas behavior and ignores possible differences in gas composition. In any real process the temperature T_2 is reduced by heat dissipation from the gas to and through the chamber walls and to any special heat exchange system introduced into the chamber. Interaction of these factors determines the time course of chamber temperature change for any given set of initial conditions and compression profile. Typically, there is a rapid temperature increase during the initial compression stages, a progressive decrease of the rate of temperature increase, and, if compression continues long enough, a broad maximum. Finally, there is a gradual decrease of chamber temperature that begins even before compression is completed. The temperature curve reflects two factors: the convexity of the curve $T_2 = f(n_2)$ for adiabatic pressurization as T_2 approaches the asymptote $T_2 = KT_{01}$ for large values of n_2, and the fact that in any real situation heat flow through vessel walls increases as intrachamber temperature rises above that of the extrachamber environment. The shape of the relationship between chamber temperature and time is determined by compression rate, temperature of added gas, initial chamber temperature, and rate of heat dissipation from chamber atmosphere to and through the chamber envelope.

When compression rates are sufficiently low, temperature equilibrium can be nearly maintained with the extrachamber environment, avoiding compressive heat stress problems even during very deep dives in large complexes. Near-equilibrium thermal conditions are maintained during descents for saturation dives following current U. S. Navy practice, which involves compression rates of only about 0.5 atm/hr. Far more rapid descent rates are often sought in industrial practice to minimize decompression time in deep intervention dives that do not attain saturation. Under these conditions, in the absence of special provisions to dissipate the heat generated, heat stress tolerance may become the factor limiting compression rate.

As indicated by equation 8, the temperature of the gas entering the chamber, T_{01}, determines both the shape and the limit of the time/temperature curve. Two effects may reduce T_{01}—cooling of the compressed gas supply by expansion and the Joule-Thomson effect. To modify the critical early ascending limb of the time/temperature curve by the former, or evaporative cooling, requires a special configuration of the gas supply conduits. The Joule-Thomson effect can very significantly cool a gas during passage through a pressure-reducing valve if the Joule-Thomson coefficient for the particular gas mixture is positive, as it is, for example, for air at room temperature:

$$\frac{\Delta T}{\Delta P} = +0.3°C/ATA$$

In cold climates this effect can "freeze up" pressure-reducing valves and conduiting and has caused fatal accidents during scuba dives. In contrast, helium or hydrogen have much lower inversion temperatures and consequently show negative Joule-Thomson effects in the temperature range of our interest:

$$\frac{\Delta T}{\Delta P} = -0.017^\circ \text{ C/ATA (for hydrogen at } 0^\circ \text{ C)}$$
$$\text{and } -0.10^\circ \text{ C/ATA (for helium at } 0^\circ \text{ C)}$$

However, such heating effects are relatively slight, and admixture of small amounts of oxygen or nitrogen will eliminate or reverse the Joule-Thomson effect.

The rate of heat dissipation to the external environment from a pressure chamber is a function of the surface area of its walls. Since the amount of heat produced for a given pressure increase is a function of the total volume of gas being compressed, increasing the surface-to-volume ratio tends to flatten the temperature curve; reducing the external temperature has a similar effect. Cooling the chamber walls before starting a compression cycle or increasing the heat capacity of the walls tends to reduce the slope of the time/temperature curve, especially during the early stages of fairly rapid compression. All these factors operate during compression in the relatively small cylindric submersible decompression chambers commonly used in current industrial practice. Thus compression during descent through the water is usually well tolerated, even when the men are completely dressed in the gear to be used in the water. Indeed, the most common complaint is that the final phase of a protracted descent from the surface exposes the men to cold rather than to heat stress. The resulting discomfort is increased, both subjectively and objectively, by a transient period of high skin temperature and sweating caused by heat stress early in the descent.

It is presently very difficult to predict the severity of physiologic heat stress to which men would be subjected in any given hyperbaric environment or to relate such predictions to empiric data concerning heat tolerance in men sitting or working at high temperatures in air at 1 ATA. Theoretic considerations of heat loss in high-pressure environments suggest that, in general, both convective and respiratory heat transfer rates are increased and that convective heat transfer, in particular, is exaggerated when helium replaces nitrogen in the composition of a hyperbaric atmosphere. For any given ambient temperature above normal skin temperature one would expect the balance of changes in heat transfer to be such that thermal tolerance time is significantly less in a high-pressure helium atmosphere than in air at 1 ATA. From a practical point of view the importance of respiratory heat exchange under these conditions places a special premium on a supply of cool respirable gas during rapid compression in hyperbaric chambers. This is readily accomplished by providing individual demand-regulated gas supplies that bring relatively cool gas from external supplies to each chamber occupant. It is also worth noting that the great difference between the densities of air and helium-oxygen mixtures can result in layering during compression, producing a piston-like compression of part of the chamber atmosphere; this effect may result in temperatures as much as 10° C higher near the bottom of the chamber than in the upper portion. This layering effect is readily avoided by assuring an adequate vertical component of forced ventilation in the chamber atmosphere.

Once the aquanaut reaches the bottom and working pressure, he encounters a new set of thermal challenges. The properties of the atmosphere surrounding him remain, as during descent, high density, high heat capacity per unit volume, and, at depths below 200 ft where changeover to helium becomes obligatory, high heat conductivity. The thermal problem now, however, is no longer that of dissipating excessive heat but rather one of conserving, or even supplying, sufficient heat for the requirement of each individual to assure comfort and maintain efficiency. In a given enclosure, having reasonably constant wall temperature and a mean atmospheric temperature of 28° to 30° C, the resting metabolic rates of subjects vary little whether in air at 1 ATA or in heliox up to 15 ATA. However, there are important differences in the partitioning of total heat dissipation among the several routes of heat loss as pressure increases (Fig. 11-15). In a relatively humid environment evaporative heat loss does not vary greatly with increasing pressure; radiative heat loss tends to decrease as a result of reduced skin temperature. There is a large increase in total convective heat loss, which may increase from less than 30% in air at 1 ATA to more than 60% of the total heat dissipated. The shapes of the curves relating the heat lost by each different route to the ambient pressure suggest that these trends continue at higher pressures. If this is true, then at a pressure of 30 ATA convective heat loss with an ambient temperature of 28% to 30% C would

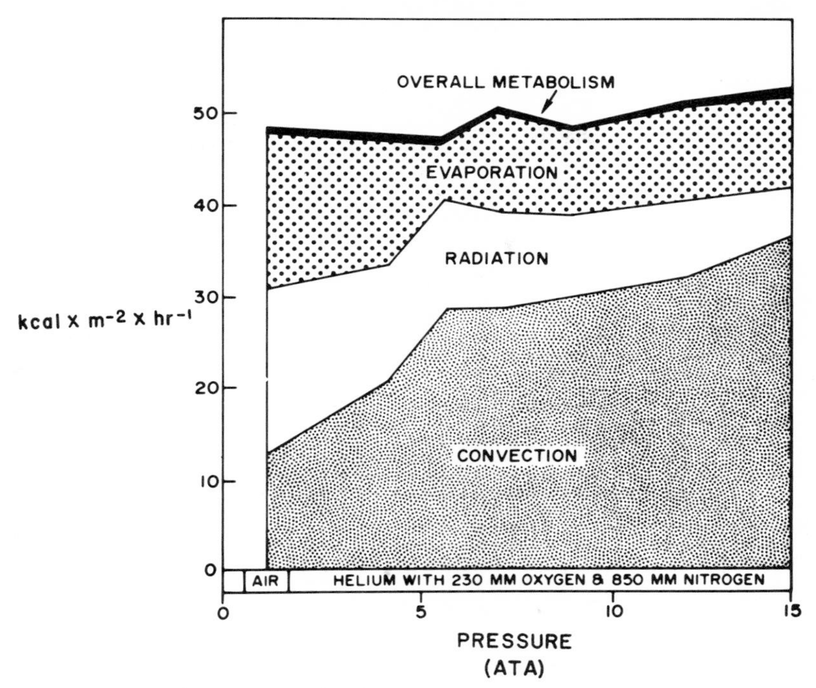

Fig. 11-15. Metabolic balance in a hyperbaric helium-oxygen atmosphere. The overall metabolic rate of seated divers is not strikingly higher than normal. Changes in the relative amounts of heat transferred by evaporation and radiation offset the large increase in convective surface cooling.

exceed the resting metabolic rate in a nude subject, causing either a net loss of body heat and fall of body core temperature or an increased metabolic rate. There is a qualitative resemblance between these effects of sojourn in high-pressure helium atmospheres and the pattern of heat loss in resting subjects immersed in water for long periods of time. In the latter subjects metabolic rates increase greatly at all water temperatures below 35° C, but heat loss often exceeds heat production until water temperature approaches 37° C. These observations suggest that the lower critical temperature for man, or the temperature below which metabolic rates begin to increase for a given species, is 28° C in air at 1 ATA but must be at least 35° C in water. High-pressure helium atmospheres may be expected to fall between these limits; extrapolation of known values suggests a critical temperature somewhat greater than 30° C for unclothed men in a heliox atmosphere at pressures between 25 and 30 ATA.

Theoretic calculations can be made using simple models to predict the effects of changes in molecular weight, pressure, and speed of external ventilation on convective heat loss in compressed atmospheres. The results for one such model are shown in Fig. 11-16. The relationship derived predicts that natural convective heat loss will be about twice as great in a heliox environment at 15 ATA as in air at the same pressure and eight times as great as in air at 1 ATA. Fig. 11-16 also shows the greatly exaggerated effect of forced ventilation in a high-pressure environment.

In environments characterized by high heat conductivity and heat capacity, the thermal behavior of man tends to be characterized by small temperature gradients between the skin surface and the surrounding medium. Thus the skin-to-ambient temperature difference in one group of subjects was as follows: in air at 1 ATA, 5° C; in heliox at 4 ATA, 3.5° C; and in heliox at 15 ATA, only 1.5° C. Such low skin temperatures with nearly constant rate of total heat dissipation implies increased thermal resistance between body core and skin, presumably on a vasomotor basis. Low skin temperature, in turn, tends to decrease radiative heat loss. Under these conditions one would expect physical exercise to be well tolerated, a sort of inverse application of the finding that in cold water physical exercise or even shivering are energetically very inefficient temperature defenses; the vasomotor response to exercise tends to over-

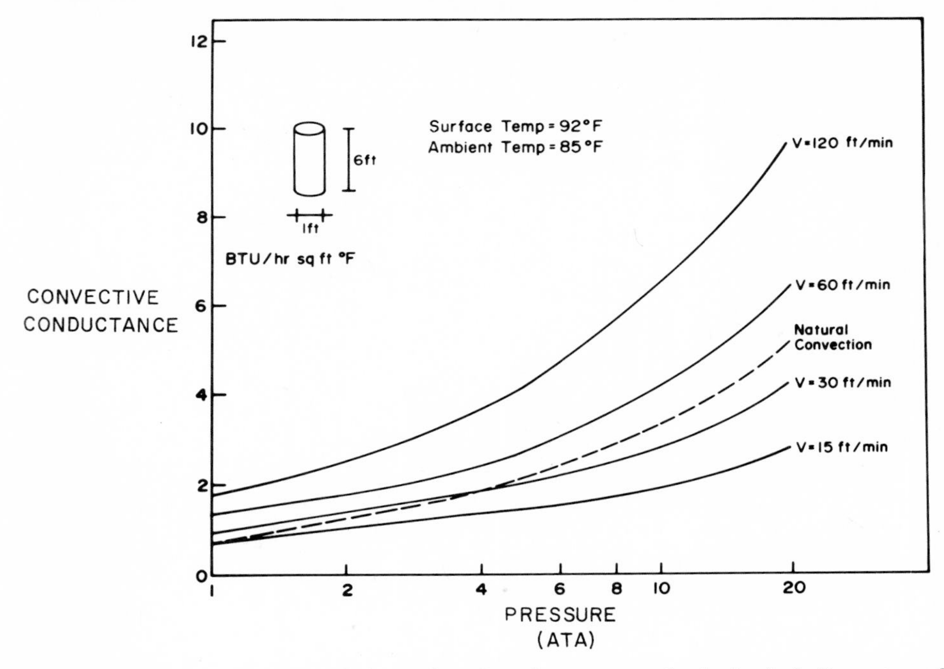

Fig. 11-16. Convective conductance (h_c) as a function of pressure and velocity in helium atmospheres.

ride peripheral vasoconstriction and cause even more rapid heat loss. Here again, hyperbaric environments elicit thermoregulatory responses intermediate in intensity between those in air at 1 ATA and those during prolonged water immersion. Thus shivering in a hyperbaric helium atmosphere may be expected to be relatively ineffective as a thermal defense and once initiated is likely to be protracted and even exaggerated. This phenomenon has been used in animal experiments to induce hibernation by deep hypothermia without the complication of water immersion.

In high-pressure environments a decreased skin-to-atmosphere temperature gradient renders evaporative heat loss by sweating less effective. Because applicable psychrometric charts are not available, the magnitude of this effect cannot be estimated. However, it is qualitatively evident that in high-pressure environments, peripheral, and especially skin, blood flow is mainly responsible for thermoregulation. With small skin-to-atmosphere temperature differences, minor skin temperature changes represent major heat flux changes and thus the effectiveness of thermoregulatory mechanisms that do not affect the skin-body core temperature gradient is very limited; under these circumstances it is hardly surprising that thermoregulation is erratic. Paradoxic thermoregulation, simultaneous shivering and sweating, is a common experience of personnel sojourning in high-pressure helium atmospheres and may be only a reflection of thermal instability resulting from the factors just discussed.

Thus far our discussion has dealt with thermal exchange across the body surface. Obviously, stresses involving this exchange can be reduced appreciably by suitable clothing and careful attention to chamber atmosphere and wall temperatures. Another remedy in the moderate pressure range is replacement of helium with a considerably heavier gas that has more favorable anesthetic properties than nitrogen. Neon is now being tested and is already known to have better thermal and acoustic properties in the moderate pressure range.

The hyperbaric chamber conditions discussed so far generally involve ambient temperatures of about 30° to 33° C. Under these conditions pulmonary heat loss accounts for only a modest proportion of the total heat budget. Even at 30 ATA in a heliox atmosphere pulmonary heat loss is not likely to account for more than 25% of total heat loss. Minor increases of metabolic rate readily compensate for this loss. Increased metabolic rate, in turn, increases minute volume of breathing, increasing heat loss; however, this increased heat loss is not of such magnitude as to seriously im-

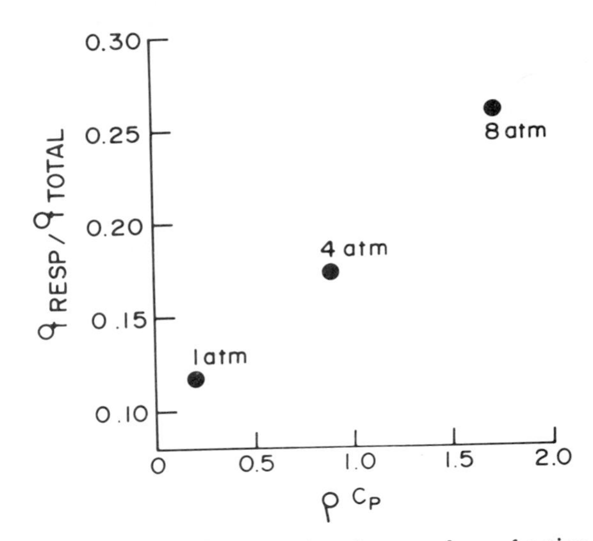

Fig. 11-17. Respiratory heat loss (expressed as a fraction of total body heat loss) as a function of heat capacity per unit volume of helium-oxygen mixture breathed.

pair the effectiveness of such metabolic thermoregulation. This situation changes rapidly, however, as the inspired-expired gas temperature difference increases. Extrapolation of presently available data suggests that at 30 ATA with a temperature difference of 16° C between inspired and expired gas, the rate of pulmonary heat loss may nearly equal the total metabolic rate (Fig. 11-17). Even if pulmonary ventilation rate is reduced by increasing P_{O_2} and by acclimatizing subjects to a higher P_{CO_2}, metabolic thermoregulation is still very ineffective under the stated conditions; the extra heat produced by increased metabolic rate will nearly all be lost by increased pulmonary ventilation.

These considerations can be treated quantitatively in terms of fractions of the maximal rate of oxygen consumption; kcal $\times$ m^{-2} $\times$ hr^{-1} $\times$ °C^{-1}; and body insulation equivalents such as the clo. In any units one must conclude that at depths of the order of 1,000 ft, even when inspired gas is at 10° or 15° C, pulmonary heat loss becomes so great that no other heat loss is tolerable nor, indeed, can increased metabolic energy production be used for mechanical work. It is thus clear that the effective function of a diver in almost any region of the world's oceans at depths in excess of 500 to 600 ft (water temperature ranges from about 10° C in the Mediterranean to less than 5° C in most other seas) requires the reduction of pulmonary heat loss. In practice, this is possible only by warming the inspired gas. Experience in the Arctic Ocean has shown that this single measure makes the difference between nearly intolerable thermal conditions and comfort at depths of 800 ft.

However, pulmonary heat loss is not the only thermoregulatory problem of divers who leave the pressure chamber environment to work at depth in the open ocean. Even in regions where surface temperatures are relatively high, the thermocline is usually at depths of 200 to 300 ft or less. Thus water temperatures encountered by man at the depths under consideration are 5° C or less. Unprotected individuals who are not excessively fat and are in excellent physical condition survive for 15 minutes or less in water of such low temperature even at the surface. Under these conditions the insulative value of peripheral tissues is only about 0.5 clo. To reduce heat loss to a tolerable level, a minimum of 1.5 clo of additional insulation must be provided. The traditional diver's dry suit, a combination of heavy woolen underwear and a water-impermeable covering inflated to ambient pressure with the atmosphere respired, provides 1 to 1.2 clo and maintains divers in reasonable comfort in cold water at a depth of 100 to 200 ft for extended periods of time as long as everything remains dry. Wet suits, made of neoprene foam, provide excellent insulation at the surface; however, because of the elasticity of the material, the volume of contained gas decreases rapidly with pressure. During descent from the surface to a depth of 60 ft the thickness of this type of fabric may decrease 60%, becoming ineffective at depths of several hundred feet. A wet suit that permits a diver to remain in water at 10° C for 75 minutes at the surface permits a stay of only 35 minutes at 60 ft and 28 minutes at 120 ft. When stored in the high-pressure environment, wet suits permit gas penetration and reexpand. Unless this is done in special containers pressurized with a suitable gas such as argon or carbon dioxide, the suits lose much of their insulating value as a result of replacement of air by helium in their gas spaces. Various solutions to the thermal problem have been considered; these include several types of heated suits. Thus far, however, the energy requirements of such suits have exceeded the amount of energy that can be provided without excessively restricting the diver's mobility.

Modified dry suits have been adapted to provide a constant internal volume with promising, but not entirely satisfactory, results. It is likely that the immediate future will see the development of some form of constant volume suit with carefully controlled respired gas temperature. This combination has given initially promising results under drastic circumstances.

EFFECTS OF HIGH PRESSURE ON EXCITABLE TISSUES—THE HIGH-PRESSURE NEUROLOGIC SYNDROME

As recently as 1966 students of the physiology of hyperbaric environments were confident that using heliox atmospheres the next limitation to diving depth would be helium narcosis. Indeed, many investigators were so certain of this conclusion that when the first neurologic symptoms were observed in England and the Soviet Union, they were immediately attributed to helium narcosis. The symptoms included tremors in men and tremors and motor disturbances in experimental animals. The severity of the neurologic syndrome was revealed in 1966 by studies of the narcotic effects of helium and hydrogen in nonhuman primates and for the first time was recognized as probably resulting from hydrostatic pressure per se. Subsequent investigation revealed that helium is almost devoid of narcotic potency, that the entire syndrome can be elicited in animals breathing fluid media containing neither helium nor any other inert gas at high pressure, and that this phenomenon is characteristic of vertebrate species ranging from fish to higher primates. A concurrent series of investigations eliminated as causes various factors that had been thought to produce this syndrome; these included high P_{O_2}, high ambient temperature, hypoxemia, hypercapnia, and excessive compression rate.

Thus today it is clear that vertebrates exposed to hydrostatic pressure in the range of 40 to 150 ATA will undergo a series of physiologic changes designated the *high-pressure neurologic syndrome* (HPNS). The sequence of events is characterized initially by tremors and motor disturbances resembling certain cerebellar lesions and not associated with any detectable electroencephalographic changes. As pressure increases, isolated spikes and eventually runs of high-voltage, high-frequency focal discharges appear. These are associated with changes of motor response characterized by involuntary jerks of isolated muscles or muscle groups, or seizures of Jacksonian type. The second phase of the syndrome reaches full development in the form of generalized electrocortical paroxysms accompanied by violent motor activity resembling in most respects various types of epileptiform seizure. The major difference between the seizures of the HPNS and true epilepsy is the striking absence of marked autonomic effects, such as changes of blood pressure, heart rate, and fecal or urinary incontinence.

The convulsion stage of the HPNS responds to prophylaxis and to therapy. Thus substituting respiratory gas mixtures with mildly narcotic properties for the commonly used heliox mixture elevates the convulsion threshold in direct proportion to the narcotic potency of the gas mixture employed. This fact is of theoretic interest. For example, identical convulsion threshold pressures are observed in hydrogen-oxygen mixtures and in helium-nitrogen-oxygen mixtures of equivalent narcotic potency despite the marked differences in gas density at the convulsion threshold pressure. Sedative doses of barbiturates, especially phenobarbital, increases the normal pressure-convulsion threshold of rhesus and squirrel monkeys by 30% to 40% without severe electric or motor seizures.

The biophysical mechanisms of the HPNS remain unknown. Studies of the effects of hydrostatic pressure on enzymatic reactions, and in particular on biosynthetic processes involving information transfer from the nucleic acids, as well as electrophysiologic studies with such model systems as the isolated frog skin preparation, have shown that pressures in excess of 50 ATA displace markedly a wide variety of cell processes and affect the properties of biologic membranes. Reviewing the information on the responses of marine organisms to hydrostatic pressure, one investigator commented that what needs explaining is not the fact that pressure effects are observed in this range but rather that such effects are on the whole relatively milder than would be expected on the basis of the biophysical and biochemical evidence.

It is interesting to compare the susceptibilities of different vertebrates to the two phases of the HPNS. In Fig. 11-18 threshold values for tremor are correlated with threshold values for convulsions for twenty-three species ranging from fish through a wide variety of different orders to such higher primates as the baboon and the rhesus monkey. Fig. 11-18 documents a range of susceptibilities from a convulsion threshold of about 50 ATA for the most susceptible species to a value slightly less than 150 ATA for one of the fish. It also strikingly illustrates the fact that all of these values conform to a common pattern; the points in the diagram define a regression line with a correlation coefficient greater than 0.85. Thus despite suggestions that the anatomic loci of the two phases of the HPNS are distinct, it appears that, for the

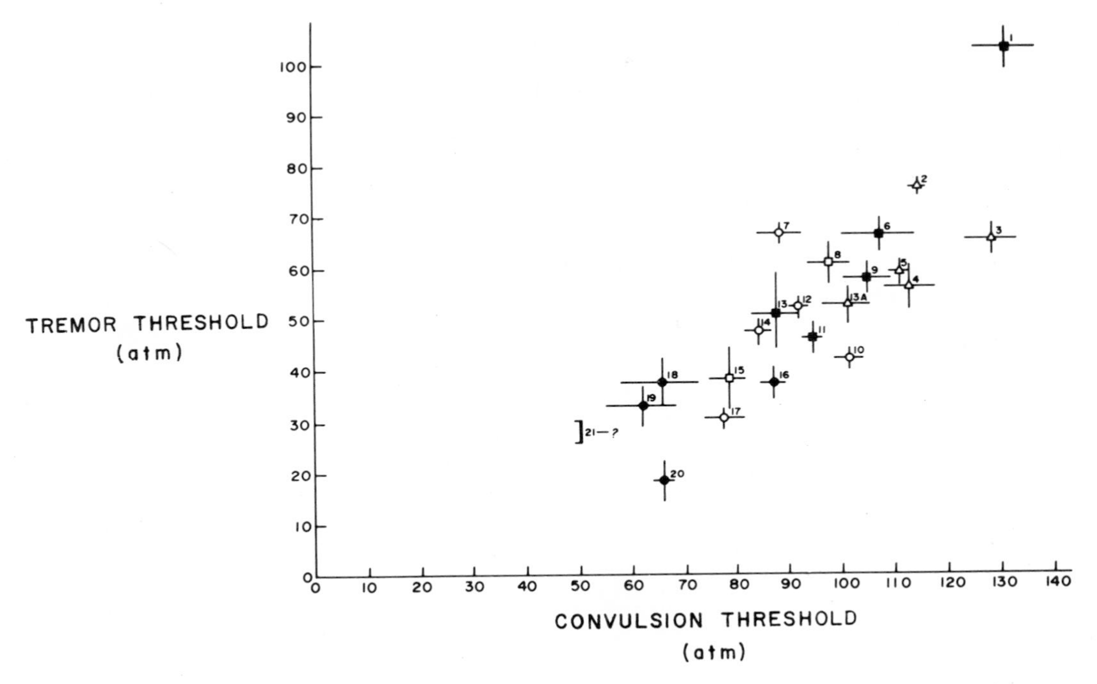

Fig. 11-18. Correlation of threshold pressures for two phases, convulsion and tremor, of the high-pressure neurologic syndrome in twenty-three species of vertebrates:

Class	Order	Species	Number
Pisces		*Achirus fasciatus*	1
		Paralichthys dentatus	6
		Anguilla rostrata	9
		Symphurus plagiusa	13
Reptilia		*Thamnophis sirtalis*	3
		Anolis carolinensis	4
Aves		*Serinus canarius*	2
		Ectopistes carolinensis	5
		Gallus domesticus	13A
Mammalian	Insectivora	*Erinaceus europaeus*	23*
	Marsupialia	*Didelphis virginiana*	15
	Edentata	*Dasypus novemcinctus*	8
	Rodentia	*Epimys rattus S.D.*	7
		Mus musculus CD-1	12
		Mus musculus AJ	14
		Cavia porcellus	17
	Lagomarpha	*Oryctolagus cuniculus*	10
	Carnivora	*Procyon lotor*	11
	Primates	*Tupaia tupaia*	20
		Saimiri sciureus	19
		Macaca mulatta	16
		Papio papio	18
		Homo sapiens	21

*Location of number 23 overlaps that of number 13A.

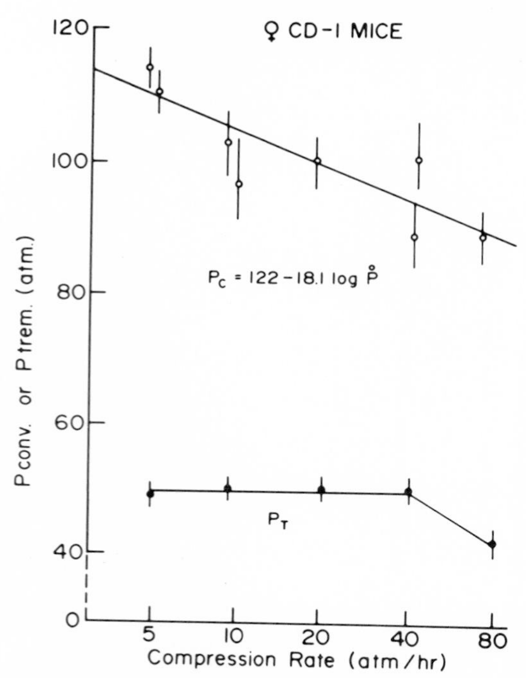

Fig. 11-19. Effect of compression rate on tremor and convulsion thresholds of the high-pressure neurologic syndrome in mice breathing a mixture of helium and oxygen, the oxygen pressure being maintained at 0.5 ATA.

entire range of vertebrates studied, there is a common pattern of development of the syndrome. This fact is of interest not merely as proof that the HPNS reflects fundamental reactions of excitable tissues of vertebrates living at or near the ocean surface but also provides a reasonable basis for predicting the onset of serious stages of the neurologic syndrome in species in which this syndrome has not yet been studied. This is pertinent to the problem of man under extreme hydrostatic pressure. The tremor phase of the neurologic syndrome has been observed in man, and the data for this have been quite consistent. At compression rates comparable to the 24 ATA/hr, as in Fig. 11-18, the tremor stage is observed in man at pressures ranging from 20 to about 25 ATA. The regression line in Fig. 11-19 suggests that for a species showing tremor in this pressure range, the convulsion phase would occur at pressures of 65 to 70 ATA with a standard deviation of approximately ± 7 ATA.

In modern practice actual diving operations are rarely conducted at compression rates as fast as those in Fig. 11-19. Thus to apply the reasoning just outlined to the probable conditions that would prevail in very deep dives requires some additional correction for the effects of compression rate. Fig. 11-19 illustrates one such set of data and suggests that while tremor threshold, but not necessarily severity of tremor, is relatively insensitive to changes of compression rate, convulsion threshold is indeed somewhat affected by compression rate. Slowing the compression rate from 25 to 1 ATA/hr shifts the convulsion threshold upward by approximately 20%. Thus the predicted mean convulsion threshold for men diving under conditions of slow compression and careful control of temperature, P_{O_2}, and breathing heliox would now fall in the range of 85 to 100 ATA, or at depths of about 2,800 to 3,000 ft of sea water.

Further expansion of this range to greater pressures by judicious use of the central nervous system depressant agents described above appears in sight. In our laboratory rhesus monkeys were pressurized to very nearly the equivalent of a depth of 1 mile of sea water. However, such penetration does not appear wholly innocuous. Indeed, there are indications that primates carried past the convulsive seizure phase of the HPNS by use of barbiturates in sedative doses develop a further stage of the syndrome characterized by marked circulatory changes. Since, even under those conditions, blood oxygenation remains well within normal limits, it is reasonable to hypothesize that this stage, like the preceding ones, is a manifestation of the effects of the HPNS on some key body functions. There is some reason to surmise that the mechanisms involved are located in various contractile tissues.

A striking characteristic of the HPNS, as of other manifestations of the effects of high hydrostatic pressures upon biologic functions, is the high degree of reversibility if exposure is stopped short of lethality. In those vertebrate species that have been studied most extensively it has been established that there is a fairly wide range of individual variation in susceptibility to the syndrome. This is evident from the statistical information in Fig. 11-18. Thus, for example, in the case of squirrel monkeys, the convulsion threshold for different individuals may vary from as little as 55 to as much as 85 ATA. If such animals are revived by careful decompression after a convulsive seizure test and allowed a period of 4 months or more to recover, a renewed test for susceptibility to the HPNS under conditions identical to the first usually yields a second convulsion threshold within less than 5 ATA of that observed during the first exposure. Similar data have been obtained in studies with mice and rhesus monkeys. Thus it appears that even quite severe exposure to the HPNS, if the experimental subject survives, results in only minimal or no residuals. The same data also suggest that selection of individuals for high resistance to HPNS may be possible and may even be one approach to increasing diving capability for very deep excursions in the future.

PERSPECTIVE

In this chapter we have described the wide variety of stresses to which men working in hyperbaric environments are exposed. We have discussed and analyzed in detail the physiologic responses to two stressors characteristic of high-pressure environments. These examples illustrate the manner in which physical, biophysical, and physiologic considerations interact, both in defining the precise nature of the strain produced and in developing the concepts with which to understand, predict, and control the adverse effects of such environmental stressors.

This mode of approach is extremely effective. At the end of World War II diving operations at depths of 500 to 600 ft were heroic ventures at the frontier of biomedical knowledge and technology. In contrast, today diving operations are often at 800 ft, ocean dives to 1,000 ft and slightly deeper have been carried out successfully, and a recent review of the subject indicates that it is now reasonable to design programs providing a capability for diving to 3,000 ft within the next decade. Although both interscience integration and biomedical application of knowledge gained from hyperbaric research has lagged, it is clear from the material presented in this chapter that promising leads have been discovered in diverse realms such as the integrative basis of thermoregulation, the biophysics of multineuronal networks, and models for important human disorders such as epilepsy and Parkinson's disease. We hope that in the future the promise of hyperbaric research will be increasingly fulfilled as technical means for the study of subjects in controlled hyperbaric environments are expanded and extended.

JUNE, 1973

GLOSSARY

acclimatization to pressure Decreasing susceptibility to decompression sickness induced by repeated exposures to decompression.

adiabatic process A thermodynamic process in which the gas within a closed system undergoes a change of volume or pressure without heat gain from an outside source or heat loss to an outside sink. In a closed system of *perfect* thermal insulation, adiabatic expansion would cool a gas, whereas adiabatic compression would heat it to the exact degree predicted by Charles' law. In practice, however, a system is never perfectly insulated and every physical process involves some heat transfer, however small.

arthralgia Joint pain; a symptom of dysbarism, or decompression sickness, and rapid compression. In the latter case it is thought to be caused by osmotically induced fluid shifts during blood and tissue uptake of inert gas.

aseptic bone necrosis Ischemic necrosis of bone resulting from occlusion of its blood supply. When associated with decompression sickness it results from embolic occlusion by gas bubbles or fat globules and primarily affects the hip and shoulder joints. Repeated often, the result is destruction of normal bone architecture and a condition that resembles osteoarthritis.

atelectasis Collapse of a lung or lung subunit following obstruction of an airway. Obliteration of pulmonary air spaces and apparent consolidation occur as gas is absorbed.

atmosphere(s) absolute (ATA) Equivalent of 760 torr, 14.7 psi, and 33 ft of sea water.

bends Type I decompression sickness manifested as a dull, throbbing pain usually localized to the extremities and relieved by recompression.

bradycardia Slow heart rate; by definition a heart rate less than 60 beats/min in a resting adult human subject.

capsule environment The microenvironment within a diver's suit or a submarine chamber.

chokes Type II decompression sickness manifested as burning substernal pain and paroxysmal coughing. It is aggravated by deep inhalation or by breathing cigarette smoke and tends to produce acute asphyxia.

continental shelf Arbitrarily defined offshore land submerged to depths not greater than 600 ft. The total global aggregate comprises an area approximately the size of Africa.

decompression sickness A syndrome of protean manifestations initiated by gas bubbles arising within tissues and body fluids. It results from excessively rapid decompression from diving depths or subambient decompression from sea-level atmospheric pressure to altitude. Synonym is dysbarism.

emphysema, interstitial A condition resulting from overinflation and tearing of the lung during excessively rapid ascent from a dive. Lung gas is forced along perivascular sheaths to the mediastinum, thoracic spaces, and sometimes subcutaneous tissues of the neck.

focal seizure A convulsive seizure originating in a localized brain region.

gas density Mass of gas per unit volume.

gas embolism Obstruction of blood vessels by bubbles of gas that form in or gain access to the bloodstream (usually systemic veins) through wounds or during decompression; distinguished from thromboembolism. Air embolism arises chiefly from regional or general lung overinflation and rupture and mainly affects the central nervous system.

heat loss Loss of heat considered primarily as a function of the thermodynamic properties of gas mixtures subjected to hyperbaric pressures.

helium voice The child-like voice of a subject breathing helium; a characteristic speech distortion during helium breathing produced by an upward shift in the frequency of energy concentrations in the speech signal at specific frequencies above the fundamental.

Henry's law At constant temperature the mass of a gas that dissolves in unit volume of a solvent is proportional to the partial pressure of the gas in the gas phase that is in equilibrium with the solvent. Thus M = KP

where M = mass of gas that dissolves in unit volume of solvent
P = equilibrium partial pressure
K = a proportionality constant

Henry's law is most applicable to the behavior of those gases that are only slightly soluble and do not react chemically with the solvent.

high-pressure environment Arbitrarily defined for the purpose of this chapter as 10 to 100 atmospheres (ATA).

hydrostatic pressure effects Physical and physiologic effects of pressure considered as a discrete environmental variable.

hyperbaric Pressure greater than 1 atmosphere (ATA).

microlung A lobe or subunit of the lung greatly reduced in size as a result of intense bronchiolar constriction.

nitrogen narcosis Psychologic and physiologic changes, including euphoria, intoxication, loss of ability to concentrate, stupefaction, and unconsciousness, associated with exposure to high-pressure air atmospheres. It is produced by P_{N_2} in excess of approximately 2 atmospheres.

osmotic pressure If solutions of different concentrations are separated by a membrane selectively permeable to the solvent molecules but impermeable to the solute molecules (semipermeable membrane), the solvent molecules tend to diffuse from the more dilute to the more concentrated solution until concentrations on both sides of the membrane are equal. The hydrostatic pressure required to prevent this process of solvent diffusion equals the osmotic pressure and is often used to measure it. Closely related to gas pressure, osmotic pressure is proportional to absolute temperature and to molecular concentration; it equals what the pressure of the dissolved substance would be if it were a gas at the same temperature occupying a volume equal to that of the pure solvent.

pneumothorax The presence of gas within either or both pleural cavities producing a corresponding degree of lung collapse. It may result from thoracic trauma or pulmonary overinflation and rupture.

psychrometry A method using wet and dry bulb thermometers to determine water vapor tension, or humidity.

recompression Compression for the purpose of treating decompression sickness and gas embolism.

respiratory heat loss The loss of body heat associated with pulmonary ventilation—warming inspired air and evaporation from moist respiratory surfaces. It is determined primarily by the parameters of humidity, gas density, and basic gas thermodynamics as a function of high pressure.

respiratory resistance Resistance to breathing; the ratio of driving pressure to the resultant volumetric flow rate. Resistance to breathing consists of inertial forces, viscous resistance, and airway resistance. Total respiratory resistance is the sum of (1) airway resistance plus (2) pulmonary tissue resistance plus (3) chest wall resistance. Gas density increases with gas pressure, increasing inertia, extent and degree of turbulence, and turbulent flow resistance.

scuba Self-contained underwater breathing apparatus; an open-circuit breathing system supplying compressed air to a diver through a mouthpiece.

surface decompression Repressurization in a dry chamber to complete the process of decompression after rapid ascent to the surface from a dive; also termed decanting.

tachycardia Rapid heart rate; by definition a heart rate faster than 100 beats/min in a resting adult human subject.

thermal properties of gases at high pressure Those properties or gases at hyperbaric pressures described by the direct parameter of temperature and the indirect parameters of internal energy, enthalpy, entropy, and free energy.

vertigo The sensation, or false perception, that the external environment is rotating around the subject (objective vertigo) or that the subject himself is rotating in space (subjective vertigo); a symptom of stimulation or dysfunction of the vestibular apparatus. Vertigo should not be confused with dizziness. Unfortunately, this term has also been used to mean spatial disorientation.

REFERENCES

1. Piatt, V. R.: Table 84. In Altman, P. L., and Dittmer, D. S., editors: Environmental biology, Bethesda, Md., 1966, Federation of American Societies for Experimental Biology, p. 328.
2. Consolazio, W. V., and others: Effects on man of high concentrations of carbon dioxide in relation to various oxygen pressures during exposures as long as 72 hours, Am. J. Physiol. **151**:479-503, 1947.
3. Raymond, L. W.: The thermal environment for undersea habitats, Second Symposium on Human Factors, San Francisco, 1970, American Society of Heating, Refrigerating, and Air Conditioning Engineers (ASHRAE).
4. Pugh, L. G. C. E.: Temperature regulation in swimmers. In Rahn, H., and Yokoyama, T.: Physiology of breath-hold diving and the Ama of Japan, Publ. 1341, Washington, D. C., 1965, National Academy of Science–National Research Council, pp. 325-348.
5. Kylstra, J. A., Longmuir, I. S., and Grace, M.: Dysbarism: osmosis caused by dissolved gas? Science **161**:289, 1968.
6. Bühlmann, A. A., and others: Saturation exposures at 31 ATA in an oxygen-helium atmosphere with excursions to 36 ATA, Aerospace Med. **41**:394, 1970.
7. Craig, A. B., Jr.: Effects of submersion and pulmonary mechanics on cardiovascular function in man. In Rahn, H., and Yokoyama, T.: Physiology of breath-hold diving and the Ama of Japan, Publ. 1341, Washington, D. C., 1965, National Academy of Science–National Research Council, pp. 295-302.
8. Scholander, P. F., and others: Circulatory adjustment in pearl divers, J. Appl. Physiol. **17**:184-190, 1962.
9. Strømme, S. B., Kerem, D., and Elsner, R.: Diving bradycardia during rest and exercise and its relation to physical fitness, J. Appl. Physiol. **28**:614, 1970.

10. Schaefer, K. E., and others: Pulmonary and circulatory adjustments determining the limits in breath-hold diving, Science **162**:1020-1023, 1968.
11. McCally, M.: Body fluid volumes and renal response to immersion. In Rahn, H., and Yokoyama, T:. Physiology of breath-hold diving and the Ama of Japan, Publ. 1341, Washington, D. C., 1965, National Academy of Science–National Research Council, pp. 253-270.
12. Graveline, D. E., and Jackson, M. M.: Diuresis associated with prolonged water immersion, J. Appl. Physiol. **17**:519-524, 1962.
13. Lanphier, E. H.: Pulmonary function. In Bennett, P. B., and Elliott, D. H., editors: The physiology and medicine of diving and compressed air work, Baltimore, 1969, The Williams & Wilkins Co., pp. 58-112.
14. Chouteau, J.: Physiological aspect of prolonged exposure at extreme ambient pressure in an oxygen-helium atmosphere, Abstract 24, Fourth International Congress on Hyperbaric Medicine, Sapporo, Japan, 1969, p. 23.
15. Behnke, A. R., Thomson, R. M., and Motley, E. P.: Psychologic effects from breathing air at 4 atmospheres, Am. J. Physiol. **112**:554-558, 1934.
16. Behnke, A. R.: Inert gas narcosis. In Fenn, W. O., and Rahn, H., editors: Handbook of physiology, sect. 3, chap. 41, Baltimore, 1965, The Williams & Wilkins Co., pp. 1059-1065.
17. Featherstone, R. M., and Muehlbaecher, C. A.: The current role of inert gases in the search for anesthetic mechanisms, Pharmacol. Rev. **15**:97-121, 1963.
18. Brauer, R. W., and Way, R. O.: Relative narcotic potencies of hydrogen, helium, nitrogen, and of their mixtures, J. Appl. Physiol. **29**:23-31, 1970.
19. Noell, W. K.: Effects of high and low oxygen tensions on the visual system. In Schaefer, K. E., editor: Environmental effects on consciousness, New York, 1962, The Macmillan Co., pp. 3-18.
20. Bennett, G. A., and Smith, F. J. C.: Pulmonary hypertension in rats living under compressed air conditions, J. Exp. Med. **59**:181, 1934.
21. Bean, J. W.: Hormonal aspects of oxygen toxicity. In Goff, L. G., editor: Underwater physiology symposium, Publ. 377, Washington, D. C., 1955, National Academy of Science–National Research Council, pp. 13-24.
22. Lambertsen, C. J.: Effects of oxygen at high partial pressure. In Fenn, W. O., and Rahn, H., editors: Handbook of physiology, sect. 3, chap. 39, Baltimore, 1965, The Williams & Wilkins Co., pp. 1027-1046.
23. Rivera, J. C.: Decompression sickness among divers: an analysis of 935 cases, Milit. Med. **129**:314-334, 1964.
24. McCallum, R. I., and others: Bone lesions in compressed air workers with special reference to men who worked on the Clyde tunnels, 1958 to 1963, J. Bone Joint Surg. **48B**:207, 1966.
25. Keays, F. L.: Compressed air illness with a report of 3,692 cases, Publ. Cornell Univ. Med. Coll. **2**:1-55, 1909.
26. Walder, D. N.: Some problems in working in a hyperbaric environment, Ann. Roy. Coll. Surg. Eng. **38**:288-307, 1966.
27. U. S. Navy Diving Manual, Washington, D. C., 1970, U. S. Government Printing Office.
28. Goodman, M. W.: Minimal-recompression, oxygen-breathing method for the therapy of decompression sickness. In Lambertsen, C. J., editor: Proceedings of the third symposium on underwater physiology, Baltimore, 1967, The Williams & Wilkins Co., pp. 165-182.
29. Lambertsen, C. J.: Modern aspects of treatment of decompression sickness, symposium on undersea-aerospace medicine, Aerospace Med. **39**:1057-1093, 1968.
30. Behnke, A. R.: Medical aspects of pressurized tunnel operations, J. Occup. Med. **12**:101-112, 1970.
31. Cockett, A. T. K., and Nakamura, R. M.: Newer concepts in the pathophysiology of experimental dysbarism-decompression sickness, Am. Surg. **30**: 447, 1964.
32. Groom, A. C., Morin, R., and Farhi, L. E.: Determination of dissolved N_2 in blood and investigation of N_2 washout from the body, J. Appl. Physiol. **23**:706, 1967.
33. Bühlmann, A. A., Frey, P., and Keller, H.: Saturation and desaturation with N_2 and He at 4 atm., J. Appl. Physiol. **23**:458, 1967.
34. Groom, A. C., and Farhi, L. E.: Cutaneous diffusion of atmospheric N_2 during N_2 washout in the dog, J. Appl. Physiol. **22**:740, 1967.
35. Hills, B. A.: Relevant phase conditions for predicting occurrence of decompression sickness, J. Appl. Physiol. **25**:310, 1969.
36. Bornmann, R. C.: SEA-LAB and the Man-in-the-Sea Programs; helium-oxygen saturation-excursion diving. In Lambertsen, C. J., editor: Proceedings of the fourth symposium on underwater physiology, New York, 1971, Academic Press, Inc.
37. Workman, R. D.: American decompression theory and practice. In Bennett, P. B., and Elliott, D. H., editors: The physiology and medicine of diving and compressed air work, Baltimore, 1969, The Williams & Wilkins Co., pp. 252-290.
38. Glasstone, S.: The gaseous state. In Glasstone, S.: Textbook of physical chemistry, New York, 1946, D. Van Nostrand Co., pp. 244-339.
39. Anthonisen, N. R., Bradley, M. E., Vorosmarti, J., and Linaweaver, P. G.: Mechanics of breathing with helium-oxygen and neon-oxygen mixtures in deep saturation diving. In Lambertsen, C. J., editor: Proceedings of the fourth symposium on underwater physiology, New York, 1971, Academic Press, Inc., pp. 339-345.
40. Kunkle, J. S., Wilson, S. D., and Cota, R. A.: Compressed gas handbook, properties of gases, NASA publ. 475, Washington, D. C., 1969, U. S. Government Printing Office, pp. 23-50.
41. Raymond, L. W., Bell, W. H., II, Bondi, K. R., and Lindberg, C. R.: Body temperature and metabolism in hyperbaric helium atmospheres, J. Appl. Physiol. **24**:678-684, 1968.
42. Raymond, L. W.: Temperature problems in multi-day exposures to high pressures in the sea: thermal balance in hyperbaric atmospheres. In Lambertsen, C. J., editor: Proceedings of the third symposium on underwater physiology, Baltimore, 1967, The Williams & Wilkins Co., pp. 138-147.

43. Landsbaum, E. M., Dodds, W. S., and Stutzman, L. F.: Humidity of compressed air, Ind. Eng. Chem. **47**:101-103, 1955.
44. Bennett, P. B., and Elliott, D. H., editors: The physiology and medicine of diving and compressed air work, Baltimore, 1969, The Williams & Wilkins Co.
45. Brauer, R. W., Way, R. O., and Perry, R. A.: Narcotic effects of helium and hydrogen in mice and hyperexcitability phenomena at simulated depths of 1,500 to 4,000 ft of sea water. In Fink, B. R.: Toxicity of anesthetics, Baltimore, 1968, The Williams & Wilkins Co., p. 241.
46. Fant, G., and Lindquist, J.: Pressure and gas mixture effects on divers' speech. In Speech Transmission Lab. Quart. Progr. Sta. Rep. 1, pp. 7-17, 1968.
47. Brauer, R. W., Way, R. O., Jordan, M. R., and Parrish, D. E.: Experimental studies on the high pressure hyperexcitability syndrome in various mammalian species. In Lambertsen, C. J., editor: Proceedings of the fourth symposium on underwater physiology, New York, 1971, Academic Press, Inc., p. 487.
48. Zimmerman, S. B., and Zimmerman, A. M.: Biostructural, cytokinetic, and biochemical aspects of hydrostatic pressure on protozoa. In Buetow, D. E., and Cameron, I. L., editors: Cell biology, New York, 1970, Academic Press, Inc., p. 179.

SUGGESTED READINGS

1. Altman, P. L., and Dittmer, D. S., editors: Environmental biology, Bethesda, Md., 1966, Federation of American Societies for Experimental Biology.
2. Barnard, E. E. P.: The treatment of decompression sickness developing at extreme pressures. In Lambertsen, C. J., editor: Proceedings of the third symposium on underwater physiology, Baltimore, 1967, The Williams & Wilkins Co., pp. 156-164.
3. Behnke, A. R., and Yaglou, C. P.: Responses of human subjects to immersion in ice water and to slow and fast rewarming, Report No. 11, Bethesda, Md., 1950, Naval Medical Research Institute; J. Appl. Physiol. **3**:591-602, 1951.
4. Fenn, W. O., and Rahn, H., editors: Handbook of physiology, sect. 3, vol. 2, Washington, D. C., 1965, American Physiological Society; Baltimore, 1965, The Williams & Wilkins Co.
5. Fulton, J. F., editor: Decompression sickness, Philadelphia, 1951, W. B. Saunders Co.
6. Goff, L. G., editor: Underwater physiology symposium, Publ. 377, Washington, D. C., 1955, National Academy of Science–National Research Council.
7. Lambertsen, C. J., editor: Proceedings of the third symposium on underwater physiology, Baltimore, 1967, The Williams & Wilkins Co.
8. Lambertsen, C. J., editor: Proceedings of the fourth symposium on underwater physiology, New York, 1971, Academic Press, Inc.
9. Lambertsen, C. J., and Greenbaum, L. J., Jr., editors: Underwater physiology symposium, Publ. 1181, Washington, D. C., 1963, National Academy of Science–National Research Council.
10. Rahn, H., and Tokoyama, T., editors: Physiology of breath-hold diving and the Ama of Japan, Publ. 1341, Washington, D. C., 1965, National Academy of Science–National Research Council.
11. Schaefer, K. E., editor: Man's dependence on the earthly atmosphere, Proceedings of the First International Symposium on Submarine and Space Medicine, New London, Conn., 1958, New York, 1962, The Macmillan Co.
12. Yarbrough, O. D., and Behnke, A. R.: The treatment of compressed air illness utilizing oxygen, J. Ind. Hyg. Toxicol. **21**:213-218, 1939.

12 RESPONSES TO CARBON DIOXIDE–CONTAINING ATMOSPHERES

N. BALFOUR SLONIM
N. KAREN BENDER

INTRODUCTION

History

The history of carbon dioxide began with the studies of Jean Baptiste van Helmont (1577-1644), the founder of the iatrochemical school. Adding acids to potash, or limestone, he collected the "air" thus produced and observed that it extinguished flame. He also knew that this "air" is the same as that produced in the process of fermentation and that present in the Grotto del Cane in Italy, a cave in which dogs died whereas their erect masters survived. Van Helmont coined the word "gas" and named this "air" the "gas sylvestre." In this sense he discovered carbonic acid gas, or carbon dioxide.

Joseph Black (1728-1799), a physician, discovered that when limestone or chalk are heated a gas evolves as the substances lose weight and that the same volume of gas effervesces during treatment of these substances with strong acid. He named this gas "fixed air." Black also found that when limewater is exposed to air, a white precipitate of chalk slowly forms, suggesting that "fixed air" is a natural component of the atmosphere. Again, using limewater as a test, he proved that "fixed air" is produced when charcoal burns, in the fermentation of beer, and by the process of metabolism. In 1764 Black confirmed the metabolic production of "fixed air" by an experiment in Glascow, where he was professor of chemistry. In an air duct in the ceiling of a church in which a congregation of 1,500 persons remained for 10 hours, he dripped a solution of limewater over rags, producing a considerable quantity of crystalline lime ($CaCO_3$). Black knew that "fixed air" would extinguish both flame and life. Although from a chemical point of view Black did not characterize "fixed air" completely, he is generally given credit for the discovery of carbon dioxide.

Cavendish (1731-1810) measured the solubility and density of "factitious airs," including "fixed air." Lavoisier (1743-1794) observed that animals die when they are confined in a sealed atmosphere as soon as they have absorbed or converted to "aeriform calcic acid" (carbon dioxide) the greater part of the respirable portion of the air. Spallanzani (1729-1799) placed small animals in sealed tubes and measured the rates of oxygen consumption, carbon dioxide production, and the change of nitrogen volume. He also found that tissues excised from freshly killed animals, and the skin and muscle of a recently deceased human being, take up oxygen and give off carbon dioxide, indicating that oxidation occurs in the tissues. In 1809 Allen and Pepys found that the volume of carbon dioxide produced is approximately equal to the amount of oxygen consumed. In 1837 Gustav Magnus improved the methods for analyzing the gas content of blood. He found that both venous and arterial blood contain carbonic acid, oxygen, and nitrogen and that there is more carbonic acid in venous than in arterial blood.

In 1867 Alexander Schmidt and Eduard Pflüger discovered that shed blood consumes oxygen and produces carbon dioxide. In the same year, Strassburg, Pflüger's student, measured the partial pressure of carbon dioxide in the tissues.

In 1868 Pflüger began a study of the control of pulmonary ventilation by oxygen and carbon dioxide in dogs. Having perfected methods of blood-gas analysis, he found that the arterial blood oxygen content of dogs breathing nitrogen decreased from a control level of 14 to 18 vol% to a value of 1 or 2 vol%, with resulting marked dyspnea. Breathing a mixture of 30% carbon dioxide and 70% oxygen increased the arterial blood carbon dioxide content of other dogs from a control level of 25 to 28 vol% to a value of 50 to 60 vol%, with only moderate dyspnea resulting. Pflüger concluded that either carbon dioxide excess or oxygen lack stimulates breathing, but he considered oxygen lack much the quicker and stronger stimulus. He quoted Dohmen, who observed enormous increases of tidal volume and moderate increases of breathing frequency during 10% carbon dioxide inhalation, but failed to realize that inhalation of 30% carbon dioxide depresses breathing. However, he concluded that the normal carbonic acid content of blood excites the medulla oblongata.

In 1885 F. Miescher-Rüsch obtained the first quantitative evidence that the resting pulmonary ventilation rate is primarily regulated by carbon dioxide. Analyzing lung gas obtained by deep exhalation, he found an average value of 5.43% carbon dioxide (on a dry gas basis) in resting human subjects. During dyspnea produced by breathing carbon dioxide–containing gas mixtures, he found to his surprise that the carbon dioxide concentration of lung gas had risen to only 6.0% to 6.4%. He concluded that an increase of lung gas carbon dioxide concentration of considerably less than 1% produces a dyspneic acceleration of breathing. Because decreasing lung gas oxygen concentration by an amount greater than the carbon dioxide concentration increase that produces dyspnea had no observable effect, Miescher-Rüsch deduced that carbon dioxide is the normal chemical stimulus for breathing.

John Scott Haldane (1860-1936) investigated the role of carbon dioxide in the regulation of pulmonary ventilation. In 1893 Haldane and Lorrain Smith observed that dyspnea occurs when the inspired carbon dioxide concentration in a closed chamber increases to a level of only 3%, whereas if carbon dioxide was removed, no effects were observed until the oxygen concentration decreased to 14%. Working with J. G. Priestley, Haldane developed a practical method of sampling alveolar gas, facilitating study of alveolar carbon dioxide concentration under a variety of experimental conditions; their work showed that the resting pulmonary ventilation rate is normally regulated by carbon dioxide rather than by oxygen. Haldane found that the P_{CO_2} in alveolar gas remains relatively constant as barometric pressure varies from 646 to 1,260 torr, despite wide variation of ambient and alveolar P_{O_2}. He also observed that during inhalation of carbon dioxide–containing air, an alveolar carbon dioxide concentration increase of only 0.2% doubles alveolar ventilation rate and that breathing 5.5% carbon dioxide prevents posthyperventilation apnea. In addition to his many other contributions, Haldane developed the science of experimental human physiology.

Since the turn of the century scientific knowledge has increased exponentially; knowledge of the physiology of carbon dioxide is no exception. In 1903 Krogh showed that most of the oxygen used by the frog enters through the lung (short diffusion path), whereas most of the more readily diffusible carbon dioxide leaves through the moist skin. In 1904 Bohr, Hasselbalch, and Krogh observed that the addition of carbon dioxide to blood drives oxygen out. This important discovery, the Bohr shift, linked the processes of oxygen and carbon dioxide transport.

Advances in the acid-base chemistry of blood clarified the relationship between carbon dioxide and blood acidity. In 1908 and 1909 L. J. Henderson applied the law of mass action to the carbon dioxide–bicarbonate system of blood, and in 1909 Sorensen introduced the logarithmic pH notation for expressing hydrogen ion concentration. In 1917 Hasselbalch introduced the logarithmic version of the mass law expression known as the Henderson-Hasselbalch equation, clarifying the relationship among blood pH, carbon dioxide pressure, total carbon dioxide content, and the bicarbonate ion concentration of the blood. In 1920 M. H. Jacobs presented conclusive evidence that molecular carbon dioxide can diffuse into the interior of tadpole cells, *Arbacia* eggs, and the petals of certain carnations, rendering them acid, whereas the charged hydrogen ion associated with acids enters the cell very slowly.

During the 1920's and 1930's there was intensive study of the acid-base chemistry of blood, the biochemistry of carbon dioxide, the interaction of carbon dioxide with oxyhemoglobin dissociation equilibria, the reaction of carbon dioxide with hemoglobin to form carbaminohemoglobin, blood carbon dioxide transport, and carbon dioxide hydration-dehydration reactions. The last studies led to the discovery of the zinc-containing enzyme *carbonic anhydrase*. The investigations of these two decades produced fundamental data on the biochemistry of blood, laid the foundation of clinical chemistry, and developed methods that were used in the following decade of clinical investigation.

The discovery by Linus Pauling of hemoglobin S, the abnormal hemoglobin of sickle cell disease, stimulated interest in the molecular biology of hemoglobin. Following this discovery many genetically determined spontaneous hemoglobinopathies were identified, some of which are associated with abnormalities of blood-gas transport. Experimental modification of the hemoglobin molecule revealed that removal of a total of only five amino acid residues from the alpha and beta chains of the globin moiety of the hemoglobin molecule increases affinity for oxygen by a factor of 50 and eliminates the Bohr shift. Other studies showed that blocking the terminal α-amino groups of hemoglobin prevents carbamate formation at physiologic pH.

The biochemical processes of carbon dioxide production and fixation in mammalian tissues were elucidated. The initial enzyme-catalyzed carbon dioxide–fixation reaction of photosynthesis was shown to be carboxylation of ribulose-1,5-diphosphate to yield two molecules of 3-phosphoglycerate. Progress was made in the biochemistry of photosynthesis, the prime source of biospheric energy.

The properties of tromethamine, an organic alcohol amine first used to delay the respiratory acidosis of fish during transport, were investigated in both clinical and laboratory research. This soluble diffusible buffer combines with hydrogen ion in aqueous environments and biologic systems favoring the formation of bicarbonate ion from carbon dioxide, thus reducing carbon dioxide concentration and pressure. After injection into experimental subjects, the charged tromethamine ion is eliminated by the kidney.

More recently there has been a resurgence of interest in certain aspects of carbon dioxide chemistry and physiology. New technics have been used to measure kinetic and thermodynamic constants for the reactions of carbon dioxide with buffer systems. The effects of oxygenation on hemoglobin and the velocity and equilibrium constants for the carbamates of hemoglobin and simple peptides have been reinvestigated. Carbonic anhydrase has been intensively studied; the amino acid sequences and structure of different forms of the enzyme are being determined and physicochemical mechanisms of its catalysis have been proposed. The mechanism of enzymic carboxylation has also been studied, including the problem of whether carbon dioxide or bicarbonate ion is the active species.

As a result of research on the chemistry and physiology of carbon dioxide, accelerating since its beginning 375 years ago with the work of van Helmont, it is now possible to understand the carbon dioxide cycle in nature in some detail and to venture predictions about the future.

Carbon dioxide cycle in nature

As living organisms and producers of carbon dioxide, we are part of the vast cycle of production, consumption, and exchange of this compound among the plants and animals of the earth, ocean, and atmosphere. Carbon dioxide is produced in nature by animal metabolism and by the decomposition, fermentation, oxidation, and combustion of organic matter.

Carbon dioxide is removed from the atmosphere by green plants. The first chemical step of photosynthesis is the enzyme-catalyzed fixation reaction, the carboxylation of ribulose-1,5-diphosphate to yield two molecules of 3-phosphoglycerate. Although small, the present atmospheric concentration of carbon dioxide is vital to the photosynthetic process and is the ultimate source of carbon in the organic chemical compounds of our bodies. Green plants produce carbohydrates, polysaccharides, and a wide range of other organic substances; indeed, it has been said that green plants hold the basic patents for chemical synthesis. Forests leave carbonaceous deposits that may undergo nonbiologic oxidation, returning carbon dioxide to the atmosphere. After ingestion by animals, plant material is digested, absorbed, and metabolized to carbon dioxide, which is eliminated, completing the natural carbon dioxide cycle.

Atmosphere and ocean constantly exchange carbon dioxide. Atmospheric carbon dioxide dissolves in water to form carbonic acid, bicarbonate ions, and carbonate ions that may remain in solution or precipitate as the insoluble carbonates familiar to the geologist. In the ocean carbon dioxide is a component of the world's largest buffer solution, which contains about 100 bicarbonate ions and an appreciable number of carbonate ions per molecule of carbon dioxide. This vast ocean reservoir stabilizes the steady-state atmospheric carbon dioxide concentration.

Carbon dioxide introduced into the atmosphere is uniformly mixed by convection up to an altitude of 90 to 100 km; above this altitude theory predicts dissociation of carbon dioxide molecules by solar ultraviolet radiation. Our atmosphere now

Table 12-1. Physical constants for carbon dioxide

Molecular weight	44.00995
Density, gas at 70° F and 1 ATA	0.1146 lb/ft³
Specific gravity, gas (air = 1)	1.5289
Vapor pressure at 70° F	830 psig
Specific volume at 70° F and 1 ATA	8.76 ft³/lb
Sublimation point at 1 ATA	−78.5° C
Triple point at 5.11 ATA or 3,885.2 torr	−56.602° ± 0.005° C
Critical temperature	31.0° C
Critical pressure	1,071.6 psia, or 72.9 ATA
Critical density	0.468 gm/ml
Latent heat of vaporization	
At triple point	149.6 BTU/lb
At 0° C	101.03 BTU/lb
Specific heat at 60° F	
Gas, c_P	0.1988 BTU × lb^{-1} × $°F^{-1}$
Gas, c_V	0.1525 BTU × lb^{-1} × $°F^{-1}$
Ratio c_P/c_V	1.303
Thermal conductivity	
At 0° C	0.0085 BTU × hr^{-1} × sq ft^{-1} × $°F^{-1}$ × ft^{-1}
at 100° C	0.0133 BTU × hr^{-1} × sq ft^{-1} × $°F^{-1}$ × ft^{-1}
Viscosity of gas at 70° F and 1 ATA	0.0148 centipoise

contains 0.0314 percent by volume of carbon dioxide[1]; however, the concentration may vary significantly relative to this average value from time to time and from place to place. Although there is reason to believe that at one time the atmosphere contained much more carbon dioxide than it presently does, atmospheric carbon dioxide concentration is now *relatively* constant despite sources and sinks. Constancy is maintained by the opposing and nearly balanced processes of production—metabolism and combustion—and consumption—photosynthesis and sedimentary deposition.

Physical and chemical properties of carbon dioxide

Carbon dioxide, also called carbonic anhydride, carbonic acid gas, "black damp," and "afterdamp," is a colorless, odorless gas that neither burns nor supports combustion. The physical properties of carbon dioxide are given in Table 12-1.

The molecular weight of carbon dioxide is 44.00995 on the basis of the carbon 12 isotope scale for which ^{12}C = 12.000; by comparison, the mean molecular weight of atmospheric air at sea level is 28.9644. The ratio is thus 1.519. The carbon dioxide molecule is linear and symmetric; it has a net dipole moment of zero. Resonance decreases the length of the two equal C═O interatomic double bonds to 1.159 Å, as compared to the typical C═O double bond length of 1.22 Å. The resonating molecular structure of carbon dioxide is represented as follows:

$$^{-}\!:\ddot{\underset{..}{O}}:C:::O:^{+} \rightleftarrows :\ddot{O}::C::\ddot{O}: \rightleftarrows {}^{+}\!:O:::C:\ddot{\underset{..}{O}}:^{-} \qquad (1)$$

Carbon has two stable isotopes of masses 12 and 13 and oxygen has three stable isotopes of masses 16, 17, and 18. Thus carbon dioxide is a mixture of twelve different isotopic species. Although the chemical and physical properties of the different isotopic species are very similar, the effect of differing masses on the vibrational frequencies alter the thermodynamic and kinetic properties of the carbon dioxide molecule. As a result of these differences, measurable isotopic fractionations are associated with many natural and laboratory processes.

Carbon dioxide is unique among gases in distributing itself in almost equal amounts per unit volume between air and water; approximately 1 volume of carbon dioxide dissolves in 1 volume of water at a carbon dioxide pressure of 1 ATA and a temperature of 15° C. The solubility coefficient* for carbon dioxide in water is exactly 1 at a little below 15° C and decreases as temperature increases. The solubility of carbon dioxide in water as a function of temperature is shown in Table 12-2.

Although the carbon dioxide molecule dissolves much more readily in water than most other gases do, it is even more soluble in hydrophobic solvents and most soluble of all in solvents that contain polar groups with C^{+}—O^{-} dipoles adjacent to nonpolar residues. The solubility of carbon dioxide in water, benzene, and ethanol at 20° C and 1 ATA partial pressure is, respectively, 0.872, 2.37, and 2.94 vol (STP)/vol liquid. Carbon dioxide is more than three times as soluble in the nonpolar solvents benzene, toluene, and heptane as it is in water. It is even more soluble in ethanol, and especially so in acetone, molecules that contain both hydrophobic and strongly polar groups. At any given temperature carbon dioxide is about thirty times as soluble as oxygen in water, a fact that accounts for its much greater diffusibility across the alveolocapillary membrane in

*The volume in cubic centimeters (STPD) of gas in solution per gram of solvent at a pressure of 1 ATA at equilibrium.

Table 12-2. Carbon dioxide dissolved in water as a function of temperature*

Temperature (°C)	Volume† (STPD)	Weight‡ (gm)
0	1.713	0.3346
5	1.424	0.2774
10	1.194	0.2318
15	1.019	0.1970
20	0.878	0.1688
25	0.759	0.1449
30	0.665	0.1257
35	0.592	0.1105
40	0.530	0.0973
45	0.479	0.0860
50	0.436	0.0761

*Data from Lange.[2]
†Absorption coefficient, or volume of gas dissolved in 1 volume of water at P_{CO_2} = 760 torr.
‡Weight of gas in grams dissolved in 100 gm of water at total pressure $P_{H_2O} + P_{CO_2}$ = 760 torr.

the lung. However, the ratio of carbon dioxide affinity for hydrophobic solvents to its affinity for water is much less than the same ratio for oxygen affinities.

When carbon dioxide dissolves in water, it is present almost entirely as unhydrated carbon dioxide. Although the carbon dioxide molecule as a whole is nonpolar, each of the two C–O bonds may act as a small dipole, with the oxygen atom at the negative end. Such dipoles may attract and orient the surrounding water molecules. Some dissolved carbon dioxide reacts with water (hydration) to form the unstable and relatively weak carbonic acid (H_2CO_3) or the bicarbonate ion (HCO_3^-); the initial product of this reversible reaction may be either carbonic acid or bicarbonate and hydrogen ions. Carbonic acid, in turn, dissociates reversibly to yield a bicarbonate ion and a proton. This means a cycle of three reversible reactions.*

In the hydration reaction the two C=O bonds of carbon dioxide must lengthen and bend toward each other, decreasing the angle between them from 180 to about 120 degrees. The electronic rearrangements associated with these changes account for the relative slowness of the hydration and dehydration reactions. The slowness of these uncatalyzed reactions is a fact of considerable biologic importance; the widely distributed enzyme *carbonic anhydrase* greatly accelerates this reaction.

Dissolved carbon dioxide reacts not only with water but also with hydroxyl groups:

*For simplicity, these three reactions are usually written sequentially.

$$CO_2 + OH^- \rightleftarrows HCO_3^- \quad (2)$$

This reaction makes a significant contribution to the rate of disappearance of carbon dioxide at a pH above 7.5. Above pH 10 it dominates the hydration reaction. It is also of theoretic interest in relation to the mechanism of the reaction catalyzed by *carbonic anhydrase,* which probably involves the donation of the hydroxyl ion to carbon dioxide as hydration proceeds and the removal of the hydroxyl ion from the bicarbonate ion as dehydration proceeds.

Atmospheric carbon dioxide can thus affect the physicochemical properties of aqueous systems. For example, the electric resistance of pure water (calculated maximum) is 26 megohms/cm compared to that of water in equilibrium with atmospheric carbon dioxide, which is 0.7 megohm/cm. The pH of absolutely pure water at 25° C is exactly 7.000. However, as a result of carbon dioxide absorption from the air, the pH of distilled water is usually from 5 to 6.

The partial pressure of carbon dioxide, or P_{CO_2}, is a measure of the tendency of carbon dioxide to escape from a system, or fugacity, and is usually expressed in millimeters of mercury (mm Hg), or torr. Gas pressure is one of the factors that determines the availability, or effective local concentration, of gas molecules for physicochemical reaction. The physiologic effects of carbon dioxide or, for that matter, of any gas are thus a direct function of its *partial pressure,* or *tension,* not its concentration. Of course, within a given system, pressure and concentration are *related* parameters and must both be considered in analyzing the *dynamics* of the physiologic effects of a gas.

The significance of P_{CO_2}, as contrasted with carbon dioxide content, depends further on the fact that this intensive factor* can be detected by

*An intensive factor is a thermodynamic property of a physicochemical system not dependent on the size of the system. Pressure is the intensive, or potential, factor of the mechanical energy of a system. As a potential, it has direction. An extensive factor is a thermodynamic property of a physicochemical system that is dependent on the size of the system. Volume is the extensive, or capacity, factor of the mechanical energy of a system. Concentration (content divided by volume) is a quantity-related measure and thus also an extensive, or capacity, factor.

The concept of pressure as a potential factor in the delivery of gas molecules to a reactive site in a physicochemical system must not be confused with the thermodynamic concept of chemical potential. Chemical potential, the intensive, or potential, factor of chemical energy, is also independent of the size of a system but measures the driving force of a chemical reaction.

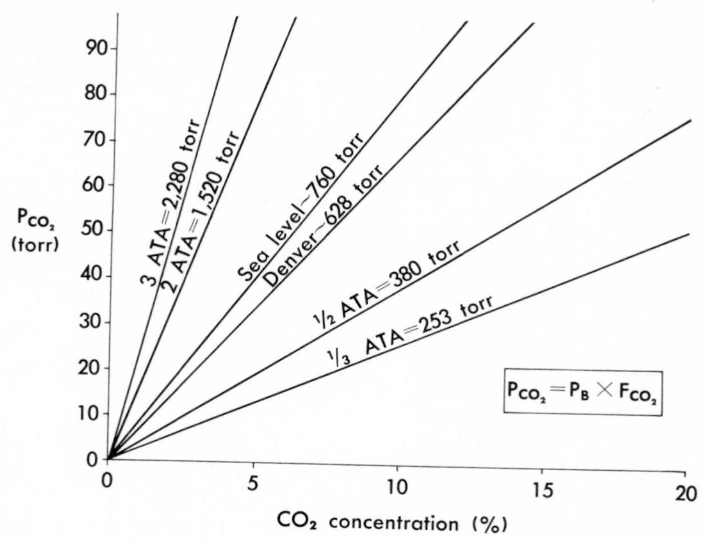

Fig. 12-1. Carbon dioxide tension as a function of carbon dioxide concentration in gas mixtures at various ambient pressures.

either physical or physiologic sensors anywhere within a system and that the information thus obtained can be used for feedback control. Carbon dioxide concentration alone without simultaneous data for total gas pressure or P_{CO_2} is generally insufficient for physiologic purposes. A useful description of environmental carbon dioxide must include either P_{CO_2} or the information necessary to calculate this factor.

The P_{CO_2} of a gas is calculated using Dalton's law. This law relates the partial pressure of a gas to its fractional concentration and to the total pressure of the gas mixture of which it is a part:

$$\text{partial pressure} = \text{fractional concentration} \times \text{total gas pressure} \quad (3)$$

For carbon dioxide this equation becomes:

$$P_{CO_2} = F_{CO_2} \times P_B \quad (4)$$

Fig. 12-1 shows this relationship for selected total gas pressure levels, including sea-level pressure, 1-mile elevation, and simple multiples and fractions of standard sea level pressure.

For example, a concentration of 1% carbon dioxide, which has slight effects when inhaled at sea level (760 torr), is physiologically equivalent to 10% carbon dioxide when inhaled by a diver 300 ft below the ocean surface, where total ambient pressure is 10 atmospheres. When experimental animals are exposed to helium-oxygen mixtures at a pressure equivalent to a depth of 4,000 ft in sea water (a total ambient pressure of 122 atmospheres), the ordinary atmospheric carbon dioxide concentration of 0.03% results in an inspired P_{CO_2} of about 30 torr. Thus even the 0.03% carbon dioxide present in atmospheric air must be reduced or removed to prevent compressed air hypercapnia at extremely high ambient pressure.

The opposite situation may exist in aerospace environments, where ambient pressure is only one half or even one third of an atmosphere while the astronaut breathes a gas mixture containing oxygen in high concentration. Under these hypobaric conditions a higher fractional concentration, or percentage, of carbon dioxide is acceptable because 3% carbon dioxide exerts the same partial pressure at one-third atmosphere as does 1% carbon dioxide at sea level (Fig. 12-1).

Carbon dioxide is a relatively stable chemical compound. Dissolved carbon dioxide reacts with water to form carbonic acid, which in turn dissociates to yield a proton and a bicarbonate ion. Aqueous solutions are thus acidic.

$$CO_2 + H_2O \rightleftarrows H_2CO_3 \rightleftarrows H^+ + HCO_3^- \quad (5)$$

Carbon dioxide is thus the anhydride of carbonic acid. The bicarbonate ion can also dissociate as follows:

$$H^+ + HCO_3^- \rightleftarrows 2H^+ + CO_3^= \quad (6)$$

Hydrogen reduces carbon dioxide as follows:

$$CO_2 + H_2 \rightarrow CO + H_2O \quad (7)$$

Above 1,700° C carbon dioxide dissociates into oxygen and carbon monoxide:

$$2\ CO_2 \rightarrow 2\ CO + O_2 \qquad (8)$$

Carbon dioxide is 15.8% dissociated at 2,227° C.

The well-known carbon dioxide permeability of both natural and synthetic rubbers, as well as certain plastics, is at once a hazard to users of physiologic equipment and a boon to designers of artificial lungs and gills.

Uses of carbon dioxide

Carbon dioxide has extensive industrial, commercial, scientific, medical, and research applications.[3] It is used as a refrigerant gas, to carbonate beverages, to prevent oxidation and combustion of inflammable materials, to cure cement, to control pH, in fire extinguishers, in inert atmospheres, in food canning, in shielded arc welding, in mining operations, and in the chemical industry. It is also used as a therapeutic respiratory stimulant and in cardiopulmonary laboratory tests. Compressed carbon dioxide is used as a source of energy for pressure spraying, for spray painting, and as an aerosol propellant; carbon dioxide cartridges are used to inflate life rafts, life vests, and other floatation devices. Solid carbon dioxide ("Dry Ice") is used to cool and preserve food as well as other perishable products during storage and transport.

Carbon dioxide is produced by animal metabolism and in processes of human technology, such as combustion of organic fuels, internal combustion engines, the kilning of limestone (heating calcium carbonate), the reaction of acids with carbonates, the burning of natural gas, and the explosion of methane ("firedamp") in mines.

There are at least seven commercial methods of producing and purifying carbon dioxide. Compressed carbon dioxide is commercially available in several grades of chemical purity. Supplied in steel cylinders, it exists as a liquid under its own vapor at a pressure of approximately 830 psig when the temperature is 70° F.[4]

Carbon dioxide analysis

The information we use to understand and interpret the responses to carbon dioxide–containing atmospheres is obtained by analysis of the content and partial pressure of carbon dioxide in samples of gases and liquids from environmental and biologic sources. Methods of carbon dioxide analysis are based on its physical and chemical properties.

Most quantitative chemical methods of carbon dioxide analysis, whether gravimetric or volumetric, are based on the principle that alkalies, such as sodium hydroxide, potassium hydroxide, lithium hydroxide, and calcium hydroxide, combine with carbon dioxide to form the respective carbonates. The Scholander Micrometer Gas Analyzer is widely used for volumetric analysis of carbon dioxide in gas samples of less than 1 cc.[5] The Van Slyke Manometric Apparatus is used for combined determination of carbon dioxide and oxygen content of blood samples.[5]

Physical methods for carbon dioxide analysis include thermal conductivity, infrared absorption, gas chromatography, mass spectrometry, microwave spectrometry, and the carbon dioxide electrode. In the carbon dioxide electrode, carbon dioxide molecules diffuse across a membrane into a precisely constituted buffer solution; there the carbon dioxide reacts with water to produce carbonic acid, and the resulting acidity, or pH, of the buffer solution is measured with a glass electrode. Using this pH, carbon dioxide pressure is then calculated from the Henderson-Hasselbalch buffer equation.

A common source of error in physiologic calculations is the simplifying assumption that inspired carbon dioxide is zero. Recirculated air in occupied closed spaces may contain considerably more carbon dioxide than fresh atmospheric air. Accumulated carbon dioxide may also affect the physiologic state of subjects as well as invalidate simplified calculations. (For a discussion of problems and calculations involving carbon dioxide see references 5 and 6).

The technic and precautions necessary for collection and storage of blood samples to be analyzed for carbon dioxide, oxygen, and pH are important.[5] If blood flowing through the pulmonary capillaries exchanges oxygen and carbon dioxide with alveolar gas through the alveolocapillary membrane in less than 1 second, it is clear that exposure of blood samples to the air, or to air bubbles, changes the interdependent blood-gas and pH parameters.

In general, the calibration of quantitative *physical* methods of analysis requires *chemical* methods. This is true regardless of whether the physical principles involved are those of the glass electrode, gas chromatography, infrared absorption, microwave spectrometry, or thermal conductivity. All too often this basic principle of methodology is not sufficiently appreciated or is overlooked in the design and operation of analytic systems.

PHYSIOLOGY OF CARBON DIOXIDE

Biologic significance of carbon dioxide

Produced by oxidation of carbon-containing substances in the processes of decay and combustion or biologically as an end-product of metabolism, carbon dioxide plays an important physiologic role in body fluids as a regulator of pulmonary ventilation, a regulator of cerebral blood flow, a component of a buffer defense system, and a substrate for carbamate formation.

Mixing uniformly in the atmosphere with other constituent gases, carbon dioxide strongly absorbs infrared radiation to play a role in the geothermal environment. Dissolving in water, carbon dioxide forms carbonic acid, bicarbonates, and carbonates, which may remain dissolved in the increasingly saline, increasingly acidic buffer solution that is the ocean, precipitate in underwater sedimentary deposits, or be fixed as calcite or aragonite carbonates in the shells and skeletons of marine biota, to become fossilized letters in the calcareous pages of geologic history.

Trapped by an enzyme in the photosynthetic process of green plants, carbon dioxide is a substrate for carboxylation, forming a wide variety of biologic organic compounds that may be stored or remain in carbonaceous deposits of wood, peat, lignite, and coal. In segments of the cycle, the natural history of carbon dioxide is reduced to that of elemental carbon or bound to that of hydrogen and nitrogen in a vast array of carbon compounds.

Animals borrow organic plant materials, digesting, absorbing, and assimilating them to semisynthesize their own species-specific brands of soluble organic compounds. Excess carbon dioxide, whether endogenous or exogenous, is deposited for storage as carbonate ions "contaminating" the hydroxyapatite mineral that hardens the organic matrix of bone; the mineral content of human bone is about 6% carbonate. Thus carbon dioxide, like water and the oxygen they both contain, is a fundamental ingredient of the biospheric broth.

Biochemical reactions involving carbon dioxide

Under certain conditions carbon dioxide, which is the anhydride of carbonic acid, combines with the amino groups of amino acids, peptides, or proteins to form carbamino compounds:

$$CO_2 + R\text{–}NH_2 \rightleftarrows R\text{–}NH\text{–}COO^- + H^+ \qquad (9)$$

Hemoglobin-carbamate is one of several forms in which the circulation transports carbon dioxide. The reaction of carbon dioxide with amino groups is rapid and does not involve an enzyme catalyst. The carbamino linkage resembles an amide or peptide bond.

Metabolic carbon dioxide production

Of the 688,500 calories of free energy liberated during the complete bio-oxidation of 1 mole of glucose to carbon dioxide and water, more than 90% is released in the aerobic phase of metabolism. The biochemical systems that oxidize pyruvate, oxidize the acetyl moiety of acetyl-coenzyme A, and effect the oxidative chain phosphorylations are mostly within subcellular organelles called mitochondria. Anaerobic glycolysis in the cell cytoplasm of one molecule of hexose yields two molecules of pyruvate. Since each of the three carbon atoms of pyruvate may be oxidized to one molecule of carbon dioxide, the complex oxidation of pyruvate yields three molecules of carbon dioxide as shown in the following overall equation:

$$CH_3COCOOH + 2\tfrac{1}{2}\ O_2 \rightarrow 3\ CO_2 + 2\ H_2O \qquad (10)$$

The first molecule of carbon dioxide is produced from pyruvate in the formation of acetyl-coenzyme A; the second and third are produced in the tricarboxylic acid cycle proper.

Pyruvate, which is derived from anaerobic glycolysis of hexose, dehydrogenation of lactate, or the oxidative deamination of the amino acid alanine, enters a mitochondrion where it is oxidatively decarboxylated by a rather complex enzyme system to acetyl-coenzyme A, producing the first carbon dioxide molecule:

$$CH_3COCOOH + HS\text{–}CoA + NAD^+ \rightarrow$$
$$CO_2 + CH_3CO\text{–}S\text{–}CoA + NADH + H^+ \qquad (11)$$

The pyruvate dehydrogenase system consists of four cofactors and an aggregate of three distinct enzymes that operate sequentially: *pyruvate decarboxylase, lipoate reductase-transacetylase,* and *dihydrolipoate dehydrogenase.* In the initial reaction catalyzed by *pyruvate decarboxylase,* pyruvate reacts in the presence of the magnesium ion with enzyme-bound thiamine pyrophosphate*; the α-carbon of the keto acid attaches to the second carbon of the thiazole ring of the vitamin to form enzyme-bound lactyl-thiamine pyrophosphate. Carbon dioxide evolves, leaving enzyme-bound α-hydroxyethyl thiamine pyrophosphate, or "active acetaldehyde." This intermediate is oxidized from the aldehyde to the acetic acid, or carboxylate,

*The pyrophosphate ester of vitamin B_1.

level by interaction with enzyme-bound lipoate, a cofactor that is linked to the ε-amino group of a lysine residue of the enzyme *lipoate reductase-transacetylase,* or *dihydrolipoyl transacetylase.* As the aldehyde is oxidized to acetic acid, lipoamide disulfide is reduced to a disulfhydryl compound. Simultaneously, lipoate accepts the acetyl group, which now becomes an acyl thioester; the formation of high-energy thioester bond conserves some of the energy liberated in the oxidation. Acetyl-lipoamide then reacts with the sulfhydryl group of coenzyme A in a thioacyl exchange, producing acetyl-coenzyme A and the disulfhydryl form of lipoamide. The latter is reconverted to the disulfide form by transfer of hydrogen atoms to the flavin prosthetic group of *dihydrolipoate dehydrogenase,* which, in turn, transfers the hydrogen atoms to nicotinamide adenine dinucleotide (NAD).* The resulting NADH is reoxidized via the chain of aerobic oxidative catalysts.

The oxidation of one molecule of pyruvate to acetyl-coenzyme A thus produces three high-energy bonds as the NADH produced in the reoxidation of the disulfhydryl form of lipoamide by the flavoprotein enzyme *dihydrolipoyl dehydrogenase* is reoxidized. However, one high-energy thioester bond is spent in the citrate condensation. The free energy change for the entire reaction sequence is about –8 kcal.

In yeast, pyruvate is irreversibly decarboxylated to acetaldehyde by a *pyruvate decarboxylase* not found in animal cells:

$$CH_3COCOOH \xrightarrow{Mg^{++}} CO_2 + CH_3CHO \quad (12)$$

Thiamine pyrophosphate is an essential coenzyme for this reaction. As in pyruvate oxidation in animal cells, α-hydroxyethyl thiamine pyrophosphate is the initial product. Here, however, the acetaldehyde moiety is not transferred and the complex decomposes, producing free acetaldehyde and regenerating thiamine pyrophosphate. Fortunately, *alcohol dehydrogenase* then catalyzes the reduction of acetaldehyde by NADH:

$$CH_3CHO + NADH + H^+ \rightleftarrows CH_3CH_2OH + NAD^+ \quad (13)$$

The complete oxidation of the acetyl moiety of acetyl-coenzyme A to carbon dioxide and water is accomplished by a series of biochemical reactions termed the tricarboxylic acid cycle. This integrated reaction sequence is the final common pathway not only for oxidation of the products of glycolysis but also for the oxidation of fatty acids and amino acids. The tricarboxylic acid cycle is the major reaction sequence supplying electrons to the transport system that reduces molecular oxygen as it generates the high-energy phosphate bonds of adenosine triphosphate (ATP) and, as such, is the major source of the free energy of higher organisms. ATP is the common currency of bioenergetics and thus the form of energy used in the endergonic processes of living cells. Of the four reactions in the tricarboxylic acid cycle that are dehydrogenations, three reduce pyridine nucleotides, forming NADH or nicotinamide adenine dinucleotide phosphate (NADPH), oxidation of which via the electron transport chain transfers most of the energy released in the tricarboxylic acid cycle.

Each revolution of the tricarboxylic acid cycle consumes one molecule of acetyl-coenzyme A and produces two molecules of carbon dioxide. The net effect of a single revolution of the tricarboxylic acid cycle is as follows:

$$CH_3COOH \longrightarrow 2\ CO_2 + 4\ H^+ + 8\ e + \text{energy} \quad (14)$$

The acetyl moiety of acetyl-coenzyme A that is oxidized in the tricarboxylic acid cycle comes mainly from glucose and fatty acid metabolism.

The first reaction of the tricarboxylic acid cycle is a condensation of acetyl-coenzyme A—a pivotal metabolite derived from (1) hexoses via pyruvate, (2) fatty acids via coenzyme A esters, or (3) amino acids via α-keto acids, pyruvate, or ketone bodies—with oxaloacetate, regenerating coenzyme A and forming the six-carbon tricarboxylic compound, citric acid. Consecutive reactions then form cis-aconitate, isocitrate, α-ketoglutarate, succinate, fumarate, malate, finally regenerating oxaloacetate, which was consumed in the initial condensation reaction. Thus the tricarboxylic acid cycle operates continuously as long as a supply of acetyl-coenzyme A is available and the resulting hydrogen ions and carbon dioxide are removed.

The oxidative decarboxylation of isocitrate produces the second molecule of carbon dioxide and a molecule of α-ketoglutarate. Isocitrate is oxidized by transfer of electrons to NAD. The reaction is catalyzed by either of two isocitrate dehydrogenases. One of these, within the mitochondrion, is Mg^{++}-dependent and specific for NAD^+; the velocity of reaction is a function of substrate, NAD^+, and Mg^{++} concentration and is also subject to regulation by a positive modifier, adenosine monophosphate. The other, located in the cell cytoplasm is Mn^{++}-dependent and specific for

*The pyridine nucleotide coenzyme of electron transport.

$NADP^+$; it produces an intermediate product, oxalosuccinate, which is bound to the enzyme. Under physiologic conditions the overall reaction involves a free energy change of about –5 kcal.

α-Ketoglutarate, a member compound of the tricarboxylic acid cycle, is a point of convergence of the metabolic pathways of carbohydrates, lipids, and certain amino acids. Oxidative deamination or transamination of glutamate produces α-ketoglutarate. Thus any amino acid that can be metabolized to glutamate—including glutamine, proline, ornithine, and histidine—is a potential source of α-ketoglutarate. Conversely, in the presence of ammonia and reduced coenzyme, α-ketoglutarate can be reductively aminated to glutamate.

The oxidative decarboxylation of α-ketoglutarate is catalyzed by a multienzyme complex composed of three-enzyme units. α-Ketoglutarate combines with enzyme-bound thiamine pyrophosphate in the presence of Mg^{++}, evolving the third molecule of carbon dioxide, which leaves enzyme-bound α-hydroxy-γ-carboxypropyl thiamine pyrophosphate, or "active succinic semialdehyde." This intermediary compound is oxidized to the succinyl level and the four-carbon chain transferred to enzyme-bound lipoate, forming the thioester succinyl lipoyl enzyme. The latter then reacts with coenzyme A to form succinyl-coenzyme A. Dihydrolipoamide is reoxidized by NAD^+. The reaction sequence also appears to involve flavin adenine dinucleotide (FAD).* The free energy change for the reaction is –8 kcal/mole.

Although this oxidative decarboxylation is analogous to that of pyruvate, it differs in the disposition of the high-energy thioester bond; in this case, catalyzed by the enzyme *succinate thiokinase,* most of the succinyl-coenzyme A reacts with guanosine diphosphate and inorganic phosphate to form succinate, coenzyme A, and guanosine triphosphate. The high-energy phosphate linkage of guanosine triphosphate can then form adenosine triphosphate.

What are the fates of the methyl and carboxyl carbon atoms of acetic acid as they go through the tricarboxylic acid cycle? Although the consumption of one molecule of acetyl-coenzyme A produces two molecules of carbon dioxide on the first revolution of the cycle, neither of the two carbon atoms in this carbon dioxide can derive from acetyl-coenzyme A. This results from the fact that citrate, because it attaches asymmetrically to aconitase, does not randomize about its center of symmetry, and the asymmetry is such that the two acetyl-derived carbon atoms are not subject to metabolic change. However, randomization does occur subsequently in the cycle with the formation of succinic acid and the carbon atoms are then distributed symmetrically about the plane of symmetry. Thus on subsequent revolutions of the tricarboxylic acid cycle, these acetyl-derived carbons are oxidized to carbon dioxide in accordance with probability.

*A cofactor derived from the vitamin riboflavin. Unlike NAD and NADP, which dissociate readily from dehydrogenases, FAD is tightly bound to an enzyme in a complex called a flavoprotein.

Metabolic carbon dioxide fixation

The following discussion is a part of our general review of the production and assimilation of carbon dioxide by living organisms and of the biocatalysts, or enzymes, that accelerate these exchanges. Carbon dioxide is usually thought of as a metabolic waste product; however, animal as well as plant cells contain enzyme systems that catalyze the incorporation of carbon dioxide into biochemical compounds. Although most of these reactions are reversals of the decarboxylations of β-keto acids such as oxaloacetate,* carbon dioxide is also involved in several important anabolic reaction sequences, including the synthesis of fatty acids, purines, and pyrimidines. The incorporation of carbon dioxide into other biochemical compounds is termed carbon dioxide fixation, or carbon dioxide assimilation.

When carbon dioxide is metabolically produced, chiefly in the tricarboxylic acid cycle, it arises from a carboxyl group; when it is incorporated, either in photosynthesis or in other assimilation reactions, it joins an organic molecule as a carboxyl group. Nearly all carboxyl groups are ionized at physiologic pH; the two C—O bonds are equal in length because of resonance, they are slightly longer (1.26 Å) than the typical C=O double bond (1.22 Å), and the bond angle is 125.6°. If the carbon dioxide molecule, rather than the bicarbonate ion, is the reactive species, the biochemical processes of carboxylation and decarboxylation involve not only the breakage or formation of a covalent bond, usually a carbon-carbon bond, but also change the angle between the two C—O bonds to an extent about equal to that occurring in hydration and dehydration reactions. The carboxyl group may be transiently

*The term "anaplerosis" means renewal of the supply of keto acids.

protonated during certain biochemical carboxylations and decarboxylations.

The vitamin biotin, widely distributed in plants and animals, is an essential prosthetic group of certain enzymes that catalyze carbon dioxide–fixing reactions. In general, acyl-coenzyme A carboxylases and α-keto acid carboxylases require biotin. Biotin is covalently bound by amide linkages to the ε-amino groups of lysine residues on the apoenzyme. The biotinyl prosthetic group is the carbon dioxide carrier. The bicarbonate ion, not molecular carbon dioxide, is the reactive species in most carbon dioxide fixation reactions catalyzed by biotin enzymes. The significance of this fact is seen in equation 5.

The biochemical mechanism involves two steps, the first of which is the ATP-dependent carboxylation of biotin by bicarbonate ion, forming an intermediate complex, *N*-carboxybiotin-enzyme. The carboxyl group is carried on the ureide N atom on the side of the biotin ring opposite to the fatty acid side chain. Biotin-requiring enzymes are inhibited by avidin, a basic biotin-binding protein found in raw egg white.

Oxaloacetate, malate, and succinate are intermediary compounds of the tricarboxylic acid cycle. Oxaloacetate may condense with acetyl-coenzyme A to form citrate, form the amino acid aspartate by transamination, undergo reduction by NADH to malate, or, in the cell cytoplasm, form phosphoenolpyruvate. Malate is oxidized to oxaloacetate. Succinate is also metabolized in the tricarboxylic acid cycle. The carbon dioxide–fixing reactions in animal cells that produce oxaloacetate, malate, and succinate are as follows:

1. Several carbon dioxide–fixing reactions form oxaloacetate by carboxylation of pyruvate. This phenomenon occurs in heterotrophic, nonphotosynthetic bacteria as well as in animal cells. The mitochondrial enzyme pyruvate carboxylase is the principal catalyst of oxaloacetate formation from carbon dioxide and pyruvate in yeast as well as in animal cells. The reaction requires biotin, ATP, Mg^{++}, and acetyl-coenzyme A; the last acts as a positive modifier but does not participate in the reaction. The bicarbonate ion is the reactive species:

$$HCO_3^- + CH_3COCOOH + ATP \xrightarrow[Mg^{++}]{\text{acetyl-coenzyme A}} \text{oxaloacetate} + ADP + P_i \quad (15)$$

Because physiologic equilibrium conditions favor decarboxylation, the quantity of carbon dioxide fixed by this reaction is usually small. However, the low K_m* for carbon dioxide, the free energy change of –5 kcal, the allosteric effect of acetyl-coenzyme A when cell demand for oxaloacetate is maximal, and impaired formation of oxaloacetate from carbon dioxide and pyruvate in biotin-deficient animals indicate that this is the chief *reserve* mechanism for oxaloacetate formation.

2. Another carbon dioxide–fixing reaction is catalyzed by *phosphoenolpyruvate carboxykinase*, an enzyme that does not require biotin:

$$CO_2 + \text{phosphoenolypyruvate} + \text{inosine diphosphate} \underset{}{\overset{Mg^{++}}{\rightleftarrows}} \text{oxaloacetate} + \text{inosine triphosphate} \quad (16)$$

The relatively high K_m for carbon dioxide and the free energy change of about +4,000 cal/mole indicate that this reaction is not a physiologically important carbon dioxide fixation pathway. On the contrary, the reverse reaction is a significant pathway of gluconeogenesis.

3. In certain plants oxaloacetate formation is catalyzed by the enzyme *phosphoenolpyruvate carboxylase*. Biotin is not required. Bicarbonate is the active species:

$$HCO_3^- + \text{phosphoenolpyruvate} \rightarrow \text{oxaloacetate} + P_i \quad (17)$$

The reaction proceeds well despite low carbon dioxide pressure; the free energy change is about –6.5 kcal/mole.

4. A transphosphorylase enzyme system is the main pathway for carbon dioxide fixation in propionibacteria. The carbon dioxide–fixing reactions that produce oxaloacetate are continuously opposed by the action of an active nucleotide-independent oxaloacetate decarboxylase in liver mitochondria.

5. *Malic enzyme*, or *malate dehydrogenase (decarboxylating)*, catalyzes the reductive carboxylation of pyruvate to malate, bypassing oxaloacetate:

$$CO_2 + CH_3COCOOH + NADPH + H^+ \overset{Mn^{++}}{\rightleftarrows} HOOCCH_2CHOHCOOH + NADP^+ \quad (18)$$

Biotin is not required. The high K_m for carbon dioxide and the free energy change of +500 cal/mole for malate formation by *malic enzyme* indicate that this reaction is probably more important as a source of NADPH for reductive biosynthesis than as a source of malate for the tricarboxylic

*The Michaelis-Menten constant; a characteristic unique for each different enzyme.

acid cycle. In the presence of NAD^+, malate is then oxidized to oxaloacetate by *malate dehydrogenase,* an enzyme distinct from *malic enzyme.*

6. The three-carbon compound propionate is a product of the oxidative metabolism of the amino acids valine, isoleucine, and methionine or catabolism of the relatively uncommon fatty acids that contain an odd number of carbon atoms. Propionate must first react with coenzyme A and ATP, forming propionyl-coenzyme A. Then, in a reaction biochemically analogous to the carboxylation of acetyl-coenzyme A, the biotin-requiring enzyme *propionyl-coenzyme A carboxylase* catalyzes the carboxylation of propionyl-coenzyme A to D-methylmalonyl-coenzyme A. Bicarbonate ion is the reactive species:

$$HCO_3^- + \text{propionyl-CoA} + ATP \overset{Mg^{++}}{\rightleftarrows} \text{D-methylmalonyl-CoA} + ADP + P_i \quad (19)$$

Certain other propionyl-coenzyme A carboxylation systems do not require biotin. The D-form of methylmalonyl-coenzyme A thus produced is next converted to its enantiomorph, the L-form, by the enzyme *methylmalonyl-coenzyme A racemase.* The thioester L-compound is then isomerized to succinyl-coenzyme A by the enzyme *methylmalonyl-coenzyme A isomerase* in a reaction that requires vitamin B_{12} as a coenzyme; the process involves intramolecular migration of the entire thioester group, including the carbonyl carbon, from its initial position to the side chain methyl group.

The succinyl-coenzyme A thus produced may now follow any of several different metabolic pathways: (1) it will most likely react with guanosine diphosphate (GDP) and inorganic phosphate (P_i), conserving the thioester bond energy in guanosine triphosphate (GTP), regenerating coenzyme A, and freeing succinate for metabolism in the tricarboxylic acid cycle; (2) it may condense with the amino acid glycine to initiate porphyrin synthesis; or (3) its thioester bond energy may be used in acylation reactions.

Patients who have pernicious anemia or other vitamin B_{12} deficiency states accumulate both propionate and methylmalonate. Increased urinary excretion of propionate and methylmalonate is used as a clinical index of vitamin B_{12} deficiency; however, increased urinary excretion of methylmalonate also occurs in children who have defective or congenitally absent *methylmalonyl-coenzyme A isomerase.*

The animal cell contains three independent systems for the biosynthesis of fatty acids—the mitochondrial, microsomal, and cytoplasmic systems—for each of which carbohydrate is the major raw material. The cytoplasmic, or palmitate, system synthesizes long-chain fatty acids de novo from the pivotal metabolite acetyl-coenzyme A; ATP supplies the energy required to form the carbon-carbon linkages of the fatty acid chain. Acetyl-coenzyme A, the starting compound of this synthetic sequence, derives mainly from the oxidative decarboxylation of pyruvate. Pyruvate, in turn, derives mainly from hexose, lactate, and glucogenic amino acids and, to a lesser extent, from malate and oxaloacetate. Ketogenic amino acids also supply acetyl-coenzyme A via acetate.

7. Carbon dioxide fixation is an essential first step in the biosynthesis of fatty acids by the cytoplasmic system of the animal cell. The biotin-requiring enzyme *acetyl-coenzyme A carboxylase* catalyzes the carboxylation of acetyl-coenzyme A, forming malonyl-coenzyme A. The reaction requires ATP, carbon dioxide, and Mn^{++}. Bicarbonate ion is the active species:

$$HCO_3^- + \text{acetyl-coenzyme A} + ATP \overset{Mn^{++}}{\rightleftarrows} \text{malonyl-coenzyme A} + ADP + P_i \quad (20)$$

Carbon dioxide fixation is also involved in the synthesis of purines—such as inosine, adenosine, guanosine, and xanthosine—and pyrimidines, such as cytidine, thymidine, and uridine. Interestingly, the mammalian cell uses relatively small amounts of preformed purines and pyrimidines; thus de novo synthesis is necessary for subsequent incorporation of these vital compounds into nucleic acids. Most purine synthesis occurs in the liver, thymus, and intestinal mucosa, where the compounds are made in discrete steps from small fragments. Purine biosynthesis begins with the formation of 5-phosphoribosyl-1-pyrophosphate, which is also a key intermediate in pyrimidine nucleotide biosynthesis.

8. Carbon-6 of the purine structure derives from carbon dioxide. The closure of formylglycineamidine ribonucleotide forms the ring structure of 5-aminoimidazole ribonucleotide, this heterocyclic compound reacts with a molecule of carbon dioxide, producing 5-aminoimidazole-4-carboxylate ribonucleotide:

$$CO_2 + \text{5-aminoimidazole ribonucleotide} \rightleftarrows \text{5-aminoimidazole-4-carboxylate ribonucleotide} \quad (21)$$

The carboxylase that catalyzes this reaction requires biotin and high concentrations of bicarbonate ion. Subsequent biochemical steps produce the parent purine compound, inosinate.

Carbon dioxide fixation is the first step in two other important metabolic processes—pyrimidine synthesis and urea formation.

9. Carbon dioxide combines with ammonia, derived from deamination of amino acids, in a reaction catalyzed by the biotin-requiring enzyme *carbamylphosphate synthetase* to form carbamylphosphate. The reaction uses two molecules of ATP and requires *N*-acetylglutamate as a cofactor, possibly for an allosteric effect. Bicarbonate ion is the reactive species:

$$HCO_3^- + NH_3 + 2\ ATP \xrightarrow{N\text{-acetylglutamate}} H_2N\text{–}CO\text{–}OPO_3H_2 + 2\ ADP + P_i \quad (22)$$

The bacillus *Escherichia coli* also contains a biotin-requiring carbamylphosphate synthetase, which catalyzes the carboxylation of ammonia and amide nitrogen.

Carbon–2 and nitrogen–1 of the pyrimidine ring structure derive simultaneously from carbon dioxide and ammonia respectively as carbamylphosphate condenses with the amino acid aspartate, forming carbamylaspartate, or ureidosuccinate; this reaction is a transcarbamylation of the amino group. Successive ring closure, oxidation, and decarboxylation produce uridylate (UMP), which is then converted to the triphosphate by kinase reactions before amination to cytidine triphosphate.

Urea is formed in a biochemical sequence called the ornithine cycle. In another transcarbamylation reaction, carbamylphosphate reacts with ornithine to form citrulline. In a subsequent step, the enzyme *arginase* hydrolyzes arginine, forming urea and regenerating ornithine; the cycle is thus completed. Carbon dioxide fixation is involved in the catabolism of an amino acid.

10. The carboxylation of β-methylcrotonyl-coenzyme A by the biotin-requiring enzyme *β-methylcrotonyl-coenzyme A carboxylase* is a step in the oxidative sequence that produces the ketone acetoacetate and acetyl-coenzyme A from the amino acid leucine:

$$CO_2 + \beta\text{-methylcrotonyl-coenzyme A} \rightarrow \beta\text{-methylglutaconyl-coenzyme A} \quad (23)$$

What is the biologic significance of carbon dioxide fixation in animal cells? Carbon dioxide fixation is one of several reserve biochemical mechanisms that assures the presence of catalytic amounts of the dicarboxylic acids required for operation of the tricarboxylic acid cycle. Another such mechanism is transamination, which yields oxaloacetate from the amino acid aspartate and α-ketoglutarate from the amino acid glutamate. Failure of the compensatory mechanisms to produce tricarboxylic acid cycle intermediates when carbohydrate deprivation accelerates fatty acid catabolism may be a factor in the development of ketosis. The reaction sequence: pyruvate → (malate) → oxaloacetate → phosphoenolpyruvate, which bypasses the pyruvate kinase reaction, may play a role in gluconeogenesis. Carbon dioxide fixation is involved in the oxidative catabolism of the amino acid L-leucine; it is also the initial step in both de novo synthesis of fatty acids by the palmitate system of animal cell cytoplasm and urea formation in the ornithine cycle of animal liver. Animal cells depend upon carbon dioxide fixation for formation of the ring structure of purines and pyrimidines and thus the synthesis of nucleic acids. Despite all these, carbon dioxide–fixing reactions in animal cells are of limited capacity and cannot substitute, even temporarily, for pulmonary carbon dioxide elimination.

Carbon dioxide transport

This section concerns the transport of carbon dioxide from its metabolic source in the metabolizing tissues to the alveolar depot in the lungs from which it is washed out into the atmospheric sink. The processes of carbon dioxide and oxygen transport are closely related.

During each minute the metabolizing cells of an average healthy man at rest consume about 250 cc (STPD) of oxygen and produce about 200 cc (STPD) of carbon dioxide, which is about 35 gm of carbon dioxide per hour, or about 288 L (STPD)/day. During heavy physical exertion the rate of carbon dioxide production and transport increases about twenty times. How is this large and variable amount of carbon dioxide, an acid anhydride, transported in the blood with relatively little pH change? Before proceeding to answer this question, we will review the relevant definitions and concepts.

Carbon dioxide production is the rate at which carbon dioxide is formed by the metabolizing cells of the body. In contrast, *carbon dioxide output* is the rate at which carbon dioxide is exhaled from the lungs. Both are expressed in milliliters (STPD) per minute. Carbon dioxide output is determined by measurement and analysis of expired gas, with correction being made for any inspired carbon dioxide. During a steady state of respiration and circulation carbon dioxide output equals carbon dioxide production, and, indeed, these two physiologic parameters must of necessity be nearly equal during any considerable period of time. However, they may differ greatly during brief unsteady, or transient, states.

Consistent with the foregoing definitions the ratio of the rate of carbon dioxide production to that of oxygen consumption is the *metabolic respiratory quotient* (R_M); in contrast, the ratio of the rate of carbon dioxide output to the rate of oxygen uptake is the *respiratory exchange ratio* (R_E). Respiratory exchange ratio is calculated from measurement and analysis of expired gas whereas metabolic respiratory quotient is assumed, inferred, or estimated by calculation of respiratory exchange ratio during a steady state at which time the two are theoretically equal:

$$\frac{\dot{V}_{CO_2}}{\dot{V}_{O_2}} = R_M = R_E \qquad (24)$$

Unfortunately, there are limitations to the steady state method for estimation of carbon dioxide production, oxygen consumption, and metabolic respiratory quotient. Even during rest a steady state is not readily achieved, and during exercise a healthy human subject is unable to achieve a steady state at oxygen uptake rates greater than 70% of maximum.

The resting metabolic respiratory quotient is about 0.82 but may vary from 0.80 to 0.85 in resting postabsorptive subjects. The theoretic value of the metabolic respiratory quotient for metabolism of carbohydrate is 1.00, fat is about 0.7, and protein is about 0.8. The value at any given time actually depends upon the particular mixture of carbohydrate, protein, and fat being metabolized. The metabolic mixture, in turn, varies with diet, physical fitness, substrate availability, and metabolic rate.

The Fick equation is a steady state relationship that can be written for carbon dioxide as well as for oxygen. It relates cardiac output rate to carbon dioxide production rate and to venoarterial blood carbon dioxide content difference:

$$\dot{Q} = \frac{\dot{V}_{CO_2}}{(C_{\bar{v}CO_2} - C_{aCO_2}) \times 10} \qquad (25)$$

where $\dot{Q}$ = cardiac output rate of blood (L/min)
$\dot{V}_{CO_2}$ = carbon dioxide production rate (ml [STPD]/min)
$C_{\bar{v}CO_2}$ = carbon dioxide concentration of mixed venous blood (vol%)
C_{aCO_2} = carbon dioxide concentration of arterial blood (vol%)

Multiplication by 10 converts milliliters (STPD) of carbon dioxide per 100 ml of blood (vol %) to milliliters (STPD) of carbon dioxide per liter of blood.

After production in metabolizing cells, carbon dioxide diffuses steadily from these intracellular regions of higher partial pressure into the surrounding interstitial fluid and from there into the systemic capillary bloodstream. The P_{CO_2} gradient determines the direction of carbon dioxide diffusion. In metabolizing tissues P_{CO_2} varies from 50 to 70 torr. It rises from 40 torr in arterial blood to 46 torr during the passage of blood through the systemic capillaries and is 40 torr in alveolar gas. The diffusion constant for carbon dioxide is much higher than that for oxygen because carbon dioxide is much more soluble in the aqueous body fluids. Thus carbon dioxide diffuses readily from venous blood across the alveolocapillary membrane into the alveolar gas despite the rather small carbon dioxide pressure gradient.

Clean atmospheric air contains only trace amounts of carbon dioxide. Thus unless carbon dioxide is added to inspired gas, the carbon dioxide present in venous blood, alveolar gas, and arterial blood originates almost entirely in metabolizing body cells. Carbon dioxide exists in the blood in several forms: (1) in physical solution as dissolved carbon dioxide and (2) bound in chemical combination as carbonic acid, bicarbonate ions, hemoglobin and other carbamates, and, in small amounts, carbonate ions. It is important to distinguish between the concentration of the various forms in which carbon dioxide exists in the blood and the contribution of each form to the process of carbon dioxide transport.

Chemical analysis of whole blood for total carbon dioxide content includes all these molecular and ionic species. The total carbon dioxide content of whole blood is measured by adding a strong acid to a sample, after which the carbon dioxide is extracted as a gas at reduced pressure.[5] Because the carbon dioxide content of erythrocytes is considerably less than that of plasma, the hematocrit is a variable affecting whole blood total carbon dioxide content. Thus it is essential to specify whether data derive from analysis of whole blood or from analysis of plasma (Table 12-3).

The transport of carbon dioxide from its many sites of production in the metabolizing tissues to

Table 12-3. Typical values for carbon dioxide and pH of human blood*

	Arterial blood			Mixed venous blood			Arteriovenous difference		
	Plasma	RBC	Whole blood	Plasma	RBC	Whole blood	Plasma	RBC	Whole blood
P_{CO_2} (torr)	40	40	40	46	46	46	6	6	6
Total carbon dioxide (vol%)	58.3			62.1			3.8		
Content (mM/L)	26.2			27.9			1.7		
Combined carbon dioxide (vol%)	55.6			59.0			3.4		
Dissolved carbon dioxide (vol%)	2.78			3.13			0.35		
Combined carbon dioxide/ dissolved carbon dioxide ratio	20.0/1			18.8/1			1.2/1		
Total carbon dioxide (vol%)			48.5			52.2			3.7
Content† (mM/L)			21.8			23.5			1.7
pH	7.40			7.376			0.024		

*Healthy sea-level–acclimatized man at rest breathing clean atmospheric air.
†Hemoglobin = 14.3 gm%.

the lungs for elimination is accomplished by circulating blood, an inhomogeneous two-phase system. Carbon dioxide diffuses first into the plasma, where transport begins, and then into the erythrocytes. From the systemic capillaries carbon dioxide is carried by the blood through the systemic venules and veins into the right heart and pulmonary arteries. Here venous bloodstreams of different origin and composition mix, flowing next into the pulmonary alveolar capillaries, where a quantity of carbon dioxide equal to the increment loaded in the tissues diffuses across the alveolocapillary membrane into the alveolar gas to be eliminated in the exhalatory stream by the intermittent washout process of alveolar ventilation.

The erythrocyte plays a key role in carbon dioxide, as well as oxygen, transport. The respiratory gas transport system within the erythrocyte involves interaction among oxygen, carbon dioxide, water, hydrogen ions, hydroxyl ions, organic phosphates, chloride ions, bicarbonate ions, the enzyme *carbonic anhydrase,* and the remarkable conjugated protein hemoglobin. The efficiency of carbon dioxide transport by the blood depends largely upon the buffer action of hemoglobin.

The hemoglobin molecule consists of globin, a protein moiety, and four heme prosthetic groups. Each heme, a ferrous-protoporphyrin complex, can combine with one molecule of oxygen. As with any protein, the buffer capacity of the hemoglobin molecule depends upon the number, kind, and degree of dissociation, or strength, of acidic and basic groups—carboxyl, amino, imidazole, and guanidino. The degree of dissociation of these groups is a function of the pH of the solution and is also affected by the physicochemical state of adjacent functional groups in the molecule. According to classic theory, imidazole groups of the amino acid histidine in the globin moiety interact with the four iron atoms in the heme prosthetic groups. As a heme combines with or releases oxygen, the dissociation of adjacent imidazole groups is affected. When heme releases oxygen, imidazole groups become less acidic (more basic); they dissociate less, tending to remove a hydrogen ion from solution to become positively charged. The reverse occurs when a heme combines with oxygen. We now know that the situation is much more complex, involving conformational changes of the hemoglobin molecule as a result of ionic bonds (salt bridges) between certain amino acid residues on both α and β chains. Thus the extent of oxygenation of hemoglobin affects its strength as an acid and its pK as a buffer. Within the physiologic pH range of 7.0 to 7.8, the imidazole groups of histidine account for most of the buffering by hemoglobin.

Conversely, as would be expected from the reciprocal nature of interaction, the acidity of the medium affects the affinity of hemoglobin for oxygen. Decreased acidity enhances imidazole dissociation, facilitating hemoglobin oxygenation, whereas increased acidity suppresses dissociation, facilitating oxygen release. By this mechanism pH changes produce characteristic modification of the shape and position of the oxyhemoglobin dissociation curve.

Thus oxygenation of hemoglobin increases its acidity, releasing hydrogen ions into the solution; deoxygenation of oxyhemoglobin decreases its acidity, removing hydrogen ions from the solution; addition of acid to a solution containing oxyhemoglobin facilitates release of oxygen; and deoxygenated hemoglobin is a considerably weaker acid than oxyhemoglobin, has greater hydrogen ion–binding capacity, and neutralizes more of the added acid.

The processes of respiratory gas exchange release significant quantities of hydrogen ions. As carbon dioxide enters blood in the systemic capillaries, hydrogen ions result from the hydration and hydroxylation of carbon dioxide. As oxygen enters blood in the lungs hydrogen ions are generated by the oxygenation of hemoglobin. Physiologically, then, hemoglobin is a hydrogen ion acceptor when carbonic acid is a hydrogen ion donor, whereas bicarbonate is a hydrogen ion acceptor when oxyhemoglobin is a hydrogen ion donor.

The following quantitative relationships exist among oxygen, carbon dioxide, hydrogen ions, and pH at the intraerythrocytic pH of 7.25. One millimole (mM) of oxyhemoglobin yields 1.88 mEq of hydrogen ions, whereas 1 mM of deoxygenated hemoglobin yields only 1.28 mEq. In systemic capillaries, the release of each millimole of oxygen (22.4 ml STPD) binds 0.6 mEq of hydrogen ions, forming 0.6 mM of bicarbonate ions from 0.6 mM of carbon dioxide (13.4 ml STPD) without pH change. The transfer of carbon dioxide without pH change is termed *isohydric* carbon dioxide transport. If the metabolic respiratory quotient were only 0.6, then 0.6 mM of carbon dioxide would be produced for each millimole of oxygen consumed. In this hypothetic circumstance, the 0.6 mM of hydrogen ions formed from 0.6 mM of carbon dioxide would be buffered completely by the concomitant deoxygenation of oxyhemoglobin. At the usual metabolic respiratory quotient of 0.82, only 0.22 mM of hydrogen ions per millimole of oxygen consumed must be buffered by other systems. Under average physiologic conditions of hemoglobin concentration, oxyhemoglobin saturation, and metabolic respiratory quotient, about 73% of the total amount of carbon dioxide that enters the blood in the tissues is buffered by the isohydric shift of hemoglobin.

Of the 3.7 ml (STPD) of carbon dioxide that enter each 100 ml of blood perfusing the metabolizing tissues of a healthy resting human subject, about 68% is transported as bicarbonate ions. For every bicarbonate ion formed 1 hydrogen ion is released, adding more than 1 mM of hydrogen ions to each liter of blood. So effective are the blood buffers that plasma pH falls only about 0.03 unit, although a hydrogen ion increment of 1 mM/L would decrease the pH of pure unbuffered water from 7.0 to 3.0. The rapid removal of hydrogen ions by deoxygenated hemoglobin permits the conversion of more than 90% of the carbon dioxide increment to bicarbonate ions. If bicarbonate ion formation were blocked by hydrogen ion accumulation, the quantity of carbon dioxide produced in the tissues could not be transported to the lungs without marked increase of tissue and venous blood P_{CO_2}.

In plasma carbon dioxide reacts with water as follows:

$$CO_2 + H_2O \rightleftarrows H_2CO_3 \rightleftarrows H^+ + HCO_3^- \qquad (26)$$

This reaction proceeds slowly; the equilibrium is far to the left. The carbon dioxide increment that occurs during passage of blood through the systemic capillaries in the metabolizing tissues drives the reaction to the right only slightly. The resulting hydrogen ion increment is buffered by the relatively weak plasma buffer systems, including proteins, and plasma pH falls slightly.

From plasma, carbon dioxide diffuses readily into erythrocytes. Within the erythrocyte the following reaction, catalyzed by the enzyme *carbonic anhydrase,* proceeds very rapidly:

$$CO_2 + OH^- + H^+ \rightleftarrows H^+ + HCO_3^- \qquad (27)$$

This reaction is facilitated by the simultaneous release of oxygen from oxyhemoglobin, forming deoxygenated hemoglobin, a stronger base. Because the end products, hydrogen ions and bicarbonate ions, are removed promptly, the fraction of the carbon dioxide increment that diffuses into the erythrocytes is much larger than that remaining in the plasma. Erythrocytes generate about 90% of the plasma bicarbonate ion increment.

Carbon dioxide can combine with any protein that bears free amino groups to form carbamates, or carbamino compounds. However, plasma proteins contain relatively few amino groups that are capable of combining with carbon dioxide under the conditions that exist in blood. The concentration of protein carbamates in plasma is low. Because the relatively small capacity of plasma proteins to form carbamino compounds is not significantly changed by the biochemical events that occur as blood flows through the systemic or pulmonary capillaries, almost none of the carbon dioxide increment loaded in the tissues is transported as plasma protein-carbamates to the lungs for elimination. Thus although some carbon dioxide reacts with plasma proteins to form carbamates, these compounds contribute very little to carbon dioxide loading in the systemic capillaries or to carbon dioxide dumping in the pulmonary capillaries.

However, a considerable fraction of the carbon dioxide produced in the tissues is bound to hemoglobin as carbamate within the erythrocyte. The reaction of carbon dioxide with hemoglobin to form carbamates is strongly affected by the concentration of 2,3-diphosphoglyceric acid (2,3-DPG), an organic phosphate present in relatively high concentration within the erythrocyte. Hemoglobin carbamate results from the combination of carbon dioxide with the four terminal α-amino groups of the globin moiety. In terms of kinetics, the rate of release of carbon dioxide from hemoglobin-carbamate is an exponential function of time ($K = 6.5/sec$, $t_{1/2} = 0.109$ sec) and is independent of extracellular pH. By contrast, the conversion of bicarbonate ions to carbon dioxide, also an exponential process, is highly dependent on extracellular pH.

Deoxygenated hemoglobin combines with more carbon dioxide to form carbamates than does oxyhemoglobin; conversely, oxygenation of hemoglobin decreases its carbon dioxide capacity. This remarkable adaptation results from a change of the number of sites available for uptake of the hydrogen ions produced by carbamate formation. Hemoglobin carbamate formation is also a function of P_{CO_2}. In vitro, the capacity of completely deoxygenated hemoglobin to form carbamino compounds is more than three times that of fully oxygenated hemoglobin. Almost all carbon dioxide transported as carbamate is combined with hemoglobin, which transports as carbamate about 20% to 25% of the total amount of carbon dioxide produced by tissue metabolism to the lungs for elimination.

The arterial blood of a healthy man resting at sea level contains from 45 to 55 ml (STPD) of carbon dioxide per 100 ml (Table 12-4). Of this total amount about 75% is in the plasma and about 25% is in the erythrocytes. The 35.6 ml in the plasma is present in three forms: (1) 1.6 ml in physical solution; (2) 34 ml as bicarbonate ions; and (3) less than 0.7 ml as plasma protein carbamates. The 12.6 ml in the erythrocytes is present in the same three forms: (1) 0.8 ml in physical solution; (2) 9.6 ml as bicarbonate ions; and (3) 2.2 ml as hemoglobin carbamates. About 90% of the total amount of carbon dioxide in arterial blood is in the form of bicarbonate ions. About 70% of the carbon dioxide that enters the blood in the metabolizing tissues is transported as bicarbonate ions, about 60% in the plasma, and the remainder within the erythrocytes.

As oxyhemoglobin releases oxygen to form deoxygenated hemoglobin, its acidic groups dissociate less and the net negative charge on the hemoglobin molecule decreases. The potassium ions that previously balanced the negatively charged acidic groups are freed to this extent to balance the concomitant bicarbonate ion increment:

$$K^+ + HbO_2^- + H^+ + HCO_3^- \rightleftarrows HCO_3^- + HHb + O_2 + K^+ \quad (28)$$

Because of the great rapidity of the carbonic anhydrase–catalyzed hydroxylation of carbon dioxide to bicarbonate ions within the erythrocyte as compared to the slow rate at which carbon dioxide is hydrated in the plasma to carbonic acid, bicarbonate ion concentration increases in the erythrocyte much faster than in the plasma. Bicarbonate ions, like carbon dioxide, diffuse readily through the erythrocyte membrane into the plasma. The law of membrane equilibrium states that the

Table 12-4. Average distribution of carbon dioxide in 100 ml of whole blood*

	Arterial blood		Mixed venous blood		Arteriovenous difference	
	ml STPD	Percent of total	ml STPD	Percent of total	ml STPD	Percent of total
Total in plasma (60 ml)	35.6	74	38.0	73	2.4	65
Dissolved carbon dioxide	1.6	3	1.8	3	0.2	5
As bicarbonate ions	34.0	71	36.2	70	2.2	60
Total in erythrocytes (40 ml)	12.6	26	13.9	27	1.3	35
Dissolved carbon dioxide	0.8	1.5	0.9	1	0.1	3
As bicarbonate ions	9.6	20	9.9	20	0.3	8
Carbamino–carbon dioxide	2.2	4.5	3.1	6	0.9	24
Total	48.2	100	51.9	100	3.7	100

*Healthy sea-level–acclimatized man at rest. Hematocrit = 40%. Modified with permission from Cantarow, A., and Schepartz, B.: Biochemistry, Philadelphia, 1967, W. B. Saunders Co.

ratios of the concentration of individual diffusible monovalent ions within the cells and in the plasma must be equal:

$$\frac{[HCO_3^-] \text{ in cells}}{[HCO_3^-] \text{ in plasma}} = \frac{[Cl^-] \text{ in cells}}{[Cl^-] \text{ in plasma}} \quad (29)$$

As bicarbonate ions diffuse from erythrocyte into plasma, chloride ions diffuse in the opposite direction. An equal number of negatively charged chloride and bicarbonate ions exchange, preserving the electric neutrality of both erythrocyte and plasma fluids. The chloride ions that diffuse from plasma into erythrocytes are balanced there by the potassium ions that were previously balanced by the intracellular bicarbonate ions. The bicarbonate ions that diffuse from erythrocytes into plasma are balanced there by the sodium ions that were previously counterbalanced by the chloride ions that shifted into the erythrocytes.

As mixed venous blood flows through the pulmonary alveolar capillaries, the loading processes just described reverse. A quantity of carbon dioxide equal to that which entered the blood in the systemic capillaries now diffuses out into the pulmonary alveoli, and the quantity of oxygen that was unloaded in the metabolizing tissues is replaced by diffusion of oxygen from the alveolar gas into the blood.

The zinc metalloenzyme *carbonic anhydrase* catalyzes the hydroxylation of carbon dioxide and, the reverse process, the dehydroxylation of the bicarbonate ion:

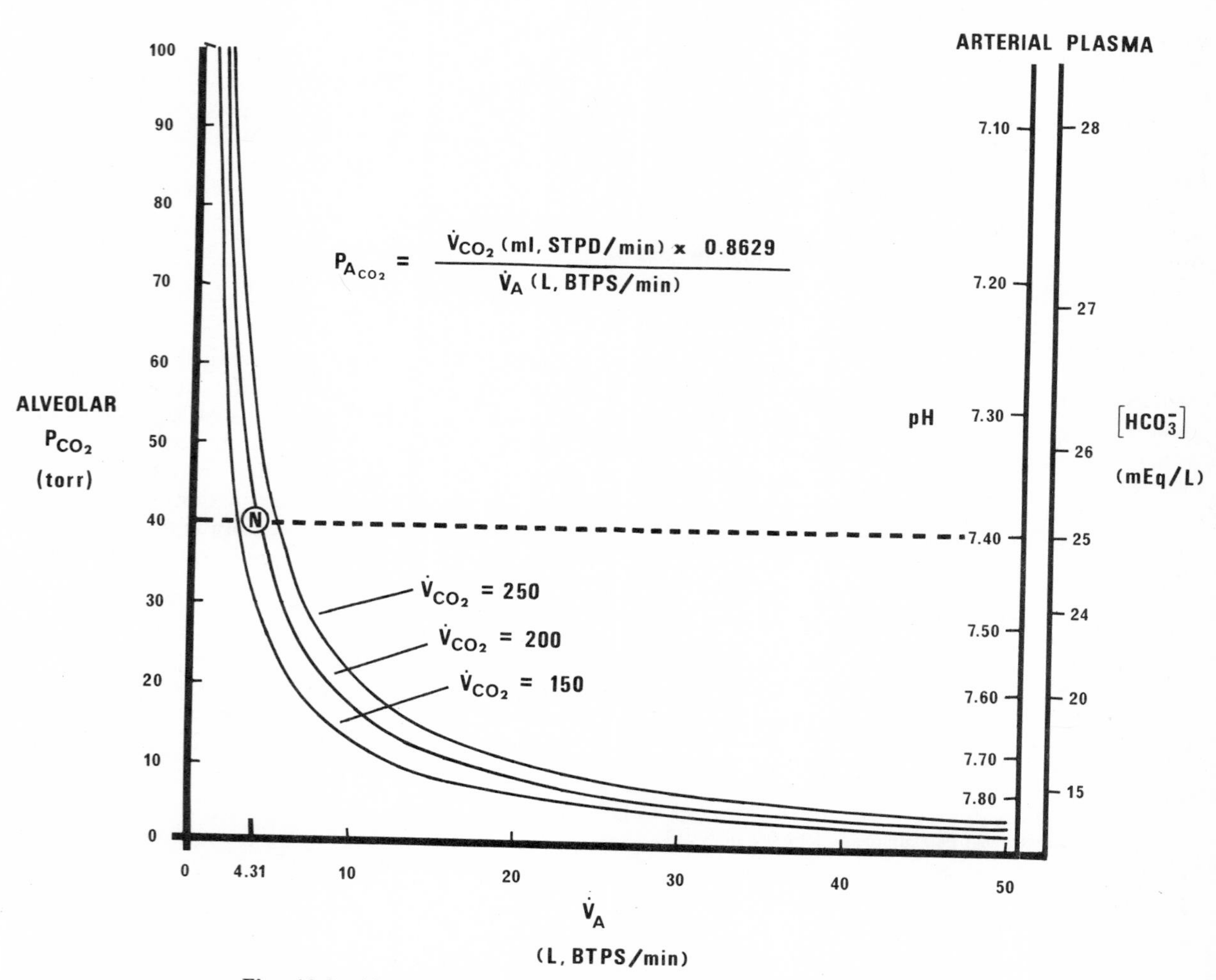

Fig. 12-2. Alveolar P_{CO_2} as a function of alveolar ventilation rate.

$$CO_2 + OH^- \underset{}{\overset{\text{carbonic anhydrase}}{\rightleftarrows}} HCO_3^- \qquad (30)$$

The enzyme contains zinc, which participates in the catalytic process and is thus essential for enzyme activity; however, substitution of cobalt ions for zinc ions yields an active enzyme.

Three distinct isoenzymes have been identified in human erythrocytes—carbonic anhydrases A, B, and C; the B isoenzyme is most abundant. Each enzyme molecule consists of a single polypeptide chain of about 265 amino acids and one zinc atom and has a molecular weight of about 30,000. The B and C isoenzymes differ both with respect to amino acid sequence and content, especially the basic amino acids. Carbonic anhydrase C has a specific activity three times greater than that of the B isoenzyme but is present in smaller amounts; the two isoenzymes thus contribute about equally to total catalytic activity within the erythrocyte.

A property widely used in research is the specific inhibition of carbonic anhydrase by sulfonamides, especially aromatic heterocyclic sulfonamides such as acetazolamide. In vitro, human erythrocyte carbonic anhydrase catalyzes not only the hydration of the carbonyl of carbon dioxide, but also the hydration of aldehydic carbonyls to the corresponding aldehydrols, the hydrolysis of 1-fluoro-2,4-dinitrobenzene, and the hydrolysis of various ester substrates, including certain naphthols, phenols, and cyclic sulfonates. As in the case of the carbon dioxide hydrase activity of the enzyme, zinc ions are also essential for its esterase activity.

Although carbonic anhydrase is widely distributed in animal tissues, its concentration may vary greatly from cell type to cell type within the same organism; for example, it is present in high concentration in neuroglial cells but practically absent from the adjacent neurons. Genetically determined variants of carbonic anhydrase exhibit subnormal carbon dioxide hydrase activity. The clinical implications of such an inherited enzyme defect remain unknown. Plants and bacteria also contain carbonic anhydrases. The carbonic anhydrase of spinach has a molecular weight of about 140,000 and occurs in the mitochondria and chloroplasts.

Carbon dioxide and pulmonary ventilation

The breathing mechanism of higher animals is a negative feedback chemostatic control system that precisely regulates its controlled variable–P_{CO_2}. Alveolar ventilation rate regulates alveolar and arterial P_{CO_2}, which in turn supplies a feedback input to the central nervous system, where it is integrated with other inputs to provide the resultant neural output that drives the breathing mechanism and ventilates the lungs. The normal range of P_{CO_2} in the body is quite narrow; the arterial blood P_{CO_2} of healthy resting subjects acclimatized at sea level averages 40 torr with a standard deviation of only $\pm$ 2 torr. In the process this remarkable regulatory mechanism also assures the body an adequate supply of oxygen.

When any air-breathing animal breathes clean atmospheric air, the quantity of carbon dioxide leaving the gas-exchanging regions of the lung (alveolar volume) in the expired gas stream per unit time is directly proportional to the product of alveolar ventilation rate and the fractional concentration of carbon dioxide in the alveolar gas.*

$$F_{A_{CO_2}}\dagger \times \dot{V}_A \text{ (L BTPS}\ddagger\text{/min)} \times 0.8262 = \dot{V}_{CO_2} \text{ (L STPD/min)} \qquad (31)$$

However,

$$F_{A_{CO_2}}\dagger = \frac{P_{A_{CO_2}}}{(P_B - P_{H_2O})} \qquad (32)$$

Substituting $P_{A_{CO_2}}$ for $F_{A_{CO_2}}$:

$$P_{A_{CO_2}} \text{ (torr)} = \frac{\dot{V}_{CO_2} \text{ (ml STPD/min)} \times 0.8629}{\dot{V}_A \text{ (L BTBS}\ddagger\text{/min)}} \qquad (33)$$

where $P_{A_{CO_2}}$ = alveolar gas carbon dioxide pressure
$\dot{V}_{CO_2}$ = carbon dioxide output rate
$\dot{V}_A$ = alveolar ventilation rate

We thus have an expression for alveolar P_{CO_2} as a function of alveolar ventilation rate and carbon dioxide output rate.

When carbon dioxide output rate is constant the equation is of the general form $X \times Y = C$. The relationship between alveolar ventilation rate and alveolar carbon dioxide pressure is now represented graphically by a rectangular hyperbola (Fig. 12-2).

Carbon dioxide production and, in a steady

*For the purpose of this particular discussion we ignore the trace amounts of carbon dioxide naturally present in clean atmospheric air and assume that all carbon dioxide that enters the lungs comes from the metabolizing cells of the body. Although mathematically convenient, the simplifying assumption that inspired $P_{CO_2} = 0$ is sometimes quite wrong or insufficiently precise and, if actually true, would represent a biologic catastrophe.

†Fractional concentration on a dry gas basis.

‡At sea level.

state, carbon dioxide output are parameters that reflect metabolic rate. Each different value of carbon dioxide output corresponds to a different curve. The three parametric curves in Fig. 12-2 bracket the range of carbon dioxide output values for healthy man at rest. The dashed horizontal line indicates the normal level of alveolar gas and arterial blood P_{CO_2}. The point Ⓝ indicates the average relationship among alveolar ventilation rate, alveolar and arterial P_{CO_2}, and arterial plasma pH and bicarbonate ion concentration for healthy sea-level–acclimatized man at rest.

$\dot{V}_A$ and P_{CO_2} are mutually reciprocal. If $\dot{V}_A$ doubles, P_{CO_2} is halved; if $\dot{V}_A$ is halved, P_{CO_2} doubles. If $\dot{V}_A$ triples, P_{CO_2} decreases to one third; if $\dot{V}_A$ decreases to one third, P_{CO_2} triples. Indeed, for any multiple of a given initial alveolar ventilation rate, the resulting alveolar P_{CO_2} is the product of the reciprocal fraction of the initial $\dot{V}_A$ $(1/\dot{V}_A)$ and the initial P_{CO_2}. P_{CO_2} is extremely sensitive to decreases of $\dot{V}_A$ from normal. Thus if $\dot{V}_A$ decreases to half its average normal resting value (from 4.31 to 2.155 L BTPS/min); P_{CO_2} rises from 40 to 80 torr!

The lung is such an efficient gas exchanger that arterial blood P_{CO_2} is almost precisely equal to alveolar gas P_{CO_2} over a wide range of physiologic and pathologic conditions:

$$\text{alveolar gas } P_{CO_2} = \text{arterial blood } P_{CO_2} \qquad (34)$$

The values for arterial plasma pH and bicarbonate ion concentration corresponding to any acutely induced arterial P_{CO_2} before renal compensation occurs are mapped on the vertical axis at the right of Fig. 12-2; these are the data of the acute steady-state, in vivo, whole-body carbon dioxide titration curve for resting sea-level–acclimatized man with normal buffer capacity. The horizontal relationship between P_{CO_2} and pH is an empirically determined set of values satisfying the Henderson-Hasselbalch equation. This curve is the result, in terms of arterial plasma pH and bicarbonate ion concentration, of acute uncompensated respiratory acidosis (hypercapnia) and alkalosis (hypocapnia). Hyperventilation promptly decreases both alveolar gas and arterial blood P_{CO_2}. Values below 25 torr are associated with psychomotor impairment, and tetany may occur.

Although the equilibration of P_{CO_2} between alveolar gas and pulmonary end-capillary blood is an important principle of respiratory physiology, the two may differ appreciably in healthy subjects under certain conditions such as rebreathing and exercise. Ventilation-perfusion abnormality may also increase arterial P_{CO_2} above the alveolar P_{CO_2} level in patients who have advanced chronic diffuse obstructive bronchopulmonary disease or cardiopulmonary venous-to-arterial shunts.

Carbon dioxide and pH homeostasis

pH homeostasis is a topic of great importance for both basic and applied biologic sciences. This significant subject is also called acid-base balance, acid-base chemistry, and neutrality regulation. Although each of these terms refers to an important aspect of the subject, we prefer the term "pH homeostasis."

The mammalian organism maintains a stable extracellular fluid hydrogen ion concentration despite continuous acid-producing metabolic processes and fluctuating environmental conditions. The homeostatic mechanism responsible for this vital stability involves a multicomponent buffer system that defends the organism against pH changes of any origin, including those that result from carbon dioxide pressure variations. Carbon dioxide is both a common cause of acid-base change and a physiologic component of an important body buffer system.

There are two important characteristics of the carbonic acid–bicarbonate buffer system: it is quantitatively the most abundant buffer system in extracellular fluid, and the pK_a of carbonic acid is 6.1. If the body were a closed system within which buffer components vary dependently at the physiologic extracellular fluid pH of 7.4, this system would be far from ideal; however, the carbonic acid–bicarbonate buffer system is an open system. Bicarbonate ion concentration is regulated primarily by the kidney, whereas carbonic acid concentration is a function of P_{CO_2}, which is regulated by the lung. A primary acid-base disturbance may involve either the carbonic acid component or the bicarbonate ion component. Thus each component of this buffer system is regulated separately and either is subject to primary disturbance.

Extracellular fluid bicarbonate ion concentration is regulated primarily by the kidney. Bicarbonate ions filter freely through the glomerulus, after which about 90% are reabsorbed in the proximal renal tubule. Bicarbonate ion reabsorption is coupled to active hydrogen ion secretion; as each hydrogen ion is produced for secretion, a bicarbonate ion is also produced and returned to the plasma. Bicarbonate ions that escape reabsorption in the proximal renal tubule are reabsorbed by a similar mechanism in the distal renal tubule. Thus urine is normally almost free of bicarbonate ions.

Ordinarily, the kidney reabsorbs more than 4,000 mEq of bicarbonate ions per day. However, to maintain acid-base balance, the kidney must also generate about 60 to 80 mEq of new bicarbonate ions per day. This newly generated bicarbonate ion replaces that used to buffer the nonvolatile acids that result from metabolism of dietary substrates, and the generation is associated with renal excretion of 60 to 80 mEq of hydrogen ions as ammonium ions (NH_4^+) and titratable acids. Buffering by ammonia (NH_3) and other buffers, chiefly phosphates, permits excretion of large quantities of hydrogen ions without decreasing urine pH below 4.5, the lowest urine pH that the healthy kidney can produce.

If more than 60 to 80 mEq of hydrogen ions per day must be eliminated, renal ammonium ion excretion increases and, to a lesser extent, so do titratable acids. Although ammonia production is quantitatively the most important mechanism of hydrogen ion excretion in response to acidosis, there is a lag of 48 to 72 hours before ammonium ion excretion becomes maximum. Ammonia is produced by deamination of glutamine and other amino acids. Increasing ammonia production involves enzyme induction.

In healthy man with a plasma bicarbonate ion concentration of 24 mEq/L, almost all bicarbonate ions that filter through the glomerulus are reabsorbed. If plasma bicarbonate ion concentration is increased by intravenous infusion of a bicarbonate ion–containing solution, the maximum renal tubular reabsorptive capacity for bicarbonate ions (T_m, HCO_3^-) is exceeded and all excess bicarbonate ions—normally that amount in excess of 24 mEq/L—are excreted in the urine. The maximum rate of renal tubular bicarbonate ion reabsorption is a regulated physiologic variable.

The kidney regulates extracellular fluid volume by controlling sodium ion reabsorption. Sodium ions are mainly reabsorbed in the proximal renal tubule as are chloride and bicarbonate ions. Chloride ion reabsorption is passive whereas bicarbonate ion reabsorption is coupled to active hydrogen ion secretion. Contraction of extracellular fluid volume increases the tubular maximum for bicarbonate ions; expansion of extracellular fluid volume tends to decrease the rate of bicarbonate ion reabsorption. Such an increase or decrease of the tubular maximum for bicarbonate ions is probably related to parallel changes of the rate of sodium ion reabsorption in the proximal renal tubule.

The rate of renal bicarbonate ion reabsorption is a linear function of blood P_{CO_2}. This relationship is independent of extracellular fluid pH. Hypercapnia increases the tubular maximum for bicarbonate ions, permitting a gradual increase of plasma bicarbonate ion concentration; conversely, hypocapnia decreases the tubular maximum for bicarbonate ions, causing urinary bicarbonate ion excretion and decreasing plasma bicarbonate ion concentration. This important mechanism is probably mediated by changes of intracellular pH. Carbon dioxide diffuses readily across all cell membranes. Increased P_{CO_2} within renal tubular cells makes more hydrogen ions available for secretion into the tubular lumen, and this increased hydrogen ion secretion is associated with an increased rate of bicarbonate ion reabsorption.

The chloride ion is the chief anion reabsorbed with the sodium ion in the proximal renal tubule. In subjects depleted of chloride ions (hypochloremia), the tubular maximum for bicarbonate ions increases and more is actively reabsorbed with sodium ions, a response that tends to maintain extracellular fluid volume. This mechanism may explain the delayed recovery of patients who have chronic hypercapnia complicated by hypochloremia. Hypochloremia stimulates sodium ion retention and the kidney is unable to excrete the excess bicarbonate ions.

The tubular maximum for bicarbonate ions is related inversely to total body potassium ion content. Potassium ion depletion increases, whereas potassium ion excess decreases, the rate of bicarbonate ion reabsorption. The mechanism by which potassium ions affect the tubular maximum for bicarbonate ions remains unknown. Adrenocorticosteroid administration increases plasma bicarbonate ion concentration. Patients who have an increased rate of endogenous mineralocorticosteroid production tend to develop metabolic alkalosis. Although this alkalosis may be secondary to the potassium ion depletion associated with adrenocorticosteroid administration or hypermineralocorticosteroidism, a direct renal action of corticosteroids has not been excluded. The physiologic role of adrenal corticosteroids in the regulation of extracellular fluid bicarbonate ion concentration remains unknown.

The stimulus for compensation in a primary acid-base disturbance is the change of pH produced by the initial shift of P_{CO_2} or bicarbonate ion concentration. Compensatory processes are secondary responses by buffers, lungs, and kidneys, which return the ratio, and thus pH, toward normal. Hyperventilation is a major compensatory mechanism in metabolic acidosis and hypoventilation occurs in metabolic alkalosis.

Henderson-Hasselbalch buffer equation

A buffer system is an aqueous solution of a weak acid (proton donor) and its conjugate base (proton acceptor). Such a system resists (buffers) changes of hydrogen ion concentration when other acids or bases are added to the solution. The dissociation constant* of the acid is an important characteristic of any buffer system:

$$K_{HB} = \frac{[H^+] \times [B^-]}{[HB]} \qquad (35)$$

where K_{HB} = dissociation constant of the acid HB
$[H^+]$ = hydrogen ion concentration
$[B^-]$ = concentration of the conjugate base
$[HB]$ = concentration of the undissociated acid

The value of K for a strong acid is larger than that for a weak acid.

The Henderson-Hasselbalch equation is derived algebraically from equation 35, which is a statement of the law of mass action. Rearranging:

$$[H^+] = \frac{K_{HB} \times [HB]}{[B^-]} \qquad (36)$$

Taking the logarithm to the base 10 of both sides:

$$\log [H^+] = \log K_{HB} + \log [HB] - \log [B^-] \qquad (37)$$

Multiplying both sides by −1:

$$-\log [H^+] = -\log K_{HB} - \log [HB] + \log [B^-] \qquad (38)$$

However, by definition:

$$-\log [H^+] = pH \quad \text{and} \quad -\log K_{HB} = pK_{HB} \qquad (39)$$

Substituting:

$$pH = pK_{HB} - \log [HB] + \log [B^-] \qquad (40)$$

$$pH = pK_{HB} + \log_{10} \frac{[B^-]}{[HB]} \qquad (41)$$

This equation defines the relationship of pH to the concentrations of the acidic and basic components of a buffer system and to pK_{HB}, the negative logarithm of the dissociation constant, which is a measure of the strength of the weak acid.

For the acid-base reaction $HB \rightleftarrows H^+ + B^-$, the precise relationship of pH to pK'_{HB} is as follows:

$$pH = pK'_{HB} + \log \frac{[B^-]}{[HB]} = pK_{HB} + \log \frac{[B^-]}{[HB]} + \log \frac{\lambda_{B^-}}{\lambda_{HB}} \qquad (42)$$

where HB = undissociated acid
B^- = conjugate base
λ_{B^-} and λ_{HB} = activity coefficients of B^- and HB, respectively

The body contains many physiologic buffer systems, including plasma proteins, hemoglobin, the bicarbonate–carbonic acid system, and cell proteins. The Henderson-Hasselbalch equation can be written separately for each of these buffer systems, but a hydrogen ion concentration* change in the intact organism is reflected in corresponding changes in all body buffer systems. Acid-base physiology is generally viewed in terms of the carbonic acid–bicarbonate buffer system for the following reasons: (1) both buffer components are present in high concentration in extracellular fluid; (2) the body is an open system with respect to this buffer; (3) both buffer components are actively regulated by lungs and kidneys; and (4) both buffer components are products of metabolism.

The Henderson-Hasselbalch equation is thus the fundamental relationship of acid-base physiology. This equilibrium equation is used to express acid-base status in terms of the important carbonic acid–bicarbonate blood buffer system. Concerning ourselves only with the first† dissociation of carbonic acid, which forms bicarbonate ions and hydrogen ions, substituting $0.0301 \times P_{CO_2}$ for the concentration of undissociated carbonic acid and using the value of 6.10 for pK_1', we have the commonly used form of the equation:

$$pH = 6.10 + \log_{10} \frac{[HCO_3^-]_{plasma}}{0.0301 \times P_{CO_2}} \qquad (43)$$

At any given time the Henderson-Hasselbalch equation contains two independent variables (two degrees of freedom), the third variable depending upon the other two. Physiologically, pH is a totally dependent variable because it can only be altered by means of changing P_{CO_2} or bicarbonate ion concentration. However, from time to time, therapeutically or experimentally, any pair of the three variables can vary independently. In this sense the Henderson-Hasselbalch equation contains (at various times) three independent continuous variables.

The pK_{HB} of an acid is the pH at which that acid is 50% ionized, because the ratio of the concentrations of base to acid at this pH is 1 and the logarithm of 1 is 0. When this ratio is 1, ad-

*Also called ionization constant.

*Blood pH means plasma pH.
†Hence, pK subscript 1.

dition of a given amount of base or acid produces less change of the ratio, and thus pH, than if the ratio is not 1. Thus a buffer solution is more effective the closer the pH is to the pK of its acid. Compare the effect of adding 10 mEq of strong acid to the buffer system when:

$$\frac{[B^-]}{[HB]} = \frac{50}{50} = 1 \qquad (44)$$

and when:

$$\frac{[B^-]}{[HB]} = \frac{90}{10} = 9 \qquad (45)$$

In the first case the ratio changes from a value of 1 to 40/60, or about 0.67, a decrease of about 33%; in the second case the ratio changes from 9 to 80/20, or 4, a decrease of about 56%.

The pK_1' of carbonic acid in body fluids is quite constant*; hence, the equation implies that the pH at any given time is determined by the numerical value of the *ratio* of bicarbonate ion concentration to carbonic acid concentration, not the *concentrations per se*. The initial insult of a primary acid-base disturbance changes the normal concentration of either the numerator (metabolic component) or the denominator (respiratory component) of the ratio. In contrast, buffer capacity, the capacity to resist pH changes, does depend upon the absolute concentrations of the acidic and basic components of the buffer system; the greater the initial concentrations, the less a given change of either buffer component will affect the ratio, and thus the pH.

The Henderson-Hasselbalch buffer equation is the most useful general relationship of acid-base physiology. As an *equilibrium* expression, it does not apply to the dynamic transient states of acid-base processes. Body responses to exposure to carbon dioxide–containing atmospheres and fluctuations of P_{CO_2} exemplify the concept of pH homeostasis. Full characterization of these responses includes description of their time-course, determination of the degree of completeness of compensation, and discrimination of the primary process of acid-base disturbance from both prompt and delayed secondary responses. Primary processes often temporally overlap and are simultaneous with the processes of recovery (response to transient stress) or compensation (response to constant stress).

*The value of pK′ is slightly affected by temperature change and other factors that modify the thermodynamic *activities* of body fluid constituents.

The Henderson-Hasselbalch buffer equation is used to calculate the P_{CO_2} of blood and other body fluids from pH and carbon dioxide content data. Carbon dioxide pressure can be estimated with a carbon dioxide electrode. Within this device bicarbonate ion concentration is known, and carbon dioxide diffuses from a fluid test sample across a thin plastic membrane into a glass electrode system that measures pH. The Henderson-Hasselbalch equation is then used to calculate P_{CO_2}.

In another method the Henderson-Hasselbalch relationship is used indirectly to estimate blood P_{CO_2} from pH measurements. A blood sample is divided into three aliquots. Two aliquots are equilibrated in tonometers with gas streams of known P_{CO_2}, one higher and the other lower than the P_{CO_2} of the original blood sample. The pH and P_{CO_2} values of these two aliquots are then plotted as points on a log-linear graph and a straight line is drawn through them; this line approximates the carbon dioxide titration curve of the particular blood sample. The pH of the original blood sample is then plotted on the straight line and its P_{CO_2} determined by interpolation on the log P_{CO_2} axis.

A three-dimensional model is necessary for simultaneous visual display of three continuous variables. Although it can be said that a three-dimensional surface is inherently no more *accurate* than the projection of that surface onto a two-dimensional coordinate system, such a projection results in a complex distortion of the three-variable relationship and is absolutely incapable of showing the true *continuous* nature of the third variable. Three-dimensional visual presentation of a three-variable problem is conceptually simpler and thus more easily comprehended.

If we assign suitable scale values within the physiologic ranges of the three variables—pH, P_{CO_2}, and bicarbonate ion concentration—then three respective rectangular Cartesian axes form a relevant acid-base cube. For any constant temperature, the Henderson-Hasselbalch equation can now be plotted, or represented geometrically, in this triaxial coordinate system, inscribing a three-dimensional surface within the physiologically relevant cube. This representation, the Acid-Base Surface* model, is thus a three-dimensional plot of the Henderson-Hasselbalch equation in a rectangular Cartesian coordinate system. The chem-

*The Acid-Base Surface model is available from Acid-Base, Inc., 2045 Franklin St., Denver, Colo. 80205.

istry and solid analytic geometry of the Henderson-Hasselbalch buffer equation provide the reference system in terms of which we can observe and analyze the dynamics of acid-base physiology. The Acid-Base Surface model is also a geometric solution of the Henderson-Hasselbalch equation. Plotting the measured or assigned values for any two of the three variables in the surface automatically implies the value of the third. The normal point, Ⓝ, for arterial blood of healthy resting man at sea level is indicated in the surface at pH = 7.40, P_{CO_2} = 40 torr, and bicarbonate ion concentration = 24 mM/L. Other reference points for blood, including the normal blood buffer curve and the four areas of primary acid-base disturbance, are also shown. The Acid-Base Surface model is available as a visual model for analysis of acid-base information.[7, 8]

Carbon dioxide titration curve

Just as the amount of oxygen in blood is a function of P_{O_2}, so the amount of carbon dioxide in blood is a function of P_{CO_2}. Within the physiologic range the average solubility coefficient (C_s) for carbon dioxide in blood at 37° C is 0.063 vol%/torr. After carbon dioxide dissolves in blood, some remains in physical solution; some reacts with proteins, mostly hemoglobin, forming carbamates; some reacts slowly with plasma water, forming carbonic acid, which then dissociates to form bicarbonate ions; some reacts with hydroxyl ions in the presence of *carbonic anhydrase* within erythrocytes, forming bicarbonate ions; and some bicarbonate ions of either origin dissociate to form a relatively small number of carbonate ions. All these forms and products of carbon dioxide in the blood are included in the blanket term *carbon dioxide content.*

As P_{CO_2} changes, both blood carbon dioxide content and pH change. Thus every respiratory acid-base change is a three-variable problem. The Henderson-Hasselbalch buffer equation implies that any given value of P_{CO_2} corresponds to a whole set of pH–bicarbonate ion concentration coordinate value combinations, constituting a P_{CO_2} isobar. The effect of changing P_{CO_2} on pH and bicarbonate ion concentration is an empiric biologic fact. Physiologically, blood buffer capacity (the slope of the carbon dioxide titration curve) imposes a second constraint upon the three-variable relationship, defining the particular manner in which plasma pH and bicarbonate ion concentration values change as blood is titrated with carbon dioxide. Such a titration curve, showing the effects of increasing or decreasing P_{CO_2} on plasma pH and bicarbonate ion concentration, is called the *carbon dioxide titration curve, carbon dioxide dissociation curve,* or *blood buffer curve;* it is a plot of the carbon dioxide content and pH of blood as a function of blood P_{CO_2}. The departure from linearity, or curvature, of such curves is caused by the chemical reactions of carbon dioxide with water and blood buffers and reflects the buffer capacity of the blood. In vitro carbon dioxide titration curves are determined experimentally by analysis of the carbon dioxide content of liquid shed blood that has been equilibrated in a tonometer with gas containing carbon dioxide at various predetermined partial pressures.

Changing hemoglobin concentration—polycythemia or anemia—change blood buffer capacity and thus the *slope* of the carbon dioxide titration curve. A simple increase of hemoglobin concentration steepens the curve, which still passes, however, through the normal arterial blood point. Anemia decreases the slope. Either change may be visualized as a rotation of the carbon dioxide titration curve about its pivotal point of origin, the normal arterial blood point. Because of hemoglobin content, the buffer capacity of whole blood is always greater than that of its plasma; although having a common initial acid-base point, a buffer curve resulting from titration of whole blood with carbon dioxide is always steeper than that for its plasma.

Varying P_{O_2} and thus oxyhemoglobin saturation shifts the carbon dioxide titration curve. This effect of P_{O_2} on the position of the carbon dioxide titration curve is called the *Haldane shift*. Deoxygenated hemoglobin is a somewhat weaker acid than oxyhemoglobin. Deoxygenation of blood, a change of greater clinical importance than the reverse, shifts any given acid-base point in the carbon dioxide titration curve to the right and higher, adding a slight metabolic, or nonrespiratory, alkalosis to any initial condition. The new curve thus produced is parallel to the normal curve. In vitro, increasing the P_{O_2} of an equilibrating gas mixture shifts the carbon dioxide titration curve to the right; conversely, decreasing P_{O_2} shifts the curve to the left. In vivo, the curve of the more oxygenated arterial blood is somewhat to the right of that for the less oxygenated venous blood; the physiologic loading and unloading of oxygen facilitates the opposite process for carbon dioxide. About 70% of the physiologic Haldane shift results from the increased affinity for carbon dioxide and correspondingly increased carbamate formation of deoxygenated hemoglobin.

We can determine the carbon dioxide titration

curve for the whole body (in vivo curve) as well as that for liquid shed blood (in vitro curve). In vivo whole-body carbon dioxide titration curves reflect significantly greater buffer capacity than in vitro curves as a result of the physiologic processes and buffer systems that contribute to the in vivo buffer capacity of the intact organism. It is important to distinguish *acute* from *chronic* and *steady state* from *transient state* in vivo whole-body carbon dioxide titration curves.

Hypercapnic respiratory acidosis results from exposure to carbon dioxide–containing atmospheres or from lung failure (pulmonary insufficiency); hypocapnic respiratory alkalosis results from any of the many causes of alveolar hyperventilation. The acid-base effects of fever illustrate well the complex interaction of respiratory and metabolic variables. Fever increases the rate of metabolic carbon dioxide production and increases pulmonary ventilation; a temperature rise increases the P_{CO_2} within any closed carbon dioxide–containing system. The net result of these three simultaneous partial effects is decreased alveolar and arterial P_{CO_2}, or respiratory alkalosis.

Body stores of carbon dioxide

The body stores of a gas are the reservoirs of that gas existing at any given time anywhere within the body—in gas-containing spaces, such as lungs, paranasal sinuses, stomach, and intestines, and dissolved in or combined with body constituents. For example, about 300 ml (STPD) of oxygen are present in body spaces, whereas about 1 L of oxygen is combined with hemoglobin within the cardiovascular system. Of the total body carbon dioxide stores of approximately 120 L (STPD), only about 100 ml is found in gas-containing spaces.

The largest body gas stores in a man breathing air at sea level are those of carbon dioxide (120 L), nitrogen (4 L), and oxygen (1.5 L). The oxygen stores readjust more frequently and more rapidly than those of carbon dioxide. Total body stores of nitrogen, a gas that is neither consumed nor produced in the body, change appreciably only in response to certain environmental changes. Thus the body stores of oxygen and nitrogen are usually in a nearly steady state.

The body stores of carbon dioxide are the largest and the most complex of the normal body gas reservoirs. The problem of defining the carbon dioxide stores is complicated by the reversible hydration and hydroxylation reactions of carbon dioxide, forming carbonic acid, bicarbonate ions, and carbonate ions in body fluids and in bone. Because carbon dioxide stores are affected by a wide variety of physiologic and pathologic processes and because P_{CO_2} has a potent effect on carbon dioxide content, body carbon dioxide stores are continually in a process of relatively slow readjustment caused by fluctuations of alveolar P_{CO_2}.

Body fluids, which comprise about 70% of body mass, normally hold about 50 ml (STPD) of carbon dioxide per 100 ml of fluid, which amounts to a carbon dioxide concentration of about 50 vol%. Human bone usually contains more than 100 vol% of carbon dioxide in the form of calcium carbonate, a contaminant of the fundamental hydroxyapatite mineral structure. The entire body of a 70-kgm man contains about 120 L (STPD) of carbon dioxide, about 20 L of which are found in the blood and soft tissues. The total carbon dioxide stored in the body is equivalent to the amount of carbon dioxide metabolically produced during a 10-hour period of rest.

Anything that increases alveolar P_{CO_2} increases body carbon dioxide stores. This includes alveolar hypoventilation, breath-holding, breathing through an external dead space, rebreathing in a closed system, and exposure to carbon dioxide-containing atmospheres. During breath-holding no carbon dioxide is eliminated and the carbon dioxide content of the body increases by an amount precisely equal to its production during that period of time. Sustained alveolar hyperventilation decreases alveolar P_{CO_2}, depleting body carbon dioxide stores.

The level of body carbon dioxide reservoirs is strongly affected by the pulmonary ventilation rate. When alveolar ventilation is eliminating carbon dioxide at precisely the same rate at which it is being produced, carbon dioxide output equals its metabolic production. Such a steady state also implies that carbon dioxide production rate is constant. The steady state relationships between alveolar gas P_{CO_2} and alveolar ventilation rate represented by the rectangular hyperbola in Fig. 12-2 are precisely valid only when (1) the subject is breathing air at sea level; (2) a steady state exists; (3) carbon dioxide production is constant and equals carbon dioxide output, which is also constant; and (4) alveolar ventilation rate, cardiac output, and body carbon dioxide reservoir levels are all constant.

Any variation of alveolar ventilation rate changes alveolar and arterial P_{CO_2}, producing a transient, or unsteady, state during which the levels of body carbon dioxide stores change. If the alveolar ventilation rate is greater than that required to eliminate current metabolic carbon dioxide production, body carbon dioxide reservoirs decrease; during alveolar hypoventilation the metabolically produced carbon dioxide is stored, filling body reservoirs. After the abrupt onset of a constant level of alveolar hyperventilation, carbon dioxide output increases rapidly. After a few minutes it begins a gradual decrease, returning toward the control value of current metabolic carbon dioxide production rate to achieve a new steady state. The difference between carbon dioxide output and carbon dioxide production during a period of alveolar hyper- or hypoventilation is the extent of depletion or accretion of body carbon dioxide stores. During such a transient state it is impossible to measure the rate of metabolic carbon dioxide production.

When alveolar ventilation rate changes abruptly from one level to another, the resulting readjustment of alveolar gas P_{CO_2} is nearly an exponential function of time. The rate constant of this exponential relationship is determined chiefly by the alveolar gas volume and the alveolar ventilation rate. After such an abrupt change of alveolar ventilation rate, alveolar P_{CO_2} changes about 50% during each 5-minute interval. The process of readjustment of body oxygen stores after an abrupt change of alveolar gas P_{O_2} is also approximately exponential. However, the rate of readjustment of body oxygen stores during alveolar hyper- or hypoventilation while breathing air is about eight times as fast as that for carbon dioxide.

Because carbon dioxide is produced in the metabolizing tissues and transported to the lungs by the systemic venous circulation, venous blood P_{CO_2} normally exceeds arterial blood P_{CO_2}. The magnitude of this arteriovenous P_{CO_2} difference depends not only upon the rates of carbon dioxide production and output but also upon the rate of cardiac output. If cardiac output increases, a given quantity of carbon dioxide produced by the metabolizing tissues is distributed throughout a larger volume of blood, decreasing venous blood P_{CO_2} and peripheral carbon dioxide stores. Conversely, if cardiac output decreases, the same number of carbon dioxide molecules are now transported in a smaller volume of blood, increasing venous blood P_{CO_2} and peripheral carbon dioxide stores. Thus variations of cardiac output rate would affect systemic venous blood P_{CO_2} even if carbon dioxide production rate, alveolar ventilation rate, alveolar gas P_{CO_2}, and arterial blood P_{CO_2} all remained constant.

Whether venous blood P_{CO_2} equals tissue P_{CO_2} remains unknown. Although no difference between the two has been found during the absorption of subcutaneous gas pockets, studies of lymph suggest that a tissue-to-venous blood P_{CO_2} difference does exist. In either case, changes of venous blood P_{CO_2} affect tissue P_{CO_2}. Thus variations of cardiac output continually readjust carbon dioxide stores everywhere in the body except in the alveolar gas and arterial blood. We are unable to measure accurately the extent of these perfusion-induced fluctuations of body carbon dioxide stores and can only calculate the net effect from indirect evidence.

If alveolar ventilation rate is maintained artificially constant, an abrupt increase of cardiac output rate would produce the following changes: within 1 minute, as peripheral reservoirs unloaded carbon dioxide, alveolar P_{CO_2} would begin a gradual increase, peaking after a few minutes, and then, as the rate of carbon dioxide loss from the peripheral stores decreases, alveolar P_{CO_2} would decrease, returning slowly to the control value. If cardiac output rate decreases abruptly while alveolar ventilation rate is held artificially constant, the resulting pattern of change of alveolar P_{CO_2} is the opposite of that just described. Initially, the changes that occur in peripheral carbon dioxide stores are partly offset by an opposite change in the alveolar gas–arterial blood compartment. Subsequently, the direction of change in the latter compartment reverses and proceeds in the same direction as the readjustments occurring in the peripheral carbon dioxide stores. In this artificial circumstance, the alveolar gas–arterial blood carbon dioxide reservoir is a buffer interposed between the peripheral carbon dioxide stores and the atmosphere. However, if alveolar ventilation rate is not held artificially constant but is allowed to change in response to the rate at which carbon dioxide is unloaded in the lungs, the changes that result from the initial transient increase or decrease of alveolar P_{CO_2} are minimized or even absent.

The body does not behave as a large simple tonometer equilibrated with carbon dioxide at alveolar carbon dioxide pressure. On the contrary, alveolar P_{CO_2} prevails only in the alveolar gas and arterial blood. Throughout the rest of the body P_{CO_2} is variably higher and the carbon dioxide content of each different tissue is a function of the P_{CO_2} to which it is exposed. Physiologically, the P_{CO_2} of a given tissue is a function of its carbon dioxide production rate, blood perfusion rate, and certain other specific local tissue factors. Experimentally, a plot of the amount of carbon dioxide stored in a tissue as a function of P_{CO_2} is the carbon dioxide dissociation curve of that particular tissue. The best known of these is that of blood. The capacity of a given tissue to store additional carbon dioxide as P_{CO_2} increases is its *storage capacity* and is defined as the slope of the carbon dioxide dissociation curve. Storage capacity is expressed as the volume (STPD) of carbon dioxide per volume of tissue, or per 100 volumes of tissue, per unit of carbon dioxide pressure in torr.

Only a small fraction of the total carbon dioxide in a tissue is present as dissolved gas. This quantity is a simple function of the P_{CO_2} and of the solubility of carbon dioxide in the particular tissue. Most of the carbon dioxide in a tissue is chemically bound, the exact amount depending upon the capacity of the given tissue to buffer carbon dioxide, forming bicarbonate or carbonate ions.

Three factors affect the storage capacity of an organ and thus the amount of carbon dioxide stored at a given P_{CO_2}: solubility, buffer capacity, and organ volume. These factors seldom change and when they do they usually change very slowly. Most available data for changes of stored carbon dioxide were obtained by varying only P_{CO_2} while the storage capacity of the tissue or organ remained constant.

The storage capacity of various tissues differs greatly. For example, fat, which does not bind carbon dioxide, stores only dissolved carbon dioxide, the fat solubility of which differs from its solubility in body fluids. The carbon dioxide storage capacity of bone is far greater than that of any other tissue. Skeletal muscle stores nearly 120 ml (STPD) of carbon dioxide per torr P_{CO_2}, whereas the myocardium stores only 0.7 ml/torr P_{CO_2}.

The actual participation of the various body carbon dioxide reservoirs in the dynamic processes of change is limited by their respective regional blood perfusion, and thus carbon dioxide transport, rates. Although the carbon dioxide storage capacity of bone is very great, it is hardly involved at all in the acute phase of carbon dioxide stores change because of its relatively low blood perfusion rate and because of the time required to deposit carbon dioxide as carbonate in bone. For example, during the first minute of breath-holding only alveolar gas and blood carbon dioxide stores are affected. If alveolar hypercapnia continues, skeletal muscle and visceral stores change, and if the inciting change of alveolar P_{CO_2} persists for several weeks, bone stores of carbon dioxide may increase markedly. Thus the change of body carbon dioxide stores is a complex function of P_{CO_2} and time.

By far the greatest carbon dioxide pressure drop between the sites of carbon dioxide production in metabolizing tissues and the atmosphere is the decrease that occurs between alveolar gas and the external environment; most of the resistance to carbon dioxide transfer resides in the process of alveolar ventilation. Because resistance between the two carbon dioxide reservoirs is low, changes of alveolar and arterial P_{CO_2} are quickly and accurately reflected in changes of mixed (mean) venous blood P_{CO_2}, the arteriovenous P_{CO_2} difference remaining almost constant. When alveolar P_{CO_2} changes, the resulting change of carbon dioxide content of the alveolar gas–arterial blood compartment is divided about evenly between alveolar gas and arterial blood.

Several different experimental methods have been used to study changes of body carbon dioxide stores. The acute carbon dioxide storage capacity of the body has been calculated as the ratio of carbon dioxide retained to the measured change of alveolar P_{CO_2} before and at the end of a breath-holding period. Alveolar P_{CO_2} has been varied by changing either alveolar ventilation rate or by adding carbon dioxide to the inspired gas mixture; the change of total body carbon dioxide stores is then computed by subtracting the metabolically produced carbon dioxide, whose production rate is assumed to remain constant, from the carbon dioxide output during the experimental period. A positive value indicates a loss from carbon dioxide stores, whereas a negative value indicates that body stores took up part of the metabolically produced carbon dioxide. Dividing the change of total body carbon dioxide stores by the corresponding change of alveolar P_{CO_2} yields an estimate of whole-body carbon dioxide storage capacity. We can also determine the carbon dioxide dissociation curve for the whole body or certain parts of experimental animals by analysis for carbon dioxide content at various time intervals after the onset of a known carbon dioxide exposure. Predictably, such analyses show that whole-body carbon dioxide content increases as the duration of exposure.

With respect to carbon dioxide storage, the body actually behaves as a multicompartment system, each component of which is a reservoir comprised of one or more body regions. Each parallel component is governed by its own variable P_{CO_2}, each has a certain aggregate carbon dioxide content, each has its own blood perfusion rate, each has its own carbon dioxide production rate, and each is described by its own time-dependent carbon dioxide dissociation curve, implying a certain carbon dioxide storage capacity. Analysis of such a multicompartment system is somewhat complex for a steady state and even more so for the dynamics of a transient state. Both states require data that remain unavailable.

For simplicity, we approximate the real situation by means of a two-compartment model. One compartment is the alveolar gas–arterial blood carbon dioxide reservoir governed by alveolar gas P_{CO_2}; the other is the venous blood–body tissue carbon dioxide reservoir governed by mixed venous blood P_{CO_2}. In this model the time-course of readjustment of the carbon dioxide stores of the two compartments is the time-course of readjustment of the respective P_{CO_2}'s that govern them.

Exposure to carbon dioxide–containing atmospheres increases alveolar P_{CO_2}, causes retention of metabolically produced carbon dioxide, and increases body carbon dioxide stores. Consider the problem of estimating the amount of carbon dioxide that is stored in the body carbon dioxide reservoirs as a result of such an exposure. Although we would like to know how total body carbon dioxide stores and distribution vary in man as a function of both P_{CO_2} and time, it is obviously impossible to determine the total amount of carbon dioxide in the human body under any conditions. However, it is possible to estimate the *change* of total body carbon dioxide stores that results from the transition from one steady state to another.

The equation used to calculate the change of total body carbon dioxide stores in the human subject is based in part on experimental data but also involves certain assumptions. Alveolar gas P_{CO_2}, which is a function of both carbon dioxide output rate and alveolar ventilation rate, governs the carbon dioxide content of alveolar gas and approximately 25% of the total blood volume—blood in the pulmonary veins, left atrium, left ventricle, and arterial segment of the systemic circulation. We assume that the carbon dioxide dissociation curve of blood is a straight line between a P_{CO_2} of 30 and 80 torr with a slope of 4.7 ml (STPD) of carbon dioxide per liter of blood per torr P_{CO_2}. Because tissues elsewhere in the body are subject to various higher P_{CO_2}'s and equilibrate at various rates, it is much more difficult to estimate the changes that occur in these other body carbon dioxide stores. We assume that the P_{CO_2} of mixed venous blood prevails throughout the entire remainder of the body—the venous blood–body tissue reservoir. The P_{CO_2} of mixed venous blood is a function of alveolar P_{CO_2}, cardiac output, and metabolic respiratory quotient. The slope of the whole-body carbon dioxide dissociation curve for experiments of ½ to 2 hours' duration varies from 1.3 to 2.1 ml (STPD) of carbon dioxide per kilogram of body weight per torr P_{CO_2}, with an average value of about 1.7. The initial total body carbon dioxide content is expressed as a constant term, K, plus or minus an increment that is a function of P_{CO_2}. The change of total body carbon dioxide stores is estimated by subtracting the initial from the final total body carbon dioxide content, an operation that removes K to give the difference. The change of total body carbon dioxide stores in liters (STPD) is estimated using either of the two following steady state equations:

$$\text{total body carbon dioxide stores} = K + BW \times [0.15 \times P_{A_{CO_2}} + (S - 0.15) \times P_{\bar{V}_{CO_2}}] \quad (46)$$

where K = initial total body carbon dioxide stores (L STPD)
BW = body weight (kgm)
0.15 = contribution of the alveolar gas–arterial blood compartment (ml STPD CO_2 × [kgm body weight]$^{-1}$ × [torr P_{CO_2}]$^{-1}$)
$P_{A_{CO_2}}$ = alveolar gas carbon dioxide pressure in torr
S = average slope of the whole-body carbon dioxide dissociation curve (ml STPD CO_2 × [kgm body weight]$^{-1}$ × [torr P_{CO_2}]$^{-1}$)
(S–0.15) = contribution of the compartment governed by mixed venous blood P_{CO_2}
$P_{\bar{V}_{CO_2}}$ = mixed venous blood carbon dioxide pressure in torr

The following equation is obtained from equation 46 by substitution. It expresses total body carbon dioxide stores as a function of related basic physiologic variables as follows:

$$\text{total body carbon dioxide stores} = K + BW \times \left[\frac{S \times 0.8629 \times \dot{V}_{CO_2}}{\dot{V}_A} + (S - 0.15) \times \frac{R - 0.32}{R \times 4.7} \times \frac{\dot{V}_{CO_2}}{\dot{Q}}\right] \quad (47)$$

where S = average slope of the whole-body carbon dioxide dissociation curve (ml STPD CO_2 × [kgm body weight]$^{-1}$ × [torr P_{CO_2}]$^{-1}$)
$\dot{V}_{CO_2}$ = carbon dioxide output and production rate (ml STPD/min)
$\dot{V}_A$ = alveolar ventilation rate (L BTPS/min)
R = metabolic respiratory quotient
0.32 = constant indicating the volume of carbon dioxide released for each volume of oxygen that combines wtih hemoglobin without changes of P_{CO_2}

Table 12-5. Responses to hypocapnia and hypercapnia*

Subject	Condition	Inspired P_{CO_2} (torr)	Arterial P_{CO_2} (torr)	Effect
Man	Hyperventilation		< 10	Tetany
Man	Hyperventilation Acclimatization to high altitude		< 15	Respiratory alkalosis producing compensatory metabolic acidosis
Man	Hyperventilation		< 35	Respiratory exchange ratio increases; psychomotor impairment; dizziness; hypotension; numbness and tingling; cerebral blood flow decreases
Man	Normal	< 1	40 ± 2	Health
Man	Rest	7.6 (1% carbon dioxide†)		Slightly increased breathing frequency and tidal volume; no subjective effects
Man	Exercise	15		Peak oxygen uptake decrease of 13%; increased ratio of pulmonary ventilation to oxygen uptake
Man	Rest	15.2 (2% carbon dioxide†)		Further increase in breathing frequency and tidal volume; pulmonary ventilation increase of 50%
Man	Rest	22.8 (3% carbon dioxide†)	45	Breathing frequency and tidal volume increase of two- to three-fold; slight respiratory acidosis producing compensatory metabolic alkalosis; decreased respiratory exchange ratio; slight transient throbbing headache; body heat loss increases; insomnia?
	Exercise			Decreased physical work capacity; for a given work rate, increase in pulmonary ventilation and oxygen uptake
Man	Rest	26.6 (3.5% carbon dioxide†)		Cerebral blood flow increased 10%; intracranial pressure increase
Man	Rest	30.4 (4% carbon dioxide†)	47	Throbbing headache; flushed face; nausea; sweating; tachycardia; palpitation; transient blood serotonin decrease
	Exercise			Decrease in physical work capacity
Man	Rest	38.0 (5% carbon dioxide†)		Four-fold increase of pulmonary ventilation; breathing becoming laborious; cerebral blood flow increased 50%; removal to fresh air produces "off-effect": headache, nausea, and vomiting
Man	Rest	45.6 (6% carbon dioxide†)		Maximum carbon dioxide concentration voluntarily tolerated for several hours; heavy physical work impossible
Dog		60.8 (8% carbon dioxide†)		Transient glucose tolerance decrease; normal eating; appear well
Man	Carbon dioxide inhalation; hypoventilation; lung failure		60 to 75	Lethargy; sympathoadrenal activation inhibited by concomitant acidosis
Monkey		65		Above this inspired P_{CO_2} cardiac and breathing frequency begin to decrease
Man	Rest	76 (10% carbon dioxide†)		Intolerable to breathe for more than a few minutes
Man	Terminal lung failure		75 to 100	Narcosis; body temperature decrease
Man	Rest	91.2 to 114.0 (12% to 15% carbon dioxide†)		Unconsciousness within minutes
Monkey		106		Narcosis

*These data are for initially healthy, sea-level–acclimatized normoxic subjects unless otherwise stated. Important variables include initial acclimatization status of test subject, the resulting arterial P_{CO_2} and the concentration-versus-time pattern of exposure. Note marked species differences. Reliable data for experimental human hypercapnia are difficult to obtain because of intolerance to breathing high carbon dioxide mixtures. The combined effects of altered P_{O_2} with altered P_{CO_2} are not presented here. The marked effects of long-term acclimatization to high carbon dioxide atmospheres are also not presented. However, note the arterial P_{CO_2} of subjects acclimatized to high altitude.

†Inspired carbon dioxide concentration in percent on a dry gas basis at a total ambient pressure of 760 torr (sea level).

Continued.

Table 12-5. Responses to hypocapnia and hypercapnia—cont'd

Subject	Condition	Inspired P_{CO_2} (torr)	Arterial P_{CO_2} (torr)	Effect
Man	Anesthesia with hypoventilation		120	Arterial blood pH = 6.99; inhibition of hyperbaric oxygen convulsions
Man		190.0 (25% carbon dioxide†)		Tunnel vision; narcosis; may cause death during hours-long exposure
Monkey		227 to 260		Maximum bradycardia (40% of control)
Man		228.0 (30% carbon dioxide†)		Anesthesia
Barley		228.0 (30% carbon dioxide†)		Death
Monkey		289		Minimum breathing frequency (one third of control)
Dog		228.0 to 304.0 (30% to 40% carbon dioxide†)		Rapid removal to air produces ventricular fibrillation and death
Monkey		389		Survival possible
Monkey	Carbon dioxide increase of 49.2 torr/hr	417		Death

4.7 = average slope of the carbon dioxide dissociation curve of blood (ml STPD CO_2 × [L of blood]$^{-1}$ × [torr P_{CO_2}]$^{-1}$)

$\dot{Q}$ = cardiac output rate (L blood/min)

Note that if R were equal to 0.32, mixed venous blood P_{CO_2} would be equal to alveolar and arterial P_{CO_2}. Using equation 47, changes of total body carbon dioxide stores can be estimated as a function of carbon dioxide output and production rate, alveolar ventilation rate, cardiac output rate, and metabolic respiratory quotient.

RESPONSES TO CARBON DIOXIDE–CONTAINING ATMOSPHERES

Effects of hypercapnia

The first observable physiologic effect of increasing inspired P_{CO_2} is increased respiratory rate and depth (Table 12-5). These two ventilatory parameters are slightly increased at a P_{CO_2} of 7.6 torr (1% carbon dioxide*) and more so at 15.2 torr (2% carbon dioxide*), but a subject is often unaware of this change. Alveolar P_{CO_2}, minute volume of breathing, and pulse rate show little change until inspired P_{CO_2} approaches the 22.8 torr (3% carbon dioxide*) level. As P_{CO_2} rises beyond this level and compensatory hyperventilation becomes inadequate to prevent a rise of alveolar P_{CO_2}, these measurements increase rather sharply. At an inspired carbon dioxide pressure of 22.8 torr, increased respiratory rate and depth are subjectively noticeable and physical work capacity is measurably diminished. Subjects report difficulty sleeping in such a carbon dioxide–enriched atmosphere. A two- to three-fold increase of pulmonary ventilation rate together with nasopharyngeal congestion neccesitates mouth breathing, causing a dry sore throat. At an ambient temperature of 15° C accelerated heat loss together with increased cutaneous blood flow create a sensation of cold that is not felt in a carbon dioxide–free environment.

An inspired P_{CO_2} of 30.4 torr (4% carbon dioxide*) produces a flushed face, headache, palpitation, and sweating. If a subject breathes a P_{CO_2} of 38 torr (5% carbon dioxide*) for a relatively short time and then switches to fresh air, he may experience "off effects"—headache, nausea, and sometimes vomiting. At 45.6 torr (6% carbon dioxide*) physical activity is difficult and heavy physical work is virtually impossible. Immediately after an abrupt shift from a high carbon dioxide to a carbon dioxide–free atmosphere, subjects report a transient smell of ammonia. A P_{CO_2} of 76 torr (10% carbon dioxide*) is subjectively intolerable to breathe and slightly higher carbon dioxide levels produce unconsciousness and, eventually, death.

The foregoing suggests that environmental P_{CO_2} should not be allowed to exceed an upper limit of 22.8 torr (3% carbon dioxide*) and, for prolonged

*On a dry gas basis at sea level.

*On a dry gas basis at sea level.

exposures, 7.6 torr (1% carbon dioxide*). However, recent evidence suggests that even this latter figure is too high.

The carbon dioxide response curve is a useful experimental and diagnostic tool. This curve is made by measuring the prompt, or transient-state, ventilatory response to breathing carbon dioxide–containing gas mixtures. The resulting minute volume of breathing is plotted as a function of inspired, alveolar, or arterial P_{CO_2}. Breathing a gas mixture containing 4.3% to 5.4% carbon dioxide* (33 to 41 torr) doubles the pulmonary ventilation rate of healthy human subjects; an inspired carbon dioxide pressure of only 15 to 23 torr (2% to 3% carbon dioxide*) doubles the pulmonary ventilation rate of patients with acidosis. Under transient-state conditions the slope of the carbon dioxide response curve in healthy subjects is approximately 1 L (BTPS) $\times$ min^{-1} $\times$ (torr arterial P_{CO_2})$^{-1}$. The curve reveals a circadian variation, shifting to the left as the day progresses.[9] Various drugs shift the carbon dioxide response curve or change its slope. For example, during incomplete curarization, reducing vital capacity to a value between 0.5 and 1 L (BTPS) decreased the amplitude of the carbon dioxide response curve about 15%.

Carbon dioxide also affects the cardiovascular system; acting directly on the cerebral vasculature, it is the principal physiologic regulator of cerebral blood flow. Unlike cerebral blood flow, systemic blood flow in other organs is strongly affected by the autonomic nervous system. Cerebral blood flow is physiologically quite insensitive to fluctuations of systemic arterial blood pressure; however, hypercapnia, dilating cerebral blood vessels and reducing cerebrovascular resistance, renders cerebral blood flow susceptible to fluctuations of the perfusing systemic arterial blood pressure. Although cerebral blood flow rate correlates better with cerebrospinal fluid pH than with arterial P_{CO_2}, P_{CO_2} is the prime variable. Ambient carbon dioxide concentrations of 3% to 4% increase cerebral blood flow 10% or more.[10]

During acute hyperventilation, cerebral blood flow rate decreases as a function of arterial P_{CO_2}, not jugular venous P_{CO_2}. Hyperventilation-induced hypocapnia also produces slow high-amplitude waves in the electroencephalogram and decreases visual critical fusion frequency (CFF).† Hypocapnia produces lactic acidosis in the cerebrospinal fluid, which is improved somewhat by increasing P_{O_2}. Experimentally, it is often difficult to separate the effects of hypocapnia from those of hypoxia. As a result of extreme hypocapnic cerebral vasoconstriction, some effects of severe hyperventilation resemble those of experimental decapitation.

Hypercapnia increases intracranial pressure as well as cerebral blood flow. Both create problems for the neurosurgeon. Thus hyperventilation is sometimes used to prepare patients for neurosurgery. Cerebral blood flow remains sensitive to P_{CO_2} during the anesthesia. Does increasing inspired carbon dioxide concentration in patients with cerebrovascular disease produce a cerebrovascular "steal syndrome," in which blood flow to normal brain regions increases while that to diseased regions decreases?

Carbon dioxide produces local vasodilation in denervated areas, but its central action is opposite

Table 12-6. Respiratory responses to an atmosphere containing carbon dioxide at 7.5 torr*

	Breathing air	Breathing carbon dioxide	Difference (percent)
Pulmonary ventilation (L BTPS/min)	7.18	8.30	+16
Breathing frequency (breaths/min)	8.7	10.0	+16
Tidal volume (L BTPS)	0.940	0.940	0
Oxygen uptake (L STPD/min)	0.262	0.265	+ 1
Carbon dioxide output (L STPD/min)	0.204	0.198	− 3
Respiratory exchange ratio	0.78	0.75	− 4
P_{CO_2} end-tidal (torr)	36.2	37.6	+ 4
P_{O_2}, end-tidal (torr)	79.6	84.6	+ 6
Ventilation equivalent for oxygen	27.3	31.3	+15
Alveolar ventilation (L BTPS/min)	4.92	5.71	+16
Alveolar ventilation/pulmonary ventilation	0.69	0.70	+ 1

*Data of Luft, U. C.: Presented at the conference on carbon dioxide levels in the atmosphere of Skylab A, Manned Spacecraft Center, Houston, Texas, August 13-14, 1970. Five healthy human subjects were studied in the basal state, two control measurements while breathing air, and two test studies while breathing carbon dioxide. Comparison shows the response of subjects acclimatized to an altitude of 5,300 ft in Albuquerque, New Mexico. The response of healthy man acclimatized at sea level is qualitatively the same and quantitatively similar.

*On a dry gas basis at sea level.

†Critical fusion frequency, or flicker fusion frequency, is a measure of the time-resolving capacity of the retina.

and dominant. A wide range of high carbon dioxide breathing mixtures blanch skin; redistribution of circulation by this mechanism may explain, in part, how carbon dioxide increases cerebral and coronary blood flow.

Acclimatization to altitude involves adjustments to hypocapnia as well as hypoxemia. A summary of the physiologic changes that occur in healthy man during acclimatization to an altitude of 1 mile provides a data base for interpretation of experimental studies conducted at this elevation.[6] Table 12-6 shows the respiratory responses of five healthy human subjects acclimatized at an altitude of 5,300 ft to an atmosphere containing carbon dioxide at 7.5 torr.

Carbon dioxide intoxication

The most obvious toxic effects of hypercapnia are signs and symptoms relating to the respiratory, cardiovascular, and central nervous systems. Sudden, extreme P_{CO_2} increase, such as that caused by inhalation of 30% carbon dioxide, simultaneously stimulates and depresses the central nervous system. Violent respiratory movements and convulsions are followed by unconsciousness and apnea. Convulsions are the paradoxic consequence of a depressant effect of carbon dioxide on the cerebral cortex, releasing the subcortical brain centers from the normally powerful cortical inhibition. Thus what is actually a depressant effect of carbon dioxide is manifested as a stimulation. Suppression is an important function of certain brain areas. Depressants often appear to stimulate by impairing the inhibitory function of higher brain centers.

Carbon dioxide narcosis is the impairment or loss of consciousness associated with extreme hypercapnia. It results from either of two primary causes —breathing high carbon dioxide gas mixtures or lung failure. In either case narcosis always involves retention and accumulation of endogenous carbon dioxide; the failing lung, diseased or drug-depressed, is unable to eliminate it, whereas carbon dioxide inhalation creates a diffusion barrier that dams it back. During general anesthesia the pulmonary process of carbon dioxide elimination is at the mercy of the anesthetist, who controls the composition of the breathing mixture, the degree of lung inflation, the tidal volume, breathing frequency, and the alveolar ventilation rate.

Although both causative agent (carbon dioxide) and cell effect (carbon dioxide intoxication) are the same in each instance of carbon dioxide narcosis, the precise clinical appearance depends on the rate at which hypercapnia develops and on the associated clinical conditions. Although no longer used, carbon dioxide narcosis was at one time deliberately induced by administration of 30% carbon dioxide for treatment of certain psychotic conditions.

When an oxygen-enriched breathing mixture is administered to patients whose ventilatory drive depends upon hypoxia as a result of carbon dioxide retention and decreased responsiveness to carbon dioxide, alveolar ventilation rate may decrease seriously, or apnea may even occur. This clinical situation is also termed *carbon dioxide narcosis.* The cerebral complications of stupor, coma, and increased cerebrospinal fluid pressure are manifestations of increased P_{CO_2}, rather than oxygen, and are relieved by increasing alveolar ventilation rate to remove carbon dioxide. Unfortunately, the term "oxygen poisoning" has been used to describe this clinical condition; however, it is inaccurate and should be abandoned. Carbon dioxide narcosis is never a contraindication for oxygen but, rather, an indication for providing *both* oxygen and the required ventilatory assistance.

The toxic effects of carbon dioxide are related to increased P_{CO_2}, increased concentration of molecular carbon dioxide, increased concentration of bicarbonate ions, acidosis, or some combination of these. Although carbon dioxide intoxication is associated with increased P_{CO_2} in alveolar gas, blood, and other body fluids, the mechanism is not fully understood. Different cell types and tissues vary in susceptibility, accounting for the pattern of developing carbon dioxide intoxication. Species differences are appreciable. At sufficiently high partial pressure, carbon dioxide can intoxicate any cell. Progressive hypercapnia is biphasic; stimulation precedes depression. Atmospheres containing very high carbon dioxide concentrations are lethal if breathed for hours or days; the higher the carbon dioxide concentration, the shorter the survival time. Hypercapnia also synergizes with either hypoxia or hyperoxia to shorten survival time. Barley, grown in oxygen in a closed, dark container, dies when the carbon dioxide concentration reaches 30%. High carbon dioxide atmospheres prevent the decay of meat at room temperature.

As carbon dioxide intoxication proceeds, metabolic rate, and thus heat production, decreases; body core temperature falls sharply. Shivering is inhibited while evaporative heat loss from the respiratory tract increases because of increased pulmonary ventilation. Hypercapnia decreases the rectal temperature of rats even when ambient temperature is high. Peripheral vasoconstriction retards heat gain from the hot environment while heat from the slowly metabolizing body core brought to the lungs by the pulmonary circulation is lost by evaporation of water in the respiratory tract.

Carbon dioxide is a relatively small uncharged molecule that dissolves in water, diffuses readily, and freely penetrates cell membranes. It enters the biochemical arena by way of the following reaction:

$$CO_2 + H_2O \rightleftarrows H_2CO_3 \rightleftarrows H^+ + HCO_3^- \quad (5)$$

This sequence is catalyzed by *carbonic anhydrase* —an enzyme found in erythrocytes, kidney, stomach, and eye. Note that this reaction produces protons and hence acid.

Acidosis is the most dangerous consequence of acute carbon dioxide intoxication. Increasing P_{CO_2} eventually acidifies the intracellular environment, impairing enzymatic processes. There is evidence that rats develop a transient intracellular metabolic acidosis from 8 to 24 hours after being placed in a 10% carbon dioxide atmosphere.[11] In man whole blood serotonin level during exposure to a 4% carbon dioxide atmosphere showed a transient 34% decrease.[12] In any biologic system each biochemical step is catalyzed by an enzyme—a specific protein catalyst that speeds the achievement of chemical equilibrium. Each of the multitude of enzymes is optimally active at a particular pH. Normal acidity of body fluids and cells is thus absolutely essential for maintenance of the dynamic biochemical equilibria that collectively generate cell structure and comprise cell function. Acidity is kept within relatively narrow limits within the body by a sensitive, well-integrated mechanism involving pH-sensing receptors, lungs, kidneys, and the effective, but limited, buffer systems. Imagine the biochemical chaos that results when pH shifts appreciably, affecting the multitude of enzymes, each to a different degree, and some actually in opposite directions. These changes throttle cell metabolism and threaten life. Carbon dioxide intoxication may involve biophysical membrane phenomena as well as biochemical alterations.

It is difficult to compare meaningfully physiologic tolerance for changes of P_{CO_2} with that for changes of P_{O_2}. Any such comparison requires standardization of conditions such as initial acclimatization status, individual as well as genetic factors, concentration-versus-time pattern of exposure, and selection of criteria for comparison, such as discomfort, disability, reversibility, or death.

Exercise in carbon dioxide–containing atmospheres

Physical work in carbon dioxide–containing atmospheres is a stress of considerable physiologic interest, involving most, if not all, body systems. While breathing air, the increased metabolic rate induced by light to moderate muscular exercise produces proportionate increases of pulmonary ventilation, oxygen uptake, and carbon dioxide output, maintaining alveolar gas and arterial blood P_{O_2} and P_{CO_2} at essentially resting levels despite the increased rates of oxygen consumption and carbon dioxide production. During heavy, or exhausting, exercise the greatly accelerated anaerobic metabolism produces a substantial metabolic acidosis; pulmonary ventilation rate rises sharply out of proportion to oxygen uptake.

Although physiologically unimportant at rest, even a low concentration of carbon dioxide in the inspired gas places a significant additional load on the carbon dioxide transport and elimination system during physical exertion; the inspired-to-alveolar carbon dioxide pressure difference decreases by an amount equal to the inspired carbon dioxide pressure. Compared to the same rate of exercise while breathing air, submaximal exercise in a carbon dioxide–containing atmosphere is characterized by a slightly higher heart rate, slightly higher systemic arterial systolic blood pressure, higher pulmonary ventilation rate, very slightly higher oxygen uptake rate, lower carbon dioxide output rate, and a lower arterial blood plasma pH.

Decreased alveolar-to-inspired carbon dioxide pressure difference impedes the increased pulmonary discharge of carbon dioxide, which is necessary to maintain homeostasis during muscular exercise. As inspired P_{CO_2} increases, the decreasing alveolar-to-inspired carbon dioxide pressure difference renders the process of carbon dioxide elimination progressively less efficient and carbon dioxide output rate falls correspondingly. Inadequate carbon dioxide elimination superimposes respiratory acidosis on the metabolic acidosis normally produced during physical exercise. Body stores of carbon dioxide increase. When the metabolic acidosis generated by accelerated anaerobic muscle metabolism can no longer be sufficiently attenuated by the process of pulmonary carbon dioxide elimination, a critical terminal rise of hydrogen ion concentration precipitates the exercise tolerance end-point.

Finkelstein and associates[13] studied healthy human subjects exercising on a bicycle ergometer while breathing an inspired P_{CO_2} of 15 torr. Compared to maximal exercise while breathing air, maximal exercise while breathing 15 torr P_{CO_2} produced the following cardiopulmonary changes*: slightly lower heart rate, slightly higher systolic blood pressure, lower peak oxygen uptake (2.692 compared to 3.102 L STPD/min), lower peak carbon dioxide output (2.842 compared to 3.444 L STPD/min), higher arterial P_{CO_2} (41 compared to 30 torr), a greater decrease of arterial blood oxygen saturation, and a smaller decrease of arterial blood P_{O_2}, the last two changes resulting from a lower pH that shifted the oxyhemoglobin dissociation curve to the right. However, pulmonary ventilation rate was unchanged.

Thus an inspired P_{CO_2} of 15 torr decreases aerobic capacity, or peak oxygen uptake rate, and limits maximal physical work capacity. Peak carbon dioxide output rate decreased even more than peak oxygen uptake rate. In contrast to the decrease of arterial P_{CO_2} during maximal exercise while breathing air, arterial P_{CO_2} rose during maximal exercise while breathing air containing carbon dioxide at 15 torr. Pulmonary ventilation was inadequate for even this relatively low inspired carbon dioxide pressure when carbon dioxide production by exercising muscles was maximal and acid metabolites were driving carbon dioxide out of the bicarbonate buffers in body fluids. Despite performance decrement, some subjects reported little or no subjective difference; others felt suddenly suffocated at the exercise tolerance endpoint.

In another study[14] eight healthy trained human subjects performed steady-state exercise at one half and at two thirds of maximum oxygen uptake rate for 30 minutes on a bicycle ergometer at inspired carbon dioxide pressures of 0, 8, 15, 21, and 30 torr. Maximal exercise was also studied at inspired carbon dioxide pressures of 0, 8, and 21 torr. Although during submaximal exertion pulmonary ventilation increased with increasing inspired carbon dioxide pressure, the pulmonary ventilation rate during maximum exercise did not increase with the level of inspired carbon dioxide. Despite pulmonary ventilation rates of 50% to 70% of maximal voluntary ventilation, carbon dioxide output rate fell progressively as inspired P_{CO_2} increased.

At inspired carbon dioxide pressures of 8 and 15 torr, no symptoms were reported at any exercise level. At an inspired carbon dioxide pressure of 21 torr, subjects felt a sensation of "air hunger" and intercostal muscle pain during maximal exertion. At an inspired carbon dioxide pressure of 30 torr, "air hunger" and headache occurred at two-thirds maximum exercise. However, symptoms did not prevent any subject from completing an exercise stint.

The maximum acceptable carbon dioxide pressure in the atmosphere of a manned closed system is presently considered to be about 8 torr. At this level of inspired carbon dioxide a healthy trained human subject can perform steady-state heavy exercise for at least 30 minutes and probably much longer without symptoms of hypercapnia or even awareness of the contamination.

There is some evidence that hypercapnia and/or exercise may increase alveolar-arterial P_{CO_2} gradient by as much as 10 torr; however, this heresy against a pillar of physiologic dogma remains to be confirmed and then explained.

Occurrence of hypercapnia

The density of carbon dioxide and the fact that it is produced by oxidation of organic matter suggest the conditions under which it accumulates. Carbon dioxide may collect at the bottom of wells, in caves or grottos, such as the Grotto del Cane, in burning buildings, and in confined spaces after explosions. Miners encounter carbon dioxide ("blackdamp") in unventilated mine passages. Fatalities have occurred in breweries, where fermentation produces carbon dioxide. Using carbon dioxide fire extinguishers in closed spaces can extinguish occupants as well as fires, a principle that was known to Joseph Black. The use of solid carbon dioxide ("Dry Ice") as a refrigerant creates another risk; handlers have been overcome by accumulation of the gas in poorly ventilated spaces.

Hypercapnia is also encountered under conditions of hyperbaria, or increased ambient pressure. There are two reasons for this: first, increased total gas pressure increases the partial pressure of the carbon dioxide fraction; and second, increased gas density tends to produce alveolar hypoventilation and carbon dioxide retention. Haldane dis-

*Studies were done in Albuquerque, New Mexico, at an elevation of 5,300 ft. At this elevation acclimatized subjects, adjusted to the hypoxemia and slight hypocapnia, show cardiopulmonary changes at rest as well as during exercise. For example, the alveolar and arterial P_{CO_2} at rest is 36, not, 40 torr and pulmonary ventilation rates are higher at rest and during all levels of exercise. Changes at sea level would be qualitatively the same but quantitatively somewhat different.

covered that divers are unable to work at depths greater than 12 fathoms (72 ft) when given only enough air flow to maintain a constant fractional concentration, or percentage, of carbon dioxide in the helmet at that depth. Increasing helmet ventilation rate in proportion to the ambient pressure at diving depth solves this problem.

Carbon dioxide production, accumulation, and removal are critical considerations in the design of manned aerospace environments, submarine habitats, and self-contained underwater breathing apparatus (scuba) of the closed-circuit type. Progressive hypercapnia is a potential consequence of accidental mechanical failure of sealed systems, such as aerospace cabins or undersea habitations.

In an occupied sealed environment that initially contains air at sea level pressure, the oxygen level falls somewhat faster than the carbon dioxide level rises. However, increasing carbon dioxide produces symptoms before hypoxia does. Hypercapnia thus warns of the deteriorating environment, provided that hypoxia has not dulled the subject too much to appreciate its significance. Death eventually results from the combined effects of hypercapnia and hypoxia. In such a situation, simple removal of carbon dioxide relieves the subject, who then will probably not detect the effects of decreasing oxygen pressure. Thus as carbon dioxide is removed from an occupied sealed environment, oxygen should always be added, and the partial pressures of both of these respiratory gases should be monitored. An equation for calculating the composition of the atmosphere within a sealed system as a function of carbon dioxide clearance rate C without the addition of oxygen was derived by Rahn and Fenn[15]:

$$\frac{C}{\dot{V}_{O_2}}\left(0.2095 - \frac{P_{O_2}}{P_B}\right) = \ln_e \frac{\dot{V}_{CO_2}}{\dot{V}_{CO_2} - \dfrac{C \times P_{CO_2}}{P_B}} \qquad (48)$$

where the initial atmosphere is fresh air

C = rate of removal of carbon dioxide expressed as clearance in liters of air cleared of CO_2 ($L \times min^{-1} \times man^{-1}$)

$\dot{V}_{O_2}$ = oxygen uptake rate ($L \times min^{-1} \times man^{-1}$)

$\dot{V}_{CO_2}$ = carbon dioxide output rate ($L \times min^{-1} \times man^{-1}$)

P_B = ambient pressure (torr)

P_{CO_2} = ambient carbon dioxide pressure (torr)

P_{O_2} = ambient oxygen pressure (torr)

Rahn and Fenn[15] analyzed the possible changes in the composition of the air with a closed system under a variety of conditions.

Chronic hypercapnia results from carbon dioxide retention in the course of lung failure in conditions such as bronchitis-emphysema syndrome. This hypercapnia produces respiratory acidosis, a complex syndrome associated with increased intracellular, as well as blood, acidity. An arterial blood pH value of 7.25 or less is considered a clinical indication for treatment of acidosis.

Homeostatic responses to hypercapnia

To defend the physiologic hydrogen ion concentration of body fluids against the stress of changing P_{CO_2} the organism must vary plasma bicarbonate ion concentration appropriately. Cell buffers are the first line of defense in the response to acute changes of P_{CO_2}. Although the carbonic acid–bicarbonate buffer system plays a key role in the defense against *metabolic* acidosis and alkalosis, and although this system is important in the response to *chronic* respiratory acidosis and alkalosis, it is unimportant in acute respiratory acid-base disturbance. The acid-base impact of acute P_{CO_2} changes is buffered primarily by intracellular buffers, which account for 97% of the buffering of acute respiratory acidosis and 99% of the buffering of acute respiratory alkalosis. Hypercapnia increases carbonic acid concentration and hydrogen ion concentration. Hydrogen ions thus created diffuse into cells in exchange for sodium and potassium ions, where they are buffered by cell proteins. This process leaves bicarbonate ions in the extracellular fluid.

Intracellular buffers account for approximately half of the prompt increase of plasma bicarbonate concentration. Some carbon dioxide molecules diffuse into erythrocytes, where they are hydroxylated in the presence of *carbonic anhydrase,* forming bicarbonate ions, or hydrated to carbonic acid, which dissociates, forming hydrogen and bicarbonate ions. The hydrogen ions thus created are buffered by hemoglobin, while bicarbonate ions diffuse into the extracellular fluid in exchange for chloride ions. This mechanism accounts for about 30% of the prompt increase of plasma bicarbonate ion concentration. Plasma proteins, which are poor buffers, contribute minimally. In man the bicarbonate ion concentration increase is small, amounting to less than 5 mEq/L as P_{CO_2} increases acutely from 40 to 80 torr.

During acute hypocapnia the process reverses; intracellular buffers release hydrogen ions while chloride and bicarbonate ions exchange in the opposite direction across the erythrocyte membrane. These processes decrease extracellular bicarbonate ion concentration by about 7 to 8 mEq/L as P_{CO_2} falls from 40 to 15 torr.

Acute steady state studies of dogs in environmental chambers show that at each ambient P_{CO_2} level within the range of 30 to 130 torr, plasma bicarbonate ion concentration achieves a plateau within about 1 hour. Within the more limited range of 40 to 80 torr P_{CO_2}, a plateau occurs in man within about 10 minutes. The plasma bicarbonate ion concentration response to increasing hypercapnia is curvilinear with smaller increments at higher P_{CO_2} levels. However, the change of hydrogen ion concentration per unit change of P_{CO_2} is linear. The slope of this curve, $\Delta[H^+]/\Delta P_{CO_2}$, is 0.77 nM $\times$ L^{-1} $\times$ $torr^{-1}$ at all levels of P_{CO_2}. Thus in acute hypercapnia the defense of hydrogen ion concentration is constant over a wide range of P_{CO_2} values. However, the defense is incomplete; hydrogen ion concentration does not return to the control level.

Human subjects made acutely hypocapnic achieve a steady state within about 10 minutes at each P_{CO_2} level within the range of 15 to 40 torr. As in acute hypercapnia, hydrogen ion concentration varies directly with P_{CO_2} and the slope, $\Delta[H^+]/\Delta P_{CO_2}$, is 0.74 nM $\times$ L^{-1} $\times$ $torr^{-1}$. The slope of this curve is about the same as that during acute hypercapnia.

In healthy man the hydrogen ion concentration response to acute hypo- and hypercapnia is linear; within the P_{CO_2} range from 15 to 80 torr the response is 0.74 to 0.77 nM of hydrogen ions $\times$ L^{-1} $\times$ $torr^{-1}$. Within the same P_{CO_2} range the bicarbonate ion response, which occurs within 10 minutes, is curvilinear, falling off at the higher P_{CO_2} levels. The bicarbonate ion response to variations of P_{CO_2} is both rapid and small. In human subjects increasing P_{CO_2} from 40 to 80 torr increases plasma bicarbonate ion from 24 mEq/L to only 27 mEq/L. Within the P_{CO_2} range from 30 to 130 torr, canine response is very similar to that of man. Does this similarity justify extrapolation to man from data derived in studies of dogs at P_{CO_2} levels to which man cannot be safely exposed?

During acute hypercapnia the plasma bicarbonate ion concentration increase results almost entirely from the action of buffer systems that already exist within the body. Persistent hypercapnia rapidly exhausts this readily available buffer capacity. To compensate for decreased pulmonary elimination of carbon dioxide during chronic hypercapnia, the kidney increases hydrogen ion excretion and bicarbonate ion production and reabsorption. Thus the kidney creates a metabolic alkalosis to compensate for persistent hypercapnic, or respiratory, acidosis.

If dogs are exposed in an environmental chamber to chronic stable hypercapnia at carbon dioxide pressures ranging from 35 to 135 torr until a steady state is achieved, a chronic steady state whole-body titration curve is observed. During the first 24 hours of exposure plasma bicarbonate ion concentration increases rapidly. This is secondary to hydrogen ion buffering by intra- and extracellular buffers and is not associated with increased urine hydrogen ion excretion. This initial rapid rise accounts for about 50% of the total bicarbonate ion concentration increase that occurs during steady state chronic hypercapnia. Plasma bicarbonate ion concentration continues to rise steadily and reaches a new stable steady state level within the next 3 to 6 days. The latter concentration increase is associated with increased urine hydrogen ion excretion in the form of ammonium ions. The increased urine hydrogen ion excretion during this second phase is sufficient to account for the total increase of plasma bicarbonate ions. Thus during the second phase the kidney replenishes all the intra- and extracellular buffers that were used during the acute phase of hypercapnia. In this mechanism of hydrogen ion homeostasis, increased blood P_{CO_2} increases renal hydrogen ion excretion while adding newly generated bicarbonate ions to extracellular fluid. Simultaneously, elevated P_{CO_2} raises the renal tubular maximum (T_m) for bicarbonate ions, permitting the kidney to maintain bicarbonate ion concentration at the new higher level.

Despite these compensatory mechanisms, the defense of extracellular hydrogen ion concentration in chronic hypercapnia is incomplete, just as it is in acute hypercapnia. Hydrogen ion concentration increases as a linear function of P_{CO_2}, but the slope of the curve for chronic hypercapnia is only 0.32 nM $\times$ L^{-1} $\times$ $torr^{-1}$, compared to a value of 0.74 for acute hypercapnia. Plasma bicarbonate ion concentration increases curvilinearly as in acute hypercapnia; however, the steady state bicarbonate ion concentration for a given level of chronic hypercapnia is significantly greater than that for the same level of acute hypercapnia.

Studies of extreme hypercapnia are obviously impossible in man. Studies of patients who have chronic hypercapnia caused by cardiopulmonary disease are difficult to interpret because other respiratory and metabolic disturbances, such as hypoxemia, are always present. The human response to chronic hypercapnia is qualitatively similar to that of the dog; there is a linear relationship between P_{CO_2} and hydrogen ion concentration and a curvilinear relationship between P_{CO_2} and bicarbonate ion concentration. The slope, $\Delta[H^+]/\Delta P_{CO_2}$, differs in published studies of human patients, ranging from 0.24 to 0.126 nM of hydrogen ions $\times L^{-1} \times torr^{-1}$. The homeostatic responses of chronic hypercapnic man do not completely restore extracellular fluid hydrogen ion concentration to normal; the compensatory metabolic alkalosis is incomplete.

What is the effect of acute P_{CO_2} changes superimposed on chronic stable hypercapnia? In dogs adapted to chronic hypercapnia, the capacity to prevent changes of hydrogen ion concentration in response to superimposed acute hypercapnia increases as a direct function of the level of the chronic hypercapnia. Dogs with more severe chronic hypercapnia defend pH better against acute P_{CO_2} elevation. However, this is not caused by a greater efficiency of intra- and extracellular buffers. On the contrary, it results at least in part from an arithmetic consideration that can be seen in the Henderson-Hasselbalch equation, where pH is a function of the bicarbonate ion concentration/P_{CO_2} ratio. If bicarbonate ion concentration is greater to begin with, as it is in chronic hypercapnia, a greater acute P_{CO_2} change is required to produce a given pH change.

Acclimatization

Chronic exposure to an atmosphere containing a low concentration of carbon dioxide results in certain responses that increase tolerance to the effects of this gas. The ventilatory response to carbon dioxide decreases with such exposure; breath-holding time and alveolar P_{CO_2} at the breath-holding breaking point are both increased. The initial respiratory acidosis is more rapidly compensated if inspired carbon dioxide concentration is high than if it is low! Initial euphoria is followed by lassitude and impaired performance, which tend to persist despite acclimatization; thus subjects can acclimatize to moderate hypercapnia, but the penalty is decreased respiratory efficiency and diminished physical work capacity.

As in altitude acclimatization, an interplay between central and peripheral chemoreceptors during carbon dioxide acclimatization gradually decreases ventilation and carbon dioxide sensitivity. Initially, increased P_{CO_2} produces respiratory acidosis in blood and in the central chemoreceptors. Hydrogen ions and, perhaps, carbon dioxide produce a sustained increase in ventilation by stimulation of peripheral chemoreceptors. This is augmented by the respiratory acidosis in the central chemoreceptors. However, as bicarbonate ions accumulate in the cerebrospinal fluid, the ventilatory drive of the central chemoreceptors decreases. Carbon dioxide sensitivity is reduced, and the breath-holding, breaking-point curve is elevated because P_{CO_2} increments along the new high carbon dioxide dissociation curve now produce less pH change than similar increments produced along the normal carbon dioxide dissociation curve.

Upon emerging from a carbon dioxide–containing atmosphere, subjects report a smell of ammonia. Return to air breathing produces respiratory alkalosis in the central chemoreceptors and reduces peripheral chemoreceptor drive. Total ventilation is thus depressed until bicarbonate ion loss restores normal central chemoreceptor pH.

PERSPECTIVE

The primeval atmosphere of the earth contained much more carbon dioxide than our present atmosphere. As the earth cooled and life evolved, the deposition of sediments and the process of photosynthesis removed carbon dioxide from the atmosphere, fixing it in carbonates and in organic compounds. Green plants have been removing carbon dioxide and adding oxygen to the atmosphere for about 1 billion years. It is calculated that chlorophyll-containing plants could completely replace the atmospheric oxygen in about 40,000 years, a relatively short time in geologic perspective.

Atmospheric carbon dioxide concentration has been increasing since the beginning of the industrial revolution because of the burning of fossil fuels. This increase has global implications because carbon dioxide traps some of the infrared radiation emitted by the earth's surface, tending to increase the average surface temperature. However, this warming "greenhouse effect" is small and may actually by overshadowed by the cooling effect of increasing atmospheric turbidity. The presently increasing atmospheric carbon dioxide concentration probably represents only a transient shift in the global carbon dioxide production-fixation balance; the supply of fossil fuels is limited, the process of photosynthesis can accelerate, and the buffer capacity of the oceans is great. Indeed, as food production becomes less adequate to meet the world demands, carbon dioxide may come to be considered a beneficial pollutant; increased carbon dioxide will increase the rate of production of organic plant material.

Prediction of the future levels of atmospheric carbon dioxide requires definition of its sources and sinks. Although the long-term trend of global carbon dioxide concentration can be adequately monitored by a limited number of ground-based stations, measurements of global carbon dioxide distribution will be needed to trace its movement from sources to sinks. Satellite-based downward looking remote sensors measuring absorption in scattered sunlight may provide the required information.

Although atmospheric carbon dioxide concentration will probably increase only slightly and transiently in the foreseeable future, man will encounter high carbon dioxide atmospheres in artificial closed environmental systems, in experimental research situations, and in the exploration of other planets such as Venus. It is likely that a compromise will be made between the physiologist's wish to maintain a natural carbon dioxide pressure in the atmosphere of sealed systems and the normal administrative wish to minimize the weight, volume, cost, and thus efficiency of carbon dioxide-removing systems. An occasional unfortunate individual will experience the combined hypercapnia and hypoxia of asphyxia as a result of accidental exposure in underventilated spaces.

Serious hypercapnia is being experienced by an increasing number of patients in the terminal stage of pulmonary insufficiency; such hypercapnia, the result of carbon dioxide retention, is associated with hypoxemia, which always develops first.[6] Bronchitis-emphysema syndrome, the most common cause of lung failure, is rapidly becoming more prevalent in the human population as a result of widespread exposure to smoke and other polluted air, the cumulative damage of many insults to the bronchopulmonary system, the genetic predisposition of certain individuals, and aging. The carbon dioxide levels that occur in the final stages of chronic lung failure, after a period of acclimatization to gradually increasing carbon dioxide, are the highest levels spontaneously experienced by man.

The cause of arterial hypoxemia and carbon dioxide retention in chronic diffuse obstructive lung disease is the mismatching of the distribution of alveolar ventilation with the distribution of pulmonary blood flow.[6] Gravity significantly affects the distribution of both inspired gas and pulmonary blood flow within the diseased, as well as the healthy, lung. Indeed, matching of alveolar ventilation with pulmonary blood flow is optimal in the weightless state. Furthermore, in the patient with bronchitis-emphysema syndrome the headward shift of the abdominal viscera and flattened hemidiaphragms back into the thoracic cavity in weightlessness would facilitate breathing. Thus it is conceivable that certain patients dying of hypoxemia and hypercapnic acidosis caused by the ventilation-perfusion abnormality of bronchitis-emphysema syndrome could survive in the weightlessness of space. Certainly, one of the dividends of space flight will be the opportunity to study cardiopulmonary physiology during weightlessness, a condition impossible to duplicate on earth.

Carbon dioxide is an inevitable product of the oxidation of organic materials. For plants, this uncommon molecule is an indispensable building block for the vital photosynthetic process. For animals, carbon dioxide is a stimulus for normal breathing but also a potential intoxicant. Indeed, the higher forms of animal life live on the brink of carbon dioxide intoxication, defended only by the constant action of their lungs and by their effective, but limited, body buffer systems. Human skin has a small, but measurable, permeability to carbon dioxide; thus far no practical use has been made of this fact. In the future methods may be developed to control the rate of transdermic (diadermic) carbon dioxide transfer, to control the rate of enzyme-catalyzed carbon dioxide fixation, or to eliminate carbon dioxide from the body by a practical nonpulmonary means.

GLOSSARY

alveolar gas The gas mixture present within gas-exchanging regions of the lungs reflecting the effects of respiratory gas exchange. The definition of alveolar gas composition is complicated by the discontinuous nature of lung ventilation, the regional effects of lung perfusion, the imperfect matching of these two aspects of lung function, and the technical problems of sampling. Alveolar gas is assumed to be saturated with water vapor at 37° C. Contrast **dead space gas.**

alveolar gas volume (V_A) The volume of the regions of the lung within which respiratory gas exchange occurs. Contrast **bronchopulmonary dead space.**

ambient Prevailing, environmental.

apnea The cessation or absence of the movements of breathing.

biosphere The region of the earth's surface, including air and water, that comprises the terrestrial life-support system; the region of the natural terrestrial ecologic system.

carbon dioxide narcosis The impaired state of awareness, ranging to unconsciousness, produced by hypercapnia.

carbon dioxide output The volumetric rate at which carbon dioxide is exhaled from the lungs. It is determined by measurement and analysis of expired gas, with correction for inspired carbon dioxide, if present, and expressed in volume units per unit time. Distinguish from carbon dioxide *production,* which is the rate at which carbon dioxide is produced by the metabolizing tissues. When a steady state of respiration and circulation exists, carbon dioxide *output* equals carbon dioxide *production*.

carbon dioxide pressure (P_{CO_2}) The tension, or partial pressure, of carbon dioxide in a defined region, system, or space; a measure of carbon dioxide fugacity, or escaping tendency.

endogenous carbon dioxide Carbon dioxide produced within the body by metabolic processes.

hydrogen ion concentration $[H^+]$ Expressed in terms of nanomoles of hydrogen ion per liter (nM/L) or as pH, the negative logarithm of the hydrogen ion concentration.

hyperbaria Increased ambient pressure. An antonym is hypobaria.

hypercapnia Increased carbon dioxide pressure within a biologic system. An antonym is hypocapnia.

hyperoxia Increased oxygen pressure. An antonym is hypoxia.

hyperventilation Generally, increased pulmonary ventilation rate beyond the actual requirement for respiratory gas exchange. It may result from increased breathing frequency, increased tidal volume, or any combination of these two. Hyperventilation is best expressed in terms of *alveolar* ventilation rate. It *increases* alveolar and arterial blood oxygen tension and *decreases* alveolar and arterial blood carbon dioxide tension. The hypocapnia produces dizziness, numbness, tingling, arterial hypotension, respiratory alkalosis, and, if hyperventilation continues, significant psychomotor impairment. An antonym is hypoventilation.

pH A measure of the acidity of an aqueous solution; the negative common logarithm of the hydrogen ion concentration (activity) or, identically, the common logarithm of the reciprocal of the hydrogen ion concentration.

partial pressure of a gas The pressure exerted by one component of a gas mixture.

respiratory acidosis The acidotic condition of the body produced by hypercapnia.

tetany A condition characterized by increased neuromuscular excitability with muscular spasm. It is usually associated with alkalosis and deficiency of calcium ions in extracellular body fluids.

REFERENCES

1. National Aeronautics and Space Administration: U. S. standard atmosphere, 1962, Table 1.2.7.: Normal composition of clean, dry atmospheric air near sea level, Washington, D. C., 1962, U. S. Government Printing Office, p. 9.
2. Lange, N. A., editor: Handbook of chemistry, ed. 10, New York, 1967, McGraw-Hill Book Co., p. 1100.
3. Leonard, J. D.: CO_2—a steadily growing giant, Chem. Eng. News **36:**116-117, 1958.
4. Braker, W., and Mossman, A. L.: Matheson gas data book, ed. 5, East Rutherford, N. J., 1971, Matheson Gas Products.
5. Slonim, N. B., Bell, B. P., and Christensen, S. E.: Cardiopulmonary laboratory basic methods and calculations: a manual of cardiopulmonary technology, Springfield, Ill., 1967, Charles C Thomas, Publisher.
6. Slonim, N. B., and Hamilton, L. H.: Respiratory physiology, ed. 2, St. Louis, 1971, The C. V. Mosby Co.
7. Slonim, N. B., and Estridge, N. K.: The acid-base surface: a 3-dimensional visual model, Rocky Mountain Med. J. **67:**59-63, 1970.
8. Slonim, N. B., and Estridge, N. K.: The acid-base surface: a three-dimensional visual model for analysis of acid-base information, J. Med. Educ. **45:**826-829, 1970.
9. Eger, E. I., II, and others: Influence of CO_2 on ventilatory acclimatization to altitude, J. Appl. Physiol. **24:**607-615, 1968.
10. Patterson, J. L., Jr., Heyman, A., Battey, L. L., and Ferguson, R. W.: Threshold response of cerebral vessels of man to increase in blood carbon dioxide, J. Clin. Invest. **34:**1857, 1955.
11. Martin, E. D., Scamman, F. L., Attebery, B. A., and Brown, E. B., Jr.: Time-related adjustments in acid-base status of extracellular and intracellular fluid in chronic respiratory acidosis, SAM-TR-67-116, Brooks Air Force Base, Texas, December, 1967, U.S.A.F. School of Aerospace Medicine.
12. Gordon, E. D., Jr.: The effect of high concentrations of carbon dioxide on blood serotonin levels in man: chronic hypercapnia, SAM-TR-68-107, Brooks Air Force Base, Texas, September 1968, U.S.A.F. School of Aerospace Medicine.
13. Finkelstein, S., Elliott, J. C., and Luft, U. C.: The effects of breathing low concentrations of CO_2 on exercise tolerance, presented at the 39th Annual Scientific Meeting of the Aerospace Medical Association, May, 1968.
14. Menn, S. J., Sinclair, R. D., and Welch, B. E.: Response of normal man to graded exercise in progressive elevations of CO_2, SAM-TR-68-116, Brooks Air Force Base, Texas, December, 1968, U.S.A.F. School of Aerospace Medicine.
15. Rahn, H., and Fenn, W. O.: A graphical analysis of the respiratory gas exchange: the O_2-CO_2 diagram, Washington, D. C., 1955, American Physiological Society.

SUGGESTED READINGS

1. American Society of Refrigerating Engineers: Refrigerating data book, New York, 1942, American Society of Refrigerating Engineers.
2. Fenn, W. O., and Rahn, H., editors: Handbook of physiology, sect. 3, vols. 1 and 2, Washington, D. C., 1965, American Physiological Society.
3. Forster, R. E., Edsall, J. T., Otis, A. B., and Roughton, F. J. W., editors: CO_2: chemical, biochemical, and physiological aspects, a symposium held at Haverford College, Haverford, Pennsylvania, August 20-21, 1968, Washington, D. C., 1969, National Aeronautics and Space Administration.
4. Granet, I., and Kass, P.: The viscosity, thermal conductivity, and specific heat of carbon dioxide at elevated pressures and temperatures, Petrol. Refiner. **31:**137-138, 1952.
5. Michels, A., and Kleerekoper, L.: Measurements of dielectric constant of CO_2 at 25°, 50°, and 100° up to 1700 atmospheres, Physica **6:**586, 1939.
6. Michels, A., and Stryland, J.: The specific heat at

constant volume of compressed carbon dioxide, Physica **18**:613, 1952.

7. Sherrat, G. G., and Griffiths, E.: A hot wire method for the thermal conductivities of gases, Phil. Mag. **27**:68, 1939.
8. Slonim, N. B., and Estridge, N. K.: Carbon dioxide—environmental health aspects, J. Environ. Health **33**:171-178, 1970.
9. Sweigert, R. L., Weber, P., and Allen, R. L.: Thermodynamic properties of gases—carbon dioxide, Ind. Eng. Chem. **38**:185-200, 1946.
10. Van Itterbeek, A., and DeClippeleir, K.: Measurement on the dielectric constant of carbon dioxide as a function of pressure and temperature, Physica **13**:459, 1947.

13 ENVIRONMENTAL POLLUTION

AUSTIN F. HENSCHEL

Through sophisticated technology modern industry and commerce produce and distribute a wide range of products and services; we generate a vast amount of power, produce food and manufactured goods on an enormous scale, provide rapid transportation, facilitate prevention, diagnosis, and treatment of disease, and create increasing leisure. Never before has man had such wealth and power at his disposal. We have come to expect and demand these luxuries as necessities of our contemporary way of life; comfort has become the object of worship.

Unfortunately, this sophisticated technology offends the earth. The price of maintaining and improving our standard of living is the increasingly rapid pollution of our environment. Many noxious and toxic chemical substances, natural and synthetic, are involved in processing raw materials, in manufacturing finished goods, and in providing essential services. Pollutants are thus direct products or indirect by-products of our technologic processes. Some of these substances escape inadvertently, some are released indiscriminately, and others are overtly dumped. Toxic substances, in both kind and quantity, are entering our environment from diverse sources at an ever-increasing rate. Disposal of waste, including human sewage, nonbiodegradable materials, noxious substances, hazardous chemicals, and radioactive residues, poses a monumental problem. Polluted air is a growing health hazard. Its impact on the total equilibrium of our ecologic system is not yet understood. Although damage to human and other life cannot be measured in dollar cost, air pollution is already enormously expensive in terms of wasted fuel and property damage. Polluted air rots and soils clothing, dissolves paint, corrodes metal, mars statues, monuments, and buildings, decreases visibility, and may injure or kill plants and animals.

Environmental degradation is certainly not new. From his beginning man has exploited earth's natural resources, fouled water, and polluted air. Progressive environmental debasement—accelerated by a rapidly increasing population mass that expects, demands, produces, and consumes ever more—is now a plainly visible menace.

Pollution knows no national boundaries; it is a critical global problem urgently demanding solutions if we are to prevent the collapse of our biospheric ecologic systems and avoid the destruction of life as we know it on this planet. Does the rate of increase in variety and concentration of environmental pollutants already exceed the relatively slow pace at which biologic adaptive capacity can develop tolerance for noxious stressors? Have we already permanently damaged our biospheric ecologic systems? What are the strain limits of the weakest links in the ecologic chain? What components of our ecologic system dare we destroy? Somehow the indispensable benefits of modern technology must be preserved and extended while environment and the quality of life are restored.

The subject of noxious exposures actually consists of three parts—occupational exposure, exposure to environmental pollution, and exposure to natural allergenic particulates. Although each topic is important in its own right, this chapter is mainly concerned with the second.

SOURCES AND CLASSIFICATION OF POLLUTANTS

In its most general sense a pollutant is an undesirable substance or agent that is present in the air, water, or soil. However, for the purpose of our discussion, we define a pollutant as a foreign or extraneous substance, material, or agent that impairs the quality or diminishes the life-supporting capacity of an environment. Pollutants may be atoms, molecules, or compounds; organic or inorganic; ions or free radicals. They may occur as simple or complex single substances, mixtures, or aggregates. They may be in the air we breathe, the food we eat, the water we drink, or the soil in which we grow our crops. Indeed, they may occur in any materials with which we come into physical contact.

Pollutants arise from natural as well as artificial sources. Volcanoes spew sulfur oxides and mineral dust. Lightning produces nitrogen oxides and ozone as it sets forest fires that generate carbon monoxide and smoke. The biologic processes themselves produce wastes that, if not removed, accumulate to extinguish life; these natural wastes include carbon dioxide and hydrogen ions.

Generally, pollutants are classified as particulates or as emulsions. Examples of particulates are dusts, fumes, and mists, the precise term depending on the size, physical state, and chemical nature of the substance involved. Unfortunately, our pulmonary clearance mechanisms are not equal to the task of eliminating the barrage of artificial particulates responsible for a number of diseases that occur in occupational environments. These include "black lung" of coal miners, granulomatous disease of beryllium workers, pulmonary fibrosis and increased lung cancer risk of asbestos workers, and a lung condition called byssinosis produced by cotton dust. Metal fumes, depending on their chemical nature, may also produce illness.

Dusts are usually fine dry particles of solid material that vary in size, shape, and physicochemical nature. When inhaled, particles 5 μ or less in diameter can reach the pulmonary alveoli; larger particles are removed from the healthy respiratory tract by physiologic clearance mechanisms but may irritate, damage, or even infect before removal. Alveolar retention of inhaled particles is actually a function of particle size and alveolar ventilation rate. For a discussion of physiologic lung clearance mechanisms, see Slonim and Hamilton.[1]

Fumes are vapors or smokes; the latter are a condensate of vaporized materials or substances such as metals suspended as fine particles in the air. Mists, or fogs, are suspensions of fine liquid droplets.

Operationally, air pollution may be defined as contamination of the atmosphere by waste products that result from human activity. Burning, common to most processes of power generation, is the major source of air pollution. Some pollutants originate in the processing and combusion of oil and gasoline; others come from industrial, chemical, and agricultural processes that produce synthetic fabrics and fertilizers. For practical purposes sources of air pollution are classified either as stationary emitters, such as factories, power plants, furnaces, and refuse burners, or as moving vehicular emitters.

Pollutants from various sources mix in the atmosphere. There they may interact with water, each other, or sunlight, producing an even more complex mixture called a photochemical smog. Thus air pollution varies in nature and intensity from region to region, depending on the physicochemical nature of the materials involved, climate, altitude, topography, and efficiency of the systems that buffer and the sinks that remove them; these same factors determine how long a given pollutant remains in the atmosphere. It is important to distinguish between emission standards and regional air quality standards.

Increased illness and higher death rates from respiratory diseases in badly polluted areas indicate that chronic or repeated exposure to polluted air may damage health. Although polluted air is obnoxious, noxious, and known to aggravate chronic bronchopulmonary disease and increase the risk of lung cancer, it is not yet possible to point to a specific disease entity caused by urban air pollution alone. However, oxidant pollutants may age tissues; free radicals may increase the rate of genetic mutation, accelerate aging, and predispose to cancer; and transient massive pollution, held in place by atmospheric inversion, may kill individuals who are elderly or debilitated or who have cardiopulmonary diseases. Notorious killer smogs occurred in Belgium's Meuse Valley, London, New York, Tokyo, and Donora, Pennsylvania. Certain types of air pollution may aggravate asthma or produce an asthma-like bronchitis.

It is a paradox that many individuals are vitally concerned about air pollution at the urban level but remain blissfully unconcerned that tobacco smoking, a self-induced personal air pollution, is a more serious health menace to themselves and even to those around them who are forced to breathe their secondhand smoke in enclosed places. Beyond doubt cigarette smoking is a cause of bronchitis, emphysema, lung cancer, and cardiovascular disease.

The pollution of water by sewage is another example of a serious pollution problem. The estuary is already known as the septic tank of the megalopolis.[2] Compounding the sewage pollution of water is disposal of chemical wastes by dumping them into rivers, lakes, and oceans. Ninety-five percent of the photosynthesis on earth takes place in the upper layers of the ocean. In the near future nuclear power plants and the processing of nuclear fuel and residues will pose problems of radiation pollution.

Despite publicity given to the problem of environmental pollution in our media, we are still not sufficiently aware of the wide variety of pollution sources or of all its actual and potential hazards. Without doubt the danger from some pollutants has been hysterically exaggerated while others receive inadequate emphasis. Many potentially hazardous substances are already present in our environment, some unrecognized, and many new ones may be present in the environments of tomorrow. It is encouraging to know that it is theoretically possible to eliminate all environmental pollution except heat, discussion of which is beyond the scope of this chapter.

Even more pervasive is the noise problem. From the ubiquitous hum of a million motors to the occasional sonic boom, we are surrounded and beleaguered by noise of ever more disturbing quality and ever greater amplitude. There is nowhere to hide. An occasional sensitive individual suffers keenly from this unremitting assault. Noise is discussed in Chapter 5, Sound, Vibration, and Impact.

COMBUSTION PRODUCTS OF CARBONACEOUS FUEL

The national surveillance network that monitors air quality is comprised of more than 7,000 sampling stations throughout the United States, reporting data to the Division of Air Surveillance in the National Environmental Research Center at Durham, North Carolina. In addition many states and localities have their own air monitoring programs. Even this extensive network affords only an incomplete picture of air quality, which varies greatly with time and place.

The most significant air pollutants resulting from combustion of carbonaceous fuels (coal, oil, gasoline, natural gas, and wood) are carbon monoxide, sulfur oxides, nitrogen oxides, hydrocarbons, particulates, and, secondarily, photochemical oxidants.* In the United States automobile exhaust is the source of nearly two thirds of such pollution, whereas industry and power-generating plants contribute about 15% each. Thus reduction or elimination of the pollutants in automotive exhaust offers the greatest potential for alleviating air pollution.

Carbon monoxide

The biologic effects of acute and chronic exposure to carbon monoxide were recently reviewed by two groups of experts—the National Air Quality Criteria Advisory Committee for the National Air Pollution Control Administration[3] and the Committee on Effects of Atmospheric Contaminants on Human Health and Welfare for the National Academy of Science.[4]

Small amounts of carbon monoxide are produced physiologically,† but the major source is incomplete combustion of carbon-containing fuels. Carbon monoxide combines avidly with hemoglobin—about 210 times as readily as oxygen—to form carboxyhemoglobin in the red blood cells; hence its effect on vertebrates is very potent. Even low concentrations of carbon monoxide effectively reduce the availability of hemoglobin to oxygen. As carbon monoxide combines with hemoglobin, making it unavailable, arterial oxygen *pressure* remains essentially unchanged despite reduced arterial blood oxygen *content*. Fortunately, the reaction between hemoglobin and carbon monoxide is reversible; when the carbon monoxide content of inspired air decreases, erythrocyte carboxyhemoglobin decreases as a function of time, releasing hemoglobin to transport oxygen again.

A healthy nonsmoker breathing fresh air essentially devoid of carbon monoxide has, because of endogenous carbon monoxide production, a carboxyhemoglobin saturation of 0.4% to 0.5%. A person who smokes twenty or more cigarettes per day and breathing fresh air has a carboxyhemoglobin saturation of about 5%. Abnormal lung

*Table A-1 in the Appendix indicates the natural presence and concentration of some of these gaseous pollutants.

†By certain plants such as seaweed and algae and in vertebrate hemoglobin metabolism.

function, a common result of cigarette smoking,* probably contributes to this carboxyhemoglobin level by producing arterial hypoxemia.

The effect of breathing carbon monoxide is a function of the carboxyhemoglobin saturation achieved; this, in turn, at equilibrium is a function of both inspired carbon monoxide level and duration of exposure. The time required to reach any given carboxyhemoglobin saturation depends on inspired carbon monoxide concentration, rate of carbon monoxide uptake, and mass of circulating hemoglobin. When inspired carbon monoxide concentration increases or decreases, a new carboxyhemoglobin equilibrium level is achieved within about 12 hours. A carboxyhemoglobin equilibrium value of about 2% is reached in 12 hours when breathing 10 parts per million (ppm) carbon monoxide, whereas the same carboxyhemoglobin saturation is reached in 2 hours when the inspired carbon monoxide concentration is 50 ppm. Breathing a 10-ppm carbon monoxide–air mixture for 12 hours increases the carboxyhemoglobin saturation of a nonsmoker about 2% above his baseline value; 35 to 50 ppm increases the saturation 5%; and 100 ppm about 10%.

Exercise, by increasing alveolar ventilation rate, increases the rate at which a new carboxyhemoglobin equilibrium value is achieved. Setting a maximum acceptable inspired carbon monoxide concentration thus involves consideration of physical exertion level as well as exposure time. It would seem that inspired carbon monoxide concentration should be held below 0.001%, the value that is in equilibrium with the normal carboxyhemoglobin content of the blood of a nonsmoker. The generally accepted threshold limit value for an 8-hour per day exposure is now set at 50 ppm. This inspired carbon monoxide concentration results in a carboxyhemoglobin equilibrium saturation of about 5% above baseline values.

The toxic effect of carbon monoxide is a result of its impairment of oxygen transport; it thus aggravates hypoxemia of any cause—ventilation-perfusion mismatching, reduced inspired oxygen pressure (hypoxia), alveolar hypoventilation, alveolocapillary diffusion defect, or venous-arterial shunt. It also aggravates the effect of anemia, creating in addition a reversible functional anemia. It aggravates the effects of changes in blood pH or temperature, facts to be understood in terms of the oxyhemoglobin dissociation curve, which it shifts to the left, interfering with oxygen unloading in the systemic capillaries. Altitude hypoxia places the victim of carbon monoxide intoxication at a special disadvantage; carbon monoxide production increases because of hypoxic combustion of organic materials, and decreased oxygen pressure enhances carbon monoxide competition with oxygen for hemoglobin.

Physical work capacity is not grossly impaired except at high carboxyhemoglobin levels. Muscle functions at lower arterial blood oxygen pressure than most other tissues. However, the myocardium is an actively metabolizing muscle with a high oxygen extraction. The central nervous system and sensory organs are also particularly vulnerable to arterial hypoxemia and thus to the effect of carbon monoxide.

Carbon monoxide is eliminated almost entirely via the lungs. There are interindividual differences, but on the average the blood carboxyhemoglobin level in an individual breathing air at sea level falls to half its original value in about 250 minutes; thus during each hour the blood loses 15% of the carbon monoxide that was present at the beginning of that hour. Under most conditions the factor limiting carbon monoxide elimination appears to be the slow rate of diffusion of the gas at the very low pressure gradients encountered while breathing air at sea level. Elimination rate decreases with increasing age of the subject; the half-time for carboxyhemoglobin desaturation increases about 1% for each year over 40.

Inhalation of pure oxygen increases the rate of carbon monoxide elimination in man and dog; the rate in man increases five- to six-fold, reducing the half-time for blood carboxyhemoglobin desaturation from 250 minutes to a range between 40 and 50 minutes. Further acceleration is attained by adding 5% or 6% carbon dioxide to the inspired oxygen. Carbon dioxide induces hyperpnea, increasing alveolar ventilation rate, and thus washes carbon monoxide out of the blood more rapidly. Carbon dioxide also decreases pH, reducing the affinity of hemoglobin for both carbon monoxide and oxygen, an effect that is greater for the former.

Healthy individuals tolerate chronic low-level elevation of carboxyhemoglobin without apparent ill effect. Even the smoker appears to function without handicap so long as carboxyhemoglobin levels remain at 5% or below. However, experimentally, mental performance deteriorates at

*There is evidence that cigarette smoking impairs every aspect of lung function.

carboxyhemoglobin levels of 5%, a saturation about equal to that in the blood of heavy smokers; performance involving estimation of time intervals is significantly impaired at 2% carboxyhemoglobin; and it is suspected that the more complex intellectual functions are actually impaired at even lower carboxyhemoglobin saturations. Indeed, a carboxyhemoglobin threshold value below which there is no effect may not even exist. Carboxyhemoglobin saturations of 3% to 5%, equivalent to inspired carbon monoxide pressures of about 0.0126 to 0.0210 torr, affect visual threshold.

The mechanism by which increased tolerance for carbon monoxide is achieved remains unknown but is believed to involve increased total hemoglobin, increased blood volume, or increased cardiac output. It is well known that chronically exposed asymptomatic individuals have carboxyhemoglobin levels that would produce measurable physiologic changes if acutely produced. Despite this fact experimental data have not yielded a clear-cut picture of the nature and mechanism of carbon monoxide adaptation.

There are large interindividual differences in ability to compensate for the effects of low concentrations of inspired carbon monoxide. Besides inherent differences in carbon monoxide tolerance, some diseases increase susceptibility to carbon monoxide. As indicated previously, any impairment or stress of the oxygen transport system such as decreased oxyhemoglobin saturation or increased oxygen demand aggravates the effect of a given carboxyhemoglobin level. Possibly, even low-level chronic carbon monoxide exposure seriously compromises the function of a diseased cardiovascular system. High carbon monoxide levels resulting in carboxyhemoglobin saturations of 15% or higher produce vascular changes resembling atherosclerosis in experimental animals. However, such high carboxyhemoglobin levels are observed in man only under unusual conditions such as acute severe carbon monoxide exposure in firemen.

It has been suggested that chronic carbon monoxide exposure accelerates degenerative vascular disease. Furthermore, in coronary artery disease decreased oxygen content of coronary artery blood decreases an already limited myocardial oxygen supply. Symptoms of myocardial ischemia sometimes occur in patients who have coronary artery disease when they breathe low levels of carbon monoxide. Statistically, it has not yet been shown that the incidence of myocardial infarction increases during periods of sustained increase of atmospheric carbon monoxide concentration; however, fatal myocardial infarction is significantly more frequent when carbon monoxide levels are 10 ppm or higher for 12 hours or more than when carbon monoxide concentration is less than 10 ppm. These observations are sufficient reason for concern and emphasize the urgent need for further controlled laboratory and epidemiologic studies.

Despite ever-increasing effluence from internal combustion engines, carbon monoxide concentration remains constant in the atmosphere from ground level up to the tropopause (junction of troposphere and stratosphere). The half-life of carbon monoxide in the atmosphere is about 4 to 5 months; the mean life (time required for carbon monoxide concentration to decay to a value of $\frac{1}{e} \times$ initial concentration) is 6 months. There are two sinks: microbial activity in soil constitutes a natural biologic sink,* whereas above the tropopause in the stratosphere carbon monoxide is oxidized to carbon dioxide in the following natural chain reaction:

$$\underset{\text{hydroxyl free radical}}{\cdot OH} + CO \rightarrow CO_2 + \underset{\text{free radical (atomic) hydrogen}}{H\cdot} \quad (1)$$

$$H\cdot + O_2 + \underset{\text{third body energy- and momentum-absorbing agent}}{M} \rightarrow \underset{\text{perhydroxyl (hydroperoxyl) free radical}}{HO_2\cdot} \quad (2)\dagger$$

$$HO_2\cdot + \text{reducing agent} \rightarrow \cdot OH \quad (3)$$

Nitrogen oxides

Two of the five oxides of nitrogen, nitrous oxide (N_2O) and nitrogen dioxide (NO_2), are naturally present in the atmosphere in trace amounts. Of the several nitrogen oxides that pollute air, nitrogen dioxide causes the most concern because of its toxicity and concentration in urban smog. The major sources of nitrogen dioxide are automotive and aircraft exhaust, industrial processes, and fumes from burning oil and coal. Other sources of exposure include the combustion of nitrogen-containing rocket propellants (missile fuel oxi-

*Certain tissues, such as myocardium, striated muscle, stomach, liver, and spleen, can oxidize a small but detectable amount of carbon monoxide to carbon dioxide at a pressure of 0.8 atmosphere.

†An exergonic termolecular recombination reaction, involving transfer of momentum and chemical energy.

dizers) and microbial action on organic material stored in agricultural silos.

Nitrogen dioxide and nitrogen tetroxide exist in a temperature-dependent equilibrium. Below 22° C the reaction 2 $NO_2 \rightleftarrows N_2O_4$ is completely to the right; at 22° C boiling occurs. Above 135° C the mixture is entirely nitrogen dioxide. After inhalation the proportions are a function of lung temperature, 37° C—approximately 30% nitrogen dioxide and 70% nitrogen tetroxide. On contact with the moist mucosa of the respiratory tract or other membranes nitrogen dioxide readily dissolves to form a mixture of nitric and nitrous acids. As with other toxic substances, the reaction to this oxidant gas depends on concentration, duration, and pattern of exposure. Other important variables are the age, sex, physical health status, and individual tolerance of the subject.

There is evidence that families living in a high nitrogen dioxide pollution area have a significantly higher relative respiratory illness rate than control groups. The odor of nitrogen dioxide is just detectable at 0.5 to 1.0 ppm. Eye and nose irritation may occur at about 5 to 10 ppm. Repeated intermittent exposure to nitrogen dioxide at concentrations of 10 to 40 ppm may cause pulmonary fibrosis and emphysema. Bronchiolitis and focal pneumonitis, lasting for 2 to 3 months, may result from exposure to concentrations of 50 to 100 ppm. Exposure to concentrations of only about 200 ppm may cause bronchopneumonia, bronchiolitis fibrosa, and death within a few weeks. Exposures of less than 1 hour to nitrogen dioxide concentrations of 500 ppm or higher may cause chemical bronchopneumonia, acute pulmonary edema, and death within a few days. The threshold limit value for an 8-hour per day exposure is set at 5 ppm.

Acute severe exposure to nitrogen dioxide or nitric oxide (NO) extensively damages or totally destroys the respiratory epithelium, especially in small bronchioles; necrotic epithelium is subsequently shed. Alveolar capillaries are also damaged and alveoli rapidly fill with edema fluid, which wells up as a frothy, watery, mucoid fluid into the air passages. A latent period of several hours may elapse between exposure and the development of pulmonary edema. If the patient survives this initial edema, bronchopneumonia is likely to occur, followed, after many days, by bronchiolitis obliterans.

Silage gassing, characterized by acute respiratory distress and bronchiolitis, is the result of nitrogen dioxide inhalation at a concentration of 100 ppm or higher. The chronic lung condition, resulting from repeated exposures, is known as silo-filler's disease; it is characterized by bronchiolitis obliterans, pulmonary fibrosis, and emphysema. Study of accidental exposures to high nitrogen dioxide concentrations, fatal and nonfatal, have provided some data for the human dose-response relationship and for maximum exposure guidelines.

Acute nitrogen dioxide exposure thus produces immediate respiratory distress followed by pulmonary edema within a period of time that depends on the concentration of the gas. Clinically, pulmonary edema reduces alveolar ventilation and impedes diffusion of respiratory gases across the alveolocapillary membrane, causing arterial hypoxemia (reduced oxygen loading pressure), carbon dioxide retention, respiratory acidosis, and, because of a rightward shift of the oxyhemoglobin dissociation curve, impairment of the oxygen transport mechanism. Death from acute nitrogen dioxide–nitrogen tetroxide exposure thus results clinically from pulmonary edema and pathophysiologically from lung failure—a combination of arterial hypoxemia, tissue hypoxia, and respiratory acidosis.

Dogs exposed acutely to 180 ppm nitrogen dioxide–nitrogen tetroxide show decreased oxyhemoglobin saturation and a rightward shift of the oxyhemoglobin dissociation curve prior to death. With cardiac output held constant, this rightward shift would produce a 10% to 15% decrease in blood oxygen transport within the physiologic range of blood oxygen pressure. Methemoglobin concentration also increases, contributing further to decreased arterial blood oxygen content. Does long-term inhalation of air polluted with nitrogen dioxide produce a similar rightward shift of the oxyhemoglobin dissociation curve and thus impair the oxygen transport mechanism?

Sulfur oxides

The bulk of sulfur oxides derive from burning coal (60%), industrial processes (22%), and burning oil (14%). Sulfur dioxide (SO_2) is the most abundant pollutant oxide of sulfur. It is largely produced in the burning of sulfur-containing coal and oil. It is a pungent, irritating, corrosive gas capable of producing significant acute and chronic bronchopulmonary reactions. Because of its high water solubility, toxic effects occur predominantly in the upper respiratory tract.

The odor of sulfur dioxide is detectable even at concentrations of 1 to 5 ppm but is not objectionable below 10 ppm. Exposure to sulfur dioxide decreases the odor threshold concentration, and continued exposure to low levels results in adaptive loss of olfactory sensitivity. There is a large interindividual difference in the ability to detect low concentrations of sulfur dioxide.

The high water solubility of sulfur dioxide facilitates irritation of the eyes and upper respiratory tract. Exposure to levels of 10 ppm and above may produce lacrimation, rhinorrhea, and throat irritation. Irritation occurs at significantly lower concentrations if particulate materials such as salt, carbon black, or oil mist are present in the air. The concentration at which sulfur dioxide irritates is higher than presently prevailing levels in urban smog. However, increased use of high-sulfur fuels may change this situation in the future.

Since exposure to various levels of sulfur dioxide is common in many industries, studies have been made to determine whether occupational exposure has health implications. Sulfur dioxide toxicity is influenced by climatic environment. Higher levels are tolerated better in a dry environment than in a humid one. One industry studied was in an arid region. There was no evidence of adverse health effects in these workers exposed to occupational sulfur dioxide concentrations ranging from 0 to 25 ppm, with peak concentrations of 100 ppm, for as long as 19 years. Evidence of sulfur dioxide acclimation was observed; concentrations of about 25 ppm in air no longer produced the typical reactions. In hot and humid regions a higher incidence of nose and throat irritation, shortness of breath, and cough is observed among workers chronically exposed to sulfur dioxide. The threshold limit value for an industrial environment is set at 5 ppm.

Sulfur dioxide exposure occasionally evokes allergic dermatitis in exposed workers. It may also provoke attacks in asthmatic patients. However, increased tolerance occurs much more often than sensitization as a result of exposure. Sulfuric acid mist is another sulfur-containing air pollutant of some importance.

The deleterious effects of sulfur dioxide on plants have been known for many years. The death of trees, especially conifers, in the vicinity of sulfuric acid factories is an example of its toxicity.

OZONE

Ozone (O_3), the most hazardous gaseous pollutant, is an important natural constituent of the upper atmosphere and, in trace concentration, of the lower atmosphere. Thanks to modern technology, exposure to ozone is increasing. Equipment producing sparks, arcs, or static discharge, as well as ultraviolet or other ionizing radiation, produces ozone from molecular oxygen. Commercially available "air purifiers" and "deodorizers" in homes, hospitals, offices, elevators, and meat storage plants generate sufficient ozone to be hazardous *under certain conditions of use and ventilation.* Some producers of distilled water add this poisonous gas to their product without informing the purchaser and without regard to whether it is for use in baby formula or the chemical laboratory. Ozone is a potential hazard in closed environmental systems such as aerospace cabins and submarine chambers in which electric discharge from equipment or ionizing radiation can interact with the oxygen-containing atmosphere. It is without doubt the most important toxicologic consistuent of photochemical smog.

Ozone is an extremely active free radical oxidant and can thus interfere in many metabolic processes at an intracellular level.[5] The biochemical mechanism of ozone toxicity is under active investigation. Free radicals are probably the basic biochemical mechanism of ozone-induced cell damage. Ozone destroys sulfhydryl-containing compounds, and in this process free radicals may be produced. However, a more important theory of ozone toxicity involves an oxidative attack on the carbon-carbon double bond of unsaturated fatty acids (UFA)—*lipid peroxidation.*

In man ozone affects mainly the lungs and the vision. The most prominent reaction is pulmonary congestion and edema. Symptoms of severe dyspnea, chest pain, and cough occurred in workers exposed to ozone at a concentration of 9 ppm. A concentration of 8 ppm decreases acetylcholinesterase activity of mouse erythrocytes in vivo. Volunteers reported substernal pain, reduced ability to think, dry mouth, cough, and the odor of ozone at a concentration of 1.5 to 2.0 ppm after a 2-hour exposure. Spirometric measurements indicated reduced vital capacity. Even during intense pollution ozone concentration rarely reaches 1 ppm in the lower atmosphere. Lung edema would be expected to interfere with alveolar gas exchange. Indeed, breathing ozone for 2 hours at a level of 0.6 to 0.8 ppm produced a diffusion defect and a reduced forced expired vital capacity. The nonrespiratory effects of ozone toxicity include decreased visual acuity and extraocular muscle imbalance. The major concern about chronic exposure to low levels of ozone is its possible role in the development of pulmonary emphysema and in hastening the general aging process.

Ozone toxicity, at least in animals, is influenced by several factors, including age, ambient temperature, and exertion. Young animals are two to three times as susceptible as older ones; at 32° C the toxicity is twice that at 21° C; and concentrations of ozone that are without effect at rest may become lethal during short periods of physical exercise. Respiratory infections decrease tolerance to ozone.

Interruption of ozone exposure by brief periods of air breathing reduces both edema formation and mortality. Some substances, such as the antioxidant vitamin ascorbic acid, reduce the effects of ozone on the lung. The phenomenon of ozone tolerance may result from reduction of lung unsaturated fatty acids after an initial exposure, preventing the accumulation of fatal free radical concentrations during subsequent exposure.

Since no adverse effects were reported at ozone concentrations below 0.6 ppm in a careful study, the present maximum acceptable concentration in industrial environments of 0.1 ppm for an 8-hour day exposure appears to provide a reasonable margin of safety. However, because of its extreme reactivity it has been suggested that there is no threshold below which the toxicologic effects of ozone are absent.

Ozone also inhibits plant growth. The chloroplast is the primary site of ozone toxicity. The relative susceptibility of chloroplasts to lipid peroxidation is presumably caused by their substantial UFA content. Growth inhibition probably results from oxidant-induced inhibition of enzymes necessary for polysaccharide synthesis, affecting formation of cell wall polysaccharides, and cellulose in particular.

PESTICIDES

The most commonly used pesticides can be classified in five major categories. These categories with representative examples are shown in the outline below; however, the categories should not be considered rigid nor is the list of examples complete. Chlorinated hydrocarbons and organophosphates are the pesticides that cause the greatest concern and are used in the greatest quantity around the world. The reactions of human beings and higher animals to chlorinated hydrocarbons, organophosphates, and carbamates are generally similar and depend on such factors as total dose, exposure pattern, route of entry, climate, age, and probably sex.

CLASSIFICATION OF SOME COMMONLY USED PESTICIDES

Chlorinated hydrocarbons
- DDT
- Chlordane
- Aldrin
- Dieldrin
- Lindane
- Methoxychlor

Organophosphates
- Parathion
- Malathion
- Dipterex
- Diazinon

Carbamates
- Carbaryl
- Banol
- Carbofuran

Inorganic compounds
- Arsenates
- Fluorides
- Thallium
- Phosphorus

Other organic compounds
- Strychnine
- Pyrethrin
- Rotenone
- Allethrin
- Piperonyl butoxide

Acute massive exposure to chlorinated hydrocarbons or organophosphates readily results in sickness and may even cause death. Such massive exposures, both fatal and nonfatal, are mainly of three types: (1) children who accidentally ingest pesticide, (2) adults who are occupationally exposed, and (3) adults who ingest pesticide with suicidal intent. All pesticides are toxic to man and should be treated accordingly. Acute massive exposure to any pesticide is a medical emergency. Cholinesterase activity, including that of plasma, is dramatically reduced, producing the familiar clinical signs and symptoms. Cholinesterase inhibition is proportional to pesticide dose.

In otherwise healthy human subjects chronic exposure to pesticides elicits a somewhat different syndrome than acute exposure. Evidence suggests that continuous low-level exposure is not a serious threat to human health. Contact dermatitis is common in workers who handle these pesticides or fruit and vegetable produce that has been sprayed with them. However, it is not a major problem and is readily controlled by appropriate safety and health precautions. Individuals have worked in the manufacture of pesticides for 10 to 20 years with pesticide exposure levels greatly exceeding the recommended safe level and suffered no apparent ill effects. Such individuals, however, have a high plasma cholinesterase and erythrocyte acetylcholinesterase activity. It is not known whether this represents increased tolerance to the pesticide. The blood level of pesticides and their metabolic products in pesticide workers who show no signs or symptoms of intoxication is about 200 to 250 times that of the general population. It is also encouraging that the pesticide level in the tissues of individuals of several different countries, including the United States, has not increased significantly during the past several years. This fact may reflect more rational use and better control programs. Human volunteers have ingested pesticides at several times the maximum acceptable exposure daily for as long as 2 years without evidence of any adverse health effects.

ROUTES OF ENTRY

Exposure to pollutants, acute or chronic, is usually an accidental occurrence; the subject is unaware that a potential hazard exists or that there is danger of contamination. In other instances exposure occurs accidentally, although the individual is aware of the hazardous nature of the material involved. Occasionally, exposure is deliberate; an individual is aware of the danger and possible consequences but chooses to accept the risks for other gain.

Environmental pollutants may affect or enter the body (1) by contact with or penetration of broken or even intact skin or exposed mucous membranes; (2) by inhalation and subsequent absorption through the remarkably permeable surfaces of the bronchopulmonary system; (3) by ingestion and subsequent absorption in the gastrointestinal tract; or (4) by any combination of these three primary routes.

Mercury poisoning can occur by absorption through intact skin or by inhalation of mercury vapor. The absorption of hexachlorophene, an antiseptic component of some medicated soaps, is another example of potentially hazardous skin absorption. Plumbism, or lead poisoning, can occur by absorption of lead fumes through the lungs as well as by ingestion of lead-containing paint.

Contact pollutants may act as primary irritants on skin or mucous membranes, as sensitizers within the skin, or, after absorption into blood or lymphatic channels, as systemic toxicants. Most pollutants that contact the skin surface do not penetrate. Some that remain on the skin or mucous membranes are neutralized by oils or fluids on the surface; others dissolve and penetrate cutaneous layers where they are neutralized or produce a dermal reaction. Substances that reach the blood or lymphatic circulations are distributed throughout the body, where they may act as systemic toxicants or be neutralized or encapsulated.

Inhaled pollutants in any physical form may, depending on their nature, irritate or inflame the respiratory tract. Particulates, by mechanical stimulation, induce bronchoconstriction and cough, expelling some inhaled particulates; others are deposited in distal airways of the bronchial tree where they are engulfed by phagocytes, inactivated by secretions, or eliminated by mucociliary transport. The mucus escalator transports material toward the pharynx where it is swallowed and usually inactivated by acid digestive juices in the stomach. Particulates smaller than 5μ in diameter may be retained in the pulmonary alveoli. Some pollutant substances are absorbed into the pulmonary interstitial tissue or into the blood or lymphatic circulations, where they are distributed throughout the body to other organ systems. Acute or chronic bronchopulmonary inflammation may produce temporary or permanent abnormalities of structure and function.

Most ingested pollutants are inactivated by digestive processes or traverse the gastrointestinal tract unchanged without causing adverse effect. Some act on the mucosal lining of the gastrointestinal tract, where they irritate, inflame, or damage. After ingestion other pollutants are absorbed into blood or lymphatic channels where they are distributed to other organs, tissues, or cells.

REACTIONS TO POLLUTANTS

Although man is responsible for most environmental pollutants now, in a certain sense it may be said that natural environmental pollution has existed since the earth's beginning. In biologic terms the earth began as a hostile, uninhabitable place—fiercely hot, studded with belching volcanoes, and covered with a poisonous atmosphere. There are some who wonder whether the earth will again become uninhabitable.

The three categories of environmental pollutants that have evoked the greatest popular concern are the combustion products of fossil fuels, pesticides, and food additives. Within each of these three categories are compounds that are chemically and toxicologically similar to each other and others that have little in common.

Mechanisms of action

Environmental pollutants of varying potency and intensity threaten life at all levels—organismic, tissue, cellular, subcellular, and molecular. The extent of displacement of physiologic processes and the nature and intensity of the reaction caused by a particular pollutant substance depends on many factors: the physicochemical nature and concentration of the pollutant substance; its distribution within the body; the availability of the substance at receptor sites; the defense mechanisms to eliminate or detoxify it, including drug-metabolizing enzymes and biochemical capacity for metabolizing the substance; the simultaneous presence of other pollutant substances; climatic environment, including ambient temperature and humidity; species differences; individual characteristics, such as age, sex, and size; individual tolerance, susceptibility or sensitivity; and even the route of entry into the body. Small wonder a useful model of the pollutant-organism interaction has not yet been developed!

When an environmental pollutant interacts with the body, it may displace the steady state equilibria of biochemical and physiologic processes at levels ranging from the molecular to the organismic. These displacements trigger a wide variety of reactions. Biochemical processes change some pollutants to physiologically inactive substances (detoxification) but increase the toxicity of others. In every interaction the pollutant compound affects the organism while the organism affects the compound. Displacements may be completely or only partially reversible. A model for the action of an environmental stressor on a biologic system in a steady state is presented in Fig. 1-2.

We know far too little about the mechanisms of action of most environmental pollutants. In some instances we cannot even identify a perturbing substance, much less discuss its nature and the mechanism of its action on a biologic system. In other instances recent observations have cast doubt on the validity of a hypothesis or called into question theories about mechanisms we thought were well understood. The possible causative role of carbon monoxide in myocardial degeneration is a case in point. Environmental pollutants and their metabolic derivatives may either act as sensitizers or exert direct effects. They may initiate defense reactions that increase tolerance to a pollutant substance or they may induce adaptive genetic changes.

There is no comprehensive system for classifying the diverse mechanisms of action of the wide variety of environmental pollutants at the many levels of physiologic organization. For no two pollutant agents are the mechanisms of interaction exactly the same.

The complexity of the interactions between man and environmental pollutants was emphasized at the National Academy of Sciences' Symposium on Physiological Characterization of Health Hazards in Man's Environment in 1966. In the proceedings of this symposium, a publication of some 450 pages, only "examples (which) would serve as models" were presented. These examples, hopefully, would stimulate scientists to probe more deeply into the nature of environmental agents and into the mechanisms of physiologic reaction to these environmental hazards.

Tissue irritation

A cell reacts in specific characteristic ways to a wide variety of environmental pollutants. The nature and extent of a given reaction pattern depend on both the nature and the concentration of the particular substance involved and on the biologic characteristics, potentialities, and condition of the cell. Reactions range from minor reversible displacement or irritation, through cell proliferation, to tissue necrosis; thus environmental pollutants may not only injure but also destroy cells and tissues. The vulnerability or susceptibility of a cell to injury by a toxicant has both genetic and environmental components. The problem of cell injury and recovery has quantitative energetic aspects that we are only beginning to comprehend. For example, there is evidence that recovery depends on an adequate, uninterrupted supply of oxygen, glucose, and energy.

Cells expend a considerable fraction of metabolic energy just to prevent superhydration. Indeed, damage from perfusion failure may result not so much from anoxic necrosis of parenchymal cells as from capillary swelling and occlusion. Pathoanatomic changes in an organ depend on the nature and distribution of the toxicant and on the structure and biologic properties of the tissues affected. Thus no single general model has been developed to describe the many mechanisms and interactions involved in cell, much less tissue, reaction to injury. An example illustrating the complexity of this problem is the reaction of the bronchopulmonary system to airborne irritants and toxicants.

The lung is basically a network of capillaries covered by alveolar epithelial cells that comprise the alveolar membrane. This continuous membrane lines the alveoli and alveolar ducts and sacs, which join the bronchioles, bronchi, and finally the trachea. Embryologically, pulmonary capillaries are of mesodermal origin, whereas the remainder of the structure derives from endoderm. It is not surprising that these tissues, which differ ontogenetically, react differently to irritants and toxicants. The reactions of all pulmonary tissues, regardless of origin, are dose-related and reflect the relative toxicity of the pollutant substance in question.

Pulmonary capillaries exhibit two kinds of reaction to airborne irritants and toxicants. If the irritant effect is not overwhelming, capillary permeability increases and blood plasma leaks into alveolar spaces, producing pulmonary edema. Severe damage or prolonged exposure produces edema, fibrin precipitation, and chemical pneumonitis. Edema and pneumonitis interfere seriously with the distribution of inspired gas, alveolar ventilation, and respiratory gas exchange. If the reaction has not progressed too far, removal of the subject from the harmful environment permits recovery processes to restore normal structure and function.

If, on the other hand, the reaction is too severe, necrosis of pulmonary capillary endothelial cells results, diminishing and impairing capillary blood supply to the involved lung tissue, and connective tissue forms scars. The result of this process may be permanent ventilation-perfusion mismatching and pulmonary emphysema of varying degree.

Generally, the alveolar membrane effectively handles solid, liquid, or gaseous inhalants that have low damage potential. Macrophages on the alveolar surface and phagocytes from interstitial tissue and blood engulf materials, inactivate some, and remove them from further contact with the alveolar surface epithelium.

Continued exposure to the pollutant substance stimulates proliferation of alveolar epithelial cells, thickening the alveolar membrane, and increasing the connective tissue component. As pulmonary capillaries are destroyed and alveolar walls thicken, the lung becomes relatively avascular. Increased membrane thickness and relatively decreased vascularity impede diffusion in two ways: by decreasing available surface area and by increasing diffusion distance between alveolar gas and pulmonary capillary blood.

The final result of such chronic or repeated insults to the lung by inhaled irritants or toxicants is ventilation-perfusion mismatching. The degree of pulmonary impairment reflects the duration and intensity of exposure, the nature of the pollutant, and the susceptibility and tolerance-increasing capability of the individual. Some evidence suggests racial differences in susceptibility to chronic bronchitis. Allergic reactions to inhaled substances are important in some conditions such as byssinosis, silicosis, and farmer's lung.

Acute bronchopulmonary reactions to inhaled particles, aerosols, and irritant gases include bronchoconstriction, cough, increased breathing frequency and depth, wheezing, dyspnea, chest pain, tracheal burning, and transient apnea. Severe acute exposure may produce acute bronchitis and bronchopneumonia. Airway reactivity varies greatly from subject to subject. In some individuals inhalation of a given irritant may produce relatively little reaction, whereas in others cough, bronchoconstriction, bronchial inflammation, and secretions may be serious or even fatal.

Allergic reactions

Both natural and artificial environments are replete with substances that are actual or potential allergens. Allergic reactions to environmental pollutants may be an important aspect of the health problem associated with pollution; however, this remains to be proved. In any case, New Orleans asthma is an example of an allergic respiratory syndrome caused by organic chemical pollutants of industrial origin.

Two factors are required for an allergic reaction, an allergen and a sensitized individual; environments provide allergens while genetic factors and exposure determine sensitization. Owing to genetic differences individuals vary greatly in the ease with which they can become sensitized to allergens.

Allergic reactions manifest themselves in many ways. The most common reactions involve the skin, the bronchopulmonary system, and the gastrointestinal system. Allergic skin reactions range in severity from mild itching to necrosis and sloughing of large skin areas. Allergic dermatitis, no respecter of socioeconomic status, is common; it is also difficult to prevent because it is difficult to isolate and protect sensitive individuals from exposure to the wide variety of environmental allergens. Although allergic dermatitis is uncomfortable with its itching and pain and may even cause disability, it is rarely fatal. Skin allergy is particularly prevalent among housewives and industrial workers, because soaps, detergents, solvents, greases, and water, which alter normal skin barrier characteristics, are among the common predisposing substances.

The air we breathe may contain many potential allergens; when inhaled these may produce mild to severe allergic respiratory symptoms. Most respiratory allergens are natural organic materials of plant or animal origin, although artificial chemical substances such as pesticides can also produce respiratory allergy. Molds, pollens, plant, animal, and mineral dusts, drugs, and other chemical substances are the major categories of respiratory allergens. The symptoms of respiratory allergy include rhinitis, sneezing, itching eyes, lacrimation, sinusitis, sore throat, breathlessness, and pulmonary congestion; asthma and hayfever, the respiratory allergies par excellence, are common; it is estimated that 3% to 4% of the population have asthma while 13% to 14% have hayfever.

Respiratory allergic reactions release histamine; antihistaminic drug therapy usually provides temporary symptomatic relief. Acute reactions are clearly recognizable and often readily associated with the particular causative airborne allergen. During humid spells accompanied by atmospheric inversion, asymptomatic hayfever and asthmatic patients may develop dyspnea and pulmonary congestion; latent sensitivities to airborne dust, molds, and pollen, dormant for years, manifest themselves in overt clinical symptoms.

Allergic sensitization via the respiratory tract is common. Respiratory infection and even anatomic changes may result from chronic or intense exposure. There is some evidence to suggest that surgery of the upper respiratory tract, such as tonsillectomy, may predispose to respiratory allergy if the operation is performed during a time of high airborne allergen concentration. Sensitization via the respiratory tract may produce not only bronchopulmonary symptoms such as asthma but may also cause allergic reactions in other organs.

Food allergens usually cause sensitization via the gastrointestinal tract but, again, the reaction is not always limited to this organ system. Many substances that contaminate water or food are potential allergens; foods themselves are common potential allergens. Gastrointestinal sensitization is particularly frequent in infants but may occur at any age if the integrity of the gastrointestinal system is compromised or if digestive processes are incomplete. Thus substances to which an individual is not sensitive can become potent allergens as a result of acute or chronic gastrointestinal disease or dysfunction.

The shock organ of an allergic reaction need not be the site of entry of the allergen into the body; for allergens that enter through the gastrointestinal tract the site of reaction and the site of entry are frequently different. Paroxysmal atrial tachycardia, migraine, angioedema of the skin, and scotomata are examples of clinical manifestations that may be triggered by ingestion of allergens in a sensitive individual.

Evidence mounts that cigarettes affect not only smokers but those around them. Petitions to ban smoking in enclosed public places and on public conveyances multiply because the accumulation of smoke in these enclosed atmospheres is offensive and may be dangerous even when smokers are segregated. Allergy to cigarette smoke may produce distressing upper respiratory tract symptoms in nonsmokers with allergic backgrounds. In a recent study 8% of a series of nonsmoking patients were hypersensitive to cigarette smoke. Symptoms appeared 30 to 60 minutes after exposure to cigarette smoke and persisted for 8 to 12 hours. Several hypersensitive patients developed headaches the day after exposure; these were either of migraine or sinus type. In another study 10% of nonsmoking allergic patients developed respiratory distress after exposure to cigarette smoke created by other persons in their immediate vicinity. Other studies reveal that only 30 minutes in a smoking environment has measurable effects on children, including increases in heart rate, systolic and diastolic blood pressure, and concentration of carbon monoxide in the blood. The effects, to a reduced degree, are the same as on smokers themselves. Other effects of exposure similar to those on smokers include eye and nose irritation, headache, cough, wheeze, sore throat, hoarseness, nausea, and dizziness. In addition mucous spirals in sputum are changed in nonsmokers who work near smokers and who are forced to breathe their exhaled smoke. The mucous spiral changes are the same as those in smokers. Thus those who smoke in enclosed public spaces are not only rude but dangerous.

TOXICOLOGIC INTERACTION OF POLLUTANTS

Seldom are we exposed to a single pollutant. Significantly, exposure to combinations of pollutants may produce different effects from those that would result from exposure to each pollutant separately. The combined effects may be purely additive, synergistic, antagonistic, or potentiative—interactions that determine the consequences of simultaneous exposure and thus the maximum permissible concentration of pollutants coexisting in a particular environment.

Fortunately, simultaneous exposure to two or more pollutant substances does not mean a priori that the combined effect will be greater than the effect of exposure to either alone. In some instances the interaction between two or more toxic substances or between a toxic substance and a drug reduces the toxicity of a pollutant or enhances the metabolism of a drug.

A well-known example of toxic substances that counteract each other is the case of arsenic and selenium. The antagonism between these two metals may account for the observation that in some areas arsenic does not appear to be a carcinogen. Equally interesting is the apparent antagonism between cadmium and either zinc or selenium; cadmium-induced anemia is prevented at least in part by zinc. Cadmium teratogenesis in animals is antagonized by selenium. Certain human cases of nephrogenic hypertension are associated with increased renal cadmium or increased cadmium/zinc ratio. This renal retention of cadmium is inhibited by zinc or selenium.

Some substances, although themselves toxic, stimulate the production of enzymes that metabolize other toxic substances. Phenobarbital and the pesticides chlordane and DDT dramatically reduce warfarin toxicity in rats. In dogs, small oral doses of chlordane stimulate the metabolism of antipyrine and phenylbutazone; the persistent effects continue to stimulate drug metabolism for months after discontinuation of chlordane ingestion. The continued chlordane effect is presumed to indicate that the pesticide is stored in body fat. Whether pesticides also enhance drug-metabolizing enzymes in man remains unknown.

In man several drugs are known to affect the rate of steroid metabolism; for example, phenobarbital markedly increases cortisol metabolism. Some halogenated hydrocarbon insecticides stimulate drug- and steroid-metabolizing enzymes in liver microsomes, enhance liver growth, promote liver protein synthesis, and accelerate glucose metabolism. Organochlorine insecticides decrease the toxicity of organophosphorus insecticides. Thus DDT increases the rate at which dieldrin metabolites are eliminated. However, a practical application of these observations remains to be discovered.

Pollutants may counteract each other by purely physical means. Oil mists diminish the toxic effects of the oxidant pollutant gases ozone and nitrogen dioxide. This effect may be caused by the capacity of the oil mist to prevent contact with the noxious gas by coating the pulmonary alveolar wall, although it is difficult to understand how this could be accomplished without impeding respiratory gas exchange. Perhaps the oil dissolves the gas and holds it in solution or contains neutralizing antioxidants.

Table 13-1. Toxicologic interaction of pesticides

Type of interaction	Compounds involved	
Antagonism	Aldrin	+ Malathion
	DDT	+ Malathion
	Endrin	+ DDT
	Aldrin	+ Parathion
	Aldrin or chlordane	+ Banol (carbamate)
Potentiation	Endrin	+ Chlordane
	Endrin	+ Aldrin
	Methoxychlor	+ Chlordane
	Methoxychlor	+ Dieldrin
	Aldrin	+ Chlordane
	Chlordane	+ Parathion
	Chlordane	+ Malathion

However, in most instances interactions between toxic substances do not lessen the effects. Examples of synergism and potentiation are far more common than those of antagonism. Combinations of some pesticides greatly increase insect-kill potential (Table 13-1). Piperonyl butoxide increases the potency of some carbamate, organophosphorus, and organochlorine pesticides through its capacity to inhibit the action of metabolizing enzymes. Toxicity of these insecticides for mammals is potentiated by substances structurally similar to piperonyl butoxide. In high doses this substance also potentiates the tumorigenic properties of several known carcinogens.

A recent observation of particular interest is that the toxicity of lindane is twice as great for rats eating a low-protein diet as for those on a standard diet. The implications, if this phenomenon is also valid for man, are far reaching and would require reconsideration of permissible levels of all pollutants in relation to the nutritional status of an exposed population. Equally confusing is the significance of genetic predisposition in determining the reaction to a pollutant. At organismic level, genetic predisposition may be the dominant factor in determining the reaction to any existing concentration of a pollutant.

More complex interactions are involved in tumorigenesis. Carcinogenesis is usually a gradual process, requiring repeated exposures to the agent involved. Skin tumors, for example, are very successfully induced by repeated application of certain aromatic hydrocarbons. In at least one instance, croton oil–hydrocarbon induction of skin tumors in mice, the process is not continuous but seems rather to progress stepwise. In this instance, the carcinogenetic process appears to be started by an "initiator" substance and completed by a "promoter" substance. To be effective, exposure to the initiator substance must precede exposure to the promoter substance. A single exposure to the initiator substance was effective for many weeks, suggesting that the process of tumor initiation is additive and irreversible. The action of the promoter substance, by contrast, was nonadditive, reversible, and affected by the time interval between exposures. Reversal of the sequence of application of the two substances vitiates the process.

Tumorigenic substances can be effective as initiators and promoters when applied topically (surface application) or when taken internally. Similarly, environmental pollutants, regardless of route of entry, may be either initiators or promoters and, therefore, potential health hazards. The fact that an environmental pollutant is not dangerous to health when acting alone does not mean that it cannot be harmful when present together with other seemingly harmless substances.

FOOD ADDITIVES

Assessment of the hazards of chemicals intentionally added to foods is a complex matter. Data for human subjects are inadequate in most instances and extrapolation or prediction from animal studies is risky at best. The truth regarding the health effects of food additives doubtless lies somewhere between the view of the hysterical, who believe that civilization is doomed by slow food additive poisoning, and that of the totally unconcerned. The artificial distinction between food additives and food residues is confusing. There are more than 10,000 compounds in the combined category of food additives and food residues. Most food on the market has been altered by addition or removal of one or more substances. We each consume about 3 pounds of food additives per year.

Most of the additives in food are not deleterious to health in the concentrations present. Indeed, additives may enhance the flavor, texture, appearance, or nutritive value of the natural food item. They may lengthen the shelf life of foods, inhibit bacterial and fungal spoilage, or delay vitamin destruction. However, in the past, certain chemical food additives were used that were capable of inducing cancer, causing allergic reactions, or destroying blood cells. The complexity and magnitude of the problem is emphasized by a task force report on the toxicology of food additives and residues.

TOWARD THE FUTURE

Doomsday predictions are always popular; but the fancied sources of doom change. Famine, war, and pestilence are traditional favorites. The media and the public embrace the opinion of anyone who is a scientist, providing it makes ominous news. Interestingly, the possibility of destruction of life on earth by the impact of a large asteroid, or by the radiation effects of a nearby supernova or solar superflare, has not yet caught the popular fancy, despite the fact that the probability of such a natural catastrophe is indeed finite, though small.

Prediction of ecologic disasters is now in vogue. There was great concern that we would exhaust the atmospheric oxygen supply by burning fossil fuels and, more recently, through destroying marine phytoplankton with DDT. Both of these prophecies have been proved false.

New technologies need not always produce potent clear-cut effects, much less wreak global havoc. For example, operation of a fleet of supersonic transport aircraft might decrease stratospheric temperature through water vapor emission; on the other hand, it might increase stratospheric temperature through particle production. In either case there is no evidence to support predictions of sea-level change or adverse biologic effects as a result of increased ultraviolet transmission.

This does not mean that we can forget about ecologic disasters. On the contrary, each and every side effect of modern technology must be conclusively investigated to determine its possible influence on global climate and on the ocean. Nothing is more important to our future on this planet than to understand the long-range effects of human activity. Geology affords abundant evidence of cataclysms. Climatic stability is unknown, nor do we know when or how the next ice age will be triggered. We must be constantly alert to the possibility of inadvertently setting off an irreversible biospheric chain reaction.

Although we could not survive without the rest of the ecosystem, history suggests that it did without us. However, man is not an intruder on the terrestrial scene; on the contrary he has been an integral part of the ecologic system for a million years. With luck our descendents can look forward to another 5 billion years of coexistence with the terrestrial ecosystem.

A discussion of environmental pollution is not complete without mention of the esthetic aspects. The psychologic reaction of some individuals to the defloration of the environment may have little health impact; nevertheless, fresh air, clean water, wholesome food, and the beauty of nature are essential to the quality of a good life; indeed, they constitute a vital natural heritage. Doubtless man can adjust to an esthetically deprived environment just as he accommodates to other sublethal deprivations; however, the price is too great and the loss too grievous. Communion with the beauty of unspoiled nature—a cloud, a sunset, a star-filled sky, the ocean, a mountain, a tree, or even a flower—pacifies, restores, and inspires us.

Goldberg introduced his 1967 Milroy lectures on The Amelioration of Food[6] with a philosophy that puts the problem of man and environmental pollution in proper perspective; it bears careful reading and serious consideration by all concerned with the pollution problem.

> Life and its environment, primarily the chemical environment of air, water, and food, are an indissoluble continuum. The normal healthy organism is poised in a state of dynamic, and perhaps uneasy, equilibrium between exogenous chemicals and nutrients and the body's endogenous metabolites. Changes in the world around us have presented all organisms with an increasing host of new compounds with which they have to contend. The extent of the chemical environment is so wide that the all-pervading exposure has to be borne in mind when considering any element within the environment, or the interactions between elements, in relation to man. The world's increasing population attests to a successful biologic adaptation to our environment, including the chemicals in it, at any rate, for the time being. While presenting the complexities of chemical exposure, therefore, equal stress must be laid on man's protective mechanisms and powers of adaptation; intra- and extracellular adjustments coupled with hormonal regulation can bring about dramatic changes in the capacity to deal with foreign compounds.

GLOSSARY

acclimatization The physiologic process or state of adjustment of an individual that occurs in response to repeated or prolonged exposure to an unaccustomed stressor. Acclimatization usually results in increased tolerance to the given stressor. Contrast **adaptation.**

adaptation Genetic changes that increase the tolerance of a species to a stressor during generations of exposure. Adaptation favors survival in the presence of the given stressor. Contrast **acclimatization.**

additive biologic effect The total biologic effect of two or more toxicants that equals the algebraic sum of the separate effects.

allergen A substance or material capable of inducing an allergic reaction or state in a susceptible individual. Allergic reactions are characteristically very specific and may involve only minute amounts of the allergen. Although often proteins, allergens may also be of a nonprotein nature.

antagonists A pair of interacting agents each of which tends to nullify the effect of the other.

carcinogen A substance or material capable of inducing cancer.

contaminant A foreign or extraneous substance or material; often one that impairs the quality or reduces the life-supporting capacity of an environment. Compare **pollutant.**

dust Fine dry particles of a substance; often small enough to be easily suspended in the atmosphere.

ecology The branch of biology that deals with the complex network of mutual interactions among living organisms and their environment; the science of organisms as affected by their interrelations and environment.

effluent A liquid, solid, or gaseous emission; the discharge or waste outflow from a machine or industrial process.

environment The totality of all elements—matter, energy, and force fields—interacting directly or indirectly with an organism at any level of physiologic organization.

exposure, acute A brief or transient exposure; in practice, often a brief exposure to a high concentration or intense level of some agent.

exposure, chronic Continuous or frequently recurring long-term exposure; in practice, often a long-term exposure to low concentrations or levels of some agent.

fume The gaseous state of a substance or material that is usually liquid or solid, especially one that is offensive or noxious. Fumes are vapors or smokes; the latter are a condensate of vaporized materials or substances such as metals suspended as fine particles in the air. Compare **vapor.**

gas A state of matter in which the atoms or molecules are almost unrestricted by cohesive forces. A gas has neither definite shape nor volume, its molecules tending to disperse uniformly throughout the entire available space.

initiator A substance that renders cells susceptible to the effects of another substance, for example, an initiator in the sequence of tumorigenesis.

mist Fine droplets of a liquid suspended in an atmosphere.

ontogenesis (ontogeny) A history of the process of growth and development in the early life of an organism.

organism Any living plant or animal; a living biologic system.

particulates Minute discrete particles or fragments of a substance or material.

perfusion, pulmonary Pulmonary capillary blood flow, supplying terminal airways and alveoli.

pollutant A foreign or extraneous substance, material, or agent that impairs the quality or diminishes the life-supporting capacity of an environment. In its most general sense, a pollutant is an undesired substance, not naturally present, that is added to an environment. It may be present in air, water, or soil.

potentiation The combined action of two substances, being greater than the simple algebraic sum of the separate effects of each; increases physiologic activity of a substance. Compare **synergistic biologic effect.**

predisposition The condition of being unusually susceptible to the action of a stressor, substance, or agent.

promoter A substance or material capable of inducing neoplasia after the action of an initiator.

receptor A cell, tissue, or organ that is the site of reaction with a toxicant, substance, or agent. Compare **shock organ.**

sensitizer A substance or material that renders an organism sensitive, hypersensitive, or unusually susceptible to its action or to the action of another substance upon subsequent exposure; also called *immune body, intermediary body, preparator, fixator.* Compare **allergen.**

shock organ The organ or organ system that manifests the greatest reaction to a toxicant, allergen, or stressor.

synergistic biologic effect The total biologic effect of two or more substances being greater than the simple algebraic sum of the separate effects of each; the action of one substance enhancing the action of another. Compare **potentiation.**

threshold concentration The lowest concentration of a foreign or abnormally accumulated compound or substance that has a detectable and usually adverse biologic effect on an organism.

tolerance The inherited or acquired capacity of an organism to endure stressors with minimal displacement of its physiologic regulatory processes.

toxicant Any substance or material that has an adverse effect on a living organism.

tumorigenic The capacity of a substance to cause or induce tumors or neoplasia.

vapor The gaseous state of an element or compound that is liquid or solid under ordinary conditions, for example, water vapor. Compare **fume.**

ventilation, pulmonary The volume of air moved into or out of the lungs per unit time; an aspect of the supply and distribution of inspired air to the gas-exchanging regions of the lung. It is calculated as the product of breathing frequency and tidal volume.

REFERENCES

1. Slonim, N. B., and Hamilton, L. H.: Respiratory physiology, ed. 2, St. Louis, 1971, The C. V. Mosby Co.
2. DeFalcio, P., Jr.: The estuary—septic tank of the megalopolis. In Lauff, G. H., editor: Estuaries, Publ. No. 83, Washington, D. C., 1967, American Association for the Advancement of Science, pp. 701-703.
3. U. S. Department of Health, Education, and Welfare: Air quality criteria for carbon monoxide, Public Health Service, NAPCA, Washington, D. C., 1970.
4. National Academy of Science–National Research Council: Effects of chronic exposure to low levels of carbon monoxide in human health, behavior, and performance, 309-01735, Washington, D. C., 1969, The Council.
5. Slonim, N. B., and Estridge, N. K.: Ozone—an underestimated environmental hazard, J. Environ. Health **31**:577-578, 1969.
6. Goldberg, L.: The amelioration of food, J. Roy. Coll. Phys. **1**:385-425, 1967.

SUGGESTED READINGS

1. American Conference of Governmental Industrial Hygienists: Threshold limit values of airborne contaminants and intended changes, 1014 Broadway, Cincinnati, Ohio, 1970.
2. Cooper, W. C., and Tabershaw, I. R.: Biologic effects of nitrogen dioxide in relation to air quality standards, Arch. Environ. Health **12**:522-530, 1964.
3. Council on Environmental Quality: First annual report: environmental quality, Washington, D. C., 1970, U. S. Government Printing Office.
4. Drury, W. H.: Environmental hazards. Are conservation values limited to esthetics? New Eng. J. Med. **275**:1168-1172, 1966.
5. Ferris, B. G.: Chronic low-level air pollution, Environ. Res. **2**:79-87, 1969.
6. Ferris, B. G., and Whittenberger, J. L.: Environmental hazards. Effects of community air pollution on prevalence of respiratory disease, New Eng. J. Med. **275**:1413-1419, 1966.
7. Jaffee, L. S.: Photochemical air pollutants and their effects on man and animals, Arch. Environ. Health **16**:241-255, 1968.
8. Kern, R. A.: Environment in relation to allergic disease, Arch. Environ. Health **4**:28-49, 1962.
9. Lee, D. H. K., and Minard, D., editors: Physiology, environment, and man, New York, 1970, Academic Press, Inc.
10. Morgan, G. B., Ozolins, G., and Tabor, E. C.: Air pollution surveillance systems, Science **170**:289-296, 1970.
11. Root, W. S.: Carbon monoxide. In Fenn, W. O., and Rahn, H., editors: Handbook of physiology, sect. 3, chap. 43, Washington, D. C., 1965, American Physiological Society.
12. Rossano, A. T., editor: Air pollution control. Guidebook for management, Stamford, Conn., 1969, Environmental Service Division.
13. Stokinger, H. E.: Effects of air pollution on animals. In Stern, A. C., editor: Air pollution, vol. 1, New York, 1962, Academic Press, Inc., p. 282.
14. Stokinger, H. E.: Pollutant gases. In Fenn, W. O., and Rahn, H., editors: Handbook of physiology, sect. 3, chap. 42, Washington, D. C., 1965, American Physiological Society.
15. Stokinger, H. E.: The spectre of today's environmental pollution—U. S. A. brand: new perspectives from an old scout, Am. Indust. Hyg. Assoc. J. **30**: 195-217, 1969.
16. Stokinger, H. E., and Coffin, D. L.: Biologic effects of air pollutants. In Stern, A. C., editor: Air pollution, New York, 1968, Academic Press, Inc., pp. 446-533.
17. Upholt, W. M., and Kearney, P. C.: Pesticides, New Eng. J. Med. **275**:1419-1426, 1966.

18. World Health Organization: Information circular on the toxicity of pesticides in man, VBC/TOX/68, 69, 70.
19. World Health Organization Regional Office: Health effects of air pollution, EURO-1143, Copenhagen, 1967, pp. 264-274.
20. Zavon, M. R.: Modern pesticides—hazard or hope, Cincinnati J. Med. **39**:157-160, 1958.

14 ARTIFICIAL CLOSED ECOLOGIC SYSTEMS

N. BALFOUR SLONIM

For many years physiologists and bioengineers have been fascinated with the challenging problems involved in the design and construction of an artificial closed ecologic system (ACES). A beginning has been made with a search for relevant principles, formulation of questions and problems, and attempts to achieve conceptual integration of the whole. Despite this study and some research, no artificial closed ecologic system has yet been constructed. Indeed, such systems are still in the stage of theory and planning. Our discussion is necessarily prospective, theoretic, and, to some extent, speculative.

The problems of ACES are multi- and interdisciplinary. Contributions will be necessary from both biologic and physical sciences, including physiology, nutrition, microbiology, psychology, bioengineering, toxicology, biochemistry, engineering, physics, and chemistry. Sacred artificial barriers between adjacent fields of science will have to be penetrated. In addition to unsolved engineering problems there are large gaps in our knowledge of the fundamental biologic aspects of artificial closed ecologic systems as illustrated by the following questions.

What, if any, components of the natural terrestrial environment are dispensable? If man is to inhabit an ACES, what precisely are the human nutritional requirements? What contaminants does man produce? What is the human tolerance for such contaminants? What biologic agents could most efficiently and safely convert human wastes to foodstuffs or fuels? What biologic agents could serve as efficient and safe chemosynthesizers? Alternatively, what physicochemical processes could accomplish these same ends? What are the psychologic considerations? What is an optimal environment? What is the measure of the quality of an ACES as a human environment? Is it the capacity to sustain human life in a state of optimal health? Does optimal health imply maximum longevity? What are the essential components, characteristics, constraints, and limitations of an artificial closed ecologic system? What is the minimum volume of an ACES? What is the minimum mass? These questions suggest the nature and extent of the gap between our present understanding of the ACES problem and the successful construction and operation of such a system. Indeed, we have more questions than answers.

An artificial closed ecologic system must contain, at minimum, all essential environmental components in a quantitative, qualitative, and functionally integrated dynamic balance. Furthermore, present knowledge of biorhythmicity suggests that rhythmic fluctuation of some, or perhaps most, environmental component intensities and concentrations will be necessary. A total environmental control system, utilizing monitor input, will be required. Without doubt, the study of artificial closed ecologic systems will accelerate the development and increase understanding of the science of ecology as a whole.

The problems, study, and design of ACES are a tour de force of environmental physiology. As any science, environmental physiology involves first description, next understanding, then prediction, and finally control. In this sense the successful design, construction, and operation of ACES is the ultimate stage in the development of the science of environmental physiology and the ultimate challenge to the environmental physiologist.

The study of artificial closed ecologic systems

began as an exercise in applied environmental physiology. The future will see the creative design and construction of a variety of experimental artificial closed ecologic systems. The subject will develop along various lines of specialization to meet specific research, biomedical, and military needs. Although we first think of *man*-oriented artificial closed ecologic systems, we can envision customized *other*-oriented systems, for example, a system in which the primary organism is a contemporary or even artificial nonhuman species. Finally, the artificial closed ecologic system will become a necessity for the survival of the descendents of the human race. Hopefully, man will develop ACES more rapidly than nature evolved hers.

The subject of ACES is so large, and still so theoretic, that we can only sketch its general outline. Considerations range from structural, thermodynamic, physiologic, and psychologic to philosophic. We will discuss selected problems of ACES in some detail.

PRINCIPLES OF NATURAL ECOLOGIC SYSTEMS

The planet Earth is an *enclosed* ecologic system. It exchanges radiant energy freely with its space environment, receiving more from the sun than it reradiates, and from time to time it receives matter from outer space in the form of meteoroids. Atoms, molecules, and energy cycle through a complex web of interconnecting biospheric channels and loops in the process of which they are used and reused. Within its ecosphere the earth contains a wide variety of natural ecologic subsystems.

A natural ecologic system is a complex network of interrelated, interacting, often interdependent, cyclic processes in a state of dynamic balance. Natural ecologic systems can be divided into two general phases: a physicochemical environmental phase and a biotic community. The source of energy for a natural ecosystem is sunlight, used by photosynthetic plants together with carbon dioxide and water to produce carbohydrates. Each cyclic subsystem involves two or more living organisms. As a result of its organizational structure and functional complexity, a natural ecologic system has certain inherent properties and exhibits characteristic behavior. The presently known principles of natural ecosystems are as follows.

1. As the definition of ecology implies, every ecologic process affects every other process. It is said that every strand of the ecologic web is in dynamic balance with every other strand. However, such physical analogies cannot do justice to the complexity, dynamic behavior, and adaptive responsiveness of a natural ecosystem. Displacement of even a single component process in a balanced ecologic system tends to produce widespread effects; these may be delayed, distant, and large. Even slight perturbation of an ecologic cycle may have profound consequences—a kind of ecologic amplification. The behavior of a balanced ecologic system is crudely analogous to the properties of an open system in a steady state; an ecologic system is intrinsically stable, flexible, self-regulating, and self-compensating. It exhibits adaptive dynamic responsiveness and feedback characteristics.

(a) Each component cycle of a natural ecosystem in its dynamic steady state oscillates naturally with characteristic frequency and amplitude. For example, the population densities of a biotic community fluctuate periodically.

(b) Complexity confers dynamic stability. Branching and alternative pathways increase the resistance of an ecologic cycle to external stress.

(c) Each component cycle has a characteristic reaction time and response rate with respect to external stressors, and each cycle has a characteristic overall turnover rate. For example, because of the relatively slow release of nutrients from humus, the turnover rate of soil ecosystems is generally less rapid than that of aquatic ecosystems.

(d) Each biologic component of an ecologic cycle has characteristic metabolic, reproductive, and growth rates; for homeothermic animals (per unit mass) metabolic rate is inversely related to size.

(e) The rates of the separate processes of an ecologic cycle are in a natural state of dynamic balance. Each cycle contains a rate-limiting process.

(f) The dynamics of any cycle can be perturbed by a wide variety of external stressors. Among these stressors are both relative deficiency (poverty) and relative excess (affluence) of essential environmental components. External factors, acting at sufficient intensity for sufficient duration, can overstress an ecologic subsystem, collapsing the entire cycle.

2. Under ordinary biospheric conditions, matter is neither created nor destroyed; it is only transformed or translocated. Everything used by a living organism is taken from its environment (source), held for a length of time depending on dynamic turnover rates, and subsequently re-

leased in some form into the environment (sink) again. In natural ecosystems, biologic substances are recycled, accumulations of material are generally compatible with life, and pollution is relatively infrequent and of limited extent.

3. Every living organism is the product of 2 to 3 billion years of evolutionary development. In the course of this process biologic systems have accumulated a complex organization of compatible components. A natural ecologic system contains two or more compatible biologic subsystems. As is true of a single living organism, the existing organization of a natural ecologic system is also likely to be better than that resulting from any given random change; indeed, a random change is almost certain to damage a natural ecologic system. Rarely, a random change improves a system.

4. *Food chains* are natural ecologic cycles that begin with autotrophic organisms, such as green plants, termed *producers,* and end with *decomposers,* organisms that decay the substance of *consumers.* Decomposers are usually bacteria or fungi. Food chains have the potentiality of concentrating pollutant substances in the higher animal organisms at the top of the chain.

5. *Energy pyramids* are characteristic of natural ecosystems. In accordance with the second law of thermodynamics, energy is required to raise materials from one *trophic level* to the next higher trophic level of a system.

6. In most natural ecosystems the mass of organisms, or *biomass,* that can be supported at successively higher trophic levels decreases. This decrement involves such factors as population density, organism size, metabolic rate, and life span. The progression of diminishing mass with increasing trophic level is termed the *pyramid of biomass.*

7. Cyclic transformations of mass and energy occur in natural ecosystems. Atoms circulate through the ecosphere in great interwoven cycles. Some move from the environment into living organisms and out again as waste products, into water, and into mineral deposits or fossil remains. Four chemical elements comprise the bulk of living matter—carbon, oxygen, nitrogen, and hydrogen. A few of the many examples of these cyclic transformations are as follows:

Carbon cycle. Green plants assimilate atmospheric carbon dioxide, synthesizing carbohydrates, fats, and, using nitrogen and sulfur, proteins. Animals ingest, digest, and assimilate plant materials, synthesizing their own proteins, fats, and carbohydrates. Organic substrates thus produced are metabolized via the aerobic tricarboxylic acid cycle, producing carbon dioxide, which returns to the atmosphere. Animals and plants die, decomposing or forming carboniferous deposits. Oxidation returns this organic carbon to the atmosphere as carbon dioxide. Some carbon is deposited in mineral carbonates.

Nitrogen cycle. The element nitrogen, which is indispensable to life, is abundant in the biosphere. Nitrogen occurs in the environment in a relatively few simple chemical forms. About 80% of the total biospheric nitrogen is in the atmosphere in the form of diatomic molecular nitrogen, a chemically inert gas, diluting atmospheric oxygen. Much of the remainder is in *humus,* a complex organic mixture in soil that is essential for plant growth. A small, but important, fraction of biospheric nitrogen resides in biochemical compounds within living and dead organisms. These compounds include amino acids, peptides, proteins, nucleic acids, vitamins, hormones, and porphyrins.

Nitrogen enters the soil from decaying biota, animal wastes, and the atmosphere. In the soil it becomes a part of the complex organic mixture called humus. The release from humus of the nitrate and ammonium salts produced by soil microorganisms is the rate-limiting step of the soil nitrogen cycle. These simple nitrogen compounds are absorbed by plant roots and used to synthesize a wide variety of nitrogenous biochemical compounds essential to plant growth and metabolism. Animals ingest, digest, absorb, assimilate, and metabolize plant material, using the nitrogenous compounds. To complete the cycle, animals and plants return nitrogen to the soil as wastes and by decay after death. Thus nitrogen returns to the soil via several pathways and in a variety of biochemical forms, ranging in complexity from urea and uric acid to porphyrins, nucleic acids, and proteins. In the soil these nitrogenous organic compounds are degraded to simple substances such as nitrates and ammonia.

Atmospheric nitrogen is fixed in the soil by certain bacteria, blue-green algae, soil yeasts, and fungi. Some of these nitrogen-fixing microorganisms, such as free-living bacteria of the genera *Azotobacter* and *Clostridia,* exist free in the soil; others, such as bacteria of the genus *Rhizobium,* live in the root nodules of legumes, such as clover; and some are associated with the leaves of certain tropical plants. A small amount of atmospheric nitrogen is fixed by natural physicochemical processes, such as the electric discharge of lightning.

A similar nitrogen cycle occurs in natural waters. Fish eliminate nitrogenous wastes; microorganisms oxidize the nitrogen to nitrate; algae reconvert this nitrate nitrogen to a variety of biochemical compounds; small aquatic animals ingest algal organic matter; and fish in turn eat these small animals.

Water cycle. The biospheric circulation of water is another example of the cyclic transformations that occur in natural ecosystems. The major cycle involves evaporation, diffusion, convective translocation, condensation, and precipitation with minor cycles of ingestion and absorption by animals and plants that use and return it to the environment by various means, including transpiration, evaporation, urination, sweating, and defecation.

We still do not know enough about natural ecologic systems to develop a comprehensive theory or to devise a satisfactory model. These await the accumulation of more information, the identification of other intrinsic properties and general principles, and the further development of biologic systems science, which thus far is inadequate to deal with even a single living organism. Hence, it appears likely that the first successful ACES will result from a combination of design theory and trial-and-error empiricism as a part of which tracer substances will be used to map ecologic pathways.

CLASSIFICATION OF ARTIFICIAL CLOSED ECOLOGIC SYSTEMS

Artificial closed ecologic systems may be classified according to purpose:

1. Research
2. Medical—climatotherapeutic
3. Aerospace—for prolonged lunar and interplanetary spaceflight; for exploration and/or colonization of space
4. Marine—for prolonged exploration and/or colonization of marine environments
5. For storage under total environmental control
6. Civil defense—for use in the event of natural or artificial catastrophe
7. Military

Artificial closed ecologic systems may also be classified by type:

Sealed or totally closed system. Such a system would have minimal outleakage of mass and minimal intake of required solar, nuclear, or chemical energy.

Enclosed or partially closed system. Such a system would have a regulated outleakage of mass and energy and a controlled intake of the required solar, nuclear, or chemical energy and the required materials, such as water. In an enclosed system the restrictions on mass and energy exchange are very specific. For interplanetary travel, the enclosed system may need to extract certain basic materials from extraterrestrial sources for successful operation of its ecologic life-support system.

THERMODYNAMIC CONSIDERATIONS

Thermodynamics—the science of energy, its transfer, and its transformations—is basic to the biosciences. Fundamental thermodynamic principles define the energetic boundaries, setting the limits within which a closed ecologic system must operate. The science of thermodynamics describes gross physical and chemical processes without regard to events at the molecular level, an independence that provides a method of wide generality. Nevertheless, thermodynamic concepts can often be applied to events at the molecular level. Artificial closed ecologic systems will involve reg-

Table 14-1. Classification of energy and work by form and factors

Form of energy	Intensive factor	Extensive factor	Element of work (dW)
Thermal (TS)	Temperature (T)	Entropy (S)	TdS
Mechanical (PV)	Pressure (P)	Volume (V)	PdV
	Force (F)	Distance (L)	FdL
	Surface tension (γ)	Surface area (A)	γdA
Chemical ($\mu_i n_i$)	Chemical potential (μ_i)	Mass or mole numbers (n_i)	$\mu_i dn_i$*
Electric (EQ)	Electric potential† (E)	Electric charge (Q)	EdQ
Magnetic (HM)	Magnetic field strength (H)	Magnetization (M)	HdM‡

*At constant temperature and pressure for calculations involving Gibbs free energy.
†Also called electromotive force (emf).
‡At constant entropy and pressure.

ulated balanced cycling and recycling of mass and energy through selected biologic and physicochemical subsystems. Hence, the laws of thermodynamics that govern the gross behavior of energy are fundamental to the design of such systems.

Some thermodynamic parameters, such as mass and volume, are proportional to the size of a system. Such parameters are called *extensive,* or *capacity,* factors. Other parameters, such as temperature and pressure, are independent of system size; these are called *intensive,* or *potential,* factors. As shown in Table 14-1, *energy* is formulated as the product of corresponding pairs of energy factors. Because one of these two factors is always a capacity factor, energy itself is a function of the size of a system. If two systems equal in all thermodynamic respects combine to form a single system, without the performance of work, the value of the capacity factor doubles, whereas that of the potential factor remains unchanged.

In thermodynamic terms *work* is the transfer of energy from a thermodynamic system to its environment, producing a displacement that is often, but not always, macroscopic. Each of the many forms of work is the product of a thermodynamic *force* and a resulting *displacement* variable. For example, the mechanical work of expansion is the work performed by a gas as it expands against an external pressure:

$$\text{work of expansion} = P\Delta V \quad (1)$$

where P = pressure
ΔV = difference between the initial and the final volumes

Other forms of work relevant to physiology include:

1. The external mechanical work done by a contracting muscle as it shortens through a distance, ΔL, against a load, or force, F:

$$\text{mechanical work} = F\Delta L \quad (2)$$

2. Surface work, for example, the work done by a cell at its growing surface as surface area increases, ΔA, against the surface tension, γ:

$$\text{surface work} = \gamma\Delta A \quad (3)$$

3. Electric work, for example, the work done as the kidney excretes sodium ions carrying a total charge, ΔQ, across the electric potential, E, of the renal tubular cells:

$$\text{electric work} = E\Delta Q \quad (4)$$

Chemical work is fundamental to biologic systems. As shown in Table 14-1, *chemical energy* is the product of a potential factor (chemical potential) and a capacity factor (mass); the unit of mass is the gram-molecule (mole), which differs for various chemical species. Biochemically, electric charges, potentials, and currents are considered to be properties of matter; electric charge is a subclass of chemical mass and electric potential is a subclass of chemical potential. At molecular level chemical energy is indistinguishable from mechanical energy.

As is true for other forms of work, chemical work is also the product of a force (chemical potential) and a displacement (change of mass of chemical species). Chemical potential, μ_i, is the thermodynamic force that drives chemical work as well as certain physical processes, such as diffusion; the direction of change is always such that chemical potential tends to equalize throughout a system. In both chemical and physical processes units of quantity transfer spontaneously from higher to lower chemical potential. Any change of the quantity of a chemical substance within a system represents work done. For example, if Δn_i* moles of a chemical species are removed from a system by either chemical reaction or transport to the surroundings, an energy change of $\mu_i \Delta n_i$ occurs within the system; this energy change is *chemical work.*

Thus chemical potential drives changes of the chemical composition of a system and determines the direction of the change. A given chemical species always tends to enter into a chemical reaction or phase transition that decreases chemical potential to a minimum or to zero, chemical reactions tending to proceed until a state of equilibrium is achieved. At equilibrium the chemical potentials within a system are equal. Unfortunately, there is no simple instrumental measurement of chemical potential. Thermodynamics describes the direction, but not the rate, of a chemical process and applies quantitatively only to closed systems at equilibrium. Nevertheless, the thermodynamic concept of chemical work has many important biologic applications.

Work is defined mathematically as the integral of the element of work:

$$W = \int dW \quad (5)$$

When work is performed the extensive factor of energy becomes a generalized displacement, whereas the intensive factor of energy becomes a gen-

*The subscript i is the entire set of atoms, molecules, and compounds.

eralized force. Thus for mechanical work equation 5 becomes:

$$W = \int PdV \quad (6)$$

In electric work, the generalized force is the electric potential E of the system, and the generalized displacement is the charge dQ that is transferred as the source discharges; the element of this work is EdQ. If a magnetic field of strength H acts on a susceptible substance to produce a magnetization of extent dM in the direction of the field, the element of the work done on the substance is HdM.

The mathematical concept of *entropy* is as important as its consequences are inevitable. *Thermal energy,* or *heat,* is disordered energy. To describe heat energy meaningfully, we must use two numbers, one for the quantity of energy and the other for the quantity of disorder. The quantity of heat energy is measured in terms of calories; the quantity of disorder is calculated in terms of entropy. As shown in Table 14-1, entropy is the capacity factor of thermal energy and thus increases with the size of a system. *Decreasing* entropy is associated with *increasing* organization and availability of energy. Entropy, disorder, and probability are related concepts; the least entropy of a system is associated with its least probable organization, order, or arrangement. If it has more entropy, a large system at low temperature could have more thermal energy than a small system at high temperature. Entropy transfers spontaneously from higher to lower thermal potentials. The value of entropy is calculated and expressed variously in joules/°K, cal/°K, or cal $\times$ mole^{-1} $\times$ °C^{-1}. There is no direct instrumental measurement of entropy.*

Thermodynamic *equilibrium* is a time-invariant state of a closed, or isolated, system in which all variables remain constant. At complete equilibrium a system exhibits no gradient with respect to any energy potential; all energy potentials are equal throughout the system. If energy potentials are uniform throughout a system, there is no tendency within the system toward change and thus no manifestation of energy. If, however, differences of energy potential exist between points in a system, energy transfer tends to occur. When such differences of energy potential exist, a system contains *potential energy.*

In a state of complete equilibrium a closed system has zero capacity to perform work, maximum entropy, minimum free energy, and maximum stability and is statistically in its most probable state; it can undergo no further change, except redistribution of internal energy and entropy, unless energy is added to the system. In an isolated system the fundamental thermodynamic concept of entropy and the basic statistical concept of probability are related as follows:

$$S = K \times \ln w \quad (7)$$

where S = entropy
K = Boltzmann's constant
ln w = natural logarithm of the number of possible molecular configurations

Thermodynamic principles are most easily applied to closed systems at equilibrium.*

Although both are time-invariant states, the steady state of an open system differs importantly from the equilibrium state of a closed system. An open chemical system takes in material from a source, modifies it, and eliminates the end-products of its internal processes; this continuous operation requires energy. The steady state concentrations of compounds within an open system depend only on the concentrations of the compounds in the source and sink and on the rate constants of the chemical reactions involved.

The composition of the system remains constant despite the fact that material passes continuously through the system. A whole organism and its individual component parts all approximate open systems in a steady state. As a result of simplicity and continuous interaction with the internal environment of the organism, the individual organs, cells, and intracellular biochemical reaction systems approximate open chemical systems most closely.

Comparison of the thermodynamic and kinetic properties of closed versus open systems is instructive. The composition of an equilibrium chemical mixture in a closed system depends on the initial concentrations of the reactants; in open chemical systems, steady state reactant concentrations are independent of the initial concentration values. Adding a reactant to a closed system at equilibrium changes the concentrations of all the reactants in the equilibrium mixture; after such an initial displacement, an open system returns to its original steady state.

*There is no direct instrumental measurement of temperature either. A thermometer measures the change of density of mercury.

*Unfortunately, a biologic system at complete thermodynamic equilibrium is dead.

The kinetic properties of open chemical systems present interesting analogies to the behavior of living biologic systems. Open chemical systems exhibit self-regulation and compensation; after a stress applied to such a system has produced an initial displacement, the system tends to reestablish the initial steady state conditions or, if the stress continues to act, the open system assumes a new steady state. Under certain conditions open systems exhibit the phenomenon of "false start": the concentration of a reactant undergoing a transition from one steady state level to another initially moves in the opposite direction away from the new steady state level before subsequently correcting itself. Open systems may also exhibit the phenomenon of "overshoot," transiently exceeding the reactant concentration of a new steady state level. Does the phenomenon of biologic oscillation result from the coupling of two or more open systems within the living organism?

The first law of thermodynamics implies that (1) any given closed system contains a certain quantity of *internal energy,* which is a function of the state of that system; (2) internal energy can be transferred from place to place or converted to other forms of energy within the system; and (3) the sum total of mass and energy within a given closed system remains constant. The first law applies to every thermodynamic process that proceeds between equilibrium states:

$$\Delta E = \Delta Q - \Delta W \tag{8}$$

where ΔE = change of internal energy within the system
ΔQ = heat gained by the system
ΔW = work done by the system

When mechanical work is transformed into heat without a change of internal energy, the amount of work done is always exactly equivalent to the quantity of heat produced. Because all forms of energy are interconvertible, the internal energy of a system may derive from or may produce thermal, radiant, chemical, or electric energy or mechanical work. The concept that energy consists of capacity factors at certain potentials accounts for all energetic changes that occur within a closed system.

A process that occurs at constant temperature is an *isothermal* process; one that occurs at constant pressure is an *isobaric* process; and a process during which a system neither gains nor loses heat is an *adiabatic* process.

A thermodynamic quantity whose integral is zero around any closed path on a pressure-volume diagram is called a *state variable*. The value of a state variable characterizes only the existing state of a system and does not depend upon how that state was arrived at. As the term implies, a set of state variables, all of which can be measured on a macroscopic scale, define the thermodynamic state of a system. Examples of state variables include entropy, internal energy, pressure, temperature, Gibbs free energy, and gravitational potential energy. By contrast, heat and work are not state variables. Unlike the second and third laws of thermodynamics, the first law is applicable to single atoms and molecules. In this sense it is more general.

The second law of thermodynamics states that entropy, the capacity factor of heat energy, always increases during spontaneous processes. A spontaneous process always goes in the direction that increases the entropy of the system-plus-environment. Hence, spontaneous processes have a natural direction. The second law also states that it is impossible to transfer heat from a colder to a hotter body by any continuous self-sustaining process. We can calculate the change of entropy for irreversible processes that begin and end in equilibrium states. We can always convert a given quantity of work completely into heat energy, but we can never convert a given amount of heat completely into work. The limit of our ability to convert heat into work is defined by a theoretically important abstraction—the reversible Carnot cycle. We might wish to achieve a reversible thermodynamic cycle in which a system would return to its original equilibrium state through a sequence of reversible processes. However, all real spontaneous processes are irreversible as a result of phenomena such as friction, fluid turbulence, and electric resistance, which render energy transformations less than 100% efficient; the theoretic maximum of work is never actually achieved. Since, in reality, no perfectly reversible change is possible, every change within a closed system increases entropy. This increase represents a portion of the internal energy change that is not available for useful work. The energy of an isolated system tends to become less and less available. Entropy content increases inexorably toward a maximum —a kind of thermodynamic nirvana.

Living organisms take available energy from a high-energy reservoir, use some to perform useful work, and transfer the remainder to a lower energy environmental reservoir as heat. Animals ingest and degrade complex biochemical compounds of relatively low entropy, such as carbohydrates and proteins, to simple compounds of higher entropy, such as carbon dioxide, water, and urea. Many of these reactions are highly irreversible, increase entropy, and produce heat. By continuously transferring heat to the environment, living organisms are able to maintain their entropy content at a low level for long periods of time. The reversal of a process that is not spontaneously reversible—a common biologic phenomenon—must always be "paid for" by an increase of environmental entropy.

Although entropy change is a test of the spontaneity of a reaction, it is a difficult test to apply because (1) entropy is difficult to measure experimentally and (2) entropy change indicates neither the maximum amount of work obtainable from a process nor the extent to which a chemical reaction will proceed, except under the special circumstance that the internal energy of a system remains unchanged. In contrast, the *Gibbs free energy* of a system is a thermodynamic function that reflects the change of both entropy and internal energy and is thus more useful in determining the spontaneity of a chemical reaction or physical transformation:

$$G = E + PV - TS \tag{9}$$

where for the system G = Gibbs free energy
E = internal energy
P = pressure
V = volume
T = absolute temperature
S = entropy

Gibbs free energy is that quantity of energy that can be removed from a system at constant pressure and temperature and used to perform work. The theoretic maximum amount of nonpressure-volume work that could be obtained as a system undergoes a given change is defined as $-\Delta G$.* For example, a chemical reaction that is the basis for a galvanic cell can perform electric work. As the cell discharges at constant pressure and temperature, the theoretic maximum amount of non-pressure-volume work, in this case electric work, that the chemical system could perform is equivalent to $-\Delta G$.

Because $H = E + PV$, equation 9 may be rewritten as follows:

$$G = H - TS \tag{10}$$

where G = Gibbs free energy
H = *enthalpy*

*The free energy change during a spontaneous process at constant pressure and temperature is negative.

For a change of free energy at constant temperature:

$$\Delta G = \Delta H - T\Delta S \tag{11}$$

where ΔH = change of enthalpy and also the heat of reaction at constant pressure and temperature

For isobaric, isothermal processes ΔH equals the ΔQ of equation 8, a statement of the first law of thermodynamics. Equation 10 shows that total enthalpy, *H*, is the sum of two terms: Gibbs free energy, *G*, and a function of entropy, *TS*. *G* is the part of the total energy that is available to perform work. *TS* is the unavailable, or bound, energy, that part of the total energy represented by the random, disordered thermal motion of the atoms and molecules.

The sign of the value of ΔG predicts whether or not a given change can occur in a system under conditions of constant pressure and temperature. If the calculated ΔG for a postulated reaction is negative, the reaction is possible; conversely, a positive value for ΔG indicates that a reaction cannot occur spontaneously, except in the reverse direction. If ΔG is zero, the system is already at equilibrium. Spontaneous processes that release free energy are *exergonic* ("downhill") and often, although not always, release heat. Nonspontaneous *endergonic* ("uphill") processes require an input of free energy to drive them.

Every chemical compound is characterized by a *standard free energy of formation*, ΔG°_f. This quantity is defined as the free energy change associated with the formation of the compound in its standard state from its elements in their standard states. The standard state of a substance is defined as its most stable form at a pressure of 1 ATA and a temperature of 0° C. The *standard free energy change*, ΔG°, for any chemical reaction can be calculated from the standard free energies of formation of the reactants and products, using the following relationship:

$$\Delta G^\circ = \Sigma\, \Delta G^\circ_f\ (\text{products}) - \Sigma\, \Delta G^\circ_f\ (\text{reactants}) \tag{12}$$

The standard free energy change of a chemical reaction is used to predict whether or not a reaction will occur spontaneously, how far the reaction will proceed, and the theoretic maximum amount of work that can be obtained from it. Free energy change, ΔG, is explicitly related to the equilibrium constant, K,* of a reversible chem-

*In this case K is not only the equilibrium constant of the reaction but also the mass law expression, or ratio of product to reactant concentrations.

ical reaction occurring at any pressure and temperature as follows:

$$\Delta G = \Delta G^\circ + RT \times \ln K \qquad (13)$$

This calculation of the free energy change predicts the direction of a chemical reaction and the extent of the reaction at equilibrium.

As mentioned previously, the Gibbs free energy of a system is a state variable. The free energy changes of a series of reactions are additive. Thus the net free energy change for a transformation that results from a series of reactions is the sum of the free energy changes of the individual reactions. Because of the additive nature of any intermediate steps, the net free energy change can be calculated from the free energy of the initial reactants and the final products alone, without reference to intervening steps and energy changes in the reaction sequence that leads from the initial reactants to the final products.

Gibbs free energy can be evaluated in isobaric, isothermal systems that are doing mainly expansion work and is thus useful in many chemical problems. However, biologic systems do kinds of work other than expansion work; indeed, they are always doing several kinds of work simultaneously. To facilitate analysis of biologic systems, a different free energy function, *total free energy,* has been defined:

$$\theta = E - TS - \Sigma(X_i Y_i) \qquad (14)$$

where θ = total free energy
X_i = set of thermodynamic forces producing displacements
Y_i = various corresponding displacements in the system on which E depends explicitly

For processes occurring at constant temperature in which all the forces in the system remain constant, each term $X_i \Delta Y_i$ is the product of a displacement and its conjugate force and represents one form of work. The sum $-\Sigma X_i \Delta Y_i$ is the net total work of all forms done by the system on the surroundings.

If they are to proceed, *endergonic* chemical reactions require a source of energy to drive them. One source of such chemical work is higher energy compounds. A process is said to be driven "uphill" if it increases free energy, increases internal energy, and decreases entropy. Within a given system, a component "uphill" process must be linked thermodynamically to, and proceed at the expense of, a simultaneously occurring "downhill" process. Except in certain reactions involving radiant energy, energy is transferred from one chemical reaction to another via an intermediate compound that is common to consecutive reactions. Energy transfer by such an arrangement is *thermodynamic coupling* and the reactions involved are *coupled* reactions.

In the process of producing essential compounds and achieving necessary syntheses, biologic systems commonly reverse the spontaneous direction of certain biochemical reactions. Two general types of endergonic reactions occur in biologic systems. An example of one type is the relatively rapid reaction sequence by which ammonia is converted to the metabolic end-product urea. The second type is exemplified by the biosynthesis of large molecules, such as glycogen and proteins. The living organism synthesizes a wide range of such complex biochemical macromolecules from a few kinds of small metabolic fragments by a sequence of chemical transference reactions involving small free energy increments. Within the biologic system, macromolecules are relatively stable, as indicated by their slow turnover rates. Many important biologic processes are endergonic, requiring an energy input to drive them. This is supplied by the common biologic device of thermodynamic coupling of these processes with exergonic biochemical reactions. The most important biologic coupling agents are organic phosphates.

As an open system takes in material from its source, some free energy is lost in uncoupled processes such as digestion and at each spontaneous step in the process as entropy, the grim reaper, takes its toll. A living organism uses much of the free energy that it derives from the metabolic oxidation of substrates to perform biologically useful work, such as the biosynthesis of large molecules (chemical work), muscular contraction (mechanical work), active transport of solutes against concentration gradients across cell membranes (osmotic work), neuronal impulse transmission (electric work), and bioluminescence (photochemical work).

Quantitative analysis of the energy transformations in living biologic systems answers important physiologic questions: How much energy is absorbed or released in a given biologic process? How much of the energy released in a given process is available for useful work? Will a given process occur spontaneously, or does it require the input of energy to drive it? Is an energy-releasing process coupled to an energy-absorbing process and, if so, what is the efficiency of the coupling? What are the overall energy transformations in a given biologic system? How efficient is a given biologic mechanism? Thermodynamic analysis helps us to understand the properties of biologic fluids and macromolecules and to quantify physiologic processes such as active transport across cell membranes.

The thermodynamic state of a closed system is determined by a certain set of parameters. The precise number of such parameters needed varies according to which aspects of the description are unimportant for a particular problem. For example, gravitational field can be ignored in most chemical problems. For the purpose of chemical thermodynamics a system is usually defined in terms of chemical content and two other variables, such as pressure and temperature. The choice of these other two variables often depends on convenience of measurement.

Is it useful to consider an artificial closed ecologic system in terms of its chemical composition, volume, and entropy? Can all other relevant thermodynamic parameters be defined as functions of these three? If so, we may consider the internal energy as a function of the state of a system and, hence, a function of its entropy, volume, and chemical composition:

$$E = f(S, V, n_i) \tag{15}$$

where E = internal energy of the system
S = entropy of the system
V = volume of the system
n_i = chemical composition expressed as the number of moles of independently variable chemical species

An *ideal* artificial closed ecologic system would be a perfectly sealed system in a perpetual state of general dynamic balance from which neither matter nor energy would escape. Each of the living organisms within the closed system is an open subsystem that is usually in a steady, but sometimes in a transient, state; molecules and energy enter and exit the living open subsystems. These biologic units comprise an interacting hierarchy of interdependent open subsystems, forming a network of interconnected channels and circuits through which matter and energy would cycle with perfect efficiency within the closed system. Thus an *ideal* artificial closed ecologic system would be completely self-sufficient and perpetually self-sustaining.

Although it is instructive to consider an *ideal* model of an artificial closed ecologic system, it is immediately apparent that some matter and energy will leak from any real "closed" or "sealed" system. The processes within a real artificial closed ecologic system will never be completely efficient. Thus to operate for a considerable length of time, any real system will require an external source of matter and energy to be supplied at intervals whose length will depend upon both the leakage rate and the overall efficiency of the processes within the system. Energy—for example, radiant solar energy, chemical energy, or nuclear energy—must enter a system to prevent the inevitable increase of entropy. Thus an artificial closed ecologic system, like a living organism, will "feed on negative entropy."

Thermodynamic considerations are fundamental to the design of artificial closed ecologic systems. However, there are significant functional aspects of both biologic and nonbiologic systems that cannot be described in terms of the equilibrium thermodynamic properties of their components. For example, there is no relationship between the internal order of a mechanism and its entropy content. Biologically, the entropy changes associated wth permutations of the monomer sequence in coded polymers—such as proteins, DNA, and RNA—are unrelated to the resulting changes of functional capacity of the polymer. Such a lack of correlation is also true for active transport by cell membranes, contractile structures that perform mechanical work, and nerve structures that generate and recognize patterns.

PHYSIOLOGIC CONSIDERATIONS

Environmental components and control subsystems

Because we do not yet know which, if any, components of the natural environment are dispensable and what physiologic penalties are incurred by their elimination, an artificial closed ecologic system will contain an as yet undetermined number of components regulated by a set of integrated subsystems. Some of these components are represented in Fig. 14-1. At present, it appears that the following subsystems are necessary to monitor and control environmental components:

1. Pressure control subsystem
2. Thermal control subsystem
   a. Temperature
   b. Heat transfer (such as by fluid circuits)
   c. Ventilation
   d. Humidity
3. Atmosphere composition control subsystem
   a. Oxygen
   b. Carbon dioxide
   c. Inert diluent gas
   d. Humidity
4. Water control subsystem

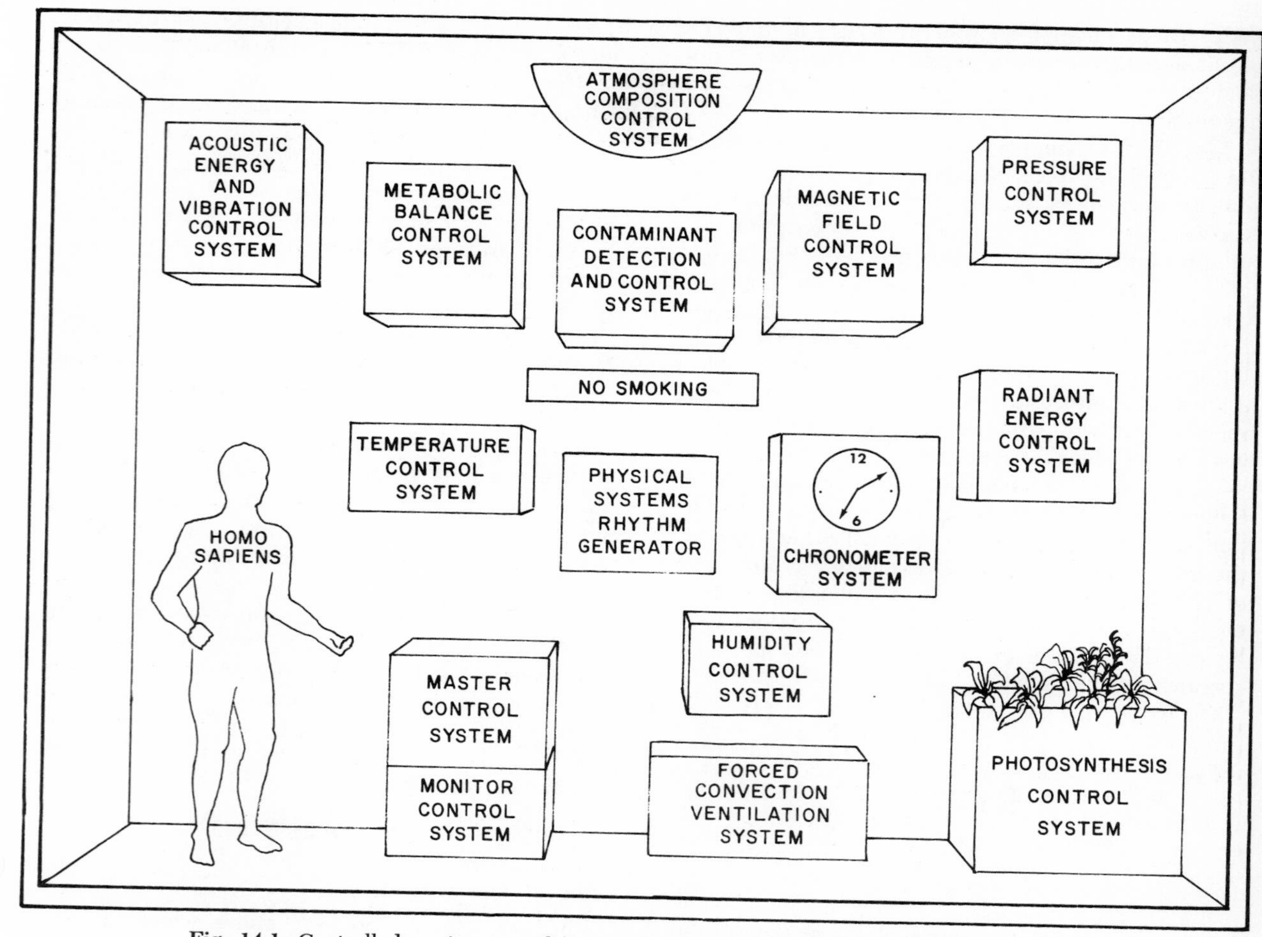

Fig. 14-1. Controlled environmental components of an artificial closed ecologic system.

5. Animal excreta control subsystem
6. Radiant energy control subsystem
7. Chronometer control subsystem
8. Metabolic balance control subsystem
   a. Photosynthesis
   b. Animal metabolism
9. Magnetic field control subsystem
10. Acoustic energy and vibration control subsystem
11. Contaminant detection control subsystem
    a. Microbial
    b. Particulate
    c. Trace gas
12. Physical components rhythm generator
13. Master control subsystem
    a. Monitor input subsystem
    b. Corrective action output subsystem

Thus the successful operation of an artificial closed ecologic system involves the cycling of essential biologic substances such as oxygen, carbon dioxide, water, a variety of nutrients, and animal waste materials; the maintenance of environmental components within physiologic limits; and the detection and elimination of certain contaminants.

Atmosphere

The atmosphere of the first artificial closed ecologic systems will probably be very similar in chemical composition and physical properties to that of the earth's natural atmosphere. This would be the simplest type of atmosphere with which to begin because present life on earth is adapted to it. The major gaseous components of the natural atmosphere—oxygen, nitrogen, carbon dioxide, and water vapor—all serve essential physiologic functions. However, a physiologic necessity for the rare gases—argon, neon, helium, xenon, krypton, and radon—has not been established. The physical properties of an atmospheric gas mixture, including density, viscosity, and thermal conductivity, affect thermoregulation and respiratory resistance, work, and flow patterns. The partial pressures of the atmospheric gases are the physiologically effective parameter.

Living organisms readily tolerate a wide range of total ambient pressures. Total gas pressure should be high enough to permit inclusion of an appreciable concentration of a physiologically inert diluent gas, such as nitrogen, in addition to oxygen and carbon dioxide. The density of a given gas mixture is a direct linear function of ambient pressure.

The gaseous composition of the atmosphere should include oxygen at a P_{O_2} of approximately 150 torr and a sufficient concentration of nitrogen, or other inert diluent gas, to minimize fire hazard and to protect against atelectasis caused by complete absorption of gas sequestered within unventilated lung regions.

Carbon dioxide, present in the natural atmosphere in the low but vital concentration of 0.0314% (0.24 torr), may rise to a P_{CO_2} of about 10 torr without impairing the health or ordinary performance of the human occupants. Because the P_{CO_2} of natural atmospheric air is the rate-limiting factor of natural photosynthesis, a slight P_{CO_2} increase would increase the rate of photosynthesis in systems involving photosynthetic plants without causing more than a slight loss of maximal physical work capacity in the human occupants. Responses to carbon dioxide–containing atmospheres are discussed in detail in Chapter 12.

Water vapor tension and relative humidity can vary widely depending on the comfort and acclimatization of the human occupants of the ACES. However, humidity should be high enough at all times to maintain the moistness of the respiratory tract essential to normal function of pulmonary clearance mechanisms. On the other hand, humidity should not be so high as to interfere with thermoregulation or induce unnecessary sweating.

An example of how an environmental component, the water vapor content of an artificial atmosphere, can affect an important physiologic parameter, alveolar oxygen pressure, is as follows. Specifically, how does the water vapor pressure of an artificial atmosphere affect the alveolar oxygen pressure of a human occupant if total ambient pressure and atmospheric oxygen pressure are held constant by a control subsystem at 760 torr and 159 torr respectively? Dalton's law of partial pressures indicates that as inspired P_{H_2O} increases, the partial pressure of the inspired diluent gas P_{N_2} decreases and the fraction of inspired oxygen* (F_{IO_2}) increases; conversely, as inspired P_{N_2} increases, F_{IO_2}* decreases. As inspired P_{H_2O} decreases, F_{IO_2} decreases. Using the alveolar gas equation, in which alveolar P_{O_2} is a function of F_{IO_2},* calculations indicate that, under the stated conditions, alveolar P_{O_2} increases about 1 torr for each 5-torr increase of ambient P_{H_2O}.

The chemical composition and physical characteristics of an artificial atmosphere may thus vary considerably within presently definable limits. The artificial atmosphere must mesh precisely with the network of cyclic processes in an ACES. Identification and elimination of pollutant substances of animal, plant, and nonbiologic origin is a significant problem, familiar in everyday life as the problem of air quality. An atmosphere contamination control subsystem will monitor microbial, particulate, and trace gas contamination of biologic and nonbiologic origin. Human occupants will produce metabolic carbon monoxide at a fixed slow rate. Certain mutant strains of algae also produce carbon monoxide as well as oxygen. Can soil bacteria, which ordinarily remove carbon monoxide from the natural atmosphere, perform this function in an ACES, or is a physicochemical carbon monoxide removal system preferable? The list of gaseous contaminants to be encountered also includes methane, ammonia, and ozone.

Nutrition and metabolism

Considerations of human nutritional requirements provide a good example of one type of problem involved. Table 14-2 shows the average daily overall metabolic requirements for one man. What precisely are the human nutritional requirements? What is optimal nutrition? The subject is discussed in terms of macronutrients, the energy-yielding dietary constituents, and micronutrients, which are necessary for metabolic processing of the fuel substances.[2] Considerations range from animal coprophagy to diet acceptability, a practical subject that has important psychologic implications.

Nitrogen balance is important. Because nitrogen is lost in urine, sweat, and feces, there is a dietary nitrogen requirement. On the average proteins contain about 16% nitrogen; however, the protein of the ordinary diet is not fully utilized. Is the specific dynamic action of protein important in computing energy balance?

*Dry gas basis.

Table 14-2. Average daily metabolic requirements of a 70-kgm, 25-year-old astronaut,* ordinary spacecrew activity

Parameter	Measured value
Oxygen uptake	0.862 kgm
Carbon dioxide output	1.056 kgm
Drinking water	2.5 L
Food-rehydrating water	1.0 L
Food	3,000 kcal
Water output	
Urine	1.6 L
Respiration and perspiration	2.13 L
Feces	0.09 kgm
Total heat output	11,100 BTU

*Data from Commoner, B.: The closing circle; nature, man, and technology, New York, 1971, Bantam Books.

The essential trace elements are iron, iodine, copper, manganese, zinc, cobalt, molybdenum, selenium, and chromium. What are the daily dietary requirements? What is the ratio of minimal intake sufficiency to the toxic intake levels? Are there other essential trace elements that remain unknown? Interrelationships and balance among the micronutrients are important. They may compete, inhibit, or enhance each other. The valence state of the ions may determine their nutritional or metabolic availability. For example, chromium must be trivalent. Chemical availability is important for both nutritional value and toxicity of trace elements. Availability of the trace elements varies greatly with the particular chemical compound in which they occur; they may be readily available or they may be sequestered in unusable form. In considering both availability and toxicity, one must think in terms of trace-element compounds.

Other nutritional problems involve the requirements for vitamins and essential amino acids. There is also the problem of lipid peroxidation, antioxidants, free radicals, and free radical scavengers in relation to aging, biologic membranes, and the stability of fat-containing foods. Compounding these problems are species differences of every type.

Although many metabolic pathways are known, the natural physiologic autoregulation of metabolism in both animals and plants is not fully understood. In the intact organism a multitude of biochemical reactions and processes are integrated into a regulated unified network. The individual biochemical reaction rates are affected by temperature, pressure, reactant concentrations, and enzyme catalysis. The effect of temperature on the metabolic rates of biologic systems has been extensively studied and is rather precisely predictable.

Plant photosynthesis, metabolism, and productivity are affected by temperature, light intensity, P_{CO_2}, and P_{O_2}. There are two major pathways of carbon dioxide fixation in plants: the reaction of carbon dioxide with phosphoenolpyruvate (PEP) in the presence of the enzyme *PEP carboxylase* to form oxaloacetate, and the reaction of carbon dioxide with ribulose-1,5-diphosphate (RuDP) in the presence of the enzyme RuDP carboxylase to form 3-phosphoglycerate. These carbon dioxide–fixing reaction sequences are both affected by ambient P_{O_2} and P_{CO_2}.

Photorespiration is a metabolic process that occurs in the peroxisomes (subcellular organelles of plant cells), releasing carbon dioxide. By contrast, dark respiration occurs in the mitochondria. Photosynthetic carbon dioxide fixation minus the carbon dioxide released by both forms of plant respiration is *net photosynthesis.* Plant species vary greatly with respect to efficiency of photosynthetic carbon dioxide fixation, and a variety have been examined for possible use in artificial closed ecologic systems. Drake and associates[3] considered the relative advantages and disadvantages of algae such as Chlorella, hydrogen bacteria such as Hydrogenomonas, other plants, and possible food-producing biologic subsystems.

Precise knowledge and complete understanding of the natural physiologic control mechanisms will eventually permit design of a metabolism control subsystem that can regulate individual metabolic processes and metabolic rates and balance them in relation to each other and to the overall requirements of an artificial closed ecologic system.

Biorhythmicity

Biorhythmicity is discussed in detail in Chapter 2. The endogenous biologic clock of an organism is composed of many oscillating subsystems that are synchronized by coupling. Environmental periodisms, some pervasive, are clues that act as synchronizers, or Zeitgeber. Environmental rhythms do not force oscillation upon a living biologic system; they entrain existing endogenous biorhythms. Oscillations of atmospheric pressure, as well as those of the more obvious Zeitgeber, can entrain endogenous circadian biorhythms.[4] The mode of action of environmental factors is not always obvious. For example, the environmental light cycle affects at least some animal species via an extraoptic photoreceptor pathway. Precise phase relationships exist between manifest biologic rhythms and natural environmental periodisms, and these relationships can be altered by certain phase-shifting influences.

What are the physiologic consequences of desynchronization of the endogenous circadian system from its environmental Zeitgeber? What are the effects of uncoupling the endogenous oscillating subsystems from each other on psychomotor function and physical performance? What are the long-term effects of deprivation of rhythmic environmental light, temperature, or pressure fluctuations? At present we may say that cyclic fluctuation of environmental component intensities within an artificial closed ecologic system may be desirable, or even necessary. The full physiologic significance of rhythmic fluctuations of environmental component intensities for an artificial closed ecologic system remains a fascinating question for the future.

FUNCTIONS OF THE CONTROL SYSTEM

A complex analytic system will make a wide variety of chemical and physical measurements. Monitors will provide information on all environmental components, including temperature, pressure, humidity, atmospheric composition, and contaminants. These monitors will determine the location and measure the concentrations and the rates at which constituent chemical components cycle within the artificial closed ecologic system. Using monitor information the master control system* will determine the quantitative and qualitative performance of the artificial ecology, assess its functional balance, and, when certain limits are exceeded, select and institute appropriate preventive, corrective, or remedial actions to maintain its health.

Will all the subsystems be integrated by function so that the mass-product outputs of one subsystem are the inputs of another? Will the subsystems also be integrated by common energy sources and sinks? Will subsystem energy demands be met by thermal transport fluids and electricity? Maximizing the efficiency of individual processes and the overall operation as well as solving the related problem of environmental thermal control are challenging problems for the future. Certainly, much remains to be learned about the properties and behavior of closed complex cyclic systems.

*An ecologic big brother.

PERSPECTIVE

Sometime in the year 5 billion A.D. the last group of scientific observers on the planet Earth slowly and solemnly boards a large interstellar spacecraft to seek a new home in a distant solar system. The earth, as predicted, is beginning to roast in the final agony of the senescent star we called the sun; the sun, having exhausted its hydrogen fuel supply, is becoming a red giant star. Discussion has long since ceased pro and con the wasteful expenditure of resources on the exploration of space. As the last highly evolved descendant of *Homo sapiens* extinguishes his cigarette to board the spacecraft, he pauses for a long last look at the heart-rending devastation of what has been the home of the human race since its evolutionary origin little more than 5 billion years earlier. This is no ordinary leave-taking; this is no au revoir.

The polar ice and snow have melted. The oceans have risen to cover all of the earth's surface except the highlands. Increasing inexorable heat has made much of the remaining dry land uninhabitable. The earth is covered with dead and dying plants and the prostrate rotting carcasses of uncounted animals. Those animals that still survive swarm toward the cooler regions of the earth. Except for a relative few, the plants and animals cannot be saved; they will be cremated with the rest of the biosphere; this is the final and absolute holocaustic sterilization of life on earth. The survival of the human race now depends on the *artificial closed ecologic system* within the spacecraft.

GLOSSARY

artificial closed ecologic system (**ACES**) A sealed self-sustaining balanced system containing an artificially designed environment within which interdependent biologic units cycle matter and energy; a system capable of independent operation for a prolonged, but not indefinite, period of time. Such a system is analogous to the natural ecologic system with its interdependent fauna and flora and crudely analogous to a "balanced" aquarium with its plants and animals.

bioregenerative life-support system A system capable of sustaining life for a limited period of time within which *living organisms* regenerate atmosphere and food. Compare **closed system** and **regenerative life-support system.**

closed system A system that exchanges neither matter nor energy with its surroundings. Compare **open system.**

ecology A branch of the science of biology; the study of the complex web of interactions, interrelationships, interdependencies, and processes that link living organisms to each other and to their natural environment.

ecosystem Ecologic system.

enclosed ecologic system A partially closed ecologic system that takes in certain matter and energy, transforms them by means of physicochemical processes to achieve an internal dynamic balance, or steady state, and releases or expels matter and energy as waste.

open system A system that takes in matter and energy, transforms them by means of physicochemical processes, and releases or expels matter and energy as waste. For example, a living organism is an open system in a steady state. Compare **closed system.**

regenerative life-support system A system capable of sustaining life for a limited period of time within which *physicochemical processes* regenerate atmosphere and food. Compare **closed system** and **bioregenerative life-support system.**

REFERENCES

1. Talbott, J. M.: Life support in space operations, Air Univ. Rev. **16:**42-52, 1965.
2. Calloway, D. H., editor: Human ecology in space flight, vol. III, New York, 1968, The New York Academy of Sciences.
3. Drake, G. L., King, C. D., Johnson, W. A., and Zuraw, E. A.: Study of life-support systems for space missions exceeding one year in duration. In Ames Research Center: The closed life-support system, NASA SP-134, Washington, D. C., 1967, National Aeronautics and Space Administration.
4. Hayden, P., and Lindberg, R. G.: Circadian rhythm in mammalian body temperature entrained by cyclic pressure changes, Science **164:**1288-1289, 1969.

SUGGESTED READINGS

1. Commoner, B.: The closing circle; nature, man, and technology, New York, 1971, Bantam Books.
2. Conkle, J. P.: Contaminant studies in closed ecological systems at the U.S.A.F. School of Aerospace Medicine, AMRL-TR-66-120, U.S.A.F. Medical Research Laboratory, 31-52, Dec., 1966.
3. Conkle, J. P., Mabson, W. E., and Adams, J. D.: A detailed study of contaminants produced by man in a space cabin simulator at 760 mm Hg, SAM-TR-67-16, U.S.A.F. School of Aerospace Medicine 1-142, March, 1967.
4. Conkle, J. P., Mabson, W. E., and Adams, J. D.: Detailed study of contaminant production in a space cabin simulator at 760 mm of mercury, Aerospace Med. **38:**491-499, 1967.
5. Cragg, J. B., editor: Advances in ecological research, vols. 1-6, New York, 1962-1969, Academic Press, Inc.
6. Darnell, R. M.: Organism and environment: a manual of quantitative ecology, San Francisco, 1971, W. H. Freeman and Company, Publishers.
7. Department of the Navy: Physiological evaluation of Sealab II. Deep submergence systems project office, Department of the Navy, Med. Ann. D. C. **37:**313-315, 1968.
8. Earls, J. H.: Human adjustment to an exotic environment: the nuclear submarine, Arch. Gen. Psychiat. (Chicago) **20:**117-123, 1969.
9. Gall, L. S., and Riely, P. E.: Microbial interactions of men and their environment inside a closed system, Contam. Contr. **6:**20-21, 1967.
10. Gall, L. S., and Riely, P. E.: Effect of diet and atmosphere on intestinal and skin flora, I. Experimental data, NASA CR-661, Washington, D. C., April, 1967, National Aeronautics and Space Administration.
11. Hamilton Standard, Division of United Aircraft Corporation: Alternate mission studies (AILSS), NASA CR-66876, Windsor Locks, Conn., July, 1969, United Aircraft Corp.
12. Holm-Hansen, O.: Ecology, physiology, and biochemistry of blue-green algae, Ann. Rev. Microbiol. **22:**47-70, 1968.
13. Katchman, B. J., Murphy, J. P. F., Linder, C. A., and Must, V. R.: The effect of cabin temperature on the nutritional, biochemical, and physiological parameters of man in a life support systems evaluator, AMRL-TR-67-107, Aerospace Medical Research Laboratory, Wright-Patterson Air Force Base, Ohio, Dec., 1969.
14. Kosmolinsky, F., and Dushkov, B.: Specific features of adaptation of a human organism to prolonged stay in sealed chambers, Aerospace Med. **39:**508-511, 1968.
15. MacNamara, W. D., and Nicholson, A. N.: Study of the effect of cabin environment on insensible water loss, Aerospace Med. **40:**657-659, 1969.
16. Margalef, R.: Perspectives in ecological theory, Chicago, 1968, University of Chicago Press.
17. McDonnell Douglas Astronautics Company: 60-day manned test of a regenerative life support system with oxygen and water recovery. Part II: Aerospace medicine and man-machine test results, NASA CR-98501, Washington, D. C., 1969, U. S. Government Printing Office.
18. Miller, R. L., and others: Design and preliminary evaluation of a man-rated photosynthetic exchanger, SAM-TR-69-64, U.S.A.F. School of Aerospace Medicine, Brooks Air Force Base, Oct., 1969.
19. Mohlman, H. T., Katchman, B. J., and Slonim, A. R.: Human water consumption and excretion data for aerospace systems, Aerospace Med. **39:**396-402, 1968.
20. National Aeronautics and Space Administration and American Institute of Biological Sciences: Bioregenerative systems, NASA SP-165, Washington, D. C., 1968, U. S. Government Printing Office.
21. Olcott, T. M., Conner, W. J., and Helvey, W. M.: Manned test of a regenerative life support system, Aerospace Med. **40:**153-160, 1969.
22. Parker, J. F., Jr., and West, V. R., editors: Bioastronautics data book, ed. 2, NASA SP-3006, Washington, D. C., 1973, National Aeronautics and Space Administration.
23. Pearson, A. O., and Grana, D. C., editors: Preliminary results from an operational 90-day manned test of a regenerative life support system, Langley Research Center, NASA SP-261, Washington, D. C., 1971, National Aeronautics and Space Administration.
24. Pielou, E. C.: Introduction to mathematical ecology, New York, 1969, John Wiley & Sons.
25. Pitts, J. N., Jr., and Metcalf, R. L.: Advances in environmental sciences and technology, vol. 1, New York, 1969, John Wiley & Sons.
26. Roth, E. M.: Physiological effects of space cabin atmospheres, Radiat. Res. **7:**413-422, 1967.

27. Saunders, R. A.: A dangerous closed atmosphere toxicant, its source and identity, AMRL-TR-66-120, U.S.A.F. Med. Res. Lab. 53-9, Dec., 1966.
28. Saunders, R. A.: A new hazard in closed environmental atmospheres, Arch. Environ. Health (Chicago) **14**:380-384, 1967.
29. Scientific American: Biosphere: A Scientific American book, San Francisco, 1970, W. H. Freeman and Company Publishers.
30. Slonim, A. R.: Waste management and personal hygiene under controlled environmental conditions, Aerospace Med. **37**:1105-1114, 1966.
31. Slonim, A. R.: Effects of minimal personal hygiene and related procedures during prolonged confinement, AMRL-TR-66-146, U.S.A.F. Med. Res. Lab. 1-25, Oct., 1966.
32. United Aircraft Corporation: Trade-off study and conceptual designs of regenerative advanced integrated life support systems (AILSS), NASA CR-1458, Washington, D. C., Jan., 1970, National Aeronautics and Space Administration.
33. Walters, J. D.: Physiological and hygiene problems involved in the study of enclosed and sealed environments, Ann. Occup. Hyg. **11**:309-320, 1968.
34. Weiss, H. S., Pitt, J. F., and Kreglow, E. S.: Three-week exposure of rodents to a neon enriched atmosphere, Aerospace Med. **39**:1215-1217, 1968.

15 AN EVOLUTIONARY PERSPECTIVE ON ENVIRONMENTAL PHYSIOLOGY

PAUL T. BAKER

In 1969 C. Ladd Prosser described the goals of environmental physiology as (1) description of adaptive variation in organisms, (2) discovery of the origins of the variations, and (3) explanation of adaptive variation by organizational level.[1] In this case he defines the word "adaptive" as any response that permits physiologic functioning and survival in a specific environment. Inevitably any attempt at perspective on the subject of environmental physiology will involve severe terminologic problems, but accepting Prosser's definition for the moment, his description of the goals of environmental physiology still fails to encompass many of the views expressed by authors in this book and is even at variance with the definition offered by the editor in his preface. Such a conflict in definition does not mean that one definition is correct while the others are in error. Instead it reflects the constant conflict of opinion between men desiring to organize the rather diffuse state of human knowledge. In this final chapter I will, therefore, try to avoid definitional issues as much as possible by attempting simply to give an overview of the information generated by environmental physiologists.

No matter how much one wishes to avoid the terminologic problems, unintended information is communicated unless authors define their use of key words very carefully. In this chapter there is no intent to set forth a group of "correct" definitions, but it is obviously necessary to state how some terms will be used and how this usage differs from that of some of the other contributors, and to suggest why I feel that eventual agreement on a standard terminology is necessary.

If we examine all of the contributions to this book and other publications in environmental physiology, we find that Prosser's definition of the field would accommodate all the activities under this heading only if his definition of adaptation included all physiologic *responses* to environmental stresses. Certainly Behnke's description in Chapter 11 of what happens to men in hyperbaric environments does not purport to include only those responses that indicate physiologic improvement during such exposure. It thus seems that environmental physiology is concerned with the total range of physiologic responses shown by organisms when exposed to environmental stresses. Having described the responses, the next goal is obviously to describe how these responses occurred. It is at this stage that the investigator often resorts to the concept of adaptation. In general this concept is used to define those response changes that tend to return physiologic parameters to those values found in the nonstress condition. It is usually assumed that with a passage of time return toward nonstress values reflects improved functional capacity of the organism. In the search for general explanatory systems the investigator may also invoke such holistic concepts as homeostasis, feedback mechanisms, and rate functions.[2]

The preceding definition of adaptation satisfies the research needs of many environmental physiologists, but as the subject progresses and more basic causal factors are sought, this simple use of the word *adaptation* loses precision and comes into conflict with other uses. Some shifts toward nonstress values occur with very brief exposure to stress, others require exposure during critical times in the ontogenetic process, and yet others appear to be based on taxonomic or individual differences in gene structure. This need for more careful description of "adaptive" timing has led to the use of a variety of words such as "acclimatization" or "acclimation" in reference to short-term changes, while long-term changes are often labeled "ontogenetic adaptation" and "genetic adaptation."[3] Even this kind of terminology can frustrate communication across interdisciplinary boundaries since, to the evolutionary geneticists, genetic adaptation has a slightly different meaning than the one implied by the physiologists. The evolutionist would consider a gene frequency change in a population to be an adaptation unless it were caused by mutational, chance, or gene flow factors.[4] Thus a gene causing blindness in man could be considered an adaptive one if the social situation led to higher reproduction and survival among the blind. On the other hand, the evolutionist might not accept that a gene increasing cold tolerance was an adaptation unless it could also be shown that this gene had increased in frequency over generations. The major differences in definition arise because the evolutionist is measuring the effect of a gene on the total performance (particularly reproductive) of all the gene bearers while the environmental physiologist is primarily concerned with how a given genetic structure affects the physiologic functioning of the particular gene bearers in the presence of specific environmental stresses. Despite these differences in interests and definition the two disciplines share the common goal of trying to understand the causes of physiologic variation. Acceptance of a common terminology would reduce mutual misunderstanding and consequently open new paths for the development of knowledge about the nature of adaptation.

Although the terminology of the field remains disparate, it is still possible to evaluate the available knowledge. The details of the physiologic responses vary considerably, according to the type of stress and taxonomic unit involved. Thus data on new stresses such as low gravity environments are primarily descriptive as compared to human responses to temperature, where description is more complete and most of the studies are concerned with response mechanisms. In contrast our knowledge of temperature responses in nonhuman primates is much less complete than it is for man.

With such diversity in the level of information, it may be more informative to review the basic descriptive goals of the subject before examining the mechanisms.

INDICES OF STRESS AND ADAPTIVE LIMITS

Indices of stress

In measuring the response of a given animal species to its environment, the environmental physiologist often begins with what is termed the neutral state. For example, an air temperature between 26.7° and 29.4° C is considered the neutral state for man. Within this temperature range a resting man without clothing or exposure to external radiation shows none of the physiologic reactions that characterize his responses to cold or heat. The neutral zone for temperature, air pressure, vapor pressure, and the like seems to represent the external conditions to which the particular species is best adapted in an evolutionary sense of the term.[5] Thus neutral zones vary significantly from one taxonomic group to another. The neutral zone concept is best applied to systems and species that respond homeostatically to environmental variation, but even so-called environmental conformers such as poikilotherms tend to have an environmental range within which they show a central response. In a statistical sense the neutral zone conforms to the mode of a response curve.[6]

When environmental conditions vary from the neutral zone they are usually considered to constitute a stress, since the organism responds functionally to maintain the conditions found during the neutral environment. If external temperature rises or falls, responses tend to maintain the body core at a constant temperature. If air pressure decreases, oxygen transport to the cells is enhanced, and even at the cell level, responses tend to maintain cell integrity.

Stress level is thus an index measuring the extent of a physiologic displacement from the neutral state, and the physiologic measurement is often termed an index of strain on the assumption that the greater the physiologic response the nearer the system is to collapse.[7] Collapse is the adaptive limit of the homeostatic system and often the adaptive limit of the total organism since organismic death may be the consequence. Fig. 15-1 illustrates this concept and indicates how individuals may vary through ontogenetic or genetic change.

For comparative environmental physiology the only measure taken is often the median lethal limit of a species, and a taxonomic comparison of the effects of acclimatization may be measured entirely by median lethal conditions. A definition of adaptive limit by this method is not totally satisfactory because it is usually based only on adult organisms. Population maintenance requires that the total life span of the species or population be studied since survival in an evolutionary sense involves all stages of the life span. For example, a definition of the adaptive limits of adult salmon would tell virtually nothing about the prob-

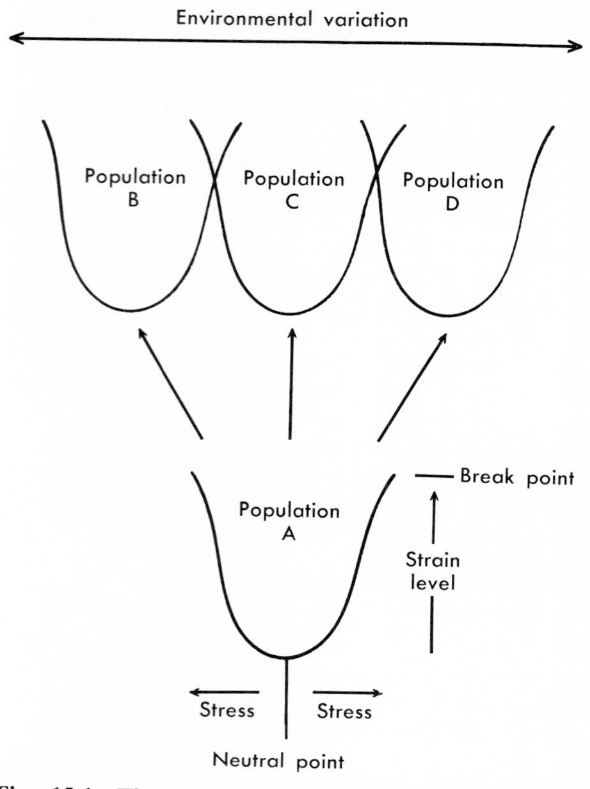

Fig. 15-1. The curves represent the response of the organism to environmental departure from neutrality. As the organism approaches its strain limits there is no further response until the breaking point; at this point the homeostatic mechanisms often collapse. Populations *B, C,* and *D* may vary from *A* as a product of either irreversible acclimatizational response (ontogenetic change) or by the process of natural selection, creating genetically different populations.

ability of salmon survival as a species unless adaptive limits were also known for the reproductive period and the adaptive limits of the fry were specified. In a related sense we know that adult man easily survives environmental conditions such as *E. coli* infections that kill a high percentage of human beings between the ages of 1 and 2.[8]

The time lag factor complicates the use of median lethal level as a measure of adaptive limit still further. For a stress such as temperature the immediate death level may offer a meaningful measure, but for other stresses such as exposure to ionizing radiation such a measure may be of little value. Ionizing radiation may produce not only reproductive death without corporeal death but it may also produce delayed somatic death. Thus adaptive limits become a rather meaningless measure, and as shown in Chapter 7 a series of stress indices must be devised to measure the joint effect of (1) immediate death, (2) life span reduction, and (3) reproductive decline.

These three measures probably offer sufficient indices for most of the purposes of comparative environmental physiology, and sound conclusions comparing species or subspecies can be drawn from such comparisons. Nevertheless, the amount of information needed for a complete description of the adaptive limits of any species remains prodigious. Not only are there a wide variety of environmental stressors that require study, but one must also consider stress interactions. For example, a description of the water temperature tolerance of lobsters is meaningful only if the oxygen content and salinity of the water are also specified.[9] By the same token the adaptive limit of a human infant to temperature change is bound to other environmental factors such as caloric intake and level of parasitic infestation.[10, 11]

Research on man creates special problems as well as opportunities in environmental physiology. The standard measures of adaptive limits cannot be determined except by compilation of data relating to naturally occurring death, and as a consequence even the temperature tolerance limits for man can be only roughly judged. Most research must be based on stress indicators instead. In the development of stress indices the assumption must be made that the indicators are related to the various survival measures. This assumption has been generally validated but the relationship is not precise. If one group of men has a rectal temperature of 37.8° C and another group 38.3° C while working in the heat, it is generally assumed that the first group has a greater tolerance for ambient heat.[12] This is probably true, but at our present level of knowledge we must state that it remains an unproved assumption because we cannot establish that the 38.3° C group will die faster if driven to higher temperatures.

Descriptive information

A reading of the chapters in this book quickly indicates a wide variation in the knowledge of adaptive limits of organisms to various facets of the physical environment. Perhaps responses to ambient temperature change can be singled out as the environmental stressor to which the response limits of organisms is best defined. This does not mean that the information is complete, but for some species such as man the limits of tolerance and their physiologic consequence can be described in general terms for short-term and even some generational effects.[13-17] On this subject the investigator has sufficient information to proceed into the nature and cause of species differences or to examine the causes of individual and population differences. At the other extreme the tolerance of organisms for magnetic fields or pollutants remains poorly defined even for acute exposures. What will happen to man after long-term exposure to relatively new environmental components such as pollutants constitutes a pressing problem in the modern world. One reason for a large difference in the available information is, of course, related to the length of time that an environmental factor has been known to scientists or has impinged on human awareness. In 1950 mercury poisoning was an oddity while the response of man to weightlessness or intense magnetic forces was a subject of untestable curiosity.

Multiple stress and cross adaptation

While considerable further research is necessary to define properly the action of specific environmental factors on physiologic function, a still less well-explored descriptive area is that of stressor interaction. As pointed out earlier, some environmental stressors tend to act in synergistic or potentiating fashion. Numerous examples are cited in the preceding chapters but a particularly clear one is provided by men at high altitude. As altitude increases both oxygen pressure and temperature decrease. For a technologically unprotected man this means an increased oxygen consumption at the same time that oxygen availability is reduced. Under these conditions both temperature and hypoxia tolerance are reduced.[18-20]

While environmental physiologists are intensely aware of the interacting nature of environmental stressors, it has proved difficult to develop indices for multiple simultaneous stressors.[21] Thermoregulatory physiology has probably been the most successful in developing indices such as those combining temperature, wind velocity, and water vapor pressure. Even for thermoregulatory physiology it has been difficult to find satisfactory values for the effects of radiation and insulation.[14]

The interacting nature of environmental stressors also raises the question of cross adaptation. In this instance the word "adaptation" is used in the sense of short-term acclimatizational processes. It is not surprising to find that men who are distance runners tend to show increased physiologic heat tolerance and perhaps vice versa. A rather straightforward explanation may be offered since the metabolic heat produced by a runner may have the same effect as an external heat load.[22] In the terms used by Adolph[23, 24] the adaptagent is the same. It has also been suggested that cross adaptation or acclimatization may occur between heat and hypoxia.[25-27] If this is confirmed by further research, the explanation will not be obvious. Most investigators would not consider these two stressors to be the same adaptagent and therefore one must assume either that the apparent cross adaptation is the product of a generalized response to stress as postulated by Selye[7] or that there exist a number of specific physiologic interactions that are unknown. Clearly cross acclimatization remains a relatively unexplored but important area of environmental physiology.

ORIGINS OF ADAPTIVE CAPACITY

Mechanisms

When the environmental physiologist goes beyond the descriptive aspect of his subject matter, he inevitably enters into the more philosophic questions of cause and effect. At the how-does-it-happen level he becomes concerned with mechanisms. A rather straightforward question may be posed such as, "How does an increase in environmental temperature trigger the sweating response in man?" To answer even such a simple question has not proved easy. Experimentally, it has been shown that local skin heating produces sweating.[28-31] It appears that a rather specific answer might be that the skin temperature sensors are linked directly to the sweat gland innervation system and work on a thermostatic principle. However, this explanation proves very inadequate since it has also been demonstrated that sweating can be induced without skin heating if body core temperature is raised.[32] Thus a dual innervation system based on multiple sensor thermostats must be postulated. Even this hypothesis must be abandoned because appropriate electric or other stimulation of the hypothalamus can produce sweating without heating any part of the body.[33, 34] Furthermore, a thermostat operating in an on-off manner appears improbable because sweat rate varies and is closely related to total heat load. The subtlety of the sweating response is further affirmed by the fact that when men work in a hot environment with high water vapor pressure, the sweat loss is less for a given rectal temperature than it is for the same rectal temperature during work

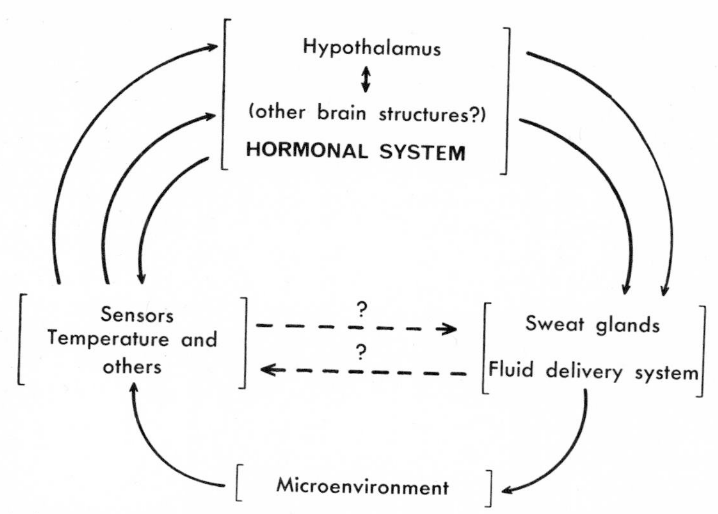

Fig. 15-2. Although incomplete, this simplified view of sweat control does indicate that there are numerous sensors involved in the regulation of sweating. The innervative interrelationships among the hypothalamus, other brain segments, and the hormonal system have not yet been thoroughly explored.

in a hot dry environment.[12, 35-37] We must then postulate a further sensor and integrative system that regulates the sweating response according to its cooling effectiveness.

The only model that accounts for all presently available descriptive and experimental evidence is the generally accepted view presented in Fig. 15-2. As indicated, many aspects remain unknown. While some specific details such as the funneling of sensory information through the hypothalamus are clear, the total network of sweat control is not completely understood.

Using the same example we may ask what happens during acclimatization or short-term "adaptation" to heat. If the individual has not regularly been exposed to high environmental temperature and is then exposed for a period of 2 weeks, sweat loss is least at the beginning of the exposure period and then increases in J-curve fashion until it reaches a virtually stable response at the end of the 2-week period. Sweat composition also changes, the nonwater components decreasing, particularly salt, until composition also stabilizes after 2 weeks. Paralleling sweat loss increase, body core temperature remains more nearly normal during physical work and the pulse rate remains closer to the nonstress level.[38-40] In context these acclimatizational changes may be considered to indicate that the organism now suffers less strain from the temperature stress. But how does sweat loss increase and why does sweat composition change? Since body temperature is lower, one might expect less sweat loss, but how does body temperature decrease when sweat evaporation is the major mechanism for heat loss, and how does the sweat gland change to conserve salt? Many suggestions have been offered such as better sweat distribution,[41] hypothalamic thermostatic changes,[42] and improved superficial blood flow, but there is still no single satisfactory unified explanation.

We examined the sweat response to heat because it appears to be one of the simpler "how" questions in physiology. Problems in altitude or hyperbaric environments demand an even more extensive examination of all body systems. In which directions then should the environmental physiologist search for causes or basic mechanisms? In a sense two divergent directions of scientific inquiry have been followed and appear occasionally to be in conflict. On one hand, research proceeds along a reductionist line based on the belief common to all science that biologic phenomena are reducible to physics and chemistry. Since life originated from a process of physical and chemical evolution this assumption seems sound. Yet another equally unassailable tenet of biologic science is that in the process of evolution new properties arise with each new level of biologic complexity and that these properties cannot be predicted from study of the individual parts. Thus examination of the separate parts of a physiologic response may provide little explanation of mechanism unless each is also studied in relationship to the other functional components. When we add to these a third basic premise—that all living organisms are in dynamic equilibrium with their environment—it becomes apparent that the environmental physiologist must assume a variety of research approaches if he hopes to arrive at a causal level of explanation.

Genes and environment

Basically, all characteristics of a living organism—whether anatomic structure, physiologic function, or even behavior—must be explained by reference to its genetic substrate and how this structure has been and is interacting with the external environment.[43] The traditional dichotomy between environmentally versus genetically produced variation proved largely unproductive as well as an untenable scientific categorization. All organisms begin as single cells, the information content packed almost entirely within the chromosomes of the nucleus. This single cell then interacts with the environment to produce a sequentially changing mass of cells ending inevitably in death for all but specific germinal cells. All aspects of this process are controlled by information contained in the original genes because environment cannot produce anything without the structuring information provided by them. On the other hand, genes act only in the presence of an appropriate environment.

Probably because the field of genetics developed from study of discontinuous organismic variables, a particular structure or function was initially thought to be of genetic origin only if it were manifest as either present or absent and not if manifest as a continuum. Thus sweat response failure of an individual lacking sweat glands was considered to be genetic, particularly if it conformed to a Mendelian inheritance pattern. All other sweat response variation was believed to be of environmental causation. The discovery of additive genes modified this view. Until recently we might have said that the difference between black and white skin color is genetically determined and that actual skin color differences are a product of this genetic difference and environmental factors, such as recent exposure to ultraviolet radiation. However, even this view is inadequate because the ability of skin to tan is also determined by genetic information.[44]

How then do genes and environment interact to produce a structure or function? The answer to this question is not known, but biochemical and developmental genetics have progressed to the point that some general mechanisms are now more clear.[45, 46] The general information template is the deoxyribonucleic acid (DNA) of the parent nucleus. From this template messenger ribonucleic acid (mRNA) is formed, which then passes to the ribosome of the cell. Through the action of other forms of RNA, mRNA then produces enzymes. These proteins, in turn, react with available environmental components, producing molecular products. At this stage several feedback mechanisms can be established. Depending upon the environmental components, enzyme production can be affected in several possible ways. Metabolite products may act directly upon the enzyme itself or may feed back to affect the various steps of RNA synthesis. Thus both RNA and enzyme production may be repressed or induced. At this most elementary molecular level the first steps of a continuing environment-gene interaction are demonstrable. Even here the process demonstrates the interdependence of the two factors in the end product. Variations of DNA structure determine mRNA and consequently enzymatic protein structure. While environment controls enzymatic production to some extent, we must also assume that the DNA and RNA structure of cells responds differently to the feedback metabolites, depending upon their original protein structure. Besides the effects of the molecular components of the environment, other physical aspects such as temperature and ambient oxygen pressure can affect both rates of production and kinds of products. Fig. 15-3 presents some of these interactions in simplified form. The extent to which environmental input can affect these processes remains open to investigation. Quite clearly a cell can function only within certain environmental limits, and, in the same sense, DNA and RNA must have very narrow limits within which product feedback can control them. If this were not the case, somatic cells would exhibit a functional variability beyond the known boundaries of biologic conservatism.

If we extend the process just described to a developing many-celled organism, the reasons for structural and functional variation are apparent. Thus as a new group of cells differentiates it feeds back new products to other cells so that still different structural and functional organization is promoted. By the time this developmental sequence has produced a complex organism such as a man, the interaction becomes subtle indeed, and

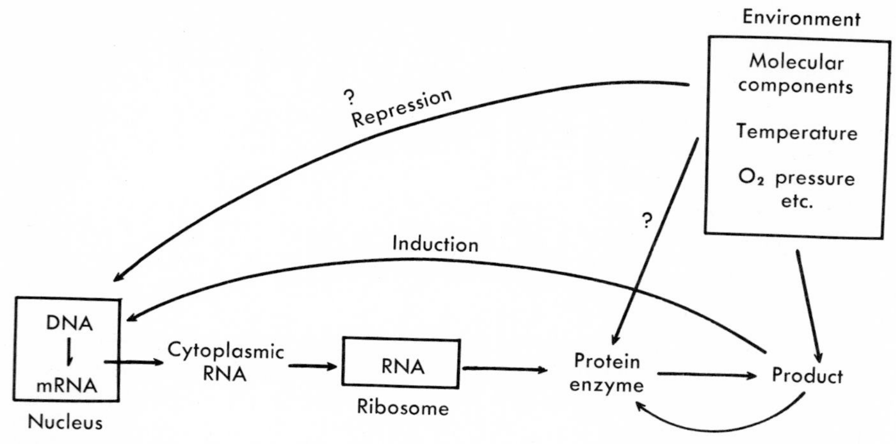

Fig. 15-3. Most of the present evidence suggests that the environment-mRNA interaction occurs at a regulatory (quantity control) level. However, research suggests that certain viruses act directly to alter the nature of RNA information and thereby qualitatively alter enzyme production.

the structure may depend on very minute differences in environmental input. A rather simple example of this is the brain that is located within the functioning structure so that its temperature is maintained constant. The brain is highly dependent on this narrow temperature range if it is to maintain the necessary survival functions.[47-49] The male gonad, on the other hand, does not function normally if maintained at brain temperature and is dependent on its anatomic location to keep it at a lower temperature.[50]

For those who are inclined to the reductionist approach in physiology, the rapidly developing field of biochemical genetics has many attractions. It offers an opportunity to ask "how?" at a chemical and physical level of explanation. In some instances research of this nature has provided examples of how acclimatization occurs. As might be expected, most of the examples concern the use of animals of a lower level of complexity and of these the greatest number involve single-cell organisms.[51] Poikilotherms also provide some examples. Fish increase protein production after cold acclimatization, and this may be a direct cell response to temperature stress.[52] Similarly, among genetically specific isozymes fish increase certain lactic dehydrogenases selectively when maintained within specific ambient temperature ranges.[53] Although no such examples are yet known for man, the finding of greater ADP content in the hearts of hypoxia-"adapted" mammals may prove to be one.[54]

In a sense such demonstrations reaffirm the belief that the search for causes can eventually reach a chemical level of explanation; however, when we recall that each evolving level of biologic structure produces new functions that are not predictable from its separate parts, it is obvious that most levels of explanation will involve causes that can only be found in structural-functional relationships. The existence of these complexes is an expression of biologic evolution and it may, therefore, be useful to review some physiologic problems within an evolutionary framework.

Evolution and adaptation

In its basic form the modern synthetic theory of evolution is deceptively simple. If we assume that genes are discrete biochemical units that match up in pairs on the chromosomes, then a mathematically straightforward conclusion can be drawn that in a very large population gene frequency will remain constant from generation to generation unless disturbed by any of a few known factors. Of these factors, point mutation and chromosomal aberration are important. Mistakes in cell reproduction mean that a second-generation cell may not be a perfect copy of the first. As far as we know this is the only way in which the genetic structure of a cell can be altered. For the present we must consider these alterations to be without pattern since no environmental force is known to alter the molecular structure of a gene or chromosome in a predetermined direction. Environmental factors such as ionizing radiation or high temperature are known to increase mutation rate in germinal cells. However, this increased rate is not specific; it only increases the probability that mutations will occur. Mutation may thus be considered as a semirandom phenomenon that is probably common enough to produce at least one mutation in each germinal cell of a complex animal like man.[55] In addition to mutation, chance processes such as genetic drift and founder's principle can alter the frequencies of existing genes from one generation to the next in small populations. Gene migration into or out of a population may also change gene frequencies, but natural selection is the evolutionary force that changes biologic organization—structure and function.[6]

Within present concepts natural selection might be visualized as any environmental force that causes a given gene systematically to increase or decrease through generations. Such a definition is, of course, oversimplified since selection may also act only to maintain a gene frequency at a previous level. Thus recurring mutations that are harmful to the reproductive performance of an organism in a given environment are constantly being eliminated by the process of selection. In a more subtle situation termed *balanced polymorphism* a gene may aid reproductive survival when present at one allele but reduce survival when present at both alleles. In this instance selection operates against both homozygous forms, thereby maintaining the gene frequency at some intermediate level in the population. In all instances the process operates through the reproductive performance of a given gene bearer. Since the process of selection maintains or improves the adaptation of a population to its environment, the theory of evolution automatically defines a gene frequency change as an adaptation. Under some conditions this definition conforms to general usage in the field of environmental physiology. Both the evolutionist and the environmental physiologist would agree that the heterozygous state of hemoglobin S was an adaptation to falciparum malaria, but the agreement would stem from divergent criteria. The physiologist would note that a hemoglobin S heterozygote has a different adaptive limit to malaria during childhood, while the evolutionist would note that under specific conditions the gene increases in frequency.[56, 57]

Despite different viewpoints these two disciplines rely inherently on each other for progress toward their respective goals. If we examine first the goals of evolutionists, the potential contribution of the environmental physiologist is quite apparent. The evolutionist's basic goal is the explanation of how selection operates. Some insight into this problem can be obtained by experimental procedures in which populations with variable genetic structures are exposed to particular environments and the differential survival is noted.[58, 59] In such experiments it is not really necessary to know how particular genes enhanced or reduced survival rates. But if the mechanisms are not known, then the evolutionist is in essence only describing what happens.[60] He is not discovering causes. When the evolutionist searches for cause he must turn to environmental physiology and to ecology. Selection does not operate directly on a gene but operates instead on functioning phenotypes. Therefore to understand how particular genes are selected, one must eventually measure the effects of the gene on phenotypic functioning in particular ecologic settings.

Many instances could be cited to show how environmental physiology has contributed to the modern concept of selection. I find two problems particularly interesting because evolutionary theory has not to date provided a thoroughly satisfactory explanation. Selection theory as presently formulated by most evolutionists involves the concept of single gene selection and, in general, it must be postulated that selection operates for or against single genes. In such a system novelty arises only by point mutation or chromosomal aberration. As Waddington[61] pointed out, no alternative to this concept exists; however, it is very difficult to explain how the enormously complex morphologic and physiologic genetic adaptations of living forms arose. A second point, which has concerned biologists since the time of Lamarck, is the fact that evolution often followed the course shown by acclimatizational processes. These changes are often so directional that they misled biologists into the idea of the inheritance of acquired characteristics. Waddington showed experimentally that in the presence of a particular environmental stressor, selection was indeed likely to act through differential mortality to select and to fix mutations in the direction of the acclimatizational syndrome. In abbreviated form his explanation for these findings[61] is that (1) developmental acclimatization introduces an organism to selective stressors that it would not encounter if it could not acclimatize; (2) selection is most likely to modify structures and functions that are already in existence; and therefore (3) the selected gene or mutation will genetically fix a trait of the same nature as the acclimatizational change. Although this formulation provides an acceptable theory, the variety of feedback systems that fix adaptive syndromes remains poorly explored and inadequately applied to both evolutionary theory and environmental physiology.

Even more difficult to explain in evolutionary theory is the evolution of acclimatizational capacity. In this instance I am using the word to encompass all of the "adaptational processes" that cannot be tied to specific genes. At the cell level one can at least conceive of the selective mechanisms that would fix regulatory genes. Extrapolating from what is known about biochemical genetics, it is easy to see how a gene might be selected that had appropriate regulatory responses to an environmental substrate; however, at the level of the complex acclimatizational capacity found in some populations the selective mechanism becomes obscure. It has been postulated that many physiologic attributes, such as aging and death, can only be explained by interpopulation or group selection.[62] Group selection is in many ways an attractive concept to explain the variety and complexity of acclimatizational capability; however, some aspects of the evolutionary record argue against accepting this mechanism. Interpopulation selection should operate more effectively the larger the number of closely related but reproductively isolated populations. Thus insects such as *Drosophila* might offer a good opportunity for such selection. At the other extreme a small number of only semiisolated populations with long generation times, such as primates, appear to offer very unlikely situations for interpopulation selection. As a general principle it might be said that the most complex and recently evolved forms are least likely to be shaped by interpopulation selection. A paradox arises from the equally sound observation that the more complex the animal species, the greater the acclimatizational capacity. This paradox suggests that either the present conception of intergroup selection is inadequate or that an alternative mode of selection remains to be discovered.

In this section I have described primarily how the theory of evolution depends upon environmental physiology, but all too often the environmental physiologist has failed to utilize the contribution that evolutionary theory could make to his problems. Evolution is the basic unifying explanatory system of biology, and whether one attempts to describe the physiologic responses of a population to its environment or search for underlying mechanisms, the findings of evolution have pertinence to the inquiry. In a sense, knowledge of the evolutionary process can form the prime base for the hypothesis of any study and should always be a consideration in the interpretation of results. As shown earlier, the concept of stress itself depends on a definition of a "neutral zone" in the environment. The neutral zones of species and subpopulations have arisen through the process of evolution and knowledge of the preceding form of animal, in conjunction with how selection modified it, constitutes a data base on which any hypothesis about adaptive limits or mechanisms of adaptation should be constructed.

Evolutionary findings provide a number of interpretive guidelines. Thus one expects greater acclimatizational capacity in more complex species, greater variability in the response of complex organisms, and, because of increasing levels of heterozygosity, greater variability in individual response in more complex organisms. Finally, the high degree of evolved genetic variability in a species such as man should warn environmental physiologists against the error of translating too freely from results on taxonomically distant species or assuming that the results obtained on one human population are valid for other human populations.

MAN—SPECIES, POPULATIONS, AND THE FUTURE

Taxonomic position

Man is one of the evolutionary end products of vertebrate evolution. His heritage can be traced with some confidence from fish through amphibians to generalized mammals. Quite early in mammalian evolution a primate line separated and in fairly rapid order diversified into separate family lines. The line leading to man appears to have been that of a quite generalized monkey form, leading to later divergences of which the present survivors are the apes and man.[63]

Considering some of the evolutionary principles previously suggested, human physiologic responses to environmental change should resemble those of apes more than any other animal. Indeed, it can be specified that we are more akin to the chimpanzee with respect to morphology and protein structure than to any other living species.[64] In general, the degree of predicted similarity decreases as evolutionary distance increases, although, of course, we share many commonalities with all living matter. If modern science had developed in tropical areas where large numbers of nonhuman primates are easily available, it is possible that primates would have been the basic experimental laboratory mammals. However, most research has involved the use of rodents and carnivores. Both rat and dog, which constitute the basic laboratory species, are fairly generalized mammals; unfortunately, their evolutionary lines diverged from that of *Homo sapiens* early in the mammalian radiation. Thus it may be said that when the study of a human problem in environmental physiology required the use of a laboratory animal, the one chosen was a form likely to demonstrate only the most general of mammalian responses and not one likely to respond in a manner specific to man. This deficiency in research methodology has been partially rectified in the past 20 years by the development of primate research centers. However, the physiologic responses and stress limits of our near relatives remain so poorly examined that the rat or dog are still often preferred as the experimental animal by the environmental physiologist searching for mechanisms.[65]

Man's differences from and similarities to closely related forms are not known in detail, and a complete description of what is known is not appropriate to this chapter. In general man and chimpanzee have very similar physiologic responses to a broad spectrum of environmental stressors. Some differences are also known, such as man's greater adaptability to hot dry conditions.[5] Such differences may provide clues to our particular evolutionary history. The outstanding difference is, of course, man's greatly superior capacity for learning and the comparatively enormous amount of information that he passes from generation to generation by nongenetic means (tradition).

Most physiologists would include nongenetically transmitted behavior as part of the total adaptive response of a species to environmental stress. Man is then the most adaptable of all animals since as a single interfertile species he lives in a wider variety of environmental conditions than any other form. The unique adaptive mechanism that allows man to survive this diversity of stressors is termed *culture* by anthropologists, but like other natural phenomena it does not lack evolutionary antecedents. Its uniqueness derives only from the fact that evolution, at each new level of complexity, produces phenomena not totally explicable from examination of the separate previous components.

Population differences

As human populations vary in genetic structure, morphology, and behavior, they also differ in their physiologic responses to environment. Man is a single species and therefore differences are usually not of a dramatic nature; furthermore, cultural adaptive mechanisms are so successful that they tend to obscure differences that may exist in the adaptive limits of various populations.[66, 67] Eskimos differ from American whites in their physiologic responses to total body or even local cooling, but using modern technology the American white can survive in far colder conditions than the Eskimo with his traditional culture. It is, of course, possible to test many of the physiologic responses under similar levels of cultural adaptation. Men can be stripped of clothing, kept inactive, and exposed to cold. Controlled tests of this nature are the only method by which noncultural or biologic adaptations can be studied, but at the same time we must remember that adaptive responses are usually syndromes and not unitary responses. Thus testing populations outside of their usual syndrome may fail to uncover the adaptive or population-specific component of a response.[3, 68] The constant interaction between cultural and biologic characteristics of human populations is probably the most important single explanation of why population differences in physiologic response and adaptation are so difficult to demonstrate.[12, 69, 70]

Culture, possible only because of man's enormous learning capacity, is such a successful method of adapting to environmental stress that many anthropologists and even a few environmental physiologists have been tempted to conclude that there are no physiologic differences in the varying responses of human populations to environmental stress. Stress physiology research of the past 20 years has clearly demonstrated differently. The response to cold of adult Australian aborigines is different from that of adult Europeans.[71] Eskimos do not respond like Bushmen, and tests on a variety of populations show a wide variation in their immediate response to a standardized heat stress. Responses to hypoxia underlie the fact that sea-level peoples suffer a drastic reduction in work capacity compared to high-altitude natives, and while the evidence is less conclusive it also appears that populations may vary widely in their responses to various infectious diseases.[72]

The demonstration of population differences does not automatically provide equally satisfactory answers as to why populations differ. In several instances the differences may be explained by reference to differences in the immediate past environment. Wyndham[36] and others have shown that the difference in response to a standardized hot-wet heat tolerance test is often an expression of the immediately past heat exposure and can often be eliminated by a brief acclimatization period. In other instances, such as differences in response to hypoxia, it has proved impossible for adult lowlanders to achieve the same physical work (aerobic) capacity as native highlanders despite prolonged exposure.[73] In the latter case, developmental acclimatization must be considered as a cause of the difference. A recent study indicates that if genetically similar lowlanders are brought to high altitude as infants they will generally have an aerobic capacity similar to that of a highland population.[74] We still do not know whether a genetically different population would show a similar developmental adaptation. Indeed, we know of very few proved instances in which response differences in human populations have a genetic basis. Even so, the theory of evolutionary adaptation is sufficiently strong that the environmental physiologist should not rule out the possibility of a genetic basis for response variation without conclusive evidence to the contrary.

The future and environmental physiology

Man has passed through a series of rapid environmental changes since the time he developed agriculture. At times and for some groups there were lulls that allowed men to become part of relatively stable ecosystems, but the present rate of change is increasingly rapid and almost worldwide. This rapidly changing environment poses major challenges to the field of environmental physiology. Man is modifying not only his own environment but, to some extent, the whole biosphere. The rapidity of change creates a certain urgency in many aspects of research. If we are to understand the responses of animals to their former natural environment, research must be completed before new selective forces modify systems in unknown ways. The last of the human populations living in traditional hunting and gathering patterns will be gone within a very few years and with them perhaps the last chance to understand some of the physiologic modes of response that characterized our formative evolution as a species.[75, 76]

A scientific desire to understand the past might be assigned a low priority in man's need for knowledge but, as much of the scientific community now realizes, the questions are more than academic. With rapid changes now occurring in our biosphere, the problems of adaptive limits and mechanisms may be critical to the survival of human civilization as we know it.

OCTOBER, 1971

REFERENCES

1. Prosser, C. L.: Principles and general concepts of adaptation, Environ. Res. **2**:404-416, 1969.
2. Prosser, C. L.: Perspectives of adaptation: theoretical aspects. In Dill, D. B., editor: Handbook of physiology, sect. 4, Washington, D. C., 1964, American Physiological Society.
3. Baker, P. T.: Multidisciplinary studies of human adaptability: theoretical justification and method. In Weiner, J. S., editor: A guide to the human adaptability proposals, ed. 2, Oxford, 1969, Blackwell Scientific Publications.
4. Wallace, B., and Srb, A.: Adaptation, Englewood Cliffs, N. J., 1964, Prentice-Hall, Inc.
5. Newman, R. W.: Why man is such a sweaty and thirsty naked animal: a speculative review, Hum. Biol. **42**:12-27, 1970.
6. Mayr, E.: Animal species and evolution, Cambridge, Mass., 1966, Harvard University Press.
7. Selye, H.: The physiology and pathology of exposure to stress, a treatise based on the concept of the general-adaptation-syndrome and the diseases of adaptation, Montreal, 1950, Acta, Inc.
8. Morgan, H. R.: The enteric bacteria. In Dubos, R. J., and Hirsch, J. G., editors: Bacterial and mycotic infections of man, ed. 4, Philadelphia, 1965, J. B. Lippincott Co.
9. McLeese, D. W.: Effects of temperature, salinity, and oxygen on the survival of the American lobster, J. Fisheries Res. Board Canad. **13**:247-272, 1956.
10. Scrimshaw, N. S., Taylor, C. E., and Gordon, J. E.: Interactions of nutrition and infection, Geneva, 1968, World Health Organization.
11. Sinclair, J. C.: Heat production and thermoregulation in the small-for-date infant, Pediat. Clin. N. Am. **17**:147-158, 1970.
12. Wyndham, C. H.: Adaptation to heat and cold, Environ. Res. **2**:442-469, 1969.
13. Baker, P. T.: Climate, culture, and evolution, Hum. Biol. **32**:3-16, 1960.
14. Carlson, L. D., and Hsieh, A. C. L.: Control of energy exchange, New York, 1970, The Macmillan Co.
15. Hammel, H. T.: Terrestrial animals in cold: recent studies of primitive man. In Dill, D. B., editor: Handbook of physiology, sect. 4, Washington, D. C., 1964, American Physiological Society.
16. Hiernaux, J.: La diversité humaine en Afrique subsaharienne. Récherches biologiques. Etudes ethnologiques, Editions de l'Institut de Sociologie, Bruxsels, 1968, Université Libre.
17. Schreider, E.: Ecological rules, body-heat regulation, and human evolution, Evolution **18**:1-9, 1964.
18. Balke, B.: Experimental studies on the conditioning of man for space flight, Air Univ. Quart. Rev. **11**: 61-75, 1959.
19. Fregly, M. J.: Cross acclimatization between cold and altitude in rats, Am. J. Physiol. **176**:267-274, 1954.
20. Mefferd, R. B., Jr., and Hale, H. B.: Effects of thermal conditioning on metabolic responses of rats to altitude, Am. J. Physiol. **195**:735-738, 1958.
21. Hale, H. B.: Cross-adaptation, Environ. Res. **2**:423-434, 1969.
22. Saltin, B., Gagge, A. P., and Stolwyk, J. A. J.: Body temperatures and sweating during thermal

transient caused by exercise, J. Appl. Physiol. **28:** 318-327, 1970.
23. Adolph, E. F.: General and specific characteristics of physiological adaptations, Am. J. Phys. Anthrop. **184:**18-28, 1956.
24. Adolph, E. F.: Perspectives of adaptation: some general properties. In Dill, D. B., editor: Handbook of physiology, sect. 4, Washington, D. C., 1964, American Physiological Society.
25. Brimer, H., Jovy, D., and Klein, K. E.: Hypoxia as a stressor, Aerospace Med. **32:**1009-1018, 1961.
26. Hale, H. B., and Mefferd, R. B., Jr.: Metabolic responses to thermal stressors of altitude-acclimated rats, Am. J. Physiol. **195:**739-743, 1958.
27. Hiestand, W. A., Stemler, F. W., and Jasper, R. L.: Increased anoxic resistance resulting from short period heat adaptation, Proc. Soc. Exp. Biol. Med. **88:**94-95, 1955.
28. Benjamin, F. B.: Sweating response to local heat application, J. Appl. Physiol. **5:**594-598, 1953.
29. Bullard, R. W., Banerjee, M. R., and MacIntyre, B. A.: The role of the skin in negative feedback regulation of eccrine sweating, Int. J. Biometeorol. **11:**93-104, 1967.
30. MacIntyre, B. A., Bullard, R. W., Banerjee, M. R., and Elizondo, R.: Mechanism of enhancement of eccrine sweating by localized heating, J. Appl. Physiol. **25:**255-260, 1968.
31. Ogawa, T.: Local effect of skin temperature on threshold concentration of sudorific agents, J. Appl. Physiol. **28:**18-22, 1970.
32. Nadel, E. R., Horvath, S. H., Dawson, C. A., and Tucker, A.: Sensitivity to central and peripheral thermal stimulation in man, J. Appl. Physiol. **29:** 603-609, 1970.
33. Fusco, M. M., Hardy, J. D., and Hammel, H. T.: Interaction of central and peripheral factors in physiological temperature regulation, Am. J. Physiol. **200:**572-580, 1961.
34. Hammel, H. T.: Neurons and temperature regulation. In Yamamoto, W. S., and Brobeck, J. R., editors: Physiological controls and regulations, Philadelphia, 1965, W. B. Saunders Co.
35. Adolph, E. F., and others: Physiology of man in the desert, New York, 1947, Interscience Publishers, Inc.
36. Wyndham, C. H.: South African ethnic adaptation to temperature and exercise. In Baker, P. T., and Weiner, J. S., editors: The biology of human adaptability, Oxford, 1966, Clarendon Press.
37. Lee, D. H. K.: Terrestrial animals in dry heat: man in the desert. In Dill, D. B., editor: Handbook of physiology, sect. 4, Washington, D. C., 1964, American Physiological Society.
38. Bass, E. E.: Thermoregulatory and circulatory adjustments during acclimatization to heat in man. In Hardy, J. S., editor: Temperature: its measurement and control in science and industry, vol. 3, New York, 1963, Van Nostrand Reinhold Co.
39. Kuno, Y.: Human perspiration, Springfield, Ill., 1956, Charles C Thomas, Publisher.
40. Wyndham, C. H.: Effect of acclimatization on the sweat rate/rectal temperature relationship, J. Appl. Physiol. **22:**27-30, 1967.
41. Thompson, M. L.: A comparison between number and distribution of functioning eccrine sweat glands in Europeans and Africans, J. Physiol. **123:**225-233, 1954.
42. Belding, H. S., and Hatch, T. F.: Relation of skin temperature to acclimatization and tolerance to heat, Fed. Proc. **22**(1):881-883, 1963.
43. Dobzhansky, T. G.: Genetics of the evolutionary process, New York, 1971, Columbia University Press.
44. Daniels, F., Jr.: Man and radiant energy: solar radiation. In Dill, D. B., editor: Handbook of physiology, sect. 4, Washington, D. C., 1964, American Physiological Society.
45. Harris, H.: The principles of human biochemical genetics, New York, 1971, American Elsevier Publishing Co., Inc.
46. Hsia, D. Y.: Human developmental genetics, Chicago, 1968, Year Book Medical Publishers, Inc.
47. Ingram, D. L., and Smith, R. E.: Brain temperature and cutaneous blood flow in the anesthetized pig, J. Appl. Physiol. **29:**698-704, 1970.
48. Kaplan, H. A., and Ford, D. H.: The brain vascular system, New York, 1966, American Elsevier Publishing Co., Inc.
49. Lassen, N. A.: Cerebral circulation, Proc. Int. Union Physiol. Sci. **6:**173, 1968.
50. Guyton, A. C.: Textbook of medical physiology, Philadelphia, 1971, W. B. Saunders Co.
51. Koffler, H., Mallett, G. E., and Adye, J.: Molecular basis of biological stability to high temperature, Proc. Nat. Acad. Sci. **43:**464, 1957.
52. Das, A., and Prosser, C. L.: Biochemical changes in goldfish acclimated to high and low temperatures. II. Protein synthesis, Comp. Biochem. Physiol. **21:** 449, 1967.
53. Hochachka, P. W.: Lactate dehydrogenases in poikilotherms: definition of a complex enzyme system, Comp. Biochem. Physiol. **18:**261, 1966.
54. Tenney, S. M., and Ou, T. C.: Some tissue factors in acclimatization to high altitude. In Hegnauer, A. H., editor: Biomedicine of high terrestrial elevations, 1969, U. S. Army Institute of Environmental Medicine, Natick, Mass., and U. S. Army Medical Research and Development Command, Washington, D. C.
55. Cavalli-Sforza, L. L., and Bodmer, W. F.: The genetics of human populations, San Francisco, 1971, W. H. Freeman and Company, Publishers.
56. Wiesenfeld, S. L.: Sickle-cell trait in human biological and cultural evolution, Science **157:**1134-1140, 1967.
57. Hexter, A.: Selective advantage of the sickle-cell trait, Science **160:**436-437, 1968.
58. Ayala, F. J.: Population fitness of geographic strains of *Drosophila serrata* as measured by interspecific competition, Evolution **24:**483-494, 1969.
59. Lewontin, R. C.: Adaptations of population to varying environments. In Cold Spring Harbor symposia on quantitative biology, vol. 22, Cold Spring Harbor, N. Y., 1957, The Biological Laboratory.
60. Morton, N. E.: Problems and methods in the genetics of primitive groups, Am. J. Phys. Anthrop. **28:** 191-202, 1968.
61. Waddington, C. H.: Evolutionary adaptation. In Tax, S., editor: Evolution after Darwin, vol. 1, Chicago, 1960, University of Chicago Press.
62. Emerson, A. E.: The evolution of adaptation in

population systems. In Tax, S., editor: Evolution after Darwin, Chicago, 1960, University of Chicago Press.

63. Pfeiffer, J.: The emergence of man, New York, 1969, Harper and Row, Publishers.
64. Goodman, M.: Serological analysis of the systematics of recent hominoids, Hum. Biol. **35**:377-436, 1963.
65. Funkhouser, G. E., Higgins, E. A., Adams, T., and Snow, C. C.: The response of the Savannah baboon to thermal stress, Life Sciences **6**:1615-1620, 1967.
66. Hildes, J. A.: Ecologic and ethnic adaptations, Environ. Res. **2**:417-422, 1969.
67. Weiner, J. S.: Human ecology. In Harrison, G. A., Weiner, J. S., Tanner, J. M., and Barnicot, N. A., editors: Human biology, New York, 1964, Oxford University Press.
68. Baker, P. T.: Micro-environment cold in a high altitude Peruvian population. In Yoshimura, H., and Weiner, J. S., editors: Human adaptability and its methodology, Tokyo, 1966, Japan Society for the Promotion of the Sciences.
69. Ladell, W. S. S.: Terrestrial animals in humid heat: man. In Dill, D. B., editor: Handbook of physiology, sect. 4, Washington, D. C., 1964, American Physiological Society.
70. Wyndham, C. H., and others: Heat reaction of Caucasian and Bantu in South Africa, J. Appl. Physiol. **19**:598-606, 1964.
71. Hammel, H. T.: Summary of thermal patterns in man, Fed. Proc. **22**:846-847, 1963.
72. Taylor, I.: Principles of epidemiology, Boston, 1964, Little, Brown and Co.
73. Velasquez, T.: Acquired acclimatization: to sea level. In Life at high altitudes, Washington, D. C., 1966, Pan American Health Organization.
74. Frisancho, R., and others: Adaptacion fisiologica acquerida durante el periodo de desarrollo. In Adaptacion biologica a la altura, Ann Arbor, 1971, The University of Michigan Center for Human Growth and Development and Department of Anthropology.
75. Baker, P. T., and Weiner, J. S.: The biology of human adaptability, Oxford, 1966, Clarendon Press.
76. Neel, J. V.: Lessons from a "primitive" people, Science **170**:815-822, 1970.

APPENDIX

Table A-1. Normal composition of clean, dry atmospheric air near sea level*

Constituent gas and formula	Content (percent by volume)	Content variable relative to its normal	Molecular weight†
Nitrogen (N_2)	78.084	—	28.0134
Oxygen (O_2)	20.9476	—	31.9988
Argon (Ar)	0.934	—	39.948
Carbon dioxide (CO_2)	0.0314	‡	44.00995
Neon (Ne)	0.001818	—	20.183
Helium (He)	0.000524	—	4.0026
Krypton (Kr)	0.000114	—	83.80
Xenon (Xe)	0.0000087	—	131.30
Hydrogen (H_2)	0.00005	?	2.01594
Methane (CH_4)	0.0002	‡	16.04303
Nitrous oxide (N_2O)	0.00005	—	44.0128
Ozone (O_3)	Summer: 0 to 0.000007	‡	47.9982
	Winter: 0 to 0.000002	‡	47.9982
Sulfur dioxide (SO_2)	0 to 0.0001	‡	64.0628
Nitrogen dioxide (NO_2)	0 to 0.000002	‡	46.0055
Ammonia (NH_3)	0 to trace	‡	17.03061
Carbon monoxide (CO)	0 to trace	‡	28.01055
Iodine (I_2)	0 to 0.000001	‡	253.8088

*From U. S. standard atmosphere, Washington, D. C., 1962, U. S. Government Printing Office, p. 9.
†On basis of carbon 12 isotope scale for which $^{12}C = 12$.
‡The content of the gases marked with a double dagger may undergo significant variations from time to time or from place to place relative to the normal indicated for these gases.

Table A-2. Altitude-pressure table

Altitude (m)	Altitude (ft)	P_B* (torr)	$(P_B - 47)$† (torr)	$0.2095 \times (P_B - 47)$‡ (torr)	Altitude (m)	Altitude (ft)	P_B* (torr)	$(P_B - 47)$† (torr)	$0.2095 \times (P_B - 47)$‡ (torr)
0	0	760	713	149	8,540	28,000	247	200	42
610	2,000	707	660	138	9,150	30,000	226	179	38
1,220	4,000	656	609	128	9,760	32,000	206	159	33
1,830	6,000	609	562	118	10,370	34,000	187	140	29
2,440	8,000	564	517	108	10,980	36,000	170	123	26
3,050	10,000	523	476	100	11,590	38,000	155	108	23
3,660	12,000	483	436	91	12,200	40,000	141	94	20
4,270	14,000	446	399	84	12,810	42,000	128	81	17
4,880	16,000	412	365	76	13,420	44,000	116	69	14
5,490	18,000	379	332	70	14,030	46,000	106	59	12
6,100	20,000	349	302	63	14,640	48,000	96	49	10
6,710	22,000	321	274	57	15,250	50,000	87	40	8
7,320	24,000	294	247	52	19,215	63,000	47	0	0
7,930	26,000	270	223	47					

*Barometric pressure.
†$(P_B - 47)$ = (1) the total pressure of the dry gases after the inspired air has been saturated with water vapor at 37° C; (2) also inspired oxygen pressure, $P_{I_{O_2}}$, when pure oxygen is breathed.
‡$0.2095 \times (P_B - 47)$ = inspired oxygen pressure, $P_{I_{O_2}}$, when air is breathed.

Table A-3. Acid-base aspects of environmental physiology

Environmental parameter	Mechanism	Primary disturbance	Compensation
Altitude or hypobaria	Decreased inspired, alveolar, and arterial P_{O_2} Decreased gas density Increased pulmonary and alveolar ventilation rate Decreased alveolar and arterial P_{CO_2}	Respiratory alkalosis	Metabolic acidosis
Thermal stress	Hyperthermia	Respiratory alkalosis Dehydration acidosis	Metabolic acidosis
Hyperbaria	Increased gas density Decreased alveolar ventilation rate Hypercapnia	Respiratory acidosis	Metabolic alkalosis
Hyperoxic atmosphere	Decreased pulmonary and alveolar ventilation rate Impaired carbon dioxide transport	Respiratory acidosis	Metabolic alkalosis
Carbon dioxide–containing atmosphere	Increased inspired, alveolar, and arterial P_{CO_2}	Respiratory acidosis	Metabolic alkalosis

Table A-4. Conversion factors for units of length

	Meter	Feet	Kilometer	Mile
Meter (m)	1	3.2808	0.001	0.000621
Feet (ft)	0.3048	1	0.0003048	0.0001893
Kilometer (km)	1,000	3,280.8	1	0.62137
Mile (mi)	1,609.34	5,280.00	1.60934	1

Table A-5. Conversion factors for units of pressure

	ATA	psi	torr
Atmosphere (ATA)	1	14.696	760
Pound/inch² (psi)	0.06805	1	51.715
Millimeter of mercury (torr)	760	0.01933	1

Table A-6. Thermal capacity of various substances*

Substance	Thermal capacity $J \times kgm^{-1} \times °C^{-1}$†	Thermal capacity $kcal \times kgm^{-1} \times °C^{-1}$
Water	4,187	1.0000
Mean body	3,350	0.80
Skin	3,324	0.77
Fat	2,303	0.55
Muscle	3,810	0.91
Blood	3,852	0.92
Rubber, synthetic	1,884	0.45
Wood	1,758	0.42
Air	1,006	0.2404
Aluminum	896	0.214
Marble	879	0.21
Asbestos	816	0.195
Copper	389	0.093
Silver	234	0.0558
Platinum	136	0.0324

*The thermal capacity of a substance is the quantity of heat necessary to produce unit change of temperature in unit mass. It is numerically equivalent to specific heat. Specific heat is the ratio of the thermal capacity of a substance to that of water at 15° C. Like specific gravity it is dimensionless.
†Also $W \times s \times kgm^{-1} \times °C^{-1}$.

Table A-7. Latent heat of evaporation of water, aqueous vapor pressure, and saturated density* in air at 1 atmosphere pressure

Temperature (°C)	Latent heat (J/gm)	Latent heat (cal/gm)	Vapor pressure (mm Hg)	Density* (gm/L)
0	2,495	595.9	4.58	0.00484
5	2,483	593.2	6.54	0.00678
10	2,472	590.4	9.21	0.00938
15	2,460	587.7	12.79	0.01280
20	2,449	584.9	17.54	0.01725
25	2,437	582.2	23.76	0.02298
30	2,426	579.5	31.82	0.03027
31	2,424	579.1	33.70	0.03195
32	2,422	578.6	35.66	0.03370
33	2,419	578.0	37.73	0.03554
35	2,415	576.9	42.18	0.03947
37	2,410	575.8	47.07	0.04376
40	2,403	574.2	55.32	0.05094

*Saturated density in air is the weight of water vapor contained in 1 L of air. It is calculated from the vapor pressure (P_w), the absolute temperature (T), and the gas constant for water vapor (R_w) as follows:
density = $(P_w/P) \times (\text{volume of air}) \times (R_w)^{-1} \times (T/P)^{-1} \times (\text{volume of air})^{-1} = 0.2882 \times P_w/T$.

Table A-8. Thermal conductivity of various substances

Substance	Thermal conductivity* A $W \times m^{-1} \times °C^{-1}$	Thermal conductivity* B $cal \times cm \times cm^{-2} \times s^{-1} \times °C^{-1}$
Silver	418.7000	1.0000000
Copper	384.4000	0.9180000
Aluminum	210.2000	0.5040000
Platinum	69.7000	0.1664000
Marble	2.9700	0.0071000
Water	0.5980	0.0014290
Animal muscle	0.3350	0.0008000
Animal fat	0.2050	0.0004900
Paper, wood	0.1260	0.0003000
Flannel	0.0963	0.0002300
Asbestos paper	0.0796	0.0001900
Cork	0.0544	0.0001300
Silk	0.0398	0.0000980
Air	0.0253	0.0000604
Eiderdown	0.0193	0.0000460
Cotton wool	0.0180	0.0000430

*Conductivity is defined as the heat transmitted per second through a plate 1 cm thick across an area of 1 cm^2 when the temperature difference is 1° C. The values in column A are obtained by multiplying those in column B by 418.7.

A STANDARD SYSTEM OF SYMBOLS AND UNITS FOR THERMAL PHYSIOLOGY

There is a continuing effort by all scientific societies to standardize the definitions and symbols used in their publications. The tables that follow are based on the proposals of the Subcommittee on Thermal Physiology of the International Union of Physiological Sciences, Dr. J. D. Hardy, chairman. (J. Appl. Physiol. **27**:439-445, 1969.)

Table A-9. Basic units for use in thermal physiology

Name	Unit*	Abbreviation of unit* (and alternatives)
Plane angle	radian	rad
Solid angle	steradin	sr
Area	square meter	m^2
Volume†	cubic meter	m^3
Frequency	hertz	Hz (l/s)
Density	kilogram per cubic meter	kgm/m^3
Velocity	meter per second	m/s
Force	newton	N ($kgm \times m \times s^{-2}$)
Pressure‡	newton per square meter	N/m^2
Work, energy, quantity of heat	joule	J (N × m)
Power	watt	W (J/s)

*Units when written out do not take a plural form and are written in lowercase letters; abbreviations of units named after persons begin with single capital letters.
†The liter (L) is a unit approved for use in conjunction with the International System of Units (SI).
‡Millimeters of mercury or torr will continue to be an acceptable unit for pressure with the SI until steam tables and other related vapor pressure tables are available more generally in N/m^2.

Table A-10. Special quantities for body heat balance equation

$$M = E \pm R \pm C \pm K \pm W \pm S$$

Quantity per unit time	Symbol
Metabolic heat production	*M*
Evaporative heat loss	*E*
Radiant heat exchange (+ for net loss)	*R*
Convective heat exchange (+ for net loss)	*C*
Conductive heat exchange (+ for net loss)	*K*
Useful work accomplished (+ for work against external forces)	*W*
Storage of body heat (+ for net heat gain by body)	*S*

Each term should be printed in upper-case italic type. The unit for each symbol is watt per square meter (W/m^2). Each term of the heat balance equation above is expressed as heat exchange rate per unit body surface area; for man, this is the DuBois area (A_D). This convention is necessary whenever the basic physiologic quantities (*M, E,* and *S*) are related to the fundamental physical coefficients involved in the process of evaporation of water from the body surface and the transfer of heat by radiation, convection, and conduction. In nutritional studies or where the overall energy equilibrium is of primary interest the energy terms may be expressed in W/kgm^n, in which n may be selected. The resulting expressions then apply only to the particular body shape or weight of the experimental animal used. Alternate indexes, now in use for physiologic activity related to thermoregulatory processes, such as ml (STPD) $O_2 \times min^{-1} \times kgm^{-n}$ for metabolic rate or ml $H_2O \times hr^{-1} \times kgm^{-n}$ for water loss from skin and respiratory tract, cannot be used directly in heat balance equations.

Table A-11. Quantities used in describing heat exchange

Principal physical quantities

Quantity	Symbol	Unit	Abbreviation of unit
Absorptance (radiation)	α	ND	
Area	A	square meter	m^2
Conductivity, thermal	k	watt per meter and degree celsius	$W \times m^{-1} \times °C^{-1}$
Density	ρ	kilogram per cubic meter	kgm/m^3
Emittance (radiation)	ε	ND	
Heat			
rate of exchange	H	watt per square meter	W/m^2
latent	λ	joule per kilogram	J/kgm
quantity (energy)	J	joule	J
Heat transfer coefficient (total surface conductance)	h	watt per square meter and degree celsius	$W \times m^{-2} \times °C^{-1}$
Irradiance (incident radiant flux density)	I	watt per square meter	W/m^2
Length	L	meter	m
Mass	m	kilogram	kgm
Mass transfer rate	$\dot{m}$	kilogram per second	kgm/s
Pressure	P	newton per square meter or millimeter of mercury	N/m^2 mm Hg
Reflectance (radiation)	ρ	ND	
Resistance, thermal (insulation)	R; I	degree celsius and square meter per watt	$°C \times m^2 \times W^{-1}$
Specific heat	c	joule per kilogram and degree celsius	$J \times kgm^{-1} \times °C^{-1}$
Time	t	second	s
Temperature	T	degree kelvin degree celsius	°K °C
Transmittance (radiation)	τ	ND	
Ventilation rate	$\dot{V}$	cubic meter per second; liter per second	m^3/s; L/s
Velocity, linear	v	meter per second	m/s
Volume	V	cubic meter, liter, milliliter	m^3, L, ml
Work rate	W	watt, joule per second	W, J/s

Special quantities useful for describing heat exchange

Quantity	Symbol	Unit
Ambient air temperature	T_a	°C
Mean radiant temperature (MRT)	$\overline{T}_r$	°K or °C
Operative temperature	T_o	°C
Effective radiant field (or flux) (+ for body warming)	H_r	W/m^2
Dimensionless factor (such as shape factor)	F	Always use with identifying subscript
Combined heat transfer coefficient	h	$W \times m^{-2} \times °C^{-1}$
Linear radiation heat transfer coefficient	h_r	$W \times m^{-2} \times °C^{-1}$
Convective heat transfer coefficient	h_c	$W \times m^{-2} \times °C^{-1}$
Insulation from the skin or clothing surface to a uniform environment at T_a	I_a	$°C \times m^2 \times W^{-1}$
Insulation of clothing	I_{cl}	1 clo = $0.155°C \times m^2 \times W^{-1}$
Relative humidity of ambient air (fraction)	ϕ_a	ND
Percent relative humidity	rh	rh = $\phi_a \times 100$
Useful quantities that cannot be used in the heat balance equation are:		
Effective temperature	T_{eff}	Defined in terms of T_a and T_{wb} (or ϕ_a)
Globe temperature	T_g	Temperature of a Vernon black 6-inch diameter globe

Table A-12. Frequently used physical subscripts

Significance	Symbol	Example
Ambient	a	T_a = ambient or dry bulb temperature
Conductive	k	h_k = coefficient of conductive heat transfer
Convective	c	h_c = coefficient of convective heat transfer
Diffusion	D	h_D = coefficient of mass transfer by diffusion
Emitting source	i	T_i = temperature of emitting source
Evaporative	e	h_e = coefficient of evaporative heat transfer
Pressure (constant)	P	c_P = specific heat at constant pressure
Projected	p	A_p = projected area
Radiation (radiative)	r	h_r = coefficient of heat transfer by linear radiation
Receiving surface	j	A_j = irradiated area
Spectral wavelength	λ	F_λ = spectral radiant flux intensity in wavelength intervals, $\Delta\lambda$
Volume (constant)	V	c_V = specific heat at constant volume
Water vapor	w	P_w = partial pressure of water vapor
Wet bulb	wb	T_{wb} = wet bulb temperature

Table A-13. Frequently used physiologic subscripts

Significance	Symbol	Example
Arterial	ar	T_{ar} = arterial blood temperature
Body	b	$\overline{T}_b$ = mean body temperature
Blood	bl	$\dot{m}_{bl}$ = blood flow rate
Expired	ex	T_{ex} = temperature of expired gas
Hypothalamic	hy	T_{hy} = hypothalamic temperature
Inspired	in	T_{in} = temperature of inspired air
Muscle	m	T_m = muscle temperature
Esophageal	es	T_{es} = esophageal temperature
Oral	or	T_{or} = oral temperature
Rectal	re	T_{re} = rectal temperature
Skin	s	$\overline{T}_s$ = mean skin temperature
Tympanic	ty	T_{ty} = tympanic temperature
Venous	ve	T_{ve} = venous blood temperature

Table A-14. Universal constants*

Name	Symbol	Value
Avogadro constant	N_A	6.0225×10^{23} mol^{-1}
Boltzmann constant	k	1.3806×10^{-23} J/°K (SI) 1.3806×10^{-16} erg/°K (cgs)
Gas constant	R	8.3143 J × °K^{-1} × mol^{-1}
Gravitational constant	G	6.670×10^{-11} N × m^2 × kgm^{-2}
Stefan-Boltzmann constant	σ	5.67×10^{-8} W × m^{-2} × °K^{-4}

*Always written in italics.

Table A-15. Conversion factors for selected units in current usage to International System of Units

To convert from	Abbreviation	To	Abbreviation	Multiply by
Energy				
British thermal unit	Btu	joule	J	1,055.9
Calorie	cal	joule	J	4.187
Foot-pound	ft-lb	joule	J	1.3558
Kilocalorie	kcal	kilojoule	kJ	4.187
		joule	J	4,187
Energy × area^{-1} × time^{-1}				
Btu per square foot and hour	Btu × ft^{-2} × hr^{-1}	watt per square meter	W/m^2	3.1525
Kilocalorie per square meter and hour	kcal × m^{-2} × hr^{-1}	watt per square meter	W/m^2	1.163
Force				
Kilogram force	kgmf	newton	N	9.807
Kilopond	kp	newton	N	9.807
Pound force (avoirdupois)	lbf	newton	N	4.4482
Power				
Btu per minute	Btu/min	watt	W	17.572
Calorie per second	cal/s	watt	W	4.187
Horsepower	1 hp = 550 ft-lbf/s	watt	W	745
Kilocalorie per hour	kcal/hr	watt	W	1.162
Kilopond meter per minute	kpm/min	watt	W	0.1634
Pressure				
Inch of mercury	in Hg	newton per square meter	N/m^2	3,386.4
Inch of water (39.2° F)	in H_2O	newton per square meter	N/m^2	248.8
Millimeter of mercury (0°C)	mm Hg	newton per square meter	N/m^2	133.3
Millimeter of water (4°C)	mm H_2O	newton per square meter	N/m^2	9.806
Speed				
Feet per minute	ft/min	meter per second	m/s	0.00508
Feet per second	ft/s	meter per second	m/s	0.3048
Kilometer per hour	km/hr	meter per second	m/s	0.2778
Miles per hour	mph	meter per second	m/s	0.447
Transfer coefficients				
Kilocalorie per square meter, hour, and °C	kcal × m^{-2} × hr^{-1} × °C^{-1}	watt per square meter and °C	W × m^{-2} × °C^{-1}	1.163
British thermal unit per square foot, hour, and °F	Btu × ft^{-2} × hr^{-1} × °F^{-1}	watt per square meter and °C	W × m^{-2} × °C^{-1}	5.67

Table A-16. Classification of living mammals

	Common name
Phylum Chordata	
Subphylum Vertebrata	
Class Mammalia	
Subclass Prototheria	Egg-laying mammals
Order Monotremata	Platypus, echidna
Subclass Theria	Live-bearing mammals
Infraclass Metatheria	Pouched mammals
Order Marsupialia	Kangaroos, koala, opossum, phalangers
Infraclass Eutheria	Placental mammals
Cohort Unguiculata	
Order Insectivora	Shrews, moles, hedgehogs
Order Dermoptera	Colugo or flying lemur
Order Chiroptera	Bats
Suborder Megachiroptera	
Suborder Microchiroptera	
Order Edentata	Sloths, armadillos, anteaters
Order Pholidota	Scaly anteaters or pangolins
Order Primates	
Suborder Prosimii	Tree shrews, lemurs, aye-ayes, tarsiers
Suborder Anthropoidea	Monkeys, marmosets, baboons, gibbons, apes, man
Cohort Glires	
Order Rodentia	Rodents
Suborder Hystricomorpha	Porcupines
Suborder Myomorpha	Mice, rats
Suborder Sciuromorpha	Squirrels, chipmunks
Order Lagomorpha	Rabbits, hares
Cohort Mutica	
Order Cetacea	Mammals of the open ocean
Suborder Odontoceti	Toothed whales, porpoises, dolphins
Suborder Mysticeti	Whale-bone whales
Cohort Ferungulata	
Order Carnivora	Flesh-eaters
Suborder Fissipedia	Bears, dogs, cats, weasels, hyenas
Suborder Pinnipedia	Walruses, seals
Order Tubulidentata	Aardvark
Order Proboscidea	Elephants
Order Hyracoidea	Hyraxes
Order Sirenia	Manatee, dugong
Order Perissodactyla	Odd-toed ungulates
Suborder Ceratomorpha	Rhinoceroses
Suborder Hippomorpha	Horses, tapirs
Order Artiodactyla	Even-toed ungulates
Suborder Suiformes	Pigs, hippopotamuses
Suborder Tylopoda	Camels
Suborder Ruminantia	Cattle, buffalo, deer, giraffes, goats, antelope, llama, sheep

GENERAL GLOSSARY OF TERMS AND CONCEPTS

Please note the special glossary of key terms following each chapter of this book. The terminology and concepts of environmental physiology continue to evolve. Definitions must incorporate the experience gained from the past, provide a precise vocabulary for current communication, and try to anticipate the needs of the future. Some of the following definitions are arbitrary; some are frankly controversial. However, we hope they are all sufficiently precise to be useful or, at the least, to encourage constructive criticism.

acclimatization The physiologic process or state of adjustment of an individual that occurs in response to repeated or prolonged exposure to an unaccustomed stressor. Acclimatization usually results in increased tolerance to the given stressor. Contrast **adaptation.**

adaptation Genetic changes that increase the tolerance of a species to a stressor during generations of exposure. Adaptation favors survival in the presence of the given stressor. Contrast **acclimatization.**

ambient A prevailing environmental condition.

alveolar gas The gas mixture present within the lungs reflecting the effects of respiratory gas exchange, as distinguished from *dead space gas.* The definition of alveolar gas composition is complicated by the discontinuous nature of lung ventilation, by the variability of lung perfusion, and by the imperfect distribution and matching of these two aspects of lung function. Alveolar gas is assumed to be saturated with water vapor at 37° C.

anaerobic Of or pertaining to the exclusion or absence of atmospheric air; growing only in the absence of molecular oxygen.

anoxia A total lack of oxygen.

apnea Cessation of breathing movements, as in anesthetized animals following hyperventilation.

basal metabolic rate The rate of energy exchange of human beings recorded under conditions of rest and thermal neutrality. The results are expressed as percentage above or below a standard established for persons of the same age and sex.

biosphere The region of the earth's surface, including air and water, that comprises the life-support system; the natural terrestrial ecosystem.

biotope A region that is uniform with respect to environmental conditions and the populations of animals and plants for which it is the habitat.

carcinogen A chemical substance or physical agent capable of inducing cancer.

carbon dioxide output The rate at which carbon dioxide is exhaled from the lungs as determined by measurement and analysis of expired gas, with correction for any inspired carbon dioxide. Expressed as a rate in volume units per unit time. Distinguish **carbon dioxide production,** which is the rate at which carbon dioxide is produced by the metabolizing tissues. When a steady state of respiration and circulation exists, *carbon dioxide output* equals *carbon dioxide production.*

contaminant A foreign or extraneous substance or material; often one that impairs the quality or reduces the life-supporting capacity of an environment. Compare **pollutant.**

dissolved gas Gas that is in simple physical solution, as distinguished from that which is chemically combined or which has reacted chemically with solutes or solvent.

dysbarism Any morbid condition or disease resulting from exposure to a change of ambient pressure and, therefore, usually sudden exposure or great change.

dyspnea The uncomfortable awareness, or consciousness, of the need for increased breathing. Dyspnea is subjective by definition; it is a symptom, not a sign. It is usually related to decreased ventilatory capacity and increased work of breathing. This sensation of dyspnea is probably not experienced by healthy, trained individuals.

ecology The branch of biology that deals with the complex network of interactions among living organisms and their environment; the science of organisms as affected by their interrelations and environment.

ecosystem An ecologic system.

effluent A liquid, solid, or gaseous emission; the discharge or waste outflow from a machine or industrial process.

environment The totality of all elements—matter, energy, and force fields—that interact directly or indirectly with an organism at any level of physiologic organization.

biotolerant Tolerant of, or compatible with, life. Neologisms include bioephetic, bioanectic.

hostile Physiologically adverse.

of a living organism An arbitrarily limited region of space containing the matter, energy, and force fields that interact with the organism directly or indirectly at any level of organization—physicochemical, biologic, or psychologic.

optimal The most favorable real environment.

environmental physiology The branch of the science of physiology concerned with the physiologic responses of presently existing forms of life to environmental change and environmental stressors.

environmental stressor An environmental change sufficient to evoke a regulatory counterresponse.

enzyme A highly specific biologic catalyst of protein nature produced by living cells and necessary for in vivo catalysis of biochemical reactions under ordinary conditions of life.

equilibrium state of a closed system A dynamic physicochemical balance according to thermodynamic principles of the matter and energy within a system that exchanges neither matter nor energy with its surroundings. Compare **steady state of an open system.**

exposure

acute A brief or transient exposure; often a brief exposure to a high concentration or intense level of some agent.

chronic Continuous or frequently recurring long-term exposure; often a long-term exposure to low concentrations or levels of some agent.

homeostatic system A physiologic regulatory system or hierarchy of systems and subsystems; generally, a feedback control mechanism.

homeotherm An animal that has relatively constant normal body temperature that is maintained despite environmental temperature change within limits.

hyperbaric Of or pertaining to pressure in excess of ambient or atmospheric pressure, for example, hyperbaric oxygenation.

hyperbarism Any condition, especially morbid, or disease resulting from exposure to increased ambient pressure; usually the result of sudden exposure or a significant increase of pressure.

hypercapnia Abnormally high carbon dioxide pressure, usually within an organism.

hyperoxia Abnormally high oxygen pressure.

hyperpnea Increased rate and/or depth of breathing, for example, hyperpnea of exercise.

hyperventilation Generally, an increased pulmonary ventilation rate beyond the actual requirement for respiratory gas exchange. It may result from increased rate or depth of breathing or any combination of these. Usually, hyperventilation is best expressed in terms of *alveolar* ventilation rate. It results in *increased* alveolar and arterial blood oxygen tension and *decreased* alveolar and arterial blood carbon dioxide tension. Hyperventilation may produce dizziness, numbness, tingling, and, if continued, significant psychomotor impairment and *respiratory alkalosis.*

hypobarogenous aerobullosis A neologism; generalized emergence and growth of gas bubbles from solution in fluids and tissues throughout the body as a result of decompression. *Aeroembolism* has been used in this sense, but as the term suggests, it should mean the *intravascular* emergence, growth, and *transport* of gas bubbles as a result of decompression.

hypobarism Any condition, especially morbid, or disease resulting from exposure to decreased ambient pressure. Hypobarism usually results from sudden exposure or a significant decrease of pressure.

hypocapnia Abnormally low carbon dioxide pressure within an organism.

hypoventilation Generally, a reduced pulmonary ventilation rate below that actually required for respiratory gas exchange. It may result from decreased rate or depth of breathing or any combination of these. Usually, hypoventilation is best expressed in terms of *alveolar* ventilation rate. It results in *decreased* alveolar and arterial blood oxygen tension and *increased* alveolar and arterial blood carbon dioxide tension. If continued, hypoventilation produces hypoxemia, hypercapnia, carbon dioxide retention, and *respiratory acidosis. Chronic alveolar hypoventilation* is an important clinical syndrome. The combination of hypoxemia and respiratory acidosis constricts the pulmonary vascular bed producing pulmonary arterial hypertension.

hypoxemia Low blood oxygen pressure and low oxyhemoglobin saturation.

hypoxia Low or reduced oxygen concentration or tension at any specified point in the transfer system from inspired gas to the metabolizing tissues; also, insufficient oxygen tension or insufficient concentration of free oxygen molecules to meet the requirements of aerobic metabolism.

histohypoxia Reduced or insufficient P_{O_2} at the tissue or cell level.

hypoxidation A state or condition of reduced aerobic metabolism in association with the reduced energy requirements of hypothermia, hibernation, hypothyroidism, or the effect of certain drugs.

hypoxidosis A state or condition of impaired aerobic metabolism in hypoxia, enzyme deficiency or dysfunction, substrate lack, or excessive accumulation of metabolites. Paradoxically, hypoxidosis may result from hyperbaric oxygenation.

isocapnia Of or having equal carbon dioxide pressure.

minute volume of breathing The volume of gas inspired or expired per minute under any given conditions. Usually, *expiratory minute volume* expressed as expired gas volume per minute at BTPS. Note that *expired* gas usually differs from *inspired* gas with regard to temperature, and water vapor, carbon dioxide, and oxygen contents.

oxygen uptake The rate at which oxygen is *removed* by the blood from alveolar gas; also termed *oxygen intake.* Distinguish the rate at which oxygen is *used* by the metabolizing tissues, termed *oxygen consumption.* When a steady state of the respiration and circulation exists, *oxygen uptake* equals *oxygen consumption.* Compare **carbon dioxide output** as opposed to **carbon dioxide production.**

ontogenesis Ontogeny; a history of the process of growth and development in the early life of an organism.

organism Any living plant or animal; a living biologic system.

facultative An organism capable of living under conditions other than the usual.

partial pressure The pressure or tension exerted by any constituent gas in a mixture; the most significant measurement with respect to the physicochemical and physiologic behavior of a gas. As described by Dalton's law, the *total* pressure of a gas mixture is the arithmetic sum of all the *individual* partial pressures of the constituent gases.

particulates Minute discrete particles or fragments of a substance or material.

pollutant A foreign or extraneous substance, material, or agent that impairs the quality or diminishes the life-supporting capacity of an environment. In its most general sense, a pollutant is an undesired substance, not naturally present, that is added to an environment. It may be present in air, water, or soil.

potentiation The combined action of two substances, being greater than the simple algebraic sum of the separate effects of each; increases the physiologic activity of a substance. Compare **synergism.**

pulmonary perfusion Pulmonary capillary blood flow, supplying terminal airways and alveoli.

pulmonary ventilation The volume of air moved into or out of the lungs per unit time; an aspect of the supply and distribution of inspired air to the gas-exchanging regions of the lungs. It is calculated as the product of breathing frequency and tidal volume.

sensitivity The quantitative extent of the primary or prompt effect or response evoked by a unit change of concentration, pressure, or intensity of a given substance or stimulus. Distinguish **threshold.**

shock organ The organ or organ system that manifests the greatest reaction to a toxicant or stressor.

steady state of an open system The state of dynamic physicochemical balance of a system that takes in matter and energy, subjects them to physicochemical transformations and expels matter and energy as waste. A living organism is an open system in a steady state. Compare **equilibrium state of a closed system.**

synergism The total biologic effect of two or more chemical substances or physical agents that is greater than the simple algebraic sum of the separate effects of each; the action of one substance or agent that enhances the action of another. Compare **potentiation.**

threshold The critical level of concentration, pressure, or intensity at which a substance or stimulus begins to exert an effect upon, or evoke a response in, an organism acutely exposed to it. Threshold may be altered in the process of acclimatization or adaptation. Distinguish **sensitivity.**

threshold concentration The lowest concentration of a foreign or abnormally accumulated compound or substance that has a detectable and usually adverse biologic effect on an organism.

tolerance The inherited or acquired capacity of an organism to endure the effects of chemical substances, physical agents, or environmental stressors with minimal displacement of its physiologic regulatory processes.

toxicant Any substance or material that has an adverse effect on a living organism.

transient state Unsteady state; a transitory condition of dynamic imbalance produced by displacement of a steady state; a condition intermediate between successive steady states. Contrast **steady state.**

INDEX

An asterisk following a page number means that a table, graph, or illustration appears on that page.

B

C

D

G

H

I

J

K

L

M

Q

R

U

W